The GALE
ENCYCLOPEDIA
of SCIENCE

THIRD EDITION

The GALE ENCYCLOPEDIA *of* SCIENCE

THIRD EDITION

VOLUME 1
Aardvark - Chaos

K. Lee Lerner and
Brenda Wilmoth Lerner,
Editors

GALE®

Detroit • New York • San Diego • San Francisco • Cleveland • New Haven, Conn. • Waterville, Maine • London • Munich

Gale Encyclopedia of Science, Third Edition

K. Lee Lerner and Brenda Wilmoth Lerner, Editors

Project Editor
Kimberley A. McGrath

Editorial
Deirdre S. Blanchfield, Chris Jeryan, Jacqueline Longe, Mark Springer

Editorial Support Services
Andrea Lopeman

Indexing Services
Synapse

Permissions
Shalice Shah-Caldwell

Imaging and Multimedia
Leitha Etheridge-Sims, Lezlie Light, Dave Oblender, Christine O'Brien, Robyn V. Young

Product Design
Michelle DiMercurio

Manufacturing
Wendy Blurton, Evi Seoud

LIBRARY OF CONGRESS CATALOGING-IN-PUBLICATION DATA

Gale encyclopedia of science / K. Lee Lerner & Brenda Wilmoth Lerner, editors.— 3rd ed.
p. cm.
Includes index.
ISBN 0-7876-7554-7 (set) — ISBN 0-7876-7555-5 (v. 1) — ISBN 0-7876-7556-3 (v. 2) — ISBN 0-7876-7557-1 (v. 3) — ISBN 0-7876-7558-X (v. 4) — ISBN 0-7876-7559-8 (v. 5) — ISBN 0-7876-7560-1 (v. 6)
1. Science—Encyclopedias. I. Lerner, K. Lee. II. Lerner, Brenda Wilmoth.

Q121.G37 2004
503—dc22

2003015731

CONTENTS

TOPIC LIST

A

Aardvark
Abacus
Abrasives
Abscess
Absolute zero
Abyssal plain
Acceleration
Accelerators
Accretion disk
Accuracy
Acetic acid
Acetone
Acetylcholine
Acetylsalicylic acid
Acid rain
Acids and bases
Acne
Acorn worm
Acoustics
Actinides
Action potential
Activated complex
Active galactic nuclei
Acupressure
Acupuncture
ADA (adenosine deaminase) deficiency
Adaptation
Addiction
Addison's disease
Addition
Adenosine diphosphate
Adenosine triphosphate
Adhesive

Adrenals
Aerobic
Aerodynamics
Aerosols
Africa
Age of the universe
Agent Orange
Aging and death
Agouti
Agricultural machines
Agrochemicals
Agronomy
AIDS
AIDS therapies and vaccines
Air masses and fronts
Air pollution
Aircraft
Airship
Albatrosses
Albedo
Albinism
Alchemy
Alcohol
Alcoholism
Aldehydes
Algae
Algebra
Algorithm
Alkali metals
Alkaline earth metals
Alkaloid
Alkyl group
Alleles
Allergy
Allotrope
Alloy

Alluvial systems
Alpha particle
Alternative energy sources
Alternative medicine
Altruism
Aluminum
Aluminum hydroxide
Alzheimer disease
Amaranth family (Amaranthaceae)
Amaryllis family (Amaryllidaceae)
American Standard Code for Information Interchange
Ames test
Amicable numbers
Amides
Amino acid
Ammonia
Ammonification
Amnesia
Amniocentesis
Amoeba
Amphetamines
Amphibians
Amplifier
Amputation
Anabolism
Anaerobic
Analemma
Analgesia
Analog signals and digital signals
Analytic geometry
Anaphylaxis
Anatomy
Anatomy, comparative
Anchovy
Anemia

Anesthesia
Aneurism
Angelfish
Angiography
Angiosperm
Angle
Anglerfish
Animal
Animal breeding
Animal cancer tests
Anion
Anode
Anoles
Ant-pipits
Antarctica
Antbirds and gnat-eaters
Anteaters
Antelopes and gazelles
Antenna
Anthrax
Anthropocentrism
Anti-inflammatory agents
Antibiotics
Antibody and antigen
Anticoagulants
Anticonvulsants
Antidepressant drugs
Antihelmintics
Antihistamines
Antimatter
Antimetabolites
Antioxidants
Antiparticle
Antipsychotic drugs
Antisepsis
Antlions
Ants
Anxiety
Apes
Apgar score
Aphasia
Aphids
Approximation
Apraxia
Aqueduct
Aquifer
Arachnids
Arapaima

Arc
ARC LAMP
Archaebacteria
Archaeoastronomy
Archaeogenetics
Archaeology
Archaeometallurgy
Archaeometry
Archeological mapping
Archeological sites
Arithmetic
Armadillos
Arrow worms
Arrowgrass
Arrowroot
Arteries
Arteriosclerosis
Arthritis
Arthropods
Arthroscopic surgery
Artifacts and artifact classification
Artificial fibers
Artificial heart and heart valve
Artificial intelligence
Artificial vision
Arum family (Araceae)
Asbestos
Asexual reproduction
Asia
Assembly line
Asses
Associative property
Asteroid 2002AA29
Asthenosphere
Asthma
Astrobiology
Astroblemes
Astrolabe
Astrometry
Astronomical unit
Astronomy
Astrophysics
Atmosphere, composition and
    structure
Atmosphere observation
Atmospheric circulation
Atmospheric optical phenomena
Atmospheric pressure

Atmospheric temperature
Atomic clock
Atomic models
Atomic number
Atomic spectroscopy
Atomic theory
Atomic weight
Atoms
Attention-deficit/Hyperactivity
    disorder (ADHD)
Auks
Australia
Autism
Autoimmune disorders
Automatic pilot
Automation
Automobile
Autotroph
Avogadro's number
Aye-ayes

B

Babblers
Baboons
Bacteria
Bacteriophage
Badgers
Ball bearing
Ballistic missiles
Ballistics
Balloon
Banana
Bandicoots
Bar code
Barberry
Barbets
Barbiturates
Bariatrics
Barium
Barium sulfate
Bark
Barley
Barnacles
Barometer
Barracuda

Barrier islands
Basin
Bass
Basswood
Bathysphere
Bats
Battery
Beach nourishment
Beardworms
Bears
Beavers
Bedrock
Bee-eaters
Beech family (Fagaceae)
Bees
Beet
Beetles
Begonia
Behavior
Bennettites
Benzene
Benzoic acid
Bernoulli's principle
Beta-blockers
Big bang theory
Binary star
Binocular
Binomial theorem
Bioaccumulation
Bioassay
Biochemical oxygen demand
Biochemistry
Biodegradable substances
Biodiversity
Bioenergy
Biofeedback
Biofilms
Bioinformatics and computational
  biology
Biological community
Biological rhythms
Biological warfare
Biology
Bioluminescence
Biomagnification
Biomass
Biome
Biophysics

Bioremediation
Biosphere
Biosphere Project
Biotechnology
Bioterrorism
Birch family (Betulaceae)
Birds
Birds of paradise
Birds of prey
Birth
Birth defects
Bison
Bitterns
Bivalves
BL Lacertae object
Black hole
Blackbirds
Blackbody radiation
Bleach
Blennies
Blindness and visual impairments
Blindsnakes
Blood
Blood gas analysis
Blood supply
Blotting analysis
Blue revolution (aquaculture)
Bluebirds
Boarfish
Boas
Bohr Model
Boiling point
Bond energy
Bony fish
Boobies and gannets
Boolean algebra
Boric acid
Botany
Botulism
Bowen's reaction series
Bowerbirds
Bowfin
Boxfish
Brachiopods
Brackish
Brain
Brewing
Brick

Bridges
Bristletails
Brittle star
Bromeliad family (Bromeliaceae)
Bronchitis
Brown dwarf
Brownian motion
Brucellosis
Bryophyte
Bubonic plague
Buckminsterfullerene
Buckthorn
Buckwheat
Buds and budding
Buffer
Building design/architecture
Bulbuls
Bunsen burner
Buoyancy, principle of
Buret
Burn
Bustards
Buttercup
Butterflies
Butterfly fish
Butyl group
Butylated hydroxyanisole
Butylated hydroxytoluene
Buzzards

C

Cactus
CAD/CAM/CIM
Caddisflies
Caecilians
Caffeine
Caisson
Calcium
Calcium carbonate
Calcium oxide
Calcium propionate
Calcium sulfate
Calculator
Calculus
Calendars

Calibration
Caliper
Calorie
Calorimetry
Camels
Canal
Cancel
Cancer
Canines
Cantilever
Capacitance
Capacitor
Capillaries
Capillary action
Caprimulgids
Captive breeding and
    reintroduction
Capuchins
Capybaras
Carbohydrate
Carbon
Carbon cycle
Carbon dioxide
Carbon monoxide
Carbon tetrachloride
Carbonyl group
Carboxyl group
Carboxylic acids
Carcinogen
Cardiac cycle
Cardinal number
Cardinals and grosbeaks
Caribou
Carnivore
Carnivorous plants
Carp
Carpal tunnel syndrome
Carrier (genetics)
Carrot family (Apiaceae)
Carrying capacity
Cartesian coordinate plane
Cartilaginous fish
Cartography
Cashew family (Anacardiaceae)
Cassini Spacecraft
Catabolism
Catalyst and catalysis
Catastrophism

Catfish
Catheters
Cathode
Cathode ray tube
Cation
Cats
Cattails
Cattle family (Bovidae)
Cauterization
Cave
Cave fish
Celestial coordinates
Celestial mechanics
Celestial sphere: The apparent
    motions of the Sun, Moon,
    planets, and stars
Cell
Cell death
Cell division
Cell, electrochemical
Cell membrane transport
Cell staining
Cellular respiration
Cellular telephone
Cellulose
Centipedes
Centrifuge
Ceramics
Cerenkov effect
Cetaceans
Chachalacas
Chameleons
Chaos
Charge-coupled device
Chelate
Chemical bond
Chemical evolution
Chemical oxygen demand
Chemical reactions
Chemical warfare
Chemistry
Chemoreception
Chestnut
Chi-square test
Chickenpox
Childhood diseases
Chimaeras
Chimpanzees

Chinchilla
Chipmunks
Chitons
Chlordane
Chlorinated hydrocarbons
Chlorination
Chlorine
Chlorofluorocarbons (CFCs)
Chloroform
Chlorophyll
Chloroplast
Cholera
Cholesterol
Chordates
Chorionic villus sampling (CVS)
Chromatin
Chromatography
Chromosomal abnormalities
Chromosome
Chromosome mapping
Cicadas
Cigarette smoke
Circle
Circulatory system
Circumscribed and inscribed
Cirrhosis
Citric acid
Citrus trees
Civets
Climax (ecological)
Clingfish
Clone and cloning
Closed curves
Closure property
Clouds
Club mosses
Coal
Coast and beach
Coatis
Coca
Cocaine
Cockatoos
Cockroaches
Codeine
Codfishes
Codons
Coefficient
Coelacanth

Coffee plant
Cogeneration
Cognition
Cold, common
Collagen
Colloid
Colobus monkeys
Color
Color blindness
Colugos
Coma
Combinatorics
Combustion
Comet Hale-Bopp
Comets
Commensalism
Community ecology
Commutative property
Compact disc
Competition
Complementary DNA
Complex
Complex numbers
Composite family
Composite materials
Composting
Compound, chemical
Compton effect
Compulsion
Computer, analog
Computer, digital
Computer languages
Computer memory, physical and
    virtual memory
Computer software
Computer virus
Computerized axial tomography
Concentration
Concrete
Conditioning
Condors
Congenital
Congruence (triangle)
Conic sections
Conifer
Connective tissue
Conservation
Conservation laws

Constellation
Constructions
Contaminated soil
Contamination
Continent
Continental drift
Continental margin
Continental shelf
Continuity
Contour plowing
Contraception
Convection
Coordination compound
Copepods
Copper
Coral and coral reef
Coriolis effect
Cork
Corm
Cormorants
Corn (maize)
Coronal ejections and magnetic
    storms
Correlation (geology)
Correlation (mathematics)
Corrosion
Cosmic background radiation
Cosmic ray
Cosmology
Cotingas
Cotton
Coulomb
Countable
Coursers and pratincoles
Courtship
Coypu
Crabs
Crane
Cranes
Crayfish
Crestfish
Creutzfeldt-Jakob disease
Crickets
Critical habitat
Crocodiles
Crop rotation
Crops
Cross multiply

Cross section
Crows and jays
Crustacea
Cryobiology
Cryogenics
Cryptography, encryption, and
    number theory
Crystal
Cubic equations
Cuckoos
Curare
Curlews
Currents
Curve
Cushing syndrome
Cuttlefish
Cybernetics
Cycads
Cyclamate
Cyclone and anticyclone
Cyclosporine
Cyclotron
Cystic fibrosis
Cytochrome
Cytology

D

Dams
Damselflies
Dark matter
Dating techniques
DDT (Dichlorodiphenyl-
    trichloroacetic acid)
Deafness and inherited hearing loss
Decimal fraction
Decomposition
Deer
Deer mouse
Deforestation
Degree
Dehydroepiandrosterone (DHEA)
Delta
Dementia
Dengue fever
Denitrification

Density
Dentistry
Deoxyribonucleic acid (DNA)
Deposit
Depression
Depth perception
Derivative
Desalination
Desert
Desertification
Determinants
Deuterium
Developmental processes
Dew point
Diabetes mellitus
Diagnosis
Dialysis
Diamond
Diatoms
Dielectric materials
Diesel engine
Diethylstilbestrol (DES)
Diffraction
Diffraction grating
Diffusion
Digestive system
Digital Recording
Digitalis
Dik-diks
Dinosaur
Diode
Dioxin
Diphtheria
Dipole
Direct variation
Disease
Dissociation
Distance
Distillation
Distributive property
Disturbance, ecological
Diurnal cycles
Division
DNA fingerprinting
DNA replication
DNA synthesis
DNA technology
DNA vaccine

Dobsonflies
Dogwood tree
Domain
Donkeys
Dopamine
Doppler effect
Dories
Dormouse
Double-blind study
Double helix
Down syndrome
Dragonflies
Drift net
Drongos
Drosophila melanogaster
Drought
Ducks
Duckweed
Duikers
Dune
Duplication of the cube
Dust devil
DVD
Dwarf antelopes
Dyes and pigments
Dysentery
Dyslexia
Dysplasia
Dystrophinopathies

E

e (number)
Eagles
Ear
Earth
Earth science
Earth's interior
Earth's magnetic field
Earth's rotation
Earthquake
Earwigs
Eating disorders
Ebola virus
Ebony
Echiuroid worms

Echolocation
Eclipses
Ecological economics
Ecological integrity
Ecological monitoring
Ecological productivity
Ecological pyramids
Ecology
Ecosystem
Ecotone
Ecotourism
Edema
Eel grass
El Niño and La Niña
Eland
Elapid snakes
Elasticity
Electric arc
Electric charge
Electric circuit
Electric conductor
Electric current
Electric motor
Electric vehicles
Electrical conductivity
Electrical power supply
Electrical resistance
Electricity
Electrocardiogram (ECG)
Electroencephalogram (EEG)
Electrolysis
Electrolyte
Electromagnetic field
Electromagnetic induction
Electromagnetic spectrum
Electromagnetism
Electromotive force
Electron
Electron cloud
Electronics
Electrophoresis
Electrostatic devices
Element, chemical
Element, families of
Element, transuranium
Elements, formation of
Elephant
Elephant shrews

Formula, structural
Fossa
Fossil and fossilization
Fossil fuels
Fractal
Fraction, common
Fraunhofer lines
Freeway
Frequency
Freshwater
Friction
Frigate birds
Frog's-bit family
Frogs
Frostbite
Fruits
Fuel cells
Function
Fundamental theorems
Fungi
Fungicide

G

Gaia hypothesis
Galaxy
Game theory
Gamete
Gametogenesis
Gamma-ray astronomy
Gamma ray burst
Gangrene
Garpike
Gases, liquefaction of
Gases, properties of
Gazelles
Gears
Geckos
Geese
Gelatin
Gene
Gene chips and microarrays
Gene mutation
Gene splicing
Gene therapy
Generator

Genetic disorders
Genetic engineering
Genetic identification of
    microorganisms
Genetic testing
Genetically modified foods and
    organisms
Genetics
Genets
Genome
Genomics (comparative)
Genotype and phenotype
Geocentric theory
Geochemical analysis
Geochemistry
Geode
Geodesic
Geodesic dome
Geographic and magnetic poles
Geologic map
Geologic time
Geology
Geometry
Geomicrobiology
Geophysics
Geotropism
Gerbils
Germ cells and the germ cell line
Germ theory
Germination
Gerontology
Gesnerias
Geyser
Gibbons and siamangs
Gila monster
Ginger
Ginkgo
Ginseng
Giraffes and okapi
GIS
Glaciers
Glands
Glass
Global climate
Global Positioning System
Global warming
Glycerol
Glycol

Glycolysis
Goats
Goatsuckers
Gobies
Goldenseal
Gophers
Gorillas
Gourd family (Cucurbitaceae)
Graft
Grand unified theory
Grapes
Graphs and graphing
Grasses
Grasshoppers
Grasslands
Gravitational lens
Gravity and gravitation
Great Barrier Reef
Greatest common factor
Grebes
Greenhouse effect
Groundhog
Groundwater
Group
Grouse
Growth and decay
Growth hormones
Guenons
Guillain-Barre syndrome
Guinea fowl
Guinea pigs and cavies
Gulls
Guppy
Gutenberg discontinuity
Gutta percha
Gymnosperm
Gynecology
Gyroscope

H

Habitat
Hagfish
Half-life
Halide, organic
Hall effect

Halley's comet
Hallucinogens
Halogenated hydrocarbons
Halogens
Halosaurs
Hamsters
Hand tools
Hantavirus infections
Hard water
Harmonics
Hartebeests
Hawks
Hazardous wastes
Hazel
Hearing
Heart
Heart diseases
Heart, embryonic development and
    changes at birth
Heart-lung machine
Heat
Heat capacity
Heat index
Heat transfer
Heath family (Ericaceae)
Hedgehogs
Heisenberg uncertainty principle
Heliocentric theory
Hematology
Hemophilia
Hemorrhagic fevers and diseases
Hemp
Henna
Hepatitis
Herb
Herbal medicine
Herbicides
Herbivore
Hermaphrodite
Hernia
Herons
Herpetology
Herrings
Hertzsprung-Russell diagram
Heterotroph
Hibernation
Himalayas, geology of
Hippopotamuses

Histamine
Historical geology
Hoatzin
Hodgkin's disease
Holly family (Aquifoliaceae)
Hologram and holography
Homeostasis
Honeycreepers
Honeyeaters
Hoopoe
Horizon
Hormones
Hornbills
Horse chestnut
Horsehair worms
Horses
Horseshoe crabs
Horsetails
Horticulture
Hot spot
Hovercraft
Hubble Space Telescope
Human artificial chromosomes
Human chorionic gonadotropin
Human cloning
Human ecology
Human evolution
Human Genome Project
Humidity
Hummingbirds
Humus
Huntington disease
Hybrid
Hydra
Hydrocarbon
Hydrocephalus
Hydrochlorofluorocarbons
Hydrofoil
Hydrogen
Hydrogen chloride
Hydrogen peroxide
Hydrogenation
Hydrologic cycle
Hydrology
Hydrolysis
Hydroponics
Hydrosphere
Hydrothermal vents

Hydrozoa
Hyena
Hyperbola
Hypertension
Hypothermia
Hyraxes

I

Ibises
Ice
Ice age refuges
Ice ages
Icebergs
Iceman
Identity element
Identity property
Igneous rocks
Iguanas
Imaginary number
Immune system
Immunology
Impact crater
Imprinting
In vitro fertilization (IVF)
In vitro and in vivo
Incandescent light
Incineration
Indicator, acid-base
Indicator species
Individual
Indoor air quality
Industrial minerals
Industrial Revolution
Inequality
Inertial guidance
Infection
Infertility
Infinity
Inflammation
Inflection point
Influenza
Infrared astronomy
Inherited disorders
Insecticides
Insectivore

Lorises
Luminescence
Lungfish
Lycophytes
Lyme disease
Lymphatic system
Lyrebirds

M

Macaques
Mach number
Machine tools
Machine vision
Machines, simple
Mackerel
Magic square
Magma
Magnesium
Magnesium sulfate
Magnetic levitation
Magnetic recording/audiocassette
Magnetic resonance imaging (MRI)
Magnetism
Magnetosphere
Magnolia
Mahogany
Maidenhair fern
Malaria
Malnutrition
Mammals
Manakins
Mangrove tree
Mania
Manic depression
Map
Maples
Marfan syndrome
Marijuana
Marlins
Marmosets and tamarins
Marmots
Mars
Mars Pathfinder
Marsupial cats
Marsupial rats and mice

Marsupials
Marten, sable, and fisher
Maser
Mass
Mass extinction
Mass number
Mass production
Mass spectrometry
Mass transportation
Mass wasting
Mathematics
Matrix
Matter
Maunder minimum
Maxima and minima
Mayflies
Mean
Median
Medical genetics
Meiosis
Membrane
Memory
Mendelian genetics
Meningitis
Menopause
Menstrual cycle
Mercurous chloride
Mercury (element)
Mercury (planet)
Mesoscopic systems
Mesozoa
Metabolic disorders
Metabolism
Metal
Metal fatigue
Metal production
Metallurgy
Metamorphic grade
Metamorphic rock
Metamorphism
Metamorphosis
Meteorology
Meteors and meteorites
Methyl group
Metric system
Mice
Michelson-Morley experiment
Microbial genetics

Microclimate
Microorganisms
Microscope
Microscopy
Microtechnology
Microwave communication
Migraine headache
Migration
Mildew
Milkweeds
Milky Way
Miller-Urey Experiment
Millipedes
Mimicry
Mineralogy
Minerals
Mining
Mink
Minnows
Minor planets
Mint family
Mir Space Station
Mirrors
Miscibility
Mistletoe
Mites
Mitosis
Mixture, chemical
Möbius strip
Mockingbirds and thrashers
Mode
Modular arithmetic
Mohs' scale
Mold
Mole
Mole-rats
Molecular biology
Molecular formula
Molecular geometry
Molecular weight
Molecule
Moles
Mollusks
Momentum
Monarch flycatchers
Mongooses
Monitor lizards
Monkeys

Organic farming
Organism
Organogenesis
Organs and organ systems
Origin of life
Orioles
Ornithology
Orthopedics
Oryx
Oscillating reactions
Oscillations
Oscilloscope
Osmosis
Osmosis (cellular)
Ossification
Osteoporosis
Otter shrews
Otters
Outcrop
Ovarian cycle and hormonal regulation
Ovenbirds
Oviparous
Ovoviviparous
Owls
Oxalic acid
Oxidation-reduction reaction
Oxidation state
Oxygen
Oystercatchers
Ozone
Ozone layer depletion

P

Pacemaker
Pain
Paleobotany
Paleoclimate
Paleoecology
Paleomagnetism
Paleontology
Paleopathology
Palindrome
Palms
Palynology

Pandas
Pangolins
Papaya
Paper
Parabola
Parallax
Parallel
Parallelogram
Parasites
Parity
Parkinson disease
Parrots
Parthenogenesis
Particle detectors
Partridges
Pascal's triangle
Passion flower
Paternity and parentage testing
Pathogens
Pathology
PCR
Peafowl
Peanut worms
Peccaries
Pedigree analysis
Pelicans
Penguins
Peninsula
Pentyl group
Peony
Pepper
Peptide linkage
Percent
Perception
Perch
Peregrine falcon
Perfect numbers
Periodic functions
Periodic table
Permafrost
Perpendicular
Pesticides
Pests
Petrels and shearwaters
Petroglyphs and pictographs
Petroleum
pH
Phalangers

Pharmacogenetics
Pheasants
Phenyl group
Phenylketonuria
Pheromones
Phlox
Phobias
Phonograph
Phoronids
Phosphoric acid
Phosphorus
Phosphorus cycle
Phosphorus removal
Photic zone
Photochemistry
Photocopying
Photoelectric cell
Photoelectric effect
Photography
Photography, electronic
Photon
Photosynthesis
Phototropism
Photovoltaic cell
Phylogeny
Physical therapy
Physics
Physiology
Physiology, comparative
Phytoplankton
Pi
Pigeons and doves
Pigs
Pike
Piltdown hoax
Pinecone fish
Pines
Pipefish
Placebo
Planck's constant
Plane
Plane family
Planet
Planet X
Planetary atmospheres
Planetary geology
Planetary nebulae
Planetary ring systems

Plankton
Plant
Plant breeding
Plant diseases
Plant pigment
Plasma
Plastic surgery
Plastics
Plate tectonics
Platonic solids
Platypus
Plovers
Pluto
Pneumonia
Podiatry
Point
Point source
Poisons and toxins
Polar coordinates
Polar ice caps
Poliomyelitis
Pollen analysis
Pollination
Pollution
Pollution control
Polybrominated biphenyls (PBBs)
Polychlorinated biphenyls (PCBs)
Polycyclic aromatic hydrocarbons
Polygons
Polyhedron
Polymer
Polynomials
Poppies
Population growth and control (human)
Population, human
Porcupines
Positive number
Positron emission tomography (PET)
Postulate
Potassium aluminum sulfate
Potassium hydrogen tartrate
Potassium nitrate
Potato
Pottery analysis
Prairie
Prairie chicken

Prairie dog
Prairie falcon
Praying mantis
Precession of the equinoxes
Precious metals
Precipitation
Predator
Prenatal surgery
Prescribed burn
Pressure
Prey
Primates
Prime numbers
Primroses
Printing
Prions
Prism
Probability theory
Proboscis monkey
Projective geometry
Prokaryote
Pronghorn
Proof
Propyl group
Prosimians
Prosthetics
Proteas
Protected area
Proteins
Proteomics
Protista
Proton
Protozoa
Psychiatry
Psychoanalysis
Psychology
Psychometry
Psychosis
Psychosurgery
Puberty
Puffbirds
Puffer fish
Pulsar
Punctuated equilibrium
Pyramid
Pythagorean theorem
Pythons

Q

Quadrilateral
Quail
Qualitative analysis
Quantitative analysis
Quantum computing
Quantum electrodynamics (QED)
Quantum mechanics
Quantum number
Quarks
Quasar
Quetzal
Quinine

R

Rabies
Raccoons
Radar
Radial keratotomy
Radiation
Radiation detectors
Radiation exposure
Radical (atomic)
Radical (math)
Radio
Radio astronomy
Radio waves
Radioactive dating
Radioactive decay
Radioactive fallout
Radioactive pollution
Radioactive tracers
Radioactive waste
Radioisotopes in medicine
Radiology
Radon
Rails
Rainbows
Rainforest
Random
Rangeland
Raptors
Rare gases
Rare genotype advantage

Rate

Ratio

Rational number

Rationalization

Rats

Rayleigh scattering

Rays

Real numbers

Reciprocal

Recombinant DNA

Rectangle

Recycling

Red giant star

Red tide

Redshift

Reflections

Reflex

Refrigerated trucks and railway cars

Rehabilitation

Reinforcement, positive and
negative

Relation

Relativity, general

Relativity, special

Remote sensing

Reproductive system

Reproductive toxicant

Reptiles

Resins

Resonance

Resources, natural

Respiration

Respiration, cellular

Respirator

Respiratory diseases

Respiratory system

Restoration ecology

Retrograde motion

Retrovirus

Reye's syndrome

Rh factor

Rhesus monkeys

Rheumatic fever

Rhinoceros

Rhizome

Rhubarb

Ribbon worms

Ribonuclease

Ribonucleic acid (RNA)

Ribosomes

Rice

Ricin

Rickettsia

Rivers

RNA function

RNA splicing

Robins

Robotics

Rockets and missiles

Rocks

Rodents

Rollers

Root system

Rose family (Rosaceae)

Rotation

Roundworms

Rumination

Rushes

Rusts and smuts

S

Saiga antelope

Salamanders

Salmon

Salmonella

Salt

Saltwater

Sample

Sand

Sand dollars

Sandfish

Sandpipers

Sapodilla tree

Sardines

Sarin gas

Satellite

Saturn

Savanna

Savant

Sawfish

Saxifrage family

Scalar

Scale insects

Scanners, digital

Scarlet fever

Scavenger

Schizophrenia

Scientific method

Scorpion flies

Scorpionfish

Screamers

Screwpines

Sculpins

Sea anemones

Sea cucumbers

Sea horses

Sea level

Sea lily

Sea lions

Sea moths

Sea spiders

Sea squirts and salps

Sea urchins

Seals

Seamounts

Seasonal winds

Seasons

Secondary pollutants

Secretary bird

Sedges

Sediment and sedimentation

Sedimentary environment

Sedimentary rock

Seed ferns

Seeds

Segmented worms

Seismograph

Selection

Sequences

Sequencing

Sequoia

Servomechanisms

Sesame

Set theory

SETI

Severe acute respiratory syndrome
(SARS)

Sewage treatment

Sewing machine

Sex change

Sextant

Sexual reproduction
Sexually transmitted diseases
Sharks
Sheep
Shell midden analysis
Shingles
Shore birds
Shoreline protection
Shotgun cloning
Shrews
Shrikes
Shrimp
Sickle cell anemia
Sieve of Eratosthenes
Silicon
Silk cotton family (Bombacaceae)
Sinkholes
Skates
Skeletal system
Skinks
Skuas
Skunks
Slash-and-burn agriculture
Sleep
Sleep disorders
Sleeping sickness
Slime molds
Sloths
Slugs
Smallpox
Smallpox vaccine
Smell
Smog
Snails
Snakeflies
Snakes
Snapdragon family
Soap
Sociobiology
Sodium
Sodium benzoate
Sodium bicarbonate
Sodium carbonate
Sodium chloride
Sodium hydroxide
Sodium hypochlorite
Soil
Soil conservation

Solar activity cycle
Solar flare
Solar illumination: Seasonal and
    diurnal patterns
Solar prominence
Solar system
Solar wind
Solder and soldering iron
Solstice
Solubility
Solution
Solution of equation
Sonar
Song birds
Sonoluminescence
Sorghum
Sound waves
South America
Soybean
Space
Space probe
Space shuttle
Spacecraft, manned
Sparrows and buntings
Species
Spectral classification of stars
Spectral lines
Spectroscope
Spectroscopy
Spectrum
Speech
Sphere
Spider monkeys
Spiderwort family
Spin of subatomic particles
Spina bifida
Spinach
Spiny anteaters
Spiny eels
Spiny-headed worms
Spiral
Spirometer
Split-brain functioning
Sponges
Spontaneous generation
Spore
Springtails
Spruce

Spurge family
Square
Square root
Squid
Squirrel fish
Squirrels
Stalactites and stalagmites
Standard model
Star
Star cluster
Star formation
Starburst galaxy
Starfish
Starlings
States of matter
Statistical mechanics
Statistics
Steady-state theory
Steam engine
Steam pressure sterilizer
Stearic acid
Steel
Stellar evolution
Stellar magnetic fields
Stellar magnitudes
Stellar populations
Stellar structure
Stellar wind
Stem cells
Stereochemistry
Sticklebacks
Stilts and avocets
Stimulus
Stone and masonry
Stoneflies
Storks
Storm
Storm surge
Strata
Stratigraphy
Stratigraphy (archeology)
Stream capacity and competence
Stream valleys, channels, and
    floodplains
Strepsiptera
Stress
Stress, ecological
String theory

Stroke
Stromatolite
Sturgeons
Subatomic particles
Submarine
Subsidence
Subsurface detection
Subtraction
Succession
Suckers
Sudden infant death syndrome
    (SIDS)
Sugar beet
Sugarcane
Sulfur
Sulfur cycle
Sulfur dioxide
Sulfuric acid
Sun
Sunbirds
Sunspots
Superclusters
Superconductor
Supernova
Surface tension
Surgery
Surveying instruments
Survival of the fittest
Sustainable development
Swallows and martins
Swamp cypress family
    (Taxodiaceae)
Swamp eels
Swans
Sweet gale family (Myricaceae)
Sweet potato
Swifts
Swordfish
Symbiosis
Symbol, chemical
Symbolic logic
Symmetry
Synapse
Syndrome
Synthesis, chemical
Synthesizer, music
Synthesizer, voice
Systems of equations

T

T cells
Tanagers
Taphonomy
Tapirs
Tarpons
Tarsiers
Tartaric acid
Tasmanian devil
Taste
Taxonomy
Tay-Sachs disease
Tea plant
Tectonics
Telegraph
Telemetry
Telephone
Telescope
Television
Temperature
Temperature regulation
Tenrecs
Teratogen
Term
Termites
Terns
Terracing
Territoriality
Tetanus
Tetrahedron
Textiles
Thalidomide
Theorem
Thermal expansion
Thermochemistry
Thermocouple
Thermodynamics
Thermometer
Thermostat
Thistle
Thoracic surgery
Thrips
Thrombosis
Thrushes
Thunderstorm
Tides

Time
Tinamous
Tissue
Tit family
Titanium
Toadfish
Toads
Tomato family
Tongue worms
Tonsillitis
Topology
Tornado
Torque
Torus
Total solar irradiance
Toucans
Touch
Towers of Hanoi
Toxic shock syndrome
Toxicology
Trace elements
Tragopans
Trains and railroads
Tranquilizers
Transcendental numbers
Transducer
Transformer
Transgenics
Transistor
Transitive
Translations
Transpiration
Transplant, surgical
Trapezoid
Tree
Tree shrews
Trichinosis
Triggerfish
Triglycerides
Trigonometry
Tritium
Trogons
Trophic levels
Tropic birds
Tropical cyclone
Tropical diseases
Trout-perch
True bugs

True eels
True flies
Trumpetfish
Tsunami
Tuatara lizard
Tuber
Tuberculosis
Tumbleweed
Tumor
Tuna
Tundra
Tunneling
Turacos
Turbine
Turbulence
Turkeys
Turner syndrome
Turtles
Typhoid fever
Typhus
Tyrannosaurus rex
Tyrant flycatchers

U

Ulcers
Ultracentrifuge
Ultrasonics
Ultraviolet astronomy
Unconformity
Underwater exploration
Ungulates
Uniformitarianism
Units and standards
Uplift
Upwelling
Uranium
Uranus
Urea
Urology

V

Vaccine

Vacuum
Vacuum tube
Valence
Van Allen belts
Van der Waals forces
Vapor pressure
Variable
Variable stars
Variance
Varicella zoster virus
Variola virus
Vegetables
Veins
Velocity
Venus
Verbena family (Verbenaceae)
Vertebrates
Video recording
Violet family (Violaceae)
Vipers
Viral genetics
Vireos
Virtual particles
Virtual reality
Virus
Viscosity
Vision
Vision disorders
Vitamin
Viviparity
Vivisection
Volatility
Volcano
Voles
Volume
Voyager spacecraft
Vulcanization
Vultures
VX agent

W

Wagtails and pipits
Walkingsticks
Walnut family
Walruses

Warblers
Wasps
Waste management
Waste, toxic
Water
Water bears
Water conservation
Water lilies
Water microbiology
Water pollution
Water treatment
Waterbuck
Watershed
Waterwheel
Wave motion
Waxbills
Waxwings
Weasels
Weather
Weather forecasting
Weather mapping
Weather modification
Weathering
Weaver finches
Weevils
Welding
West Nile virus
Wetlands
Wheat
Whisk fern
White dwarf
White-eyes
Whooping cough
Wild type
Wildfire
Wildlife
Wildlife trade (illegal)
Willow family (Salicaceae)
Wind
Wind chill
Wind shear
Wintergreen
Wolverine
Wombats
Wood
Woodpeckers
Woolly mammoth
Work

Wren-warblers
Wrens
Wrynecks

X

X-ray astronomy
X-ray crystallography
X rays
Xenogamy

Y

Y2K
Yak
Yam
Yeast
Yellow fever
Yew
Yttrium

Z

Zebras
Zero
Zodiacal light
Zoonoses
Zooplankton

ORGANIZATION OF THE ENCYCLOPEDIA

The *Gale Encyclopedia of Science, Third Edition* has been designed with ease of use and ready reference in mind.

- Entries are alphabetically arranged across six volumes, in a single sequence, rather than by scientific field

- Length of entries varies from short definitions of one or two paragraphs, to longer, more detailed entries on more complex subjects.

- Longer entries are arranged so that an overview of the subject appears first, followed by a detailed discussion conveniently arranged under subheadings.

- A list of key terms is provided where appropriate to define unfamiliar terms or concepts.

- Bold-faced terms direct the reader to related articles.

- Longer entries conclude with a "Resources" section, which points readers to other helpful materials (including books, periodicals, and Web sites).

- The author's name appears at the end of longer entries. His or her affiliation can be found in the "Contributors" section at the front of each volume.

- "See also" references appear at the end of entries to point readers to related entries.

- Cross references placed throughout the encyclopedia direct readers to where information on subjects without their own entries can be found.

- A comprehensive, two-level General Index guides readers to all topics, illustrations, tables, and persons mentioned in the book.

AVAILABLE IN ELECTRONIC FORMATS

Licensing. *The Gale Encyclopedia of Science, Third Edition* is available for licensing. The complete database is provided in a fielded format and is deliverable on such media as disk or CD-ROM. For more information, contact Gale's Business Development Group at 1-800-877-GALE, or visit our website at www.gale.com/bizdev.

ADVISORY BOARD

A number of experts in the scientific and libary communities provided invaluable assistance in the formulation of this encyclopedia. Our advisory board performed a myriad of duties, from defining the scope of coverage to reviewing individual entries for accuracy and accessibility, and in many cases, writing entries. We would therefore like to express our appreciation to them:

ACADEMIC ADVISORS

Marcelo Amar, M.D.
Senior Fellow, Molecular Disease Branch
National Institutes of Health (NIH)
Bethesda, Maryland

Robert G. Best, Ph.D.
Director
Divison of Genetics, Department of Obstetrics and
    Gynecology
University of South Carolina School of Medicine
Columbia, South Carolina

Bryan Bunch
Adjunct Instructor
Department of Mathematics
Pace University
New York, New York

Cynthia V. Burek, Ph.D.
Environment Research Group, Biology Department
Chester College
England, UK

David Campbell
Head
Department of Physics
University of Illinois at Urbana Champaign
Urbana, Illinois

Morris Chafetz
Health Education Foundation
Washington, DC

Brian Cobb, Ph.D.
Institute for Molecular and Human Genetics
Georgetown University
Washington, DC

Neil Cumberlidge
Professor
Department of Biology

Northern Michigan University
Marquette, Michigan

Nicholas Dittert, Ph.D.
Institut Universitaire Européen de la Mer
University of Western Brittany
France

William J. Engle. P.E.
Exxon-Mobil Oil Corporation (Rt.)
New Orleans, Louisiana

Bill Freedman
Professor
Department of Biology and School for Resource and
    Environmental Studies
Dalhousie University
Halifax, Nova Scotia, Canada

Antonio Farina, M.D., Ph.D.
Department of Embryology, Obstetrics, and
    Gynecology
University of Bologna
Bologna, Italy

G. Thomas Farmer, Ph.D., R.G.
Earth & Environmental Sciences Division
Los Alamos National Laboratory
Los Alamos, New Mexico

Jeffrey C. Hall
Lowell Observatory
Flagstaff, Arizona

Clayton Harris
Associate Professor
Department of Geography and Geology
Middle Tennessee State University
Murfreesboro, Tennesses

Lyal Harris, Ph.D.
Tectonics Special Research Centre
Department of Geology & Geophysics

CONTRIBUTORS

Nasrine Adibe
Professor Emeritus
Department of Education
Long Island University
Westbury, New York

Mary D. Albanese
Department of English
University of Alaska
Juneau, Alaska

Margaret Alic
Science Writer
Eastsound, Washington

James L. Anderson
Soil Science Department
University of Minnesota
St. Paul, Minnesota

Monica Anderson
Science Writer
Hoffman Estates, Illinois

Susan Andrew
Teaching Assistant
University of Maryland
Washington, DC

John Appel
Director
Fundación Museo de Ciencia y
    Tecnología
Popayán, Colombia

David Ball
Assistant Professor
Department of Chemistry
Cleveland State University
Cleveland, Ohio

Dana M. Barry
Editor and Technical Writer
Center for Advanced Materials
    Processing
Clarkston University
Potsdam, New York

Puja Batra
Department of Zoology
Michigan State University
East Lansing, Michigan

Donald Beaty
Professor Emeritus
College of San Mateo
San Mateo, California

Eugene C. Beckham
Department of Mathematics and
    Science
Northwood Institute
Midland, Michigan

Martin Beech
Research Associate
Department of Astronomy
University of Western Ontario
London, Ontario, Canada

**Julie Berwald, Ph.D. (Ocean
    Sciences)**
Austin, Texas

Massimo D. Bezoari
Associate Professor
Department of Chemistry
Huntingdon College
Montgomery, Alabama

John M. Bishop III
Translator
New York, New York

T. Parker Bishop
Professor
Middle Grades and Secondary
    Education
Georgia Southern University
Statesboro, Georgia

Carolyn Black
Professor
Incarnate Word College
San Antonio, Texas

Larry Blaser
Science Writer
Lebanon, Tennessee

Jean F. Blashfield
Science Writer
Walworth, Wisconsin

Richard L. Branham Jr.
Director
Centro Rigional de
    Investigaciones Científicas y
    Tecnológicas
Mendoza, Argentina

Patricia Braus
Editor
American Demographics
Rochester, New York

David L. Brock
Biology Instructor
St. Louis, Missouri

Leona B. Bronstein
Chemistry Teacher (retired)
East Lansing High School
Okemos, Michigan

Brandon R. Brown
Graduate Research Assistant
Oregon State University
Corvallis, Oregon

Lenonard C. Bruno
Senior Science Specialist
Library of Congress
Chevy Chase, Maryland

Janet Buchanan, Ph.D.
Microbiologist
Independent Scholar
Toronto, Ontario, Canada.

Scott Christian Cahall
Researcher
World Precision Instruments, Inc.
Bradenton, Florida

G. Lynn Carlson
Senior Lecturer
School of Science and
    Technology
University of Wisconsin—
    Parkside
Kenosha, Wisconsin

James J. Carroll
Center for Quantum Mechanics
The University of Texas at Dallas
Dallas, Texas

Steven B. Carroll
Assistant Professor
Division of Biology
Northeast Missouri State
    University
Kirksville, Missouri

Rosalyn Carson-DeWitt
Physician and Medical Writer
Durham, North Carolina

Yvonne Carts-Powell
Editor
Laser Focus World
Belmont, Massachustts

Chris Cavette
Technical Writer
Fremont, California

Lata Cherath
Science Writer
Franklin Park, New York

Kenneth B. Chiacchia
Medical Editor
University of Pittsburgh Medical
    Center
Pittsburgh, Pennsylvania

M. L. Cohen
Science Writer
Chicago, Illinois

Robert Cohen
Reporter
KPFA Radio News
Berkeley, California

Sally Cole-Misch
Assistant Director
International Joint Commission
Detroit, Michigan

George W. Collins II
Professor Emeritus
Case Western Reserve
Chesterland, Ohio

Jeffrey R. Corney
Science Writer
Thermopolis, Wyoming

Tom Crawford
Assistant Director
Division of Publication and
    Development
University of Pittsburgh Medical
    Center
Pittsburgh, Pennsylvania

Pamela Crowe
Medical and Science Writer
Oxon, England

Clinton Crowley
On-site Geologist
Selman and Associates
Fort Worth, Texas

Edward Cruetz
Physicist
Rancho Santa Fe, California

Frederick Culp
Chairman
Department of Physics
Tennessee Technical
Cookeville, Tennessee

Neil Cumberlidge
Professor
Department of Biology
Northern Michigan University
Marquette, Michigan

Mary Ann Cunningham
Environmental Writer
St. Paul, Minnesota

Les C. Cwynar
Associate Professor
Department of Biology
University of New Brunswick
Fredericton, New Brunswick

Paul Cypher
Provisional Interpreter
Lake Erie Metropark
Trenton, Michigan

Stanley J. Czyzak
Professor Emeritus
Ohio State University
Columbus, Ohio

Rosi Dagit
Conservation Biologist
Topanga-Las Virgenes Resource
    Conservation District
Topanga, California

David Dalby
President
Bruce Tool Company, Inc.
Taylors, South Carolina

Lou D'Amore
Chemistry Teacher
Father Redmund High School
Toronto, Ontario, Canada

Douglas Darnowski
Postdoctoral Fellow
Department of Plant Biology
Cornell University
Ithaca, New York

Sreela Datta
Associate Writer
Aztec Publications
Northville, Michigan

Sarah K. Dean
Science Writer
Philadelphia, Pennsylvania

Sarah de Forest
Research Assistant
Theoretical Physical Chemistry
  Lab
University of Pittsburgh
Pittsburgh, Pennsylvania

Louise Dickerson
Medical and Science Writer
Greenbelt, Maryland

Marie Doorey
Editorial Assistant
Illinois Masonic Medical Center
Chicago, Illinois

Herndon G. Dowling
Professor Emeritus
Department of Biology
New York University
New York, New York

Marion Dresner
Natural Resources Educator
Berkeley, California

John Henry Dreyfuss
Science Writer
Brooklyn, New York

Roy Dubisch
Professor Emeritus
Department of Mathematics
New York University
New York, New York

Russel Dubisch
Department of Physics
Sienna College
Loudonville, New York

Carolyn Duckworth
Science Writer
Missoula, Montana

Laurie Duncan, Ph.D.
  (Geology)
Geologist
Austin, Texas

Peter A. Ensminger
Research Associate
Cornell University
Syracuse, New York

Bernice Essenfeld
Biology Writer
Warren, New Jersey

Mary Eubanks
Instructor of Biology
The North Carolina School of
  Science and Mathematics
Durham, North Carolina

Kathryn M. C. Evans
Science Writer
Madison, Wisconsin

William G. Fastie
Department of Astronomy and
  Physics
Bloomberg Center
Baltimore, Maryland

Barbara Finkelstein
Science Writer
Riverdale, New York

Mary Finley
Supervisor of Science Curriculum
  (retired)
Pittsburgh Secondary Schools
Clairton, Pennsylvania

Gaston Fischer
Institut de Géologie
Université de Neuchâtel
Peseux, Switzerland

Sara G. B. Fishman
Professor
Quinsigamond Community
  College
Worcester, Massachusetts

David Fontes
Senior Instructor
Lloyd Center for Environmental
  Studies
Westport, Maryland

Barry Wayne Fox
Extension Specialist,
  Marine/Aquatic Education
Virginia State University
Petersburg, Virginia

Ed Fox
Charlotte Latin School
Charlotte, North Carolina

Kenneth L. Frazier
Science Teacher (retired)
North Olmstead High School
North Olmstead, Ohio

Bill Freedman
Professor
Department of Biology and
  School for Resource and
  Environmental Studies
Dalhousie University
Halifax, Nova Scotia

T. A. Freeman
Consulting Archaeologist
Quail Valley, California

Elaine Friebele
Science Writer
Cheverly, Maryland

Randall Frost
Documentation Engineering
Pleasanton, California

Agnes Galambosi, M.S.
Climatologist
Eotvos Lorand University
Budapest, Hungary

Robert Gardner
Science Education Consultant
North Eastham, Massachusetts

Gretchen M. Gillis
Senior Geologist
Maxus Exploration
Dallas, Texas

Larry Gilman, Ph.D. (Electrical
  Engineering)
Engineer
Sharon, Vermont

Kathryn Glynn
Audiologist
Portland, Oregon

David Goings, Ph.D. (Geology)
Geologist
Las Vegas, Nevada

Natalie Goldstein
Educational Environmental
  Writing
Phoenicia, New York

David Gorish
TARDEC
U.S. Army
Warren, Michigan

Louis Gotlib
South Granville High School
Durham, North Carolina

Hans G. Graetzer
Professor
Department of Physics
South Dakota State University
Brookings, South Dakota

Jim Guinn
Assistant Professor
Department of Physics
Berea College
Berea, Kentucky

Steve Gutterman
Psychology Research Assistant
University of Michigan
Ann Arbor, Michigan

Johanna Haaxma-Jurek
Educator
Nataki Tabibah Schoolhouse of
   Detroit
Detroit, Michigan

Monica H. Halka
Research Associate
Department of Physics and
   Astronomy
University of Tennessee
Knoxville, Tennessee

Brooke Hall, Ph.D.
Professor
Department of Biology
California State University at
   Sacramento
Sacramento, California

Jeffrey C. Hall
Astronomer
Lowell Observatory
Flagstaff, Arizona

C. S. Hammen
Professor Emeritus
Department of Zoology
University of Rhode Island

Lawrence Hammar, Ph.D.
Senior Research Fellow
Institute of Medical Research
Papua, New Guinea

William Haneberg, Ph.D.
   (Geology)
Geologist
Portland, Oregon

Beth Hanson
Editor
The Amicus Journal
Brooklyn, New York

Clay Harris
Associate Professor
Department of Geography and
   Geology
Middle Tennessee State
   University
Murfreesboro, Tennessee

Clinton W. Hatchett
Director Science and Space
   Theater
Pensacola Junior College
Pensacola, Florida

Catherine Hinga Haustein
Associate Professor
Department of Chemistry
Central College
Pella, Iowa

Dean Allen Haycock
Science Writer
Salem, New York

Paul A. Heckert
Professor
Department of Chemistry and
   Physics
Western Carolina University
Cullowhee, North Carolina

Darrel B. Hoff
Department of Physics
Luther College
Calmar, Iowa

Dennis Holley
Science Educator
Shelton, Nebraska

Leonard Darr Holmes
Department of Physical Science
Pembroke State University
Pembroke, North Carolina

Rita Hoots
Instructor of Biology, Anatomy,
   Chemistry
Yuba College
Woodland, California

Selma Hughes
Department of Psychology and
   Special Education
East Texas State University
Mesquite, Texas

Mara W. Cohen Ioannides
Science Writer
Springfield, Missouri

Zafer Iqbal
Allied Signal Inc.
Morristown, New Jersey

Sophie Jakowska
Pathobiologist, Environmental
   Educator
Santo Domingo, Dominican
   Republic

Richard A. Jeryan
Senior Technical Specialist
Ford Motor Company
Dearborn, Michigan

Stephen R. Johnson
Biology Writer
Richmond, Virginia

Kathleen A. Jones
School of Medicine
Southern Illinois University
Carbondale, Illinois

Harold M. Kaplan
Professor
School of Medicine
Southern Illinois University
Carbondale, Illinois

Anthony Kelly
Science Writer
Pittsburgh, Pennsylvania

Amy Kenyon-Campbell
Ecology, Evolution and
   Organismal Biology Program
University of Michigan
Ann Arbor, Michigan

Judson Knight
Science Writer
Knight Agency
Atlanta, Georgia

Eileen M. Korenic
Institute of Optics
University of Rochester
Rochester, New York

Jennifer Kramer
Science Writer
Kearny, New Jersey

Pang-Jen Kung
Los Alamos National Laboratory
Los Alamos, New Mexico

Marc Kusinitz
Assistant Director Media
   Relations
John Hopkins Medical Institution
Towsen, Maryland

Arthur M. Last
Head
Department of Chemistry
University College of the Fraser
   Valley
Abbotsford, British Columbia

Nathan Lavenda
Zoologist
Skokie, Illinios

Jennifer LeBlanc
Environmental Consultant
London, Ontario, Canada

Nicole LeBrasseur, Ph.D.
Associate News Editor
Journal of Cell Biology
New York, New York

Benedict A. Leerburger
Science Writer
Scarsdale, New York

Betsy A. Leonard
Education Facilitator

Reuben H. Fleet Space Theater
   and Science Center
San Diego, California

Adrienne Wilmoth Lerner
Graduate School of Arts &
   Science
Vanderbilt University
Nashville, Tennessee

Lee Wilmoth Lerner
Science Writer
NASA
Kennedy Space Center, Florida

Scott Lewis
Science Writer
Chicago, Illinois

Frank Lewotsky
Aerospace Engineer (retired)
Nipomo, California

Karen Lewotsky
Director of Water Programs
Oregon Environmental Council
Portland, Oregon

Kristin Lewotsky
Editor
Laser Focus World
Nashua, New Hamphire

Stephen K. Lewotsky
Architect
Grants Pass, Oregon

Agnieszka Lichanska, Ph.D.
Department of Microbiology &
   Parasitology
University of Queensland
Brisbane, Australia

Sarah Lee Lippincott
Professor Emeritus
Swarthmore College
Swarthmore, Pennsylvania

Jill Liske, M.Ed.
Wilmington, North Carolina

David Lunney
Research Scientist
Centre de Spectrométrie
   Nucléaire et de Spectrométrie
   de Masse
Orsay, France

Steven MacKenzie
Ecologist
Spring Lake, Michigan

J. R. Maddocks
Consulting Scientist
DeSoto, Texas

Gail B. C. Marsella
Technical Writer
Allentown, Pennsylvania

Karen Marshall
Research Associate
Council of State Governments
   and Centers for Environment
   and Safety
Lexington, Kentucky

Liz Marshall
Science Writer
Columbus, Ohio

James Marti
Research Scientist
Department of Mechanical
   Engineering
University of Minnesota
Minneapolis, Minnesota

Elaine L. Martin
Science Writer
Pensacola, Florida

Lilyan Mastrolla
Professor Emeritus
San Juan Unified School
Sacramento, California

Iain A. McIntyre
Manager
Electro-optic Department
Energy Compression Research
   Corporation
Vista, California

Jennifer L. McGrath
Chemistry Teacher
Northwood High School
Nappanee, Indiana

Margaret Meyers, M.D.
Physician, Medical Writer
Fairhope, Alabama

G. H. Miller
Director
Studies on Smoking
Edinboro, Pennsylvania

J. Gordon Miller
Botanist
Corvallis, Oregon

Kelli Miller
Science Writer
NewScience
Atlanta, Georgia

Christine Miner Minderovic
Nuclear Medicine Technologist
Franklin Medical Consulters
Ann Arbor, Michigan

David Mintzer
Professor Emeritus
Department of Mechanical
    Engineering
Northwestern University
Evanston, Illinois

Christine Molinari
Science Editor
University of Chicago Press
Chicago, Illinois

Frank Mooney
Professor Emeritus
Fingerlake Community College
Canandaigua, New York

Partick Moore
Department of English
University of Arkansas at Little
    Rock
Little Rock, Arkansas

Robbin Moran
Department of Systematic Botany
Institute of Biological Sciences
University of Aarhus
Risskou, Denmark

J. Paul Moulton
Department of Mathematics
Episcopal Academy
Glenside, Pennsylvania

Otto H. Muller
Geology Department

Alfred University
Alfred, New York

Angie Mullig
Publication and Development
University of Pittsburgh Medical
    Center
Trafford, Pennsylvania

David R. Murray
Senior Associate
Sydney University
Sydney, New South Wales,
    Australia

Sutharchana Murugan
Scientist
Three Boehringer Mannheim
    Corp.
Indianapolis, Indiana

Muthena Naseri
Moorpark College
Moorpark, California

David Newton
Science Writer and Educator
Ashland, Oregon

F. C. Nicholson
Science Writer
Lynn, Massachusetts

James O'Connell
Department of Physical Sciences
Frederick Community College
Gaithersburg, Maryland

Dúnal P. O'Mathúna
Associate Professor
Mount Carmel College of
    Nursing
Columbus, Ohio

Marjorie Pannell
Managing Editor, Scientific
    Publications
Field Museum of Natural History
Chicago, Illinois

Gordon A. Parker
Lecturer
Department of Natural Sciences
University of Michigan-Dearborn
Dearborn, Michigan

David Petechuk
Science Writer
Ben Avon, Pennsylvania

Borut Peterlin, M.D.
Consultant Clinical Geneticist,
    Neurologist, Head Division of
    Medical Genetics
Department of Obstetrics and
    Gynecology
University Medical Centre
    Ljubljana
Ljubljana, Slovenia

John R. Phillips
Department of Chemistry
Purdue University, Calumet
Hammond, Indiana

Kay Marie Porterfield
Science Writer
Englewood, Colorado

Paul Poskozim
Chair
Department of Chemistry, Earth
    Science and Physics
Northeastern Illinois University
Chicago, Illinois

Andrew Poss
Senior Research Chemist
Allied Signal Inc.
Buffalo, New York

Satyam Priyadarshy
Department of Chemistry
University of Pittsburgh
Pittsburgh, Pennsylvania

Patricia V. Racenis
Science Writer
Livonia, Michigan

Cynthia Twohy Ragni
Atmospheric Scientist
National Center for Atmospheric
    Research
Westminster, Colorado

Jordan P. Richman
Science Writer
Phoenix, Arizona

Kitty Richman
Science Writer
Phoenix, Arizona

Vita Richman
Science Writer
Phoenix, Arizona

Michael G. Roepel
Researcher
Department of Chemistry
University of Pittsburgh
Pittsburgh, Pennsylvania

Perry Romanowski
Science Writer
Chicago, Illinois

Nancy Ross-Flanigan
Science Writer
Belleville, Michigan

Belinda Rowland
Science Writer
Voorheesville, New York

Gordon Rutter
Royal Botanic Gardens
Edinburgh, Great Britain

Elena V. Ryzhov
Polytechnic Institute
Troy, New York

David Sahnow
Associate Research Scientist
John Hopkins University
Baltimore, Maryland

Peter Salmansohn
Educational Consultant
New York State Parks
Cold Spring, New York

Peter K. Schoch
Instructor
Department of Physics and
    Computer Science
Sussex County Community
    College
Augusta, New Jersey

Patricia G. Schroeder
Instructor
Science, Healthcare, and Math
    Division
Johnson County Community
    College
Overland Park, Kansas

Randy Schueller
Science Writer
Chicago, Illinois

Kathleen Scogna
Science Writer
Baltimore, Maryland

William Shapbell Jr.
Launch and Flight Systems
    Manager
Kennedy Space Center
KSC, Florida

Kenneth Shepherd
Science Writer
Wyandotte, Michigan

Anwar Yuna Shiekh
International Centre for
    Theoretical Physics
Trieste, Italy

Raul A. Simon
Chile Departmento de Física
Universidad de Tarapacá
Arica, Chile

Michael G. Slaughter
Science Specialist
Ingham ISD
East Lansing, Michigan

Billy W. Sloope
Professor Emeritus
Department of Physics
Virginia Commonwealth
    University
Richmond, Virginia

Douglas Smith
Science Writer
Milton, Massachusetts

Lesley L. Smith
Department of Physics and
    Astronomy
University of Kansas
Lawrence, Kansas

Kathryn D. Snavely
Policy Analyst, Air Quality Issues
U.S. General Accounting Office
Raleigh, North Carolina

Charles H. Southwick
Professor
Environmental, Population, and
    Organismic Biology
University of Colorado at Boulder
Boulder, Colorado

John Spizzirri
Science Writer
Chicago, Illinois

Frieda A. Stahl
Professor Emeritus
Department of Physics
California State University, Los
    Angeles
Los Angeles, California

Robert L. Stearns
Department of Physics
Vassar College
Poughkeepsie, New York

Ilana Steinhorn
Science Writer
Boalsburg, Pennsylvania

David Stone
Conservation Advisory Services
Gai Soleil
Chemin Des Clyettes
Le Muids, Switzerland

Eric R. Swanson
Associate Professor
Department of Earth and Physical
    Sciences
University of Texas
San Antonio, Texas

Cheryl Taylor
Science Educator
Kailua, Hawaii

Nicholas C. Thomas
Department of Physical Sciences
Auburn University at
    Montgomery
Montgomery, Alabama

W. A. Thomasson
Science and Medical Writer
Oak Park, Illinois

Marie L. Thompson
Science Writer
Ben Avon, Pennsylvania

Laurie Toupin
Science Writer
Pepperell, Massachusetts

Melvin Tracy
Science Educator
Appleton, Wisconsin

Karen Trentelman
Research Associate
Archaeometric Laboratory
University of Toronto
Toronto, Ontario, Canada

Robert K. Tyson
Senior Scientist
W. J. Schafer Assoc.
Jupiter, Florida

James Van Allen
Professor Emeritus
Department of Physics and
    Astronomy
University of Iowa
Iowa City, Iowa

Julia M. Van Denack
Biology Instructor
Silver Lake College
Manitowoc, Wisconsin

Kurt Vandervoort
Department of Chemistry and
    Physics
West Carolina University
Cullowhee, North Carolina

Chester Vander Zee
Naturalist, Science Educator
Volga, South Dakota

Rashmi Venkateswaran
Undergraduate Lab Coordinator
Department of Chemistry
University of Ottawa
Ottawa, Ontario, Canada

R. A. Virkar
Chair
Department of Biological
    Sciences
Kean College
Iselin, New Jersey

Kurt C. Wagner
Instructor
South Carolina Governor's
    School for Science and
    Technology
Hartsville, South Carolina

Cynthia Washam
Science Writer
Jensen Beach, Florida

Terry Watkins
Science Writer
Indianapolis, Indiana

Joseph D. Wassersug
Physician
Boca Raton, Florida

Tom Watson
Environmental Writer
Seattle, Washington

Jeffrey Weld
Instructor, Science Department
    Chair
Pella High School

Pella, Iowa

Frederick R. West
Astronomer
Hanover, Pennsylvania

Glenn Whiteside
Science Writer
Wichita, Kansas

John C. Whitmer
Professor
Department of Chemistry
Western Washington University
Bellingham, Washington

Donald H. Williams
Department of Chemistry
Hope College
Holland, Michigan

Robert L. Wolke
Professor Emeritus
Department of Chemistry
University of Pittsburgh
Pittsburgh, Pennsylvania

Xiaomei Zhu, Ph.D.
Postdoctoral research associate
Immunology Department
Chicago Children's Memorial
    Hospital, Northwestern
    University Medical School
Chicago, Illinois

Jim Zurasky
Optical Physicist
Nichols Research Corporation
Huntsville, Alabama

Aardvark

Aardvarks are nocturnal, secretive, termite- and ant-eating **mammals**, and are one of Africa's strangest animals. Despite superficial appearances, aardvarks are not classified as true **anteaters**; they have no close relatives and are the only living **species** of the order Tubulidentata and family Orycteropodidae. Aardvarks are large piglike animals weighing from 88-143 lb (40-65 kg) and measuring nearly 6 ft (1.8 m) from nose to tip of tail. They have an arched body with a tapering piglike snout at one end and a long tapering tail at the other. Their legs are powerful and equipped with long, strong claws for digging. The

An immature aardvark standing in the grass. *Photograph by Eric & David Hosking. The National Audubon Society Collection/Photo Researchers, Inc. Reproduced by permission.*

first white settlers in South Africa named these peculiar animals aardvarks, which means earth **pigs** in Afrikaans.

Aardvarks are found throughout Africa south of the Sahara Desert. They spend the daylight hours in burrows and forage for food at night. Grunting, shuffling, and occasionally pressing their nose to the ground, aardvarks zigzag about in search of insect **prey**. Fleshy tentacles around the nostrils may be chemical receptors that help locate prey. Their favorite food is **termites**. Using their powerful limbs and claws, aardvarks tear apart concrete-hard termite mounds and lick up the inhabitants with their sticky foot-long tongue. Aardvarks also eat **ants**, locusts, and the fruit of wild gourds. Adapted for eating termites and ants, the teeth of aardvarks are found only in the cheeks, and have almost no enamel or roots.

Female aardvarks bear one offspring per year. A young aardvark weighs approximately 4 lb (2 kg) when born, and is moved to a new burrow by its mother about every eight days. After two weeks the young aardvark accompanies its mother as she forages, and after about six months it can dig its own burrow.

Hyenas, lions, cheetahs, wild dogs, and humans prey on aardvarks. Many Africans regard aardvark meat as a delicacy, and some parts of the **animal** are valued by many tribes for their supposed magical powers. If caught in the open, aardvarks leap and bound away with surprising speed; if cornered, they roll over and lash out with their clawed feet. An aardvark's best defense is digging, which it does with astonishing speed even in sun-baked, rock-hard soil. In fact, aardvarks can penetrate soft earth faster than several men digging frantically with shovels.

Abacus

The abacus is an ancient calculating machine. This simple apparatus is about 5,000 years old and is thought to have originated in Babylon. As the concepts of **zero** and Arabic number notation became widespread, basic math functions became simpler, and the use of the abacus diminished. Most of the world employs adding machines, calculators, and computers for mathematical calculations, but today Japan, China, the Middle East, and Russia still use the abacus, and school children in these countries are often taught to use the abacus. In China, the abacus is called a suan pan, meaning counting tray. In Japan the abacus is called a soroban. The Japanese have yearly examinations and competitions in computations on the soroban.

Before the invention of counting machines, people used their fingers and toes, made marks in mud or sand, put notches in bones and wood, or used stones to count, calculate, and keep track of quantities. The first abaci

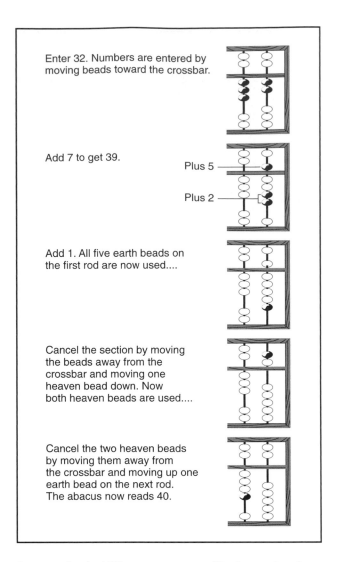

Enter 32. Numbers are entered by moving beads toward the crossbar.

Add 7 to get 39. Plus 5 — Plus 2 —

Add 1. All five earth beads on the first rod are now used....

Cancel the section by moving the beads away from the crossbar and moving one heaven bead down. Now both heaven beads are used....

Cancel the two heaven beads by moving them away from the crossbar and moving up one earth bead on the next rod. The abacus now reads 40.

An example of addition on a *suan pan*. The heaven beads have five times the value of the earth beads below them. *Illustration by Hans & Cassidy. Courtesy of Gale Group.*

were shallow trays filled with a layer of fine sand or dust. Number symbols were marked and erased easily with a finger. Some scientists think that the term *abacus* comes from the Semitic word for dust, *abq*.

A modern abacus is made of wood or plastic. It is rectangular, often about the size of a shoe-box lid. Within the rectangle, there are at least nine vertical rods strung with movable beads. The abacus is based on the decimal system. Each rod represents columns of written numbers. For example, starting from the right and moving left, the first rod represents ones, the second rod represents tens, the third rod represents hundreds, and so forth. A horizontal crossbar is **perpendicular** to the rods, separating the abacus into two unequal parts. The moveable beads are located either above or below the crossbar. Beads above the crossbar are called heaven

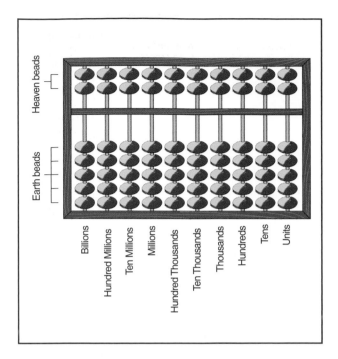

A Chinese abacus called a *suan pan* (reckoning board).
Illustration by Hans & Cassidy. Courtesy of Gale Group.

beads, and beads below are called earth beads. Each heaven bead has a value of five units and each earth bead has a value of one unit. A Chinese suan pan has two heaven and five earth beads, and the Japanese soroban has one heaven and four earth beads. These two abaci are slightly different from one another, but they are manipulated and used in the same manner. The Russian version of the abacus has many horizontal rods with moveable, undivided beads, nine to a column.

To operate, the soroban or suan pan is placed flat, and all the beads are pushed to the outer edges, away from the crossbar. Usually the heaven beads are moved with the forefinger and the earth beads are moved with the thumb. For the number one, one earth bead would be pushed up to the crossbar. Number two would require two earth beads. For number five, only one heaven bead would to be pushed to the crossbar. The number six would require one heaven (five units) plus one earth (one unit) bead. The number 24 would use four earth beads on the first rod and two earth beads on the second rod. The number 26 then, would use one heaven and one earth bead on the first rod, and two earth beads on the second rod. **Addition**, **subtraction**, **multiplication**, and **division** can be performed on an abacus. Advanced abacus users can do lengthy multiplication and division problems, and even find the **square root** or cube root of any number.

See also Arithmetic; Mathematics.

Abrasives

Abrasive materials are hard crystals that are either found in nature or manufactured. The most commonly used of such materials are **aluminum** oxide, silicon carbide, cubic boron nitride, and **diamond**.

Other materials such as garnet, zirconia, **glass**, and even walnut shells are used for special applications.

Abrasives are primarily used in metalworking because their grains can penetrate even the hardest metals and alloys. However, their great hardness also makes them suitable for working with such other hard materials as stones, glass, and certain types of **plastics**. Abrasives are also used with relatively soft materials, including wood and rubber, because their use permits high stock removal, long-lasting cutting ability, good form control, and fine finishing.

Applications for abrasives generally fall in the following categories: 1) cleaning of surfaces and the coarse removal of excess material, such as rough off-hand grinding in foundries; 2) shaping, as in form grinding and tool sharpening; 3) sizing, primarily in precision grinding; and 4) separating, as in cut-off or slicing operations.

TABLE 1. COMMON INDUSTRIAL ABRASIVES	
Abrasive	*Used for*
aluminum oxide	grinding plain and alloyed steel in a soft or hardened condition
silicon carbide	cast iron, nonferrous metals, and nonmetallic materials
diamond	grinding cemented carbides, and for grinding glass, ceramics, and hardened tool steel
cubic boron nitride	grinding hardened steels and wear-resistant superalloys

TABLE 2. MOHS HARDNESSES OF SELECTED MATERIALS

Abrasive	Mohs Hardness
wax (0 deg C)	0.2
graphite	0.5 to 1
talc	1
copper	2.5 to 3
gypsum	2
aluminum	2 to 2.9
gold	2.5 to 3
silver	2.5 to 4
calcite	3
brass	3 to 4
fluorite	4
glass	4.5 to 6.5
asbestos	5
apatite	5
steel	5 to 8.5
cerium oxide	6
orthoclase	6
vitreous silica	7
beryl	7.8
quartz	8
topaz	9
aluminum oxide	9
silicon carbide (beta type)	9.2
boron carbide	9.3
boron	9.5
diamond	10

For the past 100 years or so, manufactured abrasives such as silicon carbide and aluminum oxide have largely replaced natural abrasives-even natural diamonds have nearly been supplanted by synthetic diamonds. The success of manufactured abrasives arises from their superior, controllable properties as well as their dependable uniformity.

Both silicon carbide and aluminum oxide abrasives are very hard and brittle, and as a result they tend to form sharp edges. These edges help the abrasive to penetrate the work material and reduce the amount of **heat** generated during the abrasion. This type of abrasive is used in precision and finish grinding. Tough abrasives, which resist fracture and last longer, are used for rough grinding.

Industry uses abrasives in three basic forms: 1) bonded to form solid tools such as grinding wheels, cylinders, rings, cups, segments, or sticks; 2) coated on backings made of **paper** or cloth in the form of sheets (such as sandpaper), strips or belts; 3) loose, held in some liquid or solid carrier as for polishing or tumbling, or propelled by **force** of air or water **pressure** against a work surface (such as sandblasting for buildings).

How do abrasives work?

Abrasion most frequently results from scratching a surface. As a general rule, a substance is only seriously scratched by a material that is harder than itself. This is the basis for the Mohs scale of hardness (see Table 2) in which materials are ranked according to their ability to scratch materials of lesser hardness.

Abrasives are therefore usually considered to be refractory materials with hardness values ranging from 6 to 10 on the Mohs scale that can be used to reduce, smooth, clean, or polish the surfaces of other, less hard substances such as **metal**, glass, plastic, stone, or wood.

During abrasion, abrasive particles first penetrate the abraded material and then cause a tearing off of particles from the abraded surface. The ease with which the abrasive particles dig into the surface depends on the hardness of the abraded surface; the ease with which the deformed surface is torn off depends on the strength and, in some cases, on the toughness of the material. Between hardness, strength, and toughness, hardness is usually the most important factor determining a material's resistance to abrasion.

When two surfaces move across each other, peaks of microscopic irregularities must either shift position, increase in hardness, or break. If local stresses are sufficiently great, failure of a tiny volume of abraded material will result, and a small particle will be detached. This type of abrasion occurs regardless of whether contact of the two surfaces is due to sliding, rolling, or impact.

Some forms of abrasion involve little or no impact, but in others the **energy** of impact is a deciding factor in determining the effectiveness of the abrasive. Brittle materials, for example, tend to shatter when impacted, and their abrasion may resemble **erosion** more than fracture.

See also Crystal.

Resources

Books

Gao, Yongsheng, ed. *Advances in Abrasive Technology*. 5th ed. Enfield, NH: Trans Tech, 2003.

Gill, Arthur, Steve Krar, and Peter Smid. *Machine Tool Technology Basics*. New York: Industrial Press, 2002.

Green, Robert E., ed. *Machinery's Handbook*. New York: Industrial Press, 1992.

Riggle, Arthur L. *How to Use Diamond Abrasives*. Mentone, CA: Gembooks, 2001.

Randall Frost

Abscess

An abscess is a circumscribed collection of pus usually caused by **microorganisms**. Abscesses can occur anywhere in the body—in hard or soft **tissue**, organs or

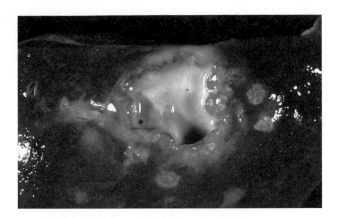

An amoebic abscess caused by *Entameoba histolytica*. Phototake (CN) /Phototake NYC. Reproduced by permission.

confined spaces. Due to their fluid content, abscesses can assume various shapes. Their internal **pressure** can cause compression and displacement of surrounding tissue, resulting in **pain**. An abscess is part of the body's natural defense mechanism; it localizes **infection** to prevent the spread of **bacteria**.

Any trauma such as injury, bacterial or amoebic infection, or **surgery** can result in an abscess. Microorganisms causing an abscess may enter tissue following penetration (e.g., a cut or puncture) with an unsterile object or be spread from an adjacent infection. These microorganisms also are disseminated by the lymph and circulatory systems.

Abscesses are more likely to occur if the urinary, biliary, respiratory, or immune systems have impaired function. A foreign object such as a splinter or stitch can predispose an area to an abscess. The body's inflammatory response mechanism reacts to trauma. The area involved has increased **blood** flow; leukocytes (mostly neutrophils) and exudates (fluid, typically serum and cellular debris) escape from blood vessels at the early stage of **inflammation** and collect in any available space. Neutrophils release enzymes which are thought to help establish the abscess cavity. The exudate attracts water, causing swelling in the affected area. Usually the body removes various exudates with its circulatory and lymphatic systems. When the body's immune response is altered by **disease**, extreme fatigue, or other predisposing factors as mentioned above, resolution of the inflamed area is slow to occur. If the affected area does not heal properly, an abscess can form.

Symptoms of an abscess vary according to location. Fever and pain can be present while dysfunction of an **organ** system sometimes is the symptom. An abscess can rupture and drain to the outside of the body or into surrounding tissue where the fluid and debris can be reabsorbed into the blood stream. Occasionally surgical drainage or **antibiotics** are needed to resolve an abscess.

See also Immune system.

Absolute dating *see* **Dating techniques**

Absolute temperature *see* **Temperature**

Absolute zero

Absolute **zero**, 0 Kelvin, −459.67° Fahrenheit, or −273.15° Celsius, is the minimum possible **temperature**: the state in which all **motion** of the particles in a substance has minimum motion. Equivalently, when the **entropy** of a substance has been reduced to zero, the substance is at absolute zero. Although the third law of **thermodynamics** declares that it is impossible to cool a substance all the way to absolute zero, temperatures of only a few billionths of a degree Kelvin have been achieved in the laboratory in the last few years.

The motions of particles near absolute zero are so slow that their behavior, even in large groups, is governed by quantum-mechanical laws that otherwise tend to be swamped by the chaotic atomic- and molecular-scale motions that are perceive as **heat**. As a result, various special phenomena (e.g., Bose-Einstein condensation, superfluids such as helium II) can only be observed in materials cooled nearly to absolute zero.

Atoms may be cooled by many methods, but **laser** cooling and trapping have proved essential achieving the lowest possible temperatures. A laser beam can cool atoms that are fired in a direction contrary to the beam because when the atoms encounter photons, they absorb them if their **energy** is at a value acceptable to the atom (atoms can only absorb and emit photons of certain energies). If a **photon** is absorbed, its **momentum** is transferred to the atom; if the atom and photon were originally traveling in opposite directions, this slows the atom down, which is equivalent to cooling it.

The third law of thermodynamics, however, dictates that absolute zero can never be achieved. The third states that the entropy of a perfect **crystal** is zero at absolute zero. If the particles comprising a substance are not ordered as a perfect crystal, then their entropy cannot be zero. At any temperature above zero, however, imperfections in the crystal lattice will be present (induced by thermal motion), and to remove them requires compensatory motion, which itself leaves a residue of imperfection. Another way of stating this dilemma is that as the temperature of a substance approaches absolute zero, it becomes increasingly more difficult to remove heat from

the substance while decreasing its entropy. Consequently, absolute zero can be approached but never attained.

When atoms have been cooled to within millionths or billionths of a degree of absolute zero, a number of important phenomena appear, such as the creation of Bose-Einstein condensates, so called because they were predicted in 1924 by German physicist Albert Einstein (1879–1955) and Indian physicist Satyendranath Bose (1894–1974). According to Bose and Einstein, bosons—particles having an **integral** value of the property termed "spin"—are allowed to coexist locally in the same quantum energy state. (Fermions, particles that have half-integer spin values, cannot coexist locally in the same energy state; electrons are fermions, and so cannot share **electron** orbitals in atoms.) At temperatures far above absolute zero, large collections of bosons (e.g., rubidium atoms) are excited by thermal energy to occupy a wide variety of energy states, but near absolute zero, some or all of the bosons will lapse into an identical, low-energy state. A collection of bosons in this condition is a Bose-Einstein condensate. Bose-Einstein condensates were first produced, with the help of laser cooling and trapping, in 1995. Since that time, numerous researchers have produced them and investigated their properties.

A Bose-Einstein condensate can emit "atom lasers," beams of fast-moving atoms analogous to the beams of photons that comprise conventional lasers. Furthermore, the speed of **light** in a Bose-Einstein condensate can be controlled by a laser beam. Researchers have succeeded in reducing the speed of light in a Bose-Einstein condensate to 38 MPH (61 km/h) and even to zero, effectively stopping a pulse of light for approximately a thousandth of a second and then restarting it. This does not contradict the famous statement that nothing can exceed the speed of light *in a vacuum*, i.e., 186,000 MPH [300,000 km/h]. Light is slowed in any transparent medium, such as **water** or **glass**, but its **vacuum** speed remains the limiting speed everywhere in the Universe.

Temperatures near absolute zero permit the study not only of Bose-Einstein condensates, but of large, fragile molecules that cannot exist at higher temperatures, of superfluids, of the orderly arrangements of electrons termed Wigner crystals, and of other phenomena.

See also Atomic theory; Matter; Physics; Quantum mechanics; Subatomic particles.

Resources

Books

Schachtman, Tom. *Absolute Zero and the Conquest of Cold.* New York: Houghton Mifflin, 1999.

KEY TERMS

. . . .

Absolute zero—Absolute zero is the lowest temperature possible. It is equal to 0K (-459°F [-273°C]).

Boson—A type of subatomic particle that has an integral value of spin and obeys the laws of Bose-Einstein statistics.

Fermion—A type of subatomic particle with fractional spin.

Periodicals

Glanz, James, "The Subtle Flirtation of Ultracold Atoms." *Science.* 5361 (April 10, 1998): 200–201.
Seife, Charles, "Laurels for a New Type of Matter." *Science.* 5542 (October 19, 2001): 503.

Larry Gilman

Abyssal plain

Abyssal plains are the vast, flat, sediment-covered areas of the deep **ocean** floor. They are the flattest, most featureless areas on the **Earth**, and have a slope of less than one foot of elevation difference for each thousand feet of **distance**. The lack of features is due to a thick blanket of sediment that covers most of the surface.

These flat abyssal plains occur at depths of over 6,500 ft (1,980 m) below **sea level**. They are underlain by the oceanic crust, which is predominantly basalt—a dark, fine-grained volcanic rock. Typically, the basalt is covered by layers of sediments, much of which is deposited by deep ocean turbidity **currents** (caused by the greater density of sediment-laden **water**), or biological materials, such as minute shells of marine plants and animals, that have "rained" down from the ocean's upper levels, or a mixture of both.

Other components of abyssal plain sediment include wind-blown dust, volcanic ash, chemical precipitates, and occasional meteorite fragments. Abyssal plains are often littered with nodules of manganese containing varying amounts of **iron**, nickel, cobalt, and **copper**. These pea- to potato-sized nodules form by direct precipitation of **minerals** from the seawater onto a bone or rock fragment. Currently, deposits of manganese nodules are not being mined from the sea bed, but it is possible that they could be collected and used in the future.

Of the 15 billion tons of river-carried clay, **sand**, and gravel that is washed into the oceans each year, only a fraction of this amount reaches the abyssal plains. The amount of biological sediments that reaches the bottom is similarly small. Thus, the **rate** of sediment accumulation on the abyssal plains is very slow, and in many areas, less than an inch of sediment accumulates per thousand years. Because of the slow rate of accumulation and the monotony of the topography, abyssal plains were once believed to be a stable, unchanging environment. However, deep ocean currents have been discovered that scour the ocean floor in places. Some currents have damaged trans-oceanic communication cables laid on these plains.

Although they are more common and widespread in the Atlantic and Indian ocean basins than in the Pacific, abyssal plains are found in all major ocean basins. Approximately 40% of our planet's ocean floor is covered by abyssal plains. The remainder of the ocean floor topography consists of hills, cone-shaped or flat-topped **mountains**, deep trenches, and mountain chains such as the mid-oceanic ridge systems.

The abyssal plains do not support a great abundance of aquatic life, though some **species** do survive in this relatively barren environment. Deep sea dredges have collected specimens of unusual-looking **fish**, worms, and clam-like creatures from these depths.

Acceleration

The term acceleration, used in **physics**, is a vector quantity. This means that acceleration contains both a number (its magnitude) and a specific direction. An object is said to be accelerating if its **rate** of change of **velocity** is increasing or decreasing over a period of **time** and/or if its direction of **motion** is changing. The units for acceleration include a **distance** unit and two time units. Examples are m/s^2 and mi/hr/s. Sir Isaac Newton (1642-1727) in his second law of motion defined acceleration as the **ratio** of an unbalanced **force** acting on an object to the **mass** of the object.

History

The study of motion by Galileo Galilei (1564-1642) in the late sixteenth and early seventeenth centuries and by Sir Isaac Newton in the mid-seventeenth century was one of the major cornerstones of modern Western experimental science. Over a period of 20 years, Galileo observed the motions of objects rolling down various inclines and attempted to time these events. He discovered that the distance an object traveled was proportional to the square of the time that it was in motion. From these experiments came the first correct concept of accelerated motion. Newton wanted to know why acceleration occurred. In order to produce a model that would help explain how the known universe of the seventeenth century worked, Newton had to give to science and physics the concept of a force which was mostly unknown at that time. With his second law of motion, he clearly demonstrated that acceleration is caused by an unbalanced force (commonly called a push or a pull) acting on an object. What we call gravity, Newton showed was nothing more than a special type of acceleration. The interaction of the acceleration of gravity on the mass of our body produces the force which is called weight. A general definition of mass is that it refers to the quantity of **matter** in a body.

Linear acceleration

An object that is moving in a straight line is accelerating if its velocity (sometimes incorrectly referred to as speed) is increasing or decreasing during a given period of time. Acceleration (a) can be either positive or **negative** depending on whether the velocity is increasing (+a) or decreasing (-a). An automobile's motion can help explain linear acceleration. The speedometer measures the velocity. If the auto starts from rest and accelerates to 60 MPH in 10 seconds, what is the acceleration? The auto's velocity changed 60 MPH in 10 seconds. Therefore, its acceleration is 60 MPH/10 s = +6 mi/hr/s. That means its acceleration changed six miles per hour every second it was moving. Notice there are one distance unit and two time units in the answer. If the auto had started at 60 MPH and then stopped in 10 seconds after the brakes were applied, the acceleration would be = -6 mi/hr/s. If this **automobile** changes direction while moving at this constant acceleration, it will have a different acceleration because the new vector will be different from the original vector. The **mathematics** of vectors is quite complex.

Circular acceleration

In circular motion, the velocity may remain constant but the direction of motion will change. If our automobile is going down the road at a constant 60 MPH and it goes around a **curve** in the road, the auto undergoes acceleration because its direction is constantly changing while it is in the curve. Roller coasters and other amusement park rides produce rapid changes in acceleration (sometimes called centripetal acceleration) which will cause such effects as "g" forces, "weightlessness" and other real or imaginary forces to act on the body, causing dramatic experiences to occur. Astronauts experience as much as 7

"gs" during lift-off of the **space shuttle** but once in **orbit** it appears that they have lost all their weight. The concept of "weightlessness" in **space** is a highly misunderstood phenomena. It is not caused by the fact that the shuttle is so far from the **Earth**; it is produced because the space shuttle is in free fall under the influence of gravity. The shuttle is traveling 17,400 MPH around Earth and it is continually falling toward Earth, but the Earth falls away from the shuttle at exactly the same rate.

Force and acceleration

Before the time of Sir Isaac Newton, the concept of force was unknown. Newton's second law was a simple equation and an insight that significantly affected physics in the seventeenth century as well as today. In the second law, given any object of mass (m), the acceleration (a) given to that object is directly proportional to the net force (F) acting on the object and inversely proportional to the mass of the object. Symbolically, this means a =

F/m or in its more familiar form F = ma. In order for acceleration to occur, a net force must act on an object.

See also Accelerators; Gravity and gravitation; Laws of motion; Velocity.

Resources

Books

Cohen, I. Bernard. *Introduction to Newton's Principia.* Lincoln, NE: iUniverse, 1999.

Galilei, Galileo. *Dialogues Concerning Two New Sciences.* Translated by H. Crew and A. DiSalvo. Glendale, CA: Prometheus Books, 1991.

Goldstein, Herbert, Charles P. Poole, and John L. Safko. *Classical Mechanics.* 3rd ed. New York, Prentice Hall, 2002.

Hewitt, Paul. *Conceptual Physics.* Englewood Cliffs, NJ: Prentice Hall, 2001.

Meriam, J.L., and L.G. Kraige. *Engineering Mechanics, Dynamics.* 5th ed. New York: John Wiley & Sons, 2002.

Methods of Motion: An Introduction to Mechanics. Washington, DC: National Science Teachers Association, 1992.

Serway, Raymond, Jerry S. Faughn, and Clement J. Moses. *College Physics.* 6th ed. Pacific Grove, CA: Brooks/Cole, 2002.

Kenneth L. Frazier

Accelerators

The term accelerators most commonly refers to particle accelerators, devices for increasing the **velocity** of **subatomic particles** such as protons, electrons, and positrons. Particle accelerators were originally invented for the purpose of studying the basic structure of **matter**, although they later found a number of practical applications. Particle accelerators can be subdivided into two large sub-groups: linear and circular accelerators. Machines of the first type accelerate particles as they travel in a straight line, sometimes over very great distances. Circular accelerators move particles along a circular or **spiral** path in machines that vary in size from less than a few feet to many miles in diameter.

The simplest particle accelerator was invented by Alabama-born physicist Robert Jemison Van de Graaff (1901-1967) in about 1929. The machine that now bears his name illustrates the fundamental principles on which all particle accelerators are based.

In the Van de Graaff accelerator, a silk conveyor belt collects positive charges from a high-voltage source at one end of the belt and transfers those charges to the outside of a hollow dome at the other end of the belt located at the top of the machine. The original Van de Graaff accelerator operated at a potential difference of 80,000

volts, although later improvements raised that value to 5,000,000 volts.

The Van de Graaff accelerator can be converted to a particle accelerator by attaching a source of positively charged ions, such as protons or He+ ions, to the hollow dome. These ions feel an increasingly strong **force** of repulsion as positive charges accumulate on the dome. At some point, the ions are released from their source, and they travel away from the dome with high **energy** and at high velocities. If this beam of rapidly-moving particles is directed at a target, the ions of which it consists may collide with **atoms** in the target and break them apart. An analysis of ion-atom collisions such as these can provide a great deal of information about the structure of the target atoms, about the ion "bullets" and about the nature of matter in general.

Linear accelerators

In a Van de Graaff **generator**, the velocity of an electrically charged particle is increased by exposing that particle to an electric field. The velocity of a **proton**, for example, may go from **zero** to 100,000 mi per second (160,000 km per second) as the particle feels a strong force of repulsion from the positive charge on the generator dome. Linear accelerators (linacs) operate on the same general principle except that a particle is exposed to a series of electrical fields, each of which increases the velocity of the particle.

A typical linac consists of a few hundred or a few thousand cylindrical **metal** tubes arranged one in front of another. The tubes are electrically charged so that each carries a charge opposite that of the tube on either side of it. Tubes 1, 3, 5, 7, 9, etc., might, for example, be charged positively, and tubes 2, 4, 6, 7, 10, etc., charged negatively.

Imagine that a negatively charged **electron** is introduced into a linac just in front of the first tube. In the circumstances described above, the electron is attracted by and accelerated toward the first tube. The electron passes toward and then into that tube. Once inside the tube, the electron no longer feels any force of attraction or repulsion and merely drifts through the tube until it reaches the opposite end. It is because of this behavior that the cylindrical tubes in a linac are generally referred to as drift tubes.

At the moment that the electron leaves the first drift tube, the charge on all drift tubes is reversed. Plates 1, 3, 5, 7, 9, etc. are now negatively charged, and plates 2, 4, 6, 8, 10, etc. are positively charged. The electron exiting the first tube now finds itself repelled by the tube it has just left and attracted to the second tube. These forces of attraction and repulsion provide a kind of "kick" that accelerates the electron in a forward direction. It passes through the space between tubes 1 and 2 and into tube 2.

The End Station A experimental hall at the Stanford Linear Accelerator Center (SLAC) in California contains three giant particle spectrometers that detect particles of various energies and angles of scatter. The particles are created when electrons from SLAC's 1.8-mi (3-km) long linear accelerator collide with a target in front of the spectrometers. The large spectrometer dominating the picture is about 98 ft (30 m) long and weighs 550 tons (500 metric tons); a man below it can be used for size comparison. A smaller, circular spectrometer is to its left, and the third, even larger, is mostly hidden by the central one. Experiments at End Station A in 1968-72 confirmed the existence of quarks. *Photograph by David Parler. National Audubon Society Collection/Photo Researchers Inc. Reproduced by permission.*

Once again, the electron drifts through this tube until it exits at the opposite end.

The electrical charge on all drift tubes reverses, and the electron is repelled by the second tube and attracted to the third tube. The added energy it receives is manifested in a greater velocity. As a result, the electron is moving faster in the third tube than in the second and can cover a greater **distance** in the same amount of **time**. To make sure that the electron exits a tube at just the right moment, the tubes must be of different lengths. Each one is slightly longer than the one before it.

The largest linac in the world is the Stanford Linear Accelerator, located at the Stanford Linear Accelerator Center (SLAC) in Stanford, California. An underground tunnel 2 mi (3 km) in length passes beneath U.S. highway 101 and holds 82,650 drift tubes along with the magnetic, electrical, and auxiliary equipment needed for the machine's operation. Electrons accelerated in the SLAC linac leave the end of the machine traveling at nearly the speed of **light** with a maximum energy of about 32 GeV (gigaelectron volts).

The term electron volt (ev) is the standard unit of energy measurement in accelerators. It is defined as the energy lost or gained by an electron as it passes through a potential difference of one volt. Most accelerators operate in the megaelectron volt (million electron volt; MeV), gigaelectron volt (billion electron volt; GeV), or teraelectron volt (trillion electron volt; TeV) range.

This particle-beam fusion accelerator can direct 36 beams of charged atomic particles at a single target simultaneously. Scientists use this technology to study the structure of matter. © *Alexander Tsiaras, National Audubon Society Collection/Photo Researchers, Inc. Reproduced with permission.*

Circular accelerators

The development of linear accelerators is limited by some obvious physical constraints. For example, the SLAC linac is so long that engineers had to take into consideration the Earth's curvature when they laid out the drift tube sequence. One way of avoiding the problems associated with the construction of a linac is to accelerate particles in a circle. Machines that operate on this principle are known, in general, as circular accelerators.

The earliest circular accelerator, the **cyclotron**, was invented by University of California professor of **physics** Ernest Orlando Lawrence in the early 1930s. Lawrence's cyclotron added to the design of the linac one new fundamental principle from physics: a charged particle that passes through a magnetic field travels in a curved path. The shape of the curved path depends on the velocity of the particle and the strength of the magnetic field.

The cyclotron consists of two hollow metal containers that look as if a tuna fish can had been cut in half vertically. Each half resembles a uppercase letter D, so the two parts of the cyclotron are known as dees. At any one time, one dee in the cyclotron is charged positively and the other negatively. But the dees are connected to a source of alternating current so that the signs on both dees change back and forth many times per second.

The second major component of a cyclotron is a large magnet that is situated above and below the dees. The presence of the magnet means that any charged particles moving within the dees will travel not in straight paths, but in curves.

Imagine that an electron is introduced into the narrow space between the two dees. The electron is accelerated into one of the dees, the one carrying a positive charge. As it moves, however, the electron travels toward the dee in a curved path.

After a fraction of a second, the current in the dees changes signs. The electron is then repelled by the dee toward which it first moved, reverses direction, and heads toward the opposite dee with an increased velocity. Again, the electron's return path is curved because of the magnetic field surrounding the dees.

Just as a particle in a linac passes through one drift tube after another, always gaining energy, so does a particle in a cyclotron. As the particle gains energy, it picks up speed and spirals outward from the center of the machine. Eventually, the particle reaches the outer circumference of the machine, passes out through a window, and strikes a target.

Lawrence's original cyclotron was a modest piece of equipment—only 4.5 in (11 cm) in diameter—capable of accelerating protons to an energy of 80,000 electron volts (80 kiloelectron volts). It was assembled from coffee cans, sealing wax, and leftover laboratory equipment. The largest accelerators of this design ever built were the 86 in (218 cm) and 87 in (225 cm) cyclotrons at the Oak Ridge National Laboratory and the Nobel Institute in Stockholm, Sweden, respectively.

Cyclotron modifications

At first, improvements in cyclotron design were directed at the construction of larger machines that could accelerate particles to greater velocities. Soon, however, a new problem arose. Physical laws state that nothing can travel faster than the speed of light. Thus, adding more and more energy to a particle will not make that particle's speed increase indefinitely. Instead, as the particle's velocity approaches the speed of light, additional energy supplied to it appears in the form of increased **mass**. A particle whose mass is constantly increasing, however, begins to travel in a path different from that of a particle with constant mass. The practical significance of this fact is that, as the velocity of particles in a cyclotron begins to approach the speed of light, those particles start to fall "out of sync" with the current change that drives them back and forth between dees.

Two different modifications-or a combination of the two-can be made in the basic cyclotron design to deal with this problem. One approach is to gradually change the **rate** at which the electrical field alternates between the dees. The goal here is to have the sign change occur at exactly the moment that particles have reached a certain point within the dees. As the particles speed up and gain weight, the rate at which electrical current alternates between the two dees slows down to "catch up" with the particles.

In the 1950s, a number of machines containing this design element were built in various countries. Those machines were known as **frequency** modulated (FM) cyclotrons, synchrocyclotrons, or, in the Soviet Union, phasotrons. The maximum particle energy attained with machines of this design ranged from about 100 MeV to about 1 GeV.

A second solution for the mass increase problem is to alter the magnetic field of the machine in such a way as to maintain precise control over the particles' paths.

This principle has been incorporated into the machines that are now the most powerful cyclotrons in the world, the synchrotrons.

A synchrotron consists essentially of a hollow circular tube (the ring) through which particles are accelerated. The particles are actually accelerated to velocities close to the speed of light in smaller machines before they are injected into the main ring. Once they are within the main ring, particles receive additional jolts of energy from accelerating chambers placed at various locations around the ring. At other locations around the ring, very strong magnets control the path followed by the particles. As particles pick up energy and tend to spiral outward, the magnetic fields are increased, pushing particles back into a circular path. The most powerful synchrotrons now in operation can produce particles with energies of at least 400 GeV.

In the 1970s, nuclear physicists proposed the design and construction of the most powerful synchrotron of all, the superconducting super collider (SSC). The SSC was expected to have an accelerating ring 51 mi (82.9 km) in circumference with the ability to produce particles having an energy of 20 TeV. Estimated cost of the SSC was originally set at about $4 billion. Shortly after construction of the machine at Waxahachie, Texas began, however, the United States congress decided to discontinue funding for the project.

Applications

By far the most common use of particle accelerators is basic research on the composition of matter. The quantities of energy released in such machines are unmatched anywhere on **Earth**. At these energy levels, new forms of matter are produced that do not exist under ordinary conditions. These forms of matter provide clues about the ultimate structure of matter.

Accelerators have also found some important applications in medical and industrial settings. As particles travel through an accelerator, they give off a form of **radiation** known as synchrotron radiation. This form of radiation is somewhat similar to **x rays** and has been used for similar purposes.

Resources

Books

Gribbin, John. *Q is for Quantum: An Encyclopedia of Particle Physics.* New York: The Free Press, 1998.
Livingston, M. Stanley, and John P. Blewett. *Particle Accelerators.* New York: McGraw-Hill, 1962.
Newton, David E. *Particle Accelerators: From the Cyclotron to the Superconducting Super Collider.* New York: Franklin Watts, 1989.

KEY TERMS

Electron—A fundamental particle of matter carrying a single unit of negative electrical charge.

Ion—An atom or molecule which has acquired electrical charge by either losing electrons (positively charged ion) or gaining electrons (negatively charged ion).

Positron—A positively charged electron.

Potential difference—The work that must be done to move a unit charge between two points.

Proton—A fundamental particle matter carrying a single unit of positive electrical charge.

Wilson, E.J. N. *An Introduction to Particle Accelerators*. Oxford:Oxford University Press, 2001.

Periodicals

Glashow, Sheldon L., and Leon M. Lederman, "The SSC: A Machine for the Nineties." *Physics Today* (March 1985): 28-37.

LaPorta, A. "Fluid Particle Accelerations in Fully Developed Turbulence." *Nature* 409, no. 6823 (2001):1017-1019.

"Particle Acceleration and Kinematics in Solar Flares." *Space Science Reviews* 101, nos. 1-2 (2002): 1-227.

Winick, Herman, "Synchrotron Radiation." *Scientific American* (November 1987): 88-99.

David E. Newton

Accretion disk

An accretion disk is an astronomical term that refers to the rapidly spiraling **matter** that is in the process of falling into an astronomical object. In principle, any **star** could have an accretion disk, but in practice, accretion disks are often associated with highly collapsed stars such as black holes or **neutron** stars.

The matter that serves as the base of the accretion disk can be obtained when a star passes through a region where the **interstellar matter** is thicker than normal. Normally, however, a star gets an accretion disk from a companion star. When two stars **orbit** each other, there is an invisible figure eight around the two stars, called the *Roche lobes*. The Roche lobes represent all the points in **space** where the gravitational potential from each star is equal. Therefore any matter on the Roche lobes could just as easily fall into either star. If one star in a binary system becomes larger than the Roche lobes, matter will fall from it onto the other star, forming an accretion disk.

The matter falling into a collapsing star hole tends to form a disk because a spherical **mass** of gas that is spinning will tend to flatten out. The faster it is spinning, the flatter it gets. So, if the falling material is orbiting the central mass, the spinning flattens the matter into an accretion disk.

Black holes are objects that have collapsed to the point that nothing, not even **light**, can escape the incredible **force** of their gravity. Because no light can escape, however, there is no way to directly observe it. However, if the **black hole** has an accretion disk, we can observe the black hole indirectly by observing the accretion disk, which will emit **x rays**. Without accretion disks there would be little hope of astronomers ever observing black hole.

Accretion disks can also occur with a **white dwarf** in a binary system. A white dwarf is a collapsed star that is the final stage in the evolution of stars similar to the **Sun**. White dwarfs contain as much mass as the Sun, compressed to about the size of **Earth**. Normally the nuclear reactions in a white dwarf have run out of fuel, but the **hydrogen** from the accretion disk falling onto a white dwarf fuels additional nuclear reactions. White dwarfs have some unusual properties that do not allow them to expand slowly to release the **heat pressure** generated by these nuclear reactions. This heat pressure therefore builds up until the surface of the whited dwarf explodes. This type of explosion is called a **nova** (not to be confused with a **supernova**), and typically releases as much **energy** in the form of protons in less than a year as the Sun does in 100,000 years.

Accuracy

Accuracy is how close an experimental reading or calculation is to the true value. Lack of accuracy may be due to **error** or due to **approximation**. The less total error in an experiment or calculation, the more accurate the results. Error analysis can provide information about the accuracy of a result.

Accuracy in measurements

Errors in experiments stem from incorrect design, inexact equipment, and approximations in measurement. Imperfections in equipment are a fact of life, and sometimes design imperfections are unavoidable as well. Approximation is also unavoidable and depends on the fineness and correctness of the measuring equipment. For example, if the finest marks on a ruler are in centimeters, then the final measurement is not likely to have an accuracy of more than half a centimeter. But if the ruler includes millimeter markings, then the measurements can be an order of magnitude more accurate.

Accuracy differs from precision. Precision deals with how repeatable measurements are—if multiple measurements return numbers that are close to each other, then the experimental results are precise. The results may be far from accurate, but they will be precise.

In calculations

Approximations are also unavoidable in calculations. Neither people nor computers can provide a totally accurate number for 1/3 or **pi** or any of several other numbers, and often the desired accuracy does not require it. The person calculating does, however make sure that the approximations are sufficiently small that they do not endanger the useful accuracy of the result. Accuracy becomes an issue in computations because rounding errors accumulate, series expansions are attenuated, and other methods that are not analytical tend to include errors.

Rounding

If you buy several items, all of which are subject to a sales tax, then you can calculate the total tax by summing the tax on each item. However, for the total tax to be accurate to the penny, you must do all the calculations to an accuracy of tenths of a penny (in other words, to three significant digits), then round the sum to the nearest penny (two significant digits). If you calculated only to the penny, then each measurement might be off by as much as half a penny ($0.005) and the total possible error would be this amount multiplied by the number of items bought. If you bought three items, then the error could be large as 1.5 cents; if you bought 10 items then your total tax could be off by as much as five cents.

As another example, if you want to know the value of pi to an accuracy of two decimal places, then you could express it as 3.14. This could also be expressed as 3.14 +/- 0.005 since any number from 3.135 to 3.145 could be expressed the same way—to two significant decimal points. Any calculations using a number accurate to two decimal places are only accurate to one decimal place. In a similar example, the accuracy of a table can either refer to the number of significant digits of the numbers in a table or the number of significant digits in computations made from the table.

Acetic acid

Acetic acid is an organic acid with the chemical formula CH_3COOH. It is found most commonly in vinegar.

In the form of vinegar, acetic acid is one of the earliest chemical compounds known to and used by humans. It is mentioned in the Bible as a condiment and was used even earlier in the manufacture of white **lead** and the extraction of mercury **metal** from its ores. The first reasonably precise chemical description of the acid was provided by the German natural philosopher Johann Rudolf Glauber in about 1648.

Acetic acid is a colorless liquid with a sharp, distinctive odor and the characteristic **taste** associated with vinegar. In its pure form it is referred to as *glacial* acetic acid because of its tendency to crystallize as it is cooled. Glacial acetic acid has a melting point of 62°F (16.7°C) and a **boiling point** of 244.4°F (118°C). The acid mixes readily with **water**, ethyl **alcohol**, and many other liquids. Its water solutions display typical acid behaviors such as **neutralization** of oxides and bases and reactions with carbonates. Glacial acetic acid is an extremely caustic substance with a tendency to **burn** the skin. This tendency is utilized by the medical profession for wart removal.

Originally acetic acid was manufactured from pyroligneous acid which, in turn, was obtained from the destructive **distillation** of wood. Today the compound is produced commercially by the oxidation of butane, ethylene, or methanol (wood alcohol). Acetic acid forms naturally during the **aerobic fermentation** of sugar or alcoholic solutions such as beer, cider, fruit juice, and wine. This process is catalyzed by the bacterium *Acetobacter*, a process from which the **species** gets its name.

Although acetic acid is best known to the average person in the form of vinegar, its primary commercial use is in the production of **cellulose** acetate, vinyl acetate, and terephthalic acid. The first of these compounds is widely used as a rubber substitute and in photographic and cinematic film, while the latter two compounds are starting points for the production of polymers such as adhesives, latex paints, and plastic film and sheeting.

A promising new use for acetic acid is in the manufacture of calcium-magnesium acetate (CMA), a highly effective and biodegradable deicer. CMA has had limited use in the past because it is 50 times more expensive than **salt**. In 1992, however, Shang-Tian Yang, an engineer at Ohio State University, announced a new method for making acetic acid from wastes produced during cheese making.

Most commonly vinegar is prepared commercially by the fermentation of apple cider, malt, or **barley**. The fermentation product is a brownish or yellow liquid consisting of 4-8% acetic acid. It is then distilled to produce a clear colorless liquid known as *white vinegar*.

Acetone

Acetone is a colorless, flammable, and volatile liquid with a characteristic odor that can be detected at very low concentrations. It is used in consumer goods such as nail polish remover, model airplane glue, lacquers, and paints. Industrially, it is used mainly as a solvent and an ingredient to make other chemicals.

Acetone is the common name for the simplest of the ketones. The formula of acetone is CH_3COCH_3.

The International Union of Pure and Applied Chemistry's (IUPAC) systematic name for acetone is 2-propanone; it is also called dimethyl ketone. The **molecular weight** is 58.08. Its **boiling point** is 133°F (56°C) and the melting point is -139.63°F (-95.4°C). The specific gravity is 0.7899.

Acetone is the simplest and most important of the ketones. It is a polar organic solvent and therefore dissolves a wide variety of substances. It has low chemical reactivity. These traits, and its relatively low cost, make it the solvent of choice for many processes. About 25% of the acetone produced is used directly as a solvent.

About 20% is used in the manufacture of methyl methacrylate to make **plastics** such as acrylic plastic, which can be used in place of **glass**. Another 20% is used to manufacture methyl isobutyl ketone, which serves as a solvent in surface coatings. Acetone is important in the manufacture of **artificial fibers**, **explosives**, and polycarbonate **resins**.

Because of its importance as a solvent and as a starting material for so many chemical processes, acetone is produced in the United States in great quantities. Acetone was 42nd in industrial volume in 1993 when 2.46 billion lb (1 billion kg) were produced. Today, acetone is available at low cost and high purity to laboratories, so it is rarely synthesized outside of industry.

Acetone is normally present in low concentrations in human **blood** and urine. Diabetic patients produce it in larger amounts. Sometimes "acetone breath" is detected on the breath of diabetics by others and wrongly attributed to the drinking of liquor. If acetone is splashed in the eyes, irritation or damage to the cornea will result. Excessive breathing of fumes causes headache, weariness, and irritation of the nose and throat. Drying results from contact with the skin.

Acetylcholine

Acetylcholine is a highly active **neurotransmitter** acting as a chemical connection between nerves (neurons). Acetylcholine diffuses across the narrow gap between nerve cells, known as the **synapse** and thus, plays an important role in connecting nerves to each other.

By the early 1900s, scientists had a reasonably clear idea of the **anatomy** of the **nervous system**. They knew that individual nerve cells—neurons—formed the basis of that system. They also knew that nerve messages traveled in the form of minute electrical signals along the length of a **neuron** and then passed from the axon of one **cell** to the dendrites of a nearby cell.

One major problem remained, however, to understand the mechanism by which the nerve message travels across the narrow gap—the synapse—between two adjacent neurons. The British neurologist, Thomas R. Elliott (1877–1961), suggested in 1903 that the nerve message is carried from one cell to another by means of a chemical compound. Elliott assumed that adrenalin might be this chemical messenger or, neurotransmitter, as it is known today.

Nearly two decades passed before evidence relating to Elliott's hypothesis was obtained. Then, in 1921, the German-American pharmacologist, Otto Loewi (1873–1961), devised a method for testing the idea. Born in Frankfurt-am-Main, Germany, in 1873, Loewi received his medical degree from the University of Strasbourg in 1896 and then taught and did research in London, England, Vienna, Austria, and Graz, Austria. With the rise of Adolf Hitler (1889–1945), Loewi left Germany first for England and then, in 1940, the United States where he became a faculty member at the New York University College of Medicine.

In his 1921 experiment, Loewi found that when he stimulated the nerves attached to a frog's **heart**, they secreted at least two chemical substances. One substance he thought was adrenalin, while the second he named vagusstoffe, after the vagus nerve in the heart.

Soon news of Loewi's discovery reached other scientists in the field, among them the English physiologist Henry Dale (1875–1968). Dale earned a medical degree from Cambridge in 1909. After a short academic career at St. Bartholomew's Hospital in London and at University College, London, Dale joined the Physiological Research Laboratories at the pharmaceutical firm of Burroughs Wellcome. Except for the war years, Dale remained at Burroughs Wellcome until 1960. He died in Cambridge on July 23, 1968.

While attending a conference in Heidelberg, Germany, in 1907, Dale became interested in the fungus ergot and the chemicals it secretes. By 1914, Dale had isolated a compound from ergot that produces effects on organs similar to those produced by nerves. He called the compound *acetylcholine*. When Dale heard of Loewi's discovery of vagusstoffe seven years later, he suggested that it was identical to the acetylcholine he had discovered earlier. For their discoveries, Loewi and Dale shared the 1936 Nobel Prize for **physiology** or medicine.

Unraveling the exact mechanism by which acetylcholine carries messages across the synapse has occupied the energies of countless neurologists since the Loewi-Dale discovery. Some of the most important work has been done by the Australian physiologist, John Carew Eccles (1903–1997), and the German-British physiologist, Bernard Katz (1911-). Eccles developed a method for inserting microelectrodes into adjacent cells and then studying the chemical and physical changes that occur when a neurotransmitter passes through the synapse. Katz discovered that neurotransmitters like acetylcholine are released in tiny packages of a few thousand molecules each. He also characterized the release of these packages in resting and active neurons. For their work on neurotransmitters, Eccles and Katz each received a Nobel Prize for physiology or medicine in 1963 and 1970, respectively.

The biochemical action of acetylcholine is now well understood. Depending on its **concentration**, it exerts two different physiological effects. Injection of small amounts into a human patient produces a fall in **blood pressure** (due to the dilation of blood vessels, or vasodilation), slowing of the heartbeat, increased contraction of smooth muscle in many organs and copious secretion from **exocrine glands**. These effects are collectively known as the "muscarinic effects" of acetylcholine, as they parallel the physiological effects of the mushroom amanita toxin, Muscarin. The rise in acetylcholine following atropine administration causes a rise in blood pressure similar to that produced by **nicotine**. This effect is therefore known as the "nicotinic effect" of acetylcholine.

See also Nerve impulses and conduction of impulses; Neuroscience

Acetylene *see* **Hydrocarbon**

Acetylsalicylic acid

Acetylsalicylic acid, commonly known as aspirin, is the most popular therapeutic drug in the world. It is an analgesic (pain-killing), antipyretic (fever-reducing), and anti-inflammatory sold without a prescription as tablets, capsules, powders, or suppositories. The drug reduces **pain** and fever, is believed to decrease the risk of **heart** attacks and strokes, and may deter colon **cancer** and help prevent premature **birth**. Often called the wonder drug, aspirin can have serious side effects, and its use results in more accidental poisoning deaths in children under five years of age than any other drug.

History

In the mid- to late-1700s, English clergyman Edward Stone chewed on a piece of willow **bark** and discovered its analgesic property after **hearing** a story that declared a brew from the bark was "good for pain and whatever else ails you." The bark's active ingredient was isolated in 1827 and named salicin for the Greek word *salix*, meaning willow. Salicylic acid, first produced from salicin in 1838 and synthetically from phenol in 1860, was effective in treating **rheumatic fever** and gout but caused severe nausea and intestinal discomfort. In 1898, a chemist named Hoffmann, working at Bayer Laboratories in Germany and whose father suffered from severe rheumatoid **arthritis**, synthesized acetylsalicylic acid in a successful attempt to eliminate the side effects of salicylic acid, which, until then, was the only drug that eased his father's pain. Soon the process for making large quantities of acetylsalicylic acid was patented, and aspirin—named for its ingredients acetyl and spiralic (salicylic) acid—became available by prescription. Its popularity was immediate and worldwide. Huge demand in the United States brought manufacture of aspirin to that country in 1915 when it also became available without a prescription.

Mechanism of action

Analgesic/anti-inflammatory action

Aspirin's recommended therapeutic adult dosage ranges from 600-1,000 mg and works best against "tolerable" pain; extreme pain is virtually unaffected, as is pain in internal organs. Aspirin inhibits (blocks) production of **hormones** (chemical substances formed by the body) called prostaglandins that may be released by an injured **cell**, triggering release of two other hormones that sensitize nerves to pain. The blocking action prevents this response and is believed to work in a similar way to prevent **tissue inflammation**. Remarkably, aspirin only acts on cells producing prostaglandins—for instance, injured cells. Its effect lasts approximately four hours.

Antipyretic action

This action is believed to occur at the anterior (frontal) hypothalamus, a portion of the **brain** that regulates such functions as heart **rate** and body **temperature**.

The body naturally reduces its **heat** through perspiration and the dilation (expansion) of **blood** vessels. Prostaglandins released in the hypothalamus inhibit the body's natural heat-reducing mechanism. As aspirin blocks these prostaglandins, the hypothalamus is free to regulate body temperature. Aspirin lowers abnormally high body temperatures while normal body temperature remains unaffected.

Blood-thinning action

One prostaglandin, thromboxane A_2, aids platelet aggregation (accumulation of blood cells). Because aspirin inhibits thromboxane production, thus "thinning the blood," it is frequently prescribed in low doses over long periods for at-risk patients to help prevent heart attacks and strokes.

Adverse affects

Poisoning

Aspirin's availability and presence in many prescription and non-prescription medications makes the risk of accidental overdose relatively high. Children and the elderly are particularly susceptible, as their toxicity thresholds are much lower than adults. About 10% of all accidental or suicidal episodes reported by hospitals are related to aspirin.

Bleeding

As aspirin slows down platelet accumulation, its use increases risk of bleeding, a particular concern during **surgery** and childbirth. Aspirin's irritant effect on the stomach lining may cause internal bleeding, sometimes resulting in **anemia**.

Reye syndrome

Reye syndrome is an extremely rare **disease**, primarily striking children between the ages of three and 15 years after they have been treated with aspirin for a viral **infection**. Reye **syndrome** manifests as severe vomiting, seizures, disorientation, and sometimes **coma**, which can result in permanent brain damage or death. The cause of Reye is unknown, but the onset strongly correlates to the treatment of viral infections with aspirin, and incidents of Reye in children on aspirin therapy for chronic arthritis is significant. In 1985, these observations were widely publicized and warning labels placed on all aspirin medications, resulting in a decline in the number of children with viruses being treated with aspirin and a corresponding decline in cases of Reye's syndrome.

KEY TERMS

. . . .

Analgesic—A compound that relieves pain without loss of consciousness.

Antipyretic—Anything that reduces fever.

Hypothalamus—A small area near the base of the brain where release of hormones influence such involuntary bodily functions as temperature, sexual behavior, sweating, heart rate, and moods.

Placenta—An organ that develops in the uterus during pregnancy to which the fetus is connected by the umbilical cord and through which the fetus receives nourishment and eliminates waste.

Platelets—Irregularly shaped disks found in the blood of mammals that aid in clotting the blood.

Prostaglandins—Groups of hormones and active substances produced by body tissue that regulate important bodily functions, such as blood pressure.

Suppository—Medication placed in a body cavity, usually the vagina or rectum, that melts and is absorbed by the body.

Other adverse affects

Aspirin can adversely affect breathing in people with sinusitis or **asthma**, and long-term use may cause kidney cancer or liver disease. There is some evidence that it delays the onset of labor in full-term pregnancies and, as it crosses the placenta, may be harmful to the fetus.

See also Analgesia; Anti-inflammatory agents; Anticoagulants

Resources

Books

Feinman, Susan E., ed. *Beneficial and Toxic Effects of Aspirin.* Boca Raton, FL: CRC Press, LLC, 1994.

O'Neil, Maryadele J. *Merck Index: An Encyclopedia of Chemicals, Drugs, & Biologicals.* 13th ed. Whitehouse Station, NJ: Merck & Co., 2001.

Ray, Oakley and Charles Ksir. *Drugs, Society & Human Behavior.* 8th ed. New York: McGraw-Hill Co., 1998.

Periodicals

"Aspirin's Next Conquest: Does it Prevent Colon Cancer?" *Journal of the National Cancer Institute* (February 2, 1994): 166-68.

Kiefer, D.M. "Chemistry Chronicles: Miracle Medicines." *Today's Chemist* 10, no. 6 (June 2001): 59-60.

Marie L. Thompson

Acid see **Acids and bases**

Acid rain

"Acid rain" is a popularly used phrase that refers to the deposition of acidifying substances from the atmosphere and the environmental damage that this causes. Acid rain became a prominent issue around 1970, and since then research has demonstrated that the deposition of atmospheric chemicals is causing widespread acidification of lakes and streams, and possibly **soil**. The resulting biological effects include the extirpation (or local **extinction**) of many populations of **fish**. Scientific understanding of the causes and consequences of acid rain, in conjunction with lobbying of government by environmental organizations, has resulted in large reductions in the atmospheric emissions of pollutants in **North America** and parts of **Europe**. If these reductions prove to be large enough, acid rain will be less of an environmental problem in those regions.

Atmospheric deposition

Strictly speaking, the term "acid rain" should only refer to rainfall, or so-called wet **precipitation**. However, the proper meaning of acid rain is "the deposition of acidifying substances from the atmosphere." This is because acidification is not just caused by acidic rain, but also by chemicals in snow and **fog**, and by inputs of gases and particulates when precipitation is not occurring.

Of the many chemicals that are deposited from the atmosphere, the most important in terms of causing acidity in soil and surface waters (such as lakes and streams) are: (1) dilute solutions of sulfuric and nitric acids (H_2SO_4 and HNO_3, respectively) deposited as acidic rain or snow, (2) the gases **sulfur dioxide** (SO_2) and oxides of **nitrogen** (NO and NO_2, together called NO_x), and (3) tiny particulates, such as ammonium sulfate ($[NH_4]_2SO_4$) and ammonium nitrate (NH_4NO_3).

The depositions of these gases and particulates primarily occur when it is not raining or snowing. This type of atmospheric input is known as "dry deposition." Large regions of Europe and North America are exposed to these acidifying depositions. However, only certain types of ecosystems are vulnerable to becoming acidified by these atmospheric inputs. These usually have a thin cover of soil that contains little **calcium**, and sits upon a **bedrock** of hard **minerals** such as granite or quartz. There is convincing evidence that atmospheric depositions have caused an acidification of **freshwater** ecosystems in such areas. Many lakes, streams, and **rivers** have become acidic, resulting in declining or locally extirpated populations of some plants and animals. However, there is not yet conclusive evidence that terrestrial ecosystems have been degraded by acidic deposition (except for cases of severe **pollution** by toxic SO_2).

Chemistry of precipitation

The acidity of an aqueous **solution** is measured as its **concentration** of **hydrogen** ions (H^+). The **pH** scale expresses this concentration in logarithmic units to the base 10, ranging from very acidic solutions of pH 0, through the neutral value of pH 7, to very alkaline (or basic) solutions of pH 14. It is important to recognize that a one-unit difference in pH (for example, from pH 3 to pH 4) implies a 10-fold difference in the concentration of hydrogen ions. The pHs of some common solutions include: lemon juice, pH 2; table vinegar, pH 3; milk, pH 6.6; milk of magnesia, pH 10.5.

As just noted, an acidic solution, strictly speaking, has a pH less than 7.0. However, in environmental science the operational definition of acidic precipitation is a pH less than 5.65. This is the pH associated with the weak solution of carbonic acid (H_2CO_3) that forms when **water** droplets in **clouds** are in chemical equilibrium with **carbon dioxide** (CO_2), an atmospheric gas with a concentration of about 360 ppm (parts per million; this is a unit of concentration).

Water in precipitation contains a mixture of positively charged ions (or cations) and negatively charged ions (or anions). The most abundant cations are usually hydrogen (H^+), ammonium (NH_4^+), calcium (Ca^{2+}), **magnesium** (Mg^{2+}), and **sodium** (Na^+), while the major anions are sulfate (SO_4^{2-}), chloride (Cl^-), and nitrate (NO_3^-). The principle of **conservation** of electrochemical neutrality of aqueous solutions states that the total number of **cation** charges must equal that of anions, so the net electrical charge is **zero**. Following from this principle, the quantity of H^+ in an aqueous solution is related to the difference in concentration of the sum of all anions, and the sum of all cations other than H^+.

Data for the **chemistry** of precipitation in a region experiencing severe acid rain are available from Hubbard Brook, New Hampshire, where one of the world's best long-term studies of this phenomenon has been undertaken. The average pH of precipitation at Hubbard Brook is 4.2, and H^+ accounts for 71% of the total amount of cations, and SO_4^{2-} and NO_3^- for 87% of the anions. Therefore, most of the acidity of precipitation at Hubbard Brook occurs as dilute sulfuric and nitric acids. The SO_4^{2-} is believed to originate from SO_2 emitted from power plants and industries, and oxidized by photochemical reactions in the atmosphere to SO_4^{2-}. The NO_3^- originates with emissions of NO_x (i.e., NO and NO_2) gases from these sources and automobiles. Not surprisingly, air masses that pass over the large **emission** sources of Boston and New York produce storms with the highest concentrations of H^+, SO_4^{2-}, and NO_3^- at Hubbard Brook.

Regions differ greatly in their precipitation chemistry. This can be demonstrated using data for precipitation chemistry monitored during a study in eastern Canada. The village of Dorset in southern Ontario is close to large sources of emission of SO_2 and NO_x. On average, the precipitation at Dorset is highly acidic at pH 4.1, and the large concentrations of SO_4^{2-} and NO_3^- suggest that the acidity is caused by dilute sulfuric and nitric acids. In comparison, the Experimental Lakes Area (ELA) is in a remote landscape in northwestern Ontario that is infrequently affected by polluted air masses. The ELA site has a less acidic precipitation (average pH 4.7) and smaller concentrations of SO_4^{2-} and NO_3^- than at Dorset. Another site near the Atlantic Ocean in Nova Scotia receives air masses that pass over large sources of emissions in New England and southeastern Canada. However, by the time Nova Scotia is reached much of the acidic SO_4^{2-} and NO_3^- have been removed by prior rain-out, and the precipitation is only moderately acidic (pH 4.6). Also, because Nova Scotia is influenced by the ocean, its precipitation chemistry is characterized by high concentrations of Na^+ and Cl^-. Finally, Lethbridge in southern Alberta is in a **prairie** landscape, and its precipitation is not acidic (average pH 6.0) because of the influence of calcium-rich, acid-neutralizing dusts blown into the atmosphere from agricultural fields.

In some places, fog moisture can be especially acidic. For example, fogwater at coastal locations in New England can be as acidic as pH 3.0-3.5. At high-elevation locations where fog is frequent there can be large depositions of cloudwater and acidity. At a site in New Hampshire where fog occurs 40% of the time, cloudwater deposition to a **conifer** forest is equivalent to 47% of the water input by rain and snow, and because of its large concentrations of some chemicals, fog deposition accounted for 62% of the total inputs of H^+, and 81% of those of SO_4^{2-} and NO_3^-.

Spatial patterns of acidic precipitation

Large regions are affected by acidic precipitation in North America, Europe, and elsewhere. A relatively small region of eastern North America is known to have experienced acidic precipitation before 1955, but this has since expanded so that most of the eastern United States and southeastern Canada is now affected.

Interestingly, the acidity of precipitation is not usually greater close to large point-sources of emission of important gaseous precursors of acidity, such as smelters or power plants that emit SO_2 and NO_x. This observation emphasizes the fact that acid rain is a regional phenomenon, and not a local one. For instance, the acidity of precipitation is not appreciably influenced by **distance** from

the world's largest point-source of SO_2 emissions, a smelter in Sudbury, Ontario. Furthermore, when that smelter was temporarily shut down by a labor dispute, the precipitation averaged pH 4.49, not significantly different from the pH 4.52 when there were large emissions of SO_2.

Dry deposition of acidifying substances

Dry deposition occurs in the intervals of time between precipitation events. Dry deposition includes inputs of tiny particulates from the atmosphere, as well as the uptake of gaseous SO_2 and NO_x by plants, soil, and water. Unlike wet deposition, the rates of dry deposition can be much larger close to point-sources of emission, compared with further away.

Once they are dry deposited, certain chemicals can generate important quantities of acidity when they are chemically transformed in the receiving **ecosystem**. For example, SO_2 gas can dissolve into the water of lakes or streams, or it can be absorbed by the foliage of plants. This dry-deposited SO_2 is then oxidized to SO_4^{2-}, which is electrochemically balanced by H^+, so that acidity results. Dry-deposited NO_x gas can similarly be oxidized to NO_3^- and also balanced by H^+.

In relatively polluted environments close to emissions sources, the total input of acidifying substances (i.e., wet + dry depositions) is dominated by the dry deposition of acidic substances and their acid-forming precursors. The dry deposition is mostly associated with gaseous SO_2 and NO_x, because wet deposition is little influenced by distance from sources of emission.

For example, within a 25 mi (40 km) radius of the large smelter at Sudbury, about 55% of the total input of **sulfur** from the atmosphere is due to dry deposition, especially SO_2. However, less than 1% of the SO_2 emission from the smelter is deposited in that area, because the tall smokestack is so effective at widely dispersing the emissions.

Because they have such a large surface area of foliage and **bark**, **forests** are especially effective at absorbing atmospheric gases and particles. Consequently, dry inputs accounted for about 33% of the total sulfur deposition to a hardwood forest in New Hampshire, 56-63% of the inputs of S and N to a hardwood forest in Tennessee, and 55% of their inputs to a conifer forest in Sweden.

Chemical changes in the forest canopy

In any forest, leaves and bark are usually the first surfaces encountered by precipitation. Most rainwater penetrates the foliar canopy and then reaches the forest floor as so-called throughfall, while a smaller amount runs down **tree** trunks as stemflow. Throughfall and stemflow have a different chemistry than the original precipitation. Because potassium is easily leached out of leaves, its concentration is especially changed. In a study of several types of forest in Nova Scotia, the concentration of potassium (K^+) was about 10 times larger in throughfall and stemflow than in rain, while calcium (Ca^{2+}) and magnesium (Mg^{2+}) were three to four times more concentrated. There was less of a change in the concentration of H^+; the rainwater pH was 4.4, but in throughfall and stemflow of hardwood stands pH averaged 4.7, and it was 4.4-4.5 in conifer stands. The decreases in acidity were associated with ion-exchange reactions occurring on foliage and bark surfaces, in which H^+ is removed from solution in exchange for Ca^{2+}, Mg^{2+}, and K^+. Overall, the "consumption" of hydrogen ions accounted for 42-66% of the input of H^+ by precipitation to these forests. Similarly, H^+ consumption by the tree canopy was 91% in a hardwood forest at Hubbard Brook, New Hampshire, 21-80% among seven stands in New Brunswick, and 14-43% in stands in upstate New York.

In areas polluted by SO_2 there can be large increases in the sulfate concentration of throughfall and stemflow, compared with ambient precipitation. This is caused by the washoff of SO_2 and SO_4 that had been previously dry-deposited to the canopy. At Hubbard Brook this SO_4 enhancement is about four times larger than ambient precipitation, while in central Germany it is about two to three times greater. These are both regions with relatively large concentrations of particulate SO_4 and gaseous SO_2 in the atmosphere.

Chemical changes in soil

Once precipitation reaches the forest floor, it percolates into the soil. Important chemical changes take place as: (1) microbes and plants selectively absorb, release, and metabolize chemicals; (2) ions are exchanged at the surfaces of particles of clay and organic **matter**; (3) minerals are made soluble by so-called acid-weathering reactions; and (4) secondary minerals such as certain clays and **metal** oxides are formed through chemical precipitation of soluble ions of **aluminum**, **iron**, and other metals. These various chemical changes can contribute to: soil acidification, the **leaching** of important chemicals such as calcium and magnesium, and the mobilization of toxic ions of aluminum, especially Al^{3+}. These are all natural, closely linked processes, occurring wherever there is well-established vegetation, and where water inputs by precipitation are greater than **evapotranspiration** (i.e., **evaporation** from vegetation and non-living surfaces). A potential influence of acid rain is to increase the rates of some of these processes, such as the leaching of toxic H^+ and Al^{3+} to lakes and other surface waters.

Some of these effects have been examined by experiments in which simulated "rainwater" of various pHs was added to soil contained in plastic tubes. These experiments have shown that very acidic solutions can cause: (1) an acidification of the soil; (2) increased leaching of the so-called "basic cations" Ca, Mg, and K, resulting in nutrient loss, decreased base saturation of cation exchange capacity, and increased vulnerability of soil to acidification; (3) increased solubilization of toxic ions of metals such as aluminum, iron, manganese, **lead**, and zinc; and (4) saturation of the ability of soil to absorb sulfate, after which sulfate leaches at about the **rate** of input. The leaching of sulfate has a secondary influence on soil acidification if it is accompanied by the loss of base cations, and it can cause acidifying and toxic effects in surface waters if accompanied by Al^{3+} and H^+.

Soil acidification can occur naturally. This fact can be illustrated by studies of ecological **succession** on newly exposed parent materials of soil. At Glacier Bay, Alaska, the melting of **glaciers** exposes a mineral substrate with a pH of about 8.0, with up to 7-10% carbonate minerals. As this material is colonized and modified by vegetation and climate, its acidity increases, reaching about pH 4.8 after 70 years when a conifer forest has established. Accompanying this acidification is a reduction of carbonates to less than 1%, caused by leaching and uptake by plants.

Several studies have attempted to determine whether naturally occurring soil acidification has been intensified as a result of acid rain and associated atmospheric depositions. So far, there is no conclusive evidence that this has occurred on a wide scale. It appears that soil acidification is a potential, longer-term risk associated with acid rain.

Chemistry of surface waters

Compared with the water of precipitation, that of lakes, ponds, streams, and rivers is relatively concentrated in ions, especially in calcium, magnesium, potassium, sodium, sulfate, and chloride. These chemicals have been mobilized from the terrestrial part of the watersheds of the surface waters. In addition, some surface waters are brown-colored because of their high concentrations of dissolved organic compounds, usually leached out of nearby bogs. Brown-water lakes are often naturally acidic, with a pH of about 4 to 5.

Seasonal variations in the chemistry of surface waters are important. Where a snowpack accumulates, meltwater in the springtime can be quite acidic. This happens because soils are frozen and/or saturated during snowmelt, so there is little possibility to neutralize the acidity of meltwater. So-called "acid shock" events in streams have

Acid rain damage done to a piece of architecture in Chicago, Illinois. *Photograph by Richard P. Jacobs. JLM Visuals. Reproduced by permission.*

been linked to the first meltwaters of the snowpack, which are generally more acidic than later fractions.

A widespread acidification of weakly-buffered waters has affected the northeastern United States, eastern Canada, Scandinavia, and elsewhere. In 1941, for example, the average pH of 21 lakes in central Norway was 7.5, but only 5.4-6.3 in the 1970s. Before 1950 the average pH of 14 Swedish water bodies was 6.6, but 5.5 in 1971. In New York's Adirondack Mountains, 4% of 320 lakes had pH less than 5 in the 1930s, compared with 51% of 217 lakes in that area in 1975 (90% were also devoid of fish). The Environmental Protection Agency sampled a large number of lakes and streams in the United States in the early 1990s. Out of 10,400 lakes, 11% were acidic, mostly in the eastern United States. Atmospheric deposition was attributed as the cause of acidification of 75% of the lakes, while 3% had been affected by acidic drainage from **coal** mines, and 22% by organic acids from bogs. Of the 4,670 streams considered acidic, 47% had been acidified by atmospheric deposition, 26% by acid-mine drainage, and 27% by bogs.

Surface waters that are vulnerable to acidification generally have a small acid-neutralizing capacity. Usual-

ly, H^+ is absorbed until a buffering threshold is exceeded, and there is then a rapid decrease in pH until another buffering system comes into play. Within the pH range of 6 to 8, bicarbonate alkalinity is the natural buffering system that can be depleted by acidic deposition. The amount of bicarbonate in water is determined by geochemical factors, especially the presence of mineral carbonates such as calcite ($CaCO_3$) or dolomite ($Ca,MgCO_3$) in the soil, bedrock, or aquatic sediment of the **watershed**. Small pockets of these minerals are sufficient to supply enough acid-neutralizing capacity to prevent acidification, even in regions where acid rain is severe. In contrast, where bedrock, soil, and sediment are composed of hard minerals such as granite and quartz, the acid-neutralizing capacity is small and acidification can occur readily. Vulnerable watersheds have little alkalinity and are subject to large depositions of acidifying substances; these are especially common in glaciated regions of eastern North America and Scandinavia, and at high altitude in more southern **mountains** (such as the Appalachians) where crustal granite has been exposed by **erosion**.

High-altitude, headwater lakes and streams are often at risk because they usually have a small watershed. Because there is little opportunity for rainwater to interact with the thin soil and bedrock typical of headwater systems, little of the acidity of precipitation is neutralized before it reaches surface water.

In overview, the acidification of freshwaters can be described as a titration of a dilute bicarbonate solution with sulfuric and nitric acids derived from atmospheric deposition. In waters with little alkalinity, and where the watershed provides large fluxes of sulfate accompanied by hydrogen and aluminum ions, the waterbody is vulnerable to acidification.

Effects of acidification on terrestrial plants

Few studies have demonstrated injury to terrestrial plants caused by an exposure to ambient acid rain. Although many experiments have demonstrated injury to plants after treatment with artificial "acid rain" solutions, the toxic thresholds are usually at substantially more acidic pHs than normally occur in nature.

For example, some Norwegian experiments involved the treating of young forests with simulated acid rain. Lodgepole pine watered for three years grew 15-20% more quickly at pHs 4 and 3, compared with a "control" treatment of pH 5.6-6.1. The height growth of **spruce** was not affected over the pH range 5.6 to 2.5, while Scotch pine was stimulated by up to 15% at pHs of 2.5 to 3.0, compared with pH 5.6-6.1. Birch trees were also stimulated by the acid treatments. However, the feather mosses that dominated the ground vegetation were negatively affected by acid treatments.

Because laboratory experiments can be well controlled, they are useful for the determination of dose-response effects of acidic solutions on plants. In general, growth reductions are not observed unless treatment pHs are more acidic than about 3.0, and some **species** are stimulated by more acidic pHs than this. In one experiment, the growth of white pine seedlings was greater after treatment at pHs of 2.3 to 4.0 than at pH 5.6. In another experiment, seedlings of 11 tree species were treated over the pH range 2.6 to 5.6. Injuries to foliage occurred at pH 2.6, but only after a week of treatment with this very acidic pH.

Overall, it appears that trees and other vascular plants are rather tolerant of acidic rain, and they may not be at risk of suffering direct, short-term injury from ambient acidic precipitation. It remains possible, however, that even in the absence of obvious injuries, stresses associated with acid rain could decrease **plant** growth. Because acid rain is regional in character, these yield decreases could occur over large areas, and this would have important economic implications. This potential problem is most relevant to forests and other natural vegetation.

This is because agricultural land is regularly treated with liming agents to reduce soil acidity, and because acid production by cropping and fertilization is much larger than that caused by atmospheric depositions.

Studies in western Europe and eastern North America have examined the possible effects of acid rain on forest productivity. Recent decreases in productivity have been shown for various tree species and in various areas. However, progressive decreases in productivity are natural as the canopy closes and **competition** intensifies in developing forests. So far, research has not separated clear effects of regional acid rain from those caused by ecological succession, insect defoliation, or climate change.

Effects of acidification on freshwater organisms

The community of microscopic **algae** (or **phytoplankton**) of lakes is quite diverse in species. Non-acidic, oligotrophic (i.e., unproductive) lakes in a temperate climate are usually dominated by golden-brown algae and **diatoms**, while acidic lakes are typically dominated by dinoflagellates, cryptomonads, and green algae.

An important experiment was performed in a remote **lake** in Ontario, in which **sulfuric acid** was added to slowly acidify the entire lake, ultimately to about pH 5.0 from the original pH of 6.5. During this whole-lake acidification, the phytoplankton community changed from an initial domination by golden-brown algae to dominance by green algae. There was no change in the total number of species, but there was a small increase in algal **biomass** after acidification because of an increased clarity of the water.

In some acidified lakes the abundance of larger plants (called macrophytes) has decreased, sometimes accompanied by increased abundance of a **moss** known as *Sphagnum*. In itself, proliferation of *Sphagnum* can cause acidification, because these plants efficiently remove cations from the water in exchange for H^+, and their mats interfere with acid neutralizing processes in the sediment.

Zooplankton are small crustaceans living in the water column of lakes. These animals can be affected by acidification through: (1) the toxicity of H^+ and associated metals ions, especially Al^{3+}; (2) changes in their phytoplankton food; and (3) changes in predation, especially if plankton-eating fish become extirpated by acidification. Surveys have demonstrated that some zooplankton species are sensitive to acidity, while others are more tolerant. In general, higher-pH lakes are richer in zooplankton species. For example, a survey of lakes in Ontario found 9-16 species with three to four dominants at pH greater than pH 5, but only 1-7 species with one to two dominants at more acidic pHs.

In the whole-lake experiment mentioned previously, the abundance of zooplankton increased by 66-93% after acidification, a change attributed to an increase in algal biomass. Although there was little change in dominant species, some less common species were extirpated.

Fish are the best-known victims of acidification. Loss of populations of trout, **salmon**, and other species have occurred in many acidified freshwaters. A survey of 700 Norwegian lakes, for example, found that brown trout were absent from 40% of the water bodies and sparse in another 40%, even though almost all of the lakes had supported healthy fish populations prior to the 1950s. Surveys during the 1930s in the Adirondack Mountains of New York found brook trout in 82% of the lakes. However, in the 1970s fish did not occur in 43% of 215 lakes in the same area, including 26 definite extirpations of brook trout in re-surveyed lakes. This dramatic change paralleled the known acidification of these lakes. Other studies documented the loss of fish populations from lakes in the Killarney region of Ontario, where there are known extirpations of lake trout in 17 lakes, while smallmouth **bass** have disappeared from 12 lakes, largemouth bass and walleye from four, and yellow **perch** and rock bass from two.

Many studies have been made of the physiological effects of acidification on fish. Younger life-history stages are generally more sensitive than adults, and most losses of fish populations can be attributed to reproductive failure, rather than mortality of adults (although adults have sometimes been killed by acid-shock episodes in the springtime).

There are large increases in concentration of certain toxic metals in acidic waters, most notably ions of aluminum. In many acidic waters aluminium ions can be sufficient to kill fish, regardless of any direct effect of H^+. In general, survival and growth of larvae and older stages of fish are reduced if dissolved aluminium concentrations are larger than 0.1 ppm, an exposure regularly exceeded in acidic waters. The most toxic ions of aluminium are Al^{3+} and $AlOH^{2+}$.

Although direct effects of acidification on aquatic **birds** have not been demonstrated, changes in their **habitat** could indirectly affect their populations. Losses of fish populations would be detrimental to fish-eating waterbirds such as **loons**, mergansers, and osprey. In contrast, an increased abundance of aquatic **insects** and zooplankton, resulting from decreased predation by fish, could be beneficial to diving **ducks** such as common goldeneye and hooded merganser, and to dabbling ducks such as the mallard and black duck.

Trees killed by acid rain in the Great Smoky Mountains.
JLM Visuals. Reproduced by permission.

Reclamation of acidified water bodies

Fishery biologists especially are interested in liming acidic lakes to create habitat for sportfish. Usually, acidic waters are treated by adding limestone ($CaCO_3$) or lime ($Ca[OH]_2$), a process analogous to a whole-lake titration to raise pH. In some parts of Scandinavia liming has been used extensively to mitigate the biological damages of acidification. By 1988 about 5,000 water bodies had been limed in Sweden, mostly with limestone, along with another several hundred lakes in southern Norway. In the early 1980s there was a program to lime 800 acidic lakes in the Adirondack region of New York.

Although liming rapidly decreases the acidity of a lake, the water later re-acidifies at a rate determined by size of the drainage **basin**, the rate of flushing of the lake, and continued atmospheric inputs. Therefore, small headwater lakes have to be re-limed more frequently. In addition, liming initially stresses the acid-adapted biota of the lake, causing changes in species dominance until a

KEY TERMS

Acid mine drainage—Surface water or groundwater that has been acidified by the oxidation of pyrite and other reduced-sulfur minerals that occur in coal and metal mines and their wastes.

Acid shock—A short-term event of great acidity. This phenomenon regularly occurs in freshwater systems that receive intense pulses of acidic water when an accumulated snowpack melts rapidly in the spring.

Acidic rain (acidic precipitation)—(1) Rain, snow, sleet or fog water having a pH less than 5.65. (2) The deposition of acidifying substances from the atmosphere during a precipitation event.

Acidification—An increase over time in the content of acidity in a system, accompanied by a decrease in acid-neutralizing capacity.

Acidifying substance—Any substance that causes acidification. The substance may have an acidic character and therefore act directly, or it may initially be non-acidic but generate acidity as a result of its chemical transformation, as happens when ammonium is nitrified to nitrate, and when sulfides are oxidized to sulfate.

Acidity—The ability of a solution to neutralize an input of hydroxide ion (OH^-). Acidity is usually measured as the concentration of hydrogen ion (H^+), in logarithmic pH units (see also pH). Strictly speaking, an acidic solution has a pH less than 7.0.

Acidophilous—Refers to organisms that only occur in acidic habitats, and are tolerant of the chemical stresses of acidity.

Conservation of electrochemical neutrality—Refers to an aqueous solution, in which the number of cation equivalents equals the number of anion equivalents, so that the solution does not have a net electrical charge.

Equivalent—Abbreviation for mole-equivalent, and calculated as the molecular or atomic weight multiplied times the number of charges of the ion. Equivalent units are necessary for a charge-balance calculation, related to the conservation of electrochemical neutrality (above).

Leaching—The movement of dissolved chemicals with water percolating through soil.

pH—The negative logarithm to the base 10 of the aqueous concentration of hydrogen ions in units of moles per liter. An acidic solution has pH less than 7, while an alkaline solution has pH greater than 7. Note that a one-unit difference in pH implies a 10-fold difference in the concentration of hydrogen ions.

new, steady-state ecosystem is achieved. It is important to recognize that liming is a temporary management strategy, and not a long-term solution to acidification.

Avoiding acid rain

Neutralization of acidic ecosystems treats the symptoms, but not the sources of acidification. Clearly, large reductions in emissions of the acid-forming gases SO_2 and NO_x are the ultimate solution to this widespread environmental problem. However, there is controversy over the amount that the emissions must be reduced in order to alleviate acidic deposition, and about how to pursue the reduction of emissions. For example, should large point sources such as power plants and smelters be targeted, with less attention paid to smaller sources such as automobiles and residential furnaces? Not surprisingly, industries and regions that are copious emitters of these gases lobby against emission controls, for which they argue the scientific justification is not yet adequate.

In spite of many uncertainties about the causes and magnitudes of the damage associated with acid rain and related atmospheric depositions, it is intuitively clear that what goes up (that is, the acid-precursor gases) must come down (as acidifying depositions). This common-sense notion is supported by a great deal of scientific evidence, and because of public awareness and concerns about acid rain in many countries, politicians have began to act effectively. Emissions of sulfur dioxide and oxides of nitrogen are being reduced, especially in western Europe and North America. For example, in 1992 the governments of the United States and Canada signed an air-quality agreement aimed at reducing acidifying depositions in both countries. This agreement calls for large expenditures by government and industry to achieve substantial reductions in the emissions of air pollutants during the 1990s. Eventually, these actions should improve environmental conditions related to damage caused by acid rain.

However, so far the actions to reduce emissions of the precursor gases of acidifying deposition have only been vigorous in western Europe and North America. Actions are also needed in other, less wealthy regions where the political focus is on industrial growth, and not on control of **air pollution** and other environmental damages that are used to subsidize that growth. In the coming years, much more attention will have to be paid to acid rain and other pollution problems in eastern Europe and the former USSR, China, India, southeast **Asia**, Mexico, and other so-called "developing" nations. Emissions of important air pollutants are rampant in these places, and are increasing rapidly.

See also Sulfur dioxide.

Resources

Books

Edmonds, A. *Acid Rain.* Sussex, England: Copper Beech Books, Ltd., 1997.

Ellerman, Danny. *Markets for Clean Air: The U. S. Acid Rain Program.* Cambridge: Cambridge University Press, 2000.

Freedman, B. *Environmental Ecology.* 2nd ed. San Diego: Academic Press, 1995.

Hancock P. L. and Skinner B. J., eds. *The Oxford Companion to the Earth.* Oxford: Oxford University Press, 2000.

Periodicals

Anonymous. *National Acid Precipitation Assessment Program* Integrated Assessment Report. Washington, DC: Superintendent of Documents, U.S. Government Printing Office, 1989.

Brimblecombe, P. "Acid Rain 2000." *Water, Air, and Soil Pollution* 130, 1-4 (2001): 25-30.

Galloway, James N. "Acidification of the World: Natural and Anthropogenic." *Water, Air, and Soil Pollution* 130, no. 1-4 (2001): 17-24.

Krajick, K. "Acid Rain: Long-term Data Show Lingering Effects from Acid Rain." *Science* 292, no. 5515 (2001): 195-196.

Milius, S. "Red Snow, Green Snow." *Science News* no. 157 (May 2000): 328-333.

Other

The United Nations. "The Conference and Kyoto Protocol," homepage [cited March 2003]. <http://unfccc.int/resource/convkp.html>.

United Stated Geological Survey. "What is Acid Rain?" [cited March 2003]. <http://pubs.usgs.gov/gip/acidrain/2.html>.

Bill Freedman

Acids and bases

Acids and bases are chemical compounds that have certain specific properties in aqueous solutions. In most chemical circumstances, acids are chemicals that produce positively-charged **hydrogen** ions, H^+, in **water**, while bases are chemicals that produce negatively-charged hydroxide ions, OH^-, in water. Bases are sometimes called *alkalis*. Acids and bases react with each other in a reaction called *neutralization*. In a **neutralization** reaction, the hydrogen ion and the hydroxide ion react to form a **molecule** of water:

$$H^+ + OH^- \rightarrow H_2O$$

Chemically, acids and bases may be considered opposites of each other. The concept of acids and bases is so important in **chemistry** that there are several useful definitions of "acid" and "base" that pertain to different chemical environments, although the definition above is the most common one.

Acids and bases have some general properties. Many acids have a sour **taste**. **Citric acid**, found in oranges and lemons, is one example where the sour taste is related to the fact that the chemical is an acid. Molecules that are bases usually have a bitter taste, like **caffeine**. Bases make solutions that are slippery. Many acids will react with metals to dissolve the **metal** and at the same time generate hydrogen gas, H_2. Perhaps the most obvious behavior of acids and bases is their abilities to change colors of certain other chemicals. Historically, an extract of **lichens** (*V. lecanora* and *V. rocella*) called *litmus* has been used since it turns blue in the presence of bases and red in the presence of acids. Litmus **paper** is still commonly used to indicate whether a compound is an acid or a base. Extracts made from red onions, red cabbage, and many other **fruits** and **vegetables** change colors in the presence of acids and bases. Such materials are called indicators.

Classic definition of acids and bases

Although acids and bases have been known since prehistoric times (vinegar, for example, is an acid), the first attempt to define what makes a compound an acid or a base was made by the Swedish chemist Svante Arrhenius (1859-1927), who proposed the definition that an acid was any compound that produced hydrogen ions, H^+, when dissolved in water, and a base was any compound that produced hydroxide ions, OH^-, when dissolved in water. Although this was and still is a very useful definition, it has two major limitations. First, it was limited to water, or aqueous, solutions. Second, it practically limited acids and bases to ionic compounds that contained the H^+ ion or the OH^- ion (compounds like hydrochloric acid, HCl, or **sodium hydroxide**, NaOH). Limited though it might be, it was an important step in the understanding of chemistry in solutions, and for his work on **solution** chemistry Arrhenius was awarded the 1903 Nobel Prize in chemistry.

Many common acids and bases are consistent with the Arrhenius definition. The following table shows a few common acids and bases and their uses. In all cases it is assumed that the acid or base is dissolved in water.

Many acids release only a single hydrogen ion per molecule into solution. Such acids are called *monoprotic*. Examples include hydrochloric acid, HCl, and **nitric acid**, HNO_3. *Diprotic* acids can release two hydrogen ions per molecule. H_2SO_4 is an example. *Triprotic* acids, like H_3PO_4, can release three hydrogen ions into solution. **Acetic acid** has the formula $HC_2H_3O_2$ and is a monoprotic acid because it is composed of one H^+ ion and one acetate ion, $C_2H_3O_2-$. The three hydrogen **atoms** in the acetate ion do not act as acids.

Strong and weak acids and bases

An important consideration when dealing with acids and bases is their *strength*; that is, how chemically reactive they act as acids and bases. The strength of an acid or base is determined by the degree of ionization of the acid or base in solution—that is, the percentage of dissolved acid or base molecules that release hydrogen or hydroxide ions. If all of the dissolved acid or base separates into ions, it is called a *strong acid* or *strong base*. Otherwise, it is a *weak acid* or *weak base*. There are only a few strong acids: hydrochloric acid (HCl), hydrobromic acid (HBr), hydriodic acid (HI), perchloric acid ($HClO_4$), nitric acid (HNO_3), and **sulfuric acid** (H_2SO_4). Similarly, there are only a few strong bases: **lithium** hydroxide (LiOH), **sodium** hydroxide (NaOH), potassium hydroxide (KOH), **calcium** hydroxide (Ca[OH]$_2$), strontium hydroxide (Sr[OH]$_2$), and **barium** hydroxide (Ba[OH]$_2$).

These strong acids and bases are 100% ionized in aqueous solution. All other Arrhenius acids and bases are weak acids and bases. For example, acetic acid ($HC_2H_3O_2$) and **oxalic acid** ($H_2C_2O_4$) are weak acids, while **iron** hydroxide, $Fe(OH)_3$, and ammonium hydroxide, NH_4OH (which is actually just **ammonia**, NH_3, dissolved in water), are examples of weak bases. The percentage of the acid and base molecules that are ionized in solution varies and depends on the **concentration** of the acid. For example, a 2% solution of acetic acid in water, which is about the concentration found in vinegar, is only 0.7% ionized. This means that fully 99.3% of the acetic acid molecules are unionized and exist in solution as the complete acetic acid molecule.

Brønsted-Lowry definition of acids and bases

Although the Arrhenius definitions of acids and bases are simplest and most useful, they are not the most widely applicable. Some compounds, like ammonia, NH_3, act like bases in aqueous solution even though they are not hydroxide-containing compounds. Also, the Arrhenius definition assumes that the acid-base reactions are occurring in aqueous solution. In many other cases,

water is indeed the solvent. In many cases, however, water is not the solvent. What was necessary was to formulate a definition of acid and base that were independent of the solvent and the presence of H^+ and OH^- ions.

Such a definition was proposed in 1923 by English chemist Thomas Lowry (1874-1936) and Danish chemists J. N. Brønsted (1879-1947) and N. Bjerrum (1879-1958) and is called the Brønsted-Lowry definition of acids and bases. (Bjerrum seems to have been forgotten.) The central chemical species of this definition is H^+, which consists merely of a **proton**. By the Brønsted-Lowry definition, an acid is any chemical species that donates a proton to another chemical species. Conversely, a base is any chemical species that accepts a proton from another chemical species. Simply put, a Brønsted-Lowry acid is a proton donor and a Brønsted-Lowry base is a proton acceptor.

The Brønsted-Lowry definition includes all Arrhenius acids and bases, since the hydrogen ion is a proton donor (in fact, it is a proton) and a hydroxide ion accepts a proton to form water:

$$H^+ \quad + \quad OH^- \quad \rightarrow \quad H_2O$$
$$\text{proton} \qquad \text{proton}$$
$$\text{donor} \qquad \text{acceptor}$$

But the Brønsted-Lowry definition also includes chemical species that are not Arrhenius-type acids or bases. The classic example is ammonia, NH_3. Ammonia dissolves in water to make a slightly basic solution even though ammonia does not contain OH^- ions. What is happening is that an ammonia molecule is accepting a proton from a water molecule to make an ammonium ion (NH_4^+) and a hydroxide ion:

$$NH_3 \quad + \quad H_2O \rightarrow NH_4^+ \quad + \quad OH^-$$
$$\text{B-L base} \qquad \text{B-L acid}$$

In essence, the water molecule is donating a proton to the ammonia molecule. The water molecule is therefore acting as the Brønsted-Lowry acid and the ammonia molecule is acting as the Brønsted-Lowry base.

In order to better understand the Brønsted-Lowry definition, it needs to be understood what is meant by a proton. The descriptions proton donor and proton acceptor are easy to remember. But are there actually bare protons floating around in solution? Not really. In aqueous solution, the protons are attached to the **oxygen** atoms of water molecules, giving them a positive charge. This species is called the *hydronium ion* and has the chemical formula H_3O^+. It is more accurate to use the hydronium ion instead of the bare hydrogen ion when writing equations for **chemical reactions** between acids and bases in aqueous solution. For example, the reaction between the hydronium ion and the hydroxide ion, the typical Arrhenius acid-base reaction, would produce two molecules of water.

Acid	Name	Use	Base	Name	Use
HCl	hydrochloric acid	cleaning, drugs, plastics	NaOH	sodium hydroxide	drain cleaner, soap
H_2SO_4	sulfuric acid	chemical synthesis, batteries	KOH	potassium hydroxide	soaps
$HC_2H_3O_2$	acetic acid	vinegar	$Mg(OH)_2$	magnesium hydroxide	antacids

Chemical reactions can go forward or backward; when the rates of the reverse reactions are equal, it is at *chemical equilibrium*. It can be shown that each side of the equilibrium has a Brønsted-Lowry acid and base. For example:

$$NH_3 \ + \ H_2O \longleftrightarrow NH_4^+ \ + \ OH^-$$
B-L base   B-L acid   B-L acid   B-L base

On each side of the reaction there is an acid and a base. The NH_4^+ ion is an acid because in the reverse reaction it donates a proton (H^+) to the OH^- ion to form NH_3 and H_2O. With respect to the reaction above, the H_2O and OH^- species make up an acid-base pair, called a conjugate acid-base pair, while the NH_3 and NH_{4+} species make up another conjugate acid-base pair. All Brønsted-Lowry acid-base reactions can be separated into reactions between two conjugate acid-base pairs. The conjugate acid always has one more H^+ than the conjugate base.

Lewis definition of acids and bases

The Brønsted-Lowry acid-base definition, while broader than the Arrhenius definition, is still limited to hydrogen-containing compounds, and is dependent on a hydrogen ion (that is, a proton) transferring from one molecule to another. Ultimately, a definition of acid and base that is completely independent of the presence of a hydrogen atom is necessary.

Such a definition was provided in 1923 by American chemist Gilbert N. Lewis (1875-1946). Instead of focusing on protons, Lewis's definition focuses on **electron** pairs. Since all compounds contain electron pairs, the Lewis definition is applicable to a wide range of chemical reactions.

A Lewis acid is defined as the reactant in a chemical reaction that accepts an electron pair from another reactant. A Lewis base is defined as the reactant in a chemical reaction that donates an electron pair to another reactant. Like the Brønsted-Lowry definition of acids and bases, the Lewis definition is reaction-dependent. A compound is not an acid or base in its own right; rather,

how that compound reacts with another compound is what determines whether it is an acid or a base.

To show that the Lewis definition is not in conflict with previous definitions of acid and base, consider the fundamental acid-base reaction of H^+ with OH^- to give H_2O. The oxygen atom in the hydroxide ion has three unbonded electron pairs around it, and during the course of the reaction one of those electron pairs is "donated" to the hydrogen ion, making a **chemical bond**. Thus, OH^- is the electron pair donor and the Lewis base, whereas H^+ is the electron pair acceptor and, therefore, the Lewis acid. These assignments are consistent with both the Arrhenius definition and the Brønsted-Lowry definitions of acid and base.

However, the Lewis acid/base definition is much broader than the previous two definitions. Consider the reaction of BF_3 and NH_3 in the gas phase, in which NH_3 is donating an electron pair to the BF_3 molecule:

Lewis acid   Lewis base

Compounds like F_3BNH_3 are stable and can be purchased as solutions in organic solvents or even as pure compounds. In the above chemical reaction, BF_3 is accepting an electron pair and therefore is the Lewis acid; NH_3 is donating the electron pair and so is the Lewis base. However, in this case neither the Arrhenius definition nor the Brønsted-Lowry definition are applicable. Therefore, while the Lewis acid/base definition includes acids and bases from the other two definitions, it expands the definitions to include compounds that are not otherwise considered "classic" acids and bases.

Organic acids

Organic chemistry is the study of compounds of the element **carbon**. Organic chemistry uses the ideas of acids and bases in two ways. The more general way is that the concept of Lewis acids and bases is used to classify organic chemical reactions as acid/base reactions because the donation of electron pairs is quite common.

The second way that organic chemistry uses the concepts of acids and bases is in the definition of certain groupings of atoms within an organic molecule called *functional groups* as acidic or basic. An organic base is, in the true Lewis base style, any molecule with electron pairs that can be donated. The most common organic base involves a **nitrogen** atom, N, bonded to carbon-containing groups. One important class of such compounds is known as amines. In these compounds, the nitrogen atom has an unbonded electron pair that it can donate as it reacts as a Lewis base. Several of these compounds are gases and have a somewhat putrid, fish-like odor. These compounds are relatively simple molecules; there are larger organic molecules, including many of natural origin, that contain a nitrogen atom and so have certain base-like properties. These compounds are called *alkaloids*. Examples include **quinine**, caffeine, strychnine, **nicotine**, **morphine**, and **cocaine**.

Organic chemistry uses the acid concept not only in the definition of the Lewis acid but also by defining a particular collection of atoms as an acid functional group. Any organic molecule containing a **carboxyl group**, -COOH, is called a *carboxylic acid*. (Non-organic acids are sometimes called *mineral acids*). Examples include formic acid, which has the formula HCOOH and is produced by some **ants** and causes their bites to sting. Another example is acetic acid, CH_3COOH, which is the acid in vinegar.

Uses of acids and bases

Many specific uses of acids and bases have been discussed above. Generally, strong acids and bases are used for cleaning and, most importantly, for synthesizing other compounds. Their utility is illustrated by the fact that three of the top 10 chemicals produced in the US in 1994 are acids or bases: sulfuric acid (#1, 89 billion lbs/40 billion kg produced), sodium hydroxide (#8, 26 billion lbs/12 billion kg produced), and **phosphoric acid** (#9, 25 billion lbs/11 billion kg produced). Weak acids and bases have specific uses in society which are so variable that the specific compound entry should be consulted.

See also Acetic acid; Alkaloid; Carboxylic acids; Citric acid; Neutralization; Nitric acid; Sodium hydroxide; Sulfuric acid.

Resources

Books

Oxtoby, David W., et al. *The Principles of Modern Chemistry.* 5th ed. Pacific Grove, CA: Brooks/Cole, 2002.

Scorpio, Ralph. *Fundamental of Acids, Bases, Buffers & Their Application to Biochemical Systems.* Falls Church, VA: Kendall/Hunt, 2000.

Snyder, C.H. *The Extraordinary Chemistry of Ordinary Things.* 4th ed. New York: John Wiley and Sons, 2002.

David W. Ball

Acne

Acne, also called *acne vulgaris*, is a chronic **inflammation** of the sebaceous **glands** embedded in the skin. These glands secrete sebum, an oily lubricant.

Although it may occur at any age, acne is most frequently associated with the maturation of young adult males. Part of the normal maturation process involves the production of—or altered expression of—hormones. During adolescence, **hormones** termed androgens are produced. Androgens stimulate the enlargement of the sebaceous glands and result in the increased production of dermal oils designed to facilitate the growth of facial hair. In females, androgen production is greater around the time of menstruation. Estrogen in females also reduces sebum production. As a result, acne often appears in young women at the time of their monthly menstrual period.

In most cases, acne resolves itself by the time the **individual** is 20-30 years old.

Contrary to popular myth, acne is not caused or aggravated by eating greasy foods or chocolate. **Bacteria** play a critical role in the development of acne. The principal bacterial **species** associated with acne is *Proprionibacterium acnes*, the other is *Staphyloccccus epidermidis*. These **microorganisms** normally reside on the skin and inside hair follicles.

The outward flow of oil forces the bacteria to the surface where it can be removed with washing. However, in the androgen-altered hair follicles, the cells lining the cavity shed more frequently, stick together, mix with the excess oil that is being produced, and pile up in clumps

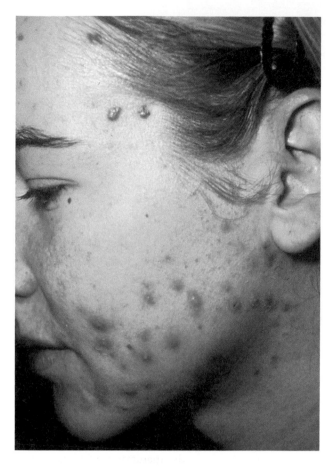

Acne vulgaris **affecting a woman's face. Acne is the general name given to a skin disorder in which the sebaceous glands become inflamed.** *Photograph by Biophoto Associates. National Audubon Society Collection/Photo Researchers, Inc. Reproduced by permission.*

inside the cavity. The accumulated material is a ready nutrient source for the *Proprionibacterium acnes* in the cavity. The bacteria grow and multiply rapidly to produce an acne sore or pustule.

As the numbers of bacteria increase, by-products of their metabolic activities cause additional inflammation. The bacteria also contain enzymes that can degrade the oil from the oil glands into free **fatty acids** that are irritating to the skin. Various other bacterial enzymes contribute to inflammation (e.g., proteases and phosphatases).

The damage caused by bacteria in acne ranges from mild to severe. In a mild case of acne, only so-called blackheads or whiteheads are evident on the skin. More severe cases are associated with more blackheads, whiteheads and pimples, and also with inflammation. The most severe form, called cystic acne, may produce marked inflammation over the entire upper body, and requires a physician's attention to reduce the bacterial populations.

Manipulating (e.g., squeezing, scratching, or picking) acne pustules can cause deep and permanent scarring. Normally, simply washing the affected area with **soap** will help dislodge the material plugging the duct. Because estrogen inhibits the development of acne, taking birth-control pills may help alleviate acne in young women. A topical antibiotic may also prove helpful. For deeper acne, injected **antibiotics** may be necessary.

Although the tendency to develop acne can be passed from parent to child, certain behaviors can aggravate acne outbreak. Acne can be caused by mechanical irritation, including pulling or stretching the skin, as often happens in athletic activities. Because steroid drugs contain androgens, taking steroids can also aggravate acne. Adolescent women who use oil-based cosmetics and moisturizers may develop an aggravated case of acne. Because the bacteria active in acne are normal residents of the skin, there is no "cure" for acne. Rather, the condition is lessened until biochemical or lifestyle changes in the individual lessen or eliminate the conditions that promote bacterial overgrowth.

See also Menstrual cycle.

Acorn worm

Acorn worms are fragile tube worms that live in **sand** or mud burrows in the intertidal areas of the world's oceans. Acorn worms are members of the phylum Hemichordata, which includes two classes—the Enteropneusta (acorn worms) and the Pterobranchia (pterobranchs). Acorn worms, also known as tongue worms, belong to one of four genera, *Balanoglossus*, *Glossobalanus*, *Ptychodera*, and *Saccoglossus*. They are mostly burrowing animals that vary in size from 1 to 39 in (1 to 100 cm) in length (*Balanoglossus gigas*). The body of acorn worms consists of proboscis, collar, and trunk. The proboscis is a digging **organ** and together with the collar (and a lot of imagination) it resembles an acorn, hence its name.

The embryos of the hemichordates show affinities with both the phylum Echinodermata (**starfish** and **sand dollars**) and with the phylum Chordata (which includes the **vertebrates**). The relationships between these phyla are tenuous and are not demonstrable in all forms. The larvae of the Chordata subphylum Cephalochordata, which includes *Amphioxus*, resemble the larvae of the Hemichordata, indicating that the Hemichordata may have given rise to the Chordata, and therefore the vertebrates.

The phylum Chordata is characterized by a dorsal, hollow nerve cord, a notochord, pharyngeal "gill" slits or pouches, and a coelom, the fluid-filled main body cavity.

Acorn worms resemble **chordates** in that these worms have pharyngeal gill slits, a nerve cord, and a coelom. A small structure in the anterior trunk was once thought to be a notochord, but it has been shown to be an extension of the gut.

Acoustics

Acoustics is the science that deals with the production, transmission, and reception of sound. Sound may be produced when a material body vibrates; it is transmitted only when there is some material body, called the medium, that can carry the vibrations away from the producing body; it is received when a third material body, attached to some indicating device, is set into vibratory **motion** by that intervening medium. However, the only vibrations that are considered sound (or sonic vibrations) are those in which the medium vibrates in the same direction as the sound travels, and for which the vibrations are very small. When the **rate** of vibration is below the range of human **hearing**, the sound is termed infrasonic; when it is above that range, it is called ultrasonic. The term supersonic refers to bodies moving at speeds greater than the speed of sound, and is not normally involved in the study of acoustics.

Production of sound

There are many examples of vibrating bodies producing sounds. Some are as simple as a string in a violin or piano, or a column of air in an organ pipe or in a clarinet; some are as complex as the vocal chords of a human. Sound may also be caused by a large disturbance which causes parts of a body to vibrate, such as sounds caused by a falling tree.

Vibrations of a string

To understand some of the fundamentals of sound production and propagation it is instructive to first consider the small vibrations of a stretched string held at both ends under tension. While these vibrations are not an example of sound, they do illustrate many of the properties of importance in acoustics as well as in the production of sound. The string may vibrate in a variety of different ways, depending upon whether it is struck or rubbed to set it in motion, and where on the string the action took place. However, its motion can be analyzed into a combination of a large number of simple motions. The simplest, called the fundamental (or the first harmonic), appears in Figure 1, which shows the outermost extensions of the string carrying out this vibration.

The second harmonic is shown in Figure 2; the third harmonic in Figure 3; and so forth (the whole set of **harmonics** beyond the first are called the overtones). The rate at which these vibrations take place (number of times per second the motion is repeated) is called the **frequency**, denoted by f (the reciprocal of the frequency, which is the **time** for one cycle to be competed, is called the period). A single complete vibration is normally termed a cycle, so that the frequency is usually given in cycles per second, or the equivalent modern unit, the hertz (abbreviated Hz). It is characteristic of the stretched string that the second harmonic has a frequency twice that of the fundamental; the third harmonic has a frequency three times that of the fundamental; and so forth. This is true for only a few very simple systems, with most sound-producing systems having a far more complex relationship among the harmonics.

Those points on the string which do not move are called the nodes; the maximum extension of the string (from the horizontal in the Figures) is called the amplitude, and is denoted by A in Figures 1-3. The **distance** one must go along the string at any instant of time to reach a section having the identical motion is called the wavelength, and is denoted by L in Figures 1-3. It can be seen that the string only contains one-half wavelength of the fundamental, that is, the wavelength of the fundamental is twice the string length. The wavelength of the second harmonic is the length of the string. The string contains one-and-one-half (3/2) wavelengths of the third harmonic, so that its wavelength is two-thirds (2/3) of the length of the string. Similar relationships hold for all the other harmonics.

If the fundamental frequency of the string is called f_0, and the length of the string is l, it can be seen from the above that the product of the frequency and the wavelength of each harmonic is equal to $2f_0 l$. The dimension of this product is a **velocity** (e.g., feet per second or centimeters per second); detailed analysis of the motion of the stretched string shows that this is the velocity with which a small disturbance on the string would travel down the string.

Vibrations of an air column

When air is blown across the entrance to an organ pipe, it causes the air in the pipe to vibrate, so that there are alternate small increases and decreases of the **density** of the air (condensations and rarefactions). These alternate in **space**, with the distance between successive condensations (or rarefactions) being the wavelength; they alternate in time, with the frequency of the vibration. One major difference here is that the string vibrates transversely (**perpendicular** to the length of the string), while the air vibrates longitudinally (in the direction of the column of air). If the pipe is open at both ends, then the density of the air at the ends must be the same as that of the air outside the pipe, while the density inside the pipe can vary above or below that value. Again, as for the vibrations of the string, the density of the air in the pipe can be analyzed into a fundamental and overtones. If the density of the air vibrating in the fundamental mode (of the open pipe) is plotted across the pipe length, the graph is as in Figure 4.

The "zero value" at the ends denotes the fact that the density at the ends of the pipe must be the same as outside the pipe (the ambient density), while inside the pipe the density varies above and below that value with the frequency of the fundamental, with a maximum (and minimum) at the center. The density plot for the fundamental looks just like that for the fundamental of the vibrating stretched string (Figure 1). In the same manner, plots of the density for the various overtones would look like those of the string overtones. The frequency of the fundamental can be calculated from the fact that the velocity, which is analogous to that found for vibrations of the string, is the velocity with which sound travels in the air, usually denoted by c. Since the wavelength of the fundamental is twice the pipe length, its frequency is $(c/2)l$, where l is the length of the organ pipe. (While the discussion here is in terms of the density variations in the air, these are accompanied by small variations in the air **pressure**, and small motions of the air itself. At places of increased density the pressure is increased; where the pressure is changing rapidly, the air motion is greatest.) When a musician blows into the mouthpiece of a clarinet, the air rushing past the reed causes it to vibrate which then causes the column of air in the clarinet to vibrate in a manner similar to, but more complicated than, the motion of the organ pipe. These vibrations (as for all vibrations) can also be analyzed into harmonics. By opening and closing the keyholes in the clarinet, different harmonics of the clarinet are made to grow louder or softer causing different tones to be heard.

Sound production in general

Thus, the production of sound depends upon the vibration of a material body, with the vibration being transmitted to the medium that carries the sound away from the sound producer. The vibrating violin string, for example, causes the body of the violin to vibrate; the "back-and-forth" motion of the parts of the body of the violin causes the air in contact with it to vibrate. That is, small variations in the density of the air are produced by the motion of the violin body, and these are carried forth into the air surrounding the violin. As the sound is carried away, the small variations in air density are propagated in the direction of travel of the sound.

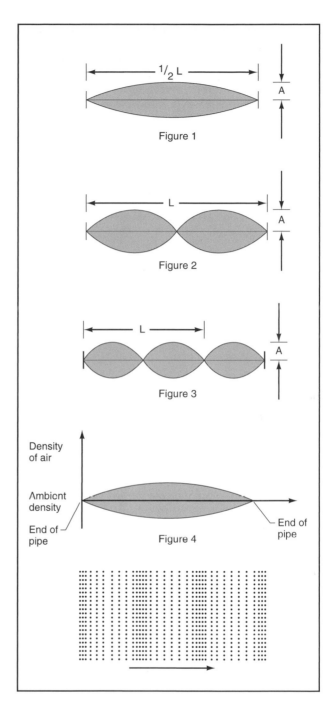

Figures 1 through 5. *Courtesy of Gale Research.*

Sounds from humans, of course, are produced by forcing air across the vocal cords, which causes them to vibrate. The various overtones are enhanced or diminished by the size and shape of the various cavities in the head (the sinuses, for example), as well as the placement of the tongue and the shape of the mouth. These factors cause specific wavelengths, of all that are produced by the vocal cords, to be amplified differently so that different people have their own characteristic voice sounds.

These sounds can be then controlled by changing the placement of the tongue and the shape of the mouth, producing **speech**. The frequencies usually involved in speech are from about 100 to 10,000 Hz. However, humans can hear sounds in the frequency range from about 20 to 18,000 Hz. These outer limits vary from person to person, with age, and with the loudness of the sound. The density variations (and corresponding pressure variations) produced in ordinary speech are extremely small, with ordinary speech producing less than one-millionth the power of a 100 watt light bulb! In the sonic range of frequencies (those produced by humans), sounds are often produced by loudspeakers, devices using electronic and mechanical components to produce sounds. The sounds to be transmitted are first changed to electrical signals by a microphone (see Reception of sounds, later), for example, or from an audio tape or **compact disc**; the frequencies carried by the electrical signals are those to be produced as the sound signals. In the simplest case, the wires carrying the electrical signals are used to form an electromagnet which attracts and releases a **metal** diaphragm. This, in turn, causes the variations in the density in the air adjacent to the diaphragm. These variations in density will have the same frequencies as were in the original electrical signals.

Ultrasonic vibrations are of great importance in industry and medicine, as well as in investigations in pure science. They are usually produced by applying an alternating electric voltage across certain types of crystals (quartz is a typical one) that expand and contract slightly as the voltage varies; the frequency of the voltage then determines the frequency of the sounds produced.

Transmission of sound

In order for sound to travel between the source and the receiver there must be some material between them that can vibrate in the direction of travel (called the propagation direction). (The fact that sound can only be transmitted by a material medium means that an explosion outside a spaceship would not be heard by its occupants!) The motion of the sound-producing body causes density variations in the medium (see Figure 5, which schematically shows the density variations associated with a sound wave), which move along in the direction of propagation. The transmission of sounds in the form of these density variations is termed a wave since these variations are carried forward without significant change, although eventually **friction** in the air itself causes the wave to dissipate. (This is analogous to a water wave in which the particles of water vibrate up and down, while the "wave" propagates forward.) Since the motion of the medium at any point is a small vibration back and forth in the direction in which the wave is proceeding, sound is termed a longitudinal wave. (The water wave, like the violin string, is an example of a transverse wave.) The most usual medium of sound transmission is air, but any substance that can be compressed can act as a medium for sound propagation. A fundamental characteristic of a wave is that it carries **energy** and **momentum** away from a source without transporting **matter** from the source.

Since the speed of sound in air is about about 1,088 ft/sec (331 m/sec), human speech involves wavelengths from about 1.3 in to 11 ft (3.3 cm to 3.3 m). Thus, the wavelengths of speech are of the size of ordinary objects, unlike light, whose wavelengths are extremely small compared to items that are part of everyday life. Because of this, sound does not ordinarily cast "acoustic shadows" but, because its wavelengths are so large, can be transmitted around ordinary objects. For example, if a light is shining on a person, and a book is placed directly between them, the person will no longer be able to see the light (a shadow is cast by the book on the eyes of the observer). However, if one person is speaking to another, then placing a book between them will hardly affect the sounds heard at all; the **sound waves** are able to go around the book to the observer's ears. On the other hand, placing a high wall between a highway and houses can greatly decrease the sounds of the traffic noises if the dimensions of the wall (height and length) are large compared with the wavelength of the traffic sounds. Thus, sound waves (as for all waves) tend to "go around" (e.g., ignore the presence of) obstacles which are small compared with the wavelength of the wave; and are reflected by obstacles which are large compared with the wavelength. For obstacles of approximately the same size as the wavelength, waves exhibit a very complex behavior known as **diffraction**, in which there are enhanced and diminished values of the wave amplitude, but which is too complicated to be described here in detail.

The speed of sound in a gas is proportional to the **square root** of the pressure divided by the density. Thus, helium, which has a much lower density than air, transmits sound at a greater speed than air. If a person breathes some helium, the characteristic wavelengths are still determined by the shape of the mouth, but the greater sound speed causes the speech to be emitted at a higher frequency—thus the "Donald Duck" sounds from someone who speaks after taking a breath of helium from a **balloon**.

In general, the speed of sound in liquids is greater than in gases, and greater still in solids. In sea water, for example, the speed is about 4,750 ft/sec (1,447 m/sec); in a gas, the speed increases as the pressure increases, and as the density decreases. Typical speeds of sound in solids are 5,450 yd/sec (5,000 m/sec), but vary considerably from one solid to another.

KEY TERMS

Amplitude—The maximum displacement of the material that is vibrating.

Condensations—When air is the vibrating medium there are alternate small increases and decreases of the density of the air; the increases are called condensations.

Cycle—A single complete vibration.

Cycles per second—The number of complete vibrations per second.

Frequency—The rate at which vibrations take place (number of times per second the motion is repeated), denoted here by f and given in cycles per second or in hertz.

Fundamental—The lowest frequency of vibration of a sound-producing body (also called the first harmonic).

Harmonics (first, second, etc.)—The various frequencies of vibration of a sound-producing body, numbered from the one of lowest frequency to higher frequencies.

Hertz—A hertz (abbreviated as Hz) is one cycle per second.

Infrasonic vibrations—When the rate of vibration is below the range of human hearing, e.g., below about 10 cycles per second.

Longitudinal wave—The case where the motion of the vibrating body is in the wave propagation direction.

Loudspeaker—A device to produce sounds from an electric current, by electrical and mechanical means, in the range of frequencies around the sonic range (that is produced by humans).

Medium—A material body that carries the acoustic vibrations away from the body producing them.

Microphone—A device to change sound waves (pressure waves), by electrical and mechanical means, into an electric current having the same frequencies as the sound, in the range of frequencies around the sonic range (that is produced by humans).

Nodes—Places where the amplitude of vibration is zero.

Overtones—The set of harmonics, beyond the first, of a soundproducing body.

Period—The length of time for one cycle to be completed; the reciprocal of the frequency.

Propagation direction—The direction in which the wave is traveling.

Rarefactions—When air is the vibrating medium there are alternate small increases and decreases of the density of the air; the decreases are called rarefactions.

Sonar—A device utilizing sound to determine the range and direction to an underwater object.

Supersonic—Refers to bodies moving at speeds greater than the speed of sound (not normally involved in the study of acoustics).

Transverse wave—The case where the motion of the vibrating body is perpendicular to the wave propagation direction.

Ultrasonic vibrations—When the rate of vibration is above the range of human hearing, e.g., above about 20,000 cycles per second.

Wave—A motion, in which energy changes are propagated, is carried away from some source, which repeats itself in space and time with little or no change.

Wavelength—The distance, at any instant of time, between parts of a vibrating body having the identical motion, denoted here by L.

Reception of sound

Physiological acoustics is the study of the transmission of sound and how it is heard by the human **ear**. Sound travels in waves, vibrations that cause compression and rarefaction of molecules in the air. The organ of hearing, the ear, has three basic parts that collect and transmit these vibrations: the outer, middle and inner ear. The outer ear is made of the pinna, the external part of the ear that can be seen, which acts to funnel sound through the ear canal toward the eardrum or tympanic membrane. The membrane is highly sensitive to vibrations and also protects the middle and inner ear. When the eardrum vibrates it sets up vibrations in the three tiny bones of the middle ear, the malleus, incus and stapes, which are often called the hammer, anvil and stirrup because of their resemblance to those objects. These bones amplify the sound. The stapes is connected to the oval window, the entrance to the inner ear, which contains a spiral-shaped, fluid-filled chamber called the cochlea. When vibrations are transmitted from the stapes to the oval window, the fluid within the cochlea is put into motion. Tiny hairs that line the basilar membrane of the cochlea, a membrane that divides the cochlea lengthwise, move in accordance with the wave pattern. The hair cells convert the mechanical energy of the waveform into nerve signals that reach the auditory nerve and then the **brain**. In the brain, sound is interpreted.

In the sonic range of frequencies, the microphone, a device using electrical and mechanical components, is the common method of receiving sounds. One simple form is to have a diaphragm as one plate of an electrical condenser. When the diaphragm vibrates under the action of a sound wave, the current in the circuit varies due to the varying **capacitance** of the condenser. This varying current can then be used to activate a meter or **oscilloscope** or, after suitable processing, make an audio tape or some such permanent record.

Applications

The applications of acoustical devices are far too numerous to describe; one only has to look around our homes to see some of them: telephones, radios and **television** sets, compact disc players and tape recorders; even clocks that "speak" the time. Probably one of the most important from the human point of view is the hearing aid, a miniature microphone-amplifier-loudspeaker that is designed to enhance whatever range of frequencies a person finds difficulty hearing.

However, one of the first large-scale industrial uses of sound propagation was by the military in World War I, in the detection of enemy submarines by means of sonar (for *sound navigation and ranging*). This was further developed during the period between then and World War II, and since then. The ship-hunting **submarine** has a sound source and receiver projecting from the ship's hull that can be used for either listening or in an echo-ranging mode; the source and receiver are directional, so that they can send and receive an acoustic signal from only a small range of directions at one time. In the listening mode of operation, the operator tries to determine what are the sources of any noise that might be heard: the regular beat of an engine heard underwater can tell that an enemy might be in the vicinity. In the echo-ranging mode, a series of short bursts of sound is sent out, and the time for the echo to return is noted; that time interval multiplied by the speed of sound in water indicates (twice) the distance to the reflecting object. Since the sound source is directional, the direction in which the object lies is also known. This is now such a well developed method of finding underwater objects that commercial versions are available for fishermen to hunt for schools of **fish**.

Ultrasonic sources, utilizing pulses of frequencies in the many millions of cycles per second (and higher), are now used for inspecting metals for flaws. The small wavelengths make the pulses liable to reflection from any imperfections in a piece of metal. Castings may have internal cracks which will weaken the structure; welds may be imperfect, possibly leading to failure of a metal-to-metal joint; **metal fatigue** may produce cracks in areas impossible to inspect by **eye**. The use of ultrasonic inspection techniques is increasingly important for failure prevention in **bridges**, **aircraft**, and pipelines, to name just a few.

The use of **ultrasonics** in medicine is also of growing importance. The detection of kidney stones or gallstones is routine, as is the imaging of fetuses to detect suspected **birth defects**, cardiac imaging, **blood** flow measurements, and so forth.

Thus, the field of acoustics covers a vast array of different areas of use, and they are constantly expanding. Acoustics in the communications industry, in various phases of the construction industries, in oil field exploration, in medicine, in the military, and in the entertainment industry, all attest to the growth of this field and to its continuing importance in the future.

Resources

Books

Deutsch, Diana. *Ear and Brain: How We Make Sense of Sounds* New York: Copernicus Books, 2003.

Kinsler, Lawrence E., et. al. *Fundamentals of Acoustics,* 4th ed. New York: John Wiley & Sons, 1999.

Organizations

The Acoustical Society of America. 2 Huntington Quadrangle, Suite 1NO1, Melville, NY 11747–4502 Phone: (516) 576–2360. <http://asa.aip.org/index.html>

Other

The University of New South Wales. "Music Acoustics" [cited March 10, 2003]. <http://www.phys.unsw.edu.au/music/>.

David Mintzer

Acrylic *see* **Artificial fibers**

Actinides

Actinides or actinoids is a generic term that refers to a series of 15 chemical elements. Denoted by the generic symbol An, these elements are all radioactive heavy metals, positioned in the seventh period and elaborated upon at the bottom of the **periodic table**.

Occurrence

Only actinium (atomic symbol, Ac), thorium (Th), protactinium (Pa), and **uranium** (U) are extracted from deposits in nature. In Canada, the United States, South Africa, and Namibia, thorium and protactinium are available in large quantities. All other actinides are synthetic or man-made.

To understand the physical and chemical properties of actinides, a basic foundation of atomic structure, radioactivity, and the periodic table is required. The atomic structure can be pictured like a solar system. In the middle of the atom is the nucleus, composed of neutrons (no charge) and protons (positively charged). Around the nucleus, electrons (negatively charged) are rotating on their own axis, as well as circulating in definite **energy** levels. Each energy level (or shell) is designated by a principal **quantum number** (n) as K, L, M, N, O, etc. or 1, 2, 3, 4, 5, etc. respectively. Each shell has sub-shells, or orbitals. The first energy level consists of one orbital (*s*); the second level consists of two orbitals (s and p); the third level consists of three orbitals (*s*, *p*, and *d*), and from the fourth level on up there are four orbitals (*s*, *p*, *d*, and *f*). The orbitals closer to the nucleus are lower in energy level than the orbitals further away from the nucleus.

The electrons are distributed according to Pauli's exclusion principle. In any atom, the number of protons is equal to the number of electrons, thus bringing the neutrality in charge. These stable and abundant **atoms** exist in nature only. In the unstable and less abundant atoms, the number of neutrons is more than the number of electrons (one element with the same **atomic number** but with a different atomic **mass**). These unstable atoms are known as isotopes, some of which are radioactive.

Radioactive isotopes become nonradioactive by the decaying process. The decaying process may involve an **emission** of: (1) electrons or **negative** beta particles; (2) helium nuclei or alpha particles; (3) gamma rays or very high **frequency** electromagnetic waves; (4) positrons or positively charged electrons or positive beta particles.

The decaying process may also be due to K-capture (an orbital **electron** of a radioactive atom that may be captured by the nucleus and taken into it). Each of the above mentioned decay processes results in a **isotope** of a different element (an element with a different atomic number). The emission of alpha particles also results in elements with different atomic weights.

The most important decay process in actinides is K-capture, followed by the splitting, or fission, of the nucleus. This fission results in enormous amounts of energy and two or more extra neutrons. These newly formed neutrons can further start K-capture, with the subsequent reactions going on like a chain reaction. Atomic reactors and atomic bombs depend on the chain reactions.

A scheme of the classification of all known (both discovered and man-made) elements is represented in the modern periodic table. The periodical table is divided into vertical columns and horizontal rows representative of the periods with increasing atomic numbers. Each box contains one element and is represented by its symbol, a single or double letter, with the atomic number as a superscript, and the **atomic weight** as a subscript. Note that in the sixth and seventh periods, there are breaks between atomic numbers 57 and 72 (**lanthanides**) and 89 and 104 (actinides). Fourteen elements are present between the atomic numbers 89 and 104, and are elaborated upon at the bottom of the periodic table. These 14 elements, plus actinium, are known as actinides.

General preparation

All actinide metals are prepared on a scale by the reduction of AnF_3 or AnF_4 with vapors of **lithium** (Li), **magnesium** (Mg), **calcium** (Ca), or **barium** (Ba) at 2,102–2,552°F (1,150–1,400°C); sometimes chlorides or oxides are used. The Van Arkel-de Boer process is a special preparation method used for thorium and protactinium.

Physical and chemical properties

A common feature of actinides is the possession of multiple oxidation states. The term **oxidation state** refers to the number of electron(s) that are involved or that can possibly become involved in the formation of chemical bond(s) in that compound, when one element combines with another element during a chemical reaction. The oxidation state is designated by a plus sign (when an electron is donated or electro-positive) or by a minus sign (when the electron is accepted or electronegative). An element can have more than one oxidation state. An electron configuration can provide the information about the oxidation state of that element. The most predominant oxidation state among actinides is +3, which is similar to lanthanides. The **crystal** structure (**geometry**), solubility property, and the formation of chemical compounds are based on the oxidation state of the given element.

Actinides ions in an aqueous **solution** are colorful, containing colors such as red purple (U^{3+}), purple (Np^{3+}), pink (Am^{3+}), green (U^{4+}), yellow green (Np^{4+}), and pink red (Am^{4+}). Actinides ions U, Np, Pu, and Am undergo **hydrolysis**, disproportionation, or formation of polymeric ions in aqueous solutions with a low **pH**.

All actinides are characterized by partially filled 5f, 6d, and 7s orbitals. Actinides form complexes easily with certain ligands as well as with halide, sulfate, and other ions. Organometallic compounds (compounds with a sign bond between the meta and **carbon** atom of organic moiety) of uranium and thorium have been prepared and are useful in organic synthesis. Several alloys of protactinium with uranium have been prepared.

Uses of actinides

Even though hazards are associated with radioactivity of actinides, many beneficial applications exist as well. Radioactive nuclides are used in **cancer** therapy, analytical **chemistry**, and in basic research in the study of chemical structures and mechanisms. The explosive power of uranium and plutonium are well exploited in making atom bombs. In fact, the uranium enriched atom bomb that exploded over Japan was the first uranium bomb released. Nuclear reactions of uranium-235 and plutonium-239 are currently utilized in atomic energy powerplants to generate electric power. Thorium is economically useful for the reason that fissionable uranium-233 can be produced from thorium-232. Plutonium-238 is used in implants in the human body to power the **heart pacemaker**, which is does not need to be replaced for at least 10 years. Curium-244 and plutonium-238 emit **heat** at 2.9 watts and 0.57 watts per gram, respectively. Therefore, curium and plutonium are used as power sources on the **Moon** to provide electrical energy for transmitting messages to **Earth**.

Sutharchanadevi Murugen

Actinium *see* **Actinides**

Action potential

Action potentials are the electrical pulses that allow the transmission of information within nerves. An action potential represents a change in electrical potential from the resting potential of the neuronal **cell membrane**, and involves a series of electrical and underlying chemical changes that travel down the length of a neural cell (**neuron**). The neural impulse is created by the controlled development of action potentials that sweep down the body (axon) of a neural cell.

There are two major control and communication systems in the human body, the **endocrine system** and the **nervous system**. In many respects, the two systems compliment each other. Although long duration effects are achieved through endocrine hormonal regulation, the nervous system allows nearly immediate control, especially regulation of homeostatic mechanisms (e.g., **blood pressure** regulation).

The neuron cell structure is specialized so that at one end, there is a flared structure termed the dendrite.

At the dendrite, the neuron is able to process chemical signals from other neurons and endocrine **hormones**. If the signals received at the dendritic end of the neuron are of a sufficient strength and properly timed, they are transformed into action potentials that are then transmitted in a "one-way" direction (unidirectional propagation) down the axon.

In neural cells, electrical potentials are created by the separation of positive and **negative** electrical charges that are carried on ions (charged **atoms**) across the cell membrane. There are a greater number of negatively charged **proteins** on the inside of the cell, and unequal distribution of cations (positively charged ions) on both sides of the cell membrane. **Sodium** ions (Na+) are, for example, much more numerous on the outside of the cell than on the inside. The normal distribution of charge represents the resting membrane potential (RMP) of a cell. Even in the rest state there is a standing potential across the membrane and, therefore, the membrane is polarized (contains an unequal distribution of charge). The inner cell membrane is negatively charged relative to the outer shell membrane. This potential difference can be measured in millivolts (mv or mvolts). Measurements of the resting potential in a normal cell average about 70 mv.

The standing potential is maintained because, although there are both electrical and **concentration** gradients (a range of high to low concentration) that induce the excess sodium ions to attempt to try to enter the cell, the channels for passage are closed and the membrane remains almost impermeable to sodium ion passage in the rest state.

The situation is reversed with regard to potassium ion (K+) concentration. The concentration of potassium ions is approximately 30 times greater on the inside of the cell than on the outside. The potassium concentration and electrical gradient forces trying to move potassium out of the cell are approximately twice the strength of the sodium ion gradient forces trying to move sodium ions into the cell. Because, however, the membrane is more permeable to potassium passage, the potassium ions leak through he membrane at a greater **rate** than sodium enters. Accordingly, there is a net loss of positively charges ions from the inner part of the cell membrane, and the inner part of the membrane carries a relatively more negative charge than the outer part of the cell membrane. These differences result in the net RMP of −70mv.

The structure of the cell membrane, and a process termed the sodium-potassium pump maintains the neural cell RMP. Driven by an ATPase **enzyme**, the sodium potassium pump moves three sodium ions from the inside of the cell for every two potassium ions that it brings back in. The ATPase is necessary because this movement or pump of ions is an active process that moves sodium and potassium ions against the standing concentration and electrical gradients. Equivalent to moving water uphill against a gravitational gradient, such action requires the expenditure of **energy** to drive the appropriate pumping mechanism.

When a neuron is subjected to sufficient electrical, chemical, or in some cases physical or mechanical **stimulus** that is greater than or equal to a threshold stimulus, there is a rapid movement of ions, and the resting membrane potential changes from $-70mv$ to $+30mv$. This change of approximately $100mv$ is an action potential that then travels down the neuron like a wave, altering the RMP as it passes.

The creation of an action potential is an "all or none" event. Accordingly, there are no partial action potentials. The stimulus must be sufficient and properly timed to create an action potential. Only when the stimulus is of sufficient strength will the sodium and potassium ions begin to migrate done their concentration gradients to reach what is termed threshold stimulus and then generate an action potential.

The action potential is characterized by three specialized phases described as depolarization, repolarization, and hyperpolarization. During depolarization, the $100mv$ electrical potential change occurs. During depolarization, the neuron cannot react to additional stimuli and this inability is termed the absolute refractory period. Also during depolarization, the RMP of $-70mv$ is reestablished. When the RMP becomes more negative than usual, this phase is termed hyperpolarization. As repolarization proceeds, the neuron achieves an increasing ability to respond to stimuli that are greater than the threshold stimulus, and so undergoes a relative refractory period.

The opening of selected channels in the cell membrane allows the rapid movement of ions down their respective electrical and concentration gradients. This movement continues until the change in charge is sufficient to close the respective channels. Because the potassium ion channels in the cell membrane are slower to close than the sodium ion channels, however, there is a continues loss of potassium ion form the inner cell that leads to hyperpolarization.

The sodium-potassium pump then restores and maintains the normal RMP.

In demyelinated nerve fibers, the depolarization induces further depolarization in adjacent areas of the membrane. In myelinated fibers, a process termed salutatory conduction allows transmission of an action potential, despite the insulating effect of the myelin sheath. Because of the sheath, ion movement takes place only at the Nodes of Ranvier. The action potential jumps from node to node along the myelinated axon. Differing types of nerve fibers exhibit different speed of action potential conduction. Larger fibers (also with decreased **electrical resistance**) exhibit faster transmission than smaller diameter fibers).

The action potential ultimately reaches the presynaptic portion of the neuron, the terminal part of the neuron adjacent to the next **synapse** in the neural pathway). The synapse is the gap or intercellular **space** between neurons. The arrival of the action potential causes the release of ions and chemicals (neurotransmitters) that travel across the synapse and act as the stimulus to create another action potential in the next neuron.

See also Adenosine triphosphate; Nerve impulses and conduction of impulses; Neuromuscular diseases; Reflex; Touch.

Resources

Books

Guyton, Arthur C., and Hall, John E. *Textbook of Medical Physiology,* 10th ed. Philadelphia: W.B. Saunders Co., 2000.

Kandel, E.R., J.H. Schwartz, and T.M. Jessell. (eds.) *Principles of Neural Science,* 4th ed. Boston: Elsevier, 2000.

Thibodeau, Gary A., and Kevin T. Patton. *Anatomy & Physiology,* 5th ed. St. Louis: Mosby, 2002.

Periodicals

Cowan, W.M., D.H. Harter, and E.R. Kandel. "The Emergence of Modern Neuroscience: Some Implications for Neurology and Psychiatry." *Annual Review of Neuroscience* 23:343–39.

Sah R., R.J. Ramirez, G.Y. Oudit, et al. "Regulation of Cardiac Excitation-Contraction Coupling By Action Potential Repolarization: Role of the Transient Outward Potassium Current." *J. Physiology* (Jan. 2003):5–18.

Organizations

National Alzheimer's Association, 919 North Michigan Avenue, Suite 1100, Chicago, IL 60611–1676. (800) 272–3900. (August 21, 2000) [cited January 18, 2003]. <http://www.alz.org>.

K. Lee Lerner

Activated complex

The term activated **complex** refers to the molecular compound or compounds that exist in the highest **energy** state, or *activated stage*, during a chemical reaction. An activated complex acts as an intermediary between the reactants and the products of the reaction.

A chemical reaction is the reorganization of **atoms** of chemically compatible and chemically reactive molecular compounds, called *reactants*. A chemical reaction goes through three stages, the initial stage consisting of the reactants, the transition stage of the activated complex, and the final stage, in which the products are formed.

For example, consider the chemical reaction

$$A + B \longleftrightarrow A—B \rightarrow C + D$$

where A and B are reactants, A-B is the activated complex, and C and D are the products. For a chemical reaction to occur, the reactant molecules should collide. Collisions between molecules are facilitated by an increase in the **concentration** of reactant, an increase in **temperature**, or the presence of a catalyst. Not every collision is successful, that is, produces a chemical reaction. For a successful collision to occur, reactants require a minimum amount of energy, called the *activation energy*. Once the reactant reaches the energy level, it enters the transition stage and forms the activated complex.

The energy of the activated complex is higher than that of reactants or the products, and the state is temporary. If there is not sufficient energy to sustain the chemical reaction, the activated complex can reform into the reactants in a backward reaction. With proper energy, though, the activated complex forms the products in a forward reaction.

See also Compound, chemical; Element, chemical.

Active galactic nuclei

Active galactic nuclei (AGNs) are perhaps the most violently energetic objects in the universe. AGNs are located at the centers of some galaxies—perhaps most galaxies—and emit a tremendous amount of **energy**, sometimes on the order of trillion times the output of the **Sun**. An AGN may outshine all the stars in its **galaxy** by a factor of 100. The energy of a typical AGN is generated in a volume smaller in diameter than our **solar system**, leading astronomers to conclude that AGNs are probably powered by supermassive black holes, that is, black holes containing a million to a billion or more times the **mass** of the Sun. The **event horizon** of such a **black hole** would be a small object by astronomical standards, with a diameter equal to perhaps one-twentieth of the **distance** from **Earth** to the Sun. Gas attracted by the black hole's gravity spirals inward, forming a rotating "accretion disk" (so-called because as **matter** approaches the black hole it accretes, or adds to, this disk). As the **accretion disk** spirals toward the event horizon of the black hole it is accelerated, compressed, and heated, causing much of its gravitational potential energy to be released in the form of **radiation**. This converted gravi-

tational potential energy is the source of the tremendous outpourings of radiation that are observed from AGNs.

Much of the energy from AGNs is emitted as **radio waves** rather than as visible **light**. These waves are emitted by electrons moving in a helical path in a strong magnetic field at speeds near the speed of light. This is known as *synchrotron radiation* (after the machine called a synchrotron, a type of **cyclotron** that confines high-speed charged particles by using a magnetic field to **force** them to move in curved paths). AGNs also emit visible light, **x rays**, and gamma rays.

About 10% of AGNs have mirror-image jets of material streaming out from the nucleus in opposite directions and at right angles to the accretion disk, moving at nearly the speed of light. Near their sources, these jets tend to vary in brightness on rapid cycles of days to months in length. Such rapid variations indicate that the energy-producing nucleus is small, ranging in size from a few light days to a few light months in diameter. Size can be deduced from the time-scale of brightness variations. Coordinated changes across a jet's source imply that some sort of coherent physical process is affecting the jet from one side of its aperture to another at least as rapidly as the observed variation, and this cannot happen faster than the speed of light. Thus a brightness change of, say, one day implies a source no more than one light-day in diameter. (A light-day is the distance traveled by light in one day, 16 billion mi. [2.54×10^{10} km].)

There are several varieties of active galactic nuclei, including compact radio galaxies, Seyfert galaxies, BL Lacertae objects, and quasars. Compact radio galaxies appear as giant elliptical galaxies. Radio telescopes, however, reveal a very energetic compact nucleus at the center, which is the source of most of the energy emitted by the galaxy. Perhaps the best-known compact radio galaxy is M87. Recent observations provide strong evidence that this core contains a supermassive black hole. Seyfert galaxies look like **spiral** galaxies with a hyperactive nucleus; that is, a set of normal-looking spiral arms surround an abnormally bright nucleus. BL Lacertae objects look like stars, but in reality are most likely very active galactic nuclei. BL Lacertae objects exhibit unusual behaviors, including extremely rapid and erratic variations in observed properties, and their exact nature is not known. Quasars also look like stars, but they are now known to be simply the most distant and energetic type of active galactic nuclei. They may be more active than nearer AGNs because they are observed in a younger condition (i.e., being distant, their light has taken longer to reach us), and so are seen in a particularly vigorous stage of accretion.

AGNs may merely be unusually active galactic nuclei, as evidence is accumulating that many or even most galaxies have black holes at their centers. If this is true, AGNs may simply have unusually active central black holes. Recent observations have confirmed that our own galaxy has a black hole at its core that is about three million times as massive as the Sun. It emits far less energy than an AGN; this may be simply because there is less matter falling into it. Black holes do not emit energy on their own, but become visible by squeezing energy out of the matter that they are swallowing. If the **space** in the near vicinity of a galaxy's central black hole is relatively free of matter, then the galaxy's core will be relatively quiet, that is, emit little energy; if a sufficient amount of matter is available for consumption by the black hole, then an AGN results.

AGNs are a particularly active area of astronomical research; about one fifth of all research astronomers are presently engaged in investigating AGNs.

Resources

Periodicals

Glanz, James, "Evidence Points to Black Hole At Center of the Milky Way." *New York Times* (October 17, 2002).

Other

Laboratory for High Energy Astrophysics. "Active Galaxies and Quasars." National Aeronautics and Space Administration. 1997; updated 2002. [cited October 20, 2002]. <http://imagine.gsfc.nasa.gov/cgi-bin/print.pl>.

Acupressure

Acupressure is an ancient method of improving a person's health by applying **pressure** to specific sites on the body. Acupressure is similar to **acupuncture**, but does not break the skin. Instead, the acupressure practitioner relies on pressure invoked by fingertip or knuckle to accomplish his purpose.

Also called Shiatzu, acupressure originated in ancient China approximately 500 years B.C. and spread throughout the Orient. It is the oldest form of **physical therapy** for which instructions are written. A basic level of acupressure can be practiced by anyone for the relief of **pain** or tension, and the practice is in active use by those who practice alternative forms of medicine.

Like acupuncture, acupressure recognizes certain pressure points located along meridians that extend the length of the body. Certain meridians and their connectors are associated with given organs or muscles, and pressure points on the meridian will affect the pain level in the **organ**. The pressure points are often located far from the organ they affect. This is a reflection of the be-

lief that **energy** flows through the body along the meridians and that pain develops in an area when the energy flow through the corresponding meridian is stopped or reduced. Acupressure opens the energy and eases pain or discomfort.

Anyone who would practice acupressure must first learn the location of the meridians and their connectors. More than a thousand pressure points have been mapped along the meridians, but the amateur practitioner need not know them all. Generally the **individual** with a recurrent or chronic pain can learn the point that best eases his pain and learn how much pressure to apply to accomplish his purpose.

Reports from various Asian and American institutions claim that acupressure can be an effective way to ease pain and relax stressed muscles without the aid of medications. It has even been employed to provide **anesthesia** for certain types of **surgery**.

See also Alternative medicine.

Larry Blaser

Acupuncture

Acupuncture is an ancient method of therapy that originated in China more than 2,000 years ago. It consists of inserting solid, hair-thin needles through the skin at very specific sites to achieve a cure of a **disease** or to relieve **pain**. Although it is not part of conventional medical treatment in most of the Western world, a 1998 consensus statement released by the National Institutes of Health (NIH) in the United States said evidence clearly shows acupuncture helps relieve many types of chronic and acute pain; nausea and vomiting associated chemotherapy, **anesthesia**, and pregnancy; and alters **immune system** functions. The World Health Organization, in conjunction with the International Acupuncture Training Center at Shanghai College of Traditional Chinese Medicine, declares acupuncture can be effective for dozens of problems—from bed wetting and allergies to chronic fatigue **syndrome** and **anxiety** disorders. Although many American physicians remain skeptical about its use as an anesthetic for **surgery** or cure for grave diseases, by 1999, approximately 10,000 acupuncturists held licenses in the United States, including some 3,000 physicians.

In the Far East, acupuncture is used extensively. It is considered one part of a total regimen that includes **herbal medicine**, closely guided dietetics, and psychological counseling. In ancient Chinese philosophy, good health de-

stand rigorous scientific scrutiny." Some health insurance programs cover treatment.

See also Acupressure; Alternative medicine.

Resources

Periodicals

"Acupuncture." (fact sheet). National Institutes of Health Office of Alternative Medicine, 1993.
"Acupuncture Illustrated." *Consumer Reports* 59 (January 1994): 54-57.
Bonta, I.L. "Acupuncture Beyond the Endorphine Concept?" *Medical Hypotheses* 58, 3(2002): 221-224.
Botello, J.G. "Acupuncture: Getting the Point." *Lears* 6 (November 1993): 43- 44.

ADA (adenosine deaminase) deficiency

ADA deficiency is an inherited condition that occurs in fewer than one in 100,000 live births worldwide. Individuals with ADA deficiency inherit defective ADA genes and are unable to produce the **enzyme** adenosine deaminase in their cells. The ADA **gene** consists of a single 32 kb locus containing 12 exons and is located on the long arm of **chromosome** 20. The enzyme adenosine deaminase is needed to break down metabolic byproducts that become toxic to T-**cell** lymphocytes, and is essential to the proper functioning of the **immune system**. Most of the body's cells have other means of removing the metabolic byproducts that ADA helps break down and remain unaffected by ADA deficiency. However, T-cell lymphocytes, white **blood** cells that help fight **infection**, are not able to remove the byproducts in the absence of ADA.

Without ADA, the toxins derived from the metabolic byproducts kill the **T cells** shortly after they are produced in the bone marrow. Instead of having a normal life span of a few months, T cells of individuals with ADA deficiency live only a few days. Consequently, their numbers are greatly reduced, and the body's entire immune system is weakened. ADA deficiency is the first known cause of a condition known as severe combined immunodeficiency (SCID).

The body's immune system includes T-cell lymphocytes and B-cell lymphocytes; these lymphocytes play different roles in fighting infections. B cells produce antibodies that lock on to disease-causing viruses and **bacteria**, thereby marking the **pathogens** for destruction. Unlike B cells, T cells cannot produce antibodies, but they do control B cell activity. T-cell helpers enable antibody production, whereas T-cell suppressors turn off an-

tibody production. Another T-cell subtype kills **cancer** cells and virus-infected cells.

Because T cells control B cell activity, the reduction of T cells results in an absence of both T cell and B cell function called severe combined immunodeficiency (SCID). Individuals with SCID are unable to mount an effective immune response to any infection. Therefore, exposures to organisms that normal, healthy individuals easily overcome become deadly infections in SCID patients. Prior to present-day treatments, most ADA-deficient SCID victims died from infections before reaching the age of two. Although SCID is usually diagnosed in the first year of life, approximately one-fifth of ADA deficient patients have delayed onset SCID, which is only diagnosed later in childhood. There are also a few cases of ADA deficiency diagnosed in adulthood.

Treatments for ADA deficiency

The treatment of choice for ADA deficiency is bone marrow transplantation from a matched sibling donor. Successful bone marrow transplants can relieve ADA deficiency. Unfortunately, only 20–30% of patients with ADA deficiency have a matched sibling donor. Another treatment involves injecting the patient with PEG-ADA, polyethylene glycol-coated bovine ADA derived from cows. The PEG coating helps keep the ADA from being prematurely degraded. Supplying the missing enzyme in this way helps some patients fight infections, while others are helped very little.

The latest treatment for ADA deficiency is **gene therapy**. Gene therapy provides victims with their own T cells into which a normal copy of the human ADA gene has been inserted. ADA deficiency is the first **disease** to be treated with human gene therapy.

The first person to receive gene therapy for ADA deficiency was four-year-old Ashanthi DeSilva. The treatment was developed by three physicians—W. French Anderson, Michael Blaese, and Kenneth Culver. DeSilva received her first treatment, an infusion of her own T cells implanted with normal ADA genes, on September 14, 1990 at the National Institutes of Health in Bethesda, Maryland.

How did DeSilva's T cells acquire the normal ADA genes? A. Dusty Miller of the Fred Hutchinson Research Center in Seattle, Washington, made the vectors for carrying the normal ADA genes into the T cells. These vectors were made from a **retrovirus**, a type of **virus** that inserts its genetic material into the cell it infects. By replacing harmful retroviral genes with normal ADA genes, Miller created the retrovirus vectors to deliver the normal ADA genes into DeSilva's T cells. The retrovirus vectors—carrying normal ADA genes—were mixed with

KEY TERMS

B cell lymphocyte—Immune system white blood cell that produces antibodies.

PEG-ADA—A drug, polyethylene-coated bovine ADA, used for treating ADA-deficiency. The polyethylene coating prevents rapid elimination of the ADA from the blood.

Retrovirus—A type of virus that inserts its genetic material into the chromosomes of the cells it infects.

Stem cells—Undifferentiated cells capable of self-replication and able to give rise to diverse types of differentiated or specialized cell lines.

T cells—Immune-system white blood cells that enable antibody production, suppress antibody production, or kill other cells.

T cells that had been extracted from DeSilva's blood and grown in culture dishes. The retrovirus vectors entered the T cells and implanted the normal ADA genes into the T-cell chromosomes. The T cells were then infused back into DeSilva's blood where the normal ADA genes in them produced ADA.

When doctors saw that DeSilva benefited and suffered no harmful effects from gene therapy, they repeated the same treatment on nine-year-old Cynthia Cutshall on January 30, 1991. Both girls developed functioning immune systems. However, since T cells have a limited life span, DeSilva and Cutshall needed to receive periodic infusions of their genetically-corrected T cells, and they both continued with PEG-ADA injections.

Subsequent research is focusing on developing a permanent cure for ADA deficiency using gene therapy. In May and June of 1993, Cutshall and three newborns with ADA deficiency received their own **stem cells** that had been implanted with normal ADA genes. Unlike T cells which only live for a few months, stem cells live throughout the patient's life, and thus the patient should have a lifetime supply of ADA without requiring further treatment.

See also Genetic disorders; Genetic engineering; Immunology.

Resources

Books

Lemoine, Nicholas R., and Richard G. Vile. *Understanding Gene Therapy* New York: Springer-Verlag, 2000.

Hershfield, M.S., Mitchell, B.S. "Immunodeficiency Diseases Caused by Adenosine Deaminase Deficiency and Purine Nucleoside Phosphorylase Deficiency." In: Scriver, C.R.;

Beaudet, A.L.; Sly, W.S.; Valle, D., eds) *The Metabolic and Molecular Bases of Inherited Disease,* 7th ed. Vol. 2. New York: McGraw-Hill 1995.

Periodicals

Blaese, Michael R. "Development of Gene Therapy for Immunodeficiency: Adenosine Deaminase Deficiency." *Pediatric Research* 33 (1993): S49–S55.

Thompson, Larry. "The First Kids With New Genes." *Time* (7 June 1993): 50–51.

Pamela Crowe

Adaptation

An adaptation is any developmental, behavioral, physiological, or anatomical change in an **organism** that gives that organism a better chance to survive and reproduce. The word "adaptation" also refers to the fitting of a whole **species**, over **time**, to function in its particular environment, and to those specific features of a species that make it better-adapted. Adaptations acquired by individuals during their lifetime, such as muscles strengthened by **exercise** or behaviors honed by experience, make an **individual** organism better-adapted; species as a whole, however, generally become better adapted to their environments only by the process of natural **selection**. Except in the form of learned **behavior**, adaptations achieved by individual organisms cannot be passed on to offspring.

Less-adapted species are less perfectly attuned to a particular environment but may be better-suited to survive changes in that environment or to colonize new areas. Highly adapted species are well-suited to their particular environment, but being more specialized, are less likely to survive changes to that environment or to spread to other environments. An example of a highly adapted species would be a **flower** that depends on a specific insect that exists only or primarily in its present environment for **pollination**. The **plant** may achieve highly reliable pollination by these means, but if its target species of insect becomes extinct, the plant will also become extinct—unless the species can adapt to make use of another pollinator.

At the level of the individual organism, an adaptation is a change in response to conditions. This is a short-term change with a short-term benefit. An example of an adaptation of this type is the production of sweat to increase cooling on a hot day.

Another type of adaptation is sensory adaptation. If a receptor or sense **organ** is over-stimulated, its excitability is reduced. For example, continually applied

A boojum tree in Mexico during the dry season. The plant has adapted to seasonal changes in precipitation by restricting the growth of its foliage during the dry season. *JLM Visuals. Reproduced by permission.*

pressure to an area of skin eventually causes the area to become numb to feeling and a considerably larger pressure has to be applied to the area subsequently to elicit a similar response. This form of adaptation enables animals to ignore most of their skin most of the time, freeing their attention for more pressing concerns.

Whether occurring within a span of minutes, over an organism's lifetime, or over thousands or millions of years, adaptation serves to increase the efficiency of organisms and thus, ultimately, their chances of survival.

Resources

Books

Gould, Stephen Jay. *The Structure of Evolutionary Theory.* Cambridge, MA: Harvard University Press, 2002.

Adder *see* **Vipers**

Addiction

Addiction is a **compulsion** to engage in unhealthy or detrimental **behavior**. Human beings can become addicted to many forms of behaviors such as gambling, overeating, sex, or reckless behavior, but the term "addiction" is most commonly used to refer to a physiological state of dependence caused by the habitual use of drugs, **alcohol**, or other substances. Addiction is characterized by uncontrolled craving, increased tolerance, and withdrawal symptoms when deprived of access to the addictive substance. Addictions afflict millions of people in the United States alone.

Addiction results from an incessant need to combat the negative side effects of a substance or situation by returning to that substance or situation for the initial enhancing effect. The desire for drugs such as heroin, **cocaine**, or alcohol all result from a need to suppress the low that follows the high. Other forms of addiction occur where seemingly harmless behaviors such as eating, running, or working become the focus of the addict's life.

Addiction and addictive substances have long been a part of human culture. The use of alcoholic beverages, such as beer, was recorded by the ancient Egyptians. The Romans and other early civilizations fermented, drank, and traded in wine. The infamous "opium dens" of the Far East offered crude opium. The discovery of America was accompanied by the discovery of tobacco, grown by the indigenous population.

Addiction today, especially addiction to illegal drugs, takes a heavy toll on modern society. Illegal drugs are easy enough to obtain, but they have a high price. In order to get money to feed their addiction, some addicts resort to theft or prostitution. Aside from criminal damage, addiction disrupts families and other social institutions in the form of divorce, abuse (mental and physical), and neglect.

Addictions

There are two classifications for addiction: chemical and nonchemical. While dependency on substances that are ingested or injected is more commonly discussed, there are a number of nonchemical addictions that can lead to equally devastating lifestyles.

Chemical addictions

Chemical addiction is the general description for an addiction to a substance that must be injected or ingested. Alcohol, opiates, and cocaine are the most common of these chemicals. Though each of them is addictive, they have different effects on the body.

Addiction to alcohol, for example, may be the result of heavy drinking coupled with a malfunctioning type of **cell** in the liver of the alcoholic. Many adults can drink large quantities of alcoholic beverages and suffer only a "hangover"—headache and nausea. The malfunctioning liver in the alcoholic, however, does not detoxify the byproducts of alcohol ingestion rapidly. The resultant accumulation of a chemical called *acetaldehyde* causes several symptoms, including **pain**, which can be relieved by the intake of more alcohol. The consumption of ever-increasing amounts of alcohol with greater frequency can lead to **organ** failure and death if the alcoholic is left untreated.

Opium, produced by a **species** of poppy, is an ancient addictive substance that is still produced for its cash value. Although raw opium is not the form most addicts encounter, purified, powdered opium has been used in many forms for hundreds of years. Tincture of opium, or laudanum, was introduced about 1500. Paregoric, a familiar household remedy today, dates from the early 1700s.

Heroin, a derivative of opium, has become a common addictive drug. Heroin is a powder dissolved in water and injected into the user's vein, giving an immediate sensation of warmth and relaxation. Physical or mental pain is relieved, and the user enters a deeply relaxed state for a few hours. The powder can also be inhaled for a milder effect. Heroin is extremely addictive and with only a few doses the user is "hooked." **Morphine**, a refinement of opium, was discovered in the early 1800s. It was first used as an effective analgesic, or painkiller, and it is still used for that purpose. Its fast action makes it a drug of choice to ease the pain of wounded soldiers during wartime. Morphine has one-fifth the addictive power of heroin.

Cocaine in its various forms is another class of addictive compounds. In fact, it is the most addictive of these drugs; some people need only a single exposure to the drug to become addicted. Cocaine is processed from the **coca plant** and is used in the form of a white powder. It can be inhaled, ingested, injected, or mixed with **marijuana** and smoked. It is also further processed into a solid crystalline substance marketed as "crack." Unlike the opiates, which bring on a warm feeling and immobility, cocaine makes its users energetic. This strong stimulation and period of hyperactivity (usually no more than half an hour) is quickly followed by a period of intense **depression**, fatigue, and paranoia. In order to relieve these harsh side effects, the user will typically retreat to taking more cocaine or using another drug, for example alcohol or heroin. Suicide is a common occurrence among cocaine addicts.

Any of these chemical substances can become the object of intense addiction. Addicts of the opiates and cocaine must have increasingly frequent doses to main-

Crack users. Crack, a form of cocaine, is one of the most addictive drugs. *Photograph by Roy Morsch Stock Market. Reproduced with permission.*

tain their desired physiological effects. Soon the addict has difficulty focusing on anything else, making it nearly impossible to hold a job or maintain a normal lifestyle. These drugs are of economic importance not only because of their high cost, but also because of the crimes committed to obtain the cash necessary to buy the drugs. The drug enforcement resources dedicated to policing those crimes and the **rehabilitation** programs provided to the drug addicts are costly.

Some experts consider drinking large amounts of coffee or cola beverages evidence of an addiction to **caffeine**. In fact, these substances do provide a short-term mood lift to the user. The first cup of coffee in the morning, the midmorning coffee break, the cola at lunch, and the dinner coffee are habitual. Withdrawal from caffeine, which is a stimulant, can cause certain mood changes and a longing for additional caffeine.

Tobacco use is also addictive (due to the **nicotine** found in tobacco). Cigarette smoking, for example, is one of the most difficult habits to stop. Withdrawal symptoms are more pronounced in the smoker than in the coffee drinker. Reforming smokers are subject to swift mood swings and intense cravings for a cigarette. A long-time smoker may never overcome the desire for cigarettes.

Withdrawal symptoms are caused by psychological, physical, and chemical reactions in the body. As the amount of addictive chemical in the **blood** begins to fall, the urge to acquire the next dose is strong. The hard drugs such as heroin and cocaine produce intense withdrawal symptoms that, if not eased by another dose of the addictive substance or an appropriate medication, can leave the user in painful helplessness. Strong muscle contractions, nausea, vomiting, sweating, and pain increase in strength until it becomes extremely difficult for the user to stay away from the drug.

The nonchemical addictions

Addictions can involve substances or actions not including addictive chemicals. Some of these addictions are difficult to define and may seem harmless enough, but they can destroy the lives of those who cannot escape them.

Gambling is one such form of addiction, affecting 6-10% of the American population, according to some experts. Gamblers begin as most others do, by placing small bets on **horses** or engaging in low-stakes card games or craps. Their successes are a form of ego enhancement, so they strive to repeat them. Their bets become larger, more frequent, and more irrational. Gamblers have been known to lose their jobs, homes, and families as a result of their activities. Their pattern is to place ever-larger bets to make up for their losses. Gamblers are difficult to treat because they refuse to recognize that they have an abnormal condition. After all, nearly everyone gambles in some form: on the lottery, horses, home poker games, or sporting events. Once a compulsive gambler is convinced that his or her problem is serious, an addiction program may be successful in treating the condition.

Food addiction can be a difficult condition to diagnose. Food addicts find comfort in eating. The physical sensations that accompany eating can become addictive, although an addict may not **taste** the food. Food addicts may indulge in binge eating—consuming prodigious quantities of food in one sitting, or they may consume smaller quantities of food over a longer period of time, but eat constantly during that time.

A food addict can become grossly overweight, leading to extremely low self-esteem, which becomes more pronounced as he or she gains weight. The addict then seeks comfort by eating more food, setting up a cycle that probably will lead to a premature death if not interrupted.

The opposite of addiction to eating is addiction to not eating. This addiction often starts as an attempt to lose weight and ends in **malnutrition**. Two common forms of this type of addiction, anorexia and bulimia, are typically associated with young females, although males and females of all ages may develop this disorder. Anorexia is a condition in which food is almost completely rejected. These addicts literally starve their bodies, consuming as little food as possible. Bulimia on the other hand, involves consuming large amounts of food uncontrollably until satisfied and then purging the food they took in as soon after eating as possible. Some experts claim that nearly 100 people a year die of malnutrition resulting from anorexia or bulimia. Others believe the number is much larger because the deaths are not recorded as anorexia or bulimia, but as **heart** failure or kidney failure, either of which can result from malnutrition.

Anorexia and bulimia are difficult to treat. In the minds of victims, they are bloated and obese even though they may be on the brink of starvation, and so they often resist treatment. Hospitalization may be required even before the official **diagnosis** is made. Treatment includes a long, slow process of psychiatric counseling.

The sex addict also is difficult to diagnose because "normal" sex behavior is not well defined. Generally, any sex act between two consenting adults is condoned if neither suffers harm. Frequency of sexual activity is not used as a deciding factor in diagnosis. More likely the sex addict is diagnosed by his or her attitude toward sex partners.

Other compulsions or addictions include **exercise**, especially running. Running releases certain **hormones** called endorphins in the **brain**, giving a feeling of euphoria or happiness. This is the "high" that runners often describe. They achieve it when they have run sufficiently to release endorphins and have felt their effects. So good is this feeling that the compulsive runner may practice his hobby in spite of bad weather, injury, or social obligation. Because running is considered a healthful hobby, it is difficult to convince an addict that he is overdoing it and must temper his activity.

Codependency could also be regarded as an addiction, although not of the classical kind. It is a form of psychological addiction to another human being. While the term codependency may sound like a mutual dependency, in reality, it is very one-sided. A person who is codependent gives up their rights, individuality, wants, and needs to another person. The other person's likes and wants become their own desires and the codependent person begins to live vicariously through the other person, totally abandoning their own life. Codependency is often the reason that women remain in abusive relationships. Codependent people tend to trust people who are untrustworthy. Self-help groups and counseling is available for codependents and provide full recovery.

Another form of addiction is addiction to work. No other addiction is so willingly embraced than that of a workaholic. Traits of workaholics are often the same traits used to identify hard workers and loyal employees. So, when does working hard become working too hard? When work becomes an addiction, it can lead to harmful effects in other areas of life, such as family neglect or deteriorating health. The individual drowns himself/herself in work to the point of shunning all societal obligations. Their parental duties and responsibilities are often handed over to the other spouse. The children are neglected by the parent and consequently end up having a poor relationship with the workaholic parent. Identifying the reason for becoming a workaholic and getting help, such as counseling, are key for overcoming this addiction.

Internet addictions are a new illness in our society. The Internet is an amazing information resource, especially for students, teachers, researchers, and physicians. People all over the globe use it to connect with individuals from other countries and cultures. However, when the computer world rivals the real world, it becomes an addiction. Some people choose to commune with the computer rather than with their spouses and children. They insulate themselves from intimate settings and relationships. Internet abuse has been cited as a contributing factor in the disintegration of many marriages and families and even in the collapse of many promising careers. Since it is a relatively new disorder, few self-help resources are available. Ironically, there are some on-line support groups designed to wean people from the Internet.

The addict

Because addictive behavior has such serious effects on the health and social well being of the addict and those around him or her, why would anyone start? One characteristic that marks addicts, whether to chemicals or nonchemical practices, is a low sense of self esteem. The addict may arise from any social or economic situation, and there is no way to discern among a group of people who will become an addict and who will not.

It has been a basic tenet that the individual who uses drugs heavily will become addicted. However, soldiers who served in Vietnam reported heavy use of marijuana and heroin while they were in the combat zone, yet the vast majority gave up the habit upon returning home. There are reports, however, of people becoming addicted to a drug with exposure only once or a few times.

Some experts believe people are born with the predisposition to become addicted. Children of addicts have a greater probability of becoming addicts themselves

KEY TERMS
. . . .

Detoxification—The process of removing a poison or toxin from the body. The liver is the primary organ of detoxification in the body.

Endorphins—A group of natural substances in the brain that are activated with exercise. They bind to opiate receptors and ease pain by raising the pain threshold. Of the three types, alpha, beta, and gamma, beta is the most potent.

Opiate—Any derivative of opium (e.g., heroin).

Opium—A natural product of the opium poppy, *Papaver somniferum*. Incising the immature pods of the plant allows the milky exudate to seep out and be collected. Air-dried, this is crude opium.

than do children whose parents are not. Thus, the potential for addiction may be hereditary. On the other hand, a psychological problem may lead the individual into addiction. The need for instant gratification, a feeling of being socially ostracized, and an inability to cope with the downfalls of life have all been cited as possible springboards to addiction.

Treatment of addiction

Habitual use of an addictive substance can produces changes in body **chemistry** and any treatment must be geared to a gradual reduction in dosage. Initially, only opium and its derivatives (morphine, heroin, **codeine**) were recognized as addictive, but many other drugs, whether therapeutic (for example, **tranquilizers**) or recreational (such as cocaine and alcohol), are now known to be addictive. Research points to a genetic predisposition to addiction; although environment and psychological make-up are other important factors and a solely genetic basis for addiction is too simplistic. Although physical addiction always has a psychological element, not all psychological dependence is accompanied by physical dependence.

Addiction of any form is difficult to treat. Many programs instituted to break the grip of addictive substances have had limited success. The "cure" depends upon the resolve of the addict, and he or she often struggles with the addiction even after treatment.

A careful medically controlled withdrawal program can reverse the chemical changes of habituation Trying to stop chemical intake without the benefit of medical help is a difficult task for the addict because of intense physical withdrawal symptoms. Pain, nausea, vomiting, sweating, and hallucinations must be endured for several days.

Most addicts are not able to cope with these symptoms, and they will relieve them by indulging in their addiction.

The standard therapy for chemical addiction is medically supervised withdrawal, along with a 12-step program, which provides physical and emotional support during withdrawal and recovery. The addict is also educated about drug and alcohol addiction. "Kicking" a habit, though, is difficult, and backsliding is frequent. Many former addicts have enough determination to avoid drugs for the remainder of their lives, but research shows that an equal number will take up the habit again.

See also Alcoholism; Amphetamines; Barbiturates.

Resources

Books

Bender, D. and Leone, B. *Drug Abuse: Opposing Viewpoints.* San Diego: Greenhaven Press, Inc., 1994.

Kuhn, C., Swartzwelder, S., and Wilson, W. *Just Say Know.* New York: W.W. Norton & Co., 2002.

Silverstein, A., V. Silverstein, and R. Silverstein. *The Addictions Handbook.* Hillside, NJ: Enslow Publishers, 1991.

Young, Kimberley. *Caught in the Net: How to Recognize the Signs of Internet Addiction—and a Winning Strategy for Recovery.* New York: John Wiley and Sons, 1998.

Other

Substance Abuse and Mental Health Services (SAMSHA), an agency of the United States Department of Health and Human Services (301)443-8956. <www.samsha.gov> (March 10, 2003).

Larry Blaser

Addison's disease

Addison's **disease**, also called adrenocortical deficiency or primary adrenal hypofunction, is a rare condition caused by destruction of the cortex of the adrenal gland, one of several **glands** the **endocrine system**. Because Addison's disease is treatable, those who develop the illness can expect to have a normal life span.

The adrenal glands

The adrenal glands, also called suprarenal glands, sit like flat, triangular caps atop each kidney. They are divided into two distinct areas-the medulla at the center and cortex surrounding the outside. The cortex, which makes up about 80% of the adrenal gland, secretes three types of hormones—sex **hormones**, mineralocorticoids (principally aldosterone), and glucocorticoids (primarily cortisol or hydrocortisone). Scientists believe these hormones perform hundreds of regulatory functions in the body, including helping to regulate **metabolism**, **blood pressure**, the effects of **insulin** in the breakdown of sugars, and the inflammatory response of the **immune system**. Addison's disease results from an injury or disease that slowly destroys the adrenal cortex, therefore shutting down the production of these hormones.

The production of cortisol by the adrenal cortex is precisely metered by a control loop that begins in an area in the **brain** called the *hypothalamus*, a collection of specialized cells that control many of the functions of the body. When necessary, the hypothalamus secretes a releasing factor that tells the pituitary gland to secrete another hormone to stimulate the adrenal gland to release more cortisol. The increased cortisol levels signal the pituitary to stop producing the adrenal stimulant. This is a finely tuned loop, and if it is interrupted or shut down, as in Addison's disease, profound changes occur in the body.

History of Addison's disease

The disease is named for its discoverer, Dr. Thomas Addison, a British surgeon who described adrenal insufficiency in 1849, though endocrine functions had yet to be explained. Addison described the condition from autopsies he performed. At the time, there was no cure for adrenal insufficiency, so victims died after contracting it. Addison also noted that 70-90% of patients with adrenal insufficiency had **tuberculosis** as well.

Addison's disease is no longer a fatal illness if it is properly diagnosed. Today, doctors note that up to 70% of cases are the result of the adrenal cortex being destroyed by the body's own immune system, so Addison's is called an *autoimmune disease*. Those who have sustained an injury to the adrenal gland and people who have diabetes are at increased risk of Addison's disease. Tuberculosis is also linked to the disease, but since this disease can now be cured, Addison's disease is rarely caused by tuberculosis today.

Addison's disease

The effects of adrenal insufficiency do not manifest themselves until more than 90% of the adrenal cortex has been lost. Then weakness and dizziness occur, and the skin darkens, especially on or near the elbows, knees, knuckles, lips, scars, and skin folds. These symptoms begin gradually and worsen over time.

The patient becomes irritable and depressed and often craves salty foods. Some people do not experience these progressive symptoms, but become aware of the disease during what is called an *addisonian crisis*. In this case, the symptoms appear suddenly and require immediate medical attention. Severe **pain** develops in the lower back, abdomen, or legs; vomiting and diarrhea leave the patient dehydrated. A person may become unconscious and may even die.

A doctor's examination reveals low blood pressure that becomes even lower when the patient rises from a sitting or lying position to a standing position. A blood test shows low blood sugar (hypoglycemia), low blood **sodium** (hyponatremia), and low levels of cortisol. Other tests are carried out to determine whether the condition is the result of adrenal insufficiency or if the low levels of cortisol are the result of problems with the hypothalamus or pituitary.

Treatment

Once diagnosed, Addison's disease is treated by replacing the natural cortisol with an oral medication. The medicine is adjusted by a doctor to bring cortisol levels in the blood up to normal and maintain them. A patient also is advised to eat salty foods, not skip any meals, and carry a packet containing a syringe with cortisone to be injected in case of an emergency.

With the loss of the ability to secrete cortisol under **stress**, a patient must take extra medication when he undergoes dental treatments or **surgery**. Even though Addison's disease is not curable, a patient with this condition can expect to live a full life span.

See also Adrenals; Diabetes mellitus.

Resources

Books

Larson, David E., ed. *Mayo Clinic Family Health Book*. New York: William Morrow, 1996.

The Official Patient's Sourcebook on Addison's Disease: A Revised and Updated Directory for the Internet Age. San Diego: ICON Health Publications, 2002.

Periodicals

Erickson, Q.L. "Addison's Disease: The Potentially Life-threatening Tan." *Cutis* 66, no. 1 (2001): 72-74.

Kessler, Christine A. "Adrenal Gland (Adrenal disorders and other problems)." In *Endocrine Problems: Nurse Review* Springhouse, PA: Springhouse Corp., 1988.

National Institute of Diabetes and Digestive and Kidney Diseases. "Addison's Disease." (fact sheet). National Institutes of Health Publication No. 90-3054.

Ten, S. "Addison's Disease 2001." *The Journal of Clinical Endocrinology & Metabolism* 86, no. 7 (2001): 2909-2922.

Larry Blaser

Addition

Addition, indicated by a + sign, is a method of combining numbers. The result of adding two numbers is called their sum.

Adding natural numbers

Consider the natural, or counting, numbers 1, 2, 3, 4,... Each natural number can be defined in terms of sets. The number 1 is the name of the collection containing every conceivable set with one element, such as the set containing 0 or the set containing the Washington Monument. The number 2 is the name of the collection containing every conceivable set with two elements, and so on. The sum of two **natural numbers** is determined by counting the number of elements in the union of two sets chosen to represent them. For example, let the set {A, B, C} represent 3 and the set {W, X, Y, Z} represent 4. Then $3 + 4$ is determined by counting the elements in {A, B, C, W, X, Y, Z}, which is the union of {A, B, C} and {W, X, Y, Z}. The result is seven, and we write $3 + 4 = 7$. In this way, the operation of addition is carried out by counting.

The addition algorithm

Addition of natural numbers is independent of the numerals used to represent the numbers being added. However, some forms of notation make addition of large numbers easier than other forms. In particular, the Hindu-Arabic positional notation (in general use today) facilitates addition of large numbers, while the use of Roman numerals, for instance, is quite cumbersome. In the Hindu-Arabic positional notation, numerals are arranged in columns, each column corresponding to numbers that are 10 times larger than those in the column to the immediate right. For example, 724 consists of 4 ones, 2 tens, and 7 hundreds. The addition **algorithm** amounts to counting by ones in the right hand column, counting by tens in the next column left, counting by hundreds in the next column left and so on. When the sum of two numbers in any column exceeds nine, the amount over 10 is retained and the rest transferred or "carried" to the next column left. Suppose it is desired to add 724 and 897. Adding each column gives 11 ones, 11 tens, and 15 hundreds. But 11 ones is equal to 1 ten and 1 one so we have 1 one, 12 tens and 15 hundreds. Checking the tens column we find 12 tens equals 2 tens and 1 hundred, so we actually have 1 one, 2 tens and 16 hundreds. Finally, 16 hundreds is 6 hundreds and 1 thousand, so the end result is 1 thousand, 6 hundreds, 2 tens, and 1 one, or 1,621.

Adding common fractions

Historically, the number system expanded as it became apparent that certain problems of interest had no solution in the then-current system. Fractions were included to deal with the problem of dividing a whole thing into a number of parts. Common fractions are numbers expressed as a **ratio**, such as 2/3, 7/9, and 3/2.

When both parts of the fraction are **integers**, the result is a **rational number**. Each rational number may be thought of as representing a number of pieces; the numerator (top number) tells how many pieces the fraction represents; the denominator (bottom number) tells us how many pieces the whole was divided into. Suppose a cake is divided into two pieces, after which one half is further divided into six pieces and the other half into three pieces, making a total of nine pieces. If you take one piece from each half, what part of the whole cake do you get? This amounts to a simple counting problem if both halves are cut into the same number of pieces, because then there are a total of six or 12 equal pieces, of which you take two. You get either 2/6 or 2/12 of the cake. The essence of adding rational numbers, then, is to turn the problem into one of counting equal size pieces. This is done by rewriting one or both of the fractions to be added so that each has the same denominator (called a common denominator). In this way, each fraction represents a number of equal size pieces. A general formula for the sum of two fractions is $a/b + c/d = (ad + bc)/bd$.

Adding decimal fractions

Together, the rational and irrational numbers constitute the set of **real numbers**. Addition of real numbers is facilitated by extending the positional notation used for integers to decimal fractions. Place a period (called a decimal point) to the right of the ones column, and let each column to its right contain numbers that are successively smaller by a factor of ten. Thus, columns to the right of the decimal point represent numbers less than one, in particular, "tenths," "hundredths," "thousandths," and so on. Addition of real numbers, then, continues to be defined in terms of counting and carrying, in the manner described above.

Adding signed numbers

Real numbers can be positive, **negative**, or **zero**. Addition of two negative numbers always results in a negative number and is carried out in the same fashion that positive numbers are added, after which a negative sign is placed in front of the result, such as $-4 + (-21) = -25$. Adding a positive and a negative number is the equivalent of **subtraction**, and, while it also proceeds by counting, the sum does not correspond to counting the members in the union of two sets, but to counting the members not in the intersection of two sets.

Addition in algebra

In **algebra**, which is a generalization of **arithmetic**, addition is also carried out by counting. For example, to sum the expressions 5x and 6x we notice that 5x means we have five xs and 6x means we have six xs, making a total of 11 xs. Thus $5x + 6x = (5 + 6)x = 11x$, which is usually established on the basis of the distributive law, an important property that the real numbers obey. In general, only like variables or powers can be added algebraically. In adding two polynomial expressions, only similar terms are combined; thus, $(3x^2 + 2x + 7y + z) + (x^3 + 3x + 4z + 2yz) = (x^3 + 3x^2 + 5x + 7y + 5z + 2yz)$.

See also Fraction, common.

Resources

Books

Eves, Howard Whitley. *Foundations and Fundamental Concepts of Mathematics.* New York: Dover, 1997.

Grahm, Alan. *Teach Yourself Basic Mathematics.* Chicago, IL: McGraw-Hill Contemporary, 2001.

Gullberg, Jan, and Peter Hilton. *Mathematics: From the Birth of Numbers.* W.W. Norton & Company, 1997.

Paulos, John Allen. *Beyond Numeracy, Ruminations of a Numbers Man.* New York: Alfred A. Knopf, 1991.

Tobey, John, and Jeffrey Slater. *Beginning Algebra.* 4th ed. NY: Prentice Hall, 1997.

Weisstein, Eric W. *The CRC Concise Encyclopedia of Mathematics.* New York: CRC Press, 1998.

J. R. Maddocks

Adenosine diphosphate

Adenosine diphosphate (ADP) is a key intermediate in the body's **energy** metabolism—it serves as the "base" to which energy-producing reactions attach an additional phosphate group, forming **adenosine triphosphate** (ATP). ATP then diffuses throughout the **cell** to drive reactions that require energy.

Structurally, ADP consists of the purine base adenine (a complex, double-ring **molecule** containing five **nitrogen atoms**) attached to the five-carbon sugar ribose; this combination is known as adenosine. Attaching two connected phosphate groups to the ribose produces ADP. Schematically, the structure may be depicted as Ad-Ph-Ph, where Ad is adenosine and **Ph** is a phosphate group.

See also Metabolism.

Adenosine triphosphate

Adenosine triphosphate (ATP) is often described as the body's "energy currency"—energy-producing meta-

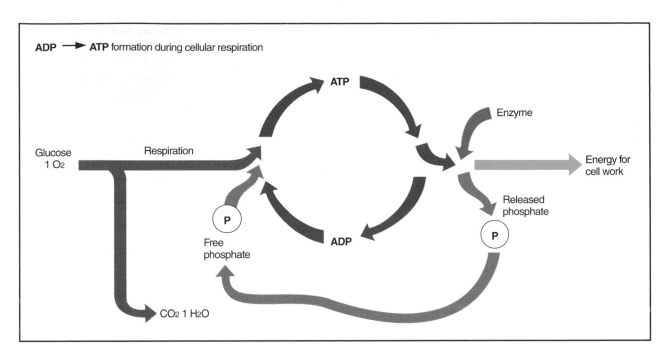

ADP ⟶ ATP formation during cellular respiration

ADP is formed during cellular respiration with energy released by the breakdown of glucose molecules. *Illustration by Hans & Cassidy. Courtesy of Gale Group.*

bolic reactions store their **energy** in the form of ATP, which can then drive energy-requiring syntheses and other reactions anywhere in the **cell**.

Structurally ATP consists of the purine base adenine (a complex, double-ring **molecule** containing five **nitrogen atoms**) attached to the five-carbon sugar ribose; this combination is known as adenosine. Attaching a string of three connected phosphate groups to the ribose produces ATP. Schematically, one may depict the structure of ATP as Ad-Ph-Ph-Ph, where Ad is adenosine and **Ph** is a phosphate group. If only two phosphate groups are attached, the resulting compound is **adenosine diphosphate** (ADP).

The final step in almost all the body's energy-producing mechanisms is attachment of the third phosphate group to ADP. This new phosphate-phosphate bond, known as a high-energy bond, effectively stores the energy that has been produced. The ATP then diffuses throughout the cell, eventually reaching sites where energy is needed for such processes as protein synthesis or muscle cell contraction. At these sites, **enzyme** mechanisms couple the energy-requiring processes to the breakdown of ATP's high-energy bond. This regenerates ADP and free phosphate, both of which diffuse back to the cell's energy-producing sites and serve as raw materials for production of more ATP.

The ATP-ADP couple is thus analogous to a rechargeable storage **battery**, with energy production sites representing the battery charger. ATP is the fully charged battery that can supply energy to a flashlight or transistor radio. ADP is the used battery that is returned for charging.

ADP is not a fully drained battery, however. It still possesses one high-energy phosphate-phosphate bond. When energy is short and ATP is scarce, the second phosphate can be transferred from one ADP to another. This creates a new ATP molecule, along with one of adenosine monophosphate (AMP). Since the "fully drained" AMP will probably be broken down and disposed of, however, this mechanism represents an emergency response that is inhibited when ATP is plentiful.

ATP is also a building block in **DNA synthesis**, with the adenosine and one phosphate being incorporated into the growing helix. (The "A" in ATP is the same as in the A-C-G-T "alphabet" of DNA.) This process differs from most other ATP-using reactions, since it releases *two* phosphate groups—initially still joined, but soon separated. With very little pyrophosphate (Ph-Ph) available in the cell, the chance that it will break the DNA chain and again form—though all enzyme reactions are theoretically reversible—is effectively infinitesimal. Since breaking the DNA chain would probably kill the cell, what at first might appear to be energy wastage turns out to be quite worthwhile. The cell also converts ATP to AMP and pyrophosphate in a few other cases where the reaction must always go only in a single direction.

See also Metabolism.

Adhesive

Adhesives bond two or more materials at their surface, and may be classified as structural or nonstructural. Structural adhesives can support heavy loads, while nonstructural adhesives cannot. Most adhesives exist in liquid, paste, or granular form, although film and fabric-backed tape varieties are also commercially available.

Adhesives have been used since ancient times. The first adhesives were probably made from boiled-down **animal** products such as hides or bones. Organic, i.e., carbon-based, adhesives have also been derived from **plant** products for use with **paper** products. While many of these organic glues have proven effective in the adhesion of furniture and other indoor products, they have not been effective in outdoor use where they are exposed to harsher environmental conditions.

Although inorganic adhesives, which are based on materials not containing **carbon**, such as the **sodium** silicates (water glasses) for bonding paper board, are sold commercially, most adhesives in common use are made of synthetic, organic materials. By far, the most widely used adhesives today are synthetic, polymer-based adhesives.

Synthetic adhesives may be made of amorphous thermoplastics above their **glass** transition temperatures; thermosetting monomers as in the case of epoxy glues and cyanoacrylates; low **molecular weight** reactive species as in the case of urethane adhesives; or block copolymers, suspensions, or latexes.

Types of adhesive bonding

Adhesive bonding may originate in a variety of ways. It may be the result of mechanical interlocking of the adhesive with the bonded surface, covalent bonding between bonded surfaces, or secondary electronic interactions between the bonded materials.

In mechanical adhesion, the adhesive flows around the substrate surface roughness so that interlocking of the two materials takes place. The adhesive may penetrate the substrate surface. Surface interpenetration often involves **polymer diffusion**; this type of bonding depends on the ability of the polymer adhesive to diffuse into the bonded surface.

Secondary electronic bonding may result from **hydrogen** bonds between the adhesive and substrate, from the interactions of overlapping polymer chains, or from such nonspecific forces as Van der Waals interactions.

In the case of covalent bonding, actual primary chemical bonds are formed between the bonded materials. For example, graft or block copolymers may bond different phases of a multicomponent polymeric material together.

Bonding applications

Adhesives are characterized by their shelf life, which is defined as the time that an adhesive can be stored after manufacture and still remain usable, and by their working life, defined as the time between mixing or making the adhesive and when the adhesive is no longer usable. The best choice of adhesive depends on the materials to be bonded.

Bonding metals

Epoxy resin adhesives perform well in the structural bonding of **metal** parts to each other. Nonstructural adhesives such as polysulfides, neoprene, or rubber-based adhesives are also available for bonding metal foils. Ethylene **cellulose** cements are used for filling recesses in metal surfaces.

When bonding metals to non-metals, the choices of adhesives are more extensive. In the case of structural bonding, for example, polyester-based adhesives may be used to bond plastic laminates to metal surfaces; low-density epoxy adhesives may be used to adhere light **plastics** such as polyurethane foam to various metals; and liquid adhesives made of neoprene and synthetic **resins** may be used to bond metals to wood. General purpose rubber, cellulose, and vinyl adhesives may be used to nonstructurally bond metals to other materials such as glass and leather.

Bonding plastics

Thermoplastic materials including nylon, polyethylene, acetal, polycarbonate, polyvinyl chloride, cellulose nitrate, and cellulose acetate are easily dissolved by solvents and softened by **heat**. These limitations restrict the use of adhesives with such materials, and solvent or heat **welding** may prove better bonding alternatives for adhering these materials.

Solvent cements can frequently be used to bond thermoplastics together. These cements combine a solvent with a base material that is the same as the thermoplastic to be adhered. In view of environmental considerations, however, many adhesives manufacturers are now reformulating their solvent-based adhesives. General purpose adhesives such as cellulosics, vinyls, rubber cements, and epoxies have also been used successfully with thermoplastics.

Thermosetting plastics, including phenolics, epoxies, and alkyds, are easily bonded with epoxy-based adhesives, neoprene, nitrile rubber, and polyester-based cements. These adhesives have been used to bond both thermosets and thermoplastics to other materials, including **ceramics**, fabric, wood, and metal.

Bonding wood

Animal glues, available in liquid and powder form, are frequently used in wood bonding. But animal glues are very sensitive to variations in **temperature** and moisture. Casein-type adhesives offer moderate resistance to moisture and high temperature, as do **urea** resin adhesives, which can be used to bond wood to wood, or wood to plastic.

Vinyl-acetate emulsions are excellent for bonding wood to materials that are especially porous, such as metal and some plastic laminates, but these adhesives also tend to be sensitive to temperature and moisture. Rubber, acrylic, and epoxy general-purpose adhesives also perform well with wood and other materials.

Fabric and paper bonding

General purpose adhesives including rubber cements and epoxies are capable of bonding fabrics together, as well as fabrics to other materials. When coated fabrics must be joined, the base adhesive material must be the same as the fabric coating. Rubber cements, gum mucilages, **wheat** pastes, and wood rosin adhesives can be used to join paper or fabric assemblies.

Resources

Books

Green, Robert E. *Machinery's Handbook.* 24th ed. New York: Industrial Press, 1992.

Petrie, Edward M. *Handbook of Adhesives & Sealants.* New York: McGraw-Hill, 1999.

Pocius, A.V. *Adhesion and Adhesives Technology.* Cincinnati, OH: Hanser Gardner Publications, 2002.

Sperling, L. H. *Introduction to Physical Polymer Science.* New York: John Wiley & Sons, 1992.

Veselovskii, R.A., Vladimir N. Kestelman, Roman A. Veselovsky. *Adhesion of Polymers.* 1st ed. New York: McGraw-Hill, 2001.

Wu, S. *Polymer Interfaces and Adhesion.* New York: Marcel Dekker, 1982.

Periodicals

Amis, E.J. "Combinatorial Investigations of Polymer Adhesion." *Polymer Preprints, American Chemical Society, Division* 42, no. 2 (2001): 645-646.

McCafferty, E. "Acid-base Effects in Polymer Adhesion at Metal Surfaces." *Journal of Adhesion Science and Technology* 16, no. 3 (2002): 239-256.

Randall Frost

Adolescence *see* **Puberty**

KEY TERMS

Composite—A mixture or mechanical combination (on a macroscopic level) of materials that are solid in their finished state, that are mutually insoluble, and that have different chemistries.

Inorganic—Not containing compounds of carbon.

Monomer—A substance composed of molecules that are capable of reacting together to form a polymer.

Organic—Containing carbon atoms, when used in the conventional chemical sense.

Polymer—A substance, usually organic, composed of very large molecular chains that consist of recurring structural units.

Synthetic—Referring to a substance that either reproduces a natural product or that produces a unique material not found in nature, and that is produced by means of chemical reactions.

Thermoplastic—A high molecular weight polymer that softens when heated and that returns to its original condition when cooled to ordinary temperatures.

Thermoset—A high molecular weight polymer that solidifies irreversibly when heated.

ADP *see* **Adenosine diphosphate**

Adrenals

The adrenal **glands** are a pair of endocrine glands that sit atop the kidneys and that release their **hormones** directly into the bloodstream. The adrenals are flattened, somewhat triangular bodies that, like other endocrine glands, receive a rich blood supply. The phrenic (from the diaphragm) and renal (from the kidney) **arteries** send many small branches to the adrenals, while a single large adrenal vein drains **blood** from the gland.

Each adrenal gland is actually two organs in one. The inner portion of the adrenal gland, the adrenal medulla, releases substances called catecholamines, specifically epinephrine, adrenaline, norepinephrine, noradrenaline, and **dopamine**. The outer portion of the adrenal gland, the adrenal cortex, releases steroids, which are hormones derived from **cholesterol**.

There are three somewhat distinct zones in the adrenal cortex: the outer part, the zona glomerulosa

(15% of cortical mass) made up of whorls of cells; the middle part, the zona fasciculata (50% of cortical mass) made up of columns of cells and that are continuous with the whorls; and an innermost area called the zona reticularis (7% of cortical mass), which is separated from the zona fasciculata by venous sinuses.

The cells of the zona glomerulosa secrete steroid hormones known as mineralocorticoids, which affect the fluid balance in the body, principally aldosterone, while the zona fasiculata and zona reticularis secrete glucocorticoids, notably cortisol and the androgen testosterone, which are involved in **carbohydrate**, protein, and **fat metabolism**.

The secretion of the adrenal cortical hormones is controlled by a region of the **brain** called the hypothalamus, which releases a corticotropin-releasing hormone. This hormone targets the anterior part of the pituitary gland, situated directly below the hypothalamus. The corticotropin-releasing hormone stimulates the release from the anterior pituitary of adreno-corticotropin (ACTH), which, in turn, enters the blood and targets the adrenal cortex. There, it binds to receptors on the surface of the gland's cells and stimulates them to produce the steroid hormones.

Steroids contain as their basic structure three 6-carbon (hexane) rings and a single 5-carbon (pentane) ring. The adrenal steroids have either 19 or 21 **carbon atoms**. These important hormones are collectively called corticoids. The 21-carbon steroids include glucocorticoids and mineralocorticoids, while the 19-carbon steroids are the androgens. Over 30 steroid hormones are made by the cortex, but only a few are secreted in physiologically significant amounts. These hormones can be classified into three main classes, glucocorticoids, mineralocorticoids, and corticosterone.

Cortisol (hydrocortisone) is the most important glucocorticoid. Its effect is the opposite to that of **insulin**. It causes the production of the sugar glucose from amino acids and glycogen stored in the liver, called gluconeogenesis, so increasing blood glucose. Cortisol also decreases the use of glucose in the body (except for the brain, spinal cord, and **heart**), and it stimulates the use of **fatty acids** for **energy**.

Glucocorticoids also have anti-inflammatory and antiallergenic action, so they are often used in the treatment of rheumatoid **arthritis**. The excessive release of glucocorticoids causes Cushing's **disease**, which is characterized by fatigue and loss of muscle mass due to the excessive conversion of amino acids into glucose. In addition, there is the redistribution of body fat to the face, causing the condition known as "moon face." The mineralocorticoids are essential for maintaining the balance of

sodium in the blood and body tissues and the **volume** of the extracellular fluid in the body. Aldosterone, the principal mineralocorticoid produced by the zona glomerulosa, enhances the uptake and retention of sodium in cells, as well as the cells' release of potassium. This steroid also causes the tubules of the kidneys to retain sodium, thus maintaining levels of this ion in the blood, while increasing the excretion of potassium into the urine. Simultaneously, aldosterone increases reabsorption of bicarbonate by the kidney, thereby decreasing the acidity of body fluids.

A deficiency of adrenal cortical hormone secretion causes **Addison's disease**, characterized by fatigue, weakness, skin pigmentation, a craving for **salt**, extreme sensitivity to **stress**, and increased vulnerability to **infection**.

The adrenal androgens are weaker than testosterone, the male hormone produced by the testes. However, some of these androgens, including androstenedione, **dehydroepiandrosterone (DHEA)**, and dehydroepiandrosterone sulfate can be converted by other tissues to stronger androgens, such as testosterone. The cortical output of androgens increases dramatically after **puberty**, giving the adrenal gland a major role in the developmental changes in both sexes. The cortex also secretes insignificant amounts of estrogen.

The steroid hormones are bound to steroid-binding **proteins** in the bloodstream, from which they are released at the surface of target cells. From there they move into the nucleus of the **cell**, where they may either stimulate or inhibit **gene** activity.

The release of the cortical hormones is controlled by adrenocorticotropic (ACTH) from the anterior pituitary gland. The level of ACTH has a diurnal periodicity, that is, it undergoes a regular, periodic change during the 24-hour **time** period. ACTH **concentration** in the blood rises in the early morning, peaks just before awaking, and reaches its lowest level shortly before **sleep**.

Several factors control the release of ACTH from the pituitary, including corticotropin-releasing hormone from the hypothalamus, free cortisol concentration in the **plasma**, stress (e.g., **surgery**, hypoglycemia, **exercise**, emotional trauma), and the sleep-wake cycle.

Mineralocorticoid release is also influenced by factors circulating in the blood. The most important of these factors is angiotensin II, the end-product of a series of steps starting in the kidney. When the body's blood **pressure** declines, this change is sensed by a special structure in the kidney called the juxtaglomerular apparatus. In response to this decreased pressure in kidney arterioles the juxtaglomerular apparatus releases an **enzyme** called renin into the kidney's blood vessels. There, the renin is converted to angiotensin I, which undergoes a further enzymatic change in the bloodstream outside the kidney to angiotensin II. Angiotensin II stimulates the adrenal cortex to release aldosterone, which causes the kidney to retain sodium. The increased concentration of sodium in the blood-filtering tubules of the kidney causes an osmotic movement of **water** into the blood, thereby increasing the blood pressure.

The adrenal medulla, which makes up 28% of the mass of the adrenal glands, is composed of irregular strands and masses of cells that are separated by venous sinuses. These cells contain many dense vesicles, which contain granules of catecholamines.

The cells of the medulla are modified ganglion (nerve) cells that are in contact with preganglionic fibers of the sympathetic **nervous system**. There are two types of medullary secretory cells, called chromaffin cells: the epinephrine (adrenalin)-secreting cells, which have large, less dense granules, and the norepinephrine (noradrenalin)-secreting cells, which contain smaller, very dense granules that do not fill their vesicles. Most chromaffin cells are the epinephrine-secreting type. These substances are released following stimulation by the acetylcholine-releasing sympathetic nerves that form synapses on the cells. Dopamine, a **neurotransmitter**, is secreted by a third type of adrenal medullar cell, different from those that secrete the other amines.

The extensive nerve connections of the medulla essentially mean that this part of the adrenal gland is a sympathetic ganglion, that is, a collection of sympathetic nerve cell bodies located outside the central nervous system. Unlike normal nerve cells, the cells of the medulla lack axons and instead, have become secretory cells.

The catecholamines released by the medulla include epinephrine, norepinephrine and dopamine. While not essential to life, they help to prepare the body to respond to short-lived but intense emergencies.

Most of the catecholamine output in the adrenal vein is epinephrine. Epinephrine stimulates the nervous system and also stimulates glycogenolysis (the breakdown of glycogen to glucose) in the liver and in skeletal muscle. The free glucose is used for energy production or to maintain the level of glucose in the blood. In addition, it stimulates lipolysis (the breakdown of fats to release energy-rich free fatty acids) and stimulates metabolism in general. Epinephrine also increases the force and **rate** of heart muscle contraction, which results in an increase in cardiac output.

The only significant disease associated with the adrenal medulla is pheochromocytoma. This **tumor** is highly vascular and secretes its hormones in large amounts. The symptoms of this disease include **hyper-tension**, sweating, headaches, excessive metabolism, **inflammation** of the heart, and palpitations.

See also Endocrine system.

Marc Kusinitz

Aerobic

Aerobic means that an **organism** needs **oxygen** to live. Some **microorganisms** can live without oxygen and they are called *anaerobic*. **Bacteria** are not dependent on oxygen to burn food for **energy**, but most other living organisms do need oxygen. Fats, **proteins**, and sugars in the diet of organisms are chemically broken down in the process of digestion to release energy to drive life activities. If oxygen is present, maximum energy is released from the food, and the process is referred to as aerobic **respiration**. The analogy of a bonfire with the energy **metabolism** of living organisms is appropriate up to the point that both processes require fuel and oxygen to produce energy and yield simpler compounds as a result of the oxidation process. There are, however, a number of important differences between the energy produced by the fire and the energy that comes from organism metabolism. The fire burns all at once and gives off large quantities of **heat** and **light**. Aerobic oxidation in an organism, on the other hand, proceeds in a series of small and controlled steps. Much of the energy released in each step is recaptured in the high-energy bonds of a chemical called **adenosine triphosphate** (ATP), a compound found in all cells and serving as an energy storage site. Part of the energy released is given off as heat.

Energy metabolism begins with an **anaerobic** sequence known as **glycolysis**. Since the reactions of glycolysis do not require the presence of oxygen, it is termed the anaerobic pathway. This pathway does not produce very much energy for the body, but it establishes a base for further aerobic steps that do have a much higher yield of energy. It is believed that **cancer** cells do not have the necessary enzymes to utilize the aerobic pathway. Since these cells rely on glycolysis for their energy metabolism, they place a heavy burden on the rest of the body.

The aerobic pathway is also known as the Krebs **citric acid** cycle and the **cytochrome** chain. In these two steps the by-products of the initial anaerobic glycolysis step are oxidized to produce **carbon dioxide**, **water**, and many energy-rich ATP molecules. All together, all these steps are referred to as **cell** respiration. Forty **percent** of

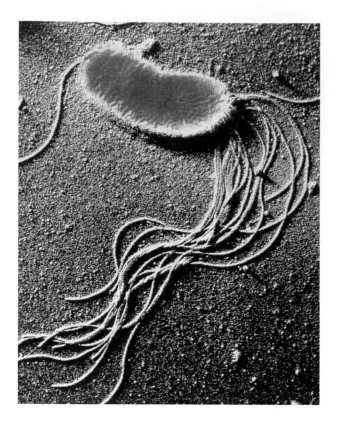

A scanning electron micrograph (SEM) of the aerobic soil bacterium *Pseudomonas fluorescens.* The bacterium uses its long, whip-like flagellae to propel itself through the water layer that surrounds soil particles. *Photograph by Dr. Tony Brain. Science Photo Library, National Audubon Society Collection/Photo Resarchers, Inc. Reproduced by permission.*

the glucose "burned" in cell respiration provides the organism with energy to drive its activities, while 60% of the oxidized glucose is dissipated as heat. This **ratio** of heat and energy is about the same as a power plant that produces **electricity** from **coal**.

See also Adenosine diphosphate; Krebs cycle.

Aerodynamics

Aerodynamics is the science of air flow over airplanes, cars, buildings, and other objects. Aerodynamic principles are used to find the best ways in which airplanes can get lift, reduce drag, and remain stable by controlling the shape and size of the wing, the **angle** at which it is positioned with respect to the airstream, and the flight speed. The flight characteristics change at higher altitudes as the surrounding air becomes colder and thinner. The behavior of the air flow also changes dramatically at flight speeds close to, and beyond, the

speed of sound. The explosion in computational capability has made it possible to understand and exploit the concepts of aerodynamics and to design improved wings for airplanes. Increasingly sophisticated wind tunnels are also available to test new models.

Basic air flow principles

Air properties that influence flow

Air flow is governed by the principles of **fluid dynamics** that deal with the **motion** of liquids and gases in and around solid surfaces. The **viscosity**, **density**, compressibility, and **temperature** of the air determine how the air will flow around a building or a plane. The viscosity of a fluid is its resistance to flow. Even though air is 55 times less viscous than **water**, viscosity is important near a solid surface since air, like all other fluids, tends to stick to the surface and slow down the flow. A fluid is compressible if its density can be increased by squeezing it into a smaller **volume**. At flow speeds less than 220 MPH (354 km/h), a third the speed of sound, we can assume that air is incompressible for all practical purposes. At speeds closer to that of sound (660 MPH [1,622 km/h]), however, the variation in the density of the air must be taken into account. The effects of temperature change also become important at these speeds. A regular commercial airplane, after landing, will feel cool to the **touch**. The Concorde jet, which flew at twice the speed of sound, felt hotter than boiling water.

Laminar and turbulent flow

Flow patterns of the air may be laminar or turbulent. In laminar or streamlined flow, air, at any point in the flow, moves with the same speed in the same direction at all times so that the flow appears to be smooth and regular. The smoke then changes to turbulent flow, which is cloudy and irregular, with the air continually changing speed and direction.

Laminar flow, without viscosity, is governed by **Bernoulli's principle**: the sum of the static and dynamic pressures in a fluid remains the same. A fluid at rest in a pipe exerts static **pressure** on the walls. If the fluid now starts moving, some of the static pressure is converted to dynamic pressure, which is proportional to the square of the speed of the fluid. The faster a fluid moves, the greater its dynamic pressure and the smaller the static pressure it exerts on the sides.

Bernoulli's principle works very well far from the surface. Near the surface, however, the effects of viscosity must be considered since the air tends to stick to the surface, slowing down the flow nearby. Thus, a boundary layer of slow-moving air is formed on the surface of an airplane or **automobile**. This boundary layer is laminar at the beginning of the flow, but it gets thicker as the air moves along the surface and becomes turbulent after a point.

Numbers used to characterize flow

Air flow is determined by many factors, all of which work together in complicated ways to influence flow. Very often, the effects of factors such as viscosity, speed, and **turbulence** cannot be separated. Engineers have found smart ways to get around the difficulty of treating such complex situations. They have defined some characteristic numbers, each of which tells us something useful about the nature of the flow by taking several different factors into account.

One such number is the Reynolds number, which is greater for faster flows and denser fluids and smaller for more viscous fluids. The Reynolds number is also higher for flow around larger objects. Flows at lower Reynolds numbers tend to be slow, viscous, and laminar. As the Reynolds number increases, there is a transition from laminar to turbulent flow. The Reynolds number is a useful similarity parameter. This means that flows in completely different situations will behave in the same way as long as the Reynolds number and the shape of the solid surface are the same. If the Reynolds number is kept the same, water moving around a small stationary airplane model will create exactly the same flow patterns as a full-scale airplane of the same shape, flying through the air. This principle makes it possible to test airplane and automobile designs using small-scale models in wind tunnels.

At speeds greater than 220 MPH (354 km/h), the compressibility of air cannot be ignored. At these speeds, two different flows may not be equivalent even if they have the same Reynolds number. Another similarity parameter, the **Mach number**, is needed to make them similar. The Mach number of an airplane is its flight speed divided by the speed of sound at the same altitude and temperature. This means that a plane flying at the speed of sound has a Mach number of one.

The drag **coefficient** and the lift coefficient are two numbers that are used to compare the forces in different flow situations. Aerodynamic drag is the **force** that opposes the motion of a car or an airplane. Lift is the upward force that keeps an airplane afloat against gravity. The drag or lift coefficient is defined as the drag or lift force divided by the dynamic pressure, and also by the area over which the force acts. Two objects with similar drag or lift coefficients experience comparable forces, even when the actual values of the drag or lift force, dynamic pressure, area, and shape are different in the two cases.

Wind tunnel testing of an aircraft model. © Dr. Gary Settles/ Science Source, National Audubon Society Collection/Photo Researchers, Inc. Reproduced with permission.

Skin friction and pressure drag

There are several sources of drag. The air that sticks to the surface of a car creates a drag force due to skin **friction**. Pressure drag is created when the shape of the surface changes abruptly, as at the point where the roof of an automobile ends. The drop from the roof increases the **space** through which the air stream flows. This slows down the flow and, by Bernoulli's principle, increases the static pressure. The air stream is unable to flow against this sudden increase in pressure and the boundary layer gets detached from the surface creating an area of low-pressure turbulent wake or flow. Since the pressure in the wake is much lower than the pressure in front of the car, a net backward drag or force is exerted on the car. Pressure drag is the major source of drag on blunt bodies. Car manufacturers experiment with vehicle shapes to minimize the drag. For smooth or "streamlined" shapes, the boundary layer remains attached longer, producing only a small wake. For such bodies, skin friction is the major source of drag, especially if they have large surface areas. Skin friction comprises almost 60% of the drag on a modern airliner.

Airfoil

An airfoil is the two-dimensional cross-section of the wing of an airplane as one looks at it from the side. It is designed to maximize lift and minimize drag. The upper surface of a typical airfoil has a curvature greater than that of the lower surface. This extra curvature is known as camber. The straight line, joining the front tip or the leading edge of the airfoil to the rear tip or the trailing edge, is known as the chord line. The angle of attack is the angle that the chord line forms with the direction of the air stream.

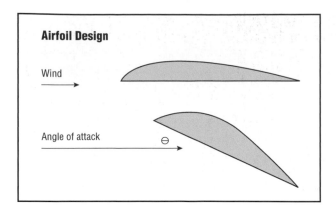

Airfoil Design

Wind

Angle of attack

⊖

The angle of attack that generates the most lift depends on many factors including the shape of the airfoil, the velocity of the airflow, and the atmospheric conditions. *Illustration by K. Lee Lerner with Argosy. The Gale Group.*

Lift

The stagnation point is the point at which the stream of air moving toward the wing divides into two streams, one flowing above and the other flowing below the wing. Air flows faster above a wing with greater camber since the same amount of air has to flow through a narrower space. According to Bernoulli's principle, the faster flowing air exerts less pressure on the top surface, so that the pressure on the lower surface is higher, and there is a net upward force on the wing, creating lift. The camber is varied, using flaps and slats on the wing in order to achieve different degrees of lift during take-off, cruise, and landing.

Since the air flows at different speeds above and below the wing, a large jump in speed will tend to arise when the two flows meet at the trailing edge, leading to a rearward stagnation point on top of the wing. Wilhelm Kutta (1867-1944) realized that a circulation of air around the wing would ensure smooth flow at the trailing edge. According to the Kutta condition, the strength of the circulation, or the speed of the air around the wing, is exactly as much as is needed to keep the flow smooth at the trailing edge.

Increasing the angle of attack moves the stagnation point down from the leading edge along the lower surface so that the effective area of the upper surface is increased. This results in a higher lift force on the wing. If the angle is increased too much, however, the boundary layer is detached from the surface, causing a sudden loss of lift. This is known as a stall and the angle at which this occurs for an airfoil of a particular shape, is known as the stall angle.

Induced drag

The airfoil is a two-dimensional section of the wing. The length of the wing in the third dimension, out to the

side, is known as the span of the wing. At the wing tip at the end of the span, the high-pressure flow below the wing meets the low-pressure flow above the wing, causing air to move up and around in wing-tip vortices. These vortices are shed as the plane moves forward, creating a downward force or downwash behind it. The downwash makes the airstream tilt downward and the resulting lift force tilt backward so that a net backward force or drag is created on the wing. This is known as induced drag or drag due to lift. About a third of the drag on a modern airliner is induced drag.

Stability and control

In addition to lift and drag, the stability and control of an **aircraft** in all three dimensions is important since an aircraft, unlike a car, is completely surrounded by air. Various control devices on the tail and wing are used to achieve this. Ailerons, for instance, control rolling motion by increasing lift on one wing and decreasing lift on the other.

Supersonic flight

Flight at speeds greater than that of sound are supersonic. Near a Mach number of one, some portions of the flow are at speeds below that of sound, while other portions move faster than sound. The range of speeds from Mach number 0.8 to 1.2 is known as transonic. Flight at Mach numbers greater than five is hypersonic.

The compressibility of air becomes an important aerodynamic factor at these high speeds. The reason for this is that **sound waves** are transmitted through the successive compression and expansion of air. The compression due to a sound wave from a supersonic aircraft does not have a chance to get away before the next compression begins. This pile up of compression creates a shock wave, which is an abrupt change in pressure, density, and temperature. The shock wave causes a steep increase in the drag and loss of stability of the aircraft. Drag due to the shock wave is known as wave drag. The familiar "sonic boom" is heard when the shock wave touches the surface of **Earth**.

Temperature effects also become important at transonic speeds. At hypersonic speeds above a Mach number of five, the **heat** causes **nitrogen** and **oxygen** molecules in the air to break up into **atoms** and form new compounds by **chemical reactions**. This changes the behavior of the air and the simple laws relating pressure, density, and temperature become invalid.

The need to overcome the effects of shock waves has been a formidable problem. Swept-back wings have helped to reduce the effects of shock. The supersonic

KEY TERMS

Airfoil—The cross-section of an airplane wing parallel to the length of the plane.

Angle of attack—The angle that the length of the airfoil forms with the oncoming airstream.

Camber—The additional curvature of the upper surface of the airfoil relative to the lower surface.

Induced drag or drag due to lift—The drag on the airplane due to vortices on the wingtips created by the same mechanism that produces lift.

Similarity parameter—A number used to characterize a flow and compare flows in different situations.

Stall—A sudden loss of lift on the airplane wing when the angle of attack increases beyond a certain value known as the stall angle.

Supersonic—Refers to bodies moving at speeds greater than the speed of sound (not normally involved in the study of acoustics).

Wave drag—Drag on the airplane due to shock waves that are produced at speeds greater than sound.

Concorde that cruises at Mach 2 and several military airplanes have delta or triangular wings. The supercritical airfoil designed by Richard Whitcomb of the NASA Langley Laboratory has made air flow around the wing much smoother and has greatly improved both the lift and drag at transonic speeds. It has only a slight curvature at the top and a thin trailing edge. The proposed hypersonic aerospace plane is expected to fly partly in air and partly in space and to travel from Washington to Tokyo within two hours. The challenge for aerodynamicists is to control the flight of the aircraft so that it does not burn up like a meteor as it enters the atmosphere at several times the speed of sound.

See also Airship; Balloon.

Resources

Books

Anderson, John D. Jr. *Introduction to Flight.* New York: McGraw-Hill, 1989.

Craig, Gale. *Introduction to Aerodynamics.* New York: Regenerative Press, 2003.

Leishman, J. Gordon. *Principles of Helicopter Aerodynamics.* Cambridge: Cambridge University Press, 2003.

Smith, H. C. *The Illustrated Guide to Aerodynamics.* Blue Ridge Summit, PA: Tab Books, 1992.

Wegener, Peter P. *What Makes Airplanes Fly?* New York: Springer-Verlag, 1991.

Periodicals

Hucho, Wolf-Heinrich. "Aerodynamics of Road Vehicles." *Annual Review of Fluid Mechanics* (1993): 485.

Vuillermoz, P. "Importance of Turbulence for Space Launchers." *Journal of Turbulence* 3, no. 1 (2002): 56.

Wesson, John. "On the Eve of the 2002 World Cup, John Wesson Examines the Aerodynamics of a Football and Explains how the Ball Can Bend as It Travels Through the Air." *Physics World* 15, no.5 (2002): 41-46.

Sreela Datta

Aerosols

Aerosols are collections of tiny particles of solid and/or liquid suspended in a gas. The size of these particles can range from about 0.001 to about 100 microns. While a number of naturally occurring aerosols exist, the most familiar form of an aerosol is the pressurized spray can. Aerosols are produced by a number of natural processes and are now manufactured in large quantities for a variety of commercial uses. They are also involved in a number of environmental problems, including **air pollution** and destruction of **ozone** in the atmosphere.

Classification

Aerosols are commonly classified into various subgroups based on the nature and size of the particles of which they are composed and, to some extent, the manner in which the aerosol is formed. Although relatively strict scientific definitions are available for each subgroup, these distinctions may become blurred in actual practical applications. The most important of these subgroups are the following:

Fumes

Fumes consist of solid particles ranging in size from 0.001 to 1 micron. Some typical fumes are those produced by the dispersion of **carbon** black, rosin, **petroleum** solids, and tobacco solids in air. Probably the most familiar form of a fume is smoke. Smoke is formed from the incomplete **combustion** of fuels such as **coal**, oil, or **natural gas**. Its particles are smaller than 10 microns in size.

Dusts

Dusts also contain solid particles suspended in a gas, usually air, but the particles are larger in size than those in a fume. They range from about 1 to 100 microns (and even larger) in size. Dust is formed by the release of materials such as **soil** and **sand**, **fertilizers**, coal dust, cement dust, pollen, and fly ash into the atmosphere. Because of their larger particle size, dusts tend to be more unstable and settle out more rapidly than is the case with fumes, which do not settle out at all.

Mists

Mists are dispersions in a gas of liquid particles less than about 10 microns in size. The most common type of mist is that formed by tiny **water** droplets suspended in the air, as on a cool summer morning. If the **concentration** of liquid particles becomes high enough to affect visibility, it is then called a **fog**. A particular form of fog that has become significant in the last half century is **smog**. Smog forms when natural moisture in the air interacts with human-produced components, such as smoke and other combustion products, to form chemically active materials.

Sprays

Sprays form when relatively large (10+ microns) droplets of a liquid are suspended in a gas. Sprays can be formed naturally, as along an **ocean** beach, but are also produced as the result of some human invention such as aerosol can dispensers of paints, deodorants, and other household products.

Sources

About three-quarters of all aerosols found in the Earth's atmosphere come from natural sources. The most important of these natural components are sea **salt**, soil and rock debris, products of volcanic emissions, smoke from forest fires, and solid and liquid particles formed by **chemical reactions** in the atmosphere. As an example of the last category, gaseous organic compounds released by plants are converted by solar **energy** in the atmosphere to liquid and solid compounds that may then become components of an aerosol. A number of **nitrogen** and **sulfur** compounds released into the atmosphere as the result of living and non-living changes undergo similar transformations.

Volcanic eruptions are major, if highly irregular, sources of atmospheric aerosols. The eruptions of Mount Hudson in Chile in August 1991, and Mount Pinatubo in the Philippines in June 1991, produced huge volumes of aerosols that had measurable effects on Earth's atmosphere.

The remaining atmospheric aerosols result from human actions. Some, such as the aerosols released from spray-can products, go directly to form aerosols in the atmosphere. Others undergo chemical changes similar to those associated with natural products. For example, oxides of nitrogen and sulfur produced during the combustion of **fossil fuels** may be converted to liquid or solid nitrates and sulfates, which are then incorporated into atmospheric aerosols.

Physical properties

The physical and chemical properties of an aerosol depend to a large extent on the size of the particles that make it up. When those particles are very large, they tend to have the same properties as a macroscopic (large size) **sample** of the same material. The smaller the particles are, however, the more likely they are to take on new characteristics different from those of the same material in bulk. Aerosols tend to coagulate, or to collide and combine with each other to form larger bodies. A cloud, for example, consists of tiny droplets of water and tiny **ice** crystals. These particles move about randomly within the cloud, colliding with each other from time to time. As a result of a collision, two water particles may adhere (stick) to each other and form a larger, heavier particle. This process results in the formation of droplets of water or crystals of ice heavy enough to fall to **Earth** as rain, snow, or some other form of **precipitation**.

Synthetic production

The synthetic production of aerosols for various commercial purposes has become such a large industry that the term aerosol itself has taken on a new meaning. Average citizens who know little or nothing about the scientific aspects of aerosols recognize the term as referring to devices for dispensing a wide variety of products.

Aerosol technology is relatively simple in concept. A spray can is filled with a product to be delivered (such as paint), a propellant, and, sometimes, a carrier to help disperse the product. Pressing a button on the can releases a mixture of these components in the form of an aerosol.

The simplicity of this concept, however, masks some difficult technological problems involved in the manufacture of certain "spray" (aerosol) products. An aerosol pesticide, for example, must be formulated in such a way that a precise amount of poison is released, enough to kill **pests**, but not so much as to produce an environmental hazard. Similarly, a therapeutic spray such as a throat spray must deliver a carefully measured quantity of medication. In cases such as these, efforts must be taken to determine the optimal particle size and concentration in the aerosol by monitoring the CFC propellants, which destroy the ozone layer.

The production of commercial aerosols fell slightly in the late 1980s because of concerns about the ozone and other environmental effects. By 1992, however, their manufacture had rebounded. In that year 990 million container units (bottles and cans) of personal aerosol products and 695 million container units of household products were manufactured. In the early 1990s many states passed legislation limiting the volatile organic compounds (or VOCs) used in consumer product aerosols such as hairspray and spray paint. These limitations has forced the aerosol industry to seek alternate propellants and solvents. In many cases this substitution has resulted in inferior products from the standpoint of drying time and spray characteristics. The industry continued to struggle with these issues into the year 2000.

Combustion aerosols

Aerosol technology has made possible vastly improved combustion systems, such as those used in fossil-fueled power **generator** plants and in rocket engines. The fundamental principle involved is that any solid or liquid fuel burns only at its surface. The combustion of a lump of coal proceeds relatively slowly because inner parts of the coal can not begin to burn until the outer layers are burned off first.

The **rate** of combustion can be increased by dividing a lump of coal or a barrel of fuel oil into very small particles, the smaller the better. Power-generating plants today often run on coal that has been pulverized to a dust, or oil that has been converted to a mist. The dust or mist is then thoroughly mixed with an oxidizing agent, such as air or pure **oxygen**, and fed into the combustion chamber. The rate of combustion of such aerosols is many times greater than would be the case for coal or oil in bulk.

Environmental factors

A number of environmental problems are associated with aerosols, the vast majority of them associated with aerosols produced by human activities. For example, smoke released during the incomplete combustion of fossil fuels results in the formation of at least two major types of aerosols that may be harmful to **plant** and **animal** life. One type consists of finely divided carbon released from unburned fuel. This soot can damage plants by coating their leaves and reducing their ability to carry out **photosynthesis**. It can also clog the alveoli, air sacs in human lungs, and interfere with a person's **respiration**.

A second type of harmful aerosol is formed when stack gases, such as **sulfur dioxide** and nitrogen oxides, react with oxygen and water vapor in the air to form sulfuric and nitric acids, respectively. Mists containing these acids may be carried hundreds of miles from their original source before conglomeration occurs and the acids fall to Earth as "acid rain." Considerable disagreement exists about the precise nature and extent of the damage caused by **acid rain**. But there seems to be little doubt that in some locations it has caused severe harm to plant and aquatic life.

Ozone depletion

A particularly serious environmental effect of aerosol technology has been damage to the Earth's ozone layer. This damage appears to be caused by a group of compounds known as **chlorofluorocarbons (CFCs)** which, for more than a half century, were by far the most popular of all propellants used in aerosol cans.

Scientists originally felt little concern about the use of CFCs in aerosol products because they are highly stable compounds at conditions encountered on the Earth's surface. They have since learned, however, that CFCs behave very differently when they diffuse into the upper atmosphere and are exposed to the intense solar **radiation** present there.

In those circumstances, CFCs decompose and release **chlorine atoms** that, in turn, react with ozone in the stratosphere. The result of this sequence of events is that the concentration of ozone in portions of the atmosphere has been decreasing over at least the past decade, and probably for much longer. This change is not a purely academic concern since Earth's ozone layer absorbs ultraviolet radiation from the **Sun** and protects animals on Earth's surface from the harmful effects of that radiation. For these reasons, CFCs have been banned from consumer product aerosols since the late 1970s. They are still employed for certain medical applications, but by and large they have been eliminated from aerosol use. The aerosol industry has replaced CFCs with other propellants such as **hydrocarbon** gases (e.g., butane and propane), compressed gases (e.g., nitrogen and **carbon dioxide**), and **hydrochlorofluorocarbons** (which are much less damaging to the ozone layer.)

Technological solutions

Methods for reducing the harmful environmental effects of aerosols such as those described above have received the serious attention of scientists for many years. As a result, a number of techniques have been invented for reducing the aerosol components of things like stack gases. One device, the electrostatic precipitator, is based on the principle that the particles of which an aerosol consists (such as unburned carbon in stack gases) carry small electrical charges. By lining a smokestack with charged **metal** grids, the charged aerosol particles can be attracted to the grids and precipitated out of the emitted smoke.

Aerosol sniffing

Another risk associated with commercial aerosols is their use as recreational drugs. Inhalation of some consumer aerosol preparations may produce a wide variety of effects, including euphoria, excitement, delusions, and hallucinations. Repeated sniffing of aerosols can result in **addiction** that can cause intoxication, damaged **vision**, slurred **speech**, and diminished mental capacity.

See also Emission; Ozone layer depletion.

Resources

Books

Baron, Paul A., and Klaus Willeke. *Aerosol Measurement: Principles, Techniques, and Applications.* 2nd ed. Hoboken, NJ: Wiley-Interscience, 2001.

Friedlander, S. K. *Smoke, Dust and Haze: Fundamentals of Aerosol Behavior.* New York: John Wiley & Sons, 1977.

Hidy, G. M. "Aerosols." In *Encyclopedia of Physical Science and Technology.* Edited by Robert A. Meyers. San Diego: Academic Press, 1987.

Hinds, William C. *Aerosol Technology: Properties, Behavior, and Measurement of Airborne Particles.* 2nd ed. Hoboken, NJ: Wiley-Interscience,1999.

Hobbs, Peter V., and M. Patrick McCormick, eds. *Aerosols and Climate.* Hampton, VA: A. Deepak, 1988.

Reist, Parker C. *Introduction to Aerosol Science.* New York: Macmillan, 1989.

Smoke, Dust, and Haze: Fundamentals of Aerosol Dynamics (Topics in Friedlander, Sheldon K. *Chemical Engineering.* 2nd ed. Oxford: Oxford University Press, 2000.

Periodicals

Browell, Edward V., et al. "Ozone and Aerosol Changes during the 1991-1992 Airborne Arctic Stratospheric Expedition." *Science* (1993): 1155-158.

Charlson, R. J., et al. "Climate Forcing by Anthropogenic Aerosols." *Science* (1992): 423-30.

Charlson, Robert J., and Tom M. L. Wigley. "Sulfate Aerosol and Climatic Change." *Scientific American* (1994): 48-55.

Haggin, Joseph. "Pressure to Market CFC Substitutes Challenges Chemical Industry." *Chemical & Engineering News* 69 (1991): 27-8.

Miller, Norman S., and Mark S. Gold. "Organic Solvent and Aerosol Abuse." *American Family Physician* 44 (1991): 183-89.

Osborne, Elizabeth G. "Administering Aerosol Therapy." *Nursing* 23 (1993): 24C-24E.

Penner, J.E., et al. "Unraveling the Role of Aerosols in Climate Change." *Environmental Science & Technology* 35, no. 15 (2001): 332a-340a.

Ramanathan, V. "Aerosols, Climate, and the Hydrological Cycle." *Science* 249, no. 5549 (2001): 2119-2114.

"The Role of Atmospheric Aerosols in the Origin Of Life." *Surveys In Geophysics* 23, no.5-5 (2002): 379-409.

"War Spurs Aerosol Research." *Geotimes* 37 (1992): 10-11.

David E. Newton

KEY TERMS

Acid rain—A form of precipitation that is significantly more acidic than neutral water, often produced as the result of industrial processes.

Chlorofluorocarbons (CFCs)—A group of organic compounds once used widely as propellants in commercial sprays, but outlawed in the United States in 1978 because of their harmful environmental effects.

Dust—An aerosol consisting of solid particles in the range of 1 to 100 microns suspended in a gas.

Electrostatic precipitator—A device for removing pollutants from a smokestack.

Fume—A type of aerosol consisting of solid particles in the range 0.001 to 1 micron suspended in a gas.

Mist—A type of aerosol consisting of droplets of liquid less than 10 microns in size suspended in a gas.

Ozone layer—A region of the upper atmosphere in which the concentration of ozone is significantly higher than in other parts of the atmosphere.

Smog—An aerosol form of air pollution produced when moisture in the air combines and reacts with the products of fossil fuel combustion.

Smoke—A form of smoke formed by the incomplete combustion of fossil fuels such as coal, oil, and natural gas.

Spray—A type of aerosol consisting of droplets of liquid greater than 10 microns in size suspended in a gas.

Stack gases—Gases released through a smokestack as the result of some power-generating or manufacturing process.

Africa

Africa is the world's second largest **continent**. From the perspective of geologists and paleontologists (scientists studying ancient life forms), Africa also takes center stage in the physical history and development of life on **Earth**. Africa possesses the world's richest and most concentrated deposits of **minerals** such as gold, diamonds, **uranium**, chromium, cobalt, and platinum. It is also the cradle of **human evolution** and the birthplace of many other **animal** and **plant species**, and has the earliest evidence of **reptiles**, dinosaurs, and **mammals**.

Origin of Africa

Present-day Africa, occupying one-fifth of Earth's land surface, is the central remnant of the ancient southern supercontinent called Gondwanaland, a landmass once made up of **South America**, **Australia**, **Antarctica**, India, and Africa. This massive supercontinent broke apart between 195 million and 135 million years ago, cleaved by the same geological forces that continue to transform Earth's crust today.

Plate tectonics are responsible for the rise of mountain ranges, the gradual drift of continents, earthquakes, and volcanic eruptions. The fracturing of Gondwanaland took place during the Jurassic period, the middle segment of the Mesozoic era when dinosaurs flourished on earth. It was during the Jurassic that flowers made their first appearance, and dinosaurs like the carnivorous Allasaurus and plant eating Stegasaurus lived.

Geologically, Africa is 3.8 billion years old, which means that in its present form or joined with other continents as it was in the past, Africa has existed for four-fifths of Earth's 4.6 billion years. Africa's age and geological continuity are unique among continents. Structurally, Africa is composed of five *cratons* (structurally stable, undeformed regions of Earth's crust). These cratons, in south, central, and west Africa are mostly igneous granite, gneiss, and basalt, and formed separately between 3.6 and 2 billion years ago, during the Precambrian era.

The Precambrian, an era which comprises more than 85% of the planet's history, was when life first evolved and Earth's atmosphere and continents developed. **Geochemical analysis** of undisturbed African **rocks** dating back 2 billion years has enabled paleoclimatologists to determine that Earth's atmosphere contained much higher levels of **oxygen** than today.

Continental drift

Africa, like other continents, "floats" on a plastic layer of the earth's upper mantle called the **asthenosphere**. The overlying rigid crust or **lithosphere**, as it is known, can be as thick as 150 mi (240 km) or under 10 mi (16 km), depending on location. The continent of Africa sits on the African plate, a section of the earth's crust bounded by mid-oceanic ridges in the Atlantic and Indian Oceans. The entire plate is creeping slowly toward the northwest at a **rate** of about 0.75 in (2 cm) per year.

The African plate is also spreading or moving outward in all directions, and therefore Africa is growing in size. Geologists say that sometime in the next 50 million years, East Africa will split off from the rest of the continent along the East African rift which stretches 4,000 miles (6,400 km) from the Red Sea in the north to Mozambique in the south.

General features

Considering its vast size, Africa has few extensive mountain ranges and fewer high peaks than any other continent. The major ranges are the Atlas Mountains along the northwest coast and the Cape ranges in South Africa. Lowland plains are also less common than on other continents.

Geologists characterize Africa's topography as an assemblage of swells and basins. Swells are rock **strata** warped upward by **heat** and **pressure** while basins are masses of lower lying crustal surfaces between swells. The swells are highest in East and central West Africa where they are capped by volcanic flows originating from the seismically active East African rift system. The continent can be visualized as an uneven tilted plateau, one that slants down toward the north and east from higher elevations in the east and south.

During much of the Cretaceous period, from 130 million to 65 million years ago, when dinosaurs like tyrannosaurus, brontosaurus, and triceratops walked the earth, Africa's coastal areas and most of the Sahara Desert were submerged underwater. **Global warming** during the Cretaceous period melted polar **ice** and caused **ocean** levels to rise. Oceanic organic sediments from this period were transformed into the **petroleum** and **natural gas** deposits now exploited by Libya, Algeria, Nigeria and Gabon. Today, oil and natural gas drilling is conducted both on land and offshore on the **continental shelf**.

The continent's considerable geological age has allowed more than enough time for widespread and repeated **erosion**, yielding soils leached of organic **nutrients** but rich in **iron** and **aluminum** oxides. Such soils are high in mineral deposits such as bauxite (aluminum **ore**), manganese, iron, and gold, but they are very poor for agriculture. Nutrient-poor **soil**, along with **deforestation** and **desertification** (expansion of deserts) are just some of the daunting challenges facing African agriculture in modern times.

East African rift system

The most distinctive and dramatic geological feature in Africa is undoubtedly the East African rift system. The rift opened up in the Tertiary period, approximately 65 million years ago, shortly after the dinosaurs became extinct. The same tectonic forces that formed the rift valley and which threaten to eventually split East Africa from the rest of the continent have caused the northeast drifting of the Arabian plate, the opening of the Red Sea to the Indian Ocean, and the volcanic uplifting of Africa's highest peaks including its highest, Kilimanjaro in Tanzania. Mount Kibo, the higher of Kilimanjaro's two peaks, soars 19,320 ft (5,796 m) and is permanently snowcapped despite its location near the equator.

Both Kilimanjaro and Africa's second highest peak, Mount Kenya (17,058 ft; 5,117 m) sitting astride the equator, are actually composite volcanos, part of the vast volcanic field associated with the East African rift valley. The rift valley is also punctuated by a string of lakes, the deepest being Lake Tanganyika with a maximum depth of 4,708 ft (1,412 m). Only Lake Baikal in Eastern Russia is deeper at 5,712 ft (1,714 m).

Seismically the rift valley is very much alive. Lava flows and volcanic eruptions occur about once a decade in the Virunga Mountains north of Lake Kivu along the western stretch of the rift valley. One **volcano** in the Virunga area in eastern Zaire which borders Rwanda and Uganda actually dammed a portion of the valley formerly drained by a tributary of the Nile River, forming Lake Kivu as a result.

On its northern reach, the 4,000-mi (6,400-km) long rift valley separates Africa from **Asia**. The rift's eastern arm can be traced from the Gulf of Aqaba separating Arabia from the Sinai Peninsula, down along the Red Sea which divides Africa from Arabia. The East African rift's grabens (basins of crust bounded by **fault** lines) stretch through the extensive highlands of central Ethiopia which range up to 15,000 ft (4,500 m) and then along the Awash River. Proceeding south, the rift valley is dotted by a series of small lakes from Lake Azai to Lake Abaya and then into Kenya by way of Lake Turkana.

Slicing through Kenya, the rift's grabens are studded by another series of small lakes from Lake Baringo to Lake Magadi. The valley's trough or **basin** is disguised by layers of volcanic ash and other sediments as it threads through Tanzania via Lake Natron. However, the rift can be clearly discerned again in the elongated shape of Lake Malawi and the Shire River Valley, where it finally terminates along the lower Zambezi River and the Indian Ocean near Beira in Mozambique.

The rift valley also has a western arm which begins north of Lake Albert (Lake Mobutu) along the Zaire-Uganda border and continues to Lake Edward. It then curves south along Zaire's eastern borders forming that country's boundaries with Burundi as it passes through Lake Kivu and Tanzania by way of Lake Tanganyika. Lake Tanganyika is not only the second deepest lake in the world but also at 420 mi (672 km) the second longest, second in length and depth only to Lake Baikal in Eastern Russia.

The rift's western arm then extends toward Lake Nysasa (Lake Malawi). Shallow but vast Lake Victoria sits in a trough between the rift's two arms. Although the surface altitude of the rift valley lakes like Nyasa and Tanganyika are hundreds of feet above **sea level**, their floors are hundreds of feet below due to their great depths. In that sense they resemble the deep fjords found in Norway.

The eastern arm of the rift valley is much more active than the western branch, volcanically and seismically. There are more volcanic eruptions in the crust of the eastern arm with intrusions of **magma** (subterranean molten rock) in the middle and lower crustal depths. Geologists consider the geological forces driving the eastern arm to be those associated with the origin of the entire rift valley and deem the eastern arm to be the older of the two.

Human evolution

It was in the great African rift valley that hominids, or human ancestors, arose. Hominid fossils of the genus *Australopithicus* dating 3-4 million years ago have been unearthed in Ethiopia and Tanzania. And the remains of a more direct ancestor of man, *Homo erectus,* who was using fire 500,000 years ago, have been found in Olduvai Gorge in Tanzania as well as in Morocco, Algeria, and Chad.

Paleontologists, who study fossil remains, employ radioisotope **dating techniques** to determine the age of hominid and other species' fossil remains. This technique measures the decay of short-lived radioactive isotopes like **carbon** and argon to determine a fossil's age. This is based on the radioscope's atomic **half-life**, or the time required for half of a **sample** of a radioisotope to undergo **radioactive decay**. Dating is typically done on volcanic ash layers and charred wood associated with hominid fossils rather than the fossils themselves, which usually do not contain significant amounts of radioactive isotopes.

Volcanic activity

Present-day volcanic activity in Africa is centered in and around the East African rift valley. Volcanos are found in Tanzania at Oldoinyo Lengai and in the Virunga range on the Zaire-Uganda border at Nyamlagira and Nyiragongo. But there is also volcanism in West Africa. Mount Cameroon (13,350 ft; 4,005 m) along with smaller volcanos in its vicinity, stand on the bend of Africa's West Coast in the Gulf of Guinea, and are the exception. They are the only active volcanos on the African mainland not in the rift valley.

However, extinct volcanos and evidence of their activity are widespread on the continent. The Ahaggar Mountains in the central Sahara contain more than 300 volcanic necks that rise above their surroundings in vertical columns of 1,000 ft or more. Also in the central Sahara, several hundred miles to the east in the Tibesti

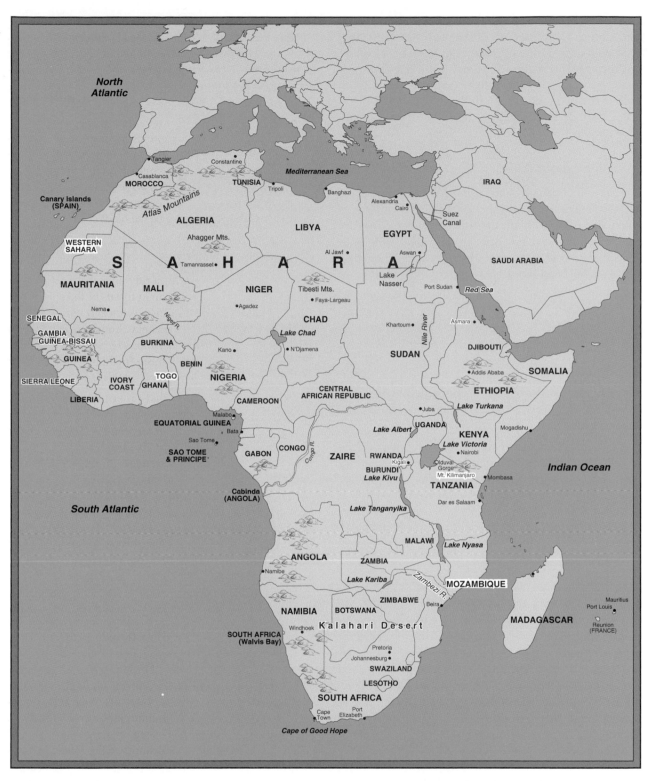

Africa.

Mountains, there exist huge volcanic craters or calderas. The Trou au Natron is 5 mi (8 km) wide and over 3,000 ft (900 m) deep. In the rift valley, the Ngorongoro Crater in Tanzania, surrounded by teeming **wildlife** and spectacular scenery, is a popular tourist attraction. Volcanism formed the diamonds found in South Africa and Zaire. The Kimberley **diamond** mine in South Africa is actually an ancient volcanic neck.

Folded mountains

The only folded mountains in Africa are found at the northern and southern reaches of the continent. Folded mountains result from the deformation and **uplift** of the earth's crust, followed by deep erosion. Over millions of years this process built ranges like the Atlas Mountains, which stretch from Morocco to Algeria and Tunisia.

Geologically, the Atlas Mountains are the southern tangent of the European Alps, geographically separated by the Strait of Gibraltar in the west and the Strait of Sicily in the east. The Atlas are strung across northwest Africa in three **parallel** arrays, the coastal, central, and Saharan ranges. By trapping moisture, the Atlas Mountains carve out an oasis along a strip of northwest Africa compared with the dry and inhospitable Sahara Desert just to the south.

The Atlas Mountains are relatively complex folded mountains featuring horizontal thrust faults and ancient crystalline cores. The Cape ranges on the other hand are older, simpler structures, analogous in age and erosion to the Appalachian mountains of the eastern United States. The Cape ranges rise in a series of steps from the ocean to the interior, flattening out in plateaus and rising again to the next ripple of mountains.

Islands

For a continent of its size Africa has very few islands lying off its coast. The major Mediterranean islands of Corsica, Sardinia, Sicily, Crete, and Cyprus owe their origins to the events that formed Europe's Alps, and are a part of the Eurasian plate, not Africa. Islands lying off Africa's Atlantic Coast like the Canaries, Azores, and even the Cape Verde Islands near North Africa are considered Atlantic structures. Two islands in the middle of the South Atlantic, Ascension and St. Helena, also belong to the Atlantic. Islands belonging to Equatorial Guinea as well as the **island** country of Sao Tome and Principe at the sharp bend of Africa off of Cameroon and Gabon are related to volcanic peaks of the Cameroon Mountains, the principal one being Mount Cameroon.

Madagascar, the world's fourth largest island after Greenland, New Guinea, and Borneo, is a geological part of ancient Gondwanaland. The island's eastern two-thirds are composed of crystalline **igneous rocks**, while the western third is largely sedimentary. Although volcanism is now quiescent on the island, vast lava flows indicate widespread past volcanic activity. Madagascar's unique plant and animal species testify to the island's long separation from the mainland.

Ocean inundations in North Africa

Marine fossils, notably tribolites dating from the Cambrian period (505-570 million years ago; the first period of the Paleozoic era) have been found in southern Morocco and Mauritania. Rocks from the succeeding period, the Ordovician (500-425 million years ago) consist of sandstones with a variety of fossilized marine organisms; these rocks occur throughout northern and western Africa, including the Sahara.

The Ordovician was characterized by the development of **brachiopods** (shellfish similar to clams), corals, **starfish**, and some organisms that have no modern counterparts, called sea scorpions, conodonts, and graptolites. At the same time the African crust was extensively deformed. The continental table of the central and western Sahara was lifted up almost a mile (1.6 km). The uplifting alternated with crustal subsidings, forming valleys that were periodically flooded.

Glaciation

During the Ordovician period, Africa, then part of Gondwanaland, was situated in the southern hemisphere on or near the South Pole. It was toward the end of this period that huge **glaciers** formed across the present-day Sahara and the valleys were filled by sandstone and glacial deposits. Although Africa today sits astride the tropics, it was once the theater of the Earth's most spectacular glacial activity. In the next period, the Silurian (425-395 million years ago), further marine sediments were deposited.

Tectonics in North Africa

The Silurian was followed by the Devonian, Mississippian, and Pennsylvanian periods (408-286 million years ago), the time interval when **insects**, reptiles, **amphibians**, and **forests** first appeared. A continental collision between Africa (Gondwanaland) and the North American plate formed a super-supercontinent (Pangaea) and raised the ancient Mauritanide mountain chain which once stretched from Morocco to Senegal. During the late Pennsylvanian period, layer upon layer of fossilized plants were deposited, forming seams of **coal** in Morocco and Algeria.

When Pangaea and later Gondwanaland split apart in the Cretaceous period (144-66 million years

Zebras grazing at the foot of Mount Kilimanjaro, the highest point in Africa at 19,340 ft (5,895 m). © Gallo Images/Corbis. Reproduced by permission.

ago), a shallow sea covered much of the northern Sahara and Egypt as far south as the Sudan. Arabia, subjected to many of the same geological and climatic influences as northern Africa, was thrust northward by tectonic movements at the end of the Oligocene and beginning of the Miocene epochs (around 30 million years ago). During the Oligocene and Miocene (5-35 million years ago; segments of the modern Cenozoic Era) **bears**, **monkeys**, **deer**, **pigs**, dolphins, and early **apes** first appeared.

Arabia at this time nearly broke away from Africa. The Mediterranean swept into the resulting rift, forming a gulf that was plugged by an **isthmus** at present-day Aden on the Arabian peninsula and Djibouti near Ethiopia. This gulf had the exact opposite configuration of today's Red Sea, which is filled by waters of the Indian Ocean.

As the Miocene epoch drew to a close about five million years ago, the isthmus of Suez was formed and the gulf (today's Red Sea) became a saline (salty) lake. During the Pliocene (5-1.6 million years ago) the Djibouti-Aden isthmus subsided, permitting the Indian Ocean to flow into the rift that is now the Red Sea.

Origin of Sahara desert

In the Pleistocene epoch (1.6-11,000 years ago), the Sahara was subjected to humid and then to dry and arid phases, spreading the Sahara Desert into adjacent forests and green areas. About 5,000-6,000 years ago in the post glacial period of our modern epoch, the Holocene, a further succession of dry and humid stages further promoted desertification in the Sahara as well as the Kalahari in southern Africa.

Earth scientists say the expansion of the Sahara is still very much in evidence today, causing the desertification of farm and grazing land and presenting the omnipresent specter of famine in the Sahel (Saharan) region.

Minerals and resources

Africa has the world's richest concentration of minerals and gems. In South Africa, the Bushveld Complex, one of the largest masses of igneous rock on Earth, contains major deposits of strategic metals such as platinum, chromium, and vanadium—metals that are indispensable in tool making and high tech industrial processes. The Bushveld complex is about 2 billion years old.

Another spectacular intrusion of magmatic rocks composed of olivine, augite, and hypersthene occurred in the Archean Eon over 2.5 billion years ago in Zimbabwe. Called the Great Dyke, it contains substantial deposits of chromium, **asbestos**, and nickel. Almost all of the world's chromium reserves are found in Africa. Chromium is used to harden alloys, to produce stainless steels, as an industrial catalyst, and to provide **corrosion** resistance.

Unique eruptions that occurred during the Cretaceous in southern and central Africa formed kimberlite pipes—vertical, near-cylindrical rock bodies caused by deep melting in the upper mantle. Kimberlite pipes are the main source of gem and industrial diamonds in Africa. Africa contains 40% of the world's diamond reserves, which occur in South Africa, Botswana, Namibia, Angola, and Zaire.

In South Africa uranium is to be found side-by-side with gold, thus decreasing costs of production. Uranium deposits are also found in Niger, Gabon, Zaire, and Namibia. South Africa alone contains half the world's gold reserves. Mineral deposits of gold also occur in Zimbabwe, Zaire, and Ghana. Alluvial gold (eroded from soils and rock strata by **rivers**) can be found in Burundi, Côte d'Ivoire, and Gabon.

As for other minerals, half of the world's cobalt is in Zaire and a continuation into Zimbabwe of Zairian cobalt-bearing geological formations gives the former country sizable reserves of cobalt as well. One quarter of the world's aluminum ore is found in a coastal belt of West Africa stretching 1,200 mi (1,920 km) from Guinea to Togo, with the largest reserves in Guinea.

Major coal deposits exist in southern Africa, North Africa, Zaire, and Nigeria. And North Africa is awash in petroleum reserves, particularly in Libya, Algeria, Egypt, and Tunisia. Nigeria is the biggest petroleum producer in West Africa, but Cameroon, Gabon, and the Congo also contain oil reserves. There are also petroleum reserves in southern Africa, chiefly in Angola.

Most of Africa's iron reserves are in western Africa, with the most significant deposits in and around Liberia, Guinea, Gabon, Nigeria, and Mauritania. In West Africa as well as in South Africa where iron deposits are also found, the ore is bound up in Precambrian rock strata.

Modern-day climatic and environmental factors

Africa, like other continents, has been subjected to gyrating swings in climate during the Quartenary period of the last 2 million years. These climatic changes have had dramatic affects on landforms and vegetation. Some

KEY TERMS

Composite volcano—A large, steep-sided volcano made of alternating sequences of lava and pyroclastic debris. Sometimes called a stratovolcano.

Craton—A piece of a continent that has remained intact since Earth's earliest history, and which functions as a foundation, or basement, for more recent pieces of a continent.

Gondwanaland—An ancestral supercontinent that broke into the present continents of Africa, South America, Antarctica, and Australia as well as the Indian subcontinent.

Graben—A block of land that has dropped down between the two sides of a fault to form a deep valley.

Lava domes—Small dome-shaped masses of volcanic rock formed in the vent of a volcano.

Paleoclimatologist—A geologist who studies climates of the earth's geologic past.

Swells—Rock strata warped upward by heat and pressure.

Volcanic neck—A usually tall, steep mountain of lava rock that solidified in the volcano's throat, stopping up the volcano as it became extinct.

of these cyclical changes may have been driven by cosmic or astronomical phenomena including asteroid and comet collisions.

But the impact of humankind upon the African environment has been radical and undeniable. Beginning 2,000 years ago and accelerating to our present day, African woodland belts have been deforested. Such environmental degradation has been exacerbated by overgrazing, agricultural abuse, and man-made climatic change, including possible global warming caused by the buildup of man-made **carbon dioxide**, **chlorofluorocarbons (CFCs)** and other greenhouse gases.

Deforestation, desertification, and soil erosion pose threats to Africa's man-made lakes and thereby Africa's hydroelectric capacity. Africa's multiplying and undernourished populations exert ever greater demands on irrigated agriculture but the continent's water resources are increasingly taxed beyond their limits. To stabilize Africa's **ecology** and safeguard its resources and mineral wealth, many earth scientists say greater use must be made of sustainable agricultural and pastoral practices. Progress in environmental and resource management, as well as population control is also vital.

Resources

Books

Hancock P. L. and Skinner B. J., eds. *The Oxford Companion to the Earth.* Oxford: Oxford University Press, 2000.

Petters, Sunday W. Lecture *Notes in Earth Sciences Series.* Vol 40, *Regional Geology of Africa.* New York: Springer-Verlag New York, Inc., 1991.

Periodicals

Leroux, M. "The Meteorology And Climate Of Tropical Africa." *Journal of Meteorology* 27, no. 271 (2002): 274.

Robert Cohen

African violet *see* **Gesnerias**

Age of the Universe

The Universe is approximately 14 billion (14,000,000,000) years old. Its age is measured from the event known as the big bang—an explosion filling all **space** and generating all of the **matter** and **energy** that exist today.

Although only in the last 50 years have astronomers been able to estimate the age of the Universe, they have long argued that the Universe must be of finite age, finite size, or both. This conclusion follows from the fact that the night sky is mostly dark. German astronomer Wilhelm Olber (1758–1840) noted in 1823 that if the Universe consisted of identical stars sprinkled through infinite space, and if it had existed for an infinitely long **time**, starlight would have had time to illuminate every point in the universe from every possible direction; in other words, no matter where you were and what direction you looked in, the sky would be a solid **mass** of **light** as bright as the surface of a **star**. Because the night sky is, in fact, dark, either the Universe does not contain an infinitely extensive population of stars, has not existed for an infinite time, or both.

Twentieth-century cosmologists have proved that "both" are true. Although space has no edges or boundaries, it does contain a finite number of cubic miles. Furthermore, time did have a beginning, some 1.4×10^{10} years ago. This figure is determined primarily by using the Doppler shift of light from distant galaxies. Doppler shift is the apparent change in **frequency** of a wave emitted by a source that is approaching or receding from an observer. If a wave source is receding from an observer, the waves detected by that observer are compressed—that is, their peaks and troughs arrive at longer intervals than they would if the source were sta-

tionary (or approaching). More widely spaced peaks and troughs correspond to lower frequency. Therefore, light from celestial sources that are receding from **Earth** is redshifted (shifted to lower frequencies, in the direction of the red end of the visible **spectrum**), while light from sources that are approaching is blueshifted (shifted to higher frequencies, in the direction of the blue end of the visible spectrum). In the 1920s, U.S. astronomer Edwin Hubble (1889–1953) observed that every distant **galaxy**, regardless of its position in the sky, is, as judged by **redshift**, receding rapidly from Earth; furthermore, more-distant galaxies are receding more rapidly than closer galaxies, and the speed of a galaxy's recession is approximately proportional to its **distance** (i.e., if galaxy B is twice as far from Earth as Galaxy A, it is receding about twice as fast).

Astronomers did not seriously consider the possibility that the whole Universe was expanding outward from a central point, with Earth located, by chance, at that central point, even though this would have explained Hubble's data. Even if the Universe had a central point (which seemed unlikely), the chances against finding Earth there by luck seemed large. Rather, Hubble's observations were interpreted as proving that space itself was expanding. However, if space was expanding at a constant rate—a concept which many scientists, including Hubble himself, resisted for years as too fantastic—it could not have been expanding forever. To know age of the Universe, one needed only to measure its present-day **rate** of expansion and calculate how long such an expansion could have been going on. If the expanding Universe were played backward like a film, how long before all the galaxies came together again?

This calculation turned out to be more difficult than it sounds, due to difficulties in measuring the rate of expansion precisely. It is easy to measure the Doppler shift of light from a star in a distant galaxy, but how does one know how far away that star is. All stars in distant galaxies are so far away as to appear as points of light without width, so their size and intrinsic (true) brightness cannot be directly measured. This problem was solved by discovering a class of stars, Cepheid variables, whose absolute brightness can be determined from the rapidity of their brightness variations. Since the absolute brightness of a Cepheid variable is known, its absolute distance can be calculated by measuring how dim it is. A Cepheid variable in a distant galaxy thus reveals that galaxy's distance from the Earth. By observing as many Cepheid variables as possible, astronomers have continually refined their estimate of the Hubble constant and thus of the age of the Universe.

Because of various uncertainties in measuring the characteristics of Cepheid variables, there is still some

KEY TERMS

. . . .

Cepheid variable star—A class of young stars that cyclically brighten and dim. From the period of its brightness variation, the absolute brightness of a Cepheid variable can be determined. Cepheid variables in distant galaxies give a measure of the absolute distance to those galaxies.

observational doubt about the Universe's rate of expansion. An independent method of calculating the age of the Universe relies on observing the types of stars making up globular clusters (relatively small, spherical-shaped groups of stars found in the vicinity of galaxies). By comparing the characteristics of clusters to knowledge about the evolution of individual stars, the age of the Universe can be estimated. The value estimated from globular cluster data—14 to 18 billion years—agrees fairly well with that estimated from the Hubble constant.

Except for **hydrogen**, all the elements of which we and Earth are composed were formed by nuclear reactions in the cores of stars billions of years after the big bang. About 4.5 billion years ago (i.e., when the Universe was about two-thirds its present age) the **solar system** condensed from the debris of exploded older stars containing such heavy elements.

Starting in the late 1990s, data have indicated that the expansion of the Universe initiated by the big bang is, contrary to what cosmologists long thought, accelerating. Several observational tests since the late 1990s have confirmed this result. If it continues to hold up, we can predict that, barring some bizarre quantum-mechanical reversal of cosmic history (as speculated by some physicists), the Universe will continue to expand forever. As it does so, its protons and neutrons will slowly break down into **radiation**, until, eventually, the entire Universe consists of a dilute, ever-expanding, ever-cooler gas of photons, neutrinos, and other fundamental particles.

It should be noted that all references to a "beginning of time" or a "zero moment" for the Universe—and thus of the "age" of the Universe itself—are a simplification. The young Universe cannot be meaningfully described in terms of "space" and "time" until its **density** drops below the Planck density, approximately 10^{94} gm/cm³; below this threshold, our commonsense concept of "time" does not apply. Therefore, if one could watch time run backwards toward the big bang one would not encounter a zero-time moment—a "beginning" of time—but rather a set of conditions under which the notion of "time" itself loses its meaning.

Resources

Books

Hawking, Stephen. *A Brief History of Time: From the Big Bang to Black Holes.* New York: Bantam, 1988.

Hawking, Stephen. *Universe in a Nutshell/Illustrated Brief History of Time.* Random House, 2002.

Livio, Mario. *The Accelerating Universe.* New York: John Wiley & Sons, 2000.

Other

University of Cambridge. "Our Own Galaxy: The Milky Way." *Cambridge Cosmology* May 16, 2000 [cited February 3, 2003]. <http://www.damtp.cam.ac.uk/user/gr/public/gal_milky.htm>.

Larry Gilman

Agent Orange

Agent Orange is a defoliant that kills plants and causes the leaves to fall off the dying plants. The name was a code devised by the United States military during the development of the chemical mixture. The name arose from the orange band that marked the containers storing the defoliant.

Agent Orange was an equal mixture of two chemicals; 2, 4–D (2,4, dichlorophenoxyl **acetic acid**) and 2, 4, 5–T (2, 4, 5-trichlorophenoxy acetic acid). Another compound designated TCDD (2, 3, 7, 8-tetrachlorodibenzo-para-dioxin) is a by-product of the manufacturing process and remains as a contaminant of the Agent Orange mixture. It is this **dioxin** contaminant that has proven to be damaging to human health.

Agent Orange was devised in the 1940s, but became infamous during the 1960s in the Vietnam War. The dispersal of a massive amount of Agent Orange throughout the tropical jungles of Vietnam (an estimated 19 million gallons were dispersed) was intended to deprive the Viet Cong of jungle cover in which they hid.

Agent Orange defoliation damage

By 1971, the use of Agent Orange in Vietnam had ended. However, the damage caused to the vegetation of the region by the spraying of Agent Orange is still visible today. Agent Orange applications affected foliage of a diversity of tropical ecosystems of Vietnam, but the most severe damage occurred in the mangrove forest of coastal areas. About 306,280 acres (1,240 km²) of the coastal mangrove forest of South Vietnam was sprayed at least once. This area comprised about 40% of the South Vietnamese coastal mangrove **ecosystem**.

The spraying killed extensive stands of the dominant mangrove **species**, *Rhizophora apiculata*. Barren, badly eroded coastal **habitat** remained, which had devastating effects on the local economy. The harvesting of dead mangroves for fuel would sustain fewer people than the living forest once did, according to a 1970 report commissioned by the United States National Academy of Sciences, because the supply of mangrove wood was not being renewed. Unless a vigorous replanting program was undertaken, the report warned of a future economic loss when the dead mangroves were harvested.

The destruction of mature seed-bearing trees has made regeneration of mangroves slow and sporadic. Weed species have become dominant. Indeed, the Academy of Sciences has estimated that full recovery of the mangrove forest might take 100 years or more.

Agent Orange was sprayed over 14 million acres of inland tropical forest. A single spray treatment killed about 10% of the tall trees comprising the forest canopy. Defoliation of the trees was much more extensive, but many of the defoliated trees continued to live. Smaller shrubs, protected from the herbicide by the high canopy, comprised the majority of vegetation in sprayed areas. The total loss of commercially useful timber caused by the military application of herbicide in South Vietnam is estimated to be 26–61 million yd^3 (20–47 million m^3). In areas that were sprayed repeatedly, the valuable **tree** species were replaced by a few resistant, commercially unimportant trees (such as *Irvingia malayana* and *Parinari annamense*), along with tussock grass and bamboo. In the dry season, the stands of grass easily catch fire, and if burned repeatedly, the land is less likely to return quickly to forest. It will take many decades before the tropical forest recovers and attains its former productivity.

Because Agent Orange herbicide remains in the **soil** for some time, there is concern that these residues might inhibit the growth of **crops** and other plants. Soil **bacteria** break down the **herbicides** into smaller molecules, but complete **decomposition** takes years to occur. Studies performed 15 years after the spraying in South Vietnam still found degradation products of Agent Orange in the soil. These byproducts, which can be toxic, can be passed through the food web. How much Agent Orange actually reached the soil is subject to question. A large proportion of the herbicide falling onto the forest was trapped by the canopy. Few drops reached the soil directly, but much of the herbicide was eventually delivered to the forest floor by a rain of dead foliage and woody **tissue**. In open areas, much more of the application reached the soil directly. The contaminant TCDD is quite persistent in soil, with a **half-life** of three years. (In that period of time, one half of the dioxin originally applied would still be present in the soil.) In studies conducted in the

United States, samples of inland soil and sediment from mangrove areas treated with herbicide still had substantial levels of TCDD after ten years. An indirect effect of the Agent Orange spraying is the poor fertility of soil in many areas, due to **erosion** following the destruction of soil-binding vegetation.

Reduction of animal habitat

As the rich, biodiverse, tropical **forests** disappeared, so did habitat for indigenous animals. Uniform grassland has poor habitat diversity compared to the complex, multi-layered tropical forest. As a result, the number of bird and mammal species living in sprayed areas declined dramatically. Most of the forest animals are adapted to living in a specific habitat and are unable to survive in the post-herbicide grassland. Wild boar, wild goat, water buffalo, tiger, and various species of **deer** became less common once the cover and food resources of the forest were removed. Domestic animals such as water buffalo, zebu, **pigs**, chickens, and **ducks** were also reported to become ill after the spraying of Agent Orange.

The defoliation and destruction of the mangrove forests had other consequences to **wildlife**. The number of coastal **birds** declined dramatically, since their habitat had vanished. **Fish** and crustacean populations also suffered, since their former breeding and nursery habitats in the web of channels winding beneath the mangrove trees were destroyed. Additionally, the wartime spraying of mangrove forest is thought to have contributed to the post-war decline in South Vietnam's offshore fishery.

Possible human health threat

Agent Orange has unquestionably been a disaster for the **ecology** of Vietnam. But evidence also suggests that the defoliant, and in particular the TCCD dioxin component, is a health threat to soldiers who were exposed to Agent Orange during their tour of duty in Vietnam. Tests using animals have identified TCCD as the cause of a wide variety of maladies. In the mid-1990s, the "Pointman" project was begun in New Jersey, which scientifically assessed select veterans in order to ascertain if their exposure to Agent Orange had damaged them. The project is ongoing. In the meantime, veterans organizations continue to lobby for financial compensation for the suffering they feel has been inflicted on some soldiers by Agent Orange.

See also Immune system; Poisons and toxins.

Resources

Books

Gough, M. *Agent Orange: The Facts.* New York: Perseus Books, 1986.

National Academy of Sciences. *Veterans and Agent Orange: Health Effects of Herbicides Used in Vietnam.* Washington, DC: National Academy Press, 1994.

Schuck, P.H.H. *Agent Orange on Trial: Mass Toxic Disasters in the Courts.* Boston: Harvard University Press, 1990.

Other

Department of Vetrans Affairs. "Where Can I Get Information on Agent Orange?" (June 23, 2000) [cited October 25, 2002]. <www.VA.gov>.

Brian Hoyle

Aging and death

Aging is the natural effect of **time** and the environment on living organisms, and death is its end result. **Gerontology** is the study of all aspects of aging. No single theory on how and why people age is able to account for all facets of aging. Although great strides have been made to postpone death as the result of certain illnesses, less headway has been made in delaying aging.

Life span is species-specific. Members of the same **species** have similar life expectancies. In most species, death occurs not long after the reproductive phase of life ends. This is obviously not the case for humans. However, there are some changes that occur in women with the onset of **menopause** when estrogen levels drop. Post-menopausal women produce less facial skin oil (which serves to delay wrinkling) and are at greater risk of developing **osteoporosis** (brittle bones). Men continue to produce comparable levels of facial oils and are thus less prone to early wrinkling. Osteoporosis occurs as **calcium** leaves bones and is used elsewhere; hence, sufficient calcium intake in older women is important because bones which are brittle break more easily.

Theories on aging

The relationship between aging and death is complex. The results from many studies indicate that aging decreases the efficiency of the body to operate, defeat infections, and to repair damage. Comparison of people aged 30–75 has demonstrated that the efficiency of lung function decreases by 50% that bones become more brittle, and that the **immune system** that safeguards the body from infections generally becomes less efficient as we age.

Why this deterioration in the functioning of the body with age occurs is still not clear. Several theories have been proposed to explain this decline. One theory proposes that after the active years of reproduction have passed, chemical changes in the body cause the gradual malfunctioning of organs and other body components. The accumulation of damage to components that are necessary for the formation of new cells of the body leads to death. For example, it has been discovered that the formation of the genetic material **deoxyribonucleic acid (DNA)** is more subject to mistakes as time goes on. Other theories relating aging with death include the negative effect of stresses to the body, and a theory that proposes that the build-up of non-functional material in the body over time lessens the ability of the body to function correctly.

The strongest arguments on the aging process favor involvement of one of, or a combination of the following: hormonal control, limited **cell division**, **gene** theory, **gene mutation** theory, protein cross-linkage theory, and free radical action. In support of hormonal control, there is the observation that the thymus gland (under the sternum) begins to shrink at adolescence, and aging is more rapid in people without a thymus. Another hormonal approach focuses on the hypothalamus (at the base of the **brain**), which controls the production of **growth hormones** in the pituitary gland. It is thought that the hypothalamus either slows down normal hormonal function or that it becomes more error-prone with time, eventually leading to physiological aging.

More recent theories on aging come from **cell biology** and **molecular biology**. Cells in culture in the laboratory keep dividing only up to a point, and then they die. Cells taken from embryos or infants divide more than those taken from adults. Hence, it is thought that this is the underlying mechanism of aging—once cells can no longer divide to replenish themselves, a person will begin to die. However, most scientists now accept that most cells (other than brain and muscle cells) are capable of division for a longer time than the normal human lifespan.

Gene theory and gene **mutation** theory both offer explanations for aging at the level of DNA. Gene theory suggests that genes are somehow altered over time, such that they naturally cause aging. Gene mutation theory is based on the observation that mutations accumulate over time, and it is mutations that cause aging and **disease**. This view is supported by the fact that samples of cells from older people do generally have more genetic mutations than cells taken from younger people. In addition, some diseases associated with age result from genetic mutations. **Cancer** is often the result of multiple mutations and some mutations reveal underlying genetic weaknesses, which cause disease in some people. Gene mutation theory also notes that for mutations to accumulate, normal DNA-repair mechanisms must have weakened. All cells have inherent repair mechanisms that routinely fix DNA errors. For these errors to accumulate, the repair system must have gone awry, and DNA-repair failure is thought to be a factor in cancer.

Protein cross-linkage and free radicals are also thought to contribute to aging. Faulty bonds (cross-linkages) can form in **proteins** with important structural and functional roles. **Collagen** makes up 25-30% of the body's protein and provides support to organs and **elasticity** to **blood** vessels. Cross-linkage in collagen molecules alters the shape and function of the organs it supports and decreases vessel elasticity. Free radicals are normal chemical byproducts resulting from the body's use of **oxygen**. However, free radicals bind unsaturated fats into cell membranes, alter the permeability of membranes, bind chromosomes, and generally alter cellular function, causing damage. **Antioxidants**, such as vitamins C and E, block free radicals and are suggested for prolonging life.

Diseases associated with aging

Some consequences of aging are age-related changes in **vision**, **hearing**, muscular strength, bone strength, immunity, and nerve function. Glaucoma and cataracts are ocular problems associated with aging that can be treated to restore failing vision in older people. Hearing loss is often noticeable by age 50, and the range of sounds heard decreases. Muscle mass and **nervous system** efficiency decrease, causing slower **reflex** times and less physical strength, and the immune system weakens, making older people more susceptible to infections.

More serious diseases of aging include Alzheimer and Huntington diseases. Patients with **Alzheimer disease**, also called primary **dementia**, exhibit loss and diminished function of a vast number of brain cells responsible for higher functions; **learning**, **memory**, and judgment are all affected. The condition primarily affects individuals over 65 years of age. Some current figures estimate that as many as 10% of people within this age group are affected by Alzheimer disease. A rapidly expanding disease whose numbers increase as the proportion of elderly Americans continues to rise, it is predicted that 14 million people will have Alzheimer disease by the year 2050. Huntington's disease is a severely degenerative malady inherited as a dominant gene. Although

As they age, men's bodies continue to produce facial skin oils at levels comparable to earlier years, and thus are less prone to early wrinkling. © *Vince Streano/Corbis. Reproduced by permission.*

its symptoms do not appear until after age 30, it is fatal, attacking major brain regions. There is no treatment for either of these age-related diseases.

Death

Death is marked by the end of blood circulation, the end of oxygen transport to organs and tissues, the end of brain function, and overall **organ** failure. The **diagnosis** of death can occur legally after breathing and the heartbeat have stopped and when the pupils are unresponsive to **light**. The two major causes of death in the United States are **heart** disease and cancer.

Other causes of death include **stroke**, accidents, infectious diseases, murder, and suicide. While most of these phenomena are understood, the concept of stroke may be unclear. A stroke occurs when blood supply to part of the brain is impaired or stopped, severely diminishing some neurological function. Some cases of dementia result from several small strokes that may not have been detected.

Some people seek to thwart aging and death through technologies such as the transplantation of organs, cosmetic **surgery**, and cryopreservation (deep-freezing) of

KEY TERMS

. . . .

Gerontology—The scientific study of aging with regard to its social, physical, and psychological aspects.

Life span—The duration of life.

the recently deceased in the hope that a future society will have found the means to revitalize the body and sustain life.

See also Artificial heart and heart valve; Autoimmune disorders; Cell death; Cryobiology; Stress.

Resources

Books

Hazzard, William. *Principles of Geriatric Medicine and Gerontology.* 4th ed. New York: McGraw-Hill, 1999.

Nuland, S. *How We Die: Reflection on Life's Final Chapter* New York: Alfred A. Knopf, 1993.

Organizations

National Institute on Aging. Building 31, Room 5C27, 31 Center Drive, MSC 2292, Bethesda, MD 20892. (301) 496–1752. <http://www.nih.gov/nia/health/health.htm.> (March 15, 2003).

United States Department of Health and Human Services Administration on Aging. (800) 677–1116. <http://www.aoa.dhhs.gov/> (March 15, 2003).

Louise Dickerson

Agouti

The twelve **species** of agoutis are the best-known members of the family Dasyproctidae (genus *Dasyprocta*) of the order Rodentia. Agoutis are found from southern Mexico through Central America to southern Brazil, including the Lesser Antilles. They are long-legged, slender-bodied, rabbit-like **mammals** with short ears and a short tail. The body length of agoutis measures 16-24 in (41.5-62 cm), and adults weigh 3-9 lb (1.3-4 kg). The fur is quite coarse and glossy, and is longest and thickest on the back. The fur ranges from pale orange through shades of brown to black on the backside, with a whitish, yellowish, or buff-colored underside. The back may be striped.

The forelimbs each have four digits, while the hind limbs have three hoof-like claws. The cheek teeth have high crowns and short roots, a condition known as hypsodont.

Agoutis live in cool, damp lowland **forests**, grassy stream banks, thick bush, high dry hillsides, savannas, and in cultivated areas. These animals are active during the day, feeding on fruit, **vegetables**, and various succulent plants, as well as corn, plantain, and cassava root. Agoutis eat sitting erect, holding the food with their forelimbs.

Female agoutis have eight mammary **glands**, and usually two, but up to four, young are born at one time. Mating may occur twice a year, and there is a three-month gestation period. The young are born in a burrow that is dug out among limestone boulders, along river banks, or under the roots of trees. The nest is lined with leaves, roots, and hair. The life span of agoutis in captivity is 13-20 years.

Other members of the family Dasyproctidae, include pacas, Itapeizcuinte, Ihei, or Iconejos pintados of the genus *Cuniculus*, which are bigger than agoutis and have spotted stripes, and the members of the genus *Myoprocta*, known as Iacushi or Iacuchi, agouti, Icutia de rabo, Icutiaia, or Icotiara, which are much smaller than agoutis and have a longer tail. Agoutis (genus *Dasyprocta*), are also known in their geographical range as Iagutis, Inequis, Icutias, Icotias, Ikonkoni, Icotuzas, and Ipicure.

Members of the genus *Plagiodontia* inhabit the Dominican Republic (*P. aedium*) and Haiti (*P. hylaeum*), and the latter is sometimes incorrectly referred to as agouti or Izagouti. The genus *Plagiodontia* are **rodents** of the family *Capromyidae* and correctly called Hispaniolan hutias or jutias.

Agricultural machines

Agriculture is an endeavor practiced in all countries. From the earliest times, humankind has engaged in some form of planting, herding, or gathering. From about 11,000-8000 B.C. in the Middle East, where many consider civilization to have begun, early farmers used crude flint-edged wooden sickles to harvest wild grains growing on the river banks. The harvested grain was often stored in caves for use during the fall and winter. Unfortunately, there was often not enough to gather and store to feed the growing population.

Soon local tribes learned how to plant and cultivate the **seeds** of wild **grasses** and raise them as food. About 9000 B.C. these same Middle Eastern tribes learned to domesticate **sheep** and raise them for both their skins and food. Communities grew along the rich, fertile banks between the Tigris and Euphrates Rivers in Mesopotamia (now Iraq). Here in the "cradle of civilization" the first true machine, the wheel, was invented and used on animal-drawn carts in the expanding fields. For the first time canals were built linking the principal rivers with local tributaries. Pulleys were used to draw water from the canals creating the first **irrigation** system.

Early farmers quickly learned that a supply of water was essential to farming. Thus, the primary fields of grain were planted alongside the great rivers of the Middle East. However, getting water from the rivers to the fields became a problem. The invention of the shaduf, or chain-of-pots, helped solve this problem. This human-powered primitive device consisted of buckets attached to a circular rope strung over a horizontal wooden wheel with wooden teeth projecting at the rim. The buckets were lowered into the water by this revolving "chain," and water was lifted from the river and carried to the fields.

The first farming tool was a pointed stick called a digging stick. The food gatherers used it to dig roots; later farmers used it to dig holes for seeds. The spade was invented by the farmer who simply added a cross bar to his digging stick so that he could use his foot to drive it deeper into the earth. A stick with a sharp branch at one end was the first hoe. Later, a stone or shell was added to the stick to give it a more effective cutting edge. Sharp stones cut along one edge converted a stick into a sickle.

After animals were domesticated for food, they were soon trained to become beasts of burden. Sometime around 2300 B.C., along the Indus River of northern India, water buffalo and zebu cattle were used to pull crude wooden plows through the earth, thus developing the practices of plowing and cultivating.

The discovery of **metal** at the end of the Neolithic period enabled farmers to have sharper, stronger blades for hoes, plows, points, and sickles. The Romans improved the design of their agricultural implements, leading to vastly improved plows and sickles using metal parts. They raised olive and fig trees as well as cereal grains; they also kept vineyards, many of which are still bearing.

The Roman plow consisted of two wooden planks in the shape of a "V." At the tip of the "V," a metal tip was attached. At the back of the "V," an upright wooden post allowed a farmer to guide the implement. A pair of oxen was required to pull the crude plow known as an aratrum. This rather basic tool could not plow a furrow or cut a slice of **soil** and turn it over as our modern plows do. Since this two-ox aratrum merely scratched the soil, it was necessary to go back over the first-plowed earth again and cross-plow it at right angles to the initial pass.

The basic plow was one of the farmer's most important implements, yet it remained unchanged for centuries. Until the 1800s, the plow was still a heavy, pointed piece of wood that was pulled by several oxen and dug an irregular furrow. It didn't turn the soil. In 1793,

Thomas Jefferson developed a curved **iron** moldboard, made according to a mathematical plan, that would lift and turn the soil yet offer little resistance to the motion of the plow. Yet his idea was never tested. However, four years later, a New Jersey farmer, Charles Newbold, patented a plow with a cast iron curved moldboard similar to the idea Jefferson proposed. Ironically, farmers were slow to accept the device, many claiming that cast iron would poison the soil and encourage the growth of weeds. In 1819, Jethro Wood followed Newbold's design with a cast-iron plow incorporating detachable parts that could be replaced when worn. Soon, more and more farmers recognized the advantages of the new design and slowly began to accept the concept of a metal, curved blade or moldboard.

However, as men began to plow the plains, cast iron plows proved to have a major disadvantage: soft or damp soil easily clung to the blade and made a full furrow difficult to achieve. James Oliver, a Scottish-American iron founder, developed an iron plow with a face hardened by chilling in the mold when it was cast. His device solved some of the previous problems. By the time of his death in 1908, his invention had made him the richest man in Iowa. Yet, even the Oliver plow had problems with the heavy, sticky soil of the prairies. Soil still stuck to the moldboard instead of turning over.

The **steel** plow was the answer to this problem. In 1833, John Lane, a blacksmith from Lockport, Illinois, began covering moldboards with strips of saw steel. For the first time a plow was successful in turning the **prairie** soil. Then, in 1837, a blacksmith from Grand Detour, Illinois, named John Deere began making a one-piece share and moldboard of saw steel, and within 25 years the steel plow had replaced the cast iron plow on the prairies. Demand for Deere's plow was so great he had to import steel from Germany. Today, the company that bears his name is one of the world's leading manufacturers of farming implements.

The Romans are also credited with the invention of a crude machine used to cut or "reap" **wheat**. It was called Pliny's reaper, not because Pliny was the inventor but because the great Roman historian mentioned it in his writings around A.D. 60. Pliny's reaper consisted of a wooden comb affixed to the front of a wooden cart. The cart was pushed through the fields by an ox. A farm hand guided the grain into the comb's teeth and manually sliced off the heads of the wheat, allowing the grain to fall into a trailing cart and leaving the straw standing. This was a disadvantage where the straw was valuable as fodder.

Another agricultural development also occurred about this time. For thousands of years the only tool used to separate or "thresh" the grain from the straw was a stick that was used to literally beat the grain from the straw. In warm climates animals were used to walk on the wheat, or "tread out the corn." However, in the colder climates of central and northern **Europe**, where the weather was uncertain, threshing was done by hand in barns.

Then, during Roman times, a farmer decided to lash two strong pieces of wood together with a leather thong. By using one stick as a handle and whipping the grain with the other stick, he could swing his device in a circular motion rather than the up-and-down motion used previously. Thus, a great deal more grain could be threshed. The invention was called a *flagullum* after the Latin word meaning a whip. It was later simply called a flail. After the fall of Rome, no further advances in reaping machines were made for 15 centuries.

During the Middle Ages, agricultural hardware also made slow progress, yet some gains were recorded. Around A.D. 1300 a new type of harness was developed which radically changed farming. This simple device allowed a horse, rather than an ox, to be hitched to a plow. Since a horse can plow three or four times faster than on ox, **horses** gradually replaced oxen as the chief source of power.

During the early 1700s, several major changes in agricultural machines were made. In this period, known today as the "Agricultural Revolution," inventors in Great Britain and the United States introduced machines that decreased the amount of labor needed and increased productivity.

For centuries farmers had planted seeds by sowing or scattering them on the soil and trusting chance to provide for **germination** and a decent crop. Sowing seeds was both labor intensive and extremely wasteful, as many seeds never grew into plants. Next, farmers tried digging a small trench and burying their seeds. This method improved the yield, but was even more labor intensive. Then, in 1701, an English farmer named Jethro Tull invented a horse-drawn device that drilled a pre-set hole in the soil and deposited a single seed. It was the first successful farm machine with inner moving parts and the ancestor of today's modern farm machinery.

Harvesting of grain is perhaps the most difficult job for farmers, and for hundreds of years all the work was performed by human hands. Horses and oxen had been used to pull plows, but harrows and carts were of no avail at harvest time. Only hours of backbreaking toil could cut and bind the grain before it rotted on the ground, a window of about ten days. At best, bringing in all the grain was uncertain; in cases of bad weather or too few laborers, it could result in famine.

The principal implement of the harvest in the earliest recorded days of history was the sickle, a curved knife with which a strong man could cut a half acre in the course of a day. Harvesting with a sickle, however, is grueling labor. Each bunch of grain must be grasped in one hand and cut by the sweep of the blade. In 1830, the sickle was still in general use under certain crop conditions.

The scythe was an ancient tool used for cutting standing grain. With it, a man could cut two acres a day. During the eighteenth century the scythe was improved by the addition of wooden fingers. With this implement, called a cradle, grain could be cut and at the same time gathered and thrown into swaths, making it simpler for others following to bind it into sheaves.

It was during the late eighteenth and early nineteenth centuries that American ingenuity altered centuries of farming practices. In 1793 for example, a young Connecticut resident, Eli Whitney, graduated from Yale University and went to live on a **cotton** plantation in Georgia. There he observed the slaves picking cotton. Each slave slowly separated and stripped the cotton fiber from its seed. However, the cotton clung so tenaciously to the green seeds that a slave working all day could only clean a single pound of cotton. Whitney recognized the problem and designed a machine to separate the cotton fibers from their seeds.

His device, called a cotton engine (the word "engine" was soon slurred to "gin"), consisted of a cylinder from which hundreds of wires (later changed to saw-toothed disks) projected. The pieces of wires worked in slots wide enough for the cotton but not wide enough for the seeds. As the hooks pulled the cotton through the slots, a revolving brush removed the cotton from the cylinder.

The cotton gin alone altered the entire economy of the South and the nation. In the year following Whitney's invention, the cotton crop increased production from five to eight million pounds. Six years later, in 1800, 35 million pounds were produced. By the time Whitney died in 1825, more than 225 million pounds of cotton were produced each year. The invention of the cotton gin led to the expansion of the plantation system with its use of slave labor and led to the South's dependence upon a single staple crop. It also encouraged the economic development of the entire nation by providing large sums for use in foreign exchange.

The invention of the reaper was probably the most influential in the history of agriculture. It greatly reduced the threat of famine in America and released thousands of men from the farm. Before the early 1800s, 95% of the world's population was needed to work on the farm. In 1930, 91% of the 220 million Americans were living in cities and small towns, with only 4% living on farms.

Many inventors worked on animal-powered machines for harvesting grain. In 1828 alone, there were four patents issued in England for reaping machines.

A combine harvesting wheat in Oklahoma. *JLM Visuals. Reproduced with permission.*

None were successful. However, in 1826, a Scottish man named Patrick Bell developed a machine that consisted of two metal strips, one fixed and the other oscillating back and forth against it. As the machine moved forward the metal strips sliced the grain. Although Bell's machine was an effective grain harvester, he was discouraged from requesting a patent by angry farm workers who feared for their jobs.

About the same time as Bell was working on his machine, a Virginian of Scottish-Irish parentage, Cyrus H. McCormick, was also trying to produce a machine that would successfully harvest grain. In 1831 McCormick demonstrated his first reaper before a skeptical gathering near his father's farm near Steel's Tavern, Virginia. The trial failed because the field was too hilly, but when another farmer offered his acreage on a flatter field, the test was a success. Although McCormick's first device was still in a crude state, it cut as much grain as six laborers working with scythes or as much as 24 could cut with sickles.

His harvester combined a number of elements, none of which were new, but had never been combined before. He used a reciprocating blade similar to Bell's design. A reel pushed the grain against a blade. The harvested grain fell onto a wooden platform located beneath the blade and was swept into swaths by a laborer standing astride it.

Despite the fact that the machine worked, no one was interested in buying one for $50. The following year McCormick demonstrated an improved model at Lexington, California, and still found no buyers. By 1840, his total sales amounted to one. He increased the price to $100 and in 1842 sold seven. By 1845, he had sold close to 100 reapers and word of their successful achievements began to spread. McCormick moved to Chicago and formed his own company to manufacture reapers. By the time he retired as a wealthy man, his son had taken over the business, which was later merged to become the International Harvester Company, now Navistar.

Other horse-drawn machines followed the improved plows and grain reapers. In 1834, threshing machines were first brought from Scotland, where they had been used since 1788. A successful American thresher was patented in 1837. The following year the combine was introduced on the American farm, and for the first time a machine combined both the harvesting and threshing of grain. These early machines were large and bulky and required horse or mule teams to pull them. It was not until the 1920s, with the successful introduction of gasoline-engine tractors, that combines were accepted on farms. By 1935 a one-man combine was in use, and by the agricultural boom of World War II, self-propelled combines were common.

Other American patents were granted in the 1840s and 1850s for an improved grain drill (thus making obsolete the sowing of seeds by hand), a mowing machine, a disk harrow, a corn planter, and a straddle row cultivator. The Marsh harvester, patented in 1858, used a traveling apron to lift the cut grain into a receiving box where men, riding on the machine, bound it in bundles. Early in the 1870s, an automatic wire binder was perfected, but it was superseded by a twine binder late in the decade.

Although **animal** power was still required, the new agricultural implements greatly saved both time and labor. During the period between 1830 and 1840, for example, the time required to harvest an acre of wheat was reduced from 37 hours to about 11 hours.

With the introduction of steam power in the early part of the nineteenth century, the days of animal powered machines were numbered. Shortly before the Civil War the first steam engines were used on farms. Initially, these heavy, rather crude devices were used in the field to provide belt power for threshing machines or other farming jobs that could be accomplished at a stationary location. During the 1850s, self propelled steam tractors were developed. However, they required a team of men to operate them. A constant supply of water and fuel (usually wood or **coal**) was required; someone was also needed to handle the controls and monitor the boiler.

Huge steam tractors were sometimes used to pull a plow, but their use was extremely limited. Because of its cost, size, and the general unwieldiness of the machine, it could only be employed profitably on the immense acreages of the west. The first internal **combustion** tractors, built near the beginning of the twentieth century, were patterned after the steam models. And, like their steam predecessors, the first gasoline tractors embodied many of the same faults: they were still too heavy; they were unreliable and broke down about as often as they ran; and, nearly all of them had an alarming tendency to dig themselves into mud holes in wet soil.

During those early years, designers were convinced that there could be neither power nor traction without great weight. The efforts of all progressive farm tractor manufacturers have since been directed toward lessening a tractor's weight per horsepower. The early steam giants weighed more than 1,000 lb (450 kg) per horsepower. Today's farm tractor weighs about 95 lb (43 kg) per horsepower.

Gas powered tractors were tested on farms in the early 1900s; however, the first all-purpose gas tractor was not in-troduced until 1922. It had a high rear wheel drive for maximum clearance under the rear axle, narrow front wheels designed to run between rows, and a power connection to attach other implements on either the front or rear end. This power take-off was standard on all farm tractors by 1934. The use of the power take-off allowed all harvesting machinery to employ a wider cut and high speed gearing. It meant that rotary power from the engine could be transmitted through a flexible shaft to drive such field implements as mowing machines, balers, combines, etc.

In 1933, pneumatic tires were introduced and, for the first time, allowed tractors to be used on paved highways, a valuable aid in moving equipment from one field to another on farms crossed by paved roads.

The continued development of farm machinery has kept pace with the rapid expansion of technology. The modern farm, for example, can use a single multipurpose machine for precision tillage, planting, shaping the beds, and fertilizing the soil, all in a single pass. For the farmer who needs to grade his field to an exact slope for proper surface irrigation, he can use special laser-leveling equipment. A **laser** beam is transmitted at the pre-set **angle** of slope and its signal is received on a mobile scraper. The scraper blade is constantly changing pitch to assure the angle of slope is leveled accurately. With laser leveling, farmers have achieved a 20-30% savings in irrigation water compared with traditional methods.

The use of **aircraft** in farming has also revolutionized many farming techniques. **Fertilizers** and **pesticides** are now widely broadcast over large fields by low-flying crop dusting planes. In some parts of the world, **rice** is sown over flooded fields by air, which saves a tremendous amount of manual labor traditionally used to plant rice.

Harvesting techniques have also been modernized since the days when McCormick's reaper was pulled across the fields. Today, almost all small grains are harvested by self-propelled combines that cut the crop, strip it if necessary, and deliver the grain to a waiting bin. Although some corn is harvested and chopped for silage, most is grown for grain. Special harvester heads strip and shell the ears, delivering clean kernels ready for bundling or storage in waiting silos. To ease the burden of handling large hay bales, machines can now harvest a hay crop, compress the hay into compact 1.6 inch square cubes, and unload the compressed feed nuggets into a waiting cart attached to the rear of the harvester.

Harvesting **fruits** and **vegetables** has always been the most labor intensive farming operation. Traditionally, harvesters simply cross a field and slice off the head of cabbage, lettuce, or celery with a sharp disk or series of knives. Most of these harvesters are non-selective, and there can be considerable waste from harvesting under-

or over-ripe produce. Selection for **color** or grade is dependent on workers at the site or at grading stations.

Today, special electronic equipment attached to a harvester can determine the correct size, color and density of certain types of produce. Tomatoes, for example, now pass through special harvesters where an optical sensor inspects each tomato as it passes at high speed. The sensor judges whether the tomato is mature or not mature, red or green, and rejects the green fruit. Today, more than 95% of California's vast tomato crop is harvested mechanically. The tomatoes have been genetically altered for tougher skin, allowing mechanical harvesting.

Asparagus harvesters can determine the pre-set length before cutting, while a lettuce harvester electronically determines which heads meet the proper size and density standards. Yet, despite the tremendous advances in agricultural technology, many fruits and vegetables are still best harvested by hand. These include most citrus fruits, melons, **grapes**, broccoli, etc.

In dairy farming, modern equipment such as automated milking machines, coolers, clarifiers, separators, and homogenizers has allowed the farmer to increase his herds as well as the productivity of his cattle. Yet the dairy farmer still puts in more man-hours of labor per dollar return than does the field-crop farmer. According to a study by the U.S. Department of Agriculture, the average number of labor-hours used to produce an acre of corn in a Midwestern state was reduced from 19.5 in 1910 to 10.3 in 1938 and to less than seven in 1995. In contrast, no change was reported in the man-hours required in the production of beef cattle and egg-laying hens, while a 5% increase was noted in the labor associated with dairy farming.

Benedict A. Leerburger

Agrochemicals

An agrochemical is any substance that humans use to help in the management of an agricultural **ecosystem**. Agrochemicals include: (1) **fertilizers**, (2) liming and acidifying agents, (3) **soil** conditioners, (4) **pesticides**, and (5) chemicals used in **animal** husbandry, such as **antibiotics** and **hormones**.

The use of agrochemicals is an increasingly prominent aspect of modern, industrial agriculture. The use of agrochemicals has been critically important in increasing the yield of agricultural **crops**. However, some uses of agrochemicals cause important environmental and ecological damages, which detract significantly from the benefits gained by the use of these materials.

Fertilizers

Fertilizers are substances that are added to agricultural lands to alleviate nutrient deficiencies, allowing large increases in the rates of crop growth. Globally, about 152 million tons (138 million metric tons) of fertilizer are used in agriculture each year. In the United States, the average **rate** of fertilizer application is about 218 lb per 2.5 acres (99 kilograms per hectare), and about 21 million tons (19 million metric tons) are used in total each year.

The most commonly used fertilizers are inorganic compounds of **nitrogen** (N). Under conditions where agricultural plants have access to sufficient **water**, their productivity is most often constrained by the supply of available forms of nitrogen, especially nitrate (NO_3^-), and sometimes ammonium (NH_4^+). Farmers commonly increase the availability of these inorganic forms of nitrogen by applying suitable fertilizers, such as **urea** or ammonium nitrate. The rate of fertilization in intensive agricultural systems is commonly several hundred pounds of nitrogen per acre per year, but it can be as high as 440 lb per acre per year (500 kg per hectare per year).

Phosphorus (P) and potassium (K) are other commonly applied **nutrients** in agriculture. Most phosphorus fertilizers are manufactured from rock phosphate, and are known as superphosphate and triple-superphosphate. Some other phosphorus fertilizers are made from bone meal or seabird guano. Potassium fertilizers are mostly manufactured from mined potash.

Often, these three macronutrients are applied in a combined formulation that includes nitrogen, in what is known as an N-P-K fertilizer. For example, a 10-10-10 fertilizer would contain materials equivalent to 10% each of nitrogen (N), P_2O_5 (source of phosphorus [P]), and K_2O (source of potassium [K]), while a 4-8-16 fertilizer would contain these nutrients in concentrations of 4%, 8%, and 16%, respectively. The desired ratios of these three nutrients are governed by the qualities of the soil being fertilized, and by the needs of the specific crop.

Sometimes other nutrients must also be supplied to agricultural crops. **Sulfur**, **calcium**, or **magnesium**, for example, are limiting to crop productivity in some places. Rarely, micronutrients such as **copper**, molybdenum, or zinc must be applied to achieve optimum crop growth.

Liming and acidifying agents

Agricultural soils are commonly too acidic or too alkaline for the optimal growth of many crop **species**. When this is the case, chemicals may be added to the soil to adjust its **pH** to a more appropriate range.

Acidic soils are an especially common problem in agriculture. Acidic soil can be caused by various factors, including the removal of acid-neutralizing bases contained in the **biomass** of harvested crops, the use of certain types of fertilizers, **acid rain**, the oxidation of sulfide **minerals**, and the presence of certain types of organic **matter** in soil. Because soil acidification is such a common occurrence, acid-neutralizing (or liming) materials are among the most important agrochemicals used, in terms of the quantities added to soil each year.

Acidic soils are commonly neutralized by adding calcium-containing minerals, usually as calcite ($CaCO_3$) in the form of powdered limestone or crushed oyster or mussel shells. Alternatively, soil acidity may be neutralized using faster-acting lime ($Ca[OH]_2$). The rate of application of acid-neutralizing substances in agriculture can vary greatly, from several hundred pounds per acre per year to more than 1,000 pounds per acre per year. The rates used depend on the acidity of the soil, the rate at which new acidity is generated, and the needs of specific crops.

Much less commonly, soils may be alkaline in reaction, and they may have to be acidified somewhat to bring them into a pH range suitable for the growth of most crops. This problem can be especially common in soils developed from parent materials having large amounts of limestone ($CaCO_3$) or dolomite ($CaMg[CO_3]_2$). Soil can be acidified by adding sulfur compounds, which generate acidity as they are oxidized, or by adding certain types of acidic organic matter, such as peat mined from bogs.

Soil conditioners

Soil conditioners are organic-rich materials that are sometimes added to soils to improve aeration and water-holding capacity, both of which are very important aspects of soil quality. Various materials can be utilized as soil conditioners, including peat, crop residues, **livestock** manure, sewage sludge, and even shredded newspapers. However, compost is the most desirable of the soil conditioners. Compost contains large quantities of well-humified organic compounds, and also supplies the soil with nutrients in the form of slow-release organic compounds.

Pesticides

Pesticides are agrochemicals that are used to reduce the abundance of **pests**, that is, organisms that are considered to interfere with some human purpose. Many kinds of pesticides are used in agriculture, but they can be categorized into simple groups on the basis of the sorts of pests that are the targets of the use of these chemicals.

Herbicides are used to kill weeds, that is, non-desired plants that interfere with the growth of crops and thereby reduce their yield. Fungicides are used to protect agricultural plants from fungal **pathogens**, which can sometimes cause complete failure of crops. **Insecticides** are used to kill **insects** that defoliate crops, or that feed on stored grains or other agricultural products. Acaricides (or miticides) are used to kill **mites**, which are important pests of crops such as apples, and ticks, which can carry debilitating diseases of livestock. Nematicides are used to kill nematodes, which are important **parasites** of the roots of some crop species. Rodenticides are used to kill **rats**, **mice**, **gophers**, and other **rodents** that are pests in fields or that eat stored crops. Preservatives are agrochemicals added to processed foods to help prevent spoilage.

Pesticides are chemically diverse substances. About 300 different insecticides are now in use, along with about 290 herbicides, 165 fungicides, and other pesticides. However, each specific pesticidal chemical (also known as the "active ingredient") may be marketed in a variety of formulations, which contain additional substances that act to increase the efficacy of the actual pesticide. These so-called "inert" ingredients of the formulation can include solvents, detergents, emulsifiers, and chemicals that allow the active ingredient to adhere better to foliage. In total, more than 3,000 different pesticidal formulations exist.

Pesticides can also be classified according to the similarities of their chemical structures. Inorganic pesticides, for example, are simple compounds of toxic elements such as arsenic, copper, **lead**, and mercury. Inorganic pesticides were formerly used in large quantities, especially as fungicides. However, they have largely been replaced by various organic (carbon-containing) pesticides.

A few of the commonly used organic pesticides are based on substances that are synthesized naturally by plants as biochemical defenses, and can be extracted and used against pests. Pyrethrin, for example, is an insecticide based on pyrethrum, which is obtained from a species of chrysanthemum, while rotenone is a rodenticide extracted from a tropical shrub.

Most organic pesticides, however, have been synthesized by chemists. The synthetic organic pesticides include such well-known groups as the **chlorinated hydrocarbons** (including the insecticide DDT, and the herbicides 2,4-D and 2,4,5-T), organophosphates (such as parathion and malathion), carbamates (for example, carbaryl and carbofuran), and triazine herbicides (such as atrazine and simazine).

A final class of pesticides is based on the action of **bacteria**, **fungi**, or viruses that are pathogenic to specific pests and can be applied as a pesticidal formulation. The most commonly used biological insecticide is manufactured using spores of the bacterium *Bacillus thuringiensis*, also known as Bt. These spores can be mass-produced in laboratory-like factories, and then used to prepare an insecticidal **solution**. Insecticides containing Bt are mostly used against leaf-eating **moths**, biting **flies**, such as blackflies, and **mosquitoes**. Most other insects are little affected by Bt-based insecticides, so the unintended nontarget effects of their usage are relatively small.

Extremely large quantities of pesticides are used in modern agriculture. Globally, about 4.4-6.6 billion lb (2-3 billion kg) of pesticides are used each year, having a total value of about $20 billion. The United States alone accounts for about one-third of all pesticide usage, even though that country only supports about 4% of the world's population.

Agrochemicals used for animal husbandry

Contagious diseases of livestock can be an important problem in modern agriculture. This is especially true when animals are being reared at a high density, for example, in feed-lots. Various agrochemicals may be used to control infectious diseases and parasites under such conditions. Antibiotics are especially important in this respect. These chemicals may be administered by injection whenever bacterial diseases are diagnosed. However, antibiotics are sometimes administered with the feed, as a prophylactic treatment to prevent the occurrence of infections. Because of the extremely crowded conditions when livestock are reared in "factory farms," antibiotics must be administered routinely to animals raised under those circumstances.

Sometimes, hormones and other animal-growth regulators are used to increase the productivity of livestock. For example, bovine growth hormone (BGH) is routinely administered in some agricultural systems to increase the growth rates of cows and their milk production.

Environmental effects of the use of agrochemicals

Many important benefits are achieved by the use of agrochemicals. These are largely associated with increased yields of **plant** and animal crops, and less spoilage during storage. These benefits are substantial. In combination with genetically improved varieties of crop species, agrochemicals have made important contributions to the successes of the "green revolution." This has helped to increase the food supply for the rapidly increasing population of humans on **Earth**.

However, the use of certain agrochemicals has also been associated with some important environmental and

Agrochemical spraying in a Michigan orchard. © *Ken Wagner/Phototake NYC. Reproduced by permission.*

ecological damages. Excessive use of fertilizers, for example, can lead to the **contamination** of **groundwater** with nitrate, rendering it unfit for consumption by humans or livestock. Water containing large concentrations of nitrate can poison animals by immobilizing some of the hemoglobin in **blood**, reducing the ability to transport **oxygen**. In addition, the run-off of agricultural fertilizer into streams, lakes, and other surface waters can cause an increased productivity of those aquatic ecosystems, a problem known as **eutrophication**. The ecological effects of eutrophication can include an extensive mortality of **fish** and other aquatic animals, along with excessive growth of nuisance **algae**, and an off-taste of drinking water.

The use of pesticides can also result in environmental problems. As was previously noted, pesticides are used in agriculture to reduce the abundance of species of pests (that is, the "targets") to below a level of acceptable damage, which is economically determined. Unfortunately, during many uses of pesticides in agriculture, the exposure of other organisms, including humans, is not well controlled. This is especially true when entire fields are sprayed, for example, when using application equipment drawn by a tractor, or mounted on an airplane or helicopter. During these sorts of broadcast applications, many non-target organisms are exposed to the pesticide. This occurs on the treated site, and also on nearby off-sites as a result of "drift" of the sprayed agrochemical. These non-target exposures cause many unnecessary poisonings and deaths of organisms that are not agricultural pests.

In addition, there is a widespread, even global contamination of the environment with some types of persistent pesticides, especially with organochlorines such as DDT, dieldrin, and aldrin. This contamination involves the widespread presence of pesticide residues in virtually all **wildlife**, well water, food, and even in humans. Residues of some of the chemicals used in animal husbandry are also believed by some people to be a problem, for example, when traces of antibiotics and bovine **growth hormones** occur in consumer products such as meat or milk.

Some of the worst examples of environmental damage caused by pesticides have been associated with the use of relatively persistent chemicals, such as DDT. Most modern usage of pesticides involves chemicals that are less persistent than DDT and related chlorinated hydrocarbons. However, severe damages are still caused by the use of some newer pesticides. In **North America**, for example, millions of wild **birds** have been killed each year as a non-target effect of the routine use of carbofuran, an agricultural insecticide. This is a substantial ecological price to pay for the benefits associated with the use of that agrochemical.

The use of some pesticides is also risky for humans. About one million pesticide poisonings occur globally every year, resulting in 20,000 fatalities. About one-half of the human poisonings occur in poorer, less-developed countries, even though these places account for only 20% of the world's use of pesticides. This disproportionate risk is due to greater rates of illiteracy in poorer countries, and to lax enforcement of regulations concerning the use of pesticides.

There have been a few examples of pesticides causing extensive toxicity to humans. The most famous case occurred at Bhopal, India, in 1984, in the vicinity of a factory that was manufacturing an agricultural insecticide. In that case, there was an accidental release of about 45 tons (40 tonnes) of deadly methyl isocyanate vapor to the atmosphere. This agrochemical-related **emission** caused the deaths of about 3,000 people, and more than 20,000 others were seriously injured.

These and other environmental effects of the use of some agrochemicals are unfortunate consequences of the application of these chemical tools to deal with agricultural problems. Researchers are constantly searching for non-chemical ways of dealing with many of these agricultural needs. Much attention is being paid, for example, to developing "organic" methods of enhancing soil fertility and dealing with pests. Unfortunately, economically effective alternatives to most uses of agrochemicals have not yet been discovered. Consequently, modern agricultural industries will continue to rely heavily on the use of agrochemicals to manage their problems of fertility, soil quality, and pests.

See also Fungicide.

Resources

Books

Briggs, D. J. and F. M. Courtney. *Agriculture and Environment.* New York: Longman, 1989.

Freedman, B. *Environmental Ecology.* 2nd ed. San Diego: Academic Press, 1995.
Knowles, D. A. *Chemistry and Technology of Agrochemical Formulations.* Dordrecht, Netherlands: Kluwer Academic Publishers, 1998.
Muller, Franz. *Agrochemicals: Composition, Production, Toxicology, Applications.* New York: VCH Publishing, 2000.
Plimmer, Jack R. *Encyclopedia of Agrochemicals.* New York: John Wiley & Sons, 2002.
Soule, J. D. and J. K. Piper. *Farming in Nature's Image: An Ecological Approach to Agriculture.* Washington, DC: Island Press, 1991.
Spearks, Donald L. *Environmental Soil Chemistry.* 2nd ed. New York: Academic Press, 2002.
Wild, A. *Soils and the Environment.* Cambridge, UK: Cambridge University Press, 1993.

Periodicals

Delin, Geoffrey N. "Effects of Surface Run-off on the Transport of Agricultural Chemicals." *Science of the Total Environment* 295, no. 1 (2002): 143-156.
Pimentel, D., et al. "Environmental End Economic Costs of Pesticide Use." *Bioscience* 41 (1992): 402-409.

Bill Freedman

KEY TERMS

Agrochemical—Any substance used in the management of an agricultural ecosystem, including fertilizers, pH-adjusting agents, soil conditioners, pesticides, and crop-growth regulators.

Fertilizer—An agrochemical that is added to soil to reduce or eliminate nutrient-caused constraints to crop productivity.

Non-target effects—Effects on organisms other than the intended pest target of a pesticide treatment.

Pest—An organism that is considered to be undesirable, from the perspective of humans.

pH—The negative logarithm to the base 10 of the aqueous concentration of hydrogen ion in units of moles per liter. An acidic solution has a pH less than 7, while an alkaline solution has a pH greater than 7. Note that a one-unit difference in pH implies a 10-fold difference in the concentration of hydrogen ion.

Soil conditioners—Substances added to soil to improve its aeration and water-holding capacity, with great benefits in terms of crop growth. Various organic compounds can be used as soil conditioners, but compost is the best.

Agronomy

Agronomy can be defined as those branches of agricultural science that deal with the production of both **plant** and **animal crops**, and the management of **soil**. The subject matter of agronomy is quite diverse, but falls into three major categories: (1) crop breeding and the genetic improvement of varieties; (2) methods of cultivation of crops (both plants and animals); and (3) sustainability of the agricultural enterprise, especially with respect to fertility of the soil.

Crop improvement

Most varieties of agricultural crops look and grow very differently than their wild progenitors. In fact, almost all of the domesticated **species** of plants and animals that humans depend upon as sources of food, materials, or **energy** have been selectively bred for various desirable traits. This evolutionary process has resulted in the development of substantial genetic differences between domesticated varieties and their wild ancestors—differences that have arisen because of deliberate selection of desirable traits by humans.

Selective breeding of agricultural species has been very important in improving their productivity under cultivation. Enormous increases in the useful yields of cultivated plants have been attained through genetically-based improvements of growth rates. These include responses to the addition of fertilizer, along with modifications of the growth form, **anatomy**, and **chemistry** of crops. Similarly, domesticated animals have been selectively bred for growth rates, compliant **behavior**, chemical quality of their produce, and other desirable traits.

Selective breeding of some agricultural species has been so intensive and thorough that they can no longer survive without the assistance of humans, that is, in the absence of cultivation. For example, the **seeds** of maize (corn) can no longer disperse from their cob because of the tightly enclosing leaves that have evolved as a result of selective breeding. Similarly, our dairy cows are no longer capable of surviving on their own—they require humans to milk them, or they die from the complications of mastitis.

Selective breeding of existing and potential agricultural species still has a great deal to contribute to future developments in agronomy. Existing crops require continual genetic refinements to make them more suitable to the changing environmental conditions of agricultural ecosystems, to improve their resistance to diseases and **pests**, and to improve their nutritional qualities. At the same time, continuing surveys of the diversity of wild species will discover many plants and animals that are of potential benefit to humans. Selective breeding will be a critical part of the process by which those new **biodiversity** resources are domesticated.

Managing the soil

The quality of agricultural soils can be easily degraded by various cultural influences, but especially: (1) nutrient removal; (2) loss of organic matter; (3) acidification; and (4) **erosion** that is caused when soils are plowed and crops are harvested. Soil degradation is one of the most important problems associated with agricultural activities, because of the obvious implications for the longer-term sustainability of productivity and harvests. Understanding the causes of soil degradation and devising ways of preventing or mitigating this problem are among the most important objectives of agronomy.

Nutrient losses from agricultural soils are caused by a number of influences:

1. Whenever crop **biomass** is removed from the land during harvesting, **nutrients** such as **nitrogen**, **phosphorus**, potassium, **calcium**, and others are also extracted. These removals occur in the form of nutrients contained in the biomass. Depending on the crop, nutrient removals during annual harvesting are not necessarily large, but over a longer period of time the cumulative losses become significant, and the soil becomes impoverished.

2. Severe disturbance of the integrity of the soil surface, for example, by plowing, makes the soil quite susceptible to erosion by **water** and **wind**. Associated with the physical losses of soil are losses of the nutrients that it contains.

3. Nutrient losses are also encouraged when there are decreases in the concentrations of organic matter in the soil, a phenomenon that is also associated with tillage of the land. Organic matter is important because it helps to bind nutrients in relatively immobile and insoluble forms, thereby helping to ensure their continuous supply, and preventing the **leaching** of nutrients from the site.

When nutrient losses from soil have been severe, it may be possible to compensate for the losses of fertility by adding nutrients. Fertilization is an important activity in modern agriculture. However, fertilization is expensive, and it causes important environmental impacts. Therefore, many agronomists are engaged in research designed to reduce the dependence of modern agriculture on intensive fertilization, and on increasing the efficiency of nutrient uptake by crops.

Losses of soil organic matter are another important problem that agronomists must address. Soil organic matter is important because of its great influence on the tilth,

An irrigated area along a river near Yazd, Iran. Note the unirrigated desert in the background. *JLM Visuals. Reproduced with permission.*

or physical structure of soil, which is closely associated with the **concentration** of humified organic matter. Tilth is very influential on the water- and nutrient-holding capacities of the soil, and is highly beneficial to the growth of crops. The emerging field of organic agriculture is largely involved with managing and optimizing the concentration of organic matter in soils. Commonly used techniques in organic agriculture include the use of green manures and composts to maintain the concentration of organic matter in soil, as well as the use of carefully designed crop rotations and mixed-cropping systems.

Soil acidification is another important agricultural problem. Acidification is caused by the removal of calcium and **magnesium** during cropping, through erosion, leaching, and actions of certain nitrogen-containing **fertilizers**, such as ammonium nitrate and **urea**.

Acidification may also be partially caused by atmospheric **pollution**, especially in regions where the air is contaminated by **sulfur dioxide**, and where acidic **precipitation** is important. Acidification is routinely countered in agriculture by mixing limestone or lime into the soil.

Erosion is another common agricultural problem that is largely caused by disturbance of the soil surface through plowing. Erosion represents a **mass wasting** of the soil resource, with great implications for fertility and other important aspects of land capability. Erosion also causes secondary impacts in the aquatic ecosystems that typically receive the large wastage of eroded materials. Erosion is particularly severe on lands with significant slopes and coarse soils, especially if these occur in a region with abundant precipitation. Erosion can be substantially prevented by plowing along contour lines rather than down-slope, and by maximizing the amount of time during which the land has a well-established plant cover. The latter can be accomplished through the use of no-tillage agricultural systems, and wherever possible, by cultivating perennial crops on sites that are vulnerable to erosion.

Managing pests and diseases

Pests and their control are another significant problem agronomists must deal with. Pests can be defined as any organisms that interfere with some human purpose. In agriculture, the most important pests are weeds, **insects** and other defoliators, and disease-causing **pathogens**. The use of **pesticides** such as **herbicides**, **insecticides**, fungicides, and **antibiotics** is a very important aspect of modern agriculture. Unfortunately, many pesticides cause important environmental damages, and many agronomists are attempting to develop systems that would decrease the reliance on pesticides in agriculture, while not compromising yields.

Animal husbandry

Devising better systems in which to raise animals as crops is another important aspect of agronomy. Considerations in animal husbandry include optimization of the productivity of the animals, achieved through selective breeding, and careful management of diet, **disease**, and housing. Agronomists concerned with animal husbandry are also interested in improving the nutritional quality of the food products, disposal of waste materials, and humane treatment of the **livestock**.

Agricultural systems

Ultimately, the goal of agronomy is to develop agricultural systems that are sustainable over the long term. An agricultural system involves particular combinations of crop species, along with methods of tillage, seeding, pest management, and harvesting. Furthermore, agricultural systems may involve the growth of successive crops in a carefully designed rotation, or perhaps the growth of several crops at the same time, for example, by row cropping or intercropping.

The ultimate judgement of the success of agronomy will be the sustainability of the agricultural systems that agronomists develop, and then persuading agriculturalists to use them.

See also Acid rain; Agrochemicals; Animal breeding; Contour plowing; Crop rotation; Crops; Fertilizers; Genetic engineering; Integrated pest management; Organic farming; Pests; Pesticides; Soil conservation.

Resources

Books

Briggs, D.J. and F.M. Courtney. *Agriculture and Environment.* New York: Longman, 1989.

Carroll, R.C., J.H. Vandermeer, and P.M. Rossett. *Agroecology.* New York: McGraw-Hill, 1990.

Freedman, B. *Environmental Ecology.* 2nd ed. San Diego: Academic Press, 1984.

Hartmann, H.T., A.M. Kofranek, V.E. Rubatzky, and W.J. Flocker. *Plant Science: Growth, Development, and Utilization of Cultivated Plants.* Englewood Cliffs, NJ: Prentice-Hall, 1988.

Miller, R.W. and R.L. Donahue. *Soils. An Introduction to Soils and Plant Growth.* New York: Prentice-Hall, 1989.

Soule, J.D. and J.K. Piper. *Farming in Nature's Image: An Ecological Approach to Agriculture.* Washington, DC: Island Press, 1991.

KEY TERMS

Agricultural system—A combination of the choice of crop species, and the methods of tillage, seeding, pest management, and harvesting. The crop may be grown in successive monocultures, or the system may involve rotations of different crops, or polyculture systems such as row cropping and intercropping.

Agronomy—The application of agricultural science to the production of plant and animal crops, and the management of soil fertility.

Nutrient—Any chemical required for life. The most important nutrients that plants obtain from soil are compounds of nitrogen, phosphorus, potassium, calcium, magnesium, and sulfur.

Organic matter—Any biomass of plants or animals, whether living or dead. Dead organic matter is the most important form in soils, particularly when occurring as humic substances.

Tilth—The physical structure of soil, closely associated with the concentration of humified organic matter. Tilth is important in water and nutrient-holding capacity of the soil, and is generally beneficial to plant growth.

Wild, A. *Soils and the Environment.* Cambridge, UK: Cambridge University Press, 1993.

Bill Freedman

AIDS

AIDS is the abbreviation for acquired immunodeficiency **syndrome**. The syndrome is caused by several types of a **virus** that is now known as the human immunodeficiency virus (HIV). AIDS is characterized by the destruction of cells that are vital to the proper operation of the **immune system**. People afflicted with AIDS can develop opportunistic infections; life-threatening illnesses caused by viruses or **bacteria** that do not sicken those with healthy immune systems. Certain types of cancers can also develop in people with AIDS.

AIDS was first detected in the 1981. At that time, the malady was not named. The identification as AIDS came in 1982. The designation of the illness as a syndrome reflected the observations that a variety of clinical symptoms were apparent, rather than a single **disease**.

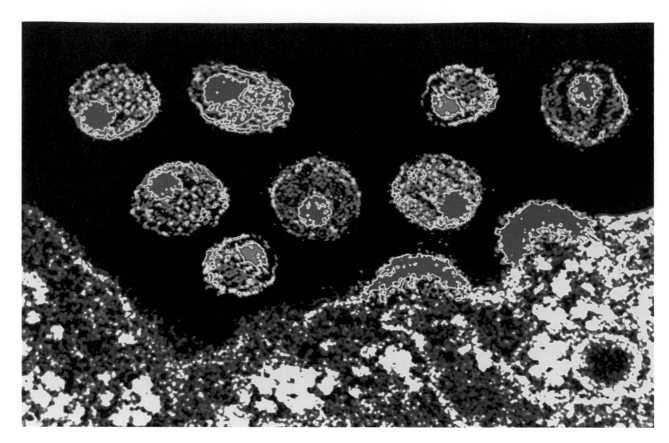

Mature HIV-1 viruses (above) and the lymphocyte from which they emerged (below). Two immature viruses can be seen budding on the surface of the lymphocyte (right of center). *Photograph by Scott Camazinr. National Audubon Society Collection/Photo Researchers, Inc. Reproduced by permission.*

Since the early 1980s, the number of cases has increased tremendously. As of 2002, more than 790,000 Americans are known to have AIDS. Globally, approximately 40 million people have AIDS, almost three million of these being children, with five million new cases being reported in 2001. The disease is considered to be an **epidemic**, especially in the world's population of black men ages 25-44, where AIDS is the leading cause of death. In 2001, three million people died of AIDS related complications and opportunistic infections.

The viral cause of AIDS was first reported in 1983 by researchers at the Pasteur Institute in France. Then, the virus was named lymphadenopathy-associated virus or LAV. The next year, researchers in the United States reported their discovery of a virus that they designated human T **cell** lymphotropic virus, type 3 (HTLV-III). After some controversy, it was later shown that the two viruses were the same.

The name of the virus was later changed to HIV, and later still to HIV-1. The origin of the virus remains unclear. The earliest known case is from **blood** sample collected in 1959 from a man in Kinshasa, Democratic Republic of Congo. A speculation, still unresolved, is that

prior to this time the virus was resident in **primates**. Indeed, an AIDS-like virus (simian immunodeficiency virus or SIV) is present in primates.

HIV-2 was discovered in 1984. The virus is similar to, but distinct from, HIV-1. AIDS caused by HIV-2 is slower to develop and the symptoms are milder than those caused by HIV-1. The designation HIV typically refers to both viral types.

Both HIV-1 and HIV-2 can be transferred from person to person during sexual contact, via infected semen. Initially, AIDS was thought to be exclusively a disease of gay males and certain racial groups. This led to a backlash against these groups. But, with time, it became clear that AIDS could infect anyone.

Maternal transfer of HIV to a developing fetus can occur, as can transmission to a nursing baby via infected breast milk. Transmission via infected blood also occurs, during a blood transfusion and the sharing of needles during the injection of illicit drugs. The possibility of blood born transmission in transfusions is far less now than in the 1970s and 1980s, because blood is now rigorously screened for the presence of the virus and is also **heat** treated to render the virus incapable of replicating.

AIDS is a progressively worsening disease. It can be difficult to diagnose during the early stages of **infection**, since symptoms that appear are similar to those of **influenza**. These symptoms include fever, headache, fatigue, and enlarged lymph nodes. Often these symptoms disappear within a week to a month.

A period follows where no symptoms are apparent. This "asymptomatic" period can be anywhere from several months to ten or more years. Although no symptoms are apparent, the virus is multiplying in cells of the immune system. This replication destroys the immune cells, in particular a class of cells called CD4 positive **T cells** (T4 cells). Indeed, the decline in T4 cells can alert a physician to the possibility of AIDS. An infection can also be detected by the presence of antibodies to the virus in the blood. However, such a test is usually done only when symptoms appear, unless a person suspects that they have been exposed to the virus.

T4 cells are one of the immune system's vital defenses against infection. So, their destruction produces a variety of symptoms. These symptoms include a loss of **energy**, loss of weight, the frequent development of fevers, frequent **yeast** infections in the mouth (thrush) or vagina, chronic skin rashes, and infections caused by the herpes virus. In infected children, growth can be slowed down.

Individuals are diagnosed with AIDS when the T cell count in their blood declines to 200 or less (healthy people have a T cell count of 1,000 or more), and when certain hallmark symptoms develop.

People with AIDS are prone to developing certain cancers. These include cancers caused by certain viruses (Kaposi's sarcoma and **cancer** of the cervix), and cancers of the immune system (lymphomas).

Treatment for AIDS is varied. Some treatments are geared toward the early stages of HIV infection, while other treatments are targeted at events that occur later in HIV infection. Still other treatments are intended to combat the opportunist microbes that cause infections.

The early stages of HIV infection involve the making of new copies of the virus. HIV is a **retrovirus**, and so contains **ribonucleic acid (RNA)** as the genetic material. Since the host cell that the virus infects contains **deoxyribonucleic acid (DNA)** as the genetic material, HIV must produce DNA. This is accomplished using a viral **enzyme** called reverse transcriptase. Drugs that inhibit the action of reverse transcriptase can prevent HIV from making new copies of itself. Some examples of reverse transcriptase inhibitors include azidothymidine (AZT; Retrovir®), stavudine, and lamivudine.

Other drugs known as protease inhibitors block the action of a protein that is vital for viral replication. Protease inhibitors act at a later step in the virus manufacture process. Examples of protease inhibitors are ritonavir, amprenavir, and lopinavir.

Often the drugs are given in combination, to overcome the development of virus resistance to any one drug. The use of reverse transcriptase inhibitors along with protease inhibitors is called highly active antiretroviral therapy, or HAART. HAART has shown positive results in people recently infected with HIV and those who have developed AIDS. Unfortunately, HAART can produce side effects, including death. As well, an intolerance to the drugs can develop, necessitating a change in the types of drugs being taken.

Treatment can also be directed at microbes that cause the opportunistic infections, rather than the HIV. Specific drugs are used to treat viral infections of the **eye**, and infections caused by yeast and **fungi**.

One of the most common and deadly opportunistic infections is that caused by the fungus *Pneumocystis carinii*. The fungus is common in the body. The lung infection caused by the fungus (*Pneumocystic carinii* **pneumonia**, or PCP) does not usually occur in a person with a healthy immune system.

As of 2002, there is no **vaccine** for AIDS. Indeed, some AIDS researchers are not optimistic that an effective vaccine will soon be developed, given that the virus replicates inside host cells where it is shielded and the often extended period between initial infection and the appearance of symptoms. The use of creams that contain HIV-inactivating chemicals, which would be applied prior to sex, is being explored. Currently, the avoidance of high risk activities, such as unprotected sex with someone whose sexual history is unknown, is the best means of minimizing the chance of developing AIDS.

Resources

Books

Duesberg, P. *Inventing the AIDS Virus.* Washington: Regnery, 1996.

Jones, C. *The Making of an Activist: Stitching a Revolution.* San Francisco: Harper, 2001.

Periodicals

Coffin, J., A. Haase, A. Levy, et al., "What to Call the AIDS Virus." *Nature* (1986): 10.

Hymes, K.B., J.B. Greene, A. Marcus, et al., "Karposi's Sarcoma in Homosexual Men: A Report of Eight Cases." *Lancet* (1981): 598–600.

Organizations

National Institute of Allergy and Infectious Disease, National Institutes of Health, Bethesda, Maryland 20892. (800) 227–8922. August 2002 [cited November 3, 2002]. <http://www.niaid.nih.gov/factsheets/hivinf.htm>.

HIV/AIDS Treatment Information Service, PO Box 6303, Rockville MD 20849–6303. (800) 448–0440. September 2002 [cited November 3, 2002]. <http://www.hivatis.org.>.

Brian Hoyle

AIDS therapies and vaccines

Acquired immunodeficiency **syndrome** (**AIDS**) is a **disease** characterized by the destruction of the **immune system**. More than 16,000 new AIDS patients are diagnosed each day. Evidence overwhelmingly supports the view that AIDS is caused by several types of a **virus** designated as the human immunodeficiency virus (HIV).

The immune system is the principle defense system of the body to a variety of infections. Thus, individuals diagnosed with AIDS are prone to illness and, in many cases, eventual death from microbiological illnesses, **organ** failures, or **cancer**.

AIDS was recognized in the early 1980s. Soon thereafter came the discovery of the various types of HIV and their association with AIDS. Since then, scientists and researchers around the world have spent billions of dollars and thousands of hours to discover how AIDS is caused and what options are available to stop the progression of the disease.

Once the scientific community became mobilized to fight AIDS, international conferences were held in the mid-1980s. Exchange of data on AIDS and HIV led to the development of **blood** tests to detect the virus or its genetic material, and to diagnose the **infection**. Only then did the severity of the illness and the extent of the worldwide **epidemic** become known.

By the late 1980s, treatments for the virus became available, and so treatment of infections that **prey** on the severely weakened immune systems of those with AIDS could be managed to some extent. The death toll internationally was still extremely high, and remained so until the mid-1990s when the research collectively began to identify the best range of treatments for the 11 or more strains of HIV.

AIDS treatment

The eleventh World AIDS conference held in Vancouver, British Columbia, Canada, in 1996 provided the turning point for treatment. Several groups of researchers presented information on a combination of drugs (also referred to as an AIDS "cocktail") that, when used together, could virtually erase any traces of the virus from the bloodstream of infected individuals. A new blood test that could detect HIV much earlier than previously available tests was also introduced.

This blood test changed early treatment options and established new ideas about how the disease works. Within a year, these novel HIV and AIDS treatments were being duplicated in medical offices and clinics around the world. By the late 1990s, HIV had moved from a progressively terminal disease to one that could be managed over the long-term, at least for those who have access to and can afford these new treatment options.

The combined drugs therapy first introduced at the 1996 conference is now the standard of care for those with HIV. Eleven different drugs that fall into three categories make up the treatment strategy. Five of the drugs are known as nucleoside analogs. That is, the drugs mimic the structure of some components of the viral genetic material. The incorporation of the analogs into viral genetic material can stop the virus from making new copies of itself. A well-known nucleoside analog is azidothymidine (AZT, which is marketed under the name Retrovir®). Other drugs block the action of protein-degrading enzymes (proteases) that are made by HIV. Still other drugs block the action of reverse transcriptase (an **enzyme** that HIV uses to duplicate its genetic material).

The differently acting drugs are used in combination (the AIDS cocktail) to block the disease's progression. All three types of drugs act to halt the manufacture of new virus inside the immune cells of the human body, where HIV has already established an active infection.

The mortality **rate** has since fallen sharply as more HIV-positive patients are prescribed this three-drug combination. The use of the improved blood test has also been expanded to measure the amount of HIV in the blood during drug treatment, which can help to pinpoint the most effective combination of drugs for each **individual**. In all cases, the goal of treatment is to keep the level of HIV in the body as low as possible, for as long as possible. Presently, a "cure" for AIDS does not exist.

Despite the treatment advances, questions remain concerning the effectiveness of such treatments over the long term. The average annual cost per patient for the three-drug combination treatment is US$15,000. This is exorbitantly expensive for the majority of AIDS patients who live in developing countries.

As well, while treatments can almost completely eliminate the virus in as few as two or more years, the immune system remains impaired. Patients may be just as susceptible to other illnesses that their immune systems cannot withstand. Furthermore, because HIV can "hide" from the immune system by occupying healthy cells, the absence of detectable HIV is no guarantee that the viral infection is truly eliminated.

A strict medicine administration schedule and diet must be maintained for the drugs to be successful. The regimen is difficult for many patients to follow, either because of privacy constraints or the side effects some of the drugs can cause.

Perhaps the most basic questions researchers are still struggling with are when to provide treatment, which drugs to begin with, how to identify when alterations are needed in the therapy, and which drugs to try next.

Vaccine development

The search for a **vaccine** that would protect people from HIV infection, by blocking the entry of the virus into the immune cells, has been ongoing almost as long as the search for AIDS treatment strategies. At least 25 experimental vaccines have been created since identification of the disease. Unfortunately, few have proved even promising enough to complete the large-scale testing on human volunteers that is required to demonstrate the vaccine's success.

Some initial vaccine trials, however, have occurred. In June 1998, a large-scale test began with 5,000 volunteers in 30 cities in the United States, and a smaller group in Thailand. Volunteers were given a series of injections that hopefully stimulate the immune system to resist the two most common strains of the AIDS virus. Previous trials using smaller numbers of people documented that 99.5% of the vaccinated volunteers produced strong levels of resistance in their immune cells, which then target and kill infections such as HIV. The trial was expected to last three years. The conclusions of the trial are expected by 2004.

Also in 1998, a group of researchers at the University of Michigan proposed to develop a vaccine that would prevent someone with HIV from passing it on to someone else. Such a vaccine would be given within the first three months after infection, when the virus is most contagious. Their theories resemble those used to create the Salk polio vaccine in the 1950s, which reduced the polio virus symptoms and drastically reduced its ability to infect others, virtually eliminating the disease within a decade. The feasibility of this approach remains to be experimentally examined.

See also Molecular biology; T cells.

Resources

Books

Bartlett, J.G., and A.K. Finkbeiner. *The Guide to Living with HIV Infection.* Baltimore: Johns Hopkins University Press, 1998.

Periodicals

Gaschen, B., J. Taylor, K. Yusin, et al. "AIDS-Diversity Considerations in HIV-1 Vaccine Selection." *Science* 296 (June 2002): 2354–2360.

Vastag, B. "HIV Vaccine Efforts Inch Forward." *Journal of the American Medical Association* 286 (October 2001): 1826–1828.

Other

United States National Institutes of Health. "National Institute of Allergy and Infectious Diseases" [cited October 19, 2002]. <http://www.niaid. nih.gov/daids/vaccine>.

Brian Hoyle

Air *see* **Atmosphere, composition and structure**

Air masses and fronts

An air mass is an extensive body of air that has a relatively homogeneous **temperature** and moisture content over a significant altitude. Air masses typically cover areas of a few hundred, thousand, or million square kilometers. A front is the boundary at which two air masses of different temperature and moisture content meet. The role of air masses and fronts in the development of **weather** systems was first appreciated by the Norwegian father and son team of Vilhelm and Jacob Bjerknes in the 1920s. Today, these two phenomena are still studied intensively as predictors of future weather patterns.

Source regions

Air masses form when a body of air comes to rest over an area large enough for it to take on the temperature and **humidity** of the land or **water** below it. Certain locations on the Earth's surface possess the topographical characteristics that favor the development of air masses. The two most important of these characteristics are topographic regularity and atmospheric stability. Deserts, plains, and oceans typically cover very wide areas with relatively few topographical irregularities. In such regions, large masses of air can accumulate without being broken apart by **mountains**, land/water interfaces, and other features that would break up the air mass.

The absence of consistent **wind** movements also favors the development of an air mass. In regions where cyclonic or anticyclonic storms are common, air masses obviously cannot develop easily.

Classification

The system by which air masses are classified reflects the fact that certain locations on the **planet** possess the topographic and atmospheric conditions that favor air mass development. That system uses two letters to designate an air mass. One letter, written in upper case, indicates the approximate latitude (and, therefore, temperature) of the region: A for arctic; P for polar; E for equatorial; T for tropical. The distinctions between arctic and polar on the one hand and equatorial and tropical on the other are relatively modest. The first two terms (arctic and polar) refer to cold air masses, and the second two (equatorial and tropical) to warm air masses.

A second letter, written in lower case, indicates whether the air mass forms over land or sea and, hence, the relative amount of moisture in the mass. The two designations are c for continental (land) air mass and m for maritime (water) air mass.

The two letters are then combined to designate both temperature and humidity of an air mass. One source region of arctic air masses, for example, is the northernmost latitudes of Alaska, upper Canada, and Greenland. Thus, air masses developing in this source region are designated as cA (cold, land) air masses. Similarly, air masses developing over the Gulf of Mexico, a source region for maritime tropical air masses, are designated as mT (warm, water) air masses.

Properties of air masses

The movement of air masses across the Earth's surface is an important component of the weather that develops in an area. For example, weather patterns in **North America** are largely dominated by the movement of about a half dozen air masses that travel across the **continent** on a regular basis.

Two of these air masses are the cP and cA systems that originate in Alaska and central Canada and sweep down over the northern United States during the winter months. These air masses bring with them very cold temperatures, strong winds, and heavy **precipitation**, such as the snowstorms commonly experienced in the Great Lakes states and New England. The name "Siberian Express" is sometimes used to describe some of the most severe storms originating from these cP and cA air masses.

From the south, mT air masses based in the Gulf of Mexico, the Caribbean, and western Atlantic Ocean move northward across the southern states, bringing hot, humid weather that is often accompanied by thunderstorms in the summer.

Weather along the western coast of North America is strongly influenced by mP air masses that flow across the region from the north Pacific Ocean. These masses actually originate as cP air over Siberia, but are modified to mP masses as they move over the broad expanse of the Pacific, where they can pick up moisture. When an mP mass strikes the west coast of North America, it releases its moisture in the form of showers and, in northern regions, snow.

Fronts

The term front was suggested by the Bjerkneses because the collisions of two air masses reminded them of a battlefront during a military operation. That collision often results in war-like weather phenomena between the two air masses.

Fronts develop when two air masses with different temperatures and, usually, different moisture content come into contact with each other. When that happens, the two bodies of air act almost as if they are made of two different materials, such as oil and water. Imagine what happens, for example, when oil is dribbled into a glass of water. The oil seems to push the water out of its way and, in return, the water pushes back on the oil. A similar shoving match takes place between warm and cold air masses along a front. The exact nature of that shoving match depends on the relative temperature and moisture content of the two air masses and the relative movement of the two masses.

Cold fronts

One possible situation is that in which a mass of cold air moving across the Earth's surface comes into contact with a warm air mass. When that happens, the cold air mass may force its way under the warm air mass like a snow shovel wedging its way under a pile of snow. The cold air moves under the warm air because the former is more dense. The boundary formed between these two air masses is a cold front.

Cold fronts are usually accompanied by a falling **barometer** and the development of large cumulonimbus **clouds** that bring rain showers and thunderstorms. During the warmer seasons, the clouds form as moisture-rich air inside the warm air mass, which is cooled and water condenses out as precipitation.

Cold fronts are represented on weather maps by means of solid lines that contain solid triangles at regular distances along them. The direction in which the triangles point shows the direction in which the cold front is moving.

Warm fronts

A situation opposite to the preceding is one in which a warm air mass approaches and then slides up and over a cold air mass. The boundary formed in this case is a warm front. As the warm air mass comes into contact with the cold air mass, it is cooled and some of the moisture held within it condenses to form clouds. In most cases, the first clouds to appear are high cirrus clouds, followed sometime later by stratus and nimbostratus clouds.

Warm fronts are designated on weather maps by means of solid lines to which are attached solid half circles. The direction in which the half circles point shows the direction in which the warm front is moving.

Occluded front

A more complex type of front is one in which a cold front overtakes a slower-moving warm front. When that happens, the cold air mass behind the cold front eventually catches up and comes into contact with the cold air mass underneath the warm front. The boundary between these two cold air masses is an occluded front.

A distinction can be made depending on whether the approaching cold air mass is colder or warmer than the second air mass beneath the warm front. The former is called a cold-type occluded front, while the latter is a warm-type occluded front.

Once again, the development of an occluded front is accompanied by the formation of clouds and, in most cases, by steady and moderate precipitation. An occluded front is represented on a weather map by means of a solid line that contains, alternatively, both triangles and half circles on the same side of the line.

Stationary fronts

In some instances, the collision of two air masses results in a stand-off. Neither mass is strong enough to displace the other, and essentially no movement occurs. The boundary between the air masses in this case is known as a stationary air mass and is designated on a weather map by a solid line with triangles and half circles on opposite sides of the line. Stationary fronts are often accompanied by fair, clear weather, although some light precipitation may occur.

See also Atmosphere, composition and structure; Storm; Weather forecasting; Weather mapping.

KEY TERMS
. . . .

Anticyclonic—Referring to an area of high pressure around which winds blow in a clockwise direction in the northern hemisphere.

Continental—Referring to very large land masses.

Cyclonic—Referring to an area of low pressure around which winds blow in a counter-clockwise direction in the northern hemisphere.

Humidity—The amount of water vapor contained in the air.

Maritime—Referring to the oceans.

Topographical—Referring to the surface features of an area.

Resources

Books

Ahrens, C. Donald. *Meteorology Today.* 2nd ed. St. Paul, MN: West Publishing Company, 1985.

Eagleman, Joe R. *Meteorology: The Atmosphere in Action.* 2nd ed. Belmont, CA: Wadsworth Publishing Company, 1985.

Lutgens, Frederick K., and Edward J. Tarbuck. *The Atmosphere: An Introduction to Meteorology.* 4th ed. Englewood Cliffs, NJ: Prentice Hall, 1989.

Lutgens, Frederick K., Edward J. Tarbuck, and Dennis Tasa. *The Atmosphere: An Intorduction to Meteorology.* 8th ed. New York: Prentice-Hall, 2000.

Moran, Joseph M., and Michael D. Morgan. *Essentials of Atmosphere and Weather.* New York: Macmillan Publishing Company, 1994.

Other

"Genesis of Fronts and Airmasses." *Weatherwise* (December 1985): 324-328.

David E. Newton

Air pollution

Air **pollution** is the presence of chemicals in the earth's atmosphere that are not a normal part of the atmosphere. In other words, air pollution is contaminated air.

Air **contamination** is divided into two broad categories: primary and secondary. Primary pollutants are those released directly into the air. Some examples include dust, smoke, and a variety of toxic chemicals, such as **lead**, mercury, vinyl chloride and **carbon monoxide**. The exhaust from vehicles and industrial smokestacks are examples of primary pollution.

A "brown cloud" (pollution) shrouds Denver, Colorado.
©Ted Spiegel/Corbis. Reproduced by permission.

Secondary pollutants are created or modified after being released into the atmosphere. In secondary pollution, a compound is released into the air. This compound is then modified into some other form, either by reaction with another chemical present in the air or by a reaction with sunlight (a photochemical reaction). The altered compound is the secondary pollutant. **Smog** that gathers above many cities is a prime example of secondary air pollution.

Pollution of the atmosphere occurs in the bulk of the atmosphere that is within 40-50 mi (64.4–80.5 km) of Earth's surface. **Nitrogen** and **oxygen** make up 99% of the atmosphere; the remaining components are argon, **carbon dioxide**, neon, helium, methane, krypton, **hydrogen**, xenon, and **ozone**. Ozone is concentrated in a band that is 12-30 mi (19–48 km) above Earth's surface.

Smog can be damaging to human health because of the formation of ozone. A complex series of **chemical reactions** involving volatile organic compounds, nitrogen oxides, sunlight, and molecular oxygen create highly reactive ozone molecules containing three oxygen **atoms**. The ozone that is present higher up in the atmosphere is beneficial. It provides an important shield against harmful ultraviolet **radiation** in sunlight. Closer to the ground, however, ozone is highly damaging to both living organisms and building materials.

Criteria pollutants

The 1970 Clean Air Act in the United States recognized seven air pollutants as being in immediate need of regulatory monitoring. These pollutants are **sulfur dioxide**, particulates (such as dust and smoke), **carbon** monoxide, volatile organic compounds, nitrogen oxides, ozone, and lead. These pollutants were regarded as the greatest danger to human health. Because criteria were established to limit their **emission**, these materials are sometimes referred to as "criteria pollutants." Major revisions to the Clean Air Act in 1990 added another 189 volatile chemical compounds from more than 250 sources to the list of regulated air pollutants in the United States.

Some major pollutants are not directly poisonous but can harm the environment over a longer period of **time**. Excess nitrogen from fertilizer use and burning of **fossil fuels** is causing widespread damage to both aquatic and terrestrial ecosystems on Earth's surface. For example, over-fertilizing of plants favors the growth of weedy **species**. Pollutants can also damage the atmosphere above Earth's surface. A well-known example of this damage is that caused by **chlorofluorocarbons (CFCs)**. CFCs were used for many years as coolant in refrigerators and as cleaning agents. While generally chemically inert and non-toxic in these settings, CFCs diffuse into the upper atmosphere where they destroy the ultraviolet-absorbing ozone shield. Ozone depletion is a concern for the health of humans, as increased exposure to the sun's ultraviolet radiation can cause genetic damage that is associated with various cancers, especially skin **cancer**.

Air pollutants can travel surprisingly far and fast. About half of the fine reddish dust visible in Miami's air during the summer is blown across the Atlantic Ocean from the Sahara Desert. **Radioactive fallout** from an explosion at the Chernobyl **nuclear reactor** in the Ukraine was detected many miles away in Sweden within two days after its release and spread around the globe in less than a week.

One of the best-known examples of long-range transport of air pollutants is **acid rain**. The acids of greatest concern in air are sulfuric and nitric acids, which are formed as secondary pollutants from **sulfur** dioxide and nitrogen oxides released by burning fossil fuels and industrial processes such as smelting ores. These acids can change the **pH** (a standard measure of the hydrogen ion **concentration** or acidity) of rain or snow from its normal, near neutral condition to an acidity that is similar to that of lemon juice. Although this acidity is not directly dangerous to humans, it damages building materials and can be lethal to sensitive aquatic organisms such as **salamanders**, **frogs**, and **fish**. Thousands of lakes in eastern Quebec, New England, and Scandinavia have been acidified to the extent that they no longer support game fish populations. Acid **precipitation** has also been implicated in forest deaths in northern **Europe**, eastern **North America**, and other places where air currents carry urban industrial pollutants.

Air pollution control

Because air pollution is visible and undesirable, most developed countries have had 50 years or more of regulations aimed at controlling this form of environmental degradation. In many cases, these regulations have had encouragingly positive effects. While urban air quality rarely matches that of pristine wilderness areas, air pollution in most of the more prosperous regions of North America, Western Europe, Japan, **Australia**, and New Zealand has been curtailed in recent years. In the United States, for example, the Environmental Protection Agency (EPA) reports that the number of days on which urban air is considered hazardous in the largest cities has decreased 93% over the past 20 years. Of the 97 metropolitan areas that failed to meet clean air standards in the 1980s, nearly half had reached compliance by the early 1990s.

Perhaps the most striking success in controlling air pollution is urban lead. Banning of leaded gasoline in the United States in 1970 resulted in a 98% decrease in atmospheric concentrations of this toxic **metal**. Similarly, particulate materials have decreased in urban air nearly 80% since the passage of the U.S. Clean Air Act, while sulfur dioxides, carbon monoxide, and ozone are down by nearly one-third.

The situation is not as encouraging in some other countries. The major metropolitan areas of developing countries often have highly elevated levels of air pollution. Rapid population growth, unregulated industrialization, local geography, and lack of enforcement have compounded the air pollution problem in cities such as Mexico City. In this city, pollution levels usually exceed World Health Organization (WHO) standards 350 days per year. More than half of all children in the city have lead levels in their **blood** sufficient to lower intelligence and retard development. The more than 5,500 metric tons of air pollutants released in Mexico City each day from the thousands of industries and millions of motor

KEY TERMS

Ecosystem—All of the organisms in a biological community interacting with the physical environment.

Ozone—A naturally occurring trace gas, having the chemical formula O_3. In the stratosphere, it serves to absorb many harmful solar UV rays.

Smog—An aerosol form of air pollution produced when moisture in the air combines and reacts with the products of fossil fuel combustion.

Volatile—Readily able to form a vapor at a relatively low temperature.

vehicles are trapped close to the surface by the **mountains** ringing the city.

Most of the developing world megacities (those with populations greater than 10 million people) have similar problems. Air quality in Cairo, Bangkok, Jakarta, Bombay, Calcutta, New Delhi, Shanghai, Beijing, and Sao Paulo regularly reach levels scientists consider dangerous to human, **animal**, and **plant** life.

See also Atmosphere, composition and structure; Global warming; Ozone layer depletion.

Resources

Books

Colls, J. *Air Pollution: An Introduction.* London: E & F.N. Spon, 1998.

Other

Environmental Protection Agency, Office of Air Quality, Planning and Standards, Information Transfer Group, Mail Code E143–03, Research Triangle Park, NC 27711. <http://www.epa.gov/airnow/> (October 18, 2002).

Lawrence Berkeley National Laboratory, 1 Cyclotron Road, Berkeley, CA 94720. <http://www.lbl.gov/Education/ELSI/pollution-main.html> (October 28, 2002).

Brian Hoyle

Aircraft

An aircraft is a machine used for traveling through the atmosphere supported either by its own buoyancy or by some sort of engine that propels the ship through the air. Aircraft of the former type are known as lighter-than-air ships, while those of the latter type tend to be heavier-than-air machines. Included in the general term aircraft are specific machines such as dirigibles, gliders, airplanes, and helicopters.

Early theories of air travel

Humans have dreamed of flying like **birds** for centuries. A Chinese myth dating to at least 1500 B.C. tells of men flying through the air in a carriage driven by something very much like a modern propeller. The Greek legend of Daedalus and Icarus is closer to the image that early humans had of flight, however. According to that legend, Daedalus constructed a set of wings that he attached to his son, Icarus, as a means of escaping from the **island** of Crete. Icarus flew so high, however, that the wax holding the wings to his body melted, and he fell to his death.

For more than 20 centuries, humans repeated Daedalus's experiment, with ever more sophisticated attempts to duplicate the flight of birds. All such attempts failed, however, as inventors failed to recognize that the power generated by a single human could never be sufficient to lift a person off the Earth's surface.

Lighter-than-air aircraft

The first real success experienced by humans in designing aircraft made use of the concept of buoyancy. Buoyancy refers to the fact that an object tends to rise if it is placed in a medium whose **density** is greater than its own. A **cork** floats in **water**, for example, because the cork is less dense than the water. Buoyant aircraft became possible when scientists discovered that certain gases—especially **hydrogen** and helium—are less dense than air. Air that has been heated is also less dense than cooler air. Thus, a container filled with one of these gases will rise in air of its own accord.

Balloons were the first aircraft to make use of this principle. The fathers of ballooning are sometimes said to be the Montgolfier brothers, Joseph and Jacques. In 1782, the brothers constructed a large **balloon** which they filled with hot air produced by an ordinary bonfire under the balloon. The balloon rose more than a mile into the air. A year later, the Montgolfiers sent up a second balloon, this one carrying a tub that held a duck, a rooster, and a sheep. Then, only two months later, they constructed yet another balloon, this one large enough to carry a human into the atmosphere.

Balloon transportation suffers from one major drawback: the balloon goes wherever the winds carry it, and passengers have almost no control over the direction or speed of their travel. The additions needed to convert a balloon into a useable aircraft are a motor to propel the balloon in any given direction and a rudder with which to steer the balloon. The modified form of a balloon with these features is known as a dirigible.

The father of the modern dirigible is generally said to be Count Ferdinand von Zeppelin. Zeppelin's dirigible consisted of a rigid **aluminum** framework supporting a fabric covering and filled with hydrogen gas. On July 2, 1900, Zeppelin's dirigible took its initial flight; his first working ship was 420 ft (125 m) long and 40 ft (12 m) in diameter. It was capable of lifting 27,000 lb (12,000 kg) and traveling at air speeds comparable to those of airplanes then available. At their peak, Zeppelin-styled dirigibles were able to carry a maximum of 72 passengers in an elaborate gondola that also held a dining room, bar, lounge, and walkways.

The end of commercial dirigible travel came in the 1930s as the result of two events. One was the continuing improvement in heavier-than-air travel which made the much slower dirigible obsolete. The other event was the dramatic explosion and destruction of the dirigible *Hindenburg* as it attempted to moor at Lakehurst, New Jersey, on May 6, 1937. The ever-present danger that the highly flammable hydrogen gas used to inflate dirigibles would ignite and burn had finally come to realization at Lakehurst. Although later dirigibles were designed to fly with non-flammable helium, they never really regained the popularity of the pre-Lakehurst period. Today, dirigibles are widely used for advertising purposes. The "Goodyear blimp" and its cousins have now become familiar sights at outdoor sporting events all over the United States.

Heavier-than-air aircraft

The father of heavier-than-air machines is said to be Sir George Cayley (1773-1857). Cayley carried out a careful study of the way birds fly and, in 1810, wrote a pioneering book on flight, *On Aerial Navigation*. Cayley's research laid the foundations of the modern science of **aerodynamics**, the study of the forces experienced by an object flying through the air. In 1853, he constructed his first working aircraft, a glider that his coachman rode above the valleys on the Cayley estate in Yorkshire, England.

Gliders

Cayley's glider is now regarded as the earliest version of the modern airplane. The glider—also known as a sailplane—differs from a modern airplane only in that it has no power source of its own. Instead, it uses updrafts and winds for propulsion and maneuvering. Cayley was well aware of the need for a powerful engine for moving a heavier-than-air machine, but only steam engines were then available, and they were much too heavy for use in an aircraft. So Cayley designed an **airship** (his glider) that could make use of natural air movements.

The B-2 Stealth Bomber. *Photograph by P. Shambroom. Photo Researchers, Inc. Reproduced by permission.*

The modern sailplane retains the characteristics that Caylcy found to be crucial in the design of his own glider. In particular, both aircrafts have very large wings, are made of the lightest possible material, and have extremely smooth surfaces. The protrusion of a single rivet can produce enough **friction** to interfere with the successful maneuvering of a sailplane or glider. Properly designed sailplanes can remain in the air for hours and can travel at speeds of up to 150 MPH (240 km/h) using only natural updrafts.

Four fundamental aerodynamic forces

The age of modern aviation can be said to have begun on December 17, 1903 at Kitty Hawk, North Carolina. On that date, Wilbur and Orville Wright, two bicycle makers from Dayton, Ohio, flew the world's first powered aircraft, a biplane (double-winged aircraft) with a wing span of 40 ft 4 in (12 m 1.5 cm) and weighing 605 lb (275 kg). The plane remained in the air a total of only 12 seconds and covered only 120 ft (37 m). But both brothers knew, as Wilbur then said, that "the age of flight had come at last." The problems that the Wright brothers had to solve in the early 1900s were essentially the same as those confronting aeronautical engineers today. In order to make a heavier-than-air machine fly, four factors have to be taken into consideration: weight, lift, drag, and thrust.

Weight is, of course, caused by the pull of the Earth's gravitational field on the airplane itself, its passengers, and its cargo. The plane will never leave the ground unless some method is found for counteracting the gravitational effect of weight. The way in which weight is counterbalanced is by means of lift, a **force** equal to and opposite in direction to the pull of gravity. Lift is provided by means of the flow of air over the airplane's wings.

Imagine that a strong **wind** blows over the wings of an airplane parked on a landing strip. Airplane wings are always designed so that the flow of air across the top surface and across the bottom surface of the wing is not identical. For example, the upper surface of the wing might have a slightly curved shape, like half of a tear drop. The lower surface of the same wing might then have a perfectly flat shape.

When air passes over a wing of this design, it moves more quickly over the top surface than it does the bottom surface. The effect produced was first observed by Swiss mathematician Daniel Bernoulli in the 1730s and is now known by his name, the Bernoulli effect. Bernoulli discovered that the faster a fluid moves over a surface, the less **pressure** it exerts on that surface.

In the case of an airplane wing, air moving over the top of the wing travels faster and exerts less pressure than air moving over the bottom of the wing. Since the pressure under the wing is greater than the pressure on top of the wing, air tends to push upward on the wing, raising the airplane off the ground.

A number of factors determine the amount of lift a wing can provide an airplane. One factor is the size of the wing. The larger the wing, the greater the total force exerted on the bottom of the wing compared to the reduced pressure on top of the wing. A second factor is speed. The faster that air moves over a wing, the less pressure it exerts and the greater the lifting force it provides to the airplane. A third factor is the relative orientation of the wing to the air flow coming toward it. This factor is known as the **angle** of attack. In general, the greater the angle of attack, the greater the lifting force provided by the wing.

A pilot has control over all three of these factors in an airplane. Speed is controlled by increasing or decreasing the **rate** at which the engine turns, thereby changing the speed at which the propeller turns. Wing size is also under the pilot's control because of flaps that are attached to the following edge of a wing. These flaps can be extended during takeoff and landing to increase the total wing area and then retracted during level flights to reduce drag. Wing flaps and flap-like sections on the tail—the elevators—change the angle at which the wing or tail meets oncoming air and, thus, the airplane's angle of attack.

Thrust and drag

An airplane does not fly, of course, simply by setting it out on the runway and waiting for a strong wind to blow over its wings. Instead, the airplane is caused to move forward, forcing it to rush through still air at a high rate of speed. The forward thrust for the aircraft comes from one of two sources: a rotating propeller blade powered by some kind of engine or a rocket engine. The propeller used to drive the Wrights' first flight at Kitty Hawk was a home built engine that weighed 180 lb (82 kg) and produced 180 horsepower.

The forward thrust provided by a propeller can be explained in exactly the same way that the lift of a wing can be explained. Think of a propeller as a short, narrow mini-wing pivoted at the center and connected to an engine that provides rotational **motion**. The mini-wing is shaped like a larger wing, with a convex forward surface and a flat back surface. As the mini-wing is caused to rotate around its pivot point, the wing sweeps through the

air, which passes over the forward surface of the mini-wing faster than it does over the back surface. As a result, the pressure on the forward surface of the mini-wing is less than it is on the back, and the mini-wing is driven forward, carrying the airplane along with it.

The pitch and angle of attack of the mini-wing—the propeller—can be changed just as the angle of attack of the airplane's main wings can be changed. During level flight, the pitch of the propeller is turned at a sharp angle to the oncoming airflow, allowing the aircraft to maintain air speed with minimal fuel consumption. During takeoff and landing, the pitch of a propeller is reduced, exposing a maximum surface area to airflow and attaining a maximum thrust. At landing, the propeller direction can actually be reversed, causing the direction of force on the propellers to shift by 180°. As a result, the propeller becomes a brake on the airplane's forward motion.

The term drag refers to a variety of factors, each of which tends to slow down the forward movement of an aircraft. The most obvious type of drag is friction drag, a retarding force that results simply from the movement of a body such as an airplane through a fluid. The amount and effect of friction drag are very much a function of the shape and design of the aircraft.

Perhaps the most complex type of drag is that caused by the very air movements that lift an aircraft off the ground. The discussion of the Bernoulli effect above assumed that air flows smoothly over a surface. Such is never the case, however. Instead, as air travels over an airplane wing, it tends to break apart and form eddies and currents. The interaction of these eddies and currents with the overall air flow and with the airplane wing itself results in a retarding force: induced drag. One of the great challenges facing aeronautical engineers is to design aircraft in which all forms of drag are reduced to a minimum.

Aircraft stability

Once it is in flight, an airplane is subject to three major types of movements: pitch, yaw, and roll. These three terms describe the possible motion of an airplane in each of three dimensions. Pitch, for example, refers to the tendency of an airplane to rotate in a forward or backward direction, tail-over-nose, or vice versa. Yaw is used to describe a horizontal motion, in which the airplane tends to rotate with the left wing forward and the right wing backward, or vice versa. Roll is the phenomenon in which an airplane twists vertically around the body, with the right wing sliding upward and the left wing downward, or vice versa.

Each of the above actions can, of course, result in an airplane's crashing, so methods must be available for preventing each. The horizontal tail at the back of an airplane body helps to prevent pitching. If the plane's nose should begin to dip or rise, the angle of attack on the tail changes, and the plane adjusts automatically. The pilot also has control over the vertical orientation of the plane's nose.

Roll is prevented by making a relatively modest adjustment in the orientation of the aircraft's wings. Instead of their being entirely horizontal to the ground, they are tipped upward at their outer edges in a very wide V shape. The shape is called a dihedral. When the airplane begins to roll over one direction or the other, the movement of air changes under each wing and the plane rights itself automatically.

Yawing is prevented partly by means of the vertical tail at the back of the airplane. As a plane's nose is pushed in one direction or the other, airflow over the tail changes, and the plane corrects its course automatically. A pilot also has control over vertical tail flaps and can adjust for yawing by shifting the plane's ailerons, flaps on the following edge of the wings and the rear horizontal tail.

Jet engines

Until the 1940s, the only system available for powering aircraft was the piston-driven propeller engine. In order to increase the speed and lifting power of an airplane, the only option that aeronautical engineers had was to try to increase the efficiency of the engine or to add more engines to the airplane. During World War II, the largest power plants consisted of as many as 28 cylinders, capable of developing 3,500 horsepower.

At the very end of World War II, German scientists produced an entirely new type of power system, the **jet engine**. As "new" as the jet airplane was in 1944, the scientific principles on which it is based had been around for more than 2,000 years. You can think of a jet engine as a very large tin can, somewhat fatter in the middle and more narrow at both ends. Both ends of the tin can have been removed so that air can pass in the front of the can (the engine) and out the back.

The center of the engine contains the elements necessary for its operation. Compressed air is mixed with a flow of fuel and ignited. As **combustion** gases are formed, they push out of the rear of the engine. As the gases leave the engine, they also turn a **turbine** which compresses the air used in the middle of the engine.

The principle on which the jet engine operates was first enunciated by Sir Isaac Newton in the seventeenth century. According to Newton's second law, for every action, there is an equal and opposite reaction. In the jet engine, the action is the surge of burned gases flowing

The first supersonic transport to fly, the Soviet Union's Tupolev-144. It flew on December 31, 1968, two months before the Concorde's inaugural flight. *U.S. National Aeronautics and Space Administration (NASA).*

out of the back of the engine. The reaction to that stream of hot gases is a forward push that moves the engine—and the wing and airplane to which it is attached—in a forward direction.

Because the exiting gases turn a turbine as well as powering the engine, a jet of this kind is also known as a turbojet. The first airplanes of this kind—the German Messerschmitt 262 and the British Gloster Meteor—were available for flight in the spring of 1944, only months before the end of the war. The first turbojet planes were capable of speeds of about 540 MPH (865 km/h), about 20 MPH (33 km/h) faster than any piston-driven aircraft then in flight.

Wind tunnels

No matter what kind of power system is being used in an aircraft, aeronautical engineers are constantly searching for new designs that will improve the aerodynamic properties of aircraft. Perhaps the single most valuable tool in that search is the wind tunnel. A wind tunnel is a closed space in which the model of a new airplane design is placed. A powerful fan at one end of the tunnel is then turned on, producing a very strong wind across the surface on the airplane model. In this respect, the wind tunnel matches the airplane-on-the-landing-strip example used above. The flow of air from the fan, across the model, and out the back of the tunnel can be followed by placing small amounts of smoke into the wind produced by the fan. Engineers can visually observe the path of smoke as it flows over the plane, and they can take photographs to obtain permanent records of these effects.

The wind tunnel was first suggested by the great Russian physicist Konstantin Tsiolkovsky. In 1897, Tsiolkovsky constructed the first wind tunnel in the town of Kaluga and investigated the effects of various aircraft bodies and wing designs on the flow of air over an aircraft. Today, wind tunnels of various sizes are in use, allowing the test of small models of airframes and wings as well as of full-size aircraft.

Specialty airplanes

Many people may think of the modern aircraft business as being dominated by military jet fighters and bombers, commercial jetliners and prop planes, and other similar large aircraft. But flying has become a popular hobby among many people in the world today, and, as a result, a number of special kinds of aircraft have been developed for use by ordinary people. One example is the bicycle drive light plane.

One of the most successful examples of this aircraft design is known as *Daedalus '88*. *Daedalus '88* was designed to be so light and so aerodynamically sound that it could be flown by means of a single person turning the pedals of a bicycle. The bicycle pedals in *Daedalus '88* are attached to a shaft that turns the aircraft's propeller, providing the thrust needed to get the plane off the ground. The wings are more than 100 ft (31 m) wide and the boom connecting the propeller to the tail of the plane is 29 ft (9 m) long.

Most of *Daedalus '88*'s success is due to the use of the lightest possible materials in its construction: polyester for wing coverings, styrofoam and balsa wood for wing ribs, and aluminum and graphite epoxy **resins** for wing spars. The final aircraft weighs no more than 70 lb (32 kg). In the late 1980s, *Daedalus' 88* was flown for a record-breaking 70 mi (110 km) at a speed of 18 mph (29 kph) by Greek bicyclist Kanellos Kanellopoulos.

Helicopters

A helicopter is an aircraft that has the capability of maneuvering in both horizontal and vertical directions. It accomplishes these maneuvers by means of a single elaborate and rather remarkable propeller-like device mounted to its top. Although the device looks like an ordinary propeller, it is much more complicated. In fact, the device is more properly thought of as a pair of wings (rotor blades) that spin around a common center. By varying the pitch, position, and angle of attack of the rotor blades, the helicopter pilot can direct the thrust upward, downward, forward, backward, or at an angle. For example, if the pilot wants the aircraft to go upward, he or she increases the pitch in each blade of the rotor, in-

creasing the upward lift that is generated. If the pilot wants to move the aircraft in a forward direction, the whole rotor shift can be tipped forward to change the lifting action to a forward thrust.

Helicopters present some difficult design problems for aeronautical engineers. One of the most serious problems is that the spinning of the rotor causes—as Newton's second law would predict—a reaction in the helicopter body itself. As the rotors spin in one direction, the aircraft has a tendency to spin at an equal speed in the opposite direction. A number of inventions have been developed to deal with this problem. Some helicopters have two sets of rotors, one turning in one direction, and the other in the opposite direction. A more common approach is to add a second propeller on the rear tail of the helicopter, mounted either horizontally or vertically. Either design helps to stabilize the helicopter and to prevent it from spinning out of control.

Navigation

Guiding the motion of aircraft through the skies is a serious problem for two reasons. First, commercial and military aircraft now fly in all kinds of weather, often under conditions that prevent them from seeing other aircraft, the ground, or the airports at which they are supposed to land. Second, there is so much traffic in the air at any one time in many parts of the world that precautions must be taken to prevent collisions.

The crucial invention that made the control of air flight possible was the development of **radar** in the 1930s by the Scottish physicist Sir Robert Watson-Watt. Radar is a system by which **radio** beams are sent out from some central location (such as an airport tower) in all directions. When those beams encounter an object in the sky—such as an aircraft—they bounce off that object and are reflected to the sending station. A controller at the sending station is able to "see" the location of all aircraft in the vicinity and, knowing this, can then direct their movements to make landings and takeoffs possible and to prevent in-air collisions of the aircraft.

Today, every movement of every aircraft is constantly monitored using radar and radio signals. The moment a commercial airliner is ready to leave its gate for a flight, for example, the pilot notifies the control tower. One controller in the tower directs the airplane's movement on the ground, telling it when to back away from the gate and what runway to use for departure. Once the airplane is airborne, the pilot switches to a new radio channel and receives instructions from departure control. Departure control guides the aircraft's movements through the busiest section of the air near the airport, a **distance** about 30 mi (50 km) in radius. Finally, depar-

KEY TERMS

Dirigible—A lighter-than-airship capable of being piloted and controlled by mechanical means.

Drag—A force of resistance caused by the movement of an aircraft through the air, such as the friction that develops between air and the aircraft body.

Glider—A motorless aircraft that remains in the air by riding on rising currents of air.

Helicopter—An aircraft with the capability of flying upward, downward, forward, backward, or in other directions.

Jet engine—An engine that obtains its power from burning a fuel within the engine, creating a backward thrust of gases that simultaneously pushes the engine forward.

Lift—The upper force on the wings of an aircraft created by differences in air pressure on top of and underneath the wings.

Pitch—The tendency of an aircraft to rotate in a forward or backward direction, tail-over-nose, or vice versa.

Roll—The tendency of an aircraft to twist vertically around its body, the right wing sliding upward and the left wing downward, or vice versa.

Thrust—The forward force on an aircraft provided by the aircraft's power system.

Wind tunnel—A closed space in which the movement of air flow over various types of aircraft bodies can be studied.

Yaw—The tendency of an aircraft to rotate in a horizontal motion, with the left wing forward and the right wing backward, or vice versa.

ture control "hands over" control of the airplane to a third system, one that guides the aircraft through various sections of the sky between its takeoff and landing points. As the plane approaches its final destination, the series of "hand offs" described above is reversed.

Resources

Books

Cumpsty, Nicholas A. *Jet Propulsion: A Simple Guide to the Aerodynamic and Thermodynamic Design and Performance of Jet Engines.* Cambridge: Cambridge University Press, 1998.

Garrison, Paul. *Lift, Thrust & Drag-A Primer of Modern Flying.* Blue Ridge, PA: TAB Books, Inc., 1981.

Leishman, J. Gordon. *Principles of Helicopter Aerodynamics.* Cambridge: Cambridge University Press, 2003.

Stever, H. Guyford, James J. Haggerty, and the editors of *Life. Flight.* New York: Time Inc., 1965.

Syson, T. *Physics of Flying Things.* Philadelphia: Institute of Physics Publishing, 2003.

David E. Newton

Airship

A technologically advanced cousin of the **balloon**, airships are streamlined vessels buoyed by gases and controlled by means of propellers, rudders, and pressurized air systems. More commonly referred to as blimps and dirigibles, the airship is comprised of non-rigid, semi-rigid, and rigid types that rely on lighter-than-air gases such as helium and **hydrogen** for lift. Since the turn of the twentieth century, they have been engaged commercially in the transport of passengers and cargo and have proven a successful means of advertising.

Airships derive their lift from forward **motion**, just as an airplane does, and all three types have long used the **internal combustion engine**, like the type used in automobiles, to propel their massive bodies through the air. These have included the earliest motorcycle engines, the diesel engines of the mammoth American ships *Akron* and *Macon*, and the beefed-up Porsche engines used to power a new generation of airships. Traditional **pressure** airships house the engine, propeller, and gear box on an outrigger that extends from the side of the car, while the modern British Skyship's 500 and 600 use inboard engines that turn long prop shafts which allow the propellers to be vectored outboard. The introduction of pivoted, or vectored, engines gave airships the ability to change the direction of thrust and afforded it such amenities as near-vertical lift-off, thus reducing the need for long runways. Capable of airborne refueling, airships can remain aloft for weeks at a time, reaching an average airspeed of 60 MPH (96.5 km/h).

Non-rigid airships

Mastery of the skies proved a dominant preoccupation with French inventors in the latter half of the eighteenth century. In 1783, Jacques and Joseph Montgolfier designed the first balloon used for manned flight, while concurrently, Jean-Baptiste-Marie Meusnier had thought to streamline the balloon and maneuver it by some mechanized means. While several airships of similar design met with limited success from 1852, Meusnier's idea did not officially get off the ground until 1898. That

A 1929 photo of the *Graf Zeppelin* airship, which used lighter-than-air hydrogen gas. *The Library of Congress.*

year, the Brazilian aeronaut Alberto Santos-Dumont became the first pilot to accurately navigate a hydrogen-filled envelope and basket by means of a propeller mounted to a motorcycle engine.

Because of their non-rigid structure, the first blimps, like the balloon, were prone to collapsing as the gas contracted during descent. To counter this, Santos-Dumont introduced the ballonet, an internal airbag that helps maintain the envelope's structure and regulates pitch as well as lift, or buoyancy. Modern blimps continue to use ballonets positioned at the front and rear of the envelope, permitting the engineer to pump air into one or the other to change the pitch **angle**. For example, by increasing the amount of air in the aft ballonet, the airship becomes tail heavy, thus raising the nose skyward. Steering is further achieved by controlling rudders affixed to one of several types of tail fin configurations, allowing for basic left, right, up, and down directions. The long standard cross-shaped fin is slowly being replaced by an X-shaped configuration. While the X is more complicated, requiring a combination of rudders to complete a maneuver, it provides better ground clearance.

The car or gondola serves as the control center, passenger quarters and cargo hold of the airship. The envelope and car are connected by a series of suspended cables attached to the envelope by different types of load-sharing surfaces. Many modern airships use a curtain structure glued or bonded along the length of the envelope which evenly distributes the weight of the car and engines.

Rigid airships

The German inventor, Count Ferdinand von Zeppelin took the guesswork out of airship **aerodynamics** by building a rigid structure of lightweight **aluminum** girders and rings that would hold the vessel's streamlined shape under varying atmospheric conditions. Unlike Dumont's single-unit envelope, Zeppelin incorporated a number of drum-shaped gasbags within compartments of the structure to maintain stabilization should one of the bags become punctured or deflate. The dirigible was then encapsulated by a fabric skin pulled tightly across its framework. Buoyancy was controlled by releasing **water** ballast to ascend or by slowly releasing the hydrogen gas through a venting system as the ship descended.

Zeppelin was among the first airship designers to realize, in practice, the functionality of greater size in

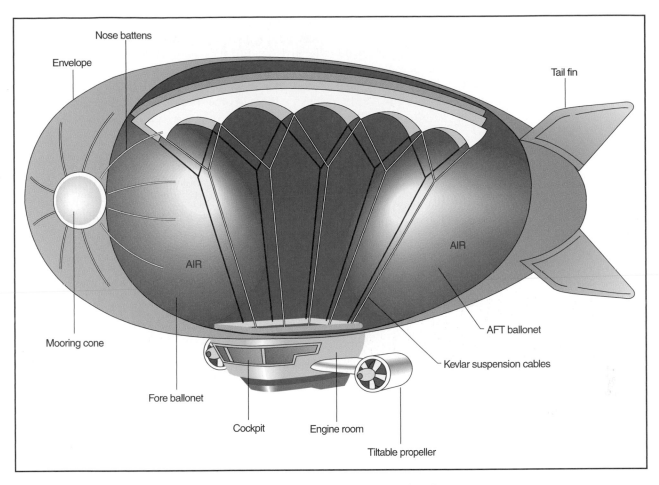

The internal structure of an airship. *Illustration by Hans & Cassidy. Courtesy of Gale Group.*

lighter-than-air vessels. As he increased the surface area of the envelope, the **volume** increased by a greater proportion. Thus, a larger volume afforded better lift to raise the aluminum hull and maximized the dirigible's cargo carrying capacity. Airships like the *Graf Zeppelin*, the *Hindenburg*, and their American counterparts, the *Macon* and the *Akron*, reached lengths of up to 800 ft (244 m) with volume capacities nearing seven million cubic feet. Hydrogen was, in part, responsible for such **engineering** feats because it is the lightest gas known to man and it is inexpensive. Its major drawback, one that would virtually close the chapter on non-rigid airship aviation, was its flammability. While passenger quarters, cargo, and fuel could be secured to or stored within the dirigible's **metal** structure, its engines were housed independently and suspended to reduce the possibility of friction-caused ignition of the gas.

Little did English writer Horace Walpole know how closely he had prophesied the future of airships when, upon the launch of the Montgolfier balloon, he wrote, "I hope these new mechanic meteors will prove only playthings for the learned or the idle, and not be converted into engines of destruction..." While zeppelins, as they had become known, were initially put into commercial service, successfully completing nearly 1,600 flights in a four-year period, they were engaged in the service of Germany during World War I as giant bombers, raiding London itself in May of 1915. After the war, the rigid dirigible was employed as both a trans-Atlantic luxury liner and airborne **aircraft** carrier, but eventually fell out of favor after the *Hindenburg* disaster in Lakehurst, New Jersey, on May 6, 1937.

Semi-rigid airships

An intermediate version of the rigid and non-rigid types, the envelope of the semi-rigid airship was fitted with a keel similar to that of a boat. The keel acted as a sort of metal spine which helped maintain the envelope's shape and supported the gondola and engines. Interest in the semi-rigid airship was short-lived, though it met with some success. In 1926, the Italian pilot Umberto Nobile navigated the airship *Norge* across the North Pole, accompanied by the Norwegian explorer Roald Amundsen.

The modern age of airships

Helium succeeded hydrogen as the gas of choice for the following generation of airships and continues as such in the early twenty-first century. Though its lifting capacity is less than that of hydrogen, helium is considered a safe resource because it is inflammable. During the 1920s, the United States discovered an abundant source of the gas in its own backyard and reinstated the blimp as a surveillance mechanism during World War II, maintaining a fleet of some of the largest non-rigid airships ever built. "It was the American monopoly of helium that made possible this Indian Summer of the small airship—long after every other country had abandoned the whole concept," wrote Patrick Abbott in his book *Airship*.

By the 1950s, the Goodyear Tire and Rubber Co. had become involved in the production of airships as part of the Navy's intended early warning defense system, and remained one of the largest manufacturers of blimps until the late 1980s. The Goodyear blimp is probably most noted as a high-flying billboard and mobile camera that has offered millions of **television** viewers a bird's-eye view of sporting events for several decades.

The early 1990s marked a resurgence in airships with the design of such models as the 222-ft (67.6-m) long *Sentinel 1000*. Built by Westinghouse Airships, Inc., its envelope is made of a lightweight, heavy-duty Dacron/Mylar/Tedlar composite that may eventually replace the traditional rubberized fabrics used by the *Sentinel*'s predecessors. According to *Aviation Week and Space Technology*, the craft "has a 345,000-cu.-ft. envelope and is powered by two modified Porsche automotive engines fitted with propellers that can be tilted through a range of plus 120 to minus 90 degrees." Its potential for transporting heavy payloads, its quietness, and relative stability, have brought the airship back to design rooms around the world. Based partly on its fuel efficiency and the fact that its shape and skin make it virtually invisible to other **radar**, the United States has plans to reintroduce the blimp as a radar platform for its Air Defense Initiative. French scientists have used the airship to navigate and study rainforests by treetop, while environmentalists have considered its usefulness as a means of monitoring coastal **pollution**. The future may see airships powered by helicopter rotor systems and solar power, as well as the return of the rigid airship.

Resources

Books

Abbott, Patrick. *Airship*. New York: Charles Scribner's Sons, 1973.

Meyer, Henry Cord. *Airshipmen, Businessmen and Politics 1890-1940*. Washington, DC: Smithsonian Institution Press, 1991.

Periodicals

Bell, Adrian. "On the Roof of the Rainforest." *New Scientist* 129 (1991): 48-51.

Garvey, William. "Rebirth of the Blimp." *Popular Mechanics* 168 (1991): 30-33+.

Hamer, Mick. "Airships Face a Military Future." *New Scientist* 115 (1987): 38-40.

Hollister, Anne. "Blimps." *Life Magazine* 4 (1988): 65-69.

Hughes, David. "New Westinghouse Airship Designed for Early Warning Surveillance." *Aviation Week and Space Technology* 135 (1991): 24-25.

John Spizzirri

Albatrosses

Albatrosses are large, long-lived seabirds in the family Diomedeidae, which contains about 13 **species**. They are found primarily in the oceans of the Southern Hemisphere. Albatrosses are superb fliers, and may be found far from land, soaring with their wings set in a characteristic bowed position. Together with petrels, shearwaters, and fulmars, albatrosses are grouped in the order Procellariiformes, which includes hook-billed sea **birds** commonly known as tubenoses. The extremely large nostrils on top of the bill lead to a pair of internal tubes, connected to a highly developed olfactory center in the **brain**. Thus, unlike most other birds, albatrosses possess a keen sense of **smell**; some biologists think that albatrosses can locate other individuals, food, and breeding and nesting areas by smell alone.

Two waved albatrosses. *JLM Visuals. Reproduced by permission.*

Flight and navigation

Albatrosses in flight are soaring birds, "floating" on the air for extended periods without flapping their wings. They have the greatest wingspan of any bird; the wingspan of the wandering albatross (*Diomedea exulans*) may reach 12 ft (3.7 m). The slender wings of albatrosses have a large aspect **ratio**, that is, a high ratio of wing length to wing width. This characteristic minimizes drag (air resistance) during flight, because the area at the tip of the wing is relatively small compared to the overall wing length. In addition, since albatrosses are large, relatively heavy birds, the amount of load on the wing is rather high. In fact, it is thought that albatrosses are close to the structural limits of wing design. In spite of this, albatrosses can soar extremely well, even in windless conditions, using slight updrafts of air created by waves on the water surface.

While albatrosses are remarkably graceful in the air, they are ungainly on land and on the surface of the water, to which they must descend to feed on **fish** and **squid**. To become airborne again, albatrosses must run into the **wind** across the surface of the water or land, until they can hoist themselves aloft. This ungraceful take-off, and their clumsy landings, are the reasons the Laysan alba-

tross (*D. immutabilis*) of Midway Island was dubbed the "Gooney Bird" by servicemen during World War II.

The navigational powers of albatrosses are impressive. They often spend many weeks at sea searching for food, well out of sight of land and obvious geographical landmarks. Some 82% of Laysan albatrosses transported experimentally to unfamiliar sites up to 4,740 mi (7,630 km) from their nesting site were able to find their way back. In contrast, only 67% of Leach's storm-petrels (*Oceanodroma leucorhoa*) were able to navigate much shorter distances back to their home area—up to 540 mi (870 km).

Salt regulation

Because albatrosses, and indeed all tubenoses, remain out at sea for days or weeks while foraging, these birds must be physiologically capable of drinking seawater without harm. The potentially serious problem of salt/fluid balance is resolved by means of specialized internal **glands** located at the base of the bill. These glands help regulate the **blood salt** content, which rises following ingestion of seawater, by producing a concentrated, salty fluid, which drips out of the tube on top of the bill. All tubenoses have the habit of sneezing and shaking their head frequently to clear this fluid.

Courtship rituals

Albatross **courtship** is unique among seabirds, both in its complexity and its duration. Males and females engage in a coordinated "dancing" display, in which the partners face one another and perform stereotyped and often synchronized behaviors such as bill "clappering" (in which the bill is quickly opened and closed repeatedly); "sky calling" (in which the bird lifts its bill to the sky, uttering a call like the "moo" of a cow); and fanning the wings while prancing in place. These displays are performed in repeating cycles for up to an hour each, numerous times per day. This **behavior** allows potential mates to evaluate each others' suitability as long-term partners.

Once formed, pair bonds in albatrosses appear to be life-long. After the initial courtship phase is over, the elaborate courtship rituals are much reduced or abandoned altogether in subsequent years. Researchers believe that these displays function more in pair formation than in the maintenance of the pair bond.

Care of the young

Albatrosses produce a single, helpless chick, which in most species requires a full year to leave the nest, longer than any other seabird. This is almost certainly because of the great effort required to collect food for the chick, who may be left alone for periods of 30-40 days while the parents forage at sea, and even up to 80 days in the case of the wandering albatross.

To compensate for their lengthy absences, adults feed their chicks a rich meal of oil, procured from their fishy diet. As a result, albatrosses have been called the "oil tankers" of the bird world: a stream of oil, produced in the adult stomach, is delivered into the hungry chick's bill. In this way the young birds are able to develop a layer of **fat** to sustain them during long parental foraging trips.

Conservation

In many parts of the range, albatrosses are suffering extensive mortality through their interaction with fishing fleets. The most intense damage is associated with so-called by-catch mortality in long-lining, in which baited hooks are strung out for several kilometers behind fishing boats. Albatrosses are attracted to the bait on the long-lines, but become caught on the hooks and drown. Because albatrosses can live to a great age, take many years to mature, and raise relatively few young, their populations are highly vulnerable to this kind of excess, by-catch mortality. The most **endangered species** are the short-tailed albatross (*Diomedea albatrus*), the Amsterdam albatross (*D. amsterdamensis*), and the wandering albatross (*D. exulans*). However, at least five additional species are also in serious decline.

Resources

Books

Ehrlich, Paul R., David S. Dobkin, and Darryl Wheye. *The Birder's Handbook: A Field Guide to the Natural History of North American Birds*. New York: Simon & Schuster, 1988.

Sibley, David Allen. *The Sibley Guide to Birds*. New York: Knopf, 2000.

Susan Andrew

Albedo

Albedo means *reflecting power* and comes from the Latin word, *albus,* for white or whiteness. The scientific meaning of albedo is the ability of a surface to reflect a certain proportion of visible **light**. A perfect mirror has an albedo of 100%; the polished surface of white metals like **aluminum** or silver comes close to that figure. Some metals like brass or **copper**, however, are colored, and they do not reflect all visible light equally well. This shows that albedo is dependent on the wavelength of the light being reflected.

In **astronomy** and **meteorology**, albedo describes the proportion of sunlight reflected back into outer **space**, for example, by a **planet** or a **satellite**. Without this reflection all the planets and their satellites would be invisible to us since, unlike the **Sun**, they are not self-luminous. The light from the Sun is white or yellowish because it is emitted from a **star** whose **temperature** is very high, close to 9,900°F (5,482°C).

The albedo of **Earth** is around 30-35%. It is higher over snow-covered surfaces, or where there is a cloud cover, and lower over clear oceans. After 30-35% of the sunlight is reflected back to space, the remaining 65-70% is first absorbed by Earth and its atmosphere and is re-emitted at the much longer wavelengths corresponding to the average temperature of our planet, 60°F (15.5°C). This re-emitted **radiation** is in the infrared part of the **spectrum**, and while we feel it as **heat**, it is not visible. Astronauts in outer space would not be able to see Earth if it had no albedo. Earth is known as the *blue planet* because the albedo reflects the particular wavelength that lies in the blue area of the spectrum. **Mars**, on the other hand, appears reddish to us, probably because its surface formations contain a large proportion of **iron** oxide, which reflects red light.

The albedo plays a crucial role in determining Earth's climate, as the average temperature at its surface is closely tied to the 65-70% of absorbed sunlight or solar **energy**. The light which is directly reflected back does not contribute to the warming of our planet. If, therefore, the albedo were to increase, the temperature at the Earth's surface would drop. If the albedo were to decrease, temperature would rise.

See also Weather.

Albinism

Albinism is a recessive inherited defect in melanin **metabolism** in which pigment is absent from hair, skin, and eyes (oculocutaneous albinism) or just from the eyes (ocular albinism). Melanin is a dark biological pigment that is formed as an end product of the metabolism of the **amino acid** tyrosine. When human skin is exposed to sunlight it gradually darkens or tans due to an increase in melanin. Tanning helps protect the underlying skin layers from the sun's harmful ultraviolet rays.

The most common examples of albinism are the white **rats**, rabbits, and **mice** found at pet stores. The characteristic white coats and pink eyes of these albino animals contrast dramatically with the brown or gray fur and dark eyes of genetically normal rats, rabbits, and mice. Domestic white chickens, **geese**, and **horses** are partial albinos. They retain pigment in their eyes, legs, and feet.

In the past, albinos were often regarded with fear or awe. Sometimes they were killed at **birth**, although albino births were common enough in some groups not to cause any excitement. For example, among the San Blas Indians of Panama, one in approximately 130 births is an albino. In the mid-nineteenth century, albinos were exhibited in carnival sideshows. Whole families were displayed at times and were described as a unique race of night people. They were said to live underground and to

KEY TERMS

Catalyst—Any agent that accelerates a chemical reaction without entering the reaction or being changed by it.

Melanin—A dark biological pigment found in the hair, skin, and eyes. Melanin absorbs harmful ultraviolet rays and is an important screen against the sun.

Melanocyte—A melanin-producing cell.

Tyrosinase—An enzyme that catalyzes the conversion of tyrosine to melanin.

Tyrosine—An amino acid that is a precursor of thyroid hormones and melanin.

come out only at night when the **light** was dim and would not hurt their eyes.

In humans, albinism is rare. One person in 17,000 has some type of albinism. Researchers have currently identified 10 different types of oculocutaneous albinism and five types of ocular albinism based on clinical appearance. Humans who have oculocutaneous albinism are unable to produce melanin; they have white, yellow, or yellow-brown hair, very light eyes (usually blue or grayish rather than pink), and very fair skin. The irises of their eyes may appear violet or pinkish because they have very little pigment and allow light to reflect back from the reddish retina in the back of the **eye**. People with albinism may also suffer from a variety optical disorders such as near- or far-sightedness, nystagmus (rapid irregular movement of the eyes back and forth), or strabismus (muscle imbalance of the eyes causing crossed eyes or lazy eye). They are very sensitive to bright light and sunburn easily. They must take great care to remain covered, wear a hat, and apply sunscreen anytime they are outdoors, since their skin is highly susceptible to precancerous and cancerous growths. People with albinism often wear sunglasses or tinted lenses even indoors to reduce light intensity to a more bearable level.

In ocular albinism only the eyes lack melanin pigment, while the skin and hair show normal or near-normal **color**. People with this condition have a variety of eye disorders because the lack of pigment impairs normal eye development. They are extremely sensitive to bright light and especially to sunlight (photophobia). Treatment of ocular albinism includes the use of visual aids and sometimes **surgery** for strabismus.

Albinism occurs when melanocytes (melanin-producing cells) fail to produce melanin. This absence of melanin production happens primarily in two ways. In tyrosinase-

An albino tiger cub and his sibling. © *Tom Brakefield/Corbis. Reproduced by permission.*

negative albinism (the most common form), the **enzyme** tyrosinase is missing from the melanocytes. Tyrosinase is a catalyst in the conversion of tyrosine to melanin. When the enzyme is missing no melanin is produced. In tyrosinase-positive albinism, a defect in the body's tyrosine transport system impairs melanin production. One in every 34,000 persons in the United States has tyrosinase-negative albinism. It is equally common among blacks and whites, while more blacks than whites are affected by tyrosinase-positive albinism. Native Americans have a high incidence to both forms of albinism.

Albinism cannot be cured, but people with this condition can expect to live a normal life span. Protection of the skin and eyes from sunlight is of primary importance for individuals with albinism. The **gene** carrying the defect that produces albinism is recessive, so both parents must carry this recessive gene in order to produce a child with the condition. When both parents carry the gene (and neither has albinism), there is a one in four chance with each pregnancy that their child will have albinism. The inheritance pattern of ocular albinism is somewhat different. This condition is X-linked, meaning that the recessive gene for ocular albinism is located on the X **chromosome**. X-linked ocular albinism appears almost exclusively in males who inherit the condition from their

mothers. Recently, a **blood** test has been developed to identify carriers of the gene that causes tyrosinase-negative albinism, and a similar test can identify this condition in the fetus by **amniocentesis**.

Vitiglio is another pigmentation disorder that resembles partial albinism. In this condition the skin exhibits stark white patches resulting from the destruction or absence of melanocytes. About 1% of the U. S. population has this disorder and it primarily affects people between the ages of 10 and 30. Unlike albinism, the specific cause of vitiglio is not known, although there seems to be a hereditary component, since about 30% of those who have vitiglio have family members with the condition. A link also exists between vitiglio and several other disorders with which it is often associated, including thyroid dysfunction, **Addison's disease**, and diabetes. Chemicals such as phenols may also cause vitiglio.

Resources

Books

Haefemeyer, J. W., R. A. King, and Bonnie LeRoy. *Facts about Albinism.* Minneapolis: International Albinism Center, 1992.
Professional Guide to Diseases. 7th ed. Springhouse, PA: Springhouse Corp., 2001

Periodicals

Haefemeyer, J. W. and J. L. Knuth. "Albinism." *Journal of Ophthalmic Nursing and Technology* 10 (1991): 55-62.

Other

"Ocular Albinism." Philadelphia: National Organization for Albinism and Hypopigmentation.
"What Is Albinism?" Philadelphia: National Organization for Albinism and Hypopigmentation.

Larry Blaser

Alchemy

Alchemy was a system of thinking about nature that preceded and contributed to the development of the modern science of **chemistry**. It was popular in ancient China, Persia, and western **Europe** throughout antiquity and the Middle Ages. A combination of philosophy, metallurgical arts, and magic, alchemy was based on a world view postulating an integral correspondence between the microcosm and the macrocosm—the smallest and largest parts of the universe. Its objectives were to find ways of accelerating the rates at which metals were thought to "grow" within the earth in their development toward perfection (gold) and of accomplishing a similar perfection in humans by achieving eternal life.

Origin

While it is not known when alchemy originated, historians agree that alchemistic ideas and practices flourished in the ancient world within several cultural traditions, as evidenced by manuscripts dating from the early centuries of the Christian era. The term "alchemy" has remained mysterious; scholars have identified "al" as an Arabic article, and proposed various etymologies for the word "chem," but a clear explanation of the term is still lacking.

Alchemy in China

Alchemical practices, namely, attempts to attain immortality, are believed to have arisen in China in the fourth century B.C. in conjunction with spread of Taoism, a mystical spiritualist doctrine which emerged in reaction to the practical spirit of Confucianism, the dominant philosophy of the period. The main emphasis in Chinese alchemy, it seems, was not on transmutation—the changing of one **metal** into another—but on the search for human immortality. In their search for an elixir of immortality, court alchemists experimented with mercury, **sulfur**, and arsenic, often creating venomous potions; several emperors died after drinking them. Such spectacular failures eventually led to the disappearance of alchemy in China.

Arabic alchemy

Alchemy flourished in the Islamic Arab caliphate of Baghdad in the eighth and ninth centuries, when court scientists, encouraged by their rulers, began studying and translating Syriac manuscripts of Greek philosophical and scientific works. The greatest representative of Arabic alchemy was ar-Razi (or Rhazes; c. 864-c.930), who worked in Baghdad. In their quest for gold, Arabic alchemists diligently studied and classified chemical elements and chemicals. Ar-Razi speculated about the possibility of using "strong waters," which were in reality corrosive **salt** solutions, as the critical ingredient for the creation of gold. Experimentation with salt solutions led to the discovery of mineral acids, but scholars are not sure if Islamic alchemy should be credited with this discovery.

Alchemy in the Western world

The history of Western alchemy probably begins in the Egyptian city of Alexandria, a great center of Greek learning during the Hellenistic period, a time of Greek cultural expansion and dominance following Alexander the Great's military conquests. Among the most prominent Alexandrian alchemists was Zosimos of Panopolis, Egypt, who may have lived in the third or fourth century A.D.

In accordance with older traditions, Zosimos that a magical ingredient was needed for the creation of gold. Greek alchemists called this ingredient *xerion*, which is Greek for powder. Through Arabic, this word came into Latin and modern European languages as *elixir*, and later became known as the elusive "philosopher's stone."

After the fall of the Western Roman Empire in the fifth century, Greek science and philosophy, as well as alchemy, sank into oblivion. In was not until the eleventh century that scholars rediscovered Greek learning, translating Syriac and Arabic manuscripts of Greek scientific and philosophical works into Latin, the universal language of educated Europeans. The pioneers of medieval science, such as Roger Bacon (c.1220-1292), viewed alchemy as a worthwhile intellectual pursuit, and alchemy continued to exert a powerful influence on intellectual life throughout the Middle Ages. However, as in ancient China, alchemists' failure to produce gold eventually provoked skepticism and led to the decline of alchemy.

In the sixteenth century, however, alchemists, frustrated by their fruitless quest for gold, turned to more practical matters, such as the use of alchemy to create medicines.

The greatest representative of this practical alchemy, which provided the basis for the development of chemistry as a science, was the German physician and alchemist Bombast von Hohenheim (known as Paracelsus;

KEY TERMS

Alcohol—Any of the large number of molecules containing a hydroxyl (OH) group bonded to a carbon atom to which only other carbon atoms or hydrogen atoms are bonded.

Azeotrope—A mixture of certain substances that distill together at the same boiling temperature, instead of separately.

Destructive distillation—An antiquated process for obtaining small amounts of alcohols, particularly methanol, from wood. The wood is heated to a high temperature in the absence of air, and gradually decomposes into a large number of chemicals.

Distillation—Collecting and condensing the vapor from a boiling solution. Each distinct, volatile chemical compound boils off individually at a specific temperature, so distillation is a way of purifying the volatile compounds in a mixture.

Ester—A molecule with a carbon both bonded to an ether linkage (carbon-oxygen-carbon), and double bonded to an oxygen.

Fermentation—The action of yeast metabolism on sugar solutions, resulting in the production of ethanol and carbon dioxide.

Glycol—An alcohol with two hydroxyl groups bonded to adjacent carbons in the molecule. Also called a diol.

Grignard synthesis—A classic laboratory method

of preparing alcohols. An alkyl halide is first reacted with magnesium, and then the addition of an aldehyde or ketone results in the formation of the alcohol.

Hydroxyl group—The -OH group attached to a carbon atom in a molecule. If the carbon atom itself is attached to only other carbon atoms or hydrogen atoms, the molecule is an alcohol.

Intermediate—In a chemical synthesis, any compound that is generated only to be used in the next step of the process.

Polyol—An alcohol with many hydroxyl groups bonded to the carbon atom backbone of the molecule.

Primary alcohol—An alcohol with the hydroxyl group at one end of the chain of carbon atoms.

Secondary alcohol—An alcohol with the hydroxyl group in the middle of a straight chain of carbon atoms.

Synthesis gas—A mixture of carbon monoxide and hydrogen gases, obtainable both from coal and natural gas, and used widely for the synthesis of alcohols and other organic compounds in the chemical industry.

Tertiary alcohol—An alcohol with the hydroxyl group in the middle of a branched chain of carbon atoms.

process known as hydration produces isopropyl alcohol when water is chemically added to propylene.

When "gasified," coal yields a mixture of hydrogen gas and **carbon monoxide** called synthesis gas, an important starting material for a variety of low- or high-molecular-weight alcohols. Methanol is now made almost entirely from synthesis gas. Before direct chemical synthesis was available, methanol was produced by the destructive distillation of wood (hence its older name, wood alcohol). Wood heated to a high **temperature** without air does not burn in the regular sense but decomposes into a very large number of different chemicals, of which methanol and some other alcohols are a small fraction.

Distilling the ethanol from fermentation products gives a mixture of 95% ethanol and 5% water, resulting in an azeotrope. This is when two chemicals distill together instead of separating at their different boiling temperatures. Most industrial ethanol is 95% alcohol unless there is specific need for very dry ethanol.

Reactions

The **plastics** industry is a major consumer of all types of alcohols, because they are intermediates in a large variety of **polymer** syntheses. The hydroxyl group is the part of an alcohol that makes the molecule relatively reactive and thus very useful in synthesis. Dozens of reactions are possible. Important esters made from ethanol include the insecticide malathion, the fragrance compound ethyl cinnamate, and the polymer building blocks ethyl acrylate and ethyl methacrylate. Examples of esters made from methanol include methyl salicylate (oil of **wintergreen**), the perfume ingredients methyl paraben and methyl benzoate, and the polymer starting material methyl acrylate. High-molecule-weight alcohols converted into esters are widely used as plasticizers in the polymer industry, and very high-molecule-weight alcohols with 12-18 carbon atoms are used to make biodegradable surfactants (detergents).

Alcohols can also be oxidized. If the alcohol's hydroxyl group is at the end of a carbon atom chain, an oxidation reaction produces either a carboxylic acid or an aldehyde. If the hydroxyl group is attached in the middle of a straight carbon atom chain, an oxidation reaction produces a ketone. An alcohol whose hydroxyl group is attached to a carbon atom that also has three other carbon branches attached to it cannot be oxidized.

The formation of double bonds in hydrocarbons can be accomplished by the dehydration of alcohols. Acid added to the alcohol removes not only the hydroxyl group, but also a hydrogen atom from an adjacent carbon atom. The reaction is called a dehydration, because H-O-H (water) is removed from the molecule and a double bond forms between the two carbon atoms.

Resources

Books

Bailey, James E. *Ullmann's Encyclopedia of Industrial Chemistry.* New York: VCH, 2003.

Bruice. Paula. *Organic Chemistry.* 3rd ed. Englewood Cliffs, NJ: Prentice-Hall, 2001.

Meyers, Robert A. *Encyclopedia of Analytical Chemistry: Applications, Theory and Instrumentation.* New York: John Wiley & Sons, 2000.

Szmant, H. H. *Organic Building Blocks of the Chemical Industry.* New York: Wiley, 1989.

Gail B. C. Marsella

Alcoholism

Alcoholism is a serious, chronic, potentially fatal condition manifested by a person's powerful **addiction** to alcoholic beverages. While experts have linked alcoholism to physiological (possibly hereditary), psychological, socioeconomic, ethnic, cultural, and other factors, there is no clear explanation of its genesis. Alcoholism occurs in all economic strata of society, in all age groups, from teenagers to the elderly, and in all races. Thus the popular stereotype of the alcoholic as a down-and-out person is misleading: alcoholics can be sufficiently functional to maintain a successful professional career. Particularly dangerous is the stereotype of alcoholism as an adult **disease**. Children and teenagers can become alcoholics—no age group is immune. While the abuse of alcoholic beverages has been known from time immemorial, the term alcoholism was coined in the nineteenth century by the Swedish physician Magnus Huss.

The psychology of alcoholism

It is commonly held that an alcoholic drinks in order to attain a euphoric state of mind. However, some re-searchers believe that it is not euphoria that the alcoholic seeks, but, more specifically, an escape from psychological **pain**. According to this view, to an alcoholic, sobriety is painful, and **alcohol** eases the pain. Most researchers, however, have not been able to explain alcoholism as simply a reaction to psychological distress. Instead, it is widely believed that physiological factors influence the alcoholic's drinking pattern.

The physiology of alcoholism

Some adults can drink alcohol-containing beverages in moderate amounts without experiencing significant side effects. There is evidence that having a glass of wine each day may be beneficial for the **heart** and digestive process.

Alcohol is a potent source of **energy** and calories. The ready availability of calories in alcohol gives an individual—alcoholic or nonalcoholic—a jolt of energy. These are, however, called empty calories, because alcohol contains no nourishment—vitamins, minerals, or other substances that the body needs. Some of the symptoms of alcoholism are the result of this phenomenon. The alcoholic can obtain the calories he needs from alcohol, but alcohol does not contain the nourishment his body needs to maintain such functions as the repair or replacement of cells.

Normally when alcohol enters the body it is rapidly absorbed from the stomach and distributed to all parts of the body in the **blood**. The alcohol is detoxified or broken down by first being changed into acetaldehyde when it flows through the liver. Acetaldehyde is a chemical that can cause painful reactions in the body. A second reaction in the liver alters acetaldehyde to form acetate, which is then changed into sugar. The liver of the alcoholic, however, is abnormally slow at the second-stage reaction.

Scientists have suggested that this conversion of acetaldehyde into acetate in the livers of alcoholics occurs at about half the speed that it does in the livers of nonalcoholics. Acetaldehyde thus accumulates in the bodies of alcoholics and causes many of the symptoms they exhibit, such as staggering gait, shaking hands, blinding headaches, and hallucinations.

Acetaldehyde is a very reactive and dominant chemical. In body cells it can block the normal chemical processes that should occur, including those in the **brain**. It can react with any other chemical in the immediate vicinity and produce byproducts of unpredictable reaction. Often the result of high levels of acetaldehyde is pain. The alcoholic may enter a cycle of drinking which leads to pain that is eased only by further drinking. The symptoms exhibited by some alcoholics when they begin to enter sobriety are collectively called delirium tremens or DTs. These symptoms can include hallucinations, illusions, trembling, and sweating.

The stages of alcoholism

The alcoholic, if he does not receive effective treatment, will progress through three stages of increasing deterioration. Alcoholism is difficult to diagnose in the early or adaptive stage. The alcoholic may drink heavily and remain functional. He does not experience any withdrawal symptoms other than the standard hangover following excessive drinking. The cells of the body adapt to large quantities of alcohol and still function. The alcohol provides a ready source of energy for **cell** functions and the cells become adept at using it. Even at this stage, however, alcohol intake will exact a penalty. The alcohol begins to attack cell structures, eroding cell membranes, altering cellular chemical balances, and otherwise upsetting a finely tuned system.

In this early stage of the condition the alcoholic can show a tremendous tolerance for alcoholic beverages. He might consume quantities that render normal adults hopelessly inebriated, yet not lose his ability to function. Only when his blood alcohol level begins to lessen does the alcoholic show symptoms of impairment. Thus, even though he does not exhibit signs of delirium tremens, the alcoholic in the early stage will know that he feels better when he drinks, functions more efficiently, and thinks more clearly. He will increase the frequency and amount of his drinking and will cross over into the middle stage of alcoholism.

No definite signpost marks the border between the early stage and middle stage of alcoholism, and the change may take years. Eventually, however, the alcoholic drinks to effect a cure, not to attain euphoria or efficiency in functioning. Deterioration of the cells of the body's organs and systems by steady infusion of alcohol begins to exert itself. The alcoholic experiences withdrawal symptoms that bring on physical and psychological pain that persists until it is eased by taking in more alcohol. These withdrawal symptoms soon worsen and require increased amounts of alcohol to erase them. The alcoholic will experience severe headaches, trembling, chills, and nausea when his blood alcohol level begins to ebb.

Full-blown DTs will eventually follow as the alcoholic continues to drink and his cellular **metabolism** becomes more and more dependent upon alcohol. He may have hallucinations, may become frightened and shrink into a corner, or may become dangerous as he lashes out to protect himself from an imaginary attack. He may manipulate his hands as if playing a game of cards or throwing dice or whittling. These symptoms are not benign, but signify a deep-seated **stress** on the body, especially the **nervous system**, and require immediate medical attention. The trauma of DTs may bring about a heart attack, **stroke**, or respiratory failure. Up to 25% of alcoholics experiencing DTs may die if not treated. At this stage, the alcoholic's body will no longer tolerate a state of low blood alcohol. His withdrawal symptoms become painful, and he can no longer limit his consumption to socially acceptable times; he must have a drink when he arises in the morning, and will probably drink on the job to alleviate his withdrawal symptoms.

But as the alcoholic's drinking increases, so does the cellular demand for alcohol, to the extent that he can no longer forestall his painful symptoms without being constantly in a state of drunkenness. This is the final, deteriorative stage of alcoholism. At this stage the alcoholic's tolerance to alcohol lessens because of widespread **organ** damage, especially in the liver and nervous system. A minority, probably about 10%, die as a result of late-stage organ damage such as **cirrhosis**. The liver simply cannot perform its functions, and the blood has a steadily increasing level of toxins. Perhaps a third of those in the late stages of alcoholism die from accidents such as falling down stairs or drowning, or by committing suicide. The physiological damage is widespread: the heart, pancreas, **digestive system**, and **respiratory system** all have characteristic changes in the late-stage alcoholic. The liver, however, suffers the most extensive damage.

Genetics of alcoholism

The contribution of **genetics** to an understanding alcoholism and other diseases having addictive **behavior** has been wrought with controversy for the past two hundred years. Because this is a politically and socially charged issue, there has been much debate regarding the true genetic contribution to alcoholism. Traditional medicine states that disease can be attributed to certain environmental conditions, specific **gene alleles** inherited from the parents or some combination of both of these factors. Most estimates of the contribution of genetics to alcoholism put the contribution of genetics about equal to that of the environment. Thus, the contribution of genetics to the disease is said to be about 50%. It should be noted, however, that various researchers have put this contribution as low as 10% or as high as 70%. The argument for heredity having a relatively strong influence on this disease rests primarily in studies involving families, adoptees and twins.

There is strong evidence that alcoholism runs in families. Most research studies in this area have demonstrated that about one fourth of the sons of alcoholics become alcoholics themselves. Daughters of alcoholics develop this disease about 5% of the time. While the estimates for the **rate** of alcoholism vary greatly for the general population, these rates are usually higher. In fact, the most consistent risk factor for developing alcoholism is a strong family history. Despite this data that shows a familial relationship for alcoholism, it could be argued that it is a learned behavior. Evidence from adoption studies further supports the contention of a genetic basis for alcoholism.

The use of adoptees is a common method to attempt to separate the effects of the environment from genetics. The rationale behind this method is that if biological children of alcoholic parents develop the disease at a greater rate than the general population when they reside with parents who do not have the disease there must be a strong genetic component to the disease. Most studies in this area have concluded that despite residing with adoptive, nonalcoholic parents, children who had alcoholic biological parents were at high risk for alcoholism. These rates reported were similar to that of children who grew up in the homes of their alcoholic biological parents. These studies strongly suggest a hereditary basis for alcoholism. This contention is further supported by twin studies which show that a second identical twin is much more likely to develop the disease if the first one developed it. Because identical twins have the same genetic instructions, this finding is consistent with the contention of a genetic component to alcoholism.

Other research models also support the assertion that heredity plays an important role in alcoholism. Researchers have turned to the science of **molecular biology** in an attempt to decipher this complex problem. One of the more compelling findings was that genetically engineered **mice** that had lacked a specific **dopamine** receptor gene in the brain were less likely to prefer alcohol and have a sensitivity to it than siblings that had the receptor. The results indicated that taking away the receptor decreased alcohol consumption by 50%. Although extensive testing needs to be completed, it is possible that mutations of these receptors in the brain may contribute to alcoholism in humans. If this avenue of research proves to be fruitful, it may be possible to treat alcoholism in the distant future through manipulation of this gene.

It is likely that alcoholism results from a combination of both genetic and environmental factors. In fact, most researchers working in this area believe that it is unlikely that science will determine the alcoholism gene. Rather, it is likely that the interaction of multiple genes contributes to the development of alcoholism. It should also be noted that those individuals with this array of genes are not predetermined to be alcoholics. While there is tremendous evidence that genes can exert influence over behavior, there is little support for the con-

Detoxification—The process of removing a poison or toxin from the body. The liver is the primary organ of detoxification in the body.

Metabolism—The physical and chemical processes that produce and maintain a living organism, including the breakdown of substances to provide energy for the organism.

Physiology—Study of the function of the organs or the body.

tention that they cause it. Thus, environment still plays a vital role in the development of this disease. While the exact alleles and their specific contribution to the development of alcoholism cannot be concluded with any certainty, it is known that genetics plays a role in the development of this disease, the actual mechanisms on how this happens has yet to be discovered by science.

Treatment

There is no cure for alcoholism. Treatment consists of bringing the alcoholic to realization of his condition and the need to avoid alcohol. It is often unsuccessful, and many alcoholics relapse into their drinking habits even after a period of abstinence.

Long-term therapeutic programs of about four weeks usually are considered necessary to arm the alcoholic to function without drinking. He is first sedated to **sleep** through the initial, painful withdrawal symptoms. After that he is subjected to an educational program to reveal to him the reasons for his condition and its inevitably fatal outcome. He is introduced to a supportive network such as Alcoholics Anonymous, the oldest such program in existence, to provide guidance for the remainder of his life. Ideally, family members participate in the treatment process to some degree, since they too can benefit from an educated understanding of alcoholism and from a support network. During his institutional stay the alcoholic receives appropriate therapy for organ damage and a concentrated nutritional regimen to restore his metabolism to normalcy rather than to an alcohol dependency.

Upon completion of therapy, the alcoholic is released into the world to resume a normal life without the physiological need for alcohol. Statistically, he has about a 50-50 chance of success. While many alcoholics return to drinking in the face of everyday stress, many others do recover, frequently finding greater peace of mind in their recovery than they possessed even before their drinking became excessive.

See also Fetal alcohol syndrome.

Resources

Books

Galanter, Mark, and Herbert D. Kleber, eds. *American Psychiatric Press Textbook of Substance Abuse Treatment.* 2nd. ed. Arlington, VA: American Psychiatric Press, 1999.

Galanter, Mark. *Recent Developments in Alcoholism: Research on Alcoholism Treatment.* New York: Plemun Publishers, 2001.

Ketcham, Katherine, et al. *Beyond the Influence: Understanding and Defeating Alcoholism Correcting the Code: Inventing the Genetic Cure for the Human Body.* New York: Bantam, 2000.

Organizations

The National Institutes of Health. "National Institute on Alcohol Abuse and Alcoholism" <http://www.niaaa.nih.gov/> (February, 4, 2003).

Larry Blaser
Zoran Minderovic
James Hoffmann

Aldehydes

Aldehydes are a class of highly reactive organic chemical compounds that contain a **carbonyl group** (in which a **carbon** atom is double-bound to an **oxygen** atom) and at least one **hydrogen** atom bound to the alpha carbon (the central carbon atom in the carbonyl group). The aldehydes are similar to the ketones, which also contain a carbonyl group. In the aldehydes, however, the carbonyl group is attached to the end of a chain of carbon **atoms**, which is not the case with the ketones. The word aldehyde is a combination of parts of the words *al*cohol and *dehyd*rogenated, because the first aldehyde was prepared by removing two hydrogen atoms (dehydrogenation) from **ethanol**. Molecules that contain an aldehyde group can be converted to alcohols by the addition of two hydrogen atoms to the central carbon oxygen double bond (reduction). Organic acids are the result of the introduction of one oxygen atom to the carbonyl group (oxidation). Aldehydes are very easy to detect by **smell**. Some are very fragrant, and others have a smell resembling that of rotten fruit.

Principal aldehydes

Formaldehyde is the simplest aldehyde. The central carbon atom in the carbonyl group is bound to two hydrogen atoms. Its chemical formula is $H_2C=O$. Formaldehyde, discovered in Russia by A. M. Butlerov in 1859, is a gas in its pure state. It is either mixed with **water** and sold as Formalin solutions or as a solid **polymer** called paraformaldehyde. The rather small formaldehyde **molecule** is very reactive and has found applications in the manufacture of many organic chemicals such as dyes and medical drugs. Formaldehyde is also a good insecticide, and it is used to kill germs in warehouses and ships. It is probably most familiar to the general public in its application as a preservative. In **biology** laboratories, animals and organs are suspended in formaldehyde solutions, which are also used as embalming fluid to preserve dead bodies from decay.

Acetaldehyde is the name of the shortest carbon chain aldehyde. It has a central carbon atom that has a double bond to an oxygen atom (the carbonyl group), a single bond to a hydrogen atom, and a single bond to another carbon atom connected to three hydrogen atoms (**methyl group**). Its chemical formula is written as CH_3CHO. Acetaldehyde is one of the oldest known aldehydes and was first made in 1774 by Carl Wilhelm Scheele. Its structure was not completely understood until 60 years later, when Justus von Liebig determined the constitution of acetaldehyde, described its preparation from ethanol, and gave the name of aldehydes to the chemical group.

The next larger aldehyde molecules have longer carbon atom chains with each carbon atom connected to two hydrogen atoms. This group of aldehydes is called aliphatic and has the general formula $CH_3(CH_2)_nCHO$, where n=1-6. When n=1, the aldehyde formula is CH_3CH_2CHO and is named propionaldehyde; when n=2, it is $CH_3(CH_2)_2CHO$ or butyraldehyde. The aliphatic aldehydes have irritating smells. For example, the smell of butyraldehyde, in low concentrations, resembles that of rotten butter. These medium-length aldehyde molecules are used as intermediates in the manufacture of other chemicals such as **acetone** and ethyl acetate used in finger nail polish remover. They are also important in the production of **plastics**.

Fatty aldehydes contain long chains of carbon atoms connected to an aldehyde group. They have between eight and 13 carbon atoms in their **molecular formula**. The fatty aldehydes have a very pleasant odor, with a fruity or a floral aroma, and can be detected in very low concentrations. Because of these characteristics, the fatty aldehydes are used in the formulation of many perfumes. The aldehyde that contains eight carbon atoms in its molecular formula is called octyl aldehyde and smells like oranges. The next longer aldehyde molecule is nonyl aldehyde, with nine carbon atoms in its structure, and has the odor of roses. A very powerful smelling compound is the 10-carbon aldehyde (decyl aldehyde), which has a scent of orange peel and is present in small **concentration** in most perfumes. Citral, a more complicated 10-carbon aldehyde, has the odor of lemons. Lauryl aldehyde, the 12-carbon

KEY TERMS

Aldehyde—A class of organic chemical compounds that contain a -CHO group.

Carbonyl group—A combination of a central carbon atom and an oxygen atom that have a double bond.

Dehydrogenation—The process of removing hydrogen atoms from a compound.

Methyl group—A terminal carbon atom connected to three hydrogen atoms.

Oxidation—The conversion of one chemical (compound) to another by the addition of oxygen atoms.

Reduction—The process by which an atom's oxidation state is decreased, by its gaining one or more electrons.

aldehyde, smells like lilacs or violets. Fatty aldehydes are also added to soaps and detergents to give them their "fresh lemon scent." The aromatic aldehydes have a **benzene** or phenyl ring connected to the aldehyde group. The aromatic aldehyde molecules have very complex structures but are probably the easiest to identify. Anisaldehyde smells like licorice. The odor of cinnamon found in various products is due to an aromatic aldehyde of complex structure named cinnamaldehyde. The aldehyde vanillin is a constituent in many vanilla-scented perfumes.

Resources

Books

Arctander, S. *Perfume and Flavor Materials of Natural Origin.* Elizabeth, NJ: S. Arctander, 1960.

Kirk-Othmer Encyclopedia of Chemical Technology. 4th ed. Suppl. New York: John Wiley & Sons, 1998.

McMurry, J. *Organic Chemistry.* 5th ed. Pacific Grove, CA: Brooks/Cole Publishing Company, 1999.

Walker, J.F. *Formaldehyde.* New York: Reinhold Publishing Corp., 1974.

Andrew Poss

Alder trees *see* **Birch family (Betulaceae)**
Alfalfa *see* **Legumes**

Algae

Algae (singular: alga) are photosynthetic, eukaryotic organisms that do not develop multicellular sex organs. Algae can be unicellular, or they may be large, multicellular organisms. Algae can occur in **salt** or fresh waters, or on the surfaces of moist **soil** or **rocks**. The multicellular algae develop specialized tissues, but they lack the true stems, leaves, or roots of the more complex, higher plants.

The algae are not a uniform group of organisms. They actually consist of seven divisions of distantly related organisms. These are considered together more as a matter of human convenience, than as a reflection of their ordered, biological, or evolutionary relationships. Therefore, the term "algae" is a common one, rather than a word that connotes a specific, scientific meaning.

Algae and their characteristics

As considered here, all of the algae are eukaryotic organisms, meaning their cells have nuclear material of **deoxyribonucleic acid (DNA)** organized within a discrete, membrane-bounded organelle, known as the nucleus. In view of this definition, the so-called blue-green algae are not discussed in this article, because those organisms are prokaryotic (that is, without an organized nucleus) and are more appropriately referred to as blue-green **bacteria**, or as cyanobacteria. The cyanobacteria are also different from the true algae in that they do not contain the photosynthetic pigment known as **chlorophyll** *a*, they do not have **cell** walls made of **cellulose**, and they do not store **energy** as starch or related polysaccharides.

Virtually all **species** of algae are photosynthetic. They have a relatively simple **anatomy**, which can range in complexity from single-celled organisms to colonial filaments, plates, or spheres, to the large, multicellular structures of the brown algae, known as thalli. Algal cell walls are generally made of cellulose, but can contain pectin, a class of hemicellulose polysaccharides that give the algae a slimy feel. The larger, multicellular algae have relatively complex tissues, which can be organized into organ-like structures that serve particular functions.

Types of algae

The actual term "algae" is not very useful in formal **biology**, because a number of disparate groups of unrelated organisms are aggregated under this broad term. The seven divisions of organisms that are considered within the algae are the Euglenophyta, Chrysophyta, Pyrrophyta, Chlorophyta, Rhodophyta, Paeophyta, and Xanthophyta. These divisions are separated on the basis of various features including their morphology and the **biochemistry** of their pigments, cell walls, and energy-storage compounds. The colors of these various algae types differ according to their particular mixtures of photosynthetic pigments, which typically include a combination of one or more chlorophylls and various accessory pigments. The latter can include carotenoids, xanthophylls, and phycobilins (these mask the green of the primary chlorophylls in various ways). The major differences among the seven divisions of algae are briefly summarized below.

Euglenophyta (euglenoids)

The Euglenophyta or euglenoids are 800 species of unicellular, protozoan-like algae, most of which occur in fresh waters. The euglenoids lack a true cell wall, and are bounded by a proteinaceous cell covering known as a pellicle. Euglenophytes have one to three flagellae for locomotion, and they store **carbohydrate** reserves as paramylon. The primary photosynthetic pigments of euglenophytes are chlorophylls *a* and *b*, while their accessory pigments are carotenoids and xanthophylls.

Most euglenoids have chloroplasts, and are photosynthetic. Some species, however, are heterotrophic, and feed on organic material suspended in the water. Even the photosynthetic species, however, are capable of surviving for some time if kept in the dark, as long as they are "fed" with suitable organic materials.

Chrysophyta (golden-brown algae)

The Chrysophyta are the golden-brown algae and **diatoms**, which respectively account for 1,100 and 40,000-100,000 species of unicellular algae. These algae occur in both marine and fresh waters, although most species are marine. The cell walls of golden-brown algae and diatoms are made of cellulose and pectic materials, a type of hemicellulose. In the diatoms especially, the cell wall is heavily impregnated with silica and is therefore quite rigid and resistant to decay. These algae store energy as a carbohydrate called leucosin, and also in oil droplets. The golden-brown algae achieve locomotion using one to two flagellae. The photosynthetic pigments of these algae are chlorophylls *a* and *c*, and the accessory pigments are carotenoids and xanthophylls, including a specialized pigment known as fucoxanthin.

Communities of diatoms (class Bacillariophyceae) can be extremely diverse, with more than 500 species commonly recorded from the **phytoplankton**, periphyton, and surface muds of individual ponds and lakes. Diatoms have double shells, or frustules, that are largely constructed of silica (SiO_2), the two halves of which (called valves) fit together like a pillbox. Diatom species are distinguished on the basis of the shape of their frustules, and the exquisite markings on the surface of these structures.

The golden-brown algae (class Chrysophyceae) are much less diverse than the diatoms. Some species of golden-brown algae lack cell walls, while others have pectin-rich walls. Golden-brown algae are especially important in open waters of the oceans, where they may dominate the productivity and **biomass** of the especially tiny size fractions of the phytoplankton. These are known as the nanoplankton, consisting of cells smaller than about 0.05 mm in diameter.

Pyrrophyta (fire algae)

The Pyrrophyta are the fire algae, including the dinoflagellates, which together account for 1,100 species of unicellular algae. Most of these species occur in marine ecosystems, but some are in fresh waters. The dinoflagellates have cell walls constructed of cellulose, and have two flagellae. These algae store energy as starch. The photosynthetic pigments of the Pyrrophyta are chlorophylls *a* and *c*, and the accessory pigments are carotenoids and xanthophyll, including fucoxanthin.

Some species of dinoflagellates can temporarily achieve a great abundance, as events that are commonly known as "red tides" because of the resulting **color** of the water. Red **tides** can be toxic to marine animals, because of the presence of poisonous chemicals that are synthesized by the dinoflagellates. Some species of dinoflagellates develop a **bioluminescence**, which can be clearly seen at night, and may cause the surface of the **ocean** to look as if it were aflame.

Chlorophyta (green algae)

The Chlorophyta or green algae consist of about 7,000 species, most of which occur in fresh water, although some others are marine. Most green algae are microscopic, but a few species, such as those in the genus *Cladophora*, are multicellular and macroscopic. The cell walls of green algae are mostly constructed of cellulose, with some incorporation of hemicellulose, and **calcium carbonate** in some species. The food reserves of green algae are starch, and their cells can have two or more organelles known as **flagella**, which are used in a whip-like fashion for locomotion. The photosynthetic pigments of green algae are chlorophylls *a* and *b*, and their accessory pigments are carotenoids and xanthophylls.

Some common examples of green algae include the unicellular genera *Chlamydomonas* and *Chlorella*, which have species dispersed in a wide range of habitats. More complex green algae include *Gonium*, which forms small, spherical colonies of four to 32 cells, and *Volvox*, which forms much larger, hollow-spherical colonies consisting of tens of thousands of cells. Some other colonial species are much larger, for example, *Cladophora*, a filamentous species that can be several meters long, and *Codium magnum*, which can be as long as 26 ft (8 m).

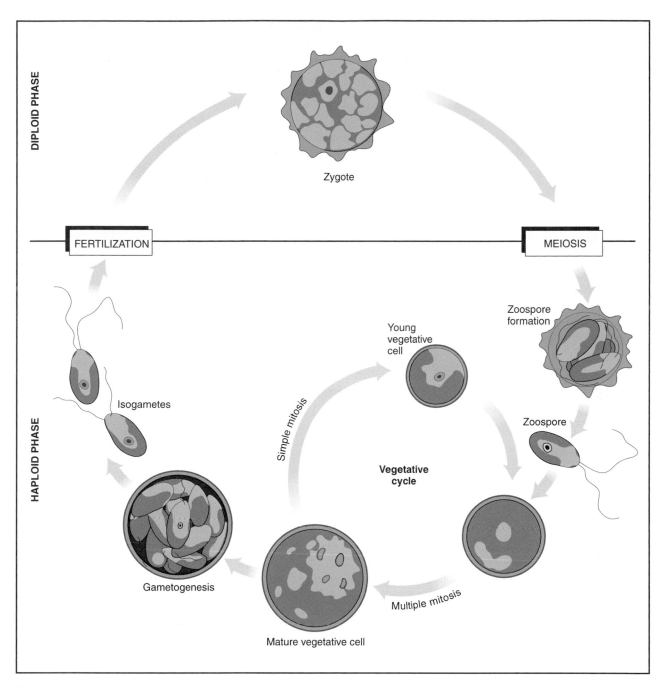

DIPLOID PHASE

HAPLOID PHASE

Zygote

FERTILIZATION

MEIOSIS

Zoospore
formation

Isogametes

Young
vegetative
cell

Zoospore

Simple mitosis

Vegetative
cycle

Gametogenesis

Multiple mitosis

Mature vegetative cell

The life cycle of *Chlorococcum*, a green alga. *Illustration by Hans & Cassidy. Courtesy of Gale Group.*

The stoneworts (class Charophyceae) are a very distinctive group of green algae that are sometimes treated as a separate division (the Charophyta). These algae can occur in fresh or **brackish** waters, and they have cell walls that contain large concentrations of **calcium** carbonate. Charophytes have relatively complex growth forms, with whorls of "branches" developing at their **tissue** nodes. Charophytes are also the only algae that develop multicellular sex organs, although these are not comparable to those of the higher plants.

Rhodophyta (red algae)

The Rhodophyta or red algae are 4,000 species of mostly marine algae, which are most diverse in tropical waters. Species of red algae range from microscopic to macroscopic in size. The larger species typically grow attached to a hard substrate, or they occur as epiphytes on other algae. The cell walls of red algae are constructed of cellulose and polysaccharides, such as agar and carrageenin. These algae lack flagellae, and they store energy as a specialized polysaccharide known as floridean starch. The photosynthetic pigments of red algae are chlorophylls *a* and *d*, and their accessory pigments are carotenoids, xanthophyll, and phycobilins.

Some examples of red algae include filamentous species such as *Pleonosporum* spp., so-called coralline algae such as *Porolithon* spp., which become heavily encrusted with calcium carbonate and contribute greatly to the building of tropical reefs, and thalloid species, such as the economically important Irish **moss** (*Chondrus crispus*).

Paeophyta (brown algae)

The Paeophyta or brown algae number about 1,500 species, almost all of which occur in marine environments. These seaweeds are especially abundant in cool waters. Species of brown algae are macroscopic in size, including the giant kelps that can routinely achieve lengths of tens of meters. Brown algae have cell walls constructed of cellulose and polysaccharides known as alginic acids. Some brown algae have relatively complex, differentiated tissues, including a holdfast that secures the **organism** to its substrate, air bladders to aid with buoyancy, a supporting stalk or stipe, wide blades that provide the major surface for nutrient exchange and **photosynthesis**, and spore-producing, reproductive tissues. The specialized, reproductive cells of brown algae are shed into the water and are motile, using two flagella to achieve locomotion. The food reserves of these algae are carbohydrate polymers known as laminarin. Their photosynthetic pigments are chlorophylls *a* and *c*, while the accessory pigments are carotenoids and xanthophylls, including fucoxanthin, a brown-colored pigment that gives these algae their characteristic dark color.

Some examples of brown algae include the sargassum weed (*Sargassum* spp.), which dominates the extensive, floating **ecosystem** in the mid-Atlantic gyre known as the Sargasso Sea. Most brown seaweeds, however, occur on hard-bottom, coastal substrates, especially in cooler waters. Examples of these include the rockweeds (*Fucus* spp. and *Ascophyllum* spp.), the kelps (*Laminaria* spp.), and the giant kelps (*Macrocystis* spp. and *Nereocystis* spp.). The giant kelps are by far the largest of the algae, achieving a length as great as 328 ft (100 m).

Xanthophyta (yellow-green algae)

The Xanthophyta or yellow-green algae are 450 species that primarily occur in fresh waters. They are unicellular or small-colonial algae, with cell walls made of cellulose and pectic compounds, and sometimes containing silica. The yellow-green algae store carbohydrate as leucosin, and they can have two or more flagellae for locomotion. The primary photosynthetic pigment of yellow-green algae is chlorophyll *a*, and the accessory pigments are carotenoids and xanthophyll.

Ecological relationships

Many types of algae are microscopic, occurring in single cells or small colonies. The usual **habitat** of many of the microscopic algae is open waters, in which case they are known as phytoplankton. Many species, however, live on the surfaces of rocks and larger plants within shallow-water habitats, and these are known as periphyton. Other microscopic algae live on the moist surfaces of soil and rocks in terrestrial environments.

Microscopic algae are at the base of their ecological food web—these are the photosynthetic, primary producers that are fed upon by herbivores. In the open waters of ponds, lakes, and especially the vast oceans, the algal phytoplankton is the only means by which diffuse solar **radiation** can be fixed into biological compounds. In these open-water (or pelagic) habitats the phytoplankton are consumed by small, grazing animals known as **zooplankton**, most of which are crustaceans. The zooplankton are in turn fed upon by larger zooplankton or by small **fish** (these predators are known as planktivores), which may then be eaten by larger fish (or piscivores). At the top of the open-water food web may be fish-eating **birds**, **seals**, whales, very large fish such as **sharks** or bluefin **tuna**, or humans. Therefore, the possibility of all of the animals occurring higher in the food webs, including the largest of the top predators, are ultimately dependent on the productivity of the microscopic phytoplankton of the pelagic marine ecosystem.

Other algae are macroscopic, meaning they can be readily observed without the aid of magnification. Some of these algae are enormous, with some species of kelps commonly reaching lengths greater than tens of meters long. Because they are primary producers, these macroscopic algae are also at the base of their ecological food webs. In most cases, however, relatively few herbivores can directly consume the biomass of macroscopic algae, and the major trophic interaction of these plants is through the decomposer, or detritivore part of the food web. In addition, because of their large size, macroscopic algae are critically important components of the physical structure of their ecosystems, providing habitat for a wide range of other organisms. The largest kelps develop a type of ecosystem that is appropriately referred to as a marine "forest" because of the scale and complexity of its habitat structure.

Some species of green algae occur as mutualistic symbionts with **fungi**, in an association of two organisms known as **lichens**. Lichens are common in many types of habitats. Other green algae occur in a **mutualism** with certain animals. In general, the host **animal** benefits from access to the photosynthetic products of the green alga, while the alga benefits from protection and access to inorganic **nutrients**. For example, species of unicellular *Chlorella* live inside of vacuoles within host cells of various species of **freshwater** protozoans, **sponges**, and **hydra**. Another species of green alga, *Platymonas convolutae*, occurs in cells of a marine flatworm, *Convoluta roscoffensis*. Various other green algae occur inside of marine **mollusks** known as nudibranchs. Similarly, various species of dinoflagellates occur as symbionts with marine corals.

Each species within an algal community has its particular ecological requirements and tolerances. Consequently, algal species tend to segregate along gradients in time and space, according to varying patterns of environmental resources, and of biological interactions, such as **competition** and predation. For example, during the growing season there is a time-series of varying abundances of phytoplankton species in open-water habitat. At certain times, particular species or closely related groups of species are abundant, but then these decline and other species of phytoplankton become dominant. This temporal dynamic is not totally predictable; it may vary significantly from year to year. The reasons for these patterns in the abundances and productivity of algal species are not understood, but they are likely associated with differences in their requirements for nutrients and other environmental factors, and perhaps with differing competitive abilities under resource-constrained conditions.

In a similar way, species of seaweeds tend to sort themselves out along stress-related environmental gradients associated with varying distances above and below the high-tide mark on rocky marine shores. The most important environmental stress for intertidal organisms is desiccation (drying), caused by exposure to the atmosphere at low tide, with the intensity of drying being related to the amount of time that is spent out of the water, and therefore to the **distance** above the high-tide line. For sub-tidal seaweeds the most important stress is the physical forces associated with waves, especially during storms. The various species of brown and red algae are arranged in rather predictable zonations along transects perpendicular to rocky shores. The largest kelps only occur in the sub-tidal habitats, because they are intolerant of desiccation. Within this near-shore habitat the species of algae are arranged in zones on the basis of their tolerance to the mechanical forces of wave action, as well as their competitive abilities in the least stressful, deeper-water habitats somewhat farther out to sea, where the tallest species grow and develop a kelp forest. In the intertidal, the various species of wracks and rockweeds dominate particular zones at various distances from the low-tide mark, with the most desiccation-tolerant species

occurring closest to the high-tide mark. Competition, however, also plays an important role in the distributions of the intertidal seaweeds.

Factors limiting the productivity of algae

Some species of algae can occur in extreme environments. For example, species of green algae have been observed in hot-water springs at Yellowstone National Park, in highly acidic volcanic lakes, in the extremely saline Great Salt Lake and Dead Sea, and on the surfaces of **glaciers** and snow. Some algae even survive suspended in the atmosphere as spores or in droplets of moisture.

These are, however, extremely stressful environmental conditions. Most algae occur in less stressful habitats, where their productivity tends to be limited by the availability of nutrients (assuming that sufficient **light** is available to support the photosynthetic process). In general, the productivity of freshwater algae is primarily limited by the availability of the nutrient phosphate (PO_4^{-3}), while that of marine algae is limited by nitrate (NO_3^-) or ammonium (NH_4^+). Some algal species, however, may have unusual nutrient requirements, and their productivity may be limited by certain micronutrients, such as silica, in the case of diatoms.

The structure of algal communities may also be greatly influenced by ecological interactions, such as competition and herbivory. For example, when herbivorous **sea urchins** are abundant, they can sometimes overgraze species of kelps in subtidal ecosystems of the west coast of **North America**, degrading the **kelp forests**. However, where sea **otters** (*Enhydra lutris*) are abundant this does not happen because the otters are effective predators of the urchins.

Another example of a biological influence on the structure of an algal community concerns the zebra mussel (*Dreissena polymorpha*). This is a bivalve mollusk that has been accidentally introduced by ocean-going ships to the Great Lakes of North America, where it has become an important pest because it clogs water pipes with its prolific growths, and can displace native species by competitively appropriating hard-substrate habitats. More to the present point, however, the zebra mussel is such an effective filter-feeder on phytoplankton, that its large populations in parts of the Great Lakes are apparently responsible for some of the clarification of the water that has occurred in recent years. Grazing by the zebra mussel has actually resulted in decreased standing **crops** of phytoplankton, even in well-fertilized waters.

Economic products obtained from algae

The most important economic products obtained from algae are associated with brown and red seaweeds, which can be utilized as food for people, and as resources for the manufacturing of industrial products. These seaweeds are mostly harvested from the wild, although increasing attention is being paid to the cultivation of large algae.

Some species of algae can be directly eaten by humans, and in eastern **Asia** they can be especially popular, with various species used as foods. An especially common food is the red alga known as nori in Japan and as laver in western **Europe** (*Porphyra* spp.), which has long been eaten by coastal peoples of China and Japan. This alga is often used as a wrapper for other foods, such as **rice** or plums, or it may be cooked into a clear soup. Nori has been cultivated for centuries in eastern Asia. Another alga known as dulse or sea kale (*Rhodymenia palmata*) is consumed dried or cooked into various stews or soups. Other commonly eaten seaweeds include the sea lettuce (*Ulva lactuca*), and murlins or edible kelp (*Alaria esculenta*).

Potentially, seaweeds are quite nutritious foods, because about 50% of their weight occurs as carbohydrates, with smaller concentrations of **proteins** and fats, and diverse micronutrients, including iodine. In practice, however, seaweeds are not very nutritious foods for humans, because we do not have the enzymes necessary to metabolize the most abundant of the complex, algal carbohydrates.

In some places, coastal **livestock** such as **sheep** and cattle, and wild **ungulates** such as **deer**, will graze algal biomass from the intertidal zone at low tide. These animals can take better advantage of the algal carbohydrates than can humans, largely because of the digestive abilities of the symbiotic **microorganisms** in their rumens and guts. Sometimes, algal biomass is harvested by humans and added to the fodder of livestock as a source of micronutrients.

The major economic importance of brown seaweeds, however, is as a natural resource for the manufacturing of a class of industrial chemicals known as alginates. These chemicals are extracted from algal biomass, and are used as thickening agents and as stabilizers for emulsions in the industrial preparation of foods and pharmaceuticals, and for other purposes.

Agar is another seaweed product, prepared from the mucilaginous components of the cell walls of certain red algae. Agar is used in the manufacturing of pharmaceuticals and cosmetics, as a culture medium for laboratory microorganisms, and for other purposes, including the preparation of jellied desserts and soups. Carrageenin is another, agar-like compound obtained from red algae that is widely used to stabilize emulsions in paints, pharmaceuticals, ice cream, and other products. Irish moss (*Chondrus crispus*) is a purplish alga that is a major source of carrageenin.

Researchers are investigating methods for the economic cultivation of red and brown seaweeds for the production of alginates, agar, and carrageenin. In California, use has been made of rafts anchored to float about 13 yd (12 m) below the surface, in shallow, less than 328-ft (100-m) deep water, to grow the highly productive, giant kelp *Macrocystis* as an industrial feedstock. Seaweeds are also cultivated on floating devices in coastal China, and research is investigating whether growth rates in dense plantings can be economically increased by enriching the seawater with nitrogen-containing **fertilizers**.

In some places, large quantities of the biomass of brown and red algae wash ashore, especially after severe storms that detach these algae from their substrates. This material, known as wrack, is an excellent substrate for **composting** into an organic-rich material that can greatly improve soil qualities in terms of aeration and water- and nutrient-holding capacity.

Over extremely long periods of time, the frustules of diatoms can accumulate in large quantities. This material is known as diatomaceous earth, and its small reserves are mined for use as a fine polishing substrate, as a fine filtering material, and for other industrial purposes.

Algae as environmental problems

Red tides are events of great abundance (or "blooms") of red, brown, or yellow-colored dinoflagellates of various species. These algae synthesize biochemicals, such as saxitoxin and domoic acid, which are extremely poisonous and can kill a wide range of marine animals, as well as humans who eat shellfish containing the toxins. The toxic syndromes of humans associated with dinoflagellate toxins are known as paralytic, diarrhetic, and amnesic shellfish poisoning.

Scientists cannot yet predict the environmental conditions that cause red tides to develop, although it seems that they are related to the availability and **ratio** of nutrients to **temperature**. Red tides are natural phenomena, but some scientists believe that human interference may have increased the frequency of these phenomena in some regions.

The dinoflagellates involved with toxic dinoflagellate blooms are commonly of the genera *Alexandrium*, *Dinophysis*, *Nitzchia*, or *Ptychodiscus*. The algal toxins can be accumulated by filter-feeding shellfish such as clams and oysters. If these are eaten while they contain red-tide toxins, they can poison humans or animals in the local ecosystem. Even creatures the size of large whales can die from eating fish containing large concentrations of dinoflagellate toxins.

In addition, freshwater algae can cause problems when they are overly abundant. Algal blooms can cause foul tastes in the water stored in reservoirs, which may be

KEY TERMS

Accessory pigments—Pigments such as the carotenoids, xanthophylls, and phycobilins, which can absorb solar radiation, and pass some of the absorbed energy to chlorophyll pigments for use in photosynthesis.

Bloom—An event of great abundance of phytoplankton, to the degree that the water is distinctly colored by the algal pigments.

Epiphyte—A plant which relies upon another plant, such as a tree, for physical support, but does not harm the host plant.

Eukaryotic cell—A cell whose genetic material is carried on chromosomes inside a nucleus encased in a membrane. Eukaryotic cells also have organelles that perform specific metabolic tasks and are supported by a cytoskeleton which runs through the cytoplasm, giving the cell form and shape.

Gyre—A zone of spirally circulating oceanic water, that tends to retain floating materials, as in the Sargasso Sea of the Atlantic Ocean.

Macroscopic—A size range that can be seen without magnification.

Microscopic—A size range that cannot be seen without magnification.

Mutualism—A symbiotic relationship between two species that is mutually beneficial.

Periphyton—Unicellular algae that occur on the surfaces of the rocks and larger plants of aquatic ecosystems.

Phytoplankton—Microscopic algae that occur suspended in the water column.

Primary photosynthetic pigment—This refers to the green-colored, chlorophyll pigments of algae and higher plants. These absorb red and blue light, to make the energy available to drive photosynthesis.

Thallus—A single plant body lacking distinct stem, leaves, and roots.

Zooplankton—Minute animal life that lives in water.

---

required by nearby towns or cities as drinking water. This can be a significant problem in naturally productive, shallow-water lakes of the prairies in North America.

Eutrophication is another major problem that is associated with algal blooms in lakes that are receiving large inputs of nutrients through sewage disposal or the runoff of agricultural fertilizers. Eutrophication can result in severe degradation of the aquatic ecosystem when large quantities of algal biomass sink to deeper waters and consume most of the **oxygen** during their **decomposition**. The anoxic (deficiency in oxygen) conditions that develop are lethal to the animals that live in the sediment and deep waters, including most species of fish. Because the primary limiting nutrient in fresh waters is usually phosphate, inputs of this nutrient can be specifically controlled by **sewage treatment**, and by the banning of detergents containing **phosphorus**. This has been done in many areas in North America, and eutrophication is now less an environmental problem than it used to be.

See also Biological community; Eukaryotae; Food chain/web; Symbiosis.

Resources

Books

Buchanan, B.B., W. Gruissem, and R.L. Jones. *Biochemistry and Molecular Biology of Plants.* Rockville, MD: American Society of Plant Physiologists, 2000.

Freedman, B. *Environmental Ecology.* 2nd ed. San Diego: Academic Press, 1995.

Pritchard, H.N. and P.T. Bradt. *Biology of Nonvascular Plants.* St. Louis: Mosby, Inc., 1984.

Raven, Peter, R.F. Evert, and Susan Eichhorn. *Biology of Plants.* 6th ed. New York: Worth Publishers Inc., 1998.

Bill Freedman

Algebra

Algebra is often referred to as a generalization of **arithmetic**. As such, it is a collection of rules: rules for translating words into the symbolic notation of **mathematics**, rules for formulating mathematical statements using symbolic notation, and rules for rewriting mathematical statements in a manner that leaves their truth unchanged.

The power of elementary algebra, which grew out of a desire to solve problems in arithmetic, stems from its use of variables to represent numbers. This allows the generalization of rules to whole sets of numbers. For example, the solution to a problem may be the **variable** x or a rule such as ab=ba can be stated for all numbers represented by the variables a and b.

Elementary algebra is concerned with expressing problems in terms of mathematical symbols and establishing general rules for the combination and manipulation of those symbols. There is another type of algebra, however, called abstract algebra, which is a further generalization of elementary algebra, and often bears little resemblance to arithmetic. Abstract algebra begins with a few basic assumptions about sets whose elements can be combined under one or more binary operations, and derives theorems that apply to all sets, satisfying the initial assumptions.

Elementary algebra

Algebra was popularized in the early ninth century by al-Khowarizmi, an Arab mathematician, and the author of the first algebra book, *Al-jabr wa'l Muqabalah*, from which the English word algebra is derived. An influential book in its day, it remained the standard text in algebra for a long time. The title translates roughly to "restoring and balancing," referring to the primary algebraic method of performing an operation on one side of an equation and restoring the balance, or equality, by doing the same thing to the other side. In his book, al-Khowarizmi did not use variables as we recognize them today, but concentrated on procedures and specific rules, presenting methods for solving numerous types of problems in arithmetic. Variables based on letters of the alphabet were first used in the late sixteenth century by the French mathematician François Viète. The idea is simply that a letter, usually from the English or Greek alphabet, stands for an element of a specific set. For example, x, y, and z are often used to represent a real number, z to represent a **complex** number, and n to stand for an integer. Variables are often used in mathematical statements to represent unknown quantities.

The rules of elementary algebra deal with the four familiar operations of **addition** $(+)$, **multiplication** $(\times)$, **subtraction** $(-)$, and **division** $(\div)$ of **real numbers**. Each operation is a rule for combining the real numbers, two at a time, in a way that gives a third real number. A combination of variables and numbers that are multiplied together, such as $64x^2$, 7yt, s/2, 32xyz, is called a monomial. The sum or difference of two monomials is referred to as a binomial, examples include $64x^2+7yt$, 13t+6x, and $12y-3ab/4$. The combination of three monomials is a trinomial (6xy+3z−2), and the combination of more than three is a polynomial. All are referred to as algebraic expressions.

One primary objective in algebra is to determine what conditions make a statement true. Statements are usually made in the form of comparisons. One expression is greater than (>), less than (<), or equal to (=) another expression, such as 6x+3 > 5, $7x^2-4 < 2$, or $5x^2+6x = 3y+4$. The application of algebraic methods then proceeds in the following way. A problem to be solved is stated in mathematical terms using symbolic notation. This results in an equation (or **inequality**). The equation contains a variable; the value of the variable that makes the equation true represents the solution to the equation, and hence the solution to the problem. Finding that solution requires manipulation of the equation in a way that leaves it essentially unchanged, that is, the two sides must remain equal at all times. The object is to select operations that will isolate the variable on one side of the equation, so that the other side will represent the solution. Thus, the most fundamental rule of algebra is the principle of al-Khowarizmi: whenever an operation is performed on one side of an equation, an equivalent operation must be performed on the other side bringing it back into balance. In this way, both sides of an equation remain equal.

Applications

Applications of algebra are found everywhere. The principles of algebra are applied in all branches of mathematics, for instance, **calculus**, **geometry**, and **topology**. They are applied every day by men and women working in all types of business. As a typical example of applying algebraic methods, consider the following problem. A painter is given the job of whitewashing three billboards along the highway. The owner of the billboards has told the painter that each is a **rectangle**, and all three are the same size, but he does not remember their exact dimensions. He does have two old drawings, one indicating the height of each billboard is two feet less than half its width, and the other indicating each has a perimeter of 68 feet. The painter is interested in determining how many gallons of paint he will need to complete the job, if a gallon of paint covers 400 square feet. To solve this problem three basic steps must be completed. First, carefully list the available information, identifying any unknown quantities. Second, translate the information into symbolic notation by assigning variables to unknown quantities and writing equations. Third, solve the equation, or equations, for the unknown quantities.

Step one, list available information: (a) three billboards of equal size and shape, (b) shape is rectangular, (c) height is 2 feet less than 1/2 the width, (d) perimeter equals 2 times sum of height plus width equals 68 feet, (e) total area, height times width times 3, is unknown, (f) height and width are unknown, (g) paint covers 400 sq ft per gallon, (h) total area divided by 400 equals required gallons of paint.

Step two, translate. Assign variables and write equations.

Let: A = area; h = height; w = width; g = number of gallons of paint needed.

Then: (1) h = 1/2w−2 (from [c] in step 1); (2) 2(h+w) = 68 (from [d] in step 1); (3) A = 3hw (from [e] in step 1); (4) g = A/400 (from [h] in step 1). Step three, solve the equations. The right hand side of equation (1) can be substituted into equation (2) for h giving 2(1/2w−2+w) = 68. By the **commutative property**, the quantity in parentheses is equal to (1/2w+w−2), which is equal to (3/2w−2). Thus, the equation 2(3/2w−2)=68 is exactly equivalent to the original. Applying the **distributive property** to the left hand side of this new equation results in another equivalent expression, 3w−4 = 68. To isolate w on one side of the equation, add 4 to both sides giving 3w−4+4 = 68+4 or 3w = 72. Finally, divide the expressions on each side of this last expression by 3 to isolate w. The result is w = 24 ft. Next, put the value 24 into equation (1) wherever w appears, h = (1/2(24)−2), and do the arithmetic to find h = (12−2) = 10ft. Then, put the values of h and w into equation (3) to find the area, A = 3×10×24 = 720 sq ft. Finally, substitute the value of A into equation (4) to find g = 720/400 = 1.8 gallons of paint.

Graphing algebraic equations

The methods of algebra are extended to geometry, and vice versa, by graphing. The value of graphing is two-fold. It can be used to describe geometric figures using the language of algebra, and it can be used to depict geometrically the algebraic relationship between two variables. For example, suppose that Fred is twice the age of his sister Irma. Since Irma's age is unknown, Fred's age is also unknown. The relationship between their ages can be expressed algebraically, though, by letting y represent Fred's age and x represent Irma's age. The result is y = 2x. Then, a graph, or picture, of the relationship can be drawn by indicating the points (x,y) in the Cartesian coordinate system for which the relationship y = 2x is always true. This is a straight line, and every **point** on it represents a possible combination of ages for Fred and Irma (of course **negative** ages have no meaning so x and y can only take on positive values). If a second relationship between their ages is given, for instance, Fred is three years older than Irma, then a second equation can be written, y = x+3, and a second graph can be drawn consisting of the ordered pairs (x,y) such that the relationship y = x+3 is always true. This second graph is also a straight line, and the point at which it intersects the line y = 2x is the point corresponding to the actual ages of Irma and Fred. For this example, the point is (3,6), meaning that Irma is three years old and Fred is six years old.

Linear algebra

Linear algebra involves the extension of techniques from elementary algebra to the solution of systems of linear equations. A linear equation is one in which no two variables are multiplied together, so that terms like xy, yz, x^2, y^2, and so on, do not appear. A system of equations is a set of two or more equations containing the same variables. **Systems of equations** arise in situations where two or more unknown quantities are present. In order for a unique solution to exist there must be as many independent conditions given as there unknowns, so that the number of equations that can be formulated equals the number of variables. Thus, we speak of two equations in two unknowns, three equations in three unknowns, and so forth. Consider the example of finding two numbers such that the first is six times larger than the second, and the second is 10 less that the first. This problem has two unknowns, and contains two independent conditions. In order to determine the two numbers, let x represent the first number and y represent the second number. Using the information provided, two equations can be formulated, x = 6y, from the first condition, and x−10 = y, from the second condition. To solve for y, replace x in the second equation with 6y from the first equation, giving 6y−10=y. Then, subtract y from both sides to obtain 5y−10=0, add 10 to both sides giving 5y=10, and divide both sides by 5 to find y=2. Finally, substitute y=2 into the first equation to obtain x=12. The first number, 12, is six times larger than the second, 2, and the second is 10 less than the first, as required. This simple example demonstrates the method of substitution. More general methods of solution involve the use of **matrix** algebra.

Matrix algebra

A matrix is a rectangular array of numbers, and matrix algebra involves the formulation of rules for manipulating matrices. The elements of a matrix are contained in square brackets and named by row and then column. For example the matrix has two rows and two columns, with the element (-6) located in row one column two. In general, a matrix can have i rows and j columns, so that an element of a matrix is denoted in double subscript notation by a_{ij}. The four elements in A are a_{11} = 1, a_{12} = -6, a_{21} = 3, a_{22} = 2. A matrix having m rows and n columns is called an "m by n" or (m × n) matrix. When the number of rows equals the number of columns the matrix is said to be square. In matrix algebra, the operations of addition and multiplication are extended to matrices and the fundamental principles for combining three or more matrices are developed. For example, two matrices are added by adding their corresponding elements. Thus, two matrices must each have the same number of rows and columns in order to be compatible for addition. When two matrices are compatible for addition, both the associative and commutative principles of elementary algebra continue to hold. One of the many applications of matrix algebra is the solution of systems of linear equations. The coefficients of a set of simultaneous equations are written in the form of a matrix, and a formula (known as Cramer's rule) is applied which provides the solution to n equations in n unknowns. The method is very powerful, especially when there are hundreds of unknowns, and a computer is available.

Abstract algebra

Abstract algebra represents a further generalization of elementary algebra. By defining such constructs as groups, based on a set of initial assumptions, called axioms, provides theorems that apply to all sets satisfying the abstract algebra axioms. A **group** is a set of elements together with a binary operation that satisfies three axioms. Because the binary operation in question may be any of a number of conceivable operations, including the familiar operations of addition, subtraction, multiplication, and division of real numbers, an asterisk or open **circle** is often used to indicate the operation. The three axioms that a set and its operation must satisfy in order to qualify as a group, are: (1) members of the set obey the associative principle [a $\times$ (b $\times$ c) = (a $\times$ b) $\times$ c]; (2) the set has an **identity element**, I, associated with the operation $\times$, such that a $\times$ I = a; (3) the set contains the inverse of each of its elements, that is, for each a in the set there is an inverse, a', such that a $\times$ a' = I. A well known group is the set of **integers**, together with the operation of addition. If it happens that the commutative principle also holds, then the group is called a commutative group. The group formed by the integers together with the operation of addition is a commutative group, but the set of integers together with the operation of subtraction is not a group, because subtraction of integers is not associative. The set of integers together with the operation of multiplication is a commutative group, but division is not strictly an operation on the integers because it does not always result in another integer, so the integers together with division do not form a group. The set of rational numbers, however, together with the operation of division is a group. The power of abstract algebra derives from its generality. The properties of groups, for instance, apply to any set and operation that satisfy the axioms of a group. It does not matter whether that set contains real numbers, **complex numbers**, vectors, matrices, functions, or probabilities, to name a few possibilities.

See also Associative property; Solution of equation.

Resources

Books

Bittinger, Marvin L, and Davic Ellenbogen. *Intermediate Algebra: Concepts and Applications.* 6th ed. Reading, MA: Addison-Wesley Publishing, 2001.
Blitzer, Robert. *Algebra and Trigonometry.* 2nd ed. Englewood Cliffs, NJ: Prentice Hall, 2003.

Gelfond, A.O. *Transcendental and Algebraic Numbers.* Dover Publications, 2003.
Immergut, Brita and Jean Burr Smith. *Arithmetic and Algebra Again.* New York: McGraw-Hill, 1994.
Stedall, Jacqueline and Timothy Edward Ward. *The Greate Invention of Algebra: Thomas Harriot's Treatise on Equations.* Oxford: Oxford University Press, 2003.
Weisstein, Eric W. *The CRC Concise Encyclopedia of Mathematics.* New York: CRC Press, 1998.

Other

Algebra Blaster 3 CD-ROM for Windows. Torrance, CA: Davidson and Associates Inc., 1994.

J.R. Maddocks

Algorithm

An algorithm is a set of instructions that indicate a method for accomplishing a task. If followed correctly, an algorithm guarantees successful completion even without the use of any intelligence. The term algorithm is derived from the name al-Khowarizmi, a ninth century Arabian mathematician who is credited with discovering **algebra**. With the advent of computers, which are particularly adept at utilizing algorithms, the creation of new and faster algorithms has become an important consideration in the study of theoretical computer science.

Algorithms can be written to solve any conceivable problem. For example, an algorithm can be developed for tying a shoe, making cookies, or determining the area of a **circle**. In an algorithm for tying a shoe, each step, from obtaining a shoe with a lace to releasing the string after it is tied, is spelled out. The individual steps are written in such a way that no judgment is ever required to successfully carry them out. The length of time required to complete an algorithm is directly dependent on the number of steps involved. The more steps, the longer it takes to complete. Consequently, algorithms are classified as fast or slow depending on the speed at which they allow a task to be completed. Typically, fast algorithms are usable while slow algorithms are unusable.

See also Computer, analog; Computer, digital.

Aliphatic hydrocarbon *see* **Hydrocarbon**

Alkali *see* **Acids and bases**

Alkali metals

The first column on the **periodic table** of the chemical elements is collectively called the alkali **metal** group: **lithium**, **sodium**, potassium, rubidium, cesium, and francium. Because their outer **electron** structure is similar, they all have somewhat similar chemical and physical properties. All are shiny, soft enough to cut with a knife, and most are white (cesium is yellow-white). All react with **water** to give **hydrogen** gas and the metal hydroxide; the heavier alkali metals react with such vigorous evolution of **heat** that the hydrogen often bursts into flame. They also react with the **oxygen** in the air to give either an oxide, peroxide, or superoxide, depending on the metal. Alkali metals almost always form ions with a positive (+1) charge, and are so reactive as elements that virtually all occur in nature only in compound form. Sodium is the most abundant, followed by potassium, rubidium, lithium, and cesium. Francium is intensely radioactive and extremely rare; only the tiniest traces occur in the earth's crust.

Most of the alkali metals glow with a characteristic **color** when placed in a flame; lithium is bright red, sodium gives off an intense yellow, potassium is violet, rubidium is a dark red, and cesium gives off blue **light**. These flame tests are useful for identifying the metals. Additionally, a striking use of sodium's characteristic emitted yellow light is in highly specialized lightbulbs, such as the very bright sodium vapor lights that appear along highways. In these bulbs, sodium **atoms** are excited with **electricity**, not a flame. Lightbulbs made with sodium use less electricity than conventional bulbs and are brighter, because the sodium gives off a larger percentage of its **energy** as light rather than heat.

Lithium

Lithium (Li) was discovered in 1817 by J.A. Arfvedson, but the free metal was not isolated until 1821, by W.T. Brande. It occurs naturally in small quantities (about 20 parts per million in the earth's crust), normally bound up with **aluminum** and silica in **minerals**. It is the smallest alkali metal, with an **atomic number** of 3 and an **atomic weight** of 6.94 amu (atomic **mass** unit). It

has a melting point of 356.9°F (180.5°C), and a **boiling point** of 2,457°F (1,347°C).

Lithium carbonate is well known for its ability to calm the mood swings of manic-depressive **psychosis**, a serious mental disorder. Industrially, however, it is used in lubricants, in batteries, in **glass** (to make it harder), and in alloys of **lead**, aluminum, and **magnesium** to make them less dense and stronger.

Sodium

Sodium (Na) is the second element in the alkali metal group, with an atomic number of 11 and an atomic weight of 22.9898 amu. Its melting point 208°F (97.8°C) and boiling point 1,621.4°F (883°C) are both lower than those of lithium, a trend that continues in the alkali metal group; as the atomic mass and size increase, the melting and boiling points decrease. Humphry Davy first isolated sodium metal by passing electricity through molten **sodium hydroxide** in 1807. It occurs naturally, in compound form, in relatively large amounts—about 20,000 parts per million in the earth's crust, plus a large **concentration** in seawater. **Sodium chloride** (or common **salt**) is one of the most common compounds on **Earth**, followed closely by **sodium carbonate** (also called soda ash or washing soda). Both of these are obtained now largely by **mining**.

Sodium compounds of various kinds are vital to industry. Sodium nitrite is a principle ingredient in gunpowder. The pulp and **paper** industry uses large amounts of sodium hydroxide, sodium carbonate, and sodium sulfate; the latter helps dissolve the lignin from wood pulp in the Kraft process so it can be made into cardboard and brown paper. In addition to paper pulping, sodium carbonate is used by power companies to absorb **sulfur dioxide**, a serious pollutant, from smokestack gases. Sodium carbonate is also important to the glass and detergent industries. Sodium hydroxide is one of the top 10 industrially produced chemicals, heavily used in manufacturing. Sodium chloride is used in foods, in water softeners, and as a de-icer for roads and sidewalks. **Sodium bicarbonate** (baking soda) is produced for the food industry as well.

Many useful chemicals and processes, particularly those of the "chlor-alkali" industry, can trace their production back to sodium chloride. Passing electricity through a concentrated salt water **solution** (the **electrolysis** of brine) produces sodium hydroxide and **chlorine** gas. Electrolysis of molten sodium chloride, on the other hand, yields elemental sodium and chlorine gas. Sodium sulfate is prepared in large quantities by reacting **sulfuric acid** with sodium chloride.

Biochemically, sodium is a vital nutrient, although excesses of it can aggravate high **blood pressure**. Sodium compounds regulate nerve transmission, alter **mem-**

brane permeability, and perform myriad other tasks for living organisms.

Potassium

Potassium (K), the third element in the alkali metal group, has an atomic number of 19 and an atomic mass of 39.0983 amu. Its melting point and boiling point are 145.9°F (63.28°C) and 1,398.2°F (759°C) respectively. Davy discovered and isolated potassium in 1807, by passing electricity through molten potassium hydroxide to obtain the free metal. Potassium is nearly as abundant as sodium in the earth's crust (21,000 parts per million). Much less potassium than sodium is present in seawater, however, partly because the **plant** life of the world absorbs potassium in large quantities. The chief minerals of potassium are sylvite, sylvinite, and carnallite.

Almost all the potassium used industrially goes into fertilizer, although small amounts of potassium hydroxide, potassium chlorate, and potassium bromide are important, respectively, in the detergent, explosive, and **photography** industries. Like sodium, potassium is a vital nutrient for organisms in a variety of ways.

Rubidium and cesium

Rubidium (Rb), the fourth element in the alkali metal group, has an atomic number of 37 and an atomic weight of 85.4678 amu. Its melting point is 102.8°F (39.31°C), and its boiling point is 1,270.4°F (688°C). Cesium (Cs), the second to last element in the group, has an atomic number of 55, an atomic weight of 132.9054 amu, a melting point of 83.12°F (28.40°C), and a boiling point of 1,239.8°F (671°C). Both were discovered in 1860-1861 by R.W. Bunsen and G.R. Kirchoff. They were the first two elements to be discovered with a **spectroscope**.

Both rubidium and cesium are rare in the earth's crust (rubidium at 90 parts per million and cesium only 3 parts per million.) The main mineral in which cesium can be found is pollucite. Rubidium can be found in pollucite, lepidolite, and carnallite. Rubidium is used almost exclusively for research, but cesium has some highly specialized industrial uses, including special glasses and **radiation** detection equipment.

Francium

Marguerite Perey discovered francium (Fr) in 1939, and named it after her homeland, France. Almost no francium occurs naturally on the earth, except very small amounts in **uranium** ores. Additionally, it is very radioactive, so the very tiny amounts produced by bombarding radium with neutrons are used almost exclusive-

ly for pure research. Presumably its **chemistry** resembles the other alkali metals, although much of that remains speculative.

Resources

Books

Emsley, John. *The Elements.* 3rd ed. New York: Oxford University Press, Inc., 1998.
Greenwood, N. N. and A. Earnshaw. *Chemistry of the Elements.* New York: Butterworth-Heinemann, 1997.

Gail B. C. Marsella

Alkaline earth metals

The second column on the **periodic table** of the chemical elements is collectively called the alkaline earth **metal** group: beryllium, **magnesium**, **calcium**, strontium, **barium**, and radium. Because the outer **electron** structure in all of these elements is similar, they all have somewhat similar chemical and physical properties. All are shiny, fairly soft—although harder than the alkali metals—and most are white or silvery colored. The steady increase in melting and boiling points with increasing molecular **mass** noticed in the alkali metal group is less pronounced in the alkaline earths; beryllium has the highest rather than the lowest melting point, for example, and the other metals do not follow a consistent pattern. All the alkaline earth metals react with **water** to produce **hydrogen** gas and the metal hydroxide, although somewhat less vigorously than the **alkali metals**. Magnesium metal can be set on fire, and burns with an extremely intense white **light** as it combines with **oxygen** in the air to form magnesium oxide. Strontium, barium, and calcium react readily with oxygen in the air to form their oxides.

Alkaline earth metals almost always form ions with a positive (+2) charge, and are sufficiently reactive as elements so they usually occur in nature only in compound form, frequently as carbonates or sulfates. Calcium is by far the most abundant, followed by magnesium, and then in much lesser amounts barium, strontium, and beryllium. Radium is radioactive and fairly rare; what exists in the earth's crust occurs almost exclusively in **uranium** deposits.

Several of the alkaline earth metals glow with a characteristic **color** when placed in a flame; calcium gives off an orange light, strontium a very bright red, and barium an apple green. These flame tests are useful for identifying the metals. Strontium compounds are often added to fireworks displays to obtain the most vivid reds.

Beryllium

Beryllium (Be), the smallest alkaline earth metal, is **atomic number** 4 on the periodic table, has an **atomic weight** of 9.01 amu (atomic mass unit), and melting and boiling points of 2,348.6°F (1,287°C) and about 4,479.8°F (2,471°C), respectively. It was discovered by N. L. Vauquelin in 1797 after a mineralogist named R. J. Hauy noticed that emeralds and beryl possessed many similar properties and might be identical substances, but it was not until 30 years later that the free metal was isolated (independently) by F. Wohler and A. Bussy. It occurs naturally in the precious stones emerald and aquamarine, which are both forms of the mineral beryl, a beryllium aluminosilicate compound.

Beryllium has no biochemical function and is extremely toxic to human beings, but small amounts (about 2%) impart superior characteristics, such as high strength, wear resistance, and **temperature** stability, to alloys. Copper-beryllium alloys make good **hand tools** in industries that use flammable solvents, because the tools do not cause sparks when struck against other objects. Nickel-beryllium alloys are used for specialized electrical connections and various high temperature applications. Beryllium is used instead of **glass** in x-ray tubes because it lets through more of the x-ray **radiation** than glass would.

Magnesium

Magnesium (Mg) is atomic number 12, has an atomic weight of 24.31 amu, and has melting and boiling points of 1,202°F (650°C) and 1,994°F (1,090°C), respectively. It was isolated (as were many other alkali and alkaline earth metals) by British chemist Humphry Davy in 1808, although its existence had been known since 1755. Magnesium, like calcium, is one of the most common elements, existing at about 23,000 parts per million in the earth's crust. Its most common mineral forms in nature are dolomite and magnesite (both carbonates), and carnallite (a chloride); relatively large amounts (about 1,200 parts per million) are also present in seawater. Asbestos is a magnesium silicate mineral, as are soapstone (or talc) and mica.

Magnesium performs a critical role in living things because it is a component of **chlorophyll**, the green pigment that captures sunlight **energy** for storage in **plant** sugars during **photosynthesis**. Chlorophyll is a large **molecule** called a porphyrin; the magnesium occupies the center of the porphyrin molecule. (In the **animal** kingdom, a similar porphyrin called heme allows hemoglobin to transport oxygen around in the bloodstream; in heme's case, however, **iron** rather than magnesium occupies the central place in the porphyrin.) Elemental magnesium is a strong, light metal, particularly when alloyed with other metals like **aluminum** or zinc, and has many

uses in construction; airplane parts are often made of magnesium alloys. Some of magnesium's rare earth alloys are so temperature resistant that they are used to make car engine parts. In organic **chemistry**, magnesium combines with alkyl halides to give Grignard reagents, vitally important in organic chemical synthesis because of their versatility. Almost any type of organic compound can be prepared by the proper selection, preparation, and reaction of a Grignard reagent.

Calcium

Calcium (Ca) is atomic number 20, has an atomic weight of 40.08 amu, and has melting and boiling points of 1,547.6 ±3°F (842 ±2°C) and 2,703.2°F (1,484°C), respectively. Davy isolated it in 1808 by electrolytic methods. It is the third most common metal on Earth, exceeded only by iron and aluminum, and the fifth most common element (41,000 parts per million in the earth's crust, and about 400 parts per million in seawater). The principle sources of calcium are limestone and dolomite (both carbonates), and gypsum (the sulfate). Other natural materials made of **calcium carbonate** include coral, chalk, and marble.

Calcium is an essential nutrient for living things, and its compounds find use in myriad industries. Both limestone and gypsum have been used in building materials since ancient times; in general, gypsum was used in drier climates. Marble is also a good building material. Limestone and dolomite are the principle sources of slaked lime (calcium hydroxide) and quick lime (**calcium oxide**) for the **steel**, glass, **paper**, dairy, and metallurgical industries. Lime can act as a flux to remove impurities from steel, as a neutralizing agent for acidic industrial waste, as a reagent for reclaiming **sodium hydroxide** from paper pulping waste, and as a "scrubbing" compound to remove pollutants from smokestack effluent. The paper industry uses calcium carbonate as an additive to give smoothness and opacity to the finished paper, and the food, cosmetic, and pharmaceutical industries use it in antacids, toothpaste, chewing gum, and vitamins.

Strontium

Strontium (Sr) is atomic number 38, has an atomic weight of 87.62 amu, and has melting and boiling points of 1,430.6°F (777°C) and 2,519.6°F (1,382°C), respectively. A. Crawford identified it as an element in 1790, and in 1808, Davy produced it as the free metal. It occurs in the earth's crust at about 370 parts per million, mostly as the **minerals** celestite (the sulfate) and strontianite (the carbonate), and occurs in the **ocean** at about eight parts per million. In addition to its use in fireworks, strontium is a glass additive.

Barium

Barium (Ba) is atomic number 56, has an atomic weight of 137.27 amu, and has melting and boiling points of 1,340.6°F (727°C) and about 3,446.6°F (1,897°C), respectively. It was isolated in 1808 by Davy, using a variation of his usual electrolytic process. Barite (the sulfate) is the main **ore** from which barium can be obtained, although witherite (the carbonate) was at one time also mined. It is not particularly plentiful, occurring in about 500 parts per million in the earth's crust, and on the average about 10 parts per billion in seawater. By far, most of the **barium sulfate** mined is used to make a sort of lubricating mud used in well-drilling operations, although small amounts of barium are alloyed with nickel for specialized uses, and some barium is used in medicine. Barium itself is toxic and has no biochemical function in living things.

Radium

Radium (Ra) is atomic number 88, has an atomic weight of 226 amu, and has melting and boiling points of about 1,292°F (700°C) and about 2,084°F (1,140°C), respectively. It was discovered by Marie and Pierre Curie in 1898; they extracted a small quantity as radium chloride by processing tons of the uranium ore called pitchblende. Radium exists in the earth's crust in only about 0.6 parts per trillion, and almost none can be found in seawater. All of the isotopes of radium are radioactive, and consequently hazardous to living things. It was formerly used in medicine to treat various kinds of **cancer** and other conditions, but its use has declined as safer radioisotopes have been discovered. Radium was also used to paint the luminous numbers on watch dials, but that use has been stopped for safety reasons.

Resources

Books

Emsley, John. *Nature's Building Blocks: An A-Z Guide to the Elements.* Oxford: Oxford University Press, 2002.

Greenwood, N. N. and A. Earnshaw. *Chemistry of the Elements.* New York: Butterworth-Heinemann, 1997.

Gail B. C. Marsella

Alkaloid

Alkaloids are chemical compounds found in plants that can react with acids to form salts. All alkaloids contain the element **nitrogen**, usually in complex, multi-ring structures.

Role in the plant

Between 10% and 15% of all plants contain some type of alkaloid. It is unclear why alkaloids are so common, and it is a matter of controversy among scientists. Some believe that plants rid themselves of excess nitrogen through the production of alkaloids just as humans and other **mammals** convert excess nitrogen into **urea** to be passed in the urine. Some modify this theory by suggesting that plants use alkaloids to temporarily store nitrogen for later use, instead of discarding altogether this difficult-to-obtain element.

Perhaps the most likely theory is that the presence of alkaloids discourages **insects** and animals from eating plants. The poisonous nature of most alkaloids supports this theory, although various alkaloids that are employed in small quantities for specific purposes can be useful to man.

Role in animals

Many alkaloids act by blocking or intensifying the actions of neurotransmitters, chemicals released by nerve cells in response to an electrical impulse called a neural signal. Neurotransmitters diffuse into neighboring cells where they produce an appropriate response, such as an electrical impulse, in another nerve **cell** or contraction in a muscle cell.

Each nerve cell produces only one type of **neurotransmitter**; **acetylcholine** and norepinephrine are the most common. Cells may respond to more than one type of neurotransmitter, however, and the response to each type may be different.

Medical use

A number of alkaloids are used as drugs. Among the oldest and best known of these is **quinine**, derived from the **bark** of the tropical cinchona **tree**. Indians of **South America** have long used cinchona bark to reduce fever, much as willow bark was used in **Europe** as a source of aspirin. In the 1600s Europeans discovered that the bark could actually cure malaria—one of the most debilitating and fatal diseases of tropical and subtropical regions.

Quinine was purified as early as 1823, and soon it replaced crude cinchona bark as the standard treatment for **malaria**. Not until the 1930s was quinine replaced by synthetic analogues that offered fewer side effects and a more reliable supply. Quinine is still used as the principal flavoring agent in tonic water—a beverage named for its ability to prevent malarial symptoms.

Cinchona bark also produces quinidine. It is used primarily to control abnormalities of **heart** rhythm such as fibrillation, a series of rapidly quivering beats that do not pump any **blood**, and heart block, a condition in which electrical currents fail to coordinate the contractions of the upper and lower chambers of the heart.

Vincaleukoblastine and vincristine, two alkaloids derived from the periwinkle **plant** (*Catharanthus roseus*), are used effectively for the treatment of white-blood-cell cancers. Vincaleukoblastine is especially useful against lymphoma (**cancer** of the lymph **glands**), while vincristine is used against the most common form of childhood **leukemia**.

Atropine is an alkaloid produced by several plants, including deadly **nightshade** (*Atropa belladonna*), Jimson weed (*Datura stramonium*), and henbane (*Hyoscyamus niger*). It has a variety of medical uses, as it is able to relax smooth muscle by blocking action of the neurotransmitter acetylcholine. Atropine is most commonly used to dilate the pupil during **eye** examinations. Atropine also relieves nasal congestion and serves as an antidote to nerve gas and insecticide poisoning.

Pilocarpine, derived from several Brazilian shrubs of the genus *Pilocarpus*, is another alkaloid used in ophthalmology, the medical specialty that treats the eye. This drug stimulates the drainage of excess fluid from the eyeball, relieving the high **pressure** in the eye caused by glaucoma. If untreated, glaucoma can lead to blindness.

Introduction of reserpine in the 1950s revolutionized high blood pressure treatment and brought new hope to those suffering from this previously untreatable and life-threatening condition. Derived from tropical trees and shrubs of the genus *Aauwolfia*, reserpine works by depleting the body's stores of the neurotransmitter norepinephrine. Among its other functions, norepinephrine contracts the **arteries** and thereby contributes to high blood pressure.

Unfortunately, reserpine also causes drowsiness and sometimes severe **depression**. Medications without these side effects have been developed in recent decades, and reserpine is rarely used.

Alkaloids for pain and pleasure

Many medically useful alkaloids act by way of the peripheral **nervous system**; others work directly on the **brain**. Prominent among the latter are the **pain** relievers **morphine** and **codeine**, derived from the opium poppy (*Papaver somniferum*). Morphine is the stronger of the two, but codeine is often prescribed for moderate pain. Codeine is also an effective cough suppressant; for years it was a standard component of cough syrups. Now, however, it has been replaced for the most part by drugs that do not have the psychological side effects of codeine.

Both morphine and codeine are addictive drugs that produce a state of relaxed, dreamy euphoria—an exag-

gerated state of "feeling good," referred to by drug addicts as a "high." The equivalent street drug is heroin, derived from morphine by a simple chemical modification. Heroin addicts typically believe that their drug is stronger and produces a more pronounced "high" than morphine; however, since heroin is rapidly converted to morphine once it enters the body, most medical scientists consider the two drugs completely equivalent.

The effects of **cocaine** are almost the opposite of those of morphine; cocaine's legal classification as a narcotic—a drug that produces stupor—is misleading from a medical standpoint. This product of the **coca** plant, native to the Andes Mountains in South America, produces a state of euphoric hyperarousal. The user feels excited, elated, and intensely aware of his or her surroundings, with an impression of enhanced physical strength and mental ability. These feelings are accompanied by the physical signs of arousal: elevated heart **rate** and blood pressure. The increased heart rate caused by a high dose may lead to fibrillation and death.

A cocaine "high," unlike the "highs" from most abused drugs, lasts less than half an hour—often much less. Once an **individual** becomes addicted, he or she needs large amounts of the expensive drug. Users' enhanced aggressiveness and physical self-confidence further increase cocaine's social dangers. Cocaine usage over time can result in paranoid **schizophrenia**, a type of insanity characterized by unfounded suspicion and fantasies of persecution; when this psychological condition is combined with continued cocaine use, the addict may perform violent acts against the supposed plotters.

Cocaine is also a local anesthetic, and was used medically for that purpose in the early part of the century. Procaine and xylocaine, synthetic local anesthetics introduced during the mid-1900s, have replaced cocaine as a medical drug.

Another pleasurable yet addictive drug is **nicotine**, usually obtained by either smoking or chewing leaves of the tobacco plant, *Nicotiana tabacum*. Ground-up leaves, known as snuff, may be placed in the nose or cheek, allowing the nicotine to diffuse through the linings of the cavities into the bloodstream.

With the possible exception of **alcohol**, nicotine is the world's most widely used addictive drug. Its attractiveness undoubtedly results from the drug's paradoxical combination of calming and stimulating properties—it can produce either relaxation or arousal, depending on the user's state. Its physical effects, however, are primarily stimulatory. By increasing the heart rate and blood pressure while constricting the arteries—including those in the heart—nicotine significantly increases the risk of a heart attack.

Some alkaloid stimulants are not addictive, however. These include **caffeine** and the related compounds theophylline and theobromine. Caffeine is found in coffee, made from beans of *Coffea arabica*; in tea, from leaves of *Camellia sinensis*; in cocoa and chocolate, from **seeds** of *Theobroma cacao*; and in cola drinks, which contain flavorings derived from nuts of *Cola* plants. In northern Argentina and southern Brazil, leaves of *Ilex paraguariensis* (a type of holly) are used to make maté, a drink more popular there than either coffee or tea.

In addition to caffeine, tea also contains small amounts of theophylline, while theobromine is the major stimulant in cocoa. Large amounts of coffee, tea, cocoa, or cola drinks (more than 6-12 cups of coffee a day, for example), can produce nervousness, shakiness (muscle tremors), and **insomnia**, and may increase the risk of heart attack. Adverse effects from smaller amounts of these beverages have been claimed but never clearly demonstrated.

Black **pepper** falls into an entirely different category of the pain/pleasure grouping. This spice derives its burning flavor primarily from the alkaloids piperine, piperidine, and chavicine.

Addiction

Addiction was originally defined by the appearance of physical symptoms—such as sweating, sniffling, and trembling—when a drug was withdrawn from an addicted person or **animal**. It was also thought that addiction was accompanied by **adaptation**, in which more and more of the drug is required to produce the same effect.

So long as the focus was on opium-derived drugs such as morphine and heroin, this definition was appropriate. Beginning in the 1960s, however, scientists realized that it did not define the properties that render cocaine and other drugs so dangerous. Cocaine does not produce adaptation, and its withdrawal does not result in physical symptoms that can be seen in laboratory animals. Cocaine withdrawal does, however, produce an intense depression that disappears when the drug is again available. Similarly, nicotine withdrawal produces psychological symptoms such as restlessness, **anxiety**, irritability, difficulty in concentrating, and a craving for the drug. Today these psychological withdrawal symptoms are recognized as valid indicators of addiction.

Poisonous alkaloids

Nearly all of the alkaloids mentioned so far are poisonous in large amounts. Some alkaloids, however, are almost solely known as poisons. One of these is strychnine, derived from the small Hawaiian tree *Strychnos nux-vomica*. Symptoms of strychnine poisoning begin with feelings of restlessness and anxiety, proceeding to muscle twitching and exaggerated reflexes. In severe poisoning, a loud sound can cause severe muscle spasms throughout the entire body. These spasms may make breathing impossible and result in death.

In the early part of the twentieth century, strychnine was widely used as a rat poison. In recent decades, however, slower-acting poisons have been used for rodent control; since **rats** can remember which foods have made them sick, one that receives a non-fatal dose of a fast-acting poison such as strychnine will never again take that type of poisonous bait.

A number of other plants also derive their lethal properties from alkaloids of one type or another. Among these are poison hemlock (*Conium maculatum*) and plants of the genus *Aconitum*, commonly called monkshood—known by devotees of werewolf stories as wolfsbane. Other examples include shrubs of the genus *Calycanthus*, known as Carolina allspice, spicebush, and sweet betty, among other names; vines of the genus *Solandra*, such as the chalice vine, cup-of-gold, silver cup, and trumpet plant; trees or shrubs of the genus *Taxus*, such as yews; plants of the lily-like genus *Veratrum*, including false hellebore; and the golden chain or bean tree (*Laburnum anagyroides*).

Although vines and non-woody plants of the genus *Solanum* pose a real danger only to children, they represent an extremely varied group. Poisonous members of this genus range from the common **potato** to the nightshade (not the same as the deadly nightshade that produces atropine). The group also includes the Jerusalem cherry, the false Jerusalem cherry, the love apple, the Carolina horse nettle, the bittersweet, the nipplefruit, the star-potato vine, and the apple of Sodom.

Poisoning by ordinary potatoes usually results from eating uncooked sprouts or sun-greened skin. These parts should be cut away and discarded. For other plants in the group, it is the immature fruit that is most likely to be poisonous. In contrast to the nervous-system effects of most alkaloids, the alkaloids found in the genus *Solanum* produce mainly fever and diarrhea.

See also Antipsychotic drugs; *Nux vomica* tree; Poisons and toxins.

Resources

Books

Lampe, Kenneth F., and Mary Ann McCann. *AMA Handbook of Poisonous & Injurious Plants*. Chicago: Chicago Review Press, 1985.

Pelletier, S. W. *Alkaloids: Chemical and Biological Perspectives*. New York: John Wiley & Sons, 1983.

W. A. Thomasson

Alkane *see* **Hydrocarbon**

Alkene *see* **Hydrocarbon**

Alkyl group

An alkyl group is a paraffinic **hydrocarbon** group that may be derived from an alkane by dropping one **hydrogen** from the structure. Such groups are often represented in chemical formulas by the letter R and have the generic name C_nH_{2n+1}.

Alkanes

Aliphatic compounds, of which the alkanes are one example, have an open chain of **carbon atoms** as a skeleton. This open chain may be straight or branched. Alkanes, also known as paraffins, are composed of carbon and hydrogen only; they have the generic formula C_nH_{2n+2}, and are the simplest and least reactive of the aliphatic compounds. Alkanes with straight chains are known as normal alkanes. (Branched chain alkanes are treated as alkyl derivatives of the straight chain compounds.) The first four members of the normal alkane series are methane, ethane, propane, and butane (see below). The names of the remaining normal alkanes are composed of a prefix that indicates the number of carbon atoms in the compound, followed by the termination -ane. Thus, n-hexane is the name given to the normal alkane having a chain of six carbon atoms.

The n-alkanes exist on a continuum that extends from simple gases to molecules of very high molecular weights. The physical properties and uses of the n-alkane change with the number of repeating CH_2 units in the chain. For example, compounds with one to four CH_2 units in the chain are simple gases (e.g. cooking gas) at room **temperature**; compounds with five to 11 CH_2 units in the chain are simple liquids (e.g. gasoline) at room temperature; compounds with 16 to 25 CH_2 units in the chain are high **viscosity** liquids (e.g. oil) at room temperature; and compounds with 1,000 to 3,000 CH_2 units in the chain are tough plastic solids (e.g. polyethylene bottles) at room temperature.

Alkyl radicals

Alkanes from which one atom of hydrogen has been removed become monovalent radicals. These radicals, which are molecular fragments having an unpaired **electron**, are known as alkyl groups. The names of the alkyl groups are formed by substituting the suffix -yl for -ane in the names of the alkanes from which they are derived.

The **methyl group** (CH_3^-) is formed from methane, CH_4. The **ethyl group**, $C_2H_5^-$, is formed from ethane, C_2H_6. Two different alkyl groups can be formed from propane, $CH_3CH_2CH_3$. Removal of a hydrogen atom from one of the carbon atoms at the end of the chain forms $C_3H_7^-$. This $CH_3CH_2CH_2^-$ group is called a normal **propyl group** (n-propyl group). Removing a hydrogen from the second carbon produces an isopropyl group (i-propyl group).

The next member of the alkanes has the formula C_4H_{10}. There are four isomers of this **molecular formula**. Removal of a hydrogen from one of the end carbons in n-butane, $CH_3CH_2CH_2CH_3$, produces the n-butyl group ($CH_3CH_2CH_2CH_2^-$). Removing a hydrogen atom from carbon 2 or 3 produces the secondary **butyl group** (sec-butyl group; $CH_3CH_2CH(CH_3)^-$). Removal of a hydrogen from carbon atoms 1, 3, or 4 forms the isobutyl group (i-butyl group; $(CH_3)_2CHCH_2^-$). Finally, removing a hydrogen from carbon 2 gives the tertiary butyl group (t-butyl group; $(CH_3)_3C^-$).

Frequently the terms primary, secondary, and tertiary are used to describe carbon atoms in a **molecule**. A primary carbon atom is one that is attached to only one other carbon atom; a secondary carbon atom is attached to two other carbon atoms; and a tertiary carbon atom is attached to three other carbon atoms in the molecule. Thus the ethyl group defined above is a primary group, since the carbon atom that has lost the hydrogen is attached to only one other carbon atom in the molecule. The n-propyl group is also a primary group for the same reason. But the i-propyl group is a secondary group, because the central carbon atom from which the hydrogen has been removed is attached to two other carbon atoms. The n-butyl and i-butyl groups are both primary groups; the secondary butyl group is a secondary group; and the tertiary butyl group is tertiary, because the central atom is attached to the three other carbon atoms in the molecule. The methyl group, however, cannot be defined using this classification scheme.

In general, the letter R is used to designate any alkyl group (R = CH_3, C_2H_5, etc.). With this convention, the alkanes are represented by the general formula R-H, and the alkyl halides by R-X, where X is a halogen.

Alkenes

The series of compounds derived from the alkanes by removing one hydrogen atom from each of two adjacent carbon atoms, thereby introducing a double bond into the molecule, bears the name olefin. The systematic names are formed by substituting the suffix -ene for -ane in the name of the alkane from which they are derived. Thus the series as a whole is called the alkenes. Some common alkenes are methylene ($-CH_2^-$), ethylene ($CH_2=CH_2$), and propylene ($CH_3CH=CH_2$).

Resources

Books

Loudon, G. Mark. *Organic Chemistry.* Oxford: Oxford University Press, 2002.

Sperling, L.H. *Introduction to Physical Polymer Science.* 3rd ed. New York: John Wiley and Sons, 2001.

Randall Frost

Alleles

Most genes exist in more than one form that, when expressed, result in different characteristics. Genes may often exist in more than one form, and these forms are termed alleles of the **gene**.

An allele is one of at least two alternative forms of a particular gene. Alleles provide the genetic instructions for products that, although similar in type, are visibly different (phenotypically different). The term allele is derived the Greek term *alleon* used to describe a difference in morphology or form. At the genetic level, alleles contain differing base sequences in their **nucleic acid** (e.g., **DNA**). As a form of a gene, an allele carries the instructions for a particular variation of the gene's protein product.

Although underlying genetic molecular complexities sometimes blur the differences in expression; nature also provides simple examples of alleles. For example, a single gene may control the **flower color** of some plants. In such cases, one allele of the gene produces one color (e.g., red flowers) while another allele of flower color gene may produce another color flower (e.g., white flowers).

Alleles reside at corresponding locations on the chromosomes that constitute a chromosomal pair. Because alleles reside in specific regions of chromosomes, they can act as markers and are subject to the laws of inheritance resulting from the apportionment of homolo-

gous chromosomes (chromosomes that match in terms of size, shape, and gene content) during **meiosis**. The alternative alleles that comprise an organism's **genome** are inherited, one allele from each parent. The allele contained on the homologous **chromosome** derived from the mother is termed the material allele. The allele located on the homologous chromosome derived from the father is termed the paternal allele.

All diploid organisms have two alleles at a given locus on a pair of homologous chromosomes. Because haploid cells (e.g., oocyte and spermatozoa in humans) contain half the chromosome compliment, such cells contain only one allele of each gene. When the diploid condition is restored following **fertilization** the pair of alleles can be described as homologous (alike) or heterologous (different). Accordingly, organisms that are homologous with respect to the alleles for a particular gene carry identical alleles for that gene. In contrast, organisms described as heterozygous for a particular gene carry alleles that differ. Alleles may also be dominant or recessive with respect to their interaction and expression.

At its most basic molecular level, an allele is an ordered sequence of bases (part of nucleotides) that code for a specific genetic product (protein, **enzyme**, **RNA molecule**, etc.).

A population with stable allele frequencies is in genetic equilibrium. Accordingly, changes in allele frequencies (the percentage of respective alleles in a population) are characteristic indicators of evolving populations. The Hardy-Weinberg **theorem** states that, in the absence of **selection** pressures, the types and frequencies of alleles in a population remain constant. The Hardy-Weinberg equation can be used to mathematically predict allele frequencies.

See also Adaptation; Chromosome mapping; Codons; Genetic engineering.

Allergy

An allergy is an excessive or hypersensitive response of the **immune system**. The allergic reaction becomes manifest as a pathological immune reaction induced either by antibodies (immediate hypersensitivity) or by lymphoid cells (delayed type allergy). Instead of fighting off a disease-causing foreign substance, the immune system launches a complex series of actions against an irritating substance, referred to as an allergen. The symptoms of an immediate hypersensitivity begin shortly after contact and decay rapidly, while the delayed type symptoms do not reach their maximum for 24–48

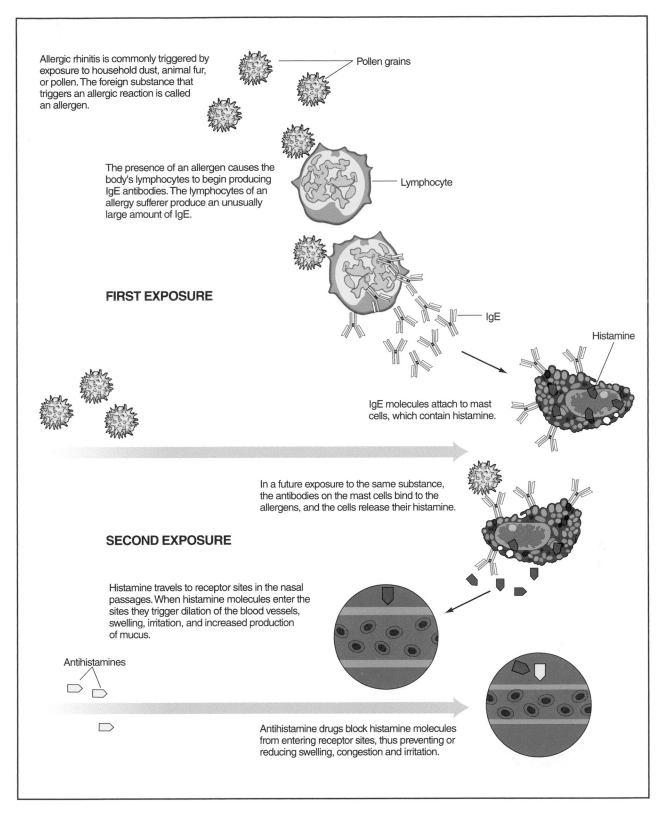

Allergic rhinitis is commonly triggered by exposure to household dust, animal fur, or pollen. The foreign substance that triggers an allergic reaction is called an allergen.

Pollen grains

The presence of an allergen causes the body's lymphocytes to begin producing IgE antibodies. The lymphocytes of an allergy sufferer produce an unusually large amount of IgE.

Lymphocyte

FIRST EXPOSURE

IgE

Histamine

IgE molecules attach to mast cells, which contain histamine.

In a future exposure to the same substance, the antibodies on the mast cells bind to the allergens, and the cells release their histamine.

SECOND EXPOSURE

Histamine travels to receptor sites in the nasal passages. When histamine molecules enter the sites they trigger dilation of the blood vessels, swelling, irritation, and increased production of mucus.

Antihistamines

Antihistamine drugs block histamine molecules from entering receptor sites, thus preventing or reducing swelling, congestion and irritation.

The allergic response. *Illustration by Hans & Cassidy. Courtesy of Gale Group.*

hours and decline slowly over a period of days or weeks. An allergic reaction may be accompanied by a number of stressful symptoms, ranging from mild to severe to life threatening. In rare cases, an allergic reaction can lead to anaphylactic shock—a condition characterized by a sudden drop in **blood pressure**, difficulty in breathing, skin irritation, collapse, and possible death.

The immune system may produce several chemical agents that cause allergic reactions. The main group of immune system substances responsible for the symptoms of allergy includes the histamines, which are produced after an exposure to an allergen. Along with other treatments and medicines, the use of **antihistamines** helps to relieve some of the symptoms of allergy by blocking out **histamine** receptor sites. The study of allergy medicine includes the identification of the different types of allergy, **immunology**, and the **diagnosis** and treatment of allergy.

Types of allergy

The most common cause of allergy is pollens that are responsible for seasonal or allergic rhinitis. The popular name for rhinitis, hay fever, a term used since the 1830s, is inaccurate because the condition is not characterized by fever. Throughout the world during every season, pollens from **grasses**, trees, and weeds affect certain individuals, producing allergic reactions like sneezing, runny nose, swollen nasal tissues, headaches, blocked sinuses, and watery, irritated eyes. Of the 46 million allergy sufferers in the United States, about 25 million have rhinitis.

Dust and the house dust mite constitute another major cause of allergies. While the mite itself is too large to be inhaled, its feces are about the size of pollen grains and can lead to allergic rhinitis. Other types of allergy can be traced to the fur of animals and pets, food, drugs, insect bites, and skin contact with chemical substances or odors. In the United States there are about 12 million people who are allergic to a variety of chemicals. In some cases an allergic reaction to an insect sting or a drug reaction can cause sudden death. Serious **asthma** attacks are associated with seasonal rhinitis and other allergies. About nine million people in the United States suffer from asthma.

Role of immune system

Some people are allergic to a wide range of allergens, while others are allergic to only a few or none. The reasons for these differences can be found in the makeup of an individual's immune system. The immune system is the body's defense against substances (antigens) that it recognizes as foreign to the body. Lymphocytes, a type of white blood **cell**, produce antibodies, which play an important role in neutralizing antigens, including **bacteria**, viruses and toxins. Those antigens specifically producing an allergic response are known as allergens When an allergen first enters the body, the lymphocytes produce an antibody called immunoglobulin E (IgE). The IgE antibodies attach to mast cells, large cells that are found in **connective tissue** and contain histamines along with a number of other chemical substances.

Studies show that allergy sufferers produce an excessive amount of IgE, indicating a hereditary factor for their allergic responses. How individuals adjust over time to allergens in their environments also determines their degree of susceptibility to allergic disorders.

The second time any given allergen enters the body, it becomes attached to the newly-formed Y-shaped IgE antibodies. These antibodies, in turn, stimulate the mast cells to discharge its histamines and other anti-allergen substances. There are two types of histamine: H_1 and H_2. H_1 histamines travel to receptor sites located in the nasal passages, **respiratory system**, and skin, dilating smaller blood vessels and constricting airways. The H_2 histamines, which constrict the larger blood vessels, travel to the receptor sites found in the salivary and tear **glands** and in the stomach's mucosal lining. H_2 histamines play a role in stimulating the release of stomach acid, thus creating a seasonal stomach ulcer condition.

Diagnosis and treatment

The patient's medical history provides the primary basis for the diagnosis of allergies. Skin patch tests are sometimes used to determine exactly which potential allergens can produce a reaction in the patient. A group of substances are placed in patches under the skin; any actual allergen will raise a weal (a red, circular swelling) at the site of the patch. The weal is a circular area of swelling that itches and is reddened. The tests at times produce false positives, so that they cannot be relied on exclusively.

The simplest form of treatment is the avoidance of the allergic substance, but that is not always possible. In such cases, desensitization to the allergen is sometimes attempted by exposing the patient to slight amounts of the allergen at regular intervals.

Antihistamines, which are now prescribed and sold over the counter as a rhinitis remedy, were discovered in the 1940s. There are a number of different ones, and they either inhibit the production of histamine or block them at receptor sites. After the administration of antihistamines, IgE receptor sites on the mast cells are blocked, thereby preventing the release of the histamines that cause the allergic reactions. The allergens are still there, but the body's "protective" actions are suspended for the

KEY TERMS

Allergen—An otherwise harmless substance that can cause a hypersensitive allergic response.

Anaphylactic shock—A violent, sometimes fatal, response to an allergen after initial contact.

Corticosteroids—Drugs that stimulate the adrenal gland and are highly effective in treating asthma and allergies but also have many side effects.

Decongestants—Drugs used for a short term to reduce mucous membrane congestion.

Histamines—A chemical released from cells in the immune system as part of an allergic reaction.

IgE—The chief immunoglobulin responsible for producing the compounds that cause allergic reactions.

Lymphocyte—A white blood cell that stimulates antibody formation.

Mast cell—A tissue white blood cell located at the site of small blood vessels.

Receptor sites—Places on the mast cell where Y-shaped antibodies stimulate histamine production.

Rhinitis—The common condition of upper respiratory tract inflammation occurring in both colds and allergy.

Weal—The reddened, itchy swelling caused by a skin patch test.

period of time that the antihistamines are active. Antihistamines also constrict the smaller blood vessels and **capillaries**, thereby removing excess fluids. Recent research has identified specific receptor sites on the mast cells for the IgE. This knowledge makes it possible to develop medicines that will be more effective in reducing the symptoms of various allergies.

Corticosteroids are sometimes prescribed to allergy sufferers as anti-inflammatories. Decongestants can also bring relief, but these can be used for a short time only, since their continued use can set up a rebound effect and intensify the allergic reaction.

See also Antibody and antigen.

Resources

Books

Adelman, Daniel C., and Thomas B. Casale. *Manual of Allergy and Immunology: Diagnosis and Therapy.* 4th ed. Philadelphia: Lippincott Williams & Wilkins Publishers, 2002.
Joneja, Janice Vickerstaff, and Leonard Bielory. *Understanding Allergy, Sensitivity, & Immunity.* New Brunswick, NJ: Rutgers University Press, 1990.
Steinman, Marion. *A Parent's Guide to Allergies and Asthma.* New York: Dell Publishing, 1992.

Organizations

American College of Allergy, Asthma & Immunology, 85 West Algonquin Rd., Suite 550, Arlington Heights, IL 60005. 1–800–842–7777. <http://allergy.mcg.edu> (October 19, 2002).

Other

National Institutes of Health. "The National Institue of Allergy and Infectious Diseases" <http://www.niaid.nih.gov/default.htm> (October 19, 2002).

Jordan P. Richman

Alligators *see* **Crocodiles**

Allotrope

Allotropes are two or more forms of the same element in the same physical state (solid, liquid, or gas) that differ from each other in their physical, and sometimes chemical, properties. The most notable examples of allotropes are found in groups 14, 15, and 16 of the **periodic table**. Gaseous **oxygen**, for example, exists in three allotropic forms: monatomic oxygen (O), a diatomic **molecule** (O_2), and in a triatomic molecule known as **ozone** (O_3).

A striking example of differing physical properties among allotropes is the case of **carbon**. Solid carbon exists in two allotropic forms: **diamond** and graphite. Diamond is the hardest naturally occurring substance and has the highest melting point (more than 6,335°F [3,502°C]) of any element. In contrast, graphite is a very soft material, the substance from which the "lead" in lead pencils is made.

The allotropes of **phosphorus** illustrate the variations in chemical properties that may occur among such forms. White phosphorus, for example, is a waxy white solid that bursts into flame spontaneously when exposed to air. It is also highly toxic. On the other hand, a second allotrope of phosphorus known as red phosphorus is far more stable, does not react with air, and is essentially nontoxic.

Allotropes differ from each other structurally depending on the number of **atoms** in a molecule of the element. There are allotropes of **sulfur**, for example, that contain 2, 6, 7, 8, 10, 12, 18, and 20 atoms per molecule (formulas S_2 to S_{20}). Several of these, however, are not very stable.

The term allotrope was first suggested by Swedish chemist J. J. Berzelius (1779-1848). He took the name from the Greek term *allotropos*, meaning other way. Berzelius was unable to explain the structure of allotropes, however. The first step in that direction was accomplished by British father and son crystallographers W. H. and L.W. Bragg in 1914. The Braggs used x-ray **diffraction** to show that diamond and graphite differ from each other in their atomic structure.

Alloy

A mixture of two or more metals is called an alloy. Alloys are distinguished from composite metals in that alloys are thoroughly mixed, creating, in effect, a synthetic **metal**. In metal composites, the introduced metal retains its identity within the matrix in the form of fibers, beads, or other shapes.

Alloys can be created by mixing the metals while in a molten state or by bonding metal powders. Various alloys have different desired properties such as strength, visual attractiveness, or malleability. The number of possible alloy combinations is almost endless since any metal can be alloyed in pairs or in multiples.

An entire period of human prehistory is named for the earliest known alloy—bronze. During the Bronze age (c.3500-1000 B.C.) humans first fashioned tools and weapons from something other than basic materials found in nature. Humans combined **copper** and tin to form a strong metal that was still easily malleable. Modern bronze contains a 25:75 **ratio** of tin to copper. The use of bronze in early times was greatest in nations where tin deposits were most plentiful, like Asia Minor, and among countries that traded with tin-mining nations.

Brass is an alloy of copper and zinc. It is valued for its light weight and rigid strength. It has a ratio of about one-third zinc to two-thirds copper. The exact ratio of metals determines the qualities of the alloy. For example, brass, having less than 63% copper, must be heated to be worked. Brass is noted for its beauty when polished. Brass was perhaps first produced in Palestine from 1400 to 1200 B.C. It was later used by the Romans for coins. Many references to brass in the Bible and other ancient documents are really mistranslations of mentions of bronze.

Pewter is an alloy of copper, tin, and antimony. It is a very soft mixture that can be worked when cold and beat repeatedly without becoming brittle. It was used in Roman times, but its greatest period of popularity began in England in the fourteenth century and continued into the eighteenth. Colonial American metalworks produced some notable pewter work. As a cheaper version of silver, it was used in plates, cups, pitchers, and candelabras.

The various types of **steel** and **iron** are all alloys classifiable by their content of other materials. For instance, wrought iron has a very small **carbon** content, while cast iron has at least 2% carbon.

Steels contain varying amounts of carbon and metals such as tungsten, molybdenum, vanadium, and cobalt, giving them the strength, durability, and anti-corrosion capabilities required by their different uses. Stainless steel, which has 18% chromium and 8% nickel alloyed to it, is valued for its anti-corrosive qualities.

Duraluminum contains one-third steel and two-thirds **aluminum**. It was developed during World War I for the superstructures of the Zeppelin airships built in Germany.

Many alloys add function to physical beauty. For example, sterling silver is made with 8% copper to add strength so that it can be made into chalices and silverware.

All American coins are made from copper alloy, sometimes sandwiched between layers of silver.

Alloys greatly enhance the versatility of metals. Without them there would be total dependency on pure metals, which would affect their cost and availability. Alloys are a very important part of humankind's past and future.

See also Metallurgy.

Allspice *see* **Myrtle family (Myrtaceae)**

Alluvial systems

An alluvial system consists of sediments eroded, transported, and deposited by **water** flowing in **rivers** or streams. The sediments, known as alluvium, can range from clay-sized particles less than 0.002 mm in diameter to boulders greater than 64 mm in diameter, depending on their source and the sediment transport capacity of streams in the system. The term alluvial is closely related to the term fluvial, which refers to flowing water. Thus, alluvial systems are the result of fluvial processes.

Modern alluvial systems can create flat and fertile valley bottoms that are attractive for farming because of their rich soils, which are replenished during frequent floods. The same floods that replenish soils, though, can become hazardous when homes are built on floodplains. Ancient alluvial systems that now lie below Earth's surface can be exceptionally good aquifers and **petroleum** reservoirs.

This delta formed downstream of a break in a natural levee. Note the marsh growth that developed along the banks of the distributaries. *Photograph by Dan Guravich. Photo Researchers, Inc. Reproduced by permission.*

Alluvium

Alluvium is the product of sediment **erosion**, transportation, and deposition. Therefore, its nature is controlled by the sediment supply and sediment transport capacity of streams in the watersheds from which it is derived. Regions subjected to high rates of tectonic **uplift** or tropical climates can supply large amounts of sediments to fluvial systems because **rocks** weather rapidly under those conditions. Tectonically stable, arid, and cold regions, in contrast, generally produce sediments at much lower rates. Global **sea level** changes influence alluvial processes in much the same way as tectonic uplift, controlling the base level to which streams erode.

The type of **bedrock** underlying a **watershed** also exerts an important influence on the nature of the alluvium deposited downstream. A drainage **basin** underlain by easily eroded sandstone may produce only **sand** and silt, whereas one underlain by hard metamorphic rocks may produce cobbles and boulders. Geologists often study sedimentary rocks that were deposited in alluvial systems in order to learn about the climate, **tectonics**, and bedrock **geology** of ancient mountain ranges that supplied the sediment.

Once sediment is produced by the physical and chemical **weathering** of rocks, it must be transported to an area in which deposition, also known as aggradation or alluviation, is possible. Although sediment transport is a complicated process, it is well known that sediment transport capacity is exponentially proportional to the stream discharge. Discharge is in turn controlled by precipitation, channel shape, channel roughness, and stream gradient. Graded streams have adjusted their longitudinal profiles (by erosion and deposition at different points) in order to transport the volume of sediment supplied to them.

Sediment particles that are small enough to be lifted and carried within the flowing water are known as the suspended load, whereas larger particles that roll, bounce, or skip along the streambed are known as the bedload. The specific sizes of sediment transported as suspended load and bedload will change as the stream discharge changes in space and **time**. Therefore, a sediment particle may be part of the bedload during times of discharge and part of the suspended load during times of high discharge such as floods. Sediment that is small enough to remain constantly suspended is often referred to as the wash load. Although a large river may be capable of transporting large boulders, its supply may be limited by the ability of watersheds boulders to produce or tributaries to deliver them to the river. In many cases, boulders that exceed the sediment transport capacity of tributary streams are supplied to rivers by landslides, rockfalls, and debris flows.

Alluvium is deposited during repeated cycles of valley incision and alluviation. Valley incision generally occurs during times of rapid tectonic uplift (when sediment transport capacity exceeds sediment supply) or dry climates (when little sediment is produced). Alluviation, in contrast, is more likely to occur during times of tectonic quiescence or wet climatic conditions. Successive cycles of incision and alluviation can produce a series of stream terraces, which are the remnants of abandoned flood plains and appear to form steps along valley walls.

Commmon components

The channel is the depression through which water flows from the head to the mouth of a stream. Some channels follow sinuous, or meandering, paths. Erosion occurs along the outside edge of meanders, where water **velocity** is the greatest, and deposition occurs along the inside edge of meanders, where water velocity is the lowest. The resulting depositional feature is known as a point bar, and alluvial channel deposits are formed by the constant migration of meanders through **geologic time**. Other channels are braided, meaning that they are composed of two or more interconnected low-sinuosity channels. Braided stream channels form when the sediment load is large compared to the sediment transport capacity of a stream. Braided streams can occur naturally in areas of high sediment supply, for example near the snouts of **glaciers** or where the stream gradient abruptly decreases. A change from a meandering stream to a braided stream over a few years or decades, however, can indicate an undesirable increase in sediment supply as the result of activities such as cattle grazing or logging.

During times of high discharge, for example shortly after heavy rainstorms or spring snowmelt, streams can rise above their banks and flood surrounding areas. The low-lying and flood-prone areas adjacent to streams are known as flood plains. The velocity of water flowing across a flood plain is much lower than that in the adjacent channel, which causes suspended sediment to be deposited on flood plain. Coarse sediment is deposited close to the channel and forms natural levees that help to control future floods, but silt and clay can be carried to the far reaches of the floodplain. Thus, alluvial systems generally consist of both coarse-grained channel deposits representing bedload and fine-grained floodplain (also known as overbank) deposits.

Meanders can be abandoned, particularly when they become extremely sinuous, to form crescent-shaped oxbow lakes within the flood plain. Abandonment occurs when it becomes more efficient for the stream to transport its sediment load by cutting a short new channel, thereby locally increasing its gradient, than by flowing through a long meander.

Another common feature of alluvial systems is the stream terrace. A stream terrace is simply a floodplain that was abandoned when the stream incised to a lower level due to a change in base level. Careful observation can often reveal several generations of step-like terraces, each of which represents the elevation of an abandoned flood plain, along a stream valley.

Coastal alluvial plains

Coastal alluvial plains form when streams transporting sediment from mountainous areas reach low elevations and **deposit** a large proportion of their sediment load. This occurs due to a decrease in stream gradient, which reduces sediment transport capacity. The extent of coastal alluvial plains is controlled in large part by sea level, and alluvium deposited during previous times of low sea level (for example, during glacial epochs) may now lay tens or hundreds of meters below sea level.

Alluvial fans

Alluvial fans form where high-gradient mountain streams flow into valleys or onto plains and deposit their sediment load. Such is the case along the foot of the Panamint Mountains bordering Death Valley in California. As with overbank floodplain deposits, the coarsest sediment is generally deposited closest to the **mountains** and finer sediment can be carried many miles. Geologists often refer to the sediment deposited near the mountains as proximal and the sediment deposited far from the mountains as distal. Alluvial fans typically form three-dimensional cones that resemble folding fans when viewed from above, hence their name. The main stream branches out into many channels that distribute sediment across the alluvial fan. Alluvial fans formed by streams that drain large watershed tend to be larger than those that drain small watersheds. They can grow in size until they begin to merge with fans created by neighboring streams, at which point they coalesce into a broad sloping surface known as a bajada.

Deltas

Deltas are formed at the mouths of streams that flow into **lakes** or oceans. They are fan-like deposits similar to alluvial fans, but located in the water rather than on dry

KEY TERMS

Bajada—A feature produced when adjacent alluvial fans overlap or coalesce and form a continuous deposit at the foot of a mountain range. Bajadas are common features in arid to semi-arid regions of the American West and Southwest.

Base level—The lowest elevation to which a fluvial system grades or adjusts itself. Local base levels can be lakes or larger rivers. The global base level is sea level, which changes through geologic time.

Bedload—The portion of sediment that is transported by rolling, skipping, and hopping along the stream bed at any given time because it is too heavy to be lifted by flowing stream water. It stands in contrast to suspended load.

Discharge—The volume of water flowing across an imaginary vertical plane perpendicular to the stream channel per unit of time. In the United

States it is customary to express stream discharge in units of cubic feet per second.

Gradient—The slope of a stream channel, measured in terms of the change in elevation per unit of channel length. Stream gradients can be expressed as percentages or using dimensionless terms such as meters of elevation change per kilometer of channel length.

Sinuosity—The degree of curvature of a stream channel as viewed from above. Highly sinuous streams contain many curves or meanders along the lengths.

Suspended load—Sediment particles transported within flowing stream water at any given time, as opposed to bedload. Suspended load is responsible for the muddiness or turbidity of river water.

land. Like alluvial fans, coarse sediments are deposited close to shore and fine-grained sediment is carried farther out to sea. The Mississippi River has formed the most prominent example of a **delta** within the United States. Other well-known examples are the Nile Delta of North **Africa** and the Amazon Delta of **South America**. When Aristotle observed the Nile Delta, he recognized it was shaped like the Greek letter delta, hence the name. Most deltas clog their channels with sediment and so must eventually abandon them. If the river then flows to the sea along a significantly different path, the delta will be abandoned and a new delta lobe will form. This process, known as delta switching, helps build the coastline outward.

See also Continental margin; Continental shelf; Earth science; Global climate; Hydrologic cycle; Hydrology; Land use; Landform; Sediment and sedimentation; Sedimentary environment; Sedimentary rock.

Resources

Books

Knighton, D. *Fluvial Forms and Processes.* London: Arnold, 1998.

Leopold, L. B. *A View of the River.* Cambridge, MA: Harvard University Press, 1994.

Leopold, L. B., M. G. Wolman, and J. P. Miller. *Fluvial Processes in Geomorphology.* New York: Dover Publications, 1995.

Periodicals

Bull, W. B. "Alluvial fans." *Journal of Geological Education* 26, no. 3 (1978): 101–06.

Other

Ableman, C. "Glossary of Hydrologic Terms." 2000 [cited October 19, 2002]. <www.srh.noaa.gov/wgrfc/glossary/>.

Pidwirny, M. J. *"Fluvial Landforms."* 2000 [cited October 19, 2002]. <www.geog.ouc.bc.ca/physgeog/contents/11j.html>.

William C. Haneberg

Aloe *see* **Lily family (Liliaceae)**

Alpacas *see* **Camels**

Alpha particle

The alpha particle is emitted by certain radioactive elements as they decay to a stable element. It consists of two protons and two neutrons; it is positively charged. The element that undergoes "alpha decay" changes into a new element whose **atomic number** is down two and atomic **mass** is down four from the original element. Alpha decay occurs when a nucleus has so many protons that the strong nuclear **force** is unable to counterbalance the strong repulsion of the electrical force between the protons. Because of its mass, the alpha particle travels relatively slowly (less than 10% the speed of **light**), and it can be stopped by a thin sheet of **aluminum** foil.

When Henri Becquerel first discovered the property of radioactivity in 1896, he did not know that the **radiation** consisted of particles as well as **energy**. Ernest

Rutherford began experimenting to determine the nature of this radiation in 1898. One experiment demonstrated that the radiation actually consisted of 3 different types: a positive particle called "alpha," a **negative** particle, and a form of electromagnetic radiation that carried high energy. By 1902 Rutherford and his colleague Frederick Soddy were proposing that a different chemical element is formed whenever a radioactive element decays, a process known as transmutation. Rutherford was awarded the Nobel Prize in **chemistry** in 1908 for discovering these basic principles of radioactivity. In Rutherford's classic "gold foil" experiment to determine the structure of an atom, his assistants Hans Geiger and E. Marsden used positively charged high speed alpha particles that were emitted from radioactive polonium to bombard a gold foil. The results of this experiment showed that an atom consisted of mostly empty **space**, with essentially all its mass concentrated in a very small, dense, positively charged center called the nucleus. In fact, the alpha particle itself was identified as the nucleus of the helium atom! Soddy, working with William Ramsay, who in 1895 had discovered that the element helium was a component of earth **minerals**, verified that helium is produced when radium is allowed to decay in a closed tube.

The first production in the laboratory of radioisotopes—as opposed to the naturally occurring radioactive isotopes—was achieved by bombarding stable isotopes with alpha particles. In 1919, Rutherford succeeded in producing oxygen-17 by bombarding ordinary nitrogen-14 with alpha particles; a **proton** was set free in the reaction. The 1935 Nobel Prize in chemistry was awarded to Irène and Frédéric Joliet-Curie for their work in producing radioisotopes through the bombardment of stable elements with alpha particles from polonium decay. After bombarding aluminum with alpha particles, they found that when the nucleus absorbed an alpha particle, it changed into a previously unknown radioisotope of **phosphorus**; an unidentified neutral particle was set free in the reaction. James Chadwick repeated their experiment using a beryllium target, then captured the particle that was emitted and used principles of classical **physics** to identify its mass as being the same as that of the proton. This particle was named the **neutron**, and Chadwick received the 1935 Nobel Prize in physics for its discovery. Subsequently, scientists began accelerating alpha particles to very high energies in specially constructed devices like the **cyclotron** and linear accelerator before shooting them at the element they wanted to transmute.

A number of radioisotopes, both natural and man made, have been identified as alpha emitters. The decay chain of the most abundant **uranium isotope**, uranium-238, to stable lead-206 involves eight different alpha decay reactions. Uranium has a very long half life (4.5 bil-

lion years) and because it is present in the earth in easily measured amounts, the **ratio** of uranium to **lead** is used to estimate the age of our **planet** Earth. More recently, Guenther Lugmair at the University of California at San Diego introduced age-dating using samarium-147, which undergoes alpha decay to produce neodymium-143.

Most of the transuranic radioactive elements undergo alpha decay. During the 1960s when an instrument was needed to analyze the lunar surface, Anthony Turkevich developed an alpha scattering instrument, using for the alpha source curium-242, a transuranic element produced by the alpha decay of americium-241. This instrument was used on the moon's surface by Surveyors 5, 6, and 7. Plutonium has been used to power more than twenty spacecraft since 1972; it is also being used to power cardiac pacemakers.

Ionization smoke detectors detect the presence of smoke using an ionization chamber and a source of **ionizing radiation** (americium-241). The alpha particles emitted by the **radioactive decay** of the americium-241 ionize the **oxygen** and **nitrogen atoms** present in the chamber, giving rise to free electrons and ions. When smoke is present in the chamber, the electrical current produced by the free electrons and ions (as they move toward positively and negatively charged plates in the detector) is neutralized by smoke particles; this results in a drop in electrical current and sets off an alarm.

Because of their low penetrability, alpha particles do not usually pose a threat to living organisms, unless they are ingested. This is the problem presented by the radioactive gas **radon**, which is formed through the natural radioactive decay of the uranium (present in rock, **soil**, and **water** throughout the world). As radon gas seeps up through the ground, it can become trapped in buildings, where it may build up to toxic levels. Radon can enter the body through the lungs, where it undergoes alpha decay to form polonium, a radioactive solid that remains in the lungs and continues to emit cancer-causing radiation.

Alternating current *see* **Electric current**

Alternative energy sources

Nonrenewable fossil fuels—coal, **petroleum**, and natural gas—provide more than 85% of the **energy** used around the world. In the United States, **fossil fuels** comprise 81.6% of the total energy supply, **nuclear power** provides 7.7%, and all renewable energy sources provide 7.3%. **Wind** power, active and passive solar systems, geothermal energy, and **biomass** are examples of renewable or alternative energy sources. Although such alternative sources make up a small fraction of total energy production today, their share is growing. Scientists estimate that easily extractable fossil fuels will be largely used up within the twenty-first century (known petroleum reserves will last less than 40 years at current rates of use). Nuclear power has several drawbacks, among which are military vulnerability, and waste disposal problems. Further, nuclear power technologies cannot be disseminated globally without disseminating at the same time all of the materials and much of the know-how for producing **nuclear weapons**. Achieving wider use of renewable sources of energy is thus, seen by many planners as key for a sustainable global economy. In 2002, the 15-nation European Union declared its intention to shift away from both fossil fuels and nuclear power, with an initial goal of generating 12% of its total energy and 22% of its **electricity** from renewable sources by 2010.

The exact contribution that alternative energy sources make to the total primary energy used around the world is not known. Conservative estimates place their share at only 3–4%, but some energy experts dispute these figures. U.S. energy analysts have argued that the **statistics** collected are based primarily on data supplied by large electric utilities and the regions they serve, and so do not fully account for areas remote from major power grids, which are more likely to use solar energy, wind energy, biomass, and other alternative sources. When these areas are taken into consideration, alternative energy sources may already contribute as much as 11% to the total primary energy used in, for example, the United States, where alternative use is lower than in most poor countries. **Animal** manure, furthermore, is widely used as an energy source in India, parts of China, and many African nations. When this is taken into account, the percentage of the worldwide contribution of alternative sources to energy production could rise as high as 10–15%.

Wind power

Although today considered an alternative energy source, wind power is one of the earliest forms of energy harvested by humans. Wind is caused by the uneven heating of Earth's surface, and its energy is equivalent to about 2% of the solar energy reaching the **planet**. The amount of energy theoretically available from wind is thus, very great, although it would be neither practical, wise, nor necessary to intercept more than a tiny percentage of the world's total windflow.

Wind is usually harvested by windmills, which may either supply mechanical energy directly to machinery or drive generators to produce electricity. (Energy must be carefully distinguished from electricity; electricity is not a source of energy, but a form of it. In processes that burn chemical or nuclear fuel to generate electricity, more energy is lost as low-grade **heat** than is delivered as electricity; a windmill, likewise, supplies less usable energy when it is used to generate electricity than when it is used to do mechanical **work**. Electricity has the positive qualities of being transmissible over long distances via powerlines and of being useful for many applications—lighting, motors, **electronics**, and so on—at its points of end-use.) The kinetic energy of wind is proportional to its **velocity**, so the ideal location for a windmill **generator** is in a place with constant and relatively fast winds and no obstacles such as tall buildings or trees. An efficient windmill can produce 175 watts of electricity per square meter of propeller-blade area at a height of 75 ft (25 m). The estimated cost of generating one kilowatt-hour (the amount of energy consumed by ten 100-watt light bulbs in one hour) by wind power is about eight cents, as compared to five cents for typical hydropower and 15 cents for nuclear power. California leads the United States in utilization of wind power, producing approximately 1.3% of its electric usage in 2000 from wind, enough to light San Francisco. Denmark leads the world in this respect, presently obtaining 21% of its electricity from windmills (and six more **percent** from other renewable sources).

Solar power

Solar energy can be utilized either directly as heat or indirectly after conversion to electrical power using photovoltaic cells or steam generators. Greenhouses and solariums are common examples of the direct use of solar energy, having **glass** surfaces that allow the passage of visible light from the **sun** but slow the escape of heat and infrared. Another direct method involves flat-plate solar collectors that can be mounted on rooftops to provide energy needed for water heating or space heating. Windows and collectors are considered passive systems for harnessing solar energy. Active solar systems use fans, pumps, or other machinery to transport heat derived from sunlight.

Photovoltaic cells are flat electronic devices that convert some of the light that falls on them directly to electricity. Typical commercial photovoltaic cells convert 10–20% of the sunlight that falls on them to electricity. In the laboratory, the highest efficiency demonstrated so far is over 30%. (Photovoltaic efficiency is important even though sunlight is free; higher-efficiency cells produce more power in a limited space, such as on a rooftop.) Photovoltaics are already economic for use in remote applications, such as highway construction signs, spacecraft, lighthouses, boats, rural villages, and isolated homes, and large-scale initiatives are under way in Cali-

Wind generators in Whitewater, California. *JLM Visuals. Reproduced by permission.*

fornia and other places to produce hundreds of megawatts of power from rooftop-mounted photovoltaic systems.

Thermal-electric solar systems have also been developed using tracking circuits that follow the sun and mirrored reflectors that concentrate its energy. These systems develop intense heat that generates steam, which in turn drives a **turbine** generator to produce electricity.

Geothermal energy

Geothermal energy is heat generated naturally in the interior of the **earth**, and can be used either directly, as heat, or indirectly, to generate electricity. Geothermal energy can be used to generate electricity by the flashed-steam method, in which high-temperature geothermal brine is used as a heat source to convert water injected from the surface into steam. The steam is used to turn a turbine and generator, which produces electricity. When geothermal wells are not hot enough to create steam, a fluid that evaporates at a lower **temperature** than water, such as isobutane or **ammonia**, can be used in a closed loop in which the geothermal heat evaporates the fluid to run a turbine and the cooled vapor is recondensed and reused. More than 20 countries utilize geothermal energy, including Iceland, Italy, Japan, Mexico, Russia, and the United States. Unlike solar energy and wind power, geothermal energy may contribute to **air pollution** and may raise dissolved salts and toxic elements such as mercury and arsenic to the surface.

Oceanic sources

Although there are several ways of utilizing energy from the oceans, the most promising ones are the harnessing of tidal power and **ocean** thermal energy conversion. The power of oceanic **tides** is based on the difference between the (usually) twice-daily high and low water levels. In order for tidal power to be effective, this difference in height must exceed about 15 ft (3 m). There are only a few places in the world where such differences exist, however, including the Bay of Fundy in Canada and a few sites in China.

Oceanic thermal energy conversion utilizes temperature differences rather than tides. Ocean temperature is commonly stratified, especially near the tropics; that is, the ocean is warmer at the surface and cooler at depth. Thermal energy conversion takes advantage of this fact by using a fluid with a low **boiling point**, such as ammonia. Vapor boiled from this fluid by warm surface water drives a turbine, with cold water from lower depths being pumped up to condense the vapor back into liquid.

The electrical power thus generated can be shipped to shore over transmission lines or used to operate a floating industrial operation such as a cannery.

It is unlikely, however, that ocean-derived energy will ever make a large contribution to world energy use; the number of suitable sites for tidal power is small, and the large-scale use of thermal energy conversion would likely cause unacceptable environmental damage.

Biomass

Biomass—wood, dried animal dung, or materials left over from agriculture—is the oldest fuel used by people, and was initially utilized for space-heating and cooking food. These uses are still major energy sources in developing countries, especially in rural areas. Biomass can also be combusted in a boiler to produce steam, which can be used to generate electricity. The biomass of trees, sugar cane, and corn can also be used to manufacture **ethanol** or methanol, which are useful liquid fuels.

Other sources of alternative energy

Other sources of alternative energy, some experimental, are also being explored. Methane gas is generated from the **anaerobic** breakdown of organic waste in landfills and in wastewater treatment plants; this methane can be collected and used as a gaseous fuel for the generation of electricity. With the cost of garbage disposal rapidly increasing, the burning of organic garbage is becoming a viable option as an energy source. Incinerators doing this are sometimes known as "waste to energy" facilities. Adequate air **pollution** controls are necessary, however, to prevent the **emission** of toxic chemicals to the environment—a "landfill in the air" effect.

Fuel cells are another rapidly developing technology. These devices oxidize **hydrogen** gas, produce electricity, and release only water as a waste product. Experimental vehicles (including buses) and medium-sized generating units are already running using this promising technology. Like electricity itself, however, hydrogen is not an energy source; hydrogen gas (H_2) does not occur naturally on Earth in significant quantities, but must be manufactured using energy from fossil fuel, solar power, or some other source. Fuel cells have the advantage of producing electricity at their point of end-use from a concentrated fuel that does not produce pollution; if their fuel can be produced by nonpolluting means, such as from solar energy,

Although not in the strictest sense an alternative source of energy, **conservation** is perhaps the most important way of reducing society's dependence on nonrenewable fossil and nuclear fuels. Improving the efficiency of energy usage is an excellent way of meeting energy demands without producing pollution or requiring changes in lifestyle (though lifestyle changes may also ultimately be necessary, as nonrenewable energy stocks decline). If a society needs, for example, to double the number of refrigerators it uses from 10 to 20, it is far cheaper at this time to engineer and manufacture 20 refrigerators with twice the efficiency of the old ones than to manufacture 10 refrigerators of the old type and double the amount of electricity produced. New electric generation facilities of any type are expensive, and all—even alternative types—impose some costs on the environment. Experts have estimated that it is still possible to double the efficiency of electric motors, triple the efficiency of light bulbs, quadruple the efficiency of refrigerators and air conditioners, and quintuple the gasoline mileage of automobiles. Several European and Japanese **automobile** manufacturers are already marketing hybrid vehicles with extremely high gasoline mileage (40–70+ miles per gallon), and these by no means reflect the upper limit of efficiency possible.

Resources

Books

Rosenberg, Paul. *Alternative Energy Handbook.* Terre Haute, IN: TWI Press, 2001.

Berger, J.J. *Charging Ahead: The Business of Renewable Energy and What It Means for America.* Berkeley: University of California Press, 1998.

Brower, M. *Cool Energy: Renewable Solutions to Environmental Problems.* Cambridge: MIT Press, 1992.

Goldemberg, J. *Energy for a Sustainable World.* New York: Wiley, 1988.

Periodicals

Dresselhaus, M.S., and Thomas, I.L., "Alternative Energy Technologies." *Nature.* (November 15, 2001): 332–337.

Meller, Paul, "Europe Pushes for Alternative Energy." *New York Times.* October 16, 2002.

Muthena Naseri
Douglas Smith

Alternative medicine

National Institutes of Health classifies alternative medicine as an unrelated group of non-orthodox therapeutic practices, often with explanatory systems that do not follow conventional biomedical explanations or more seriously, based on pseudoscience. Others more generally define it as medical interventions not taught at United States medical schools or not available at United States hospitals.

Alternative therapies include, but are not limited to the following disciplines: folk medicine, **herbal medicine**, diet fads, homeopathy, faith healing, new age healing, chiropractic, **acupuncture**, naturopathy, massage, and music therapy. Studies suggest these therapies are sought out by individuals who suffer a variety of medical problems. In general, alternative medical practice that fits three criteria: it is not taught in the standard medical school curriculum; there is not sufficient scientific evidence that the treatment is safe and effective against a specific **disease**; and insurance companies do not reimburse the patient for its cost.

Such a definition could include nearly all unproven but ineffective practices that offer little in benefit but draw billions of health care dollars from desperate patients. The use of laetrile (a derivative of apricot pits) to treat **cancer** and chelation therapy to remove **cholesterol** deposits from severely affected **arteries** are cases in point. Both are highly touted by their practitioners, both have been tested under rigid scientific research standards, and both have been found ineffective and useless. The primary harm of such treatments lies in the fact that patients who utilize them often do not seek more effective, mainstream medical care.

The first known example of alternative medicine in the United States was the introduction and patenting in 1797 of a "mechanical tractor" to pull bad **electricity**, alleged to be the source of all illnesses, from the body. A chief justice of the Supreme Court, several members of Congress, and the retired president, George Washington, all used this device.

Although some alternative medical practices are clearly ineffective and sometimes dangerous, others have achieved a degree of acceptability in the eyes of organized medicine. Among these are naturopathy, yoga, **biofeedback**, hypnotism, acupuncture, chiropractic medicine, homeopathy, and relaxation techniques.

Naturopathy

The practitioner of naturopathic medicine considers the person as a whole, treats symptoms such as fever as a natural manifestation of the body's defense mechanism that should not be interrupted, and works to heal disease by altering the patient's diet, lifestyle, or work habits. The basis of naturopathy can be traced back through Native American practices, to India, China, and ancient Greece.

In naturopathy, the body's power to heal is acknowledged to be a powerful process that the practitioner should enhance using natural remedies. Fever, **inflammation**, and other symptoms are not the underlying cause of disease, but are reflections of the body's attempt to rid itself of the underlying cause. The disease itself originates from spiritual, physical, or emotional roots, and the cause must be identified in order that effective therapy may be applied. The patient is viewed holistically and not as a collection of symptoms; the cure is gauged to be safe and not harmful to the patient. The practitioner is a teacher who is trained to recognize the underlying problems and teach the patient to adopt a healthier lifestyle, diet, or attitude to forestall disease. The naturopathic practitioner is a specialist in preventive medicine who believes prevention can best be achieved by teaching patients to live in ways that maintain good health.

In addition to advising the patient on lifestyle changes to prevent disease, the naturopathic practitioner may also call upon acupuncture, homeopathy, **physical therapy**, and other means to strengthen the patient's ability to fight disease. Herbal preparations as well as **vitamin** and mineral supplements may be used to strengthen weakened immune systems. Stressful situations must be eased so that the **digestive system** can function properly, and any spiritual disharmony is identified and corrected.

Naturopaths are trained in herbal medicine, clinical dietetics, hydrotherapy, acupuncture, and other noninvasive means to treat disease. They provide therapy for chronic as well as acute conditions, and may work beside physicians to help patients recover from major **surgery**. The naturopath does only minor surgery and depends upon natural remedies for the bulk of patient therapy. Naturopathy is not widely accepted by physicians, although some practitioners are also doctors of medicine (M.D.).

Lifestyle changes

Lifestyle changes can be as simple as getting more **exercise** or as complex as completely redesigning the diet. Exercise in moderation is a preventive measure against **heart** disease, **stroke**, and other serious conditions. It is only when such exercise programs become excessive or all consuming that they may be harmful. Some people use very high daily doses of vitamins in an attempt to forestall the aging process or to assist the body in ridding itself of cancer or the HIV **virus**. Megadoses of vitamins have not been proven effective for these purposes. Vitamins play a specific role in the **metabolism** and excess vitamins simply are stored in the **fat** or are eliminated from the body through the kidneys.

Changes to achieve a more balanced intake of **nutrients** or to reduce the amount of fat in the diet are beneficial, but dietary programs that add herbal supplements to the diet may be ineffective or even harmful. Chinese and Far Eastern cultures use herbal therapy to achieve weight loss, delay aging, or increase strength. Dietary supplements of Chinese herbs have become increasingly popular among Westerners, although most Americans do not know specifically what herbs they are consuming. Laboratory tests have raised questions about the effectiveness of many of these herbs; some may have high **lead** content and therefore are potentially toxic.

Relaxation

Many practices are included under the general term of relaxation. Relaxation techniques are generally accepted as beneficial to individuals who are otherwise unable to **sleep**, in **pain**, under ongoing job-related **stress**, or recovering from surgery. Various relaxation techniques frequently are used in hospitals to help patients deal with pain or to help them sleep.

In the simplest form of relaxation therapy, the **individual** is taught to lie quietly and to consciously relax each part of the body. Beginning with the feet and progressing through the ankles, calves, thighs, abdomen, and so forth up to the neck and forehead, each part of the body is told to relax and the individual focuses his thoughts on the body part that is being told to relax. It is possible to feel the muscles of the leg or the arm relaxing under this focused attention.

The ancient practice of yoga is also considered a relaxation technique. The practice of assuming a specified position (the lotus position, for example), clearing the mind of the sources of stress, and concentrating on one's inner being for a short time can be beneficial. Following a yoga session, the individual often is less stressed and can order his thoughts in a more organized manner. Transcendental meditation is a variation of yoga that consists of assuming specific body positions and chanting a mantra, a word or two that is repeated and serves to concentrate the mind. This practice is claimed to clear the mind of stressful thoughts and **anxiety**, and enables the practitioner to reorder his priorities in a more relaxed manner.

Yet another variation on relaxation came into widespread use in the late 1960s. Biofeedback became a popular practice that initiated an industry devoted to manufacturing the devices needed to practice it effectively. Biofeedback is a process by which an individual consciously controls certain physiologic processes. These can be processes that normally are subject to thought control, such as muscle tension, or those that are not, such as heart **rate**. To effect such control, the person is connected to a gauge or signal device that changes tone with changes in the **organ** being controlled. This visual or auditory signal provides evidence of the effectiveness of the person's effort. The heart rate can be monitored by the scale of a tone or a blip on a small screen. The tone lowers in pitch or the blip appears with decreasing frequency as the heart rate slows. The goal is for the indi-

vidual to learn to influence the signal in front of him and, having acquired this proficiency, to be able to accomplish the same physiologic changes without the visual or auditory signal. Biofeedback has been used successfully to reduce stress, eliminate headaches, control **asthma** attacks, and relieve pain.

Hypnotism, despite its use for entertainment purposes, also has a place in medical practice. Hypnotism was first introduced to the medical community in the late eighteenth century by a German physician, Franz Anton Mesmer (1734–1815), and was first called mesmerism. Mesmerism fell out of favor in France when a scientific committee failed to verify Mesmer's claims for the practice. The name was later changed to hypnotism (from the Greek word *hypnos* for sleep) by James Braid (1795–1860), an English ophthalmologist. Although technically hypnotism is not sleep, the name stuck.

Sigmund Freud (1856–1939) adopted hypnotism into his practice in the nineteenth century. Early in his use of the practice he praised its benefits, writing two scientific papers on hypnotism and employing it to treat his patients. By the early 1890s, however, Freud abandoned hypnotism in favor of his own methods of analysis. It was not until the 1950s that the British and American medical societies approved the use of hypnosis as an adjunct to pain treatment. In clinical use, hypnotism is called hypnotherapy, and it is used to treat both physical and psychological disorders. The patient is placed in a trance-like state so that the physician may delve into the deepest levels of the mind to relieve such conditions as migraine headaches, muscle aches, chronic headaches, and postoperative pain. This trance-like condition can be induced by the practitioner or by the patient. It is achieved by first relaxing the body and then by concentrating the patient's attention on a single object or idea, shifting his thoughts away from the immediate environment. In the lightest form of hypnosis, the superficial level, the patient may accept suggestions but will not always take steps to carry them out. Therapists try to reach the deeper state of hypnosis, the somnambulistic stage, in which the patient is readily susceptible to suggestion and carries out instructions while hypnotized as well as after he has come out of the trance (post-hypnotic suggestion).

While in the trance the patient can be induced to ignore pain, to fully relax, or to carry out other beneficial suggestions by the therapist. Also the therapist may suggest that the patient can hypnotize himself when he needs relief from pain or needs to blunt his appetite. The patient is given a simple ritual to follow including specific words to say to place himself in a hypnotic trance. He will then convince himself that his pain has been relieved or that he has eaten a sufficient amount. Upon recovering his normal level of consciousness he will find that his pain is less or that he has no need for additional food.

Unlike portrayals of hypnotists in films, a therapist cannot hypnotize anyone who does not want to be hypnotized. It is essential that the patient and therapist have a close rapport, that the patient fully believes the practice will be of benefit to him, and that the surroundings are devoid of distracting stimuli. Even when in a trance, the patient will not carry out any act he would find morally unacceptable in his waking state. The hypnotist cannot place someone in a trance, for example, and direct him to steal a car or rob a bank. The subject will awaken with the shock of the suggestion.

Chiropractic medicine

Chiropractic medicine is founded on the hypothesis that many human diseases and disorders stem from deviations or subluxations of the spine, which impinge on the spinal nerves, causing pain or disfunction of the affected organs. Treatment consists of determining which of the vertebrae have shifted and then realigning them properly. This may be accomplished in a single treatment or may require a series of treatments over time.

Chiropractic, derived from the Greek words for "practice by the hands," was developed by a Canadian-born Iowa grocer, Daniel David Palmer (1845–1913), in 1895. Palmer believed that the source of illness was the misalignment, or subluxation, of the spinal column in such a way that the vertebrae impinged upon the spinal nerves that passed from the spinal cord, between the vertebrae, to the various organs and muscles of the body. This constriction of the spinal nerve prevented the neural impulses from flowing properly, thus making it impossible for the **brain** to regulate body functions and leaving tissues susceptible to diseases. Correcting the subluxation would, therefore, restore the neural impulses and strengthen the body.

In 1898, Palmer established the Palmer College of Chiropractic in Davenport, Iowa. In 1910 he published a textbook on chiropractic, which outlined his theories. Since then, the number of chiropractic schools has increased to 16 with a total enrollment of approximately 10,000. To practice as a doctor of chiropractic (D.C.) an individual must complete the four years of chiropractic school and pass a licensing test. Some states specify that entry into a chiropractic college requires only a high school diploma and others require two years of college prior to entry.

Chiropractors themselves differ in the definition of their specialty—this has led to the formation of two separate professional organizations. The International Chiropractors Association advocates chiropractic therapy limited to spinal manipulation only, while members of the American Chiropractors Association endorse a wider range of therapeutics including physical therapy, diathermy (heating of body tissues with electromagnetic **radiation**, **electric current**, and ultrasonic waves), and dietary counseling in addition to the basic spinal manipulation.

Chiropractors do not prescribe medication or perform surgery. Only with great reluctance did the American Medical Association recognize chiropractic as a legitimate specialty. The practice is still approached with skepticism by many in the mainstream medical community because no chiropractic school is recognized by any accrediting body and because the practice itself is based on unsound, unscientific principles. Still, many physicians refer patients with back pain to chiropractors who are more skilled at manipulating misaligned vertebrae.

Acupuncture

Acupuncture is a form of therapy developed by the ancient Chinese and subsequently refined by Chinese practitioners. It consists of inserting needles through the skin in very specific places to alleviate pain, cure disease, or provide **anesthesia** for surgery. Acupuncture as a palliative is accepted among the medical community. Its use as a cure for serious disease or for anesthesia is not endorsed by many physicians.

For the practitioner of acupuncture, the human body is a collection of thousands of acupuncture points that lie along specific lines or meridians. Twelve pairs of meridians are plotted on the body, one of each pair on each side. An additional meridian, the Conception Vessel, courses along the midline of the front of the body and another, the Governor Vessel, along the spine. The meridians are connected by extrameridians. Additional acupuncture points lie outside the meridians on areas such as the **ear** lobes, fingers, toes, and so forth.

Each meridian is a course for the perceived flow of **energy** through the body and that flow may be connected to an organ removed from the actual location of the meridian. Needles are inserted at points along the meridian that are specific for a given organ. For example, although the liver lies in the right side of the abdomen, the acupuncture points for liver disease may include areas on the opposite side of the body as well as one of the earlobes.

Acupuncture therapy consists of first locating the source of pain, then deciding upon the appropriate meridian and acupuncture points. Needles used in acupuncture may be short, for use in less-fleshy areas, or long for use in areas with copious flesh or muscle. The needle is simply inserted into the proper acupuncture point and rotated. The needles are left in place for a given time and rotated periodically while they are in place. Some patients find dramatic relief from pain with acupuncture and a number of physicians have incorporated the procedure into their practices.

In the Far East, acupuncture is a recognized form of therapy, and it is combined with herbal medicine, diet restrictions, and exercise. Not only is it used for pain relief, but it is also used frequently as anesthesia during surgery and for the treatment of serious diseases such as cancer. No scientific proof has been offered that it is effective against serious diseases, though its anesthetic and analgesic properties have been demonstrated.

Homeopathy

Homeopathy is a term derived from the Greek words meaning "similar suffering." It is a system in which diluted **plant**, mineral, or **animal** substances are given to stimulate the body's natural healing powers. Homeopathy was developed in the late eighteenth century by Samuel Hahnemann (17551–1843), a German physician. Hahnemann conducted experiments to improve standard therapy, which then consisted of bloodletting and administering purgatives made with mercury, which were highly toxic. In one of his experiments he ingested an extract of cinchona, the **bark** from a Peruvian **tree** used by the natives to treat **malaria**. Hahnemann consumed large doses of the bark and developed the symptoms of malaria. From this he concluded that if large doses resulted in symptoms of the disease, small doses should stimulate the body's own disease-fighting mechanism.

Homeopathy is based upon three principles formulated by Hahnemann. The first is the law of similars, stating that like cures like. The second is the law of infinitesimal dose, stating that the potency of a remedy is a reflection of how much it is diluted. Third is the holistic medical model, stating that any illness is specific to the individual who has it.

The law of similars is seen in more traditional medical practice in the use of immunizations. Inoculations of attenuated or dead viruses or **bacteria** are given to stimulate the production of antibodies to resist a full-scale invasion of the same virus or bacterium. Thus, immunizing a child against **poliomyelitis** consists of administering a **solution** containing the dead polio virus; this results in the formation of antibodies that are available to repel the living polio virus if the child is exposed to it.

Homeopathy is much more accepted in **Europe**, Latin America, and India than it is in the United States. It is touted as a low-cost, nontoxic, effective means of delivering medication that can cure even chronic diseases, including those that conventional medications fail to cure. In France, pharmacies are required to stock homeopathic remedies in addition to regular pharmaceutical drugs. Hospital and outpatient clinics specializing in

homeopathy are part of the British health care system and the practice of homeopathy is a recognized postgraduate medical specialty.

In the United States homeopathy has only begun to be accepted by the mainstream medical community. Approximately 3,000 physicians or other health care providers endorse the practice. Though it excites little enthusiasm among practicing physicians in America, homeopathy is an ongoing specialty, and the production of homeopathic remedies is regulated by the Food and Drug Administration to assure their purity, and proper labeling and dispensing.

See also Acupressure.

Resources

Books

Cohen, Michael. *Complementary Medicine: Legal Boundaries and Regulatory Perspectives* Baltimore: Johns Hopkins University Press, 1999.

common aluminum-containing mineral, and then passing an **electric current** through the hot liquid. Molten aluminum metal collects at the **cathode** (**negative** electrode) in a process called **electrolysis**. Not long after the development of this process, the price of aluminum metal plummeted to around 30 cents a pound.

In the production of aluminum today by the Hall-Héroult process, the aluminum oxide is dissolved in a molten mixture of **sodium, calcium,** and aluminum fluorides, which melts at a lower **temperature** than cryolite. The aluminum oxide is in the form of bauxite, a white, brown, or red earthy clay; it was first found near Les Baux, France, in 1821 by P. Berthier, and is now the main source of all aluminum. It is mined in various parts of **Africa** and in France, Surinam, Jamaica, and the United States—mainly in Alabama, Arkansas, and Georgia. The world's supply of bauxite appears to be immense enough to last for hundreds of years at the rate it is being mined today.

Uses

In spite of the fact that aluminum is very active chemically, it does not corrode in moist air the way iron does. Instead, it quickly forms a thin, hard coating of aluminum oxide. Unlike iron oxide or rust, which flakes off, the aluminum oxide sticks tightly to the metal and protects it from further oxidation. The oxide coating is so thin that it is transparent, so the aluminum retains its silvery metallic appearance. Sea **water**, however, will corrode aluminum unless it has been given an unusually thick coating of oxide by the anodizing process.

When aluminum is heated to high temperatures in a **vacuum**, it evaporates and condenses onto any nearby cool surface such as **glass** or plastic. When evaporated onto glass, it makes a very good mirror, and aluminum has largely replaced silver for that purpose because it does not tarnish and turn black, as silver does when exposed to impure air. Many food-packaging materials and shiny plastic novelties are made of **paper** or plastic with an evaporated coating of bright aluminum. The "silver" helium balloons that we see at birthday parties are made of a tough plastic called Mylar, covered with a thin, evaporated coating of aluminum metal.

Aluminum conducts **electricity** about 60% as well as copper, which is still very good among metals. Because it is also light in weight and highly ductile (can be drawn out into thin wires), it is used instead of copper in almost all of the high-voltage electric transmission lines in the United States.

Aluminum is used to make kitchen pots and pans because of its high **heat** conductivity. It is handy as an air- and water-tight food wrapping because it is very malleable; it can be pressed between steel rollers to make foil (a thin sheet) less than a thousandth of an inch thick. Claims are occasionally made that aluminum is toxic and that aluminum cookware is therefore dangerous, but no clear evidence for this belief has ever been found. Many widely used antacids in the drug store contain thousands of times more aluminum (in the form of **aluminum hydroxide**) than a person could ever get from eating food cooked in an aluminum pot. Aluminum is the only light element that has no known physiological function in the human body.

Chemistry and compounds

Aluminum is an unusual metal in that it reacts not only with acids, but with bases as well. Like many active metals, aluminum dissolves in strong acids to evolve **hydrogen** gas and form salts. In fact, cooking even weakly acidic foods such as tomatoes in an aluminum pot can dissolve enough aluminum to give the dish a "metallic" **taste**. But aluminum also dissolves in strong bases such as **sodium hydroxide**, commonly known as lye. Most oven cleaners, which are designed to work on steel and porcelain, contain sodium or potassium hydroxide; the user must take care not to get it on any aluminum parts of the range because it will cause adverse effects. Some commercial drain cleaners contain lye mixed with shavings of aluminum metal; the aluminum dissolves in the sodium hydroxide **solution** to produce bubbles of hydrogen gas, which add a mechanical clog-breaking action to the grease-dissolving action of the lye.

Hydrated aluminum chloride, $AlCl_3 \cdot H_2O$, also called aluminum chlorohydrate, is used in antiperspirants because, like alum (**potassium aluminum sulfate**), it has an astringent effect—a tissue-shrinking effect—that closes up the sweat-gland ducts and stops perspiration.

Over one million tons of aluminum sulfate, $Al_2(SO_4)_3$, are produced in the United States each year by dissolving aluminum oxide in **sulfuric acid**, H_2SO_4. It is used in water purification because when it reacts with lime (or any base), it forms a sticky precipitate of aluminum hydroxide that sweeps out tiny particles of impurities. Sodium aluminum sulfate, $NaAl(SO_4)_2 \cdot 12H_2O$, a kind of alum, is used in "double-acting" baking powders. It acts as an acid, reacting at oven temperatures with the **sodium bicarbonate** in the powder to form bubbles of **carbon dioxide** gas.

See also Metal production; Metallurgy.

Resources

Books

"Aluminum." *Kirk-Othmer Encyclopedia of Chemical Technology.* 4th ed. Suppl. New York: John Wiley & Sons, 1998.

Braungart, Michael and William McDonough. *Cradle to Cradle: Remaking the Way We Make Things.* North Point Press, 2002.

Lide, D.R., ed. *CRC Handbook of Chemistry and Physics* Boca Raton: CRC Press, 2001.

Snyder, C.H. *The Extraordinary Chemistry of Ordinary Things.* 4th ed. New York: John Wiley and Sons, 2002.

Robert L. Wolke

Aluminum hydroxide

Aluminum hydroxide is a common compound of aluminum, **hydrogen**, and **oxygen** which can be considered either a base with the formula $Al(OH)_3$, or an acid with the formula H_3AlO_3. In addition, the compound is frequently treated as a hydrate—a water-bonded compound—of aluminum oxide and designated variously as hydrated alumina, or aluminum hydrate or trihydrate, hydrated aluminum, or hydrated aluminum oxide, with the formula $Al_2O_3(H_2O)_x$.

Properties

Aluminum hydroxide is found in nature as the mineral bayerite or gibbsite (also called hydrargillite). A mixed aluminum oxide-hydroxide mineral is known as diaspore or boehmite.

In a purified form, aluminum hydroxide is either a white bulky powder or granules with a **density** of about 2.42 g/mL. It is insoluble in **water**, but soluble in strong **acids and bases**. In water, aluminum hydroxide behaves as an amphoteric substance. That is, it acts as an acid in the presence of a strong base and as a base in the presence of a strong acid. This behavior can be represented by the following somewhat oversimplified equation.

$$3H^+ + AlO_3^{3-} \leftrightarrows Al(OH)_3 \leftrightarrows Al^{3+} + 3OH^-$$

In the presence of a strong acid such as hydrochloric acid, the above equilibrium shifts to the right, and aluminum chloride is formed.

$$Al(OH)_3 + 3\ HCl \rightarrow 3H_2O + AlCl_3$$

In the presence of a strong base such as **sodium hydroxide**, the equilibrium is driven to the left and a **salt** of the aluminate ion (AlO_{2-}) is formed.

$$NaOH + H_3AlO_3 \rightarrow NaAlO_2 + 2H_2O$$

Sodium aluminate, $NaAlO_2$, has a number of practical applications, such as in water softening, the sizing of **paper**, the manufacture of **soap** and milk **glass**, and in the printing of **textiles** and fabrics.

Uses

Aluminum hydroxide and its closely related compounds have a number of practical uses. In one process of water purification, for example, aluminum sulfate, $Al_2(SO_4)_3$, or alum (usually **potassium aluminum sulfate**, $KAl(SO_4)_2$), is mixed with lime (**calcium** hydroxide, $Ca(OH)_2$) in a container of water to be purified. The reaction between these compounds results in the formation of a gelatinous precipitate aluminum hydroxide. As the precipitate settles out of **solution**, it adsorbs on its surface particles of dirt and **bacteria** that were suspended in the impure water, which can then be removed by filtering off the aluminum hydroxide precipitate.

The ability of aluminum hydroxide to adsorb substances on its surface explains a number of its other applications. It is used in a number of chemical operations, for example, as a filtering medium and in ion-exchange and **chromatography** devices.

Aluminum hydroxide is popular as an antacid. It behaves as a base, reacting with and neutralizing excess stomach acid (hydrochloric acid) to bring relief from "heartburn." A few of the more common commercial antacids containing aluminum hydroxide are Amphojel, Di-Gel, Gelusil, and Maalox. Ingredients can change without notice, however, so the labels should always be checked.

Aluminum hydroxide is also used as a mordant in dyeing. In most cases, the compound is precipitated out of a water solution onto the fibers to be dyed. The material is then immersed into the dye bath. The **color** of the final product depends on the combination of dye and mordant used in the process. A similar process is used in the manufacture of certain paint pigments. A given dye and aluminum hydroxide are precipitated together in a reaction vessel and the insoluble compound thus formed is then filtered off.

Additional uses of aluminum hydroxide include the manufacture of aluminosilicate glass, a high melting point glass used in cooking utensils, the waterproofing of fabrics, and the production of fire clay, paper, pottery, and printing inks.

A close chemical relative of aluminum hydroxide, aluminum hydroxychloride $Al_2(OH)_5Cl$, is an ingredient in many commercial antiperspirants. The compound acts as an astringent, a substance that closes pores and stops the flow of perspiration.

Resources

Books

Brown, Theodore L., and H. Eugene LeMay, Jr. *Chemistry: The Central Science.* 8th ed. Englewood Cliffs, NJ: Prentice-Hall, 1999.

KEY TERMS

Adsorption—The process by which atoms, ions, or molecules of one substance adhere to the surface of a second substance.

Amphoterism—The property of being able to act as either an acid or a base.

Equilibrium—The conditions under which a system shows no tendency for a change in its state. At equilibrium the net rate of reaction becomes zero.

Mordant—A material that is capable of binding a dye to a fabric.

Pigment—Any substance that imparts color to another substance.

Hawley, Gessner G., ed. *The Condensed Chemical Dictionary.* 9th ed. New York: Van Nostrand Reinhold, 1977.

O'Neil, Maryadele J. *Merck Index: An Encyclopedia of Chemicals, Drugs, & Biologicals.* 13th ed. Whitehouse Station, NJ: Merck & Co., 2001.

David E. Newton

Alzheimer disease

Alzheimer **disease** is the most common form of the **brain** disorder called **dementia**. People with dementia experience difficulty in carrying out daily activities because of damage to the regions of their brains that control thought, language, and **memory**. While many older people have Alzheimer disease, it is not a normal part of aging. Currently, there is no cure.

The disease is named after the German physician Alois Alzheimer. In 1906, Alzheimer noticed changes over time in samples of brain **tissue** of a woman who died of mental illness. Specifically, Alzheimer noticed the development of clumps and tangled bundles of fibers in the brain. The clumps are now known as amyloid plaques, and the tangled regions are called neurofibrillary tangles. Both are hallmarks of Alzheimer disease.

Biology of Alzheimer disease

Other changes occur in the brain of a patient with the disease. For example, there is a loss of nerve cells in the brain, especially in the area of the brain associated with memory. In addition, there is a loss of the chemical **neurotransmitter** called **acetylcholine**. This chemical aids in the transmission of signals from one nerve to another in the brain. The loss of acetylcholine means that brain signals are not transmitted through the brain as well as in people who do not have Alzheimer disease.

Alzheimer disease is a progressive disorder. The formation of the amyloid plaques, loss of nerve cells, and loss of acetylcholine occurs over time. Thus, the deterioration in brain function occurs gradually in an Alzheimer's patient. The disease is also associated more with older people (i.e., 60 and older) than with younger people. Only about three **percent** of people ages 65 to 74 have the disease, whereas nearly half of all those 85 and older may have the disease.

The reasons for the formation of amyloid plaques and neurofibrillary tangles are not clear.

Diagnoses and treatment of Alzheimer disease

Diagnosis of the disease often results from an awareness of a change in **behavior** or memory by an **individual** or his/her physician. A definitive diagnosis requires the observation of plaques and tangled fibers in samples of brain tissue. However, such a tissue examination is possible only after death. So, in practical terms, only a "possible" or "probable" diagnosis of Alzheimer's is possible prior to death.

The pharmaceutical industry is developing drugs for a variety of therapeutic approaches to treating Alzheimer disease. These include drugs that will inhibit the inflammatory response the brain appears to mount against plaques. In this approach, it is assumed that the inflammatory response contributes to **cell death**. By inhibiting it, the researchers hope to decrease the loss of brain cells in Alzheimer disease.

Another tact is the development of drugs that increase the amounts of acetylcholine in the brains of patients with Alzheimer disease. By preventing the breakdown of remaining acetylcholine, researchers hope to alleviate some of the symptoms of the disease. The drug called Tacrine stops the destruction of acetylcholine that occurs normally, allowing acetylcholine to persist longer in the brain. Tacrine, which is currently one of a few drugs that has Food and Drug Administration (FDA) approval for the treatment of Alzheimer disease, appears to be helpful only in the early to middle stages of Alzheimer disease. Other drugs that exhibit similar effects include donepezil, rivastigmine, and galantamine.

Other efforts at preventing the onset of Alzheimer disease are focusing on the effects of **hormones** such as estrogen, drugs that reduce the inflammatory response (which might contribute to brain damage), drugs that decrease the oxidation of chemicals in the body, natural "remedies" such as gingko biloba, and increasing attention to the careful control of **hypertension** (specifically using **calcium** channel blockers).

Current research

Research into the cause of Alzheimer disease focuses on the nature of the basic **biology** of the disease and the development of drugs that will counteract or prevent the deterioration associated with the disease. These quests are difficult, as the cause of Alzheimer disease at the molecular level is still unknown. Similarly, the formation of amyloid plaques and why the connections between brain cells are destroyed are unclear. Until the exact biological causes of Alzheimer disease are identified, drug trials may be used more for research than therapy.

A key to understanding the disease and the best hope for an effective treatment may lie in a better understanding of the beta-amyloid protein that forms the plaques in the brain. The formation and biological function of beta amyloid protein is still unclear, even though its existence has been known since 1906, when Alois Alzheimer saw the protein in plaques in the brain of a patient with dementia.

Researchers are striving to decrease the production of beta-amyloid protein or of the larger amyloid precur-

sor protein, from which it is derived, in hopes of decreasing the formation of the damaging plaques.

Other researchers have developed **animal** models of Alzheimer disease. By introducing into **mice** genes that code for a changed form of human beta-amyloid protein, scientists have created animals that experience brain degeneration and plaques. Tests with these animals may one day help determine the role of beta-amyloid protein in Alzheimer disease and could help researchers find drugs for treatment.

Not all researchers agree on the significance of beta-amyloid protein in the progression of Alzheimer disease.

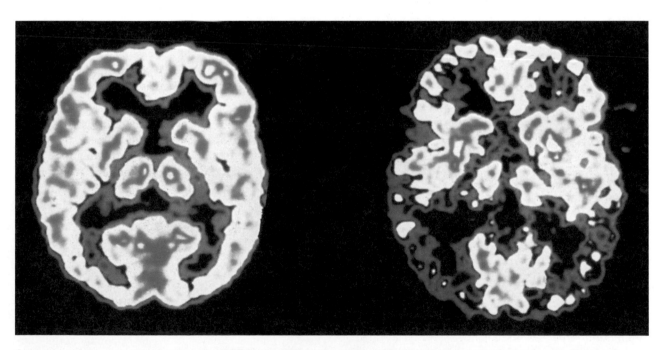

Colored positron emission tomography (PET) brain scans comparing a normal brain (left) with the brain with Alzheimer disease. © *Photo Reasearchers. Reproduced by permission.*

Indeed, some scientists doubt that the beta-amyloid protein is the key to understanding and treating the disease. Rather, they suppose that the accumulation of large amounts of beta-amyloid protein is a response to an as-yet undiscovered cause.

See also Chromosomal abnormalities; Prions.

Resources

Periodicals

Selkoe, D.J. "Alzheimer disease: Genotypes, Phenotype and Treatment." *Science* 275 (January 1997): 630–631.

Organizations

Alzheimer's Association, 919 North Michigan Avenue, Suite 1100, Chicago, IL 60611–1676. (800) 272–3900. <http://www.alz.org>.

Administration on Aging, 330 Independence Avenue SW, Washington, DC 20201. (800) 677-1116. <http://www.aoa.gov/factsheets/alz.html.>.

Brian Hoyle

AM *see* Radio waves

Amaranth family (Amaranthaceae)

The amaranth (or pigweed) family is a large group of dicotyledonous flowering plants known to botanists as the Amaranthaceae. It is a relatively large family, having about 65 genera and 900 **species**. The species in this family are mostly annual or perennial herbs, although a few species are shrubs or small trees. Botanists divide Amaranthaceae into two subfamilies: the Amaranthoideae and the Gomphrenoideae, based on certain morphological characteristics of their flowers.

The flowers of most species in the Amaranthaceae are bisexual (or monoecious), meaning they have both male and female reproductive organs. In all species, the flowers are small and have radial **symmetry**. The flowers of most species arise in a dense inflorescence, or **flower** cluster, with each flower of the inflorescence subtended by one or more small red bracts (modified leaves). The small red bracts remain present as the flower matures into a fruit. The flowers of most species produce **nectar** and are insect-pollinated. An exception is *Amaranthus*, a genus with about 50 species, whose flowers are **wind** pollinated and do not make nectar. All species have simple, non-compound leaves.

Many species in Amaranthaceae have red inflorescences, **fruits**, and vegetative parts, due to the presence

Love-lies-bleeding, a member of the amaranth family.
Photograph by Alan L. Detrick. National Audubon Society Collection/Photo Researchers, Inc. Reproduced by permission.

of betalain pigments. Betalains are a class of nitrogen-containing pigments, which only occur in 10 evolutionarily related **plant** families, known as the Centrospermae. Interestingly, none of the species with betalain pigments also have flavonoid pigments. Flavonoids and betalains can be similar in **color**, even though they have different chemical structures.

Most of the 900 species of Amaranthaceae are native to tropical and subtropical regions of **Africa**, Central America, and **South America**. The number of Amaranthaceae species declines as one approaches the northern and southern temperate zones. There are about 100 species of this family in **North America**. Many species of Amaranthaceae are considered weeds, since they invade disturbed areas, such as agricultural fields and roadsides.

Several species in the Amaranthaceae family are used by humans. Some species are important horticultural plants, such as *Amaranthus caudatus*, commonly known as "love-lies-bleeding." The **seeds** of several

Tumbleweed plants (of the amaranth family) growing in dry soil. *Photograph by Tom McHugh. Photo Researchers, Inc. Reproduced by permission.*

species of the *Amaranthus* genus were eaten by indigenous peoples of North and South America, and were cultivated over 5,000 years ago in the Tehuacan region of modern-day Mexico. Grain amaranths are still grown throughout Central America and Mexico, and also as a minor cash crop in the United States. Many health food stores currently sell amaranth grain, a flour-like substance made by grinding amaranth seeds. Amaranth grain can be used with **wheat** to make bread, or can be cooked with water to make a side dish.

Amaryllis family (Amaryllidaceae)

Species in the amaryllis family are flowering plants, and are mostly long-lived, perennial herbs arising from a bulb or, less commonly, from rhizomes (underground stems). These plants have linear or strap shaped leaves, either crowded around the base of a leafless flowering stem, or arranged in two tight rows along a short stem, as in the common houseplant *Clivia*. The leaves are usually hairless and contain mucilage cells, or cells filled with **calcium** oxalate crystals known as raphides for defense against herbivores. Silica-filled (**glass**) cells, which are typical of many other monocotyledonous **plant** families, are absent from the amaryllis family.

The flowers of amaryllids are bisexual, with six perianth parts (or tepals) that sometimes have appendages

A Joshua tree (*Yucca brevifolia*). *Photograph by Gregory Ochoki. Photo Researchers, Inc. Reproduced by permission.*

that form a corona, as in the central, protruding part of a daffodil **flower** (*Narcissus* spp.). The flowers are white, yellow, purple, or red, but never blue. The flowers are pollinated by **bees** or **moths**, but many are also adapted to bird and some to bat **pollination**. The **fruits** are usually many-seeded capsules, or sometimes berries.

Between 900 and 1,300 species of amaryllids have been recognized. Most are tropical or subtropical, with centers of distribution in South Africa, the western Mediterranean (especially Spain and Morocco), and to a lesser extent, Andean **South America**. Many species are drought-resistant xerophytes that produce leaves in the spring or when the rainy season begins, open their stomates only at night, have stomates located in the bottom of pits, and have thick waxy leaves—all to conserve water.

Many amaryllids are prized as ornamentals for home or garden because of their large, showy flowers, which are held high above the contrasting dark green leaves. The Cape belladona (*Amaryllis belladona*) is a native of dry regions of southwestern Cape Province,

On the left, a flowering century plant (*Agave shawii*), and on the right a cirio (*Idria columnaris*), Baja California, Mexico. Some plant taxonomists include agaves in the amaryllis family, while others place them in their own family, the Agavaceae. *Photograph by F. Gohier. National Audubon Society Collection/Photo Researchers, Inc. Reproduced by permission.*

South Africa, and is widely cultivated for its large, pink, bell-shaped flowers, which are moth-pollinated. The genus *Hippeastrum*, which is native to the West Indies, Mexico, and as far south as Argentina, has also been called *Amaryllis* by some taxonomists. Whatever its correct identity, many species of this genus have spectacular, large flowers that have evolved for pollination by **birds**, and are commonly grown as ornamentals.

Many members of the tribe Narcisseae are also widely cultivated, both indoors and outdoors. Most of the horticultural varieties originate from Spain, Portugal, or Morocco, and are small to medium-sized herbs, with linear leaves around a leafless stem that typically bears one to several flowers. The spring-flowering species have been most intensively bred as garden ornamentals, especially the common daffodil (*Narcissus pseudonarcissus*). Snowdrops (*Galanthus nivalis*), an early bloom-

ing ornamental, and St. John's lily (*Crinum asiaticum*), are also amaryllids. Clivia is a common houseplant, especially in **Europe**, that is prized for its deep-green, shiny leaves and its large salmon-colored flowers, and for the fact that it requires little **light**, water, or attention.

Agaves are often included in the amaryllis family. However, this taxonomic treatment is controversial, and agaves are sometimes put into their own family, the Agavaceae. About 300 species live in dry habitats from the southern United States to northern South America. These are conspicuous perennials, with a dense rosette of large, persistent sword-like leaves that bear spines at the tip, and often along the margins as well.

The scientific name *Agave* comes from the Greek *agaue*, which means noble, referring to the height of their flowering stalks. Their common name, century plant, refers to the long period of time that these plants remain in a non-sexual, vegetative state before flowering. This period generally lasts for 5-50 years, and not 100 years as the name implies. When they are ready to flower, *Agave* plants rapidly develop a thick flowering stem that may reach 20 ft (6 m) in height. Greenhouse keepers have sometimes come to work in the morning to find a flowering stem of an agave poking through the broken glass of their greenhouse. After this one episode of sexual activity, some species of agave plants die.

A few species of agaves are commercially important. *Agave americana,* commonly called American aloe or century plant, contains aloe, which is a commonly used ingredient in shampoos, moisturizers, and salves. Also important is the production of Mexico's national alcoholic beverage, pulque, which is mostly made from *Agave americana,* although a few other species of agave are also used. When the flowering stem is formed, a large amount of sap is produced. The bud at the end of the developing flowering stem is removed, and a cavity scooped out in which the sugary sap collects. As much as 238 gal (900 l) can be recovered from one plant over three to four months. The sap is fermented, resulting in a milky liquid with an **alcohol concentration** of 4-8 %. This beverage is pulque. A more potent liquor known as mescal is produced by **distillation** of pulque.

Many species of agave are valuable for the long fibers in their leaves. Aztecs used the fibers of henequen (*A. fourcroydes*) to make a twine, and the practice continues today. The stronger sisal **hemp** is derived from leaves of *A. sisalana,* native to the Americas but now grown in many parts of the tropics. Sisal is commonly used to manufacture baler twine for agriculture and a parcelling twine. The hard fibers of *Furcraea macrophylla* of Colombia have been used to make the large bags used for shipping coffee beans from that and other countries.

Resources

Books

The American Horticultural Society. *The American Horticultural Society Encyclopedia of Plants and Flowers.* New York: DK Publishing, 2002.

Heywood, Vernon H. ed. *Flowering Plants of the World.* New York: Oxford University Press, 1993.

Les C. Cwynar

Amber *see* **Fossil and fossilization**

American Standard Code for Information Interchange

The American Standard Code for Information Interchange (ASCII; pronounced "askee") was first introduced in 1968 as a method of encoding alphabetic and numeric data in digital format. Although ASCII code was originally developed for the teletypewriter industry, it has since found widespread use in computer and information-transfer technologies.

Because ASCII code is standardized, computers and other electronic devices can use it to exchange data with each other. This is true even of computers that use different operating systems, for example PCs and Macintoshes.

As originally formulated, each ASCII-encoded representation consisted of a string of seven digits, where each digit was either a 0 or a 1 (i.e., binary code). There were as a result 128 possible ways of arranging these 0s and 1s. In this representation, each alphanumeric character was uniquely assigned a number between 0 and 127, which was represented by its binary equivalent in a string of seven 0s and 1s. The ASCII notation for the capital letter A, for example, is the binary code representation (1000001) for the base-10 number 65; similarly, a blank space has the binary code for the base-10 number 32.

Inside computers, each English alphabet character is represented by a string of eight 0s and 1s. Each of the digits in the string is known as a bit, and a series of eight bits is known as a byte. Because ASCII code as originally formulated constituted only 7 bits, when 7-bit ASCII code was embedded in the eight-bit computer code, there was one bit left over.

At one time, this extra bit was used primarily for the purpose of checking errors in data transmission. But today, computers use this extra bit to encode an additional 128 characters for the purpose of representing special symbols. Note that with eight-bit encoding, the number of possible arrangements of 0s and 1s increases from 128 to 256.

As an example of the eight-bit encapsulation of seven-bit ASCII code, note that the eight-bit representation for the letter "A" (01000001) simply places a 0 in the eighth-bit position relative to the seven-bit representation (1000001). Seven-bit ASCII still has some advantages, however, as it is recognized by all computers, including PCs, Macs, UNIX or VMS mainframes, printers, and any other computer-related equipment.

Eight-bit ASCII code is known as extended ASCII code. This representation was introduced by IBM in 1981 for use in its first personal computer. Extended ASCII code quickly became a standard in the personal computer industry. Unlike the original seven-bit ASCII code, the extended code uses 32 of its 256 character representations to encode nonprinting commands such as "form feed."

Another 32 character representations are reserved for numbers and punctuation marks. Thirty-two more representations are for upper case letters and additional punctuation marks. The last 32 representations are reserved for lower case letters. Note that the upper and lower case letters have distinct representations differing by 32.

In languages other than English, where there are much larger character sets, for example Chinese and Japanese, a single byte (eight bits) is not sufficient for representing all the characters in the language. However, by representing each character by two bytes (16 bits), it is possible to assign a unique number code to each character.

In the United States, most computers require slightly modified operating systems to be able to handle two bytes at a time, and special reference tables to display the characters. It is therefore necessary to change operating systems before one can run Japanese or Chinese software here. But U.S. software applications will run without problems on computers in Japan and China equipped with operating systems that recognize two-byte characters.

The term ASCII is sometimes used imprecisely to refer to a type of text computer document. A file that contains ASCII text (also known as plain text) is one that does not contain any special embedded control characters. This encoding system not only lets a computer store a document as a series of numbers, but also lets the computer share the document with other computers that use the ASCII system.

Resources

Other

American National Standard for Information Sciences: Codes for the Representation of Languages for Information Interchange. National Information Standards Organization, 1991.

Randall Frost

Americium *see* **Element, transuranium**

Ames test

The Ames test, named for its developer, Bruce Ames, is a method to test chemicals for their cancer-causing properties. It is used by cosmetic companies, pharmaceutical manufacturers, and other industries that must prove that their products will not cause **cancer** in humans.

Ames, a cancer researcher at the University of California, began development of his method in the late 1950s. He believed an efficient, less-expensive means could be found to screen substances than the cumbersome methods in use. He hit upon the use of **bacteria**, which could be grown (cultured) cheaply, and grew rapidly so that testing could be completed quickly, yet could indicate the carcinogenic (cancer-causing) potential of many chemicals.

The bacterium used is a strain of *Salmonella typhimurium* that lacks an **enzyme** needed to form colonies. The bacterium is grown on agar culture (agar is a gelatin-like substance with **nutrients**). The substance to be tested is blotted on a bit of **paper** and placed on the agar. If the substance is a **carcinogen** it will cause mutations in the bacterium as the cells divide. The mutant cells will have the enzyme to form colonies. The test can be completed in a day.

Bacterial mutations are the result of DNA damage. Because DNA in bacteria is similar to that in the higher animals, it is assumed the substance also will damage DNA in those animals and cause **cell** mutations and possibly cancer.

The Ames test is a screening test; it is not the final test that any substance must undergo before being commercially produced. It is designed to detect a cancer-causing agent quickly and inexpensively.

Prior to development of this test the procedure to determine whether a substance was a carcinogen required feeding the test substance to or injecting it into laboratory animals such as **rats** or **mice** and then examining them for evidence of **tumor** formation. Testing took years to complete, hundreds of animals, and millions of dollars.

Any substance that causes bacterial **mutation** in the Ames test is not given further consideration for development. A substance that does not produce bacterial mutation still must undergo **animal** testing, but at least the manufacturer has a good idea that it is not a cancer-causing agent.

Amicable numbers

Two numbers are said to be amicable (i.e., friendly) if each one of them is equal to the sum of the *proper* divisors of the others, i.e., whole numbers less than the given numbers that divide the given number with no remainder. For example, 220 has proper divisors 1, 2, 4, 5, 10, 11, 20, 22, 44, 55, and 110. The sum of these divisors is 284. The proper divisors of 284 are 1, 2, 4, 71, and 142. Their sum is 220; so 220 and 284 are amicable. This is the smallest pair of amicable numbers.

The discovery of amicable numbers is attributed to the neo-Pythagorean philosopher Iamblichus (c.250-330), who credited Pythagoras with the original knowledge of their nature. The Pythagoreans believed that amicable numbers, like all special numbers, had a profound cosmic significance. A biblical reference (a gift of 220 **goats** from Jacob to Esau, Genesis 23: 14) is thought by some to indicate an earlier knowledge of amicable numbers.

No pairs of amicable numbers other than 220 and 284 were discovered by European mathematicians until 1636, when the French mathematician Pierre de Fermat (1601-1665) found the pair 18,496 and 17,296. A century later, the Swiss mathematician Leonhard Euler (1707-1783) made an extensive search and found about 60 additional pairs. Surprisingly, however, he overlooked the smallest pair, 1184 and 1210, which was subsequently discovered in 1866 by a 16-year-old boy, Nicolo Paganini.

During the medieval period, Arabian mathematicians preserved and developed the mathematical knowledge of the ancient Greeks. For example, The polymath Thabit ibn Qurra (836-901) formulated an ingenious rule for generating amicable number pairs: Let $a = 3(2^n) - 1$, $b = 3(2^{n-1}) - 1$, and $c = 9(2^{2n-1}) - 1$; then, if a, b, and c are primes, $2^n ab$ and $2^n c$ are amicable. This rule produces 220 and 284 when n is 2. When n is 3, c is not a prime, and the resulting numbers are not amicable. For n = 4, it produces Fermat's pair, 17,296 and 18,416, skipping over Paganini's pair and others.

Amicable numbers serve no practical purpose, but professionals and amateurs alike have for centuries enjoyed seeking them and exploring their properties.

Amides

An amide is a nitrogen-containing compound that can be considered a derivative of **ammonia**, NH_3. Organic amides contain the **group** (-C=O). The simplest organic amide is methamide,

$$HC=O$$
$$|$$
$$NH_2$$

Inorganic amides consist of a **metal** or some other **cation** combined with the amide ion ($-NH_2$), as in **sodium** amide ($NaNH_2$). In comparison with their better known organic cousins, the inorganic amides are less important. One member of the inorganic family, sodium amide, does find some application as a dehydrating agent and in the production of the dye indigo, the rocket fuel hydrazine, sodium cyanide, and other compounds.

Classification and properties

Like the amines, the amides can be classified as primary, secondary, or tertiary, depending on the number of **hydrogen atoms** substituted in the ammonia **molecule**. An amide containing the $-NH_2$ group is a primary amide, one containing the -NH group is a secondary amine, and one containing the -N- group is a tertiary amine.

Although amides and amines both contain an amino group ($-NH_2$, NH or N), the former are much weaker bases and much stronger acids than the latter. Amides undergo many of the same reactions as do other derivatives of organic acids. For example, they undergo **hydrolysis** to produce the parent carboxylic acid and ammonia. They can also be dehydrated with a strong agent such as diphosphorus pentoxide, P_2O_5. The product of this reaction, a nitrile, a compound containing the $-C\equiv N$ group, is widely used in the synthesis of other organic compounds.

Some familiar amides

The synthesis of a protein results in a protein in the formation of an amide bond between adjacent amino acids. **Proteins** can be considered the most common examples of amides in the natural world. A naturally occurring amide is nicotinamide, one of the B vitamins. A third familiar natural amide is **urea**, also known as carbamide. Urea is the compound by which otherwise toxic wastes are excreted from mammalian bodies.

Important synthetic amides

Perhaps the best known of all synthetic amides is the fiber known as nylon. In 1931, the American chemist Wallace Hume Carothers discovered a process for making one of the first synthetic fibers. He found that the addition of adipic acid to hexamethylene diamine resulted in the formation of a strong, fiber-like product to which he gave the name Nylon 66. The 66 part of the name reflects the fact that adipic acid and hexamethylene diamine each contain six **carbon** atoms in their molecules.

The reaction between these two substances results in the formation of a long **polymer**, somewhat similar to the structure of natural protein. As in protein, the subunits of nylon are joined by amide bonds. For this reason, both protein and nylon can be thought of as polyamides, compounds in which a large number of amide units are joined to each other in a long chain.

Other types of nylon were also developed at later dates. One form, known as Nylon 6, is produced by the polymerization of a single kind of molecule, 6-aminohexanoic acid. The bonding between sub-units in Nylon-6-amide bonds is the same as it is in Nylon 66. In all types of nylon, the fiber obtains its strength from hydrogen bonding that occurs between **oxygen** and hydrogen atoms on adjacent chains of the material.

Another type of polymer is formed when two of the simplest organic compounds, urea and formaldehyde, react with each other. In this reaction, amide bonds form between alternate urea and formaldehyde molecules, resulting in a very long polyamide chain. Urea formaldehyde polymers are in great demand by industry, where they are used as molding compounds, in the treatment of **paper** and **textiles**, and as a binder in particle board, to mention but a few uses.

An amide with which many people are familiar is acetaminophen, an analgesic (pain-killer). It is the active ingredient in products such as Amadil, Cetadol, Datril, Naprinol, Panadol, and Tylenol. Another amide analgesic is phenacetin, found in products such as APC (aspirin, phenacetin, and **caffeine**) tablets and Empirin.

Other commercially important amides include the insect repellant N,N-dimethyl-m-toluamide (Off, Deet), the local anesthetics lidocaine (Xylocaine) and dibucaine (Nupercaine), the tranquilizer meprobromate (Miltown, Equaine), and the **insecticides** Sevin and Mipcin.

See also Artificial fibers.

KEY TERMS

Amide ion—An anion with the formula -NH$_2$.

Hydrogen bond—A weak chemical bond that results from the electrical attraction between oppositely charged particles, usually a hydrogen ion and an oxygen-containing ion.

Nylon—A group of synthetic fibers that have in common the fact that they are large polymers held together by amide bonding.

Polymer—A large molecule made up of many small sub-units repeated over and over again.

Protein—Any large naturally occurring polymer made of many amino acids bonded to each other by means of amide bonds.

Synthetic detergent—A soap-like material that is manufactured from raw materials.

Resources

Books

Carey, Francis A., and Richard J. Sundberg *Advanced Organic Chemistry: Structure and Mechanisms.* 4th ed. New York: Plenum, 2001.

Carey, Francis A. *Organic Chemistry.* New York: McGraw-Hill, 2002.

Embree, Harland D. *Brief Course: Organic Chemistry.* Glenview, IL: Scott, Foresman, 1983, pp. 362-67.

Joesten, Melvin D., David O. Johnston, John T. Netterville, and James L. Wood. *World of Chemistry.* Belmont, CA: Brooks/Cole Publishing Company, 1995.

Ouellette, Robert. *Chemistry: An Introduction to General, Organic, and Biological Chemistry.* Prentice Hall, 1994.

David E. Newton

Amino acid

Amino acids are organic compounds made of **carbon**, **hydrogen**, **oxygen**, **nitrogen** and, in a few cases, **sulfur**. The basic structure of an amino acid **molecule** consists of a carbon atom that is bonded to an amino group (-NH$_2$), a **carboxyl group** (-COOH), a hydrogen atom and a fourth group that differs from one amino acid to another and is often referred to as the -R group or the side chain. The -R group can vary widely and is responsible for the differences in the chemical properties. The name, amino acid, comes from the amino group and the acid group which are the most reactive parts of the molecule. The amino acids that are important in the biological

world are referred to as a-amino acids because the amino group is bonded to the a-carbon atom, that is, the one adjacent to the carboxyl group.

A chemical reaction that is characteristic of amino acids involves the formation of a bond, called a **peptide linkage**, between the carboxyl group of one amino acid and the amino group of a second amino acid. Very long chains of amino acids can bond together in this way to form **proteins**. The importance of the amino acids in nature arises from their ability to form proteins, which are the basic building blocks of all living things.

The specific properties of each kind of protein are largely dependent on the kind and sequence of the amino acids in it. Other chemical behavior of these protein molecules is due to interactions between the amino and the carboxyl groups or between the various -R groups along the long chains of amino acids in the molecule. These chemical interactions confer a three-dimensional configuration on the protein, which is essential to its proper functioning.

Chemical structure

Although most of the known amino acids were identified and isolated (sometimes in impure form) during the nineteenth century, the chemical structures of many of them were not known until much later. Understanding their importance in the formation of proteins, the basis of the structure and function of all cells is of even more recent origin, dating to the first part of the twentieth century. Only about 20 amino acids are common in humans, with two others present in a few **animal species**. There are over 100 other lesser known ones that are found mostly in plants.

Each of the common amino acids has, in addition to its chemical name, a more familiar name and a three-letter abbreviation that is frequently used to identify it. They are often grouped by similarities in the chemical properties of the side chains. The side chain of glycine (gly) consists of a single hydrogen atom; alanine (ala), valine (val), leucine (leu), and isoleucine (ile) all have **hydrocarbon** (containing only hydrogen and carbon) side chains; proline (pro) has a hydrocarbon that is part of a ring structure; serine (ser) and threonine (thr) have an **alcohol** (-OH) side chain; cysteine (cys) and methionine (met) both have sulfur **atoms** as part of the -R group; phenylalanine (phe), tyrosine (tyr), and tryptophan (trp) all contain an aromatic ring (related to the **benzene** ring) as part of the side chain; aspartic acid (asp) and glutamic acid (glu) have a second carboxylic acid group while asparagine (asn) and glutamine (gln) have a carboxylic acid derivative (a -CONH$_2$) group; and lysine (lys), arginine (arg), and histidine (his) have an -R group that contains a second amino group.

Although amino acid molecules contain an amino group and a carboxyl group, certain chemical properties are not consistent with this structure. Unlike the behavior of molecules with amino or carboxylic acid functional groups alone, amino acids exist mostly as crystalline solids that decompose rather than melt at temperatures over 392°F (200°C). They are quite soluble in **water** but insoluble in non-polar solvents like benzene or **ether**. Their acidic and basic properties are exceptionally weak for molecules that contain an acid carboxyl group and a basic amino group.

This problem was resolved when it was realized that amino acids are better represented as dipolar ions, sometimes called zwitterions (from the German, meaning hybrid ions). Although the molecule as a whole does not have a net charge, there is a transfer of an H^+ ion from the carboxyl group to the amino group; consequently, the amino group is present as an $-NH_3^+$ and the carboxyl group is present as a $-COO^-$ (Fig. 1). This reaction is an acid-base interaction between two groups in the same molecule and occurs because the $-COOH$ group is a rather strong acid and the $-NH_2$ group is a rather strong base. As a result of this structure, amino acids can behave as acids in the presence of strong bases or they can behave as bases in the presence of strong acids.

One other property of amino acids that is important to their chemical behavior is that all of the amino acids except glycine can exist as mirror images of each other; that is, right- or left-handed versions of the molecule. Like the positions of the thumb and fingers of a glove, the right hand being the mirror image of the left hand, the positions of the functional groups on the a carbon can be mirror images of each other. Interestingly, nearly all of the amino acids occurring in nature are the left-handed versions of the molecules. Right-handed versions are not found in the proteins of higher organisms but are present in some lower forms of life such as in the **cell** walls of **bacteria**. They are also found in some **antibiotics** such as streptomycin, actinomycin, bacitracin, and tetracycline. These antibiotics can kill bacterial cells by interfering with the formation of protein necessary for maintaining life and for reproducing.

Bonding

Amino acids are extremely important in nature as the monomers, or individual units, that join together in chains to form copolymers (polymers made of more than one kind of **monomer**). The chains may contain as few as two or as many as 3,000 amino acid units. Groups of only two amino acids are called dipeptides; three amino acids bonded together are called tripeptides; if there are more than 10 in a chain, they are called polypeptides; and if there are 50 or more, they are known as proteins.

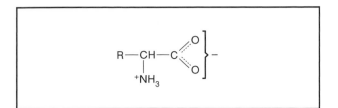

Figure 1. *Illustration by Hans & Cassidy. Courtesy of Gale Group.*

In 1902 the German organic chemist, Emil Fischer, first proposed that the amino acids in polypeptides are linked together between the carboxyl group of one amino acid and the amino group of the other. This bond forms when the -OH from the carboxyl end of one molecule and the hydrogen from the amino end of another molecule split off and form a small molecule byproduct, H_2O or water. This type of reaction is called a condensation reaction. The new bond between the carbon atom and the nitrogen atom is called a peptide bond, also known as an amide linkage. Because every amino acid molecule has a carboxyl end and an amino end, each one can join to any other amino acid by the formation of a peptide bond.

All the millions of different proteins in living things are formed by the bonding of only 20 amino acids to form long **polymer** chains. Like the 26 letters of the alphabet that join together to form different words, depending on which letters are used and what the sequence is, the 20 amino acids can join together in different combinations and sequences to form proteins. But whereas words usually have only about 10 or fewer letters per word, proteins are usually made from at least 50 amino acids to more than 3,000. Because each amino acid can be used many times along the chain and because there are no restrictions on the length of the chain, the number of possible combinations for the formation of protein is truly enormous.

The amino acids in polypeptides can be represented in three ways: by writing out the complete chemical formulas; by writing the amino acid sequence using the standard, three-letter abbreviation for each acid as in gly-ser-ala (which represents glycine, serine and alanine); or by naming the polypeptide as in glycylserylalanine. The name is derived by dropping the -ine or -ic ending of each amino acid along the chain and replacing it with a -yl ending. The last acid of the chain is given its full name. It is common practice to write polypeptides with the free, unbonded amino group on the left and the free carboxylic acid group on the right.

Order is important in the functioning of a protein; gly-ser-ala, gly-ala-ser, and ala-ser-gly, for example, are different peptides. In fact, there are 27 different tripeptides that are possible from these three amino acids. (Each may be used more than once.) There are about two quadrillion different proteins that can exist if each of the 20 amino acids present in humans is used only once. However, just as not all sequences of letters make sense, not all sequences of amino acids make functioning proteins and other sequences can cause undesirable effects. While small mistakes in the amino acid sequence can sometimes be tolerated in nature without serious problems, at other times malfunctioning proteins can be caused by a single incorrect amino acid in the polymer chain. **Sickle cell anemia** is a fatal **disease** caused by a single amino acid, glutamic acid being replaced by a different one, valine, at the sixth position from the end of the protein chain in the hemoglobin molecule. This small difference causes lower **solubility** of the sickle cell hemoglobin molecules. They precipitate out as small rods which give the cells the characteristic sickle shape and result in the often fatal disease.

The specific sequence of the amino acids along the protein chain is referred to as the primary structure of the protein. However, these chains are not rigid, but rather they are long and flexible like string. The strands of protein can twist to form helixes or fold into sheets. They can bend and fold back on themselves to form globs and several protein molecules sometimes combine into a larger molecule. All of these configurations are caused by interactions both within a single protein strand as well as between two or three separate strands of protein.

Just as proteins are formed when amino acids bond together to form long chains, they can be broken down again into their individual amino acids by a reaction called **hydrolysis**. This reaction is just the reverse of the formation of the peptide bond. In the process of digestion, proteins are once again broken down into their individual amino acid components. Special digestive enzymes are necessary to cause the peptide linkage to break and a molecule of water is added when the reaction occurs. The resulting amino acids are released into the small intestine where they can easily pass into the bloodstream and be carried to every cell of the **organism**. There, once again, each individual cell can use these amino acids to assemble the new and different proteins required for its specific functions. Life goes on by the continual breakdown of protein into the individual amino acid units followed by the buildup of new protein from these amino acids.

Of the 20 amino acids required by humans for making protein, 12 of them can be made within the body from other **nutrients**. But the other eight, called the essential amino acids, cannot be made by the body and must be obtained from the diet. These are isoleucine, leucine, lysine, methionine, phenlyalanine, threonine,

KEY TERMS

Amino group—An -NH_2 group.

BETA-amino acid—An amino acid in which the -NH_2 (amino) group is bonded to the carbon atom.

BETA-carbon atom—The carbon atom adjacent to the carboxyl group.

Carboxyl group—A -COOH group; also written -CO_2H.

Essential amino acid—Amino acids that cannot be synthesized by the body and must be obtained from the diet.

Monomers—Small, individual subunits which join together to form polymers.

Peptide—Substances made up of chains of amino acids, usually fewer than 50.

Peptide bond—The bond formed when the carboxyl group of one amino acid joins with the amino group of a second amino acid and splits off a water molecule.

tryptophan, and valine. In addition, arginine and histidine are believed to be essential to growing children but may not be essential to mature adults. An adequate protein is one that contains all of the essential amino acids in sufficient quantities for growth and repair of body **tissue**. Most proteins from animal sources (**gelatin** being the only exception), contain all the essential amino acids and are considered adequate proteins. Many **plant** proteins do not contain all of the essential amino acids. Corn, for example, does not contain the essential amino acids lysine and tryptophan. **Rice** is lacking in lysine and threonine, **wheat** is lacking in lysine, and soy beans are lacking in methionine. People who are vegetarians and do not consume animal proteins in their diets sometimes suffer from **malnutrition** because of the lack of one or more amino acids in their diets even though they may consume enough food and plenty of calories.

See also Nutrition.

Resources

Books

Durbin, Richard, et al. *Biological Sequence Analysis: Probabilistic Models of Proteins and Nucleic Acids.* Cambridge: Cambridge University Press, 1999.

Lide, D. R., ed. *CRC Handbook of Chemistry and Physics* Boca Raton: CRC Press, 2001.

Newhouse, Elizabeth L., et al., eds. *Inventors and Discoverers: Changing Our World.* Washington, DC: National Geographic Society, 1994.

White, James, and Dorothy C. White, eds. *Proteins, Peptides, and Amino Acids Sourcebooks.* Humana Press; 2002.

Periodicals

Bishop, Katherine. "Baby Boomers Fight Aging by Dropping Acid (Amino)." *New York Times* (10 June 1992): B1(N).

Venter, J.C., et al. "The Sequence of the Human Genome." *Science* 291 (2001): 1304-1351.

Leona B. Bronstein

Ammonia

Ammonia, composed of three parts **hydrogen** and one part **nitrogen**, is a sharp-smelling, flammable, and toxic gas that is very soluble in **water**, where it acts as a base in its **chemical reactions**.

Ammonia in the past

Ammonia was present in the primordial atmosphere of **Earth**, and may have been the source of nitrogen for the earliest forms of life, although much controversy exists over the details. In ancient Egypt, ammonium compounds were used in rites honoring the god Ammon, from which came the name we still use for the gas and its compounds. Early chemists learned to generate ammonia from **animal** parts such as deerhorn, and obtained ammonial preparations (spirits of hartshorn, etc.), but Joseph Priestley (1733-1804) first collected and experimented with the pure substance. C. L. Berthollet (1748-1822) proved that ammonia is composed of nitrogen and hydrogen.

In the nineteenth century, ammonia was sometimes manufactured by the action of steam on **calcium** cyanamide, called the cyanamide process, which in turn was made by reacting calcium carbide with nitrogen at high temperatures. In the early twentieth century, German chemists Fritz Haber and Carl Bosch learned how to make ammonia in large quantities by high-pressure catalytic reactions of nitrogen (from air) with hydrogen. Both men were awarded Nobel prizes—Haber in 1918 and Bosch in 1931. The Haber-Bosch process is the basis for modern ammonia production, although many improvements have been made in the details of the technology.

Physical and chemical properties of ammonia

Ammonia (**boiling point** -28.03°F [-33.35°C]) can be made in the laboratory by heating ammonium chloride with lime, and the gas collected by downward displacement of air, or displacement of mercury. Water solutions of ammonia, called ammonium hydroxides, having as much as 28% ammonia by weight, can be obtained by this method. Ammonium hydroxide exhibits the characteristics of a weak base, turning litmus **paper** blue, and neutralizing acids with the formation of ammonium salts. Transition **metal** ions are either precipitated as hydroxides (**iron** [II], iron [III]) or converted to ammonia complexes (**copper** [II], nickel [II], zinc [II], silver [I]). The copper (II) ammonia **complex**, in **solution**, is deep blue in **color**, and serves as a qualitative test for copper. It also has the ability to dissolve **cellulose**, and has been used in the process for making regenerated cellulose fibers, or rayon.

Ammonia molecules possess a pyramidal shape, with the nitrogen atom at the vertex. These molecules continually undergo a type of **motion** called inversion, in which the nitrogen atom passes through the **plane** of the three hydrogen **atoms** like an umbrella turning inside out in the wind. When ammonia acts as a base, the nitrogen atom bonds either to a **proton** (to form ammonium ion) or to a metal **cation**. Ammonium salts such as ammonium chloride, called sal ammoniac, are water soluble and volatile when heated. It is often found that considerable **heat** is absorbed when ammonium salts dissolve in water, leading to dramatic reduction in **temperature**. Ammonium salts containing anions of weak acids (carbonate, sulfide) easily liberate ammonia owing to the tendency of a proton to break off the nitrogen atom and be bound by the weak acid **anion**.

In liquid or frozen ammonia, the molecules attract one another through sharing a hydrogen atom between one **molecule** and the next, called hydrogen bonding. In this attraction, called association, compounds apparently containing free electrons can be obtained by treating sodium/ammonia solutions with complexing agents.

Ammonia is a flammable gas, and reacts with **oxygen** to form nitrogen and water, or nitrogen (II) oxide and water. Oxidation of ammonia in solution leads to hydrazine, a corrosive and volatile ingredient in fuels. Ammonium salts of oxidizing anions—nitrate, dichromate, and perchlorate—are unstable and can explode or deflagrate when heated. Ammonium nitrate is used as a high explosive; ammonium perchlorate as a component of rocket fuels. Ammonium dichromate is used in a popular artificial **volcano** demonstration, in which a conical pile of the **salt** is ignited and burns vigorously, throwing off quantities of green chromium (III) oxide-the lava.

When ammonium hydroxide is treated with iodine crystals, an explosive brown solid, nitrogen triiodide, is formed. When dry, this substance is so sensitive that the lightest touch will cause it to explode with a crackling sound and a puff of purple iodine vapor.

Sources and production of ammonia

Ammonia is manufactured by the reaction of hydrogen with nitrogen in the presence of an iron catalyst, which is known as the Haber-Bosch process. The reaction is exothermic and is accompanied by a **concentration** in **volume**. (The ammonia occupies less volume than the gases from which it is made.) High **pressure** conditions (150-250 bar) are used, and temperatures range from 752–932°F (400–500°C). The mixed gases circulate through the catalyst, ammonia is formed and removed, and the unconverted reactants are recirculated. Large ammonia plants can produce over 1,000 tons per day. Each ton of ammonia requires 3,100 cu yd (2,400 cu m) of hydrogen and 1,050 cu yd (800 cu m) of nitrogen, as well as 60 gigajoules of **energy**. Much of the energy is consumed in the compressors needed to attain the high pressure used in the synthesis, and in heating the reactants. Further energy is needed to produce the hydrogen from **hydrocarbon** feedstocks, and to separate nitrogen from air. The synthesis reaction itself produces some heat, and great attention is given to heat efficiency, and use of waste heat. The gases that enter the catalytic converter must be highly purified and free of **sulfur** compounds, which adversely affect the catalyst. The catalyst is prepared in place by hydrogen treatment of magnetite, an iron oxide containing potassium hydroxide and other oxides in small amounts as promoters. A large ammonia plant might have as much as 100 tons of catalyst.

Since the hydrogen is usually derived from a **natural gas** called methane, the price of ammonia is very sensitive to the availability or price of fuels. United States production of ammonia reached 17 million tons in 1991, and demand was even larger than U.S. production, leading to about two million tons of imports. World ammonia production is about 100 million tons per year, which amounts to about 40 lbs (18 kg) for each person on earth.

Ammonia is formed from nitrogen in air by the action of nitrogen-fixing **bacteria** that exist in the **soil** on the roots of certain plants like alfalfa. **Nitrogen fixation** can also be accomplished by blue-green **algae** in the sea. These bacteria and algae possess an **enzyme** called nitrogenase that permits them to convert nitrogen to ammonia at 77°F (25°C) and 1 bar of pressure, much milder conditions than those of the Haber-Bosch process. Nitrogenase is known to be a complex protein containing metal atoms, such as iron and molybdenum, and sulfide ions, but its structure and mode of action are imperfectly understood, even after decades of research. Recent research indicates that the nitrogen molecule may bind to iron atoms in the enzyme as a reaction step.

Ammonia can be formed in the human body, and may build up abnormally during serious illnesses such as **Reye's syndrome**. Much nitrogen is normally excreted by humans (and other **mammals**) as **urea**, a water soluble solid, but **fish** can excrete ammonia directly.

Urea eventually reacts with water to form ammonia, which therefore is usually present to some extent in waste water. Low concentrations of ammonia in water can be detected and measured using a solution called Nessler's reagent, which develops a strong color in the presence of ammonia. A recent toxic substance inventory done by the United States government estimated that in 1989, 200,000 tons of ammonia were released into the environment. This figure does not include fertilizer applications of ammonia.

Although Earth's atmosphere is free of ammonia, liquid and solid ammonia exist on other planets, such as **Jupiter**, where it may have originally formed from metal nitrides reacting with water. Ammonia has also been detected in interstellar **space** by radioastronomy.

Uses of ammonia

The largest use of ammonia is in **fertilizers**, which are applied to the soil and help provide increased yields of **crops** such as corn, **wheat**, and soybeans. Liquid ammonia, ammonia/water solutions, and chemicals made from ammonia, such as ammonium salts and urea, are all used as sources of soluble nitrogen. Urea, which is made from ammonia and **carbon dioxide**, can also be used as a feed supplement for cattle, aiding in the rapid building of protein by the animals.

All other important nitrogen chemicals are now made from ammonia. **Nitric acid** results from oxidation of ammonia in the presence of a platinum catalyst, called the Ostwald process, followed by treatment of the resulting nitrogen oxides with water. Nitric acid and nitrates are needed for the manufacture of **explosives** like TNT, nitroglycerin, gunpowder, and also for the propellants in cartridges for rifles and machine guns.

Two types of polymers needed for **artificial fibers** require the use of ammonia, polyamides (nylon) and acrylics (orlon). The original polyamide named nylon, brought out by DuPont Chemical Co., was made from two components, adipic acid and hexamethylenediamine. The nitrogen in the second named component is derived from ammonia. Acrylics are made from a three-carbon nitrogen compound, acrylonitrile. Acrylonitrile comes from the reaction of propene, ammonia, and oxygen in the presence of a catalyst.

Because of its basic properties, ammonia is able to react with acidic gases such as nitrogen oxides and sulfur oxides to form ammonium salts. Thus ammonia is useful in scrubbers that remove acidic gases before they can be released into the environment.

KEY TERMS

. . . .

Ammonia complexes—Species, usually positively charged ions, formed by linking several ammonia molecules through their nitrogen atoms to a transition metal ion.

Gigajoule—A billion joules. An amount of energy equal to 277 kilowatt-hours, or about the electrical energy used by a family in a month.

Polyamide—A polymer such as nylon, containing recurrent amide groups linking segments of the polymer chain.

Future prospects

Ammonia will continue to be important for agriculture and for the whole nitrogen chemicals industry. As countries in **Asia** and Latin America develop high standards of living and stronger economies, they will begin to need their own ammonia plants. For this reason, capacity and production will continue to grow. New uses may develop, particularly for ammonia as a relatively inexpensive base with unique properties, for liquid ammonia as a solvent, and as a storage medium for hydrogen, as the nations evolve toward alternative fuels.

See also Amides.

Resources

Books

Greenwood, N. N. and A. Earnshaw. *Chemistry of the Elements.* New York: Butterworth-Heinemann, 1997.

K.H. Buechel, et al. *Industrial Inorganic Chemistry.* New York: VCH, 2000.

Minerals Yearbook 2000. Washington, DC: Government Printing Office, 2001.

Periodicals

Seiler, N. "Ammonia and Alzheimer's Disease." *Neurochemistry International* 41, no. 2-3 (2002): 187-207.

John R. Phillips

Ammonification

Strictly speaking, ammonification refers to any chemical reaction that generates **ammonia** (NH_3) as an end product (or its ionic form, ammonium, NH_4^+). Ammonification can occur through various inorganic reactions or as a result of the metabolic functions of **microorganisms**, plants, and animals. In the ecological context, however, ammonification refers to the processes by which organically bound forms of **nitrogen** occurring in dead **biomass** (such as amino acids and **proteins**) are oxidized into ammonia and ammonium. The ecological process of ammonification is carried out in **soil** and **water** by a great diversity of microbes and is one of the many types of chemical transformations that occur during the **decomposition** of dead organic **matter**.

Ammonification is a key component in the **nitrogen cycle** of ecosystems. The nitrogen cycle consists of a complex of integrated processes by which nitrogen circulates among its major compartments in the atmosphere, water, soil, and organisms. During various phases of the nitrogen cycle, this element is transformed among its various organic and inorganic compounds.

As with all components of the nitrogen cycle, the proper functioning of ammonification is critical to the health of ecosystems. In the absence of ammonification, organic forms of nitrogen would accumulate in large quantities. Because growing plants need access to inorganic forms of nitrogen, particularly ammonium and nitrate (NO_3^-), the oxidation of organic nitrogen of dead biomass through ammonification is necessary for maintenance of the productivity of **species** and ecosystems.

Ammonification

Nitrogen is one of the most abundant elements in the tissues of all organisms and is a component of many biochemicals, particularly amino acids, proteins, and nucleic acids. Consequently, nitrogen is one of the critically important **nutrients** and is required in relatively large quantities by all organisms. Animals receive their supply of nitrogen through the foods they eat, but plants must assimilate inorganic forms of this nutrient from their environment.

However, the **rate** at which the environment can supply inorganic nitrogen is limited and usually small in relation to the metabolic demands of plants. Therefore, the availability of inorganic forms of nitrogen is frequently a **limiting factor** for the productivity of plants. This is a particularly common occurrence for plants growing in terrestrial and marine environments, and to a lesser degree, in fresh waters (where phosphate supply is usually the primary limiting nutrient, followed by nitrate).

The dead biomass of plants, animals, and microorganisms contains large concentrations of organically bound nitrogen in various forms, such as proteins and amino acids. The process of decomposition is responsible for recycling the inorganic constituents of the dead biomass and preventing it from accumulating in large unusable quantities. Decomposition is, of course, mostly carried out through the metabolic functions of a diverse array of **bacteria**, **fungi**, actinomycetes, other microorganisms, and some animals. Ammonification is a particular aspect of the more complex process of organic decay, specifically referring to the microbial conversion of organic-nitrogen into ammonia (NH_3) or ammonium (NH_4^+).

Ammonification occurs under oxidizing conditions in virtually all ecosystems and is carried out by virtually all microorganisms that are involved in the decay of dead organic matter. In situations where **oxygen** is not present, a condition referred to as **anaerobic**, different microbial decay reactions occur, and these produce nitrogen compounds known as amines.

The microbes derive some metabolically useful **energy** from the oxidation of organic-nitrogen to ammonium. In addition, much of the ammonium is assimilated and used as a nutrient for the metabolic purposes of the microbes. However, if the microbes produce ammonium in quantities that exceed their own requirements, as is usually the case, the surplus is excreted into the ambient environment (such as the soil), and is available for use as a nutrient by plants, or as a substrate for another microbial process, known as **nitrification** (see below). Animals, in contrast, mostly excrete **urea** or uric acid in their nitrogen-containing liquid wastes (such as urine), along with diverse organic-nitrogen compounds in their feces. The urea, uric acid, and organic nitrogen of feces are all substrates for microbial ammonification.

One of the most elementary of the ammonification reactions is the oxidation of the simple organic compound, urea ($CO(NH_2)_2$), to ammonia through the action of a microbial **enzyme** known as urease. (Note that two units of ammonia are produced for every unit of urea that is oxidized.) Urea is a commonly utilized agricultural fertilizer, used to supply ammonia or ammonium for direct uptake by plants, or as a substrate for the microbial production of nitrate through nitrification (see below).

Ammonium is a suitable source of nitrogen uptake for many species of plants, particularly those that live in acidic soils and waters. However, most plants that occur in non-acidic soils cannot utilize ammonium very efficiently, and they require the **anion** nitrate (NO_3^+) as their source of nitrogen uptake. The nitrate is generally derived by the bacterial oxidation of ammonium to nitrite, and then to nitrate, in an important ecological process known as nitrification. Because the species of bacteria that carry out nitrification are extremely intolerant of acidity, this process does not occur at significant rates in acidic soils or waters. This is the reason why plants growing in acidic habitats can only rely on ammonium as their source of nitrogen **nutrition**.

Because ammonium is a positively charged **cation**, it is held relatively strongly by ion-exchange reactions occurring at the surfaces of clay **minerals** and organic

KEY TERMS

Decomposition—The breakdown of the complex molecules composing dead organisms into simple nutrients that can be reutilized by living organisms.

Leaching—The process of movement of dissolved substances in soil along with percolating water.

Nutrient—Any chemical that is required for life.

matter in soils. Consequently, ammonium is not leached very effectively by water as it percolates downward through the soil. This is in contrast to nitrate, which is highly soluble in soil water and is leached readily. As a result, nitrate **pollution** can be an important problem in agricultural areas that have been heavily fertilized with nitrogen-containing **fertilizers**.

Humans and ammonification

Humans have a major influence on the nitrogen cycle, especially through the use of fertilizers in agriculture. Under nutrient-limited conditions, farmers commonly attempt to increase the availability of soil nitrogen, particularly as nitrate, and to a lesser degree, as ammonium. Rates of **fertilization** in intensive agricultural systems can exceed 446.2 lb/ac (500 kg/ha) of nitrogen per year. The nitrogen in the fertilizer may be added as ammonium nitrate (NO_4NH_4) or as urea. The latter compound must be ammonified before inorganic forms of nitrogen are present, that is, the ammonium and nitrate that can be taken up by plants. In some agricultural systems, compost or other organic materials may be added to soils as a conditioner and fertilizer. In such cases, the organic nitrogen is converted to available ammonium through microbial ammonification, and nitrate may subsequently be generated through nitrification.

In situations where the rates of fertilization are excessive, the ability of the **ecosystem** to assimilate the nitrogen input becomes satiated. Although the ammonium produced by ammonification does not leach readily, the nitrate does, and this can lead to the pollution of **groundwater** and surface waters, such as streams and **rivers**. Pollution of groundwater with nitrate poses risks for human health, while surface waters may experience an increased productivity through **eutrophication**.

Resources

Books

Atlas, R.M. and R. Bartha. *Microbial Ecology*. Menlo Park, CA: Benjamin/Cummings, 1987.

Brady, Nyle C., and Ray R. Weil. *The Nature and Properties of Soils*. 13th ed. Englewood Cliffs, NJ: Prentice Hall, 2001.

Freedman, B. *Environmental Ecology*. 2nd ed. San Diego: Academic Press, 1995.

Spearks, Donald L. *Environmental Soil Chemistry*. 2nd ed. New York: Academic Press, 2002.

Wild, Alan. *Soil Conditions and Plant Growth*. 11th ed. London: Longman, 1998.

Bill Freedman

Amnesia

Amnesia is a dissociative psychological disorder manifested by total or partial loss of **memory** and usually caused by a trauma. Unlike ordinary forgetfulness (the inability to remember a friend's **telephone** number), amnesia is a serious threat to a person's professional and social life. Amnesia, which depending on its cause can be either organic and psychogenic, has several types.

How amnesia is manifested

Global or generalized amnesia indicates the total loss of a person's identity: the **individual** has forgotten who she or he is. What makes this type of amnesia baffling is the fact that only personal memory is affected. The amnesiac does not remember who he or she is but displays no loss of general knowledge. A person suffering from retrograde amnesia cannot remember events that happened immediately before the trauma. In anterograde amnesia, all events following the trauma are forgotten. Finally, amnesia can also be selective, or categorical, manifested by a person's inability to remember events related to a specific incident.

Causes of amnesia

The causes of amnesia can be physiological and/or psychological. Amnesia caused by physical trauma is called organic amnesia, while the term psychogenic amnesia is used in reference to amnesia caused by psychological trauma.

Organic amnesia

Examples of organic amnesia include cases of memory loss following head injuries, **brain** lesions, **stroke**, substance abuse, **carbon monoxide** poisoning, **malnutrition**, electro-convulsive therapy, **surgery**, and infections. Persons suffering from any kind of organic amnesia display a number of typical characteristics. Their memory loss is anterograde: events after the trauma are

---

KEY TERMS

Dissociative disorder—Referring to psychological conditions, such as psychogenic amnesia and multiple personality disorder, in which an area of personal memory is dissociated from a person's consciousness.

Etiology—The cause or origin of a disease or condition.

---

forgotten. In addition, they can remember the distant past well, but their grasp of the immediate past is tenuous. If treatment is unsuccessful, the amnesiac's condition can worsen, leading to progressive memory loss. In such cases, memory loss is irreversible. If therapy is successful, the patient may partially regain memories blocked by retrograde amnesia, while anterograde amnesia usually remains.

Psychogenic amnesia

The causes of psychogenic amnesia are psychological, and they include career-related **stress**, economic hardship, and emotional distress. Experts have maintained that psychogenic amnesia has no physiological causes, although recent research has established that emotional trauma may alter the brain **physiology**, thus setting the stage for the interplay of psychological and physiological factors in the **etiology** of amnesia. In other words, psychogenic amnesia may have secondary causes that could be defined as organic. The most enigmatic psychogenic amnesia is identity loss; the person affected by this type of amnesia loses all personal memories, while retaining his or her general (impersonal) knowledge. For example, the amnesiac may not know his or her name, but can still be able to speak an acquired second language. Furthermore, in psychogenic amnesia, there is no anterograde memory loss. Finally, although psychogenic amnesia is reversible and can end within hours or days, it is a serious condition that can be difficult to treat.

Resources

Books

American Psychiatric Association. *Diagnostical and Statistical Manual of Mental Disorders*. 4th ed. Washington, DC: APA, 1994.

Berrios, German E., and John R. Hodges, eds. *Memory Disorders in Psychiatric Practice*. Cambridge: Cambridge University Press, 2000.

Comer, Ronald J. *Abnormal Psychology*. 2nd ed. New York: W. H. Freeman, 2000.

Feldman, R. *Understanding Psychology.* New York: McGraw-Hill, 1998.

Periodicals

Cipolotti, L. "Long-term Retrograde Amnesia." *Neurocase* 8, no. 3 (2002): 177.

Cohen, N.J. "Amnesia is A Deficit in Relational Memory." *Psychological Science* 11, no.6 (2001): 454-461.

Golden, Frederic. "Mental Illness: Probing the Chemistry of the Brain." *Time* 157 (January 2001).

Hyman, S.E. "The Genetics of Mental Illness: Implications for Practice." *Bulletin of the World Health Organization* 78 (April 2000): 455-463.

Knowlton, B.J. "Retrograde Amnesia." *Hippocampus* 11, no. 1 (2001): 50-55.

Zoran Minderovic

Amniocentesis

Amniocentesis is an invasive procedure used to obtain amniotic fluid for prenatal **diagnosis** of a fetus (e.g., assessment of fetal lung maturity).

In the 1950s the measurement of bilirubin concentrations present in amniotic fluid in monitoring the rhesus diseases was first reported. Amniocentesis for fetal **chromosome** analysis was also initiated in the 1950s. The first application was for fetal sex determination. **Down syndrome** via amniocentesis was first detected in 1968.

Cells naturally are exfoliated from the surface and from the mucosae of the fetus and some of these cells survive for a time in the fluid surrounding the fetus in the amniotic cavity. Soluble biochemical material of clinical

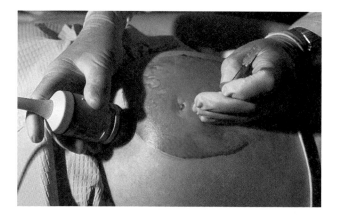

A physician uses an ultrasound monitor (left) to position the needle for insertion into the amnion when performing amniocentesis. *National Audubon Society Collection/Photo Researchers, Inc. Reproduced by permission.*

significance produced by the fetus may also accumulate in the amniotic fluid. The fluid can be analyzed for these substances directly.

For amniocentesis, the maternal abdomen is washed with antiseptic **solution**. A local anesthetic is given and a hollow needle (22 gouge) is inserted through the mother's abdominal wall into the amniotic cavity avoiding the placenta if possible and a sample of the fluid (approximately 20 mL) is withdrawn with a syringe attached to the needle. In order to insure the safety of the fetus, the procedure is monitored in real time via an ultrasound scan. In twin pregnancies, after withdrawal of amniotic fluid of the first sac, an injection of a dye is necessary to understand if the fluid of the second sac has been drawn. If the fluid of the second puncture is clear, then it does not come from the first sac.

Viable cells in the fluid are then cultured (grown) *in vitro*. At 16 weeks' gestation amniotic fluid contains 200,000 cells mL, but a very small number are capable of forming colonies. The chromosomes of the cultured cells can then be examined.

Viewing the chromosomes under a light **microscope** will reveal if a normal diploid number of chromosomes are present or if extra or fewer chromosomes are present. Additionally, structural chromosomal aberrations, as well as uniparental disomies can be detected.

FISH (**fluorescence** in situ hybridization) analysis can also be used for rapid detecting numerical anomalies involving chromosome 13, 18, 21, X and Y. Quantitative polymerase chain reaction (Q-PCR) has been also used for detecting the most common aneuploidies. More recently, it has been shown how amniocentesis can be used for detecting DNA anomalies responsible for the **etiology** of many autosomal and X-linked disorders as well as **hemophilia** A, **sickle cell anemia**, DiGeorge **syndrome**, and other diseases.

Amniocentesis is an elective procedure that can detect the presence of many types of **genetic disorders**, thus allowing doctors and prospective parents to make important decisions about early treatment and intervention. Down syndrome is a chromosomal disorder characterized by a diversity of physical abnormalities, mental retardation, and shortened life expectancy. It is by far the most common, nonhereditary, genetic **birth** defect, afflicting about one in every 1,000 babies. Since the risk of bearing a child with a nonhereditary genetic defect such as Down syndrome is directly related to a woman's age, amniocentesis is recommended for women who will be older than 35 on their due date. Thirty-five is the recommended age to begin amniocentesis because that is the age at which the risk of carrying a fetus with such a defect roughly equals the risk of miscarriage caused by the procedure—about 1 in 200.

Maternal complications of amniocentesis such as septic shock and amnionitis are rare. Rhesus isoimmunization can be prevented by prophylactic administration of anti-D immunoglobulin to Rh-negative women. Risk of abortion has been quoted of 1% about for single pregnancies and 3% about for twin pregnancies. Amniocentesis is ordinarily performed between the 14th and 16th week of pregnancy, with results usually available within three weeks. It is possible to perform amniocentesis as early as the 11th week but this is not usually recommended because there appears to be an increased risk of miscarriage when done at this time. Furthermore, the CEMAT study (Canadian Early and Mid Trimester Amniocentesis Trail) in 1998 cleared that early amniocentesis from 11 to 12 weeks is associated with significant disadvantages because of difficult or unsuccessful procedures as well as more than one needle insertion and more likely fetal cells culture failure. The advantage of early amniocentesis is the extra time for decision making if a problem is detected. Potential treatment for the fetus can begin earlier. Elective abortions are safer and less controversial the earlier they are performed.

See also Embryo and embryonic development; Embryology; Germ cells and the germ cell line; Sexual reproduction.

Resources

Periodicals

Antsaklis A., et al. "Genetic Amniocentesis in Women 20-34 Years Old: Associated Risks." *Prenat Diagn* 20(3) (2003): 247-50.

Dugoff L., and Hobbins, J.C. "Invasive Procedures to Evaluate the Fetus." *Clin Obstet Gynecol.* 45(4) (2002):1039-53.

Gordon M.C., et al. "Complications of Third-trimester Amniocentesis using Continuous Ultrasound Guidance." *Obstet Gynecol.* 99(2) (2002):255-9.

Antonio Farina
Brenda Wilmoth Lerner

Amoeba

Amoebas are single-celled protozoans of the order Amoebida. They consist of a mass of cellular fluid surrounded by a **membrane**, and containing one or more nuclei (depending upon the **species**), as well as other **cell** organelles, such as food vacuoles.

The word amoeba is derived from the Greek word ameibein (to change), which describes the amoeba's most easily distinguishable feature, the continuous changing of shape by repeated formation of *pseudopods* (Greek: false feet).

Pseudopodal movement is based on a continual change in the state of protoplasm flowing into the foot-like appendage. An interior fluid (endoplasm), under **pressure** from an exterior gel (ectoplasm), flows forward in the cell. When the endoplasm reaches the tip of a developing pseudopod, the fluid is forced backward against the ectoplasm, and is turned into a gel. After returning to the body of the cell, the newly formed ectoplasm gel is converted back to fluid endoplasm, and again flows forward under the pressure of the exterior gel.

Pseudopods serve two important functions—locomotion and food capture, activities that are often interrelated. Amoebas use their pseudopods to ingest food by a method called phagocytosis (Greek: phagein, to eat).

The streaming of protoplasm inside the pseudopods moves the amoeba forward. When the **organism** contacts a food particle, the pseudopods surround the particle. After the food is corralled by the amoeba, an opening in the membrane allows the food particle to pass into the cell. Inside the cell, the food is enclosed within food vacuoles, digested by enzymes, and assimilated by the amoeba. The amoeba expels particles that are not acceptable as food.

The organisms generally implied by the term "amoeba" belong to the phylum **Protozoa**, class Mastigophora, which includes organisms with flagellae (whip-like organs of locomotion) such as *Chlamydomonas angulosa*, as well as those with pseudopods. The class Sarcodina, which has as its principle distinguishing feature the almost universal presence of pseudopods, includes *Amoeba proteus*, the best-known protozoan.

The Rhizopoda (in some classifications a subclass of Sarcodina) contains all common "naked amoebas," which are either tubular or somewhat flattened. They move by means of protoplasmic flow, by producing pseudopodia, or by advancing as a single mass. Rhizopoda also includes sarcodinids known as giant amoebas and testaceous forms (those with tests, or shells). Some apparently "naked" amoebas have coatings of various kinds, such as scales, mucoid layers called *glycocalyces*, or complex filaments much smaller than scales.

In addition to the naked forms, many species of amoeba have tests (hard coverings), and are referred to as shelled amoebas. Most of these shelled amoebas are classified in the order Arcellinida. They have a test with a single opening, and are predominantly **freshwater** organisms. Shelled amoebas feed on a variety of organisms, such as **bacteria**, **algae**, and other protozoans.

Most members of the order of Amoebida are free-living in fresh or **salt water** or **soil**, and ingest bacteria. Larger members also feed on algae and other protozoans. Several amoebas of this group are pathogenic to humans.

The family Amoebidae includes mostly freshwater species, whose pseudopodal movement is either monopodial (the entire protoplasmic mass moves forward) or polypodial (several pseudopods advance simultaneously). One member, *Amoeba proteus* is commonly used for teaching and cell **biology** research. *Chaos carolinense,* one of the larger species, has multiple nuclei and can reach a length of 0.12 in (3 mm).

The Hartmannellidae family includes small and medium-sized amoebas that move forward monopodially, advancing by means of a steady flow. They feed on bacteria, although some species of the genus *Saccamoeba* also feed on unicellular algae.

The family Entamoebidae includes most of the obligately endozoic (parasitic inside a host) Amoebida organisms, including *Entamoeba histolytica*. Amebiasis (**infection** with *E. histolytica*) is a serious intestinal **disease** also called amoebic **dysentery**. It is characterized by diarrhea, fever, and dehydration. Although amebiasis is usually limited to the intestine, it can spread to other areas of the body, especially the liver.

E. histolytica exists as either a trophozoite or cyst. The trophozoite is motile, possesses a single nucleus, and lives in the intestine. It is passed from the body in diarrhea, but cannot survive outside the host. The cyst form, consisting of condensed protoplasm surrounded by a protective wall, is produced in the intestine, can survive outside the host, and even withstands the acid of the stomach when it is ingested with food or contaminated water. Once inside the intestine, *E. histolytica* multiplies by means of binary fission.

Another family, Acanthamoebidae (in the Amoebida suborder Acanthopodina), includes the genus *Acanthamoeba* genera, which are often isolated from fresh water and soil. *Acanthamoeba* cause primary amebic meningoencephalitis (PAM, **inflammation** of the **brain** and its protective membranes), especially in individuals who are ill and whose immune systems are weakened. *Acanthamoeba* infections have been traced to fresh water, hot tubs, soil, and homemade contact **lens** solutions. In the latter case, **contamination** of contact lens **solution** with the organism has caused keratitis, an inflammation of the cornea accompanied by **pain** and blurred **vision**. Severe cases can require a corneal transplant or even removal of the **eye**.

A member of the order Schizopyrenida, *Naegleria fowleri* is an especially dangerous human parasite, causing rapidly fatal PAM in people swimming in heated water, or warm, freshwater ponds and lakes, mainly in the southern United States. Both *Naegleria* and *Acanthamoeba* enter through the nasal mucosa and spread to the brain along nerves.

Ampere *see* **Units and standards**

Amphetamines

Amphetamines are a group of **nervous system** stimulants that includes amphetamine, dextroamphetamine, and methamphetamine. They are used to induce a state of alert wakefulness and euphoria, and since they inhibit appetite, they also serve as diet pills. After World War II, they were widely prescribed by physicians as diet pills, but they are generally no longer recommended for weight loss programs since there are too many hazards in the prolonged use of amphetamines. Prolonged exposure may result in **organ** impairment, affecting particularly the kidneys. Amphetamines are addictive and may lead to compulsive **behavior**, hallucinations, paranoia, and suicidal actions. Their medical use has currently been narrowed to treating only two disorders. One is a condition known as attention-deficit hyperactivity disorder (ADHD) in children. When used to treat overactive children, amphetamines are carefully administered under controlled situations as part of a larger program. The other condition for which amphetamines are prescribed is a **sleep** disorder known as narcolepsy, the sudden uncontrollable urge to sleep during the hours of wakefulness.

In street language, amphetamines are known as pep pills, as speed (when injected), and as ice (when smoked in a crystalline form). The popularity of amphetamines as a street drug appears to have been facilitated originally by pilfering from the drug companies manufacturing the pills. They are now also illegally manufactured in secret laboratories.

History

Amphetamines were first synthesized in 1887 by the drug company Smith, Kline and French. They were not marketed until 1932, however, as Benzedrine inhalers for relief from nasal congestion due to hay fever, colds, or **asthma**. In 1935, after noting its stimulant effects, the drug company encouraged prescription of the drug for the chronic sleep disorder narcolepsy. Clinical enthusiasm for the drug led to its misapplication for the treatment of various conditions, including **addiction** to opiates. The harmful effects of the drug were first noted by the British press, and in 1939 amphetamines were placed on a list of toxic substances for the United Kingdom.

The early abuse of Benzedrine inhalers involved the removal of the strip containing the amphetamines from the casing of the inhaler. The strips were then either chewed or placed in coffee to produce an intense stimulant reaction. Since the inhalers were inexpensive and easily obtainable at local drug stores, they were purchased by young people searching for ways of getting "high." But amphetamines became particularly popular in World War II. Soldiers on both sides were given large amounts of amphetamines as a way of fighting fatigue and boosting morale. The British issued 72 million tablets to the armed forces. Records also show that kamikaze pilots and German panzer troops were given large doses of the drug to motivate their fighting spirit. Hitler's own medical records show that he received eight injections a day of methamphetamine, a drug known to create paranoia and unpredictable behavior when administered in large dosages.

The demand for amphetamines was high in the 1950s and early 1960s. They were used by people who had to stay awake for long periods of time. Truck drivers who had to make long hauls used them to drive through the night. Those who had long tours of duty in the armed forces relied on them to stay awake. High school and college students cramming for tests took them to study through the nights before their examinations. Athletes looked to amphetamines for more **energy**, while English and American popular musicians structured their lives and music around them. The Food and Drug Administration (FDA) estimated that there were well over 200 million amphetamine pills in circulation by 1962 in the United States alone.

During that period of time about half of the quantity of amphetamines produced were used outside of the medically prescribed purposes mandated by the legal system. Of the 19 companies producing amphetamines then, nine were not required to show their registry of buyers to the FDA. It is believed that these nine companies supplied much of the illegal traffic in amphetamines for that period.

By 1975 a large number of street preparations were being passed off as amphetamine tablets. Tests indicated that only about 10% of the street drugs represented as amphetamines contained any amphetamine substance at all. The false amphetamines were in fact mixtures of **caffeine** and other drugs that resembled amphetamine, such as phenethylamine, an over-the-counter drug used to relieve coughs and asthma or to inhibit appetite. Other false amphetamine tablets contained such over-the-counter drugs as ephedrine and pseudoephedrine. These

bootleg preparations came under such names as Black Beauty, Hustler, and Penthouse, and they were promoted in magazines that catered to counterculture sentiment.

The use of amphetamines and drugs like amphetamine showed a sharp decrease in the 1980s. The decrease was probably due to the increasing use of **cocaine**, which was introduced in the mid-1970s and continues to be a major street drug at the present time. Another reason was the introduction of newer types of appetite suppressants and stimulants on the pharmacological market and then to the street trade. Still, a survey done in 1987 showed that a large number of high school seniors (12%) had used drugs of the amphetamine type during the previous year.

Ice

Illegal users of methamphetamine originally took the drug in pill form or prepared it for injection. More recently, however, a crystalline form of the drug that is smoked like crack cocaine has appeared on the market. The practice of smoking methamphetamine began in Hawaii and then spread to California. Various names are given to smokable methamphetamine, such as Ice, LA Ice, and Crank. Ice is much cheaper than crack because it is made from easily available chemicals and does not require complicated equipment for its production. An illegal drug manufacturer can produce ice at a much lower cost than cocaine and therefore realize a much greater profit margin. Like cocaine, amphetamine reaches the **brain** faster when it is smoked. Users have begun to prefer ice over crack because the high lasts much longer, persisting for well over 14 hours. The side effects of an ice high can be quite severe, however. Side effects such as paranoia, hallucinations, impulsive behavior, and other psychotic effects may last for several days after a prolonged high from ice.

Action

Amphetamines, according to recent research, act on the neurotransmitters of the brain to produce their mood-altering effects. The two main neurotransmitters affected are **dopamine** and norepinephrine, produced by cells in the brain. Amphetamines appear to stimulate the production of these two neurotransmitters and then prevent their uptake by other cells. They further increase the amount of surplus neurotransmitters by inhibiting the action of enzymes that help to absorb them into the nervous system. It is believed that the excess amount of neurotransmitters caused by the amphetamines are also responsible for the behavioral changes that follow a high.

Drugs that pose a high risk of addiction like amphetamines, opiates, and cocaine all seem to arouse the centers of the brain that control the urge to seek out pleasurable sensations. Addictive drugs overcome those centers and displace the urge to find pleasure in food, sex, or sleep, or other types of activity that motivate people not addicted to drugs. The drug addict's primary concern is to relive the pleasure of the drug high, even at the risk of "crashing" (coming down from the high in a painful way) and in the face of the social disapproval the habit inevitably entails. Laboratory experiments have shown that animals self-administering amphetamines will reject food and water in favor of the drug. They eventually perish in order to keep up their supply of the drug.

Withdrawal symptoms for chronic users include **depression**, **anxiety**, and the need for prolonged periods of sleep.

Physical and psychological effects

Amphetamines inhibit appetite and stimulate **respiration** as a result. On an oral dose of 10-15 mg daily an **individual** feels more alert and more confident in performing both physical and mental work and is able to show an increase in levels of activity. It has not been determined how the drug affects the quality of work done under its influence. The drug also results in a rise in **blood pressure** and an increased, though sometimes irregular, **heart rate**.

Psychological dependency arises from the desire to continue and heighten the euphoric effects of the drug. During an amphetamine euphoria, the individual feels an enlargement of physical, mental, and sexual powers along with the absence of the urge to eat or sleep. Those who inject the drug feel a "rush" of the euphoric effect moments after the injection. They will feel energized and focused in an unusual way.

Depending on the user's medical history, the dosage, and the manner in which the drug was delivered to the body, a number of toxic effects can accompany amphetamine abuse. Large intravenous dosages can lead to delirium, seizures, restlessness, the acting out of paranoic fantasies, and hallucinations. In hot weather there is a danger of **heat** stroke, since amphetamines raise the body **temperature**. The increased blood pressure can lead to stroke. Heart conditions such as arrhythmia (irregular heartbeat) can develop and become fatal, especially for those with heart **disease**. Since the dosage levels of street drugs are not reliable, it is possible to overdose unknowingly when using the drug intravenously. The results can be **coma** and death. Chronic users will show much weight loss and chronic skin lesions. Those who are "shooting up" (injecting) street versions of amphetamine face the further dangers

from contaminated substances, adulterations in the chemicals used, and a lack of sterilized needles. These conditions carry the same risks associated with heroin use, such as **hepatitis** and infections to vital organs, along with irreversible damage to blood vessels. Contaminated needles may also transmit the HIV **virus** that causes **AIDS**.

Treatment

It takes several days to help a person recover from an acute amphetamine reaction. It is important to control body temperature and to reassure a person undergoing the psychological effects of the drug. In order to control violent behavior, **tranquilizers** are administered to quiet the patient. Treatment of the depression which is an after-effect of heavy usage is also required. Patients will seek to deal with the fatigue that comes after the body has eliminated the drug by resuming its use. A long-term program for maintaining abstinence from the drug has to be adhered to. Just as in the case of recovery from **alcoholism** and other forms of drug abuse, recovering addicts benefit from support groups.

KEY TERMS

AIDS—Acquired immunodeficiency syndrome; a fatal viral disease contracted by a virus transmitted through the blood or body fluids.

Attention-deficit hyperactivity disorder (ADHD)—A childhood condition marked by extreme restlessness and the inability to concentrate, which is sometimes treated with amphetamines.

Crashing—Coming down from a prolonged drug high such as that produced by amphetamines.

Euphoria—Feelings of elation and well being produced by drugs like amphetamines.

HIV—Human immunodeficiency virus, which leads to AIDS.

Ice—Crystalline methamphetamine that is smoked to produce a high.

Neurotransmitters—Chemicals produced in the brain, which are responsible for different emotional states.

Paranoia—Delusions of persecution; one of the main psychotic conditions produced by an excess use of amphetamines.

Speed—An injectable form of methamphetamine.

Tranquilizers—Drugs used to pacify anxiety attacks.

Resources

Books

Clayton, Lawrence. *Amphetamines and Other Stimulants.* New York: Rosen Publishing Group, 1998.

Conolly, Sean. *Amphetamines (Just the Facts).* Oxford: Heinemann Library, 2000.

Klaassen, Curtis D. *Casarett and Doull's Toxicology.* 6th ed. Columbus: McGraw-Hill, Inc., 2001.

O'Neil, Maryadele J. *Merck Index: An Encyclopedia of Chemicals, Drugs, & Biologicals.* 13th ed. Whitehouse Station, NJ: Merck & Co., 2001.

Shapiro, Harry. *Waiting for the Man.* New York: William Morrow, 1988.

Stimmel, Barry. *The Facts About Drug Use.* New York: Haworth Medical Press, 1991.

Periodicals

Chan, Paul, et al. "Fatal and Nonfatal Methamphetamine Intoxication in the Intensive Care Unit." *Journal of Toxicology: Clinical Toxicology* 32 (June 1994): 147-56.

Kiefer, D.M. "Chemistry Chronicles: Miracle Medicines." *Today's Chemist* 10, no. 6 (June 2001): 59-60.

Steele, M.T. "Screening for Stimulant Use in Adult Emergency Department Seizure Cases." *Journal of Toxocology* 38, no.6 (2001): 609-613.

Jordan Richman

Amphibians

The vertebrate class Amphibia, to date, includes about 3,500 **species** in three orders: **frogs** and **toads** (order Anura), **salamanders** and **newts** (order Caudata), and **caecilians** (order Gymnophiona). There is, however, a much larger number of extinct species, because this ancient group of animals were the first **vertebrates** to begin exploiting terrestrial environments. Fossil amphibians are known from at least the Devonian era, about 400 million years ago. However, this group was most diverse during the late Carboniferous and Triassic eras, about 360-230 million years ago.

None of the surviving groups of amphibians can be traced back farther than about 200 million years. All of the living amphibians are predators as adults, mostly eating a wide variety of **invertebrates**, although the largest frogs and toads can also eat small **mammals**, **birds**, **fish**, and other amphibians. In contrast with adults, larval frogs and toads (tadpoles) are mostly herbivorous, feeding on **algae**, rotting or soft tissues of higher plants, and infusions of **microorganisms**.

Amphibians are poikilothermic animals—their body **temperature** is not regulated, so it conforms to the environmental temperature. Amphibians have a moist, glandular, scaleless skin, which is poorly waterproofed in most species; this skin allows gaseous exchange and actively pumps salts. Most amphibians have tails, but the tail in adult frogs and toads is vestigial, and is fused with the pelvis and sacral vertebrae into a specialized structure called a urostyle. Some species of caecilians have lost their limbs and limb-girdles, and have a wormlike appearance.

All amphibians have a complex life cycle, which begins with eggs that hatch into larvae, and eventually metamorphose into adult animals. Usually, the eggs are laid into **water** and are externally fertilized. The larvae or tadpoles have gills or gill slits and are aquatic. Adult amphibians may be either terrestrial or aquatic, and breathe either through their skin (when in water) or by their simple saclike lungs (when on land). However, these are all generalized characteristics of the amphibian lifestyle; some species have more specialized life histories, and can display attributes that differ substantially from those described above. Rare idiosyncrasies of amphibian life history can include ovoviviparity, in which fully formed, self-nourishing, developing eggs are retained inside the female's body until they hatch as tadpoles, and even **viviparity**, in which larvae develop within the female but are nourished by the parent, as well as by their incompletely formed egg, until they are released as miniature frogs.

Frogs and toads lack tails but have greatly enlarged hind legs that are well adapted for jumping and swimming. Most of the living species of amphibians are anurans, comprising about 3,000 species. Most anurans are aquatic, but some are well adapted to drier habitats. Some common anurans of **North America** include the bullfrog (*Rana catesbeiana*, family Ranidae), spring peeper (*Hyla crucifer*, family Hylidae), and the American toad (*Scaphiopus holbrooki*, family Pelobatidae). The latter species lives in arid regions, estivating (spending the summer in **hibernation**) during dry periods but emerging after rains to feed, and taking advantage of heavy but unpredictable periods of rain to engage in frenzies of breeding. The largest frogs reach 11.8 in (30 cm) in length and weigh several pounds.

There are about 250 species of newts and salamanders, ranging in size from approximately 6 in (15 cm) to more than 5 ft (1.5 m). These amphibians have a tail and similarly sized legs well adapted to walking, but are usually found in or near water. Most species lay their eggs in water, however, adults usually spend most of their time in moist habitats on land. An exception is the red-spotted newt (*Notophthalmus viridescens*, family Salamandridae) of eastern North America, which in its juvenile stage (the red eft) wanders in moist terrestrial habitats for several years before returning to water to develop into its aquatic adult stage. Some species, such as the

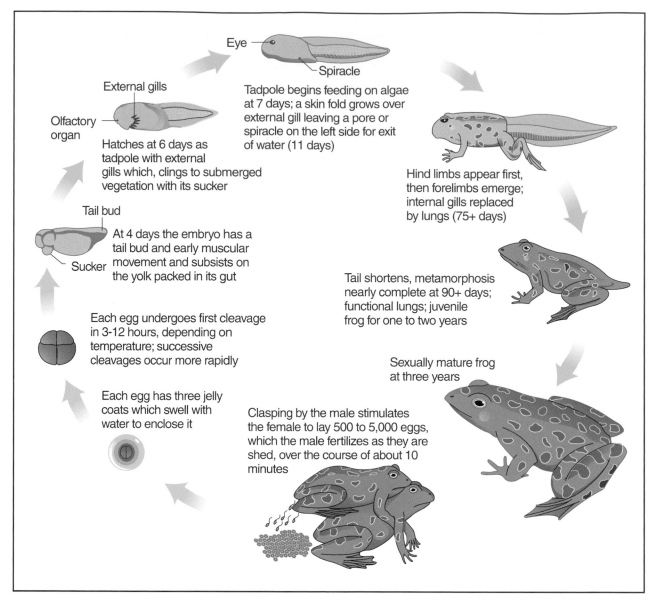

The life cycle of a frog. *Illustration by Hans & Cassidy. Courtesy of Gale Group.*

lungless red-backed salamander (*Plethodon cinereus*, family Plethodontidae) of North America, are fully terrestrial. This species lays its eggs in moist places on the forest floor, where the animals develop through embryonic and larval stages and hatch directly as tiny versions of the adult stage.

Caecilians are legless, almost tailless, wormlike, burrowing amphibians found in moist, tropical habitats. They feed on soil invertebrates. There are at least 160 species of caecilians, reaching 5 ft (1.5 m) in length, but most are rarely seen despite their size.

Recently, within the last 50 years, an alarming global decline in the gross number of amphibians and amphibian species has been documented. Because of their dependence on fresh water supplies for reproduction, and since almost all species spend their adult lives in moist environments, it is hypothesized that widespread **water pollution** is catalyzing the decline of this class of organisms. In other cases, the **extinction** and decline of species is known to be the direct consequence of human intervention (**habitat** destruction). However, scientists have noted that in untouched, pristine environments where **pollution** and human encroachment are minimal or nonexistent, amphibian populations also are declining. As such, it is speculated that other changes, such as increased UV **radiation** due to **ozone** depletion, or the introduction of competing exotic species is responsible for their collapse.

The decline is unfortunate because amphibians are important indicators of the health of ecosystems, sources of potent and medically useful chemicals, stabilizers of ecological balance in areas where they live, and contributors to the aesthetic beauty of the natural world.

See also Chordates.

Amplifier

An amplifier is a device, usually electronic, that magnifies information to a more powerful signal at the amplifier's output. Amplifiers are usually based on electronic principles but may utilize hydraulics or magnetics.

Amplifiers are used when the electrical power of a signal must be increased. Audio amplifiers can increase the microwatts developed by a microphone to more than a million watts of power required to fill a stadium during a concert. Satellites use amplifiers to strengthen **television** and **telephone** signals so they can be received easily when beamed back to **Earth**.

Long-distance telephone circuits were made possible when amplifiers magnified power that had been dissipated by the resistance of cross-country phone wires. Amplifiers were also needed to restore lost volume. Undersea telephone cables require amplifiers beneath the sea. Cable-television systems require as many as 100 sophisticated broad-band width amplifiers to serve subscribers.

Amplifiers and energy

Just as a faucet is not the source of the water it dispenses, an amplifier does not create the **energy** it controls. An amplifier draws on the power of the weak input signal and supplements it with energy provided by a power source, increasing the power of the signal.

A replacement **battery** installed in a portable cassette-tape player supplies it with all the energy that will eventually create sounds during the life of that battery. The battery contains no information about the music or **speech** the player will eventually produce. Amplifiers in the tape player make it possible for the program on a **compact disc**, cassette tape, or a **radio** signal to dispense the battery's energy at a controlled **rate** as needed to produce the desired sounds.

Efficiency

No amplifier can be 100% efficient. All amplifiers waste some of the energy supplied to them. For instance, an amplifier's efficiency may be improved, but the result may be increased distortion in the final output.

Cascading amplifiers

To process an extremely weak signal, an amplifier must be able to magnify data power by a factor of millions. To achieve this, amplifier stages are frequently connected in series to multiply their gain. Each stage in a chain provides the signal for the following stage, an arrangement called a cascade. The total amplification of a cascade is equal to the product of the individual-stage gains. If each of three amplifiers in cascade has a voltage gain of 100, the overall voltage gain will equal one million. A 10-micro volt signal processed by this cascade increases to a 10-volt signal.

Discrete and integrated amplifiers

Electronic amplifiers using separate transistors, resistors, and capacitors wired into place one by one are called discrete amplifiers. Discrete amplifiers have been all but superseded by integrated circuits (ICs) for small-

The power amplifier for the Antares fusion laser. *Photograph by Ray Nelson. Phototake NYC. Reproduced by permission.*

signal applications. Vast numbers of transistors and many supporting components are contained on an IC's single silicon **crystal** chip. Circuit boards now use just a few encapsulated chips in place of the hundreds of individual components once required. An engineer or technician normally does not need to be aware of an IC's internal circuitry, further simplifying their use.

See also Electronics.

Resources

Books

Cannon, Don L. *Understanding Solid-State Electronics.* 5th ed. SAMS division of Prentice Hall Pub. Co., 1991.

Donald Beaty

Amplitude modulation *see* **Radio waves**

Amputation

The term amputation refers to the complete or partial severance of a limb or other body part. Surgical amputations may be performed because of an injury, **congenital (birth)** defect, malignant **disease**, **infection**, or vascular disease. Approximately 80% of all surgical amputations are performed on the lower limbs, such as the leg or foot. Artificial limbs (**prosthetics**) are often used to restore complete or partial functioning, such as walking, after amputation.

History

Although surgical amputations date back at least to the time of Hippocrates (c.460-375 B.C.), amputating limbs to save lives did not become widespread until the sixteenth century. Many of the advances in amputation **surgery** were made by military surgeons during the course of wars. In 1529, French military surgeon Ambroise Paré rediscovered the use of ligation, in which a thread-like or wire material is used to tie off, or constrict, **blood** vessels. This surgical technique, which stops the flow of blood from a severed vein, greatly reduces the patient's chances of bleeding to death and helped to make amputation a viable surgical approach.

The introduction of the tourniquet in 1674 further advanced surgical amputation. Essentially, a tourniquet is a circling device that is wrapped around the limb above the area to be amputated and then twisted to apply **pressure** to stop the flow of blood.

In 1867, Lord Lister's introduction of antiseptic techniques to surgery further advanced amputation. Anti-septics, such as iodine and chloride, reduced the chances of infection by inhibiting the growth of infectious agents such as **bacteria**. Other advances at this time included the use of **chloroform** and **ether** as anesthetics to reduce **pain** and keep the patient unconscious during surgery.

Reasons for amputation

The reasons for surgical amputations can be classified under four major categories: trauma, disease, tumors, and congenital defects. Amputations resulting from trauma to the limb are usually the result of physical injury, for example, from an accident; thermal injury due to a limb being exposed to extreme hot or cold temperatures; or infections, such as **gangrene**. Certain diseases, such as **diabetes mellitus** and vascular disease, may also lead to complete or partial amputation of a limb. Vascular disease is a leading cause of amputation in people over 51 years of age. The development of either malignant or nonmalignant tumors may also lead to amputation. Finally, congenital deficiencies, such as absence of part of an arm or leg or some other deformity, may be severe enough to require amputation, particularly if the defect interferes with the individual's ability to function.

Levels and goals of amputation

In determining how much of a limb to amputate, the surgeon must take several factors into consideration. When dealing with amputation due to disease, the surgeon's first and most important goal is to remove enough of the appendage or limb to insure the elimination of the disease. For example, when amputating to stop the spread of a malignant **tumor**, the surgeon's objective is to remove any portion of the limb or **tissue** that may be infected by the malignancy.

Other considerations in determining the level of amputation include leaving enough of a stump so that an artificial limb (prosthesis) may be attached in a functional manner. As a result, whenever possible, the surgeon will try to save functioning joints like knees and elbows.

Further goals of amputation surgery include leaving a scar that will heal well and is painless, retaining as much functioning muscle as possible, successfully managing nerve ends, achieving hemostasis (stopping the flow of blood) in **veins** and **arteries** through ligation, and proper management of remaining bone.

Prosthetics and limb reattachment

The use of artificial limbs or prosthetics most likely dates back to prehistoric man's use of tree limbs and forked sticks for support or replacement of an appendage. In 1858, a **copper** and wood leg dating back to

300 B.C. was discovered in Italy. In the fifteenth century, a knight who had lost a hand in battle could acquire an **iron** replacement. Recent medical, surgical, and **engineering** advances have led to the development of state-of-the-art prosthetics, some of which can function nearly as well as the original limb. For example, some individuals who lose a leg may even be able to run again with the aid of a modern prosthetic device.

Recent advances have also led to more and more accidentally amputated limbs being successfully reattached. Depending on a number of factors, including the condition of the limb and how long it has been severed, full functional ability may be regained. In a notorious case that occurred in 1993, a man's severed penis was reattached with full functional ability.

Phantom limb

A baffling medical phenomenon associated with amputation is the amputee's perception of a phantom limb. In these cases, which are quite common among amputees, the amputees will perceive their amputated limb as though it still exists as part of their body. This phantom limb may be so real to an amputee that he or she may actually try to stand on a phantom foot or perform some task such as lifting a cup with a phantom hand. Although amputees may feel a number of sensations in a phantom limb, including numbness, coldness, and **heat**, the most troubling sensation is pain. Approximately 70% of all amputees complain of feeling pain in their phantom limbs. Such pain ranges from mild and sporadic to severe and constant.

Although it probably is related to the central **nervous system**, the exact cause of the phantom limb phenomenon is unknown. Theories on the origin of the phantom limb phenomenon include impulses from remaining nerve endings and spontaneous firing of spinal cord neurons (nerve cells). More recent studies indicate that the phenomenon may have its origin in the brain's neuronal circuitry.

Treatments for phantom limb pain include excision (cutting out) of neuromas (nodules that form at the end of cut nerves), reamputation at a higher point on the limb, or operation on the spinal cord. Although success has been achieved with these approaches in some cases, the patient usually perceives pain in the phantom limb again after a certain interval of time.

Resources

Books

Atlas of Limb Prosthetics: Surgical, Prosthetic, and Rehabilitation Principles. 2nd ed. New York: Mosby Year Book, 1992.

Barnes, Robert W., and Birck Cox. *Amputations: An Illustrated Manual.* Philadelphia: Hanley & Belfus, 2000.

Bella J., Edd, and Fapta May Pt. *Amputations and Prosthetics: A Case Study Approach.* 2nd ed. New York: F. A. Davis, 2002.

Murdoch, George, and Wilson A. Bennett, Jr. *A Primer on Amputations and Artificial Limbs.* Springfield, IL: Charles C. Thomas, 1998.

Periodicals

Johnston, J., and C. Elliott. "Healthy Limb Amputation: Ethical and Legal Aspects." *Clinical Medicine* 2, no. 5 (September/October, 2002):435-435.

McPhee, A.T. "Scientist Solves Mystery: Why a Missing Arm Itches." *Current Science* 78 (1993): 8-9.

Melzack, Ronald. "Phantom Limbs." *Scientific American* (April 1994): 120-126.

Mulvey, Martha A. "Traumatic Amputation." *RN* 54 (1991): 26-30.

Sherman, R. "To Reconstruct or Not to Reconstruct?" *New England Journal of Medicine* 12, no. 347-24 (2001):1906-7

van der Schans, C.P., et al. "Phantom Pain and Health-related Quality of Life in Lower Limb Amputees." *Journal Pain Symptom Management* 24, no.4 (October, 2002):429-36.

David Petechuk

Anabolism

Anabolism, or biosynthesis, is the process by which living organisms synthesize complex molecules of life from simpler ones. Anabolism, together with **catabolism**, are the two series of chemical processes in cells that are, together, called **metabolism**. Anabolic reactions are divergent processes. That is, relatively few types of raw materials are used to synthesize a wide variety of end products. This results in an increase in cellular size or complexity—or both.

Anabolic processes produce peptides, **proteins**, polysaccharides, lipids, and nucleic acids. These molecules comprise all the materials of living cells, such as membranes and chromosomes, as well as the specialized products of specific types of cells, such as enzymes, antibodies, **hormones**, and neurotransmitters.

Catabolism, the opposite of anabolism, produces smaller molecules used by the **cell** to synthesize larger molecules, as will be described below. Thus, in contrast to the divergent reactions of anabolism, catabolism is a convergent process, in which many different types of molecules are broken down into relatively few types of end products.

The **energy** required for anabolism is supplied by the energy-rich **molecule adenosine triphosphate** (ATP). This energy exists in the form of the high-energy **chemical bond** between the second and third molecule of phosphate on ATP. ATP's energy is released when this bond is broken, turning ATP into **adenosine diphosphate** (ADP). During anabolic reactions, the high-energy phosphate bond of ATP is transferred to a substrate (a molecule worked on by an **enzyme**) in order to energize it in preparation for the molecule's subsequent use as a raw material for the synthesis of a larger molecule. In addition to ATP, some anabolic processes also require high-energy **hydrogen atoms** that are supplied by the molecule NADPH.

Although anabolism and catabolism occur simultaneously in the cell, the rates of their **chemical reactions** are controlled independently of each other. For example, there are two enzymatic pathways for glucose metabolism. The anabolic pathway synthesizes glucose, while catabolism breaks down glucose. The two pathways share 9 of the 11 enzymatic steps of glucose metabolism, which can occur in either sequence (i.e., in the direction of anabolism or catabolism). However, two steps of glucose anabolism use an entirely different set of enzyme-catalyzed reactions.

There are two important reasons that the cell must have separate complementary anabolic and catabolic pathways. First, catabolism is a so-called "downhill" process during which energy is released, while anabolism requires the input of energy, and is therefore an energetically "uphill" process. At certain points in the anabolic pathway, the cell must put more energy into a reaction than is released during catabolism. Such anabolic steps require a different series of reaction than are used at this point during catabolism.

Second, the different pathways permit the cell to control the anabolic and catabolic pathways of specific molecules independently of each other. This is important because there are times when the cell must slow or halt a particular catabolic or anabolic pathway in order to reduce breakdown or synthesis of a particular molecule. If both anabolism and catabolism used the same pathway, the cell would not be able control the **rate** of either process independent of the other: slowing the rate of catabolism would slow the rate of anabolism.

Opposite anabolic and catabolic pathways can occur in different parts of the same cell. For example, in the liver the breakdown of **fatty acids** to the molecule acetyl-CoA takes place inside mitochondria. Mitochondria are the tiny, membrane-bound organelles that function as the cell's major site of ATP production. The buildup of fatty acids from acetyl-CoA occurs in the cytosol of the cell, that is, in the aqueous area of the cell that contains various solutes.

Although anabolic and catabolic pathways are controlled independently, both metabolic routes share an important common sequence of reactions that is known collectively as the **citric acid** cycle, or **Krebs cycle**. The Krebs cycle is part of a larger series of enzymatic reactions collectively called oxidative phosphorylation. This pathway is an important means of breaking down glucose to produce energy, which is stored in the form of ATP. But the molecules produced by the Krebs cycle can also be used as precursor molecules, or raw materials, for anabolic reactions that make proteins, fats, and carbohydrates.

Despite the independence of anabolism and catabolism, the various steps of these processes are in some ways so intimately linked that they form what might be considered an "enzymatic ecological system." In this system, a change in one part of a metabolic series of reactions can have a ripple effect throughout the linked anabolic and catabolic pathways.

This ripple effect is the cell's way of counterbalancing an increase or decrease in anabolism of a molecule with an opposite increase or decrease in catabolism. This lets the cell adjust the rate of anabolic and catabolic reactions to meet its immediate needs, and prevent imbalance of either anabolic or catabolic products.

For example, when the cell needs to produce specific proteins, it produces only enough of each of the various amino acids needed to synthesize those proteins. Moreover, certain amino acids are used by the cell to make glucose, which appears in the **blood**, or glycogen, a **carbohydrate** stored in the liver. So the products of **amino acid** catabolism do not accumulate, but rather feed the anabolic pathways of carbohydrate synthesis. Thus, while many organisms store energy-rich **nutrients** such as carbohydrates and **fat**, most do not store other biomolecules, such as proteins, or nucleic acids, the building blocks of **deoxyribonucleic acid (DNA)**.

The cell regulates the rate of anabolic reactions by means of allosteric enzymes. The activity of these enzymes increases or decreases in response to the presence or absence of the end product of the series of reactions. For example, if an anabolic series of reactions produces a particular amino acid, that amino acid inhibits the action of the allosteric enzyme, reducing the synthesis of that amino acid.

Anaconda *see* **Boas**

Anaerobic

The term anaerobic refers to living processes (usually the release of **energy** from **nutrients**) that take place in the absence of molecular **oxygen**. The earliest organisms, the prokaryotic **bacteria**, lived in an oxygen deficient atmosphere and extracted energy from organic compounds without oxygen (that is, by anaerobic **respiration**). Most organisms alive today extract their energy from nutrients aerobically (in the presence of oxygen) although a few respire. **Aerobic** respiration releases a lot of energy from nutrients, whereas anaerobic respiration releases relatively little energy.

Anaerobic organisms

Microorganisms that cannot tolerate oxygen and are killed in its presence are called obligate (or strict) anaerobes. Bacteria of the genus *Clostridium*, which cause gas **gangrene**, **tetanus**, and **botulism**, belong to this group as well as *Bacteroides gingivalis*, which thrives in anaerobic crevices between the teeth. Strict anaerobes are killed by oxygen, which is why **hydrogen peroxide** (which releases oxygen) is frequently applied to wounds. Methane-producing bacteria (methanogens), can be isolated from the anaerobic habitats of swamp sludge sewage and from the guts of certain animals. Methanogens generate the marsh gases of swamps and **sewage treatment** plants by converting **hydrogen** and **carbon dioxide** gases to methane. Some organisms, such as **yeast**, have adapted to grow in either the presence or absence of oxygen and are termed facultative anaerobes.

Anaerobic respiration

All cells carry out the process of **glycolysis**, which is an example of anaerobic respiration. Glycolysis is the initial phase of **cellular respiration** that involves the splitting of the six-carbon glucose **molecule** into two three-carbon pyruvate fragments. This is the main energy-releasing pathway that occurs in the cytoplasm of all prokaryotes and eukaryotes, and no oxygen is required. During glycolysis, oxidation (the removal of electrons and hydrogen ions) is facilitated by the coenzyme NAD$^+$ (nicotinamide adenine dinucleotide), which is then reduced to NADH. Only two ATP molecules result from this initial anaerobic reaction. This is a small amount of energy when compared to the net aerobic energy yield of 36 ATP in the complete oxidation of one molecule of glucose.

Fermentation

Under anaerobic conditions, the pyruvate molecule can follow other anaerobic pathways to regenerate the NAD$^+$ necessary for glycolysis to continue. These include alcoholic **fermentation** and lactate fermentation. In the absence of oxygen the further reduction or addition of hydrogen ions and electrons to the pyruvate molecules that were produced during glycolysis is termed fermentation. This process recycles the reduced NADH to the free NAD$^+$ coenzyme which once again serves as the hydrogen acceptor enabling glycolysis to continue. Alcoholic fermentation, characteristic of some plants and many microorganisms, yields **alcohol** and **carbon** dioxide as its products. Yeast is used by the **biotechnology** industries to generate carbon dioxide gas necessary for bread-making and in the fermentation of hops and **grapes** to produce alcoholic beverages. Depending on the yeast variety used, the different alcohol levels realized act as a form of population control by serving as the toxic element which kills the producers. **Birds** have been noted to fly erratically after they have gorged themselves on the fermenting fruit of the *Pyracantha* shrub.

Reduction of pyruvate by NADH to release the NAD$^+$ necessary for the glycolytic pathway can also result in lactate fermentation, which takes place in some **animal** tissues and in some microorganisms. Lactic acid-producing bacterial cells are responsible for the souring of milk and production of yogurt. In working animal muscle cells, lactate fermentation follows the exhaustion of the ATP stores. Fast twitch muscle fibers store little energy and rely on quick spurts of anaerobic activity, but the **lactic acid** that accumulates within the cells eventually leads to muscle fatigue and cramp.

Analemma

The Earth's **orbit** around the **Sun** is not a perfect **circle**. It is an **ellipse**, albeit not a very flattened one, and this leads to a number of interesting observational effects. One of these is the analemma, the apparent path traced by the Sun in the sky when observed at the same **time** of day over the course of a year. The path resembles a lopsided figure eight, which you sometimes see printed on a globe, usually somewhere on the Pacific Ocean where there is lots of room to show it.

Suppose you go outside and measure the Sun's position in the sky every day, precisely at noon, over the course of a year. You would expect to see the Sun appear to move higher in the sky as summer approached, and then move lower as winter approached. This occurs because the tilt of Earth's axis causes the Sun's apparent celestial latitude, or *declination*, to change over the course of the year.

However, you would also notice that at some times of the year, the Sun would appear slightly farther west in the sky than it did at other times, as if it were somehow gaining time on your watch. This results from the ellipticity of Earth's orbit. According to Kepler's second law of **motion**, planets moving in an elliptical orbit will move faster when they are closer to the Sun than when they are farther away. Therefore, Earth's speed in its orbit is constantly changing, decelerating as it moves from perihelion (its closest point to the Sun) to aphelion (its farthest point from the Sun), and accelerating as it then "falls" inward toward perihelion again.

It would be a nightmare for watchmakers to try to make a clock that kept actual solar time. The clock would have to tick at different rates each day to account for Earth's changing **velocity** about the Sun. Instead, watches keep what is called **mean** solar time, which is the average value of the advance of solar time over the course of the year. As a result, the Sun gets ahead of, and behind, mean solar time by up to 16 minutes at different times of the year. In other words, if you measured the position of the Sun at noon mean solar time at one time of year, the Sun might not reach that position until 12:16 P.M. at another time of year.

Now all the elements are in place to explain the figure eight. The tilt of Earth's orbital axis causes the Sun to appear higher and lower in the sky at different times of year; this forms the vertical axis of the eight. The ellipticity of Earth's orbit causes the actual solar time to first get ahead of, and then fall behind, mean solar time. This makes the Sun appear to slide back and forth across the vertical axis of the eight, forming the rest of the figure.

Clearly, the shape of the analemma depends upon a particular planet's orbital inclination and ellipticity. The Sun would appear to trace a unique analemma for any of the planets in the **solar system**; the analemmas thus formed are fat, thin, or even teardrop-shaped variants on the basic figure eight.

Jeffrey Hall

Analgesia

Analgesia is the loss of **pain** without the loss of consciousness.

Techniques for controlling and relieving pain include **acupuncture**, **anesthesia**, hypnosis, **biofeedback**, and the use of analgesic drugs. Acupuncture is the ancient Chinese practice of inserting fine needles along certain pathways of the body and is used to relieve pain, especially in **surgery**, and to cure **disease**. In Western medicine the discovery of **ether** was a landmark in the development of anesthesia. Other techniques for pain control include electrical stimulation of the skin, massage, and stress-management therapy.

Analgesia is of primary importance for the treatment of injury or illness. The main agents for accomplishing analgesia in medical practice are analgesic drugs. These fall into two main categories: addictive and nonaddictive. Nonaddictive analgesics are generally used for treating moderate to severe pain and can be purchased without a prescription as over-the-counter drugs. More powerful analgesics have the potential for **addiction** and other undesirable side effects. They are usually used in hospitals or prescribed for relief from severe pain.

Presently, efforts are underway to develop powerful nonaddictive pain-relieving drugs. In order to improve the effectiveness and minimize the harm of analgesic drugs, pharmacological research has focused on the mechanism of how analgesics accomplish the task of pain relief. *Mechanism* in this context means the way the drug works in the body to accomplish its results.

Nonaddictive analgesics

While sold under many different brand names, the three main nonaddictive analgesics are aspirin, acetaminophen, and ibuprofen.

Aspirin was first synthesized in 1853 from vinegar and salicylic acid (**acetylsalicylic acid**). It is a member of the salicin family, which is a bitter white chemical found in willow **bark** and leaves. The analgesic qualities of willow bark were known to the ancient Greeks and others throughout the ages. In 1898 a German company, Bayer, further developed and marketed acetylsalicylic acid from industrial dyes and one year later named it aspirin. It soon became enormously popular as a pain reliever and antifever medicine. Its use as an anti-inflammatory for the treatment of **arthritis** and rheumatism made it the "gold standard" for readily available pain relief.

One of the major drawbacks of aspirin, however, is its effect on the stomach. It acts as an irritant and can cause bleeding **ulcers** in persons who take it over a long period of time. Recent research shows that aspirin's effectiveness not only as an analgesic but also as an anticoagulant makes it useful in the treatment of **heart** attacks and **stroke**, as well as in preventive medicine for other diseases.

Acetaminophen, introduced in 1955, is an over-the-counter drug that has become a very popular alternative to aspirin, since it relieves moderate pain without irritating the stomach. For the treatment of arthritis and rheumatism, acetaminophen does not have aspirin's anti-inflammatory effect, but as a pain and fever reliever it is just as effective as aspirin.

Ibuprofen is a nonsteroidal anti-inflammatory drug (NSAID) first introduced as a prescription drug in 1974. The Food and Drug Administration allowed it to be marketed as an over-the-counter drug in 1984. As an analgesic for minor pain and fever, it has about the same performance as aspirin and acetaminophen. Unlike acetaminophen, however, it can be irritating to the stomach.

Mechanism of nonaddictive analgesics

Pain that is caused by trauma (injury) in the body sets off the creation of a chemical called prostaglandin. The initial pain is caused by the nerve impulse that relays the injury message to the **brain**. According to the prostaglandin theory, the pain that is felt afterward is due to the prostaglandin at the site of the injury sending pain messages to the spinal cord, which then transmits these messages to the brain.

In 1971, a Nobel prize-winning pharmacologist, Sir John R. Vane, theorized that by blocking the formation of certain prostaglandins, aspirin was able to relieve the **inflammation** and pain that accompanies the trauma of damaged cells. Even though analgesics like aspirin had been used for centuries, Vane's discovery was the first real breakthrough in understanding how they work, and interest in prostaglandin research grew rapidly.

How over-the-counter medicines like aspirin, acetaminophen, and ibuprofen are able to temporarily relieve mild to moderate pain and even some severe pain is not completely understood. The blocking of prostaglandin production throughout the body, not just at the site of the pain, is the major theory that is currently accepted among medical scientists, although there are other theories to explain how these drugs work. Recent research suggests that aspirin has the ability to shut off communication of pain-transmitting nerves in the spinal cord.

The effectiveness of over-the-counter pain medicine depends on the kind of pain being experienced. Some pain originates in the skin and mucous areas of the body, while other pain comes from the smooth muscles and organs within the body. This latter type of pain is often difficult to pinpoint; it may feel like a generalized dull ache or throb, or may be referred pain, meaning pain that is felt in a part of the body away from its actual source.

Addictive analgesics

The treatment of severe pain, like pain that accompanies heart attack, kidney stones, gallstones, or terminal **cancer**, requires the use of prescription medicines that are far more potent than nonprescription drugs. **Morphine**, a drug derived from opium, has a long history of effective relief of severe pain, but it also is addictive and has dangerous side effects. Morphine and other drugs like it are called opiates.

Unlike nonaddictive drugs, increasing dosages of opiates also increases their analgesic effects. Thus, when they are self-administered, as pain increases, there is a danger of an overdose. Morphine is both a depressant and a stimulant. As a stimulant, it may cause nausea and vomiting, constriction of the pupils of the **eye**, and stimulation of the vagus nerve, which regulates heartbeat. This stimulation of the vagus nerve interferes with the treatment of pain in coronary **thrombosis**. Its main side effects—addictive potential, tolerance (dosages must be increased to get the same effect), constipation, and a marked **depression** of the respiratory system—restrict morphine's use as an analgesic.

Mechanism of addictive analgesics

The first line of defense against pain in the body is the endorphins. These chemicals are peptides (compounds of amino acids) found in the brain and parts of the central **nervous system**. Like opiates, they produce a sense of well being (euphoria) and relieve pain. Endorphins are released from a transmitting nerve **cell** and then bind to the receptor sites of a receiving cell. After an endorphin sends the message to block pain signals to the receptor site, it is annihilated, thus allowing other endorphins to be produced.

Morphine molecules flow through the spaces (synapses) between these sending and receiving cells and position themselves on the receptor sites, locking out endorphins. The morphine **molecule** sends the same message as the endorphin to block the pain signal. With long-term use of morphine, the endorphins decline in number and so do the receptor sites, leading to the twin problems of addiction and drug tolerance.

Development of new analgesics

The first phase in the development of new analgesics came with the development of **narcotic** blockers used to help drug addicts who overdose on a narcotic. Since these drugs have the ability to block the effects of morphine, they are called antagonists. They do not, however, provide any pain relief.

KEY TERMS

Addictive analgesics—Opiate drugs, like morphine, that provide relief from severe pain.

Agonist-antagonist analgesics—Drugs that provide pain relief in addition to blocking the effects of narcotics; these may be more effective and safer than opiates.

Antagonists—Drugs that block the effects of narcotics.

Endorphins—Biochemicals produced by the brain that act as opiates and reduce pain.

Nonaddictive analgesics—Drugs that relieve minor to moderate and, in some cases, severe pain.

Prostaglandins—Hormone-like substances found throughout the body, some of which are responsible for the pain and inflammation of an injury or illness.

Referred pain—Pain that is felt away from the part of the body in which it originates.

A second generation of narcotic antagonists are called agonist-antagonist analgesics. An agonist drug does provide pain relief by occupying receptor sites that block pain signals to the brain. The new group of agonist-antagonist medications show improved performance over morphine for providing pain relief, with fewer side effects and less chance of addiction.

The present challenge for medical science is to find medicines that are as effective as the opiates and morphine for relieving pain but do not have their side effects. Recent research on the brain has uncovered how endorphins and other related brain chemicals work, and provides hope for improved analgesic drugs.

See also Novocain.

Resources

Books

Berkow, Robert and Andrew J. Fletcher. *The Merck Manual of Diagnosis and Therapy.* Rahway: Merck Research Laboratories, 1992.

Gold, Mark S., and Michael Boyett. *Wonder Drugs: How They Work.* New York: Simon & Schuster, 1987.

Kehrer, James P., and Daniel M. Kehrer. *Pills and Potions.* New York: Arco, 1984.

McCaffery, Margo, and Alexandra Beebe. *Pain: Clinical Manual for Nursing Practice.* St. Louis: Mosby, 1989.

McKenry, Leda M., and Evelyn Salerno. *Mosby's Pharmacology in Nursing.* St. Louis: Mosby, 1989.

O'Neil, Maryadele J. *Merck Index: An Encyclopedia of Chemicals, Drugs, & Biologicals.* 13th ed. Whitehouse Station, NJ: Merck & Co., 2001.

Physicians Desk Reference 2003 with Physicians Desk Reference Family Guide. Montvale, NJ: Medical Economics, 2002.

Periodicals

Kiefer, D.M. "Chemistry Chronicles: Miracle Medicines." *Today's Chemist* 10, no. 6 (June 2001): 59-60.

Jordan P. Richman

Analog signals and digital signals

A signal is any time-varying physical quantity—voltage, **light** beam, sound wave, or other—that is used to convey information. Analog signals convey information by analogy (i.e., by mimicking the behavior of some other quantity). Digital signals convey information by assuming a series of distinct states symbolizing numbers (digits). Both analog and digital signals are essential to modern communications and computing, but the greater simplicity and generality of digital signaling has encouraged an increasing reliance on digital devices in recent decades.

Analog signals

An analog signal varies in step some other physical phenomenon, acting as an analog to or model of it. For example, the electrical signal produced by a microphone is an analog of the **sound waves** impinging on the mike.

The term analog is also commonly used to denote any smoothly varying waveform, even one (e.g., the voltage available from an AC power outlet) that does not convey information. Any waveform that is continuous in both **time** and amplitude is an analog waveform, while an analog waveform that happens to convey information is an analog signal.

The elemental or archetypal analog signal is a sinusoidal wave or sinusoid, because any analog signal can be viewed as a sum of sinusoids of different frequencies that have been variously shifted in time and magnified in amplitude. The rapidity with which a sinusoid repeats its cycle (one crest plus one dip) is termed its **frequency**. A plot of the frequencies and amplitudes of all the sinusoids that would be needed to build up a given analog waveform depicts its frequency content, or spectrum. Processing of analog signals consists largely of altering their spectra. For example, turning up the treble on a stereo system selectively amplifies the high-frequency part of the music's spectrum.

Digital signals

Digital signals convey discrete symbols that are usually interpreted as digits. For example, a voltage that signals the numbers 1 through N by shifting between N distinct levels is a digital signal, and so is a sinusoid that signals N digits by shifting between N distinct frequencies or amplitudes. (The latter would be analog as regards its waveform, but digital as regards its signaling strategy.)

Most digital signals are binary; that is, they signal the digits 0 and 1 by shifting between two distinct physical states (e.g., a high voltage and a low voltage). Each 0 or 1 is a bit (binary digit). Other numbers are communicated by transmitting "words," bundles of 0s and 1s, either bit by bit along a single channel or in **parallel** (as by N wires all signaling at the same time). Typical word lengths are: $2^4 = 16$, $2^5 = 32$, and $2^6 = 64$ bits.

Although the term digital emphasizes the use of a finite number of signal states to communicate digits, it is really the use of such signals to convey symbols that makes digital signals uniquely useful. Whether the two states of a binary signal represent 0 and 1, Yes and No, "first half of alphabet" and "second half of alphabet," or any other pair of meanings is entirely up to the human designer. In principle, it is possible to use analog signals the same way, but in practice it is quite awkward to do so.

Another feature of virtually all digital systems is that all signals in the system change state (low to high, high to low) at frequent, regularly spaced instants. These system-wide changes of state are governed by a central timing device, or system clock. The system clock in a modern digital device may change state millions or billions of times per second.

Analog to digital and back again

Because most physical quantities can be described by measurements, and because any measurement can be represented by a sufficiently long series of 0s and 1s, it is possible to transfer some of the information in any analog signal to a digital signal. In fact, according to the Nyquist sampling **theorem**, sufficiently precise measurements of an analog waveform made at twice or more the maximum frequency present in that waveform will preserve all its information. Digital signals can, conversely, be converted to analog signals by using them as inputs to a device whose output shifts smoothly between a series of voltages (or other physical states) as directed by a changing set of input bits. Analog-to-digital conversion is per-

formed, for example, when a digital **compact disc** (CD) is recorded, and digital-to-analog conversion is performed when a CD is played back over an audio system.

See also Computer, analog; Computer, digital.

Resources

Books

Sklar, Bernard. *Digital Communications: Fundamentals and Applications,* 2nd ed. Englewood Cliffs, NJ: Prentice-Hall, Inc., 2001.

Lathi, B.P. *Modern Digital and Analog Communications Systems,* 3rd ed. Oxford, UK: Oxford University Press, 1998.

Larry Gilman

Analytic geometry

Analytic **geometry** is a branch of **mathematics** that uses algebraic equations to describe the size and position of geometric figures on a coordinate system. Developed during the seventeenth century, it is also known as Cartesian geometry or coordinate geometry. The use of a coordinate system to relate geometric points to **real numbers** is the central idea of analytic geometry. By defining each **point** with a unique set of real numbers, geometric figures such as lines, circles, and conics can be described with algebraic equations. Analytic geometry has found important applications in science and industry alike.

Historical development of analytic geometry

During the seventeenth century, finding the solution to problems involving curves became important to industry and science. In **astronomy**, the slow acceptance of the **heliocentric theory** (Sun-centered theory) of planetary **motion** required mathematical formulas that would predict elliptical orbits. Other areas such as **optics**, navigation, and the military required formulas for things such as determining the curvature of a **lens**, the shortest route to a destination, and the trajectory of a cannon ball. Although the Greeks had developed hundreds of theorems related to curves, these did not provide quantitative values so they were not useful for practical applications. Consequently, many seventeenth-century mathematicians devoted their attention to the quantitative evaluation of curves. Two French mathematicians, Rene Descartes (1596-1650) and Pierre de Fermat (1601-1665) independently developed the foundations for analytic geometry. Descartes was first to publish his methods in an appendix titled *La geometrie* of his book *Discours de la methode* (1637).

Cartesian coordinate system

The link between **algebra** and geometry was made possible by the development of a coordinate system which allowed geometric ideas, such as point and line, to be described in algebraic terms like real numbers and equations. In the system developed by Descartes, called the rectangular cartesian coordinate system, points on a geometric **plane** are associated with an ordered pair of real numbers known as coordinates. Each coordinate describes the location of a single point relative to a fixed point, the origin, which is created by the intersection of a horizontal and a vertical line known as the x-axis and y-axis respectively. The relationship between a point and its coordinates is called one-to-one since each point corresponds to only one set of coordinates.

The x and y axes divide the plane into four quadrants. The sign of the coordinates is either positive or **negative** depending in which quadrant the point is located. Starting in the upper right quadrant and working clockwise, a point in the first quadrant would have a positive value for the abscissa and the ordinate. A point in the fourth quadrant (lower right hand corner) would have negative values for each coordinate.

The notation P (x,y) describes a point P having coordinates x and y. The x value, called the abscissa, represents the horizontal **distance** of a point away from the origin. The y value, known as the ordinate, represents the vertical distance of a point away from the origin.

Distance between two points

Using the ideas of analytic geometry, it is possible to calculate the distance between the two points A and B, represented by the line segment AB that connects the points. If two points have the same ordinate but different abscissas, the distance between them is $AB = x_2 - x_1$. Similarly, if both points have the same abscissa but different ordinates, the distance is $AB = y_2 - y_1$. For points that have neither a common abscissa or ordinate, the **Pythagorean theorem** is used to determine distance. By drawing horizontal and vertical lines through points A and B to form a right triangle, it can be shown using the distance formula that $AB = -(x_2 - x_1)^2 + (y_2 - y_1)^2$. The distance between the two points is equal to the length of the line segment AB.

In addition to length, it is often desirable to find the coordinates of the midpoint of a line segment. These coordinates can be determined by taking the average of the x and y coordinates of each point. For example, the coordinates for the midpoint M (x,y) between points P (2,5) and Q (4,3) are x = (2 + 4)/2 = 3 and y = (5 + 3)/2 = 4.

Algebraic equations of lines

One of the most important aspects of analytic geometry is the idea that an algebraic equation can relate to a geometric figure. Consider the equation 2x + 3y = 44. The solution to this equation is an ordered pair (x,y) which represents a point. If the set of every point that makes the equation true (called the *locus*) were plotted, the resulting graph would be a straight line. For this reason, equations such as these are known as linear equations. The standard form of a linear equation is Ax + By = C, where A, B, and C are constants and A and B are not both 0. It is interesting to note that an equation such as x = 4 is a linear equation. The graph of this equation is made up of the set of all ordered pairs in which x = 4.

The steepness of a line relative to the x-axis can be determined by using the concept of the slope. The slope of a line is defined by the equation

$$m = \frac{Y_2 - Y_1}{x_2 - x_1}$$

The value of the slope can be used to describe a line geometrically. If the slope is positive, the line is said to be rising. For a negative slope, the line is falling. A slope of **zero** is a horizontal line and an undefined slope is a vertical line. If two lines have the same slope, then these lines are **parallel**.

The slope gives us another common form for a linear equation. The slope-intercept form of a linear equation is written y = mx + b, where m is the slope of the line and b is the y intercept. The y intercept is defined as the value for y when x is zero and represents a point on the line that intersects the y axis. Similarly, the x intercept represents a point where the line crosses the x axis and is equal to the value of x when y is zero. Yet another form of a linear equation is the point-slope form, $y - y_1 = m(x - x_1)$. This form is useful because it allows us to determine the equation for a line if we know the slope and the coordinates of a point.

Calculating area using coordinates

One of the most frequent activities in geometry is determining the area of a polygon such as a triangle or **square**. By using coordinates to represent the vertices, the areas of any polygon can be determined. The area of triangle OPQ, where O lies at (0,0), P at (a,b), and Q at (c,d), is found by first calculating the area of the entire **rectangle** and subtracting the areas of the three right triangles. Thus the area of the triangle formed by points OPQ is = da − (dc/2) − (ab/2) − [(d−b)(a−c)]/2. Through the use of a determinant, it can be shown that the area of this triangle is:

$$\text{Area of triangle} = \ ^{1}/_{2} \ \begin{vmatrix} a & b \\ c & d \end{vmatrix}$$

This specific case was made easier by the fact that one of the points used for a vertex was the origin.

$$\text{Area of triangle} = \ ^{1}/_{2} \ \begin{vmatrix} x_1 & y_1 \\ x_2 & y_2 \end{vmatrix} + \ ^{1}/_{2} \ \begin{vmatrix} x_2 & y_2 \\ x_3 & y_3 \end{vmatrix} + \ ^{1}/_{2} \ \begin{vmatrix} x_3 & y_3 \\ x_1 & y_1 \end{vmatrix}$$

The general equation for the area of a triangle defined by coordinates is represented by the previous equation.

In a similar manner, the area for any other polygon can be determined if the coordinates of its points are known.

Equations for geometric figures

In addition to lines and the figures that are made with them, algebraic equations exist for other types of geometric figures. One of the most common examples is the **circle**. A circle is defined as a figure created by the set of all points in a plane that are a constant distance from a central point. If the center of the circle is at the origin, the formula for the circle is $r^2 = x^2 + y^2$ where r is the distance of each point from the center and called the radius. For example, if a radius of 4 is chosen, a plot of all the x and y pairs that satisfy the equation $4^2 = x^2 + y^2$ would create a circle. Note, this equation, which is similar to the distance formula, is called the center-radius form of the equation. When the radius of the circle is at the point (a,b) the formula, known as the general form, becomes $r^2 = (x - a)^2 + (y - b)^2$.

The circle is one kind of a broader type of geometric figures known as **conic sections**. Conic sections are formed by the intersection of a geometric plane and a double-napped cone. After the circle, the most common conics are parabolas, ellipses, and hyperbolas.

Curves known as parabolas are found all around us. For example, they are the shape formed by the sagging of **telephone** wires or the path a ball travels when it is thrown in the air. Mathematically, these figures are described as a **curve** created by the set of all points in a plane at a constant distance from a fixed point, known as the focus, and a fixed line, called the directrix. This means that if we take any point on the **parabola**, the distance of the point from the focus is the same as the distance from the directrix. A line can be drawn through the focus **perpendicular** to the directrix. This line is called the axis of **symmetry** of the parabola. The midpoint between the focus and the directrix is the vertex.

The equation for a parabola is derived from the distance formula. Consider a parabola that has a vertex at point (h,k). The linear equation for the directrix can be represented by y = k − p, where p is the distance of the focus from the vertex. The standard form of the equation of the parabola is then $(x - h)^2 = 4p(y - k)$. In this case, the axis of symmetry is a vertical line. In the case of a horizontal axis of symmetry, the equation becomes $(y - k)^2 = 4p(x - h)$ where the equation for the directrix is x = h − p. This formula can be expanded to give the more common quadratic formula which is $y = Ax^2 + Bx + C$, such that A does not equal 0.

Another widely used conic is an **ellipse**, which looks like a flattened circle. An ellipse is formed by the graph of the set of points, the sum of whose distances from two fixed points (foci) is constant. To visualize this definition, imagine two tacks placed at the foci. A string is knotted into a circle and placed around the two tacks. The string is pulled taut with a pencil and an ellipse is formed by the path traced. Certain parts of the ellipse are given various names. The two points on an ellipse intersected by a line passing through both foci are called the vertices. The chord connecting both vertices is the major axis and the chord perpendicular to it is known as the minor axis. The point at which the chords meet is known as the center.

Again by using the distance formula, the equation for an ellipse can be derived. If the center of the ellipse is at point (h,k) and the major and minor axes have lengths of 2a and 2b respectively, the standard equation is

$$\frac{(x - h)^2}{a^2} + \frac{(y - k)^2}{b^2} = 1$$

Major axis is horizontal

$$\frac{(x - h)^2}{b^2} + \frac{(y - k)^2}{a^2} = 1$$

Major axis is vertical

If the center of the ellipse is at the origin, the equation simplifies to $(x^2/a^2) + (y^2/b^2) = 1$.

The "flatness" of an ellipse depends on a number called the eccentricity. This number is given by the **ratio** of the distance from the center to the focus divided by the distance from the center to the vertex. The greater the eccentricity value, the flatter the ellipse.

Another conic section is a **hyperbola**, which looks like two facing parabolas. Mathematically, it is similar in definition to an ellipse. It is formed by the graph of the set of points, the difference of whose distances from two fixed points (foci) is constant. Notice that in the case of a hyperbola, the difference between the two distances

KEY TERMS

. . . .

Abscissa—The x-coordinate of a point representing its horizontal distance away from the origin.

Conic—A geometric figure created by a plane passing through a right circular cone.

Coordinate system—A system that relates geometric point s to real numbers based on their location in space relative to a fixed point called the origin.

Directrix—A line which, together with a focus, determines the shape of a conic section.

Ellipse—An eccentric or elongated circle, or oval.

Focus—A point, or one of a pair of points, whose position determines the shape of a conic section.

Hyperbola—A conic section created by a plane passing through the base of two cones.

Intercept—The point at which a curve meets the x or y axes.

Linear equations—A mathematical equation which represents a line.

Locus—The set of all points that make an equation true.

Ordinate—The y-coordinate of a point representing its vertical distance away from the origin.

Pythagorean theorem—An idea suggesting that the sum of the squares of the sides of a right triangle is equal to the square of the hypotenuse. It is used to find the distance between two points.

Slope—Slope is the ratio of the vertical distance separating any two points on a line, to the horizontal distance separating the same two points.

from fixed points is plotted and not the sum of this value as was done with the ellipse.

As with other conics, the hyperbola has various characteristics. It has vertices, the points at which a line passing through the foci intersects the graph, and a center. The line segment connecting the two vertices is called the transverse axis. The simplified equation for a hyperbola with its center at the origin is $(x^2/a^2) - (y^2/b^2) = 1$. In this case, a is the distance between the center and a vertex, b is the difference of the distance between the focus and the center and the vertex and the center.

Three-dimensional coordinate systems and beyond

Geometric figures such as points, lines, and conics are two-dimensional because they are confined to a sin-gle plane. The term two-dimensional is used because each point in this plane is represented by two real numbers. Other geometric shapes like spheres and cubes do not exist in a single plane. These shapes, called surfaces, require a third dimension to describe their location in **space**. To create this needed dimension, a third axis (traditionally called the z-axis) is added to the coordinate system. Consequently, the location of each point is defined by three real numbers instead of two. For example, a point defined by the coordinates (2,3,4) would be located 2 units away from the x axis, 3 units from the y axis, and 4 units from the z axis.

The algebraic equations for three-dimensional figures are determined in a way similar to their two-dimensional counterparts. For example, the equation for a **sphere** is $x^2 + y^2 + z^2 = r^2$. As can be seen, this is slightly more complicated than the equation for its two-dimensional cousin, the circle, because of the additional **variable** z^2.

It is interesting to note that just as the creation of a third dimension was possible, more dimensions can be added to our coordinate system. Mathematically, these dimensions can exist, and valid equations have been developed to describe figures in these dimensions. However, it should be noted that this does not mean that these figures physically exist and in fact, at present they only exist in the minds of people who study this type of multi-dimensional analytic geometry.

Resources

Books

Larson, Ron. *Calculus With Analytic Geometry.* Boston: Houghton Mifflin College, 2002.

Paulos, John Allen. *Beyond Numeracy.* New York: Alfred A. Knopf Inc., 1991.

Weisstein, Eric W. *The CRC Concise Encyclopedia of Mathematics.* New York: CRC Press, 1998.

Perry Romanowski

Anaphylaxis

Anaphylaxis is a severe, sudden, often fatal bodily reaction to a foreign substance or antigen. C. R. Richet first coined the term to define the puzzling reactions that occurred in dogs following injection of an eel toxin. Instead of acquiring immunity from the toxin as expected, the dogs experienced acute reactions, including often fatal respiratory difficulties, shock, and internal hemorrhaging. In humans, anaphylaxis is a rare event usually triggered by an antiserum (to treat snake or insect bites), **antibiotics** (especially immunoglobulin), or after wasp or bee stings. Certain foods can also trigger these severe reactions, including seafood, **rice**, potatoes, egg-whites, raw milk, and pinto beans.

In systemic or system-wide cases, symptoms occur just minutes (or in rare cases weeks) after introduction of the foreign substance and include flushed skin, itching of the scalp and tongue, breathing difficulties caused by bronchial spasms or swollen tissues, vomiting, diarrhea, a sudden drop in **blood pressure**, shock, and loss of consciousness. Less severe cases, usually caused by nonimmunologic mechanisms, may produce widespread hives or severe headache. These less severe cases are called anaphylactoid reactions.

While the exact biological process is poorly understood, anaphylaxis is thought to result from antigen-antibody interactions on the surface of mast cells, **connective tissue** cells believed to contain a number of regulatory chemicals. This interaction damages **cell** membranes, causing a sudden release of chemicals, including **histamine**, heparin, serotonin, bradykinin, and other pharmacologic mediators. Once released, these mediators produce the frightening bodily reactions that characterize anaphylaxis.

Because of the severity of these reactions, treatment must begin as soon as possible. The most common emergency treatment involves injection of epinephrine (adrenaline), followed by administration of cortisone, **antihistamines**, and other drugs that can reduce the effects of the unleashed chemical mediators. For people with known reactions to antibiotics, foods, insect and snake bites, or other factors, avoidance of the symptom-inducing agent is the best form of prevention.

See also Antibody and antigen.

Anatomy

Anatomy, a subfield of **biology**, is the study of the structure of living things. There are three main areas of anatomy: **cytology** studies the structure of **cell**; histology examines the structure of tissues; and gross anatomy deals with organs and **organ** groupings called systems. **Comparative anatomy**, which strives to identify general structural patterns in families of plants and animals, provided the basis for the classification of **species**. Human anatomy is a crucial element of the modern medical curriculum.

History

Modern anatomy, as a branch of Western science, was founded by the Flemish scientist Andreas Vesalius

(1514–1564), who in 1543 published *De humani corporis fabrica* (Structure of the human body). In addition to correcting numerous misconceptions about the human body, Vesalius's book was the first description of human anatomy that organized the organs into systems. Although initially rejected by many followers of classical anatomical doctrines, Vesalius's systematic conception of anatomy soon became the foundation of anatomical research and education throughout the world; anatomists still use his systematic approach.

Human anatomy

Human anatomy divides the body into the following distinct functional systems: cutaneous, muscular, skeletal, circulatory, nervous, digestive, urinary, endocrine, respiratory, and reproductive. This division helps the student understand the organs, their relationships, and the relations of individual organs to the body as a whole.

The cutaneous system consists of the integument— the covering of the body, including the skin, hair, and nails. The skin is the largest organ in the body, and its most important function is to act as a barrier between the body and the outside world. The skin's minute openings (pores) also provide an outlet for sweat, which regulates the body **temperature**. Melanin, a dark pigment found in the skin, provides protection from sunburn. The skin also contains oil-producing cells.

The muscles of the **muscular system** enable the body to move and provide power to the hands and fingers. There are two basic types of muscles. Voluntary (skeletal) muscles enable movements under conscious direction (e.g., to walk, move an arm, or smile). Involuntary (smooth) muscles are not consciously controlled, and operate independent of conscious direction. For example, they play an important role in digestion. The third type of muscle, cardiac muscle is involuntary, but also is striated, as in skeletal muscles. Because cardiac muscle is self-contractile it allows the **heart** to pumps **blood** throughout the body, without pause, from early in embryogenesis to death.

The **skeletal system**, or the skeleton, is the general supportive structure of the body. In addition, the skeletal system is the site of many important and complex physiological and immunological processes. The skeletal frame provides the support that muscles need in order to function. Of the 206 bones in the human body, the largest is the femur, or thigh bone. The smallest are the tiny **ear** ossicles, three in each ear, named the hammer (malleus), anvil (incus), and stirrup (stapes). Often included in the skeletal system are the ligaments, which connect bone to bone; the joints, which allow the connected bones to move; and the tendons, which connect muscle to bone.

The **circulatory system** comprises the heart, **arteries**, **veins**, **capillaries**, blood and blood-forming organs, and the lymphatic sub-system. The four chambers of the heart allow the heart to act as a dual pump to propel blood to the lungs for oxygenation (pulmonary system) and to pump blood throughout the body (systemic circulation). From the heart, the blood circulates through arteries. The blood is distributed through smaller and smaller tubes until it passes into the microscopic capillaries which bathe every cell. The veins collect the "used" blood from the capillaries and return it to the heart.

The **nervous system** consists of the **brain**, the spinal cord, and the sensory organs that provide information to them. For example, our eyes, ears, nose, tongue, and skin receive stimuli and send signals that travel both electrically and chemically to the brain. The brain is an intricate system of complicated neurons (nerve cells) that allow us to process sensory information, visceral signals (e.g. regulating breathing, body temperature, etc.), and perform cognitive thought.

The **digestive system** is essentially a long tube extending from the mouth to the anus. Food entering the mouth is conducted through the stomach, small intestine, and large intestine, where accessory organs contribute digestive juices to break down the food, extracting the molecules that can be used to nourish the body. The unusable parts of the ingested food are expelled through the anus as fecal **matter**. The salivary **glands** (in the mouth), the liver, and the pancreas are the primary digestive glands.

The urinary system consists of the kidneys, the bladder, and the connecting tubules. The kidneys filter **water** and waste products from the blood and pass them into the bladder. At intervals, the bladder is emptied through the urinary tract, ridding the body of unneeded waste.

The **endocrine system** consists of ductless (endocrine) glands that produce **hormones** that regulate various bodily functions. The pancreas secretes **insulin** to regulate sugar **metabolism**, for example. The pituitary gland in the brain is the principal or "master" gland that regulates many other glands and endocrine functions.

The **respiratory system** includes the lungs, the diaphragm, and the tubes that connect them to the outside atmosphere. **Respiration** is the process whereby an **organism** absorbs **oxygen** from the air and returns **carbon dioxide**. The diaphragm is the muscle that enables the lungs to work.

Finally, the **reproductive system** enables sperm and egg to unite and the egg to remain in the uterus or womb to develop into a functional human.

Anatomical nomenclature

Over the centuries, anatomists developed a standard nomenclature, or method of naming anatomical structures. Terms such as "up" or "down" obviously have no meaning unless the orientation of the body is clear. When a body is lying on it's back, the thorax and abdomen are at the same level. The upright sense of up and down is lost. Further, because anatomical studies and particularly embryological studies were often carried out in animals, the development of the nomenclature relative to comparative anatomy had an enormous impact on the development of human anatomical nomenclature. There were obvious difficulties in relating terms from quadrupeds (animals that walk on four legs) who have abdominal and thoracic regions at the same level as opposed to human bipeds in whom an upward and downward orientation might seem more obvious.

In order to standardize nomenclature, anatomical terms relate to the *standard anatomical position*. When the human body is in the standard anatomical position it is upright, erect on two legs, facing frontward, with the arms at the sides each rotated so that the palms of the hands turn forward.

In the standard anatomical position, *superior* means toward the head or the *cranial* end of the body.

The term *inferior* means toward the feet or the *caudal* end of the body.

The frontal surface of the body is the *anterior* or *ventral* surface of the body. Accordingly, the terms "anteriorly" and "ventrally" specify a position closer to—or toward—the frontal surface of the body. The back surface of the body is the *posterior* or *dorsal* surface and the terms "posteriorly" and "dorsally" specify a position closer to—or toward—the posterior surface of the body.

The terms *superficial* and *deep* relate to the **distance** from the exterior surface of the body. Cavities such as the thoracic cavity have internal and external regions that correspond to deep and superficial relationships in the midsagittal **plane**.

The bones of the skull are fused by sutures that form important anatomical landmarks. Sutures are joints that run jaggedly along the interface between the bones. At **birth**, the sutures are soft, broad, and cartilaginous. The sutures eventually fuse and become rigid and ossified near the end of **puberty** or early in adulthood.

The sagittal suture unties the parietal bones of the skull along the midline of the body. The suture is used as an anatomical landmark in anatomical nomenclature to establish what are termed *sagittal planes* of the body. The primary sagittal plane is the sagittal plane that runs through the length of the sagittal suture. Planes that are

parallel to the sagittal plane, but that are offset from the midsagittal plane are termed *parasagittal planes*. Sagittal planes run anteriorly and posteriorly, are always at right angles to the coronal planes. The *medial plane* or *midsagittal plane* divides the body vertically into superficially symmetrical *right* and *left* halves.

The medial plane also establishes a centerline axis for the body. The terms *medial* and *lateral* relate positions relative to the medial axis. If a structure is medial to another structure, the medial structure is closer to the medial or center axis. If a structure is lateral to another structure, the lateral structure is farther way from the medial axis. For example, the lungs are lateral to the heart.

The coronal suture unites the frontal bone with the parietal bones. In anatomical nomenclature, the primary *coronal plane* designates the plane that runs through the length of the coronal suture. The primary coronal plane is also termed the *frontal plane* because it divides the body into frontal and back halves.

Planes that divide the body into superior and inferior portions, and that are at right angles to both the sagittal and coronal planes are termed *transverse planes*. Anatomical planes that are not parallel to sagittal, coronal, or transverse planes are termed *oblique planes*.

The body is also divided into several regional areas. The most superior area is the *cephalic region* that includes the head. The *thoracic region* is commonly known as the chest region. Although the *celiac region* more specifically refers to the center of the *abdominal region*, celiac is sometimes used to designate a wider area of abdominal structures. At the inferior end of the abdominal region lies the *pelvic region* or *pelvis*. The posterior or dorsal side of the body has its own special regions, named for the underlying vertebrae. From superior to inferior along the midline of the dorsal surface lie the *cervical, thoracic, lumbar* and *sacral* regions. The buttocks is the most prominent feature of the *gluteal region*.

The term *upper limbs* or *upper extremities* refers to the arms. The term *lower limbs* or *lower extremities* refers to the legs.

The *proximal* end of an extremity is at the junction of the extremity (i.e., arm or leg) with the trunk of the body. The *distal* end of an extremity is the point on the extremity farthest away from the trunk (e.g., fingers and toes). Accordingly, if a structure is proximate to another structure it is closer to the trunk (e.g., the elbow is proximate to the wrist). If a structure is distal to another, it is farther from the trunk (e.g., the fingers are distal to the wrist).

Structures may also be described as being medial or lateral to the midline axis of each extremity. Within the upper limbs, the terms radial and ulnar may be used synonymous with lateral and medial. In the lower extremities, the terms fibular and tibial may be used as synonyms for lateral and medial.

Rotations of the extremities may de described as medial rotations (toward the midline) or lateral rotations (away from the midline).

Many structural relationships are described by combined anatomical terms (e.g. the eyes are anterio-medial to the ears).

There are also terms of movement that are standardized by anatomical nomenclature. Starting from the anatomical position, *abduction* indicates the movement of an arm or leg away from the midline or midsagittal plane. *Adduction* indicates movement of an extremity toward the midline.

The opening of the hands into the anatomical position is *supination* of the hands. Rotation so the dorsal side of the hands face forward is termed *pronation*.

The term *flexion* means movement toward the flexor or anterior surface. In contrast, *extension* may be generally regarded as movement toward the extensor or posterior surface. Flexion occurs when the arm brings the hand from the anatomical position toward the shoulder (a curl) or when the arm is raised over the head from the anatomical position. Extension returns the upper arm and or lower to the anatomical position. Because of the embryological rotation of the lower limbs that rotates the primitive dorsal side to the adult form ventral side, flexion occurs as the thigh is raised anteriorly and superiorly toward the anterior portion of the pelvis. Extension occurs when the thigh is returned to anatomical position. Specifically, due to the embryological rotation, flexion of the lower leg occurs as the foot is raised toward the back of the thigh and extension of the lower leg occurs with the kicking **motion** that returns the lower leg to anatomical position.

The term *palmar surface* (palm side) is applied to the flexion side of the hand. The term *plantar surface* is applied to the bottom sole of the foot. From the anatomical position, extension occurs when the toes are curled back and the foot arches upward and flexion occurs as the foot is returned to anatomical position.

Rolling motions of the foot are described as *inversion* (rolling with the big toe initially lifting upward) and *eversion* (rolling with the big toe initially moving downward).

See also Anatomy, comparative; Forensic science; Human evolution; Physiology, comparative; Physiology; Surgery.

Resources

Books

Gray, Henry. *Gray's Anatomy.* Philadelphia: Running Press, 1999.

Marieb, Elaine Nicpon. *Human Anatomy & Physiology.* 5th Edition. San Francisco: Benjamin/Cummings, 2000.

Netter, Frank H., and Sharon Colacino. *Atlas of Human Anatomy.* Yardley, PA: Icon Learning Systems, 2003.

K. Lee Lerner
Larry Blaser

Anatomy, comparative

There are many forms of evidence for **evolution**. One of the strongest forms of evidence is comparative **anatomy**; comparing structural similarities of organisms to determine their evolutionary relationships. Organisms with similar anatomical features are assumed to be relatively closely related evolutionarily, and they are assumed to share a common ancestor. As a result of the study of evolutionary relationships, anatomical similarities and differences are important factors in determining and establishing classification of organisms.

Some organisms have anatomical structures that are very similar in embryological development and form, but very different in function. These are called homologous structures. Since these structures are so similar, they indicate an evolutionary relationship and a common ancestor of the **species** that possess them. A clear example of homologous structures is the forelimb of **mammals**. When examined closely, the forelimbs of humans, whales, dogs, and **bats** all are very similar in structure. Each possesses the same number of bones, arranged in almost the same way. While they have different external features and they function in different ways, the embryological development and anatomical similarities in form are striking. By comparing the anatomy of these organisms, scientists have determined that they share a common evolutionary ancestor and in an evolutionary sense, they are relatively closely related.

Other organisms have anatomical structures that function in very similar ways, however, morphologically and developmentally these structures are very different. These are called analogous structures. Since these structures are so different, even though they have the same function, they do not indicate an evolutionary relationship nor that two species share a common ancestor. For example, the wings of a bird and dragonfly both serve the same function; they help the **organism** to fly. However, when comparing the anatomy of these wings, they are very different. The bird wing has bones inside and is covered with feathers, while the dragonfly wing is missing both of these structures. They are analogous structures. Thus, by comparing the anatomy of these organisms, scientists have determined that **birds** and **dragonflies** do not share a common evolutionary ancestor, nor that, in an evolutionary sense, they are closely related. Analogous structures are evidence that these organisms evolved along separate lines.

Vestigial structures are anatomical features that are still present in an organism (although often reduced in size) even though they no longer serve a function. When comparing anatomy of two organisms, presence of a structure in one and a related, although vestigial structure in the other is evidence that the organisms share a common evolutionary ancestor and that, in an evolutionary sense, they are relatively closely related. Whales, which evolved from land mammals, have vestigial hind leg bones in their bodies. While they no longer use these bones in their marine **habitat**, they do indicate that whales share an evolutionary relationship with land mammals. Humans have more than 100 vestigial structures in their bodies.

Comparative anatomy is an important tool that helps determine evolutionary relationships between organisms and whether or not they share common ancestors. However, it is also important evidence for evolution. Anatomical similarities between organisms support the idea that these organisms evolved from a common ancestor. Thus, the fact that all **vertebrates** have four limbs and gill pouches at some part of their development indicates that evolutionary changes have occurred over time resulting in the diversity we have today.

Anchovy

Anchovies are small, **bony fish** in the order Clupeiformes, a large group that also includes herring, **salmon**, and trout. Anchovies are in the family Engraulidae, and all of the more than 100 **species** are in the genus *Engraulis*. Anchovies are predominantly marine **fish**, but are occasionally found in **brackish** waters and even in **freshwater**. Species of anchovies are found in the Mediterranean Sea, Black Sea, European Atlantic coastal waters, and the Pacific coasts of Peru and Chile.

Anchovies are about 4-8 in (10-20 cm) long, with smooth scales, and soft fins. They live in schools made up of many thousands of individuals, often grouped according to size. Anchovies come to surface waters during the spawning season (May-July). Their eggs float on the surface of the **water** and hatch in 3-4 days. Anchovies swim with their mouth open, and feed on **plankton**, small crustaceans, and fish larvae. When food is scarce, anchovies take turns swimming at the front of their school, where they are more likely to encounter the best food. When a school of anchovies sense danger, it swims together to make a tight ball in which the fish on the inside are more protected, while those on the outer part have a greater chance of being consumed. Anchovies are usually caught by fishermen at night—lights on the boat serve as a lure.

In Peru, the anchovy known as anchovetta (*Engraulis ringens*), is economically important as a source of fish meal and fertilizer. In 1970, the Peruvian anchovetta fishery yielded about 12 million tons of fish, but this crashed to less than 2 million tons in most years between 1973-1987. The collapse of the fishery was likely due to excessive harvesting, although oceanographic and climatic changes associated with warm-water El Niño events may have also played a role. More recently, there has been somewhat of a recovery of the anchovetta stocks, and the fishery has yielded about 4 million tons a year. Other species of anchovy (such as *E. encrasicolus*) are fished in smaller numbers, canned in oil and salt, and sold as a delicacy (these are the tiny fish on pizza, and are also used to make the dressing for Caesar salad). Other species of anchovy are used as fish bait or are added to pet food and **livestock** feed.

Anemia

Anemia means literally lack of **blood**. In fact it is a reduction in the number of red blood cells, **plasma**, or packed red blood cells to a level that is lower than necessary for normal functioning. This is the result of the inability to replace lost cells or plasma **volume** at the **rate** they are being lost. The underlying cause for anemia may be one of several conditions.

Causes of anemia

Anemia can result from a rapid loss of blood cells as with trauma, from a reduced rate of replacement, or from an abnormally rapid destruction of blood cells within the body that outstrips the replacement ability of bone marrow.

Normally the red blood **cell** count for an adult male is 5.4 million per cubic mm of blood at sea level. The count for the adult female is 4.8 million per cubic mm. An adult male is considered anemic if his red cell count falls below 4.5 million per cubic mm, and the female if her count is less than 4 million per cubic mm. The normal rate of replacement for red blood cells is 40,000 to 50,000 new cells per cubic mm per day. Blood cells are generated within the bone marrow.

Trauma and surgery

Any time the integrity of the body is violated blood can be lost. An **automobile** accident, a wound, a fall, or a surgical procedure all can open the body to blood loss. A trauma or wound that opens a major blood vessel allows blood to be pumped from the closed **circulatory system** and depletes the blood volume. Certainly the body is not able to replace such a rapid loss of blood cells or plasma.

Surgical procedures have become more and more "bloodless." Refinements in techniques as well as in instrumentation have rendered modern surgical procedures blood-conserving transactions.

A great deal of **surgery** is carried out using flexible scopes and instruments that can be inserted into the body through very small incisions. The surgical procedure is carried out by the physician who can see the surgical field by means of a miniature camera on the tip of the instrument. Long incisions followed by clamping of bleeding **arteries** and **veins** have been replaced. The need for blood transfusions during surgery has been reduced markedly over the past decade. The patient who may need blood replacement is urged to have his own blood collected prior to surgery, so there will be no subsequent problems from the transfusion.

Low red blood cell production

A deficiency in red blood cell production is the most common cause of anemia, and a lack of **iron** is the most common reason for low cell production. Iron is the basic atom of hemoglobin, the substance that gives blood its red **color** and is responsible for carrying **oxygen** and **carbon dioxide** through the body. The normal level of iron in the body is 0.1-0.2 oz (3-5 g), depending upon one's sex and size.

Iron deficiency occurs as the result of inadequate intake of iron, loss of iron in young women of childbearing age through menstruation and child bearing, impaired absorption of iron, which may occur following removal of part of the stomach and small intestine, or because of a chronic loss of low volumes of blood within the **digestive system** through polyps, **cancer**, or ulcerations brought on by excessive aspirin intake.

Old and nonfunctional red blood cells are culled from the blood stream by the spleen. The cells are destroyed and the iron **molecule** within the hemoglobin is transferred back to the bone marrow for reuse, a classic example of recycling. Red blood cells that escape the body, however, take iron with them. Of the usual iron intake of 10-20 mg daily in the diet, about one or two mg is absorbed to replace lost iron. Absorption is increased in women who are menstruating or are pregnant.

People with iron deficiency anemia will have abnormally small red blood cells and low hemoglobin levels even though the total red blood cell count may be normal. Supplementary iron intake may resolve the problem, but surgery may be necessary if the underlying cause is a polyp or other area bleeding internally.

Aplastic anemia

Aplastic anemia is a life-threatening form of anemia resulting from insufficient production of blood cells. The underlying cause of the **disease** is in the bone marrow itself. Aplastic anemia can be brought about by exposure to toxic chemicals or **radiation** or by excessive intake of certain medications. Removal of the toxic agent or getting away from the source of radiation usually will allow the patient to recover without further events. However, a form of aplastic anemia known as idiopathic, which means a disease of unknown cause, may well result in death.

Aplastic anemia affects all cells in the blood including the white blood cells and blood platelets. Loss of white blood cells, the core of the **immune system**, leaves one susceptible to **infection**. Loss of blood platelets, which are functional in the clotting process, means bleeding into the skin, digestive system, urine, or **nervous system** may occur, as may repeated nosebleeds.

Treatment for this form of anemia is to remove the agent causing it, if known, and to provide supportive treatment until the bone marrow recovers. In extreme cases bone marrow transplantation from a compatible **individual** may be attempted and is sometimes successful.

Megaloblastic anemia

Megaloblastic refers to the large, immature blood cells seen with this kind of anemia. One form called pernicious anemia is the result of a digestive inadequacy in which **vitamin** B_{12} is not absorbed through the intestine. The condition also is caused by a parasitic infection, as with a tapeworm, certain digestive diseases, cancer, anticancer chemotherapeutic agents, and most commonly by lack of folate. It is an anemia often seen in elderly people

---

KEY TERMS

Bone marrow—A spongy tissue located in the hollow centers of certain bones, such as the skull and hip bones. Bone marrow is the site of blood cell generation.

Hemolytic anemia—Hemolytic means destruction of the blood cell, an abnormal rate of which may lead to lowered levels of these cells.

Menstruation—The normal, nearly monthly bleeding cycle experienced by women of childbearing age. It is the result of lack of a fertilized egg that triggers the loss of the lining of the uterus (womb) and subsequent bleeding.

Spleen—An organ lying just under the diaphragm in the abdomen that removes old or damaged blood cells from circulation. The spleen is not a critical organ and can be removed if necessary.

---

who suffer from **malnutrition**, alcoholics, teenagers (possibly also related to malnutrition), and pregnant women.

This form of anemia may be corrected easily by placing the individual on a balanced, adequate diet or by replacing lowered levels of folate or vitamins.

Sickle cell anemia

Although blood cell levels are not lowered in **sickle cell anemia**, the blood cells may be nonfunctional at times, resulting in oxygen starvation of some body tissues.

Others

There are many other kinds of anemias, though they are rarely seen. Those discussed are the most common types. Others include a form called spherocytosis, which, as the name implies, results in a spherical form of red blood cells resulting from an abnormality in the cell **membrane**; a similar form called elliptocytosis; and others resulting from decreased or abnormal hemoglobin production.

Resources

Periodicals

Friedland, I. "The Anemia Epidemic." *Working Mother* 16 (July, 1993): 18.

Organizations

National Heart, Lung, and Blood Institute, Bethesda, MD. Publications on anemia and sickle cell anemia.

Larry Blaser

Anesthesia

Anesthesia is the loss of feeling or sensation. It may be accomplished without the loss of consciousness, or with partial or total loss of consciousness.

Anesthesiology is a branch of medical science that relates to anesthesia and anesthetics. The anesthetist is a specialized physician in charge of supervising and administering anesthesia in the course of a surgical operation. Depending on the type of operation and procedures used, there are two types of anesthesia: general anesthesia, which causes a loss of consciousness, and local anesthesia, where the anesthetic "freezes" the nerves in the area covered by the operation. In local anesthesia, the patient may be conscious during the course of the operation or given a sedative, a drug that induces sleep.

History of anesthesia

While the search for pain control during surgery dates back to the ancient world, it was not until 1846 that it went on record that a patient was successfully rendered unconscious during a surgical procedure. Performed in a Boston hospital, the operation used a gas called ether to anesthetize the patient while a neck tumor was removed. In Western medicine, the development of anesthesia has made possible complex operations like open heart surgery and organ transplants. Medical tests that would otherwise be impossible to perform are routinely carried out with the use of anesthesia.

Before the landmark discovery of ether as an anesthetic, patients who needed surgery for either illness or injury had to face the surgeon's knife with only the help of alcohol, opium, or other narcotics. Often a group of men held the patient down during the operation in case the narcotic or alcohol wore off before it was over. Under these conditions many patients died just from the pain of the operation.

Nitrous oxide

In 1776 Joseph Priestley, a British chemist, discovered the gas nitrous oxide. Another British chemist, Humphry Davy, proposed nitrous oxide as a means for pain-free surgery, but his views were dismissed by other physicians of the day. In the next century, Horace Wells, a Connecticut dentist, began to experiment with nitrous oxide, and in 1845 attempted to demonstrate its anesthetic qualities to a public audience. However, the patient woke before the operation was over and began to scream in pain. Because of this spectacle, it took another 20 years before nitrous oxide again gained attention. By 1870, nitrous oxide was a commonplace dental anesthetic.

Ether

The gas ether was discovered in 1540 and given its name in 1730. The gas was first successfully used by an American physician, Crawford W. Long, in an operation in 1842. The operation, however, was unrecorded, so official credit went instead to William Morton for his 1846 demonstration of an operation with the use of ether.

Chloroform

The credit for discovering the third major anesthetic of this period in medical history goes to James Young Simpson, a Scottish gynecologist and obstetrician. Simpson used ether in his practice but searched for an anesthetic that would make bearing children less painful for women. He tested several gases until he came upon chloroform in 1847 and began to use it on women in labor. Chloroform use, though, had higher risks than those associated with ether, and it called for greater skill from the physician. Neither ether nor chloroform are used in surgery today.

Emergence of anesthesiology

Anesthesiology as a medical specialty was slow to develop. By the end of the nineteenth century, ether, which was considered safer than chloroform, was administered by *etherizers* who had little medical experience, including students, new physicians, non-medical specialists, nurses, and caretakers. Eventually, nurses began to be used for this job, becoming the first anesthetists by the end of the nineteenth century.

While the practice of surgery began to make considerable progress by the turn of the century, anesthesiology lagged behind. In the twentieth century, though, the need for specialists in anesthesia was sparked by two world wars and advanced surgical techniques. To meet these demands, the American Society of Anesthetists was formed in 1931 and specialists were then certified by the American Board of Anesthesiology in 1937. By 1986, the Board certified 13,145 specialists—physicians and nurses, called nurse anesthetists—in the field of anesthesiology.

Types of anesthesia

Modern anesthesiology can be divided into two types, pharmacological and non-pharmacological. Pharmacological anesthesia uses a wide variety of anesthetic agents to obtain varying degrees of sedation and pain control. The anesthesia is administered orally, by injection, or with a gas mask for inhalation. Examples of non-pharmacological anesthesia are the use of breathing techniques during conscious childbirth (Lamaze method

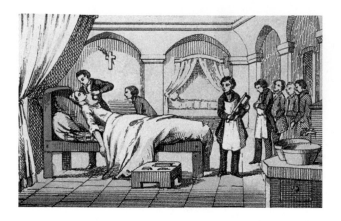

A nineteenth-century physician administering chloroform prior to surgery (probably an amputation). Ether was one of the earliest anesthetics to be used, but was difficult to administer, as it made the patient choke. For a time it was replaced by chloroform, but made a comeback when chloroform proved even more dangerous. *National Audubon Society Collection/Photo Researchers, Inc. Reproduced by*

of natural childbirth) and the ancient art of Chinese **acupuncture**. Non-pharmacological anesthesia requires special skills on the part of its practitioners, and its effects are not as reliable as pharmacological techniques.

Pharmacological anesthesia is described as either general or local.

General anesthesia

There are three phases to general anesthesia. The anesthetist must first induce the state of unconsciousness (induction), keep the patient unconscious while the procedure is performed (maintenance), then allow the patient to emerge back into consciousness (emergence).

A drug commonly used to induce unconsciousness is thiopentone **sodium**. It is a barbiturate that produces unconsciousness within 30 seconds after being injected intravenously. Thiopentone does not reduce pain; it actually lowers the threshold of pain. It is used in the induction stage to bring about a quick state of unconsciousness before using other drugs to maintain the anesthetic condition during surgery.

Other agents used for the induction and maintenance of anesthesia are gases or volatile liquids such as nitrous oxide, halothane, enflurane, methoxyflurane and cyclopropane.

Nitrous oxide is still commonly used in **dentistry**, minor surgery, and major surgery when it is accompanied by other anesthetics. Though the gas has been used for many years, it is still uncertain how nitrous oxide accomplishes its anesthetic effect. Mixtures of **oxygen** and nitrous oxide appear to enhance its effect. Unlike other

agents used today, it appears to have no toxic side effects on the body.

Halothane is a colorless liquid with a very low **boiling point**. Its use, though, may be connected to liver toxicity. Enflurane and methoxyflurane are also liquids that are useful as analgesics (pain relievers) and **muscle relaxants**, but they also may have undesirable side effects. Cyclopropane, which is an expensive and explosive gas used for rapid induction and quick recovery, has over the years been replaced with the use of halothane.

The anesthesiologist interviews the patient before the operation and examines his or her medical records to determine which of the many anesthetic agents available will be used. Cyclopropane or atropine may be given before the operation to relieve pain and **anxiety**. When a muscle relaxant is given for the surgical procedure, the anesthesiologist monitors the respiratory equipment to ensure the patient is breathing properly.

Administration of the anesthetic is usually accomplished by the insertion of a cannula (small tube) into a vein. Sometimes a gas anesthetic may be introduced through a mask. If a muscle relaxant is used, the patient may not be able to breathe on his own, and a breathing tube is passed into the windpipe (trachea). The tube then serves either to deliver the anesthetic gases or to ventilate (oxygenate) the lungs.

During the course of the surgery, the anesthesiologist maintains the level of anesthetic needed to keep up the patient's level of anesthesia to the necessary state of unawareness while monitoring vital functions, such as heart beat, breathing, and blood/gas exchange.

COMPLICATIONS OF GENERAL ANESTHESIA. There are a number of possible complications that can occur under general anesthesia. They include loss of **blood pressure**, irregular heart beat, heart attack, vomiting and then inhaling the vomit into the lungs, **coma**, and death. Although mishaps do occur, the chance of a serious complication is extremely low. Avoidance of complications depends on a recognition of the condition of the patient before the operation, the choice of the appropriate anesthetic procedure, and the nature of the surgery itself.

Local anesthesia

Local anesthetics block pain in regions of the body without affecting other functions of the body or overall consciousness. They are used for medical examinations, diagnoses, minor surgical and dental procedures, and for relieving symptoms of minor distress, such as itching, toothaches, and hemorrhoids. They can be taken as creams, ointments, sprays, gels, or liquid; or they can be given by injection and in **eye** drops.

Some local anesthetics are benzocaine, bupivacaine, **cocaine**, lidocaine, procaine, and tetracaine. Some act rapidly and have a short duration of effect, while others may have a slow action and a short duration. They act by blocking nerve impulses from the immediate area to the higher pain centers. Regional anesthetics allow for pain control along a wider area of the body by blocking the action of a large nerve (nerve block). Sprays can be used on the throat and related areas for a bronchoscopy, and gels can be used for the urethra to numb the area for a catherization or cystoscopy.

Spinal anesthesia is used for surgery of the abdomen, lower back and legs. Spinal or epidural anesthesia is also used for surgery on the prostate gland and hip. A fine needle is inserted between two vertebrae in the lumbar (lower part) of the spine and the anesthetic flows into the fluid surrounding the spinal cord. The nerves absorb the anesthetic as they emerge from the spinal fluid. The area anesthetized is controlled by the location of the injection and the amount of absorption of the anesthetic by the spinal fluid.

COMPLICATIONS OF LOCAL ANESTHESIA. It is possible to have adverse reactions to local anesthetics, such as dizziness, hypotension (low blood pressure), convulsions, and even death. These effects are rare but can occur if the dose is too high or if the drug has been absorbed too rapidly. A small percentage of patients (1-5%) may develop headaches with spinal anesthesia.

Theory of the mechanism of anesthesia

Although scientists are not sure exactly how anesthesia works, there are many theories that have been proposed. In addition, different anesthetics may have different mechanisms of action. One theory proposes a relationship between the **solubility** of the anesthetic agent into the **fat** cells of the body (**lipid** solubility) as determining the degree of its potency as an anesthetic agent. Since nerve **cell** membranes are highly lipid, the **brain**, with its high nerve cell content, soaks up the anesthetic. Not all lipid soluble substances, however, are anesthetics. Lipid solubility, therefore, is only a partial explanation of the anesthetic's mechanism.

Another feature of anesthetic absorption is the way it is passed from the lungs to other cells in the body. At first there is a quick transmission from the lungs to the rest of the body, but as an equilibrium is reached, the anesthetic begins to quickly pass out from the lungs. However, fat cells retain the anesthetic longer than other cells.

Studies have shown that some inhaled anesthetics are metabolized by the liver (hepatic **metabolism**). Here is where the skill of the anesthetist is needed to control the amounts administered in order to avoid the problem of toxicity.

KEY TERMS

Chloroform—A simple chlorinated hydrocarbon compound consisting of one carbon atom, one hydrogen atom, and three chlorine atoms. One of several early gases, along with ether and nitrous oxide, that opened the way for the successful pharmacological use of anesthesia.

Consciousness—A mental state involving awareness of the self and the environment..

Halothane—An important, current volatile liquid anesthetic.

Lipid solubility—The ability of a drug to dissolve in fatty tissue.

The future of anesthesia

Since World War II, many changes have taken place in anesthesiology. Important discoveries have been made with such volatile liquids as halothane and synthetic opiates. The technology of delivery systems has been greatly improved. But with all these changes, the basic goal of anesthesia has been the same—the control of a motionless surgical field in the patient. In the next 50 years it is possible that the goals of anesthesia will be widened. The role of anesthesia will broaden as newer surgical techniques develop in the area of organ transplants. Anesthesia may also be used in the future to treat acute infectious illness, mental disorders, and different types of heart conditions. There may be a wide range of new therapeutic applications for anesthesia.

Anesthesiologists compete strongly for research funds. Better trained anesthesiologists need to do research to gain further knowledge on the effects and mechanisms of anesthesia. Since understanding and controlling pain is the central problem of anesthesiology, it will be necessary to gain more knowledge about the mechanism of pain and pain control. New anesthetics, delivery, and monitoring systems will need to be developed to keep up with the pace of medical development as it moves closer to noninvasive surgical techniques.

See also Analgesia; Novocain.

Resources

Books

Barash, Paul G., Bruce F. Cullen, and Robert K. Stoelting. *Clinical Anesthesia.* Philadelphia: Lippincott, 1992.

McKenry, Leda M. and Evelyn Salerno. *Mosby's Pharmacology in Nursing*. Philadelphia: Mosby, 1989.

Jordan P. Richman

Aneurism

An aneurism is a weak spot in the wall of an artery or a vein that dilates or balloons out, forming a blood-filled sack or pouch. Aneurisms can occur almost anywhere in the body and are found in all age groups, although they occur primarily in the elderly. The foremost cause of aneurisms is atherosclerosis, or fatty deposits in the **arteries**. If an aneurism bursts, a massive amount of **blood** is released, which results in an almost instantaneous drop in blood **pressure** and can cause death. **Surgery** can be a successful treatment for unruptured aneurisms.

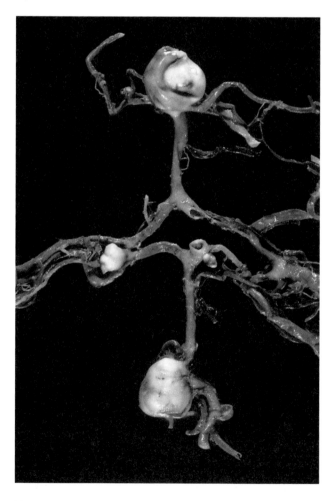

Three aneurisms can be seen in this section of cerebral artery removed from a human brain. © *Martin Rotker/Phototake NYC Reproduced with permission.*

Atherosclerotic aneurisms are the most common aneurisms. They are found primarily in the abdominal aorta (a larger elastic artery), typically occur after the age of 50, and are more common in men than women. Syphilitic (luetic) aneurisms occur primarily in the thoracic aorta and are due to syphilis, a sexually transmitted **disease**. Dissecting aneurisms occur when blood penetrates the arterial or venous wall, causing dissections between the wall's various layers. Aneurisms occurring from injury to the artery are known as false aneurisms.

Aneurisms are also classified by their size and shape. A berry aneurism is spherical, or circular, and usually 0.4-0.6 in (1-1.5 cm) in diameter. A saccular aneurism is a larger berry aneurism and may reach a size of 6-8 in (15-20 cm). Fusiform aneurisms are shaped like spindles, and cylindroid (or tubular) aneurisms are cylinder dilations that occur over a considerable length of the artery.

Although recent studies show that **genetics** can be a factor in the development of aneurisms, environmental factors, such as a bad diet resulting in high blood pressure, are a major cause of aneurisms. Cigarette smoking is considered to be the highest risk factor for developing an aneurism. As a result, treatment of **hypertension** through proper diet and prescription drugs and by refraining from smoking are approaches to preventing the likelihood of an aneurism forming.

A key to the successful treatment of aneurisms is to discover them before they burst. Diagnostic procedures for aneurisms include the use of ultrasound, computed tomography, and magnetic **resonance angiography**. Surgical "clipping" or "grafting" is the treatment of choice for most aneurisms. This process involves cutting out the aneurism and then reconnecting the artery or vein with a metallic clip or synthetic tube **graft**.

Angelfish

The word angelfish is a general term that refers to many different kinds of **fish**. Typically, angelfish have thin bodies that are flattened laterally. They tend to have elongated dorsal, anal, and pelvic fins, and display a wide variety of colors, making them popular aquarium **species**. The association of these fish with angels may be due to the fins resembling wings.

All angelfish belong to the taxonomic order Perciformes. This varied group contains two families of angelfish. The so-called true angelfish belong to the family Pomacanthidae. **Butterfly fish**, another group of angelfish species, belong to the family Chaetodontidae. True angelfish are distinct from butterfly fish by the presence of a spine near the bottom margin of the *operculum*, the outer gill covering. There are eight scientific genera of angelfish found worldwide.

All species of angelfish are tropical, but surprisingly, some species inhabit **freshwater** while others prefer **salt water**. Angelfish are typically small, reaching only a few centimeters in length. However, some species are much larger, growing up to 2 ft (61 cm) in length. Most angelfish feed on small aquatic **invertebrates**. Several species are bred and cultivated as aquarium pets, others are taken from coral reefs and sold to enthusiasts.

Angelfish varieties have colorful names that reflect their appearance. Examples include Flame, Coral Beauty, Lemon Peel, and Dwarf Angel. Members of the shark genus *Squatina* are called angelfish because their pectoral fins resemble wings. These animals are not closely related to the angelfish kept as pets in aquariums.

Terry Watkins

Angina *see* **Heart diseases**

Angiography

Angiography is a medical diagnostic test in which a fluid that is visible on **x rays** is used to take photographs of the **arteries** of the **heart** or other organs.

First used in the early 1950s, angiography is now a standard procedure to locate areas where an artery is closed or constricted and interfering with the circulation of **blood**.

Angiography applied to the heart is called coronary angiography. A constriction of one of the arteries feeding the heart can be serious enough that a person will experience chest pains when he or she exercises because the heart muscle has an insufficient supply of blood.

The heart is a special muscular **organ** that must continue beating at all times. When you are awake or asleep, resting or exercising, the heart must supply the body with blood. In times of extra need, as when you run or bicycle rapidly, the heart must work harder to supply oxygen-laden blood to the muscles. To accomplish its work, the heart is given the first arteries that branch off the main artery leaving the heart, the aorta. In this way the heart has a source of blood at high **pressure** and with the most **oxygen**.

The arteries that supply the heart muscle are called the coronary arteries, and they course around the outside of the heart to carry blood to all parts of the hard-working muscle. As a person ages, however, these coronary arteries may begin to close down from **cholesterol** deposits, or they may have spasms in the muscle of the arteries, or a blood clot that has been circulating in the blood stream may lodge in one of the arteries. Any of these situations can result in **pain** or even death because the heart muscle is receiving too little oxygen. Pain in the chest caused by an oxygen-starved heart is called angina. Angina is a serious condition requiring medical attention. It may indicate a blockage that is easily controlled by medications, or it may be a life-threatening blockage requiring bypass **surgery** to restore circulation. To treat angina, the physician must know the exact location of the blockage, whether only one artery is blocked or if several are affected, and how severe the blockage is—whether it is only partially obscuring the artery or entirely plugging the blood passage.

To locate a blockage and discern its severity the doctor usually must resort to an angiogram. This is called an invasive study because it requires a catheter to be inserted into one of the patient's arteries.

To perform an angiogram the cardiologist inserts a long, thin tube (a catheter) into an artery usually in the thigh. The patient is fully awake during an angiogram, which is performed under local **anesthesia**. In this way the patient can turn over or from side to side if the doctor needs a different view. The arteries do not sense pain or **touch**, so the patient cannot feel the catheter as it progresses through the arteries.

The catheter is radio-opaque, that is, it can be seen on an x ray so the doctor can follow the progress of the catheter. He feeds the catheter into the artery and from there into the main artery of the body, the aorta. The tip of the catheter is slightly bent so that it can be steered from one artery into another by turning the catheter. The doctor follows the progress of the catheter on a fluoroscope. The fluoroscope is a screen onto which x rays are projected after they pass through the patient. The radio-opaque catheter shows up on the screen as a long, moving shadow.

The physician guides the tip of the catheter from the aorta into the main coronary artery and injects a contrast medium. This is a liquid that also is visible on x rays. As the contrast medium floods through the coronary arteries a videotape is made of the progress of the fluid into and through the arterial tree. The contrast medium is visible only for a few seconds on the fluoroscope and then is pumped out of the arteries, but the videotape can be reviewed at a slower pace. Angiography provides a method to visualize vessels in a given time frame following the distribution or dispersal of the dye through arterial or venous phases. Any constriction or stoppage in the artery

KEY TERMS

Diagnostic—A means to find the source of a condition or disease, enabling the physician to apply the appropriate therapy.

Fluoroscope—An instrument consisting of an x-ray machine and a television-screen. The x rays are passed through the patient and focused on the screen so the physician can observe the structure being studied or the progress of a contrast medium.

Heart attack—Myocardial infarction, or damage of the heart muscle in a locale fed by an artery that has become blocked. Without blood for only a short while, the heart muscle can cause chest pain called angina pectoris.

Spasm—A sudden flexing of the arterial wall that constricts the artery and slows blood flow. In a coronary artery, a spasm can cause a heart attack.

will be evident by looking at the pattern of the medium. It will show the artery coursing over the heart until the contrast medium reaches a constriction, where it is pinched into a small stream, or a stoppage, which the medium is not able to pass at all. The physician can move the tip of the catheter to position it at another artery as needed. Computerized enhancement techniques are utilized to improve the resolution of angiograms.

Once the troublesome area or areas have been located the doctor can decide what form of treatment is most appropriate.

A coronary angiogram is a form of test generally called an arteriogram, which means literally a picture of an artery. Arteries are long tubes with muscular walls that carry blood away from the heart. The arterial muscle enables the size of the artery to be enlarged or reduced in accordance with the demand for blood. Immediately after a meal, for example, the arteries to the digestive organs are enlarged to carry away the digested **nutrients**. The arteries in the arms and legs will be constricted to carry less blood. On the other hand, when a person runs or plays actively the arteries to the arms and legs and other muscles used in the activity are dilated to carry a full load of oxygen-rich blood to the muscles; the arteries to the **digestive system** are constricted.

Arteriograms are used to see the arteries in organs other than the heart. This diagnostic study can be carried out with the arteries in the **brain**, to find the location of a ruptured artery that has caused a **stroke**, for example; or in the kidneys and in the legs. There are variations of arteriograms, a splenoportograph, involves the injection of

contrast medium directly into the spleen to view the splenic and portal **veins**. In all cases the test consists of injecting a radio-opaque contrast medium into a suitable blood vessel and then capturing the image of the arteries made visible by the medium.

See also Circulatory system.

Resources

Books

PM Medical News *21st Century Complete Medical Guide to Heart Disease, Heart Attack, Cholesterol, Coronary Artery Disease, Bypass Surgery, Angioplasty* New York: Progressive Management Medical News, 2002.

Larry Blaser

Angiosperm

Angiosperm is the name given to those plants that produce flowers during **sexual reproduction**. The term literally means "vessel seed" and refers to the fact that **seeds** are contained in a highly specialized **organ** called an ovary.

Flowering plants are the most recently evolved of the major groups of plants, arising only about 130 million years ago. Despite their geological youthfulness, angiosperms are the dominant plants of the world today: about 80% of all living **plant species** are flowering plants. Furthermore, they occupy a greater variety of habitats than any other group of plants. The ancestors of flowering plants are the gymnosperms (e.g., pine and fir), which are the other major group of plants that produce seeds. The gymnosperms, however, produce their seeds on the surface of leaf-like structures, which makes the seeds vulnerable to mechanical damage when winds whip the branches back and forth, and to drying out. Most importantly, **conifer** seeds are vulnerable to **insects** and other animals, which view seeds as nutritious, **energy** packed treats. In angiosperms, the margins of the seed-bearing leaves have become inrolled and fused, so the seeds are no longer exposed but are more safely tucked inside the newly evolved "vessel," which is the ovary.

The other major advance of the angiosperms over the gymnosperms was the **evolution** of the **flower**, which is the structure responsible for sexual reproduction in these plants. The function of sexual reproduction is to bring together genetic material from two individuals of differing ancestry, so that the offspring will have a new genetic makeup. The gymnosperms dealt with their immobility by packaging their male component into tiny pollen grains, which can be released into the **wind** to be blown to the female component of another **individual** of the same species. Although this method of **pollination** succeeds, it is wasteful and inefficient because most of the pollen grains land somewhere other than on a female, such as in one's nose, where they may cause hay fever. Furthermore, pollen grains are rich in fixed energy and **nutrients** such as **nitrogen**, so they are costly to make (native North Americans used to make pancakes out of the pollen of cattail). By evolving bright colors, scents, and **nectar**, the flowers of angiosperms served to attract animals. By traveling from one flower to another, these animals would accidentally move pollen as well, enabling sexual reproduction to take place. Because flower-seeking animals such as **bees**, **butterflies**, and **hummingbirds** can learn to recognize different types of flowers, they can move pollen from flower to flower quite efficiently. Therefore, animal-pollinated species of flowering plants do not need to produce as much pollen as gymnosperms, and the resources they save can be put into other important functions, such as growth and greater seed production. Therefore, the flower and its ovary have provided angiosperms with tremendous advantages, and have enabled them to become rapidly dominant over their **gymnosperm** ancestors.

Les C. Cwynar

Angle

An angle is a geometric figure created by two line segments that extend from a single **point** or two planes which extend from a single line. The size of an angle, measured in units of degrees or radians, is related to the amount of **rotation** required to superimpose one of its sides on the other. First used by ancient civilizations, angles continue to be an important tool to science and industry today.

The study of angles has been known since the time of the ancient Babylonians (4,000-300 B.C.). These people used angles for measurement in many areas such as construction, commerce, and **astronomy**. The ancient Greeks developed the idea of an angle further and were even able to use them to calculate the circumference of **Earth** and the **distance** to the **moon**.

A geometric angle is formed by two lines (rays) that intersect at a common endpoint called the vertex. The two rays are known as the sides of the angle. An angle can be specified in various ways. If the vertex of an angle is at point P, then the angle could be denoted by ∠P. It can be further described by using a point from each ray. For example, the angle ∠OPQ would have the point O on one

ray, a vertex at point P, and the point Q on the remaining ray. An angle can also be denoted by a single number or character which is placed on it. The most common character used is the Greek letter θ (theta).

Units of measurement of an angle

An angle is commonly given an **arithmetic** value which describes its size. To specify its this value, an angle is drawn in a standard position on a coordinate system, with its vertex at the center and one side, called the initial side, along the x axis. The value of the angle then represents the amount of rotation needed to get from the initial side to the other side, called the terminal side. The direction of rotation indicates the sign of the angle. Traditionally, a counterclockwise rotation gives a positive value and a clockwise rotation gives a **negative** value. The three terms which are typically used to express the value of an angle include revolutions, degrees, or radians.

The revolution is the most natural unit of measurement for an angle. It is defined as the amount of rotation required to go from the initial side of the angle all the way around back to the initial side. One way to visualize a revolution is to imagine spinning a wheel around one time. The distance traveled by any point on the wheel is equal to one revolution. An angle can then be given a value based on the fraction of the distance a point travels divided by the distance traveled in one rotation. For example, an angle represented by a quarter turn of the wheel is equal to .25 rotations.

A more common unit of measurement for an angle is the degree. This unit was used by the Babylonians as early as 1,000 B.C. At that time, they used a number system based on the number 60, so it was natural for mathematicians of the day to divide the angles of an equilateral triangle into 60 individual units. These units became known as degrees. Since six equilateral triangles can be evenly arranged in a **circle**, the number of degrees in one revolution became $6 \times 60 = 360$. The unit of degrees was subdivided into 60 smaller units called minutes and in turn, these minutes were subdivided into 60 smaller units called seconds. Consequently, the notation for an angle which has a value of 44 degrees, 15 minutes, and 25 seconds would be 44° 15' 25".

An angle may be measured with a protractor, which is a flat instrument in the shape of a semi-circle. There are marks on its outer edges which subdivide it into 180 evenly spaced units, or degrees. Measurements are taken by placing the midpoint of the flat edge over the vertex of the angle and lining the 0° mark up with the initial side. The number of degrees can be read off at the point where the terminal side intersects the **curve** of the protractor.

Another unit of angle measurement, used extensively in **trigonometry**, is the radian. This unit relates a unique angle to each real number. Consider a circle with its center at the origin of a graph and its radius along the x-axis. One radian is defined as the angle created by a counterclockwise rotation of the radius around the circle such that the length of the **arc** traveled is equal to the length of the radius. Using the formula for the circumference of a circle, it can be shown that the total number of radians in a complete revolution of 360° is 2π. Given this relationship, it is possible to convert between a degree and a radian measurement.

Geometric characteristics of angles

An angle is typically classified into four categories including acute, right, obtuse, and straight. An acute angle is one which has a degree measurement greater than 0° but less than 90°. A right angle has a 90° angle measurement. An obtuse angle has a measurement greater than 90° but less than 180°, and a straight angle, which looks like a straight line, has a 180° angle measurement.

Two angles are known as congruent angles if they have the same measurement. If their sum is 90°, then they are said to be complementary angles. If their sum is 180°, they are supplementary angles. Angles can be bisected (divided in half) or trisected (divided in thirds) by rays protruding from the vertex.

When two lines intersect, they form four angles. The angles directly across from each other are known as ver-

tical angles and are congruent. The neighboring angles are called adjacent because they share a common side. If the lines intersect such that each angle measures 90°, the lines are then considered **perpendicular** or orthogonal.

In addition to size, angles also have trigonometric values associated with them such as sine, cosine, and tangent. These values relate the size of an angle to a given length of its sides. These values are particularly important in areas such as navigation, astronomy, and architecture.

See also Geometry.

Perry Romanowski

Anglerfish

Anglerfish are marine **fish** that attract **prey** by dangling a fleshy, bait-like appendage (the esca) in front of their heads. The appendage, which resembles a fishing pole, is attached to the end of the dorsal fin's foremost spine (the illicium), which is separated from the rest of the fin.

Anglerfish belong to the order Lophiiformes, which includes three suborders, 15 families, and about 215 **species**. The order Lophiiformes is in the class Osteichthyes, the bony fishes, which, in turn, is in the subphylum Vertebrata, phylum Chordata.

Anglerfish are distributed throughout the world and include both free-swimming (pelagic) species and seabed-dwelling (benthic) species. Most species of anglerfish live on the **ocean** floor and have bizarre body forms. They have small gill openings located at or behind the base of the pectoral fins, which are limb-like in structure; some species also have limb-like pelvic fins. The swim bladder of anglerfish is physoclistic, that is, it has no connection to the digestive tract.

The common names of many anglerfish are particularly colorful: for example, the flattened goosefish, the warty angler, and the spherical frogfish. The flattened goosefish inhabits the muddy bottom of the continental slopes of the Atlantic, Indian, and western Pacific Oceans and remains motionless as it angles for prey with its esca. The balloon-shaped warty angler is found in South **Australia** and New South Wales, and the spherical frogfish, which propels itself in hopping motions with pectoral fins modified into special "jumping" limbs, is found in tropical and subtropical seas.

The diet of anglerfish consists mostly of other fish, although some species consume marine **invertebrates**. For example, batfish, anglerfish with free flaps of skin, also eat marine **snails**, clams, **crustacea**, and worms; deep-sea anglers of the family Melanocetidae consume **copepods** and **arrow worms** in addition to fish.

Deep-sea anglerfish, which belong to the suborder Ceratioidei, live at depths between 4,920 and 8,200 ft (1,500 and 2,500 m). The deep-sea anglerfish lack pelvic fins, are free-swimming (pelagic) and do not remain still on the bottom, as do other anglerfish. Deep-sea anglerfish are distributed widely from subarctic to subantarctic waters, but are absent from the Mediterranean Sea.

Anglerfish range in length from about 4 in to 2 ft (10–60 cm). The males of four families of the suborder Ceratioidei show a dramatic sexual dimorphism, in that males are much smaller than the females. The male *Ceratias holboelli*, for example, grows to a maximum length of only 2.5 in (6 cm) while the females can reach up to 4 ft (1.2 m) in length.

Soon after **birth**, parasitic male deep-sea anglers use their pincerlike mouths to affix themselves to a female. The **tissue** of their mouths fuses completely with the tissues of the female, so that their **blood** supplies mingle. The female is the hunter, attracting prey with a luminous esca, while the male receives nourishment via the shared blood supply. The male, assured of a life without the need to hunt for food, ceases growth, except for development of reproductive organs. In turn, the female is assured a lifetime supply of sperm from the male to fertilize her eggs.

Some anglerfish are used by humans as food; for example, the European goosefish, *Lophius piscatorius*, also called monkfish, is highly prized; and frogfish are sometimes sold for home aquariums and are occasionally eaten.

Animal

Animals are creatures in the kingdom Animalia, one of the five major divisions of organisms (the others are: Monera or **bacteria**, **Fungi**, Protists or protozoans, and Plantae or plants). Animals are multicellular, eukaryotic organisms, with cells that do not have walls made of **cellulose**. Animals are capable of voluntary, spontaneous movements, often in response to sensory perceptions. For their **nutrition**, animals require sources of biologically fixed **energy**, that is, the **biomass** of either plants or other animals. Zoology is the scientific study of animals.

The time of the origin of animals during Earth's evolutionary history is not known, because the first animals were soft-bodied, multicellular life forms that did not preserve well as fossils. By the time multicellular animals became represented in the geological record about

640-670 million years ago, there were already numerous phyla of animals, so the actual origin of the kingdom must have occurred earlier.

Most zoologists recognize the existence of 30-35 phyla of animals, some of which are extinct and only known from their fossil record. The simplest animals with the most ancient lineages are asymmetric or radially symmetric, for example, Porifera (**sponges**) and Cnidaria (**jellyfish** and **sea anemones**). The simplest of the bilaterally symmetric animals include Platyhelminthes (**flatworms**) and Nematoda (nematodes). Coelomates are a functional group of animals with an enclosed body cavity, the best known of which are Mollusca (**snails**, clams, **octopus**, and **squid**), Annelida (**segmented worms** and leeches), Arthropoda (**insects**, spiders, and **crustacea**), Echinodermata (**sea urchins** and **starfish**), and Chordata (**fish**, **amphibians**, **reptiles**, **birds**, and **mammals**).

About one million **species** of animals have been named. However, biologists estimate that a much larger number of animal species has yet to be discovered, and that the actual total could be as large as 30-50 million species. The undiscovered species mostly occur in Earth's richest ecosystems, especially old-growth tropical rain **forests**, and perhaps the deep oceans. Most of the undiscovered species of terrestrial animals are believed to be insects, especially **beetles**.

See also Arrow worms; Arthropods; Brachiopods; Chordates; Horsehair worms; Mesozoa; Mollusks; Moss animals; Phoronids; Ribbon worms; Spiny-headed worms.

Bill Freedman

Animal breeding

Animal breeding is the selective mating of animals to increase the possibility of obtaining desired traits in the offspring. It has been performed with most domesticated animals, especially **cats** and dogs, but its main use has been to breed better agricultural stock. The more modern techniques involve a wide variety of laboratory methods, including the modification of embryos, sex selection, and **genetic engineering**. These procedures are beginning to supplant traditional breeding methods, which focus on selectively combining and isolating **livestock** strains. In general, the most effective strategy for isolating traits is by selective inbreeding; but different strains are sometimes crossed to take advantage of **hybrid** vigor and to forestall the negative results of inbreeding, which include reduced fertility, low immunity, and the development of genetic abnormalities.

The genetic basis of animal breeding

Breeders engage in genetic "experiments" each time they plan a mating. The type of mating selected depends on the goals. To some breeders, determining which traits will appear in the offspring of a mating is like rolling the dice—a combination of luck and chance. For others, producing certain traits involves more skill than luck—the result of careful study and planning. Breeders have to understand how to manipulate genes within their breeding stock to produce the kinds of dogs they want. They have to first understand dogs as a **species**, then dogs as genetic individuals.

Once the optimal environment for raising an animal to maturity has been established (i.e., the proper **nutrition** and care has been determined) the only way to manipulate an animal's potential is to manipulate its genetic information. In general, the genetic information of animals is both diverse and uniform: diverse, in the sense that a population will contain many different forms of the same **gene** (for instance, the human population has 300 different forms of the protein hemoglobin); and uniform, in the sense that there is a basic physical expression of the genetic information that makes, for instance, most **goats** look similar to each other.

In order to properly understand the basis of animal breeding, it is important to distinguish between **genotype and phenotype**. Genotype refers to the information contained in an animal's DNA, or genetic material. An animal's phenotype is the physical expression of its genotype. Although every creature is born with a fixed genotype, the phenotype is a variable influenced by many factors in the animal's environment and development. For example, two cows with identical genotypes could develop quite different phenotypes if raised in different environments and fed different foods.

The close association of environment with the expression of the genetic information makes animal breeding a challenging endeavor, because the physical traits a breeder desires to selectively breed for cannot always be attributed entirely to the animal's genes. Moreover, most traits are due not just to one or two genes, but to the complex interplay of many different genes.

DNA consists of a set of chromosomes; the number of chromosomes varies between species (humans, for example, have 46 chromosomes). **Mammals** (and indeed most creatures) have two copies of each **chromosome** in the DNA (this is called diploidy). This means there are two copies of the same gene in an animal's DNA. Sometimes each of these will be partially expressed. For example, in a person having one copy of a gene that codes for normal hemoglobin and one coding for sickle-cell hemoglobin, about half of the hemoglobin will be normal and the other half will be sickle-cell. In other cases, only one of the genes can be expressed in the animal's phenotype. The gene expressed is called dominant, and the gene that is not expressed is called recessive. For instance, a human being could have two copies of the gene coding for **eye color**; one of them could code for blue, one for brown. The gene coding for brown eyes would be dominant, and the individual's eyes would be brown. But the blue-eyes gene would still exist, and could be passed on to the person's children.

Most of the traits an animal breeder might wish to select will be recessive, for the obvious reason that if the gene were always expressed in the animals, there would be no need to breed for it. If a gene is completely recessive, the animal will need to have two copies of the same gene for it to be expressed (in other words, the animal is homozygous for that particular gene). For this reason, animal breeding is usually most successful when animals are selectively inbred. If a bull has two copies of a gene for a desirable recessive trait, it will pass one copy of this gene to each of its offspring. The other copy of the gene will come from the cow, and assuming it will be normal, none of the offspring will show the desirable trait in their phenotype. However, each of the offspring will have a copy of the recessive gene. If they are then bred with each other, some of their offspring will have two copies of the recessive gene. If two animals with two copies of the recessive gene are bred with each other, all of their offspring will have the desired trait.

There are disadvantages to this method, although it is extremely effective. One of these is that for animal breeding to be performed productively, a number of animals must be involved in the process. Another problem is that undesirable traits can also mistakenly be selected for. For this reason, too much inbreeding will produce sickly or unproductive stock, and at times it is useful to breed two entirely different strains with each other. The resulting offspring are usually extremely healthy; this is referred to as "hybrid vigor." Usually hybrid vigor is only expressed for a generation or two, but crossbreeding is still a very effective means to combat some of the disadvantages of inbreeding.

Another practical disadvantage to selective inbreeding is that the DNA of the parents is altered during the production of eggs and sperm. In order to make eggs and sperm, which are called gametes, a special kind of **cell division** occurs called **meiosis**, in which cells divide so that each one has half the normal number of chromosomes (in humans, each sperm and egg contains 23 chromosomes). Before this division occurs, the two pairs of chromosomes wrap around each other, and a phenomenon known as crossing over takes place in which sections of one chromosome will be exchanged with sections of the other chromosome so that new combinations are generated. The problem with crossing over is that some unexpected results can occur. For instance, the offspring of a bull homozygous for two recessive but desirable traits and a cow with "normal" genes will all have one copy of each recessive gene. But when these offspring produce gametes, one recessive gene may migrate to a different chromosome, so that the two traits no longer appear in one **gamete**. Since most genes work in complicity with others to produce a certain trait, this can make the process of animal breeding very slow, and it requires many generations before the desired traits are obtained—if ever.

Economic considerations

There are many reasons why animal breeding is of paramount importance to those who use animals for their livelihood. Cats have been bred largely for aesthetic beauty; many people are willing to pay a great deal of money for a Siamese or Persian cat, even though the affection felt for a pet has little to do with physical appearance. But the most extensive animal breeding has occurred in those areas where animals have been used to serve specific practical purposes. For instance, most dog breeds are the result of a deliberate attempt to isolate traits that would produce better hunting and herding dogs (although some, like toy poodles, were bred for traits that would make them desirable pets). **Horses** have also been extensively bred for certain useful qualities; some for size and strength, some for speed. But farm animals, particularly food animals, have been the subject of the most intensive breeding efforts.

The physical qualities of economic importance in farm animals vary for each species, but a generalized goal is to eliminate the effects of environment and nutrition. An ideal strain of milk cow, for instance, would produce a large amount of high-quality milk despite the type of food it is fed and the environment in which it is reared. Thus, animals are generally all bred for feed efficiency, growth **rate**, and resistance to **disease**. However, a pig might be bred for lean content in its meat, while a hen would be bred for its laying potential. Many cows have been bred to be hornless, so they cannot inadvertently or deliberately gore each other.

Although maximum food production is always a major goal, modern animal breeders are also concerned about nutritional value and the ability of animals to survive in extreme environments. Many parts of the world are sparsely vegetated or have harsh climatic conditions, and a high efficiency producer able to endure these environments would be extremely useful to the people who live there. In addition, many people of industrialized countries are concerned not about food availability but about the quality of this food; so breeders seek to eliminate the qualities that make meat or milk or eggs or other animal products unhealthy, while enhancing those qualities that make them nutritious.

Modern methods in biotechnology

Although earlier animal breeders had to confine themselves to choosing which of their animals should mate, modern technological advances have altered the face of animal breeding, making it both more selective and more effective. Techniques like genetic **engineering**, embryo manipulation, artificial insemination, and cloning are becoming more and more refined. Some, like artificial insemination and the manipulation of embryos to produce twins, are now used habitually. Others, such as genetic engineering and cloning, are the subject of intense research and will probably have a great impact on future animal breeding programs.

Artificial insemination

Artificial insemination is the artificial introduction of semen from a male with desirable traits into females of the species to produce pregnancy, and is useful because a far larger number of offspring can be produced than would be possible if the animals were traditionally bred. Because of this, the value of the male as breeding stock can be determined much more rapidly, and the use of many different females will permit a more accurate evaluation of the heritability of the desirable traits. In addition, if the traits produced in the offspring do prove to be advantageous, it is easier to disperse them within an animal population in this fashion, as there is a larger breeding stock available. One reason artificial insemination has been an extremely important tool is that it allowed new strains of superior stock to be introduced into a supply of animals in an economically feasible fashion.

The process of artificial insemination requires several steps. Semen must be obtained and effectively diluted, so that the largest number of females can be inseminated consistent with a high probability of pregnancy. The semen must be properly stored so that it remains viable. The females must be tested before the sample is introduced to ensure they are fertile, and following the procedure, they must be tested for pregnancy to determine its success. All these factors make artificial insemination more expensive and more difficult than traditional breeding methods, but the processes have been improved and refined so that the economic advantages far outweigh the procedural disadvantages, and artificial insemination is the most widely applied breeding technique.

Embryo manipulation

In order to understand the techniques of embryo manipulation, it is important to understand the early stages of reproduction. When the egg and sperm unite to form a zygote, each of the parents supply the zygote with half of the chromosomes necessary for a full set. The zygote, which is a single **cell**, then begins to reproduce itself by the cellular division process called **mitosis**, in which each chromosome is duplicated before separation so that each new cell has a full set of chromosomes. This is called the morula stage, and the new cells are called blastomeres. When enough cells have been produced (the number varies from species to species), cell differentiation begins to take place. The first differentiation appears to be when the blastocyst is formed, which is an almost hollow sphere with a cluster of cells inside; and the differentiation appears to be between the cells inside, which become the fetus, and the cells outside, which become the fetal membranes and placenta. However, the process is not entirely understood at the present time and there is some variation between species; so it is difficult to pinpoint the onset of differentiation, which some scientists believe occurs during blastomere division.

During the first stages of cell division, it is possible to separate the blastomeres with the result that each one develops into a separate embryo. Blastomeres with this capability are called totipotent. The purpose of this ability of a single blastomere to produce an entire embryo is probably to safeguard the process of embryo development against the destruction of any of the blastomeres. In theory, it should be possible to produce an entire embryo from each blastomere (and blastomeres are generally totipotent from the four to eight cell stage), but in practice it is usually only possible to produce two embryos. That is why this procedure is generally referred to as embryo splitting rather than cloning, although both terms refer to the same thing (cloning is the production of genetically identical embryos, which is a direct result of embryo splitting).

Interestingly enough, although the embryos produced from separated blastomeres usually have fewer cells than a normal embryo, the resulting offspring fall within the normal range of size for the species.

It is also possible to divide an embryo at other stages of development. For instance, the time at which embryo division is most successful is after the blastocyst has formed. Great care must be taken when dividing a blastocyst, since differentiation has already occurred to some extent, and it is necessary to halve the blastocyst very precisely.

Another interesting embryonic manipulation is the creation of chimaeras. These are formed by uniting two different gametes, so that the embryo has two distinct cell lineages. Chimaeras do not combine the genetic information of both lineages in each cell. Instead, they are a patchwork of cells containing one lineage or the other. For this reason, the offspring of chimaeras are from one distinct genotype or the other, but not from both. Thus chimaeras are not useful for creating new animal populations beyond the first generation. However, they are extremely useful in other contexts. For instance, while embryo division as described above is limited in the number of viable embryos that can be produced, chimaeras can be used to increase the number. After the blastomeres are separated, they can be combined with blastomeres of a different genetic lineage. It has been found that with the additional **tissue**, the survival rate of the new embryos is more favorable. For some reason only a small percentage of the resulting embryos are chimaeric; this is thought to be because only one cell lineage develops into the cells inside the blastocyst, while the other lineage forms extra-embryonic tissue. It is believed that the more advanced cells are more likely to form the inner cells.

Another application of chimaeras could be for breeding **endangered species**. Because of the different biochemical environments in the uterus, and the different regulatory mechanisms for fetal development, only very closely related species are able to bear each other's embryos to term. For example, when a goat is implanted with a **sheep** embryo or the other way around, the embryo is unable to develop properly. This problem can perhaps be surmounted by creating chimaeras in which the placenta stems from the cell lineage of the host species. The **immune system** of an animal attacks tissue it recognizes as "non-self," but it is possible that the mature chimaeras would be compatible with both the host species and the target species, so that it could bear either embryo to term. This has already proved to be true in studies with **mice**.

A further technique being developed to manipulate embryos involves the creation of uniparental embryos and same-sex matings. In the former case, the cell from a single gamete is made to go through mitosis, so that the resulting cell is completely homozygous. In the latter case, the DNA from two females (parthogenesis) or two males (androgenesis) is combined to form cells that have only female- or male-derived DNA. These zygotes cannot be developed into live animals, as genetic information from male and female derived DNA is necessary for embryonic development. However, these cells can be used to generate chimaeras. In the case of parthogenetic cells, these chimaeras produce viable gametes. The androgenetic cells do not become incorporated in the embryo; they are used to form extra-embryonic tissue, and so no gametes are recovered.

Aside from these more ambitious embryo manipulation endeavors, multiple ovulation and **embryo transfer**

KEY TERMS

Androgenesis—Reproduction from two male parents.

Artificial insemination—The artificial introduction of semen from a male with desirable traits into females of the species to produce pregnancy.

Blastocyst—An embryo at that stage of development in which the cells have differentiated to form embryonic and extra-embryonic tissue. The blastocyst resembles a sphere with the extra-embryonic tissue making up the surface of the sphere and the future embryonic tissue appearing as a cluster of cells inside the sphere.

Blastomere—The embryonic cell during the first cellular divisions, before differentiation has occurred.

Chimera—An animal (or embryo) formed from two distinct cellular lineages that do not mingle in the cells of the animal, so that some cells contain genetic information from one lineage and some from the other.

Cloning—The production of multiple genetically identical embryos or zygotes.

Crossing over—In meiosis, a process in which adjacent chromosomes exchange pieces of genetic information.

Diploid—Nucleus or cell containing two copies of each chromosome, generated by fusion of two haploid nuclei.

Extra-embryonic tissue—That part of the developmental tissue that does not form the embryo.

Fetus—The unborn or unhatched animal during the latter stages of development.

Genome—Half a diploid set of chromosomes; the genetic information from one parent.

Hybrid vigor—The quality of increased health and fertility (superior to either parent) usually produced when two different genetic strains are crossed.

Parthogenesis—Reproduction from two female parents.

Placenta—The organ to which a fetus is attached by the umbilical cord in the womb. It provides nutrients for the fetus.

Totipotent—Embryonic cells able to produce an entire fetus.

Transgenic—Cells or species that have undergone genetic engineering.

Uniparental—Having only one genetic parent.

(MOET) could soon become a useful tool. MOET is the production of multiple embryos from a female with desirable traits, which are then implanted in the wombs of other females of the same species. This circumvents the disadvantages of breeding from a female line (which are that a female can only produce a limited number of offspring due to the time investment and physical rigors of pregnancy). At the present time, MOET is still too expensive for commercial application, but is being applied experimentally.

Genetic engineering

Genetic engineering is being implemented to create animals that have had a new gene inserted directly into their DNA. These animals are called transgenic. The procedure involves microinjection of the desired gene into the nucleus of fertilized eggs. It has been found that in many cases, but with varying rates of success, the new gene is reproduced in all developing cells, and the gene can be transcribed (which means the information contained in the gene can be read and utilized by the cell). This is a startling breakthrough in animal breeding endeavors, because it means a specific trait can be incorporated into a population

in a single generation, rather than the several generations this takes when conventional breeding techniques are used.

However, there are some serious limitations to the procedure. The first of these has to do with the manner in which many genes work together to produce most traits. In fact, there are very few traits a breeder would like to include in an animal population that involve only one or two genes. Although it might some day be possible to incorporate any number of genes into an embryo's DNA, the complex interplay of genes is not understood very well, and the process of identifying all of the genes related to a desired trait is costly and time-consuming.

Another problem in the production of transgenic animals is that they pass their modified DNA on to their offspring with varying success rates and unpredictable results. In some cases, the new gene is present in the offspring but it is not utilized. The new gene may also be altered or rearranged in some way, probably during the process of gamete production.

These factors have made it difficult to successfully produce a transgenic strain of animals. However, with further research into the mechanism by which the gene is incorporated into the **genome**, and by successfully mapping the target animal genome and identifying the genes responsible for various traits, genetic engineering will no doubt become a major tool for improving animal strains.

Sex selection

It would be extremely useful if a breeder were able to predetermine the sex of each embryo produced, because in many cases one sex is preferred. For instance, in a herd of dairy cows or a flock of laying hens, females are the only commercially useful sex. When the owner of a dairy herd has inseminated a cow at some expense, this issue becomes more crucial. In some cases, an animal is being bred specifically for use as breeding stock; in this case, it is far more useful to produce a male that can be bred with multiple females than a female, which can only produce a limited number of offspring.

Whether or not an animal is male or female is determined by its sex chromosomes, which are called X and Y chromosomes. An animal with two X chromosomes will develop into a female, while an animal with one X and one Y chromosome will become a male. In mammals, the sex of the offspring is almost always determined by the male parent, because the female can only donate an X chromosome, and it is the presence or absence of the Y chromosome that causes maleness (this is not true in, for instance, **birds**; in that case it is the female who has two different sex chromosomes). The problem in sex selection is to separate the Y-carrying sperm from the X-carrying sperm. Thus far, attempts to do so have been largely unsuccessful or too expensive for commercial application, but the economic advantages make this an area of intense research, and it is quite probable that an efficient and cost-effective method will soon be developed.

See also Biotechnology; Captive breeding and reintroduction; Genetics.

Resources

Books

Babiuk, Lorne A., and John J. Phillips, eds. *Animal Biotechnology.* New York: Pergamon Press, 1989.

Dawkins, Richard. *The Selfish Gene.* New York: Oxford University Press, 1989.

Hill, William G. and Trudy F.C. Mackay, eds. *Evolution and Animal Breeding.* Wallingford, UK: CAB International, 1989.

Periodicals

Blasco, A. "The Bayesian Controversy in Animal Breeding." *Journal of Animal Science* 79, no.8 (2002): 2023-2046.

Sarah A. de Forest

Animal cancer tests

The chemical causation of **cancer** is not a simple process. Many, perhaps most, chemical carcinogens do not have the potency to cause cancer in their usual condition. The non-cancer causing form of the chemical is called a procarcinogen. Procarcinogens are frequently complex organic compounds that the human body attempts to dispose of when ingested. Hepatic enzymes chemically change the procarcinogen in several steps to yield a chemical that is more easily excreted. These chemical changes result in modification of the procarcinogen (with no cancer forming ability) to the ultimate **carcinogen** (with cancer causing competence). Ultimate carcinogens have been shown to have a great affinity for DNA, RNA, and cellular **proteins**, and it is the interaction of the ultimate carcinogen with the **cell** macromolecules that causes cancer. It is unfortunate indeed that one cannot look at the chemical structure of a potential carcinogen and predict whether or not it will cause cancer. There is no computer program that can predict what hepatic enzymes will do to procarcinogens and how the metabolized end product(s) will interact with cells.

Great strides have been made in the development of chemotherapeutic agents designed to cure cancer. The drugs have significant efficacy with certain cancers (these include, but are not limited to, pediatric acute lymphocytic **leukemia**, choriocarcinoma, **Hodgkin's disease**, and testicular cancer) and some treated patients attain a normal life span. While this development is heartening, the cancers listed are, for the most part, relatively infrequent. More common cancers such as colorectal carcinoma, lung cancer, breast cancer, and ovarian cancer remain intractable with regard to treatment.

These several reasons are why **animal** testing is used in cancer research. The majority of Americans support the effort of the biomedical community to use animals to identify potential carcinogens with the hope that such knowledge will lead to a reduction of cancer prevalence. Similarly, they support efforts to develop more effective chemotherapy. Animals are used under terms of the Animal Welfare Act of 1966 and its several amendments. The act designates that the U. S. Department of Agriculture is responsible for the humane care and handling of warm-blooded and other animals used for biomedical research. The act also calls for inspection of research facilities to insure that adequate food, housing, and care are provided. It is the belief of many that the constraints of the current law have enhanced the quality of biomedical research. Poorly maintained animals do not provide quality research.

Mice—the best animal model for cancer research

In trying to evaluate whether mouse cancer research will apply to humans, the first question that crops up is why use **mice** when one is interested in curing human cancer? There are several reasons. Mice share many common features with humans and develop all the types of cancer that humans develop. They also have the same genes involved in a lot of these cancers.

Mice are readily available. Candidate drugs cannot only be tested on people. It would be unethical, immoral, and probably illegal. However, before a new drug can be tested on humans, there have to be some guidelines about what it can do. These guidelines come from mice and other animal models.

Mice can develop human cancers. Furthermore, human tumors can be implanted in mice. In addition, mice have simpler **genetics**. Genetically identical mice simplify experiments by minimizing the confusion that arises when testing drugs on mixed populations. In theory, genetically identical mice should all respond the same to a given form of treatment. By adding or deleting genes from mice, scientists learn how that gene's products influence a treatment and thus, obtain valuable clues to the **biochemistry** of cancer.

The "nude" mouse lacks a thymus gland, needed for the development of the **immune system**. Since its immune systems is so defective, this mouse does not reject transplanted human cancers and is widely used to test cancer drugs.

Despite the numerous advantages in using mice as models for studying human cancers, it has to be emphasized that mice are not men and mouse results may not apply to people. No generalizations can be made about the individual drugs. Each new drug is new—and one has to go through the process of testing its toxicity and effectiveness. There are many reasons to be cautious when interpreting animal tests. Many mouse experiments have been proven wrong in the past and scientists have been disappointed going from mice to people. There are several cancer cures—interferon, interleukins, and cytokines—that worked much better in mice than in people.

Although much of the rationale for testing drugs on animals rests on the similarities among organisms—similar does not mean identical. Basic cellular enzymes—and the DNA code that governs life's **chemistry** and structure—are similar in widely divergent organisms. Genes with the same function in mice and men contain, on average, 85% similarity in the actual sequence of DNA subunits. But, it is possible that even a single change in the order of DNA bases can make a huge dif-

ference. **Sickle cell anemia** is a painful **disease** caused by one erroneous atom in the hemoglobin **molecule** that carries **oxygen** in red **blood** cells. This mistake, results from a single erroneous subunit among the thousands that comprise the hemoglobin **gene**.

And then there is the toxicity issue. Interleukin-2 is a blood-borne signaling molecule that produced remarkable results against cancer in animals a decade ago. But in people, it caused leaks in blood vessels, and could never be used at the dosages that worked so well in animals. Instead of becoming the magic bullet some had predicted, IL-2 is now a small member of the overall anti-cancer tool kit.

Beyond these issues is another more serious concern—the nature of the experimental tumors. The mouse experiments use tumors that grow well in laboratories, but human tumors arise spontaneously and could have quite different genetics and properties. The **biology** of the **tumor** may be different, not because it is a human tumor growing on a mouse, but because it is spontaneous and not grown in a lab.

Resources

Periodicals

Abelson, P. H. "Testing for Carcinogens With Rodents." *Science* 249 (21 September 1990): 1357.
Donnelly, S., and K. Nolan. "Animals, Science, and Ethics." *Hastings Center Report* 20 (May-June 1990 suppl.): 1-32.
Marx, J. "Animal Carcinogen Testing Challenged: Bruce Ames Has Stirred Up the Cancer Research Community." *Science* 250 (9 November 1990): 743-5.
Rider, E.L. "Housing and Care of Monkeys and Apes in Laboratories." *Laboratory Animals* 36, no. 3 (2002): 221-242.

Robert G. McKinnell

Anion

An anion is a negatively charged atom or group of **atoms**. Anions are attracted to the **anode**, or positive electrode, in an electrolytic cell. Some common anions are the hydroxide ion (OH^-), the chloride ion (Cl^-), the nitrate ion (NO_3^-), and the bicarbonate ion (HCO_3^-). The single minus signs indicate that these ions carry one electron's worth of **negative** charge. The carbonate ion (CO_3^{2-}), for example, carries two units of negative charge.

The names of anions consisting of single atoms (monatomic ions) end in the suffix -ide. Fluoride (F^-), sulfide (S^{2-}), and oxide (O^{2-}), are examples of such ions. A few polyatomic ions (ions with more than one atom) also have an -ide ending. The cyanide ion (CN^-) is an example.

The names of most polyatomic anions end in either -ate or -ite. For example, the most common polyatomic anions of **sulfur** are the sulfate (SO_4^{2-}) and sulfite (SO_3^-) ions. In pairs such as this one, the -ate suffix is used for the ion that contains sulfur in the higher oxidation number, and the -ite suffix for the ion with the lower oxidation number. The oxidation number of sulfur is six in the sulfate ion and four in the sulfite ion.

Anise *see* **Carrot family (Apiaceae)**

Anode

The word anode is used in two different sets of circumstances: with respect to **vacuum** tubes and with respect to electrochemical cells.

Vacuum tubes

A **vacuum tube** is a tube (usually made of **glass**) with most of the air pumped out, and usually containing two electrodes—two pieces of **metal** with a potential difference (a voltage difference) applied between them. Electrons, being negatively charged, are repelled out of the **negative** electrode and fly through the vacuum toward the positive electrode, to which they are simultaneously being attracted. The positive electrode is called the anode; the negative electrode is called the **cathode**.

Common examples of vacuum tubes in which electrons flow from a cathode to an anode are cathode ray tubes, such as **television** tubes and computer monitors, and x-ray tubes. In an x-ray tube, the kind of metal that the anode is made of determines the kind of **x rays** (i.e., the x-ray **energy**) that the tube emits.

Electrochemical cells

There are two kinds of electrochemical cells: those in which **chemical reactions** produce electricity—called galvanic cells or voltaic cells—and those in which **electricity** produces chemical reactions—called electrolytic cells. An example of a galvanic cell is a flashlight **battery**, and an example of an electrolytic cell is a cell used for electroplating silver or gold. In either case, there are two electrodes called the anode and the cathode.

Unfortunately, there has been much confusion about which electrode is to be called the anode in each type of cell. Chemists and physicists correctly consider electricity to be a flow of negative electrons, but for historical reasons, engineers have considered electricity to be a flow of positive charge in the opposite direction. Furthermore, even chemists have been confused because negative charge flows away from one of the electrodes inside the cell, but in the external circuit negative charge flows toward that same electrode. This has led to a variety of conflicting definitions of anodes in various textbooks and reference works.

The confusion can be cleared up by defining the anode and cathode in terms of the actual chemical reactions—the oxidation and reduction reactions—that are taking place inside the cell, whether the cell is generating electricity as a galvanic cell or consuming it as an electrolytic cell. The anode is now defined as the electrode at which an oxidation reaction is taking place in the cell. The cathode, then, is the electrode at which the corresponding reduction reaction is taking place.

Anodes in practical use

A sacrificial anode is a piece of metal that is made to act as an anode and therefore be oxidized, in order to protect another piece of metal from being oxidized. For example, to keep **iron** or **steel** from oxidizing (rusting) when in contact with air and moisture, such as when it is being used as a fence post, the post can be connected to a piece of zinc that is buried in the ground next to it. The iron and zinc in the moist **soil** constitute the two electrodes of a galvanic cell. But because zinc oxidizes more easily than iron, the zinc acts as the anode and is preferentially oxidized. It is said to be "sacrificed" because it gradually gets eaten away by oxidation instead of the iron. For this reason, a cable made of zinc was buried alongside the Alaskan pipeline, the huge steel pipe that transports **petroleum** from the Alaskan oil fields to the lower states.

Galvanized iron is iron that has been coated with zinc so that it can be used outdoors or in the ground without rusting. The zinc oxidizes in preference to the iron. Galvanized iron is widely used in making garbage cans, pails, and chain-link fencing.

Anodizing is a process in which a piece of metal is made the anode in an electrolytic cell in order to oxidize it deliberately. When **aluminum** is anodized in this way, a coating of aluminum oxide is built up on its surface. This coating, unlike the metal itself, can take dyes. Many kinds of aluminum utensils and novelties in bright blue, green, red, and gold colors are made of anodized aluminum.

See also Cell, electrochemical; Oxidation-reduction reaction.

Robert L. Wolke

Anoles

Anoles are small lizards in the genus *Anolis* (family Iguanidae), found only in the Americas, mostly in the tropical countries. Because anoles can change the **color** of their skin according to their mood, **temperature**, **humidity**, and **light** intensity, these animals are sometimes called **chameleons**. However, none of the more than 300 **species** of anoles is closely related to the true chameleons (family Chamaeleonidae) of Eurasia and **Africa**.

Aggressive encounters and defense responses to predators prompt the extension of the throat fan, or dewlap, of male anoles. This stereotyped visual display is often accompanied by a vigorous demonstration of head-bobbing, the frequency and amplitude of which are important among anoles in species recognition.

Because anoles spend a great deal of time engaged in displays and other activities associated with holding their breeding territory, the males are at a relatively greater risk of predation than the more inconspicuous females. However, the greater risks of predation are balanced by the better reproductive success a male anole may achieve during the period of his life that he is able to hold a high-quality territory.

The green anole (*Anolis carolinensis*) is the only species native to **North America**, occurring in the southeastern United States, and also in Cuba and nearby islands. The green anole does not hibernate, and is active on warm, sunny days throughout the winter, remaining inactive in a sheltered place on colder days. The usual color of these animals is brown, but male animals quickly become green when they are engaged in aggressive encounters with other males, or when they are courting a female.

Four additional species of anoles have been introduced to Florida from their natural ranges in the West Indies or Central America. These are the brown anole (*Anolis sagrei*), the large-headed anole (*A. cybotes*), the **bark** anole (*A. distichus*), and the knight anole (*A. equestris*). Although these species are not part of the native **fauna** of Florida, they are now entrenched components of that state's ecosystems as are many other **introduced species** of plants and animals.

Anorexia nervosa *see* **Eating disorders**

Ant-pipits

The ant-pipits are 10-11 **species** of **birds** that make up the family Conopophagidae. These birds are exclu-

A brown anole (*Anolis sagrei*) on Estero Island, Florida. Anoles are the largest group of reptiles in the Western Hemisphere. *Photograph by Robert J. Huffman. Field Mark Publications. Reproduced by permission.*

sively South American, occurring in tropical rain **forests** of Amazonia. The usual **habitat** of ant-pipits is thick and lush with foliage, and the birds are rather shy. Consequently, these small birds are difficult to see and demanding to study. Therefore, little is known about their **biology** and **ecology**.

Ant-pipits are small, stocky, wren-like birds, with a short tail, short wings, and long legs. The body length of these almost tail-less birds is 4-5.5 in (10-14 cm). Ant-pipits feed on the forest floor, mostly by using their strong legs and feet to scratch about in **plant** litter to expose their food of **insects** and other **arthropods**. Ant-pipits are permanent residents in their forest habitats, meaning they are not known to undertake long-distance, migratory movements.

The true ant-pipits are eight to nine species in the genus *Conopophaga*. The black-bellied ant-pipit (*Conopophaga melanogaster*) is a rather attractive species of tropical forests in Amazonian Brazil. Male individuals have chestnut back, wings, and tail, black head and breast, and a white eye-stripe that extends back into a distinctive plume at the back of the head. Coloration of the female black-bellied ant-pipit consists of more subdued hues of brown and grey.

The black-cheeked ant-pipit (*C. melanops*) of Amazonian Brazil has a chestnut cap, an olive back and wings, and a large, black cheek-patch. The chestnut-crowned ant-pipit (*C. castaneiceps*) occurs in Amazonian Colombia, Ecuador, and Peru, while the slaty ant-pipit (*C. ardesiaca*) occurs in Bolivia and southern Peru.

There are two species of *Corythopis* ant-pipits, the ringed ant-pipit (*Corythopis torquata*) and the southern ant-pipit (*C. delalandi*).

Antarctica

Of the seven continents on **planet** Earth—**North America, South America, Europe, Africa, Asia, Australia**, and Antarctica—the last lies at the southernmost tip of the world. It is the coldest, driest, and windiest **continent**. **Ice** covers 98% of the land, and its 5,100,000 sq mi (13,209,000 sq km) occupy nearly one-tenth of the Earth's land surface, or the same area as Europe and the United States combined. Despite its barren appearance, Antarctica and its surrounding waters and islands teem with life all their own, and the continent plays a significant role in the climate and health of the entire planet.

Humans have never settled on Antarctica because of its brutal climate, but, since its discovery in the early 1800s, explorers and scientists have traveled across dangerous seas to study the continent's winds, temperatures, **rocks**, **wildlife**, and ice. Scientists treasure the unequaled chance at undisturbed research; as travel to the continent improves, tourists enjoy the opportunity to visit the last "frontier" on the **earth**; environmentalists focus on Antarctica as the only continent largely unspoiled by human hands; and, in an increasingly resource-hungry world, others look at the continent as a key source of oil and mineral resources. While some countries have tried to claim parts of the continent as their own, Antarctica is an independent continent protected by international treaty from ownership by any one country.

Antarctica—an overview

Antarctica does not have a town, a **tree**, or even a blade of grass on the entire continent. That does not mean that Antarctica is not vital to life on earth. Seventy **percent** of the world's fresh **water** is frozen atop the continent. These icecaps reflect warmth from the **Sun** back into the atmosphere, preventing planet Earth from overheating. Huge **icebergs** break away from the stationary ice and flow north to mix with warm water from the equator, producing **currents**, **clouds**, and complex **weather** patterns. Creatures as small as microscopic **phytoplankton** and as large as whales live on and around the continent, including more than 40 **species** of **birds**. Thus, the continent provides habitats for vital links in the world's food chain.

Geologists believe that, millions of years ago, Antarctica was part of a larger continent called Gondwanaland, based on findings of similar fossils, rocks, and other geological features on all of the other southern continents. About 200 million years ago, Gondwanaland broke apart into the separate continents of Antarctica, Africa, Australia, South America, and India (which later collided with Asia to merge with that continent). Antarc-

tica and these other continents drifted away from each other as a result of shifting of the plates of the earth's crust, a process called **continental drift** that continues today. The continent is currently centered roughly on the geographic South Pole, the point where all south latitudinal lines meet. It is the most isolated continent on earth, 600 mi (1,000 km) from the southernmost tip of South America and more than 1,550 mi (2,494 km) away from Australia.

Geology

Antarctica is considered both an island—because it is surrounded by water—and a continent. The land itself is divided into east and west parts by the Transantarctic Mountains. The larger side, to the east, is located mainly in the eastern longitudes. West Antarctica is actually a group of islands held together by permanent ice.

Almost all of Antarctica is under ice, in some areas by as much as 2 mi (3 km). The ice has an average thickness of about 6,600 ft (2,000 m), which is higher than many **mountains** in warmer countries. This grand accumulation of ice makes Antarctica the highest continent on Earth, with an average elevation of 7,500 ft (2,286 m).

While the ice is extremely high in elevation, the actual land mass of the continent is, in most places, well below **sea level** due to the weight of the ice. If all of this ice were to melt, global sea levels would rise by about 200 ft (65 m), **flooding** the world's major coastal ports and vast areas of low-lying land. Even if only one-tenth of Antarctica's ice were to slide into the sea, sea levels would rise by 20 ft (6 m), severely damaging the world's coastlines.

Under all that ice, the Antarctic continent is made up of mountains. The Transantarctic Mountains are the longest range on the continent, stretching 3,000 mi (4,828 km) from Ross Sea to Weddell Sea. Vinson Massif, at 16,859 ft (5,140 m), is the highest mountain peak. The few areas where mountains peek through the ice are called nunataks.

Among Antarctica's many mountain ranges lie three large, moon-like valleys—the Wright, Taylor, and Victoria Valleys—which are the largest continuous areas of ice-free land on the continent. Known as the "dry valleys," geologists estimate that it has not rained or snowed there for at least one million years. Any falling snow evaporates before it reaches the ground, because the air is so dry from the ceaseless winds and brutally cold temperatures. The dryness also means that nothing decomposes, including seal carcasses found to be more than 1,000 years old. Each valley is 25 mi (40 km) long and 3 mi (5 km) wide and provides rare glimpses of the rocks that form the continent and the Transantarctic Mountains.

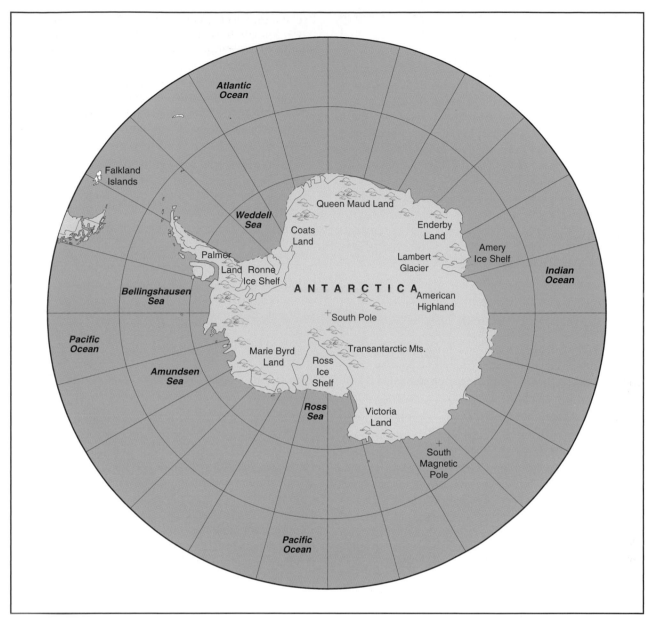

Antarctica. *Illustration by Hans & Cassidy. Courtesy of Gale Group.*

Around several parts of the continent, ice forms vast floating shelves. The largest, known as the Ross Ice Shelf, is about the same size as Texas or Spain. The shelves are fed by **glaciers** on the continent, so the resulting shelves and icebergs are made up of frozen fresh water. Antarctica hosts the largest glacier on Earth; the Lambert Glacier on the eastern half of the continent is 25 mi (40 km) wide and more than 248 mi (400 km) long.

Gigantic icebergs are a unique feature of Antarctic waters. They are created when huge chunks of ice separate from an ice shelf, a cliff, or glacier in a process known as calving. Icebergs can be amazingly huge; an

iceberg measured in 1956 was 208 mi (335 km) long by 60 mi (97 km) wide (larger than some small countries) and was estimated to contain enough fresh water to supply London, England, for 700 years. Only 10-15% of an iceberg normally appears above the water's surface, which can create great dangers to ships traveling in Antarctic waters. As these icebergs break away from the continent, new ice is added to the continent by snowfall.

Icebergs generally flow northward and, if they do not become trapped in a bay or inlet, will reach the Antarctic Convergence, the point in the **ocean** where cold Antarctic waters meet warmer waters. At this point, ocean currents usually sweep the icebergs from west to east until they melt. An average iceberg will last several years before melting.

Three oceans surround Antarctica—the Atlantic, Pacific, and Indian Oceans. Some oceanographers refer to the parts of these oceans around Antartica as the Southern Ocean. While the **saltwater** that makes up these oceans does not usually freeze, the air is so cold adjacent to the continent that even the **salt** and currents cannot keep the water from freezing. In the winter months, in fact, the ice covering the ocean waters may extend over an area almost as large as the continent. This ice forms a solid ring close to the continent and loose chunks at the northern stretches. In October (early spring) as temperatures and strong winds rise, the ice over the oceans breaks up, creating huge icebergs.

Subantarctic islands are widely scattered across the ocean around Antarctica. Some, such as Tristan da Cunha and the Falkland Islands, have human populations that live there year-round. Others, including the Marion Islands, the Crozets and Kerguelen, St. Paul, Amsterdam, Macquarie and Campbell Islands, have small scientific bases. Others are populated only by **penguins**, **seals**, and birds.

Strong winds blow constantly against the western shores of most of the northernmost islands, creating some of the stormiest seas in the world. Always wet in the summer, they are blanketed by snow in the winter and **grasses** in the summer. Further south in the colder subantarctic, the islands are covered with snow for much of the year and may have patches of mosses, **lichens**, and grasses in the summer. Seals and enormous populations of sea birds, including penguins and petrels, come to the islands' beaches and cliffs to breed in the summer.

The Falkland Islands, which are British Crown colonies, are the largest group of subantarctic islands, lying 300 mi (480 km) east of South America. Two main islands, East and West, have about 400 smaller islets surrounding them. The islands cover 6,200 sq mi (16,000 sq km). Much of the ground is covered with wet, springy turf, overlying thick beds of peat. Primary industries are farming and ranching of cattle and **sheep**.

Climate

Antarctica is the coldest and windiest place on Earth. The **wind** can gust up to 200 MPH (322 km/h), or twice as hard as the average hurricane. Surprisingly, little snow actually falls in Antarctica; because the air is so cold, the snow that does fall turns immediately to ice.

Because of the way the Earth tilts on its axis as it rotates around the Sun, both polar regions experience long winter nights and long summer days. At the South Pole itself, the sun shines around the clock during the six months of summer and virtually disappears during the cold winter months. The tilt also affects the **angle** at which the Sun's **radiation** hits the earth. When it is directly overhead at the equator, it strikes the polar regions at more indirect angles. As a result, the Sun's radiation generates much less **heat**, even though the polar regions receive as much annual daylight as the rest of the world.

Even without the **wind chill**, the continent's temperatures can be almost incomprehensible to anyone who has not visited there. In winter, temperatures may fall to −100°F (−73°C). The world's record for lowest **temperature** was recorded on Antarctica in 1960, when it fell to −126.9°F (−88.3°C).

The coastal regions are generally warmer than the interior of the continent. The Antarctic Peninsula may get as warm as 50°F (10°C), although average coastal temperatures are generally around 32°F (0°C). During the dark winter months, temperatures drop drastically, however, and the warmest temperatures range from −4 to −22°F (−20 to −30°C). In the colder interior, winter temperatures range from −40 to −94°F (−40 to −70°C).

The strong winds that constantly travel over the continent as cold air races over the high ice caps and then flows down to the coastal regions, are called katabatic winds. Winds associated with Antarctica blizzards commonly gust to more than 120 mi (193 km) per hour and are among the strongest winds on Earth. Even at its calmest, the continent's winds can average 50-90 mi (80-145 km) per hour. Cyclones occur continually from west to east around the continent. Warm, moist ocean air strikes the cold, dry polar air and swirls its way toward the coast, usually losing its force well before it reaches land. These cyclones play a vital role in the exchange of heat and moisture between the tropical and the cold polar air.

While the Antarctic sky can be clear, a white-out blizzard may be occurring at ground level because the strong winds whip up the fallen snow. The wind redesigns the snow into irregularly shaped ridges, called sastrugi, which are difficult to traverse. Blizzards are common on the continent and have hindered several exploration teams from completing their missions.

Surprisingly, with all its ice and snow, Antarctica is the driest continent on Earth based on annual **precipitation** amounts. The constantly cold temperatures have allowed each year's annual snowfall to build up over the centuries without melting. Along the polar ice cap, annual snowfall is only 1-2 in (2.5-5 cm). More precipitation falls along the coast and in the coastal mountains, where it may snow 10-20 in (25-51 cm) per year.

Plants and animals

While the Arctic region teems with life, the Antarctic continent is nearly barren due to the persistently cold and dry climate. Plants that grow in the region reflect this climate and **geology**. The pearlwort (*Colobanthus quitensis*) and grass (*Deschampsia antarctica*) are the only two flowering plants on the continent. Both grow in a small area on or near the warmest part of the continent, the Antarctic Peninsula. Larger plants include mosses and lichens (a combination of **algae** and **fungi**) found along the coast and on the peninsula. Green, nonflowering liverworts live on the western side of the peninsula. Brightly colored snow algae often form on top of the snow and ice, coloring it red, yellow, or green.

A few hardy organisms live on rocks in the dry valleys; these are primarily lichens that hide inside the porous orange sandstone. These lichens, called *cryptoendoliths* or "hidden in rock," use up more than 99.9% of their photosynthetic productivity simply to stay alive. In contrast, a typical **plant** uses 90% for survival. Ironically, the lichens found in these valleys are among the longest-living organisms on earth. The dry valleys also host pockets of algae, fungi, and **bacteria** between frozen rock crystals; these give scientists clues about how life might survive on a frozen planet like **Mars**.

Few creatures can survive Antarctica's brutal climate. Except for a few **mites** and midges, native animals do not exist on Antarctica's land. Life in the sea and along the coast of Antarctica and its islands, however, is often abundant. A wide variety of animals make the surrounding waters their home, from **zooplankton** to large birds and **mammals**. A few **fish** have developed their own form of antifreeze over the centuries to prevent ice crystals from forming in their bodies, while others have evolved into cold-blooded species to survive the cold.

The base of Antarctica's marine food chain is phytoplankton, which feed on the rich **nutrients** found in coastal waters. The zooplankton feed on the phytoplankton, which are in turn consumed by the native fish, birds, and mammals. Antarctic krill (tiny shrimplike creatures about 1.5 in [4 cm] long) are the most abundant zooplankton and are essential to almost every other life form in the region. They swim in large pools and look like red patches on the ocean. At night, their crusts shimmer like billions of fireflies beneath the sea. Because of their abundance, krill have also been explored as a potential food source for humans.

Among the whales that make the southern oceans their home for at least part of the year are the blue, fin, sei, minke, humpback, and southern right whales. Known as baleen whales, this whale group has a bristly substance called baleen located in plates in their mouths that filter food such as krill from the water. In fact, the blue whale is the largest **animal** ever known to have lived on Earth. The blue whale eats 3 tons (6,000 pounds or 2.7 metric tons) of krill each day and has been measured to weigh up to 180 tons (163,000 kg) and span 124 ft (38 m) in length. After it was discovered in the 1800s, the blue whale was heavily hunted for its blubber, which was melted into oil for fuel. While scientists believe more than 200,000 existed before whaling, there are as few as 1,000 blue whales today. All baleen and toothed whales are now protected from hunting by international agreements.

Two toothed whales also swim in Antarctic waters, the sperm and the orca or killer whale. The sperm whale is the larger of the two, measuring as long as 60 ft (18 m) and weighing as much as 70 tons (63,500 kg). It can dive down to 3,300 ft (1,006 m).

More than half the seals in the world live in the Antarctic—their blubber and dense fur insulate them from the cold. Five species of true or earless seals live in the region, the Weddell, Ross, leopard, crabeater, and elephant. The Weddell seal is the only one that lives in the Antarctic year-round, on or under the ice attached to the continent in the winter. Using their sawlike teeth to cut holes in the ice for **oxygen**, they can dive down to 2,000 ft (610 m) to catch fish and **squid**. The seals use a complex system to control their bodies' oxygen levels, which allows them to dive to such depths and stay underwater for as long as an hour.

The Ross seal, named for English explorer James Ross, is quick underwater and catches fish easily with its sharp teeth. It lives on the thickest patches of ice and is the smallest and least plentiful of the species. Leopard seals are long and sleek and are fierce predators, living on the northern edges of pack ice and in the sea or near penguin rookeries, where they eat small penguins and their eggs as well as other seals. Crabeater seals are the most plentiful species of seal on Earth, with an estimated 40 million or more in the Antarctic region alone. Elephant seals are the largest species of seal, live on the sub-antarctic islands, and eat squid and fish. Unlike most seals, the males are much larger than the females. All five seal species are now protected under international law from hunting, which almost wiped out the Ross and elephant seals in the 1800s. One other type of seal, the southern fur seal, is also plentiful on Antarctica. It has visible ears and longer flippers than the true seals, which makes it much more agile on land as well as in the water.

Several seabirds make the Antarctic their home, including 24 species of petrels, small seabirds that dart over the water and nest in rocks along the shore. Examples include the albatross (a gliding bird with narrow, long wings that may live up to 40 years), the southern giant fulmar, dove prion, and snow petrel. **Shore birds** that feed in the shallow waters near the shoreline include the blue-eyed cormorant, the Dominican gull, and the brown skua, which eats the eggs and young of other birds. The Arctic tern is the world's best at long-distance flying, because it raises its young in the Arctic but spends the rest of the year in the Antarctic, a **distance** of over 10,000 mi (16,090 km). Land birds include the wattled sheathbill, South Georgia pintail, and South Georgia pipit.

Of all the animals, penguins are the primary inhabitants of Antarctica. Believed to have evolved 40–50 million years ago, they have oily feathers that provide a waterproof coat and a thick layer of **fat** for insulation. Penguins' bones are solid, not hollow like those of most birds that allow them to fly. While solid bones prevent penguins from flying, they add weight and make it easier for penguins to dive into the water for food. Because predators cannot live in the brutally cold climate, penguins do not need to fly; thus, their wings have evolved over the centuries to resemble flippers or paddles.

Seven of the 18 known species of penguins live on the Antarctic: the Adelie and emperor (both considered true Antarctic penguins because they live on the continent), the chinstrap, gentoo, macaroni, rockhopper, and king penguins. The Adelie is the most plentiful species of penguin and can be found over the widest area of the continent. They spend their winters on the pack ice away from the continent, then return to land in October to nest in large rookeries or colonies along the rocky coasts. The emperor penguin is the largest species of penguin; it is the only Antarctic bird never to set foot on land, and it breeds on sea ice attached to the mainland. The most popular type of penguin for zoos, emperor penguins are 4 ft (1.2 m) tall and can weigh up to 80 lb (30 kg). They are the hardiest of all the animals that inhabit the Antarctic, staying throughout the year while other birds head north to escape the brutal winter. They breed on the ice surface during the winter months because their immense size requires a longer incubation period. This schedule also ensures that the chicks will hatch in July or early spring in the Antarctic, providing the most days for the chicks to put on weight before the next winter's cold arrives.

The female lays one egg on the ice, then walks up to 50 mi (80 km) to open sea for food. When she returns, filled with food for the chick, the male—who has been incubating the egg atop the ice during the coldest winter months—makes the same trek out to sea to restore its body weight, which may drop by 50% during this period. The parents take turns traveling for food after the chick has hatched.

Because the emperor penguin is one of the few species that lives on Antarctica year-round, researchers believe it could serve as an indicator to measure the health of the Antarctic **ecosystem**. The penguins travel long distances and hunt at various levels in the ocean, covering wide portions of the continent. At the same time, they are easily tracked because the emperor penguins return to their chicks and mates in predictable ways. Such indicators of the continent's health become more important as more humans travel to and explore Antarctica and as other global conditions are found to affect the southernmost part of the world.

Exploration of the continent

Greek philosopher Aristotle hypothesized more than 2,000 years ago that the earth was round and that the southern hemisphere must have a landmass large enough to balance the lands in the northern hemisphere. He called the probable but undiscovered land mass "Antarktikos," meaning the opposite of the Arctic.

The Greek geographer and astronomer Ptolemy called Antarctica "Terra Australis Incognita" or "unknown southern land" in the second century A.D. He claimed the land was fertile and populated but separated from the rest of the world by a region of torrid heat and fire around the equator. This concept was believed for several centuries, and the continent remained a mystery until James Cook crossed the Antarctic Circle and circumnavigated the continent in 1773. While he stated that the land was uninhabitable because of the ice fields surrounding the continent, he noted that the Antarctic Ocean was rich in whales and seals. For the next 50 years, hunters exploited this region for the fur and oil trade.

As hunting ships began traveling farther and farther south in the early 1800s to find fur seals, it was inevitable that the continent would be encountered and explored. Three countries claim first discovery rights to Antarctic land: Russia, due to explorer Fabian von Bellinghausen, on January 27, 1820; England as a result of English explorer Edward Bransfield, on January 30, 1820; and the United States by way of American sealer Nathaniel Palmer, on November 18, 1820. In actuality, American sealer John Davis was the first person to actually step onto the continent, on February 7, 1821.

James Weddell was a sealer who traveled to the continent on January 13, 1823. He found several previously unknown seal species, one of which later became known as the Weddell seal, then took his two ships—the *Jane* and the *Beaufoy*—farther south than any explorer had previously traveled. He reached 74° south latitude on February 20, 1823, and the vast sea he had entered became known as the Weddell Sea.

In 1895, the first landing on the continent was accomplished by the Norwegian whaling ship *Antarctic*. The

A glacier system in Antarctica. *JLM Visuals. Reproduced by permission.*

British were the first to spend a winter on Antarctica, in 1899. By 1911, a race had begun to see who would reach the South Pole first: the South Pole is an imaginary geographical center point at the bottom of the earth. Again, it was a Norwegian, Roald Amundsen, who was the first to reach the South Pole on December 14, 1911. Robert Scott of England and his four men arrived a month later. While the first team made it home safely, the Scott team ran out of food and froze to death on their way home.

Airplanes first landed on Antarctica when Australian Hubert Wilkins flew 1,300 mi (2,092 km) over the Antarctic Peninsula in the 1920s, viewing terrain never before seen by another human. American Richard Byrd, the first person to fly over the North Pole in 1926, took his first Antarctic flight in 1929 and discovered a new mountain range he named Rockefeller. Thereafter, the continent was mapped and explored primarily from the air. Byrd continued his explorations to Antarctica over the next three decades and revolutionized the use of modern vehicles and communications equipment for polar exploration.

Scientific exploration

While various countries were busy claiming rights to Antarctica, scientists were cooperating effectively on research as early as 1875. Twelve nations participated in the first International Polar Year in 1882 and 1883. While most of the research was done in the Arctic, one German station was located in the Antarctic region. A second International Polar Year occurred in 1932-33, followed by the International Geophysical Year (IGY) from July 1, 1957 to December 31, 1958. This time, all twelve nations conducted research in Antarctica and set up base camps in various locations, some of which are still used today. Topics of research included the pull of gravity, glaciology, meteorites, cosmic rays, the southern lights, dynamics of the sun, the passage of **comets** near the earth, and changes in the atmosphere. Several organizations have been formed and agreements have been signed since these cooperative research projects to ensure that political conflicts do not arise concerning research and use of Antarctica.

Current events

A wide variety of research is continuing on Antarctica, primarily during the relatively warmer summer months from October to February when temperatures may reach a balmy 30–50°F (−1–10°C). The cold temperatures and high altitude of Antarctica allow as-

KEY TERMS

Antarctic Circle—The line of latitude at 66 degrees 32 minutes South, where there are 24 hours of daylight in midsummer and 24 hours of darkness in midwinter.

Antarctic convergence—A 25-mi (40-km) region where cold Antarctic surface water meets warmer, subantarctic water and sinks below it.

Antarctic Ocean—The seas surrounding the continent, where the Atlantic, Pacific, and Indian Oceans converge.

Blubber—Whale or seal fat used to create fuel.

Calving—The process in which huge chunks of ice or icebergs break off from ice shelves and sheets or glaciers to form icebergs.

Dry valleys—Areas on the continent where no rain is known to have fallen for more than two million years, and the extremely dry katabatic winds cause any snow blown into the valleys to evaporate before hitting the ground. The Taylor, Victoria, and Wright valleys are the largest continuous areas of ice-free land on the continent.

Glacier—A river of ice that moves down a valley to the sea, where it breaks into icebergs.

Iceberg—A large piece of floating ice that has broken off a glacier, ice sheet, or ice shelf.

Katabatic winds—Fierce, dry winds that flow down along the steep slopes of the interior mountains of Antarctica, and along ice caps and glaciers on the subantarctic islands.

Krill—Tiny sea animals or zooplankton that are the main food for most larger species in the Antarctic region.

Nunataks—Mountain peaks that are visible above the ice and snow cover.

Pack ice—Ice from seawater, which forms a belt approximately 300–1,800 mi (483–2,897 km) wide around the continent in winter.

South magnetic pole—The point to which a compass is attracted and which is some distance from the geographic South Pole. It varies from year to year as the earth's magnetic field changes.

South pole—The geographically southernmost place on Earth.

Southern lights—Also known as the aurora australis, they are streamers of different colors in the sky, especially at night, and are thought to be caused by electrical disturbances and cosmic particles in the upper atmosphere that are attracted by the South Magnetic Pole.

Subantarctica—The region just north of Antarctica and the Antarctic Circle, but south of Australia, South America, and Africa.

tronomers to put their telescopes above the lower atmosphere, which lessens blurring. During the summer months, they can study the Sun around the clock, because it shines 24 hours a day. Antarctica is also the best place to study interactions between **solar wind** and **Earth's magnetic field**, temperature circulation in the oceans, unique animal life, **ozone** depletion, ice-zone ecosystems, and glacial history. Buried deep in Antarctica's ice lie clues to ancient climates, which may provide answers to whether the earth is due for **global warming** or the next ice age.

Scientists consider Antarctica to be a planetary bellwether, an early indicator of negative changes in the entire planet's health. For example, they have discovered that a hole is developing in the ozone layer over the continent, a protective layer of gas in the upper atmosphere that screens out the ultraviolet **light** that is harmful to all life on Earth. The ozone hole was first observed in 1980 during the spring and summer months, from September through November. Each year, greater destruction of the

layer has been observed during these months, and the first four years of the 1990s have produced the greatest rates of depletion thus far. The hole was measured to be about the size of the continental United States in 1994, and it lasts for longer intervals each year. Scientists have identified various chemicals created and used by humans, such as **chlorofluorocarbons (CFCs)**, as the cause of this destruction, and bans on uses of these chemicals have begun in some countries.

Researchers have also determined that a major climate change may have occurred in Antarctica in the 1980s and 1990s, based on recorded changes in ozone levels and an increase in cloudiness over the South Pole. This, coupled with a recorded weakening of the ozone shield over North America in 1991, has led scientists to conclude that the ozone layer is weakening around the entire planet.

Others are studying the ice cap on Antarctica to determine if, in fact, the earth's climate is warming due to the burning of **fossil fuels**. The global warming hypothesis is based on the atmospheric process known as the **greenhouse effect**, in which **pollution** prevents the heat **energy** of the earth from escaping into the outer atmosphere. Global warming could cause some of the ice cap to melt, flooding many cities and lowland areas. Because the polar regions are the engines that drive the world's weather system, this research is essential to identify the effect of human activity on these regions.

Most recently, a growing body of evidence is showing that the continent's ice has fluctuated dramatically in the past few million years, vanishing completely from the continent once and from its western third at least several times. These collapses in the ice structure might be triggered by climatic change, such as global warming, or by far less predictable factors, such as volcanic eruptions under the ice. While the east Antarctic ice sheet has remained relatively stable because it lies on a single tectonic plate, the western ice sheet is a jumble of small plates whose erratic behavior has been charted through **satellite** data. The west Antarctic is also dominated by two seas, the Ross and Weddell, whose landward regions are covered by thick, floating shelves of ice. Some researchers speculate that if warmer, rising oceans were to melt this ice, the entire western sheet might disintegrate quickly, pushing global sea levels up by 15-20 ft (5-6 m).

See also Gaia hypothesis; Greenhouse effect; Ozone layer depletion; Plate tectonics.

Resources

Books

Anderson, J. B. *Antarctic Marine Geology.* Cambridge: Cambridge University Press, 1999.

Billings, Henry. *Antarctica—Enchantment of the World.* Chicago: Childrens Press, 1994.

Hancock P. L. and Skinner B. J., eds. *The Oxford Companion to the Earth.* Oxford: Oxford University Press, 2000.

Hurley, Frank. *South with Endurance: Shackleton's Antarctic Expedition, 1914-1917.* New York: Simon & Schuster, 2001.

Joughin, I. "Antarctica: A Review of Recent Medical Research." *Trends in Pharmacological Sciences.* 23, no. 10 (2002): 487-490.

McGonigal, David. Lynn Woodworth, and Sir Edmund Hillary, eds. *Antarctica and the Arctic: The Complete Encyclopedia.* Westport, CT: Firefly Books, 2001.

Periodicals

Grotta, Daniel and Sally. "Antarctica: Whose Continent is it Anyway?" *Popular Science* 240, no. 1 (1992): 62-7, 90-1.

Horgan, John. "Antarctic Meltdown." *Scientific American* 266, no. 3: 19-28.

Keeling, Ralph F. "Palaeoceanography: Antarctic Stratification and Glacial CO2." *Nature* 412, no. 6847 (2001): 605-606.

Kiernan, K. "Impacts of Geoscience Research on the Physical Environment of the Antarctic." *Australian Journal of Earth Sciences* 48, no. 5 (2001): 767.

Monastersky, Richard. "Antarctic Ozone Level Reaches New Low." *Science News* 144, no. 16: 247.

Monastersky, Richard. "Science on Ice." *Science News* 143, no. 15 (1993): 232-35.

Palca, Joseph. "Poles Apart, Science Thrives on Thin Ice." *Science* 255, no. 5042 (1992): 276-78.

Sally Cole-Misch

Antbirds and gnat-eaters

The antbirds and gnat-eaters are 231 **species** of **birds** that comprise the relatively large family, Formicariidae. These birds only occur in Central and **South America**, mostly in lowland tropical **forests**.

The antbirds and gnat-eaters are variable in their body form and size. Their body length ranges from 4–14 in (10–36 cm), and they have short, rounded wings, and a rounded tail that can be very short or quite long. Most species have a rather large head with a short neck, and the bill is stout and hooked at the tip. Species that live and feed in the forest canopy have a relatively long tail and wings, while those of ground-feeding species are shorter.

The colors of the plumage of most species are rather subdued hues of browns and greys, although there are often bold patterns of white, blue, or black. Males and females of most species have different plumage. These birds generally occur in solitary pairs, which are permanent residents in a defended territory.

Species of antbirds forage widely in the forest floor or canopy for their food of **insects**, spiders, and other **invertebrates**. Species of antbirds are prominent members of the local, mixed-species foraging flocks that often occur in their tropical forest **habitat**. These flocks can contain as many as 50 species, and are thought to be adaptive because they allow better detection of **birds of prey**.

Despite their name, antbirds rarely eat **ants**. Antbirds received their common name from the habit of some species of following a column of army-ants as it moves through their tropical forest habitat. These predatory assemblages of social insects disturb many insects as they move along the forest floor. Antbirds and other species of birds often follow these columns to capture insects and other **prey** that have been disturbed by the army ants. About 27 species of antbirds have the habit of actively following army ants, while other species do this on a more casual, less focused basis.

Antbirds lay two to three eggs in a cup-shaped nest located in a low **tree** or on the ground. Both parents share in the incubation of the eggs and the nurturing of the young. Pairs of antbirds are monogamous for life, the partners remaining faithful to one another until death.

Species of antbirds are prominent elements of the avian community of the lowland tropical forests that are their usual habitat. In some cases in Amazonia, as many as 30–40 species of antbirds can occur in the same area, dividing up the habitat into subtly defined niches.

The white-faced ant-catcher (*Pithys albifrons*) is an ant-following species of tropical forests of Amazonian Brazil and Venezuela. The ocellated ant-thrush (*Phaenostictus mcclennani*) ranges from Nicaragua to Ecuador. The rufous-capped ant-thrush (*Formicarius colma*) forages on the floor of Amazonian forests of Brazil, Venezuela, Ecuador, and Peru.

Anteaters

Anteaters belong to the family Myrmecophagidae, which includes four **species** in three genera. They are found in Trinidad and range from southern Mexico to northern Argentina. The spiny anteater (echidna) of **Australia** is an egg-laying mammal and is not related to the placental anteaters of the New World. The banded anteater (or **numbat**) of Australia is a marsupial mammal, and not a close relative of the placental anteaters. The anteater's closest relatives are **sloths**, **armadillos**, and **pangolins**. All belong to the order Edentata, meaning without teeth, although only the anteaters are strictly toothless.

Anteaters feed by shooting their whiplike tongues in and out of insect nests up to 160 times per minute. **Ants,**

A giant anteater (*Myrmecophaga tridactyla*). *Photograph by Robert J. Huffman. Field Mark Publications. Reproduced by permission.*

along with sticks and gravel, stick to the sticky tongue like **flies** to flypaper. Horny papillae toward the rear of the two-foot-long tongue help this toothless mammal to grind its food, which is ground up further by the muscular stomach. It is thought that grit the anteater swallows may actually help its stomach grind up the food. In a typical day, a giant anteater, the largest of four species, will consume up to 30,000 ants.

Anteaters' bodies have a narrow head and torso and a long, slender snout that is kept close to the ground to sniff for **insects**. Anteaters have poor eyesight and **hearing** but a keen sense of **smell**. Their legs end in long, sharp claws that are used primarily to open ant and termite nests but double as defensive weapons. To protect their claws, anteaters walk on their knuckles with their claws turned inward.

A diet of ants and **termites** provides little **energy**, so anteaters have adapted by evolving an unusually low resting metabolic **rate** and low core body **temperature**, moving slowly, and spending much of the day sleeping. Female anteaters bear only one offspring per year and devote much of their energy to caring for the young.

The largest of the four New World species is the giant anteater (*Myrmecophaga tridactyla*), which is widely distributed throughout Central and **South America** east of the Andes to northern Argentina. The giant anteater is about the size of a large dog. It is covered with short, mostly gray hair, except on its tail, which has long, bushy fur. True to its name, the giant anteater subsists almost entirely on large, ground-dwelling ants. It moves between ant nests, taking just a little from each nest, thereby avoiding excessive ant bites and depletion of its food supplies.

Giant anteaters are solitary creatures, pairing up shortly before mating and parting just afterward. Females suckle their young for about six months and carry them on their back for up to a year, when the young anteaters are nearly fully grown. The giant anteater is an **endangered species** due to **habitat** destruction and hunting.

The lesser, or collared, anteater consists of two species, *Tamandua mexicana* and *T. tetradactyla*. *Tamandua mexicana* is found from southern Mexico to northwestern Venezuela and Peru, while *T. tetradactyla* lives in Trinidad and South America, east of the Andes, from Venezuela to northern Argentina and southern Brazil. Lesser anteaters are distinguished from giant anteaters by their large ears, prehensile tail, and affinity for climbing trees. They are about half the size of giant anteaters and are covered with bristly hair that varies in **color** from blond to brown. The term collared anteater refers to the band of black fur encircling the abdomen, found in *T. mexicana* and *T. tetradactyla* from the south-

KEY TERMS

Pangolin—A toothless, scaly anteater-like mammal that feeds on insects (ants and termites) and is found in Asia and Africa.

Papillae—A general term for any tiny projections on the body, including the roots of developing teeth, hair, feathers, and the bumps on the tongue.

Prehensile—A strong tail adapted for grasping or wrapping around objects.

eastern part of their range. The tree-climbing lesser anteaters feed mostly on termites. Like the giant anteater, the female tamandua carries her offspring on her back.

The most elusive of the four species of anteater is the silky anteater (*Cyclopes didactylus*). This squirrel-sized **animal** spends most of the day sleeping in trees, and comes out to forage for ants at night. Silky anteaters are distributed from southern Mexico to most of the Amazon **basin**, and west of the Andes to northern Peru. These anteaters rest on branches of the silk cotton trees, where its silky, gold and gray fur blends with the tree's soft, silver fibers. This camouflage protects the silky anteater from **owls**, **eagles**, and other predators. In this species, both parents feed the young, and both carry their offspring on their backs.

See also Monotremes; Spiny anteaters.

Resources

Books

Gould, Edwin, and Gregory McKay, eds. *The Encyclopedia of Mammals.* 2nd ed. New York: Academic Press, 1998.

Nowak, Ronald M. *Walker's Mammals of the World.* Baltimore: The Johns Hopkins University Press, 1999.

Periodicals

"Quick-snacking Anteater Avoids Attack" *Science* (8 August 1991): 88.

Cynthia Washam

Antelopes and gazelles

Antelopes and **gazelles** belong to the family Bovidae, which includes even-toed hoofed animals with hollow horns and a four-chambered stomach. **Sheep**, cattle, and **goats** are also bovids. The family Bovidae in **Africa** includes nine tribes of antelopes, one of which includes the 12 **species** of gazelles (Antilopini). Other tribes are

the **duikers** (Cephalophini), **dwarf antelopes** (Neotragini), reedbuck, kob, and **waterbuck** (Reduncini), hartebeeste, topi, and wildebeeste (Alcelaphini), impala (Aepycerotini), bushbuck, kudu, and **eland** (Tragelaphini), rhebok (Peleini), and horse antelopes (Hippotragini). Most antelopes and gazelles are found in Africa where 72 of the 84 species live. The word gazelle comes from an Arabic word that means affectionate. Antelopes and gazelles can be found throughout the **grasslands** of Africa, in **mountains**, **forests**, and deserts. Antelopes range in size from small 15 lb (7 kg) antelopes to a 1,200 lb (545 kg) **animal**, the eland, of East and West Africa.

Antelopes and gazelles are noted for the beauty of their horns. Some are **spiral** in shape, others are ringed, lyre-shaped, or S-shaped. Gazelles have black-ringed horns 10–15 in (25–38 cm) long. Depending on the size of the species, the horns can be as short as 1 in (2.5 cm) or as long as 5 ft (1.5 m). The Grant's gazelle has horns that are as long as the shoulder height of the animal. In most species the females as well as the males have horns.

The prevailing **color** of antelopes and gazelles is brown or black and white, but different species show a range of coloration and markings. All antelopes and gazelles have scent **glands** that they use to mark territory and to signal, age, sex, and social status. Glands can be preorbital (below the eyes), interdigital (between the hooves), subauricular (below the ears), or on the back, shins, and genital areas of the animals.

Smaller species of antelopes with well-developed hind quarters and coloration indicate a reliance on concealment and bounding escape runs. The larger species tend to inhabit open spaces, since they are able to escape predators by their ability to reach high speeds, some reaching 35 MPH (56 km/h).

Some species, such as the Dorcas gazelle and the **oryx** of the dry **savanna** regions, have developed effective ways to decrease the need for **water**. Some species are solitary in habit while others such as the impala live in herds, which are either single sex (all females) that mate only with the dominant male, or are herds of both males and females. They can be polygamous—one male to a number of females—or monogamous. Some groups are adolescent males. Common to all species is the **birth** of one offspring. Pregnancy ranges from four to nine months depending on the size of the species.

Antenna

An antenna is a device used to transmit and receive electromagnetic waves such as **radio waves** and mi-

Antenna for a Doppler weather radar site. Doppler radar can be used to detect wind shear and microburst weather conditions. *Photograph by Brownie Harris. Stock Market. Reproduced by permission.*

crowaves. Antennas provide the transition between a guided wave (flowing in a wire) and a free space wave (flowing in air or **vacuum**). An antenna can take high **frequency** pulses from an electrical signal **generator**, focus them, and launch them into space, like the antenna at a **radio** station. Conversely, it can pick up waves from space, focus them, and send them to a receiver, like the antenna on your car radio. You can think of an antenna as a soap bubble pipe: pulses (soap film) travel down the transmission line (pipe stem), reach the bowl (antenna), and are electrically shaped and pushed out into free space. The horn antennas used for **microwave communication** are designed to let the **radiation** spread out gradually rather than undergo an abrupt transition from the waveguide into free space. This is known as impedance matching, and contributes to the propagation of the radiation in the same way as cupping your hands around your mouth when shouting makes your voice travel further.

Basically there are two types of antennas: those that rotate and those that are stationary. Rotating antennas usually operate as search and detection systems. They are typically found on ships, airports, or weather stations. Often, an antenna will include a reflecting element to focus the radio waves, commonly parabolic or shaped something like an orange slice.

The stationary antenna type is generally found at radio or microwave transmitting sites. This antenna configuration can be a long wire between pylons, a single pylon with a long rod at the top, or include a number of unevenly spaced rods like an outdoor **television** antenna. The **satellite** dish, an antenna with a parabolic reflector, is another common type of stationary configuration.

See also Radar.

Anther *see* **Flower**

Anthrax

Anthrax is the name given to an **infection** that is caused by the bacterium *Bacillus anthracis*. The bacterium is common in cattle, **sheep**, **goats**, **camels**, antelopes, and other plant-eating animals. Humans can also become contaminated with the anthrax bacterium. In the past, such human **contamination** was only associated with farmers or sheepherders, people who worked in close contact with infected animals. Now, however, the population as a whole is more at risk of anthrax infection because the **organism** has been used as a bioterrorist weapon and weapon of war.

The use of *Bacillus anthracis* as a weapon is due to the ability of the bacterium to form a structure known as a **spore**. The spore form is able to withstand prolonged periods of **drought** and conditions that would quickly kill the growing form of the microorganism. The dust-like spores are easily spread through air or liquids. Anthrax skin infections due to the entry of spores or the growing **bacteria** into a wound or scrapped region of skin are treatable and death from superficial infection is rare. Inhalation of the spores, however, can produce a lung infection that develops rapidly and is frequently fatal.

Anthrax is relatively rare in the United States because of widespread **animal** vaccination and practices used to disinfect hides or other animal products. For those in high-risk professions, such as **livestock** workers, veterinarians, or those in the military, an anthrax **vaccine** is available. The vaccine, which does not contain living bacteria, is 93% effective in protecting against infection. To provide this immunity, an **individual** must be given

Antibody—A molecule created by the immune system in response to the presence of an antigen (a foreign substance or particle). It marks foreign microorganisms in the body for destruction by other immune cells.

Bronchitis—Inflammation of the mucous membrane of the bronchial tubes of the lung that can make it difficult to breathe.

Cutaneous—Pertaining to the skin.

Meningitis—Inflammation of the meninges; membranes covering the brain and spinal cord.

Pulmonary—Having to do with the lungs or respiratory system.

Spore—A dormant form assumed by some bacteria, such as anthrax, that enable the bacterium to survive high temperatures, dryness, and lack of nourishment for long periods of time. Under proper conditions, the spore may revert to the actively multiplying form of the bacteria.

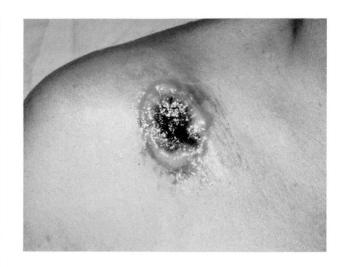

A case of cutaneous anthrax. *NMSB/Custom Medical Stock Photo. Reproduced by permission.*

Resources

Periodicals

Inglesby, T. V., D. A. Henderson, J. G. Bartlett, et al. "Anthrax as a Biological Weapon: Medical and Public Health Management." *Journal of the American Medical Association* 281 (May 1999): 1735–1745.

Pannifer, A. D., T. Y. Wong, R. Schwarzenbacher, et al. "Crystal Structure of the Anthrax Lethal Factor." *Nature* 414 (November 2001): 229–233.

Other

Centers for Disease Control and Prevention. 1600 Clifton Road, Atlanta, GA 30333. (800) 311–3435. <http://www.cdc.gov/ncidod/dbmd/diseaseinfo/anthrax_g.htm.>

Brian Hoyle

an initial course of three injections, given two weeks apart, followed by booster injections at 6, 12, and 18 months, and an annual immunization thereafter.

Approximately 30% of those who have been vaccinated against anthrax may notice mild local reactions such as a slight tenderness at the injection site. Someone who has already had anthrax might have a more severe local reaction upon vaccination. Infrequently, there may be a severe local reaction with extensive swelling of the forearm, and only a very few vaccine recipients may have a more general flu-like reaction to the shot.

Other means of preventing the spread of infection include careful handling of dead animals suspected of having the **disease** and providing good ventilation when processing hides, fur, wool, or hair. Additionally, anyone visiting a country where anthrax is common or where herd animals are not often vaccinated should avoid contact with livestock or animal products and avoid eating meat that has not been properly prepared and thoroughly cooked.

In 2001, the United States experienced several deliberate releases of anthrax spores through the mailing of contaminated letters. These incidents led to the call for a nationwide immunization program against a possible airborne bioterrorist release of the anthrax spores. However, whether the vaccine would provide complete protection against anthrax used as a biological weapon is, as yet, unclear.

See also Biological warfare.

Anthropocentrism

Anthropocentrism is a world view that considers humans to be the most important factor and value in the Universe. In contrast, the biocentric world view considers humans to be no more than a particular **species** of **animal**, without greater intrinsic value than any of the other species of organisms that occur on **Earth**. The ecocentric world view incorporates the biocentric one, while additionally proposing that humans are a natural component of Earth's **ecosystem**, and that humans have an absolute and undeniable requirement of the products and services of ecosystems in order to sustain themselves and their societies.

There are a number of important implications of the anthropocentric view, which strongly influence the ways in which humans interpret their relationships with other species and with nature and ecosystems. Some of these are discussed below:

1. The anthropocentric view suggests that humans have greater intrinsic value than other species. A result of this attitude is that any species that are of potential use to humans can be a "resource" to be exploited. This use often occurs in an unsustainable fashion that results in degradation, sometimes to the point of **extinction** of the biological resource, as has occurred with the dodo, great auk, and other animals.

2. The view that humans have greater intrinsic value than other species also influences ethical judgments about interactions with other organisms. These ethics are often used to legitimize treating other species in ways that would be considered morally unacceptable if humans were similarly treated. For example, animals are often treated very cruelly during the normal course of events in medical research and agriculture. This prejudiced treatment of other species has been labeled "speciesism" by ethicists.

3. Another implication of the anthropocentric view is the belief that humans rank at the acme of the natural evolutionary progression of species and of life. This belief is in contrast to the modern biological interpretation of **evolution**, which suggests that no species are "higher" than any others, although some clearly have a more ancient evolutionary lineage, or may occur as relatively simple life forms.

The individual, cultural, and technological skills of humans are among the attributes that make their species, *Homo sapiens*, special and different. The qualities of humans have empowered their species to a degree that no other species has achieved during the history of life on Earth, through the development of social systems and technologies that make possible an intense exploitation and management of the environment. This power has allowed humans to become the most successful species on Earth. This success is indicated by the population of humans that is now being maintained, the explosive growth of those numbers, and the increasing amounts of Earth's biological and environmental resources that are being appropriated to sustain the human species.

However, the true measure of evolutionary success, in contrast to temporary empowerment and intensity of resource exploitation, is related to the length of time that a species remains powerful—the sustainability of its enterprise. There are clear signals that the intense exploitation of the environment by humans is causing widespread ecological degradation and a diminished **carrying capacity** to sustain people, numerous other species, and many types of natural ecosystems. If this environmental deterioration proves to truly be important, and there are many indications that it will, then the recent centuries of unparalleled success of the human species will turn out to be a short-term phenomenon, and will not represent evolutionary success. This will be a clear demonstration of the fact that humans have always, and will always, require access to a continued flow of ecological goods and services to sustain themselves and their societies.

Anti-inflammatory agents

Anti-inflammatory agents are compounds that reduce the **pain** and swelling associated with **inflammation**. Inflammation is a response of the body to injuries such as a blow or a **burn**. The swelling of the affected region of the body occurs because fluid is directed to that region. The inflammatory response can aid the healing process.

In conditions such as rheumatoid **arthritis**, however, the swelling and increased tenderness that are characteristic of inflammation are undesirable. The intake of anti-inflammatory agents can ease the discomfort of arthritis, and other conditions such as **asthma**.

The relief provided by anti-inflammatory agents has been known for millennia. Hippocrates, the "father" of medicine who lived 2,500 years ago, knew infusions of willow **bark** could aid in relieving pain. Only a little more than 100 years ago was the basis of this relief identified as a family of anti-inflammatory chemicals called salicylates. A modified version of salicylic acid is aspirin. Today, anti-inflammatory agents are extremely popular as pain relievers. For example, over 80 billion tablets of aspirin are taken each year around the world.

There are two groups of anti-inflammatory agents, the corticosteroids and the nonsteroidal anti-inflammatory drugs (NSAIDs).

Corticosteroids are produced by the adrenal gland in carefully controlled amounts. Higher levels of the compounds are achieved by ingesting a pill or receiving an injection (the systematic route), or by use of a skin cream, nasal spray, or inhaler (the local route). Examples of corticosteroids include prednisone, prednislone, and hydrocortisone. These compounds are potent anti-inflammatory agents.

As their name implies, NSAIDs, are not steroids. In other words, they are not produced in the adrenal gland. They can be found naturally or are chemically made. Examples of NSAIDs include ibuprofen and **acetylsalicylic acid** (popularly known as aspirin).

Corticosteroids and NSAIDs inhibit inflammation in the same manner. Both classes of agents inhibit the production of a compound in the body called prostaglandin. Prostaglandin is made from another **molecule** called arachidonic acid in a reaction that depends on the activity of an **enzyme** called cyclooxidase. Anti-inflammatory agents prevent cyclooxidase from working properly. The shut-down of prostaglandin production curtails the inflammatory response and, because prostaglandin also aids in the passage of nerve impulses, pain is lessened.

The relief produced by anti-inflammatory agents comes with a caution, however. Side effects sometimes occur, particularly with the steroids. Corticosteroids taken over an extended period of time via the systematic route can produce fluid retention, weight gain, increased **blood pressure**, **sleep** disturbance, headaches, glaucoma, and can retard growth of children. The side effects of most NSAIDs are not so pronounced (ringing in ears, nausea, rash), and usually result from an overdose of the compound. Nonetheless, a potentially fatal condition in children called Reye **syndrome** has been linked to the use of aspirin, particularly if a child has recently had a bout of viral illness.

See also Immune system.

Resources

Books

Almond, C. *Emergency Medicine: A Comprehensive Study.* 4th ed. New York: McGraw-Hill, 1995.

Periodicals

Dubose, T. D., Jr., and D. A. Nolony, "Nephrotoxicity of Nonsteroidal Anti-inflammatory Drugs." *Lancet* 344 (August 1994): 515–518.

Singh, G., D. R. Ramey, D. Morfeld, et al. "Gastrointestinal Tract Complications of Nonsteroidal Anti-inflammatory Drug Treatment in Rheumatoid Arthritis." *Archives of Internal Medicine* 156 (1996): 1530–1536.

Vane, J. R. "Mechanism of Action of NSAIDs." *British Journal of Rheumatology* 35 (January 1996): 1–3.

Antibiotics

Antibiotics are natural or synthetic compounds that kill **bacteria**. Antibiotics are not active against viruses.

There are many different antibiotics that have different bacterial targets. Some antibiotics are specific in their activity, affecting only one or a few types (genera) of bacteria. Other antibiotics, such as penicillin, are active against a wide variety of bacteria. Such antibiotics are described as being "broad spectrum" antibiotics.

The first antibiotic discovered was penicillin. Before the discovery of penicillin by Sir Alexander Flemming (1881-1955) in 1928, bacterial infections were difficult to fight. Illnesses such as **pneumonia**, **tuberculosis**, and **typhoid fever** were untreatable, and bacterial infections that nowadays are minor inconveniences could become life threatening. Following the discovery of penicillin, many environmental sites were examined for compounds that exhibited anti-bacterial activity, resulting in the discovery of several naturally occurring antibiotics. As the molecular basis of activity of these antibiotics became known, antibiotics could be chemically synthesized with the ability to target specific sites on the bacterial surface, or inside bacteria.

Antibiotics can be produced by some bacteria and various eukaryotic organisms, such as plants. The antibiotics serve to protect the **organism** from other bacteria. Such antibiotics are typically found by screening a bacterial extract against other bacteria, and looking for inhibition in the growth of the target bacteria. Pharmaceutical companies have automated this screening process, so that thousands of samples can be examined each day.

Antibiotics can also be made by customizing a compound to a selected target on the bacterial surface or inside the bacterial **cell**. Molecular **sequencing** technology and computerized three-dimensional image simulation is extensively used in this antibiotic design process.

Antibiotic classes

There are different structures of antibiotics. Groups of antibiotics can have the same basic structure, with minor differences, such as the presence of different chemical groups protruding off of the main core structure. The different groups of antibiotics are known as classes.

Penicillin is in a class known as beta-lactam antibiotics. The name of this class is based on the beta-lactam ring that forms the core of the antibiotic **molecule**. Tetracyclines, aminoglycosides, rifamycins, quinolones, and sulphonamides are other classes of antibiotics.

Penicillin culture. *Custom Medical Stock Photo, Inc. Reproduced by permission.*

The mode of action of the different classes of antibiotics is varied. For example, beta-lactam antibiotics destroy the assembly of a bacterial structure called the peptidoglycan. The peptidoglycan is a rigid net that encircles the bacterial cell. It acts as the main stress-bearing layer of the bacterial cell wall. When the assembly of the peptidoglycan is disrupted, the ability of peptidoglycan to hold the bacterial wall together vanishes, and the cell explodes. Another class of antibiotics called aminoglycosides has a different method of killing bacteria. These antibiotics bind to a section of the bacterial structure called the ribosome. The ribosome is involved in making protein. By blocking the function of the ribosome, new protein cannot be made and the bacterial cell dies. Some aminoglycoside antibiotics also reduce the ease by which molecules can move from the outside of the cell to the inside of the cell. Once again, the result is death. In another example, the class of antibiotics known as quinolones act to disrupt an **enzyme** that unwinds the coiled **double helix** of deoxyribonucleic acid. If the DNA cannot unwind, new copies cannot be made. Without new DNA, the growth and division of the bacteria stops.

Antibiotic resistance

Following the discovery of penicillin, the many new antibiotics that were discovered or made to effectively control infectious bacteria. By the 1970s, the scientific community assumed that the battle against bacterial infections had been won. Beginning in the 1980s, however,

instances of bacterial resistance to previously effective antibiotics began to appear. The problem of resistance has accelerated throughout the 1990s to the present.

Altering an antibiotic slightly by adding or modifying a chemical side group can restore the effectiveness of the antibiotic. It is now clear, however, that such effectiveness may be short-lived. Resistance to the modified antibiotic can develop in a relatively short time.

An important contributor to the problem of antibiotic resistance is the overuse or misuse of antibiotics. Proper use of an antibiotic for the prescribed time either kills the target bacteria directly, or weakens the bacteria so that they are killed by the host's immune response. If the **concentration** of the antibiotic is too low to kill the bacteria, however, or if a patient stops taking the antibiotic before the course of the drug is complete, the surviving bacteria can then develop resistance to the drug. The resistant trait can be passed on to subsequent generations of the bacteria.

Some types of bacteria are now resistant to all but a few antibiotics. One strain of a *Staphylococcus* bacterium is resistant to every known antibiotic. Infections caused by this microbe are extremely difficult to treat. So far, this strain is rare, but clinical microbiologists expect that cases will become more frequent.

See also Infection; Membrane.

Resources

Books

Murray, P. R. *Manual of Clinical Microbiology.* 7th ed. Washington: American Society for Microbiology Press, 1999.

Reese, R. E., and R. F. Betts. *A Practical Approach to Infectious Diseases.* 4th ed. Boston: Little, Brown and Company, 1996.

Salyers, A. A., and D. D. Whitt. *Bacterial pathogenesis: A Molecular Approach.* 2nd ed. Washington: American Society for Microbiology press, 2001.

Other

Alliance for the Prudent Use of Antibiotics. 75 Kneeland Street, Boston, MA 02111–190. (617) 636-0966. <http://www.tufts.edu/med/apua/>.

Brian Hoyle

Antibody and antigen

The antibody and antigen reaction is an important protective mechanism against invading foreign substances. The antibody and antigen reaction, together with phagocytosis, constitute the immune response (humoral immune response). Invading foreign substances are antigens while the antibodies, or immunoglobulins, are specific **proteins** generated (or previously and present in **blood**, lymph or mucosal secretions) to react with a specific antigen.

Antigens, which are usually proteins or polysaccharides, stimulate the **immune system** to produce antibodies. The antibodies inactivate the antigen and help to remove it from the body. While antigens can be the source of infections from pathogenic **bacteria** and viruses, organic molecules detrimental to the body from internal or environmental sources also act as antigens.

Once the immune system has created an antibody for an antigen whose attack it has survived, it continues to produce antibodies for subsequent attacks from that antigen. This long-term **memory** of the immune system provides the basis for the practice of vaccination against **disease**. The immune system, with its production of antibodies, has the ability to recognize, remember, and destroy well over a million different antigens.

There are several types of simple proteins known as globulins in the blood: alpha, beta, and gamma. Antibodies are gamma globulins produced by B lymphocytes when antigens enter the body. The gamma globulins are referred to as immunoglobulins. In medical literature they appear in the abbreviated form as Ig. Each antigen stimulates the production of a specific antibody (Ig).

Antibodies are all in a Y-shape with differences in the upper branch of the Y. These structural differences of amino acids in each of the antibodies enable the individual antibody to recognize an antigen. An antigen has on its surface a combining site that the antibody recognizes from the combining sites on the arms of its Y-shaped structure. In response to the antigen that has called it forth, the antibody wraps its two combining sites like a "lock" around the "key" of the antigen combining sites to destroy it.

An antibody's mode of action varies with different types of antigens. With its two-armed Y-shaped structure, the antibody can attack two antigens at the same time with each arm. If the antigen is a toxin produced by pathogenic bacteria that cause an **infection** like **diphtheria** or **tetanus**, the binding process of the antibody will nullify the antigen's toxin. When an antibody surrounds a **virus**, such as one that causes **influenza**, it prevents it from entering other body cells. Another mode of action by the antibodies is to call forth the assistance of a group of immune agents which operate in what is known as the **plasma** complement system. First the antibodies will coat infectious bacteria and then white blood cells will complete the job by engulfing the bacteria, destroying them, and then removing them from the body.

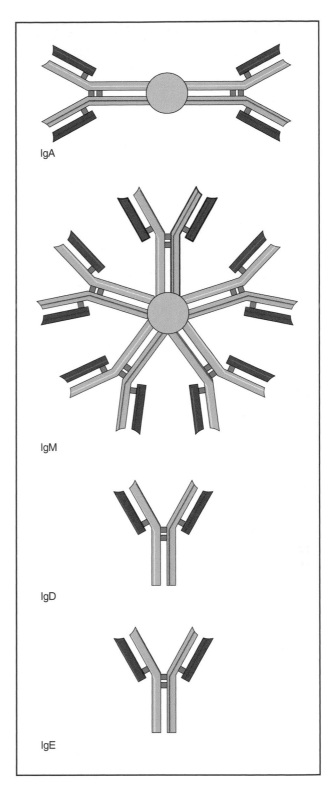

Schematic diagrams of several classes of antibody molecules. *Illustration by Hans & Cassidy. Courtesy of Gale Group.*

Functions of antibody types

There are five different antibody types, each one having a different Y-shaped configuration and function. They are the Ig G, A, M, D, and E antibodies.

IgG

IgG is the most common type of antibody. It is the most common Ig against microbes. It acts by coating the microbe to hasten its removal by other immune system cells. It gives lifetime or long-standing immunity against infectious diseases. It is highly mobile, passing out of the blood stream and between cells, going from organs to the skin where it neutralizes surface bacteria and other invading **microorganisms**. This mobility allows the antibody to pass through the placenta of the mother to her fetus, thus conferring a temporary defense to the unborn child.

After **birth**, IgG is passed along to the child through the mother's milk, assuming that she nurses the baby. But some of the Ig will still be retained in the baby from the placental transmission until it has time to develop its own antibodies. Placental transfer of antibodies does not occur in ruminant animals, such as **horses**, **pigs**, cows, and **sheep**. They pass their antibodies to their offspring only through their milk.

IgA

This antibody is found in body fluids such as tears, saliva, mucosa, and other bodily secretions. It is an antibody that provides a first line of defense against invading **pathogens** and allergens, and is the body's major defense against viruses. It is found in large quantities in the bloodstream and protects other wet surfaces of the body. While they have basic similarities, each IgA is further differentiated to deal with the specific types of invaders that are present at different openings of the body.

IgM

Since this is the largest of the antibodies, it is effective against larger microorganisms. Because of its large size (it combines 5 Y-shaped units), it remains in the bloodstream where it provides an early and diffuse protection against invading antigens, while the more specific and effective IgG antibodies are being produced by the plasma cells.

The **ratio** of IgM and IgG cells can indicate the various stages of a disease. In an early stage of a disease there are more IgM antibodies. As an typical infection progresses, many B cells shift from the IgM production to production of IgG antibodies. The presence of a greater number of IgG antibodies would indicate a later stage of the disease. IgM antibodies usually form clusters that are in the shape of a star.

IgD

This antibody appears to act in conjunction with B and **T cells** to help them in location of antigens. Research continues on establishing more precise functions of this antibody.

IgE

The antibody responsible for allergic reactions, IgE acts by attaching to cells in the skin called mast cells and basophil cells (mast cells that circulate in the body). In the presence of environmental antigens like pollens, foods, chemicals, and drugs, IgE releases histamines from the mast cells. The histamines cause the nasal **inflammation** (swollen tissues, running nose, sneezing) and the other discomforts of hay fever or other types of allergic responses, such as hives, **asthma**, and in rare cases, anaphylactic shock (a life-threatening condition brought on by an **allergy** to a drug or insect bite). An explanation for the role of IgE in allergy is that it was an antibody that was useful to early man to prepare the immune system to fight **parasites**. This function is presently overextended in reacting to environmental antigens.

The presence of antibodies can be detected whenever antigens such as bacteria or red blood cells are found to agglutinate (clump together), or where they precipitate out of **solution**, or where there has been a stimulation of the plasma complement system. Antibodies are also used in laboratory tests for blood typing when transfusions are needed and in a number of different types of clinical tests, such as the Wassermann test for syphilis and tests for **typhoid fever** and infectious mononucleosis.

Types of antigens

By definition, anything that makes the immune system respond to produce antibodies is an antigen. Antigens are living foreign bodies such as viruses, bacteria, and **fungi** that cause disease and infection. Or they can be dust, chemicals, pollen grains, or food proteins that cause allergic reactions.

Antigens that cause allergic reactions are called allergens. A large percentage of any population, in varying degrees, is allergic to animals, fabrics, drugs, foods, and products for the home and industry. Not all antigens are foreign bodies. They may be produced in the body itself. For example, **cancer** cells are antigens that the body produces. In an attempt to differentiate its

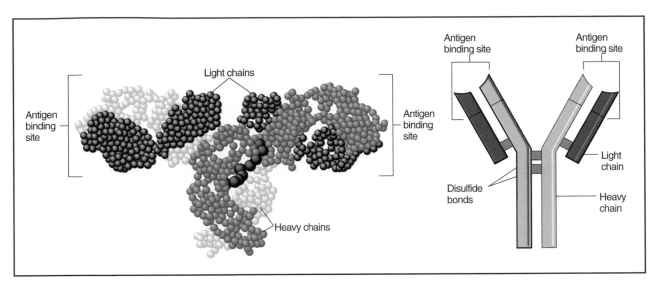

An IgG molecule (left) is shown schematically (right). *Illustration by Hans & Cassidy. Courtesy of Gale Group.*

"self" from foreign substances, the immune system will reject an **organ** transplant that is trying to maintain the body or a blood transfusion that is not of the same blood type as itself.

There are some substances such as nylon, plastic, or Teflon that rarely display antigenic properties. For that reason, nonantigenic substances are used for artificial blood vessels, component parts in **heart** pacemakers, and needles for hypodermic syringes. These substances seldom trigger an immune system response, but there are other substances that are highly antigenic and will almost certainly cause an immune system reaction. Practically everyone reacts to certain chemicals, for example, the resin from the poison ivy **plant**, the venoms from insect and reptile bites, solvents, formalin, and **asbestos**. Viral and bacterial infections also generally trigger an antibody response from the immune system. For most people penicillin is not antigenic, but for some there can be an immunological response that ranges from severe skin rashes to death.

Another type of antigen is found in the **tissue** cells of organ transplants. If, for example, a kidney is transplanted, the surface cells of the kidney contain antigens that the new host body will begin to reject. These are called human leukocyte antigens (HLA), and there are four major types of HLA subdivided into further groups. In order to avoid organ rejection, tissue samples are taken to see how well the new organ tissues match for HLA compatibility with the recipient's body. Drugs will also be used to suppress and control the production of helper/suppressor T cells and the amount of antibodies.

Red blood cells with the ABO antigens pose a problem when the need for blood transfusions arises. Before a transfusion, the blood is tested for type so that a compatible type is used. Type A blood has one kind of antigen and type B another. A person with type AB blood has both the A and B antigen. Type O blood has no antigens. A person with type A blood would require either type A or O for a successful transfusion. Type B and AB would be rejected. Type B blood would be compatible with a B donor or an O donor. Since O has no antigens, it is considered to be the universal donor. Type AB is the universal recipient because its antibodies can accept A, B, AB, or O. One way of getting around the problem of blood types in transfusion came about as a result of World War II. The great need for blood transfusions led to the development of blood plasma, blood in which the red and white cells are removed. Without the red blood cells, blood could be quickly administered to a wounded soldier without the delay of checking for the blood antigen type.

Another antigenic blood condition can affect the life of newborn babies. Rhesus disease (also called erythroblastosis fetalis) is a blood disease caused by the incompatibility of Rh factors between a fetus and a mother's red blood cells. When an Rh negative mother gives birth to an Rh positive baby, any transfer of the baby's blood to the mother will result in the production of antibodies against Rh positive red blood cells. At her next pregnancy the mother will then pass those antibodies against Rh positive blood to the fetus. If this fetus is Rh positive, it will suffer from Rh disease. Tests for Rh blood factors are routinely administered during pregnancy.

Vaccination

Western medicine's interest in the practice of vaccination began in the eighteenth century. This practice probably originated with the ancient Chinese and was adopted by Turkish doctors. A British aristocrat, Lady Mary Wortley Montagu (1689-1762), discovered a crude form of vaccination taking place in a lower-class section of the city of Constantinople while she was traveling through Turkey. She described her experience in a letter to a friend. Children who were injected with pus from a **smallpox** victim did not die from the disease but built up an immunity to it. Rejected in England by most doctors who thought the practice was barbarous, smallpox vaccination was adopted by a few English physicians of the period. They demonstrated an almost 100% **rate** of effectiveness in smallpox prevention.

By the end of the eighteenth century, Edward Jenner (1749-1823) improved the effectiveness of vaccination by injecting a subject with cowpox, then later injecting the same subject with smallpox. The experiment showed that immunity against a disease could be achieved by using a **vaccine** that did not contain the specific pathogen for the disease. In the nineteenth century, Louis Pasteur (1822-1895) proposed the **germ theory** of disease. He went on to develop a **rabies** vaccine that was made from the spinal cords of rabid rabbits. Through a series of injections starting from the weakest strain of the disease, Pasteur was able, after 13 injections, to prevent the death of a child who had been bitten by a rabid dog.

There is now greater understanding of the principles of vaccines and the immunizations they bring because of our knowledge of the role played by antibodies and antigens within the immune system. Vaccination provides active immunity because our immune systems have had the time to recognize the invading germ and then to begin production of specific antibodies for the germ. The immune system can continue producing new antibodies whenever the body is attacked again by the same **organism** or resistance can be bolstered by booster shots of the vaccine.

Monoclonal antibodies

For research purposes there were repeated efforts to obtain a laboratory specimen of one single antibody in sufficient quantities to further study the mechanisms and applications of antibody production. Success came in 1975 when two British biologists, César Milstein (1927-) and Georges Kohler (1946-) were able to clone immunoglobulin (Ig) cells of a particular type that came from multiple myeloma cells. Multiple myeloma is a rare form of cancer in which white blood cells keep turning out a specific type of Ig antibody at the expense of others, thus making the individual more susceptible to outside infection. By combining the myeloma **cell** with any se-

lected antibody-producing cell, large numbers of specific monoclonal antibodies can be produced. Researchers have used other animals, such as **mice**, to produce **hybrid** antibodies which increase the range of known antibodies.

Monoclonal antibodies are used as drug delivery vehicles in the treatment of specific diseases, and they also act as catalytic agents for protein reactions in various sites of the body. They are also used for **diagnosis** of different types of diseases and for complex analysis of a wide range of biological substances. There is hope that they will be as effective as enzymes in chemical and technological processes and that they will play a role in **genetic engineering** research.

See also Anaphylaxis; Transplant, surgical.

Resources

Books

Roitt, Ivan M., Peter J. Delves *Roitt's Essential Immunology.* 10th. ed. Oxford: Blackwell Scientific Publications, 2001.
Sompayrac, Lauren M. *How the Immune System Works.* Oxford: Blackwell Scientific Publications, 1999.

Jordan P. Richman

Anticoagulants

Anticoagulants are complex organic or synthetic compounds, often carbohydrates, that help prevent the clotting or coagulation of **blood**. The most widely used of these is heparin, which blocks the formation of thromboplastin, an important clotting factor in the blood. Most anticoagulants are used for treating existing thromboses (clots that form in blood vessels) to prevent further clotting. Oral anticoagulants, such as warfarin and dicumarol, are effective treatments for venous thromboembolisms (a blockage in a vein caused by a clot), but heparin is usually prescribed for treating the more dangerous arterial **thrombosis**.

Anticoagulants are often mistakenly referred to as blood thinners. Their real role is not to thin the blood but to inhibit the biochemical series of events that lead to the unnatural coagulation of blood inside unsevered blood vessels, a major cause of **stroke** and **heart** attack.

The coagulation process

In 1887, Russian scientist Ivan Pavlov first postulated the existence of natural anticlotting factors in animals and humans. His extensive studies of blood circulation led him to the realization that when blood reaches the lungs, it loses some of its ability to coagulate, a process aided, he believed, by the addition of some anticlotting substance. In 1892, A. Schmidt, the father of the enzymatic theory of blood coagulation, published the first data proving the existence of coagulation-inhibiting agents in liver, spleen, and lymph node cells. He later isolated this agent from liver **tissue** and demonstrated its anticoagulant properties. In 1905, P. Morawitz hypothesized that thrombosis might be effectively controlled by reducing the coagulation properties of the blood using antithrombins found in **plasma**.

Most modern theories of coagulation are scientifically complex and involve numerous substances known as clotting factors. The major mechanism of clot formation involves the conversion of fibrinogen, a highly soluble plasma protein, into fibrin, a stringy protein. There are a number of steps in this conversion process. First, when blood vessels are severed, prothrombin activator is produced. This agent's interaction with **calcium** ions causes prothrombin, an alpha globulin produced by the liver, to undergo conversion to thrombin. Next, thrombin, acting as an **enzyme**, triggers **chemical reactions** in fibrinogen, binding the molecules together end to end in long threads. Once these fibrin structures are formed, they adhere to the damaged area of the blood vessel, creating a mesh that traps blood cells and platelets. The resulting sticky mass, a clot, acts as a plug to seal the vessel and prevent further blood loss.

Thrombosis and embolism

Normally, clots form only in response to tissue injury. The natural flow of blood keeps thrombin from congregating in any one area. Clots can usually form only when the blood flows slowly or when wounds are opened. A clot that forms in a vessel abnormally is known as a thrombus; if the clot breaks free and is swept up along through the blood stream to another location, it is known as an embolus. These abnormal formations are thought to be caused by condition changes in the linings of blood vessels. Atherosclerosis and other diseases that damage arterial linings may lead directly to the formation of blood clots. Anticoagulants are used to treat these conditions.

Heparin

The first effective anticoagulant agent was discovered in 1916 by a medical student, who isolated a specific coagulation inhibitor from the liver of a dog. This substance, known as heparin because it is found in high concentrations in the liver, could not be widely produced until 1933, when Canadian scientists began extracting the substance from the lungs of cattle. In 1937, researchers began using heparin to treat and prevent surgical thrombosis and **embolism**.

Heparin is a complex organic acid found in all mammalian tissues that contain mast cells (allergic reaction mediators). It plays a direct role in all phases of blood coagulation. In animals and humans, it is produced by mast cells or heparinocytes, which are found in the connective tissues of the **capillaries**, inside blood vessels, and in the spleen, kidneys, and lymph nodes. There are several types of heparin in widescale clinical use, all of which differ in physiologic activity. Various salts of heparin have been created, including **sodium**, **barium**, benzidine, and others. The most widely used form in medical practice is heparin sulfate.

How it works

Heparin works by inhibiting or inactivating the three major clotting factors—thrombin, thromboplastin, and prothrombin. It slows the process of thromboplastin synthesis, decelerates the conversion of prothrombin to thrombin, and inhibits the effects of thrombin on fibrinogen, blocking its conversion to fibrin. The agent also causes an increase in the number of negatively charged ions in the vascular wall, which helps prevent the formation of intravascular clots.

Heparin is administered either by periodic injections or by an infusion pump. The initial dose is usually 5,000 units, followed by 1,000 units per hour, depending on the patient's weight, age, and other factors. The therapy usually lasts for seven to ten days. Heparin is also used in the treatment of deep vein thrombosis, a serious surgical complication which is also associated with traumatic injury. This condition can lead to immediate death from pulmonary embolism or produce long-term, adverse effects. Patients with pelvic or lower extremity fractures, spinal cord injuries, a previous thromboembolism, varicose **veins**, and those over age 40 are most at risk. Low-dose heparin is a proven therapy and is associated with only a minimal risk of irregular bleeding.

Oral anticoagulants

The development of oral anticoagulants can be linked directly to a widespread cattle **epidemic** in the United States and Canada during the mid-1920s. A scientist traced the cause of this outbreak to the cattle feed, a fodder containing spoiled sweet clover, which caused the cattle to bleed to death internally. Mixing alfalfa, a food rich in **vitamin** K, into the fodder seemed to prevent the **disease**. In 1941, research showed that the decaying sweet clover contained a substance that produced an anti-vitamin-K effect. The substance was isolated from the sweet clover, and called it dicumarol. During the 1940s, the agent was synthesized and widely used in the United States to treat postoperative thrombosis. In 1948, a more powerful synthetic compound was derived for use, initially as a rodentcide. This substance, known as warfarin, is now one of the most widely prescribed oral anticoagulants. There are numerous other agents in clinical use. Acenocoumarol, ethyl biscoumacetate, and phenprocoumon, which are seldom used in the United States, are widely prescribed elsewhere in the world.

Most oral anticoagulants work by suppressing the action of vitamin K in the coagulation process. These agents are extremely similar in chemical structure to vitamin K and effectively displace it from the enzymatic process, which is necessary for the synthesis of prothrombin and other clotting factors. Indeed, one of the ways to treat irregular bleeding, the most common side effect of oral anticoagulants, is vitamin K therapy.

In addition to bleeding, another side effect of oral anticoagulant use is negative interaction with numerous other drugs and substances. Even unaided, oral anticoagulants can have serious effects. For example, use of warfarin during pregnancy can cause **birth defects**, fetal hemorrhages, and miscarriage. Dicumarol, the original oral anticoagulant, is seldom used today because it causes painful intestinal problems and is clinically inferior to warfarin.

Over the past two decades, the search for new, less toxic anticoagulants has led to the development and use of a number of synthetic agents, including fibrinlysin,

thrombolytin, and urokinase. The enzyme streptokinase, developed in the early 1980s, is routinely injected into the coronary artery to stop a heart attack. Another new agent, tissue plasminogen activator, a blood protein, is being generated in large quantities using recombinant-DNA techniques. The search is on for a natural anticlotting factor that can be produced in mass quantities.

See also Acetylsalicylic acid.

Resources

Books

Hardman, J., et al. *Goodman and Gillman's Pharmacological Basis of Therapeutics.* New York: McGraw-Hill, 2001.
Rubenstein, Edward, and Daniel Federman, eds. *Scientific American Medicine.* New York: Scientific American, 1994.

Anticonvulsants

Anticonvulsants are drugs designed to prevent the seizures or convulsions typical of **epilepsy** or other convulsant disorders. Epilepsy is not a single disease—it is a set of symptoms that may have different causes in different people. There is an imbalance in the brain's electrical activity, which causes seizures. These may affect part or all of the body and may or may not cause a loss of consciousness. Anticonvulsant drugs act on the **brain** to reduce the frequency and severity of seizures.

Petit mal seizures may be so subtle that an observer will not notice that an **individual** is having one. Grand mal seizures are more dramatic and unmistakable. The patient may cry out, lose consciousness, and drop to the ground with muscle spasms in the extremities, trunk, and neck. The patient may remain unconscious after the seizure.

Anticonvulsant drugs are an important part of the treatment program for epilepsy. Anticonvulsant drugs are available only by prescription because they are so potent and toxic if taken in excess. Their consumption must be carefully monitored by **blood** tests. Once an individual has been diagnosed with epilepsy, he or she must continue taking the anticonvulsant drugs for life.

Anticonvulsant drugs include medicines such as phenobarbitol, carbamazepine (Tegretol), phenytoin (Dilantin), and valproic acid (Depakote, Depakene). The drugs are available only with a physician's prescription and come in tablet, capsule, liquid, and "sprinkle" forms. The recommended dosage depends on the type of anticonvulsant, its strength, and the type of seizures for which it is being taken.

Phenobarbital is an anticonvulsant drug that has been used since 1912 and it is still one of the better drugs for this purpose. Phenobarbital is used to treat infants (ages 0–1 year) with any type of seizure disorder, and other children with generalized, partial, or febrile seizures. It is also used for treatment of status epilepticus (seizures lasting longer than 15 minutes). The **barbiturates**, such as mephobarbital, and metharbital, are also sometimes used as anticonvulsants. Of the family of barbiturate drugs, these are the only three that are satisfactory for use over a long period of time. They act directly on the central **nervous system** and can produce effects such as drowsiness, hypnosis, deep **coma**, or death, depending upon the dose taken. Because they are habit forming drugs, the barbiturates probably are the least desirable to use as anticonvulsant drugs.

Tegretol is an antiepileptic drug. Types of seizures treated with Tegretol include: grand mal, focal, psychomotor and mixed (seizures that include complex partial or grand mal seizures). Absence seizures (petit mal) do not respond to Tegretol. Phenytoin is a close relative to the barbiturates, but differs in chemical structure. Phenytoin acts on the motor cortex of the brain to prevent the spread of the signal that initiates a seizure. Sudden withdrawal of the drug after a patient has taken it for a long period of time can have serious consequences. The patient can be plunged into a constant epileptic state. Lowering the dosage or taking the patient off phenytoin must be done gradually.

Valproic acid compounds are also antiepileptic drugs, though their mechanism of action is unknown. One of their major side effects is liver toxicity, appearing most often in young patients and in those who are taking more than one anticonvulsant drug.

Another class of anticonvulsant drugs, the succinimides, suppress the brain wave pattern leading to seizures and stabilizes the cortex against them. These are useful drugs in the treatment of petit mal epilepsy, and like phenytoin, must be withdrawn slowly.

Neurontin is another anticonvulsant medication belonging to the antiepileptic drug class. It is a medication used as add-on therapy with other antiepileptic medications to treat partial seizures common among adults over 12 years old with epilepsy.

Anticonvulsant drugs may interact with medicines used during **surgery**, dental procedures, or emergency treatment. These interactions could increase the chance of side effects. Anyone who is taking anticonvulsant drugs should be sure to tell the health care professional in charge before having any surgical or dental procedures or receiving emergency treatment. Some people feel drowsy, dizzy, lightheaded, or less alert when using these drugs, especially when they first begin taking them or when dosage is increased. Anyone who takes anticonvulsant drugs should not drive, use machines or do anything else that might be dangerous until they have found out how the drugs affect them. This medicine may increase sensitivity to sunlight. Even brief exposure to **sun** can cause a severe sunburn or a rash.

People with certain medical conditions, or who are taking certain other medicines, can have problems if they take anticonvulsant drugs. Before taking these drugs, be sure to let the physician know about any of these conditions. The physician should also be told about any allergies to foods, dyes, preservatives, or other substances.

Birth defects have been reported in babies born to mothers who took anticonvulsant drugs during pregnancy. Women who are pregnant or who may become pregnant should check with their physicians about the safety of using anticonvulsant drugs during pregnancy. Some anticonvulsant drugs pass into breast milk and may cause unwanted effects in babies whose mothers take the medicine. Women who are breast-feeding should check with their physicians about the benefits and risks of using anticonvulsant drugs.

Anticonvulsant drugs may affect blood sugar levels. Patients with diabetes who notice changes in the results of their urine or blood tests should check with their physicians. Taking anticonvulsant drugs with certain other drugs may affect the way the drugs work or may increase the chance of side effects. The most common side effects are constipation, mild nausea or vomiting, and mild dizziness, drowsiness, or lightheadedness. These problems usually go away as the body adjusts to the drug and do not require medical treatment. Less common side effects may occur and do not need medical attention unless they persist or are troublesome. Anyone who has unusual symptoms after taking anticonvulsant drugs should get in touch with his or her physician.

Antidepressant drugs

Antidepressant drugs are used to treat serious, continuing mental **depression** that interferes with a person's ability to function. Everyone feels sad, "blue," or discouraged occasionally, but usually those feelings do not interfere with everyday life and do not need treatment. However, when the feelings become overwhelming and last for weeks or months, professional treatment can help. Although depression is one of the most common and serious mental disorders, it is also one of the most treatable. According to the American Psychiatric Association, 80–90% of people with depression can be helped. If untreated, depression can lead to social withdrawal, physical complaints, such as fatigue, **sleep** problems, and aches and pains, and even suicide.

The first step in treating depression is an accurate **diagnosis** by a physician or mental health professional. The physician or mental health professional will ask questions about the person's medical and psychiatric history and will try to rule out other causes, such as thyroid problems or side effects of medicines the person is taking. Lab tests may be ordered to help rule out medical problems. Once a person has been diagnosed with depression, treatment will be tailored to the person's specific problem. The treatment may consist of drugs alone, counseling alone, or drugs in combination with counseling methods such as psychotherapy or cognitive behavioral therapy.

Antidepressant drugs help reduce the extreme sadness, hopelessness, and lack of interest in life that are typical in people with depression. These drugs also may be used to treat other conditions, such as obsessive compulsive disorder, premenstrual **syndrome**, chronic **pain**, and **eating disorders**.

Description

Antidepressant drugs, also called antidepressants, are thought to work by influencing communication between cells in the **brain**. The drugs affect chemicals called neurotransmitters, which carry signals from one nerve **cell** to another. These neurotransmitters are involved in the control of mood and in other responses and functions, such as eating, sleep, pain, and thinking.

The main types of antidepressant drugs in use today are:

- Tricyclic antidepressants, such as amitriptyline (Elavil), imipramine (Tofranil), and nortriptyline (Pamelor).
- Selective serotonin reuptake inhibitors (SSRIs or serotonin boosters), such as fluoxetine (Prozac), paroxetine (Paxil), and sertraline (Zoloft).
- Monoamine oxidase inhibitors (MAO inhibitors), such as phenelzine (Nardil), and tranylcypromine (Parnate).
- Lithium (used mainly to treat **manic depression**, but also sometimes prescribed for recurring bouts of depression).

Selective serotonin reuptake inhibitors act only on the **neurotransmitter** serotonin, while tricyclic antidepressants and MAO inhibitors act on both serotonin and another neurotransmitter, norepinephrine, and may also interact with other chemicals throughout the body. Selective serotonin reuptake inhibitors have fewer side effects than tricyclic antidepressants and MAO inhibitors, perhaps because selective serotonin reuptake inhibitors act only on one body chemical, serotonin.

Because the neurotransmitters involved in the control of moods are also involved in other processes, such as sleep, eating, and pain, drugs that affect these neurotransmitters can be used for more than just treating depression. Headache, eating disorders, bed-wetting, and other problems are now being treated with antidepressants.

All antidepressant drugs are effective, but different types work best for certain kinds of depression. For example, people who are depressed and agitated do best when they take an antidepressant drug that also calms them down. People who are depressed and withdrawn may benefit more from an antidepressant drug that has a stimulating effect.

Recommended dosage

Recommended dosage depends on the kind of antidepressant drug, the type and severity of the condition for which it is prescribed, and other factors such as the patient's age. Check with the physician who prescribed the drug or the pharmacist who filled the prescription for the correct dosage.

Precautions

While antidepressant drugs help people feel better, they cannot solve problems in people's lives. Some mental health professionals worry that people who could benefit from psychotherapy rely instead on antidepressant drugs for a "quick fix." Others point out that the drugs work gradually and do not produce instant happiness. The best approach is often a combination of counseling and medicine, but the correct treatment for a spe-

KEY TERMS

Cognitive behavioral therapy—A type of psychotherapy in which people learn to recognize and change negative and self-defeating patterns of thinking and behavior.

Depression—A mood disorder where the predominant symptoms are apathy, hopelessness, sleeping too little or too much, loss of pleasure, self-blame, and possibly suicidal thoughts.

Obsessive-compulsive disorder—An anxiety disorder in which people cannot prevent themselves from dwelling on unwanted thoughts, acting on urges, or performing repetitive rituals, such as washing their hands or checking to make sure they turned off the lights.

Premenstrual syndrome (PMS)—A set of symptoms that occur in some women 2–14 days before they begin menstruating each month. Symptoms include headache, fatigue, irritability, depression, abdominal bloating, and breast tenderness.

cific patient depends on many factors. The decision of how to treat depression or other conditions that may respond to antidepressant drugs should be made carefully and will be different for different people.

Always take antidepressant drugs exactly as directed. Never take larger or more frequent doses, and do not take the drug for longer than directed.

Most antidepressant drugs do not begin working right away. The effects may not be felt for several weeks. Continuing to take the medicine is important, even if it does not seem to be working at first.

Side effects

Side effects depend on the type of antidepressant drug.

Interactions

Antidepressant drugs may interact with a variety of other medicines. When this happens, the effects of one or both of the drugs may change or the risk of side effects may be greater. Some interactions may be life-threatening. Anyone who takes antidepressant drugs should let the physician know all other medicines he or she is taking.

Resources

Books

Breggin, Peter. *The Anti-Depressant Fact Book: What Your Doctor Won't Tell You About Prozac, Zoloft, Paxil, Celexa, and Luvox*. New York: Persus, 2001.

Periodicals

Johnson, Lois. "Daylight for Depression." *Total Health* 16 (December 1994): 14.

Sacra, Cheryl. "The New Cure-alls: Mood Lifters May Offer Handfuls Of Hope for More Than Just Depression." *Health* 22 (September 1990): 36.

"Treatment of Depression: Drugs Alone Are Not Enough." *HealthFacts* 20 (February 1995): 189.

Nancy Ross-Flanigan

Antihelmintics

Antihelmintics are drugs used to kill parasitic worms (from the Greek word *helmins*, worm). These preparations are also called vermicides.

Worm infestations are among the most common parasitic diseases of man. Often the life cycle of the worm begins when a child playing in dirt ingests the eggs of the worm. The egg hatches in the child's digestive tract and the worms begin their unending quest to reproduce, make more eggs, and infect more hosts.

Parasitic worms may be either round (called nematodes), or have a segmented, flat configuration (called cestodes). Worms are most problematic in areas where sanitation is poor.

The most common parasitic worm infestation in nontropical climates is by the pinworm (*Enterobias vermicularis*), which is highly infectious and may affect an entire family before the **infection** is diagnosed. At any time, up to 20% of the childhood population has pinworms and may jump to 90% among children who are institutionalized, as in an orphanage.

Other roundworm infections are by the hookworm and whipworm, either of which can gain entrance to the body by penetrating the skin of a bare foot. **Trichinosis** derives from eating raw or undercooked pork containing the worm larva.

Flat, segmented worms, also called tapeworms, can originate in raw beef, pork, or **fish**. These worms attach their heads to the walls of the intestine and shed square segments that are packages of eggs.

Most roundworm infestations can be cleared by taking pyrantel pamoate, a drug for sale over the counter. It was developed in 1972 as a prescription drug. Because it is given in the form of a pill or syrup in a single dose, based on body weight, without serious side effects, pyrantel was approved for nonprescription sale. This medication relaxes the muscles that the worm holds on to on the intestinal wall so the worm is passed out with a normal fecal movement.

Other, more potent drugs are available by prescription to treat worm infections. They include piperazine, tetrachloroethylene, and thiabendazole for **roundworms**, and niclosamide for tapeworms.

See also Parasites.

Larry Blaser

Antihistamines

Antihistamines are medicines that relieve or prevent the symptoms of hay fever and other kinds of **allergy**. An allergy is a condition in which the body becomes unusually sensitive to some substance, such as pollen, **mold** spores, dust particles, certain foods, or medicines. These substances, known as allergens, cause no unusual reactions in most people. But in people who are sensitive to them, exposure to allergens causes the **immune system** to overreact. The main reaction is the release of a chemical called **histamine** from specialized cells in the body tissues. Histamine causes such familiar and annoying allergy symptoms as sneezing, itching, runny nose, and watery eyes.

As their name suggests, antihistamines block the effects of histamine, reducing allergy symptoms. When used for this purpose, they work best when taken before symptoms are too severe. Antihistamine creams and ointments may be used to temporarily relieve itching. Some antihistamines are also used to treat **motion** sickness, nausea, dizziness, and vomiting. And because some cause drowsiness, they may be used as **sleep** aids.

Some antihistamine products are available only with a physician's prescription. Others can be bought without a prescription. These drugs come in many forms, including tablets, capsules, liquids, injections and suppositories. Some common antihistamines are astemizole (Hismanal), brompheniramine (Dimetane, Dimetapp), chlorpheniramine (Deconamine), clemastine (Tavist), diphenhydramine (Benadryl), doxylamine (an ingredient in sleep aids such as Unisom and Vicks NyQuil), loratadine (Claritin), and promethazine (Phenergan).

Recommended dosage

Recommended dosage depends on the type of antihistamine. Check with the physician who prescribed the drug or the pharmacist who filled the prescription for the correct dosage, and always take antihistamines exactly as directed. If using non-prescription (over-the-counter)

types, follow the directions on the package label. Never take larger or more frequent doses, and do not take the drug longer than directed.

For best effects, take antihistamines on a schedule, not just as needed. Histamine is released more or less continuously, so countering its effects requires regular use of antihistamines.

Precautions

For best effects, people who have seasonal allergies should take antihistamines before allergy season starts or immediately after being exposed to an allergen. Even then, however, antihistamines do not cure allergies or prevent histamine from being released. They also have no effect on other chemicals that the body releases when exposed to allergens. For these reasons, antihistamines can be expected to reduce allergy symptoms by only about 50%.

In some people antihistamines become less effective when used over a long time. Switching to another type of antihistamine may help.

The antihistamines Seldane and Seldane-D were taken off the market in 1997 due to concerns that the active ingredient, terfenadine, triggered life-threatening **heart** rhythm problems when taken with certain drugs. Terfenadine can also be dangerous to people with liver **disease**. The U.S. Food and Drug Administration encouraged people who were using Seldane to talk to their physicians about switching to another antihistamine such as loratadine (Claritin) or to fexofenadine (Allegra), which is similar to Seldane but has a safer active ingredient.

The antihistamine astemizole (Hismanal) may also cause life-threatening heart rhythm problems or severe allergic reactions when taken in higher-than-recommended doses or with certain drugs. Taking astemizole with food may reduce its absorption into the bloodstream. Astemizole (Hismanal) should not be combined with any of the following:

- the **antibiotics** erythromycin (E-Mycin and other brands), clarithromycin (Biaxin), or troleandomycin (TAO)
- the **blood pressure** medicine mibefradil (Posicor)
- medicines used in treating HIV **infection** such as indinavir (Crixivan), ritonavir (Norvir), and nelfinavir (Viracept)
- antidepressants such as fluoxetine (Prozac), sertraline (Zoloft), or paroxetine (Paxil)
- **asthma** medicines such as zileuton (Zyflo)
- the antifungal drugs ketoconazole (Nizoral) or itraconazole (Sporanox)
- large doses of quinine

In addition, patients with liver disease should not take Hismanal.

People with asthma, **emphysema**, chronic **bronchitis**, or other breathing problems should not use antihistamines unless directed to do so by a physician.

Some antihistamines make people drowsy, dizzy, uncoordinated, or less alert. For this reason, anyone who takes these drugs should not drive, use machines or do anything else that might be dangerous until they have found out how the drugs affect them.

Antihistamines can interfere with the results of skin and blood tests. The antihistamine promethazine (Phenergan) can interfere with pregnancy tests and can raise blood sugar. Anyone who is taking antihistamines should notify the health care provider in charge before scheduling medical tests.

People with **phenylketonuria** should be aware that some antihistamine products contain aspartame (Nutrasweet), which breaks down in the body to phenylalanine.

Anyone who has sleep apnea (periods when breathing stops during sleep) should not take the antihistamine promethazine (Phenergan). This drug may also cause people with seizure disorders to have more frequent seizures.

Because children are often more sensitive to antihistamines, they may be more likely to have side effects and to suffer from accidental overdoses. Check with a physician before giving antihistamines to children under 12 years.

Older people may also be more likely to have side effects, such as nervousness, irritability, dizziness, sleepiness, and low blood pressure from antihistamines. Older men may also have problems urinating. Unless these problems are severe, they can usually be handled by taking a lower dose or switching to a different antihistamine. However, people over 80 or older people with serious physical problems or **dementia** may become confused, disoriented, and incoherent after taking even small amounts of the antihistamine diphenhydramine (Benadryl).

Special conditions

People with certain medical conditions or who are taking certain other medicines can have problems if they take antihistamines. Before taking these drugs, be sure to let the physician or pharmacist know about any of these conditions:

Allergies

Some antihistamine products may contain the dye tartrazine, which causes allergic reactions, including bronchial asthma, in some people. People who are allergic to aspirin may also be allergic to tartrazine.

Sodium bisulfite, a preservative, is found in some antihistamine products. People who are sensitive to this chemical may have allergic-type reactions, including **anaphylaxis** and severe asthma attacks. People with asthma are especially likely to be sensitive to sodium bisulfite. Ask the pharmacist which antihistamine products are sulfite-free.

Anyone who has had unusual reactions to antihistamines in the past should let his or her physician know before taking the drugs again. The physician should also be told about any allergies to foods, dyes, preservatives, or other substances.

Pregnancy

Pregnant women should not use antihistamines unless directed to do so by a physician.

Breast-feeding

Antihistamines pass into breast milk and may cause side effects in nursing babies. Women who are breast-feeding should check with their physicians before using antihistamines.

Other medical conditions

Before using antihistamines, people with any of these medical problems should make sure their physicians are aware of their conditions:

- glaucoma
- hyperthyroidism (overactive thyroid)
- high blood pressure
- enlarged prostate
- heart disease
- ulcers or other stomach problems
- stomach or intestinal blockage
- liver disease
- kidney disease
- bladder obstruction
- diabetes

Use of certain medicines

Taking antihistamines with certain other drugs may affect the way the drugs work or may increase the chance of side effects.

Side effects

Common side effects of antihistamines include drowsiness, dizziness, poor coordination, restlessness, excitability, nervousness, and upset stomach. These problems usually go away as the body adjusts to the drug and do not require medical treatment. Less common side effects, such as dry mouth, nose, and eyes, irritability, difficulty urinating, and blurred **vision**, also may occur and do not need medical attention unless they do not go away or they interfere with normal activities.

If any of the following side effects occur, check with the physician who prescribed the medicine as soon as possible:

• rapid, irregular, pounding, or fluttering heartbeat
• convulsions
• sweating
• vomiting
• fever
• chills
• fainting
• breathing problems
• hallucinations
• unusual bleeding or bruising
• low blood pressure
• unusual sensitivity to light
• uncontrolled movements

Other rare side effects may occur. Anyone who has unusual symptoms after taking antihistamines should get in touch with his or her physician.

Interactions

Antihistamines may increase the effects of other drugs that slow down the central **nervous system** (CNS), such as **alcohol**, **tranquilizers**, **barbiturates**, and sleep aids. Avoid drinking alcohol while taking antihistamines, and check with a physician before combining antihistamines with other CNS depressants.

Certain antihistamines should not be used within two weeks of using monoamine oxidase inhibitors (MAO inhibitors), which are drugs used to treat Parkinson's disease, **depression**, and other psychiatric conditions. Examples of MAO inhibitors are Parnate and Nardil. People who have been taking MAO inhibitors or who are not sure if they have should check with their physician or pharmacist before taking antihistamines.

Although no such interactions have been reported, antihistamine loratadine (Claritin) could potentially interact with the antiulcer drugs cimetidine (Tagamet) and ranitidine (Zantac), both of which are also taken for heartburn; with antibiotics such as erythromycin and Biaxin; with the antifungal drug ketoconazole (Nizoral); and with the bronchodilator theophylline (Theo-Dur).

These interactions can cause liver problems. Persons who are taking these drugs should consult with an physician or pharmacist to see whether this is an issue for them.

Check with a physician before combining the antihistamine chlorpheniramine (Deconamine) with any of the following:

• asthma medicines such as albuterol (Proventil) and bromocriptine (Parlodel)
• blood pressure drugs such as mecamylamine (Inversine), methyldopa (Aldomet), and reserpine
• narcotic painkillers such as meperidine (Demerol), oxycodone-aspirin (Percodan) and oxycodone-acetaminophen (Percocet)
• the anticonvulsant drug phenytoin (Dilantin)
• sleep aids such as triazolam (Halcion) and secobarbital (Seconal)
• tranquilizers such as diazepam (Valium) and alprazolam (Xanax)

Hismanal should not be taken with grapefruit juice or combined with any of the following drugs:

• antibiotics such as erythromycin (E-Mycin and other brands), clarithromycin (Biaxin), or troleandomycin (TAO)
• the blood pressure medicine mibefradil (Posicor)
• medicines used in treating HIV infection such as indinavir (Crixivan), ritonavir (Norvir), and nelfinavir (Viracept)
• antidepressants such as fluoxetine (Prozac), sertraline (Zoloft), or paroxetine (Paxil)
• asthma medicines such as zileuton (Zyflo)
• the antifungal drugs ketoconazole (Nizoral) or itraconazole (Sporanox)
• large doses of quinine

Check with a physician before combining *any* antihistamine with any of the following drugs:

• Medicines for stomach or abdominal cramps or spasms, such as dicyclomine (Bentyl).
• The antibiotics azithromycin (Zithromax), clarithromycin (Biaxin), or erythromycin (E-mycin).
• The antifungal drugs itraconazole (Sporanox) or ketoconazole (Nizoral).
• The **calcium** channel blocker bepridil (Vascor).
• Drugs used to treat irregular heartbeat, such as disopyramide (Norpace), procainamide (Pronestyl), or quinidine (Quinaglute Dura-Tabs, Cardioquin).
• Antidepressants such as maprotiline (Ludiomil) or tricyclic antidepressants such as desipramine (Norpramin) or imipramine (Tofranil).

KEY TERMS

Allergen—An otherwise harmless substance that can cause a hypersensitive allergic response.

Anaphylaxis—A sudden, life-threatening allergic reaction.

Hallucination—A sensory experience of something that does not exist outside the mind. A person can experience a hallucination in any of the five senses. Hallucinations usually result from drugs or mental disorders.

Histamine—A chemical released from cells in the immune system as part of an allergic reaction.

Phenylketonuria—(PKU) A genetic disorder in which the body lacks an important enzyme. If untreated, the disorder can lead to brain damage and mental retardation.

Pollen—Dust-like grains produced by the male parts of plants and carried by wind, insects, or other methods, to the female parts of plants.

• Medicines called phenothiazines, used to treat mental, emotional, and nervous disorders. Examples are chlorpromazine (Thorazine) and prochlorperazine (Compazine).

• Pimozide (Orap), used to treat symptoms of Tourette's syndrome.

Not all possible interactions of antihistamines with other drugs are listed here. Be sure to check with a physician or pharmacist before combining antihistamines with any other prescription or nonprescription (over-the-counter) medicine.

Nancy Ross-Flanigan

Antihydrogen *see* **Antimatter**

Antimatter

Antimatter is **matter** comprising particles that are equal in **mass** to the particles comprising ordinary matter—neutrons, protons, electrons, and so forth—but with opposite electrical properties. An antiproton has the same mass as the **proton**, but **negative** charge, an antielectron (positron) has the same mass as an **electron**, but negative charge, and an antineutron has the same mass as a **neutron**, but with a magnetic moment opposite in sign

to the neutron's. Antiparticles are built up of antiquarks, which come in six varieties that mirror the six varieties of **quarks** that make up ordinary matter. In theory, antimatter has almost exactly the same properties as ordinary matter.

When particles and antiparticles approach closely, they annihilate each other in a burst of high-energy photons. A complementary or reverse process also occurs: high-energy photons can produce particle-antiparticle pairs. Antiparticles are also given off by the fissioning nuclei of certain isotopes.

The existence of antimatter was proposed by British physicist Paul Dirac (1902–1984) in 1927. The positron was first observed in 1932 and the antiproton and antineutron in the 1950s. Antimatter does yield **energy**; the annihilation process that occurs when matter and antimatter meet provides a 100% matter-to-energy conversion, as opposed to the extremely small efficiencies achieved in **nuclear fission** and fusion reactors. However, antimatter cannot be used as a practical energy source because there are no bulk sources of antimatter. Antimatter can only be collected painstakingly, a few particles at a time.

Antimatter's rarity is one of the great mysteries of **cosmology**. The production of antimatter from energy must always, so far as present-day **physics** can determine, produce matter and antimatter in equal quantities. Therefore, the big bang—the explosion in which the Universe originated—should have yielded equal quantities of matter and antimatter. However, the Universe seems to consist entirely of matter. Where did all the antimatter go?

For decades, some physicists theorized that it might still be out there—that some galaxies might consist of antimatter. Because antimatter should absorb and emit photons just like normal matter, it would be impossible to tell a matter **galaxy** from an antimatter galaxy by observing it through the **telescope**.

Theoretical physicists now argue that this is unlikely, and that the entire Universe almost certainly does consist entirely of ordinary matter—although they still don't know why. If matter and antimatter had segregated into separate galactic domains after the big bang, these domains must have gone through a process of mutual annihilation along their contacting borders; this would have produced large quantities of gamma rays, and these which would still be observable as a gamma-ray background everywhere in the sky. A cosmic gamma-ray background of uncertain origin does exist, but is only about one fifth as intense as it would be if the formation of matter and antimatter domains were responsible for it. Nevertheless, if antimatter galaxies do exist, some of the very-high-speed **atoms** from **space** that are known as cosmic rays should consist of antimatter. A device to detect antimatter cosmic rays was tested briefly on a **space shuttle** flight in 1998, and found no antimatter; it is scheduled to make a more thorough investigation, operating on the **International Space Station**, in 2003.

Antimatter, despite its scarcity, has practical uses. Radioactive isotopes that give off positrons are the basis of PET (positron-emission tomography) scanning technique. When administered to a patient, a positron-emitting **isotope** of (say) **oxygen** is utilized chemically just like ordinary oxygen; when it breaks down, it emits positrons. Each positron cannot travel far before it encounters an electron. The two particles annihilate, giving off two identical, back-to-back gamma rays. These can be picked up by a detector array, and a **map** of positron-emission activity can be created by a computer that deduces the locations of large numbers of positron-electron annihilations from these gamma-ray detections. PET produces real-time images of **brain** activity of importance to neurology, and has also been used to probe the structure of nonliving systems (engines, semiconductor crystals, etc.).

The latest goal of antimatter research is the creation of cool antihydrogen atoms (composed of an antiproton orbited by a positron). These early antihydrogen atoms, however, were formed moving at high speed, and quickly struck normal atoms in the apparatus and were destroyed. In 2002, researchers at CERN (Conseil Europée pour la Recherche Nnucléaire) produced thousands of antihydrogen atoms cooled to about 50 degrees Kelvin. The goal of such research is to isolate the antihydrogen from ordinary matter long enough to observe its properties, such as its absorption and **emission** spectra; cool atoms, which are moving more slowly, are easier to confine. The properties of antihydrogen are predicted by theory to be identical to those of ordinary **hydrogen**; therefore, if they are not, revisions to basic physics will be required. CERN researchers hope to obtain their first antihydrogen spectra in late 2003.

See also Acceleration; Accelerators; Atomic theory; Electromagnetic spectrum; Particle detectors; Quantum mechanics; Subatomic particles.

Resources

Books

Fraser, Gordon. *Antimatter: The Ultimate Mirror.* Cambridge, UK: Cambridge University Press, 2000.

Periodicals

Peskin, Michael, "The Matter with Antimatter." *Science.* 419 (October 3, 2002):24–25.

Seife, Charles, "CERN Team Produces Antimatter in Bulk." *Science.* 297 (September 20, 2002):1979–1981.

Hijmans, Tom W., "Cold Antihydrogen." *Science.* 419 (October 3, 2002):439–440.

Larry Gilman

Antimetabolites

Antimetabolites are substances that interfere with the normal **metabolism** of an **organism**, thereby causing its death. They are widely used in the medical sciences because they have the ability to kill or inactivate **microorganisms** that cause **disease**. Terms such as antibacterials, antifungals, and antivirals are used to describe antimetabolites that act on **bacteria**, **fungi**, and viruses, respectively. In most cases, an antimetabolite works by inhibiting the action of an **enzyme** that is crucial to the process of metabolism. When the enzyme is immobilized, the series of reactions by which metabolism occurs is interrupted and the microorganism dies.

One of the classic examples of antimetabolite action is that of the sulfa drugs, discovered in the 1930s. Some examples of the sulfa drugs are sulfathiazole, sulfadiazine, and sulfacetamide. All sulfa drugs affect the metabolism of microorganisms in the same way. Under normal circumstances, a bacterium makes use of a compound known as para-aminobenzoic acid (PABA) to produce a second compound, folic acid. Folic acid is then used in the manufacture of nucleic acids in the bacterium.

Sulfa drugs have chemical structures that are very similar to that of PABA. When a sulfa compound is ingested by a bacterium, the microorganism attempts to make folic acid using the sulfa drug rather than PABA. The folic acid-like compound that is produced, however, can not be used to make nucleic acids. The bacterium's normal metabolism is interrupted, and it dies.

Antimetabolites generally work in one of three ways to interrupt the metabolism of an organism. First, as in the example above, they may prevent the formation of nucleic acids, essential for the production of DNA in the organism. Second, they may interfere with the synthesis of **proteins** in the **cell** of a microorganism. Third, they may interfere with the synthesis of a cell wall, causing the cell to break apart and die.

Today, a wide variety of antimetabolite drugs are available to physicians. In addition to the sulfa drugs described above, other examples of such drugs include members of the penicillin family, the tetracyclines, chloramphenicol, streptomycin, and the anti-cancer drug known as 5-fluorouracil.

Antimony *see* **Element, chemical**

Antioxidants

Antioxidants are molecules that prevent or slow down the breakdown of other substances by **oxygen**.

In **biology**, antioxidants are scavengers of small, reactive molecules known as free radicals and include intracellular enzymes such as superoxide dismutase (SOD), catalase and glutathione peroxidase. Antioxidants can also be extracellular originating as exogenous cofactors such as vitamins. **Nutrients** functioning as antioxidants include vitamins, for example ascorbic acid (**vitamin** C), tocopherol (vitamin E) and vitamin A. **Trace elements** such as the divalent **metal** ions selenium and zinc also have antioxidant activity as does uric acid, an endogenous product of purine **metabolism**. Free radicals are molecules with one or more unpaired electrons, which can react rapidly with other molecules in processes of oxidation. They are the normal products of metabolism and are usually controlled by the antioxidants produced by the body or taken in as nutrients. However, **stress**, aging, and environmental sources such as polluted air and **cigarette smoke** can add to the number of free radicals in the body, creating an imbalance. The highly reactive free radicals can damage nucleic acids and have been linked to changes that accompany aging (such as age-related macular degeneration, an important cause of blindness in older people) and with **disease** processes that lead to **cancer**, **heart** disease, and **stroke**.

The **brain** is particularly vulnerable to oxidative stress. Free radicals play an important role in a number of neurological conditions including stroke, **Parkinson disease**, **Alzheimer disease**, **epilepsy** and **schizophrenia**. Some other diseases in which oxidative stress and depletion of antioxidant defence mechanisms are prominent features include hepatic **cirrhosis**, pre-eclampsia, pancreatitis, rheumatoid **arthritis**, mitochondrial diseases, systemic sclerosis, **malaria**, neonatal oxidative stress and renal **dialysis**.

Antioxidant simply refers to a **molecule** that protects cells from a process called oxidation, the negative effect of oxygen. Antioxidants act by neutralizing free radicals, which are **atoms** or groups of atoms that become destructive to cells when produced in large quantities. While our bodies naturally manufacture some antioxidant enzymes, it cannot manufacture selenium, the trace mineral necessary to produce these enzymes, nor can it manufacture the essential micronutrient antioxidants vitamins C, E, and beta-carotene and other carotenoids. They must be obtained by eating foods rich in these essential nutrients and, in some cases, supporting the diet with vitamin supplements.

To generate **energy**, cells must produce millions of **chemical reactions**. During this process, some oxygen molecules finish up with an odd instead of even number of electrons. Because each **electron** needs a mate, an unpaired electron makes the oxygen molecule unstable. These unstable molecules, called free radicals, then take an electron from a neighboring **cell**. The neighboring cell then becomes unstable, and affects its neighboring cell, starting a chain reaction. This domino effect becomes dangerous when the cell **membrane** or DNA is attacked, causing the cell to function poorly or die. When the body is unable to produce enough antioxidants to counteract free radicals, the resultant cell damage can cause certain types of cancer, heart disease, premature aging, cataracts, and other serious illnesses. For example, when free radicals damage artery walls, fatty deposits accumulate around the scar **tissue**, causing atherosclerosis and heart disease.

Most of the body's cells are continually exposed to free radicals, the creation of which is also stimulated by external sources such as tobacco smoke, **smog**, **radiation**, **herbicides**, and other environmental pollutants. Antioxidants are the body's mechanism for stabilizing free radicals. They work by donating one of their electrons to the unstable molecule, preventing the unpaired electron from "stealing" from another cell. Unlike oxygen molecules, antioxidants remain stable, even after donating an electron.

Vitamins as antioxidants

Vitamins play a vital role in the body's healthy functioning. Researchers now know that vitamins E, C, B_1, B_2, and B_3, the carotenoids (beta-carotene, lycopene, and lutein), selenium, and **magnesium**, are important antioxidants in the body's continuous fight against free radicals. These essential nutrients are readily available in fresh fruit and **vegetables** and many studies have linked fruit and vegetable consumption to a lowered risk of certain diseases. While vitamin supplements may be helpful in fighting free radicals, there is no comparable evidence to show their effectiveness. Most traditional health authorities recommend three to five servings of fruit and vegetables daily, along with vitamin supplements, if necessary, for a total daily intake of 6–15 mg beta-carotene, 250–500 mg vitamin C, and 200–800 IU vitamin E.

The vitamins

Carotenoids, excellent antioxidants, are available in abundance in red, orange, and deep yellow vegetables and fruit such as tomatoes, sweet potatoes, winter squash, carrots, peaches, and cantaloupe; dark green leafy vegetables such as **spinach** and broccoli; as well as liver, egg yolks, milk, and butter. Studies show that men with the lowest levels of beta-carotene are at the highest risk for prostate cancer. (Vitamin A, a fat-soluble vitamin, is not—as is sometimes suggested—an antioxidant. However, beta-carotene—a vitamin A analog, or precursor, is converted into vitamin A by the body. Because it is stored in the liver, vitamin A can be toxic when taken in large quantities.) Lycopene, another carotenoid and one of the most powerful antioxidants, is believed to be effective against many diseases including cancers of the mouth, pharynx, esophagus, stomach, colon, rectum, and cervix. Early in 1999, studies showed that tomatoes and their byproducts drastically increase **blood** levels of lycopene and reduce cellular damage from free radicals. Lutein, also a carotenoid, is believed to decrease the risk of macular degeneration.

Vitamin E, a fat-soluble vitamin, is not only essential for promoting general good health in the areas of aging, **infertility**, and athletic performance, research has also shown this antioxidant counteracts cell damage that leads to cancer, heart disease, and cataracts. It is also known to work together with other antioxidants, such as vitamin C, to help prevent some chronic illnesses. Available in vegetable oils, margarine, **wheat** germ, nuts, **seeds**, and peanut butter, supplementation of this important vitamin may be necessary to maintain adequate amounts within the body.

Vitamin C (also called ascorbic acid) is a water-soluble vitamin and is not easily stored in the body. Therefore, daily intake is required. This "antioxidant workhorse" is perhaps the most famous, and research has shown its effectiveness in lowering the risk of cataracts and other **eye** problems, lowering blood **pressure** and **cholesterol** levels, and reducing the risk of heart attacks and strokes. Researchers have also found low levels of vitamin C in people suffering from **asthma**, arthritis, cancer, and diabetes. Excellent dietary sources of vitamin C include citrus **fruits** (especially oranges and grapefruit), kiwi fruit, strawberries, red peppers, broccoli, and potatoes.

Current research on antioxidants

The last few years have witnessed an explosion of information on the role of oxidative stress in causing a number of serious diseases, and there appears to be a potential therapeutic role for antioxidants in preventing such diseases. For example, recent epidemiological studies have shown that a higher consumption of vitamin E and to a lesser extent β-carotene is associated with a large decrease in the **rate** of coronary arterial disease.

KEY TERMS

Antioxidant—A molecule that protects cells from the process of oxidation, the negative effect of oxygen.

Fat-soluble vitamin—Vitamins that can be dissolved in oil or fat and are stored in the body's fatty tissue.

Free radicals—Unstable molecules containing an odd number of electrons and, therefore, seeking an electron from another molecule.

Water-soluble vitamin—Vitamins that can be dissolved in water and are excreted from the body through the kidneys.

The most effective dose of vitamin E is apparently 400–800 mg/day. Other studies have shown that a diet rich in fruits and vegetables leads to a marked decline in the cancer rate in most organs with the exception of blood, breast and prostate. Antioxidants play a major role in this. It is now abundantly clear that toxic free radicals play an important role in carcinogenesis. In several studies high cancer rates were associated with low blood levels of antioxidants particularly vitamin E. Similarly, vitamin C is thought to protect against stomach cancer by scavenging carcinogenic nitrosamines in the stomach.

Studies have suggested that the antioxidants occurring naturally in fresh fruits and vegetables are very beneficial and protect against excessive oxidative stress. There is still some question as to whether antioxidants in the form of dietary supplements are equally beneficial. Some scientists believe that regular consumption of such supplements interferes with the body's own production of antioxidants.

See also Alternative medicine; Biochemistry; Nutrition.

Judyth Sassoon
Larry Blaser

Resources

Books

Mahan, L. Kathleen and Sylvia Escott-Stump. *Krause's Food, Nutrition, and Diet Therapy.* Philadelphia: W.B. Saunders, 2000.
Packer, Lester, et. al *Antioxidant Food Supplements in Human Health.* Academic Press, 1999.

Other

National Cancer Institute. NCI Fact Sheet: "Tea and Cancer Prevention" December 6, 2002 [cited February 5, 2003]. <http://www.cancer.gov/newscenter/content_nav.aspx?viewid=afc8f2c0-f3df-4f6c-9c30-28fc15c0054e>.

National Eye Institute. United States National Library of Medicine. Clinical Advisory: "Antioxidant Vitamins and Zinc Reduce Vision Loss from Age-Related Macular Degeneration" October 25, 2001 [cited February 5, 2003]. <http://www.nlm.nih.gov/databases/alerts/amd.html>.

Antiparticle

An antiparticle is a subatomic particle identical with more familiar **subatomic particles** such as electrons or protons, but with the opposite electrical charge or, in the case of uncharged particles, the opposite magnetic moment. For example, an antielectron (also known as a positron) is identical with the more familiar **electron**, except that the former carries a single unit of positive electrical charge rather than a single unit of **negative** electrical charge. Antiparticles are not considered to be unusual or abnormal but are as fundamental a part of the natural world as are non-antiparticles. The main difference between the two classes of particles is that the world with which humans normally deal is constituted of protons, neutrons, and electrons rather than antiprotons, antineutrons, and antielectrons. To avoid suggesting that non-antiparticles are more "normal" than antiparticles, the name koinoparticle has been suggested for "ordinary" particles such as the **proton**, electron, and **neutron**.

Dirac's Hypothesis

During the late 1920s, the British physicist Paul Dirac attempted to modify the currently accepted model of the atom by including in it the relativistic properties of electrons. As a result of his analysis, Dirac found that electrons should be expected to exist in two **energy** states, one positive and one negative. The concept of positive energy presents no problems, of course, but Dirac and other physicists were uncertain as to the meaning of a negative energy state. What did it mean to say that an electron had less than **zero** energy?

Eventually Dirac concluded that the negative energy state for an electron might imply the existence of a kind of electron that no one had yet imagined, one that is identical with the familiar negatively-charged electron in every respect except its charge. It was, Dirac suggested, an electron with a positive charge, or an antielectron.

Within five years, Dirac's hypothesis had been confirmed. In 1932, the American physicist Carl Anderson found in photographs of a **cosmic ray** shower the tracks of a particle that satisfied all the properties predicted by Dirac for the antielectron. Anderson suggested the name positron for the new particle.

Other antiparticles

The existence of the positron strongly suggested to scientists that other antiparticles might exist. If there was a positively-charged electron, they asked, why could there also not be a negatively-charged proton... the antiproton. The search for the antiproton took much longer than the search for the antielectron. In fact, it was not until 1955 that Emilio Segre and Owen Chamberlain were able to prove that antiprotons are produced when protons from a powerful **cyclotron** collide with each other.

The antineutron is a fundamentally different kind of antiparticle than the antielectron or antiproton. Since neutrons have no electrical charge, an antineutron could not differ in this respect from its mirror image. Instead, an antineutron is a particle whose direction of spin is opposite that of the neutron. Since a particle's spin is expressed in the magnetic field that it generates—its magnetic moment—the antineutron is defined as the antiparticle with a magnetic moment equal in magnitude, but opposite in sign, to that of the neutron.

Other antiparticles also exist. For example, the electron is a member of a group of fundamental particles known as the leptons. Other leptons include the mu **neutrino** (muon) and the tau neutrino (tauon), electron-like particles that exist only at very high energy levels not observed under circumstances of our ordinary everyday world. Both muons and tauons have their own antiparticles, the antimuon and the antitauon.

Antimatter

Given the existence of the antiproton, antineutron, and antielectron, one might imagine the existence of antiatoms, **atoms** that are identical to the atoms of everyday life but have mirror image electrical charges and mirror image magnetic moments. In fact, some scientists believe that a whole universe of **antimatter** made of antiatoms may actually exist in some dimensions of which we are not aware.

Locating the existence of such an antiuniverse would be very difficult, however. When an antiparticle comes into contact with its mirror image—an antielectron with an electron, for example—the two particles annihilate each other and their **mass** is converted to energy. Thus, any time **matter** comes into contact with antimatter, both are destroyed and converted into energy.

Antiparticles and cosmology

The Swedish physicist Hans Alfvén has studied in some detail the possible role of antiparticles in the creation of the universe. At first glance, one would assume that the number of koinoparticles and antiparticles pro-

KEY TERMS

Cyclotron—A machine used to accelerate particles to very high energies, making possible nuclear reactions that do not take place at the energies of everyday life.

Lepton—A particle that undergoes weak interactions. Three classes of leptons—electrons, muons, and tauons and their antiparticles— are known to exist.

Magnetic moment—A measure of the strength of a magnetic field.

Neutrino—An elementary particle with no electrical charge and virtually no mass.

Subatomic particle—An elementary particle smaller than an atom. Protons, neutrons, and electrons are examples of subatomic particles.

duced during the big bang would be equal. As it happens, however, the way in which the two classes of particles decay is very slightly different, a difference that would have become more and more important as the nascent universe aged during the first second of creation. Eventually, the very small difference in decay properties between particles might have produced a larger and larger difference, with koinomatter finally winning a predominance in terms of numbers throughout the universe. Until and if scientists can learn more about the presence of antimatter in other parts of the universe, however, questions such as these will remain unanswered.

Resources

Books

Alfvén, Hans. *Worlds-Antiworlds: Antimatter in Cosmology.* San Francisco: W. H. Freeman and Co., 1966.
Hewitt, Paul. *Conceptual Physics.* New York: Prentice Hall, 2001.
Introduction to Astronomy and Astrophysics. 4th ed. New York: Harcourt Brace, 1997.
Trefil, James. *Encyclopedia of Science and Technology.* The Reference Works, Inc., 2001.

David E. Newton

Antipsychotic drugs

An antipsychotic drug, sometimes called a neuroleptic, is a prescription medication used to treat **psychosis**. Psychosis is a major psychiatric disorder characterized by derangement or disorganization of personality and/or by the inability to tell what is real from what is not real, often with hallucinations, delusions, and thought disorders. People who are psychotic often have a difficult time communicating with or relating to others. Sometimes they become agitated and violent. They may hear voices, see things that aren't really there, and have strange or untrue thoughts, such as believing that other people can hear their thoughts or are trying to harm them. They may also neglect their appearances and may stop talking or talk only "nonsense."

Among the conditions considered to be psychoses are **schizophrenia**, major **depression**, and bipolar affective disorder (**manic depression**). Psychosis can arise from emotional or organic causes. Organic causes include **brain** tumors, drug interactions, and substance abuse. Antipsychotic drugs do not cure mental illness, but can reduce some of the symptoms or make them milder. The medicine may improve symptoms enough for the person to undergo counseling and live a more normal life. The type of antipsychotic medicine prescribed depends on the type of mental problem the patient has.

The vast majority of antipsychotics work by blocking the absorption of **dopamine**, a chemical that occurs naturally in the brain and is responsible for causing psychotic reactions, especially those that happen as a result of mental illness. Dopamine is one of the substances in the brain responsible for transmitting messages across the gaps, or synapses, of nerve cells. Too much dopamine in a person's brain speeds up nerve impulses to the point of causing hallucinations, delusions, and thought disorders. By blocking the dopamine receptors, antipsychotics reduce the severity of these symptoms. The brain has several types of dopamine receptors, and their unselective blockage by antipsychotic drugs causes the side effects.

When a patient takes an antipsychotic drug, he or she enters what is called a neuroleptic state. Impulsiveness and aggression decrease as do concern and arousal about events going on in the environment outside the person. The person taking the drug has fewer hallucinations and delusions as well. Once these symptoms are controlled by antipsychotic drugs, he or she can live a more normal life, and physicians can more easily treat the cause of the psychosis.

Antipsychotic medications were not used in the United States before 1956. The first drug used to treat psychosis was reserpine, which is made from a **plant** called rauwolfia. Reserpine was first made in India, where rauwolfia had been used to treat psychotic symptoms for centuries. Chlorpromazine, which was invented at about the same time and marketed under the name Thorazine, soon became the most favored drug. Once these and other antipsychotic medicines were introduced in the United States, they gained widespread acceptance for the treatment of schizophrenia. The use of these drugs allowed the release of many people who had been confined to mental institutions.

Antipsychotic drugs are also known as neuroleptics or major **tranquilizers**. Several types of these drugs are available, such as haloperidol (Haldol), **lithium** (Lithonate), chlorpromazine (Thorazine), and thioridazine (Mellaril). The newer antipsychotics include risperidone (Risperdal), quetiapine (Seroquel) and olanzapine (Zyprexa). These medicines are available only with a physician's prescription. The recommended dosage depends on the type of antipsychotic drug, the condition for which it is prescribed, and other factors. Despite their benefits, antipsychotic medicines have a number of strong side effects.

Although it usually takes at least two weeks for the drug to work on symptoms of psychosis, side effects often show up sooner. The most severe include muscle rigidity, muscle spasms, twitching, and constant movement. Tardive dyskinesia (TD)—a rhythmic, uncontrollable movement of the tongue, lips, jaw, or arms and legs—develops after a mental patient has taken an antipsychotic drug for a longer period of time. Twenty-six **percent** of chronically mentally ill people who are, or have been hospitalized develop TD. Often these side effects do not disappear when a person stops taking his or her medication.

Perhaps the most serious side effect of antipsychotic medications is neuroleptic malignant **syndrome** or NMS, a condition that occurs when someone taking an antipsychotic drug is ill or takes a combination of drugs. People with NMS cannot move or talk. They also have unstable **blood pressure** and **heart** rates. Often NMS is fatal. Even when the person recovers, he or she has an 80% chance of experiencing NMS again if given antipsychotic drugs.

Today a new generation of antipsychotics has been developed as a result of recent **molecular biology** discoveries about how the brain works. These new drugs have fewer side effects. Some do not completely block dopamine receptors; others are more selective, blocking only one type of dopamine receptor. A few of the newer drugs block serotonin or glutamate, two other neurotransmitters.

Resources

Books

American Psychiatric Association. *Let's Talk about Psychiatric Drugs.* Washington, DC: American Psychiatric Association, 1993.

Buelow, George, and Suzanne Hebert. *Counselor's Resource on Psychiatric Medications, Issues of Treatment and Referral.* Pacific Grove, CA: Brooks/Cole, 1995.

Shives, Louise Rebecca. *Basic Concepts of Psychiatric-Mental Health Nursing.* 4th ed. Philadelphia: Lippincott Williams & Wilkins, 1997.

Kay Marie Porterfield

Antisepsis

Antisepsis is the prevention or inhibition of an **infection** by either killing the **organism** responsible for the infection, or weakening the organism so that it is unable to cause the infection or survive. This is usually achieved by application of an antiseptic or germicidal preparation.

An antiseptic differs from an antibiotic. An antibiotic is specifically directed to a target bacterium or different types of **bacteria**. There are many different classes of **antibiotics**, some of which can kill only a few types of bacteria; others are effective against many types of bacteria.

An antiseptic is a chemical compound that is "broad spectrum" in its activity; that is, it kills a wide variety of bacteria and other **microorganisms**. Because many antiseptics are used on the skin (e.g., "swabbing" the skin with iodine before an injection), antiseptics tend to be non-irritating.

The search for antiseptics

The search for antiseptic agents is as old as humanity. In a hieroglyphic prescription dating from c. 1500 B.C., an Egyptian is depicted ordering a mixture of grease and honey for treatment of a wound. While infections and their causes were not known at that time, the relief provided by antiseptic compounds was recognized.

Other historical "cures" for infections include **plant** extracts, broths of **animal** or plant materials, and poultices of **moss**, mud, or dung, Not until the eighteenth century did progress begin to be made toward conquering everyday infections.

As late as the beginning of the nineteenth century, physicians had no knowledge of the septic (infectious) process or its prevention. Although **surgery** had developed steadily, the mortality **rate** among patients was high. Whether the patient would die from an infected **organ** or from the surgery to remove it often was a moot point. Surgeons went from one patient to the next without washing their hands or changing aprons. Thus, the bacteria from one patient were readily passed to the next and sepsis was an accepted fact.

In the middle of the nineteenth century, the Hungarian obstetrician Ignaz Semmelweiss (1818–1865) proposed that the infectious agent of puerperal fever, which was fatal to many women during childbirth (hence its other name, childbed fever), could be spread by the attending physician. Semmelweiss further suggested that washing hands between patients could prevent the infection. At first he was ridiculed, but when his rate of fatal puerperal fever infections declined rapidly with his practice of washing his hands, other obstetricians soon adopted the practice. This was the first introduction of antisepsis into medical practice.

In the latter half of the nineteenth century, British physician Joseph Lister (1827–1912) introduced the practice of spraying carbolic acid over patients during operations. This reduced the **contamination** of the open wound from airborne microorganisms and microbes on the doctor's clothing or gloves. Lister's innovation brought aseptic technique into the operating theatre.

The early antiseptics were based on mercury (mercurochrome, merthiolate), but have fallen into disuse. Although mercury poses a serious health hazard if absorbed into the body, the small amounts of mercury in the mercury-based antiseptics posed little threat. They were discontinued because they were relatively ineffective. Although mercury-based antiseptics readily stopped bacteria from reproducing and spreading, they did not kill the microorganism. Once the merthiolate was washed away, the bacteria revived and resumed their invasion of the tissues.

Modern antisepsis

Now, the innovations of Semmelweiss and Lister are an accepted part of medicine. Modern antisepsis is both

KEY TERMS

Attenuated—A bacterium that has been killed or weakened, often used as the basis of a vaccine against the disease caused by the bacterium.

Etiology—The cause or origin of a disease or condition.

Organic—Carbon based material. The word organism is derived from organic, meaning any life form.

Pathogenic—Disease causing.

Sepsis—From the Greek, meaning decay, the presence in the blood or other tissue of a microorganism that causes an infection or disease.

preventive and therapeutic. Examples of preventive measures include hand washing by the surgeon, use of sterile surgical gowns, masks, gloves and equipment, and the preparation of the patient's skin with antiseptic. Therapeutic antisepsis is the application of a bactericidal agent to an infected area to kill the infectious agent.

Antiseptics have also found their way into the home. Various powders, liquids, or ointments are applied to the surface of the skin to prevent infection of a cut, splinter, or other superficial wound. These antiseptics are for external use only and each is effective against only one type of bacterium (e.g., gram positive bacteria).

Newer antiseptics are based on the quaternary ammonium compounds (such as benzalkonium chloride and benzethonium chloride). "Quats" are longstanding and powerful antiseptics. Other common antiseptics include alcohols (e.g., ethyl or isopropyl **alcohol**), **hydrogen peroxide**, and phenol. Each is effective against a narrow range of bacterial infections, but none is effective against viruses. These antiseptics often are mixed to provide a wider range of antibacterial activity. They are applied externally on a cut or scrape to prevent infection and, when incorporated into mouthwash, can be gargled to kill bacteria that may cause a sore throat.

Resources

Books

Krasner, R.I. *The Microbial Challenge: Human-Microbe Interactions.* Washington: American Society for Microbiology Press, 2002.

Drug Facts and Comparisons, 56th ed. New York: Facts and Comparisons, 2002.

Periodicals

Purdy, C. "It's the Little Things That Count." *Current Health* 20 (March 1994): 20–22.

Brian Hoyle

Antlions

Antlions or doodlebugs are **insects** best known by their larvae, which have small, fat bodies with a huge sickle-shaped pair of mandibles. Antlions belong to the family Myrmeleonidae, of the order Neuroptera, which also includes the **lacewings**. Members of this order are named for the delicate venation on the wings of the adult, but most people are probably more familiar with the larval stage of antlions.

Some **species** of antlions simply chase down their **prey** of small insects, while others construct a pitfall trap, or **sand** trap, which they dig in the loose **soil**. This is done by pushing the sand away from the center of a **circle** while walking backwards. This trap, shaped like a funnel, can measure up to 2 in (5 cm) across with the antlion larva hidden at the bottom.

Ants and other insects stumble upon the antlion trap and lose their footing. The loose nature of the soil prevents them from regaining their balance as they slide down the side of the pit into the waiting jaws of the antlion. The prey is caught and injected with a paralyzing secretion. The body fluids are then sucked out and the empty exoskeleton is discarded. If the prey manages to regain its balance and footing, it is greeted with a shower of sand that is thrown by the antlion, causing it to lose its footing again, often with the same deadly result.

Adult antlions superficially resemble **damselflies**. However, compared to the damselfly, the antlion is a very feeble flyer, has a very complex wing venation, and has clubbed antennae almost a quarter inch (0.6 cm) long. Unfortunately, adult antlions are not easy to find, and little is known about their **behavior**. Adult antlions mate in the summer and the female lays her fertilized eggs in sandy soils, which are required by the larvae. Depending on species, from one to three years are spent as a larva that eventually pupates, usually in the spring, in a sand-encrusted silk cocoon at the bottom of the sand trap. One month later an adult antlion emerges from the pupal cocoon.

Ants

Ants are **insects** in the family Formicidae in the order Hymenoptera, which also includes **bees** and **wasps**. The body of ants is divided into three sections: head, thorax, and abdomen. The head bears two long, flexible antennae (for **touch** and chemical detection), two eyes, and a pair of powerful mandibles (jaws) for feeding and defense. Ants have three pairs of long legs that end with a claw. They are attached to the thorax, which is connected by a narrow petiole, or waist, to the segmented abdomen. At the tip of the abdomen are the reproductive organs and the stinging **organ** (in some **species**). Ants live in highly successful social communities called colonies, and are found worldwide in cool scrublands, hot deserts, inner cities, and tropical rain **forests**. Their nests are constructed underground or in tree-top **leaf** nests woven with silken thread.

Ants weigh 0.28–1.41 oz (1–5 mg), depending on the species. In 1994, 9,500 species of ants in 300 genera were recognized, and it is expected that many more species will be added to this total.

Mandibles are elongated, saw-toothed, blade-like pinchers that snap together sideways, allowing for the efficient capture of living **prey** and providing excellent defense against predators. Females of ground-dwelling species of ants secrete an antibiotic substance, which they smear throughout the nest, thus protecting the entire colony from the **fungi** and **bacteria** that thrive in damp, decaying vegetation.

Social structure, development, and behavior

Ants live in eusocial communal societies where, typically, members are clearly segregated into breeding and working castes. In the colony, several generations of adults reside together, and the young are fed, nurtured, and protected deep within the mound. A typical colony of the *Pheidole tepicana* comprises the queen, the males, and six castes of workers.

Mating, reproduction, and life span

Ants undergo complete metamorphosis—from egg, to larva, to pupa, to adult. Each ant colony begins with, and centers on, the queen, whose sole purpose is to reproduce. Although the queen may copulate with several males during her brief mating period, she never mates again. She stores sperm in an internal pouch, the spermatheca, near the tip of her abdomen, where sperm remain immobile until she opens a valve that allows them to enter her reproductive tract to fertilize the eggs.

The queen controls the sex of her offspring. Fertilized eggs produce females (either wingless workers seldom capable of reproduction, or reproductive virgin queens). Unfertilized eggs develop into winged males who do no work, and exist solely to fertilize a virgin queen. The queen produces myriads of workers by secreting a chemical that retards wing growth and ovary development in the female larvae. Virgin queens are produced only when there are sufficient workers to allow for the expansion of the colony.

Queens live long lives in comparison with their workers and are prolific breeders. A queen of *Lasius niger*, a common ant found in **Europe**, lived for 29 years in captivity, while the queen of the urban Pharaoh's ant, *Monomorium pharaonis*, lives for only three months. The queen of the leafcutter ant from **South America** produces 150 million workers during her 14-year life span.

The first phase of colony development is the founding stage, beginning with mating, when winged males and virgin queens leave the nest in massive swarms called nuptial flights, searching out a mate from another colony. In colonies with large populations, like that of the fire ant *Solenopsis,* hundreds of thousands of young queens take to the air in less than an hour, but only one or two individuals will survive long enough to reproduce. Most are taken by predators such as **birds**, **frogs**, **beetles**, **centipedes**, spiders, or by defensive workers of other ant colonies. A similar fate awaits the male ants, none of which survive after mating.

After mating, queen ants and male ants lose their wings. The queen scurries off in search of a site to start her new nest. If she survives, she digs a nest, lays eggs, and single-handedly raises her first brood that consists entirely of workers. In leafcutter ants, adults emerge 40–60 days after the eggs are laid. The young daughter ants feed, clean, and groom the queen ant. The workers enlarge the nest, excavate elaborate tunnel systems, and transport new eggs into special hatching chambers. Hatchling larvae are fed and cleaned, and pupated larvae in cocoons are protected until the young adults emerge to become workers themselves.

The colony now enters the ergonomic stage, a time entirely devoted to work and expansion. It may take a single season or five years before the colony is large enough to enter the reproductive stage, when the queen ant begins to produce virgin queens and males that leave the nest at mating time to begin the entire cycle anew.

In some species, a new queen founds a new colony alone; in others species, several queens do so together. Sometimes, groups of workers swarm from the nest with a young queen to help her establish her nest. In colonies with several already fertile queens, such as in the Costa Rican army ant *Eciton burchelli,* entire groups break away with their **individual** queens to establish individual colonies. In single queen colonies, such as those of the fire ant, the death of the queen means the death of the colony, as she leaves no successors. Colonies with multiple queens survive and thrive.

Labor management

Some species of ant develop a caste of big, strong, major workers (soldiers) responsible for milling (chewing and pulverizing hard seed food), storing liquid food, and defense. Workers gather and store food for the entire colony, lugging loads much larger than themselves back to the nest. The workers of *Myrmecocystus mimicus*, the honeypot ant of the southwestern United States, collect **nectar** from flowers, sweet moisture from fruit, and honeydew produced by sucking insects like **aphids**. The food is carried back to the nest, where it is regurgitated into the crop of storage worker ants, which become living storage barrels. When a hungry ant touches the head or mouth of a storage worker ant, the storage worker responds by regurgitating food for the hungry ant.

Worker ants remove all waste (such as body parts and feces) from the nest, or bury objects too large for removal. Different species use different methods of disposing of their dead. Many simply eat the dead. Others, such as the army ant (*Eciton*), carry the corpses out of the nest, while the fire ant (*Solenopsis invicta*) scatters the corpses about the nest's periphery. In some instances, sick and dying ants actually leave the nest to die.

Defense and offense

Ants employ diverse strategies to protect their colony, territory, or food. Ants are aggressive, often raiding other ant colonies, fighting to the death and snapping off limbs, heads, and body parts of enemies with their strong, sharp mandibles. Minor workers grab the enemy by the legs, pinning them down so majors can attack the body. In some species, soldier ants do the fighting, while the minor workers scurry to and from the battleground, dragging corpses of both enemy and kin back to the nest to feed the family. When moving colony sites, workers transport the queen, males, aged or ill workers, pupae, larvae, and eggs.

Communication

Ants secrete substances called **pheromones**, which are chemical messages detected by other ants through sense organs or the antennae. This process, called **chemoreception**, is the primary communication vehicle that facilitates mate attraction, kin, and non-kin recognition. It is also used to discriminate between egg, larva, and pupa, as warning signals, recruitment to defensive action or a new food source, the laying of odor trails from which workers or scouts find their way home or lead an entire colony to a new location, and delineation of territorial boundaries.

Chemoreception is supported by tactile (touch and feel), acoustic (**hearing** and vibration detection), and visual communication. Ants send tactile signals by touching and stroking each others' bodies with their antennae

KEY TERMS

Chemoreception—Detection of chemical substances which act as messengers.

Crop—Part of an ant's digestive tract that expands to form a sac in which liquid food is stored.

Eusocial—Truly social, with complex societal structures.

Mandibles—A pair of biting jaws in insects.

Milling—Chewing and pulverizing hard seed into a powdery texture.

Pheromones—Hormonal and chemical secretions.

Replete—Ants that receive regurgitated liquid food from many worker ants, storing it in the crop for future regurgitation back to other hungry ants.

Spermatheca—Oval pouch or sperm sac.

and forelegs. Ants produce high-pitched chirps known as stridulations by rubbing together specialized body parts on the abdomen called files and scrapers. Stridulations are sometimes heard, but most often felt, the vibrations being detected by sensitive receptors on the legs. The young queen stridulates frantically during mating season to announce a full sperm sac, deterring other would-be mates and allowing her to escape to begin nesting. Drumming and body-rapping are used primarily by tree-dwelling ants and carpenter ants, and involve banging the head or antenna on a hard surface, sending vibrational warning signals to nest mates. Some large-eyed species, such as *Gigantiops*, can see form and movement but **vision** in most ants is virtually nonexistent and the least important of all their communication senses.

Ants and the ecosystem

Earth-dwelling species of ants turn and enrich more **soil** than do earthworms; predatory species of ants control insect **pests** and spider populations, as well as **animal** litter by devouring rotting carcasses. Other species of ants spread **seeds**, thereby propagating valuable vegetation. Ants can also be serious pests; for example, leaf-cutter ants, which grow fungi gardens for food, also strip massive amounts of leaves and flowers. They haul the vegetation to their nests and pulverize it into a paste, which they feed to fungi that grows like **mold** on bread. Ant colonies are enormous: one nest of the Brazilian ant *Atta sexdens* housed about 8 million ants. A colony this size can strip as much vegetation as a cow in just one day, causing serious agricultural destruction.

Resources

Books

Holldobler, Bert and Edward O Wilson. *The Ants*. Cambridge: Harvard University Press, 1990.

Holldobler, Bert, and Edward O. Wilson. *Journey to the Ants*. Cambridge: Harvard University Press, 1994.

Williams, David F., ed. *Exotic Ants: Biology, Impact, and Control of Introduced Species*. Boulder, CO: Westview Press, 1994.

Marie L. Thompson

Anxiety

Anxiety is an unpleasant emotional state characterized by an often vague apprehension, uneasiness, or dread. Anxiety is often accompanied by physical sensations similar to those of fear such as perspiration, tightness of the chest, difficulty breathing or breathlessness, dry mouth, and headache. Unlike fear, in which the **individual** is usually aware of its cause, the cause of anxiety is often not clear.

Everyone experiences anxiety; it is a natural and healthy human response many theorists believe has evolved to warn us of impending dangers so that we might better cope with them. If the anxiety, however, seems to be excessive in strength or duration, or happens without sufficient objective reasons, it might be considered an unhealthy, possibly abnormal, response. There are numerous theories as to the causes and functions of anxiety. This entry will cover the four most extensive and influential theories: the existential, psychoanalytic, behavioral and **learning**, and cognitive.

Existential theorists generally distinguish between normal and neurotic anxiety. They believe normal anxiety is an unavoidable and natural part of being alive. It is the emotional accompaniment of the fear of death and of the immediate awareness of the meaninglessness of the world we live in. Anxiety is also felt on experiencing freedom and realizing we can create and define our lives through the choices we make. In this sense, anxiety is positive, showing us we are basically free to do whatever we choose. Neurotic anxiety is a blocking of normal anxiety which interferes with self-awareness. Rather than facing and dealing with the threat causing the normal anxiety, the individual cuts him or herself off from it.

Sigmund Freud, the Austrian physician who founded the highly influential theory and treatment method called **psychoanalysis**, distinguished three types of anxiety: reality, neurotic, and superego or moral. Reality anxiety is fear of real and possible dangers in the outside world.

Neurotic anxiety is fear of being punished by society for losing control of one's instincts, for instance by eating large amounts of food very rapidly, or openly expressing sexual desire. Moral or superego anxiety is fear of negative self-evaluation from the conscience or superego. The anxiety may be felt as guilt, and those with strong superegos may feel guilt or anxiety when they do (or even think of doing) something they were raised to believe was wrong. In Freudian theory, anxiety functions to warn individuals of impending danger, and it signals the ego to take actions to avoid or cope with the potential danger.

Learning and behavioral theories focus on how fears and anxieties can be learned through direct experience with a noxious **stimulus**, for example, touching a hot pan, or by indirect observation such as seeing someone else touching a hot pan and expressing **pain**. Most people learn to avoid the stimuli or situations that lead to anxiety, but taken too far, avoidance can be very limiting. For example, if someone avoided all tests or job interviews, he or she would never succeed in school or have many job options. Extreme avoidance can also lead to extreme behaviors such as those seen in some **phobias**, which are persistent, intense, irrational fears of a thing or situation with a strong urge to flee from the source of fear. For example, someone with a phobia of pigeons might have difficulty walking calmly down a city street when pigeons are nearby. In both instances, avoidance prevents the individual from learning that the original feared stimulus may not be dangerous.

Cognitive theories focus on the role of **cognition**, or thinking, in anxiety. They look at how interpretations or evaluations of situations affect reactions to the situations. This is based on evidence that internal mental statements or thoughts can dictate whether anxiety or other emotions are felt.

Some theorists believe anxiety plays a central role in most, if not all, mental disorders. Today, there are many classifications of anxiety disorders. Panic attacks, obsessive-compulsive disorder (OCD), post-traumatic **stress** disorder, social anxiety disorder, generalized anxiety disorder, phobias, and even certain **eating disorders** are all believed to be forms of anxiety dysfunction. Even clinical **depression** is believed to have an anxiety component. Clinical research has focused on finding ways to diagnose certain anxiety disorders so that therapists can accurately help their patients. Examples of the tools therapists use to diagnose anxiety disorders are the Social Phobia and Anxiety Inventory and the Social Interactions Anxiety Scale. Both are series of questions answered by the patient that help the therapist diagnose their condition.

There is evidence that some individuals may be biologically predisposed toward experiencing strong anxiety.

Knowledge of the chemical basis of anxiety and other psychological phenomena is rapidly increasing. Because the chemical basis of **brain** function is better understood, new medications to control inherited and acquired anxiety disorders are being developed. Paroxetine (Paxil), sertraline (Zoloft), and fluoxetine (Prozac) are examples of newer anti-anxiety medications now in wide usage.

Resources

Books

Kaplan, H.I., and B.J. Sadock. *Comprehensive Textbook of Psychiatry*. 6th ed. Baltimore: Williams and Wilkins, 1995.

Wolman, B.B., and G. Stricker, eds. *Anxiety and Related Disorders: A Handbook*. New York: John Wiley and Sons, 1994.

Periodicals

Golden, Frederic. "Mental Illness: Probing the Chemistry of the Brain." *Time* 157 (January 2001).

Hyman, S.E. "The Genetics of Mental Illness: Implications for Practice." *Bulletin of the World Health Organization* 78 (April 2000): 455-463.

Apes

Apes are a group of **primates** that includes **gorillas**, orang-utans, **chimpanzees**, and gibbons. These are the primate **species** that are the most closely related to humans. The hands, feet, and face of an ape are hairless, while the rest of its body is covered with coarse black, brown, or red hair. Apes share some characteristics that set them apart from other primates: they have an appendix, lack a tail, and their skeletal structures have certain features not found in the skeletons of other primates.

Gorilla

Gorillas (*Gorilla gorilla*) inhabit **forests** of Central **Africa** and are the largest and most powerful of all primates. Adult males stand 6 ft (1.8 m) upright (although this is an unnatural position for a gorilla) and weigh up to 450 lb (200 kg), while females are much smaller. Gorillas live to about 44 years old. Mature males (older than about 13 years), called silverbacks, are marked by a band of silver-gray hair on their back.

Gorillas live in small family groups of several females and their young, led by a dominant silverback male. The females comprise a harem for the silverback, who holds the sole mating rights among males in the troop. Female gorillas produce one infant after a gestation period of nine months. The large size and great strength of the silverback are advantages in competing with other males for leadership of the group and in defending against outside threats. Despite its ferocious image to some people, the gorilla is not an aggressive **animal**. Even in a clash between two adult males, most of the conflict consists of aggressive posturing, roaring, and chest-beating, rather than physical contact.

During the day these ground-living, vegetarian apes move slowly through the forest, selecting leaves, fruit, and stems from the vegetation as food. Their home range is about 9–14 sq mi (25–40 sq km). At night the family group sleeps in trees, resting on platform nests that they make each evening from branches; silverbacks usually **sleep** near the foot of the **tree**.

Gorilla numbers are declining rapidly and only about 50,000 remain in the wild. Other than humans, gorillas have no real predators, although leopards will occasionally take young individuals. Hunting, poaching (a live mountain gorilla can be worth $150,000), and **habitat** loss are causing gorilla populations to decline. The trade of gorillas has been banned by the countries where they occur and by the Convention on International Trade in Endangered Species of Wild Fauna and Flora (CITES), but they are nevertheless threatened by the illegal black market. The shrinking forest refuge of these great apes is being felled in order to accommodate the needs of the ever-expanding human population in Central Africa.

Orang-utan

The **orang-utan** (*Pongo pygmaeus*) is restricted to the rainforests of the islands of Sumatra and Borneo in Indonesia and Malaysia. The orang-utan is the largest living arboreal mammal. It spends most of the daylight hours moving slowly and deliberately through the forest canopy in search of food. Sixty **percent** of its diet consists of fruit, and the remainder is composed of young leaves and shoots, tree **bark**, mineral-rich **soil**, and **insects**. Orang-utans are long-lived, with many individuals reaching 50–60 years of age in the wild. These large, chestnut-colored, long-haired apes are endangered because of habitat destruction and illegal capture for the wild-animal trade.

Even though Indonesia and Malaysia have more than 400,000 sq mi (1,000,000 sq km) of **rainforest** habitat remaining, the rapid **rate** of **deforestation** threatens the continued existence of the wild orang-utan population, which is now estimated at about 25,000 individuals. The Indonesian and Malaysian governments and CITES have banned the local and international trading of orang-utans, but they are still threatened by the illegal market. In order to meet the demand for these apes as pets around the world, poachers kill mother orang-utans to secure their young. The mortality rate of these captured orphans is extremely high, with fewer than 20% of those smuggled arriving alive at their final destination. Some hope for the species rests in a global effort to manage a captive propagation program in zoos, although this is far less preferable to conserving them in their wild habitat.

Chimpanzee

The common chimpanzee (*Pan troglodytes*) is relatively widespread in the forested parts of West, Central, and East Africa. A closely related species, the pygmy chimpanzee or bonobo (*P. paniscus*), is restricted to swampy lowland forests of the Zaire **basin**. Despite their names, common chimpanzees are no longer common, and pygmy chimpanzees are no smaller than the other species.

Chimpanzees are partly arboreal and partly ground-dwelling. They feed in fruit trees by day, nest in other trees at night, and can move rapidly through treetops. On the ground, chimpanzees usually walk on all fours (this is called knuckle walking), since their arms are longer than their legs. Their hands have fully opposable thumbs and, although lacking a precision grip, they can manipulate objects dexterously. Chimpanzees make and use a variety of simple tools: they shape and strip "fishing sticks" from twigs to poke into termite mounds, and they chew the ends of shoots to fashion fly whisks. They also throw sticks and stones as offensive weapons and when they hunt and kill **monkeys**.

Chimpanzees live in small nomadic groups of 3–6 animals (common chimpanzee) or 6–15 animals (pygmy chimpanzee), which make up a larger community of 30–80 individuals that occupy a territory. Adult male chimpanzees cooperate in defending their territory against predators. Chimpanzee society consists of promiscuous, mixed-sex groups. Female common chimpanzees are sexually receptive for only a brief period in mid-month (estrous), while female pygmy chimpanzees are sexually receptive for most of the month. Ovulating females capable of being fertilized have swollen pink hindquarters and copulate with most of the males in the group. Female chimpanzees give **birth** to a single infant after a gestation period of about eight months.

Jane Goodall has studied common chimpanzees for almost 30 years in the Gombe Stream National Park of Tanzania. She has found that chimpanzee personalities are as variable as those of humans, that chimpanzees form alliances, have friendships, have personal dislikes, and run feuds. Chimpanzees also have a cultural tradition, that is, they pass learned **behavior** and skills from generation to generation. Chimpanzees have been taught complex sign language (the chimpanzee larynx will not allow **speech**), through which abstract ideas have been conveyed to people. These studies show that chimpanzees can develop a large vocabulary and that they can manipulate this vocabulary to frame original thoughts.

Humans share approximately 95% of their genes with chimpanzees, so only 1.6% of human DNA is responsible for all the differences between the two species. The DNA of gorillas differs 2.3% from that of chimpanzees, which means that the closest relatives of chimpanzees are humans, and not gorillas. The close relatedness of chimpanzees (and other apes) with humans is a key element of the ethical argument for a higher standard of care, and even granting of legal rights, for these animals in captivity. Further studies of chimpanzees will undoubtedly help us to better understand the origins of the social behavior and **evolution** of humans. Despite their status as close relatives of humans, both species of chimpanzees are threatened by the destruction of their forest habitat and by hunting and by capture for research. Both species of chimpanzees are considered endangered, and their international trade is closely regulated by CITES.

Gibbons

Gibbons (genus *Hylobates*) are the smallest members of the ape family. Gibbons are found in Southeast **Asia**, China, and India, and nine species are recognized. They spend most of their lives at the tops of trees in the jungle, eating leaves and fruit. They are extremely agile, swinging with their long arms on branches to move from tree to tree, and they often walk upright on tree branches. Gibbons are known for their loud calls and songs, which are used to announce their territory and warn away others. They are devoted parents, raising one or two offspring at a time and showing extraordinary affection in caring for them. Conservationists and biologists who have worked with gibbons describe them as extremely intelligent, sensitive, and affectionate.

Gibbons have long been hunted as food, for medical research, and for sale as pets and zoo specimens. A common method of collecting them is to shoot the mother and then capture the infant. The mortality rate in collecting and transporting gibbons to places where they can be sold is extremely high, and this coupled with the destruction of their jungle habitat has resulted in severe depletion of their numbers. Despite a ban on the international trade in gibbons (by CITES), illegal commerce, particularly of babies, continues in markets throughout Asia.

Resources

Books

Benirschke, K. *Primates: The Road to Self-Sustaining Populations*. New York: Springer-Verlag, 1986.

Goodall, Jane. *Through a Window: My Thirty Years With the Chimpanzees of Gombe*. Boston: Houghton Mifflin, 1990.
Grace, E.S. and R.D. Lawrence. *Apes*. San Francisco: Sierra Club Books for Children, 1995.
Rumbaugh, Duane M., and D.A. Shaw. *Intelligence of Apes and Other Rational Beings (Current Perspectives in Psychology)*. New Haven, CT: Yale University Press, 2003.
Schaller, G.B. *The Year of the Gorilla*. Chicago: University of Chicago Press, 1988.

Periodicals

"Profile: Ian Redmond: An 11th-Hour Rescue for Great Apes?" *Science* 297 no. 5590 (2002): 2203.
Rider, E.L. "Housing and Care of Monkeys and Apes in Laboratories." *Laboratory Animals* 36, no. 3 (2002): 221-242.
Speart, J. "Orang Odyssey." *Wildlife Conservation* 95 (1992): 18-25.

Eugene C. Beckham
Neil Cumberlidge
Lewis G. Regenstein

Apgar score

Apgar score is the assessment of a newborn baby's physical condition based on skin **color**, **heart rate**, response to stimulation, muscle tone, and respiratory effort. Each criteria is rated from **zero** to two with a total score of 10 signifying the best possible physical condition. The assessment determines the need for immediate emergency treatment, helps prevent unnecessary emergency intervention, and indicates possible **brain** damage. Because the score corresponds closely to an infant's life expectancy, it is used as a guideline to advise parents on their baby's chances of survival.

Dr. Virginia Apgar published her scoring system in 1953 during her tenure as professor of anesthesiology at Columbia-Presbyterian Medical Center, New York, where she was involved in the **birth** of more than 17,000 babies. She observed the need for a quick, accurate, scientific evaluation of the newborn, primarily to aid in diagnosing asphyxiation (suffocation) and to determine the need for resuscitation (aided breathing). The evaluations, made and recorded one, five, and ten minutes after birth, quickly became the standard by which modern medicine throughout the world measured the health of the newborn infant.

Apgar's name became an acronym for Appearance, Pulse, Grimace, Activity, and **Respiration**. Appearance scores two if the baby's skin is a healthy tone such as pink, one if extremities are bluish, and zero if the entire body is blue. Pulse (heart rate) scores two for higher than 100 per minute, one for below 100, and zero if absent.

Grimace scores two for an energetic cry (with or without the traditional slap on the bottom or soles of the feet), one for a slight wail, and zero for no response. Actively moving babies score two for muscle tone, one for some effort at movement, and zero if limp. Respiration scores two for strong efforts to breathe; one for irregular breathing, and zero for no effort. With a total five-minute score of seven to 10, the infant's chances of surviving the first month are almost 100%, approximately 80% with a score of four, and 50% with a score of zero to one.

In 1989, an article in *The Lancet* concluded the Apgar Score was outmoded in light of advanced diagnostic and treatment techniques. Magee Women's Hospital in Pittsburgh, Pennsylvania, the largest obstetrical services hospital in the United States, still uses the Apgar Score as an indicator of the newborn's chances of survival. However, immediate resuscitation needs are determined under the Neonatal Resuscitation Program, developed in 1986 by the American Council of Pediatrics and the American Heart Association, and whose guidelines are used across the United States and by modern medical centers throughout the world.

Aphasia

Aphasia is a disorder caused by damage to the areas of the **brain** that direct the ability to speak, interpret, and understand language. Usually, aphasia is caused by a head injury, a brain **tumor**, a **stroke**, or a serious **infection**.

In adults, one of the most common causes of aphasia is a cerebrovascular accident—a stroke. A stroke occurs when the **blood** and **oxygen** supply to the brain is blocked, either by a clogged blood vessel (cerebral **thrombosis**) or a burst blood vessel (cerebral hemorrhage). When an injury or stroke interferes with the blood and oxygen supply, the brain cells cut off from oxygen die.

The areas of the brain involved in communication and language—all located on the left side of the brain—include the auditory cortex, which sorts what is heard into categories that make sense; Wernicke's area, where words and word patterns are stored; and Broca's area, which receives information from Wernicke's area and sends signals to the tongue, lips, and jaw that translate brain messages into actual **speech**.

Because these areas of the brain control different language skills, the communication problems that occur depend on what parts of the brain are damaged. For example, if the Broca area is injured, one may understand what is said and be able to think of an appropriate response. But because the link between thought and the physical act of speaking is damaged, one has trouble coordinating lips, tongue, and jaw to form understandable words. Damage to the Broca area may also make it difficult to communicate in writing; one knows what to write but the connection between thought and hand movement to form words on **paper** has been damaged.

There are several different systems for classifying aphasias. Some broad areas include Wernicke's aphasia (difficulty understanding language because words spoken cannot be matched to words stored in the brain); conduction aphasia (a break in the fibers that connect the Wernicke and Broca areas of the brain; a person understands what is said, but can not repeat it); transcortical aphasia (repetition without understanding); and global aphasia (all language abilities are impaired because all portions of the brain related to language have been damaged).

Aphids

Aphids are **insects** in the order Homoptera, which are also known as **plant** lice. Some 3,800 **species** of aphids have been identified worldwide with 1,300 species occurring in **North America**, which includes some 80 species that are **pests** of **crops** and ornamental plants. Aphids have a distinctive pear-shaped body, and most are soft and green in **color**. The wings are transparent and are held in a tent-like position over the abdomen, which has a short tail, called a cauda. The legs are long and thin, and the antennae are thin and have six segments. Two tube-like structures, called cornicles, project from the fifth or sixth abdominal segments of aphids. The cornicles excrete a defensive chemical when the aphid is threatened.

Reproductive habits

Aphids have a complicated life cycle and reproduction habits that make them extremely adaptable to their host plants and environments. When aphid eggs that have overwintered on their host plants hatch in the spring, they produce females without wings. These females and are capable of reproducing asexually, a process called **parthenogenesis**. Several asexual generations of aphids may be produced during a growing season.

When it becomes necessary to move to another plant, females with wings are produced and move to another host plant. As winter approaches, both males and females are produced and their fertilized eggs again overwinter until the next spring. Sometimes winged females that produce asexually also migrate to new hosts.

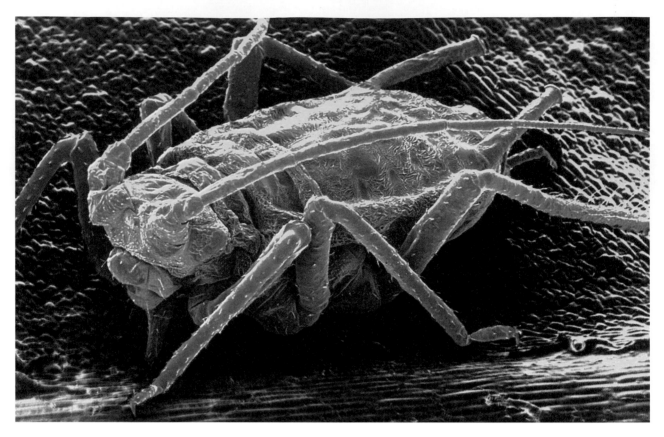

A scanning electron micrograph (SEM) of a green peach aphid (*Myzus persicae*) on the leaf of a Marvel of Peru (*Mirabilis jalapa*). This aphid winters on peach trees and migrates to other plants for the summer. It is an economic pest when it infests crops such as potatoes. The aphid is also a carrier of plant viruses, which are transmitted when it inserts its stylet into the veins of a leaf in search of sugar-carrying cells. The tubes on the aphid's back are siphons for sending pheromone signals. ©*Dr. Jeremy Burgess/Science Photo Library, National Audubon Society Collection/Photo Researchers, Inc. Reproduced with permission.*

The lack of wings among generations of aphids that have no need to migrate is seen as an adaptive advantage, since it helps keep them from being blown away in windy **weather**.

Ants and aphids

An intimate, symbiotic relationship exists between **ants** and aphids. They are often compared to cattle, with the ants acting as protectors and ranchers. What aphids have that ants want is something called honeydew, a sweet substance that is excreted by aphids through their anus and contains surplus sugar from the aphid's diet. Ants protect aphid eggs during the winter, and carry the newly hatched aphids to new host plants, where the aphids feed on the leaves and the ants get a supply of honeydew.

Because of their ability to reproduce rapidly and grow large colonies, their feeding on plants causes yellowing, stunting, mottling, browning, and curling of leaves, as well as inhibiting the ability of the host plant to produce crops. Infestations by aphids can cause plants to die, and the insects can carry other diseases, such as plant viruses, from one plant to another. Their saliva is also toxic to plant tissues. Among the biological controls of aphid infestations in agriculture and **horticulture** are **lacewings**, sometimes called "aphid lions," lady **beetles** or ladybird beetles (ladybugs), and syrphid **flies**. **Pesticides**, including diazinon, disyston, malathion, **nicotine** sulfate, and others, are also used to control aphids. On a smaller scale, some gardeners control aphids by simply washing them off with a spray of soapy **water**.

Resources

Books

Arnett, Ross H. *American Insects.* New York: CRC Publishing, 2000.

Hubbell, Sue. *Broadsides from the Other Orders: A Book of Bugs.* New York: Random House, 1993.

Imes, Rick. *The Practical Entomologist.* New York: Simon & Schuster, 1992.

McGavin, George C. *Bugs of the World.* Blandford Press, 1999.

Vita Richman
Neil Cumberlidge

KEY TERMS

Exoskeleton—A hard, shell-like structure that serves both to protect the vital organs of animals without an internal skeleton and to support their muscle systems.

Honeydew—A sweet substance excreted by aphids that ants need.

Parthenogenesis—Asexual reproduction without the fertilization of eggs.

Spiracles—Openings that lead to a system of tubes that supply air to insects.

Symbiosis—A biological relationship between two or more organisms that is mutually beneficial. The relationship is obligate, meaning that the partners cannot successfully live apart in nature.

Apnea *see* **Sleep disorders**

Apoptosis *see* **Cell death**

Apple *see* **Rose family (Rosaceae)**

Approximation

In **mathematics**, making an approximation is the act or process of finding a number acceptably close to an exact value; that number is then called an approximation or approximate value. Approximating has always been an important process in the experimental sciences and **engineering**, in part because it is impossible to make perfectly accurate measurements. Approximation also arises because some numbers can never be expressed completely in decimal notation. In these cases approximations are used. For example, irrational numbers, such as pi (π), are nonterminating, nonrepeating decimals. Every **irrational number** can be approximated by a **rational number**, simply by truncating it. Thus, π can be approximated by 3.14, or 3.1416, or 3.141593, and so on, until the desired **accuracy** is obtained.

Another application of the approximation process occurs in iterative procedures. **Iteration** is the process of solving equations by finding an approximate solution, then using that approximation to find successively better approximations, until a solution of adequate accuracy is found. Iterative methods, or formulas, exist for finding square roots, solving higher order polynomial equations, solving differential equations, and evaluating integrals.

The limiting process of making successively better approximations is also an important ingredient in defining some very important operations in mathematics. For instance, both the **derivative** and the definite **integral** come about as natural extensions of the approximation process. The derivative arises from the process of approximating the instantaneous **rate** of change of a **curve** by using short line segments. The shorter the segment the more accurate the approximation, until, in the **limit** that the length approaches **zero**, an exact value is reached. Similarly, the definite integral is the result of approximating the area under a curve using a series of rectangles. As the number of rectangles increases, the area of each **rectangle** decreases, and the sum of the areas becomes a better approximation of the total area under investigation. As in the case of the derivative, in the limit that the area of each rectangle approaches zero, the sum becomes an exact result.

Every **function** can be expressed as a series, the indicated sum of an infinite sequence. For instance, the sine function is equal to the sum: $\sin(x) = x - x^3/3! + x^5/5! - x^7/7!+...$. In this series the symbol (!) is read "factorial" and means to take the product of all positive **integers** up to and including the number preceding the symbol ($3! = 1 \times 2 \times 3$, and so on). Thus, the value of the sine function for any value of x can be approximated by keeping as many terms in the series as required to obtain the desired degree of accuracy. With the growing popularity of digital computers, the use of approximating procedures has become increasingly important. In fact, series like this one for the sine function are often the basis upon which handheld scientific calculators operate.

An approximation is often indicated by showing the limits within which the actual value will fall, such as 25 ±3, which means the actual value is in the **interval** from 22 to 28. Scientific notation is used to show the degree of approximation also. For example, 1.5×10^6 means that the approximation 1,500,000 has been measured to the nearest hundred thousand; the actual value is between 1,450,000 and 1,550,000. But 1.500×10^6 means 1,500,000 measured to the nearest thousand. The true value is between 1,499,500 and 1,500,500.

Apraxia

Apraxia is a disorder of **brain** function in which a person is unable to perform learned motor acts even though the physical ability exists and the desire to perform them is there. Brain damage to the parietal lobes, particularly in the dominant hemisphere, results in apraxia. Unlike paralysis, movements remain intact but

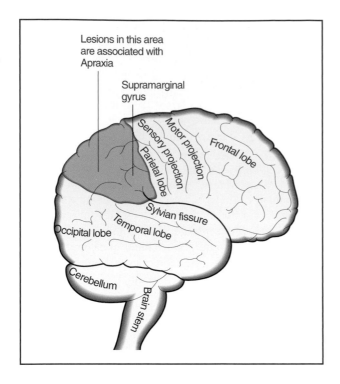

Lesions in this area are associated with Apraxia

Supramarginal gyrus

Motor projection

Sensory projection

Frontal lobe

Parietal lobe

Sylvian fissure

Occipital lobe

Temporal lobe

Cerebellum

Brain stem

The area of the brain associated with apraxia. *Illustration by Hans & Cassidy. Courtesy of Gale Group.*

the patient can no longer combine them sequentially to perform desired functions like dressing. Damage to the parietal lobes can arise from a variety of causes including metabolic diseases, **stroke**, and head injuries.

The German neurologist Hugo Liepmann (1863-1925) introduced the term apraxia in 1900 after observation of an impaired patient. Based on anatomic data, he suggested that planned or commanded actions are controlled not in the frontal lobe of the brain but in the parietal lobe of the brain's dominant hemisphere. Liepmann then postulated that damage to this portion of the brain prevents the activation of "motor programmes," learned sequences of activities that produce desired results on command. He also divided apraxia into three types: ideational, ideomotor, and kinetic.

Ideational apraxia, sometimes called object blindness, renders patients incapable of making appropriate use of familiar objects upon command, even though they can name the object and describe how to use it. Ideomotor apraxia is the inability to follow verbal commands or imitate an action, such as waving goodbye. The harder the patient tries, the more difficult execution becomes. Ironically, the patient often performs the gesture spontaneously or as an emotional response, like waving goodbye when a loved-one leaves. Kinetic apraxia refers to clumsiness in performing a skilled act that is not due to paralysis, muscle weakness, or sensory loss.

Other types of apraxia have been described since Liepmann's time. Apraxia of **speech** is the inability to program muscles used in speech, resulting in incorrect verbal output. It is frequently seen in conjunction with **aphasia**, the inability to select words and communicate via speech, writing, or signals. In dressing apraxia, patients can put clothes on but cannot program the appropriate movement sequences. Therefore, a coat goes on back-to-front, or socks over shoes. Facial apraxia leaves patients unable to move portions of their face upon command. They often, however, use other parts of their body to achieve a similar end. For example, when asked to blow out a match, the patient may step on it. Constructional apraxia refers to the inability to apply well-known and practiced skills to a new situation, like drawing a picture of a simple object from **memory**.

Although intense research has increased scientific understanding of this complex disorder, many mysteries remain.

Resources

Books

Brown, Jason W., *Aphasia, Apraxia and Agnosia, Clinical and Theoretical Aspects.* Springfield, IL: Charles C. Thomas, 1979.

Brown, Jason W., ed. *Agnosia and Apraxia: Selected Papers of Liepmann, Lange, and Potzl.* Mahwah, NJ: Lawrence Erlbaum Associates, 1988.

Hammond, Geoffrey R., ed. *Cerebral Control of Speech and Limb Movements.* Amsterdam: North-Holland, 1990.

Roy, E. A., ed. *Neuropsychological Studies of Apraxia and Related Disorders.* Amsterdam: North-Holland, 1985.

Williams, Moyra. *Brain Damage, Behavior, and the Mind.* New York: John Wiley & Sons, 1979.

Marie L. Thompson

Apricot *see* **Rose family (Rosaceae)**

Aqueduct

Aqueducts are structures used to carry **water** from a supply source to distant areas in need of water. The word aqueduct comes from two Latin words, *aqua* (water) and *ducere* (to lead). The first aqueducts were built as early as the tenth century B.C. by ancient communities. While primitive people lived very close to water, as people moved inland and away from direct water supplies, they created systems of water retrieval. Wells were dug to reach underground water supplies. Also, cisterns, underground collecting tanks, were used to store water. Eventually, **dams** were constructed to block water flow, allowing water **pressure** to increase, and run-off channels were constructed to guide water to specific regions. Early aqueducts redirected water by use of conduits (covered canals, or pipes, usually made of stone) that were often buried a few inches below ground for protection. Aqueducts were driven by the **force** of gravity pulling water downhill and extended for many miles. The use of aqueducts for drinking water, agriculture, and other uses is a part of Greek, Mexican, Roman, and Asian history. The city of ancient Rome had 11 active aqueducts traversing roughly 300 mi (485 km). Modern aqueducts use electrical power to elevate water that travels through many miles of pipes. Many modern pipes are deep beneath the ground and supply cities with water for personal and industrial use.

History

The first record of an aqueduct appeared in 691 B.C. in Assyria. This 34 mi (55 km) long aqueduct was simple, consisting of a single arch over one valley. At that time, Greeks were using wells to retrieve water from underground pools. Certain plants, such as fig trees, marked water sources because their roots grow in water. The first Greek aqueduct followed in 530 B.C. on the **island** of Samos. This aqueduct was built by an engineer named Eupalinus, who was told to supply the city with water by tunneling a pathway through a mountain. The Samos aqueduct extended for about 1 mi (1.6 km) underground, and had a diameter of 8 ft (2.4 m). These first aqueducts demonstrated an understanding of siphons and other basic hydraulic principles.

While ancient Roman aqueducts evolved into an extensive network of canals supplying the city, the first one, the Aqua Appia, was not built until 312 B.C. This aqueduct was a simple subterranean covered ditch. Roman aqueducts were usually built as open troughs, covered with a top, and then covered with **soil**. They were made from a variety of materials including masonry, lead, terra cotta, and **wood**. The Appia was about 50 ft (15 m) underground to make it inaccessible to Roman enemies on the city's outskirts. The Anio Vetus, built in 272 B.C., brought more water to the city, but both the Appia and the Vetus had sewer-like designs. The Aqua Marcia, built in 140 B.C., was made of stone and had lofty arches. The Aqua Tepula of 125 B.C. was made from poured **concrete**. Later Roman aqueducts were mainly built to meet the needs of the people or the desires of the rulers of the time. The average Roman aqueduct was 10–50 mi (16–80 km) long with a 7–15 sq ft (0.7–1.4 sq m) cross-section. Aqueducts were generally wide enough for a man to enter and clean.

Around the world, communities made advances in **irrigation** and water management. In the Mexican Tehuacan Valley, evidence of irrigation dates back to

The Pont du Gard, a Roman aqueduct in Nimes, France, dates from the first century A.D. *Photograph by John Moss. National Audubon Society Collection/Photo Researchers, Inc. Reproduced by permission.*

around 700 B.C. in the remains of the Purron Dam. The dam was used to direct water to domestic and crop regions for several hundred years. In the same valley, the Xiquila Aqueduct was built around A.D. 400. Early North American aqueducts include the Potomac aqueduct in Washington, DC. This aqueduct, which was built in 1830, extends over the Potomac River at the Key Bridge, which joins Northern Virginia and the Georgetown area of the city. It was built with support from eight piers and two stone abutments to carry water from the upper Potomac to the city.

Later aqueducts of the United States include the Colorado River Aqueduct that supplies Los Angeles and the Delaware River Aqueduct that carries water into New York. In addition, aqueducts carry water from northern to southern California. The southwestern region of the United States is particularly dry and requires water import. Water can be collected from aquifers (underground water reservoir), **rivers**, lakes, or man-made reservoirs.

Technology

While modern water pipes are much wider (20–30 ft or 6.1–9.1 m in diameter) and significantly longer (hundreds of miles long) than the first aqueducts, the hydraulic principles governing water carriage remain essentially the same. Water flows along gradients, and its **velocity** depends on a number of factors. Water flows more quickly along steeper gradients, but wear and tear on such pipes is greater, resulting in the need for more frequent repair. More gradual sloping pipes result in slower-flowing water with greater sludge deposits; hence, these pipes require more cleaning with less repair.

Water velocity along conduits is also greater in larger, smoother pipes. Pipes or canals that have rough surfaces disrupt water flow, slowing it down. In addition, larger diameter passageways provide less resistance, because a smaller percentage of the flowing water is retarded by the surface **friction** of the conduit. Thus, smaller diameter pipes slow the flow of water compared to larger diameter pipes.

The use of water to generate other forms of power is not new at all. The ancient Greeks and Romans both used water mills for work in such places as flour factories. In such mills, aqueducts were used to supply water on a relatively continuous basis. A modern application of water power is hydroelectric power.

Resources

Books

Hodge, A. *Roman Aqueducts and Water Supply.* London: Gerald Duckworth & Co., 1992.

Hodge, A., ed. *Future Currents in Aqueduct Studies.* Leeds: Francis Cairns Ltd., 1991.

Reisner, Marc. *Cadillac Desert: The American West and its Disappearing Water.* New York: Penguin Books Ltd., 1993.

Louise Dickerson

Aquifer

Although **groundwater** exists beneath most land surfaces, it is frequently limited in its availability to human users by local hydrogeologic conditions. Those portions of the water-bearing subsurface that are capable, within their hydrogeologic constraints, of yielding significant amounts of that **water** are called aquifers. Aquifers can store large amounts of water within pore spaces throughout the rock or sediment. These same voids are the means by which water is transported into and out of the aquifer, and ultimately, to the user. An aquifer might also be known as a groundwater reservoir. By contrast, aquicludes are capable of groundwater storage, but their internal structure is such that movement of the water through the rock is severely limited, making them unsatisfactory for water supply. The term aquitard is applied to a unit of rock that restricts the movement of

water through it but to a lesser degree than an aquiclude. In the extreme case, rock that neither transmits nor stores any water is called an aquifuge. This represents a rock that either contains no voids at all or the existing voids have no interconnection, thereby prohibiting both the storage and transmission of water.

The aforementioned terms are used in a relativistic manner and most have no strict definition associated with them. The hydrologic context of the aquifer, i.e., the relative abundance of water, will frequently be the determining factor as to which of the terms are used in defining a particular aquifer. For example, in an arid environment, the lack of a more productive unit might lead one to refer to a restrictive layer as an aquifer, while the same layer in an area of more plentiful groundwater and free-flowing rock types would be classified as an aquitard. This imprecise usage leads many hydrogeologists to define an aquifer as a subsurface zone capable of producing water in sufficient quantity to make it economically useful.

Aquifers can occur in a variety of forms. The classic representation is a uniform sandy **horizon** with well-sorted **sand** grains and an ample percentage of void **space** that permits substantial storage and transport. Aquifers can frequently be found in unconsolidated valley sediments, i.e., sand and gravels through which the water can readily flow. In more dense **rocks**, such as granite, groundwater might flow readily only through fractures. The most important factor in the classification of an aquifer is the presence of sufficient void **volume** and the degree to which the openings allow movement of the water.

Aquifers can be further classified on the basis of contact with the atmosphere. Water within an unconfined aquifer is in direct contact with the atmosphere through the open pore spaces of the material that overlies the aquifer. Water at the top of the saturated portion of an unconfined aquifer, known as the water table, is at **atmospheric pressure** and is free to move vertically in response to water level changes within the aquifer. When an impermeable material, such as a clay layer, separates the water within a water-bearing formation from the atmosphere, the aquifer is known as a confined aquifer. The overlying layer restricts the upward movement of the water within the aquifer and causes the **pressure** at the top of the aquifer to be at levels greater than atmospheric pressure.

See also Groundwater.

Arachnids

Arachnids (class Arachnida) form the second largest group of terrestrial **arthropods** (phylum Arthropoda) with the class Insecta being the most numerous. There are over 70,000 **species** of arachnids, which include such familiar creatures as scorpions, spiders, harvestmen or daddy longlegs, and ticks and **mites**, as well as the less common whip scorpions, pseudoscorpions, and **sun** spiders. Arachnids are members of the subphylum Chelicerata, which also includes the phylogenetically ancient **horseshoe crabs**.

Like other arthropods, arachnids have paired, jointed appendages, a hardened exoskeleton, a segmented body, and a well-developed head. They differ from other arthropods by the organization of their body into two main parts, the prosoma (equivalent to the head and thorax of **insects**) and the opisthosoma (or the abdomen). There are six pairs of appendages associated with the prosoma. The first pair are stabbing appendages near the mouth called chelicerae, used for grasping and cutting, and the second pair are called pedipalps or general purpose mouthparts. The last four pairs of appendages are the walking legs. Most arachnids are terrestrial and respire by means of book lungs, or by tracheae (air tubes from the outside to the tissues), or both. Most arachnids are terrestrial carnivorous predators. They feed by piercing the body of their **prey**, and then either directly ingesting its body fluids, or by releasing digestive secretions onto the outside of the prey to predigest the food before ingestion.

Scorpions (order Scorpiones) are distinguished by their large, pincer-like pedipalps, and a segmented abdomen consisting of a broad anterior part and a narrow posterior part ("tail"), which ends in a sharp pointed stinger. The latter contains a pair of poison **glands** whose ducts open at its tip. The venom is neurotoxic (attacking nerve functions) but, except in a handful of species, not potent enough to harm humans. Scorpions breathe by means of book lungs. Mating in scorpions is preceded by a complex courtship **behavior**. Newly-hatched scorpions are carried around by the mother on her back for one to two weeks. Scorpions are nocturnal, and feed mostly on insects. During the day they hide in crevices, under **bark**, and in other secluded places. They are distributed worldwide in tropical and subtropical regions.

In spiders (order Araneae) the abdomen is not segmented and it is separated from the cephalothorax by a narrow waist. The large and powerful chelicera of some spiders contains a poison gland at the base, while the tips serve as fangs to inject the poison into the prey. The pedipalps of spiders are long and leg-like. In male spiders, the pedipalps each contain a palpal **organ**, used to transfer sperm to the female. Some species of spider have only book lungs for **respiration**, while others have both book lungs and tracheae. Spiders possess silk producing glands whose secretion is drawn into fine threads by structures called spinnerets located on the lower side

A Cosmetid harvestman. *JLM Visuals. Reproduced by permission.*

of the abdomen. Different types of silk are produced and used for a variety of purposes, including orb-weaving, ensnaring prey, packaging sperm to be transferred to the female, and making egg sacs. Although all spiders produce silk, not all weave orbs. The **courtship** patterns of spiders are quite varied.

Spiders have a worldwide distribution and are ubiquitous, living in and around human habitations, in burrows in the ground, in **forests**, and even underwater. Spiders are predators, feeding mostly on insects. Despite their reputation as fearsome animals, spiders actually benefit humans by keeping some insect populations under control. The bite of the very few potentially harmful species of spiders is rarely fatal to humans.

Most mites and ticks (order Acari) are small, mites being microscopic, and ticks measuring only 0.2–1 in (5–25 mm) in length. The oval body of acarines consists of the fused cephalothorax and abdomen. The chelicerae and pedipalps are small, and form part of the feeding apparatus. Adult ticks and mites have four pairs of walking legs, but the larvae have only three pairs. Respiration in acarines is by tracheae. The ticks are mostly bloodsucking ectoparasites of **mammals**. In addition to sucking **blood**, and injecting poison into the host in the process, ticks transmit the agents of diseases such as Rocky Mountain spotted fever, **Lyme disease**, relapsing fever, **typhus**, and Texas cattle fever. A female tick needs to engorge by feeding on her host's blood before she can lay eggs. An engorged female is three or more times her original size. The feeding requires attachment to the host for days. The larval and nymphal stages likewise feed before they molt and progress to the next stage. Ticks have specialized sense organs which enable them to locate a host more than 25 ft (7.5 m) away.

Many mites are either ecto- or endoparasites of **birds** and mammals, feeding on the skin and underlying tissues. Many more mites are free living. Some, such as the chigger, are parasitic as larvae but free living in the nymph and adult stages. The ectoparasites live on the host's body surface, while the endoparasites excavate tunnels under the host's skin in which they live and reproduce. While some parasitic mites transmit **disease** organisms, many produce diseases such as scabies, mange, or cause an intense itch. Ticks and parasitic mites are clearly of great economic importance. Some free-living mites are also of considerable importance because they cause destruction of stored grain and other products. House dust mites cause allergies in many people.

Harvestmen (order Opiliones) look superficially like spiders but differ in many respects. Harvestmen lack a waist separating the abdomen from the cephalothorax and their abdomen is segmented. They can ingest solid food as well as fluids. They do not produce silk and are non-venomous. Harvestmen feed on insects and contribute to insect control, although they are less important in this respect than spiders.

Arapaima

The giant of **freshwater** fishes, the arapaima or pirarucu (*Arapaima gigas*) is a legend among **fish**. Weighing up to 440 lb (200 kg), this **species**, which has only been recorded in the **rivers** of Brazil and the Guianas, may reach a length of some 16.5 ft (5 m), although most specimens today are less than 10 ft (3 m) long. The origins of the arapaima, which belongs to the bony-tongued fishes (Osteoglossidae), date back to the Cretaceous period (some 65–135 million years ago); it is one of just five remaining species of this ancient group.

In appearance the arapaima exhibits many archaic characteristics, such as an asymmetrical tail fin and a swim bladder that also functions as an air-breathing **organ**. Only very young arapaima have functional gills; adult fish always come to the surface to breathe in **oxygen** and expel **carbon dioxide**, usually at intervals of 10–15 minutes.

Arapaima are active predators and often seek out fish in pools that are drying out, or backwaters where the fish are slowly being starved of oxygen. Captured **prey** are held in the jaws and the toothed tongue is then used to press and grind the hapless prey against the roof of the mouth. When prey is abundant, arapaima gorge themselves on fish, laying down rich **fat** deposits that will see them through the breeding season. Reproduction is marked by vivid changes in **color** and the pairing of adult fish. Both male and female participate in excavat-

ing a small hole in the substrate, usually in shallow **water** and well concealed by vegetation. When the nest site is completed, the female proceeds to lay her eggs, which are then fertilized by the attendant male. Both parents remain in the vicinity of the nest to ward off potential predators. Upon emerging from the eggs, the young fish remain close to one or both of the adults until such time as they are able to fend for themselves. Despite such lavish parental attention, the predation **rate** on young arapaima is thought to be considerable.

Prior to the nineteenth century, this species was seldom captured as neither the techniques nor the means of preserving such a large amount of food were available. In recent decades, however, the introduction of steel-tipped harpoons and gill nets have resulted in large catches of this giant fish. Smoking and salting techniques have also developed, enabling people to store larger quantities of fish for longer periods. As a result, this species has developed as one of the most important fish caught for food throughout the Amazon **basin**. No estimates of the population size of this species exist, but scientists have expressed concern for its future as a result of increasingly heavy hunting pressures in some regions.

Arc

An arc is a segment of a **curve**, most often a **circle**. In the strictest definition, an arc is a segment of a curve in a **plane**. Examples include segments of geometrical forms such as circles, ellipses, and parabolas, as well as irregular arcs defined by analytical functions.

Arcs of circles can be classified by size. A minor arc is one whose length is shorter than one-half of the circumference of a circle. A major arc is one whose length is longer than one half of the circumference of a circle. An arc whose length is exactly one-half of the circumference of the circle is simply called a semi-circle. The line connecting the endpoints of a major arc or minor arc is called a chord.

Angles subtended by circles can be classified by the location of the vertex. One important type of **angle** has the vertex located at the circumference. An angle whose vertex is at the center of the circle is called a central angle. Each specific central angle is subtended by only one arc, but each arc subtends infinitely many angles.

An arc of a circle can be measured by length along the circumference, or in terms of the angle subtended by the arc. A **theorem** of **geometry** states that the measure of the central angle of the circle is the measure of corresponding arc. If the arc lies on a circle of radius r and subtends a central angle (*L*A) measured in degrees, then the length of the arc is given by $b = 2\pi r(LA/360)$.

In the case of irregular arcs, the length can be determined using **calculus** and differential geometry.

Kristin Lewotsky

Arc lamp

Long before the incandescent electric **light** bulb was invented, **arc** lamps had given **birth** to the science of electric lighting. In the early 1800s, when the first large batteries were being built, researchers noticed that **electric current** would leap across a gap in a circuit, from one electrode to the other, creating a brilliant light. Sir Humphry Davy is credited with discovering this **electric arc** and inventing the first arc lamp, which used **carbon** electrodes. Yet the electric arc lamp remained a curiosity for decades. Many scientists gave public demonstrations of arc lighting, and the invention of automatic controls in the 1840s made it possible for arc lamps to be used in special applications such as lighthouses, theaters, and microscopes. But arc lamps still relied on expensive batteries or generators as their source of power.

Then a flurry of inventions brought arc lighting into widespread use. First came the development in 1871 of a relatively cheap source of **electricity**, the dynamo, a type of **generator** which produces direct current power. Public interest quickly reawakened, and people began installing arc lighting in factories, mills, and railway stations, any place light was required over a large, open **space**. France pioneered in this field, though Great Britain and America soon followed. The next step forward was the electric candle, a type of arc lamp invented in 1876 by Pavel Jablochkoff (1847-1894), a Russian engineer who later moved to Paris. This device, which could burn for two hours without adjustment, eliminated the need for expensive automatic controls. Although defects soon led to its downfall, this arc lamp greatly stim-

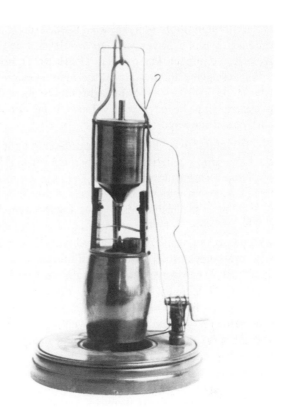

Arc lamp. *Corbis-Bettmann. Reproduced by permission.*

ulated development of electric lighting and increased the demand for better generating equipment.

By this time, American scientists were active in improving and installing arc lighting systems. In 1877, a dynamo invented earlier by William Wallace and American inventor and electrician Moses Farmer (1820-1893) was adapted for arc lighting by Wallace. This was probably the first commercial arc lamp made in the United States. Around the same time, American Charles Brush's arc lamp, which used magnets to move the electrodes, could be lit by remote control. He also invented a way to operate multiple arc lamps from a single dynamo, which greatly improved upon the European method. In 1879, Brush demonstrated his first streetlight system in Cleveland a success that led many other American and European cities to install Brush arc lighting. Finally, a team of two American electrical engineers, Edwin Houston (1847-1937) and Elihu Thomson, introduced an arc lighting system that wasted less electricity by maintaining a constant current. Two years later, in 1881, they patented automatic controls for the system.

At the turn of the century, after other improvements to arc lamps, spin-off technologies began to spring from the original concept. Scientists knew that electricity, when passed through certain gases at very low pressures, would discharge light, producing a glow instead of an arc. Although high voltage was needed to start the process, a much lower voltage would sustain it. An American engineer, Peter Hewitt (1861-1921), invented a starting device and developed the first discharge light, which used mercury vapor in a **glass** tube. Soon, higher **pressure** lamps began to be developed, using either mercury or **sodium** vapor.

In contrast to arc lamps and **incandescent light** bulbs, discharge lamps deliver nearly all their **energy** in the form of visible light or ultraviolet rays, rather than producing large amounts of useless **heat**. The **color** of the light varies depending on the gas. Mercury gives a bluish light, which can be corrected to look more natural by coating the tube with phosphors, while sodium vapor light is distinctly yellow. Both types provide excellent illumination for large areas such as roadways, shopping malls, parking lots, and exhibit halls. Mercury lamps are used where the quality of light is an aesthetic concern in a city's downtown area, for example, while sodium lights work well where visibility is more important than appearance. **Metal** halide lamps, a fairly recent development, produce a **spectrum** that is ideal for color **television** pickup, so they are often used in sport stadiums and athletic fields. Fluorescent lamps and neon lights are also variations of discharge lamps.

Meanwhile, the original arc lamp has come full circle. Ironically, today's extremely powerful versions make use of the lamp's heat, rather than its light. These high-tech arc lamps, which can simulate the heat of the **sun**, have proved useful in testing aerospace materials and hardening metal surfaces.

Archaebacteria

Life on **Earth** can divided into three large collections, or domains. These are the **Eubacteria** (or "true" **bacteria**), Eukaryota (the domain that humans belong to), and Archae. The members of this last domain are the archaebacteria.

Most archaebacteria (also called archae) look bacteria-like when viewed under the **microscope**. They have features that are quite different, however, from both bacteria and eukaryotic organisms. These differences led American microbiologist Carl Woese to propose in the 1970s that archaebacteria be classified in a separate domain of life. Indeed, because the organisms are truly separate from bacteria, Woese proposed that the designation archaebacteria be replaced by archae.

Archae are similar to eukaryotic organisms in that they lack a part of the **cell** wall called the peptidoglycan. Also, archae and eukaryotes share similarities in the way

that they make a new copy of their genetic material. However, archae are similar to bacteria in that their genetic material is not confined within a **membrane**, but instead is spread throughout the cell. Thus, archae represent a blend of bacteria and eukaryotes (some scientists call them the "missing link"), although generally they are more like eukaryotes than bacteria.

General characteristics

Archaebacteria are described as being obligate anaerobes; that is, they can only live in areas without **oxygen**. Their oxygen-free environments, and the observations that habitats of Archaebacteria can frequently be harsh (so harsh that bacteria and eukaryotic organisms such as humans cannot survive), supports the view that Archaebacteria were ones of the first life forms to evolve on Earth.

Archaebacteria are microscopic organisms with diameters ranging from 0.0002–0.0004 in (0.5–1.0 micrometer). The **volume** of their cells is only around one-thousandth that of a typical eukaryotic cell. They come in a variety of shapes, which can be characterized into three common forms. Spherical cells are called cocci, rod shaped cells are called bacilli, and **spiral** cells can either be vibrio (a short helix), spirillum (a long helix), or spirochete (a long, flexible helix). Archaebacteria, like all prokaryotes, have no membrane bound organelles. This means that the archaebacteria are without nuclei, mitochondria, endoplasmic reticula, lysosomes, Golgi complexes, or chloroplasts. The cells contain a thick cytoplasm that contains all of the molecules and compounds of **metabolism** and **nutrition**. Archaebacteria have a cell wall that contains no peptidoglycan. This rigid cell wall supports the cell, allowing an archaebacterium to maintain its shape, and protecting the cell from bursting when in a hypotonic environment. Because these organisms have no nucleus, the genetic material floats freely in the cytoplasm. The DNA consists of a single circular **molecule**. This molecule is tightly wound and compact, and if stretched out would be more than 1,000 times longer than the actual cell. Little or no protein is associated with the DNA. Plasmids may be present in the archaebacterial cell. These are small, circular pieces of DNA that can duplicate independent of the larger, genomic DNA circle. Plasmids often code for particular enzymes or for antibiotic resistance.

Groups of Archaebacteria

Archaebacteria can be divided into three groups. The first group is comprised of the methane producers (or methanogens). These archaebacteria live in environments without oxygen. Methanogens are widely distributed in nature. Habitats include swamps, deep-sea waters, **sewage treatment** facilities, and even in the stomachs of cows. Methanogens obtain their **energy** from the use of **carbon dioxide** and **hydrogen** gas.

The second group of Archaebacteria are known as the extreme halophiles. Halophile means "**salt** loving." Members of this second group live in areas with high salt **concentration**, such as the Dead Sea or the Great Salt Lake in Utah. In fact, some of the archaebacteria cannot tolerate a relatively unsalty environment such as seawater. Halophilic microbes produce a purple pigment called bacteriorhodopsin, which allows them to use sunlight as a source of photosynthetic energy, similar to plants.

The last group of archaebacteria lives in hot, acidic waters such as those found in **sulfur** springs or deep-sea thermal vents. These organisms are called the extreme thermophiles. Thermophilic means **heat** loving. They thrive at temperatures of 160°F (70°C) or higher and at **pH** levels of pH=1 or pH=2 (the same pH as concentrated **sulfuric acid**).

Archaebacteria reproduce asexually by a process called binary fission. In binary fission, the bacterial DNA replicates and the cell wall pinches off in the center of the cell. This divides the **organism** into two new cells, each with a copy of the circular DNA. This is a quick process, with some **species** dividing once every twenty minutes. **Sexual reproduction** is absent in the archaebacteria, although genetic material can be exchanged between cells by three different processes. In transformation, DNA fragments that have been released by one bacterium are taken up by another bacterium. In transduction, a bacterial phage (a **virus** that infects bacterial cells) transfers genetic material from one organism to another. In conjugation, two bacteria come together and exchange genetic material. These mechanisms give rise to genetic recombination, allowing for the continued **evolution** of the archaebacteria.

Archaebacteria are fundamentally important to the study of evolution and how life first appeared on Earth. The organisms are also proving to be useful and commercially important. For example, methanogens are used to dissolve components of sewage. The methane they give off can be harnessed as a source of power and fuel. Archaebacteria are also used to clean up environmental spills, particularly in harsher environments where most bacteria will fail to survive.

A thermophilic archaebacterium called *Thermus aquaticus* has revolutionized **molecular biology** and the **biotechnology** industry. This is because the cells contain an **enzyme** that both operates at a high **temperature** and is key to making genetic material. This enzyme has been

---

KEY TERMS

Chloroplast—Green organelle in higher plants and algae in which photosynthesis occurs.

Domain—One of the three primary divisions, Archae, Bacteria, or Eukaryota, of all living systems.

Enzyme—Biological molecule, usually a protein, which promotes a biochemical reaction but is not consumed by the reaction.

Eukaryote—A cell whose genetic material is carried on chromosomes inside a nucleus encased in a membrane. Eukaryotic cells also have organelles that perform specific metabolic tasks and are supported by a cytoskeleton which runs through the cytoplasm, giving the cell form and shape.

Golgi complex—Organelle in which newly synthesized polypeptide chains and lipids are modified and packaged.

Lysosome—The main organelle of digestion, with enzymes that can break down food into nutrients.

Mitochondria—An organelle that specializes in ATP formation, the "powerhouse" of the cell.

Nucleus—A membrane-bound organelle in a eukaryote that isolates and organizes the DNA.

Organelle—An internal, membrane-bound sac or compartment that has a specific, specialized metabolic function.

---

harnessed as the basis for a technique called the polymerase chain reaction (**PCR**). PCR is now one of the bedrocks of molecular **biology**.

See also Evolution, divergent.

Resources

Books

Howland, J.L. *The Surprising Archaea.* New York: Oxford University Press, 2000.

Periodicals

Doolittle, W.F. "What are the Archaebacteria and Why are They Important?" *Biochemical Society Symposium* 58 (1992): 1–6.

Woese, C.R. "Bacterial Evolution." *Microbiological Reviews* 51 (1987): 221–271.

Woese, C.R., O. Kandler, and M.L. Wheelis. "Towards a Natural System of Organisms: Proposal for the Domains Archae, Bacteria, and Eucaya." *Proceedings of the National Academy of Sciences USA* 87 (1990): 4576–4579.

Brian Hoyle

Archaeoastronomy

Archaeoastronomy is the study of prescientific peoples' relation to the sky as part of their natural environment. As a formal investigation, the field of archaeoastronomy is relatively young, having begun only in the 1960s. It is often known as cultural **astronomy** to indicate the multidisciplinary breadth of the field and its emphasis on cultural practices and issues rather than on the "correctness" of ancient observations. Archaeoastronomers are concerned to know what observations were made by an ancient society, who made them, and how those observations were integrated into the society's political life, agricultural or hunting-gathering practices, and civic and religious customs. Thus, the tools of modern archaeoastronomers are as likely to be those of art history, sociology, or linguistics as those of the quantitative sciences, such as computer-processing algorithms, large databases, and statistical inference.

Cosmology

Most prescientific peoples developed a **cosmology** that explained human existence as seamlessly interwoven into the workings of the universe. This relationship of the part to the whole was usually expressed through symbols and metaphors. A simple, almost universal cosmological principle was captured in the idea of mirroring: events and powers in the sky mirrored those on **Earth**; the earth was but a microcosm of the sky. In virtually all Northern Hemisphere societies, for example, earthly dwellings (the tepee, yurt, or igloo) were seen as particularized representations of the larger dwelling that arched high overhead in the heavens to create the celestial vault and which rotated around the Pole Star. An actual pole of **rotation**, extending from Earth to heaven, was a strong element of native North American cosmology. In Inuit cosmology, the superior plane (the mythological equivalent of the sky) was known as the Land Above. Other cosmologies have figured the universe as an endlessly folded ribbon, with Earth in the center fold; as a set of nested boxes; or as a series of interlocking spheres.

Prescientific societies held the celestial bodies in great reverence, yet were also on an intimate footing with them. Ancient peoples regarded the sky as inhabited by Sky People, deities, departed ancestors, or simply forces. The Sky People or powers were thought to impose order on chaotic human affairs. At the same time, the sky powers could be solicited and manipulated to serve human goals. Their authority could be invoked to justify the actions of a chief priest or ruler. A **moon** associated with important periods in the agricultural or hunting cycle could be honored to ensure better food supplies. A desire to place the sky powers in the service of the human agenda may have been the impetus that led prescientific societies to take up regular observations of the skies—in other words, astronomy.

Early observatories

Written records are missing for many prescientific societies as they turned from noting a single celestial event to making the kinds of repeated observations that could be applied predictively to events in their own lives, such as harvesting or knowing when to expect newborns in their herds. Many ancient peoples did, however, leave physical signs of their observing activities. Among the most intriguing are the sites that, to a modern **eye**, seemingly could have been used as very early observatories.

Between about 3500 and 1500 B.C., Bronze age builders in Britain and the northwestern portion of France known as Brittany erected, or marked in the existing landscape, thousands of sites that invite speculation about astronomical use. The most fundamental arrangement consisted of a natural indicator on the horizon, such as a notch in a mountain (the foresight), which was aligned with another, manmade marking, such as a standing post or stone, or a hollowed-out depression in a rock (the backsight). Because distances of up to 28 mi (45 km) have been measured between foresight and backsight, these common configurations have become known as long alignments. Statistical studies show that few long alignments could have been used to establish the date of a major celestial event, such as a **solstice**, with certainty. Even a rough **approximation**, however, would have sufficed for ceremonial reasons such as **sun** worship.

The Neolithic builders are better known for erecting stone circles, such as Stonehenge. (A henge is an earthen mound surrounded by a low bank and a ditch; wooden or stone pillars may be arranged on the top.) Astronomical opinion is divided over the uses of Stonehenge, but the evidence for its primary use as an observatory is considered weak. Stonehenge was built in three phases over a period of about 400 years, beginning around 1700 B.C., and is one of many circles built during this period. It stands on flat Salisbury Plain, in southern England. Because the horizon lacks distinctive features, short sight lines may have been incorporated into the placement of the stones. A ritual figure such as a priest or priestess silhouetted against the rising moon, and framed between two megaliths, would also have been an impressive sight.

Astronomers argue that Stonehenge could have been used to observe the winter solstice, the time when the rising of the Sun is farthest south, and the extreme rising and setting positions of the Moon. Its primary purpose seems to have been ceremonial. That a ceremonial structure might have later evolved into an observatory in prehistoric Britain has important implications for social organization, for it suggests that an educated, elite class existed to make the observations and supervise the construction and repair of sites. Those educated observers in turn might have been the forerunners of the Druids of the Iron Age in Britain.

Culturally dissimilar groups in the American Southwest and California appear to have followed observing practices like those used in prehistoric Britain. Oral histories taken from nineteenth century Pueblan informants in the Four Corners region indicate that both horizon observations (that is, long alignments) and wall **calendars** were used. The position of the Pueblan observer was not marked by a standing stone or gouge mark or wall painting; rather, it appeared to be esoteric knowledge held by the priest-astronomer, who simply walked to the same spot every time that observations were to be made. Wall calendars were created as sunlight penetrated an opening in a house or residential **cave** to fall on the opposite wall. With the use of both horizon and wall calendars, the Pueblans could track the motions of the Sun, Moon, and stars and also events that occurred in the four sacred quadrants of the sky. Similar horizon and wall calendars were used by many of the estimated 300 tribes living in California before contact with Hispanic traders and explorers, which occurred in the 1760s and 1770s. Some precontact California tribes are thought to have had two calendars: a secular one, known to all, and a secret calendar to guide the timing of sacred rituals.

Simple observing stations that made use of existing sight lines and horizon marks stand at one end of the spectrum of early observatories. At the other end is elite, corporate architecture, such as is found in the northern Yucatan site of Chichn Itz. There, the highly elaborated architecture, with its steeply ascending steps and ornately carved and painted reliefs, reflects a complex union of political power and astronomical knowledge.

Chichn Itz was built over two periods that lasted, in aggregate, from about A.D. 700 to 1263. Its people were both numerate and literate, creating written works that detailed their astronomical culture. The Maya also had a warrior class, waged war regularly, took captives as slaves, and practiced ritual human sacrifice and bloodletting.

The Mayan calendar and Mayan life were dominated by the Sun and **Venus**, whose astrophysical activities are related to each other in a 5:8 **ratio** considered sacred by the Maya. The Sun was associated with warfare; Venus, a fearful power, was associated with warfare, sacrifice, fertility, rain, and maize. In the 1960s and later, glyphs from Mayan writings were interpreted as showing that raids were undertaken during important Venus sta-

tions, such as its first appearance as the Morning Star or the Evening Star. These raids have come to be called star war events. The Caracol, a building probably designed as an observatory, and several other important ceremonial buildings at Chichn Itz, such as the Great Ball Court, the Upper Temple of the Jaguars, and the Temple of the Warriors, are fairly precisely aligned to face significant Sun and Venus positions.

Mayan interest in genealogy made calendrics important. Their basic calendar consisted of two cycles, one of 260 named days and one a year of 365 days, which ran concurrently. Fifty-two years of 365 days each formed the "Calendar Round." The Maya developed a system for uniquely identifying every one of the 18,980 days in the Calendar Round. A table in the Dresden Codex, one of three Mayan manuscripts to have survived the Spanish conquest, indicates ability to predict solar and lunar **eclipses**, as well as the behavior of certain other celestial bodies. At the time of the conquest, these predictions may have been accurate to within a day, rather than to within an hour or minute, as was then possible in **Europe** with the aid of advanced instrumentation. The Maya, however, did not have the concept of an hour.

See also Seasons.

Resources

Books

Ruggles, Clive L. N., and Nicholas J. Saunders, eds. *Astronomies and Cultures*. Chicago: University Press of Chicago, 1993.

Ruggles, Clive L. N., ed. *Records in Stone: Papers in Memory of Alexander Thom*. Cambridge: Cambridge University Press, 1988.

Smolin, Lee. *The Life of the Cosmos*. Oxford: Oxford University Press, 1999.

Zhenoao Xu, Yaotiao Jiang, and David W. Pankenier. *East-Asian Archaeoastronomy: Historical Records of Astronomical Observations of China, Japan, and Korea*. Philadelphia: Taylor & Francis, 2000.

Periodicals

Belmonte, J. A. "On the Orientation of Old Kingdom Egyptian Pyramids." *Archaeoastronomy* 26 (2001): S1-S20

Esteban, C. "Orientations of Pre-Islamic Temples in Northwest Africa." *Archaeoastronomy*. 26 (2001): S65-S84.

Henriksson, G. "Archaeoastronomy: New Trends in the Field, with Methods and Result." *Journal of Radioanalytical and Nuclear Chemistry*. 247, no. 3 (2001): 609-619.

Krupp, E. C., ed. "Archaeoastronomy and the Roots of Science." *AAAS Selected Symposium 71*. Boulder, CO: Westview Press, 1984.

Milbrath, Susan. "Astronomical Images and Orientations in the Architecture of Chichn Itz." *New Directions in American Astronomy. Proceedings of the 46th International Congress of Americanists*. Oxford, England: B.A.R., 1988.

Ruggles, Clive L. N., ed. *Archaeoastronomy in the 1990s. Papers Derived From the Third "Oxford" International Symposium on Archaeoastronomy* Loughborough, U.K.: Group D Publications Ltd., 1993.

Zeilik, Michael. "Astronomy and Ritual: The Rhythm of the Sacred Calendar in the U.S. Southwest." *New Directions in American Astronomy. Proceedings of the 46th International Congress of Americanists* Oxford, England: B.A.R., 1988.

Marjorie Pannell

Archaeogenetics

By applying modern **genetics** to population studies, **archaeology**, and anthropology, scientists are forming a new interpretation of prehistoric migrations. The initial peopling of **Europe**, **Asia**, and the Americas is usually explained by basic theories that appeal to reason. For example, scientists consider that groups of prehistoric peoples would periodically migrate into **North America** via a land bridge over the shortest span of **ocean** separating the **continent** from Asia as plausible. Answering questions such as when, and for how long, is more problematic. Archaeological field research occasionally yields new breakthroughs, but the work is painstaking and site discovery can often be a matter of chance. Scientists now address the issue in the lab rather than in the field.

Despite the density of various nations and ethnic groups in Europe, the human population of the continent is the least genetically diverse. Current anthropological interpretations of the peopling of Europe assert that the advance of agriculture was the driving force behind the **migration** into the region. Ten-thousand years ago, Neolithic farmers who migrated along the Mediterranean coast and then up through northern Europe were widely thought to have settled the land with their **livestock** and readily established farms. It was assumed that this land grab and population explosion effectively drove more primitive hunters and gathers out of the region.

Current genetic research offers a counter-thesis. Radically reinterpreting the accepted theories of a great population dispersal in Neolithic Europe, archaeogenetics suggests that Europeans are rooted in much earlier—and more primitive—peoples. The descendants of most Europeans were most likely ice age stalwarts, grouped in small, migratory hunter and gather bands. An estimated 80% of modern Europeans can potentially trace their ancestry back to one of seven female or ten male founder lineages. Such evidence suggests that not only did Neolithic farmers fail to immediately drive more primitive populations from the land, they most likely coexisted and interbred.

Archaeology and genetics can sometimes make uncomfortable neighbors—especially when more breakthrough discoveries are achieved in the lab than in the field. Over the past decade, Y **chromosome** research helped to narrow down possibilities for Europe's founding lineages. Y chromosome research works on one of the most basic concepts of genetics: the segregation of parental genes in offspring. The Y chromosome, which determinations the male sex in offspring, is usually passed on from father to son without any form of recombination. This segregation does not mean the **gene** is completely unscathed; as **evolution** dictates, the gene is imprinted by very minute mutations. Only 10 such genetic "tags" were observed, but were found in over 90% of the specimen.

Matrilineal studies yielded similar results. Mitochondria DNA (mDNA), passed on from mother to child, is analyzed in a similar method as its male chromosome counterpart. However, mDNA research has an added benefit. Since mDNA has an established value for **mutation** rates, it is useful in estimating the **time** scale of population change. Just as their cohorts predicted in their Y chromosome studies, mDNA evidence determined that the founder lineage for over three-fourths of Europe was in the Paleolithic era.

As with any evidence that purports to shatter existing knowledge in a field, use of both Y chromosome and mDNA has drawn many questions. Some critics worry that the data surrounding linkages with modern populations is based upon inadequate sampling. Still others doubt that mDNA is an effective standard of chronological dating. Regardless, both forms of research are making pioneering discoveries in archaeogenetics and **biodiversity**.

Beyond Europe, archaeogeneticists and physical anthropologists (people who study hominid fossils)

Traditional Aboriginal tribal ceremony. © *Penny Tweedie/Corbis. Reproduced by permission.*

search for increasingly primordial roots of modern populations. In 1987, scientists and mathematicians identified a theoretical 140,000–280,000 year-old source presently upheld as "Mitochondrial Eve." Though such a source is the most recent common ancestor of all humans living on **Earth** today, "Eve" less represents an actual human than a mathematical concept that illustrates the transfer of mDNA along a matrilineal line. As Earth's population changes over time, so too will the theoretical "Eve" mitochondrial donor. Thus, the "Mitochondrial Eve" is a crucial concept in the discussion of the genetic relationship of humans presently alive, but it is not the origin of the human **species** or a tangible rung in the ladder of evolution.

Studies similar to the research on the population prehistory of Europe are currently being conducted in Asia and the Americas. A skeleton found in Washington state, now known as "Kennewick Man," has yielded sever archaeogenetic clues about the peopling of North America, convincing some scientists to revisit long established population theories. However, the remains, their availability for scientific testing, and preliminary scientific studies on "Kennewick Man" remain hotly contested.

See also DNA technology.

Resources

Books

Renfrew, Colin. Katie Boyle, eds. *Archaeogenetics: DNA and the Population Prehistory of Europe.* Oakville, CT: David Brown Book Company, 2001.

Archaeology

The term archaeology refers, in part, to the study of human culture and of cultural changes that occur over **time**. In practice, archaeologists attempt to logically reconstruct human activities of the past by systematically recovering and examining artifacts or objects of human origin. However, archaeology is a multi-faceted scientific pursuit, and includes various specialized disciplines and subfields of study. Depending on the specific field of interest, archaeological artifacts can encompass anything from ancient Greek pottery vessels to disposable plastic bottles at modern dump sites. Thus, studies in archaeology can extend from the advent of human prehistory to the most recent of modern times. The two most common areas of study in archaeological research today, particularly in the United States, are prehistoric and historic archaeology.

Background

Although there has been a natural interest in the collection of antiquities for many hundreds of years, controlled scientific excavation and the systematic study of artifacts and the cultures who made them was not widely practiced until early in the twentieth century. Prior to that time, most archaeology consisted of randomly collecting artifacts that could be found lying on the surface of sites or in caves. In **Europe** and the Middle East, well-marked tombs and other ancient structures provided visual clues that valuable antiquities might be found hidden nearby.

Because most of these early expeditions were financed by private individuals and wealthy collectors, broken artifacts were often left behind, because only intact and highly-crafted items were thought to have any value. Consequently, early theories regarding the cultures from which artifacts originated were little more than speculation.

One of the first Americans to practice many of the techniques used in modern archaeology was Thomas Jefferson (1743-1826). Before serving his term as president of the United States, Jefferson directed systematic excavations in Virginia of prehistoric Indian "mounds" or earthworks that resemble low pyramids made from **soil**. His published report by the American Philosophical Society in 1799 marked the beginning of a new era in archaeological studies.

In addition to his thorough examination and recording of artifacts, Jefferson was perhaps the first researcher to observe and note a phenomenon known as stratigraphy, where soil and artifacts are deposited and layered one above the other like the skins of an onion. Jefferson's observation of site **stratigraphy** is still in use today as a basic field technique in determining the age and complexity of archaeological sites.

During the late nineteenth century, academic archaeology as a university subject sprang from a branch of anthropology, one of the social sciences that mainly concentrates on understanding the cultural traditions and activities of non-literate peoples.

American anthropologists of the time discovered that a better way to understand the social structure of living Native Americans was to try to reconstruct their prehistoric lifestyles. At the same time, archaeologists found that studies of contemporary Native Americans help in the interpretation of prehistoric Indian cultures, of which there are no written records. Thus, modern ar-

chaeology evolved as a result of the mutual effort between these two separate fields of academic discipline.

In addition to the many important discoveries made by researchers over the past century, perhaps the most sweeping event in the history of archaeology was the development of radiocarbon or C-14 dating in the late 1940s by Willard F. Libby.

Instead of relying solely on theories and hypotheses to date a site, scientists could derive, through laboratory analysis, highly accurate dates from small samples of organic material such as **wood** or bone. In many cases, radiocarbon dating proved or disproved theories regarding the postulated ages of important archaeological finds.

Today, archaeologists use a variety of techniques to unlock ancient mysteries. Some of these newly-developed techniques even include the use of **space** technology. The National Aeronautics and Space Administration's (NASA) ongoing project, Mission to Planet Earth, is one example. The primary objective of the project is to **map** global environmental changes by using specially designed SIR-C/X-SAR (Spaceborne Imaging Radar C/X-Band Synthetic Aperture Radar). However, a recent Middle East mission conducted by the **space shuttle** *Endeavor* resulted in several radar photographs of a series of ancient **desert** roadways leading to a central location. Field investigations, currently underway, have revealed that the roadways lead to the 5,000-year-old lost city of Ubar in southern Oman on the Arab Peninsula.

Disciplines

Over the past several decades, as new laboratory technologies have become available, so have the number of disciplines or fields of study that are linked to archaeology. Although they may seem unrelated, each is important, in that they contribute to the growing body of knowledge about humankind.

Physical anthropology, **ethnobotany**, DNA analysis, x-ray emission microanalysis, and **palynology** are but a few of the dozens of specialized areas of study. However, the individual field archaeologist tends to focus his or her research on a specific culture and/or era in time.

Prehistoric archaeology

Everything that is currently known about human prehistory is derived through the excavation and study of ancient materials.

Prehistoric archaeology encompasses the study of humankind prior to the advent of written languages or written history. In effect, the job of the prehistoric archaeologist is to discover and write about the histories of ancient peoples who had no written history of their own.

However, prehistory lasted for different times in different parts of the world. In Europe, primitive written languages made their appearance around 2500 B.C., thus technically ending Old World prehistory. On the other hand, American prehistory came to a close only 500 years ago when explorers such as Christopher Columbus (c. 1451-1506) visited the New World for the first time. By returning to their homelands with written reports of what they had seen, those explorers ushered in the early historic era of the Western Hemisphere.

Historic archaeology

As opposed to prehistoric archaeological studies, historic archaeology is a discipline that focuses on a more detailed understanding of the recent past. Typically, any site or building over 50 years old but younger than the regional area's prehistory is considered historic. For example, Civil War battlefields are historic landmarks. Because most historic sites and buildings were used during times when records were kept, the historic archaeologist can review written documentation pertaining to that time and place as part of the overall study.

Artifactual material discovered during an historic archaeological excavation adds specific detailed information to the knowledge already gathered from existing historic documents.

A relatively new sub-discipline of historic archaeology, known as urban archaeology, attempts to quantify our current culture and cultural trends by examining material found at modern dump sites. Urban archaeology has been primarily used as a teaching aid in helping students of cultural anthropology and archaeology develop interpretive skills.

Classical archaeology

The classical archaeologist is perhaps the most popular public stereotype—the "Indiana Jones" movie character, for example, is based on the fantastic exploits of a classical archaeologist. The roots of classical archaeology can be traced to the European fascination with Biblical studies and ancient scholarship. Classical archaeologists focus on monumental art, architecture, and ancient history. Greek mythology, Chinese dynasties, and all ancient civilizations are the domain of the classical archaeologist.

This field is perhaps one of the most complex areas of the study of human culture, for it must utilize a combination of both prehistoric and historic archaeological techniques. In addition, classical studies often employ what is referred to as underwater archaeology to recover the cargo of sunken ancient sailing vessels that carried trade goods from port to port.

Archaeological excavation, Eldon Pueblo, Coconino National Forest, Arizona. *Photograph by Tom Bean. Stock Market. Reproduced by permission.*

Cultural resource management

Beginning in the 1940s, a number of state and federal laws have been enacted in the United States to protect archaeological resources from potential destruction that could be caused by governmental developments such as highway construction. In the 1970s, additional legislation extended these laws to cover development projects in the private sector. Other countries have also passed similar laws protecting their antiquities. As a result, archaeology has moved from a purely academic study into the realm of private enterprise, spawning hundreds of consulting firms that specialize in contract archaeology or what is commonly known as cultural resource management (CRM).

CRM work has become a major economic industry in the United States, generating an estimated 250 million dollars a year in annual business through government and private contracts. Most archaeological studies conducted in the United States today are a direct result of compliance with antiquities laws.

Construction projects such as the building of **dams**, highways, power lines, and housing developments are a recognized and accepted pattern of human growth. It is also recognized that these activities sometimes have a detrimental affect on the evidences of human history. Both historic buildings and prehistoric sites are often encountered during the course of new construction planning. However, various steps can be taken to insure that such sites are acknowledged and investigated prior to any activities that might damage them.

The two most common measures that can be taken when a site lies in the path of development include (1) preservation by simple avoidance or (2) data recovery. Preservation by avoidance could mean re-routing a highway around an archaeological site, or moving the placement of a planned building to a different location. On the other hand, when avoidance is not possible, then data recovery by excavation becomes the next alternative.

In the case of historic structures, for example, data recovery usually includes a thorough recordation of the building. This might include photographs, scale interior and exterior drawings, and historic document searches to determine who designed and built the structure or who might have lived there. For prehistoric resources, sampling the contents of the subsurface **deposit** by excavation is perhaps the only method by which to determine the significance of a site.

Field methods

Archaeological survey/field reconnaissance

Prior to any archaeological excavation, whether it be prehistoric or historic, a survey or field reconnaissance must be conducted. In general terms, a survey systematically inspects the surface of a site to determine whether or not it is significant and may warrant further investigation. Although the determination of significance is somewhat subjective, most researchers agree that it hinges on a site's potential to yield useful information. Significance may include the ability of a site to produce new, previously unknown information, or additional data to add to what is currently known about the site's original inhabitants and their culture.

Surveys are also conducted on undeveloped properties where no sites have been previously recorded. These types of surveys are performed to verify that no sites are on the subject property, or if one is found, to record its presence and document its location, condition, size, and type. All records pertaining to archaeological surveys and sites are permanently stored at either government or university clearing houses for future reference by other archaeologists.

Test excavation

The term test excavation refers to the intermediate stage of an archaeological investigation between surveying and salvage excavation or full data recovery program. It generally incorporates the digging of units or **square** pits in order to **sample** the contents and depth of an archaeological site. Test units can be randomly located on a site or placed in specific locations. The use of test units helps the archaeologist determine what areas of a site will yield the highest quantity of artifacts or most useful information before committing time and resources to a more intensive study.

Typically, test units are measured metrically, such as in a one meter by one meter square, and excavated through the archaeological deposit to sterile soil, or soil which fails to produce any additional finds. Depending on the project and the type of site, various methods and techniques are used during an excavation. Sometimes the artifact-bearing soils (called *midden*) are excavated from units in 2–4 in (5–10 cm) levels, carefully peeling one layer of soil and then the next. This is known as arbitrary excavation.

In instances where natural soil layering can be observed, excavation comprises removal of each layer or **strata** regardless of its thickness. The purpose in both cases is to maintain a vertical control of where artifacts were recovered, as well as separating the shallow, younger materials from deeper, older materials.

Each layer or level of soil recovered from unit excavations is sifted through a wire mesh or screen. Soil falls through the screen while artifacts and other material are left on top, where they are then washed, sorted according to type, bagged, and labeled. In this way, archaeologists not only record from what unit a particular artifact was found but also at what depth.

Thus, archaeologists can reconstruct a three-dimensional view of a site layout.

Salvage excavation

Salvage excavations, or what is referred to as rescue archaeology outside the United States, generally represent the final data recovery program of an archaeological site. Although the methods used in salvage excavations are similar to those of test excavations, there are some distinctions with regard to objectives.

Whereas the term test implies an initial investigation to be followed by further study, the term salvage generally denotes that no additional research may be undertaken once the excavation is complete. This is particularly true for archaeological sites that will be destroyed as a result of construction activities.

KEY TERMS

. . . .

Artifact—A man-made object that has been shaped and fashioned for human use.

Classical archaeology—Archaeological research that deals with ancient history, ancient architecture, or any of the now-extinct civilizations of Greece, Egypt, Rome, Aztec, Mayan, etc.

Cultural resource management—Contract archaeology performed by privately owned and operated archaeological consulting firms.

Data recovery—An excavation intended to recover artifacts which represent the basic, raw data of any archaeological study.

Deposit—Refers to the three-dimensional, subsurface or below-ground portion of an archaeological site.

Discipline—A specialized field or sub-field of study.

Historic archaeology—Archaeological studies focusing on the eras of recorded history.

Midden—Darkened archaeological site soil caused by organic waste material such as food refuse and charcoal from ancient campfires.

Prehistoric archaeology—Archaeological studies dealing with material which dates to before the historic era, or before the advent of written languages.

Salvage/rescue excavation—The final phase of a typical two-part series of archaeological site excavations.

Site—An archaeological resource such as an ancient Indian campsite, or an old, historic building.

Survey—A systematic surface inspection conducted to examine a known archaeological site or to verify that no site exists on the property under examination.

Test excavation—A preliminary excavation conducted to determine the size, depth, and significance of an archaeological site prior to committing to a salvage or final excavation.

Unfortunately, it is an accepted fact in archaeology that recovery of 100% of the contents of a site is impractical due to either time or budget constraints. In actuality, that number is estimated from less than 1% to as much as 7%, leaving the remaining bulk of the deposit unstudied. Thus, archaeologists are forced by necessity to make determinations on the most appropriate mode of data recovery and what avenues of research would best benefit from the excavation.

Traditional methods of excavation typically include the use of picks, shovels, trowels, and hand-held shaker screens.

However, archaeologists have realized that although laboratory techniques have taken full advantage of the latest technologies, field methods have not made a corresponding advancement, and in fact have not changed dramatically since the 1930s. To tackle this problem, researchers have begun to experiment with various alternative methods of data recovery. These alternatives are designed to increase sample size, lower costs to sponsors, and at the same time maintain careful scientific control over the recovery process.

One of these alternatives includes the use of earth-moving machinery. For decades, machines such as backhoes or tractor-mounted augers have been employed during test excavations to aid in determining the boundaries and depths of archaeological deposits. Recently, however, machines have been used to actually salvage excavate sites by simulating traditional digging methods, or digging level by level. As a result, new methods of hydraulic **water** screening have been developed to process large amounts of midden soils. Although machines have been successfully utilized in salvage excavations, the use of mechanized earth-moving machinery in archaeology is not widely practiced and cannot be applied to all sites.

Current controversy

Among many North American Indian tribes, treatment of the dead has traditionally been a matter of great concern. Some modern-day Native Americans have expressed that ancestral graves should not be disturbed or, if that cannot be prevented, then any remains and artifacts recovered should be reburied with ceremony.

However, it was not until the 1970s that this issue became a nationwide concern. By then, public attitudes in the United States had become more favorable toward both Native American interests and religious values. Passage of the Native American Religious Freedom Act of 1978 was a reflection of this change in public attitude, as well as a result of the newly developed political awareness and organization of Native American activist groups.

In 1990, the Native American Grave Protection and Repatriation Act (NAGPRA) was signed into federal law.

In addition to applying penalties for the trafficking of illegally obtained Native American human remains and cultural items, the law mandated that all federally-funded institutions (museums, universities, etc.) are required to repatriate or "give back" their Native American collections to tribes who claim cultural or religious ownership over those materials. These and other recently adopted state laws have sparked a heated controversy among scientists and Native American groups.

For archaeologists, physical anthropologists, and other scholars who study humankind's past, graves have provided a very important source of knowledge about past cultures. This has been particularly true in the reconstruction of prehistoric North American cultures whose peoples left no written history but who buried their dead surrounded with material goods of the time. Repatriation of this material will prevent any further studies from being conducted in the future.

Although many researchers support repatriation of historic material that can be directly linked to living tribal descendants, others have stated that it is not possible to make such determinations on very ancient materials that date to before the pyramids of Egypt. Another argument is that ongoing medical studies of diseases found in the bones of ancient remains could lead to breakthroughs in treatments to help the living.

Archaeologists have expressed that museum materials are part of the heritage of the nation, and that these new laws fail to take into consideration the many complex factors that separate ancient human remains from modern Native American cultures.

See also Dating techniques; Ethnoarchaeology.

Resources

Books

Ashmore, Wendy, and Robert J. Sharer. *Discovering Our Past: A Brief Introduction to Archaeology.* 3rd ed. New York: McGraw Hill, 1999.

Bahn, Paul G., ed. *The Cambridge Illustrated History of Archaeology.* Cambridge: Cambridge University Press, 1999.

Fagan, Brian M. *Archaeology: A Brief Introduction.* 8th ed. Englewood Cliffs, NJ: Prentice Hall, 2002.

Feder, Kenneth L. *Frauds, Myths, and Mysteries: Science and Pseudoscience in Archaeology.* 4th ed. New York: McGraw-Hill, 2001.

Haviland, William A. *Human Evolution and Prehistory.* 2nd ed. New York: CBS College Publishing, 1983.

Joukowsky, Marth. *A Complete Field Manual of Archaeology: Tools and Techniques of Field Work for Archaeologists.* Englewood Cliffs, NJ: Prentice-Hall, 1980.

Renfrew, Colin, and Paul G. Bahn. *Archaeology: Theories, Methods, and Practice.* 3rd ed. London: Thames & Hudson, 2000.

Periodicals

Gibbins, D. "Shipwrecks and Maritime Archaeology." *Archaeology Prospection* 9, no. 2 (2002): 279-291.

Noble, Vergil E. "Nineteenth- and Early Twentieth-Century Domestic Site Archaeology." *Historical Archaeology* 35, no. 2 (2001): 142.

T. A. Freeman

Archaeometallurgy

Archaeometallurgy is the study of **metal** artifacts, the technology that was used to smelt them, and the ways ancient societies acquired ores. In addition to understanding the history of metal technology, archaeometallurgists seek to learn more about the people who made and used metal implements and gain a broader understanding of the economic and social contexts in which the people lived. Archaeometallurgy can help to answer archaeological questions concerning the rise of craft specialization, the effects new technologies have on societies, the level of interaction between cultures, and the forces required to change societies.

Archaeometallurgy is a type of **archaeometry**, which is the use of scientific methods to study archaeological materials. It incorporates many different fields of study, including **geology**, ethnography, history, **chemistry**, and materials science. Archaeometallurgists reconstruct ancient smelting (**ore** melting) furnaces, conduct experiments, and analyze metals and slag (the glassy residue left by smelting).

It is a misconception that somehow the use of metals is limited to certain ages (e.g., the Bronze Age or the **Iron** Age). For example, until relatively recently, there was no evidence of **metallurgy** in pre-Bronze Age southeast **Europe**. **Copper** artifacts had been found and there was evidence of ore **mining**, but because no slag had been found, some archaeologists believed the copper had been smelted elsewhere. In 1979, copper slag was discovered with material from the Vinca culture (5400-4000 B.C.). The pieces of slag were small and scattered, and had been overlooked by earlier investigators. Spectroscopic analysis of the slag showed it was similar to local ores. This is strong evidence for local smelting of the ore. In addition, a few tin bronze artifacts have been found with the Vinca and contemporary cultures of southeast Europe. This suggests that the Bronze Age, when it arrived, may have been a scaled-up version of a technology that already existed, rather than something fundamentally new. This is one example of how archaeometallurgy helps us understand ancient societies.

See also Archaeology; Spectroscopy.

Archaeometry

Archaeometry is the analysis of archeological materials using analytical techniques borrowed from the physical sciences and **engineering**. Examples include trace element analysis to determine the source of obsidian used to manufacture arrowheads, and chemical analysis of the growth rings of fossilized sea shells to determine seasonal variations in local **temperature** over **time**.

Modern archaeometry began with the discovery of radiocarbon dating in the 1950s. Today, artifact analyses use excavation techniques, **remote sensing**, and dating methods that all draw on archaeometry.

Archaeometricians are currently using sophisticated computer techniques to handle the masses of data this field continues to generate.

Archaeomagnetic and paleomagnetic dating

Because shifts in the molten core of the **planet** cause **Earth's magnetic field** to vary, and because this causes our planet's magnetic North Pole to change position over time, magnetic alignments in archeological specimens can be used to date specimens.

In **paleomagnetism**, **rocks** are dated based on the occurrence of reversal's in Earth's magnetic poles. These types of pole reversals have occurred with irregular frequency every hundred thousand years or so in Earth's history. Geologists collect samples to be analyzed by drilling into **bedrock**, removing a core, and noting the relative alignment to Earth's present magnetic field. The **sample** is then analyzed in the laboratory to determine its remnant magnetism—the pole's alignment when the sample crystallized. Using a compiled master chronology of pole reversals, scientists can then date the specimen. Because the time between pole reversals is so large, this technique can only be used to date objects to an **accuracy** of a few thousand to tens of thousands of years. The technique has been used to date human remains in the Siwalki Hills of India, in the Olduvai Gorge in Kenya, and in the Hadar region of Ethiopia.

Archaeomagnetism makes use of the fact that the magnetic North Pole has shifted position over time. When clay in an object is heated to a sufficiently high temperature, the **iron** particles in the clay will align to the magnetic pole. If the clay has remained undisturbed since it was fired, it will indicate the position of the pole when it was made. Archaeomagnetism can therefore be used to date fixed objects such as lined fire pits, plaster walls, and house floors. Other techniques, such as radiocarbon dating and dendrochronology, can be used to date

wood from the fire. By comparing data, a master **curve** showing the position of the magnetic North Pole over time can be generated. This master curve then provides a basis for assigning dates to undated clay samples based on where their remnant **magnetism** indicates the pole was when they were fired. Because the pole position can be determined rather exactly for the last 100,000 years or so, dates for materials of this age and younger can be quite accurate. However, disturbances occur at times in the earth's magnetic field at various geographical locations, so it has been necessary to develop separate master curves for different regions. In the southwestern United States, where dendrochronology was used to help calibrate the master curve, archaeomagnetism can yield dates with accuracy as great as +/- 50 years, a precision unmatched by radiocarbon dating.

Dendrochronology

Dendrochronology is the extraction of chronological and environmental information from the annual growth rings of trees. This technique uses well established **tree** ring sequences to date events. Reconstruction of environmental occurrences, droughts for example, which took place when the trees were growing, is also possible based on traits such as changes in tree ring thickness. Tree-ring dating allows dates to be assigned to archeological artifacts; reconstructed environmental events shed light on the ways that human societies have changed in response to environmental conditions.

Dendrochronology was developed in the early 1900s by the American astronomer Andrew Ellicott Douglas as part of his research on the effects of **sunspots** on Earth's climate. Douglas developed a continuous 450-year record of tree ring variability, which he succeeded in correlating with the winter rainfalls preceding the growth years.

The technique very quickly proved useful for dating wood and charcoal remains found in the American southwest. By 1929, dendrochronology had become the first independent dating technique to be used in archeology. Since then, approximately 50,000 tree ring dates from about 5,000 sites have yielded the finest prehistoric dating controls anywhere in the world. Tree ring dating later proved successful in other parts of **North America**, including Alaska and the Great Plains. Today, the technique is practiced in one form or another throughout the world.

The key to successful dendrochronolgical dating is cross-dating—comparing one tree's rings with other trees in the area. This may be done by looking for covariations in tree ring width, or comparing other tree ring attributes to identify overlapping sequences.

By incorporating overlapping sequences from multiple trees, it has been possible to produce chronologies that go back further than any of the individual tree ring specimens. In this way, it has been possible to extend the chronology for the southwest as far back as 322 B.C. The longest individual tree ring chronologies developed to date have been for an 8,700-year California bristlecone pine sequence, and a 10,000-year sequence in **Europe**.

Besides chronological information, archeological tree ring dating yields information about the way wood was used in an ancient culture, and about past climates.

Fission-track dating

Radioactive decay (fission) of **uranium** U-238 causes microscopic tracks of **subatomic particles** to develop in **minerals** and **glass**. By measuring the number of these present in an artifact, which is a function of the sample's age and the amount of uranium present, scientists can determine the absolute age of an artifact.

Fission-track dating has been used to determine the age of glaze coverings on 400-500 year old Japanese bowls. A glass shard dating to Gallo-Roman times was determined to date from A.D. 150, but the precision of that date was only +/- 20% (a possible date range from A.D. 120–180) Nineteenth-century glass produced in central Europe, on the other hand, was dated very precisely. The technique has occasionally proven useful for **pottery analysis** when the objects contained inclusions of materials such as obsidian in which the fission tracks had not been erased over time by the high temperatures of glazing.

Lithics

Lithics are stone tools. Stone tools are capable of revealing information about sources of raw materials and ancient trade routes, usage (through wear patterns), function (from residues such as **blood** and **plant** material), and the **evolution** of craft specializations.

To determine how a tool was made, the archeologist may attempt to reproduce the tool in the laboratory (the traditional method), or take an analytic approach using models based on **physics**, fracture mechanics, and the physical properties of various materials.

Luminescence dating

When certain materials such as quartz, feldspar, and flint are buried, they store trapped electrons that are deposited by background sources of nuclear and cosmic **radiation**. As long as the material is buried, the population of trapped electrons accumulates at a constant **rate**.

Once the material is exposed to daylight or **heat**, however, the trapped electrons are released from their traps. By monitoring the **luminescence** produced by the released electrons, it is possible to determine the length of time that an object has been underground.

Metals analysis

Archeologists analyze **metal** artifacts to determine the sources of ores used to produce the artifacts, and to learn more about trade patterns and fabrication technologies. Metal analyses are also used to authenticate artifacts. Analytical techniques frequently used to determine elemental compositions include x-ray **fluorescence** spectrometry, atomic absorption spectrometry, and **neutron activation analysis**. **Lead isotope** analysis is the technique of choice for determining sources of ores containing **copper**, silver, and tin (all of which contain trace amounts of lead).

Obsidian hydration dating

In many cultures, obsidian was the preferred material for working into stone tools. When obsidian, which is a volcanic glass, is fractured, the fresh surfaces absorb **water**. The thickness of the water-absorbing edge, or rind, increases with time. Measurement of the rind with powerful microscopes thus yields a dimension that can be correlated with the age of the tool.

Although this technique has been widely used in California and the Great Basin, it remains a relatively inaccurate technique when used alone to date artifacts.

Paleobotany and paleoethnobotany

In **paleobotany**, the remains of plants recovered from prehistoric **soil** deposits are analyzed to determine the **species** of plants that were present, the parts of the plant used, the time of year they were collected, and genetic changes in the plant species over time. In order to use this technique, the paleobotanist must have access to a complete reference collection indicating the changes in a plant species over time. Paleobotanists have, using this technique, been able to reconstruct information about prehistoric climates, patterns of plant use, seasonal patterns of site occupation, **vegetables** included in diets, and transitions from plant-gathering to plant-cultivation practices.

The paleoethnobotanist, like the paleobotanist, studies plant remains in the context of archeology, but in addition looks at the interactions between the plant materials and the people who used them. The first techniques used for plant recovery involved methods of flotation to separate organic from inorganic **matter**. Modified flotation techniques are still used to extract

KEY TERMS

Artifact—A man-made object that has been shaped and fashioned for human use.

Atomic absorption spectrometry—Method of analysis in which the specimen is placed in a flame and the light emitted is analyzed.

Cosmic radiation—Electrons and atomic nuclei that impinge upon the earth from outer space.

Flotation—A method of separating organic remains by causing them to float to the surface.

Fracture mechanics—Analysis of the way an objects breaks.

Luminescence—Light emission from a body that is not due only to that body's temperature. Luminescence is frequently produced by chemical reactions, irradiation with electrons or electromagnetic radiation, or by electric fields.

Magnetic field—The electromagnetic phenomenon produced by a magnetic force around a magnet.

Neutron activation analysis—Method of analysis in which a specimen is bombarded with neutrons, and the resultant radio isotopes are measured.

Nuclear radiation—Particles emitted from the atomic nucleus during radioactive decay or nuclear reactions.

Radioactivity—Spontaneous release of subatomic particles or gamma rays by unstable atoms as their nuclei decay.

X-ray fluorescence spectrometry—A nondestructive method of analysis in which a specimen is irradiated with x rays and the resultant spectrum is analyzed.

---

carbonized plant fragments from sediment. Modern analytical techniques for examining recovered plant materials, based on genetic and DNA research, permit the identification of plant **proteins**, isotopes, starches, and lipids. With these methods, it has been possible to determine the sequences of domestication of such plants as maize, **wheat**, **barley**, and **rice**.

Potassium-argon dating

The potassium-argon method of dating allows scientists to date rocks that were formed between 50,000 and two billion years ago. The rate at which argon 40 forms from the decay of potassium 40, one of the most common minerals in the earth's crust, is known. Therefore, it is possible to determine the age of an object based on the accumulation of argon 40 in the specimen. The first archeological site to be dated by this method (using lava samples) was at Olduvai Gorge in Tanzania.

Radiocarbon dating

Radiocarbon dating allows archeologists to date materials, formed between 300 and 40-50,000 years ago, that contain organic **carbon**. Carbon 14 is a naturally occurring radioisotope of ordinary carbon (carbon-12) that is created in the upper atmosphere when carbon-12 is bombarded by cosmic rays. On **Earth**, living organisms metabolize carbon 14 in the same percentage that it exists in the atmosphere. Once the plant or **animal** dies, however, the carbon-14 **atoms** began to decay at a known rate. Consequently, the age of a carbon-containing specimen such as charcoal, wood, shells, bone, antlers, peat, and sediments with organic matter, can be determined. One of the first applications of this technique was to assign a date to the beginning of the post-glacial period of about 10,000 years ago.

Resources

Books

Daniel, Glyn, ed. *The Illustrated Encyclopedia of Archeology.* New York: Thomas Y. Crowell: 1977.

Fagan, Brian M., ed. *The Oxford Companion to Archeology.* New York: Oxford University Press, 1996.

Maloney, Norah. *The Young Oxford Book of Archeology.* New York: Oxford University Press, 1997.

Sullivan, George. *Discover Archeology: An Introduction to the Tools and Techniques of Archeological Fieldwork.* Garden City, NY: Doubleday & Company, 1980.

Periodicals

Fowler, M.J. "Satellite Remote Sensing and Archaeology." *Archaeology Prospection* 9, no. 2 (2002): 55-70.

Randall Frost

Archeological mapping

Before any excavation is begun at a site, the archeologist must prepare a survey **map** of the site. Site mapping may be as simple as a sketch of the site boundaries, or as complex as a topographic map complete with details about vegetation, artifacts, structures, and features on the site. By recording the presence of artifacts on the site, the site map may reveal information about the way the site was used, including patterns of occupational use. Contour maps may shed light on ways in which more recent environmental activity may have changed the original patterns of use. In cases where structural remains are visible at a site, the site map can provide a basis for planning excavations.

When staking out a site to be excavated, the archeologist typically lays out a **square** grid that will serve as a reference for recording data about the site. The tools required to construct the grid may be as simple as a compass, a measuring tape, stakes, and a ball of twine. After the grid has been laid out, the archeologist draws a representation of it on graph **paper**, being careful to note the presence of any physical landmarks such as trees, **rivers**, and large **rocks**. Once the excavation is underway, each artifact recovered is mapped into the square in the grid, and layer in which it was found.

As artifacts are removed from each layer in the site, their exact positions are plotted on a map. At the end of the excavation, a record of the site will exist in the form of maps for each excavated layer at the site. Photographs are also taken of each layer for comparison with the maps.

To facilitate artifact recovery, deposited material at the site may be screened or sifted to make sure materials such as **animal** bones, **snails**, **seeds**, or chipping debris are not overlooked. When a screen is used, shovelfuls of **soil** are thrown on it so that the dirt sifts through the screen, leaving any artifacts behind. In some cases, the **deposit** may be submerged in a container filled with plain or chemically-treated **water**. When the water is agitated, light objects such as seeds, small bones, and charred **plant** material rise to the top of the container.

Prior to shipment to the laboratory for processing, artifacts are placed in a bag that is labeled with a code indicating the location and stratigraphic layer in which the artifacts were found. Relevant information about each artifact is recorded in the field notes for the site.

Many mapping techniques developed for use on land have also been adapted for underwater archeology. Grids can be laid out to assist in mapping and drawing the site, and to assist the divers who perform the excavation. In this case, however, the grids must be weighted to keep them from floating away, and all mapping, recording, and photographing must be done with special equipment designed for underwater use.

Spatial mapping and stratigraphic mapping

Most modern archeologists will attempt to place data taken from a site into archeological context by mapping the spatial and stratigraphic dimensions of the site.

Spatial dimensions include the distribution of artifacts, and other features in three dimensions. The level of detail given in the spatial description typically depends on the goals of the research project. One hundred years ago, finds were recorded much less precisely than they are today; it might have been sufficient to map an object's location to within 25 sq yd (7 sq m). Today, the location of the same artifact might be recorded to the nearest centimeter. Modern archeologists still use maps to record spatial information about a site. Such information includes the spatial distribution of artifacts, features, and deposits, all of which are recorded on the map. Measuring tools range from simple tapes and plumb bobs to highly accurate and precise **surveying instruments** called **laser** theodolites.

The **accuracy** of a map is the degree to which a recorded measurement reflects the true value; the precision of the map reflects the consistency with which a measurement can be repeated. Although the archeologist strives for accuracy in representing the site by the map, the fact that much of what is recorded represents a subjective interpretation of what is present makes any map a simplification of reality. The levels of accuracy and precision that will be deemed acceptable for the project must be determined by the archeologists directing the investigation.

The second technique involved in recording the archeological context of a site is stratigraphic mapping. Any process that contributed to the formation of a site (e.g., dumping, **flooding**, digging, **erosion**, etc.) can be expected to have left some evidence of its activity in the stratification at the site. The sequential order these processes contribute to the formation of a site must be carefully evaluated in the course of an excavation. The archeologist records evidence of ordering in any deposits and interfaces found at the site for the purposes of establishing a relative chronology of the site and interpreting the site's history. In order to document the stratification at the site, the archeologist may draw or photograph vertical sections in the course of an excavation. Specific graphing techniques have been developed to aid archeologists in recording this information. Finally, the archeologist typically notes such details as soil **color** and texture, and the presence and size of any stones, often with the aid of reference charts to standardize the descriptions.

Although all archeologists agree that keeping careful records of an excavation is essential to good practice, there is a certain amount of disagreement as to what constitutes archeological data. Many of the practices of the eighteenth century archeologist seem crude when compared to the detailed site information that is now considered vital—for example, the exact positioning and magnetic properties of fired clay. However, the practices of today will no doubt seem coarse to the archeologist of the next century.

KEY TERMS

Artifact—A man-made object that has been shaped and fashioned for human use.

Contour map—Map illustrating the elevation or depth of the land surface using lines of equal elevation; also known as a topographic map.

Stratigraphy—The study of layers of rocks or soil, based on the assumption that the oldest material will usually be found at the bottom of a sequence.

Theodolite—An optical instrument consisting of a small telescope used to measure angles in surveying, meteorology, and navigation.

Resources

Books

Daniel, Glyn, ed. *The Illustrated Encyclopedia of Archeology.* New York: Thomas Y. Crowell: 1977.

Fagan, Brian M., ed. *The Oxford Companion to Archeology.* New York: Oxford University Press, 1996.

Maloney, Norah. *The Young Oxford Book of Archeology.* New York: Oxford University Press, 1997.

Renfrew, Colin, and Paul G. Bahn. *Archaeology: Theories, Methods, and Practice.* 3rd ed. London: Thames & Hudson, 2000.

Sullivan, George. *Discover Archeology: An Introduction to the Tools and Techniques of Archeological Fieldwork.* Garden City, NY: Doubleday & Company, 1980.

Randall Frost

Archeological sites

Archeologists are concerned with the activities of people and nature that create evidence of a cultural past. Such evidence, which may include any remnant of human habitation, is referred to as the archeological record. The processes that produce this evidence are called formation processes.

There are two types of formation processes: cultural and environmental. Cultural formation processes are those that follow the actual use of an artifact, e.g., reuse, discard, disturbance, and archeological recovery. Environmental formation processes are those agents that impact cultural materials at any stage of their existence.

Cultural formation

The four cultural formation processes are reuse, cultural deposition, reclamation, and disturbance. Reuse might include **recycling**, secondary use, or use by another party. Cultural deposition processes take cultural materials from the context in which they are used in a culture and place them in an environmental context; examples include discarded dishes, burials, and abandonments. Reclamation processes are those in which archeological materials are retrieved for reuse by a culture; examples include scavenging, looting of previously deposited artifacts, and archeological recovery. Disturbance processes alter the earth's surface and alter archeological materials, deposits and sites; examples include plowing and land leveling.

Environmental formation

Environmental formation processes include the chemical, physical, and biological processes by which nature alters cultural materials. The scale at which these processes affect cultural material may be at the artifact level (e.g., the rotting of **wood** or the **corrosion** of metals), the site level (e.g., the burrowing of animals), or the regional level (the burial or **erosion** of sites).

Archeologists must sort out the contributions of the various formation processes to achieve new understandings of past human **behavior**. An artifact such as a hand axe would be expected to acquire signs of wear in the course of normal use, but could also acquire similar patterns of wear from cultural and environmental processes. By identifying ways that formation processes have altered an artifact, the archeologist can better assess the way the artifact was used in the culture that produced it.

Finding an archeological site

Besides drawing on information from large archeological sites such as Stonehenge and Angkor Wat, archeologists must also rely on data from a myriad of much smaller sites if they are to construct an accurate interpretation of the economic, environmental, and ideological factors that governed occupation of the larger sites. In many cases, the only evidence of human occupation must come from the remnants of seasonal campsites, and artifacts such as stone tools or bones.

In order to efficiently **sample** sites within a region and locate very small sites, archeologists have developed a variety of ground-survey and remote-sensing techniques.

Remote sensing and geophysical analysis

Although some archeological sites can be recognized above ground, the majority lie beneath the ground's surface. **Remote sensing** techniques allow archeologists to identify buried sites. In addition, by examining the ways human intervention have altered the

KEY TERMS

. . . .

Artifact—A man-made object that has been shaped and fashioned for human use.

Electrical resistivity—A measure of the resistance (opposition) an object poses to electrical current flowing through it.

Magnetic field—The electromagnetic phenomenon producing a magnetic force around a magnet.

Radar—A method of detecting distant objects based on the reflection of radio waves from their surfaces.

Solar radiation—Energy from the sun.

Stratification—A method of describing the ages of different strata of rocks or soil, based on the assumption that the oldest material will usually be found at the bottom layer.

Thermal detector—A device that detects heat.

surface near a site (e.g., through the construction of refuse pits or hearths), the archeologist may be able to identify patterns of previous usage.

Archeologists may employ techniques borrowed from the fields of **geophysics** and **geochemistry** to detect and **map** archeological sites and features. Many geophysical techniques, including **electrical resistance** measurements of the **soil** above a site and magnetic measurements of a pottery kiln, were first employed by archeologists in the 1940s and 1950s.

Satellite detectors have also been used to monitor the reflected solar **radiation** above a site. The characteristics of the reflected **light** allow the archeologist to identify differences in soil or vegetation covering a site, and at sufficiently high resolutions, to recognize archeological features. In this way, archeologists have been able to map out drainage canals once used by Mayan farmers, but now lying beneath the umbrella of the Yucatan **rainforest**.

Airborne thermal detectors, capable of monitoring the surface **temperature** of the soil and covering vegetation, take advantage of differences in the way materials retain **heat** to isolate archeological features from surrounding soil. With this technique, buried Egyptian villages appear to glow at night beneath thin layers of **sand**.

Electrical resistivity measurements of the soil are sensitive to the presence of **water** and dissolved salts in the water. Because constriction materials such as granite or limestone have a higher electrical resistance than the surrounding soil, electrical resistivity measurements may

be of use in determining the locations of buried structures such as stone walls.

The magnetic properties of soil depend on the presence of **iron** particles, which when heated to sufficiently high temperatures tend to align themselves with the **earth's magnetic field**. However, the earth's magnetic field and intensity change over **time**. Wherever human activity alters the iron compounds in the soil by subjecting them to high temperatures, for example by building fires in a hearth or firing pottery in a kiln, the heated soil upon cooling takes on magnetic properties that reflect the direction and intensity of the earth's magnetic field at the time of cooling. Since archeologically related changes in local magnetic fields may only amount to one part in 10,000, and because of daily fluctuations in the magnetic field due to electrical currents in the ionosphere, this technique usually requires monitoring of the earth's field at a reference point during the archeological investigation.

Other electromagnetic measurements used to probe a site may examine the soil for phosphates and heavy **minerals** often associated with past human habitation.

Three dimensional representations of a buried feature may be constructed using ground-sensing **radar** or resistivity profiling to obtain vertical geophysical cross-sections across a site. When placed beside each other, these sections create a three-dimensional image of buried objects.

Ground surveys

The techniques of ground surveying date to the 1930s and 1940s. Ground surveys require no special equipment, just an observant archeologist with some knowledge of what might be found at a site. Ground survey records may include notes about any visible cultural features and artifacts on the site, site measurements, preparing maps or sketches of the site, and sometimes gathering small collections of artifacts. Surface artifacts may be gathered either as **random** grabs or complete samplings in a given area.

Site assessments

Initially, the archeologist must determine the size, depth, and stratification of a site. Second, the age or ages of the site must be determined. Third, the types of artifacts and features present at the site must be identified. And finally, information about the environment and the way that it influenced human habitation at the site must be known.

Site assessment techniques fall into two categories: destructive and nondestructive. Surface collecting, testing with shovels, digging pits, and mechanical trenching all disturb the site, and are considered destructive. Nondestructive techniques include mapping and remote sensing.

Resources

Books

Daniel, Glyn, ed. *The Illustrated Encyclopedia of Archeology.* New York: Thomas Y. Crowell: 1977.

Fagan, Brian M., ed. *The Oxford Companion to Archaeology.* New York: Oxford University Press, 1996.

Hester, Thomas R., Harry J. Schafer, and Kenneth L. Feder. *Field Methods in Archaeology.* 7th ed. New York: McGraw-Hill, 1997.

Krivanek, R. "Specifics and Limitations of Geophysical Work on Archaeological Sites." *Archaeological Prospection* 8, no. 2 (2001): 113-134.

Maloney, Norah. *The Young Oxford Book of Archeology.* New York: Oxford University Press, 1997.

Sullivan, George. *Discover Archeology: An Introduction to the Tools and Techniques of Archeological Fieldwork.* Garden City, NY: Doubleday & Company, 1980.

Waters, Michael R. *Principles of Geoarchaeology: A North American Perspective* Tucson, AZ: University of Arizona Press, 1997.

Periodicals

Smith, Monica L. "The Archaeology of a 'Destroyed' Site." *Historical Archaeology* 35, no. 2 (2001): 31-40.

Randall Frost

Argand diagram *see* **Complex numbers**

Argon *see* **Rare gases**

Arithmetic

Arithmetic is a branch of **mathematics** concerned with the numerical manipulation of numbers using the operations of **addition, subtraction, multiplication, division**, and the extraction of roots. General arithmetic principles slowly developed over **time** from the principle of counting objects. Critical to the advancement of arithmetic was the development of a positional number system and a symbol to represent the quantity **zero**. All arithmetic knowledge is derived from the primary axioms of addition and multiplication. These axioms describe the rules which apply to all **real numbers**, including whole numbers, **integers**, rational, and irrational numbers.

Early development of arithmetic

Arithmetic developed slowly over the course of human history, primarily evolving from the operation of counting. Prior to 4000 B.C., few civilizations were even able to count up to ten. Over time however, people learned to associate objects with numbers. They also learned to think about numbers as abstract ideas. They recognized that four trees and four cows had a common quantity called four. The best evidence suggests that the ancient Sumerians of Mesopotamia were the first civilization to develop a respectable method of dealing with numbers. By far the most mathematically advanced of these ancient civilizations were the Egyptians, Babylonians, Indians, and Chinese. Each of these civilizations possessed whole numbers, fractions, and basic rules of arithmetic. They used arithmetic to solve specific problems in areas such as trade and commerce. As impressive as the knowledge that these civilizations developed was, they still did not develop a theoretical system of arithmetic.

The first significant advances in the subject of arithmetic were made by the ancient Greeks during the third century B.C. Most importantly, they realized that a sequence of numbers could be extended infinitely. They also learned to develop theorems which could be generally applied to all numbers. At this time, arithmetic was transformed from a tool of commerce to a general theory of numbers.

Numbering system

Our numbering system is of central importance in the subject of arithmetic. The system we use today called the Hindu-Arabic system, was developed by the Hindu civilization of India some 1,500 years ago. It was brought to **Europe** during the middle ages by the Arabs and fully replaced the Roman numeral system during the seventeenth century.

The Hindu-Arabic system is called a decimal system because it is based on the number 10. This means that it uses 10 distinct symbols to represent numbers. The fact that 10 is used is not important because it could have just as easily been based on another number of symbols like 14. An important feature of our system is that it is a positional system. This means that the number 532 is different from the number 325 or 253. Critical to the invention of a positional system is perhaps the most significant feature of our system: a symbol for zero. Note that zero is a number just as any other and we can perform arithmetic operations with it.

Axioms of the operations of arithmetic

Arithmetic is the study of mathematics related to the manipulation of real numbers. The two fundamental properties of arithmetic are addition and multiplication. When two numbers are added together, the resulting number is called a sum. For example, 6 is the sum of 4 + 2. Similarly, when two numbers are multiplied, the resulting number is called the product. Both of these operations have a related inverse operation which reverses or "undoes" its action. The inverse operation of addition is subtraction. The result obtained by subtracting two numbers is known as the difference. Division is the inverse operation of multiplication and results in a quotient when two numbers are divided. The operations of arithmetic on real numbers are subject to a number of basic rules, called axioms. These include axioms of addition, multiplication, distributivity, and order. For simplicity, the letters a, b, and c, denote real numbers in all of the following axioms.

There are three axioms related to the operation of addition. The first, called the commutative law, is denoted by the equation a + b = b + a. This means that the order in which you add two numbers does not change the end result. For example, 2 + 4 and 4 + 2 both mean the same thing. The next is the associative law which is written a + (b + c) = (a + b) + c. This axiom suggests that grouping numbers also does not effect the sum. The third axiom of addition is the **closure property** which states that the equation a + b is a real number.

From the axioms of addition, two other properties can be derived. One is the additive **identity property** which says that for any real number a + 0 = a. The other is the additive inverse property which suggests that for every number a, there is a number −a such that −a + a = 0.

Like addition, the operation of multiplication has three axioms related to it. There is the commutative law of multiplication stated by the equation a × b = b × a. There is also an associative law of multiplication denoted by a × (b × c) = (a × b) × c. And finally, there is the closure property of multiplication which states that a × b is a real number. Another axiom related to both addition and multiplication is the axiom of distributivity represented by the equation (a + b) × c = (a × c) + (b × c).

The axioms of multiplication also suggest two more properties. These include the multiplicative identity property which says for any real number a, 1 × a = a, and the multiplicative inverse property that states for every real number there exists a unique number (1/a) such that (1/a) × a = 1.

The axioms related to the operations of addition and multiplication indicate that real numbers form an algebraic **field**. Four additional axioms assert that within the set of real numbers there is an order. One states that for any two real numbers, one and only one of the following relations is true: either a < b, a > b or a = b. Another suggests that if a < b, and b < c, then a < c. The monotonic property of addition states that if a < b, then a + c < b + c. Finally, the monotonic property of multiplication states that if a < b and c > 0, then a × c < b × c.

Numbers and their properties

These axioms apply to all real numbers. It is important to note that real numbers is the general class of all numbers that includes whole numbers, integers, rational numbers, and irrational numbers. For each of these number types only certain axioms apply.

Whole numbers, also called **natural numbers**, include only numbers that are positive integers and zero. These numbers are typically the first ones to which a person is introduced, and they are used extensively for counting objects. Addition of whole numbers involves combining them to get a sum. Whole number multiplication is just a method of repeated addition. For example, 2 × 4 is the same as 2 + 2 + 2 + 2. Since whole numbers do not involve **negative** numbers or fractions, the two inverse properties do not apply. The smallest whole number is zero but there is no limit to the size of the largest.

Integers are whole numbers that include negative numbers. For these numbers the inverse property of addition does apply. For these numbers, zero is not the smallest number but it is the middle number with an infinite number of positive and negative integers existing before and after it. Integers are used to measure values which can increase or decrease such as the amount of money in a cash register. The standard rules for addition are followed when two positive or two negative numbers are added together and the sign stays the same. When a positive integer is added to a negative integer, the numbers are subtracted and the appropriate sign is applied. Using the axioms of multiplication it can be shown that when two negative integers are multiplied, the result is a **positive number**. Also, when a positive and negative are multiplied, a negative number is obtained.

Numbers to which both inverse properties apply are called rational numbers. Rational numbers are numbers that can be expressed as a **ratio** of two integers, for example, $\frac{1}{2}$. In this example, the number 1 is called the numerator and the 2 is called the denominator. Though rational numbers represent more numbers than whole numbers or integers, they do not represent all numbers. Another type of number exists called an **irrational number** which cannot be represented as the ratio of two integers. Examples of these types of numbers include square roots of numbers which are not perfect squares and cube roots of numbers which are not perfect cubes. Also, numbers such as the universal constants π and e are irrational numbers.

The principles of arithmetic create the foundations for all other branches of mathematics. They also represent the most practical application of mathematics in everyday life. From determining the change received from a purchase to calculating the amount of sugar in a batch of cookies, **learning** arithmetic skills is extremely important.

See also Algebra; Calculus; Function; Geometry; Trigonometry.

KEY TERMS

. . . .

Associative law—Axiom stating that grouping numbers during addition or multiplication does not change the final result.

Axiom—A basic statement of fact that is stipulated as true without being subject to proof.

Closure property—Axiom stating that the result of the addition or multiplication of two real numbers is a real number.

Commutative law—Axiom of addition and multiplication stating that the order in which numbers are added or multiplied does not change the final result.

Hindu-Arabic number system—A positional number system that uses 10 symbols to represent numbers and uses zero as a place holder. It is the number system that we use today.

Inverse operation—A mathematical operation that reverses the work of another operation. For example, subtraction is the inverse operation of addition.

Resources

Books

Paulos, John Allen. *Beyond Numeracy.* New York: Alfred A. Knopf, Inc., 1991.

Perry Romanowski

Armadillos

Armadillos are bony-skinned **mammals** native to Central and **South America**. Armadillos (family Dasypodidae) number 20 **species** in eight genera. The species include the long-nosed armadillo (six species), the naked-tailed armadillo (four species), the hairy armadillo (three species), the three-banded armadillo (two species), the fairy armadillo (two species), the six-banded or yellow armadillo (one species), the pichi (one species), and the giant armadillo (one species).

Distribution and habitat

Armadillos are found through the whole of South and Central America, from the Strait of Magellan northward to eastern Mexico.

The common long-nosed (or nine-banded) armadillo is the most widespread and is the only species found in the United States. In the 1850s several armadillos were recorded in Texas, and their descendants spread rapidly through the Gulf States toward the Atlantic in what has become the swiftest mammalian distribution ever witnessed. In 1922 a captive pair of common long-nosed armadillos escaped in Florida, and in a few decades their descendants numbered in the tens of thousands. Moving westward, the eastern population met with the Texas group only within the last decade.

Rivers and streams present no barrier to the spread of armadillos. Gulping air into their stomachs and intestines to buoy themselves, armadillos float leisurely across the **water**. Others have been observed walking into streams on one side and strolling out on the other side a few minutes later.

While water presents no barrier to armadillos, cold does, and winter temperatures have slowed their northern advance in the United States. Armadillos are poorly insulated and cannot withstand chilling. Cold also reduces the abundance of **insects** that armadillos depend on for food. Because of this, armadillos have moved northward only as far as Oklahoma and southern Kansas.

Armadillos are found in habitats ranging from pampas (**grasslands**) to arid deserts and from coastal prairies to rainforests and deciduous **forests**.

Physical appearance

Armadillos appear to be a conglomeration of other **animal** parts—the shell of a turtle, the ears of an **aardvark**, the feet of a lizard, the face of a pig, and the tail of a **dinosaur**. However, the patches and bands of coarse bristles and hairs on their bodies reveal them to be true mammals.

Extinct species of armadillo grew to enormous sizes and their bony shells were used as roofs and tombs by early South American Indians. Surviving species are nowhere near that large, ranging in size from the 99–132 lb (45–60 kg) giant armadillo to the 2.8–3.5 oz (80–100 g) lesser fairy armadillo. The familiar common long-nosed armadillo weighs in at 6–10 lb (2.7–4.5 kg).

The most obvious and unusual feature of armadillos is their bony skin armor, found in no other living mammal. Bands of a double-layered covering of horn and bone develop from the skin and cover most of the upper surfaces and sides of the body. These bony bands or plates are connected by flexible skin. The top of the head is capped with a bony shield and the tail is usually encased with bony rings or plates. Their underside is covered only with soft, hairy skin. Armadillos have a flattened and elongated head with a long, extendable tongue. Set in their jaws are numerous small, peglike

A nine-banded armadillo (*Dasypus novemcinctus*) in the Aransas National Wildlife Refuge, Texas. *Photograph by Robert J. Huffman. Field Mark Publications. Reproduced by permission.*

teeth. The teeth are not covered by enamel and grow continuously. Their hind limbs have five clawed toes while the powerful forelimbs end in three, four, or five curved digging claws.

Feeding and defense

Armadillos are predominantly nocturnal in their foraging habits and their teeth dictate their diet. Those with sturdy teeth eat insects, **snails**, worms, small lizards, carrion, tubers, and **fruits**. Those with soft teeth eat primarily insects such as **ants** and **termites**. Using their long, sticky tongue to remove insects from their nests, they have been observed to eat as many as 40,000 ants at one feeding. It is estimated that a single armadillo can eat 200 lb (90 kg) of insects in a year.

A keen sense of **smell** helps the armadillo locate **prey** as much as 6 in (10 cm) underground. Pressing its nose to the ground to keep the scent and holding its breath to avoid inhaling dust, the armadillo digs into the **soil** and litter with astonishing speed.

Armadillos also defend themselves by burrowing into the earth, disappearing completely in a few minutes. Once dug in, they expand their bony shell and wedge themselves into the burrow. They can also run surprisingly fast and, if cornered, will use their claws to fight. Once a **predator** catches an armadillo it must deal with its bony armor. The three-banded armadillo can roll up into a tight ball presenting nothing but armor to its enemies. Their armor also protects them from **cactus** spines and dense, thorny undergrowth.

While some species of armadillo are hunted by humans in Central and South America for their meat, the greatest danger to armadillos in the United States is the **automobile**. Dozens of armadillos are run down as they wander onto highways at dusk. Their habit of leaping

several feet into the air when startled contributes to many of the automobile-related deaths.

Reproduction

Armadillos are solitary creatures that seek companions only during the mating season. After mating, female armadillos can suspend their pregnancy for up to two years. Another reproductive peculiarity of armadillos that has caught the attention of geneticists is the ability to produce multiple births from a single fertilized egg: depending on the species, 4, 8, or 12 genetically identical offspring may be produced.

Young armadillos are born in nest chambers within a burrow. At **birth**, the young are pink and have a soft, leathery skin. This soft skin hardens within a few weeks. Young armadillos stay close to their mother for about two weeks before striking out on their own.

Resources

Books

Gould, Edwin, and Gregory McKay, eds. *The Encyclopedia of Mammals.* 2nd ed. New York: Academic Press, 1998.

Nowak, Ronald M. *Walker's Mammals of the World.* Baltimore: Johns Hopkins University Press, 1999.

Periodicals

Schueler, Donald G. "Armadillos Make Me Smile a Lot." *Audubon* (July 1988): 73.

Storrs, Elanor. "I'll Think about That Tomorrow." *Discover* (16 February 1990): 16.

Watson, Jim. "Rising Star." *National Wildlife* (October/ November 1989): 47.

Dennis Holley

Aromatic hydrocarbon *see* **Hydrocarbon**

Arrow worms

Arrow worms are small marine planktonic animals of the phylum Chaetognatha found in tropical seas. Most of

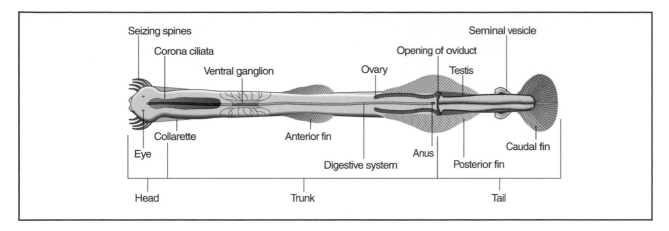

The anatomy of an arrow worm. *Illustration by Hans & Cassidy. Courtesy of Gale Group.*

the 50 **species** of arrow worms belong to the genus *Sagitta*. Arrow worms have a head with eyes and hook-like spines on their jaws that identify them as predators of smaller planktonic animals and larvae. Arrow worms have an elongated body, roughly the shape of an arrow with two pairs of lateral fins and a tail fin. They are cross-fertilizing hermaphroditic animals. Sperm from one **individual** is received by another individual in a sperm pouch, which later fertilizes the maturing egg in the ovary.

Arrow worms are thought to be distantly related to the phylum Chordata (which includes the **vertebrates**) but they lack many of the other important chordate characteristics. Nevertheless, arrow worms do have a coelom (a fluid-filled body cavity) which is a characteristic of **chordates** and the phylum Protochordata. The coelom in arrow worms forms as an out-pocketing of the larval intestine. A similar origin of the coelom is also found in the phylum Echinodermata and the subphylum Cephalochordata (*Amphioxus*) of the phylum Chordata. However, the majority of chordates (the vertebrates in the subphylum Vertebrata) have a coelom that arises in a different way—by a splitting of the tissues to form the cavity.

Arrowgrass

The arrowgrass family (Juncaginaceae) is a family of herbaceous plants whose leaves are grass-like and shaped somewhat like an arrowhead. The arrowgrass family has four genera: *Scheuchzeria* with two **species**; *Thrighlochin* with 12 species; *Maundia* with one species; and *Tetroncium* with one species.

All species in the arrowgrass family grow in wet or moist habitats in temperate and cold regions of the world. Many species grow in fresh **water** and are common in sphagnum bogs. Other species grow in **brackish** (semi-salty) water.

All plants in this family have thin, grass-like leaves that are flat, linear, and smooth. They have a specialized underground stem, referred to as a **rhizome**. The leaves and roots arise from the rhizome. The roots of some species are fat and tuberous. Most species are perennial, in that they maintain leaves all year round.

The flowers of all species are small and inconspicuous, with clusters of individual flowers arising from an erect stalk. This inflorescence is referred to as a spike or raceme. The flowers are symmetrical, and the different parts of the flowers occur in threes or in multiples of three. Some species have bisexual (monoecious) flowers, in that male and female organs occur on the same **flower**. Other species have unisexual (dioecious) flowers, with male and female organs occurring in different, separate flowers. The flowers of all species are wind-pollinated.

The fertilized flowers give rise to **fruits**, referred to as follicles. A follicle is a dry fruit that splits along a suture on one side to release the seed(s). The follicles of plants in the arrowgrass family have one or two **seeds**. Each embryonic seed has one cotyledon (seed **leaf**).

The plants of the arrowgrass family are not of great economic significance to humans. However, the leaves or rhizomes of some species have been traditionally eaten by some aboriginal peoples of **North America** and **Australia**.

Arrowroot

Arrowroot is an edible starch obtained from the underground stems, or rhizomes, of several **species** of the genus *Maranta*, family Marantaceae. The most common species of arrowroot is *Maranta arundinacea*, native to the tropical areas of Florida and the West Indies, and called true, Bermuda, or West Indian arrowroot. Several relatives of true arrowroot are also known locally as arrowroot and have roots containing edible starch. For example, Brazilian arrowroot from the cassava **plant** (*Manihot esculenta*) is the source of tapioca. Other starches also called arrowroot are obtained from the genera *Curcuma* and *Tacca*. True arrowroot and its relatives are currently cultivated in **Australia**, southeast **Asia**, Brazil, and the West Indies.

The roots of arrowroot grow underground to about 1.5 ft (46 cm) long and 2 cm in diameter. Above ground, branched stems grow up to 6 ft (2 m) tall, having big, ovate leaves and a few white flowers. The jointed, light yellow rhizomes are harvested after one year of growth, when they are full of starch. After harvesting, the roots are soaked in **water**, making their tough, fibrous covering easier to peel off, and the remaining starchy **tissue** is then beaten into a pulp. The pulp is rinsed with water many times to separate the starch from the residual fiber. The liquid pulp is allowed to dry; the powder that remains is starch. One acre (0.4 ha) of arrowroot can yield 13,200 lb (6 mt) of roots. From this amount, 2,200 lb (1 mt) of starch can be obtained.

Arrowroot starch is very pure; it has no **taste** or odor, and has minimal nutritional value, other than as a source of **energy**. It is used in cooking as a thickening agent for soups, sauces, and puddings. What makes this arrowroot unique is that when boiled with a liquid such as water or broth, the gel-like mixture remains transparent, and does not become cloudy or opaque, as is the case with other starches. Arrowroot starch digests easily and is frequently used in food products for babies (for example, arrowroot cookies) or for people who need to eat bland, low protein diets because of illness. Native Americans used this root to absorb poison out of arrow wounds, giving the plant its common name.

Arsenic *see* **Element, chemical**

Arteries

Arteries are **blood** vessels that transport oxygenated blood from the **heart** to other organs and systems throughout the body. In humans, healthy arteries are smooth, elastic structures, while diseased arteries may contain bulges due to high blood **pressure**, hard, inelastic areas, or internal blockages resulting from the accumulation of fatty plaques circulating in the blood. Atheroscle-

rosis is the hardening or narrowing of an artery after plaque formation has partially restricted blood flow through the artery. Atherosclerosis is the major contributor to coronary artery **disease** (CAD), and coronary artery disease, often resulting in heart attack or arrhythmia, is the number one cause of deaths in the United States.

In humans, a typical artery contains an elastic arterial wall that can be divided into three principal layers, although the absolute and relative thickness of each layer varies with the type or diameter of artery. The outer layer is termed the tunica adventia, the middle layer is termed the tunica media, and an inner layer is the tunica intima. These layers surround a lumen, or opening, that varies in size with the particular artery, through which blood passes.

Arteries of varying size comprise a greater arterial blood system that includes, in descending diameter, the aorta, major arteries, smaller arteries, arterioles, meta-arterioles, and **capillaries**. It is only at the level of the capillary that branches of arteries become thin enough to permit gas and nutrient exchange. As the arterial system progresses toward the smaller diameter capillaries, there is a general and corresponding increase in the number of branches and total area of lumen available for blood flow. As a result, the **rate** of flow slows as blood approaches the capillary beds. This slowing is an important feature that enables efficient exchange of gases—especially **oxygen**.

In larger arteries, the outer, middle, and inner endothelial and muscle layers are supported by elastic fibers, and serve to channel the high pressure and high rate of blood flow. A difference in the orientation of cells within the layers (e.g., the outer endothelial cells are oriented longitudinally, while the middle layer smooth muscle cells run in a circumference around the lumen) also contributes both strength and **elasticity** to arterial structure.

The aorta and major arties are highly elastic, and contain walls with high amounts of elastin. During heart systole (contraction of the heart ventricles), the arterial walls expand to accommodate the increased blood flow. Correspondingly, the vessels contract during diastole and this contraction also serves to drive blood through the arterial system.

In the systemic arterial network that supplies oxygenated blood to the body, aortas are regions of the large-lumened singular artery arising from the left ventricle of the heart. Starting with the ascending aorta that arises from the left ventricle, the aortas form the main trunk of the systemic arterial system. Before the ascending aorta curves into the aortic arch, right and left coronary arteries branch off to supply the heart with oxy-

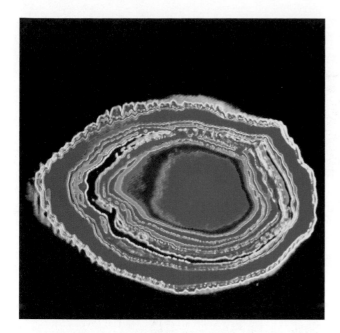

Hardening artery. © 1991 Howard Sochurek/The Stock Market. Reproduced by permission.

See also Anatomy; Blood gas analysis; Blood supply; Circulatory system; Heart diseases; Stroke; Veins.

Arteriosclerosis

Arteriosclerosis literally means "hardening of the arteries." As people age, their **blood** vessel walls naturally grow a bit stiffer and harder, with less flexibility. A common complication of arteriosclerosis is called atherosclerosis. In this condition, plaques (hardened masses composed of lipids, dead cells, fibrous **tissue**, and platelets) collect in the **arteries**. If a plaque grows large enough, it can block the flow of blood through the artery. If this blockage is severe, it can lead to **stroke**. A stroke occurs when an artery in the **brain** becomes blocked, depriving an area of the brain of **oxygen**. Similarly, a **heart** attack occurs when a blocked coronary artery deprives the heart muscle of oxygen. Damage to either the brain or heart occurs when areas of oxygen-deprived tissue die. A severe complication of atherosclerosis occurs when a piece of the plaque breaks off and migrates through the artery to other sites. This bit of circulating plaque is called an **embolism**. An embolism causes damage by blocking blood flow at its destination, resulting in oxygen-deprived tissue and tissue death.

Since atherosclerosis causes strokes and heart attacks, it is considered one of the leading causes of illness and death in the United States. Arteriosclerosis has been linked to high blood **cholesterol** levels, lack of **exercise**, and smoking. Cholesterol is a substance that is similar to **fat** and is found in fatty foods. Although a diet low in cholesterol and fat can reduce the risk of atherosclerosis, some people have a genetic predisposition to high cholesterol levels. Even on normal or low fat diets, these individuals still have high cholesterol (hypercholesterolemia), and therefore a high risk of developing atherosclerosis. Researchers are currently working on ways to help people with hypercholesteremia lower their cholesterol levels. For now, lowering cholesterol is achieved through diet and some drug therapies. In the future, it is hoped that a bioengineered **gene** that makes a cholesterol-neutralizing protein can be implanted into hypercholesteremia patients.

The cause of atherosclerosis

Atherosclerosis means "hardening of a paste-like, fatty material." This pasty, fatty material develops within arteries over a period of many years. If the condition is not treated, this material will eventually harden into atherosclerotic plaques.

Although researchers continue to speculate about how and why this fatty material gets into arteries, a con-

genated blood. Before the aortic arch turns to continue downward (inferiorly) as the descending aorta, it gives rise to a number of important arteries. Branching either directly off of—or from a trunk communicating with the aortic arch—is a brachiocephalic trunk that branches into the right subclavian and right common carotid artery that supply oxygenated blood to the right sight of the head and neck, as well as portions of the right arm.

The aortic arch also gives rise to the left common carotid artery that, along with the right common carotid artery, branches into the external and internal carotid arteries to supply oxygenated blood to the head, neck, **brain**.

The left subclavian artery branches from the aortic arch and—with the right subclavian arising from the brachiocephalic trunk—supplies blood to neck, chest (thoracic wall), central **nervous system**, and arms via axillary, brachial, and vertebral arteries.

In the chest (thoracic region), the continuation of the aortic arch—the descending aorta—is specifically referred to as the thoracic aorta. The thoracic aorta is the trunk of arterial blood supply to the thoracic region. As the thoracic aorta passes through an opening in the diaphragm (aortic hiatus) to become the abdominal aorta, parietal and visceral branches supply oxygenated blood to abdominal organs and structures. The abdominal aorta ultimately branches into left and right common iliac arteries that then branch into internal and external iliac arteries, supplying oxygenated blood to the organs and tissues of the lower abdomen, pelvis, and legs.

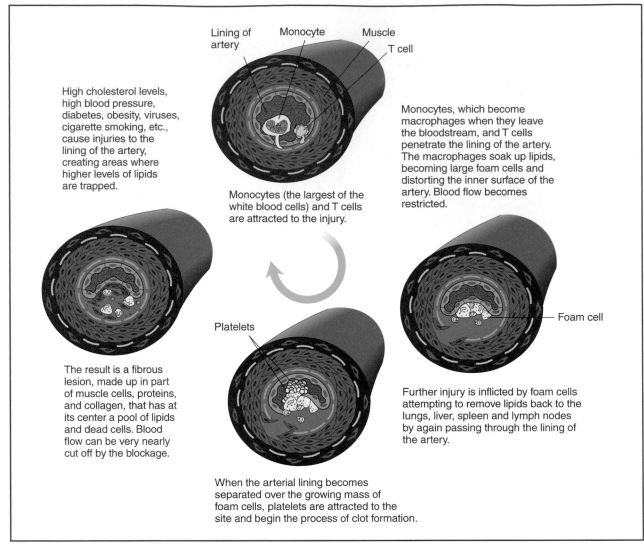

High cholesterol levels, high blood pressure, diabetes, obesity, viruses, cigarette smoking, etc., cause injuries to the lining of the artery, creating areas where higher levels of lipids are trapped.

Lining of artery   Monocyte   Muscle   T cell

Monocytes (the largest of the white blood cells) and T cells are attracted to the injury.

Monocytes, which become macrophages when they leave the bloodstream, and T cells penetrate the lining of the artery. The macrophages soak up lipids, becoming large foam cells and distorting the inner surface of the artery. Blood flow becomes restricted.

Foam cell

Platelets

The result is a fibrous lesion, made up in part of muscle cells, proteins, and collagen, that has at its center a pool of lipids and dead cells. Blood flow can be very nearly cut off by the blockage.

Further injury is inflicted by foam cells attempting to remove lipids back to the lungs, liver, spleen and lymph nodes by again passing through the lining of the artery.

When the arterial lining becomes separated over the growing mass of foam cells, platelets are attracted to the site and begin the process of clot formation.

The progression of arteriosclerosis. *Illustration by Hans & Cassidy. Courtesy of Gale Group.*

sensus has emerged in the last 20 years that points to injury of the artery walls as the underlying cause of atherosclerosis. This injury can be mechanical, such as an artery that has been nicked or cut in some way. Injury can also be in the form of exposure to various agents, such as toxins (including **cigarette smoke**), viruses, or cholesterol. When the artery wall is injured, it tries to heal itself. The response of the artery cells and the **immune system** to the injury leads, paradoxically, to the formation of plaques.

How plaques form

Interestingly, all middle-aged and older people have some form of arteriosclerosis. In fact, some experts consider arteriosclerosis part of the normal wear and tear of aging arteries. Most people will not progress beyond arterioscle-

rosis to atherosclerosis. In some people, however, such as those with hypercholesteremia or people with high-cholesterol diets, hardened plaques form that block the arteries.

Researchers have formed the injury model of atherosclerosis by studying the arteries of people of all ages. In infants, for instance, a "fatty streak" consisting of lipids and immune cells is present in the innermost layer of arteries. Lipids are molecules found in cholesterol. Even at this young age, the arteries are already responding to injury, most likely the presence of cholesterol.

In teenagers and young adults at high risk for atherosclerosis, layers of macrophages (special white cells that ingest foreign material) and smooth muscle cells overlie the interior wall of arteries. In an attempt to reduce the amount of **lipid** within the artery, macrophages ingest the lipid. Because under a **microscope** the lipid within the macrophage appears to be "bubbly," or foamy, a macrophage that has ingested lipid is called a foam **cell**.

As time passes, more and more macrophages ingest more and more lipid, and more foam cells are present within the artery. To rid the artery of the ever increasing load of lipid, foam cells migrate through the artery wall to return the lipid to the liver, spleen, and lymph nodes. As the foam cells penetrate the artery walls, they further injure the artery. This injury attracts platelets, special components of the blood that lead to the formation of blood clots. When you cut your finger, platelets rush to the site of injury and begin forming a clot to stop the flow of blood. In an artery, platelets also rush to the site of injury and begin the process of clot formation. But instead of helping the situation, the clot further complicates the growing plaque in the artery, which at this point consists of lipid-laden macrophages, lipids, muscle cells, and platelets.

The platelets, in turn, release chemicals called growth factors. Several different growth factors are released by the platelets. One causes the cells in the artery to release **proteins** and other substances that eventually lead to the formation of a matrix, a collection of tough protein fibers, that further hardens the artery. Another growth factor prompts the development of blood vessels in the plaque. At this stage, the plaque is fully formed. It is large enough to block the flow of blood through the artery. At its center is a pool of lipid and dead artery cells, and it is covered by a cap of **connective tissue**.

Diagnosis and treatment

Most people are unaware of the atherosclerotic process going on in their arteries. The condition usually goes undiagnosed until a person develops symptoms of blood vessel obstruction, such as the heart **pain** called angina. Angina is a warning sign that the atherosclerosis may become life-threatening. People with angina take medication and are usually monitored by physicians to make sure the arteries do not become completely blocked. Some angina patients are able to slow or halt the progression of atherosclerosis with proper diet and lifestyle changes, such as quitting smoking.

The atherosclerotic plaques can be visualized by a special x-ray technique. A slim tube called a catheter is inserted through an artery in the leg or other location until it reaches the plaque site. Dye that can be seen on **x rays** is shot through the catheter. When an x ray of the site is taken, the plaque can be clearly seen.

Currently, atherosclerosis in the arteries of the heart is treated with drugs, **surgery**, or a technique called angioplasty. Like the x-ray technique, angioplasty involves inserting a catheter through an artery. Instead of shooting

dye through the catheter, the catheter is used as a "roto-rooter" to open up the narrowed arteries. Some researchers have used anti-clotting drugs delivered through the catheter to dissolve clots.

Surgery is also used to treat atherosclerotic obstruction of he heart's arteries. Bypass surgery in which a section of an artery in the leg is used to "bypass" a section of a blocked coronary artery, can allow the heart muscle to again receive adequate oxygen. Because this surgery carries a risk of stroke and other serious complications, angioplasty and drug therapy, in conjunction with diet management, are tried first. If no positive results are found, surgery is performed as a last resort.

One problem with angioplasty is that the solution is often only temporary. The plaques return because angioplasty addresses only the plaques, not the process that leads to their formation. In the future, molecules that target the growth factors released by the platelets may be delivered directly to the plaques through special **catheters**. This treatment is some years away.

Prevention

Because atherosclerosis may be the result of the artery's response to cholesterol, it makes sense to reduce the intake of cholesterol. Two types of cholesterol are found in foods: cholesterol that contains high **density** lipoprotein (the HDLs) and cholesterol that contains low density lipoprotein (the LDLs). Researchers have found that LDL cholesterol is the culprit in atherosclerosis.

To keep the arteries healthy, individuals should eat no more than 300 mg of cholesterol a day. Cholesterol is found only in **animal** products; **plant** foods contain no cholesterol. Since many foods that are high in fat are also high in cholesterol, limiting fat intake can help reduce cholesterol levels. Knowing which foods are high in cholesterol and avoiding these foods (or limiting these foods) can also lower cholesterol. People should have their blood cholesterol levels checked periodically, particularly if there is a family history of arteriosclerosis. Those with hypercholesteremia or a history of heart **disease** may want to try a stricter diet that eliminates all fats and cholesterol. Before embarking on any major dietary change, however, consult your physician.

See also Circulatory system.

Resources

Books

Acierno, Louis J. *The History of Cardiology.* New York: Parthenon Publishing Group, 1994.
Filer, Lloyd J. Jr., Ronald M. Lauer, and Russell L. Leupker, eds. *Prevention of Atherosclerosis and Hypertension Beginning in Youth.* Philadelphia: Lea and Febiger, 1994.

KEY TERMS
. . . .

Angina—Warning pain that signals the progressive worsening of atherosclerosis in heart arteries.

Angioplasty—A technique in which a catheter is introduced into an artery narrowed by atherosclerosis in order to widen the artery.

Arteriosclerosis—Hardening and thickening of artery walls.

Atherosclerosis—Abnormal narrowing of the arteries of the body that generally originates from the buildup of fatty plaque on the artery wall.

Cholesterol—A fat-like substance that contains lipids; found in animal products.

Embolism—A piece of an arteriosclerotic plaque that breaks off and lodges in a distant artery.

Fatty streak—The first stage in atherosclerosis; consists of lipid, macrophages, and immune cells.

Foam cell—A macrophage that has ingested lipid.

Hypercholesteremia—A genetic condition in which the body accumulates high levels of cholesterol.

Lipid—A molecule that is a component of fats and cholesterol.

Macrophages—Special white blood cells that ingest foreign substances or materials such as lipids.

Plaque—A mass of lipid, fibrous tissue, dead cells, and platelets that collects within the artery; plaques can harden and become so large that they block the flow of blood through the artery.

Platelets—Special component of blood that contributes to clot formation.

Fuster, Valentin, ed. *Progression-Regression of Atherogenesis: Molecular, Cellular, and Clinical Bases.* Dallas: American Heart Association, 1992.

Yeagle, Philip. *Understanding Your Cholesterol.* San Diego: Academic Press, 1991.

Periodicals

Ross, Russell. "The Pathenogenesis of Atherosclerosis: A Perspective for the 1990s." *Nature* 362 (April 29, 1993): 801+.

Tunis, Sean R., et al. "The Use of Angioplasty, Bypass Surgery, and Amputation in the Treatment of Peripheral Vascular Diseases." *New England Journal of Medicine* 325, no. 8 (August 22, 1991): 556.

"Warding Off Artherosclerosis." *The Lancet* v361, i9365 (April 12, 2003).

Kathleen Scogna

Arthritis

Arthritis is a term that refers to the **inflammation** of joints (the point where the ends of two bones meet each other). Upwards of 43 million American adults and children (1 out of every 6 citizens) have some form of the more than 100 different types of arthritis.

Inflammation is a reaction of the body to injury. Excess fluid is directed to the affected area, which produces swelling. The fluid is meant to aid the healing process, and is temporary for many injuries. However, in arthritis, the constant or recurring inflammation causes tenderness and stiffness that is debilitating over long periods of time.

In a typical joint, the ends of the bones are covered with a smooth material called cartilage. The cartilage allows the bones to move smoothly against each other. A joint is also wrapped in a network called the synovium. Fluid within the synovium (synovial fluid) helps ease the **friction** of bones rubbing against each other. Finally, the joint is supported and movement is possible because of ligaments, muscles, and tendons that attach to various regions of the joint. All of these components can be subject to arthritic inflammation.

The two most common types of arthritis are osteoarthritis and rheumatoid arthritis. Osteoarthritis is the gradual wearing away of the cartilage. This commonly occurs due to overuse of the joint, or because of an injury such as a fracture. As such, osteoarthritis is associated more with adults than with children or youth.

In rheumatoid arthritis, the synovium surrounding a joint becomes inflamed. Also, the bodies' own **immune**

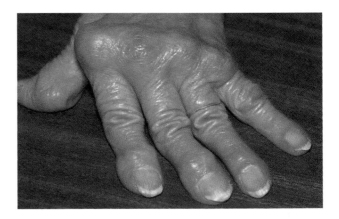

A close-up of a hand deformed by rheumatoid arthritis; the knuckles are swollen and reddened and the fingers curve away from the thumb. The ends of the middle fingers are swollen with cartilage accretion (called Heberden's nodes) that are an indication of osteoarthritis. *Science Photo Library, National Audubon Society Collection/Photo Researchers, Inc. Reproduced by permission.*

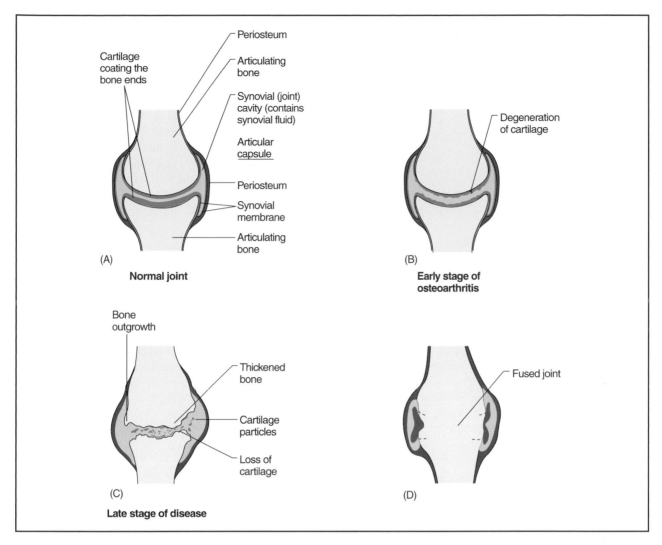

Periosteum

Cartilage
coating the
bone ends

Articulating
bone

Synovial (joint)
cavity (contains
synovial fluid)

Articular
capsule

Periosteum

Synovial
membrane

Articulating
bone

(A)

Normal joint

Degeneration
of cartilage

(B)

**Early stage of
osteoarthritis**

Bone
outgrowth

Thickened
bone

Cartilage
particles

Loss of
cartilage

(C)

Late stage of disease

Fused joint

(D)

The progression of osteoarthritis. *Illustration by Hans & Cassidy. Courtesy of Gale Group.*

system begins to attack and destroy the surface of the joint. This "self against self" immune reaction is typical of autoimmune diseases or conditions like rheumatoid arthritis. Both adults and children are susceptible to rheumatoid arthritis.

The first line of drug treatment for osteoarthritis, as well as other types of arthritis, is nonsteroidal **anti-inflammatory agents** (NSAIDs), including aspirin and medicines that are closely related to aspirin. Some NSAIDs are sold over-the-counter. But those having more potent dosages are sold only by prescription and have to be monitored carefully to avoid adverse side effects.

A second line of drug treatment involving corticosteroids is often needed for treatment of rheumatoid arthritis and other forms of the **disease**. Corticosteroids are used to reduce inflammation. These drugs, such as cortisone, simulate hydrocortisone, a natural chemical produced in the adrenal cortex. The function of com-

pounds like corticosteroids is to try to slow the disease down or make it go into remission. This is done by suppressing the immune response that is key to the damage caused by rheumatoid arthritis.

The length of treatment can range from several days to years. Taken either as a pill or through injection, dosages vary according to the type of arthritis and the needs of the **individual**. Corticosteroids are used for both osteoarthritis and rheumatoid arthritis. These medicines are injected into a specific site, such as a finger joint or the knee, for quick relief from **pain** and inflammation. In addition to blocking prostaglandin production, they also reduce the amount of white **blood** cells that enter into the damaged area of the joint. Though aspirin and the other NSAIDs all work the same way to suppress prostaglandin production in the body, there are major differences in the way individuals will respond to particular NSAIDs.

Stomach bleeding and irritation of the gastrointestinal tract are the two major drawbacks of long-term aspirin and the other NSAID therapy. A newer form of NSAIDs called COX-2 inhibitors reduce the production of an **enzyme** that stimulates the immune response, thereby relieving arthritic inflammation, without blocking the enzyme that protects the stomach lining, thereby reducing stomach irritation. Acetaminophen relieves pain without stomach irritation, but it is not an anti-inflammatory, nor does it reduce the swelling that accompanies arthritis. Because suppression of the immune system can leave someone vulnerable to other infections, and because steroid compounds can have unwanted side effects that are more severe than those produced by the NSAIDs, the use of corticosteroids and other similarly-acting compounds to treat arthritis must be monitored under a physician's care.

See also Autoimmune disorders; Carpal tunnel syndrome; Physiology.

Resources

Books

Mayo Foundation. *Mayo Clinic on Arthritis* Rochester, MN: Mayo Foundation for medical Education & Research, 1999.
Trien, S.F., and D.S. Pisetsky. *The Duke University Medical Center Book of Arthritis.* New York: Fawcett, 1995.

Organizations

Arthritis Foundation. PO Box 7669, Atlanta, GA 30357-0669. (800) 283-7800. <http://www.arthritis.org>.

Bryan Hoyle

Arthropods

Arthropods are **invertebrates** such as **insects**, spiders and other **arachnids**, and crustaceans that comprise the phylum Arthropoda. The phylum Arthropoda includes three major classes—the Insecta, Arachnida, and **Crustacea**.

Arthropods are characterized by their external skeleton, or exoskeleton, made mostly of *chitin*, a complex, rigid **carbohydrate** usually covered by a waxy, waterproof cuticle. This integument is important in reducing **water** loss in terrestrial habitats, in providing protection, and in providing a rigid skeleton against which muscles can work in order to develop **motion** of the **animal** or in its body parts. The exoskeleton is segmented, which allows for easy movement of the body, and there are numerous paired, segmented appendages, such as legs, antennae, and external mouth parts. Periodically, the entire rigid exoskeleton is shed, the temporarily soft animal swells in size, and its new, larger exoskeleton hardens.

Most arthropods have compound eyes, each with numerous lenses capable of forming complex, composite images. Arthropods have various mechanisms for the exchange of respiratory gases which, depending on the group, include gills, chambered structures known as book lungs, tracheal tubes, and various moist areas of the body surface.

Most arthropods exhibit sexual dimorphism, in that the male animals look distinctly different from the females, at least in the appearance of their external genitalia. Arthropods have internal **fertilization**, and they lay eggs. Arthropods have a complex life cycle. This generally involves eggs, a juvenile larval stage, and the adult form, with complex **metamorphosis** occurring during the transitions between these stages. In some insects, there is an additional stage between the larva and the adult, known as a pupa.

Arthropods are extremely diverse in **species** richness. Approximately 874,000 living species of arthropods have been named, comprising more than 80% of all named species of animals. However, some estimates predict large numbers of species of arthropods that have not yet been described and named by biologists. Most of these unnamed species are small **beetles** and other inconspicuous arthropods, and most of these occur in old-growth tropical rainforests, a **biome** that has not yet been well explored and studied by taxonomists and ecologists.

Species of arthropods utilize an enormous variety of Earth's habitats. Most species of crustaceans are aquatic, although a few, such as woodlice and land **crabs**, occur in moist habitats on land. The spiders, **mites**, scorpions, and other arachnids are almost entirely terrestrial animals, as are the extremely diverse insects.

Some species of arthropods are very important to humans. A few species are important as vectors in the transmission of microbial diseases, such as **malaria**, **yellow fever**, **encephalitis**, plague, Chagas **disease**, and **Lyme disease**. Some arthropods are venomous, and can hurt or kill people by single or more multiple stinging, for example, scorpions, some spiders, and **bees** and wasps. Some arthropods are a highly nutritious source of food for people, as is the case of **lobsters**, **crayfish**, **shrimp**, many species of crabs, and some insects.

However, the most critical importance of arthropods relates to the extremely diverse and beneficial ecological functions that they carry out. Arthropods play an important role in nutrient cycling and other aspects of ecological food webs. Earth's ecosystems would be in a great deal of trouble if there was any substantial decline in the myriad species of arthropods in the **biosphere**.

Arthroscopic surgery

Arthroscopic **surgery** is a type of orthopedic surgery that utilizes an instrument called an arthroscope. An arthroscope is a small fiber-optic device that allows the surgeon to view the inside of a joint without a large incision. Most arthroscopic surgery can be performed requiring only three small incisions, each about 0.25 in (6 mm) long. These incisions are called portals. The word arthroscope is from the Greek words meaning "to look at joints." The arthroscope is made up of a **lens** and a **light** source, and is connected to a video camera. The surgeon can view the inside of the joint directly through the arthroscope, or an image may be displayed on a video screen. This image gives the surgeon a clear view of the **tissue** inside the joint. The surgeon can then use other tiny instruments (on the order of only one-eighth of an inch, or 3–4 mm, in diameter) to perform necessary procedures. Arthroscopic surgery can be used as a **diagnosis** tool, or for corrective procedures ranging from easing the **pain** of **arthritis** patients to mending torn ligaments.

Benefits of arthroscopic surgery

The benefits of arthroscopic surgery over traditional open surgery are numerous. Small incisions produce little scarring and heal quickly, allowing patients to return to normal activity in less **time** than with open surgery. There is also a reduced risk of **infection** and swelling associated with small incisions. Diagnosis and surgery can both be performed in one procedure. Skeletal joints, such as the knee, hip, or shoulder, are more easily examined with an arthroscope. There is less pain associated with arthroscopic surgery, and the procedure is usually more successful than open joint surgery. The small instruments used cause less damage to surrounding tissues than larger surgical devices. A typical arthroscopic procedure takes less than one hour, and the patient can return home the same day. A surgeon may perform an arthroscopic procedure in an office, a hospital, or an outpatient surgery center.

Development of the procedure

Only recently has this type of surgery become widespread, despite the fact it has been around for many years. The Japanese physician Kengi Takagi first described the technique in the early 1930s, although it was not perfected until much later. A student of Takagi, Masaki Watanabe, developed the first arthroscope in the late 1960s. This instrument was first used to view the inside of the knee. In the 1970s, fiber-optic technology became readily available and more surgeons began to use the procedure on the hip and shoulder in addition to the knee. In the 1980s, arthroscopic surgery could be performed on smaller joints such as the ankle. Currently, there are over one million arthroscopic surgeries performed each year in the United States alone. The procedure is most often used in sports medicine.

The operation

Arthroscopic procedures are relatively simple. Before knee, ankle, or hip surgery, a patient may want to practice walking on crutches to make the post-operative recovery more comfortable. Fasting is usually required before the procedure because of the use of anesthetics. **Anesthesia** is usually local, although in more complicated surgeries general anesthesia can be given. A spinal anesthesia or epidural is given to numb the patient from the waist down if necessary. A patient may also be given a sedative to help with relaxation during the procedure. Hair will typically need to be shaved around the incision sites. Monitoring devices such as a pulse reader or an EKG may be attached to the patient in some cases. All instruments, as well as the incision sites, must be completely sterilized to reduce the risk of infection.

The surgery itself begins when a tourniquet is wrapped around the joint. The limb is then elevated to drain **blood** away from the joint. Saline **solution** is injected into the joint through a cannula, or drainage tube, first to distend the joint, or cause it to swell, and then to irrigate it, or flush it during the procedure. Swelling of the joint gives the surgeon more room to operate. The continuous flushing washes away any cartilage or other fragments that might break off into the joint during the procedure. The arthroscope is then inserted and the surgeon examines the joint using high-resolution imagery. After an initial diagnosis, a surgical procedure may be performed, such as using a cutter to remove tissue. After the procedure is finished, the instruments are removed from the joint and the saline is squeezed out. Sometimes steroids are injected into the joint to reduce **inflammation**. The incisions are closed with sutures and covered with bandages, thus completing the surgery.

Post-operative procedures include icing the joint and elevating the limb to minimize swelling. Patients are often advised not to drive for 24 hours. Painkillers may be prescribed to alleviate soreness or discomfort. Incisions should be kept clean and dry until they are completely healed. The patient may need to return to the surgeon for removal of sutures. Use of the joint is usually limited for a time after surgery, and **physical therapy** is often recommended to help restore mobility and strength. Complications are uncommon but may include infection or bleeding in the area of the incision.

Types of arthroscopic surgery

Arthroscopy can be used for many different procedures. It can be used in the knee, hip, shoulder, ankle, wrist, or elbow. Diagnoses made with arthroscopes are significantly more accurate than those made based on symptoms alone. One common type of arthroscopic surgery is a meniscectomy, the removal of torn cartilage in the knee. Other common procedures in the knee include the repair of torn cartilage or ligaments and removal of scar tissue on the patella, or kneecap. Hip arthroscopies are used to remove loose bodies, pieces of cartilage or scar tissue that have broken away from surrounding tissue. Other procedures used in the hip joint include removing bone fragments or cartilage build-up. In the shoulder, arthroscopic surgery is used to repair the rotator cuff as well as to remove loose bodies. Arthroscopic procedures can be employed to alleviate the pain of certain types of arthritis. It may also be used with an open surgical procedure, such as an ACL (anterior cruciate ligament) reconstruction.

Resources

Periodicals

Pierce, Michael A. "Arthroscopic Surgery." *Scientific American* (February 1997).

Sekiya, Jon K., et al. "Hip Arthroscopy Using a Limited Anterior Exposure: An Alternative Approach for Arthroscopic Access." *Arthroscopy: the Journal of Arthroscopic and Related Surgery* (January-February 2000).

Jennifer McGrath

Artichoke *see* **Composite family**

Artifacts and artifact classification

Artifacts are often the most intriguing part of archaeological research. Whether priceless or common, artifacts are key to deciphering the archaeological record and garnering information about how people lived in the past. However, most of the information from archaeological excavation is gathered from an artifact's context, or where an artifact is found, and with what other items it is recovered. Artifacts, and their context, help archaeologists describe and compare aspects of past cultures, as well as form a chronology of those cultures. There are also limitations on the amount of scientific information that artifacts alone can provide.

An artifact is any object that was intentionally designed and shaped through human efforts. Some artifacts are discovered by accident, for example, by a farmer plowing his field or by a construction worker digging a building foundation. However, archeological excavation and artifact retrieval always proceeds by well-established methods designed to record as much information as possible about a site and its artifact assemblage, or group of recovered objects.

When collecting artifacts from an archeological site, the archeologist endeavors to establish and document the context in which an artifact was found. To understand context, one must take care to document the artifact's exact horizontal and vertical positions, its relationship to the stratum in which it was found (that is, its stratigraphic position), and any cultural factors that contributed to its location. Each step of the excavation is recorded with detailed maps and photographs of the site. Some archeologists use specially prepared data sheets to record information about recovered artifacts that is later entered into a computer. Recovered artifacts are placed in bags (and sometimes assigned field numbers) before being sent to a laboratory for analysis.

Besides artifacts, archeologists may take sediment samples from a site back to the laboratory for fine-screening. This allows recovery of artifacts that typical field-screening techniques would miss. For example, sediments may provide microscopic pollen grains that will aid paleoclimatic reconstructions. Material from ancient hearths may contain **seeds**, hulls, and small **animal** bones that help archeologists decipher the diet of that site's occupants. Charcoal samples can be retrieved for age dating in the laboratory using carbon-14 (radiocarbon), for example.

In wet or submerged sites, the recovery of artifacts is rendered more difficult by the tendency of the artifacts to disintegrate when dried too rapidly. Even in dry caves, some recovered materials may require special treatment if they are to be preserved. It is important in these cases that an archeological conservator be present at the excavation site to assist in the recovery of artifacts. Delicate pieces may be protected in plaster, polyurethane foam, resin, or latex rubber.

After an artifact's position has been mapped and recorded in field notes, the artifact is taken to the site laboratory to be cleaned and labeled. Artifacts are then sorted according to type of material, e.g., stone, ceramic, **metal**, **glass**, or bone, and after that into subgroups based on similarities in shape, manner of decoration, or method of manufacture. By comparing these object groupings with the stratigraphic positions in which the objects were found, the archeologist has a basis for assigning relative ages (older vs. younger) to the objects.

The objects are finally wrapped for transfer to an off-site laboratory. That any off-site processing be performed quickly is desirable so that the documentation of all artifacts found at the site may proceed without delay. Each specialist involved in an excavation will usually be responsible for writing a report of his or her findings at the site. These reports will later be collectively published in scientific journal or book as the site report. Often, archaeologists are responsible for writing a report that details how the site should be managed, further excavated, or preserved.

Artifact classification

The notion that artifacts can be classified into types rests on the principles of typological analysis. Typology is the study of artifacts based on observable traits such as form, methods of manufacture, and materials. Classification should not be based on an artifact's function because this often cannot be unambiguously determined. According to this concept, within any given region, artifacts that are similar in form or style were produced at about the same

time, and stylistic changes are likely to have been gradual or evolutionary. Typologic categories are, however, only arbitrary constructions used by archeologists to come to terms with the archeological record. There is consequently no single, best way to classify artifacts into types.

A typologist first classifies artifacts in terms of attributes, for example, raw material, **color** and size. The typologist then classifies artifacts using mutually exclusive characteristics, called attribute states. Thus, ceramic pots could be sorted on the basis of shape, for example, into groups of bowls, jars, and plates.

While the typologist is generally satisfied using attribute types to describe artifacts, the goal of archeological classification is to so completely describe artifacts that they can be easily compared with objects from other sites. Unfortunately, this is much easier said than done. Because typology is largely based on an almost intuitive ability to recognize types, or requires several typological guides or other comparative resources, it has been refuted by practitioners of more rigorous analysis. By employing mathematical and statistical analyses, typologists have tried to demonstrate that attribute types are not arbitrary, and that their use provides significant, reproducible results.

When statistical methods were introduced into this field in the early 1950s, typologists claimed that artifact types in each culture were inherent, and that they could be determined by statistical analysis. Since that time, however, this assumption has been repeatedly questioned by those who doubt that the existence of absolute types can or will ever be verified.

At one time, typological analysis in conjunction with careful excavation provided archeologists with their only basis for reconstructing cultural and historical sequences. But with the advent of absolute **dating techniques**, archeologists required a less arbitrary basis for classifying artifacts.

While typology is no longer the standard means of dating the components of most artifact assemblages, typological analysis is still one of the most useful means of describing and comparing various artifacts. Absolute dating methods used for prehistoric and old historic sites cannot be applied to more modern sites. In recent decades, archaeological excavation and analysis of more recent historical sites has increased. Urban, or industrial **archaeology** focuses on artifacts produced during and after the **Industrial Revolution** of the eighteenth and nineteenth centuries, most especially those sites associated with manufacturing or related urban neighborhoods. Many artifacts recovered from these sites have decorations, maker's marks, or shapes that are easily identifiable because they are documented in the historical record or resemble something still used today. For example, historic archaeologists of the modern period sometimes rely on antique guides, old photographs, or factory archives to identify and date an object. While more definite because of the wealth of collaborative material, this type of analysis is still a form of typology.

See also Archaeology; Archaeometallurgy; Archaeometry; Stratigraphy (archeology).

Resources

Books

Fagan, Brian M., ed. *The Oxford Companion to Archeology.* New York: Oxford University Press, 1996.

Maloney, Norah. *The Young Oxford Book of Archeology.* New York: Oxford University Press, 1997.

Lyman, R. Lee. Michael J. O'Brien. *Seriation, Stratigraphy, and Index Fossils - The Backbone of Archaeological Dating* New York: Kluwer Academic Publishers, 1999.

Nash, Stephen Edward, ed. *It's about Time: A History of Archaeological Dating in North America.* Salt Lake City, UT: University of Utah Press, 2000.

Randall Frost

Artificial fibers

Polymeric fibers

Most synthetic fibers are polymer-based, and are produced by a process known as spinning. This process involves extrusion of a polymeric liquid through fine holes known as spinnerets. After the liquid has been spun, the resulting fibers are oriented by stretching or drawing. This increases the polymeric chain orientation and degree of crystallinity, and has the effect of increasing the modulus and tensile strength of the fibers. Fiber manufacture is classified according to the type of spinning that the **polymer** liquid undergoes: this may be melt spinning, dry spinning, or wet spinning.

Melt spinning is the simplest of these three methods, but it still requires that the polymer constituent be stable above its melting **temperature**. In melt spinning, the polymer is melted and forced through the spinnerets, which may contain from 50-500 holes. The diameter of the fiber immediately following extrusion exceeds the hole diameter. During the cooling process, the fiber is drawn to induce orientation. Further orientation may later be achieved by stretching the fiber to a higher draw **ratio**. Melt spinning is used with polymers such as nylon, polyethylene, polyvinyl chloride, **cellulose** triacetate, and polyethylene terephthalate, and in the multifilament extrusion of polypropylene.

In dry spinning, the polymer is first dissolved in a solvent. The polymer **solution** is extruded through the spinnerets. The solvent is evaporated with hot air and collected for reuse. The fiber then passes over rollers, and is stretched to orient the molecules and increase the fiber strength. Cellulose acetate, cellulose triacetate, acrylic, modacrylic, aromatic nylon, and polyvinyl chloride are made by dry spinning.

In wet spinning, the polymer solution is spun into a coagulating solution to precipitate the polymer. This process has been used with acrylic, modacrylic, aromatic nylon, and polyvinyl chloride fibers. Viscose rayon is produced from regenerated cellulose by a wet spinning technique.

Table 1 (Artificial Polymeric Fibers) provides detailed information about each of the important classes of spun fibers.

Other synthetic fibers

Besides the polymer-based synthetic fibers described above, there are other types of synthetic fibers that have special commercial applications. These include the fibers made of **glass**, **metal**, **ceramics**, and **carbon** described in Table 2 (Artificial, Nonpolymeric Fibers).

Resources

Books

Basta, Nicholas. *Shreve's Chemical Process Industries.* New York: McGraw-Hill Book Company. 1999.

TABLE 1. ARTIFICIAL POLYMERIC FIBERS

ACRYLIC

Compostion	At least 85% acrylonitrile units. Anidex is a cross-linked polyacrylate consisting of at least 50 wt% esters of a monohydric alcohol and acrylic acid.
Processing	Orlon® is made by dissolving acrylonitrile in an organic solvent. The solvent is filtered and dry-spun. Fibers are drawn at high temperature to 3 to 8 times their original length and the molecules are oriented into long parallel chains.
Properties	Resistant to dilute acids and alkalies, solvents, insects, mildew, weather. Damaged by alkalies and acids, heat above 356° F (180° C), acetone, ketones.
Uses	Sweaters, women's coats, men's winter suiting, carpets, blankets, outdoor fabrics, knits, fur-like fabrics, blankets. Orlon® and Acrilan® have been used as wool substitutes.
Trade Names	Orlon® (E.I. duPont de Nemours & Co., Inc.), Acrilan® (Monsanto Co.), Cantrece® (E.I. duPont de Nemours & Co., Inc.).

MODACRYLIC

Composition	35 to 85% acrylonitrile units.
Processing	Union Carbide makes Dynel®, a staple copolymer modacrylic fiber made from resin of 40% acrylonitrile and 60% vinyl chloride. It is converted into staple in a continuous wet-spinning process. The resin powder dissolved in acetone, filtered and spun. Fiber is dried, cut and crimped.
Uses	Dynel® resembles wool. Used for work clothing, water-softener bags, dye nets, filter cloth, blankets, draperies, sweaters, pile fabric.
Trade Names	Verel®-copolymer (Tennessee Eastman Co.), Dynel®-copolymer (E.I. duPont de Nemours & Co., Inc.)

POLYESTER

Composition	85% ester of a dihydric alcohol and terephthalic acid.
Processing	Melt spinning.
Properties	Resistant to weak acids and alkalies, solvents, oils, mildew, moths. Damaged by phenol, heat above 338° F (170° C).
Uses	Apparel, curtains, rope, twine, sailcloth, belting, fiberfill, tire cord, belts, blankets, blends with cotton.

TABLE 1. ARTIFICIAL POLYMERIC FIBERS (cont'd)

Trade Names	Dacron® (E.I. duPont de Nemours & Co., Inc.), Kodel® (Tennessee Eastman Co.), Fortrel® (Fiber Industries, Inc.
	RAYON
Composition	The pioneer artificial fibers viscose, cuprammonium cellulose, and cellulose acetate were originally referred to as rayon. This name now is reserved for viscose. Rayon is now a generic name for a semisynthetic fiber composed of regenerated cellulose and manufactured fibers consisting of regenerated cellulose in which substituents have replaced not more than 15% of the hydroxyl group hydrogens.
Processing	Rayon was first made by denitration of cellulose nitrate fibers, but now most is made from wood pulp by the viscose process. The viscose process produces filaments of regenerated cellulose. First, a solution of cellulose undergoes chemical reaction, ageing, or solution ripening; followed by filtration and removal of air; spinning of fiber; combining of the filaments into yarn; and finishing (bleaching, washing, oiling, and drying).
Properties	Rayon can be selectively dyed in combination with cotton. Hydroxyl groups in the cellulose molecules cause the fiber to absorb water—this causes low wet strength. In the dry state, the hydroxyl groups are hydrogen bonded, and the molecules are held together. Thus the dry fibers maintain their strength even at high temperatures.
Uses	High tenacity viscose yarn is used in cords for tires, hose, and belting. Strength is achieved by orienting the fiber molecules when they are made. Textile rayon is used primarily in women's apparel, draperies, upholstery, and in blends with wool in carpets and rugs. Surgical dressings.
	ACETATE (CELLULOSE ACETATE)
Composition	Cellulose acetate and its homologs are esters of cellulose and are not regenerated cellulose. Where not less than 92% of the hydroxyl groups are acetylated, the product is called a triacetate.
Processing	Cellulose is converted to cellulose acetate by treatment with a mix containing acetic anhydride. No process exists whereby the desired number of acetyl groups can be achieved directly. The process involves first producing the triacetate, then hydrolyzing a portion of the acetate groups. The desired material is usually about half way between triacetate and diacetate. Arnel® (cellulose triacetate) is produced by Celanese Corp. It is a machine-washable fiber, that shows low shrinkage when stretched, and has good crease and pleat resistance.
Uses	Blankets, carpets, modacrylic fibers, cigarette filters.
Trade Names	Arnel® (Celanese Corp.).

TABLE 1. ARTIFICIAL POLYMERIC FIBERS (cont'd)

VINYLS AND VINYLIDENES

Composition	Saran is a copolymer of vinyl chloride and vinylidene chloride (at least 80% by weight vinylidene chloride units). Vinyon is the trade name given to copolymers of 90% vinyl chloride and 10% vinyl acetate (at least 85% vinyl chloride units).
Processing	Saran is prepared by mixing the two monomers and heating. The copolymer is heated, extruded at 356° F (180° C), air-cooled, and stretched. The vinyon copolymer is dissolved in acetone to 22% solids and filtered. The fibers are extruded by dry spinning, left to stand, then wet-twisted and stretched. The fibers are resistant to acids and alkalies, sunlight, and aging. Vinyon is useful in heat-sealing fabrics, work clothing. Bayer chemists first spun polyvinyl chloride into this chemically resistant and rot-proof fiber in 1931.
Properties	These fibers are resistant to mildew, bacterial, and insect attack. Polyvinyl chloride is resistant to acids and alkalies, insects, mildew, alcohol, oils. Polyvinylidene chloride is resistant to acids, most alkalies, alcohol, bleaches, insects, mildew, and weather. Polyvinyl chloride is damaged by ethers, esters, aromatic hydrocarbons, ketones, hot acids, and heat above 158° F (70° C). Polyvinylidene chloride is damaged by heat above 194° F (90° C) and by many solvents.
Uses	Saran can be used for insect screens. Widest use is for automobile seat covers and home upholstery. Typical polyvinyl chloride uses include nonwoven materials, felts, filters, and blends with other fibers. Typical polyvinylidene chloride uses include outdoor fabrics, insect screens, curtains, upholstery, carpets, work clothes.
Trade Names	Proprietary polyvinylidene chloride names include Dynel®-copolymer (Uniroyal, Inc.). Saran is a generic name for polyvinylidene chloride.

NYLON

Composition	Nylon is a generic name for a family of polyamide polymers characterized by the presence of an amide group. Common types are nylon 66, nylon 6, nylon 4, nylon 9, nylon 11, and nylon 12.
Processing	Nylon 66 was developed by Carothers by reacting adipic acid and hexamethylenediamine in 1935. Nylon 6 is based on caprolactam. It was developed by I.G. Farbenindustrie in 1940.
Properties	Resistant to alkalies, molds, solvents, moths. Damaged by strong acids, phenol, bleaches, and heat above 338° F (170° C).
Uses	Typical uses include tire cord, carpets, upholstery, apparel, belting, hose, tents, toothbrush bristles, hairbrushes, fish nets and lines, tennis rackets, parachutes, and surgical sutures.
Trade Names	Chemstrand nylon® (Monsanto Co.).

TABLE 1. ARTIFICIAL POLYMERIC FIBERS (cont'd)

SPANDEX

Composition	At least 85% by weight segmented polyurethane.
Processing	Segmented polyurethanes are produced by reacting diisocyanates with long-chain glycols. The product is chain-extended or coupled, then converted to fibers by dry spinning.
Properties	Resistant to solvents, oils, alkalies, insects, oxidation. Damaged by heat above 284° F(140° C), strong acids.
Uses	Fibers are used in foundation garments, hose, swimwear, surgical hose, and other elastic products.
Trade Names	Proprietary names include Lycra® (E.I. duPont de Nemours & Co., Inc.), Spandelle® (Firestone Synthetic Fibers Co.).

OLEFIN

Composition	At least 85 wt% ethylene, propylene, or other olefin units other than amorphous rubber polyolefins.
Processing	Polymer is spun from a melt at about 212° F (100° C) above the melting point because the polymer is very viscous near its melting point.
Properties	Difficult to dye. Low melting points. Polypropylene has a very low specific gravity, making it very light and suitable for blankets. It has 3 to 4 times the resistance of nylon to snags and runs, and is softer, smoother, and lighter. Polypropylene is resistant to alkalies, acids, solvents, insects, mildew. Polyethylene is resistant to alkalies, acids (except nitric), insects, mildew. Polypropylene is damaged by heat above 230° F (110° C). Polyethylene is damaged by oil and grease, heat above 212° F (100° C), oxidizers.
Uses	Olefins make excellent ropes, laundry nets, carpets, blankets, and carpet backing. Polypropylene is typically used for rope, twine, outdoor fabrics, carpets, upholstery. High density polyethylene is typically used for rope, twine, and fishnets. Low density polyethylene is typically used for outdoor fabrics, filter fabrics, decorative coverings.
Trade Names	Proprietary names for polypropylene include Herculon® (Hercules Powder Co.), Polycrest® (Uniroyal, Inc.); proprietary polyethylene names include DLP® (W. R. Grace & Co.).

FLUOROCARBON

Composition	Long chain carbon molecules with bonds saturated by fluorine. Teflon is polytetrafluoroethylene.
Processing	Fluorocarbon sheets are made by combining polytetrafluoroethylene with another microgranular material to form thin, flexible sheets. The filler is then dissolved out, leaving a pure, porous polytetrafluroethylene sheet. The pores must be small enough to be vapor permeable but large enough to extend through the fabric. The sheet has to be very thin, and is therefore fragile. GoreTex® has to protected by being sandwiched between robust fabrics such as polyester or nylon weave.

TABLE 1. ARTIFICIAL POLYMERIC FIBERS (cont'd)	
Properties	As a fiber, Teflon is highly resistant to oxidation and the action of chemicals, including strong acids, alkalies, and oxidizing agents. It is nonflammable. It retains these properties at high temperatures 446°-554° F (230°-290° C). It is strong and tough. It exhibits low friction, which coupled with its chemical inertness, makes it suitable in pump packings and shift bearings. Polytetrafluoroethylene is resistant to almost all chemicals, solvents, insects, mildew. Polytetrafluoroethylene is damaged by heat above 482° F (250° C), and by fluorine at high temperatures.
Uses	Typical uses include corrosion-resistant packings, etc., tapes, filters, bearings, weatherproof outdoorwear.
Trade Name	Teflon® (E.I. duPont de Nemours & Co., Inc.) and GoreTex® (W. L. Gore & Associates) are proprietary names.
VINAL	
Composition	Vinal is the U.S. term for vinyl alcohol fibers. At least 50 wt% of the long synthetic polymer chain is composed of vinyl alcohol units. The total of the vinyl alcohol units and any one or more of the various acetal units is at least 85 wt% of the fiber.
Processing	Vinyl acetate is first polymerized, then saponified to polyvinyl alcohol. The fiber is spun, then treated with formalin and heat to make it insoluble in water. Production has remained largely confined to Japan.
Properties	The fiber has reasonable tensile strength, a moderately low melting point (432° F [222° C]), limited elastic recovery, good chemical resistance, and resistance to degradation by organisms. Has good chemical resistance, low affinity for water, good resistance to mildew and fungi. Combustible. Used for fishing nets, stockings, gloves, hats, rainwear, swimsuits. Polyvinyl alcohol is resistant to acids, alkalies, insects, mildew, oils; it is damaged by heat above 320° F (160° C), phenol, cresol, and formic acid.
Uses	Used in bristles, filter cloths, sewing thread, fishnets, and apparel. Polyvinyl alcohol is typically used for a wide range of industrial and apparel uses, rope, work clothes, fish nets.
AZLON	
Composition	Any regenerated naturally occurring protein. Azlon is the generic name for manufactured fiber in which the fiber-forming substance is composed of any regenerated naturally occurring protein. Proteins from corn, soybeans, peanuts, fish, and milk have been used. Azlon consists of polymeric amino acids.

TABLE 1. ARTIFICIAL POLYMERIC FIBERS (cont'd)

Processing	Vegetable matter is first crushed and the oil is extracted. Then the remaining protein matter is dissolved in a caustic solution and aged. The resulting viscous solution is extruded through a spinneret into an acid bath to coagulate the filaments. The fiber is cured and stretched to give it strength. Then the fiber is washed, dried, and used in a filament or staple form. Casein fibers are extracted from skim milk, then dissolved in water, and extruded under heat and pressure. The filaments are hardened and aged in an acid bath. Then they are washed, dried, and used in either filament or staple form. Seaweed fibers are made by extracting sodium alginate from brown seaweed using an alkali. The resulting solution is purified, filtered, and wet spun into a coagulating bath to form the fibers.
Properties	Azlon fibers have a soft hand, and blend well with other fibers. Combustible.
Uses	Used in blends to add a wool-like hand to other fibers, and to add loft, softness, and resiliency. Protein fibers resist moths and mildew, do not shrink, and impart a cashmere-like hand to blended fabrics.

TABLE 2. ARTIFICIAL, NONPOLYMERIC FIBERS

Glass	
Composition	Comprised primarily of silica.
Processing	Continuous filament process—molten glass drawn into fibers, which are wound mechanically. Winding stretches the fibers. Subsequently formed into glass fiber yarns and cords. Staple fiber process—uses jets of compressed air to attenuate or draw out molten glass into fine fibers. Used for yarns of various sizes. Wool process—molten glass attenuated into long, resilient fibers. Forms glass wool, used for thermal insulation or fabricated into other items.
Properties	Nonflammable. Can be subjected to temperatures as high as 1200° F (650° C) before they deteriorate. Non-absorbent Impervious. They resist most chemicals, and do not break when washed in water. Mothproof, mildew-proof, do not degrade in sunlight or with age. Strong. Among the strongest man-made fibers. Derived from limestone, silica sand, boric acid, clay, coal, and fluospar. Different amounts of these ingredients result in different properties.

TABLE 2. ARTIFICIAL, NONPOLYMERIC FIBERS (cont'd)

Uses	Decorative fabrics, fireproof clothing, soundproofing, plastics reinforcement, tires, upholstery, electrical and thermal insulation, air filters, insect screens, roofing, ceiling panels.
	METAL
Composition	Whiskers are single-crystal fibers up to 2 in (5 cm) long. They are made from tungsten, cobalt, tantalum, and other metals, and are used largely in composite structures for specialized functions. Metallic yarns consist of metallized polyester film.
Processing	Filaments are alloys drawn through diamond dies to diameters as small as 0.002 cm. In ancient times, gold and silver threads were widely used in royal and ecclesiastical garments. Today, metallic yarns usually consist of a core of single-ply polyester film that is metallized on one side by vacuum-depositing aluminum. The film is then lacquered on both sides with clear or tinted colors.
Properties	Metallic whiskers may have extremely high tensile strength. Modern metallic yarns are soft, lightweight, and do not tarnish.
Uses	Whiskers are used in biconstituent structures composed of a metal and a polymeric material. Examples include aluminum filaments covered with cellulose acetate butyrate. Steels for tire cord and antistatic devices have also been developed. Metallic yarns are used for draperies, fabrics, suits, dresses, coats, ribbons, tapes, and shoelaces.
Trade Names	Producers of metallic yarns include Metlon Corp.; Metal Film Co.; and Multi-tex Products Co. Dobeckman Co. produced the first widely used metallic yarn under the tradename Lurex®.
	CERAMIC
Composition	Alumina and silica.
Processing	Insertion of aluminum ions into silica.
Properties	Retains properties to 2300° F (1260° C), and under some conditions to 3000° F (1648° C), lightweight, inert to most acids and unaffected by hydrogen atmosphere, resilient.
Uses	Used for high temperature insulation of kilns and furnaces, packing expansion joints, heating elements, burner blocks, rolls for roller hearth furnaces and piping, fine filtration, insulating electrical wire and motors, insulating jet motors, sound deadening.
Trade Names	Fiberfrax® (Carborundum).
	CARBON
Composition	Carbonized rayon, polyacrylonitrile, pitch or coal tar.

TABLE 2. ARTIFICIAL, NONPOLYMERIC FIBERS (cont'd)

Processing	High modulus carbon fibers are made from rayon, polyacrylonitrile, or pitch. Rayon fibers are charred at 413°-662° F (200°-350° C), then carbonized at 1832°-3632° F (1000° to 2000° C). The resulting carbon fibers are then heat treated at 5432° F (3000° C) and stretched during the heat treatment. Carbon fibers are also obtained by heat treating polyacrylonitrile, coal tar or petroleum pitch.
Properties	Capable of withstanding high temperatures.
Uses	Carbon fibers come in three forms. Low modulus fibers used for electrically conducting surfaces. Medium modulus fibers used for fabrics. High modulus fibers used for stiff yarn. Used in the manufacture of heat shields for aerospace vehicles and for aircraft brakes. Carbon fibers are also used to reinforce plastics. These plastics may be used for sporting goods and engineering plastics.

A scanning electron micrograph (SEM) of fibers of a dacron polyester material used in sleeping bags. The core of each fiber has up to seven air cavities that increase its insulating ability. *Photograph. Science Photo Library/National Audubon Society Collection/Photo Researchers, Inc. Reproduced by permission.*

KEY TERMS

Fiber—A complex morphological unit with an extremely high ratio of length to diameter (typically several hundred to one) and a relatively high tenacity.

Nonpolymeric fibers—Fibers made of materials other than polymeric liquid, including glass, metal, ceramics, and carbon. These fibers have special commercial applications.

Polymer—A substance, usually organic, composed of very large molecular chains that consist of recurring structural units. From the Greek "poly" meaning many, and "meros" meaning parts.

Polymeric fibers—Fibers created by "spinning" and divided into the following classes: acrylic, modacrylic, rayon, acetate, vinyls and vinylidenes, nylon, spandex, olefin, fluorocarbon, vinal, and azlon.

Spinning—The process by which most polymer-based synthetic fibers are created. It involves extrusion of polymeric liquid through spinnerets. Fiber manufacture is classified by the type of spinning involved: melt spinning, dry spinning, or wet spinning.

Synthetic—Referring to a substance that either reproduces a natural product or that is a unique material not found in nature, and which is produced by means of chemical reactions.

Jerde, Judith. *The New Encyclopedia of Textiles.* New York: Facts on File, 1992.

Lynch, Charles T. *Practical Handbook of Materials Science.* Boca Raton, FL: CRC Press, Inc. 1989.

Sperling, L. H. *Introduction to Physical Polymer Science.* New York: John Wiley & Sons, Inc. 1992.

Artificial heart and heart valve

An artificial **heart** is a manmade device that is intended to replace the heart muscle that pumps approximately 2,000 gal (7,571 L) of **blood** through the body each day. The heart muscle is composed of several chambers and the blood flow into and out of the chambers is controlled by a system of valves. Valve failure can lead to congestive heart failure, pulmonary **edema**, and other serious cardiovascular illnesses. Replacement of diseased or defective heart valves with artificial valves can be a solution to restore heart function.

Four one-way valves control the movement of blood into, through, and out of the human heart. All are designed to permit the flow of blood in one direction only and to prevent its backflow. Blood returning from the body to the heart enters the right atrium and from there passes through a valve called the tricuspid (or right atrioventricular) valve into the right ventricle. The right ventricle pumps the blood through another valve (the semilunar valve) to the lungs where **carbon dioxide** is removed and the blood is infused with fresh **oxygen**. Blood returns to the left atrium and is pumped into the left ventricle through the bicuspid or mitral (or left atrioventricular) valve. When the left ventricle contracts, it forces blood through the aortic semilunar valve into the aorta and on through the body.

Heart valves are constructed of a pair of flaps. The flaps are made of strong, thin, and fibrous material that is connected by strong fibers to muscles within the main heart muscle. This construction allows the flaps to remain shut against back **pressure**, thus allowing blood to flow only one way through them.

When the valves malfunction because of **birth defects**, deposits of **cholesterol** or **calcium** on the leaflets

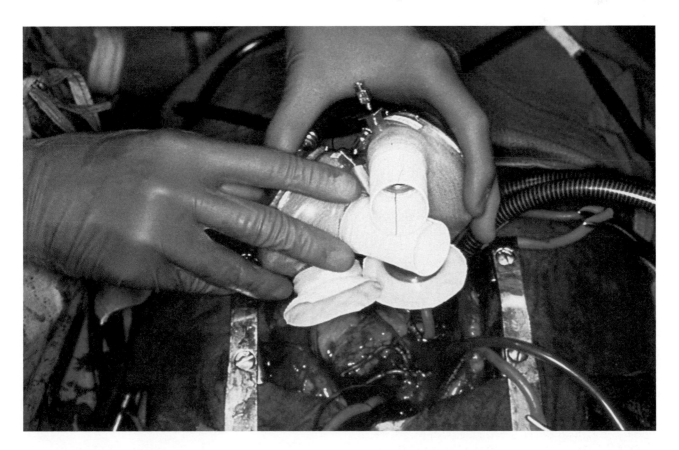

Surgical implantation of a Jarvik artificial heart. *NIH, National Audubon Society Collection/Photo Researchers, Inc. Reproduced by permission.*

of the valves, or because of a rheumatic **disease**, they do not close completely. Backflow of blood can result (a physician hears this as a heart murmur). Though a small **volume** of backflow is not harmful, a deteriorating valve that permits greater and greater amounts of blood to pass back through it can have serious consequences on the heart itself.

It is often necessary to replace the natural valve with a man-made artificial device. A one-way valve is a rather simple device to construct to function outside of the body. Inside the body, however, the valve must perform flawlessly and unceasingly for many years. The early mechanical valves, though adequate for controlling blood flow, tended to fracture after some years of use. The **metal** cage enclosing the valve broke under repeated and constant taps from the movable part of the valve. A fractured valve means a malfunctioning valve, possibly the formation of a blood clot and potential heart attack or **stroke**.

One modern design of an artificial valve consists of a ring by means of which the valve is sewn into place in the heart and some means of controlling the flow of blood. One artificial valve known as the ball-and-cage model has a three-pronged cage within which is a ball. The ball lifts to allow blood to pass through and is pressed down into the valvular opening to seal it and prevent backflow. Another design, called a disk-and-cage valve, has a similar action, except the ball is replaced by a flat disk that swivels back and forth to open and close the passage.

A more modern replacement valve uses the valve from a pig heart, treated to prevent rejection by the human **immune system**, mounted into a ring. The ring is sewn into the heart. The leaflets in the pig valve behave more like those in the human heart with less danger of breakage than occurs in the cage-type valves.

The use of artificial hearts as a replacement for an ailing human heart began in 1953. Then, the use of a **heart-lung machine** designed by a physician named John Gibbon in open-heart **surgery** demonstrated that an artificial device could, at least for a time, replace the real heart. In 1966, Michael DeBakey implanted a partial artificial heart. From 1982 to 1985, William DeVries carried out a series of implants of a device called the Jarvik-7 artificial heart. The first heart was implanted in Barney Clark, who lived for 112 days before dying of complications caused by the device. Mechanical and other problems ultimately stopped the use of the Jarvik-7. In 2000, the modified version of the Jarvik heart (the Jarvik 2000) was implanted. This was the first completely artificial heart to be installed.

In July of 2001, surgeons at Jewish Hospital in Louisville, Kentucky, implanted an artificial heart (the

KEY TERMS

Aorta—The major blood vessel leaving the heart and carrying blood to numerous other smaller vessels that branch off and deliver blood to the entire body.

Atrium (plural: atria)—One of two upper chambers of the heart. They are receiving units that hold blood to be pumped into the lower chambers, the ventricles.

Congestive heart failure—A heart condition in which the heart has enlarged because of back pressure against its pumping action. The heart will finally become so large and inefficient that it is unable to pump enough blood to maintain life.

Leaflets—Fibrous tissue flaps that open and close in the heart valves to allow the passage of blood.

Ventricles—The two lower chambers of the heart; also the main pumping chambers.

AbioCor® heart) in Robert Tools, a 59-year-old man whose own heart was damaged and failing. The AbioCor® heart was the first self-contained artificial heart to be implanted in a human. It is also the smallest artificial heart yet devised, being about the size of a softball.

Tools suffered a serious stroke and died in November, 2001. Nonetheless, his progress since the installation of the artificial heart was encouraging. As of May 2003, at least eight patients received AbioCor® hearts.

See also Anatomy, comparative; Birth defects; Cholesterol; Electrocardiogram (ECG); Pacemaker.

Resources

Periodicals

Guy, T.S. "Evolution and Current Status of the Total Artificial Heart. *American Society for Artificial Internal Organs Journal* 44 (1998): 28–32.

Organizations

American Society for Artificial Internal Organs, PO Box C, Boca Raton, FL 33429. (561) 391–8589. [cited October, 19, 2002]. <http://www. asaio.org>.

Other

"How Artificial Hearts Work." How Stuff Works [cited October 19, 2002]. <http://www. howstuffworks.com/artitificial-heart.htm.>.

Brian Hoyle

Artificial intelligence

Certain tasks can be performed faster and more accurately by traditionally programmed computers than by human beings, particularly numerical computation and the storage, retrieval, and sorting of large quantities of information. However, the ability of computers to interact flexibly with the real world—their "intelligence"—remains slight. Artificial intelligence (AI) is a subfield of computer science that seeks to remedy this situation by creating software and hardware that possess some of the behavioral flexibility shown by natural intelligences (people and animals).

In the 1940s and 1950s, the first large, electronic, digital computers were designed to accomplish specific tasks (e.g., a numerical calculation set up by a human programmer) by completing a series of clearly defined steps, an **algorithm**. Programmers wrote algorithmic software that precisely specified both the problem and how to solve it. AI programmers, in contrast, seek to program computers not with rigid algorithms but with flexible rules for seeking solutions. An AI program may even be designed to modify the rules it is given or to develop entirely new rules.

What types of problem are appropriate for traditional, algorithmic computing and what types call out for AI? Some of the tasks that are hardest for people are, fortunately, algorithmic in nature. Teachers can describe in detail the process for multiplying numbers, and accountants can state accurately the rules for completing tax forms, yet many people have difficulty performing such tasks. Straightforward, algorithmic programs can perform them easily because they can be broken down into a series of precise procedures or steps that do not vary from case to case. On the other hand, tasks that require little thought for human beings can be hard to translate into algorithms and therefore difficult for computers to perform. For example, most people know that a pot of boiling **water** requires careful handling. We identify hot pots by flexible recognition of many possible signs: steam rising, radiant **heat** felt on the skin, a glimpse of blue flame or red coils under the pot, a rattling lid, and so forth. Once we know that the pot is boiling, we plan our actions accordingly. This process seems simple, yet describing exactly to a computer how to reliably conclude "this pot is hot" and to take appropriate action turns out to be extremely difficult. The goal of AI is to create computers that can handle such complex, flexible situations. One obstacle to this goal is uncertainty or confusion about what is intelligence.

What is intelligence?

One possible definition of intelligence is the acquisition and application of knowledge. An intelligent entity, on this view, is one that learns—acquires knowledge—and is able to apply this knowledge to changing real-world situations. In this sense, a rat is intelligent, but most computers, despite their impressive number-crunching capabilities, are not. To qualify as intelligent, an AI system must use knowledge (whether acquired from databases, sensory devices, trial and error, or all of the above) to make effective choices when confronted with data that are to some extent unpredictable Insofar as a computer can do this, it may be said, for the purposes of AI, to display intelligence. Note that this definition is purely functional, and that in AI the question of consciousness, though intriguing, need not be considered.

This limited characterization of intelligence would, perhaps, have been considered overcautious by some AI researchers in the early days, when optimism ran high. For example, U.S. economist, AI pioneer, and Nobel Prize winner Herbert Simon (1916–2001) predicted in 1965 that by 1985 "machines will be capable of doing any work man can do." Yet over 35 years later, despite exponential growth in memory size and processing speed, no computer even comes close to commonplace human skills like conversing, driving a car, or diapering a baby, much less "doing any work" a human being can do. Why has progress in AI been so slow?

One answer is that while intelligence, as defined above, requires knowledge, computers are only good at handling information, which is not the same thing. Knowledge is meaningful information, and "meaning" is a nonmeasurable, multivalued variable arising in the real world of things and values. Bits—"binary digits," 1s and 0s—have no meaning, as such; they are meaningful only when people assign meanings to them. Consider a single bit, a "1": its information content is one bit regardless of what it means, yet it may mean nothing or anything, including "The circuit is connected," "We surrender," "It is more likely to rain than snow," and "I like apples." The question for AI is, how can information be made meaningful to a computer? Simply adding more bits does not work, for meaning arises not from information as such, but from relationships involving the real world. In one form or another, this basic problem has been stymieing AI research for decades. Nor is it the only such problem. Another problem is that computers, even those employing "fuzzy logic" and autonomous **learning**, function by processing symbols (e.g., 1s and 0s) according to rules (e.g., those of Boolean algebra)—yet human beings do not usually think by processing symbols according to rules. Humans are able think this way, as when we do **arithmetic**, but more commonly we interpret situations, leap to conclusions, utter sentences, and plan actions using thought processes that do not involve symbolic computation at all. What our thought processes really are—that is, what our intelligence is, preicsely—and how to translate it into (or mimic it by means of) a system of computable symbols and rules is a problem that remains unsolved, in general, by AI researchers.

In 1950, British mathematician Alan Turing (1912–1954) proposed a hypothetical game to help decide the issue of whether a given machine is truly intelligent. The "imitation game," as Turing originally called it, consisted of a human questioner in a room typing questions on a keyboard. In another room, an unseen respondents—either a human or a computer—would type back answers. The questioner could pose queries to the respondent in an attempt to determine if he or she was corresponding with a human or a computer. In Turing's opinion, if the computer could fool the questioner into believing that he or she was having a dialog with a human being then the computer could be said to be truly intelligent.

The Turing test is obviously biased toward human language prowess, which most AI programs do not even seek to emulate. Nevertheless, it is significant that even the most advanced AI programs devoted to natural language are as far as ever from passing the Turing test. "Intelligence" has proved a far tougher nut to crack than the pioneers of AI believed, half a century ago. Computers remain unintelligent.

Even so, AI has made many gains since the 1950s. AI software is now present in many devices, such as automatic teller machines (ATMs), that are part of daily life, and is finding increasing commercial application in many industries.

Overview of AI

All AI computer programs are built on two basic elements: a knowledge base and an inferencing capability. (Inferencing means drawing conclusions based on logic and prior knowledge.) A knowledge base is made up of many discrete units of information—representing facts, concepts, theories, procedures, and relationships—all relevant to a particular task or aspect of the world. Programs are written to give the computer the ability to manipulate this information and to reason, make judgments, reach conclusions, and choose solutions to the problem at hand, such as guessing whether a series of credit-card transactions involves fraud or driving an automated rover across a rocky Martian landscape. Whereas conventional, deterministic software must follow a strictly logical series of steps to reach a conclusion, AI software uses the techniques of search and pattern matching; it may also, in some cases, modify its knowledge base or its own structure ("learn"). Pattern matching may still be algorithmic; that is, the computer must be told exactly where to look in its knowledge base and what constitutes

a match. The computer searches its knowledge base for specific conditions or patterns that fit the criteria of the problem to be solved. Microchip technology has increased computational speed, allowing AI programs to quickly scan huge arrays of data. For example, computers can scan enough possible chess moves to provide a challenging opponent for even the best human players. Artificial intelligence has many other applications, including problem solving in **mathematics** and other fields, expert systems in medicine, natural language processing, **robotics**, and education.

The ability of some AI programs to solve problems based on facts rather than on a predetermined series of steps is what most closely resembles "thinking" and causes some in the AI field to argue that such devices are indeed intelligent.

General problem solving

Problem solving is thus, something AI does very well as long as the problem is narrow in focus and clearly defined. For example, mathematicians, scientists, and engineers are often called upon to prove theorems. (A **theorem** is a mathematical statement that is part of a larger theory or structure of ideas.) Because the formulas involved in such tasks may be large and complex, this can take an enormous amount of time, thought, and trial and error. A specially designed AI program can reduce and simplify such formulas in a fraction of the time needed by human workers.

Artificial intelligence can also assist with problems in planning. An effective step-by-step sequence of actions that has the lowest cost and fewest steps is very important in business and manufacturing operations. An AI program can be designed that includes all possible steps and outcomes. The programmer must also set some criteria with which to judge the outcome, such as whether speed is more important than cost in accomplishing the task, or if lowest cost is the desired result, regardless of how long it takes. The plan generated by this type of AI program will take less time to generate than by traditional methods.

Expert systems

The expert system is a major application of AI today. Also known as knowledge-based systems, expert systems act as intelligent assistants to human experts or serve as a resource to people who may not have access to an expert. The major difference between an expert system and a simple database containing information on a particular subject is that the database can only give the user discrete facts about the subject, whereas an expert system uses reasoning to draw conclusions from stored information. The purpose of this AI application is not to replace our human experts, but to make their knowledge and experience more widely available.

An expert system has three parts: knowledge base, inference engine, and user interface. The knowledge base contains both declarative (factual) and procedural (rules-of-usage) knowledge in a very narrow field. The inference engine runs the system by determining which procedural knowledge to access in order to obtain the appropriate declarative knowledge, then draws conclusions and decides when an applicable solution is found.

An interface is usually defined as the point where the machine and the human "touch." An interface is usually a keyboard, mouse, or similar devices. In an expert system, there are actually two different user interfaces: One is for the designer of the system (who is generally experienced with computers) the other is for the user (generally a computer novice). Because most users of an expert system will not be computer experts, it is important that system be easy for them to use. All user interfaces are bi-directional; that is, are able to receive information from the user and respond to the user with its recommendations. The designer's user interface must also be capable of adding new information to the knowledge base.

Natural language processing

Natural language is human language. Natural-language-processing programs use artificial intelligence to allow a user to communicate with a computer in the user's natural language. The computer can both understand and respond to commands given in a natural language.

Computer languages are artificial languages, invented for the sake of communicating instructions to computers and enabling them to communicate with each other. Most computer languages consist of a combination of symbols, numbers, and some words. These languages are complex and may take years to master. By programming computers (via computer languages) to respond to our natural languages, we make them easier to use.

However, there are many problems in trying to make a computer understand people. Four problems arise that can cause misunderstanding: (1) Ambiguity—confusion over what is meant due to multiple meanings of words and phrases. (2) Imprecision—thoughts are sometimes expressed in vague and inexact terms. (3) Incompleteness—the entire idea is not presented, and the listener is expected to "read between the lines." (4) Inaccuracy—spelling, punctuation, and grammar problems can obscure meaning. When we speak to one another, furthermore, we generally expect to be understood because our common language assumes all the meanings that we share as members of a specific cultural group. To a non-native speaker, who shares fewer of our cultural background, our meaning may not always be clear. It is even more difficult for computers, which have no share at all in the real-world relationships that confer meaning upon information, to correctly interpret natural language.

To alleviate these problems, natural language processing programs seek to analyze syntax—the way words are put together in a sentence or phrase; semantics—the derived meaning of the phrase or sentence; and context—the meaning of distinct words within a sentence. But even this is not enough. The computer must also have access to a dictionary which contains definitions of every word and phrase it is likely to encounter, and may also use keyword analysis—a pattern-matching technique in which the program scans the text, looking for words that it has been programmed to recognize. If a keyword is found, the program responds by manipulating the text to form a reasonable response.

In its simplest form, a natural language processing program works like this: a sentence is typed in on the keyboard; if the program can derive meaning—that is, if it has a reference in its knowledge base for every word and phrase—it will respond, more or less appropriately An example of a computer with a natural language processor is the computerized card catalog available in many public libraries. The main menu usually offers four choices for looking up information: search by author, search by title, search by subject, or search by keyword. If you want a list of books on a specific topic or subject you type in the appropriate phrase. You are asking the computer—in English—to tell you what is available on the topic. The computer usually responds in a very short time—in English—with a list of books along with call numbers so you can find what you need.

Computer vision

Computer vision is the use of a computer to analyze and evaluate visual information. A camera is used to collect visual data. The camera translates the image into a series of electrical signals. This data is analog in nature—that is, it is directly measurable and quantifiable. A digital computer, however, operates using numbers expressed directly as digits. It cannot read analog signals, so the image must be digitized using an analog-to-digital converter. The image becomes a very long series of binary numbers that can be stored and interpreted by the computer. Just how long the series is depends on how densely packed the pixels are in the visual image. To get an idea of pixels and digitized images, take a close look at a newspaper photograph. If you move the **paper** very close to your eyes, you will notice that the image is a sequence of black and white dots—called pixels, for picture elements—arranged in a certain pattern. When you move the picture away from your eyes, the picture becomes clearly defined and your **brain** is able to recognize the image.

Artificial intelligence works much the same way. Clues provided by the arrangement of pixels in the image give information as to the relative **color** and texture of an object, as well as the **distance** between objects. In this way, the computer can interpret and analyze visual images. In the field of robotics, visual analysis is very important.

Robotics

Robotics is the study of robots, which are machines that can be programmed to perform manual tasks. Most robots in use today perform various functions in an industrial setting. These robots typically are used in factory assembly lines, by the military and law enforcement agencies, or in hazardous waste facilities handling substances far too dangerous for humans to handle safely.

Most robots do not resemble the humanoid creations of popular science fiction. Instead, they usually consist of a manipulator (arm), an end effector (hand), and some kind of control device. Industrial use robots generally are programmed to perform repetitive tasks in a highly controlled environment. However, more research is being done in the field of intelligent robots that can learn from their environment and move about it autonomously. These robots use AI programming techniques to understand their environment and make appropriate decisions based on the information obtained. In order to learn about one's environment, one must have a means of sensing the environment. Artificial intelligence programs allow the robot to gather information about its surroundings by using one of the following techniques: contact sensing, in which a robot sensor physically touches another object; noncontact sensing, such as computer vision, in which the robot sensor does not physically **touch** the object but uses a camera to obtain and record information; and environmental sensing, in which the robot can sense external changes in the environment, such as **temperature** or **radiation**.

The most recent robotics research centers around mobile robots that can cope with environments that are hostile to humans, such as damaged nuclear-reactor cores, active volcanic craters, or the surfaces of other planets.

Computer-assisted instruction

Intelligent computer-assisted instruction (ICAI) has three basic components: problem-solving expertise, student model, and tutoring module. The student using this type of program is presented with some information from the problem-solving expertise component. This is the knowledge base of this type of AI program. The student responds in some way to the material that was presented, either by answering questions or otherwise demonstrating his or her understanding. The student model analyzes the student's responses and decides on a course of action. Typically this involves either presenting some review material or allowing the student to advance to the next level of knowledge presentation. The tutoring module may or may not be employed at this point, depending on the student's level of mastery of the material. The system does not allow the student to advance further than his or her own level of mastery.

Most ICAI programs in use today operate in a set sequence of presentation of new material, evaluation of student response, and employment of tutorial (if necessary). However, researchers at Yale University have created software that uses a more Socratic way of teaching. These programs encourage discovery and often will not respond directly to a student's questions about a specific topic. The basic premise of this type of computer-assisted learning is to present new material only when a student needs it. This is when the brain is most ready to accept and retain the information. This is exactly the scenario most teachers hope for: students who become adroit self-educators, enthusiastically seeking the wisdom and truth that is meaningful to them. The cost of these programs, however, can be far beyond the means of many school districts. For this reason, these types of ICAI are used mostly in corporate training settings.

See also Automation; Computer, analog; Computer, digital; Computer software; Cybernetics.

Resources

Books

Caudill, Maureen. *In Our Own Image: Building an Artificial Person.* New York: Oxford University Press, 1992.

Kelly, Derek. *A Layman's Introduction to Robotics.* Princeton: Petrocelli Books, 1986.

Periodicals

Feder, Barnaby J., "Artificial Intelligence For the New Millennium; A Revolution More Bland Than Kubrick's '2001'." *New York Times.* June 30, 2001.

Travis, John, "Building a Baby Brain in a Robot." *Science* (20 May 1994): 1080–1082.

"An Encounter with A.I." *Popular Science* (June 1994).

Weng, Juyang, et al. "Autonomous Mental Development by Robots and Animals." *Science.* (January 26, 2001): 599–00.

Johanna Haaxma-Jurek

Artificial limbs and joints *see* **Prosthetics**

Artificial selection *see* **Selection**

Artificial vision

Artificial **vision** refers to the technology (usually, visual implants) that allows blind people to see. The main aim of visual implants is to relay the picture to the **brain** using either cameras or photoreceptor arrays. There are different types of implants used to stimulate vision; retinal, cortical, optic nerve, and biohybrid implants. None of the currently available technologies restores full vision, but they are often able to improve one's ability to recognize shapes and movements.

Cortical implants

A number of scientific groups are working on cortical implants, which directly stimulate the visual cortex and can be used by a majority of blind patients. The main differences between implants are the way in which they interact with the brain. Surface-type implants, such as the Dobelle implant, are placed directly on the brain surface. Others, such as penetrating implants, are designed to be introduced into brain **tissue** for direct contact with the neurons responsible for relaying visual information. As of 2003, penetrating implants are under study at several universities and laboratories affiliated with the National Institutes of Health.

The first surface visual implant, an electrode array, was surgically inserted by a team led by American ophthalmologist William H. Dobelle in 1978. With the Dobelle implant, surface electrodes are connected to a camera installed on one side of special eyeglasses. During the treatment, the patient's cortex is systematically and slowly stimulated to re-learn how to see. Camera and **distance** sensor in sunglasses send the signals to a computer and it in turn, sends pulses to an electrode array. Vision is black and white with **light** appearing as phosphenes, which are visual sensations resulting from mechanical stimulation of the **eye** (similar to the visual sensations created when pressing upon the eyeball with closed eyes). Quality of vision is dependent not only on the technology, but also on the patient, as different people see varying numbers of phosphenes. The necessity to carry a **battery** and a computer on the belt can be considered a disadvantage, but the Dobelle implant allows easy upgrades to newer technology. The main drawback of this implant is the price tag (around $100,000) for the treatment. Research and scientific scrutiny of the Dobelle implant continues into the twenty-first century, with limited clinical trials showing positive results.

Retinal implants

Retinal implants were designed for people with retinal diseases such as retinitis pigmentosa or age-related macu-

Japan's Mitsubishi Electric presents the first color artificial retina in 1999. © *AFP/Corbis. Reproduced by permission.*

lar degeneration. These implants rely on intact retinal neurons to transmit the stimuli to the brain. A number of varying designs are being used in clinical trails; the main difference between them is the localization of the implant in the retina. Epiretinal implants are placed on the retinal surface and subretinal implants are placed under the retina.

The first retinal implant to move into clinical trials (in 2000) was a subretinally placed artificial silicon retina (ARS). The ARS is a very small silicon microchip of 2mm in diameter and 25 micrometers thick, composed of miniature solar cells. The arrays of these cells respond to light similar to natural photoreceptors.

The artificial retina component chip (ARCC) is an example of an epiretinal implant. The ARCC is composed of light sensors and electrode arrays that send signals to the retinal neurons. However, the system requires additional power and is dependent upon an external camera to detect a picture and induce a **laser** pulse. Incoming laser light is detected by the photo sensors in the implant to create a picture. The camera and the laser are build into special sunglasses.

vOICe system was developed as an alternative to surgical implants. Users claim that they are receiving visual sensations from using the system, which involves changing images from a video camera into corresponding sound patterns that in turn, stimulate the visual cortex. The vOICe technology, therefore, can be classified as artificial vision, although only its effectiveness as a sensory substitution has been established. The main ad-

KEY TERMS

Phosphenes—Visual sensations or impressions resulting from a mechanical pressure on an eyeball or excitation of the retina by stimulus other than light.

Retina—An extremely light-sensitive layer of cells at the back part of the eyeball. The image formed by the lens on the retina is carried to the brain by the optic nerve.

Visual acuity—Keenness of sight and the ability to focus sharply on small objects.

Visual cortex—Area of the brain (occipital lobe) concerned with vision.

vantages are no surgical intervention and cost of the device is only $2,500.

Quality of artificial vision

Resolution of the artificial systems is an important consideration in their design and usefulness. A pixel resolution of 5x5, such as with the ARCC, allows reading of some individual letters; a 10x10 pixel vision (ASR) can allow further distinction of form, but a pixel resolution of 32x32 is usually necessary to allow a person freedom of movement. Better resolutions of 64x64 pixels and up to 250x250 pixels are considered to be a matter of time. Resolution of the cortical implants is measured mostly in vision acuity because their aim is not only freedom of movement, but also ability to read.

See also Biotechnology; Blindness and visual impairments; Nanotechnology; Prosthetics.

Resources

Periodicals

Dobelle, W.H., "Artificial Vision for the Blind by Connecting a Television Camera to the Visual Cortex." *ASAIO Journal* (January 2000):3–9.

Margalit, Eyal, Maia, Mauricio, Weiland, James D., et al., "Retinal Prosthesis for the Blind." *Surv. Ophthalmol.* (July-August 2002):335–56

Meijer, Peter B.L. "An Experimental System for Auditory Image Representation." *IEEE Transactions on Biomedical Engineering* (February 1992):112–121

Organizations

The Dobelle Institute, Inc, 61 Mall Drive, Commack, NY. 631–864–1600. [cited January 12, 2003]. <http://www.artificialvision.com/index.html>.

Other

Brown University. "Artificial Vision" [cited January 14, 2003]. <http://biomed.brown.edu/Courses/BI108/BI108_1999_Groups/Vision_Team/Vision.htm>.

How Stuff Works. Kevin Bonsor. "How Artificial Vision Will Work" [cited January 14, 2003]. <http://www.howstuffworks.com/artificial-vision. htm>.

Vision Technology for the Totally Blind. February 15, 1996 [cited January 14, 2003]. <http://www.seeingwithsound.com.voice.html>.

Agnieszka Lichanska

Arum family (Araceae)

Arums, also called aroids, are flowering plants in the family Araceae. The 2,500 **species** of arums are distributed worldwide, primarily in tropical and subtropical regions, where they grow in rainforests, mostly on the ground but also commonly as epiphytes. Arums are generally absent from the arctic and deserts. Only 11 species occur in **North America** and other north temperate regions.

Most species are small to medium-sized perennial herbs, often climbing as vines, and a few are shrubs. The leaves of arums are generally broad, frequently dissected, and occasionally have natural holes. The leaves commonly contain abundant sharp-pointed crystals of **calcium** oxalate, which give the leaves an acrid **smell**. Calcium oxalate crystals are poisonous and irritating when chewed, thus protecting the leaves from herbivores. The tiny flowers are densely borne on a stucture called a spadix, which is accompanied by a large, often colorful **leaf** known as a spathe. The spathe varies in size and shape, but in the most advanced species of arums it forms a hood that encloses the spadix, as in the familiar jack-in-the-pulpit (*Arisaema triphyllum*) of North America. The **fruits** of arums are almost always brightly colored berries that are eaten and dispersed by animals.

Arums are famous for the variety of offensive odors they produce in association with a **pollination** strategy that involves deception. Arums are usually pollinated by **flies** or **beetles** that normally feed on rotting organic **matter**, such as decaying plants or **mushrooms**, dung, or **animal** carcasses. Arums mimic the odors of decay by emitting vapors of fatty compounds from their spadix and spathe, thereby luring **insects** who expect to find a tasty mass of rotting flesh or decaying plants.

Many aroids imprison the insects that they deceive. A fine example of this strategy is *Helicodiceros muscivorus,* a native of Corsica, Sardinia, and nearby Mediteranean islands. On the **island** of Calvioli, this **plant** grows in open areas between **rocks**, sometimes in gull colonies. Bird droppings, regurgitated seafood, the carcasses of chicks, and eggs broken by predators all contribute to breathtaking odors. *Helicodiceros* flowers when

A jack-in-the-pulpit near St. Mary's, Ontario. *Photograph by Alan and Linda Detrick. Photo Researchers, Inc. Reproduced by permission.*

Skunk cabbage. *Michael P. Gadomski, all rights reserved. Photo Researchers, Inc. Reproduced by permission.*

the **gulls** are breeding, and it produces an open spathe with the shape of a shortened, slightly compressed bullhorn. The spathe is a mottled grey-and-red **color**, and its appearance and odor resemble rotting meat. Excited blowflies will actually choose the stench of the arum over the gull-mess, landing on the spathe in search for food. They are eventually drawn to the dark, smelly, narrow end of the spathe, where they enter through a small opening into a chamber, which becomes their dungeon. The blowflies are unable to escape because of a dense barrier of stiff, sharp, downward-pointing hairs that guard the opening. The chamber encloses the basal portion of the spadix, on which are located the flowers. Some of the flies will have previously visited other plants and been dusted by pollen. When the flies first become trapped, only the female flowers are receptive and so they will be pollinated by the accidental stumbling of the blowflies in the dungeon. The flowers also exude a small amount of **nectar**, just enough to keep the flies alive. After a few days, the male flowers mature, and release their pollen onto the blowflies. Simultaneously, the sharp hairs wither, thus releasing the flies. Some of the flies will be duped a second time and cross-pollinate the arums.

Arums are unique among plants in possessing a remarkable ability to generate metabolic **heat**. Their spadix commonly respires **fat** rapidly to produce heat during pollination, apparently as a means of increasing the vaporization of their foul-smelling compounds. The philodendron (*Philodendron scandens*) can raise its spadix **temperature** to as high as 116°F (47°C), even when the air is close to freezing. The skunk cabbage (*Symplocarpus foetidus*), which is a native of swamps and other wet places in eastern North America, flowers early in the spring, often when the ground is still covered by snow. Its spadix can generate enough heat to attain temperatures up to 77°F (25°C) above air temperatures, melt the surrounding snow, and get a head start on attracting flies.

The arum known as jack-in-the-pulpit is a perennial plant, native to moist or wet **forests** throughout eastern North America, and it is sometimes cultivated as an interesting garden ornamental. The sex of **individual** plants depends on their size. When small, they only produce male flowers, but in later years when they are larger, they switch sex and produce only female flowers. The explanation for this phenomenon appears to be that when the plant is small, it has relatively few resources available to it, insufficient to develop the large berries that the jack-in-the-pulpit produces. Therefore, plants are male first because pollen grains are small and take relatively little **energy** to produce. When the plant becomes larger, it can afford to invest in the higher costs of producing fruits. Thus, it switches its sex to female.

Several arums are important economically. The *Monstera deliciosa* produces an edible fruit and is a popular indoor plant because of its unusual leaves, which have large holes due to arrested development of parts of its growing surface. *Colocasia esculentum*, commonly called taro or poi, is a native of **Asia** with many varieties that are widely cultivated in the tropics because of their large, starch-rich tubers. Many arums are cultivated as ornamentals for their interesting foliage, which is often intricately dissected, such as the *Philodendron* and *Monstera* described previously. Other species are prized as indoor plants because of their large, brightly colored spathes, which may range in color from pure white to bright red and even iridescent. Species of the genus *Cryptocoryne* are commonly used as aquarium plants.

Resources

Books

Brown, D. 1988. *Aroids: Plants of the Arum Family.* Portland, OR: Timber Press, 1988.

Dahlgren, R.M.T., H.T. Clifford, and P.F. Yeo. *The Families of the Monocotyledons: Structure, Evolution, and Taxonomy.* Berlin: Springer-Verlag, 1985.

Les C. Cwynar

Asbestos

Asbestos is the general name for a wide variety of silicate **minerals**, mostly silicates of **calcium**, **magnesium**, and **iron**. Their common characteristics are a fibrous structure and resistance to fire. The two most common families of asbestos minerals are called amphibole and serpentine. The mineral has been known and used by humans for centuries. The ancient Romans, for example, wove asbestos wicks for the lamps used by vestal virgins. The story is also told about Charlemagne's effort to impress barbarian visitors by throwing a table cloth woven of asbestos into the fire.

One of the first complete scientific descriptions of asbestos was provided by J. P. Tournefort in the early 1700s. He explained that the substance "softens in oil and thereby acquires suppleness enough to be spun into Threads; it makes Purses and Handkerchiefs, which not only resist the Fire, but are whiten'd and cleansed by it."

Travelers to **North America** in the 1750s also told of widespread use of asbestos among both colonists and Native Americans.

Classification and properties

The various minerals that make up the asbestos group are so diverse that they share only one major property, their fibrous character. The form known for the longest time and most widely used is chrysotile, or white asbestos, a member of the serpentine asbestos family. Its fibers are long, hollow cylinders with a diameter of about 25 nanometers (10^{-9} meter). The fibers are strong and relatively inflexible. The chemical formula assigned to chrysotile is $Mg_3Si_2O_5(OH)$. Like other forms of asbestos, chrysotile is noncombustible. The whole class of minerals was, in fact, named after the Greek word *asbeston*, for noncombustible. The amphibole asbestos minerals are:

- riebeckite ($Na_2Fe^{2+}3Fe^{3+}2Si_8O_{22}(OH)_2$)
- anthophyllite ($Mg_7Si_8O_{22}(OH)$)
- actinolite ($Ca_2(Mg,Fe^{2+})_5Si_8O_{22}(OH)_2$)
- tremolite ($Ca_2Mg_5Si_8O(OH)_2$).

Riebeckite is also called crocidolite, or blue asbestos.

The asbestos amosite is sometimes included among the amphibole asbestos minerals and sometimes placed in its own group. Amosite ($Fe_7Si_8O_{22}(OH)$) is also called grenerite. It is typically ash-gray in **color**.

In general, amphibole minerals and amosite tend to have longer, more rigid fibers with a lower melting point than that of chrysotile. This fact makes them less desirable as fireproofing materials.

Occurrence and mining

The primary sources of the asbestos minerals are Quebec and the Yukon in Canada and the Ural Mountain region of Russia (chrysotile) and southern **Africa** (the amphiboles and amosite). Some asbestos is also found in Mexico and Italy, and in the United States, in Arizona, California, North Carolina, and Vermont.

By far the greatest amount (95%) of asbestos produced today is chrysotile. An additional 3.5% consists of crocidolite, and the final 1.5% is amosite.

The largest supplier of asbestos minerals has traditionally been Russia, or the former Soviet Union, which accounted for about half of all the asbestos mined in the world. The second largest source has been Canada (about 30% of the world's output), followed by the European nations, Zimbabwe, China, South Africa, and the United States.

Asbestos occurs either in seams that run at or just beneath the earth's surface or in veins that may go as deep as 327 yd (300 m). One method of quarrying the seams is known as block caving. In this process, trenches are dug underneath an asbestos seam and the whole section is then allowed to fall and break apart. In another technique, open seams of the mineral are plowed up and allowed to dry in air.

Underground veins are mined in much the same way as is **coal**. The distinctive fibrous character of asbestos makes it relatively easy to separate from other rocky material with which it is found.

Processing

After asbestos is removed from the earth, it is processed in order to divide it into groups according to fiber length. Longer fibers are separated out for weaving into a cloth-like material. Shorter fibers, known as shingles, are combined with each other and often with other materials to make composite products. Perhaps the best known of these composites is asbestos cement which was invented in the late 1800s. Asbestos cement contains about 12.5% asbestos with the remainder consisting of portland cement; this mixture is used for a variety of construction purposes.

The first step in making asbestos cement is to form a thick, pasty mixture of cement and asbestos in **water**. That mixture is then passed along a conveyor belt, where water is removed. At the end of the belt, the damp mixture of cement and asbestos is laid down on some type of base. Layers are allowed to build up until a material of the desired thickness if obtained. It is then dried.

Uses

About two-thirds of the world production of chrysotile is used to make asbestos cement. That material can be fabricated into corrugated or flat sheets for use as a building material in industrial and agricultural structures. Altering the process by which the asbestos cement is made can improve thermal, acoustical, and other properties to make it more suitable for interior structures also. Asbestos cement can also be fabricated as cylinders, making a material that is suitable for ducts and **pressure** pipes.

Long-fiber asbestos is used in other kinds of applications. It can be woven alone or with other fibers (such as **glass** fibers) to make protective clothing for fire fighters, brake and clutch linings, electrical insulation, moldings for **automobile** components, and linings for chemical containers.

Health considerations

The deleterious health effects of asbestos have become apparent only since the end of World War II. Prior

---

KEY TERMS

. . . .

Asbestos cement—A composite material made by mixing together cement and asbestos.

Asbestosis—A disorder that affects the respiratory tract as a result of inhaling asbestos fibers, leading eventually to a variety of serious and generally fatal respiratory illnesses.

Fabricate—To shape a material into a form that has some commercial value, as in shaping asbestos fibers into a flat or corrugated board.

Fiber—A complex morphological unit with an extremely high ratio of length to diameter (typically several hundred to one) and a relatively high tenacity.

Mesothelioma—Tumors that occur in structures found in the lining of the lungs.

Quarrying—A method by which some commercially valuable substance is extracted, usually from the earth's surface.

Shingles—A name given to shorter fibers of asbestos.

---

to that time, very few measurements had been made of the **concentration** of asbestos in the air around workplaces and other settings in which asbestos was used. In addition, the connection between the mineral and its health effects was difficult to recognize since those effects typically do not manifest themselves for 20 years or more after exposure.

Today scientists know that a rather narrow range of asbestos fiber lengths (less than two microns and five to 100 microns in length) can cause a range of respiratory problems, especially asbestosis, lung **cancer**, and mesothelioma. These problems begin when asbestos fibers enter the **respiratory system** and become lodged in the interstitial areas—the areas between the alveoli—in the lungs. As the fibers continue to accumulate in the lungs, they can cause the development of fibrous scar **tissue** that reduces the flow of air through the respiratory system.

Symptoms that gradually develop include coughing and shortness of breath, weight loss, and anorexia. Other respiratory conditions, such as **pneumonia** and **bronchitis**, become more common and more difficult to cure. Eventually the fibers may initiate other anatomical and physiological changes, such as the development of tumors and carcinomas. Individuals most at risk for asbestos-related problems are those continually exposed to the mineral fibers. This includes those who work in as-

---

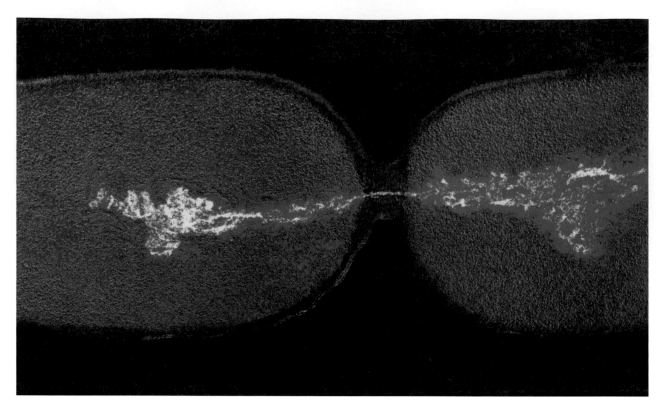

A colored transmission electron micrograph (TEM) of the *Salmonella typhimurium* bacterium reproducing by binary fission. *Photograph by Dr. Kari Lounatmaa/Photo Researchers, Inc. Reproduced by permission.*

bestos **mining** and processing as well as those who use the product in some other manufacturing line, as in the production of brake linings.

Over the past two decades, intense efforts have been made to remove asbestos-based materials from buildings where they are especially likely to pose health risks, as in school buildings and public auditoriums. Recent critics of asbestos removal maintain that if not done properly, asbestos removal spreads more asbestos fibers into the air than it actually removes. Also, there has not been a satisfactory substitute found for the asbestos materials being removed.

See also Poisons and toxins; Respiratory diseases.

Resources

Books

Brodeur, Paul. *Outrageous Misconduct: The Asbestos Industry on Trial.* New York: Pantheon Books, 1985.
Greenwood, N. N., and A. Earnshaw. *Chemistry of the Elements.* Oxford: Butterworth-Heinneman Press, 1997.

Periodicals

Yuspa, S. H. "Overview of Carcinogenesis: Past, Present and Future." *Carcinogenesis* 21 (2000): 341–344.

David E. Newton

Asexual reproduction

Sexual reproduction involves the production of new cells by the fusion of sex cells (sperm and ova) to produce a genetically different **cell**. Asexual reproduction, on the other hand, is the production of new cells by simple division of the parent cell into two daughter cells (called binary fission). Since there is no fusion of two different cells, the daughter cells produced by asexual reproduction are genetically identical to the parent cell.

The adaptive advantage of asexual reproduction is that organisms can reproduce rapidly, and so colonize favorable environments rapidly.

In nature

Bacteria, cyanobacteria, **algae**, most **protozoa**, **yeast**, dandelions, and **flatworms** all reproduce asexually. When asexual reproduction occurs, the new individuals are called clones, because they are exact duplicates of their parent cells. Mosses reproduce by forming runners that grow horizontally, produce new stalks, and then the runner decomposes, leaving a new **plant** which is a clone of the original.

Starfish can regenerate and eventually produce a whole new **organism** from one of its severed appendages.

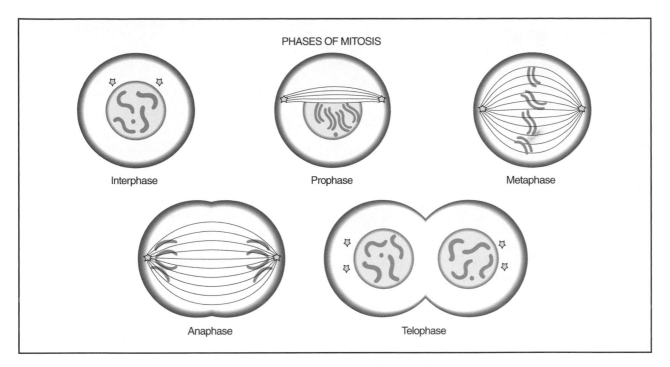

PHASES OF MITOSIS

Interphase

Prophase

Metaphase

Anaphase

Telophase

Phases of mitosis. *Illustration by Hans & Cassidy. Courtesy of Gale Group.*

Duplication of organisms, whether sexually or asexually, involves the partitioning of the genetic material (chromosomes) in the cell nucleus.

During asexual reproduction, the chromosomes divide by **mitosis**, which results in the exact duplication of the genetic material into the nuclei of the two daughter cells. Sexual reproduction involves the fusion of two **gamete** cells (the sperm and ova) which each have half the normal number of chromosomes, a result of reduction division known as **meiosis**.

Bacteria reproducing asexually double their numbers rapidly, approximately every 20 minutes. This reproduction **rate** is offset by a high death rate that may be the result of the accumulation of **alcohol** or acids that concentrate from the bacterial colonies.

Yeasts reproduce asexually by budding, as well as reproducing sexually. In the budding process, a bulge forms on the outer edge of the yeast cell as nuclear division takes place. One of these nuclei moves into the bud, which eventually breaks off completely from the parent cell. Budding also occurs in flatworms, which divide into two and then regenerate to form two new flatworms.

Bees, **ants**, **wasps**, and other **insects** can reproduce sexually or asexually. In asexual reproduction, eggs develop without **fertilization**, a process called **parthenogenesis**. In some **species** the eggs may or may not be fertilized; fertilized eggs produce females, while unfertilized eggs produce males.

There are a number of crop plants which are propagated asexually. The advantage of asexual propagation to farmers is that the **crops** will be more uniform than those produced from seed. Some plants are difficult to cultivate from seed and asexual reproduction in these plants makes it possible to produce crops that would otherwise not be available for commercial marketing.

The process of producing plants asexually is called vegetative propagation and is used for such crops as potatoes, bananas, raspberries, pineapples, and some flowering plants used as ornamentals. Farmers plant the so-called " eyes" of potatoes to produce duplicates of the parent. With **banana** plants, the suckers that grow from the root of the plant are separated and then planted as new ones. With raspberry bushes, branches are bent and covered with **soil**. They then grow into a separate plant with their own **root system** and can eventually be detached from the parent plant.

See also Buds and budding; Clone and cloning; Genetics.

Resources

Books

Leone, Francis. *Genetics: The Mystery and the Promise.* Blue Ridge Summit, PA: Tab Books, 1992.

Taylor, Martha. *Campbell's Biology Student Study Guide.* Redwood City, CA: Benjamin/Cummings, 1990.

Periodicals

Allison, Richard. "Genetic Engineering Studied." *Cancer Researcher Weekly* (21 March 1994): 13.

KEY TERMS

Binary fission—The process in which cell division occurs and two cells are produced where only one existed before.

Blastomere separation—Cloning by splitting multicelled embryos.

Gamete—A male or female sex cell capable of reproduction.

Regeneration—The ability of an organism to reproduce wholly from a part of another one.

Replication—The production of new cells like the original one.

Vegetative propagation—A type of asexual reproduction in plants involving production of a new plant from the vegetative structures—stem, leaf, or root—of the parent plant.

Nash, J. Madeleine. "Is Sex Really Necessary?" *Time* (20 January 1992): 47.

Robertson, John A. "The Question of Human Cloning." *The Hastings Center Report* (March/April 1994): 6.

Vita Richman

Ash *see* **Olive family (Oleaceae)**

Asia

Asia is the world's largest **continent**, encompassing an area of 17,177,000 sq mi (44,500,000 sq km), 29.8% of the world's land area. The Himalayan Mountains, which are the highest and youngest mountain range in the world, stretch across the continent from Afghanistan to Burma. The highest of the Himalayan peaks, called Mount Everest, reaches an altitude of 29,028 ft (8,848 m). There are many famous deserts in Asia, including the Gobi Desert, the Thar Desert, and Ar-Rub'al-Khali ("the empty quarter"). The continent has a wide range of climatic zones, from the tropical jungles of the south to the Arctic wastelands of the north in Siberia.

The continent of Asia encompasses such an enormous area and contains so many countries and islands that its exact borders remain unclear. In the broadest sense, it includes central and eastern Russia, the countries of the Arabian Peninsula, the far eastern countries, the Indian subcontinent, and numerous **island** chains. It is convenient to divide this huge region into five cate-

gories: the Middle East, South Asia, Central Asia, the Far East, and Southeast Asia.

The Middle East

The Middle Eastern countries lie on the Arabian Peninsula, southwest of Russia and northeast of **Africa**, separated from the African continent by the Red Sea and from **Europe** in the northwest by the Mediterranean Sea. This area stretches from Turkey in the northwest to Yemen in the south, which is bordered by the Arabian Sea. In general, the climate is extremely dry, and much of the area is still a desertwilderness. **Precipitation** is low, so the fertile regions of the middle east lie around the **rivers** or in valleys which drain the mountains. Much of the coastal areas are very arid, and the vegetation is mostly desert scrub.

Iran

Iran is separated somewhat from the rest of the Arabian Peninsula by a great gulf which divides it from most of Saudi Arabia. This gulf is known as the Oman Gulf where it meets the Arabian Sea and is called the Persian Gulf as it extends past the Strait of Hormuz. Most of Iran is a plateau lying about 4,000 ft (1,200 m) above **sea level**, and this plateau is crossed by the mountain ranges of Zagros and Elburz. These meet at an angle, forming an inverted V; between them the land is mostly **salt** marshes and desert. The highest elevation is Mount Damavand, which reaches 18,606 ft (5,671 m). The valleys between the mountain peaks are the main fertile regions of the country.

Iraq and Kuwait

Bordering Iran in the northeast is the country of Iraq. The west and southwest, where it borders with Syria and Saudi Arabia, is a desert region; in the northeast is a mountain range which reaches altitudes of over 10,000 ft (3,000 m). Between these two regions are fertile riverplains which are watered by the Tigris and Euphrates Rivers. In the southeast the two rivers join together, forming the broad Shatt al-'Arab, which flows between Iraq and Iran.

In the southeast corner of Iraq, along the tip of the Persian Gulf, is the tiny country of Kuwait. Its terrain is almost entirely made up of desert and mud flats, but along the southwestern part of the coast are a few low hills.

Saudi Arabia and Yemen

Saudi Arabia is the largest of the middle eastern countries. In the west it is bordered by the Red Sea, which lies between Saudi Arabia and the African continent. The Hijaz Mountains run parallel to this coast in the northwest, rising sharply from the sea to elevations ranging from 3,000–9,000 ft (910–2,740 m). In the south is another mountainous region called the Asir, stretching along the coast for about 230 mi (370 km) and inland about 180–200 mi (290–320 km). Between the two ranges lies a narrow coastal plain called the Tihamat ash-Sham. East of the Hijaz Mountains are two great plateaus called the Najd, which slopes gradually downward over a range of about 3,000 ft (910 m) from west to east, and the Hasa, which is only about 800 ft (240 m) above sea level. Between these two plateaus is a desert region called the Dahna.

About one third of Saudi Arabia is estimated to be desert. The largest of these is the Ar-Rub'al-Khali, which lies in the south and covers an area of about 250,000 sq mi (647,500 sq km). In the north is another desert, called the An-Nafud. The climate in Saudi Arabia is generally very dry; there are no lakes and only seasonally flowing rivers. Saudi Arabia, like most of the Middle Eastern countries, has large oil reserves; also found here are rich gold and silver mines which are thought to date from the time of King Solomon.

Yemen, which lies along the southern border of Saudi Arabia, is divided into South Yemen, called the People's Democratic Republic of Yemen (PDRY), and North Yemen, called the Yemen Arab Republic (YAR). YAR, which lies below the Asir region of Saudi Arabia, is very mountainous. It consists mostly of plateaus and tablelands which are also the main fertile regions; but along the coast of the Red Sea is a stretch of flat coastal plains called the Tihama. This plain continues into PDRY along the gulf of Aden, and is very arid. The west of PDRY, near the YAR border, is a mountain range; and in the north PDRY borders on the great Ar-Rub'al-Khali desert. The southern plateau region is the most fertile part of the country.

Oman, UAI, and Qatar

Oman, which is bordered by the Arabian Sea and the Gulf of Oman, has three main geographical regions. These are a coastal plain, a mountain range which borders on them, and a plateau region beyond the mountains which extends inland to the Ar-Rub'al-Khali desert.

The strip of territory called United Arab Emirates (UAI) is a country divided up into different emirates or provinces. It is bordered by the Persian Gulf and a small part of the Gulf of Oman in the southwestern tip of the country, and it consists mainly of **sand** and gravel desert regions, but includes some fertile coastal strips and many islands. In parts of the country, the coastal sand dunes are over 300 ft (90 m) high.

North of UAI and bordered on three sides by the Persian Gulf is the country of Qatar, which is believed to have been an island before it joined with the Arabian Peninsula. It is mostly flat and sandy, with some low cliffs rising on the northeastern shore and a low chain of hills on the west coast.

Israel and Jordan

Israel contains three main regions. Along the Mediterranean Sea lies a coastal plain. Inland is a hilly area that includes the hills of Galilee in the north and Samaria and Judea in the center. In the south of Israel lies the Negev desert, which covers about half of Israel's land area. The two bodies of **water** in Israel are the Sea of Galilee and the Dead Sea. The latter, which takes its name from its heavy salinity, lies 1,290 ft (393 m) below sea level, and is the lowest point on the earth's landmasses. It is also a great resource for potassium chloride, **magnesium** bromide, and many other salts.

Jordan borders on Israel in the east near the Dead Sea, whose surface is about 1,290 ft (393 m) below sea level. To the east of the Jordan river, which feeds the Dead Sea, is a plateau region. The low hills gradually slope downward to a large desert, which occupies most of the eastern part of the country.

Lebanon and Syria

Lebanon, which borders Israel in the north, is divided up by its steep mountain ranges. These have been carved by **erosion** into intricate clefts and valleys, lending the landscape an unusual rugged beauty. On the western border, which lies along the Mediterranean Sea, is the Mount Lebanon area. These mountains rise from sea level to a height of 6,600–9,800 ft (2,000–3,000 m) in less than 25 mi (40 km). On the eastern border is the Anti-Lebanon mountain range, which separates Lebanon from Syria. Between the mountains lies Bekaa Valley, Lebanon's main fertile region.

Syria has three major mountain ranges. In the southwest, the Anti-Lebanon mountain range separates the country geographically from Lebanon. In the southeast is the Jabal Ad-Duruz range, and in the northwest, running parallel to the Mediterranean coast, are the Ansariyah mountains. Between these and the sea is a thin stretch of coastal plains. The most fertile area is in the central part of the country east of the Anti-Lebanon and Ansariyah mountains; the east and northeastern part of Syria is made up of steppe and desert region.

Turkey

Turkey, at the extreme north of the Arabian Peninsula, borders on the Aegean, the Mediterranean, and the Black Seas. Much of the country is cut up by mountain ranges, and the highest peak, called Mount Ararat, reaches an altitude of 16,854 ft (5,137 m). In the northwest is the Sea of Marmara, which connects the Black Sea with the Aegean Sea. Most of this area, called Turkish Thrace, is fertile and has a temperate climate. In the south, along the Mediterranean, there are two fertile plains called the Adana and the Antalya, which are separated by the Taurus mountains.

The two largest lakes in Turkey are called Lake Van, which is close to the border with Iraq, and Lake Tuz, which lies in the center of the country. Lake Tuz has such a high level of salinity that it is actually used as a source of salt. Turkey is a country of seismic activity, and earthquakes are frequent.

The Far East

Most of the far eastern countries are rugged and mountainous, but rainfall is more plentiful than in the middle east, so there are many forested regions. Volcanic activity and **plate tectonics** have formed many island chains in this region of the world, and nearly all the countries on the coast include some of these among their territories.

China and Taiwan

China, with a land area of 3,646,448 sq mi (9,444,292 sq km), is an enormous territory. The northeastern part of the country is an area of mountains and rich forest land, and its mineral resources include **iron**, **coal**, gold, oil, **lead**, **copper**, and magnesium. In the north, most of the land is made up of fertile plains. It is here that the Yellow (Huang) River is found, which has been called "China's Sorrow" because of its great **flooding**. The northwest of China is a region of mountains and highlands, including the cold and arid steppes of Inner Mongolia. It is here that the Gobi Desert, the fifth largest desert in the world, is found. The Gobi was named by the Mongolians, and its name means "waterless place." It encompasses an area of 500,000 sq mi (1,295,000 sq km), and averages two to four inches of rainfall a year. In contrast, central China is a region of fertile land and temperate climate. Many rivers, including the great Chang (Yangtze) River, flow through this region, and there are several **freshwater** lakes. The largest of these, and the largest in China, is called the Poyang Hu. In the south of China the climate becomes tropical, and the land is very fertile; the Pearl (Zhu or Chu) River **delta**, which lies in this region, has some of the richest agricultural land in China. In the southwestern region, the land becomes mountainous in parts, and coal, iron, phosphorous, manganese, **aluminum**, tin, **natural gas**, copper, and gold are all found here. In the west, before the line of the Himalayas which divides China from India, lies Tibet,

which is about twice as large as Texas and makes up about a quarter of China's land area. This is a high plateau region, and the climate is cold and arid. A little to the north and east of Tibet lies a region of **mountains** and **grasslands** where the Yangtze and Yellow Rivers arise.

One of the largest of the islands off the China coast is called Taiwan, which consists of Taiwan proper and about 85 additional tiny islands in the region. Because Taiwan lies on the edge of the **continental shelf**, the western seas are shallow (about 300 ft; 90 m) while the eastern seas reach a depth of 13,000 ft (4,000 m) only 31 mi (50 km) from the shore. The area is prone to mild earthquakes.

Japan and Korea

Japan consists of a group of four large islands, called Honshu, Hokkaido, Kyushu, and Shikoku, and more than 3,000 smaller islands. It is a country of intense volcanic activity, with more than 60 active volcanoes, and frequent earthquakes. The terrain is rugged and mountainous, with lowlands making up only about 29% of the country. The highest of the mountain peaks is an extinct **volcano** found on Honshu called Mount Fuji. It reaches an altitude of 12,388 ft (3,776 m). Although the climate is generally mild, tropical cyclones usually strike in the fall, and can cause severe damage.

The two Korean republics lie between China and Japan, and are bordered by the Yellow Sea (Huang Hai) on one side and the Sea of Japan on the other. North Korea is a very mountainous region, with only 20% of its area consisting of lowlands and plains. Mount Paektu, an extinct volcano with a lake in the crater, is the highest point in the country at 9,003 ft (2,744 m). South Korea is also quite mountainous, but with lower elevations; the highest point on the mainland is Mount Chiri, at an altitude of 6,283 ft (1,915 m). The plains region is only slightly larger than in the north, taking up about 30% of the country's land area.

Central Asia

Central Asia includes Mongolia and central and eastern Russia. This part of Asia is mostly cold and inhospitable. While only 5% of the country is mountainous, Mongolia has an average elevation of 5,184 ft (1,580 m). Most of the country consists of plateaus. The **temperature** variation is extreme, ranging from −40 to 104°F (−40 to 40°C). The Gobi Desert takes up about 17% of Mongolia's land mass, and an additional 28% is desert steppe. The remainder of the country is forest steppe and rolling plains.

North of China and Mongolia lies Russian Siberia. This region is almost half as large as the African conti-

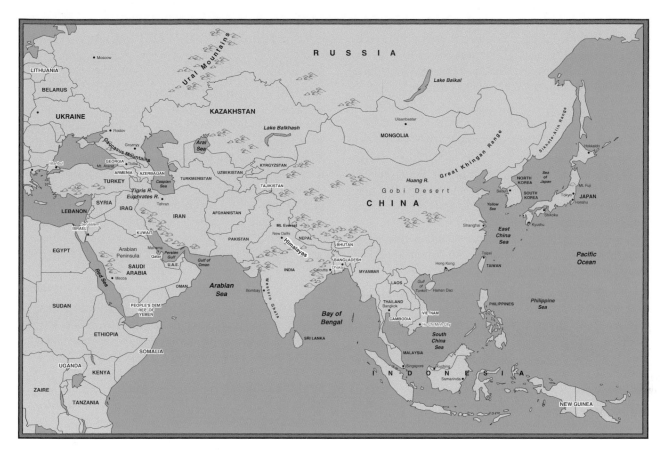

Asia. *Illustration by Hans & Cassidy. Courtesy of Gale Group.*

nent, and is usually divided into the eastern and western regions. About the top third of Siberia lies within the Arctic Circle, and the climate is very harsh. The most extreme temperatures occur in eastern Siberia, where it falls as low as −94°F (−70°C), and there are only 100 days a year when it climbs above 50°F (10°C). Most of the region along the east coast is mountainous, but in the west lies the vast West Siberian Plain.

The most important lake in this area, and one of the most important lakes in the world, is called Lake Baikal. Its surface area is about the size of Belgium, but it is a mile deep and contains about a fifth of the world's fresh water supply. The diversity of aquatic life found here is unparalleled; it is the only **habitat** of 600 kinds of plants and 1,200 kinds of animals, making it the home of two-thirds of the freshwater **species** on **earth**.

Southeast Asia

Southeast Asia includes a number of island chains as well as the countries east of India and south of China on the mainland. The area is quite tropical, and tends to be very humid. Much of the mountainous regions are extremely rugged and inaccessible; they are taken up by forest and jungle and have been left largely untouched; as a result, they provide habitat for much unusual **wildlife**.

Thailand

Thailand, which is a country almost twice the size of Colorado, has a hot and humid tropical climate. In the north, northeast, west, and southeast are highlands which surround a central lowland plain. This plain is drained by the river Chao Phraya, and is rich and fertile land. The highlands are mostly covered with **forests**, which includes tropical rainforests, deciduous forests, and coniferous pine forests. Thailand also has two coastal regions; the largest borders on the Gulf of Thailand in the east and southeast, and on the west is the shore of the Andaman Sea.

Vietnam

Vietnam, which borders on the South China Sea at the Gulf of Tonkin, consists mainly of two fertile river delta regions separated by rugged, mountainous terrain. In the south, the Mekong delta region is the largest and most fertile of the lowland areas, making up about a quarter of the total area. The northern delta region, of the Red (Hong) River, is much smaller. It is divided from the south by the Annamese Highlands, which take up the greatest part of the north. Vietnam has a moist, tropical climate, and its highlands are densely forested.

Cambodia and Laos

Between Thailand and Vietnam lies Cambodia, a country of low plains. In the center of the country is the Tonle Sap (Great Lake), and many of the rivers that water Cambodia flow into this **lake**. During the winter, when the Mekong floods, it forces the flow back from the Tonle Sap into the tributaries, flooding the surrounding area with rich silt. In the north and southwest are some mountain ranges, and the Cardamom range lies along the southern coast.

North of Cambodia, lying between Thailand in the west and Vietnam in the east, is the country of Laos. The Mekong flows along most of its western boundary with Thailand, and most of the country's rivers drain into the Mekong. On the eastern border lie the Annamese Highlands. The northern part of Laos is also very mountainous and covered with thick jungle and some coniferous forests.

Myanmar

Myanmar, formerly called Burma, lies largely between China and India, but also borders on Thailand in a strip of coast along the Andaman Sea. The country is geographically isolated by mountain ranges lying along its western and eastern borders; these run from north to south, meeting in the extreme north. Like most of the mountains in southeast Asia, these are covered with dense forest and jungle. Between the ranges is a large fertile expanse of plains watered by the Irrawaddy River; a little north of this valley below the northern mountains is a small region of dry desert.

Malaysia, Indonesia, and the Philippines

South of the mainland countries lie the island chains of Malaysia, Indonesia, and the Philippines. The latter two are both sites of much volcanic activity; Indonesia is estimated to have 100 active volcanoes. These islands, in particular Malaysia, are extremely fertile and have large regions of tropical rain forests with an enormous diversity in the native **plant** and wildlife.

South Asia

South Asia includes three main regions: the Himalayan mountains, the Ganges Plains, and the Indian Peninsula.

The Himalayas: Afghanistan, Pakistan, Nepal, and Bhutan

The Himalayas stretch about 1,860 mi (3,000 km) across Asia, from Afghanistan to Burma, and range from 150–210 mi (250–350 km) wide. They are the highest mountains in the world, and are still being pushed upward at a **rate** of about 2.3 in (6 cm) a year. This great mountain range originated when the Indian subcontinent collided with Asia, which occurred due to the subduction of the Indian plate beneath the Asian continent. The Himalayas are the youngest mountains in the world, which accounts in part for their great height. At present they are still growing as India continues to push into the Asian continent at the rate of about 2.3 in (6 cm) annually. The Indian subcontinent is believed to have penetrated at least 1,240 mi (2,000 km) into Asia thus far. The range begins in Afghanistan, which is a land of harsh climate and rugged environment.

Bordered by China, Russia, Pakistan, and Iran, Afghanistan is completely landlocked. High, barren mountains separate the northern plains of Turan from the southwestern desert region, which covers most of Afghanistan's land area. This desert is subject to violent sand storms during the winter months. The mountains of Afghanistan, which include a spur of the Himalayas called the Hindu Kush, reach an elevation of more than 20,000 ft (6,100 m), and some are snow-covered year-round and contain **glaciers**. The rivers of the country flow outward from the mountain range in the center of the country; the largest of these are the Kabul, the Helmand, the Hari Rud, and the Kunduz. Except for the Kabul, all of these dry up soon after flowing onto the dry plains.

To the east of Afghanistan and separated from it by the Hindu Kush, lies Pakistan. In the north of the country are the mountain ranges of the Himalayas and the Karakoram, the highest mountains in the world. Most of the peaks are over 15,000 ft (4,580 m) and almost 70 are higher than 22,000 ft (6,700 m). By comparison, the highest mountains in the United States, Mount McKinley in Alaska, is only 20,321 ft (6,194 m). Not surprisingly, many of the mountains in this range are covered with glaciers.

In the west of the country, bordering on Afghanistan, is the Baluchistan Plateau, which reaches an altitude of about 3,000–4,000 ft (900–1,200 m). Further south, the mountains disappear, replaced by a stony and sandy desert. The major rivers of Pakistan are the Kabul, the Jhelum, the Chenab, the Ravi, and the Sutlej; all of these drain into the Indus River, which flows into the Arabian Sea in the south of Pakistan.

Also found in the Himalayan Mountains are Nepal and the kingdom of Bhutan. Both of these countries border on the fertile Ganges Plains, so that in the south they are densely forested with tropical jungles; but most of both territories consists of high mountains. It is in Nepal that the highest peak in the world, called Mount Everest, is found; it is 29,028 ft (8,848 m) high.

The Ganges Plains: India and Bangladesh

South of the Himalayan mountains, India is divided into two major regions. In the north are the Ganges Plains, which stretch from the Indus to the Ganges river delta. This part of India is almost completely flat and immensely fertile; it is thought to have alluvium reaching a depth of 9,842 ft (3,000 m). It is fed by the snow and **ice** from the high peaks, and streams and rivers from the mountains have carved up the northern edge of the plains into rough gullies and crevices. Bangladesh, a country to the north and east of India, lies within the Ganges Plains. The Ganges and the Brahmaputra flow into Bangladesh from India, and they are fed by many tributaries, so the country is one of the most well-watered and fertile regions of Asia. However, it is also close to sea level, and plagued by frequent flooding.

The Peninsula: India, Sri Lanka, and the Maldives

South of the plains is the Peninsula, a region of low plateaus and river valleys. It is bounded on the west, parallel to the Arabian Sea, by the Ghat mountain range; further north by the border of Pakistan is the Thar desert, which encompasses an area of 100,387 sq mi (260,000 sq km). In its southern extent, the Thar borders on salt marshes and the great lava expanse called the Deccan plateau. The island of Sri Lanka, which lies south of India, is the only other country which is part of the Peninsula, although it is separated from it by the **ocean**.

Off the southwestern tip of the **peninsula** are the Maldives, a group of about 1,200 islands. At their highest point, they only reach an altitude of about 80 ft (24 m) above sea level; and their number and identity varies as old islands are constantly submerged and new ones created.

Resources

Books

Babaev, Agadzhan, and Agajan G. Babaev, eds. *Desert Problems and Desertification in Central Asia: The Researches of the Desert Institute.* Berlin: Springer Verlag, 1999.

Chapman, Graham P., and Kathleen M. Baker, eds. *The Changing Geography of Asia.* New York: Routledge, 1992.

Menzies, Gavin. *1421: The Year China Discovered America.* New York: William Morrow & Co., 2003.

Taylor, Robert H., ed. *Asia and the Pacific.* New York: Facts on File, 1991.

Ulack, Richard, and Gyula Pauer. *Atlas of Southeast Asia.* New York: Macmillan Publishing Co., 1989.

Sarah A. de Forest

Asparagus *see* **Lily family (Liliaceae)**
Aspirin *see* **Acetylsalicylic acid**

Assembly line

An assembly line is a system of **mass production** in which a product is manufactured in a step-by-step process as it moves continuously past an arrangement of workers and machines. Introduced in the nineteenth century, it provided the basis for the modern methods of mass production of quantities of standardized, relatively low-cost goods available to great numbers of consumers. As one of the most powerful productivity concepts in history, it was largely responsible for the emergence and expansion of the industrialized, consumer-based system we have today.

History

The principle of continuous movement is perhaps the simplest and most obvious fact of an assembly line, dating back to Assyrian times, where there is evidence of a

The assembly of robots at Renault in France. *Photograph by Cecilia Posada. Phototake NYC. Reproduced by permission.*

system of bucket elevators called the "chain of pots." Miners in medieval **Europe** also used these bucket elevators, and by the time of the Renaissance, engineers were becoming familiar with some form of the assembly line. In the fourteenth century, for example, the shipbuilding arsenal of Venice used moving lines of prefabricated parts to equip their war galleys. What may have been the first powered-roller conveyer system was introduced in 1804 by the British Navy's automatic production of biscuits or "hardtack." It used a **steam engine** to power its rollers. By the 1830s, the principle of continuous processing was starting to enter the consciousness of manufacturers, although it was by no means fully embraced until the 1870s in the United States. By then, the principles of division of labor and interchangeable parts had been successfully demonstrated by the American inventors Eli Whitney (1765-1825) and Samuel Colt (1814-1862).

The assembly line was first used on a large scale by the meat-packing industries of Chicago and Cincinnati during the 1870s. These slaughterhouses used monorail trolleys to move suspended carcasses past a line of stationary workers, each of whom did one specific task. Contrary to most factories' lines in which products are gradually put together step-by-step, this first assembly line was in fact more of a "dis-assembly" line, since each worker butchered a piece of a diminishing **animal**. The apparent breakthroughs in efficiency and productivity that were achieved by these meat packers were not immediately realized by any other industry until the American industrialist Henry Ford (1863-1947) designed an assembly line in 1913 to manufacture his Model T automobiles. Ford openly admitted using the meat-packing lines as a model. When the total time of assembly for a single car fell from 12.5 labor hours to 93 labor minutes, Ford was able to drastically reduce the price of his cars. His success not only brought **automobile** ownership within the grasp of the average person, but it served notice to all types of manufacturers that the assembly line was here to stay. The assembly line transformed in a revolutionary way the manner and organization of work, and by the end of World War I, the principle of continuous movement was sweeping mass-production industries of the world and was soon to become an integral part of modern industry.

The basic elements of traditional assembly line methods are nearly all the same. First, the sequence in which a product's component parts are put together must be planned and actually designed into the process. Then the first manufactured component passes from station to station, often by conveyor belt, and something is done or added to it. By the last station, the product is fully assembled and is identical to each one before and after it. This system ensures that a large quantity of uniform-quality goods are produced at a relatively low cost.

When manufacturers first implemented the idea of the assembly line, they enjoyed dramatic gains in productivity, and the consumer realized lower costs. However, the nature of work in a factory changed radically. Skilled workers were replaced by semi-skilled or even unskilled workers, since tasks had been minutely compartmentalized or broken down and each person was responsible only for assembling or adding one particular part. Manufacturers soon realized however, that not only were a great number of managers and supervisors required to oversee these laborers, but a high degree of preplanning on their part was absolutely essential. Overall operations had become much more complex and correct sequencing was essential. Thus, before actual assembly line production could begin, proper design of both the product and the assembly line itself had to be accomplished. Even the simplest tasks were critical to its overall success, and the apparently straightforward assembly line became a highly complex process when broken down and considered step-by-step.

Role of workers

Early twentieth century assembly line systems carried the concept of division of labor to an extreme and usually restricted each worker to the repetitive performance of one simple task. These individuals had few real skills, and they were not required to know any more than their basic job demanded. This human element proved to be the weakest link in the entire system. For most people, assembly line work eventually entailed a physical and mental drudgery that became seriously counterproductive. Often the work itself was detrimental to an individual's physical and mental well-being, and from a manufacturer's standpoint, this usually resulted in diminished productivity.

Henry Ford and his fellow industrialists soon discovered this phenomenon when they tried to speed up their assembly lines. Since the pact of the assembly line was dictated by machines, supervisors often accelerated them, forcing workers to try to keep up. When this constant pressure to increase production was combined with the essentially dull and repetitive nature of the job, the result was often a drop in quality as well as output, not to mention worker unrest and dissatisfaction. By the 1920s, industry leaders realized that they could not ignore the dehumanizing aspects of the assembly line. However, it was not until after World War II that the major industries made serious attempts to make the mechanical aspects of the assembly line accommodate itself to the human **physiology** and **nervous system**.

The logical evolution of the assembly line would seem to lead to one that is fully automated. Such an automated system would ideally imply the elimination of the human element and its replacement with automatic controls that guarantee a level of **accuracy** and quality that is beyond human skills. In fact, this is the case today where **automation** has completely changed the nature of the traditional assembly line. Computer advances have resulted in assembly lines that are entirely run by computers controlling industrial robots of all kinds. Increasingly, such robots not only perform the repetitive, elementary tasks, but also are sufficiently intelligent (via feedback systems) to regulate or adjust their own performance to suit a changing situation. Especially in the automobile industry, assembly lines consist of machines that are run by machines. People are still needed of course, for quality control, repair, and routine inspection, as well as for highly specialized tasks. In fact, rather than minimizing the human skills needed to oversee these systems, today's automated assembly lines require more highly skilled workers to operate and maintain the sophisticated, computer-controlled equipment.

In the 1980s, Japanese and Italian automobile manufacturers so successfully automated their assembly lines that certain of their factories consisted almost entirely of robots regularly doing their jobs. On one particular Italian Fiat, only 30 of the 2,700 welds were done by human hands. In principle, they are Henry Ford's assembly lines carried to their ultimate conclusion. Starting again with the bare chassis, major components (which themselves have been automatically assembled elsewhere) are attached by robots, and the computer keeps track of exactly what is to be added to each. Each vehicle is considered unique and the central computer assures its total assembly. On the other hand, GM found that robots could not replace human workers and had to retrench from technology and focus on retraining workers. Final product assembly and delivery to the dealer offers the consumer, if not always the most affordable product, an extremely wide array of special options.

See also Robotics.

Resources

Books

How in the World? Pleasantville, NY: Reader's Digest Association, Inc., 1990.

McNeil, Ian. *An Encyclopaedia of the History of Technology.* New York: Routledge, 1996.

KEY TERMS

Automation—The application of self-governing machines to tasks once performed by human beings.

Division of labor—The separation of a job or task into a number of parts, with each part performed by a separate individual or machine.

Industrial robots—Programmable, multi-purpose, electromechanical machines that can perform manufacturing-related tasks that were traditionally done by human beings.

Interchangeable parts—The production of high precision parts that are identical as opposed to unique, hand-made parts; this standardization of size and shape assures its quality and quantity and permits low-cost mass-production.

Mass production—A method of organizing manufacturing processes to produce more things at a lower cost that is based on specialized human labor and the use of tools and machines.

Periodicals

Rae, John B. "The Rationalization of Production." In *Technology in Western Civilization* Vol. 2, edited by Melvin Kranzberg and Carroll E. Pursell, Jr. New York: Oxford University Press, 1967.

"Who Says the Assembly-Line Age Is History?" *U.S. News & World Report* (July 16, 1984): 48-49.

Leonard C. Bruno

Asses

Asses include three of the seven genera that make up the family Equidae, which also includes **horses** and **zebras**. Wild asses are completely wary and apt to run swiftly away, so they have been difficult to study. Asses can survive in poor **habitat** such as scrub and near **desert** regions. Asses have loud voices, most notable in the raucous bray of the domestic burro and a keen sense of **hearing**. Male asses (stallions) tend to leave the herd and live solitary lives except during the mating season in late summer. Female asses tend to stay in the herd, especially when caring for their young. All of the **species** of wild asses are endangered.

The Asiatic wild ass (*Equus hemionus*) was formerly distributed in **Asia** from China to the Middle East.

The largest species of the kiang ass (*E. kiang*) lives in the high steppes of Tibet and China. This spieces is about 4.5 ft (1.4 m) tall at the shoulder, weighs up to 880 lbs (400 kg) with a red-brown to black back, and white sides and belly. The coat becomes thicker during the cold Tibetan winters.

The other Asiatic asses are smaller than the kiang, with narrower heads and longer ears. The onager of Iran was perhaps the first member of the horse family to be domesticated. These wild asses once lived in large herds in the deserts and **grasslands** of Asia, but now are limited to a few very small areas and may even be extinct in the wild, though their exact status is uncertain. The kulan is a small wild ass found in the Mongolian Desert which can run at speeds of up to 40 MPH (64 km/h). The khur, or Indian onager and the dziggetai of Mongolia are both endangered and probably exist today only in **wildlife** reserves. The small Syrian onager is the wild ass of the Bible, stands only slightly more than 3 ft (1 m) high at the shoulder, has not been seen since 1927, and is probably extinct in the wild.

The African wild ass (*E. africanus*), is the ancestor of the domesticated donkey, and is represented by a few thousand individuals in Ethiopia, Somalia, and the Sudan. The domesticated donkey is sometimes given a separate name, *E. asinus*. The African wild ass has hooves that are higher and narrower than those of other equids, allowing sure footing in its dry, hilly home. Like many desert living animals, these wild asses need little **water**, can withstand dehydration, even in temperatures of 125°F (52°C), and can survive two or three days without drinking.

There are two varieties of African wild ass. The Somali wild ass of Somalia and Ethiopia has a dark stripe along its back, light stripes on its legs, dark tips on its ears, and a dark, short mane. The animal's base coat **color** may turn yellowish or tan during the summer. The Somali wild ass is an **endangered species**, while the slightly smaller Nubian wild ass, which lacks stripes on its legs, is probably already extinct in the wild.

Domesticated asses are known as **donkeys**, jackasses, or burros. Their size varies from the tiny 2-ft (less than a meter) burro of Sicily to the Spanish donkey that stands more than 5 ft (2 m) at the shoulder. Numerous feral (wild, formerly domestic) burros live in the western United States, which are regularly rounded up and sold as pets. These sure-footed animals carry tourists on the steep narrow paths leading down into the Grand Canyon.

The **hybrid** offspring produced when a horse mare mates with a donkey stallion are called mules, which are as sure-footed as burros, and are even stronger than horses. However, mules, being hybrids, are almost always sterile. The offspring hybrid produced by the mating of a horse stallion with a donkey mare is called a hinny. Hinnies tend to resemble a horse more than a mule but are relatively rare because female donkeys do not easily become pregnant.

Resources

Books

Duncan, P., ed. *Zebras, Horses and Asses: An Action Plan for the Conservation of Wild Equids.* Washington, DC: Island Press, 1992.

Knight, Linsay. *The Sierra Club Book of Great Mammals.* San Francisco: Sierra Club Books for Children, 1992.

Patent, Dorothy Hinshaw. *Horses and Their Wild Relatives.* New York: Holiday House, 1981.

Special Publications Division. *National Geographic Book of Mammals.* Vol. 1 & 2. Washington, DC: National Geographic Society, 1981.

Stidworthy, John. *Mammals: The Large Plant-Eaters.* Encyclopedia of the Animal World. New York: Facts On File, 1988.

Wild Horses. Zoobooks series. San Diego: Wildlife Education, 1987.

Jean F. Blashfield

Associative property

In **algebra**, a binary operation is a rule for combining the elements of a set two at a time. In most important examples that combination is also another member of the same set. **Addition, subtraction, multiplication**, and **division** are familiar binary operations. A familiar example of a binary operation that is associative (obeys the associative principle) is addition (+) of **real numbers**. For example, the sum of 10, 2, and 35 is determined equally as well as (10 + 2) + 35 = 12 + 35 = 47, or 10 + (2 + 35) = 10 + 37 = 47. The parentheses on either side of the defining equation indicate which two elements are to be combined first. Thus, the associative property states that combining a with b first, and then combining the result with c, is equivalent to combining b with c first, and then combining a with that result. A binary operation (*) defined on a set S obeys the associative property if (a * b) * c = a * (b * c), for any three elements a, b, and c in S. Multiplication of real numbers is another associative operation, for example, $(5 \times 2) \times 3 = 10 \times 3 = 30$, and $5 \times (2 \times 3) = 5 \times 6 = 30$. However, not all binary operations are associative. Subtraction of real numbers is not associative since in general (a − b) −c does not equal a − (b − c), for example $(35 - 2) - 6 = 33 - 6 = 27$, while $35 - (2 - 6) = 35 - (-4) = 39$. Division of real numbers is not associative either. When the associative property holds for all the members of a set, every combination of elements must result in another element of the same set.

Astatine *see* **Halogens**

Aster *see* **Composite family (Compositaceae)**

Asteroid *see* **Minor planets**

Asteroid 2002 AA29

Late in 2002, astronomers and the International Astronomical Union (IAU) confirmed the discovery the discovery of an asteroid, designated Asteroid 2002 AA29, in a companion **orbit** to **Earth**. It is the first object ever identified to be in a companion orbit around the **Sun** (i.e., it shares at least some of the same orbital path and **space**). In another 600 years, the asteroid will technically and temporarily become an Earth **moon**.

Although in a companion orbit, computer driven mathematical estimates establish that Asteroid 2002 AA29 never comes closer than approximately 3.5 million m (5.6 million km) from Earth at its closest approach.

Asteroid 2002 AA29 was first detected by the linear automated sky survey project in January 2002. Optical and gravitational evidence indicate that Asteroid 2002 AA29 is approximately 109 yd (100 m) wide.

Although co-orbital for some of its travel around the Sun, Asteroid 2002 AA29 does not follow the exact same path as Earth. Asteroid 2002 AA29 travels a horseshoe-like path that allows it vary in relative position to the Sun and Earth (i.e., it oscillates between appearing on both sides of the Sun from Earth's perspective. Asteroid 2002 AA29 made its closest recent approach to Earth—approximately 10 to 12 times the normal Earth-Moon distance—at 1900 GMT on 8 January 2003.

Orbital dynamics projections establish that in 550 A.D. Asteroid 2002 AA29 technically became an Earth orbital satellite—technically a second moon. Because of

Asteroid 2002 AA29's odd orbit this event is due to occur again in 2600 and will last approximately 50 years.

Resources

Other

NASA Near Earth Object Program. 2002 AA29 Animations. <http://neo.jpl.nasa.gov/2002aa29.html>. (March 12, 2003).

Asthenosphere

The asthenosphere is the layer of **Earth** situated at an average depth of about 62 mi (about 100 km) beneath Earth's surface. It was first named in 1914 by the British geologist Joseph Barrell, who divided Earth's overall structure into three major sections: the **lithosphere**, or outer layer of rock-like material; the asthenosphere; and the centrosphere, or central part of the **planet**. The asthenosphere gets its name from the Greek word for weak, *asthenis,* because of the relatively fragile nature of the materials of which it is made. It lies in the upper portion of Earth's internal structure traditionally known as the mantle. Scientists have not seen the asthenosphere of Earth, but its existence has a profound effect upon the planet and the manner in which the Earth's crust behaves. For anyone living near a plate boundary on Earth, the asthenosphere contributes mightily to the uneasy geologic conditions which may plague the area.

Evidence for the existence of the asthenosphere

Geologists are somewhat limited as to the methods by which they can collect information about **Earth's interior**. For example, they may be able to study rocky material ejected from volcanoes and lava flows for hints about properties of the interior regions. Generally speaking, however, the single most dependable source of such information is the way in which seismic waves are transmitted through Earth's interior. These waves can be produced naturally as the result of earth movements, or they can be generated synthetically by means of explosions, air guns, or other techniques.

In any case, seismic studies have shown that a type of waves known as S-waves slow down significantly as they reach an average depth of about 62 mi (100 km) beneath Earth's surface. Then, at a depth of about 155 mi (250 km), their **velocity** increases once more. Geologists have taken these changes in wave velocity as indications of the boundaries for the region now known as the asthenosphere.

Properties of the asthenosphere

The material of which the asthenosphere is composed can be described as plastic-like, with much less rigidity than the lithosphere above it. This property is caused by the interaction of **temperature** and **pressure** on asthenospheric materials. Any rock will, of course, melt if its temperature is raised to a high enough temperature. However, the melting point of any rock (or of any material) is also a function of the pressure exerted on the rock (or the material). In general, as the pressure is increased on a material, its melting point increases.

Materials that make up the asthenosphere tend to be slightly cooler than their melting point. This gives them a plastic-like quality that can be compared to **glass**. As the temperature of the material increases or as the pressure exerted on the material increases, the material tends to deform and flow. If the pressure on the material is sharply reduced, so will be its melting point, and the material may begin to melt quickly. The fragile melting point/pressure balance in the asthenosphere is reflected in the estimate made by some geologists that up to 10% of the asthenospheric material may actually be molten. The rest is so close to being molten that relatively modest changes in pressure or temperature may cause further melting.

In addition to loss of pressure on the asthenosphere, another factor that can bring about melting is an increase in temperature. The asthenosphere is heated by contact with hot materials that make up the mesosphere beneath it. Obviously, the temperature of the mesosphere is not constant. It is hotter in some places than in others. In those regions where the mesosphere is warmer than average, the extra **heat** may actually increase the extent to which asthenospheric materials are heated and a more extensive melting may occur. The results of such an event are described below.

The asthenosphere in plate tectonic theory

The asthenosphere is now thought to play a critical role in the movement of plates across the face of Earth's surface. According to plate tectonic theory, the lithosphere consists of a relatively small number of very large slabs of rocky material. These plates tend to be about 60 mi (100 km) thick and in most instances many thousands of miles wide. They are thought to be very rigid themselves but capable of being moved on top of the asthenosphere. The collision of plates with each other, their lateral sliding past each other, and their separation from each other are thought to be responsible for major geologic features and events such as volcanoes, lava flows, mountain building, and deep crustal faults and rifts.

In order for plate tectonic theory to make any sense, some mechanism must be available for permitting the flow of plates. That mechanism is the semi-fluid character of the asthenosphere itself. Some observers have described the asthenosphere as the 'lubricating oil' that permits the movement of plates in the lithosphere. Others view the asthenosphere as the driving force or means of conveyance for the plates.

Geologists have now developed theories to explain the changes that take place in the asthenosphere when plates begin to diverge from or converge toward each other. For example, suppose that a region of weakness has developed in the lithosphere. In that case, the pressure exerted on the asthenosphere beneath it is reduced, melting begins to occur, and asthenospheric materials begin to flow upward. If the lithosphere has not actually broken, those asthenospheric materials cool as they approach Earth's surface and eventually become part of the lithosphere itself. On the other hand, suppose that a break in the lithosphere has actually occurred. In that case, the asthenospheric materials may escape through that break and flow outward before they have cooled. Depending on the temperature and pressure in the region, that outflow of material (**magma**) may occur rather violently, as in a **volcano**, or more moderately, as in a lave flow. Both these cases produce crustal plate divergence, or spreading apart. Pressure on the asthenosphere may also be reduced in zones of divergence, where two plates are separating from each other. Again, this reduction in pressure may allow asthenospheric materials in the asthenosphere to begin melting and to flow upward. If the two overlying plates have actually separated, asthenospheric material may flow through the separation and form a new section of lithosphere.

In zones of convergence, where two plates are moving toward each other, asthenospheric materials may also be exposed to increased pressure and begin to flow downward. In this case, the lighter of the colliding plates slides upward and over the heavier of the plates, which dives down into the asthenosphere. Since the heavier lithospheric material is more rigid than the material in the asthenosphere, the latter is pushed outward and upward. During this movement of plates, material of the downgoing plate is heated in the asthenosphere, melting occurs, and molten materials flow upward to Earth's surface. Mountain building is the result of continental collision in such situations, and great mountain chains like the Urals, Appalachian, and Himalayas have been formed in such a fashion. When oceanic plates meet one another, **island** arcs (e.g., Japan or the Aleutians) are formed. Great **ocean** trenches occur in places of plate convergence. In any one of the examples cited here, the asthenosphere supplies new material to replace lithospheric materials that have been displaced by some other tectonic or geologic mechanism.

KEY TERMS

Lithosphere—The outer layer of Earth, that extends to a depth of about 60 mi (100 km).

Magma—Molten material exuded from below Earth's surface, generally consisting of rock-like materials rich in silicon and oxygen.

Seismic wave—A disturbance produced by compression or distortion on or within the earth, which propagates through Earth materials; a seismic wave may be produced by natural (e.g. earthquakes) or artificial (e.g. explosions) means.

Therefore, whether scientists are considering the origin of compressed mountain ranges like the Himalayas, or the origin of the great ocean trenches (like the Peru-Chile trench), they also consider the activity of the asthenosphere, which keeps Earth's plates continually geologically active.

See also Continental drift; Continental margin; Continental shelf; Planetary geology; Plate tectonics.

Resources

Books

Press, Frank, and Raymond Sevier. *Understanding Earth.* San Francisco: Freeman, 2000.

Tarbuck, Edward. J., Frederick K. Lutgens, and Dennis Tassa, eds. *Earth: An Introduction to Physical Geology,* 7th ed. Upper Saddle River, NJ: Prentice Hall, 2002.

Fuchs, Karl, and Claude Froidevaux. *Composition, Structure, and Dynamics of the Lithosphere and Asthenosphere System.* Washington, DC: American Geophysical Union, 1987.

David E. Newton

Asthma

Asthma is a lung **disease** that affects approximately four million people in the United States. In people with asthma, the airways of the lungs are hypersensitive to irritants such as **cigarette smoke** or allergens. When these irritants are inhaled, the airways react by constricting, or narrowing. Some people with asthma have only mild, intermittent symptoms that can be controlled without drugs. In others, the symptoms are chronic, severe, and sometimes life threatening. Although researchers have learned more about the underlying causes of asthma in recent years, a definitive treatment is still unavailable. In fact, deaths from asthma are on the rise. In the last decade, asthma deaths worldwide rose 31%. The reasons for this increase are not clear; however, many experts believe that the lack of standard treatments and the inconsistent monitoring of asthma patients have contributed to the increased mortality **rate**.

What is asthma?

Asthma is sometimes referred to as a disease of "twitchy lungs," which means that the airways are extremely sensitive to irritants. The airways are the tubes that bring air from the windpipe, or trachea, to the lungs. These tubes are called the bronchi. Each bronchus, in turn, branches into smaller tubes called bronchioles. At the end of the bronchioles are small, balloon-like structures called alveoli. The alveoli are tiny sacs that allow **oxygen** to diffuse into the **blood** and **carbon dioxide** to diffuse from body tissues into the lungs to be exhaled.

During an asthma attack, the bronchi and bronchioles constrict and obstruct the passage of air into the alveoli. Besides constricting, the airways may secrete copious amounts of mucus in an effort to clear the irritation from the lungs. The airway walls also swell, causing **inflammation** and further obstruction. As the airways become increasingly obstructed, oxygen cannot reach the small airsacs; blood oxygen levels drop and the body's tissues and organs become oxygen-deprived. At the same time, **carbon** dioxide cannot escape the small airsacs for exhalation; blood levels of carbon dioxide increase, and exert a toxic effect on the tissues and organs of the body.

Underlying the bronchial inflammation is an immune response in which white blood cells known as type 2 helper T (Th2) cells (a type of CD4 helper T **cell**) are prominent. Th2 cells secrete chemicals known as interleukins that promote allergic inflammation and stimulate another set of cells known as B cells to produce IgE and other antibodies. In contrast, type 1 helper T (Th1) cells, another class of CD4 **T cells**, produce interferon-g and interleukin-2, which initiate the killing of viruses and other intracellular organisms by activating macrophages and cytotoxic T cells. These two subgroups of helper T cells arise in response to different immunogenic stimuli and cytokines, and they constitute an immunoregulatory loop: cytokines from Th1 cells inhibit Th2 cells, and vice versa. An imbalance in this reciprocal arrangement may be the key to asthma and there is credible evidence that, when freed from the restraining influence of interferon-g, Th2 cells can provoke airway inflammation. Recent experiments in support of this concept have focused on a newly discovered transcription factor, T-bet, which is necessary to induce helper T cells to differentiate into Th1 cells and for Th1 cells to produce interferon-g. For

these reasons, T-bet is thought to be central to the feedback loops that regulate Th1 and Th2 cells, and in this way it could be important in asthma.

One of the hallmarks of asthma is that the airway obstruction is reversible. This reversibility of the airway swelling is used to definitively diagnose asthma. If the swelling and inflammation can be brought under control with asthma drugs, the person has asthma and not some other upper respiratory tract disease.

In addition to cigarette smoke and various allergens, other triggers can cause asthma attacks. A cold or other upper respiratory **infection** may bring on an asthma attack. Strong emotions, such as excitement, tension, or **anxiety**, may trigger asthma symptoms. **Exercise** can cause symptoms of asthma. **Weather** conditions, such as extreme cold, **heat**, or **humidity** can cause an asthma attack. **Pollution** and increasing **ozone** levels are also associated with episodes of asthma. Other environmental factors include occupational exposure to certain substances like **animal** dander, **wood** particles, dusts, various industrial chemicals, and **metal** salts.

The characteristic sign of asthma is wheezing, the noisy, whistling breathing that a person makes as he or she tries to push air in and out of narrowed airways. Other symptoms of asthma include a tight chest, shortness of breath, and a cough.

Treatment of asthma

Currently, several drugs are used to treat asthma. Not all of the asthma drugs, however, should be used by every asthma patient. Some patients with mild asthma only need to use medication intermittently to control wheezing, while patients with more serious asthma need to take medication at regular intervals to avoid life-threatening attacks. It is important for asthma patients to see their doctors if the frequency or severity of their symptoms change. It has been suggested that many of the life-threatening asthma attacks are in people who once had mild asthma—with symptoms that could be treated as they occurred—which then progressed to a more severe case of the disease.

Bronchodilators dilate constricted lung airways by relaxing the muscles that line the bronchial tubes. Oral bronchodilators include theophylline; theophylline's counterpart, aminophylline, is used through a needle in the vein (intravenous or IV) for severe episodes of asthma. During severe, acute attacks of asthma, injections of epinephrine are given just under the patient's skin. Epinephrine has a quick, but short-lasting effect of bronchodilatation.

Most asthma patients are given bronchodilators such as albuterol that are used in a mist form that is in-

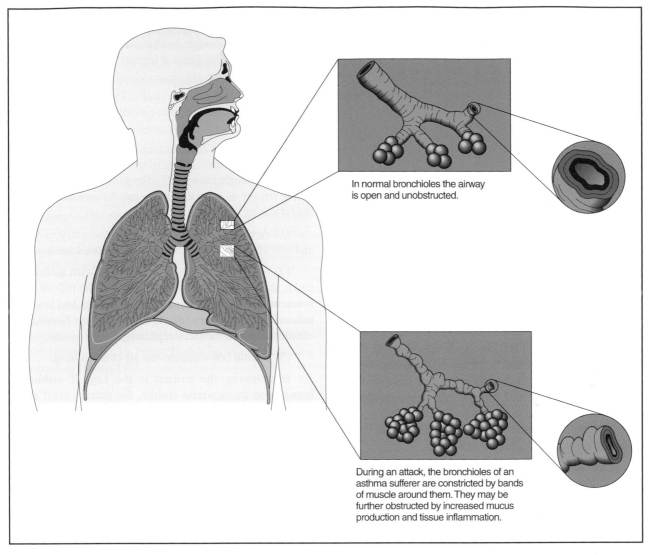

In normal bronchioles the airway is open and unobstructed.

During an attack, the bronchioles of an asthma sufferer are constricted by bands of muscle around them. They may be further obstructed by increased mucus production and tissue inflammation.

A comparison of normal bronchioles and those of an asthma sufferer. *Illustration by Hans & Cassidy. Courtesy of Gale Group.*

haled from either a special inhaler device or an aerosol machine. Some patients are instructed to use their bronchodilator at regular intervals, while others may just be told to use the inhaler if they notice the beginning of an asthma attack. The inhaled medications are quick-acting because they are directly applied to the constricted airways.

In the 1990s, some controversy about inhaled bronchodilators arose in the medical field. In a study published in 1993, doctors found an increased risk of death or near-death from asthma when patients used a type of inhaled bronchodilator commonly prescribed to control asthma. Although more information is still needed regarding the reasons behind the increase in deaths and near-deaths and their association with inhaled bronchodilators, some experts think that the association can be explained by several factors:

(1) More people who use inhaled bronchodilators die because their asthma suddenly becomes more severe and they do not see their doctors. These patients are treating severe asthma with a drug usually prescribed for milder forms.

(2) Bronchodilators may have long-term effects on **organ** systems.

(3) Bronchodilators may, over time, increase airway hyper-responsiveness.

(4) Physicians are not adequately monitoring their patients for progression from mild to severe asthma.

These factors are currently being investigated. Asthma experts stress, however, that people with asthma who use inhaled bronchodilators should continue to do so, but under medical supervision. They should also immediate-

NASA's Mars *Viking Lander* (full-scale operational model) its arm extended as it would be to collect a soil sample to be examined for chemical indicators of biogenic activity. *© Lowell Georgia/Corbis. Reproduced by permission.*

and meteors. These objects falling onto Earth were likely to bring inorganic and organic compounds used later to create **proteins** and DNA, the basis of life, as it is known today. Analysis of space dust from our **galaxy** and beyond revealed presence of **ammonia** and organic compounds such as glycine, vinyl **alcohol**, **benzene**, **polycyclic aromatic hydrocarbons**, and formaldehyde.

The crushing of large objects onto the Earth's surface not only brought new molecules to Earth, but also often had catastrophic consequences leading to sudden climate changes and mass extinctions (for example, the disappearance of the dinosaurs). Events from Earth's past were described by scientists studying fossils, **rocks**, and more recently, Greenland's and Antarctic **ice** sheets.

Signs of life on planets and moons of the solar system are investigated by analyzing meteors that landed on Earth. These, however, could have been potentially contaminated by terrestrial material making it

impossible to interpret the results. Therefore, the studies of rocks originating on **Mars** are much more interesting, provided that samples are kept in isolation. Clues to the existence of extraterrestial life are also provided by studies of the **chemistry** and **geology** of the surface of planets and moons. The presence of hydrated salts and formations similar to ice fields of the Antarctic were identified on Jupiter's moons Callisto, Europa, and Ganymeda.

Beside searching for life in and beyond our solar system, astrobiology is involved in studying the response of Earth lifeforms to weightlessness in **orbit** and possible space travel. **Space shuttle** missions have investigated the effects of lack of gravity on the **physiology** of human and **animal** bodies and their **immune system**, **plant** growth in space, and **fertilization**. Early twenty-first century planned research includes flying worms into space and studying **gene** expression in their multiple generations to see if weightlessness affects them similar-

KEY TERMS

Anoxic—Deprived of oxygen

Astrobiology—The study of living organisms on planets beyond Earth and in space.

Evolution—in biology, inheritable changes occurring over a time span greater than one generation.

Extraterrestrial—Beyond Earth

ly to **vertebrates**. So far, the main problems encountered with prolonged stay in space are muscle atrophy (shrinkage), loss of bone mass, and **radiation exposure**, especially during walks outside the shuttle.

Future of astrobiology

New challenges for astrobiology include developing methods to test the newly discovered **extrasolar planets** for life signs, which can be quite different from what is now known. Although traveling in space is still limited, it is important that scientists establish ways in which terrestrial organisms can safely live in space for prolonged periods.

See also Astronomy; Biosphere; Chemical evolution; Extrasolar planets; Life history.

Resources

Books

Horneck, Gerda, and Christa Baumstark-Khan, eds. *Astrobiology: the Quest for the Conditions of Life* Berlin: Springer, 2002.

Periodicals

Blumberg, Baruch S., "Astrobiology: an Introduction." *Anatomical Record* (November 2002):169–170.

Drake, Michael J, and Bruce M. Jakosky, "Narrow Horizons in Astrobiology." *Nature* (February 2002):733–734.

Organizations

NASA Astrobiology Institute, Ames Research Center, Moffett Field, CA 94035 [cited November 10, 2002]. <http://astrobiology.arc.nasa.gov>.

Other

"The Astrobiology Web. Your Online Guide to the Living Universe" [cited November 14, 2002]. <http://www.astrobiology.com>.

Astrobiology Web. "Astrobiology 101:Exploring the Living Universe." Mitchell K. Hobish, and Keith Cowing 1999 [cited November 14, 2002]. <http://www.astrobiology.com/adastra/astrobiology.101.html>.

Agnieszka Lichanska

Astroblemes

Astroblemes are the scars left on Earth's surface by the high **velocity** impact of large objects from outer **space**. Such colliding bodies are usually meteorites, but some may have been comet heads or asteroids. Few of these impacts are obvious today because our active **earth** tends to erode meteorite craters over short periods of **geologic time**. The term astrobleme was coined in 1961 by Robert S. Dietz from two Greek roots meaning "star wound." Most geologists were not convinced until the 1930s that a mysterious handful of huge circular depressions on the earth were caused by meteorites. The most studied astrobleme during that time was Barringer Crater, a meteor crater in northern Arizona, measuring 0.7 mi (1.2 km) across and 590 ft (180 m) deep. It is now thought to have been blown out about 25,000 years ago by a nickel-iron meteorite about the size of a large house traveling at 9 mi (15 km) per second. Over the years aerial **photography** and **satellite** imagery have revealed many other astroblemes. About 100 around the world have been confirmed by various geological methods. A number have diameters 10–60 times larger than that of the Barringer Crater and are hundreds of millions of years old. The largest astrobleme is South Africa's Vredefort Ring, whose diameter spans 24 mi (40 km).

Exploding or collapsing volcanoes can make roughly circular craters, so it is not easy to interpret such features unless there are a lot of meteorite fragments present. However, because only meteorites collide with the earth at terrific speeds, geologists also have the option of searching for the effects of tremendous **pressure** applied in an instant of time at potential astrobleme sites. Important clues along this line are a large body of shattered rock (impact breccia) radiating downward from a central focus, similar small-scale "shatter cones," very high pressure forms of the mineral silica not found anywhere else in the earth's crust (coesite and stishovite), finely cracked, "shocked" quartz particles, and bits of impact-melted silicate rock that cool into tiny balls of **glass** called "tektites."

See also Comets.

Astrolabe

An astrolabe is an astronomical instrument once used widely to measure stars or planets in order to determine latitude and **time**, primarily for navigational purposes. The original meaning of the word in Greek is

An astrobleme in Arizona. *JLM Visuals. Reproduced by permission.*

"star-taker." The astrolabe was probably invented by astronomers in the second century B.C.

At least two forms of the astrolabe have existed. The older form, known as the planispheric astrolabe, consists of two circular **metal** disks, one representing **Earth** and the other, the celestial **sphere** at some particular location (latitude) on the Earth's surface. The first of these disks, called the plate or tympan, is fixed in position on a supporting disk known as the mater. It shows the great circles of altitude at the given latitude. Any given plate can be removed and replaced by a plate for some other altitude with the appropriate markings for that latitude.

The second of the disks, called the rete or spider, is attached to the plate and the mater by a metal pin through its center. The metal disk that makes up the rete is primarily cut out so that it consists of a complex series of curved lines ending in points. The points indicate the location of particular stars in the celestial sphere. The rete can be rotated around the central pin to show the position of stars at various times of the day or night, as indicated by markings along the circumference of the mater.

To use the astrolabe, an observer hangs the instrument from a metal ring attached at the top of the mater. A sighting device on the back of the astrolabe, the alidade, is then lined up with some specific **star** in the sky. As the alidade is moved to locate the star, the rete on the front of the astrolabe is also pivoted to provide the correct setting of the celestial sphere for the given time of day. That time of day can then be read directly off the mater.

A much simpler form of the astrolabe was invented in about the fifteenth century by Portuguese navigators. It consisted only of the mater and the alidade, suspended from a ring attached to the mater. The alidade was used to determine the elevation of a star above the horizon and, thus, the latitude of the ship's position. This form of the astrolabe, known as the mariner's astrolabe, later evolved into the instrument known as the **sextant**.

More elaborate forms of the mariner's astrolabe were later developed and are still used for some specialized purposes. One of these, known as the impersonal astrolabe, was invented by the French astronomer André Danjon (1890-1967). The modern prismatic astrolabe is based on Danjon's concept. In this form of the astrolabe, two **light** rays from the same star are passed through a **prism**, one directly and one after reflection from the surface of a pool of mercury. The star is observed as it rises (or sets) in the sky. During most of this period, the two light rays passing through the prism are out of phase

Galileo's astrolabe. *Photo Researchers, Inc. Reproduced by permission.*

with each other. At some point, the specific latitude for which the astrolabe is designed is attained and the two star images coincide with each other, giving the star's precise location at that moment.

See also Celestial coordinates.

Astrometry

Astrometry literally means measuring the stars. This type of measurement determines a specific star's location in the sky with great precision. In order to establish a star's location, it is necessary to first establish a coordinate system in which the location can be specified. Traditionally, very distant stars, which show very little **motion** as viewed from **Earth**, have been used to establish that coordinate system. However, the **accuracy** of the coordinate system is dependent on the accuracy of the positions of defining stars, effort has been made to use the extremely distant point-like objects known as quasars to establish an improved standard coordinate system. Because quasars give off **radio waves**, their positions can be determined with extreme accuracy, but the implementation of this system has yet to be accomplished.

Astrometry is of fundamental importance to the study of the stars. Astronomers can use the **distance** of the **star** to help determine its other properties. The annual motion of the Earth about the **Sun** causes nearby stars to appear to move about in the sky with respect to distant background stars. The amplitude of this apparent motion determines the distance of the star from our sun, which is known as its trigonometric **parallax**. The angular **rate** of change of the star's position is called its proper motion. If the distance to the star is also known, the proper motion can be converted into a transverse **velocity** relative to the Sun, which is the apparent speed of the star across the line of sight. For most stars this motion is extremely small and may require positional determinations 50 years or longer for accurate measurement. The transverse velocity may be combined with the radial velocity determined from the star's spectra to yield the true **space** velocity with respect to the Sun.

Occasionally the proper motion will be found to vary in a periodic manner, suggesting that the target star is orbiting another object in addition to its steady motion across the sky. Such stars are called astrometric binary stars. Stars that are orbited by planets, which are too faint to be directly observed, show this motion. However, the motion is liable to be extremely small unless the star is quite small and the **planet** rather large.

Often, astronomers cannot determine the distance of the star directly from the coordinate system. In this case, a method called statistical parallax is used. For example, if one independently knew the transverse velocity of a star, one could use the proper motion to obtain a distance. While it is impossible to determine the transverse velocity of a specific star without knowledge of its distance and proper motion an estimation can be obtained by using the transverse velocity for a collection of similar stars. In addition, the radial velocity of the star can be obtained directly from its spectra without knowledge of its distance or proper motion. These can then be combined with the observed proper motions to yield distances to the similar stars of a particular type and average values for their intrinsic properties.

A similar trick can be used on a group of stars which move together through space on more-or-less **parallel** tracks. Such groups of stars are called galactic, or open, clusters. Just as the parallel tracks of a railroad appear to converge to a **point** in the distance, so the stellar motions will appear to point to a distant convergent point in the

KEY TERMS

. . . .

Moving cluster method—A method of determining the distances to stars in a cluster, which are assumed to share a similar direction of motion in space. The accuracy of the method depends largely on the extent to which the cluster is spread across the sky and therefore has only been applied successfully to the nearest clusters.

Parallax—A generic term used to denote the distance to an astronomical object.

Proper motion—Measures the angular motion of an object across the line of sight.

Quasar—Originally stood for Quasi-Stellar Radio Source. However, the term is now commonly used to refer to any quasi-stellar source exhibiting large recessional velocities associated with the expansion of the universe.

Radial velocity—Motion of an object along the line of sight generally determined from the spectrum of the object. By convention, a positive radial

velocity denotes motion away from the observer while a negative radial velocity indicates motion toward the observer.

Space velocity—The vector combination of the radial velocity (vr) and transverse velocity (vt) of an object. The magnitude of the space velocity, vs, is given by $vs^2 = vr^2 + vt^2$.

Statistical parallax—A method for determining the distance to a class of similar objects assumed to have the same random motion with respect to the Sun.

Transverse velocity—That component of the space velocity representing motion across the line of sight.

Trigonometric parallax—The parallax of a stellar object determined from the angular shift of the object with respect to the distant stars resulting from the orbital motion of the Earth about the Sun.

---

sky. The location of this point with respect to each star specifies the **angle** between the radial velocity and the space velocity for that star. Knowledge of that angle allows the tangential velocity of the star to be obtained from the directly measured radial velocity. Again, knowledge of the individual proper motion and tangential velocity allows for a determination of the distance to each star of the cluster. This scheme is known as the moving cluster method.

Adaptive **optics** has greatly improved the accuracy of stellar positions made from the ground. Since the mid-twentieth century, astronomers have known that it was possible in principle to undo the distortions of astronomical images generated by the atmosphere. First one had to measure those distortions, then construct a "lens" with characteristics that could be changed as fast as the atmosphere itself changed. Theoretically such a **lens** could then "undo" the distortions of the atmosphere, leaving the astronomer with the steady image of the star beyond the atmosphere. This seemed impossible to accomplish until very recently. Powerful lasers have now been used to produce artificial stars high in the atmosphere, enabling astronomers to measure the atmospheric distortions. Remarkable increases in computer speed have allowed the analysis of those distortions to be completed in milliseconds so that a thin mirror can be adjusted to correct for atmospheric distortions. Such systems are generally referred to as adaptive optics and in principle

they allow observations of stellar positions to be made from the ground.

However, while adaptive optics systems were being developed, several satellites and **satellite** programs addressed the fundamental problems of astrometry. The pointing accuracy of the **Hubble Space Telescope** required a greatly increased "catalogue" of stars and their positions so that guide stars could be found for all potential targets of the **telescope**. A number of ground surveys which provided positions of several million stars were undertaken expressly to provide those guide stars. Partly in response to these surveys, machines were developed that could automatically measure star positions on the thousands of photographic **glass** plates that were taken for the projects. Now the determination of stellar positions can be accomplished directly by an electronic detector much like those found in a video camera, thereby replacing the photographic plate. This development also allowed the design of satellites dedicated to the determination of stellar positions. The must notable of these is Hipparcos, developed by the European Space Agency (ESA). Hipparcos was designed to measure the positions of more than 100,000 stars with an accuracy of between two and four milliarc seconds. This is easily more than ten times the accuracy readily achievable from the ground. After nearly a decade of development, Hipparcos was launched aboard the Ariane spacecraft in 1989. Unfortunately, due to a failure of the final stage of the rocket, the satellite never achieved the geo-stationary **orbit** approximately 25,000 mi (40,250 km) above the earth. Instead, its orbit is highly elliptical with its furthest point near the desired distance, but dipping down close to the earth's atmosphere at its low point. This greatly reduced the efficiency of the satellite.

Eventually satellites like Hipparcos, along with improved ground observations, will significantly enhance the number of stars for which we have good positions and other astrometrical data. Not only will this clarify our view of the local stellar neighborhood within our **galaxy**, it will also increase our fundamental knowledge of stars themselves.

See also Celestial coordinates.

Astronomical unit

An astronomical unit (AU) is a unit of length that astronomers use for measuring distances within the **solar system**. One astronomical unit is the **mean** distance between the **Earth** and the **Sun**, called the semimajor axis, or 92,919,000 mi (149,597,870 km).

The relative distances between the Sun and the planets, in astronomical units, were known long before the actual distances were established. Kepler, in developing his third law, showed that the **ratio** of the square of a planet's period (the **time** to make one complete revolution) to the cube of the semimajor axis of its **orbit** is a constant; that is, the ratio is the same for all the planets. Kepler's law can be summarized by the formula $a^3/p^2 = K$ where a is the semimajor axis of the planet's orbit, p is its period, and K is the proportionality constant, a constant that holds for all bodies orbiting the Sun. By choosing the period of the Earth as one year and its orbital radius as one AU, the constant K has a numerical value of one.

Kepler's third law (in a more accurate form derived by Isaac Newton) can be used to calculate a precise value of the AU, if the exact **distance** between the earth and another **planet** can be measured. An early attempt took place in 1671, when Jean Cassini in Paris and Jean Richer about 5,000 mi (8,000 km) away in Cayenne, Guiana, simultaneously determined the **parallax** of **Mars**. Their measurements, which allowed them to calculate the distance from earth to Mars by triangulation, showed Mars to be about 50 million mi (80 million km) from Earth. Since the relative distance between Earth and Mars was known, it was a simple matter to determine the actual value of an AU in miles or kilometers. Today, the value of the AU is known very accurately. By measuring the time for a **radar** pulse to reach **Venus** and

return, the distance can be calculated because radar waves travel at the speed of **light**.

See also Kepler's laws.

Astronomy

Astronomy, the oldest of all the sciences, seeks to describe the structure, movements and processes of celestial bodies.

History and impact of astronomy

Ancient ruins provide evidence that the most remote ancestors observed and attempted to understand the workings of the Cosmos. Although not always fully understood, these ancient ruins demonstrate that early man attempted to mark the progression of the **seasons** as related to the changing of the apparent changing positions of the **Sun**, stars, planets and **Moon** on the celestial **sphere**. Archaeologists speculate that such observation made more reliable the determination of times for planting and harvest in developing agrarian communities and cultures.

The regularity of the heavens also profoundly affected the development of indigenous religious beliefs and cultural practices. For example, according to Aristotle (384–322 B.C.), **Earth** occupied the center of the Cosmos, and the Sun and planets orbited Earth in perfectly circular orbits at an unvarying **rate** of speed. The word astronomy is a Greek term for **star** arrangement. Although heliocentric (Sun-centered) theories were also advanced among ancient Greek and Roman scientists, the embodiment of the **geocentric theory** conformed to prevailing religious beliefs and, in the form of the Ptolemaic model subsequently embraced by the growing Christian church, dominated Western thought until the rise of empirical science and the use of the **telescope** during the Scientific Revolution of the sixteenth and seventeenth centuries.

In the East, Chinese astronomers, carefully charted the night sky, noting the appearance of "guest stars" (**comets**, novae, etc.). As early as 240 B.C., the records of Chinese astronomers record the passage of a "guest star" known now as Comet Halley, and in A.D. 1054, the records indicate that one star became bright enough to be seen in daylight. Archaeoastronmers argue that this transient brightness was a **supernova** explosion, the remnants of which now constitute the Crab Nebula. The appearance of the supernova was also recorded by the Anasazi Indians of the American Southwest.

Observations were not limited to spectacular celestial events. After decades of patient observation, the

Mayan peoples of Central America were able to accurately predict the movements of the Sun, Moon, and stars. This civilization also devised a calendar that accurately predicted the length of a year, to what would now be measured to be within six seconds.

Early in the sixteenth century, Polish astronomer Nicolaus Copernicus (1473–1543) reasserted the **heliocentric theory** abandoned by the Greeks and Romans. Although sparking a revolution in astronomy, Copernicus's system was deeply flawed by an insistence on circular orbits. Danish astronomer Tycho Brahe's (1546–1601) precise observations of the celestial movements allowed German astronomer and mathematician Johannes Kepler (1571–1630) to formulate his laws of planetary **motion** that correctly described the elliptical orbits of the planets.

Italian astronomer and physicist Galileo Galilei (1564–1642) was the first scientist to utilize a newly invented telescope to make recorded observations of celestial objects. In a prolific career, Galileo's discoveries, including phases of **Venus** and moons orbiting **Jupiter** dealt a death blow to geocentric theory.

In the seventeenth century, English physicist and mathematician Sir Isaac Newton's (1642–1727) development of the **laws of motion** and gravitation marked the beginning of Newtonian **physics** and modern **astrophysics**. In addition to developing **calculus**, Newton made tremendous advances the understanding of **light** and **optics** critical to the development of astronomy. Newton's seminal 1687 work, *Philosophiae Naturalis Principia Mathematica (Mathematical Principles of Natural Philosophy)* dominated the Western intellectual landscape for more than two centuries and proved the impetus for the advancement of celestial dynamics.

Theories surrounding **celestial mechanics** during the eighteenth century were profoundly shaped by important contributions by French mathematician Joseph-Louis Lagrange (1736–1813), French mathematician Pierre Simon de Laplace, (1749–1827) and Swiss mathematician Leonhard Euler (1707–1783) that explained small discrepancies between Newton's predicted and the observed orbits of the planets. These explanations contributed to the concept of a clockwork-like mechanistic universe that operated according to knowable physical laws.

Just as primitive astronomy influenced early religious concepts, during the eighteenth century, advancements in astronomy caused significant changes in Western scientific and theological concepts based upon an unchanging, immutable God who ruled a static universe. During the course of the eighteenth century, there developed a growing scientific disregard for understanding based upon divine revelation and a growing acceptance of an understanding of Nature based upon the development and application of scientific laws. Whether God intervened to operate the mechanisms of the universe through miracles or signs (such as comets) became a topic of lively philosophical and theological debate. Concepts of the divine became increasing identified with the assumed eternity or **infinity** of the Cosmos. Theologians argued that the assumed immutability of a static universe, a concept shaken by the discoveries of Copernicus, Kepler, Galileo and Newton, offered proof of the existence of God. The clockwork universe viewed as confirmation of the existence of a God of infinite power who was the "prime mover" or creator of the universe. For many scientists and astronomers, however, the revelations of a mechanistic universe left no place for the influence of the Divine, and they discarded their religious views. These philosophical shifts sent sweeping changes across the political and social landscape.

In contrast to the theological viewpoint, astronomers increasingly sought to explain "miracles" in terms of natural phenomena. Accordingly, by the nineteenth century, the appearance of comets was no longer viewed as direct signs from God but rather a natural, explainable and predictable result of a deterministic universe. Explanations for catastrophic events (e.g., comet impacts, extinctions, etc.) increasingly came to be viewed as the inevitable results of **time** and statistical probability.

The need for greater **accuracy** and precision in astronomical measurements, particularly those used in navigation, spurred development of improved telescopes and pendulum driven clocks that greatly increased the pace of astronomical discovery. In 1781, improved mathematical techniques combined with technological improvements along with the proper application of Newtonian laws, allowed English astronomer William Herschel to discover the **planet Uranus**.

Until the twentieth century, astronomy essentially remained concerned with the accurate description of the movements of planets and stars. Developments in electromagnetic theories of light and the formulation of quantum and relativity theories, however, allowed astronomers to probe the inner workings of the celestial objects. Influenced by German-American physicist Albert Einstein's (1879–1955) theories of relativity and the emergence of quantum theory, Indian-born American astrophysicist Subrahmanyan Chandrasekhar (1910–1995) first articulating the evolution of stars into supernova, white dwarfs, **neutron** stars and accurately predicting the conditions required for the formation of black holes subsequently found in the later half of the twentieth century. The articulation of the stellar evolutionary cycle allowed rapid advancements in cosmological theories regarding the creation of the universe. In particular, Ameri-

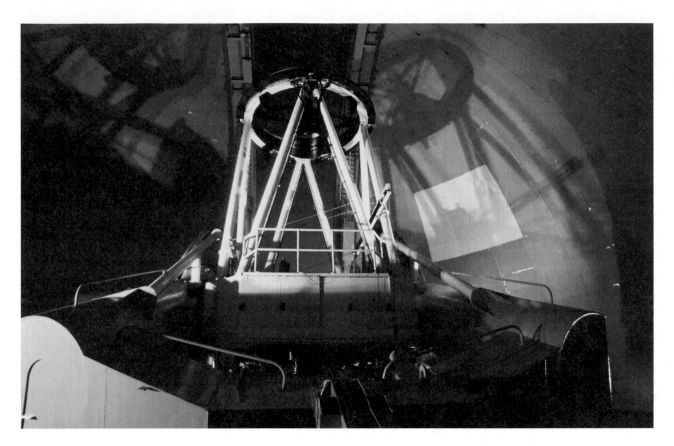

A view of the inner dome and CFHT (Canada-France-Hawaii) telescope at the Mauna Kea Observatory in Hawaii. *Stephen and Donna O'Meara/Photo Researchers, Inc. Reproduced by permission.*

can astronomer Edwin Hubble's (1889–1953) discovery of red shifted spectra from stars provided evidence of an expanding universe that, along with increased understanding of **stellar evolution**, ultimately led to the abandonment of static models of the universe and the formulation of big bang based cosmological models.

In 1932, American engineer Karl Janskey (1905–1945) discovered existence of **radio waves** of emanating from beyond the Earth. Janskey's discovery led to the birth of **radio astronomy** that ultimately became one of the most productive means of astronomical observation and spurred continuing studies of the Cosmos across all regions of the **electromagnetic spectrum**.

Profound questions regarding the birth and death of stars led to the stunning realization that, in a real sense, because the heavier **atoms** of which he was comprised were derived from neucleosynthesis in dying stars, man too was a product of stellar evolution. After millenniums of observing the Cosmos, by the dawn of the twenty-first century, advances in astronomy allowed humans to gaze into the night sky and realize that they were looking at the light from stars distant in **space** and time, and that they, also, were made from the very dust of stars.

The science of astronomy

At its most fundamental, astronomy is based on the electromagnetic **radiation** emitted by the stars. The ability to gather light is the key to acquiring useful data. The bigger the primary mirror of a telescope, the greater its light-gathering capabilities and the greater the magnification of the instrument. These two attributes allow a large telescope to image fainter, smaller objects than a telescope of lesser size. Thus, astronomers build ever-larger telescopes, such as the 33-ft-diameter (10-m) Keck telescopes in Hawaii, or escape the distorting effects of the atmosphere with orbital observatories like the **Hubble Space Telescope**.

Astronomy is not just about visible light, however. Though the visible spectral region is most familiar to us because our eyes are optimized for these wavelengths, observation in the visible region shows only a small portion of the activities and processes underway in the universe. When astronomers view the night sky in other regions of the electromagnetic **spectrum**, it presents an entirely different picture. Hot gases seethe and boil when viewed at infrared wavelengths, newly forming galaxies and stars glow with **x rays**, and mysterious objects generate explosive bursts of gamma rays. **Radio** wave and ultraviolet observations likewise bring astronomers new insights about stellar objects.

Each spectral region requires different instrumentation, and different approaches to data analysis. Radio astronomy, for example, is performed by 20- and 30-ft-diameter (6- and 9-m) antennas, or even telescopes like the one in Arecibo, Puerto Rico, in which a 1,000-ft (303-m) diameter natural bowl in the landscape has been lined to act as an enormous radio wave collector. In the Very Large Array in New Mexico, 27 antennas placed as much as a mile apart from one another are linked by computer to make simultaneous observations, effectively synthesizing a telescope with a 22-mi (35-km) aperture—a radio-frequency analog to the Keck telescope. Infrared, x-ray, and gamma-ray telescopes require special materials and designs for both the focusing optics and the detectors, and cannot be performed below the Earth's atmosphere.

Quantifying light—luminosity and spectral classes

Astronomy is based upon the information we can derive by what we observe when we gaze at the stars. One of the characteristics of a star that can be determined observationally is its luminosity—the amount of light that the star emits. When combined with other information about a star such as its size or **temperature**, luminosity can indicate the intensity of fusion reactions taking place in the stellar core. Luminosity cannot always be determined by direct observation, however, as **distance** can decrease the apparent luminosity of an object. The Sun, for example, is not excessively luminous as stars go; it only appears brighter than any other stellar object because it is so close to us.

Magnitude is another way of expressing the luminosity of a star. The Greek astronomer Hipparchus developed the magnitude scale for stars, rating their brightness on a scale of 1 to 6. According to the scale, a star of first magnitude is defined as appearing 100 times as bright as a star of sixth magnitude, so the larger the magnitude, the fainter the object. As telescopes have allowed astronomers to peer deeper into the universe, the scale has expanded: Sirius, which appears to be the brightest star in the heavens, has an apparent magnitude of −1.27, while **Pluto** has a magnitude of 14.

Apparent magnitude, like apparent luminosity, can be deceptive. To avoid invalid comparisons, astronomers have developed the concept of absolute magnitude, which is defined as the apparent magnitude the object would have when viewed at a distance of 32.6 light years. Thus, measuring the distance to various objects is an important task in astronomy and astrophysics.

The **color** of light emitted by a star indicates its temperature. At the beginning of the century, astronomers began classifying stars based on color, or spectral classes. The classes are O, B, A, F, G, K, and M. O-type stars are the hottest (63,000°F [34,632°C]) and tend to appear white or blue-white, while M-type stars are the coolest (5,400°F [2,952°C]) and tend to appear red; our yellow sun, type G, falls in the middle. Another rating—L-type, for dim, cool objects below M-type—has recently been proposed for addition to the listing.

Astronomers can glean a tremendous amount of information from **stellar magnitudes** and glasses. Between 1911–13, Danish astronomer Ejnar Hertzsprung (1873–1967) and American astronomer Henry Norris Russell (1877–1957) independently developed what is now known as the **Hertzsprung-Russell diagram** that plots the magnitude and color of stars. According to the diagram, most stars fall on a slightly curving diagonal that runs from very bright, hot stars down to very cool, red stars. Most stars follow this so-called main sequence as they gradually burn out. Some stars fall off of the main sequence, for example red giants, which are relatively cool but appear bright because of their enormous size; or white dwarfs, which are bright but so small as to appear faint.

Spectroscopy

When we think of astronomy, spectacular, colorful pictures of swirling galaxies, collapsing stars, and giant **clouds** of interstellar gas come to mind. In reality, however, some of the most useful observational data in astronomy does not involve images at all. Spectroscopic techniques are powerful tools that allow scientists detect the presence of certain elements or processes in faraway galaxies.

In **spectroscopy**, incoming light—such as that from a star—is passed through a grating or a **prism** that splits the light up into its constituent wavelengths, or colors. Normally, a very bright, hot star will emit a continuous spectrum of light that spreads like a rainbow across the electromagnetic spectrum. In the case of lower **density** gas masses such as nebulae, however, the light will be emitted only at certain specific wavelengths defined by the elements found in the nebula—hydrogen atoms, for example—generate vivid yellow lines at characteristic wavelengths. The spectra will thus consist of a collection of bright lines in an otherwise dark background; this is called an **emission** spectrum. Similarly, if a star is surrounded by a cooler atmosphere, the atoms in the atmosphere will absorb certain wavelengths, leaving dark lines

KEY TERMS

· ·

Absolute magnitude—The apparent brightness of a star, measured in units of magnitudes, at a fixed distance of 10 parsecs.

Absorption spectrum—The record of wavelengths (or frequencies) of electromagnetic radiation absorbed by a substance; the absorption spectrum of each pure substance is unique.

Apparent magnitude—The brightness of a star, measured in units of magnitudes, in the visual part of the electromagnetic spectrum, the region to which our eyes are most sensitive.

Emission spectrum—A spectrum consisting of bright lines generated by specific atoms or atomic processes.

Luminosity—The amount of light emitted from a source per unit area.

Spectroscopy—A technique for studying light by breaking it down into its constituent wavelengths.

Infrared astronomy; Relativity, general; Relativity, special; Space shuttle; Spacecraft, manned; Spectral classification of stars; Spectral lines; Spectroscope.

Resources

Books

Croswell, Ken. *Magnificent Universe*. New York: Simon & Schuster, 1999.

Crosswell, Ken. *See the Stars: Your First Guide to the Night Sky*. Boyds Mills PA: Boyds Mills Press, 2000.

Hawking, Stephen. *The Illustrated Brief History of Time, Updated and Expanded*. New York: Bantam, 2001.

Rees, Martin J. *Our Cosmic Habitat*. Princeton, NJ: Princeton University Press, 2001.

Sagan, Carl. *Cosmos* New York: Random House, 2002.

Other

Nemiroff, Robert, and Jerry Bonnell. National Air and Space Administration and Michigan Technological University. "Astronony Picture of the Day" [cited February 5, 2003]. <http://antwrp.gsfc.nasa.gov/apod/astropix.html>.

K. Lee Lerner
Kristen Lewotsky

in what would otherwise be a continuum. This is known as an absorption spectrum.

Scientists study absorption and emission spectra to discover the elements present in stars, galaxies, gas clouds, or planet-forming nebulae. By monitoring the amount by which spectroscopic lines shift toward red wavelengths or toward blue wavelengths, astronomers can determine whether objects are moving toward or away from the Earth. This technique, based on the Doppler shift, is not only used to help astronomers study the expansion of the universe, but to determine the distance or age of the object under study. By studying the Doppler shift of stellar spectra, astronomers have been able to monitor faint wobbles in the motion of stars that indicate the presence of a companion star or even of **extrasolar planets**.

Although the sophisticated instruments and analysis techniques of astrophysics assist in the understanding of universe, astronomy is essentially about the observation of light. Using the data produced by a multitude of telescopes around the world and in **orbit**, astronomers are making new discoveries on a daily basis, and just as often exposing new puzzles to solve. The basic tools described above help scientists to extract information about stellar objects, and thus about the processes at work in the Universe.

See also Astrobiology; Astroblemes; Astrolabe; Astrometry; Astronomical unit; Cosmic background radiation; Cosmic ray; Cosmology; Gravity and gravitation;

Astrophysics

Astrophysics describes the processes that give rise to the observable features of our universe in terms of previously developed physical theories. It ties together **physics** and **astronomy** by describing astronomical phenomena in terms of the physics and **chemistry** we are familiar with in our everyday life.

Background

Why do the stars shine? How did our **galaxy** form? Will the universe expand forever? These are the types of questions asked by astrophysicists in an attempt to understand the processes which cause our universe, and everything in it, to behave the way it does. From the low-energy gravitational interactions between planets and stars, to the violent, high **energy** processes occurring in the centers of galaxies, astrophysical theories are used to explain what we see, and to understand how phenomena are related.

For thousands of years, astronomy was simply an observational science—humans could observe phenomena in the sky, but had no physical explanation for what they saw. Early humans could offer only supernatural explanations for what they observed, which seemed drastically different from what they experienced in everyday life. Only in the twentieth century have scientists been able to explain many astronomical phenomena in terms of detailed physical theories, relating them to the same chemistry and physics at work in our everyday lives.

Astrophysical experiments, unlike experiments in many other sciences, cannot be done under controlled conditions or repeated in a laboratory; the energies, distances, and **time** scales involved are simply too great. As a result, astrophysicists are forced into the role of observer, watching events as they happen without being able to control the parameters of the experiment.

For centuries, humans have made such observations and attempted to understand the forces at work. But how did scientists develop our picture of the universe if they could not reproduce what they see in the laboratory? Instead of controlling the experiments, they used what data they were able to obtain in order to develop theories based on extensions of the physical laws which govern our day-to-day experiences on **Earth**.

Astrophysics often involves the creation of mathematical models as a means of interpreting observations. This theoretical is important not just for explaining what has already been seen, but also for predicting other observable effects. These models are often based on well-established physics, but often must be simplified, because real astronomical phenomena can be enormously complex.

Processes in the universe

Even looking close to the Earth, in our own **solar system** we see widely varying conditions. The properties of the rocky **planet** Mercury, very close to the **Sun**, differ dramatically from those of the gas giant **Saturn**, with its complex ring structure, and from the cold, icy **Pluto**. But the range of variations found in our solar system is minuscule when compared to that of the stars, galaxies, and more exotic objects such as quasars. The properties of all of these objects, however, can be measured by observation, and an understanding of how they work can be reached by the extension and application of the same physical laws with which we are familiar.

The first astrophysical concept or law to be recognized was the law of gravity. We are all familiar with the **force** of gravity. Although it is a very weak force compared to the other fundamental forces of nature, it is the dominant factor determining the structure and the fate of the universe. Large structures, such as galaxies, and smaller ones, such as stars and planets, coalesced due to the force of gravity, which acts over vast distances of **space**.

Much of the evolution of our universe is due to gravity's effects. However, scientists generally hold the view that the understanding of atomic processes marks the true beginning of astrophysics. Indeed, even such enormous objects as stars are governed by the interaction and behavior of **atoms**. Thus it is often said that astrophysics began in the early decades of the twentieth century, when **quantum mechanics** and atomic physics were born.

Importance of instrumentation

Scientists learn about distant objects by measuring the properties that we can observe directly—by detecting emissions from the objects. The most common measurements are of electromagnetic **radiation**, extending from **radio waves**, through visible wavelengths to high energy gamma rays. Each time a class of objects has been studied in a new wavelength region, astrophysicists gain insights into composition, structure, and properties. Emissions from each wavelength region are generated by and affected by different processes, and so provide fresh understanding of the object. For this reason, the development of new instrumentation has been crucial to the development of astrophysics.

The development of space instrumentation that can detect photons before they are obscured by the Earth's atmosphere has been critical to our understanding of the universe. Large space-based observatories, such as the **Hubble Space Telescope**, continually spawn major advances in astrophysics due to their ability to study the universe over specific regions of the **electromagnetic spectrum** with unprecedented sensitivity. In addition, probes such as the *Voyagers,* which visited most of the outer planets of our solar system, have provided detailed measurements of the physical environment throughout our solar system. The use of **spectroscopy**, which can determine the chemical composition of distant objects from their wavelength distribution, is a particularly important tool of the astrophysicist.

In addition to the photons of electromagnetic radiation, emitted particles can be detected. These can be protons and electrons, the constituents of ordinary **matter** on Earth (though often with extremely high energies), or ghostly neutrinos, which only weakly interact with matter on Earth (and are thus extremely difficult to detect), but help us learn about the nuclear reactions which power stars.

Astrophysics proceeds through hypothesis, prediction, and test (via observation), its common belief being that laws of physics are consistent throughout the universe. These laws of physics have served us well, and scientists are most skeptical of proposed explanations that violate them.

See also Cosmology; Gamma-ray astronomy; Infrared astronomy; Pulsar; Quasar; Relativity, general; Relativity, special; Spectral classification of stars; Spectral lines; Star; Telescope; Ultraviolet astronomy; X-ray astronomy.

Resources

Books

Audouze, Jean, Guy and Israël, eds. *The Cambridge Atlas of Astronomy.* Cambridge: Cambridge University Press, 1994.

Bacon, Dennis Henry, and Percy Seymour. *A Mechanical History of the Universe.* London: Philip Wilson Publishing, Ltd., 2003.

Introduction to Astronomy and Astrophysics. 4th ed. New York: Harcourt Brace, 1997.

Kaufmann, William J. III. *Discovering the Universe.* 2nd ed. New York: W. H. Freeman, 1990.

Pasachoff, Jay M. *Contemporary Astronomy.* 4th ed. Philadelphia: Saunders College Publishing, 1989.

David Sahnow

Atmosphere, composition and structure

Earth's atmosphere is composed of about 78% **nitrogen**, 21% **oxygen**, and 0.93% argon. The remainder, less than 0.1%, contains many small but important trace gases, including **water** vapor, **carbon dioxide**, and **ozone**. All of these trace gases have important effects on the earth's climate. The atmosphere can be divided into vertical layers determined by the way **temperature** changes with height. The layer closest to the surface is the troposphere, which contains over 80% of the atmospheric **mass** and nearly all the water vapor. The next layer, the stratosphere, contains most of the atmosphere's ozone, which absorbs high **energy radiation** from the **sun** and makes life on the surface possible. Above the stratosphere are the mesosphere and thermosphere. These two layers include regions of charged **atoms** and molecules, or ions. Called the ionosphere, this region is important to **radio** communications, since **radio waves** can bounce off the layer and travel great distances. It is thought that the present atmosphere developed from gases ejected by volcanoes. Oxygen, upon which all **animal** life depends, probably built up as excess emissions from plants that produce it as a waste product during **photosynthesis**. Human activities may be affecting the levels of some important atmospheric components, particularly **carbon** dioxide and ozone.

Composition of the atmosphere

Major gases

The most common atmospheric gas, nitrogen (chemical symbol N_2) accounts for about 78% of the atmosphere. Nitrogen gas is largely inert, meaning that it

does not readily react with other substances to form new chemical compounds. The next most common gas, oxygen (O_2), makes up about 21% of the atmosphere. Oxygen is required for the **respiration** (breathing) of all animal life on **Earth**, from humans to **bacteria**. In contrast to nitrogen, oxygen is extremely reactive. It participates in oxidation, a type of chemical reaction that can be observed everywhere. Some common examples of oxidation are apples turning from white to brown after being sliced, the rusting of **iron**, and the very rapid oxidation reaction we call fire. Just under 1% of the atmosphere is made up of argon (Ar), which is a very inert noble gas, meaning that it does not take part in any **chemical reactions** under normal circumstances.

Together, these three gases account for 99.96% of the atmosphere. The remaining 0.04% contains a wide variety of trace gases, several of which are crucial to life on Earth.

Important trace gases

Carbon dioxide (CO_2) affects the earth's climate and plays a large support role in the **biosphere**, the collection of living things that populate the earth's surface. Only about 0.0325% of the atmosphere is CO_2. Carbon dioxide is required by **plant** life for photosynthesis, the process of using sunlight to store energy as simple sugars, upon which all life on Earth depends. Carbon dioxide is also one of a class of compounds called greenhouse gases. These gases are made up of molecules that absorb and emit infrared radiation, which we feel as **heat**. The solar energy radiated from the sun is mostly in the visible range, within a narrow band of wavelengths. This radiation is absorbed by the earth's surface, then re-radiated back out to **space** not as visible **light**, but as longer wavelength infrared radiation. Greenhouse gas molecules absorb some of this radiation before it escapes to space, and re-emit some of it back toward the surface. In this way, these gases trap some of the escaping heat and increase the overall temperature of the atmosphere. If the atmosphere had no greenhouse gases, it is estimated that the earth's surface would be 90°F (32°C) cooler.

Water vapor (H_2O) is found in the atmosphere in small and highly variable amounts. While it is nearly absent in most of the atmosphere, its **concentration** can range up to 4% in very warm, humid areas close to the surface. Despite its relative scarcity, atmospheric water probably has more of an impact on the earth than any of the major gases, aside from oxygen. Water vapor participates in the **hydrologic cycle**, the process that moves water between the oceans, the land surface waters, the atmosphere, and the **polar ice caps**. This water cycling drives **erosion** and rock **weathering**, determines the earth's **weather**, and sets up climate conditions that make land areas dry or wet, habitable or inhospitable. When cooled sufficiently, water vapor forms **clouds** by condensing to liquid water droplets, or at lower temperatures, solid **ice** crystals. Besides creating rain or snow, clouds affect Earth's climate by reflecting some of the energy coming from the sun, making the **planet** somewhat cooler. Water vapor is also an important greenhouse gas. It is concentrated near the surface and is much more prevalent near the tropics than in the polar regions.

Ozone (O_3) is almost all found in a layer about 9–36 mi (15–60 km) in attitude. Ozone gas is irritating to peoples' eyes and skin, and chemically attacks rubber and plant **tissue**. Nevertheless, it is vital to life on Earth because it absorbs most of the high energy radiation from the sun that is harmful to plants and animals. A portion of the energy radiated by the sun lies in the ultraviolet (UV) region. This shorter wavelength radiation is responsible for suntans, and is sufficiently powerful to harm cells, cause skin **cancer**, and burn tissue, as anyone who has had a painful sunburn knows. The ozone molecules, along with molecules of O_2, absorb nearly all the high energy UV rays, protecting the earth's surface from the most damaging radiation. The first step in this process occurs high in the atmosphere, where O_2 molecules absorb very high energy UV radiation. Upon doing so, each absorbing **molecule** breaks up into two oxygen atoms. The oxygen atoms eventually collide with another O_2 molecule, forming a molecule of ozone, O_3 (a third molecule is required in the collision to carry away excess energy). Ozone in turn may absorb UV of slightly longer wavelength, which knocks off one of its oxygen atoms and leaves O_2. The free oxygen atom, being very reactive, will almost immediately recombine with another O_2, forming more ozone. The last two steps of this cycle keep repeating but do not create any new chemical compounds; they only act to absorb ultraviolet radiation. The amount of ozone in the stratosphere is minute. If it were all transported to the surface, the ozone gas would form a layer about 0.1–0.16 in (2.5–4.0 mm) thick. This layer, as thin as it is, is sufficient to shield the earth's occupants from harmful solar radiation.

Aerosols

In addition to gases, the atmosphere has a wide variety of tiny particles suspended in the air, known collectively as **aerosols**. These particles may be liquid or solid, and are so small that they may require very long times to settle out of the atmosphere by gravity. Examples of aerosols include bits of suspended **soil** or **desert sand**, tiny smoke particles from a forest fire, **salt** particles left over after a droplet of **ocean** water has evaporated, plant pollen, volcanic dust plumes, and particles formed from the **pollution** created by a **coal** burning power plant.

Aerosols significantly affect the atmospheric heat balance, cloud growth, and optical properties.

Aerosols cover a very wide size range. Raindrops suspended in a cloud are about 0.04–0.24 in (1–6 mm) in diameter. Fine desert sand and cloud droplets range in diameter down to about 0.0004 in (0.01 mm). Sea salt particles and smoke particles are 1/100th of this, about 0.0001 mm, or 0.1 micrometer, in diameter (1 micrometer = one thousandth of a millimeter). Smallest of all are the particles that form when certain gases condense; that is, when several gas molecules come together to form a stable cluster. These are the Aitkin nuclei, whose diameters can be measured down to a few nanometers (1 nanometer = one millionth of a millimeter).

Some aerosols are just the right size to efficiently scatter sunlight, making the atmosphere look hazy. Under the right conditions, aerosols act as collecting points for water vapor molecules, encouraging the growth of cloud droplets and speeding the formation of clouds. They may also play a role in Earth's climate; the aerosols are known to reflect a portion of incoming solar radiation back to space, which lowers the temperature of the earth's surface. Current research is focused on estimating how much cooling is provided by aerosols, as well as how and when aerosols form in the atmosphere.

Atmospheric structure

The atmosphere can be divided into layers based on the **atmospheric pressure** and temperature profiles (the way these quantities change with height). **Atmospheric temperature** drops steadily from its value at the surface, about 290K (63°F; 17°C), until it reaches a minimum of around 220K (−64°F; −53°C) at 6 mi (10 km) above the surface. This first layer is called the troposphere, and ranges in **pressure** from over 1,000 millibars at **sea level** to 100 millibars at the top of the layer, the tropopause. Above the tropopause, the temperature rises with increasing altitude up to about 27 mi (45 km). This region of increasing temperatures is the stratosphere, spanning a pressure range from 100 millibars at its base to about 10 millibars at the stratopause, the top of the layer. Above 30 mi (50 km), the temperature resumes its drop with altitude, reaching a very cold minimum of 180K (−135°F; −93°C) at around 48 mi (80 km). This layer is the mesosphere, which at its top (the mesopause) has an atmospheric pressure of only 0.01 millibars (that is, only 1/100,000th of the surface pressure). Above the mesosphere lies the thermosphere, extending hundreds of miles upward toward the **vacuum** of space. It is not possible to place an exact "top" of the atmosphere; air molecules simply become scarcer and more rarefied until the atmosphere blends with the material found in space.

The troposphere

The troposphere contains over 80% of the mass of the atmosphere, along with nearly all of the water vapor. This layer contains the air we breathe, the winds we observe and the clouds that bring our rain. In fact, all of what we know as "weather" occurs in the troposphere, whose name means "changing sphere." All of the cold fronts, warm fronts, high and low pressure systems, **storm** systems, and other features seen on a weather **map** occur in this lowest layer. Severe thunderstorms may penetrate the tropopause.

Within the troposphere the temperature drops with increasing height at an average **rate** of about 11.7°F per every 3,281 ft (6.5°C per every 1,000 meters). This quantity is known as the lapse rate. When air begins to rise, it will expand and cool at a faster rate determined by the laws of **thermodynamics**. This means that if a parcel of air begins to rise, it will soon find itself cooler and denser than its surroundings, and will sink back downward. This is an example of a stable atmosphere—vertical air **motion** is prevented. Due to the fact that air masses move around in the troposphere, a cold air mass may move into an area and have a higher lapse rate. That is, its temperature drops off more quickly with height. Under these weather conditions, air that begins rising and cooling will become warmer than its surroundings. It then is like a hot-air **balloon**, it is less dense than the surrounding air and is buoyant, so it will continue to rise and cool in a process called **convection**. If this is sustained, the atmosphere is said to be unstable and the rising parcel of air will cool to the point where its water vapor condenses to form cloud droplets. The air parcel is now a convective cloud. If the buoyancy is vigorous enough, a storm cloud will develop as the cloud droplets grow to the size of raindrops and begin to fall out of the cloud as rain. Thus under certain conditions, the temperature profile of the troposphere makes possible storm clouds and **precipitation**.

During a strong **thunderstorm**, cumulonimbus clouds (the type that produce heavy rain, high winds, and hail) may grow tall enough to reach or extend into the tropopause. Here they run into strong stratospheric winds, which may shear off the top of the clouds and stop their growth. One can see this effect in the "anvil" clouds associated with strong summer thunderstorms.

The stratosphere

The beginning of the stratosphere is defined as that point where the temperature reaches a minimum and the lapse rate abruptly drops to **zero**. This temperature structure has one important consequence: it inhibits rising air. Any air that begins to rise will become cooler and denser than the surrounding air. The stratosphere, then, is very stable.

Although the stratosphere has very little water, clouds of ice crystals may form at times in the lower stratosphere over the polar regions. Early Arctic explorers named these clouds nacreous or mother-of-pearl clouds because of their iridescent appearance. More recently, very thin, widespread clouds have been found to form in the polar stratosphere under extremely cold conditions. These clouds, called polar stratospheric clouds, or PSCs, appear to be small crystals of ice or frozen mixtures of ice and **nitric acid**. PSCs play a key role in the development of the ozone hole, which is described below.

The stratosphere contains most of the ozone found in the earth's atmosphere. In fact, the presence of ozone is the reason for the temperature profile found in the stratosphere. As described previously, ozone and oxygen gas both absorb short wave solar radiation. In the series of reactions that follow, heat is released. This heat warms the atmosphere in the layer at about 12–27 mi (20–45 km) and gives the stratosphere its characteristic temperature increase with height.

The ozone layer has been the subject of some concern. In 1985, scientists from the British Antarctic Survey noticed that the amount of stratospheric ozone over the South Pole was dropping sharply during the spring months, recovering somewhat as spring turned to summer. An examination of the historical records revealed that the springtime ozone losses had begun around the late 1960s and had grown much more severe by the late 1970s. By the mid-1980s virtually all the ozone was disappearing from parts of the polar stratosphere during the late winter and early spring. These ozone losses, dubbed the ozone hole, were the subject of intense research both in the field and in the laboratory. The picture that has emerged implicates **chlorine** as the chemical responsible for ozone destruction in the ozone hole. Chlorine apparently gets into the stratosphere from chlorofluorocarbons, or CFCs—industrial chemicals widely used as refrigerants, aerosol propellants, and solvents. Laboratory experiments show that after destroying an ozone molecule, chlorine is tied up in a form unable to react with any more ozone. However, it can chemically react with other chlorine compounds on the surfaces of polar stratospheric cloud particles, which frees the chlorine to attack more ozone. In other words, each chlorine molecule is recycled many times so that it can destroy thousands of ozone molecules. The realization of chlorine's role in ozone depletion brought about an international agreement in 1987, the Montreal Protocol, which committed the participating industrialized countries to begin phasing out CFCs.

The mesosphere and thermosphere

The upper mesosphere and the lower thermosphere contain charged atoms and molecules (ions in a region known as the ionosphere. The atmospheric constituents at this level include nitrogen gas, atomic oxygen and nitrogen (O and N), and nitric oxide (NO). All of these are exposed to strong solar **emission** of ultraviolet and x ray radiation, which can result in ionization, knocking off an **electron** to form an atom or molecule with a positive charge. The ionosphere is a region enriched in free electrons and positive ions. This charged particle region affects the propagation of radio waves, reflecting them as a mirror reflects light. The ionosphere makes it possible to tune in radio stations very far from the transmitter; even if the radio waves coming directly from the transmitter are blocked by **mountains** or the curvature of the earth, one can still receive the waves bounced off the ionosphere. After the sun sets, the numbers of electrons and ions in the lower layers drop drastically, since the sun's radiation is no longer available to keep them ionized. Even at night, however, the higher layers retain some ions. The result is that the ionosphere is higher at night, which allows radio waves to bounce for longer distances. This is the reason that one can frequently tune in more distant radio stations at night than during the day.

The upper thermosphere is also where the bright nighttime displays of colors and flashes known as the aurora occur. The aurora are caused by energetic particles emitted by the sun. These particles become trapped by **Earth's magnetic field** and collide with the relatively few gas atoms present above about 60 mi (100 km), mostly atomic oxygen (O) and nitrogen gas (N_2). These collisions cause the atoms and molecules to emit light, resulting in spectacular displays.

The past and future of the atmosphere

If any atmosphere was present after Earth was formed about 4.5 billion years ago it was probably much different than that of today. Most likely it resembled those of the outer planets—Jupiter, **Saturn**, **Uranus**, and Neptune—with an abundance of **hydrogen**, methane, and **ammonia** gases. The present atmosphere did not form until after this primary atmosphere was lost. One theory holds that the primary atmosphere was blasted from the earth by the sun. If the sun is like other stars of its type, it may have gone through a phase where it violently ejected material outward toward the planets. All of the inner planets, including Earth, would have lost their gaseous envelopes. A secondary atmosphere began to form when gases were released from the crust of the early Earth by volcanic activity. These gases included water vapor, carbon dioxide, nitrogen, and **sulfur** or sulfur compounds. Oxygen was absent from this early secondary atmosphere.

The large amount of water vapor released by the volcanos formed clouds that continually rained on the early Earth, forming the oceans. Since carbon dioxide dissolves easily in water, the new oceans gradually absorbed most of it. (Nitrogen, being unreactive, was left behind to become the most common gas in the atmosphere.) The carbon dioxide that remained began to be used by early plant life in the process of photosynthesis. Geologic evidence indicates this may have begun about two to three billion years ago, probably in an ocean or aquatic environment. Around this time, there appeared **aerobic** (oxygen using) bacteria and other early animal life, which consumed the products of photosynthesis and emitted CO_2. This completed the cycle for CO_2 and O_2: as long as all plant material was consumed by an oxygen breathing **organism**, the two gases stayed in balance. However, some plant material was inevitably lost or buried before it could be decomposed. This effectively removed carbon dioxide from the atmosphere and left a net increase in oxygen. Over the course of billions of years, a considerable excess built up this way, so that oxygen now makes up over 20% of the atmosphere (and carbon dioxide makes up less than 0.033%). All animal life thus depends on the oxygen accumulated gradually by the biosphere over the past two billion years.

Future changes to the atmosphere are difficult to predict. There is currently growing concern that human activity may be altering the atmosphere to the point that it may affect the earth's climate. This is particularly the case with carbon dioxide. When **fossil fuels** such as coal and oil are dug up and burned, buried carbon dioxide is released back into the air. As discussed earlier, carbon dioxide is a greenhouse gas—it acts to trap infrared (heat) energy radiated by the earth, warming up the atmosphere. What effect will this have on future temperatures? While no one has a definite answer, this is an area of active research, using computers to model the oceans, the atmosphere, and the land areas as a very complicated climate system.

See also Atmospheric circulation; Greenhouse effect.

Resources

Books

Bohren, Craig. *Clouds in a Glass of Beer.* New York: John Wiley and Sons, 1987.

Erickson, Jon. *Greenhouse Earth.* Blue Ridge Summit, PA: Tab Books, 1990

Firor, John. *The Changing Atmosphere.* New Haven, CT: Yale University Press, 1990.

KEY TERMS

Infrared radiation—Radiation similar to visible light but of slightly longer wavelength. We sense infrared radiation as heat.

Ionosphere—Region of the atmosphere above about 48 mi (80 km) with elevated concentrations of charged atoms and molecules (ions).

Lapse rate—The rate at which the atmosphere cools with increasing altitude.

Mesosphere—The third layer of the atmosphere, lying about 30–48 mi (50–80 km) in height and characterized by small lapse rate.

Ozone hole—The sharp decease in stratospheric ozone over Antarctica that occurs every spring.

Stratosphere—A layer of the upper atmosphere above an altitude of 5–10.6 mi (8–17 km) and extending to about 31 mi (50 km), depending on season and latitude. Within the stratosphere, air temperature changes little with altitude, and there are few convective air currents.

Thermosphere—The top layer of the atmosphere, starting at about 48 mi (80 km) and stretching up hundreds of miles into space. Due to bombardment by very energetic solar radiation, this layer can possess very high gas temperatures.

Troposphere—The layer of air up to 15 mi (24 km) above the surface of the Earth, also known as the lower atmosphere.

Ultraviolet radiation—Radiation similar to visible light but of shorter wavelength, and thus higher energy.

X-ray radiation—Light radiation with wavelengths shorter than the shortest ultraviolet; very energetic and harmful to living organisms.

Hamblin, W.K., and E.H. Christiansen. *Earth's Dynamic Systems.* 9th ed. Upper Saddle River: Prentice Hall, 2001.

Hancock P.L. and B.J. Skinner, eds. *The Oxford Companion to the Earth.* Oxford: Oxford University Press, 2000.

Lutgens, Frederick K., Edward J. Tarbuck, and Dennis Tasa. *The Atmosphere: An Intorduction to Meteorology.* 8th ed. New York: Prentice-Hall, 2000.

McNeill, Robert. *Understanding the Weather.* Las Vegas: Arbor Publishers, 1991.

Wallace, John M., and Peter V. Hobbs. *Atmospheric Science: An Introductory Survey.* San Diego: Academic Press, 1997.

James Marti

Atmosphere observation

The term **weather** observation refers to all of the equipment and techniques used to study the properties of the atmosphere. These include such well-known instruments as the **thermometer** and **barometer** as well as less familiar devices such as the radiosonde and devices for detecting the presence of trace gases in the atmosphere.

History

The fundamental principles on which the most common atmospheric observational instruments are based were discovered during the seventeenth and eighteenth centuries. For example, the Italian physicist Evangelista Torricelli invented the first barometer in 1643, while the air hygrometer, a device for measuring atmospheric **humidity**, was first constructed by the Swiss physicist Horace Bénédict de Saussure in about 1780.

These instruments were useful at first in studying atmospheric properties close to the ground, but not at very high altitudes. In 1648, the French physicist Florin Périer asked his brother-in-law to carry a pair of barometers to the top of Puy-de-Dôme to make measurements of air **pressure** there, but that was about the limit to which humans themselves could go.

Kites

One of the first means developed for raising scientific instruments to higher altitudes was the kite. In one of the most famous kite experiments of this kind, Benjamin Franklin used a kite in 1752 to discover that **lightning** was nothing other than a form of **electricity**. Within a short period of time, kites were being used by other scientists to carry recording thermometers into the atmosphere, where they could read temperatures at various altitudes.

Weather balloons

An important breakthrough in atmospheric observation came in the late eighteenth century with the invention of the hot-air **balloon**. Balloon flights made it possible to carry instruments thousands of feet into the atmosphere to take measurements. The English physician John Jeffries is often given credit for the first balloon ascension for the purpose of making meteorological measurements. In 1785 Jeffries carried a thermometer, barometer, and hygrometer to a height of 9,000 ft (2,700 m) in his balloon.

For the next 150 years, balloons were the primary means by which instruments were lifted into the atmosphere for purposes of observation. A number of devices were invented specifically for use in weather balloons.

Most commonly used of these devices were the meteorograph and the radiosonde, both of which are combinations of instruments for measuring **temperature**, pressure, humidity, and other atmospheric properties.

The radiosonde differs from a meteorograph in that it also includes a **radio** that can transmit the data collected back to the **earth**. When the radiosonde is used to collect data about atmospheric winds also, it is then known as a rawinsonde. In most cases, data collected with the meteorograph is recovered only when the instrument is jettisoned from the balloon or airplane carrying it. At one time, scientists paid five dollars to anyone who found and returned one of these measuring devices.

Balloons are still an important way of transporting weather instruments into the atmosphere. Today, they are often very large pieces of equipment, made of very thin plastic materials and filled with helium gas. When the balloons are first released from the ground, they look as if they are nearly empty. However, as they rise into the atmosphere and the pressure around them decreases, they fill to their full capacity. Balloons used to study the properties of the upper atmosphere are known as sounding balloons.

Rockets and aircraft

The invention of the airplane and the rocket created a multitude of new opportunities for the study of the atmosphere by making it possible to carry instruments far higher than they had ever gone before. At first, airplanes did similar work to that done by scientists traveling in balloons. Airplanes, however, performed that work much more efficiently, and at higher altitudes, with greater safety and comfort. As technology improved, **aircraft** began to take on new and more complex tasks. For example, they could fly through a cloud and collect cloud droplets on slides for future study. Today, airplanes are also used by "hurricane hunters," who fly into the middle of a hurricane to study its properties and movement.

Airplanes now used for atmospheric observation often have bizarre appearances. They may carry large platforms on their tops, oversized needles on their noses, or other attachments in which an array of observational instruments can be carried. One airplane used for atmospheric observation, a commercial DC8 aircraft, has been redesigned and outfitted to carry the equipment needed to measure levels of **ozone** and related chemicals over the Antarctic. Data collected from this airplane has been crucial in helping scientists to understand how ozone levels have been decreasing over the South Pole over the past decade or more.

Unmanned rockets can carry measuring devices to altitudes even greater than is possible with a piloted aircraft. Again, rockets can perform the same standard mea-

surements as a radiosonde, except at greater atmospheric heights. But they can also perform more complex measurements. For example, they can be designed to carry and release a variety of chemicals that can then be tracked by **radar** and other systems located on the ground. Some rockets have also released explosive devices high in the atmosphere so that scientists can study the way in which **sound waves** are transported there.

Weather satellites

The most sophisticated atmospheric observational systems of all are those that make use of artificial satellites. A weather **satellite** is a device that is lifted into Earth **orbit** by a rocket and that carries inside it a large number of instruments for measuring many properties of the atmosphere. The first weather satellite ever launched was put into orbit by the U.S. government on April 1, 1960. Its name was *TIROS 1* (Television and Infrared Observation Satellite). Over the next five years, nine more satellites of the same name were launched. One of the primary functions of the TIROS satellites was to collect and transmit photographs of the Earth's cloud patterns.

In addition to the TIROS program, the U.S. government has put into operation a number of other weather satellite systems, including Nimbus, ESSA (Environmental Sciences Service Administration), and GOES (geosynchronous environmental satellites). The former Soviet Union also had an active program of weather observation by satellite. The first Soviet satellite was known as *Kosmos 122*, followed by a series of satellites known by the code name of Meteor. Following the launches by the United States and the former Soviet Union, Japan, the European Space Agency, India and China have all launched weather satellites.

Satellites provide a variety of data about atmospheric properties that can contribute to improved **weather forecasting**. Satellites can track the development, growth, and movement of large **storm** systems, such as hurricanes and cyclones. This information can be used to warn human populations of oncoming storms and thereby save human lives and reduce some property damage.

Satellites can also take measurements using various wavelengths of **light**, thereby collecting data that would not be accessible to some other kinds of instruments. As an example, a satellite can photograph a cloud cover using both visible and infrared light and, by comparing the two, predict which cloud system is more likely to produce **precipitation**.

Atmospheric composition

The observational systems described so far can be used to measure more than just physical properties such as

KEY TERMS

. . . .

Meteorograph—An instrument designed to be sent into the atmosphere to record certain measurements, such as temperature and pressure.

Radiosonde—An instrument for collecting data in the atmosphere and then transmitting that data back to Earth by means of radio waves.

Rawinsonde—A type of radiosonde that is also capable of measuring wind patterns.

temperature, pressure, and air movements. They can also be used to determine the chemical composition of the atmosphere. Such measurements can be valuable not only in the field of **meteorology**, but in other fields as well.

One of the earliest examples of such research dates to 1804 when the French physicist Joseph Louis Gay-Lussac traveled in a balloon to a height of 23,000 ft (6,900 m). At this altitude, he collected an air **sample** that he analyzed upon his return to the ground. Gay-Lussac found that the composition of air at 23,000 ft (6,900 m) was the same as it was at **sea level**.

An example of this kind of research today involves the issue of climate change. Over the past decade there has been a great deal of interest with respect to large scale changes in the Earth's climate. Many scientists believe that an accumulation of **carbon dioxide** and other gases in the atmosphere has been contributing to a gradual increase in the Earth's overall average annual temperature. This phenomenon is referred to as **global warming**. If such a change were, in fact, to occur, it would have significant effects on **plant** and **animal** (including human) life on Earth.

Many of the questions about **global climate** change cannot adequately be answered, however, without a fairly good understanding of the gases present in the atmosphere, changes in their **concentration** over time, and **chemical reactions** that occur among those gases. Until recently, most of those questions could not have been answered. Today, however, manned aircraft, rockets, and satellites are able to collect some of the kinds of data that will allow scientists to develop a better understanding of the chemical processes that occur in the atmosphere and the effects they may have on both weather and climate.

Resources

Books

Hodgson, Michael, and Devin Wick. *Basic Essentials: Weather Forecasting.* 2nd ed. Guilford, CT: Globe Pequot Press, 1999.

Lutgens, Frederick K., Edward J. Tarbuck, and Dennis Tasa. *The Atmosphere: An Intorduction to Meteorology.* 8th ed. New York: Prentice-Hall, 2000.

Other

National Oceanic & Atmospheric Administration (NOAA) 2002. <http://www.noaa.gov>.

David E. Newton

Atmospheric circulation

Atmospheric circulation is the movement of air at all levels of the atmosphere over all parts of the **planet**. The driving force behind atmospheric circulation is solar **energy**, which heats the atmosphere with different intensities at the equator, the middle latitudes, and the poles. Differential heating causes air to rise in the atmosphere at some locations on the planet and then to sink back to the earth's surface at other locations. **Earth's rotation** on its axis and the unequal distribution of land and **water** masses on the planet also contribute to various features of atmospheric circulation.

An idealized model of atmospheric circulation

As early as the 1730s, the English lawyer and amateur scientist George Hadley described an idealized model for the movement of air in the earth's atmosphere. It is well known, Hadley pointed out, that air at the equator is heated more strongly than at any other place on **Earth**. In comparison, air above the poles is cooler than at any other location. One can hypothesize, therefore, that surface air near the equator will rise into the upper atmosphere and, above the poles, sink from the upper atmosphere to ground level. In order to balance these vertical movements of air, it was also necessary to hypothesize that air flows across the earth's surface from each pole back to the equator and, in the upper atmosphere, from above the equator to above the poles.

The movement of air described by Hadley can be called a **convection** cell. The term convection refers to the transfer of **heat** as it is carried from place to place by a moving fluid, air in this case.

Hadley knew, of course, that surface winds do not blow from north to south in the northern hemisphere and from south to north in the southern hemisphere, as his simple model would require. He explained that winds actually tend to blow from the east or west because of Earth's **rotation**. The spinning planet causes air flows that would otherwise be from the north or south to be diverted to the east or west, Hadley said.

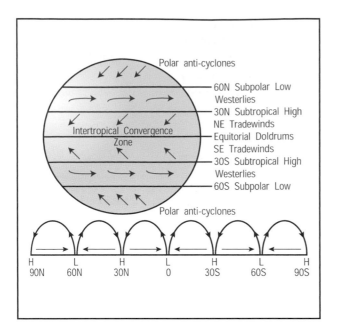

Polar anti-cyclones

Intertropical Convergence Zone

60N Subpolar Low
Westerlies
30N Subtropical High
NE Tradewinds
Equitorial Doldrums
SE Tradewinds
30S Subtropical High
Westerlies
60S Subpolar Low

Polar anti-cyclones

H	L	H	L	H	L	H
90N	60N	30N	0	30S	60S	90S

Atmospheric circulation results in prevailing global wind belts. *Illustration by K. Lee Lerner, with Argosy. The Gale Group.*

As an analogy of how this change could occur, suppose that you are sitting on a spinning merry-go-round trying to catch a ball thrown by a friend at the center of the platform. The ball will obviously travel in a straight line from the thrower to the intended catcher on the rim. But to the catcher, the ball will appear to follow a curved path, and he or she will have to reach out to catch the ball.

A century after Hadley's initial theory was proposed, a mathematical description of this "merry-go-round effect" was published by the French physicist Gaspard Gustave de Coriolis. Coriolis was able to prove mathematically that an object in **motion** on any rotating body always appears to follow a curved path in relation to any other body on the same rotating body. This discovery, now known as the **Coriolis effect**, provided a more exact explanation of the reason that surface winds are deflected to the east or west than did Hadley's original theory.

The three-cell model

At about the time that Coriolis published his studies on rotating bodies, scientists were beginning to realize that Hadley's single convection cell model was too simple. **Atmospheric pressure** and **wind** measurements taken at many locations around the planet did not fit the predictions made by the Hadley model.

Some important modifications in the Hadley model were suggested in the 1850s, therefore, by the American meteorologist William Ferrell. Ferrell had, of course,

much more data about wind patterns than had been available to Hadley. On the basis of these data, Ferrell proposed a three-cell model for atmospheric circulation.

Ferrell's model begins where Hadley's began, with the upward flow of air over the equator and its continued flow toward the poles along the upper atmosphere. At approximately 30° latitude, however, Ferrell hypothesized that this air had become sufficiently cooled so that it began to descend to the earth's surface. Once at surface level, some of this air would then flow back toward the equator, as in the Hadley model. Today this large convection current over the third of the globe above and below the equator is called a Hadley cell.

Ferrell's new idea, however, was that some of the air descending to the earth near latitude 30° would flow away from the equator and toward the poles along the earth's surface. It was this flow of air that made Ferrell's model more complex and more accurate than Hadley's. For at about 60° latitude, this surface flow of air collided with a flow of polar air to make two additional convection cells.

Ferrell had agreed with Hadley about the movement of air above the poles. That is, cool air would descend from higher altitudes and flow toward the equator along the earth's surface. At about 60° latitude, however, this flow of polar air would collide with air flowing toward it from the 30° latitude outflow.

The accumulation of air resulting from this collision along latitude 60° would produce a region of high **pressure** that could be relieved, Ferrell said, by massive updrafts that would carry air high into the atmosphere. There the air would split into two streams, one flowing toward the equator and descending to the earth's surface once more at about 30° latitude. This downward flow would complete a second convection cell covering the mid-latitudes and now known as the Ferrell cell. The second stream above 30° latitude would flow toward the poles and complete the third, or polar, cell.

One can hardly expect a model of the atmosphere developed nearly 150 years ago to be completely valid today. We know a great deal more about the atmosphere and have much more data than Ferrell knew or had. Still, his hypothesis is still valuable because it provides some general outlines about the nature of atmospheric circulation. It also explains a number of well-known circulation phenomena.

Observed patterns of circulation

One of the implications of the Ferrell hypothesis is that there should be relatively little surface wind near the equator. In this region, surface winds should be flowing toward the equator from the Hadley cells and, when they meet, rising upward into the upper atmosphere. Equatorial regions would be expected to be characterized, therefore, by relatively low pressures with weak surface winds.

But these conditions are exactly what mariners have observed for centuries. Indeed, they long ago gave the name of the doldrums to the equatorial seas. For centuries, ship captains have feared and avoided equatorial waters because winds are so weak and unreliable there that they could easily become stranded for days or weeks at a time.

A second region of calm on the earth's surface, according to the three-cell model, would be around latitude 30°. In this region, air moving downward from both the Hadley and Ferrell cells collides as it reaches the earth's surface, producing regions of high pressure. As in the doldrums, the regions around latitude 30° are characterized by weak, unpredictable winds.

Again, such regions have long been feared and avoided by sailors, who have given them the name of the horse latitudes. The origin of this name comes from the fact that ships bringing **horses** to the Americas often became becalmed in the waters around 30°N latitude. As supplies ran low, ships were forced to throw their horses overboard. Many stories are told of the waters in these latitudes being littered with the carcasses of the unfortunate animals.

The regions between the horse latitudes and the doldrums (between 0° and 30° latitude) are those in which surface winds flow toward the equator. That flow is not directly from north to south or south to north, of course, because of the Coriolis effect. Instead, winds in these regions tend to blow from the northeast to the southwest in the northern hemisphere and from the southeast to the northwest in the southern hemisphere. Since the winds tend to be strong and dependable—the sorts of wind upon which sailing ships depend—these winds have long been known as the trade winds.

The intersection of the Ferrell and polar cells around latitude 60° is another region at which surface flows of air meet. One, from the Ferrell cell, consists of relatively warm air flowing toward the poles. The other, from the polar cell, consists of much colder air flowing toward the equator. The point at which these two systems meet is called the polar front and is characterized by some of the world's most dramatic storms.

The prevailing direction of surface winds with the Ferrell and polar cells is determined by the Coriolis effect. In the former cell, winds tend to blow from the southwest to the northeast in the northern hemisphere and from the northwest to the southeast in the southern hemisphere. To residents of **North America**, these pre-

vailing westerlies are well known as the mechanism by which **weather** systems are carried across the **continent** from west to east.

In the polar cell, the predominant air movements are just the opposite of the prevailing westerlies: from northeast to southwest in the northern hemisphere and from southeast to northwest in the southern hemisphere.

Patterns of surface pressure

Any student of **meteorology** understands that conceptual models have only limited applicability to the real world. A number of factors in the real world differ from the ideal conditions used to construct a model. These factors insure that actual weather conditions will be far more complex than the general conditions described above.

For example, both the Hadley and Ferrell models assumed that the earth has a homogeneous composition and that the **sun** always shines directly over the equator. Neither condition, of course, is actually true. For example, most parts of the planet are covered with water, and land masses are distributed unequally among this watery background. The flow of air in any one cell, then, may be undisturbed for long stretches in one region (as across an **ocean**), but highly disrupted in another region (as across a mountainous area).

Useful tools for meteorologists interested in studying air movements are charts of air pressure at various locations on the earth's surface. These charts are of value because, whatever models may predict, we known that in the real world air movements tend to occur from regions of higher pressure to those of lower pressure.

Such charts indicate that certain parts of the planet tend to be characterized by unusually high or low pressure centers at various times of the year. In general, about eight semipermanent high and low pressure cells have been identified. The term semipermanent is used for such cells because they seem to reappear every year on a regular basis.

For example, a semipermanent high pressure area occurs over the Bermuda Islands and persists throughout the year. A semipermanent low pressure—the Icelandic low—is usually found somewhat to the north of the Bermuda high, although it tends to shift from east to west and back again during various parts of the year. During the winter in the northern hemisphere, a semipermanent high exists over Siberia, although by summer it has disappeared and been replaced by a semipermanent low over India. The existence of these semipermanent highs and lows accounts for fairly predictable air movements over relatively large areas of the earth's surface.

The jet streams

During World War II, an especially dramatic type of atmospheric air movement was discovered: the jet streams. On a bombing raid over Japan, a sortie of B-29 bombers found themselves being carried along with a tail wind of about 186 MPH (300 km/h). After the war, meteorologists found that these winds were part of permanent air movements now known as the jet streams. Jet streams are currents of air located at altitudes of 30,000–45,000 ft (9,100–13,700 m) that generally move with speeds ranging from about 30–75 MPH (50–120 km/h). It is not uncommon, however, for the speed of jet streams to be much greater than these average figures, as high as 300 MPH (500 km/h) having been measured.

The jet streams discovered in 1944 are formed along the polar front between the Ferrell and polar cells. For this reason, they are usually known as polar jet streams. Polar jet streams usually travel on a west to east direction between 30°N and 50°N latitude. Commercial **aircraft** often take advantage of the extra push provided by the polar **jet stream** when they travel from west to east, although the same winds slow down planes going in the opposite direction.

The pathway followed by jet streams is quite variable. They may break apart into two separate streams and then rejoin, or not. They also tend to meander north and south from a central west-east axis. The movement of the jet streams is an important factor in determining weather conditions in mid-latitude regions.

Since the end of World War II, jet streams other than those along the polar front have been discovered. For example, a tropical easterly jet stream has been found to develop during the summer months over **Africa**, India, and southeast **Asia**. Some low-level jet streams have also been identified. One of these is located over the Central Plains in the United States, where topographic and climatic conditions favor the development of unusually severe wind systems.

Other violent wind systems

A number of air movements are not large enough to be described as forms of global circulation although they do cover extensive regions of the planet. Monsoons, for example, are heavy rain systems that sweep across the Indian subcontinent for about six months of each year. They are caused by a massive movement of air from Siberia to Africa by way of India and back again.

During the winter, cold, dry air from central Asia sweeps over India, out across the Indian Ocean, and into Africa. Relatively little moisture is transported out of Siberia during this time of the year. As summer ap-

proaches, however, the Asian land mass warms up, low pressures develop, and the winter air movement pattern is reversed. Winds blow out of Africa, across the Indian Ocean and the Indian **peninsula**, and back into Siberia. These winds pick up moisture from the ocean and bring nearly constant rains—the monsoons—to India for about six months.

See also Air masses and fronts; Global climate; Monsoon.

Resources

Books

Ahrens, C. David, Rachel Alvelais, and Nina Horne. *Essentials of Meteorology: An Invitation to the Atmosphere.* Belmont, CA: Brooks/Cole, 2000.

Ahrens, C. Donald. *Meteorology Today.* 2nd ed. St. Paul, MN: West Publishing Company, 1985.

Allen, Oliver E., and the Editors of Time-Life Books. *Planet Earth: Atmosphere.* Alexandria, VA: Time-Life Books, 1983.

Eagleman, Joe R. *Meteorology: The Atmosphere in Action.* 2nd ed. Belmont, CA: Wadsworth Publishing Company, 1985.

Hamblin, W.K., and Christiansen, E.H. *Earth's Dynamic Systems.* 9th ed. Upper Saddle River: Prentice Hall, 2001.

Hodgson, Michael, and Devin Wick. *Basic Essentials: Weather Forecasting.* 2nd ed. Guilford, CT: Globe Pequot Press, 1999.

Houghton, John. *The Physics of Atmospheres.* 3rd ed. Cambridge: Cambridge University Press, 2002.

James, I. N. *Introduction to Circulating Atmospheres.* New York: Cambridge University Press, 1994.

KEY TERMS

Convection—The transfer of heat by means of a moving fluid.

Coriolis effect—An apparent force experienced by any object that is moving across the face of a rotating body.

Doldrums—A region of the equatorial ocean where winds are light and unpredictable.

Horse latitudes—A region of the oceans around 30° latitude where winds are light and unpredictable.

Jet stream—A rapidly moving band of air in the upper atmosphere.

Polar front—A relatively permanent front formed at the junction of the Ferrell and polar cells.

Trade winds—Relatively constant wind patterns that blow toward the equator at about 30° latitude.

Lorenz, Edward N. *The Nature and Theory of the General Circulation of the Atmosphere.* Geneva: World Meteorological Organization, 1967.

Lutgens, Frederick K., Edward J. Tarbuck, and Dennis Tasa. *The Atmosphere: An Intorduction to Meteorology.* 8th ed. New York: Prentice-Hall, 2000.

Wagner, A. James, "Persistent Circulation Patterns." *Weatherwise* (February 1989): 18-21.

David E. Newton

Atmospheric optical phenomena

Atmospheric optical phenomena are visual events that take place in Earth's atmosphere as a consequence of the way **light** is reflected, refracted, and diffracted by solid particles, liquids droplets, and other materials present in the atmosphere. Such phenomena include a wide variety of events ranging from the blue **color** of the sky itself to mirages and **rainbows** to sundogs and solar pillars.

Reflection and refraction

If Earth's atmosphere were a **vacuum**, the only atmospheric optical phenomenon observable would be a stream of white light from the **sun**. The fact that colors appear in the atmosphere is a consequence of the way that white light is broken up into its component parts—red, orange, yellow, green, blue, indigo, and violet: the spectrum—during its interaction with materials in the atmosphere. That interaction takes one of three general forms: reflection, refraction, and **diffraction**.

Reflection occurs when light rays strike a smooth surface and bounce off at an **angle** equal to that of the incoming rays. Reflection can explain the origin of color in some cases because certain portions of white light are more easily absorbed or reflected than are others. For example, an object that appears to have a green color does so because that object absorbs all wavelengths of white light except that of green, which is reflected.

One form of reflection—internal reflection—is often involved in the explanation of optical phenomena. During internal reflection, light enters one surface of a transparent material (such as a **water** droplet), is reflected off the inside surface of the material, and is then reflected a second time out of the material. The color of a rainbow can partially be explained in terms of internal reflection.

Refraction is the bending of light as it passes at an angle from one transparent material into a second transparent material. The process of refraction accounts for

An aurora borealis display. *Photograph by Pekka Parviatnen. © Pekka Parviatnen. Photo Researchers, Inc. Reproduced by permission.*

the fact that objects under water appear to have a different size and location than they have in air. Light waves passing through water and then through air are bent, causing the **eye** to create a visual image of the object.

Displacement phenomena

Perhaps the most common example of an atmospheric effect created by refraction is the displacement of astronomical bodies. When the sun is directly overhead, the light rays it emits pass straight through Earth's atmosphere. No refraction occurs, and no change in the sun's apparent position takes place.

As the sun approaches the horizon, that situation changes. Light from the sun now enters Earth's atmosphere at an angle and is refracted. The eye sees the path of the light as it is bent and assumes that it has come from a position in the sky somewhat higher than it really is. That is, the sun's apparent location is displaced by some angle from its true location. The same situation is true for any astronomical object. The closer a **star** is to the horizon, for example, the more its apparent position is displaced from its true position.

Green flash

One of the most dramatic examples of sunlight refraction is the green flash. That term refers to the fact that in the moment following sunset or sunrise, a flash of green light lasting no more than a second can sometimes be seen on the horizon on the upper part of the sun. The green light is the very last remnant of sunlight refracted by Earth's atmosphere, still observable after all red, orange, and yellow rays have disappeared. The green light remains at this moment because the light rays of shorter wavelength—blue and violet—are scattered by the at-

Sundogs at sunset on the frozen sea at Cape Churchill, Hudson Bay. © Dan Guravich 1987, National Audubon Society Collection/Photo Researchers, Inc. Reproduced by permission.

mosphere. The green flash is rarely seen, but when it is, it makes a remarkable impression on the observer.

Scattered light

Light that bounces off very small objects is not reflected uniformly, but is scattered in all directions. The process of scattering is responsible for the fact that humans observe the sky as blue. When white light from the sun collides with molecules of **oxygen** and **nitrogen**, it is scattered selectively. That is, light with shorter wavelengths—blue, green, indigo, and violet—is scattered more strongly than is light with longer wavelengths—red, orange, and yellow. No matter where a person stands on Earth's surface, she or he is more likely to see the bluish light scattered by air molecules than the light of other hues.

Twinkling

Stars twinkle; planets do not. This general, though not inviolable, rule can be explained in terms of refraction. Stars are so far away that their light reaches Earth's atmosphere as a single point of light. As that very narrow beam of light passes through Earth's atmosphere, it is refracted and scattered by molecules and larger particles of matter. Sometimes the light travels straight toward an observer, but sometimes its path is deflected. To the observer, the star's light appears to go on and off many times per second. That is, it twinkles.

Planets usually do not twinkle because they are closer to **Earth**. The light that reaches Earth from them consists of wider beams rather than narrow rays. The refraction or scattering of only one or two light rays out of the whole beam does not make the light seem to disappear. At any one moment, enough light rays reach Earth's surface from a **planet** to give a sense of one continuous beam of light.

Mirages

One of the most familiar optical phenomena produced by refraction is a mirage. One type of mirage—the inferior mirage—is caused when a layer of air close to the ground is heated more strongly than is the air immediately above it. When that happens, light rays pass through two transparent media—the hot, less dense air and the cooler, more dense air—and are refracted. As a result of the refraction, the blue sky appears to be present on Earth's surface; it may look like a body of water and objects such as trees appear to be reflected in that water.

A second type of mirage—the superior mirage—forms when a layer of air next to the ground is much cooler than the air above it. In this situation, light rays from an object are refracted in such a way that an object appears to be suspended in air above its true position. This phenomenon is sometimes referred to as looming.

Rainbows

The most remarkable phenomenon in the atmosphere is likely to be the rainbow. To understand how a rainbow is created, imagine a single beam of white light entering a spherical droplet of water. As the light passes from air into water, it undergoes refraction (that is, it is bent). However, each color present in the white light is bent by a different amount—the blues and violets more than the reds and yellows. The light is said to be dispersed, or separated according to color. After the dispersed rays pass into the water droplet, they reflect off the rear inner surface of the droplet and exit into the air once more. As the light rays pass out of the water into the air, they are refracted a second time. As a result of this second refraction, the separation of blues and violets from reds and yellows is made more distinct.

An observer on Earth's surface can see the net result of this sequence of events repeated over and over again by billions of individual water droplets. The rainbow that is produced consists simply of the white light of the sun separated into its component parts by each separate water droplet.

Haloes, sundogs, and sun pillars

The passage of sunlight through cirrus **clouds** can produce any one of the optical phenomena known as haloes, sundogs, or sun pillars. One explanation for phenomena of this kind is that cirrus clouds consist of tiny **ice** crystals that refract light through very specific angles, namely 22° and 46°. When sunlight shines through a cirrus cloud, each tiny ice **crystal** acts like a **glass prism**, refracting light at an angle of 22° (more commonly) or 46° (less commonly).

A halo is one example of this phenomenon. Sunlight shining through a cirrus cloud is refracted in such a way that a **circle** of light—the halo—forms around the sun. The halo may occur at 22° or 46°.

Sundogs are formed by a similar process, and occur during sunrise or sunset. When relatively large (about 30 microns) crystals of ice orient themselves horizontally in a cirrus cloud, the refraction pattern they form is not a circle (a halo), but a reflected image of the sun. This reflected image is located at a **distance** of 22° from the actual sun, often at or just above the horizon. Sundogs are also known as mock suns or parhelia.

Sun pillars are, as their name suggests, narrow columns of light that seem to grow out of the top or (less commonly) from the bottom of the sun. This phenomena is a result not of refraction, but of reflection. Sunlight reflects off the bottom of flat ice crystals as they settle slowly toward Earth's surface. The exact shape and orientation of the sun pillar depends on the position of the sun above the horizon and the exact orientation of the ice crystals to the ground.

Coronas and glories

In addition to reflection and refraction, the path of a light ray can be altered by yet a third mechanism, diffraction. Diffraction occurs when a light ray passes close to some object. For comparison, you can think of the way in which water waves are bent as they travel around a rock. Diffraction may also result in the separation of white light into its colored components.

When light rays from the **Moon** pass through a thin cloud, they may be diffracted. **Interference** of the various components of white light that generates the colors make up the corona. The pattern formed by the diffraction rays is a ring around the Moon. The ring may be fairly sharp and crisp, or it can be diffuse and hazy. The ring is known as a corona. Coronas may also form around the sun, although because the sun is much brighter, they are more difficult to observe.

A glory is similar to a corona but is most commonly observed during an airplane ride. As sunlight passes over the airplane, it may fall on water droplets in a cloud below. The light that is diffracted then forms a series of colored rings—the glory—around the airplane's shadow.

Resources

Books

Ahrens, C. Donald. *Meteorology Today.* 2d ed. St. Paul, MN: West Publishing Company, 1985.

Eagleman, Joe R. *Meteorology: The Atmosphere in Action.* 2nd ed. Belmont, CA: Wadsworth Publishing Company, 1985.

Greenler, Robert. *Rainbows, Halos, and Glories.* New York: Cambridge University Press, 1980.

Lutgens, Frederick K., and Edward J. Tarbuck. *The Atmosphere: An Introduction to Meteorology.* 4th ed. Englewood Cliffs, NJ: Prentice Hall, 1989.

Lutgens, Frederick K., Edward J. Tarbuck, and Dennis Tasa. *The Atmosphere: An Intorduction to Meteorology.* 8th ed. New York: Prentice-Hall, 2000.

David E. Newton

Atmospheric pressure

The earth's atmosphere exerts a **force** on everything within it. This force, divided by the area over which it acts, is the atmospheric **pressure**. The atmospheric pressure at **sea level** has an average value of 1,013.25 millibars. Expressed with other units, this pressure is 14.7 lb per square inch, 29.92 inches of mercury, or 1.01×10^5 pascals. Atmospheric pressure decreases with increasing altitude: it is half of the sea level value at an altitude of about 3.1 mi (5 km) and falls to only 20% of the surface pressure at the cruising altitude of a jetliner. Atmospheric pressure also changes slightly from day to day as **weather** systems move through the atmosphere.

The earth's atmosphere consists of gases that surround the surface, and like any gas, the atmosphere exerts a pressure on everything within it. A gas is made up of molecules that are constantly in **motion**. If the gas is in a container, some gas molecules are always bouncing off the container walls. When they do so, they exert a tiny force on the walls. With a sufficient number of molecules, their impacts add up to make a force that can easily be measured. Dividing the total force by the area over which it is measured gives the gas pressure. Anything else the gas touches will also have this pressure exerted on it. Thus anywhere we go within the earth's atmosphere we can detect atmospheric pressure.

Atmospheric pressure decreases as one climbs higher in the atmosphere, and increases the closer one gets to the earth's surface. The reason for this change with altitude is that atmospheric pressure at any point is really a measure of the weight, per unit area, of the atmosphere above that point. At sea level, for example, the pressure is 14.7 pounds per square inch. This means that a slice of the atmosphere in the shape of a long, thin column, with a one square inch base and as tall as the top of the atmosphere (at least 120 mi or 200 km), would have air within the column weighing 14.7 lb (6.7 kg). At a higher elevation, such as the top of a 10,000 ft (3,048 m) mountain, one is above some of the atmosphere. Here the atmospheric pressure is lower than at sea level, because there is less air weighing down from above. A person feels this sort of pressure effect when they dive to the bottom of a **lake** or deep swimming pool. As the diver descends deeper into the **water**, more and more water lies overhead. The extra water exerts an increasing pressure that the diver can feel on his or her skin (and especially on the eardrums).

Atmospheric pressure is closely related to weather. Regions of pressure that are slightly higher or slightly lower than the **mean** atmospheric pressure develop as air circulates around the **earth**. The air rushes from regions of high pressure to low pressure, causing winds. The properties of the moving air (cool or warm, dry or humid) will determine the weather for the areas through which it passes. Knowing the location of high and low pressure areas is vital to **weather forecasting**, which is why they are shown on the weather maps printed in newspapers and shown on **television**.

Atmospheric pressure is measured by a **barometer**, of which there are several designs. The first barometer was made by Evangelista Torricelli in 1643, using a column closed at one end and partially filled with mercury. The column was placed vertically in a small pool of mercury with the open end downward. In this arrangement, the mercury does not run out the open end. Rather, it stays at a height such that the pressure exerted by the suspended mercury upon the pool will equal the atmospheric pressure on the pool. The mercury barometer is still in common use today (this is the reason pressure is still given the units "inches of mercury" on weather reports). Modern barometers include the aneroid barometer, which substitutes a sealed container of air for the mercury column, and the electronic **capacitance** manometer, which senses pressure electronically.

James Marti

Atmospheric temperature

The **temperature** of the atmosphere varies with the **distance** from the equator (latitude) and height above the surface (altitude). It also changes in **time**, varying from season to season, from day to night and irregularly due to passing **weather** systems. If these variations are averaged out on a global basis, a pattern of average temperatures emerges for the atmosphere. The vertical temperature profile (the way temperature changes with height) divides the atmosphere into four layers: the troposphere, the stratosphere, the mesosphere, and the thermosphere.

The vertical temperature profile

Averaging atmospheric temperatures over all latitudes and across an entire year gives us the average vertical temperature profile. This plot is sometimes called a standard atmosphere. The average vertical temperature profile suggests four distinct layers (Figure 1). In the first layer, called the troposphere, average atmospheric temperature drops steadily from its value at the surface, about 290K (63°F; 17°C) until it reaches of minimum of around 220K (−64°F; −53°C) at a level about 6.2 mi (10 km) high. This level, known as the tropopause, is just above the cruising altitude of large commercial jet **aircraft**. The drop of temperature with height, called the lapse **rate**, is nearly steady throughout the troposphere at 43.7°F (6.5°C) per 0.6 mi (1 km). At the tropopause, the lapse rate abruptly shrinks to very low values. Atmospheric temperature is roughly constant over the next 12 mi (20 km), then begins to rise with increasing altitude up to about 31 mi (50 km). This region of increasing temperatures is the stratosphere. At the top of the layer, called the stratopause, temperatures are nearly as warm as the surface values. Between about 31–50 mi (50–80 km) lies the mesosphere, where atmospheric temperature resumes its drop with altitude and reaches a very cold minimum of 180K (−136°F; −93°C) at the top of the layer (the mesopause), around 50 mi (80 km). Above the mesopause is the thermosphere, which as its name implies is a zone of high gas temperatures. In the very high thermosphere (about 311 mi (500 km) above Earth's surface) gas temperatures can reach from 500–2,000K (441–3,141°F; 227–1,727°C), depending on how active the **sun** is. However, these figures are somewhat misleading. Temperature is a measure of the **energy** of the gas molecules' **motion**. Although they have high energies, the molecules in the thermosphere are present in very low numbers, less than one millionth of the amount present on average at Earth's surface. If a person were in the thermosphere, it would feel to them much more like the icy cold of **space** because such a small number of energetic gas molecules would be unable to transfer much of their **heat** energy.

To add more information to the temperature graph, one can plot atmospheric temperature as a function of both latitude and altitude. Figures 2 and 3 show such plots, with latitude as the x coordinate and altitude as the y.

The sun's role in atmospheric temperature

The reason that temperature is distributed as shown in the figures is mostly due to the sun and the way solar energy is deposited in the atmosphere. Most of the solar **radiation** is emitted as visible **light**, with smaller portions at shorter wavelengths (ultraviolet radiation)

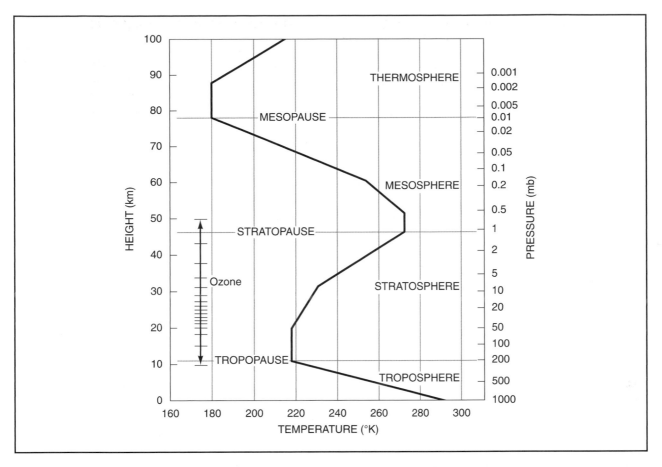

Figure 1. The temperature of Earth's atmosphere is broadly determined by the deposition of solar energy and by the absorption of infrared (heat) radiation by "greenhouse" gases. Concern has recently been growing about increases in greenhouse warming due to human activities. This possible global warming is the subject of active research. *Illustration by Hans & Cassidy. Courtesy of Gale Group.*

and longer wavelengths (infrared radiation, or heat). Little of the visible light is absorbed by the atmosphere (although some is reflected back into space by **clouds**), so most of this energy is absorbed by the Earth's surface. **Earth** is warmed in the process and radiates heat (infrared radiation) back upward. This warms the atmosphere, and just as one will be warmer when standing closer to a fire, the layers of air closest to the surface are the warmest.

According to this explanation, the temperature should continually drop as one goes higher in atmosphere. Yet Figure 1 shows that temperature rises with height in the stratosphere. The reason for this apparent contradiction is another case of solar energy deposition in the atmosphere. The stratosphere contains nearly all the atmosphere's **ozone**. Ozone (O_3) and molecular **oxygen** (O_2) absorb most of the sun's short wavelength ultraviolet radiation. In the process they are broken apart and reform again and again. The net result is that the ozone molecules transform the ultraviolet radiation to heat en-

ergy, heating up the layer and causing the increasing temperature profile observed in the stratosphere.

The mesosphere resumes the temperature drop with height. The thermosphere however is subject to very high energy, short wavelength ultraviolet and x-ray solar radiation. As the **atoms** or molecules present at this level absorb some of this energy, they are ionized (have an **electron** knocked off) or dissociated (molecules are split into their component atoms). The gas layer is strongly heated by this energy bombardment, especially during periods when the sun is active, that is, emitting elevated amounts of short wavelength radiation.

The greenhouse effect

Solar energy is not the only determinant of atmospheric temperature. As noted above, Earth's surface, after absorbing solar radiation in the visible region, emits infrared radiation back to space. Several atmospheric gases absorb this heat radiation and re-radiate it in all direc-

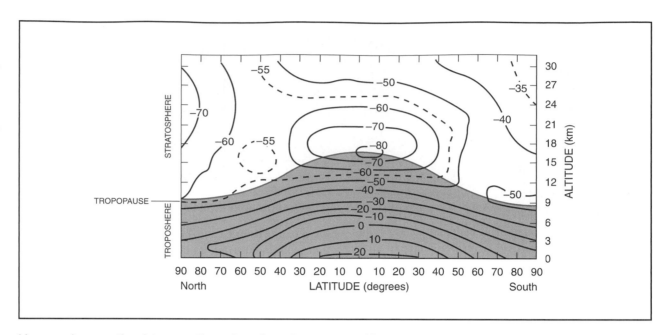

Lines are drawn on the plot connecting points of equal temperature (like contour lines on a map), given in degrees C. Figure 2 is for December though February, which is winter in the northern hemisphere and summer in the southern. As one might expect, the warmest temperature is found at the surface near the equator, and drops as one travels toward either pole and/or as one increases in altitude. Surprisingly, however, the coldest spot in the lower atmosphere is at the tropopause over the equator, which is colder than even over polar regions. *Illustration by Hans & Cassidy. Courtesy of Gale Group.*

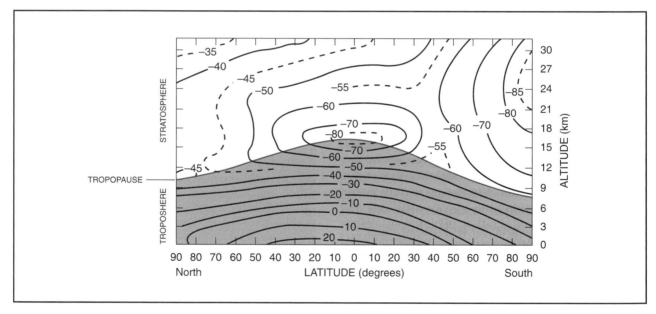

The temperature plot (Figure 3) for June through August (Southern hemisphere winter, Northern summer) shows that the equatorial temperature does not change much with the seasons. The middle and high latitudes have experienced much more change, as the temperature contours have shifted northward. The tropopause over the equator is still extremely cold, surpassed only by the stratosphere over the Antarctic. *Illustration by Hans & Cassidy. Courtesy of Gale Group.*

tions, including back toward the surface. These so-called greenhouse gases thus trap infrared radiation within the atmosphere, raising its temperature. Important greenhouse gases include **water** vapor (H_2O), **carbon dioxide**

(CO_2), and methane (CH_4). It is estimated that the Earth's surface temperature would average about 32°C (90°F) cooler in the absence of greenhouse gases. Since this temperature is well below the freezing point of water, it is

KEY TERMS

. . . .

Greenhouse effect—The warming of the Earth's atmosphere as a result of the capture of heat re-radiated from the Earth by certain gases present in the atmosphere.

Infrared radiation—Radiation similar to visible light but of slightly longer wavelength. We sense infrared radiation as heat.

Lapse rate—The rate at which the atmosphere cools with increasing altitude, given in units of degrees C per kilometer.

Mesosphere—The third layer of the atmosphere, lying between about 50 and 80 kilometers in height and characterized by a small lapse rate.

Stratosphere—A layer of the upper atmosphere above an altitude of 5–10.6 mi (8–17 km) and extending to about 31 mi (50 km), depending on season and latitude. Within the stratosphere, air tem-

perature changes little with altitude, and there are few convective air currents.

Thermosphere—The top layer of the atmosphere, starting at about 50 mi (80 km) and stretching up hundreds of miles or kilometers in to space. Due to bombardment by very energetic solar radiation, this layer can possess very high gas temperatures.

Troposphere—The layer of air up to 15 mi (24 km) above the surface of the Earth, also known as the lower atmosphere.

Ultraviolet radiation—Radiation similar to visible light but of shorter wavelength, and thus higher energy.

X-ray radiation—Light radiation with wavelengths shorter than the shortest ultraviolet; very energetic and harmful to living organisms.

apparent that the **planet** would be much less hospitable to life in the absence of the **greenhouse effect**.

While greenhouse gases are essential to supporting life on the planet, more is not necessarily better. Since the beginning of the **industrial revolution** in the mid-nineteenth century, humans have released increasing amounts of **carbon** dioxide to the atmosphere through the burning of **fossil fuels**. The level of carbon dioxide measured in the remote atmosphere has shown a continuous increase since record keeping began in 1958. If this increase translates into a like rise in atmospheric temperature, the results would be dire indeed: melting **polar ice caps** and swelling seas, resulting in coastal cities being covered by the **ocean**; radical shifts in climate, dooming plants and animals that could not adapt quickly enough; and unpredictable changes in **wind** and weather patterns, posing significant challenges for agriculture. The problem in forecasting the changes that increasing greenhouse gases may bring is that the Earth's climate is a very complicated, interconnected system. The interplay of the atmosphere, the oceans, the continents and the **ice** caps is not completely understood. While it is known that some of the emitted carbon dioxide is absorbed by the oceans and eventually deposited as carbonate rock (such as limestone), we do not know if this is a steady process or if it can keep pace with our constant releases. Computer models designed to mimic the Earth's climate must make many approximations. Nonetheless, calculations by these less-than-perfect models suggest that a doubling of carbon dioxide levels

would mean an increase in the average Northern hemisphere surface temperatures of 39–43°F (4–6°C). While this may not sound like much, note that during the last ice age, when large ice sheets covered much of the northern hemisphere, the Earth's average temperature was only 41°F (5°C) below current levels.

Resources

Books

Ahrens, C. David, Rachel Alvelais, and Nina Horne. *Essentials of Meteorology: An Invitation to the Atmosphere.* Belmont, CA: Brooks/Cole, 2000.

Dessler, A. *The Chemistry and Physics of Stratospheric Ozone* Cornwall, UK: Academic Press, 2000.

Fisher, David E. *Fire and Ice: The Greenhouse Effect, Ozone Depletion, and Nuclear Winter* New York: Harper & Row, 1999.

Lutgens, Frederick K., Edward J. Tarbuck, and Dennis Tasa. *The Atmosphere: An Intorduction to Meteorology.* 8th ed. New York: Prentice-Hall, 2000.

McNeill, Robert. *Understanding the Weather.* Las Vegas: Arbor Publishers, 1991.

Newton, David E. *Global Warming.* Santa Barbara, CA: ABC-CLIO Pub., Inc., 1993.

Periodicals

Bard, E. "Ice Age Temperatures and Geochemistry." *Science* no. 284 (May 1999): 1133-1134.

James Marti

Atoll *see* **Coral and coral reef**

Atomic clock

Atomic clocks are the world's most accurate **time** keepers—more accurate than astronomical time or quartz clocks. Originally, a second was defined as 1/86,400 of a **mean** solar day. Today it is defined as 9,192,631,770 periods or wavelengths of the **radiation** absorbed by the cesium-133 atom as it changes between two hyperfine **energy** levels. The change in definition was the result of the atomic clocks' ability to accurately measure these very short periods. Atomic clocks do not resemble ordinary clocks or watches. The **atoms** serve the same purpose as did pendulums or quartz crystals in earlier clocks. They have no "hands" to turn or liquid **crystal** displays for number read-outs. They simply produce electrical pulses that serve as a standard for calibrating other less accurate clocks.

In 1945, Isidor Rabi, a **physics** professor at Columbia University, first suggested that a clock could be made from a technique he developed in the 1930s called atomic beam magnetic **resonance**. Using Rabi's technique, the Commerce Department's National Institute of Standards and Technology (NIST) (then the National Bureau of Standards), developed the world's first atomic clock in 1949 using the **ammonia molecule** as the source of vibrations. In 1952, NIST revised its design using cesium atoms as the vibration source. This clock was named NBS-1.

Like all atoms, the cesium atoms used in an atomic clock are quantized; that is, they can absorb or give up only discrete quantities of energy. It is the quantum nature of the atom that is the underlying principle of atomic clocks. An atom of cesium can exist at a minimum energy level of, for example, E1, which is called its ground

state. It may absorb a certain amount of energy and reach a somewhat greater energy level—E2, E3, E4, and so on. Thus, an atom in its ground state can accept a quantity of energy equal to E2 − E1, E3 − E1, E4 − E1, and so on, but it cannot exist at an energy level that lies between these values. It cannot, for example, absorb a quantity of energy equal to 1/2(E2 − E1). Once an atom is at an energy level greater than its ground state, it can only release energy quantities equal to the difference in energy levels: E2 − E1, E3 − E1, E4 − E1, E4 − E3, E3 − E2, and so on. When energy is emitted by an atom, it is in the form of electromagnetic radiation, such as **light**. Only radiation with frequencies between 4.3×10^{14} and 7.5×10^{14} Hz can be seen because those frequencies mark the ends of the range visible to the human **eye**. The greater the **frequency** of the radiation, the greater its energy. In fact, the energy, E, of the radiation is given by the equation $E = hf$, where f is the frequency and h is **Planck's constant** (6.626×10^{-34} J•s).

The energy of the radiation absorbed by the cesium atoms used in most atomic clocks is very small. It is absorbed (or released) when cesium atoms pass between two so-called hyperfine energy levels. These energy levels, which are very close together, are the result of magnetic forces that arise because of the spin of the atom's nucleus and the electrons that surround it. The frequency of the radiation absorbed or released as atoms oscillate between two hyperfine energy states can be used as a standard for time. Such frequencies make ideal standards because they are very stable—they are not affected by **temperature**, air **pressure**, light, or other common factors that often affect ordinary **chemical reactions**.

On December 30, 1999, NIST started a new atomic clock, the NIST F-1, which is estimated to neither gain nor lose a second for 20 million years. Located at NIST's Boulder, Colorado, laboratories, the NIST F-1, currently the United States's primary frequency standard, shares the distinction of being the world's most accurate clock with a similar device in Paris.

The NIST F-1 uses a fountain-like movement of atoms to keep time. First, a gas of cesium atoms is introduced into the clock's **vacuum** chamber. Six infrared **laser** beams then are directed at right angles to each other at the center of the chamber. The lasers gently push the cesium atoms together into a ball, thus slowing the movement of the atoms and cooling them to near **absolute zero**. The fountain action is created when two vertical lasers gently toss the ball upward. This little push is just enough to propel the ball through a microwave-filled cavity for about one meter. Gravity pulls the ball downward again toward the lasers.

Repairs completed to an atomic clock. This clock is timed to the resonance frequency of cesium. *Royal Greenwhich Observatory, National Audubon Society Collection/Photo Researchers, Inc. Reproduced by permission.*

While inside the cavity, the atoms interact with the microwave signal and their atomic states are altered in relation to the frequency of the signal. The entire round trip takes about a second. When finished, another laser, directed at the ball, causes the altered cesium atoms to emit light, or fluoresce. Photons emitted during **fluorescence** are measured by a detector. This procedure is repeated many times while the microwave energy in the cavity is tuned to different frequencies. Eventually, a microwave frequency is achieved that alters the states of most of the cesium atoms and maximizes their fluorescence. This frequency is the natural resonance frequency for the cesium atom; the characteristic that defines the second and, in turn, makes ultraprecise timekeeping possible.

The NIST F-1's predecessor, the NIST-7, as well as many versions before it, fired heated cesium atoms horizontally through a microwave cavity at a high speed. NIST F-1's cooler and slower atoms allow more time for the microwaves to "interrogate" the atoms and determine their characteristic frequency, thus providing a more sharply defined signal. NIST F-1, along with an international pool of atomic clocks, define the official world time called Coordinated Universal Time.

Atomic clocks have been used on jet planes and satellites to verify Einstein's theory of relativity, which states that time slows down as the **velocity** of one object relative to another increases. Until the advent of atomic clocks that can measure time to within one second in a million years, there was no direct way of accurately measuring the time dilation predicted by Einstein, even at the velocities of **space** probes. Although these space vehicles reach speeds of 25,000 MPH (40,000 km/h), such a speed is only 0.004% of the speed of light, and it is only at velocities close to the speed of light that time dilation becomes significant. Atomic clocks on satellites are used in navigation. The signals sent by the atomic clocks in satellites travel at the speed of light (186,000 mi/s or 300,000 km/s). Signals from different satellites reach a ship or a plane at slightly different times because their distances from the plane or vessel are not the same. For example, by the time simultaneous signals sent from two satellites at distances of 3,100 mi (5,000 km) and 4,970 mi (8,000 km) from a ship reach the vessel, they will be separated by a time interval of 0.01 second. By knowing the position of several such satellites and the time delay between their signals, the longitude and latitude of a ship or plane can be established to within several feet.

International Atomic Time, based on cesium clocks, is periodically compared with mean solar time. Because the Earth's **rate** of **rotation** is slowly decreasing, the length of a solar day is increasing. Today's day is about three milliseconds longer than it was in 1900. The change is too small for us to notice; however, it is readily

KEY TERMS

Frequency—Number of oscillations or waves emitted per second.

Ionized atoms—Atoms that have acquired a charge by gaining or losing an electron.

Mean solar day—The average solar day; that is, the average time for the earth to make one complete rotation relative to the sun.

Photons—The smallest units or bundles of light energy. The energy of a photon is equal to the frequency (f) of the light times Planck's constant (h); thus, E = hf.

detected by atomic clocks. Whenever the difference between International Atomic Time and astronomical time is more than 0.9 seconds, a "leap second" is added to the mean solar time. To keep these two time systems synchronized, leap seconds have been added about every year and a half.

Resources

Books

Gribbin, John. *Q is for Quantum: An Encyclopedia of Particle Physics.* New York: The Free Press, 1998.

Itano, Wayne M., and Norman F. Ramsey. "Accurate Measurement of Time." *Scientific American* 269 (July 1993): 56-65.

Morrison, Leslie. "The Day Time Stands Still." *New Scientist* (27 June 1985).

Periodicals

NIST. "NIST F-1 Cesium Fountain Clock." *NIST* (29 December 1999).

Wineland, D. J. "Trapped Ions, Laser Cooling, and Better Clocks." *Science* (26 October 1984).

Robert Gardner

Atomic force microscope *see* **Microscopy**

Atomic mass unit *see* **Atomic weight**

Atomic models

The atom is defined as the smallest part of an element that retains the chemical properties of that element. The existence of **atoms** was first proposed as early as 400 B.C. Greek philosophers debated whether one could divide a substance into infinitely smaller pieces, or if eventually the smallest, indivisible particle would be

was composed of particles of approximately the same mass as protons. In addition, magnetic or electric fields could not deflect this beam. Chadwick concluded that the beam must be made up of neutral particles of approximately the same size as protons, which he called neutrons. Neutrons, together with protons, make up the nucleus of the atom and contribute to the majority of its mass. Electrons are located in the empty space surrounding the atom, which makes up most of its **volume**. The mass of the electron is insignificant when compared to the mass of the protons and neutrons.

The dual nature of matter

Even with the discovery of the proton, Rutherford's atomic model still did not explain how electrons could have stable orbits around the nucleus. The development of a mathematical constant by the German physicist Max Planck (1858–1947) served as the basis for the next atomic model. Planck developed his constant in 1900 when explaining how **light** was emitted from hot objects. He hypothesized that electromagnetic **radiation** could only be associated with specific amounts of energy, which he called quanta. The energy lost or gained by an atom must occur in a quantum, which can be thought of as a "packet" containing a minimum amount of energy. He described the relationship between a quantum of energy and the **frequency** of the radiation emitted mathematically by the equation $E=\lambda$ (where E is the energy, in joules, of one quantum of radiation, and λ is the frequency of the radiation). The letter h symbolizes **Planck's constant**.

In 1905 Albert Einstein (1879-1955) developed a theory stating that light has a dual nature. Light acts not only as a wave, but also as a particle. Each particle of light has a quantum of energy associated with it and is called a **photon**. The energy of a photon can be expressed using Planck's equation. Einstein's hypothesis helped explain the light emitted when current is passed through a gas in a cathode ray tube. An atom that has the lowest potential energy possible is said to be in the ground state. When a current passes through a gas at low pressure, the potential energy of the atoms increase. An atom having a higher potential energy than its ground state it is said to be in an excited state. The excited state atom is unstable and will return to the ground state. When it does, it gives off the lost energy as electromagnetic radiation (a photon). When an electric current is passed through an elemental gas, a characteristic **color** of light is emitted. This light can be passed through a **prism** where it splits into various bands of light at specific wavelengths. These bands are known as the line-emission **spectrum** for that element. The line-emission spectrum for hydrogen was the first to be described

mathematically. Scientists now faced the task of developing a model of the atom that could account for this mathematical relationship.

The Bohr model of the atom

In 1913, the Danish theorist Niels Bohr (1885–1962) developed his quantized shell model of the atom. Bohr modified Rutherford's model by hypothesizing that the electrons **orbit** the nucleus in specific regions of fixed size and energy. The energy of the electron depends on the size of the orbit. Electrons in the smallest orbits have the least energy. An atom is stable when its electrons occupy orbits of the lowest possible energy. The energy of an electron increases as it occupies orbits farther and farther from the nucleus.

These orbits can be thought of as the rungs of a ladder. As a person climbs up a ladder, they step on one rung or another, but not in between rungs, because a person cannot stand on air. Likewise, the electrons of an atom can occupy one orbit or another, but cannot exist in between orbits. While in an orbit, the electron has a fixed amount of energy. The electron gains or loses energy by moving to a new orbit, either further from or closer to the nucleus.

When an electron falls from the excited state to the ground state, a photon is emitted with a specific energy. The energy of the photon is equal to the energy difference between the two orbits. The energy of each photon corresponds to a particular frequency of radiation given by Planck's equation, $E = h\lambda$. Bohr was able to calculate the energy of the electron in a hydrogen atom by measuring the wavelengths of the light emitted in its line-emission spectrum. Bohr's atomic model was very stable because the electron could not lose any more energy than it had in the smallest orbit. One major problem with Bohr's model was that it could not explain the properties of atoms with more than one electron, and by the early 1920s, the search for a new atomic model had begun.

The modern atomic model

The development of **quantum mechanics** served as the foundation of the modern atomic theory. In 1922, the American physicist Arthur H. Compton (1892–1962) conduced x-ray scattering experiments that confirmed and advanced Einstein's theory on the dual nature of light. In 1923, the French physicist Louis-Victor de Broglie (1892–1987) expanded on this theory by proposing that all matter, as well as radiation, behaves both as a particle and a wave. Until this time, scientists had viewed matter and energy as distinct phenomena that followed different laws.

Broglie's proposal was not supported by experimental or mathematical evidence until 1926 when the Austrian physicist Erwin Schrödinger (1887–1961) developed his mathematical wave equation. Schrödinger proposed that electrons also behaved like waves. His wave equation could be used to find the frequency of the electrons and then Planck's equation could be used to find the corresponding energy. Schrödinger's equation gave a much more precise description of an electron's location and energy than Bohr's model could. It could also be used for atoms with more than one electron. Furthermore, only waves of specific frequencies could be solved using his equation. This demonstrated that only certain energies are possible for the electrons in an atom. Further experiments demonstrated that Broglie was correct in his assertion that matter could behave as waves, as electrons were diffracted and exhibited **interference**.

In 1927, German physicist Werner Heisenberg (1901–1976) developed what is now known as the **Heisenberg uncertainty principle**. This hypothesis states that the position and **velocity** of an electron, or any moving particle, cannot both be known at the same time. This meant that the solutions to the Schrödinger wave equation, known as wave functions, could describe only the probability of finding an electron in a given orbit. Therefore, the electrons are not located in discrete orbits, as hypothesized in the **Bohr model**, but instead occupy a hazier region, called an orbital. An orbital indicates a probable location of the electrons in an atom instead of a definite path that they follow. The probable location of the electrons in an orbital is described by a series of numbers called quantum numbers.

The quantum model of the atom uses four quantum numbers to describe the arrangement of electrons in an atom, much like an address describes the locations of houses on a street. This arrangement is known as the electron configuration. The atoms for each element have their own distinct electron configuration. The ground state electron configuration of an atom represents the lowest energy arrangement of the electrons in an atom. The placement of electrons in a particular configuration is based on three principles. The first, the Aufbau principle, states that an electron will occupy the lowest possible energy orbital available. The **Pauli exclusion principle** states that each electron in an atom has its own distinct set of four quantum numbers. No two electrons in an atom will have the same set.

Lastly, Hund's rule states that even though each orbital can hold two electrons, the electrons will occupy the orbitals such that there are a maximum number of orbitals with only one electron. Developed by the German scientist, Friedrich Hund (1896–1997), Hund's rule allows scientists to predict the order in which electrons fill an atom's suborbital shells.

Hund's rule is based on the Aufbau principle that electrons are added to the lowest available energy level (shell) of an atom. Around each atomic nucleus, electrons occupy energy levels termed shells. Each shell has a spherical s orbital and, starting with the second shell, orbitals (p, d, f, etc.) and suborbitals (e.g., 2px,2py, 2pz) with differing size, shapes and orientation (i.e., direction in space).

Although each suborbital can hold two electrons, the electrons all carry negative charges and, because like charges repel, electrons repel each other. In accord with Hund's rule, electrons space themselves as far apart as possible by occupying all available vacant suborbitals before pairing up with another electron. The unpaired electrons all have the same spin **quantum number** (represented in electron configuration diagrams with arrows all pointing either upward or downward).

In accord with the Pauli exclusion principle that states that each electron must have its own unique set of quantum numbers that specify its energy and because all electrons have a spin of 1/2, each suborbital can hold up to two electrons only if their spins are paired +1/2 with -1/2. In electron configuration diagrams, paired electrons with opposite spins are represented by paired arrows pointing up and down.

Although Hund's rule accurately predicts the electron configuration of most elements, exceptions exist, especially when atoms and ions have the opportunity to gain additional stability by having filled s shells or half-filled d or f orbitals.

In 1928, the English physicist P.A.M. Dirac (1902–1984) formulated a new equation to describe the electron. Schrödinger's equation did not allow for the principles of relativity, and could only be used to describe movement of particles that are slower than the speed of light. Because electrons move at a much greater velocity, Dirac introduced four new wave functions to describe the behavior of electrons. These functions described electrons in various states. Two of the states corresponded to their spin orientations in the atom, but the other two could not be explained. In 1932, the American physicist Carl David Anderson (1905–1991) discovered the positron, which explained the two mystery states described by Dirac.

Modern physics has expanded the atomic model by introducing new particles that can be created in **vacuum** tubes. Such particles are called antiparticles, because they can be "destroyed" or converted into other forms of energy. Antiparticles include positrons, muons, pions, hadrons, baryons, mesons, and quarks. These particles

KEY TERMS

....

Alpha particle—Two protons and two neutrons bound together and emitted from the nucleus during some kinds of radioactive decay.

Anode—A positively charged electrode.

Cathode—A negatively charged electrode.

Electromagnetic radiation—The energy of photons, having properties of both particles and waves. The major wavelength bands are, from short to long: cosmic, ultraviolet, visible or "light," infrared, and radio.

Law of conservation of mass—Mass is neither created nor destroyed during ordinary chemical or physical reactions.

Law of definite proportions—A chemical compound contains the same elements in exactly the same proportions by mass regardless of the size of the sample or the source of the compound.

Law of multiple proportions—If two or more different compounds are composed of the same two elements, then the ratio of the masses of the second element combined with a certain mass of the first element is always a ratio of small whole numbers.

Positron—A positively charged particle having the same mass and magnitude of charge as the electron.

Potential energy—Stored energy.

Quarks—Believed to be the most fundamental units of protons and neutrons.

Subatomic particle—An elementary particle smaller than an atom. Protons, neutrons, and electrons are examples of subatomic particles.

can combine with each other or split to form new and different particles. For example, three quarks can combine to form a proton, a **neutron**, or a baryon. Many of these particles have such high energies, they have never actually been observed in the laboratory. The quantum field theory is the study of the behavior of these antiparticles and how they relate to the three subatomic particles (the proton, neutron, and electron). According to the quantum field theory, the atom can be subdivided not only into protons, neutrons, and electrons, but into antiparticles as well.

See also Atomic spectroscopy; Atomic weight; Electromagnetic spectrum; Particle detectors; Quantum mechanics.

Resources

Books

Chattergee, Lali. *The Exotic Lifestyles of Subatomic Particles* Dubuque, OH: Kendall/Hunt Publishers, 2000.

Davis and Metcalf. *Modern Chemistry.* New York: HBJ School, 2000.

Gribbin, John R. and Mary Gribbin. *Q is for Quantum: An Encyclopedia of Particle Physics* New York: Free Press, 2000.

LaMay, H. Eugene. *Chemistry: Connections to Our Changing World.* Upper Saddle River, NJ: Prentice Hall, 2000.

Myers, Oldham, and Tocci. *Chemistry: Visualizing Matter.* New York: Holt Rinehart & Winston, 2000.

Other

Particle Data Group. Lawrence Berkeley National Laboratory. "The Particle Adventure: The Fundamentals of Matter and Force" [cited February 5, 2003]. <http://particleadventure. org/particleadventure/>.

K. Lee Lerner
Jennifer McGrath

Atomic number

The atomic number of an element is equal to the number of protons in the nucleus of its atom. For example, the nucleus of an **oxygen** atom contains eight protons and eight neutrons. Oxygen's atomic number is, therefore, eight. Since each **proton** carries a single positive charge, the atomic number is also equal to the total positive charge of the atomic nucleus of an element.

The atomic number of an element can be read directly from any **periodic table**. It is always the smaller whole number found in association with an element's symbol in the table. In nuclear **chemistry**, an element's atomic number is written to the left and below the element's symbol; since an element's atomic number can always be determined simply by knowing its symbol, however, the former is often omitted from a nuclear symbol, as in ^{16}O, where the superscript represents the atomic **mass**.

The concept of atomic number evolved from the historic research of Henry Gwyn-Jeffreys Moseley in the 1910s. Moseley bombarded a number of chemical elements with **x rays** and observed the pattern formed by the reflected rays. He discovered that the wavelength of the reflected x rays decreases in a regular predictable pattern with increasing atomic mass. Moseley hypothesized that the regular change in wavelength from element to element was caused by an increase in the positive charge on atomic nuclei in going from one element to the next heavier element.

Moseley's discovery made possible a new understanding of the periodic law first proposed by Dmitri Mendeleev in the late 1850s. Mendeleev had said that the properties of the elements vary in a regular, predictable pattern when the elements are arranged according to their atomic masses. Although he was essentially correct, the periodic table constructed on this basis had three major flaws. Certain pairs of elements (tellurium and iodine constitute one example) appear to be misplaced when arranged according to their masses.

When atomic number, rather than atomic mass, is used to construct a periodic table, these problems disappear. The reason is that an element's chemical properties depend on the number and arrangement of electrons in its **atoms**. The number of electrons in an atom, in turn, is determined by the nuclear charge. It is obvious, then, that the number of protons in a nucleus (or, the nuclear charge, or the atomic number) determines the chemical properties of an element.

See also Element, chemical.

Atomic physics *see* **Physics**

Atomic spectroscopy

Atomic **spectroscopy** is the technique of analyzing the **energy** emitted by **atoms** in order to determine the energy levels of the atom's electrons.

Electrons can have only certain discrete energies. These energies are characteristic of each element; that is, every atom of an element has the same set of available energies. Normally, electrons in atoms are distributed in the lowest energy levels. This is called the "ground state" of the atom. If energy is added to the atom in the form of **light**, **heat**, or **electricity**, the electrons can move to a higher energy level. The electrons are said to be in an "excited" state. When the electrons return to their ground state distribution, they emit the excess energy in the form of light. The light carries the energy that is the difference

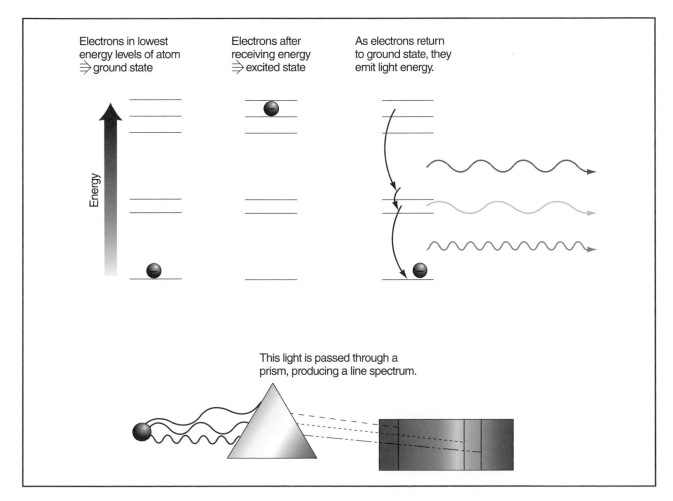

Atomic spectroscopy. *Illustration by Hans & Cassidy. Courtesy of Gale Group.*

between one energy level and another. The distribution of light energies is called a **spectrum** (plural spectra). The study of spectra is called spectroscopy. When light is emitted from an atom, the different colors of light can be seen after they are separated by a device like a **prism** or a **diffraction grating**. If white light passed through a prism or grating, you would see a full rainbow. But when the light emitted by an element passes through, not every **color** is there—only those specific colors corresponding to energy level differences. These correspond to lines in the spectrum.

A line spectrum is very useful in identifying an element because no two elements have the same line spectrum. This is how elements can be identified even on far away stars. The spectrum of light from the **star** is analyzed for the lines of color in its spectrum. These lines can then be matched to known line spectra of elements on **Earth**. The element helium was discovered in this way. Its line spectrum was seen when sunlight was passed through a prism.

See also Electron.

Atomic theory

Atomic theory is the description of **atoms**, the smallest units of elements. The scientific evidence for the existence of atoms and its even smaller constituents is so vast that most people now consider the existence of atoms to be a fact and not just a theory.

History

Beginning in about 600 B.C., many Greek philosophers struggled to understand the nature of **matter**. Some said everything was made of **water**, which comes in three forms (solid **ice**, liquid water, and gaseous steam). Others believed that matter was made entirely of fire in ever-changing forms. Still others believed that whatever comprised matter, it must be something that could not be destroyed but only recombined into new forms. If they could see small enough things, they would find that the same "building blocks" they started with were still there. One of these philosophers was named Democritus. He imagined starting with a large piece of matter and gradually cutting it into smaller and smaller pieces, finally reaching the smallest piece. This tiniest building block that could no longer be cut he named *atomos*, Greek for "no-cut." *Atomos* has been changed in modern times to "atom." The atoms Democritus envisioned differed only in shape and size. In his theory, different objects looked different because of the way the atoms were arranged.

Aristotle, one of the most influential philosophers of that time, believed in some kind of "smallest part" of matter but not with Democritus's descriptions. Aristotle said there were only four elements (earth, air, fire, water) and that these had some smallest unit that made up all matter. Aristotle's teachings against the idea of Democritus's atom were so powerful that the idea of the atom fell out of philosophical fashion for the next 2,000 years.

Although atomic theory was abandoned for this long period, scientific experimentation, especially in **chemistry**, flourished. From the Middle Ages (C. 1100) onward, many **chemical reactions** were studied. By the seventeenth century, some of these chemists began thinking about the reactions they were seeing in terms of smallest parts. They even began using the word atom again. One of the most famous chemists of the end of the eighteenth century was Antoine Lavoisier. His chemical experiments involved very careful weighing of all the chemicals. He reacted various substances until they were in their simplest state. He found two important factors: (1) the simplest substances, which he called elements, could not be broken down any further, and (2) these elements always reacted with each other in the same proportions. The same more complex substances he called compounds. For example, two volumes of **hydrogen** reacted exactly with one **volume** of **oxygen** to produce water. Water could be broken down to always give exactly two volumes of hydrogen and one volume of oxygen. Lavoisier had no explanation for these amazingly consistent results. However, his numerous and careful measurements provided the clue to another chemist named John Dalton.

Dalton realized that if elements were made up of atoms, a different atom for each different element, atomic theory could explain Lavoisier's results. If two atoms of hydrogen always combined with one atom of oxygen, the resulting combination of atoms, called a **molecule**, would be water. Dalton published his explanation in 1803. This year is considered the beginning of modern atomic theory. Scientific experiments that followed Dalton were attempts to characterize how many elements there were, what the atoms of each element were like, how the atoms of each element were the same and how they differed, and, ultimately, whether there was anything smaller than an atom.

Describing characteristics of atoms

One of the first attributes of atoms to be described was relative **atomic weight**. Although a single atom was too small to weigh, atoms could be compared to each other. The chemist Jons Berzelius assumed that equal volumes of gases at the same **temperature** and **pressure** contained equal numbers of atoms. He used this idea to

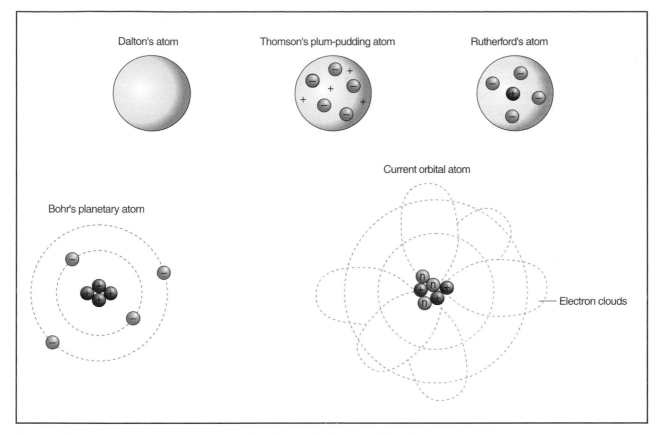

The evolution of atomic theory. *Illustration by Hans & Cassidy. Courtesy of Gale Group.*

compare the weights of reacting gases. He was able to determine that, for example, oxygen atoms were 16 times heavier than hydrogen atoms. He made a list of these relative atomic weights for as many elements as he knew. He devised symbols for the elements by using the first letter or first two letters of their Latin names, a system still in use today. The symbol for hydrogen is H, for oxygen is O, for **sodium** (natrium, in Latin) is Na, and so on. The symbols also proved useful in describing how many atoms combine to form a molecule of a particular compound. For example, to show that water is made of two atoms of hydrogen and one atom of oxygen, the symbol for water is H_2O. One oxygen atom can even combine with one other oxygen atom to produce a molecule of oxygen with the symbol O_2.

As more and more elements continued to be discovered, it became convenient to begin listing them in symbol form in a chart. In 1869, Dmitri Mendeleev listed the elements in order of increasing atomic weight and grouped elements that seemed to have similar chemical reactions. For example, **lithium** (Li), sodium (Na), and potassium (K) are all metallic elements that burst into flame if they get wet. Similar elements were placed in the same column of his chart. Mendeleev began to see a pattern among the elements, where every eighth element

on the atomic weight listing would belong to the same column. Because of this periodicity or repeating pattern, Mendeleev's chart is called the "Periodic table of the elements." The table was so regular, in fact, that when there was a "hole" in the table, Mendeleev predicted that an element would eventually be discovered to fill the place. For instance, there was a **space** for an element with an atomic weight of about 72 (72 times heavier than hydrogen) but no known element. In 1886, 15 years after its prediction, the element Germanium (Ge) was isolated and found to have an atomic weight of 72.3. Many more elements continued to be predicted and found in this way. However, as more elements were added to the **periodic table**, it was found that if some elements were placed in the correct column because of similar reactions, they did not follow the right order of increasing atomic weight. Some other atomic characteristic was needed to order the elements properly. Many years passed before the correct property was found.

As chemistry experiments were searching for and characterizing more elements, other branches of science were making discoveries about **electricity** and **light** that were to contribute to the development of atomic theory. Michael Faraday had done much work to characterize electricity; James Clerk Maxwell characterized light. In the 1870s, William Crookes built an apparatus, now called a Crookes tube, to examine "rays" being given off by metals. He wanted to determine whether the rays were light or electricity based on Faraday's and Maxwell's descriptions of both. Crookes's tube consisted of a **glass** bulb, from which most of the air had been removed, encasing two **metal** plates called electrodes. One electrode was called the **anode** and the other was called the **cathode**. The plates each had a wire leading outside the bulb to a source of electricity. When electricity was applied to the electrodes, rays appeared to come from the cathode. Crookes determined that these cathode rays were particles with a **negative** electrical charge that were being given off by the metal of the cathode plate. In 1897, J. J. Thomson discovered that these negatively charged particles were coming out of the atoms and must have been present in the metal atoms to begin with. He called these negatively charged **subatomic particles** "electrons." Since the electrons were negatively charged, the rest of the atom had to be positively charged. Thomson believed that the electrons were scattered in the atom like raisins in a positively-charged bread dough, or like plums in a pudding. Although Thomson's "plum-pudding" model was not correct, it was the first attempt to show that atoms were more complex than just homogeneous spheres.

At the same time, scientists were examining other kinds of mysterious rays that were coming from the Crookes tube that did not originate at its cathode. In 1895, Wilhelm Roentgen noticed that photographic plates held near a Crookes tube would become fogged by some invisible, unknown rays. Roentgen called these rays "x rays," using "x" for unknown as in **mathematics**. Roentgen also established the use of photographic plates as a way to take pictures of mysterious rays. He found that by blocking the **x rays** with his hand, for instance, bones would block the x rays but skin and **tissue** would not. Doctors still use Roentgen's x rays for imaging the human body.

Photographic plates became standard equipment for scientists of Roentgen's time. One of these scientists, Henri Becquerel, left some photographic plates in a drawer with **uranium**, a new element he was studying. When he removed the plates, he found that they had become fogged. Since there was nothing else in the drawer, he concluded that the uranium must have been giving off some type of ray. Becquerel showed that this **radiation** was not as penetrating as x rays since it could be blocked by **paper**. The element itself was actively producing radiation, a property referred to as radioactivity. Largely through the work of Pierre and Marie Curie, more radioactive elements were found. The attempts to characterize the different types of radioactivity led to the next great chapter in the development of atomic theory.

In 1896, Ernest Rutherford, a student of J. J. Thomson, began studying radioactivity. By testing various elements and determining what kinds of materials could block the radiation from reaching a photographic plate, Rutherford concluded that there were two types of radioactivity coming from elements. He named them using the first two letters of the Greek alphabet, alpha and beta. Alpha radiation was made of positively charged particles about four times as heavy as a hydrogen atom. Beta radiation was made of negatively charged particles that seemed to be just like electrons. Rutherford decided to try an experiment using the alpha particles. He set up a piece of thin gold foil with photographic plates encircling it. He then allowed alpha particles to hit the gold. Most of the alpha particles went right through the gold foil. But a few of them did not. A few alpha particles were deflected from their straight course. A few even came straight backward. Rutherford wrote that it was as surprising as if one had fired a bullet at a piece of tissue paper only to have it bounce back. Rutherford concluded that since most of the alpha particles went through, the atoms of the gold must be mostly empty space, not Thomson's space-filling plum-pudding. Since a few of the alpha particles were deflected, there must be a densely packed positive region in each atom that he called the nucleus. With all the positive charge in the nucleus, the next question was the arrangement of the electrons in the atom.

In 1900, physicist Max Planck had been studying processes of light and **heat**, specifically trying to understand the light radiation given off by a "black-body," an ideal cavity made by perfectly reflecting walls. This cavity was imagined as containing objects called oscillators which absorbed and emitted light and heat. Given enough time, the radiation from such a black-body would produce a colored-light distribution called a **spectrum** that depended only on the temperature of the black-body and not on what it was made of. Many scientists attempted to find a mathematical relationship that would predict how the oscillators of a black-body could produce a particular spectral distribution. Max Planck found that correct mathematical relationship. He assumed that the **energy** absorbed or emitted by the oscillators was always a multiple of some fundamental "packet of energy" he called a quantum. Objects that emit or absorb energy do it in discrete amounts, called quanta.

At this same time, there was a physicist working with Thomson and Rutherford named Niels Bohr. Bohr realized that the idea of a quantum of energy could explain how the electrons in the atom are arranged. He described the electrons as being "in orbit" around the nucleus like planets around the **sun**. Like oscillators in a black-body could not have just any energy, electrons in the atom could not have just any **orbit**. There were only

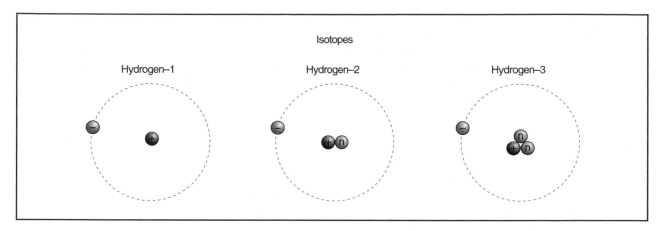

Isotopes

Hydrogen–1   Hydrogen–2   Hydrogen–3

Hydrogen's isotopes: hydrogen, deuterium, and tritium. *Illustration by Hans & Cassidy. Courtesy of Gale Group.*

certain distances that were allowed by the energy that an **electron** had. If an electron of a particular atom absorbed the precisely right quantum of energy, it could move farther away from the nucleus. If an electron farther from the nucleus emitted the precisely right quantum of energy, it could move closer to the nucleus. What the precisely right values were differed for every element. These values could be determined by a process called **atomic spectroscopy**, an experimental technique that looked at the light spectrum produced by atoms. An atom was heated so that all of its electrons were moved far away from the nucleus. As they moved closer to the nucleus, the electrons would begin emitting their quanta of energy as light. The spectrum of light produced could be examined using a **prism**. The spectrum produced in this way did not show every possible **color**, but only those few that matched the energies corresponding to the electron orbit differences. Although later refined, Bohr's "planetary model" of the atom explained atomic **spectroscopy** data well enough that scientists turned their attention back to the nucleus of the atom.

Rutherford, along with Frederick Soddy, continued work with radioactive elements. Soddy, in particular, noticed that as alpha and beta particles were emitted from atoms, the atoms changed in one of two ways: (1) the element became a totally different element with completely new chemical reactions, or (2) the element maintained the same chemical reactions and the same atomic spectrum but only changed in atomic weight.

He called atoms of the second group isotopes, atoms of the same element with different atomic weights. In any natural **sample** of an element, there may be several types of isotopes. As a result, the atomic weight of an element that was calculated by Berzelius was actually an average of all the **isotope** weights for that element. This was the reason that some elements did not fall into the correct order on Mendeleev's periodic table—the average atomic weight depended on how much of each kind of isotope was present. Soddy suggested placing the elements in the periodic table by similarity of chemical reactions and then numbering them in order. The number assigned to each element in this way is called the **atomic number**. The atomic numbers were convenient ways to refer to elements.

Meanwhile, Thomson had continued his work with the Crookes tube. He found that, not only were cathode rays of electrons produced, but so were positive particles. After much painstaking work, he was able to separate the many different kinds of positive particles by weight. Based on these measurements, he was able to determine a fundamental particle, the smallest positive particle produced, called a **proton**. Since these were being produced by the atoms of the cathode and since Rutherford showed that the nucleus of the atom was positive, Thomson realized that the nucleus of an atom must contain protons. A young scientist named Henry Moseley experimented with bombarding atoms of different elements with x rays. Just as in atomic spectroscopy, where heat gives electrons more energy, x rays give protons in the nucleus more energy. And just as electrons give out light of specific energies when they cool, the nucleus emits x rays of a specific energy when it "de-excites." Moseley discovered that the energy of the emitted x rays for every element followed a simple mathematical relationship. The energy depended on the atomic number for that element, and the atomic number corresponded to the number of positive charges in the nucleus. So the correct ordering of the periodic table is by increasing number of protons in the atomic nucleus. The number of protons equals the number of electrons in a neutral atom. The electrons are responsible for the chemical reactions. Elements in the same column of the periodic table have similar arrangements of electrons with the highest energies, and this is why their reactions are similar.

Only one problem remained. Electrons had very little weight, 1/1,836 the weight of a proton. Yet the protons did not account for all of the atomic weight of an atom. It was not until 1932 that James Chadwick discovered the existence of a particle in the nucleus with no electrical charge but with a weight slightly greater than a proton. He named this particle the **neutron**. Neutrons are responsible for the existence of isotopes. Two atoms of the same element will have the same number of protons and electrons but they might have different numbers of neutrons and therefore different atomic weights. Isotopes are named by stating the name of the element and then the number of protons plus neutrons in the nucleus. The sum of the protons and neutrons is called the **mass number**. For example, uranium-235 has 235 protons and neutrons. We can look on a periodic table to find uranium's atomic number (92) which tells us the number of protons. Then by subtracting, we know that this isotope has 143 neutrons. There is another isotope of uranium, ^{238}U, with 92 protons and 146 neutrons. Some combinations of protons and neutrons are less stable than others. Picture trying to hold 10 bowling balls in your arms. There will be some arrangement where you might be able to manage it. Now try holding 11 or only nine. There might not be a stable arrangement and you would drop the bowling balls. The same thing happens with protons and neutrons. Unstable arrangements spontaneously fall apart, emitting particles, until a stable structure is reached. This is how radioactivity like alpha particles is produced. Alpha particles are made of two protons and two neutrons tumbling out of an unstable nucleus.

Hydrogen has three kinds of isotopes: hydrogen, ^2H (deuterium), and ^3H (tritium).

The atomic weights of the other elements were originally compared to hydrogen without specifying which isotope. It is also difficult to get single atoms of hydrogen because it usually reacts with other atoms to form molecules like H_2 or H_2O. So a different element's isotope was chosen for comparison. The atomic weights are now based on ^{12}C (carbon-12). This isotope has six protons and six neutrons in its nucleus. Carbon-12 was defined to be 12 atomic **mass** units. (Atomic mass units, abbreviated amu, are units used to compare the relative weights of atoms. One amu is less than 200 sextillionths of a gram.) Every other isotope of every other element is compared to this. Then the weights of a given element's isotopes are averaged to give the atomic weights found on the periodic table.

Until this point in the story of the atom, all of the particles comprising the atom were thought of as hard, uniform spheres. Beginning in 1920 with the work of Louis de Broglie, this image changed. De Broglie showed that particles like electrons could sometimes have properties of waves. For instance, if water waves are produced by two sources, like dropping two pebbles

into a pond, the waves can interfere with each other. This means that high spots add to make even higher spots. Low spots add to make even lower regions. When electrons were made to travel through a double slit, with some electrons going through one slit and some through the other, they effectively created two sources. The electrons showed this same kind of **interference**, producing a pattern on a collection plate. The ability of electrons and other particles to sometimes show properties of particles and sometimes of waves is called wave-particle duality. This complication to the nature of the electron meant that Bohr's idea of a planetary atom was not quite right. The electrons do have different discrete energies, but they do not follow circular orbits. In 1925, Werner Heisenberg stated that the precise speed and location of an electron cannot both be known at the same time. This **"Heisenberg uncertainty principle"** inspired Erwin Schrödinger to devise an equation to calculate how an electron with a certain energy moves. Schrödinger's equation describes regions in an atom where an electron with a certain energy is likely to be but not exactly where it is. This region of probability is called an orbital. Electrons move so fast within these orbitals that we can think of them as blurring into an **electron cloud**. Electrons move from one orbital into another by absorbing or emitting a quantum of energy, just as Bohr explained.

Applications of atomic theory

Early studies of radioactivity revealed that certain atomic nuclei were naturally radioactive. Some scientists wondered that if particles could come out of the nucleus, would it also be possible to force particles into the nucleus? In 1932, Cockcroft and Walton succeeded in building a particle accelerator, a device that could make streams of charged particles move faster and faster. These fast particles, protons for example, were then aimed at a thin plate of a lighter element like lithium (Li). If a lithium atom nucleus "captures" a proton, the nucleus becomes unstable and breaks apart into two alpha particles. This technique of inducing radioactivity by bombardment with accelerated particles is still the most used method of studying nuclear structure and subatomic particles. Today, **accelerators** race the particles in straight lines or, to save land space, in ringed paths several miles in diameter.

The spontaneous rearrangement of the atomic nucleus always results in a release of energy in the form of kinetic **motion** in fast-moving neutrons. When a large nucleus falls apart to form smaller atoms, the process is called fission. When lighter atoms are forced together to produce a heavier atom, the process is called fusion. In either case, fast neutrons are released. These can transfer their kinetic energy to the surroundings, heating it. This heat can be used to boil water, producing steam to run a

KEY TERMS

Accelerator—A device that causes particles to move faster and faster.

Alpha particle—Two protons and two neutrons bound together and emitted from the nucleus during some kinds of radioactive decay.

Atomic mass—The mass of an atom relative to carbon-12, ^{12}C (which has a mass of exactly 12 atomic mass units); also the mass, in grams, of an element that contains one mole of atoms.

Atomic mass unit (u or amu)—A unit used to express the mass of atoms equal to exactly one-twelfth of the mass of carbon-12.

Beta particle—One type of radioactive decay particle emitted from radioactive atomic nuclei. A beta particle is the same thing as an electron.

Electrode—A metal plate that carries electrical current.

Electron cloud—The image of an electron moving so fast that it seems to fill a region of space.

Interference—The combination of waves in which high spots combine to give even higher spots and low spots combine to give even lower spots.

Kinetic energy—The energy of a moving object.

Mass number—The sum of protons and neutrons.

Nucleus—The dense central part of an atom containing the protons and neutrons; plural is nuclei.

Orbital—The region of probability within an atom where an electron with a particular energy is likely to be.

Oscillators—Objects that can absorb or emit energy and convert it into kinetic energy.

Periodicity—Repeatability of a pattern.

Quantum—The amount of radiant energy in the different orbits of an electron around the nucleus of an atom.

Quarks—Believed to be the most fundamental units of protons and neutrons.

Uncertainty principle—Heisenberg's statement that both the position and velocity of a particle cannot be known with equal precision at the same time.

Wave-particle duality—The ability of objects to show characteristics of both waves and particles.

---

turbine that turns an electric **generator**. Fusion is the process occurring in the center of the Sun and other stars. So much energy can be released quickly that the process has also been used for the hydrogen bomb. However, fusion is not yet controlled enough for running a power plant. Research continues to find a controlled method of using fusion energy.

On the other hand, fission reactions have also been used for very powerful weapons. The first atomic bomb was detonated in 1945. Since then, however, fission energy has also been controlled enough to operate the many **nuclear power** plants around the world.

While an atom is the smallest part of an element which still is that element, atoms are not the smallest particles that exist. Even the protons and neutrons in the atomic nucleus are believed to made of even smaller particles called **quarks**. Current research in atomic **physics** focuses on describing the internal structure of atoms. By using particle accelerators, scientists are trying to characterize quarks which may combine in a number of ways to produce other types of subatomic particles.

No one has ever seen a single atom even with the best optical microscopes. Special types of microscopes called scanning **tunneling** microscopes and atomic force microscopes make use of the forces produced by the electrons to obtain images of the electron **clouds**. These clouds indicate how atoms are arranged but we cannot "see" through the cloud to the nucleus. Because of the limitations of size, we will never see an atom with our own eyes. Everything we know about atoms must be deduced from larger-scale experiments. As a result, the description of atoms is still called a theory. However, this theory explains atomic experiments so well that we usually think of the existence of atoms as a fact.

See also Element, chemical; Nuclear fission; Nuclear fusion.

Eileen Korenic

Atomic weight

Atoms are exceedingly small, so small that actual weights of atoms were not able to be determined until early in the twentieth century. The weight of an atom of oxygen-16 (an **oxygen** atom with eight neutrons in the nucleus) was found to be 2.657×10^{-23} grams and an atom of carbon-12 (a **carbon** atom with six neutrons in the nucleus) was found to weigh 1.99×10^{-23} grams. Because these units are so very small, they are not practical and are seldom used in everyday laboratory work. Rather, the weight of an atom is usually calculated in units other than grams, one that is closer to the size of the particle being weighed and is therefore more practical.

The table of atomic weights is based on a unit called an atomic **mass** unit, abbreviated u, or in older notation, amu. This unit is defined as 1/12 the mass of carbon-12 (^{12}C) and is equal to 1.6606×10^{-24} grams. On this scale, carbon-12 weighs exactly 12 atomic mass units. But because even the smallest amount of **matter** contains enormous numbers of atoms, atomic weights are usually interpreted to mean grams of an element rather than atomic mass units. When interpreted in grams, the atomic weight of an element represents 6.02×10^{23} atoms, which is defined as one **mole**. Thus, the atomic weight in grams is the mass of an element that contains one mole or 6.02×10^{23} atoms.

Atomic weights are actually atomic masses but historically they were called atomic weights because the method used to determine them was called weighing. This terminology has persisted and is more familiar to most people even though the values obtained are actually atomic masses.

History

Although the **atomic theory** of matter, in its various forms, existed a good two thousand years before the time of John Dalton, he was the first to propose, in his 1808 book *A New System of Chemical Philosophy*, that atoms had weight. Atoms, as Dalton defined them, were hard, solid, indivisible particles with no inner spaces, rather than something that could not be seen, touched, or tasted. They were indestructible and preserved their identities in all **chemical reactions**. Furthermore, each kind of element had its own specific kind of atom different from the atoms of other elements. These assumptions led him to propose that atoms were tangible matter and therefore had weight.

Because atoms were much too small to be seen or measured by any common methods, absolute weights of atoms could not be determined. Rather, these first measurements were made by comparing weights of various atoms to **hydrogen**. Hydrogen was chosen as the unit of comparison because it was the lightest substance known and the weights of the other elements would be very close to whole numbers.

The weight of oxygen could then be calculated because of earlier work by Humboldt and Gay-Lussac, who

found that **water** consisted of only two elements, hydrogen and oxygen, and that there were eight parts of oxygen for every one part of hydrogen. Lacking any knowledge about how many atoms of hydrogen and oxygen combine in a **molecule** of water, Dalton again had to make some assumptions. He assumed that nature is basically very simple and, therefore, one atom of hydrogen combines with only one atom of oxygen. Using this hypothesis and the fact that hydrogen was assigned a weight of one unit, it follows that oxygen, which is eight times heavier than hydrogen, would have a weight of eight units. Of course, if the **ratio** between hydrogen and oxygen in water were not one to one, but some other ratio, the weight of oxygen would have to be adjusted accordingly. Dalton used experimental results and similar reasoning to prepare the very first Table of Atomic Weights, but because of the lack of knowledge about the real formulas for substances, many of the weights were incorrect and had to be modified later.

Knowledge about absolute formulas of substances came mainly from the work of two chemists. In 1809, Gay-Lussac observed that gases react with each other in very simple proportions. For example, at the same **temperature** and **pressure**, two volumes of hydrogen react with one **volume** of oxygen and form two volumes of water. Then in 1811, Amedeo Avogadro proposed that equal volumes of gases have the same number of particles if measured at the same temperature and pressure. The difficulty of explaining how one volume of oxygen could form two volumes of water without violating the current theory that atoms were indivisible was not resolved until the 1850s when Avogadro's explanation that molecules of gases, such as hydrogen and oxygen, existed as diatomic molecules (molecules with two atoms joined together) was finally accepted. If each oxygen molecule was composed of two oxygen atoms, then it was the molecule and not the atom that split apart to form two volumes of water.

Although other scientists contributed to knowledge about atomic weights, much of the experimental work that was used to improve the Table of Atomic Weights was done by J. J. Berzelius who published his list of the weights of 54 elements in 1828. Unlike Dalton's atomic weights, the weights published by Berzelius match quite well the atomic weights used today.

So far, all the knowledge about atomic weights was relative to the weight of hydrogen as one unit. The first of the experiments to uncover knowledge about the absolute weight of parts of the atom were done early in the twentieth century by J. J. Thomson and Robert Millikan. Thomson studied rays of **negative** particles (later discovered to be electrons) in partially evacuated tubes. He measured how the beam deflected or bent when placed in a magnetic field and used this information to calculate mathematically the ratio of the charge on the **electron** to the mass of the electron.

Millikan devised a clever experiment in which he produced a very fine spray of charged oil droplets and allowed them to fall between two charged plates. By adjusting the charge on the plates as he observed the droplets under a **microscope**, he was able to suspend the droplets midway between the two plates and with this information calculate the charge of the electron. Now, along with the charge/mass ratio calculated by Thomson, the mass of the electron could be calculated and was found to equal 9.11×10^{-28} grams (a decimal with 27 zeros before the 9).

About five years later, the charge and mass of the **proton** were calculated. The proton was found to weigh 1.6726×10^{-24} grams or about 1,836 times as much as an electron. Because most atoms (hydrogen being the only exception) were heavier than would be expected from the number of protons they had, it was known that there must be another neutral particle in the atom. Because of the difficulty in observing neutral particles, the **neutron** was not discovered until 1932 by James Chadwick. The mass was found to be 1.6749×10^{-24} grams, about the same as the mass of the proton.

Isotopes

The atomic weight represents the sum of the masses of the particles that make up the atom, protons, neutrons, and electrons. But since the mass of the electron is so small and essentially all the weight of the atom comes from the protons and neutrons, the atomic weight is considered to represent the sum of the masses of the protons and neutrons present in the atom. These weights were given in relative units called atomic mass units (abbreviated u or, in older notation, amu) in which the protons and neutrons have nearly equal masses. Consequently, the sum of the protons and neutrons in the nucleus would be the same as the atomic weight of the atom.

Today, a very sophisticated instrument, called a mass spectrometer, is used to obtain accurate measurements of atomic masses. In this instrument, atoms are vaporized and then changed to positively charged particles by knocking off electrons. These charged particles are passed through a magnetic field which causes them to be deflected different amounts, depending on the size of the charge and mass. The particles are eventually deposited on a detector plate where the amount of deflection can be measured and compared with the charge. Very accurate relative masses are determined in this way.

When atoms of various elements were analyzed with the mass spectrometer, scientists were surprised to find that not all atoms of the same element had exactly the same mass. Oxygen, for example, was found to exist in three different forms, each differing by one atomic mass unit or about the mass of one proton or one neutron. Since the number of protons in the nucleus was known because of their association with a +1 charge, the three different masses for oxygen had to be caused by different numbers of neutrons in the nucleus. Atoms of this type were called isotopes. Both the identity of the element (since the number of protons remains the same) and the chemical properties (since the electrons remain unchanged) are identical in isotopes of the same element. However, the mass is different because of the different number of neutrons in the nucleus, and this sometimes makes the atom unstable and radioactive. Radioactive isotopes are frequently used in research because the radioactivity can be followed using a Geiger counter. They can be administered to living systems like plants or animals and the **isotope** is observed as it moves and reacts throughout the system. Oxygen has three isotopes with masses of 16, 17, and 18 (often written as oxygen-16, oxygen-17, and oxygen-18 [^{16}O, ^{17}O, ^{18}O]). Similarly, carbon exists as carbon-12, carbon-13, and carbon-14 and hydrogen as hydrogen-1, hydrogen-2, and hydrogen(^{12}C, ^{13}C, ^{14}C)-3(H, ^{2}H, ^{3}H). Each of these successive isotopes have one more neutron in the nucleus than the preceding one.

Interpretation of atomic weights

Early work on atomic weights used naturally occurring oxygen, with an assigned atomic weight of exactly 16 as the basis for the scale of atomic weights. All other atomic weights were found in relation to it. Confusion arose when, in 1929, the three isotopes of oxygen were discovered. In 1961, it was finally decided to adopt carbon-12 as the basis for all other atomic weights. Under this system still in use today, the atomic weight of carbon-12 is taken to be exactly 12 and the atomic mass unit is defined as exactly one-twelfth the mass of carbon-12. All other atomic weights are measured in relation to this unit.

When examining the table of atomic weights, it is found that the weight of carbon is not given as exactly 12 as would be expected, but rather 12.01. The reason is that the weights used in the table represent the average weight of the isotopes of carbon that are found in a naturally occurring **sample**. For example, most of the carbon found in nature, 98.89% of it to be exact, is carbon-12 and has a weight of exactly 12. The rest of it (1.11%) is carbon-13 (with an atomic weight of 13.00) and carbon-14 (which exists in quantities too minute to affect this calculation). The atomic weight of carbon is calculated by taking 98.89% of the weight of carbon-12 and 1.11% of the weight of carbon-13 to give 12.0112. All weights in the table of atomic weights are calculated by using the percentage of each isotope in a naturally occurring sample.

Because atoms are so small, making it impossible for chemists to observe or weigh them, the weights of individual atoms are not very useful for experimentation. Very large numbers of atoms are involved in even the tiniest samples of matter. It is important to match the unit that is used to make a measurement to the size of the thing being measured. For example, it is useful to measure the length of a room in feet rather than miles because the unit, foot, corresponds to the length of a room. One would not measure the **distance** to London or Paris or to the **sun** in inches or feet because the distance is so large in relation to the size of the unit. Miles would be a much more appropriate unit.

A new unit, called a mole, was created as a more useful unit for working with atoms. A mole is a counting number much like a dozen. A dozen involves 12 of anything, 12 books, 12 cookies, 12 pencils, etc. Similarly, a mole involves 6.02×10^{23} (602 with 21 zeros after it) particles of anything. The mole is such a large number that it is not a useful measurement for anything except counting very, very small particles, too small to even imagine. For example, if a mole of dollars were divided evenly among all the people of the world (5.5 billion), every single person alive would receive 1.09×10^{14} dollars! That is enough money to last nearly 300 years if a billion dollars were spent every single day of the year. Yet a mole of carbon atoms is contained in a chunk of **coal** about as big as a marble.

Needless to say, atoms cannot be counted in the same way that cookies or books are counted. But they can be counted by weighing, and the mole is the unit that can express this quantity. If a ping-pong ball weighs one ounce, then 12 ounces of ping-pong balls would contain 12 balls. Twenty ounces would contain 20 balls. If golf balls weigh four ounces each, then 48 ounces are needed in order to obtain 12 balls, and 80 ounces are needed in order to obtain 20 balls. Since the golf ball weighs four times as much as the ping-pong ball, it is easy to obtain equal numbers of these two balls by weighing four times as much for the golf balls as you weigh for the ping-pong balls. Actually, it is easier to count small numbers of ping-pong balls or golf balls than to weigh them. But if 10 or 20 or 30 thousand of them were needed, it would be much easier figure the weight of the balls and weigh them than it would be to count them.

Likewise, the weighing method is more useful and, in fact, is the only method by which atoms can be counted. It was discovered in the early 1800s, mostly through the work of Amadeo Avogadro, that when the atomic weight of an atom is interpreted in grams rather than atomic mass units, the number of atoms in the sample is always 6.02×10^{23} atoms or a mole of atoms. Thus, 12 grams of carbon contain one mole of carbon atoms and 16 grams of oxygen contain one mole of oxygen atoms. One mole of the lightest atom, hydrogen, weighs just one gram and one mole of the heaviest of the naturally occurring elements, **uranium**, weighs 238 grams.

Molecules are particles made up of more than one atom. The weight of the molecule, called the **molecular weight**, can be found by adding the atomic weights of each of the atoms that make up the molecule. Water is a molecule with a formula, H_2O. It is composed of two atoms of hydrogen, with an atomic weight of one, and one atom of oxygen with an atomic weight of 16. Water, therefore has a molecular weight of 18. When this molecular weight is interpreted as 18 atomic mass units, it represents the weight of one molecule in relation to one-twelfth of carbon-12. When the molecular weight is interpreted as 18 grams (less than 400 drops of water), it represents the weight of one mole or 6.02×10^{23} molecules of water. Similarly, the molecular weight of **carbon dioxide** (CO_2) is 44 atomic mass units or 44 grams. A chunk of solid carbon dioxide (known as dry **ice**) about the size of a baseball contains one mole of molecules. If this chunk were allowed to change to a gas at room conditions of temperature and pressure, this mole of carbon dioxide would take up slightly over a cubic foot.

Uses

When new substances are found in nature or are produced in the laboratory, the first thing chemists try to determine is the chemical formula for the substance. This new compound, a substance made of two or more kinds of atoms, is analyzed to find what elements it is composed of. This is usually done by chemically separating the compound into its elements and then determining how much of each element was present. Chemical formulas tell how many atoms are in a compound, not the amount of mass. So the mass of each element must be expressed as a part of a mole by comparing it to the atomic weight. When expressed in this manner, the quantity is a way of representing how many atoms are present in the compound. These numbers of moles are expressed as ratios, reduced to the lowest whole numbers and then combined with the symbols for the elements to represent the simplest chemical formula.

Companies that produce raw materials or manufacture goods use atomic and molecular weights to help determine the amounts of reactants needed to produce a given amount of product. Or they can determine how much product they can produce from a given amount of reactant. Once again, the quantities involved in chemical reactions depend on how many atoms or molecules react, not on the amount of mass of each. So the known amount of reactant or product must be expressed as a part of a mole by comparing it to the molecular weight.

KEY TERMS

. . . .

Atomic mass—The mass of an atom relative to carbon-12 (which has a mass of exactly 12 atomic mass units); also the mass, in grams, of an element that contains one mole of atoms.

Atomic mass unit (u or amu)—A unit used to express the mass of atoms equal to exactly one-twelfth of the mass of carbon-12.

Molecule—The smallest particle of a compound that can exist, formed when two or more atoms join together to form a substance.

Although other factors are involved in these determinations, this quantity, along with the balanced equation for the chemical reaction, allows chemists to figure out how much of any other reactant or product is involved in the reaction. Calculations of this type can save manufacturers many dollars because the amounts of chemicals needed to manufacture a product can be accurately determined. If a billion tires are produced in one year and one penny can be saved on each tire by not using more of a substance than can be reacted, it would be a substantial savings to the company of $10,000,000 per year.

See also Avogadro's number; Mass spectrometry; Periodic table.

Resources

Books

Brock, William H. *The Norton History of Chemistry.* New York: W. W. Norton & Company, 1993.

Feather, Ralph M. et al. *Science Connections.* Columbus, OH: Merrill Publishing Company, 1990.

Leona B. Bronstein

Atoms

Atoms are the smallest particles of **matter** that have distinct physical and chemical properties. Each different type of atom makes up an element which is characterized by an **atomic weight** and an atomic symbol. Since the **atomic theory** was first proposed in the early nineteenth century, scientists have discovered a number of **subatomic particles**.

The development of the atomic theory traces its history to early human civilizations. To these people, change was a concept to ponder. Ancient Greek philosophers tried to explain the causes of changes in their environment, typically, chemical changes. This led them to propose a variety to ideas about the nature of matter. By 400 B.C., it was believed that all matter was made up of four elements including earth, fire, air and **water**. At around this time, Democritus proposed the idea of matter being made up of small indivisible particles. He called these particles *atomos*, or atoms. While Democritus may have suggested the theory of atoms, the Greeks had no experimental method for testing his theory.

This experimental method was suggested by Robert Boyle in the seventeenth century. At this time, he advanced the idea that matter existed as elements which could not be broken down further. Scientists built on Boyle's ideas, and in the early nineteenth century, John Dalton proposed the atomic theory.

Dalton's theory had four primary postulates. First, he suggested that all elements are made up of tiny particles called atoms. Second, all atoms of the same element are identical. Atoms of different elements are different in some fundamental way. Third, chemical compounds are formed when atoms from different elements combine with each other. Finally, **chemical reactions** involve the reorganization of the way atoms are bound. Atoms themselves do not change.

Using Dalton's theory, scientists investigated the atom more closely. They wanted to determine the structure of these atoms. The first subatomic particle was discovered by J. J. Thomson (1856-1940) in the late nineteenth century. Using a **cathode ray tube** he discovered negatively charged particles called electrons. Around this same time, scientists began to find that certain atoms produced radioactivity. In 1911, Ernest Rutherford (1871-1937) proposed the idea that atoms had a nucleus which the electrons orbited around. This led to the discovery of positively charged protons and neutral particles called neutrons.

Over time, scientists developed a chart known as the **periodic table** of elements to list all known elements. Atoms on this chart are symbolized by abbreviations called the atomic symbol. For example, **oxygen** atoms are denoted by the letter O. Each atom also has a unique **mass** denoted by its atomic weight. The **atomic number** is also distinct to each type of atom denoting the number of protons in their nucleus.

While atoms generally contain the same number of protons as neutrons, this is not always the case. Atoms which have more or less neutrons than protons are known as isotopes. For example, **carbon** atoms can have 12, 13 or 14 neutrons. When a nucleus has too many neutrons, as in the case of carbon-14, it is unstable and gives off **radiation** which can be measured. Radioactive isotopes have found many useful applications in **biology**. Scientists have used them in **radioactive dating** to determine the age of fossils. They have also used them as tracer atoms to follow a chemical as it goes through metabolic processes in an **organism**. This has made them an important tool in medicine.

ATP *see* **Adenosine triphosphate**

Attention-deficit/ hyperactivity disorder (ADHD)

Attention deficit/hyperactivity disorder (ADHD), also known as hyperkinetic disorder (HKD) outside of the United States, is estimated to affect 3–9% of children, and afflicts boys more often than girls. Although difficult to assess in infancy and toddlerhood, signs of ADHD may begin to appear as early as age two or three, but the symptom picture changes as adolescence approaches. Many symptoms, particularly hyperactivity, diminish in early adulthood, but impulsivity and inattention problems remain with up to 50% of ADHD individuals throughout their adult life.

Children with ADHD have short attention spans, becoming easily bored or frustrated with tasks. Although they may be quite intelligent, their lack of focus frequently results in poor grades and difficulties in school. ADHD children act impulsively, taking action first and thinking later. They are constantly moving, running, climbing, squirming, and fidgeting, but often have trouble with gross and fine motor skills and, as a result, may be physically clumsy and awkward. Their clumsiness may extend to the social arena, where they are sometimes shunned due to their impulsive and intrusive **behavior**.

Causes and symptoms

The causes of ADHD are not known. However, it appears that heredity plays a major role in the development of ADHD, with many researchers assuming that ADHD is due to a genetic defect that results in altered **brain biochemistry**. Children with an ADHD parent or sibling are more likely to develop the disorder themselves. Before **birth**, ADHD children may have been exposed to poor maternal **nutrition**, viral infections, or maternal substance abuse. In early childhood, exposure to lead or other toxins can cause ADHD-like symptoms. Traumatic brain injury or neurological disorders may also trigger ADHD symptoms. Although the exact cause of ADHD is not known, an imbalance of certain neurotransmitters, the chemicals in the brain that transmit messages between nerve cells, is believed to be the mechanism behind ADHD symptoms. In 1990, a study by researchers at the National Institute for Mental Health documented the neurobiological effects of ADHD through brain imaging. The results showed that the **rate** at which the brain uses glucose, its main **energy** source, was shown to be lower in persons with ADHD, especially in the portion of the brain that is responsible for attention, handwriting, motor control and inhibition responses.

A widely publicized study conducted in the early 1970s suggested that allergies to certain foods and food additives caused the characteristic hyperactivity of ADHD children. Although some children may have adverse reactions to certain foods that can affect their behavior (for example, a rash might temporarily cause a child to be distracted from other tasks), carefully controlled follow-up studies have uncovered no link between food allergies and ADHD. Another popularly held misconception about food and ADHD is that the consumption of sugar causes hyperactive behavior. Again, studies have shown no link between sugar intake and ADHD. It is important to note, however, that a nutritionally balanced diet is important for normal development in *all* children.

Diagnosis is based on a collaborative process that involves affected children, psychiatrists or other physicians, the child's family and school. Deciding what treatment will best benefit the affected child requires a careful diagnostic assessment after a comprehensive evaluation of psychiatric, social, cognitive, educational, family and medical/neurological factors. A thorough evaluation can take several hours and may require more than one visit to a physician. Treatment follows only after the evaluation is made.

Psychologists and other mental health professionals typically use the criteria listed in the *Diagnostic and Statistical Manual of Mental Disorders, Fourth Edition, Text Revision (DSM-IV-TR)* as a guideline for determining the presence of ADHD. For a diagnosis of ADHD, *DSM-IV-TR* requires the presence of at least six of the following symptoms of inattention, or six or more symptoms of hyperactivity and impulsivity combined:

Inattention:

• Fails to pay close attention to detail or makes careless mistakes in schoolwork or other activities.

• Has difficulty sustaining attention in tasks or activities.

• Does not appear to listen when spoken to.

• Does not follow through on instructions and does not finish tasks.

• Has difficulty organizing tasks and activities.

• Avoids or dislikes tasks that require sustained mental effort (e.g., homework).

- Is easily distracted.
- Is forgetful in daily activities.

    Hyperactivity:

- Fidgets with hands or feet or squirms in seat.
- Does not remain seated when expected to.
- Runs or climbs excessively when inappropriate (in adolescence and adults, feelings of restlessness).
- Has difficulty playing quietly.
- Is constantly on the move.
- Talks excessively.

    Impulsivity:

- Blurts out answers before the question has been completed.
- Has difficulty waiting for his or her turn.
- Interrupts and/or intrudes on others.

DSM-IV-TR also requires that some symptoms develop before age seven, and that they significantly impair functioning in two or more settings (e.g., home and school) for a period of at least six months. Children who meet the symptom criteria for inattention, but not for hyperactivity/impulsivity are diagnosed with attention-deficit/ hyperactivity disorder, predominantly inattentive type, commonly called ADD. (Young girls with ADHD may not be diagnosed because they have mainly this subtype of the disorder.)

Diagnosis

The first step in determining if a child has ADHD is to consult with a pediatrician. The pediatrician can make an initial evaluation of the child's developmental maturity compared to other children in his or her age group. The physician should also perform a comprehensive physical examination to rule out any organic causes of ADHD symptoms, such as an overactive thyroid or **vision** or **hearing** problems.

If no organic problem can be found, a psychologist, psychiatrist, neurologist, neuropsychologist, or **learning** specialist is typically consulted to perform a comprehensive ADHD assessment. A complete medical, family, social, psychiatric, and educational history is compiled from existing medical and school records and from interviews with parents and teachers. Interviews may also be conducted with the child, depending on his or her age. Along with these interviews, several clinical inventories may also be used, such as the Conners Rating Scales (Teacher's Questionnaire and Parent's Questionnaire), Child Behavior Checklist (CBCL), and the Achenbach Child Behavior Rating Scales. These inventories provide valuable information on the child's behavior in different settings and situations. In addition, the Wender Utah Rating Scale has been adapted for use in diagnosing ADHD in adults.

It is important to note that mental disorders such as **depression** and **anxiety** disorder can cause symptoms similar to ADHD. A complete and comprehensive psychiatric assessment is critical to differentiate ADHD from other possible mood and behavioral disorders. Bipolar disorder, for example, may be misdiagnosed as ADHD.

Public schools are required by federal law to offer free ADHD testing upon request. A pediatrician can also provide a referral to a psychologist or pediatric specialist for ADHD assessment. Parents should check with their insurance plans to see if these services are covered.

Treatment

Psychosocial therapy, usually combined with medications, is the treatment approach of choice to alleviate ADHD symptoms. Psychostimulants, such as dextroamphetamine (Dexedrine), pemoline (Cylert), and methylphenidate (Ritalin) are commonly prescribed to control hyperactive and impulsive behavior and increase attention span. They work by stimulating the production of certain neurotransmitters in the brain. Possible side effects of stimulants include nervous tics, irregular heartbeat, loss of appetite, and **insomnia**. However, the medications are usually well-tolerated and safe in most cases.

In children who don't respond well to stimulant therapy, tricyclic antidepressants such as desipramine (Norpramin, Pertofane) and amitriptyline (Elavil) are frequently recommended. Reported side effects of these drugs include persistent dry mouth, sedation, disorientation, and cardiac arrhythmia (particularly with desipramine). Other medications prescribed for ADHD therapy include buproprion (Wellbutrin), an antidepressant; fluoxetine (Prozac), an SSRI antidepressant; and carbamazepine (Tegretol, Atretol), an anticonvulsant drug. Clonidine (Catapres), an antihypertensive medication, has also been used to control aggression and hyperactivity in some ADHD children, although it should not be used with Ritalin. A child's response to medication will change with age and maturation, so ADHD symptoms should be monitored closely and prescriptions adjusted accordingly.

Behavior modification therapy uses a reward system to reinforce good behavior and task completion and can be implemented both in the classroom and at home. A variation on this is cognitive-behavioral therapy. This decreases impulsive behavior by getting the child to recognize the connection between thoughts and behavior, and to change behavior by changing negative thinking patterns. **Individual** psychotherapy can help an ADHD child build self-esteem, give them a place to discuss their worries and anxieties, and help them gain insight into their behavior and feelings. Family therapy may also be beneficial in helping family members develop coping skills and in working through feelings of guilt or anger parents may be experiencing.

ADHD children perform better within a familiar, consistent, and structured routine with positive reinforcements for good behavior and real consequences for bad. Family, friends, and caretakers should all be educated on the special needs and behaviors of the ADHD child. Communication between parents and teachers is especially critical to ensuring an ADHD child has an appropriate learning environment.

Alternative treatment

A number of alternative treatments exist for ADHD. Although there is a lack of controlled studies to prove their efficacy, proponents report that they are successful in controlling symptoms in some ADHD patients. Some of the more popular alternative treatments include:

- EEG (electroencephalograph) **biofeedback**. By measuring brainwave activity and teaching the ADHD patient which type of brainwave is associated with attention, EEG biofeedback attempts to train patients to generate the desired brainwave activity.

- Dietary therapy. Based in part on the Feingold food **allergy** diet, dietary therapy focuses on a nutritional plan that is high in protein and complex carbohydrates and free of white sugar and salicylate-containing foods such as strawberries, tomatoes, and **grapes**.

- Herbal therapy. Herbal therapy uses a variety of natural remedies to address the symptoms of ADHD, such as **ginkgo** (*Gingko biloba*) for **memory** and mental sharpness and chamomile (*Matricaria recutita*) extract for calming. The safety of herbal remedies has not been demonstrated in controlled studies. For example, it is known that gingko may affect **blood** coagulation, but controlled studies have not yet evaluated the risk of the effect.

- Homeopathic medicine. This is probably the most effective alternative therapy for ADD and ADHD because it treats the whole person at a core level. Constitutional homeopathic care is most appropriate and requires consulting with a well-trained homeopath who has experience working with ADD and ADHD individuals.

Prognosis

Untreated, ADHD can negatively affect a child's social and educational performance and can seriously damage his or her sense of self-esteem. ADHD children have

---

KEY TERMS

Conduct disorder—A behavioral and emotional disorder of childhood and adolescence. Children with a conduct disorder act inappropriately, infringe on the rights of others, and violate societal norms.

Nervous tic—A repetitive, involuntary action, such as the twitching of a muscle or repeated blinking.

Oppositional defiant disorder—A disorder characterized by hostile, deliberately argumentative, and defiant behavior towards authority figures.

---

impaired relationships with their peers, and may be looked upon as social outcasts. They may be perceived as slow learners or troublemakers in the classroom. Siblings and even parents may develop resentful feelings towards the ADHD child.

Some ADHD children also develop a conduct disorder problem. For those adolescents who have both ADHD and a conduct disorder, up to 25% go on to develop antisocial personality disorder and the criminal behavior, substance abuse, and high rate of suicide attempts that are symptomatic of it. Children diagnosed with ADHD are also more likely to have a learning disorder, a mood disorder such as depression, or an anxiety disorder.

Approximately 70–80% of ADHD patients treated with stimulant medication experience significant relief from symptoms, at least in the short-term. Approximately half of ADHD children seem to "outgrow" the disorder in adolescence or early adulthood; the other half will retain some or all symptoms of ADHD as adults. With early identification and intervention, careful compliance with a treatment program, and a supportive and nurturing home and school environment, ADHD children can flourish socially and academically.

See also Nervous system; Neuroscience; Psychiatry; Psychoanalysis; Psychology.

Resources

Books

Barkley, Russell A. *Taking Charge of ADHD.* New York: Guilford Press, 2000.

Hallowell, Edward M. and John J. Ratey. *Driven to Distraction.* New York: Pantheon Books, 1995.

Wender, Paul H. *ADHD: Attention-Deficit Hyperactivity Disorder in Children and Adults.* Oxford: Oxford University Press, 2001.

Periodicals

Hallowell, Edward M. "What I've Learned from \A.D.D." *Psychology Today* 30, no. 3 (May-June 1997): 40–6.

Osman, Betty B. *Learning Disabilities and ADHD: A Family Guide to Living and Learning Together.* New York: John Wiley & Sons, 1997.

Swanson, J.M., et al. "Attention-Deficit Hyperactivity Disorder and Hyperkinetic Disorder." *The Lancet* 351 (Feb 7, 1997): 429–33.

Other

National Institutes of Mental Health. "Attention-Deficit Hyperactivity Disorder" [cited January, 10, 2003]. <http://www.nimh.nih.gov/publicat/adhd.cfm>.

Paula Anne Ford-Martin

Auks

Auks are penguinlike seabirds found in the Northern Hemisphere. These **birds** spend most of their lives in the coastal waters north of 25°N latitude, coming ashore only to lay their eggs and raise their young. There are 22 **species** of auks, including the Atlantic puffin, the common murre, the dovekie or lesser auk, and the extinct great auk.

Called alcids, the members of the auk family fill an ecological **niche** similar to that filled by the **penguins** in the southern hemisphere. However similar their role, the penguins and alcids are not closely related, the alcids being more closely related to the **gulls**.

Like penguins, alcids have waterproof feathers and swim and dive for their **prey**. Unlike penguins, auks can fly. Their wings are relatively small, and while auks are not especially graceful in flight, they are able to "fly" gracefully underwater in pursuit of prey. Auks obtain all of their food from the sea. Some of the smaller species subsist on **plankton** alone, but most eat **fish**. Like penguins, auks have deceptive coloration; when in

Razor-billed auks (*Alca torda*). *Photograph by J.L. Lepore. Photo Researchers, Inc. Reproduced by permission.*

water, their white fronts make them nearly invisible to fish below.

Auks' legs are near the rear of their bodies, giving them an upright, penguinlike posture. In some species the feet and bill are brightly colored-most notably in the Atlantic puffin, whose blue, yellow, red, and white striped bill is important to the species during the mating season. The colorful plates that make up the harlequinesque bill are shed when the bird molts. Other species, such as the **rhinoceros** auklet, grow special tufts of feathers during mating season. Auks mate for life and are generally monogamous, although males will attempt to copulate with a female if she is not attended by her mate.

Most auks lay their eggs on bare stone ledges, or scoop out a nest in a burrow. An exception is the marbled murrelet, which builds a simple nest in the branches of seaside **pines**. Depending on the species, one or two eggs are laid and incubated for 29-42 days.

A few auk species breed in solitary pairs, but most congregate in large colonies. One of the most densely populated auk colonies on record included 70 pairs of common murres in a **space** of 7.5 sq ft (0.65 sq m). In species that congregate in such large rookeries, each bird's egg is uniquely colored and/or patterned, allowing for easy identification. Chicks, too, are recognized individually by their voice; chicks and parents start getting to know each other's voice even before hatching. Such recognition ensures that each auk feeds only its own offspring.

Chick development varies greatly among the auks. The young of the tiny ancient murrelet take to sea with their parents just a day after they hatch. Other species brood their chicks for 20-50 days. In general, smaller auks lay proportionally larger eggs, from which hatch more precocious chicks. For example, the ancient murrelet weights just over 0.7 oz (200 g), but lays an egg that is approximately one-fourth of the adult's body weight. The chicks reach sexual maturity in about three years.

Auks are long-lived; some birds banded as adults have been found in breeding colonies 20 years later. Their natural enemies include the great skua, the gyrfalcon, and the **peregrine falcon**.

Although auks are protected by law in **North America**, humans remain their greatest threat. In the 1500s, sailors slaughtered huge numbers of great auks for food on long sea voyages. Flightless and 2 ft (0.6 m) tall, the great auk was helpless when caught on land. Tens of millions of these birds were killed until the species became extinct in 1844, when the last great auk was killed on Eldez Island, off the coast of Iceland. More recently, hunting of other auk species has become a popular sport in Greenland.

Oil pollution—both from spills and from tanker maintenance—also kills countless birds each year. Oil destroys the waterproof quality of the auks' feathers and is swallowed by the birds when they attempt to clean themselves. Auk drownings in fishermen's gill nets have decreased in recent years. However, the decline in food-fish species such as cod and haddock have led fishermen to turn their attention to the fish species auks eat, such as sprats. Such **competition** does not bode well for the auks.

Resources

Books

Brooke, M., and T. Birkhead, eds. *The Cambridge Encyclopedia of Ornithology.* Cambridge: Cambridge University Press, 1991.

Forshaw, Joseph. *Encyclopedia of Birds.* New York: Academic Press, 1998.

Terres, John K. *The Audubon Encyclopedia of North American Birds.* Avenel, NJ: Wings Books, 1991.

F.C. Nicholson

Aurora *see* **Solar wind**

Australia

Of the seven continents, Australia is the flattest, smallest, and except for **Antarctica**, the most arid. Including the southeastern **island** of Tasmania, the island **continent** is roughly equal in area to the United States, excluding Alaska and Hawaii. Millions of years of geographic isolation from other landmasses accounts for Australia's unique **animal species**, notably marsupial **mammals** like the kangaroo, egg laying mammals like the **platypus**, and the flightless emu bird. Excluding folded structures (areas warped by geologic forces) along Australia's east coast, patches of the northern coastline and the relatively lush island of Tasmania, the continent is mostly dry, bleak, and inhospitable.

Topography and origin of Australia

Australia has been less affected by seismic and orogenic (mountain building) forces than other continents during the past 400 million years. Although seismic (**earthquake**) activity persists in the eastern and western highlands, Australia is the most stable of all continents. In the recent geological past, it has experienced none of the massive upheavals responsible for uplifting the Andes in **South America**, the Himalayas in south **Asia** or the European Alps. Instead Australia's topography is the end result of gradual changes over millions of years.

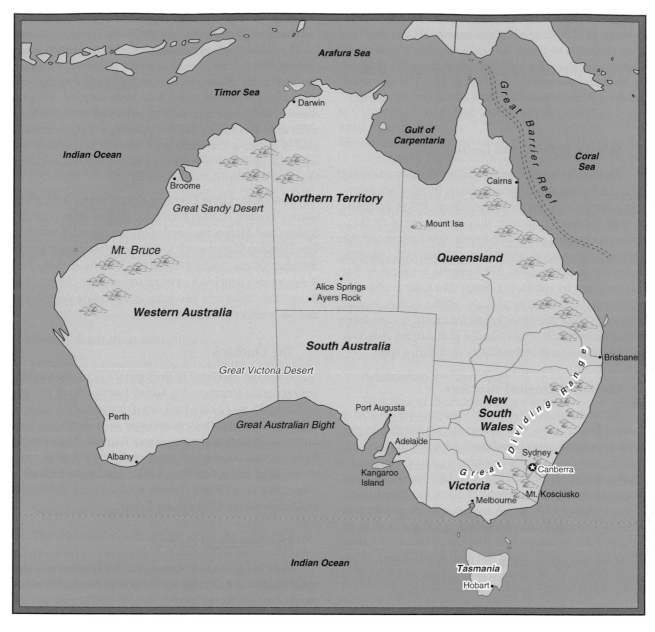

Australia. *Illustration by Hans & Cassidy. Courtesy of Gale Group.*

Australia is not the oldest continent, a common misconception arising from the continent's flat, seemingly unchanged expanse. Geologically it is the same age as the Americas, Asia, **Africa**, **Europe**, and Antarctica. But Australia's crust has escaped strong **earth** forces in recent geological history, accounting for its relatively uniform appearance. As a result, the continent serves as a window to early geological ages.

Splitting of Australia from Antarctica

About 95 million years ago, tectonic forces (movements and pressures of the earth's crust) split Australia from Antarctica and the southern supercontinent of Gondwanaland. Geologists estimate that the continent is drifting northward at a **rate** of approximately 18 inches (28 cm) per year. They theorize that south Australia was joined to Antarctica at the Antarctic regions of Wilkes Land, including Commonwealth Bay. Over a period of 65 million years, beginning 160 million years ago, Australia's crust was stretched hundreds of miles by **tectonics** before it finally cleaved from Antarctica.

Testimony to the continental stretching and splitting includes Kangaroo Island off South Australia, made up of volcanic basalts, as well as thick layers of sediment

logical period is characterized by specific types of fossilized organisms (**plant** and animal) the record for which has been formalized into the **geologic time** scale.

Such fossil testimony is spotty in eastern Australia except for a few areas including Broken Hill in western New South Wales, central Victoria and Tasmania. So while it is difficult to reconstruct an authoritative overall history of Australia in Cambrian times, some general conclusions can be withdrawn. One is that western Australia has a more stable geological history relatively speaking than the eastern section of the continent. There is much greater mobility and folding generally in the eastern half than in the west.

Geography of Victoria

Victoria is also characterized by a belt of old rocks upon which sediments have been deposited called the Lachlan geosyncline. Marine rocks were deposited in quiet **water** to great thicknesses in Victoria, forming black shales. Some of the sediment was built up by mud-laden **currents** from higher areas on the sea floor. These current-borne sediments have produced muddy sandstones called *graywackes*.

At the end of the Ordovician and early Silurian periods (about 425 million years ago) there was widespread folding of the Lachlan geosyncline called the Benambran orogeny. The folding was accompanied by granite intrusions and is thought to be responsible for the composition and texture of the rocks of the Snowy Mountains in Victoria, including Mt. Kosciusko, Australia's tallest peak at 7,310 ft (2,193 m).

The Melbourne trough in Victoria is full of Silurian graywackes and mudstones featuring graptolite fossils, cup or tube-shaped organisms with durable exoskeletons (shells) that congregated in colonies. Mid-Devonian strata (375 million years old) are abundant with armored fishes which are also found in central Australia, **North America**, Eurasia, and Antarctica. The sandstones of the Grampian Mountains in western Victoria were formed from the erosion of uplifted areas which were deposited in lakes and by rivers during the late Devonian, early Carboniferous periods (about 350 million years ago).

Mountain-building in eastern Australia

In eastern Australia, Paleozoic era volcanic activity built up much of the rock strata. Mountain glaciation during the late Carboniferous period when **insects**, **amphibians**, and early **reptiles** first evolved, also transformed the landscape. Mountain building in eastern Australia culminated during the middle and later Permian period (about 250 million years ago) when a huge mass of **magma** (un-

derground molten rock) was emplaced in older rocks in the New England area of northeastern New South Wales. This huge mass or batholith, caused extensive folding to the west and ended the sedimentation phase of the Tasman geosyncline. It was also the last major episode of orogeny (mountain building) on the continent.

Glaciers and ocean inundations

In parts of Western Australia, particularly the Carnarvon Basin at the mouth of the Gascoyne River, glacial sediments are as thick as three miles. Western Australia, particularly along the coast, has been inundated repeatedly by the sea and has been described by geologists as a mobile shelf area. This is reflected in the alternating strata of deposited marine and non-marine layers.

Sedimentary features of Australia

In the center of Australi a is a large sedimentary basin or depression spanning 450 mi (720 km) from east to west and 160 mi (256 km) north to south at its widest point. Sedimentary rocks of all varieties can be found in the basin rocks which erosion shaped into spectacular scenery including Ayres Rock and Mt. Olga. These deposits are mostly of pre-Cambrian age (over 570 million years old), while sediment along the present-day coastline including those in the Eucla Basin off the Great Australian Bight are less than 70 million years old. North of the Eucla Basin is the Nullarbor (meaning treeless) Plain which contains many unexplored limestone caves.

Dominating interior southern Queensland is the Great Artesian Basin which features non-marine sands built up during the Jurassic period (190 million to 130 million years ago), sands which contain much of the basin's Artesian water. Thousands of holes have been bored in the Great Artesian Basin to extract the water resources underneath but the salt content of water from the basin is relatively high and the water supplies have been used for **livestock** only.

The Sydney basin formed over the folded rocks of the Tasman geosyncline and is also considered to be an extension of the Great Artesian Basins. Composed of sediments from the Permian and Triassic periods (290 million to 190 million years old) it extends south and eastward along the **continental shelf**. The sandstone cliffs around Sydney Harbor, often exploited for building stones, date from Triassic sediments.

Geology of Tasmania

During the early to mid-Jurassic period, there was an intrusion of 2,000 cubic miles of dark, layered volcanic rocks in Tasmania, similar to the magmatic forma-

Ayers Rock, central Australia, 1,143 ft (349 m) high. *JLM Visuals. Reproduced by permission.*

tions of the Karroo region in South Africa and the Palisades in New York. Tasmania separated from mainland Australia only 10,000 years ago, when sea levels rose after the thawing of the last **ice** age.

Mountain-building and glaciation in Victoria

The Glass House Mountains in southeastern Queensland and the Warbungle Mountains in northern New South Wales were formed by volcanism in recent times, during the Miocene Epoch, around 20 million years ago. Gold bearing deep leads can be found in basaltic lava flows along ancient valleys. The Flinders ranges in South Australia were uplifted in the modern geological era, the Cenozoic. About the same time the sea retreated from the Murray Basin in South Australia. During the Pleistocene (less than two million years ago), a 400 sq mi (103,600 ha) area around Mount Kosciusko was covered with **glaciers**.

Climate

The climatological record of Australia shows a pronounced **temperature** drop on the continent in the late

Miocene and early Pliocene epochs between 26 and five million years ago when **monkeys** and early apes first evolved and saber toothed **cats** prowled the earth. On the Eyre Peninsula in South Australia and in Gippsland in Victoria, eucalyptus and acacia trees supplanted the previously dominant beech trees that had thrived in the warmer climate of the Miocene era.

Natural resources

Of course geology is inextricably intertwined with natural resources and mineral exploitation. **Minerals** in Australia have had a tremendous impact on the country's human history and patterns of settlement. Alluvial gold (gold sediments deposited by rivers and streams) spurred several gold fevers and set the stage for Australia's present demographic patterns. During the post-World War II period there has been almost a continuous run of mineral discoveries, including gold, bauxite, iron, and manganese reserves as well as opals, sapphires, and other precious stones.

It is estimated that Australia has 24 billion tons (22 billion tonnes) of **coal** reserves, over one-quarter of which (7 billion tons/6 billion ton) is anthracite or black coal deposited in Permian sediments in the Sydney Basin of New South Wales and in Queensland. Brown coal suitable for electricity production in found in Victoria. Australia meets its domestic coal consumption needs with its own reserves and exports the surplus.

Natural gas fields are liberally distributed throughout the country and now supply most of Australia's domestic needs. There are commercial gas fields in every state and pipelines connecting those fields to major cities. Within three years, Australian natural gas production leapt almost 14-fold from 8.6 billion cu ft (258 million cu m) in 1969, the first year of production, to 110 billion cu ft (3.3 billion cu m) in 1972. All in all, Australia has trillions of tons of estimated natural gas reserves trapped in sedimentary strata distributed around the continent.

Australia supplies much of its oil consumption needs domestically. The first Australian oil discoveries were in southern Queensland near Moonie. Australian oil production now amounts to about 25 million barrels per year and includes pumping from oil fields off northwestern Australia near Barrow Island, Mereenie in the southern Northern Territory, and fields in the Bass Strait. The Barrow Islands, Mereenie, and Bass Strait fields are also sites of natural gas production.

Australia has rich deposits of **uranium** ore, which is refined for use for fuel for the **nuclear power** industry. Western Queensland, near Mount Isa and Cloncurry contains three billion tons (2.7 billion tonnes) of uranium ore reserves. There are also uranium deposits in Arnhem Land in far northern Australia, as well as in Queensland and Victoria.

Most of Australia's substantial iron ore reserves are in Western Australia in and around the Hammersley Range. Australia has billions of tons of iron ore reserves, exporting magnetite iron from mines in Tasmania to Japan while still extracting ore from older mines on the Eyre Peninsula of South Australia and in the Koolyanobbing Range of southern Western Australia.

The Western Australian shield is rich in nickel deposits that were first discovered at Kambalda near Kalgoorlie in south Western Australia in 1964. Other nickel deposits have been found in old goldmine areas in Western Australia. Small quantities of platinum and palladium have been extracted side-by-side with nickel reserves.

Australia is also extremely rich in zinc reserves, the principal sources for which are Mt. Isa and Mt. Morgan in Queensland. The Northern Territory also has **lead** and zinc mines as well as vast reserves of bauxite (**aluminum** ore), namely at Weipa on the Gulf of Carpenteria and at Gove in Arnhem Land.

Gold production in Australia, which was substantial earlier in the century, has declined from a peak production of four million fine ounces in 1904 to several hundred thousand fine ounces. Most gold is extracted from the Kalgoorlie-Norseman area of Western Australia. The continent is also well-known for its precious stones, particularly white and black opals from South Australia and western New South Wales. There are sapphires and topaz in Queensland and in the New England District of northeastern New South Wales.

Because of its aridity, Australia suffers from leached, sandy, and salty soils. The continent's largely arid land and marginal water resources represent challenges for **conservation** and prudent environmental management. The challenge is to maximize the use of these resources for human beings while preserving ecosystems for animal and plant life.

Resources

Books

Brizga, Sandra, and Brian L. Finlayson, eds. *River Management: The Australasian Experience.* John Wiley & Sons, 2000.

Drummond, Barry J., ed. *The Australian Lithosphere.* 1991.

The Geology of South Australia. South Australia: State Print, 1993.

Hamblin, W.K., and E.H. Christiansen. *Earth's Dynamic Systems.* 9th ed. Upper Saddle River: Prentice Hall, 2001.

Hancock P. L. and Skinner B. J., eds. *The Oxford Companion to the Earth.* Oxford: Oxford University Press, 2000.

Morrison, Reg. *Australia: Land Beyond Time.* Ithica, NY: Cornell University Press, 2002.

Periodicals

"Underground Current Electrifies Australia." *New Scientist* (March 30, 1991): 10.

Robert Cohen

Autism

Autism is a profound mental disorder marked by an inability to communicate and interact with others. The condition's characteristics include language abnormalities, restricted and repetitive interests, and the appearance of these characteristics in early childhood. The disorder begins in infancy, but typically is not diagnosed until the ages two to five. Although individuals with autism are more likely to be mentally retarded than other individuals some people with the disorder have a high intelligence level. The cause of autism is unknown although it is probably biological in origin.

A singular world view

Children with autism were described as early as 1908 by Heller, a Viennese educator. Autism was not named and identified as a distinct condition until 1943. That year American psychiatrist Leo Kanner wrote about what he called "infantile autism." Kanner derived the term autism from the Latin word *aut*, meaning self. Kanner described a group of children who looked normal but showed limited communication skills and were drawn to repetitive **behavior**. Researchers have since discovered that the disorder is not common, occurring in two to five of every 10,000 births. The disorder is more common in males than in females.

As many as two-thirds of children with autistic symptoms are mentally deficient. But individuals with autism can be highly intelligent. Some achieve great levels of success at school and at work. The term autism is more a description of a range of behavioral traits than a term to describe a single type of person with a single level of potential.

Autistic individuals generally share a defect in "the theory of mind." This term is used to describe the way normal individuals develop a sense of what others are thinking and feeling. This sense is usually developed by about age four. Autistic individuals typically are limited in their ability to communicate nonverbally and verbally. About half of all autistic people never learn to speak. They are likely to fail in developing social relationships with peers, have limited ability to initiate conversation if they do learn how to talk, and show a need for routines and rituals.

The range of ability and intelligence among autistic individuals is great. Some individuals are profoundly withdrawn. These aloof individuals generally do not greet parents when they enter the house or seek comfort when in **pain**. Others can conduct a conversation but may be obsessed with strange or unusual behaviors, such as a fascination with **calendars** or timetables. Still others use their ability to focus on particular bodies of fact to master a job or a profession. Temple Grandin is a Ph.D. and a successful **animal** behavior expert who is autistic and an international expert in her field.

Certain autistic people have areas of expertise in which they are superior to normal individuals. These skills are called **savant** abilities and have been well documented in music, drawing, and areas where calculation is involved. The vibrant drawings of buildings by autistic British artist Stephen Wiltshire have been published to great acclaim.

Abundance of theories

The precise cause of autism is not known and many theories have emerged concerning its origin. Initially autism was believed to be caused by destructive parents. Kanner observed in 1943 that there are few "really warmhearted fathers and mothers" among parents of autistic children (Groden 1988). Other experts suggested that parents of autistic children are more likely to be cold and unsupportive of their children than parents of normal children. In the 1950s and 1960s it was still generally believed that this parental behavior caused autism. Children were therefore advised to have psychotherapy for autism. This therapy was generally unsuccessful. Some experts suggested that autistic children be removed from their parents.

In the mid-1960s experts began to challenge the assumption that parents cause autism. Evidence emerged that while autistic children look normal they have particular physical abnormalities. These include a higher-than-normal likelihood of epilepsy—it occurs in as many as 30% of children with autism. Researchers also looked at the way parents interacted with autistic children. Their findings showed that parents of autistic children are equally as skilled as parents of normal children on average.

The general belief today is that autism is a biological disorder and has nothing to do with parenting skill. Clues to what causes autism include a wealth of abnormalities documented to occur in higher percentages among autistic people than normal people. These include certain genetic conditions, **epilepsy**, mental disabilities, and some **birth defects**.

Genetics appears to play a role in autism but its role is not completely understood. Brothers or sisters of individuals with autism are slightly more likely than others to be autistic. Approximately 2–3% of siblings of autistic people have the disorder. The twin of an autistic **individual** is also more likely to have autism.

The occurrence of fragile X **syndrome**, a genetic disorder, in about 10% of autistic people has presented researchers with a documented cause of the disorder. Fragile X victims have a gap on their X **chromosome**. The condition is generally linked to mental disabilities and a characteristic facial appearance (a high forehead and long ears) among other traits. Brothers or sisters of individuals with fragile X syndrome are nearly 50% more likely to be autistic than are brothers and sisters of normal children. It is not clear what causes the syndrome.

Another genetic trait more common among autistic individuals than others is neurofibromatosis, a genetic condition which affects the skin and the nerves and sometimes causes **brain** damage. Researchers also have noted that mothers of autistic children are more likely to receive medication during pregnancy. They have also found that autistic children are more likely to have been born with meconium (the first stool of the infant) in the amniotic fluid during labor. However these events also commonly occur among normal children. Other possible causes of autism include rubella **infection** in early pregnancy, herpes **encephalitis** (which can cause **inflammation** of the brain), and cytomegalovirus infection.

Whatever the cause of the damage, various abnormalities in the autistic brain have been documented. These include variations in the frontal lobes of the brain which focus on control and planning, and in the limbic system. The limbic system is a group of structures in the brain which are linked to emotion, behavior, **smell**, and other functions. Autistic individuals may suffer from a limited development of the limbic system. This would explain some of the difficulties faced by autistic individuals in processing information.

Studies using MRI (**magnetic resonance imaging**) scanners have found abnormalities in the cerebellum. Some researchers suggest that the section where abnormalities have been documented is the part of the brain concerned with attention. Other tests using electroencephalograms have found that autistic children show abnormalities in the way the brain processes sound.

Teaching and learning

As theories about the cause of autism have changed so have approaches to teaching autistic individuals. Individuals with autism were once considered unteachable and were often institutionalized. Experts currently recommend early education for autistic individuals using approaches geared specifically for them. Those who cannot speak may learn sign language. Often some form of behavioral modification is suggested including offering positive reinforcement for good behavior.

Given the need for organization and repetitive behavior among autistic people, many experts suggest a structured environment with a clearly defined schedule. Some experts advocate special schools for autistic children while others recommend including them in a general school program with appropriate help.

Controversy exists concerning the best way to teach and communicate with autistic children who do not speak. In the 1980s some parents and educators claimed great success using so-called facilitated communication. This technique involves the use of a keyboard or letter board and a facilitator to help the autistic individual use the device. Some have said the device allows autistic individuals to break through the barriers of communication. Others have criticized facilitated communication as a contrived and false method of communication. The technique remains controversial.

KEY TERMS

Limbic system—A group of structures in the brain including the hippocampus and dentate gyrus. These structures are associated with smell, emotion, behavior, and other brain activity.

Psychotherapy—A broad term that usually refers to interpersonal verbal treatment of disease or disorder that addresses psychological and social factors.

With no cure and no prenatal test for autism, there is no prospect of eliminating this condition in the near future. Continued educational research concerning the best way to teach autistic children and continued scientific research concerning possible treatments for autism are the greatest hope for dealing with the effects of this profound developmental problem.

Resources

Books

Baron-Cohen, Simon, and Patrick Bolton. *Autism: The Facts.* Oxford: Oxford University Press, 1993.

Bauman, Margaret L., and Thomas L. Kemper. *The Neurobiology of Autism.* Baltimore: Johns Hopkins University Press, 1994.

Brill, Marlene Targ. *Keys to Parenting the Child With Autism.* Hauppauge, NY: Barron's, 1994.

Cohen, Donald J., and Anne M. Donnellan. *Handbook of Autism and Pervasive Developmental Disorders.* Silver Spring, MD: V. H. Winston and Sons, 1987.

Hart, Charles A. *A Parent's Guide to Autism.* New York: Pocket Books, 2001.

Isselbacher, Kurt J., et al., eds. *Harrison's Principles of Internal Medicine.* 13th ed. New York: McGraw-Hill, 1994.

Periodicals

Eberlin, Michael, et al. "Facilitated Communication: A Failure to Replicate the Phenomenon." *Journal of Autism and Communication Disorders* 23 (1993): 507-528.

"Interventions To Facilitate Social Interaction For Young Children." *Journal of Autism and Developmental Disorders.* 32, no. 5-5 (2002): 351-372.

Patricia Braus

Autoimmune disorders

Autoimmunity is a condition where the **immune system** mistakenly recognizes host **tissue** or cells as foreign. (The word "auto" is the Greek word for self.) Because of this false recognition, the immune system reacts against the host components. There are a variety of autoimmune disorders (also called autoimmune diseases).

An autoimmune **disease** can be very specific, involving a single **organ**. Three examples are Crohn's disease (where the intestinal tract is the target), multiple sclerosis (where tissues of the **brain** are the target), and **diabetes mellitus** Type I (where the insulin-producing cells of the pancreas are the target).

Other autoimmune disorders are more general, and involve multiple sites in the body. One example is rheumatoid **arthritis**.

Each autoimmune disorder occurs rarely. However, when all the disorders are tallied together, autoimmunity is found to be a disorder that affects millions of people in the United States alone.

The causes of autoimmune disease are not clearly known. However, there are indications that some disorders, or at least the potential to develop the disorder, can be genetically passed on from one generation to the next. Psoriasis is one such example. The environment also has an influence. The way the immune system responds to environmental factors and to infections (e.g., those caused by some viruses) can trigger the development of a disorder. Other factors, which are less understood, include aging, **hormones**, physiological changes during pregnancy, and chronic **stress**.

General autoimmune disorders

Systemic lupus erythematosus

In systemic lupus erythematosus (often known as LUPUS), antibodies attack a number of the body's own different tissues. The disease recurs periodically and is seen mainly in young and middle-aged women. Symptoms include fever, chills, fatigue, weight loss, skin rashes (particularly the classic "butterfly" rash on the face), vasculitis, joint **pain**, patchy hair loss, sores in the mouth or nose, lymph-node enlargement, gastric problems, and, in women, irregular menstrual periods. About half of those who suffer from lupus develop cardiopulmonary problems, and some may also develop urinary problems. Lupus can also affect the central **nervous system**, causing seizures, **depression**, and **psychosis**.

Rheumatoid arthritis

Rheumatoid arthritis occurs when the immune system attacks and destroys the tissues that line bone joints and cartilage. The disorder occurs throughout the body, although some joints may be more affected than others. Initially, the disorder is characterized by a low-grade fever, loss of appetite, weight loss, and a generalized pain in the joints. The joint pain then becomes more specific, usually beginning in the fingers then spreading to other areas (i.e., wrists, elbows, knees, ankles). As the disease progresses, joint function diminishes sharply and deformities occur. A particularly distinctive feature is the "swan's neck" curling of the fingers.

Scleroderma

This disorder, which affects **connective tissue**, is also called CREST **syndrome** or progressive systemic sclerosis. Symptoms include pain, swelling, and stiffness of the joints. As well, the skin takes on a tight, shiny appearance. The **digestive system** becomes involved, resulting in weight loss, appetite loss, diarrhea, constipation, and distention of the abdomen. As the disease progresses, the **heart**, lungs, and kidneys become involved, and malignant **hypertension** causes death in approximately 30% of cases.

Goodpasture's syndrome

Goodpasture's syndrome occurs when antibodies are deposited in the membranes of both the lung and kidneys, causing both **inflammation** of kidney glomerulus (glomerulonephritis) and lung bleeding. It is typically a disease of young males. Symptoms are similar to that of **iron** deficiency **anemia**, including fatigue and pallor. Symptoms involving the lungs may range from a cough that produces bloody sputum to outright hemorrhaging. Symptoms involving the urinary system include **blood** in the urine and/or swelling.

Polymyositis and dermatomyositis

These immune disorders affect the neuromuscular system. In polymyositis, symptoms include muscle weakness, particularly in the shoulders or pelvis that prevents the patient from performing everyday activities. In dermatomyositis, this muscle weakness is accompanied by a rash that appears on the upper body, arms, fingertips, and sometimes on the eyelids.

Ankylosing spondylitis

Immune-system-induced degeneration of the joints and soft tissue of the spine is the hallmarks of ankylosing spondylitis. The disease generally begins with lower back pain that progresses up the spine. The pain may eventually become crippling.

Sjogren's syndrome

Exocrine glands are attacked, resulting in excessive dryness of the mouth and eyes.

Autoimmune disorders of the endocrine glands

Type I (immune-mediated) diabetes mellitus

This disorder is considered to be caused by an antibody that attacks and destroys the insulin-producing islet cells of the pancreas. Type I diabetes mellitus is characterized by fatigue and an abnormally high level of glucose in the blood (a condition called hyperglycemia).

Grave's disease

The disorder is caused by an antibody that binds to specific cells in the thyroid gland, causing them to make excessive amounts of thyroid hormone. This disease is characterized by an enlarged thyroid gland, weight loss without loss of appetite, sweating, heart palpitations, nervousness, and an inability to tolerate **heat**.

Hashimoto's thyroiditis

An antibody that binds to cells in the thyroid gland causes the disorder known as Hashimoto's thyroiditis. This disorder generally displays no symptoms, but patients can exhibit weight gain, fatigue, dry skin, and hair loss. Unlike Grave's disease, however, less thyroid hormone is made.

Autoimmune disorders of the blood and blood vessels

Pernicious anemia

Pernicious anemia is a disorder in which the immune system attacks the lining of the stomach, destroying the ability to utilize **vitamin** B_{12}. Signs of pernicious anemia include weakness, sore tongue, bleeding gums, and tingling in the extremities. Because the disease causes a decrease in stomach acid, nausea, vomiting, loss of appetite, weight loss, diarrhea, and constipation are possible. Also, because Vitamin B_{12} is essential for the nervous system function, the deficiency of it brought on by the disease can result in a host of neurological problems, including weakness, lack of coordination, blurred **vision**, loss of fine motor skills, loss of the sense of **taste**, ringing in the ears, and loss of bladder control.

Autoimmune thrombocytopenic purpura

In this disorder, the immune system targets and destroys blood platelets. It is characterized by pinhead-size red dots on the skin, unexplained bruises, bleeding from the nose and gums, and blood in the stool.

Autoimmune hemolytic anemia

Antibodies coat and lead to the destruction of red blood cells in autoimmune hemolytic anemia. Symptoms include fatigue and abdominal tenderness due to an enlarged spleen.

Vasculitis

A group of autoimmune disorders in which the immune system attacks and destroys blood vessels can cause vasculitis. The symptoms vary, and depend upon the group of **veins** affected.

Autoimmune disorders of the skin

Pemphigus vulgaris

Pemphigus vulgaris involves a group of autoimmune disorders that affect the skin. This disease is characterized by blisters and deep lesions on the skin.

Autoimmune disorders of the nervous system

Myasthenia gravis

In myasthenia gravis, the immune system attacks a receptor on the surface of muscle cells, preventing the muscle from receiving nerve impulses and resulting in severe muscle weakness. The disease is characterized by fatigue and muscle weakness that at first may be confined to certain muscle groups, but then may progress to the point of paralysis. Patients often have expressionless faces as well as difficulty chewing and swallowing. If the disease progresses to the **respiratory system**, artificial **respiration** may be required.

The above are only some of the more than 30 autoimmune disorders.

The immune system

The immune system defends the body against attack by infectious **microorganisms** and inanimate foreign objects. Immune recognition and attack of an invader is a highly specific process. A particular immune system **cell** will only recognize and target one type of invader. The immune system must develop this specialized knowledge of individual invaders, and learn to recognize and not destroy cells that belong to the body itself.

Immune recognition depends upon the **chemistry** of the surface of cells and tissues. Every cell carries protein markers on its surface. The markers—called major histocompatability complexes (MHCs)—identify the cell as to its type (e.g. nerve cell, muscle cell, blood cell, etc.), and also to which organ or tissue the cell comprises. In a properly functioning immune system, the class of immune cells called **T cells** recognizes the host MHCs. Conversely, if the T cells encounter a MHC that is not recognized as that belongs to the host, another class of immune cell called B cells will be stimulated to produce antibodies. There are a myriad of B cells, each of which produces a single characteristic antibody directed toward one specific antigen. The binding of an antibody to the antigen on the invading cell or particle initiates a process that destroys the invader.

In autoimmune disorders, the immune system cannot distinguish between "self" cells and invader cells. As a result, the same destructive operation is carried out on the body's own cells that would normally be carried out on **bacteria**, viruses, and other invaders. The reasons why the immune systems become dysfunctional are still not clearly understood. It is conceded by the majority of immunologists that a combination of genetic, environmental, and hormonal factors contribute to the development of autoimmunity.

A number of other mechanisms may also trigger autoimmunity. A substance that is normally restricted to one part of the body—and so not usually exposed to the immune system—is released into other areas of the body. The substance is vulnerable to immune attack in these other areas. In a second mechanism, the antigenic similarity between a host **molecule** and a molecule on an invader may fool the immune system into mistaking the host's component as foreign. Additionally, drugs, **infection**, or some other environmental factor can alter host cells. The altered cells are no longer recognizable as "self" to the immune system. Finally, the immune system can become damaged and malfunction by, for example, a genetic **mutation**.

Diagnosis of autoimmune disorders

A number of different tests can help diagnose autoimmune diseases. A common feature of the tests is the detection of antibodies that react with host antigens. Such tests involve measuring the level of antibodies found in the blood. An elevated amount of antibodies indicates that a humoral immune reaction is occurring. Antibody production is, of course, a normal response of the immune system to an infection. The normal operation of the immune system must be ruled out as the cause for the increased antibody levels. A useful approach is to determine the class of antibody that is present. There are five classes of antibodies. IgG antibody is the class that is usually associated with autoimmune diseases. Unfortunately, IgG is also dominant in normal immune responses. The most useful antibody tests involve introducing the patient's antibodies to samples of his or her own tissue. If the antibodies bind to the host tissue it is diagnostic for an autoimmune disorder. Antibodies from a person without an autoimmune disorder

---

KEY TERMS

Autoantibody—An antibody made by a person that reacts with their own tissues.

Complement system—A series of 20 proteins that "complement" the immune system; complement proteins destroy virus-infected cells and enhance the phagocytic activity of macrophages.

Macrophage—A white blood cell that engulfs invading cells or material and dissolves the invader.

---

would not react to "self" tissue. Tissues from the thyroid, stomach, liver, and kidney are used most frequently in this type of testing.

Treatment of autoimmune disorders

Treatment is specific to the disease, and usually concentrates on lessening the discomfort of the symptoms rather than correcting the underlying cause. Treatment also involves controlling the physiological aspects of the immune response, such as inflammation. This is typically achieved using two types of drugs. Steroids are used to control inflammation. There are many different steroids, each having side effects. The use of steroids is determined by the benefits gained by their use versus the side effects produced. Another form of treatment uses immunosuppressive drugs, which inhibit the replication of cells. By stopping **cell division**, non-immune cells are also suppressed. This can lead to, for example, side effects such as anemia.

See also Cell death; Immunology.

Resources

Books

Abbus, Abdul K., and Andrew H. Lichtman. *Basic Immunology: Function and Disorders of the Immune System* Philadelphia: W. B. Saunders, 2001.

Baron-Faust, Rita, Jill P. Buyon and Virginia Ladd. *The Autoimmune Connection: Essential Information for Women on Diagnosis, Treatment, and Getting On with Your Life* New York: McGraw-Hill, 2003.

Moore, Elaine A., Lisa Moore, and Kelly R. Hale. *Graves' Disease: A Practical Guide.* Jefferson, NC: McFarland & Company, 2001.

Santamaria, Pere. *Cytokines and Chemokines in Autoimmune Diseases (Medical Intelligence Unit, 30)* Georgetown, TX: Eurekah.com, Inc., 2002.

Other

National Institutes of Health. National Institute of Allergy and Infectious Diseases, 31 Center Drive, MSC 2520, Bethesda, MD 20892–2520 [cited January 10, 2003]. <http://www.niaid.nih.gov/publications/autoimmune/autoimmune.tm>.

John Thomas Lohr

Automatic pilot

The automatic pilot has it roots in the **gyroscope**, a weighted, balanced wheel mounted in bearings and spinning at high **velocity**. As early as 1852 the French scientist Jean-Bernard-Léon Foucault had experimented with the gyroscope and found that it tended to stay aligned with its original position and also tended to orient itself **parallel** to Earth's axis in a north-south direction. Thus, he reasoned, the gyroscope could be used as a compass because it designated true, or geographic, north rather than magnetic north, as traditional compasses did, which varied according to their location.

By the early 1900s the gyrocompass was a crucial part of navigation. A German manufacturer, Hermann Anschutz-Kaempfe, and an American, inventor, Elmer Sperry, had produced two gyrocompasses for use on board ships. Sperry also invented the first automatic pilot for ships, named " Metal Mike," which used the information from the ship's gyrocompass to steer the vessel.

Soon interest arose in applying this method to control **aircraft**. Sperry again led the way when one of his devices was used aboard a Curtiss flying boat in 1912. It used a single gyroscope which, like all spinning masses, tended to resist any change in the plane's axis of **rotation**. Whenever the airplane departed from its original altitude, a small **force** was applied to a spring connected to one end of the gyro axis, and this, magnified mechanically, was used to restore movement of the aircraft controls. In 1914, Sperry's son, Lawrence, competed in Paris with fifty-three other entrants to win a prize of fifty thousand francs for the most stable airplane. He demonstrated his plane's stability by flying low over the judges while he took his hands off the controls and a mechanic walked out on the wing.

All simple autopilots since that time have used similar principles. However, for airplane control in three directions (left/right, up/down, wing up/wing down) three gyros are needed. By 1930 the British Royal Aircraft Establishment and several private companies had developed refined autopilot systems that were gradually introduced to both military and civilian aircraft. During World War II much more complicated autopilots were produced. By a simple movement of a control knob, the aircraft could be held on a steady heading at a constant altitude, made to turn at a steady **rate** in either direction, or even change altitude in a precise manner.

A German, Irmgard Flugge-Lotz (1903-1974) played a key role in developing these automatic controls. As a child in Germany she spent her time at the movies watching **engineering** documentaries rather than Charlie Chaplin comedies. During World War II she developed methods of controlling aircraft during **acceleration** and in flying curves; with faster planes, pilots had no chance to correct for any miscalculations. She worked on what she called "discontinuous automatic control," which laid the foundation for automatic on-off aircraft control systems in jets.

After World War II the United States armed forces became interested in developing an **inertial guidance** system that used autopilots for rockets, submarines and manned aircraft. Such a system would not rely on any information from the outside, such as **radio waves** or celestial bodies, for "fixes." Instead, it would plot its course from information via gyroscopes and from calculations accounting for the rotation of **Earth**. The United States Air Force turned to a Massachusetts Institute of Technology professor, Charles Stark Draper, to develop such a system. He worked on improving gyro units, specific force receivers, amplifiers, servodrives, and other elements. On February 8, 1953, his equipment was aboard a B-29 bomber that left Bedford, Massachusetts, for Los Angeles, California, on a twelve-hour flight. It kept course without deviation, corrected for **wind** and currents, rose to clear the Rocky Mountains—all with no input from the ground or from the pilot and co-pilot. The inertial guidance system sensed every change in the forward velocity, every move right and left and up and down. By continually digesting these changes, and remembering the plane's take-off point, the system was able to determine inflight position with great **accuracy**.

Today, much of the burden of flight has been transferred to autopilots. They can control attitude and altitude, speed, heading, and flight path selection. One great advance has been in landings: signals from an instrument landing system control an aircraft automatically during approach, to land the plane down the glide slope beam, keep it on the runway, and even turn it onto the taxiway, all in totally blind conditions.

Automation

Automation is the use of scientific and technological principles in the manufacture of machines that take over work normally done by humans. This definition has been disputed by professional scientists and engineers, but in any case, the term is derived from the longer term automatization or from the phrase automatic operation. Delmar S. Harder, a plant manager for General Motors, is credited with first having used the term in 1935.

History

Ideas for ways of automating tasks have been in existence since the time of the ancient Greeks. The Greek inventor Hero (fl. about A.D. 50), for example, is credited with having developed an automated system that would open a temple door when a priest lit a fire on the temple altar. The real impetus for the development of automation came, however, during the **Industrial Revolution** of the early eighteenth century. Many of the steam-powered devices built by James Watt, Richard Trevithick, Richard Arkwright, Thomas Savery, Thomas Newcomen, and their contemporaries were simple examples of machines capable of taking over the work of humans. One of the most elaborate examples of automated machinery developed during this period was the drawloom designed by the French inventor Basile Bouchon in 1725. The instructions for the operation of the Bouchon loom were recorded on sheets of **paper** in the form of holes. The needles that carried thread through the loom to make cloth were guided by the presence or absence of those holes. The manual process of weaving a pattern into a piece of cloth through the work of an **individual** was transformed by the Bouchon process into an operation that could be performed mindlessly by merely stepping on a pedal.

Types of automation

Automated machines can be subdivided into two large categories—open-loop and closed-loop machines, which can then be subdivided into even smaller categories. Open-loop machines are devices that, once started, go through a cycle and then stop. A common example is the automatic dishwashing machine. Once dishes are loaded into the machine and a button pushed, the machine goes through a predetermined cycle of operations: pre-rinse, wash, rinse, and dry, for example. A human operator may have choices as to which sequence the machine should follow—heavy wash, light wash, warm and cold, and so on—but each of these operations is alike in that the machine simply does the task and then stops. Many of the most familiar appliances in homes today operate on this basis. A microwave oven, a coffee maker, and a CD player are examples.

Larger, more complex industrial operations also use open-cycle operations. For example, in the production of a car, a single machine may be programmed to place a side panel in place on the car and then weld it in a dozen or more locations. Each of the steps involved in this process—from placing the door properly to each of the different welds—takes place according to instructions programmed into the machine.

Closed-loop machines

Closed-loop machines are devices capable of responding to new instructions at some point in their operation. The instructions may come from the operation being performed itself or from a human operator. The ability of a machine to change its operation based on new information is known as feedback. One example of a closed-loop operation is the machine used in the manufacture of paper. Paper is formed when a suspension of pulpy fibers in **water** is emptied onto a conveyer belt whose surface is a sieve. Water drains out of the suspension, leaving the pulp on the belt. As the pulp dries, paper is formed. The **rate** at which the pulpy suspension is added to the conveyer belt can be automatically controlled by a machine. A sensing device at the end of the conveyor belt is capable of measuring the thickness of the paper and reporting back to the pouring machine on the condition of the product. If the paper becomes too thick, the sensor can tell the pouring machine to slow the rate at which pulpy suspension is added to the belt. If the paper becomes too thin, the sensor can tell the machine to increase the rate at which the raw material is added to the conveyor belt.

Many types of closed-loop machines exist, such as the papermaking machine. Some contain sensors, but are unable to make necessary adjustments on their own. Instead, sensor readings are sent to human operators who monitor the machine's operation and input any changes it may need to make in its functioning. Other closed-loop machines contain feedback mechanisms, like the papermaking machine described above. The results of the operation determine what changes, if any, the machine has to make

Still other closed-loop machines have feedforward mechanisms. That is, the first step they perform is to examine the raw materials that come to them and then decide what operations to perform. Letter-sorting machines are of this type. The first step such a machine takes in sorting letters is to read the zip code on the address and then send the letter to the appropriate sub-system.

The role of computers in automation

Since the 1960s, the nature of automation has undergone dramatic changes as a result of the availability of computers. For many years, automated machines were limited by the amount of feedback data they could collect and interpret. Thus, their operation was limited to a relatively small number of alternatives. When an automated machine is placed under the control of a computer, however, that disadvantage disappears. The computer can analyze a vast number of sensory inputs from a system and decide which of many responses it should make.

Artificial intelligence

The availability of computers has also made possible a revolution in the most advanced of all forms of automation, operations that are designed to replicate human thought processes. An automated machine is said to be "thinking" if the term is used for only the simplest of mental processes: "Should I do A or B," for example. The enormous capability of a computer, however, makes it possible for a machine to analyze many more options, compare options with each other, consider possible outcomes for various options, and perform basic reasoning and problem-solving steps not contained within the machine's programmed memory. At this point, the automated machine is approaching the types of mental functions normally associated with human beings, and is therefore said to have **artificial intelligence**.

If not quite human-level intelligence, at least animal-level. For example, researchers at Genobyte, a company in Boulder, CO, developed Robokoneko, a robot kitten (ko=child, neko=cat, in Japanese). The cat's **brain**, the Cellular Automata Machine (CAM) designed by Hugo de Garis of Advanced Telecommunications Research (Kyoto, Japan), contains nearly 40 million artificial neurons. The CAM-Brain's neurons are real electronic devices, rather than the software simulations used in most artificial intelligence research. The neural net circuit modules, with a maximum of 1,000 neurons each, are evolved at speeds of less than a second using a special computer chip, called a field programmable gate array (FPGA). Researchers hope the CAM-Brain will, for the first time, allow a robot to interact with stimuli in its environment to develop the sort of intelligence seen in animals, and that it will allow the kitten to walk, turn, jump, arch its back, and sit up.

Applications

Manufacturing companies in virtually every industry are achieving rapid increases in productivity by taking advantage of automation technologies. When one thinks of automation in manufacturing, robots usually come to mind. The automotive industry was the early adopter of **robotics**, using these automated machines for material handling, processing operations, and assembly and inspection. Donald A. Vincent, executive vice president, Robotic Industries Association, predicts a greater use of robots for assembly, paint systems, final trim, and parts transfer will be seen in the near future.

Vincent expects other industries to heavily invest in robotics as well. Industries such as the **electronics** industry, with its need for mass customization of electronic goods, the miniaturization of electronics goods and their internal components, and the re-standardization of the semiconductor industry, which, he says, will completely retool itself by 2004. Robotics will continue to expand into the food and beverage industry where they will perform such tasks as packaging, palletizing, and filling; as well as the aerospace, appliance, and non-manufacturing markets.

One can break down automation in production into basically three categories: fixed automation, programmable automation, and flexible automation. The automotive industry primarily uses fixed automation. Also known as "hard automation," this refers to an automated production facility in which the sequence of processing operations is fixed by the equipment layout. A good example of this would be an automated production line where a series of workstations are connected by a transfer system to move parts between the stations. What starts as a piece of sheet **metal** in the beginning of the process, becomes a car at the end.

Programmable automation is a form of automation for producing products in batches. The products are made in batch quantities ranging from several dozen to several thousand units at a time. For each new batch, the production equipment must be reprogrammed and changed over to accommodate the new product style.

Flexible automation is an extension of programmable automation. Here, the variety of products is sufficiently limited so that the changeover of the equipment can be done very quickly and automatically. The reprogramming of the equipment in flexible automation is done off-line; that is, the programming is accomplished at a computer terminal without using the production equipment itself.

Computer numerical control (CNC) is a form of programmable automation in which a machine is controlled by numbers (and other symbols) that have been coded into a computer. The program is actuated from the computer's memory. The machine tool industry was the first to use numerical control to control the position of a cutting tool relative to the work part being machined. The CNC part program represents the set of machining instructions for the particular part, while the coded numbers in the sequenced program specifies x-y-z coordinates in a Cartesian axis system, defining the various positions of the cutting tool in relation to the work part.

In the home

There has been a lot of talk about the "automated home" where everything in the house is networked and controlled via the computer through digital technology. Experts predict that eventually a homeowner will be able to set the **thermostat**, start the oven, and program a **DVD** player from the office or the home study thanks to computers.

The human impact of automation

The impact of automation on individuals and societies has been profound. On one level, many otherwise unpleasant and/or time-consuming tasks are now being performed by machines: dishwashing being one of the obvious examples. The transformation of the communications industry is another example of how automation has enhanced the lives of people worldwide. Today, millions of **telephone** calls that once would have passed through human operators are now handled by automatic switching machines.

Other applications of automation in communications systems include local area networks (LAN) and communications satellites. A LAN operates like an automated telephone company, however, they can transmit not only voice, but also digital data between terminals in the system. Satellites, necessary for transmitting telephone or video signals throughout the world, depend on automated guidance systems to place and retain the satellites in predetermined orbits.

For banking, automatic tellers are ubiquitous. The medical industry employs robots to aid the doctor in analyzing and treating patients. Automatic reservation, navigation, and instrument landing systems, not to mention automatic pilots have revolutionized the travel industry.

However, automation has also resulted in drastic dislocations in employment patterns. When one machine can do the work of ten workers, most or all of those people must be relocated or retrained to learn newer and higher skills. Whether or not this is a wholly negative impact has been strongly debated. As population and consumer demand for the products of automation increases, new jobs have been created.

Positive impacts on employment patterns include computerized programs that help designers in many fields develop and test new concepts quickly, without ever building a physical prototype. Automated systems also make it much easier for people to carry out the work they do in non-traditional places. They may be able to stay home, for example, and do their jobs by communicating with other individuals and machines by means of highly automated communications systems.

Resources

Books

"Automatic Control in Industry." In *The Illustrated Science and Invention Encyclopedia*. Vol. 2, pp. 192-96. Westport, CT: H. S. Stuttman, Inc., Publishers, 1982.

Braungart, Michael and William McDonough. *Cradle to Cradle: Remaking the Way We Make Things*. North Point Press, 2002.

Decelle, Linda S. "Automation." In *McGraw-Hill Encyclopedia of Science & Technology*. 7th ed. Vol. 1, pp. 300-04. New York: McGraw-Hill Book Company, 1992.

Dunlop, John. *Automation and Technological Change*. Englewood Cliffs, NJ: Prentice-Hall, 1962.

O'Brien, Robert, and the Editors of Life magazine. *Machines*. New York: Time Incorporated, 1964.

Stringer, Howard. *Opening Keynote Consumer Electronics Show in Las Vegas*. 1999.

Trefil, James. *Encyclopedia of Science and Technology*. The Reference Works, Inc., 2001.

Periodicals

"Technology Bulletin." *Design News* (February 15, 1999).

"Time-Triggered Control Network For Industrial Automation." *Assembly Automation* Author/s: Heffernan Volume: 22, no. 1 (2002): 60-68.

Vincent, Donald. "The North American Robotics Industry: Leading the Charge to a Productive 21st Century." *RAS Robotics and Automation Society* (Sept. 1999).

David E. Newton

Automobile

Few inventions in modern times have had as much impact on human life and on the global environment as the automobile. Automobiles and trucks have had a strong influence on the history, economy, and social life of much of the world.

Entire societies, especially those of the industrialized countries, have been restructured around the power of rapid, long-distance movement that the automobile confers on individuals and around the flexible distribution patterns made possible by trucks. Automobiles have given great freedom of movement to their owners, but encourage sprawl (i.e., straggling, low-density urban development). Sprawl degrades landscapes and produces traffic congestion that tends to immobilize the automobiles that make sprawl itself possible. Furthermore, the dependence on **petroleum** fuel of automobiles and trucks, and thus of the economies based on these machines, imposes strong patterns on global politics, moving industrial societies such as the United States (which consumes approximately 25% of the world's oil production) to be deeply concerned with the power struggles of the Persian Gulf, where approximately 70% of the world's oil reserves are located.

The automobile is also a significant health hazard, both directly and indirectly. According to the United Nations, over a million people (both vehicle occupants and pedestrians) die every year on the world's roads; the United States alone loses over 40,000 lives annually to car crashes. Meanwhile, automobile exhausts are the largest single source of **air pollution** (in the United States). Air **pollution** from the manufacture and operation of automobiles contributes importantly to crop damage, **acid rain**, destruction of the **ozone** layer, lung **disease**, and early death by a variety of health problems. Sixteen **percent** of the world's annual greenhouse-gas production comes from automobiles; greenhouse gasses contribute to **global climate** change, which may disrupt food production and flood coasts and low islands worldwide.

The first automobiles, constructed in the late nineteenth century, were essentially horse-drawn carriages with the **horses** removed and engines installed. After more than a century of development, the modern automobile is a sophisticated system. It combines fuel efficiency and speed to offer the mobility and flexibility of use demanded by an enormous variety of lifestyles and industries. Automobiles affect every aspect of society, from the design of our cities, to police, ambulance, fire, and utility services, to such personal uses as vacation travel, dining, and shopping. **Mass production** techniques, first developed for the automobile in the early twentieth century, have been adapted for use in nearly every industry, providing for the efficient and inexpensive production of products.

Trucks, especially 18-wheel tractor-trailer trucks, have become the major form of transporting goods across the country, allowing, for example, produce to be quickly transported to markets while still fresh. The use of automotive technology—tractors, combines, pickers, sprayers, and other self-propelled machines—in agriculture has en-

abled farmers to increase the quantity, quality, and variety of our foods, albeit at the price of increased **soil erosion** rates, agricultural dependence on petroleum-derived fuels and **fertilizers**, and increased runoff pollution. Meanwhile, dozens of industries depend, directly or indirectly, on the automobile. These industries include producers of **steel** and other metals, **plastics**, rubber, **glass**, fabrics, petroleum products, and electronic components.

Structure of the automobile

Thousands of individual parts make up the modern automobile. Much like the human body, these parts are arranged into several semi-independent systems, each with a different function. For example, the human **circulatory system** comprises the **heart**, **blood** vessels, and blood. The automobile contains analogous circulatory systems for coolant fluid (mostly **water**), for lubricating oil, and for fuel. The engine—the "heart" of the automobile—is comprised of pistons, cylinders, tubes to deliver fuel to the cylinders, and other components. Each system is necessary for making the automobile run and reducing noise and pollution.

The major systems of an automobile are the engine, fuel system, transmission, electrical system, cooling and lubrication system, and the chassis, which includes the suspension system, braking system, wheels and tires, and the body. These systems will be found in every form of motor vehicle and are designed to interact with and support each other.

Design factors

When an automobile is designed, the arrangement, choice, and type of components depend on various factors. The use of the automobile is one factor. Some cars are required only for local driving; these cars may be capable of achieving good fuel economy on short trips, but they may be less comfortable to drive at high speeds. A sports car, built for speed, will have enhanced steering and handling abilities, but requires a stronger engine, more fuel, and a more sophisticated suspension system. Yet, an automobile must also be flexible enough to perform in every situation and use.

Other factors in the design of automobiles include the requirements for pollution-control components that have been placed on the modern automobile. Safety features are also a factor in the automobile's design, affecting everything from the braking and steering systems to the materials used to construct the body. The design of the body must incorporate standards of safety, size and weight, **aerodynamics** or ways to reduce the **friction** of airflow, and appearance.

Henry Ford and his first automobile. *Bettmann/Corbis. Reproduced by permission.*

The choice of front-wheel drive allows for a smaller, more fuel-efficient car. But the arrangement of the engine and its relationship to other automobile systems will be different from a rear-wheel-driven car. Independent suspension for all four wheels improves the automobile's handling, safety, and comfort, but requires a more complex arrangement. The use of computer technology, the most recently added system to the automobile, requires changes in many of the car's other systems. Lastly, cost is an important factor in the design of a car. Many features useful for improving the various systems and characteristics of an automobile may make it too costly to produce and too expensive for many people to buy.

The design of an automobile, therefore, is a balance of many factors. Each must be taken into consideration, and compromises among features satisfy as many factors as possible. Yet, for all the variety among automobiles, the basic systems remain essentially the same.

Interaction of systems

Before examining the components of each system, it is useful to understand how the systems interact. The human body is again a good example of the interaction of systems. The heart pumps blood, which feeds the tissues of the body while at the same time helping to remove impurities. The tissues, fed by blood, are able to perform their tasks, and often are required to support the action of the heart. Muscle **tissue**, for example, depends on the availability of oxygen-rich blood from the heart in order to move the body.

The **internal combustion engine** is the heart of the automobile. The engine produces **energy** from fuel and converts that energy into the power to move the different components that will move the car. The engine converts the chemical energy produced by the burning of the fuel into mechanical energy. This energy is used to spin a shaft. The spinning shaft, through the interaction of the transmission and other components, causes the wheels to turn and the car to move. A similar transfer of energy to **motion** can be seen when bicycling. The up and down motion of the feet and legs is converted to the turning motion of the pedals, which in turn pulls the chain that causes the rear wheel to spin.

Just as it is more difficult to pedal a bicycle from a standstill than it is while already rolling, the engine requires the electrical system to give it the push to move on its own. The electric starter motor of an automobile provides a powerful **force** to give the engine its initial movement. The **battery** supplies energy for the engine to use when burning the fuel needed to make it run. The alternator is driven by a belt attached to the engine, recharging the battery so there will be a constant supply of energy. The sensors of the computer control system, which governs many of the processes in an automobile, also require **electricity**.

The burning of fuel is a hot, noisy process that also produces pollutants in the form of exhaust. This exhaust must be carried away from the engine and away from the automobile. The exhaust system, with its muffler, also acts to reduce the noise produced by the vehicle. Burning fuel in the cylinders produces two other results: friction and extremely high temperatures. In order to protect the parts from being worn down from the friction and from melting with the **heat**, they must be properly lubricated and cooled. These systems depend on the engine and the electrical system for the power to perform their tasks.

The engine's power is used to turn the automobile's wheels. Because the tires are the only parts of the automobile that are actually in contact with the road, they must rest on a system of supports that will carry the weight of the car and respond to conditions of the road surface. At the same time, the driver must be capable of guiding the direction of the automobile. Once an automobile is moving, it will continue to move until some sort of friction, the brake, is applied to stop it.

The wheels, suspension, steering, and braking systems are all attached to the car's chassis, as is the rest of the automobile. The chassis and body, analogous to the skeletal structure in the human body, provide support for all the various systems and components, while also providing safety, comfort, and protection from the elements for the automobile's passengers.

Engine

The engine operates on internal **combustion**; that is, the fuel used for its power is burned inside the engine. This burning occurs inside cylinders. Within the cylinder is a piston. When the fuel is burned, it creates an explosive force that causes the piston to move up and down.

The piston is attached, via a connecting rod, to a crankshaft, where the up and down movement of the piston converts to a circular motion. When bicycling, the upper part of a person's leg is akin to the piston. Power from the leg is passed through the pedal in order to turn the crank.

Gasoline is the most common automobile fuel. It is pulled into the cylinder by the **vacuum** created as the piston moves down through the cylinder. The gasoline is then compressed up into the cylinder by the next movement of the piston. A spark is introduced through a spark plug placed at the end of the cylinder. The spark causes the gasoline to explode, and the explosion drives the piston down again into the cylinder. This movement, called the power stroke, turns the crankshaft. A final movement of the piston upward again forces the exhaust gases, the byproducts of the fuel's combustion, from the cylinder. These four movements—intake, compression, power, and exhaust—are called strokes. The four-stroke engine is the most common type of automobile engine.

Most automobiles have from four to eight cylinders, although there are also two-cylinder and 12-cylinder automobiles. The cylinders work together in a sequence to turn the crankshaft, so that while one cylinder is in its intake stroke, another is in the compression stroke, and so forth. Generally, the more cylinders, the more smoothly the engine will run. The size of the automobile will affect the number of cylinders the engine uses. Smaller cars generally have the smaller four-cylinder engine. Mid-sized cars will generally require a six-cylinder engine, while larger cars need the power of an eight-cylinder engine.

The number of cylinders, however, is less important to the level of an engine's power than is its displacement. Displacement is a measure of the total **volume** of fuel mixture moved by all the pistons working together. The more fuel burned at one time, the more explosive the force, and thus, the power will be. Displacement is often expressed as cubic centimeters (cc) or as liters. A smaller engine will displace 1,200 cc (1.2 L) for 60 horsepower, while a larger engine may displace as much as 4,000 cc (4 L), generating more than 100 horsepower. Horsepower is the measurement of the engine's ability to perform work. The size and weight of the car also affects its power. It takes less work to propel a lighter car than a heavier car, even if they have the same engine, just as a horse carrying a single rider can go faster with less effort than a horse drawing a cart.

Fuel system

Gasoline must be properly mixed with air before it can be introduced into the cylinder. The combination of gasoline and air creates a more volatile explosion. The fuel pump draws the gasoline from the gas tank mounted toward the rear of the car. The gasoline is drawn into a carburetor on some cars, while it is fuel-injected on others; both devices mix the gasoline with air (approximately 14 parts of air to one part of gasoline) and spray this mixture as a fine mist into the cylinders. Other parts of the fuel system include the air cleaner, which is a filter to ensure that the air mixed into the fuel is free of impurities; and the intake manifold, which distributes the fuel mixture to the cylinders.

Exhaust system

After the fuel is burned in the pistons, the gases and heat created must be discharged from the cylinder to make room for the next infusion of fuel. The exhaust system is also responsible for reducing the noise caused by the explosion of the fuel.

Exhaust gases are discharged from the cylinder through an exhaust valve. The exhaust gathers in an exhaust manifold before eventually being channeled through the exhaust pipe and muffler and finally out the tailpipe and away from the car. The muffler is constructed with a maze of what are called baffles, specially developed walls that absorb energy, in the form of heat, force, and sound, as the exhaust passes through the muffler.

The burning of fuel creates additional byproducts of hazardous gases—hydrocarbons, **carbon monoxide**, and **nitrogen** oxide—which are harmful both to the engine's components and to the environment. The **emission** control system of a car is linked to the exhaust system, and functions in two primary ways. The first is to reduce the levels of unburned fuel. This is achieved by returning the exhaust to the fuel-air mixture injected into the cylinders to burn as much of the exhaust as possible. The second method is through a catalytic converter. Fitted before the muffler, the catalytic converter contains **precious metals** that act as catalysts. That is, they increase the **rate** of conversion of the harmful gases to less harmful forms.

Cooling system

The automobile uses an additional system to reduce the level of heat created by the engine. The cooling system also maintains the engine at a **temperature** that will allow it to run most efficiently. A liquid-cooled system is most commonly used.

The explosion of fuel in the cylinders can produce temperatures as high as 4,000°F (2,204°C); the temperature of exhaust gases, while cooler, still reach to 1,500°F (816°C). Liquid-cooling systems use water (mixed with an antifreeze that lowers the freezing point and raises the **boiling point** of water) guided through a series of jackets attached around the engine. As the water **solution** circu-

lates through the jackets, it absorbs the heat from the engine. It is then pumped to the radiator at the front of the car, which is constructed of many small pipes and thin **metal** fins. These allow a large surface area to draw the heat from the water solution. A fan attached to the radiator uses the **wind** created by the movement of the car to cool the water solution further. Temperature sensors in the engine control the operation of the cooling system, so that the engine remains in its optimal temperature range.

Lubrication

Without proper lubrication, the heat and friction created by the rapid movements of the engine's parts would quickly cause it to fail. The lubrication system of an automobile acts to reduce engine wear caused by the friction of its metal parts, as well as to carry off heat. At the bottom of the engine is the crankcase, which holds a supply of oil. A pump, powered by the engine, carries oil from the crankcase and through a series of passages and holes to all the various parts of the engine. As the oil flows through the engine, it forms a thin layer between the moving parts, so that they do not actually touch. The heated oil drains back into the crankcase, where it cools. The fumes given off by the crankcase are circulated by the PCV (positive crankcase ventilation) valve back to the cylinders, where they are burned off, further reducing the level of pollution given off by the automobile.

Electrical system

Electricity is used for many parts of the car, from the headlights to the **radio**, but its chief function is to provide the electrical spark needed to ignite the fuel in the cylinders. This is performed by an electrical system comprises a battery, starter motor, alternator, distributor, ignition coil, and ignition switch. As discussed above, the starter motor is necessary for generating the power to carry the engine through its initial movements. Initial voltage is supplied by the battery, which is kept charged by the alternator. The alternator creates electrical current from the movement of the engine, much as windmills and watermills generate current from the movement of air or water.

Turning the key in the ignition switch draws current from the battery. This current, however, is not strong enough to provide spark to the spark plugs so it is drawn through the ignition coil, which is comprised of the tight primary winding and the looser secondary winding. The introduction of current between these windings creates a powerful magnetic field. Interrupting the current flow, which happens many times a second, causes the magnetic field to collapse. The collapsing of the magnetic field produces a powerful electrical surge. In this way, the 12-

volt current from the battery is converted to the 20,000 volts needed as spark to ignite the gasoline.

Because there are two or more cylinders, and therefore as many spark plugs, this powerful current must be distributed—by the distributor—to each spark plug in a carefully controlled sequence. This sequence must be timed so that the cylinders and the pistons powering the crankshaft work smoothly together. For this reason, most automobiles manufactured today utilize an electronic ignition, in which a computer precisely controls the timing and distribution of current to the spark plugs.

Transmission

Once the pistons are firing and the crankshaft is spinning, this energy must be converted, or transmitted, to drive the wheels. But the crankshaft spins only within a limited range, usually between 1,000 to 6,000 revolutions per minute (rpm), and this is not enough power to cause the wheel to turn when applied directly. The transmission accomplishes the task of bringing the engine's **torque** (the amount of twisting force the crankshaft has as it spins) to a range that will turn the wheels. One way to experience the effect of torque is by using two wrenches, one with a short handle, the other with a long handle. It may be difficult to turn a nut with the shorter wrench; but the nut turns much more easily with the longer wrench, because it allows a more powerful twisting force. The transmission has two other functions: allowing reverse movement, and braking the engine.

There are two types of transmission: manual and automatic. With a manual transmission, the driver controls the shifting of the **gears**. In an automatic transmission, as its name implies, gears are engaged automatically. Both types of transmission make use of a clutch, which allows the gears to be engaged and disengaged.

Automobiles generally have at least three forward gears plus a reverse gear, although many manual transmissions have four or even five gears. Each gear provides a different **ratio** of the number of revolutions per minute (rpms) of the crankshaft (the power input) to the number of revolutions per minute of the output of the transmission, directed to the wheels. In first gear, for example, which is needed to move the automobile from a standstill, the ratio of input to output is 3.5 to one or even higher. The greater the ratio, the more torque will be achieved in the output. Each successively higher gear has a lower input to output ratio. This is because once the automobile is rolling, progressively less torque is needed to maintain its movement. The fourth and fifth gears found on most cars are used when the engine has achieved higher speeds; often called overdrive gears, these gears allow the input to output ratio to sink lower than one to one. In other words,

the wheels are spinning faster than the crankshaft. This allows for higher speeds and greater fuel efficiency.

Chassis

The chassis is the framework to which the various parts of the automobile are mounted. The chassis must be strong enough to bear the weight of the car, yet somewhat flexible in order to sustain the shocks and tension caused by turning and road conditions. Attached to the chassis are the wheels and steering assembly, the suspension, the brakes, and the body.

The suspension system enables the automobile to absorb the shocks and variations in the road surface, keeping the automobile stable. Most cars feature independent front suspension, that is, the two wheels in front are supported independently of each other. In this way, if one wheel hits a bump while the other wheel is in a dip, both wheels will maintain contact with the road. This is especially important because steering the automobile is performed with the front wheels. More and more cars also feature independent rear suspension, improving handling and the smoothness of the ride.

The main components of the suspension system are the springs and the shock absorbers. The springs suspend the automobile above the wheel, absorbing the shocks and bumps in the road surface. As the chassis bounces on the springs, the shock absorbers act to dampen, or quiet, the movement of the springs, using tubes and chambers filled with hydraulic fluid.

The steering system is another part of the suspension system. It allows the front wheels to guide the automobile. The steering wheel is attached to the steering column, which in turn is fitted to a gear assembly that allows the circular movement of the steering wheel to be converted to the more linear, or straight, movement of the front wheels. The gear assembly is attached to the front axle by tie rods. The axle is connected to the hubs of the wheels.

Wheels and the tires around them are the only contact the automobile has with the road. Tires are generally made of layers of rubber or synthetic rubber around steel fibers that greatly increase the rubber's strength and ability to resist puncture. Tires are inflated with air at a level that balances the greatest contact with the road surface with the ability to offer puncture resistance. Proper inflation of the tires will decrease wear on the tires and improve fuel efficiency.

Body

The body of a car is usually composed of steel or **aluminum**, although fiberglass and plastic are also used. The body is actually a part of the chassis, the whole formed by **welding** stamped components into a single unit. While the body forms the passenger compartment, offers storage **space**, and houses the automobile's systems, it has other important functions as well. Passenger safety is achieved by providing structural support strong enough to withstand the force of an accident. Other parts of the car, such as the front and hood, are designed to crumple in a crash, thereby absorbing much of the impact. A firewall between the engine and the interior of the car protects the passengers in case of an engine fire. Lastly, the body's shape contributes to reducing the level of wind resistance as the car moves, allowing the driver better handling and improving fuel efficiency.

Hybrids

In the last few years several makes of affordable hybrid cars have appeared on the U.S. market. Hybrid cars burn gasoline in an efficient engine to produce electricity. This power is then stored in batteries and used to run an **electric motor**, and power from the electric motor is combined with power from the gasoline motor to move the car. Today's commercially-available hybrid cars produce a small fraction of the air pollution per mile traveled that is produced by conventional cars of comparable size, and get significantly better mileage as well.

Resources

Books

Duffy, James E. *Modern Automotive Mechanics.* Tinley Park, IL: Goodheart-Willcox Publisher, 1990.

Lewis, David L., and Laurence Goldstein, eds. *The Automobile and American Culture.* Ann Arbor: University of Michigan Press, 1986.

Magliozzi, Tom and Ray. *Car Talk.* New York: Dell Publishing, 1991.

Thiessen, Frank J., and David N. Dales. *Automotive Principles and Service.* 4th ed. Upper Saddle River, NJ: Prentice Hall, 1993.

Other

Fowler, Jonathan. "Traffic Deaths on Rise Globally." Associated Press. August 29, 2002 [cited October 19, 2002]. <http://www.twincities.com/mld/twincities/3962244.htm>.

Hermance, David, and Shoichi Sasaki. "Hybrid Vehicles Take to the Streets." IEEE Spectrum. November, 1998 [cited October 19, 2002]. <http:// www.spectrum.ieee.org/select/1198/hyb.html>.

International Center for Technology Assessment. "The Real Price of Gasoline" 2002 [cited October 17, 2002]. <http://www.icta.org/projects/ trans/realpricegas.pdf>.

U.S. Census Bureau. "2001 Statistical Abstract of the United States" [cited Oct. 19, 2002]. <http://www.census.gov/prod/2002pubs/01statab/trans.pdf>.

M.L. Cohen

Autonomic nervous system *see* **Nervous system**

KEY TERMS

Catalyst—Any agent that accelerates a chemical reaction without entering the reaction or being changed by it.

Combustion—A form of oxidation that occurs so rapidly that noticeable heat and light are produced.

Friction—A force caused by the movement of an object through liquid, gas, or against a second object that works to oppose the first object's movement.

Gear—A wheel arrayed with teeth that meshes with the teeth of a second wheel to move it.

Ratio—A measurement in quantity, size, or speed of the relationship between two or more things.

Shaft—A rod that, when spun, can be used to move other parts of a machine.

Torque—The ability or force needed to turn or twist a shaft or other object.

Voltage—Measured in volts, the amount of electrons moved by an electric current or charge.

Autotroph

An autotroph is an **organism** able to make its own food. Autotrophic organisms take inorganic substances into their bodies and transform them into organic nourishment. Autotrophs are essential to all life because they are the primary producers at the base of all food chains. There are two categories of autotrophs, distinguished by the **energy** each uses to synthesize food. Photoautotrophs use **light** energy; chemoautotrophs use chemical energy.

Photoautotrophs

Plants are the most abundant and recognizable autotrophs on **Earth**. If you have noticed a houseplant on a windowsill imperceptibly turn its leaves toward the **sun**, you have probably guessed that plants are photoautotrophs. **Plant** leaves soak up the energy in sunlight and use it to make food. Plants take in **water** through their roots and atmospheric **carbon dioxide** through their leaves. Plant cells absorb light energy to fuel the synthesis of inorganic **hydrogen**, **oxygen**, and **carbon** into a sugar that nourishes them. This process is known as **photosynthesis**.

Because plants, as autotrophs, make living **tissue** solely out of nonliving material, they form the foundation of all food chains. Can you think of one thing you eat that does not, ultimately, come from plants? Plants are called primary producers because they create themselves out of transformed inorganic **matter** and, thus, are the "original food" that sustains all living things.

Chemoautotrophs

Until recently, scientists believed there existed only a few kinds of **bacteria** that used chemical energy to create their own food. Some of these bacteria were found living near vents and active volcanos on the lightless **ocean** floor. The bacteria create their food using inorganic **sulfur** compounds gushing out of the vents from the hot interior of the **planet**.

In 1993, scientists found many new **species** of chemoautotrophic bacteria living in fissured rock far below the ocean floor. These bacteria take carbon dioxide and water into their bodies and use the chemical energy in sulfur compounds to create nourishing carbohydrates and sugars. A unique characteristic of these chemoautotrophic bacteria is that they thrive at temperatures high enough to kill other organisms. Some scientists believe these unique bacteria should be classified in their own new taxonomic kingdom.

Avalanche *see* **Mass wasting**

Aviation meteorology *see* **Meteorology**

Avocado *see* **Laurel family (Lauraceae)**

Avocets *see* **Stilts and avocets**

Avogadro's number

Avogadro's number is the number of particles in one **mole** of any substance. Its numerical value is 6.02225×10^{23}. One mole of **oxygen** gas contains 6.02×10^{23} molecules of oxygen, while one mole of **sodium chloride** contains 6.02×10^{23} sodium ions and 6.02×10^{23} chloride ions. Avogadro's number is used extensively in calculating the volumes, masses, and numbers of particles involved in chemical changes.

The concept that a mole of any substance contains the same number of particles arose out of research conducted in the early 1800s by the Italian physicist Amedeo Avogadro (1776-1856). Avogadro based his work on the earlier discovery by Joseph Gay-Lussac that gases combine with each other in simple, whole-number ratios of volumes. For example, one liter of oxygen combines with two liters of **hydrogen** to make two liters of **water** vapor.

Avogadro argued that the only way Gay-Lussac's discovery could be explained was to assume that one liter of any gas contains the same number of particles as one liter of any other gas. To explain the water example above, he further hypothesized that the particles of at least some gases consist of two particles bound together, a structure to which he gave the name **molecule**.

The question then becomes, "What is this number of particles in a liter of any gas?" Avogadro himself never attempted to calculate this. Other scientists did make that effort, however. In 1865, for example, the German physicist J. Loschmidt estimated the number of molecules in a liter of gas to be 2.7×10^{22}. The accepted value today is 2.69×10^{22}.

For all elements and compounds, not just gases, a given weight must contain a certain number of **atoms** or molecules. A weight (in grams) equal to the atomic or **molecular weight** of the substance-that is, one mole of any element or compound-must contain the same number of atoms or molecules, because there is always a constant relationship between atomic weights and grams. (One atomic **mass** unit = 1.66×10^{-22} g.) The number of atoms or molecules in one mole of an element or compound has been named Avogadro's number, in honor of his realization about the numbers of particles in gases. As stated above, that number has been determined to be 6.0225×10^{23}.

See also Atomic weight.

Axolotyl *see* **Salamanders**

Aye-ayes

The aye-aye (*Daubentonia madagascariensis*) is a rare tree-dwelling **animal** that is found only at a few localities along the eastern half of Madagascar, off eastern **Africa**. It is a member of a group of primitive **primates** known as **Prosimians**, most of which are **lemurs**. The aye-aye is the only surviving member of the family Daubentoniidae; a slightly larger relative (*D. robusta*) became extinct about 1,000 years ago.

Only a handful of aye-ayes survive in the wild, and it is a critically **endangered species**. It has suffered from extensive loss of its natural **habitat** of humid tropical forest, which now exists only in fragmented remnants, including 16 protected areas. The aye-aye is also killed by local people, who believe that it is an evil omen.

Because of its strange appearance and **behavior**, the aye-aye is sometimes referred to as the most bizarre of all primates. When it was first encountered by Europeans, it was incorrectly classified as a rodent. This was because of its large ears, squirrel-like bushy tail, and large, continuously growing incisor teeth. This primate has large eyes set in a cat-like face, which give it good night **vision**, and dark brown to black fur with long guard-hairs. Adults are approximately 3.3 ft (1 m) long and weigh up to 6.5 lb (3 kg).

The aye-aye's digits end in claws, except for the big toe and thumb, which have flat nails. The middle finger of each hand is extremely long and thin, and has a long claw that is used to dig insect larvae out of **tree bark** and crevices. When an aye-aye hears the movement of a beetle larva under the bark of a tree, it gnaws through the **wood** to uncover the tunnel of the grub, and then inserts its long finger to catch it. Research indicates that by tapping a branch, aye-ayes can locate grubs using a sort of **echolocation**. Aye-ayes also eat fruit, bamboo shoots, and free-ranging **insects**.

Aye-ayes forage at night, mostly in trees. They **sleep** during the day in a globe-shaped nest woven of leaves and twigs. They may spend several days in one nest before moving on to another. Several animals may share the same home range, but they are solitary animals, except for mothers with young. They have lived for as long 26 years in captivity.

The nocturnal habits of aye-ayes make it difficult to observe them in the wild, so little is known about their breeding and behavior. They are thought to breed every two or three years. A single infant is born after a gesta-

A young aye-aye (*Daubentonia madagascariensis*). *Will McIntyre/Photo Researchers, Inc. Reproduced by permission.*

tion period of about 158 days (according to observations in captivity). In April 1992, an aye-aye named "Blue Devil" was born at Duke University's primate center (it was named after the university mascot). It was the first aye-aye born in captivity in the Western Hemisphere.

roots, and bulbs. Females and young hamadryas baboons have a brown coat, while adult males have a silver-gray mane over their shoulders and brilliant red bare skin on their face and around their genitals. This baboon was sacred to the ancient Egyptians; it is depicted in Egyptian art as an attendant or representative of the god Thoth, the god of letters and scribe of the gods. This species has been exterminated in Egypt and its numbers are reduced elsewhere in its range.

The Guinea baboon (*P. papio*) also feeds on grass, fruit, seeds, insects, and small animals and is found in the savanna woodlands of Senegal and Sierra Leone. These baboons have a brown coat, bare red skin around their rump, and a reddish brown face.

The drill (*Mandrillus leucophaeus*) lives in the rainforests of southeast Nigeria, western Cameroon, and Gabon, and feeds on fruit, seeds, **fungi**, roots, insects, and small animals. Drills have a brown-black coat and a naked rump ranging from blue to purple. A fringe of white hair surrounds a black face and the long muzzle has ridges along its sides.

The mandrill (*Mandrillus sphinx*) is the largest monkey. It lives in southern Cameroon, Gabon, and Congo in rainforests. Its diet includes fruit, seeds, fungi, roots, insects, and small animals. Mandrill males have a blue-purple bare rump, and a bright red stripe running down the middle of their muzzle with blue ridges on the sides and a yellow beard. Female and the young mandrills are similarly colored, but less brilliantly.

Gelada baboons (*Theropithecus gelada*) are found in the **grasslands** of Ethiopia, where they eat grass, roots, bulbs, seeds, fruit, and insects. Geladas have a long brown coat, hair that is cream colored at the tips and a long mantle of hair covering their shoulders and giving them the appearance of wearing a cape. The bare rump of both sexes is red and somewhat fat. Both sexes also have an area of bare red skin around their necks and the females have small white blister-like lumps in this area of their bodies that swell during menstruation.

Social behavior

Baboon social **behavior** is matrilineal, in which a network of social relationships are sustained over three generations from the female members of the species. A troop of baboons can range in number from 30 to over 200 members, depending upon the availability of food. The baboon troop consists of related bands composed of several clans, where each clan may have a number of smaller harem families made up of mothers, their children, and a male. Female baboons remain with the group into which they are born for the duration of their lives,

while the males leave to join other troops as they become mature.

Ranking within the group of females begins with the mother, with female offspring ranking below their mothers. Adult females are either nursing or pregnant for most of their lives, and they spend a great deal of time with other female friends, avoiding the males. During a daytime rest period, the females gather around the oldest female in the troop and lie close together. The way baboons huddle together while they are resting and the other movements in their troops are defense measures against outsiders and predators. The dominant males travel in the center of the troop to keep order among the females and the juveniles, while the younger males travel around the outer fringes of the group.

During their rest periods, baboons spend considerable time grooming one another, which helps to reinforce their social bonds. Females have male baboons that assist them in caring for their infants and protecting them from danger. These males may not be the fathers, but they may later mate with the female.

Baboon friendships

When a young male baboon matures, he leaves the family group to join a new troop. His first gestures are toward an adult female who may make friends with him. His gestures of friendship include lip-smacking, grunting, and grooming. It will take several months of this kind of friendly behavior before a more permanent bond is established between them. A female may have friendships with more than one male, and when she is ready to mate she may do so with all of her male friends. Social grooming between females and males is always between friends. A female benefits from this relationship by gaining protection and help in caring for her offspring. A male benefits by having a female with which to mate. These friendships do not insure the male paternity of infants, but the social benefits seem to outweigh paternity rights in baboon societies.

There are often fights among males for the right to mate with females in estrus. Males do not have the strong ranking order that the females have among their family relatives, and males must compete for mating rights. Among some species of baboon, the older males mate more than younger dominant males because the older males have more friendships with females, and these bonds are of a longer duration.

Baboons, like people, will often trick their peers into getting something they want. Researchers once witnessed a young male baboon tricking an adult female into relinquishing the roots she was digging by signaling that he was being attacked. When his mother came to the

A savanna baboon (*Papio cynocephalus*) in South Luangwa National Park, Zambia. *Photograph by Malcolm Boulton. National Audubon Society Collection/Photo Researchers, Inc. Reproduced by permission.*

rescue and saw only the other female, she chased the other baboon into the woods. As soon as the root digger was chased away, the young male took the tubers for himself. In another example, a female baboon groomed an adult male until he became so relaxed he fell asleep and forgot his antelope meat. Once he was asleep, she took the meat for herself.

Food and foraging habits

Baboons have the same number of teeth and dental pattern as human beings. Baboons, like other members of the subfamily Cercopithecinae, have cheek pouches that can hold a stomach's worth of food. This enables them to literally eat on the run, and is helpful to them when they have to compete for food or avoid danger. Baboons can quickly fill up their pouches, then retreat to safety to eat at their leisure.

Baboons walk on all four limbs and their rear feet are plantigrade, meaning they walk on the whole foot, not just on their toes. The walking surface of their hands is the complete surface of their four fingers. When feeding, baboons tend to stand on three of their limbs and pluck food while eating with one hand. When baboons

are walking, their shoulders are higher than their hips and they are able to easily see what is going on around them as they forage for food.

Baboons are basically fruit-eaters, but they also eat seeds, flowers, buds, leaves, **bark**, roots, bulbs, rhizomes, insects, **snails**, **crabs**, **fish**, lizards, **birds**, and small mammals. Young baboons learn what to eat and what not to eat through trial and error. Adults monitor their choices and intervene to prevent younger baboons from eating unusual food. When **water** is not available, baboons dig up roots and tubers to find liquid and often dig holes in dry river beds to find water.

It has been observed that baboons adapt their food choices to what is available in their habitats. In some regions they have developed group hunting techniques.

Communication

Baboons have a complex system of communication that includes vocalizations, facial expressions, posturing, and gesturing. These vocalizations, which baboons use to express emotions, include grunts, lip-smacking, screams, and alarm calls. The intensity of the emotion is conveyed by repetition of the sounds in association with other forms of communication.

Baboons communicate with each other primarily through body gestures and facial expressions. The most noticeable facial expression is an open-mouth threat where the baboon bares the canine teeth. Preceding this may be an eyelid signal, raising the eyebrows and showing the whites of the eyes, that is used to show displeasure. If a baboon really becomes aggressive, the hair may also stand on end, threatening sounds will be made, and the ground will be slapped.

In response to aggressive facial expressions and body gestures, other baboons usually exhibit submissive gestures. A fear-face, a response to aggression, involves pulling the mouth back in what looks like a wide grin.

Presenting among baboons takes place in both sexual and nonsexual contexts. A female will approach a male and turn her rump for him to show that she is receptive. This type of presentation can lead to mating or to a special relationship between the pair. A female may present in the same way to an infant to let the infant know it may come close to her. She may use this body gesture as a simple greeting to a male, indicating that she respects his position. A female will also present to another female when she want to fondle the other female's infant. Males present to other males as a greeting signal. Their tails, however, are not raised as high as those of females when they present.

Presentation is also used as an invitation or request for grooming, and for protection. Baboons freely engage in embracing to show affection to infants and juveniles. The frontal embrace has also been seen as a gesture of reassurance between baboons when they are upset. Infants have their own forms of communication that involves a looping kind of walk, wrestling, and a play-face. The play-face is an open mouth gesture they use to try to bite one another.

Baboon models

Baboons were studied for a long time as models of primate behavior and used to help construct the **evolution** of human behavior. More recently, **chimpanzees** have been used as a model because of their genetically close relationship to humans, and because they exhibit toolmaking, some language-like ability, and some mathematical **cognition**. Until recently, baboon troops were thought to be male-dominated. Studies have now demonstrated the cohesive nature of the matrilineal structure of baboon society.

What continues to interest researchers about baboons is how adaptable they are, even when their habitats are threatened. Baboons have become skillful crop

KEY TERMS

Estrus—A condition marking ovulation and sexual receptiveness in female mammals.

Foraging—The act of hunting for food.

Matrilineal—Social relationships built around the mother and her children.

Plantigrade—Walking on the whole foot, not just the toes as cats and dogs do.

Presenting—Body gestures that signal intentions or emotions among baboons.

raiders in areas where their terrains have been taken up by humans for agriculture. In spite of this adaptability, some species are threatened or endangered. For example, the drill is highly endangered due to **habitat** destruction and commercial hunting. In 1984, in a unique study designed to save three troops of baboons from human encroachment, baboon biologist Shirley C. Strum translocated the 132 animals 125 mi (200 km) away from their original home to a harsh outpost of 15,000 acres (6,075 hectares) near the Ndorobo Reserve on the Laikipia Plateau north of Mt. Kenya. Every baboon of the three different groups were captured to ensure social integrity.

After release, Strum found that the translocated groups watched and followed local troops of baboons to find food and water. Within six weeks, indigenous males joined the translocated troops, providing a good source of intelligence about the area. Strum found that the translocated animals learned by trial and error. The topography of the area meant that some portions of the range had good food in the wet seasons and other portions had good food in the dry seasons. To feed efficiently, the baboons had to learn this difference and to switch their ranging seasonally. Today, there are still a few differences between indigenous and translocated baboons in behavior and diet, but on the whole, it is difficult to tell that the groups had different origins.

Resources

Books

Cheney, Dorothy L., and Robert M. Seyfarth. *How Monkeys See the World.* Chicago: University of Chicago Press, 1990.

Grzimek, Bernhard. *Encyclopedia of Mammals.* Vol. 2. New York: McGraw-Hill, 1990.

Loy, James, and Calvin B. Peters. *Understanding Behavior: What Primate Studies Tell Us about Human Behavior.* New York: Oxford University Press, 1991.

MacDonald, David, and Sasha Norris, eds. *Encyclopedia of Mammals.* New York: Facts on File, 2001.

Mason, William A., and Sally P. Mendoza. *Primate Social Conflict*. Albany: State University of New York Press, 1993.

Strum, Shirley C. *Almost Human: A Journey into the World of Baboons*. New York: Random House, 1987.

Periodicals

"Abnormal, Abusive, And Stress-Related Behaviors In Baboon Mothers." *Biological Psychiatry* 52, no. 11 (2002): 1047-1056.

Maestripieri, Dario. "Evolutionary Theory And Primate Behavior." *International Journal of Primatology* 23, no. 4 (2002): 703-705.

Strum, Shirley C. "Moving The Pumphouse Gang." *International Wildlife* (May/June 1998).

Vita Richman

Bacteria

Bacteria are mostly unicellular organisms that lack **chlorophyll** and are among the smallest living things on earth—only viruses are smaller. Multiplying rapidly under favorable conditions, bacteria can aggregate into colonies of millions or even billions of organisms within a **space** as small as a drop of **water**.

The Dutch merchant and amateur scientist Anton van Leeuwenhoek was the first to observe bacteria and other **microorganisms**. Using single-lens microscopes of his own design, he described bacteria and other microorganisms (calling them "animacules") in a series of letters to the Royal Society of London between 1674 and 1723.

Bacteria are classified as prokaryotes. Broadly, this taxonomic ranking reflects the fact that the genetic material of bacteria is contained in a single, circular chain of **deoxyribonucleic acid (DNA)** that is not enclosed within a nuclear **membrane**. The word **prokaryote** is derived from Greek meaning "prenucleus." Moreover, the DNA of prokaryotes is not associated with the special **chromosome proteins** called histones, which are found in higher organisms. In addition, prokaryotic cells lack other membrane-bounded organelles, such as mitochondria. Prokaryotes belong to the kingdom Monera. Some scientists have proposed splitting this designation into the kingdoms **Eubacteria** and **Archaebacteria**. Eubacteria, or true bacteria, consist of more common **species**, while Archaebacteria (with the prefix archae—meaning ancient) represent strange bacteria that inhabit very hostile environments. Scientists believe these bacteria are most closely related to the bacteria which lived when the **earth** was very young. Examples of archaebacteria are those bacteria which currently live in extremely salty environments or extremely hot environments, like geothermal vents of the **ocean** floor.

Characteristics of bacteria

Although all bacteria share certain structural, genetic, and metabolic characteristics, important biochemical differences exist among the many species of bacteria. These differences permit bacteria to live in many different, and sometimes extreme, environments. For example, some bacteria recycle **nitrogen** and **carbon** from decaying organic **matter**, then release these gases into the atmosphere to be reused by other living things. Other bacteria cause diseases in humans and animals, help digest sewage in treatment plants, or produce the **alcohol** in wine, beer and liquors. Still others are used by humans to break down toxic waste chemicals in the environment, a process called **bioremediation**.

The cytoplasm of all bacteria is enclosed within a **cell** membrane surrounded by a rigid cell wall whose polymers, with few exceptions, include peptidoglycans—large, structural molecules made of protein **carbohydrate**.

Bacteria also secrete a viscous, gelatinous **polymer** (called the glycocalyx) on their cell surfaces. This polymer, composed either of polysaccharide, polypeptide, or both, is called a capsule when it occurs as an organized layer firmly attached to the cell wall. Capsules increase the disease-causing ability (virulence) of bacteria by inhibiting **immune system** cells called phagocytes from engulfing them. One such bacterium, *Streptococcus pneumoniae*, is the cause of **pneumonia**.

The shape of bacterial cells are classified as spherical (coccus), rodlike (bacillus), **spiral** (spirochete), helical (spirilla) and comma-shaped (vibrio) cells. Many bacilli and vibrio bacteria have whiplike appendages (called **flagella**) protruding from the cell surface. Flagella are composed of tight, helical rotors made of chains of globular protein called flagellin, and act as tiny propellers, making the bacteria very mobile.

Flagella may be arranged in any of four ways, depending on the species of bacteria. There is the monotrichous condition (single flagellum at one end), the amphitrichous (single flagellum at each end of the bacterium), the lophotrichous (two or more flagella at either or both ends of the bacterium), and the peritrichous condition (flagella distributed over the entire cell).

Spirochetes are spiral-shaped bacteria that live in contaminated water, sewage, **soil** and decaying organic matter, as well as inside humans and animals. Spirochetes move by means of axial filaments, which consist of bundles of fibrils arising at each end of the cell beneath an outer sheath. The fibrils, which spiral around the cell, rotate, causing an opposite movement of the outer sheath that propels the spirochetes forward, like a turning corkscrew. The best known spirochete is *Treponema pallidum*, the **organism** that causes syphilis.

On the surface of some bacteria are short, hairlike, proteinaceous projections that may arise at the ends of the cell or over the entire surface. These projections, called fimbriae, let the bacteria adhere to surfaces. For example, fimbriae on the bacterium *Neisseria gonorrhoea*, which causes gonorrhea, allow these organisms to attach to mucous membranes.

Other proteinaceous projections, called pili, occur singly or in pairs, and join pairs of bacteria together, facilitating transfer of DNA between them.

During periods of harsh environmental conditions some bacteria, such as those of the genera *Clostridium* and *Bacillus,* produce within themselves a dehydrated, thick-walled endospore. These endospores can survive extreme temperatures, dryness, and exposure to many toxic chemicals and to **radiation**. Endospores can remain dormant for long periods (hundreds of years in some cases) before being reactivated by the return of favorable conditions.

A primitive form of exchange of genetic material between bacteria involving plasmids does occur. Plasmids are small, circular, extrachromosomal DNA molecules that are capable of replication and are known to be capable of transferring genes among bacteria. For example, resistance plasmids carry genes for resistance to **antibiotics** from one bacterium to another, while other plasmids carry genes that confer pathogenicity. In addition, the transfer of genes via bacteriophages—viruses that specifically parasitize bacteria—also serves as a means of genetic recombination. *Corynebacterium diphtheriae*, for example, produces the **diphtheria** toxin only when infected by a phage that carries the diphtherotoxin **gene**.

The above examples of genetic information exchange between bacterial cells occurs regularly in nature. This natural exchange, or lateral gene transfer, can be mimicked artificially in the laboratory. Bioengineering uses sophisticated techniques to purposely transfer DNA from one organism to another in order to give the second organism some new characteristic it did not have previously. For example, in a process called transformation, antibiotic susceptible bacteria that are induced to absorb manipulated plasmids placed in their environment, can acquire resistance to that antibiotic substance due to the new genes they have incorporated. Similarly, in a process called transfection, specially constructed viruses are used to artificially inject bioengineered DNA into bacteria, giving infected cells some new characteristic.

Bacteria synthesize special DNA-cutting enzymes (known as restriction enzymes) that destroy the DNA of phages that do not normally infect them. Purified restriction enzymes are used in the laboratory to slice pieces of DNA from one organism and insert them into the genetic material of another organism as mentioned above.

Bacterial growth

The term "bacterial growth" generally refers to growth of a population of bacteria, rather than of an individual cell. Individual cells usually reproduce asexually by means of binary fission, in which one cell divides into two cells. Thus, bacterial growth of the population is a geometric progression of numbers of cells, with division occurring in regular intervals, called generation **time**, ranging from 15 minutes to 16 hours, depending upon the type of bacterium. In addition, some filamentous bacteria (actinomycetes) reproduce by producing chains of spores at their tips, while other filamentous species fragment into new cells.

Stages of bacterial growth

Under ideal conditions, the growth of a population of bacteria occurs in several stages termed lag, log, stationary, and death.

During the lag phase, active metabolic activity occurs involving synthesis of DNA and enzymes, but no growth. Geometric population growth occurs during the log, or exponential phase, when metabolic activity is most intense and cell reproduction exceeds **cell death**. Following the log phase, the growth **rate** slows and the production of new cells equals the rate of cell death. This period, known as the stationary phase, involves the establishment of an equilibrium in population numbers and a slowing of the metabolic activities of individual cells. The stationary phase reflects a change in growing condition—for example, a lack of **nutrients** and/or the accumulation of waste products.

When the rate of cell deaths exceeds the number of new cells formed, the population equilibrium shifts to a net reduction in numbers and the population enters the death phase, or logarithmic decline phase. The population may diminish until only a few cells remain, or the population may die out entirely.

Physical and chemical requirements for bacterial growth

The physical and chemical requirements for growth can vary widely among different species of bacteria, and some are found in environments as extreme as cold polar regions and hot, acid springs.

In general, the physical requirements for bacteria include proper **temperature**, **pH** and osmotic **pressure**.

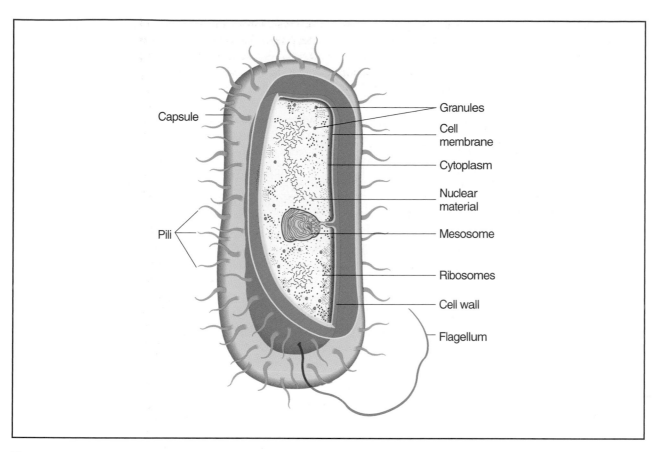

The anatomy of a typical bacterium. *Illustration by Hans & Cassidy. Courtesy of Gale Group.*

Most bacteria thrive only within narrow ranges of these conditions, however extreme those ranges may be.

Temperature and bacteria

The lowest temperature at which a particular species will grow is the minimum growth temperature, while the maximum growth temperature is the highest temperature at which they will grow. The temperature at which their growth is optimal is called the optimum growth temperature. In general, the maximum and minimum growth temperatures of any particular type of bacteria are about 30°F (-1°C) apart.

Most bacteria thrive at temperatures at or around that of the human body 98.6°F (37°C), and some, such as *Escherichia coli*, are normal parts of the human intestinal **flora**. These organisms are mesophiles (moderate-temperature-loving), with an optimum growth temperature between 77°F (25°C) and 104°F (40°C). Mesophiles have adapted to thrive in temperatures close to that of their host.

Psychrophiles, which prefer cold temperatures, are divided into two groups. One group has an optimal growth temperature of about 59°F (15°C), but can grow at temperatures as low as 32°F (0°C). These organisms live in ocean depths or Arctic regions. Other psychrophiles that can also grow at 32°F (0°C) have an optimal growth temperature between 68°F (20°C) and 86°F (30°C). These organisms, sometimes called psychrotrophs, are often those associated with food spoilage under refrigeration.

Thermophiles thrive in very hot environments, many having an optimum growth temperature between 122°F (50°C) and 140°F (60°C), similar to that of hot springs in Yellowstone National Park. Such organisms thrive in compost piles, where temperatures can rise as high as 140°F (60°C). Extreme thermophiles grow at temperatures above 195°F (91°C). Along the sides of **hydrothermal vents** on the ocean bottom 217 mi (350 km) north of the Galapagos Islands, for example, bacteria grow in temperatures that can reach 662°F (350°C).

pH and bacteria

Like temperature, pH also plays a role in determining the ability of bacteria to grow or thrive in particular environments. Most commonly, bacteria grow optimally within a narrow range of pH between 6.7 and 7.5.

Acidophiles, however, prefer acidic conditions. For example, *Thiobacillus ferrooxidans*, which occurs in drainage water from **coal** mines, can survive at pH 1. Other bacteria, such as *Vibrio cholera*, the cause of **cholera**, can thrive at a pH as high as 9.0.

Osmotic pressure and bacteria

Osmotic pressure is another **limiting factor** in the growth of bacteria. Bacteria are about 80-90% water; they require moisture to grow because they obtain most of their nutrients from their aqueous environment.

Cell walls protect prokaryotes against changes in osmotic pressure over a wide range. However, sufficiently hypertonic media at concentrations greater than those inside the cell (such as 20% sucrose) cause water loss from the cell by **osmosis**. Fluid leaves the bacteria causing the cell to contract, which, in turn, causes the cell membrane to separate from the overlying cell wall. This process of cell shrinkage is called plasmolysis.

Because plasmolysis inhibits bacterial cell growth, the addition of salts or other solutes to a **solution** inhibits food spoilage by bacteria, as occurs when meats or **fish** is salted.

Some types of bacteria, called extreme or obligate halophiles, are adapted to—and require—high **salt** concentrations, such as found in the Dead Sea, where salt concentrations can reach 30%. Facultative halophiles do not require high salt environments to survive, but are capable of tolerating these conditions. Halophiles can grow in salt concentrations up to 2%, a level that would inhibit the growth of other bacteria. However, some facultative halophiles, such as *Halobacterium halobium* grow in salt lakes, salt flats, and other environments where the **concentration** of salts is up to seven times greater than that of the oceans.

When bacteria are placed in hypotonic media with concentrations weaker than the inside of the cell, water tends to enter by osmosis. The accumulation of this water causes the cell to swell and then to burst, a process called osmotic lysis.

Carbon, nitrogen and other growth factors

In addition to water and the correct salt balance, bacteria also require a wide variety of elements, especially carbon, **hydrogen**, and nitrogen, **sulfur** and **phosphorus**, potassium, **iron**, **magnesium** and **calcium**. Growth factors, such as vitamins and pyrimidines and purines (the building blocks of DNA), are also necessary.

Carbon is the fundamental building block of all the organic compounds needed by living things, including nucleic acids, carbohydrates, proteins and fats.

Chemoheterotrophs are bacteria that use organic compounds such as proteins, carbohydrates and lipids as their carbon source, and which use electrons from organic compounds as their **energy** source. Most bacteria (as well as all **fungi**, protozoans and animals) are chemoheterotrophs. Chemoautotrophs (for example hydrogen, sulfur, iron, and nitrifying bacteria) use **carbon dioxide** as their carbon source and electrons from inorganic compounds as their energy source.

Saprophytes are heterotrophs that obtain their carbon from decaying dead organic matter. Many different soil bacteria release **plant** nitrogen as **ammonia (ammonification)**. Other bacteria, the *Nitrosomonas*, convert ammonia to nitrite, while *Nitrobacter* convert nitrite to nitrate. Other bacteria, especially *Pseudomonas*, convert nitrate to nitrogen gas. These bacteria complement the activity of nitrogen-fixing bacteria (for example, *Rhizobium),* which fix nitrogen from the atmosphere and make it available to leguminous plants, and *Azotobacter*, which are also found in fresh and marine waters. Together, the activity of these bacteria underlies the **nitrogen cycle**, by which the gas is taken up by living organisms, used to make proteins and other organic compounds, returned to the soil during decay, then released into the atmosphere to be reused by living things.

Phototrophs use **light** as their primary source of energy, but may differ in their carbon sources. Photoheterotrophs (purple nonsulfur and green nonsulfur bacteria) use organic compounds as their carbon source, while photoautotrophs (for example, photosynthetic green sulfur and purple sulfur bacteria) use carbon dioxide as a source of carbon.

Aerobic and anaerobic bacteria

Oxygen may or may not be a requirement for a particular species of bacteria, depending on the type of **metabolism** used to extract energy from food (**aerobic or anaerobic**). In all cases, the initial breakdown of glucose to pyruvic acid occurs during **glycolysis**, which produces a net gain of two molecules of the energy-rich **molecule adenosine triphosphate** (ATP).

Aerobic bacteria

Aerobic bacteria use oxygen to break down pyruvic acid, releasing much more ATP than is produced during glycolysis during the process known as aerobic **respiration**. In addition, aerobic bacteria have enzymes such as superoxide dismutase capable of breaking down toxic forms of oxygen, such as superoxide free radicals, which are also formed by aerobic respiration.

During aerobic respiration, enzymes remove electrons from the organic substrate and transfer them to the electron transport chain, which is located in the membrane of the mitochondrion. The electrons are transferred along a chain of electron carrier molecules. At the final transfer position, the electrons combine with **atoms** of oxygen—the final electron acceptor—which in turn combines with protons (H⁺) to produce water molecules. Energy, in the form of ATP, is also made here. Along the chain of electron carriers, protons that are pumped across the mitochondrial membrane re-enter the mitochondrion. This flow of electrons across the membrane fuels oxidative phosphorylation, the chemical reaction that adds a phosphate group to **adenosine diphosphate** (ADP) to produce ATP.

Obligate aerobes must have oxygen in order to live. Facultative aerobes can also exist in the absence of oxygen by using **fermentation** or anaerobic respiration. Anaerobic respiration and fermentation occur in the absence of oxygen, and produce substantially less ATP than aerobic respiration.

Anaerobic bacteria

Anaerobic bacteria use inorganic substances other than oxygen as a final electron acceptor. For example, *Pseudomonas* and *Bacillus* reduce nitrate ion (NO_3^-) to nitrite ion (NO_2^-), nitrous oxide (N_2O) or nitrogen gas (N_2). *Clostridium* species, which include those that cause **tetanus** and **botulism**, are obligate anaerobes. That is, they are not only unable to use molecular oxygen to produce ATP, but are harmed by toxic forms of oxygen formed during aerobic respiration. Unlike aerobic bacteria, obligate anaerobes lack the ability synthesize enzymes to neutralize these toxic forms of oxygen.

The role of bacteria in fermentation

Fermentation bacteria are anaerobic, but use organic molecules as their final electron acceptor to produce fermentation end-products. *Streptococcus*, *Lactobacillus*, and *Bacillus*, for example, produce **lactic acid**, while *Escherichia* and *Salmonella* produce **ethanol**, lactic acid, succinic acid, **acetic acid**, CO_2, and H_2.

Fermenting bacteria have characteristic sugar fermentation patterns, i.e., they can metabolize some sugars but not others. For example, *Neisseria meningitidis* ferments glucose and maltose, but not sucrose and lactose, while *Neisseria gonorrhoea* ferments glucose, but not maltose, sucrose or lactose. Such fermentation patterns can be used to identify and classify bacteria.

During the 1860s, the French microbiologist Louis Pasteur studied fermenting bacteria. He demonstrated that fermenting bacteria could contaminate wine and beer during manufacturing, turning the alcohol produced by yeast into acetic acid (vinegar). Pasteur also showed that heating the beer and wine to kill the bacteria preserved the flavor of these beverages. The process of heating, now called pasteurization in his honor, is still used to kill bacteria in some alcoholic beverages, as well as milk.

Pasteur described the spoilage by bacteria of alcohol during fermentation as being a "disease" of wine and beer. His work was thus vital to the later idea that human diseases could also be caused by microorganisms, and that heating can destroy them.

Identifying and classifying bacteria

The most fundamental technique for classifying bacteria is the gram stain, developed in 1884 by Danish scientist Christian Gram. It is called a differential stain because it differentiates among bacteria and can be used to distinguish among them, based on differences in their cell wall.

In this procedure, bacteria are first stained with crystal violet, then treated with a mordant—a solution that fixes the stain inside the cell (e.g., iodine-KI mixture). The bacteria are then washed with a decolorizing agent, such as alcohol, and counterstained with safranin, a light red dye.

The walls of gram positive bacteria (for example, *Staphylococcus aureus*) have more peptidoglycans (the large molecular network of repeating disaccharides attached to chains of four or five amino acids) than do gram-negative bacteria. Thus, gram-positive bacteria retain the original violet dye and cannot be counterstained.

Gram negative bacteria (e.g., *Escherichia coli*) have thinner walls, containing an outer layer of lipopolysaccharide, which is disrupted by the alcohol wash. This permits the original dye to escape, allowing the cell to take up the second dye, or counterstain. Thus, gram-positive bacteria stain violet, and gram-negative bacteria stain pink.

The gram stain works best on young, growing populations of bacteria, and can be inconsistent in older populations maintained in the laboratory.

Microbiologists have accumulated and organized the known characteristics of different bacteria in a reference book called *Bergey's Manual of Systematic Bacteriology* (the first edition of which was written primarily by David Hendricks Bergey of the University of Pennsylvania in 1923).

The identification schemes of *Bergey's Manual* are based on morphology (e.g., coccus, bacillus), staining (gram-positive or negative), cell wall composition (e.g., presence or absence of peptidoglycan), oxygen requirements (e.g., aerobic, facultatively anaerobic) and biochemical tests (e.g., which sugars are aerobically metabolized or fermented).

In addition to the gram stain, other stains include the acid-fast stain, used to distinguish *Mycobacterium* species (for example, *Mycobacterium tuberculosis*, the cause of **tuberculosis**); endospore stain, used to detect the presence of endospores; negative stain, used to demonstrate the presence of capsules; and flagella stain, used to demonstrate the presence of flagella.

Another important identification technique is based on the principles of antigenicity—the ability to stimulate the formation of antibodies by the immune system. Commercially available solutions of antibodies against specific bacteria (antisera) are used to identify unknown organisms in a procedure called a slide agglutination test. A **sample** of unknown bacteria in a drop of saline is mixed with antisera that has been raised against a known species of bacteria. If the antisera causes the unknown bacteria to clump (agglutinate), then the test positively identifies the bacteria as being identical to that against which the antisera was raised. The test can also be used to distinguish between strains, slightly different bacteria belonging to the same species.

Phage typing, like serological testing, identifies bacteria according to their response to the test agent, in this case viruses. Phages are viruses that infect specific bacteria. Bacterial susceptibility to phages is determined by growing bacteria on an agar plate, to which solutions of phages that infect only a specific species of bacteria are added. Areas that are devoid of visible bacterial growth following incubation of the plate represent organisms susceptible to the specific phages.

Because a specific bacterium might be susceptible to **infection** by two or more different phages, it may be necessary to perform several tests to definitively identify a specific bacterium.

The evolutionary relatedness of different species can also be determined by laboratory analysis. For example, analysis of the **amino acid** sequences of proteins from different bacteria disclose how similar the proteins are. In turn, this reflects the similarity of the genes coding for these proteins.

Protein analysis compares the similarity or extent of differences between the entire set of protein products of each bacterium. Using a technique called **electrophoresis**, the entire set of proteins of each bacterium is separated according to size by an electrical charge applied across gel. The patterns produced when the gel is stained to show the separate bands of proteins reflects the genetic makeup, and relatedness, of the bacteria.

The powerful techniques of **molecular biology** have given bacteriologists other tools to determine the identity and relatedness of bacteria.

Taxonomists interested in studying the relatedness of bacteria compare the **ratio** of **nucleic acid** base pairs in the DNA of microorganisms, that is, the number of guanosine-cytosine pairs in the DNA. Because each guanosine on a double-stranded molecule of DNA has a complementary cytosine on the opposite strand, comparing the number of G-C pairs in one bacterium, with that in another bacterium, provides evidence for the extent of their relatedness.

Determining the percentage of G-C pairs making up the DNA also discloses the percentage of adenosine-thymine (A-T)—the other pair of complementary nucleic acids making up DNA (100% - [% G-C] = % A-T).

The closer the two percentages are, the more closely related the bacteria may be, although other lines of evidence are needed to make a definitive determination regarding relationships.

The principle of complementarity is also used to identify bacteria by means of nucleic acid hybridization. The technique assumes that if two bacteria are closely related, they will have long stretches of identical DNA. First, one bacterium's DNA is isolated and gently heated to break the bonds between the two complementary strands. Specially prepared DNA probes representing short segments of the other organism's DNA are added to this solution of single-stranded DNA. The greater degree to which the probes combine with (hybridize) complementary stretches of the single stranded DNA, the greater the relatedness of the two organisms.

In addition to helping bacteriologists better classify bacteria, the various laboratory tests are valuable tools for identifying disease-causing organisms. This is especially important when physicians must determine which antibiotic or other medication to use to treat an infection.

Bacteria and disease

The medical community did not accept the concept that bacteria can cause **disease** until well into the nineteenth century. Joseph Lister, an English surgeon, applied the so-called "germ theory" to medical practice in the 1860s. Lister soaked surgical dressing in carbolic acid (phenol), which reduced the rate of post-surgical infections so dramatically that the practice spread.

In 1876, the German physician Robert Koch identified *Bacillus anthracis* as the cause of **anthrax**, and in so doing, developed a series of laboratory procedures for proving that a specific organism can cause a specific disease. These procedures, called Koch's postulates, are still generally valid. Briefly, they state that, to prove an organism causes a specific disease, the investigator must:

1. Find the same pathogenic microorganism in every case of the disease.

2. Isolate the pathogen from the diseased patient or experimental **animal** and grow it in pure culture.

3. Demonstrate that the pathogen from the pure culture causes the disease when it is injected into a healthy laboratory animal.

4. Isolate the pathogen from the inoculated animal and demonstrate that it is the original organism injected into the animal.

The ability to isolate, study, and identify bacteria has greatly enhanced the understanding of their disease-causing role in humans and animals, and the subsequent development of treatments. Part of that understanding derives from the realization that since bacteria are ubiquitous and are found in large numbers in and on humans, they can cause a wide variety of diseases.

The skin and the nervous, cardiovascular, respiratory, digestive and genitourinary systems are common sites of bacterial infections, as are the eyes and ears.

The skin is the body's first line of defense against infection by bacteria and other microorganisms, although it supports enormous numbers of bacteria itself, especially *Staphylococcus* and *Streptococcus* species. Sometimes these bacteria are only dangerous if they enter a break in the skin or invade a wound, for example, the potentially fatal staphylococcal **toxic shock syndrome**. Among other common bacterial skin ailments are **acne**, caused by *Propionibacterium acnes* and superficial infection of the outer **ear** canal, caused by *Pseudomonas aeruginosa*.

Among the neurological diseases are **meningitis**, an **inflammation** of the brain's membranes caused by *Neisseria meningitidis* and *Hemophilus influenzae*.

Many medically important bacteria produce toxins, poisonous substances that have effects in specific areas of the body. Exotoxins are proteins produced during bacterial growth and metabolism and released into the environment. Most of these toxin-producing bacteria are gram positive.

Among the gram positive toxin-producing bacteria are *Clostridium tetani*, which causes tetanus, an often fatal paralytic disease of muscles; *Clostridium botulinum*, which causes botulism, a form of potentially lethal **food poisoning**; and *Staphylococcus aureus*, which also causes a form of food poisoning (gastroenteritis).

Most gram negative bacteria (for example, *Salmonella typhi*, the cause of **typhoid fever**) produce endotoxins, toxins that are part of the bacterial cell wall.

As the role of bacteria in causing disease became understood, entire industries developed that addressed the public health issues of these diseases.

As far back as 1810, the French confectioner Nicholas Appert proved that food stored in **glass** bottles and heated to high temperatures could be stored for long periods of time without spoiling. Appert developed tables that instructed how long such containers should be boiled, depending upon the type of food and size of the container. Today, the **food preservation** industry includes not only canning, but also freezing and freeze-drying. An important benefit to food preservation is the ability to destroy potentially lethal **contamination** by *Clostridium botulinum* spores.

Even as concepts of prevention of bacterial diseases were being developed, scientists were looking for specific treatments. Early in the twentieth century, the German medical researcher Paul Ehrlich theorized about producing a "magic bullet" that would destroy pathogenic organisms without harming the host.

In 1928, the discovery by Scottish bacteriologist Alexander Fleming that the **mold** *Penicillium notatum* inhibited growth of *Staphylococcus aureus* ushered in the age of antibiotics. Subsequently, English scientists Howard Florey and Ernst Chain, working at Oxford University in England, demonstrated the usefulness of penicillin, the anti-bacterial substance isolated from *P. notatum* in halting growth of this bacterium. This inhibitory effect of penicillin on bacteria is an example of antibiosis, and from this term is derived the word antibiotic, which refers to a substance produced by microorganisms that inhibits other microorganisms.

Beginning in the 1930s, the development of synthetic anti-bacterial compounds called sulfa drugs further stimulated the field of anti-bacterial drug research. The many different anti-bacterial drugs available today work in a variety of ways, such as the inhibition of synthesis of cell walls, of proteins, or of DNA or RNA.

Today, medical science and the multi-billion dollar pharmaceutical industry are facing the problem of bacterial resistance to drugs, even as genetically engineered bacteria are being used to produce important medications for humans.

Bacterial adaptation

Bacteria have been designed to be adaptable. Their surrounding layers and the genetic information for these and other structures associated with a bacterium are capable of alteration. Some alterations are reversible, disappearing when the particular pressure is lifted. Other alterations are maintained and can even be passed on to succeeding generations of bacteria.

KEY TERMS

. . . .

Capsule—A viscous, gelatinous polymer composed either of polysaccharide, polypeptide, or both, that surrounds the surface of some bacteria cells. Capsules increase the disease-causing ability (virulence) of bacteria by inhibiting immune system cells called phagocytes from engulfing them.

Death phase—Stage of bacterial growth when the rate of cell deaths exceeds the number of new cells formed and the population equilibrium shifts to a net reduction in numbers. The population may diminish until only a few cells remain, or the population may die out entirely.

Exotoxins—Toxic proteins produced during bacterial growth and metabolism and released into the environment.

Fimbriae—Short, hairlike, proteinaceous projections that may arise at the ends of the bacterial cell or over the entire surface. These projections let the bacteria adhere to surfaces.

Gram staining—A method for classifying bacteria, developed in 1884 by Danish scientist Christian Gram, which is based upon a bacterium's ability or inability to retain a purple dye.

Koch's postulates—A series of laboratory procedures, developed by German physician Robert Koch in the late nineteenth century, for proving that a specific organism cause a specific disease.

Lag phase—Stage of bacterial growth in which metabolic activity occurs but no growth.

Log phase—Stage of bacterial growth when metabolic activity is most intense and cell reproduction exceeds cell death. Also known as exponential phase.

Phage typing—A method for identifying bacteria according to their response to bacteriophages, which are viruses that infect specific bacteria.

Pili—Proteinaceous projections that occur singly or in pairs and join pairs of bacteria together, facilitating transfer of DNA between them.

Spirochetes—Spiral-shaped bacteria which live in contaminated water, sewage, soil and decaying organic matter, as well as inside humans and animals.

Stationary phase—Stage of bacterial growth in which the growth rate slows and the production of new cells equals the rate of cell death.

The first antibiotic was discovered in 1929. Since then, a myriad of naturally occurring and chemically synthesized antibiotics have been used to control bacteria. Introduction of an antibiotic is frequently followed by the development of resistance to the agent. Resistance is an example of the **adaptation** of the bacteria to the antibacterial agent.

Antibiotic resistance can develop swiftly. For example, resistance to penicillin (the first antibiotic discovered) was recognized almost immediately after introduction of the drug. As of the mid 1990s, almost 80% of all strains of *Staphylococcus aureus* were resistant to penicillin. Meanwhile, other bacteria remain susceptible to penicillin. An example is provided by Group A *Streptococcus pyogenes*, another Gram-positive bacteria.

The adaptation of bacteria to an antibacterial agent such as an antibiotic can occur in two ways. The first method is known as inherent (or natural) resistance. Gram-negative bacteria are often naturally resistant to penicillin, for example. This is because these bacteria have another outer membrane, which makes the penetration of penicillin to its target more difficult. Sometimes when bacteria acquire resistance to an antibacterial

agent, the cause is a membrane alteration that has made the passage of the molecule into the cell more difficult.

The second category of adaptive resistance is called acquired resistance. This resistance is almost always due to a change in the genetic make-up of the bacterial **genome**. Acquired resistance can occur because of **mutation** or as a response by the bacteria to the selective pressure imposed by the antibacterial agent. Once the genetic alteration that confers resistance is present, it can be passed on to subsequent generations. Acquired adaptation and resistance of bacteria to some clinically important antibiotics has become a great problem in the last decade of the twentieth century.

Bacteria adapt to other environmental conditions as well. These include adaptations to changes in temperature, pH, concentrations of ions such as **sodium**, and the nature of the surrounding support. An example of the latter is the response shown by *Vibrio parahaemolyticus* to growth in a watery environment versus a more viscous environment. In the more viscous setting, the bacteria adapt by forming what are called swarmer cells. These cells adopt a different means of movement, which is more efficient for moving over a more solid surface. This adaptation is under tight genetic control, involving the expression of multiple genes.

Bacteria react to a sudden change in their environment by expressing or repressing the expression of a whole lost of genes. This response changes the properties of both the interior of the organism and its surface **chemistry**. A well-known example of this adaptation is the so-called **heat** shock response of *Escherichia coli*. The name derives from the fact that the response was first observed in bacteria suddenly shifted to a higher growth temperature.

One of the adaptations in the surface chemistry of Gram-negative bacteria is the alteration of a molecule called lipopolysaccharide. Depending on the growth conditions or whether the bacteria are growing on an artificial growth medium or inside a human, as examples, the lipopolysaccharide chemistry can become more or less water-repellent. These changes can profoundly affect the ability of antibacterial agents or immune components to kill the bacteria.

Another adaptation exhibited by *Vibrio parahaemolyticus*, and a great many other bacteria as well, is the formation of adherent populations on solid surfaces. This mode of growth is called a biofilm. Adoption of a biofilm mode of growth induces a myriad of changes, many involving the expression of previously unexpressed genes. In addition,l de-activation of actively expressing genes can occur. Furthermore, the pattern of gene expression may not be uniform throughout the biofilm. Bacteria within a biofilm and bacteria found in other niches, such as in a wound where oxygen is limited, grow and divide at a far slower speed than the bacteria found in the test tube in the laboratory. Such bacteria are able to adapt to the slower growth rate, once again by changing their chemistry and gene expression pattern.

A further example of adaptation is the phenomenon of chemotaxis, whereby a bacterium can sense the chemical composition of the environment and either moves toward an attractive compound, or shifts direction and moves away from a compound sensed as being detrimental. Chemotaxis is controlled by more than 40 genes that code for the production of components of the flagella that propels the bacterium along, for sensory receptor proteins in the membrane, and for components that are involved in signaling a bacterium to move toward or away from a compound. The adaptation involved in the chemotactic response must have a **memory** component, because the concentration of a compound at one moment in time must be compared to the concentration a few moments later.

See also Antisepsis; Biodegradable substances; Biodiversity; Composting; Microbial genetics; Origin of life; Water microbiology; Nitrogen fixation.

Resources

Books

Alberts, et al. *Molecular Biology of the Cell,* 4th. ed New York: Garland Science, 2002.

Cullimore, Roy D. *Practical Atlas for Bacterial Determination* Boca Raton, FL: CRC Press, 2000.

Dyer, Betsey Dexter. *A Field Guide to Bacteria.* Ithaca, NY: Cornell University Press, 2003.

Groisman, Eduardo A. *Principles of Bacterial Pathogenesis.* Burlington, MA: Academic Press, 2000.

Koehler, T.M. *Anthrax* New York: Springer Verlag, 2002.

Walsh, Christopher. *Antibiotics: Actions, Origins, Resistance.* Washington, DC: American Society for Microbiology Press, 2003.

Other

The Foundation for Bacteriology, New York University. "Virtual Museum of Bacteria" [cited February 5, 2003]. <http://www.bacteriamuseum. org/main1.shtml>.

Marc Kusinitz
Brian Hoyle

Bacteriophage

Bacteriophage (also known as phages) are viruses that target and infect only bacterial cells. The first observation of what since turned out to be bacteriophage was made in 1896. Almost twenty years later, the British bacteriologist Frederick Twort demonstrated that an unknown microorganism that could pass through a filter that excluded **bacteria** was capable of destroying bacteria. He did not explore this finding in detail, however. In 1915, the French Canadian microbiologist Felix d'Herelle observed the same result, and named the microorganism bacteriophage (bacteria eater, from the Greek *phago,* meaning to eat).

Many types of bacteriophage have been identified since their discovery in 1915, and they are named according to the type of bacteria they infect. For example, staphylophages are specific viruses of the staphylococcal bacteria, and coliphages specifically infect coliform bacteria.

Bacteriophage are the most thoroughly studied and well-understood viruses. They occur frequently in nature, carry out similar biological functions as other viruses, yet do not target human cells for **infection**. Phages have proven to be a valuable scientific research tool for a variety of applications: as models for the study of viral infectious mechanisms, as tools of **biotechnology** that introduce new genes into bacterial cells, and as potential treatments for human bacterial infection. For example, the experiments that lead to the discovery of messenger

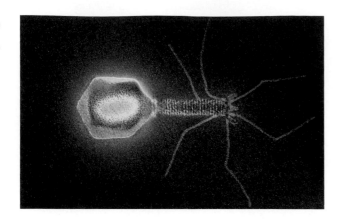

Colored transmission electron micrograph of a T4 bacteriophage virus. © *Department of Microbiology, Biozentrum/Science Photo Library/Photo Researchers, Inc.*

ribonucleic acid, one of the keys to the manufacture of protein in bacteria, viruses, and even cells found in humans, were accomplished using a bacteriophage. Another example is the bacteriophage designated T4, which specifically infects the bacterium *Escherichia coli*. T4 has been a cornerstone of **molecular biology**; studies of the way T4 makes new copies of itself has revealed a great deal of information about bacteriophage **genetics** and the regulation of the expression the **gene** viral genetic material. Additionally, another bacteriophage, called lambda, has been fundamentally important to molecular **biology** as a model system for gene regulation and as a means of moving genetic material from one bacterium to another.

Bacteriophage structure

Bacteriophage have different three-dimensional shapes (or morphologies). Those that are known as T-even phages (i.e., T2, T4, and T6) have a shape similar to the *Apollo* spacecraft that landed on the **Moon** in the 1960s. These phages have a head that has a slightly spherical shape called an icosahedron. A tube connects the head to spider-like supporting legs. This overall shape is vital to the way T-even bacteriophage deliver their payload of genetic material into a bacterial **cell**. Once on the surface on a bacterium, the tube portion of the phage contracts, and the phage acts like microscopic hypodermic needles, literally injecting the genetic material into the bacterium.

Other bacteriophage can be spherical (e.g. PhiX174, S13) or long and thread-like (e.g., M13).

Structurally, bacteriophage, like most viruses, are composed of a protein coat surrounding a core containing DNA (**deoxyribonucleic acid**) or RNA (ribonucleic acid), though many variations on this basic design exist. The bacteria these viruses infect often measure about one

micron in diameter (a micron is one thousandth of a millimeter) and the phages themselves may be as small as twenty-five thousandths of a micron. Bacteriophage infect their hosts by binding to the wall of the bacterial cell. The wall is perforated by **enzyme** action and phage DNA is injected into the cell. The genetic machinery of the cell is altered to make more bacteriophage DNA. Ultimately, the host cell dies when phage copies accumulate to the point of lysing (bursting) the cell **membrane** and releasing the phages, which go forth and continue the cycle.

Phages as valuable molecular tools

Much of what has been learned about the mechanisms of viral infection in general has been discerned through the study of bacteriophage. They have proved to be valuable molecular tools for biotechnology, as they can be used as vehicles to move genetic material from one **organism** into another organism.

It is through this revolutionary use of phages to introduce foreign DNA into new cells that human **insulin** was first safely and cheaply produced. In a process called lateral gene transfer, genes from one source are transplanted into a different living cell so that they will give the different cell a new characteristic (found in the first cell). For example, specific human genes are implanted in bacterial cells with the aid of phages that allow bacterial cells to produce human insulin and other valuable protein products in great purity and quantity. Lateral gene transfer has given a new, human characteristic to bacterial cells. Bacteriophage act as the deliverers of transferred genes.

Today, bacteriophage used to inject DNA into host cells for research or biotechnology can be "manufactured" in test tubes. Kits containing bacteriophage **proteins** and structural components are used to create intact phages from pieces that spontaneously self-assemble under the right chemical conditions. In this way, scientists can customize bacteriphage and the DNA they contain for many uses.

Additionally, bacteriophage are only now beginning to fulfill the dream of Felix d'Herelle, in combating infection in humans and animals. The medical potential of many bacteriophage is great as a treatment for **blood** infection and **meningitis** for example, along with a host of bacterial infections increasingly resistant to **antibiotics**.

See also Ebola virus; Epstein-Barr virus; Retrovirus.

Resources

Books

Flint, S.J., L.W. Enquist, R.M. Krug, et al. *Principles of Virology: Molecular Biology, Pathogenesis, and Control.*

Washington, DC: American Society for Microbiology Press, 1999.

Stahl, F.W. *We Can Sleep Later: Alfred D. Hershey and the Origins of Molecular Biology.* Cold Spring Harbor, NY: Cold Spring Harbor Press, 2000.

Summers, W.C. *Felix d'Herelle and the Origins of Molecular Biology.* New Haven: Yale University Press, 2000.

Brian Hoyle

Badgers

Badgers are eight **species** of robust, burrowing carnivores in the subfamily Melinae of the family Mustelidae, which also includes **weasels**, **mink**, marten, **otters**, and **skunks**. There are eight species of badgers, distributed among six genera.

Badgers have a strong, sturdy body, with short, powerful, strong-clawed legs, and a short tail. The head is slender and triangular-shaped. The fur of badgers is largely composed of long, stiff, rather thin guard hairs, with relatively little underfur. Badgers are fossorial animals, meaning they are enthusiastic diggers, often constructing substantial den-complexes, usually in sandy ground. Badgers are generally crepuscular, being active at dusk, night, and dawn. Badgers are strong, tough animals, and can readily defend themselves against all but the largest predators.

The American badger

The American or **prairie** badger (*Taxidea taxus*) is widespread in the prairies and savannas of western **North America**. The American badger is a stout-bodied and stubby-tailed **animal**. It has strong front legs armed with long, sturdy claws useful for digging and shorter-clawed hind legs. The primary **color** of the fur is grayish to brownish-red, with a black or dark-brown snout and feet, and bright white markings on the face and top of the head, extending over the front of the back. The hair is longest on the sides of the animal, which accentuates the rather compressed appearance of this species. Males are somewhat larger than females, and can be as long as about 3 ft ([1 m] body plus the tail) and weigh as much as 22 lb (10 kg).

The American badger is a solitary animal during most of the year, coming together only during the breeding season. This species is highly fossorial, and it digs numerous dens. The dens may be used for breeding by successive generations of animals and can be complicated assemblages of tunnels, access holes, and chambers. The sleeping chambers are comfortably lined with grassy hay, which is renewed frequently for cleanliness. However, **individual** badgers may change their dens rather frequently, sometimes moving around and digging new holes over a rather extensive area. Defecation occurs in holes dug aboveground, which are then covered up. Badgers in northern and alpine parts of the range of the species will hibernate during the winter, but more southerly populations are active throughout the year. American badgers scent-mark their territory, using secretions from a pair of anal **glands**.

The American badger is primarily a **carnivore**, catching its own **prey** or scavenging dead animals. However, it also feeds on **plant** materials. Prey species include rabbits, ground **squirrels**, small **mammals** such as **mice** and **voles**, ground-nesting **birds**, earthworms, **snails**, and **insects**. Foraging can occur at any time of day, but most commonly around the late afternoon to dusk. Burrowing prey are excavated by vigorous, extensive digging. Baby badgers are born in the early springtime, and they disperse from the natal den in the following autumn.

The American badger has a relatively dense and lustrous fur, which in the past was of commercial value, mostly for use as a fur trim, and also for the manufacturing of shaving brushes. The American badger is considered by some farmers and ranchers to be a pest, primarily because its access holes can represent a leg-breaking hazard to large **livestock**. Consequently, this species has been excessively trapped and poisoned in many areas, greatly reducing the extent and abundance of its wild populations. Attempts have been made to cultivate American badgers on fur farms, but these did not prove successful.

Other species of badgers

The natural range of the Eurasian badger (*Meles meles*) extends south of the **tundra** throughout most of **Europe**, Russia, Mongolia, Tibet, China, and Japan. The Eurasian badger is primarily a species of **forests** and thick scrub, although it also occurs in relatively disturbed habitats, such as parks. The Eurasian badger can reach a length of 3 ft (1 m) and a weight of 35 lb (16 kg). It has pronounced white stripes running along the head and the forepart of its back, overlying a grayish basal color. The feet are colored dark brown or black.

The Eurasian badger digs its den or "set" in open sites with sandy **soil**, using its strong forelegs and stout, sharp claws. The den may be used continuously by many

A badger at its den hole. *Photograph by Jess R. Lee. The National Audubon Society Collection/Photo Researchers, Inc. Reproduced by permission.*

generations of animals, and is a complex of tunnels, with numerous exits, entrances, ventilation holes, sleeping chambers, and even an underground toilet area for use by young (adults defecate in outside pits). The sleeping areas are lined with a bedding of plant materials, which are kept clean by frequent renewals.

The Eurasian badger is monogamous and pairs for life, which can be as long as 15 years. This species is somewhat gregarious, and several pairs will live in harmony in the same complex of burrows. Once the young badgers have matured, however, they are driven from the natal den. This usually occurs after the end of their first winter, when the animals are almost one-year old.

European badgers forage at dusk and during the night, although they may also be seen basking during the day. These animals are omnivorous, with plant materials comprising about three quarters of the food consumed, and hunted and scavenged animals the remainder. If they are hand-raised as babies, European badgers will become quite tame, but wild-caught adults are not tamable.

The hog-badger (*Arctonyx collaris*) occurs in hilly, tropical and subtropical forests of northern India, Nepal, southern China, and Southeast **Asia**, as far south as Indonesian Sumatra. The hog-badger can weigh as much as 31 lb (14 kg), and is an omnivorous, nocturnal animal with a pig-like snout. This species digs enthusiastically, and climbs well.

The teludu or Malayan stink badger (*Mydaus javanensis*) is a brown animal with a broad, white stripe running along its back from the head to the tail. The teludu has well-developed anal glands, which can be used in a skunk-like fashion to deter potential predators by squirting a smelly secretion as far as 5 ft (1.5 m). This species occurs on the Malay Peninsula, Borneo, Sumatra, and Java. The palawan or calamian stink badger (*Suillotaxus marchei*) occurs on some Philippine Islands.

The ferret badgers are various species of relatively slender, ferret-like animals, with a long, bushy tail, a face mask, and an active and inquisitive demeanor. The Chinese ferret badger (*Melogale moschata*) occurs in China and northern Southeast Asia. The Bornean or Javan ferret badger (*M. orientalis*) occurs on the Southeast Asian islands of Borneo and Sumatra. The Indian ferret badger (*M. personata*) occurs from eastern India and Nepal to Thailand and Vietnam. Ferret badgers live in open forests and savannas, and they den in holes dug in the ground or in hollow trees. Ferret badgers are predators of small mammals, birds, and **invertebrates**.

Resources

Books

Banfield, A.W.F. *The Mammals of Canada.* Toronto: University of Toronto Press, 1974.

Grzimek, B., ed. *Grzimek's Encyclopedia of Mammals.* London: McGraw Hill, 1990.

Hall, E. R. *The Mammals of North America.* 2nd ed. New York: Wiley & Sons, 1981.

Nowak, R. M., ed. *Walker's Mammals of the World.* 5th ed. Baltimore: John Hopkins University Press, 1991.

Wilson, D. E., and D. Reeder. *Mammal Species of the World.* 2nd ed. Washington, DC: Smithsonian Institution Press, 1993.

Bill Freedman

Baguios *see* **Tropical cyclone**

Baking soda *see* **Sodium bicarbonate**

Ball bearing

Ball bearings help reduce **friction** and improve efficiency by minimizing the frictional contact between machine parts through bearings and lubrication. Ball bearings allow rotary or linear movement between two surfaces. As the name indicates, a ball bearing involves a number of balls, typically **steel**, sandwiched between a spinning inner race (a small steel ring with a rounded grove on its outer surface) and a stationary outer race (a larger steel ring with a rounded grove on its inner surface). These balls are held in place by a cage or a retainer. The sides often have shield or seal rings to retain the grease or some other lubrication and to keep the bearing as clean as possible. The area of contact between the balls and the moving parts is small, therefore friction and **heat** is low. Bearings based on rolling action are called rolling-element bearings and contain cylindrical rollers instead of balls, but operate on the same principle.

A bearing can carry loads along its axis of **rotation** or **perpendicular** to its axis of rotation. Those carrying loads along the axis of rotation are referred to as thrust bearings. Rolling-element bearings carrying loads perpendicular to the rotational axis are called radially loaded bearings.

Ball bearings are generally used for small to medium size loads and are suitable for high speed operation. The type of bearing used is dependent on the application. An engineer will evaluate a bearing based on load, speed that they can carry this load, and required life expectancy under specified conditions. Friction, start-up **torque**, ability to withstand impact or harsh environments, rigidity, size, cost, and complexity also are important design considerations.

Some of the different types of bearing include:

- The rigid single row or radial ball bearing, probably the most widely used, is designed so that balls run in comparatively deep grooved tracks, which make the bearing suitable for both radial and axial loads. Sealed versions are lubricated for life and maintenance-free.

- A rigid single row bearing with filling slots for balls contain more balls than the standard. This allows it to withstand heavier radial loads but only limited thrust.

- Rigid double row bearings are good for heavy radial loads and provide greater rigidity.

- Double row bearings appear in a wide variety of applications, ranging from electric motors and centrifugal pumps to electromagnetic clutches.

- The self-aligning bearing has two rows of balls with a common sphered raceway in the outer ring. This feature gives the bearings their self-aligning property, permitting angular misalignment of the shaft with respect to the housing.

- The angular-contact bearing has one side of the outer-race groove cut away to allow the insertion of more balls, which enables the bearing to carry large axial loads in one direction only. Such bearings are usually used in pairs so that high axial loads can be carried in both directions.

Chair castor showing ball bearings made of palladium alloys. *Photograph by Robert J. Huffman. FieldMark Publications. Reproduced by permission.*

Miniature ball bearings are particular useful for the automotive industry for anti-lock braking systems. These systems use a variety of precision small bearings in electric motors. Other applications include throttle assembly, cooling fans, and idle control motors.

A team of scientists from the the Weizmann Institute of Science in Rehovot, Israel, are developing molecular ball bearings from tungsten disulfide.

Resources

Books

Macaulay, David. *The New Way Things Work*. Boston: Houghton Mifflin Company, 1998.

Periodicals

Rapoport, L., Yu. Bilik, Y. Feldman, M. Homyonfer, S. R. Cohen, and R. Tenne. "Hollow Nanoparticles of WS2 as Potential Solid-State Lubricants." *Nature* (June 19, 1997):791-793.

Laurie Toupin

Ballistic missiles

Any missile that lofts an explosive payload which descends to its target as a ballistic projectile—that is, solely under the influence of gravity and air resistance—is a ballistic missile. Missiles that do not deliver a free-falling payload, such as engine powered-cruise missiles (which fly to their targets as robotic airplanes), are not "ballistic."

A ballistic missile has two basic components: a package contains guidance systems and **explosives** (the payload), and the rocket that lofts the payload into the upper atmosphere or into **space** (the booster). Ballistic missiles traverse **distance** rapidly; a long-range ballistic missile can to the other side of the world in 30 minutes. Because they give so little advance warning and deliver small, fast-moving payloads that may contain **nuclear weapons** capable of destroying entire cities, ballistic weapons are highly destructive and difficult to defend against.

Categories of ballistic missile

With the exception of submarine-launched ballistic missiles (SLBMs), ballistic missiles are categorized according to range. Five commonly-accepted categories of ballistic missile, with their associated ranges, are as follows: (1) battlefield short range ballistic missiles (BSRMBs: <93 mi [150 km]); (2) short range ballistic missiles (SRBMs: 93–497 mi [150–800 km]), (3) medium range ballistic missiles (MRBMs: 497–1490 mi [800–2400 km]), (4) intermediate range ballistic missiles (IRBMs: 1490–3416 mi [2400–5500 km]), and (5) intercontinental range ballistic missiles (ICBMs: >3416 mi [> 5500 km]).

Alternatively, the U.S. Department of Defense defines ballistic missiles with ranges less than 683 mi (1100 km) as SRBMs, those with ranges between 683 and 1708 mi (1100–2750 km) as MRBMs, those with ranges between 1708 and 3416 mi (1100–5500 km) as IRBMs.

Ballistic missiles can be launched from submarines, silos (i.e., vertical underground tubes), ships, or trailers. All ballistic missiles launched from submarines, regardless of range, are categorized as SLBMs; modern SLBMs have ranges comparable to those of ICBMs. The purpose of mounting ballistic missiles on submarines is to make them secure from attack. Modern missile submarines, such as those in the U.S. Trident class, are difficult to locate and can launch their missiles without surfacing.

Ballistic missile function

The flight of a ballistic missile can be divided into three phases: boost phase, cruise phase, and descent (terminal) phase. Boost phase begins with the ignition of the missile's booster rocket. The booster lofts the missile at a steep angle, imparting a high speed to the payload before burning out. The payload and booster then separate, beginning the cruise phase. The spent booster falls back to **Earth** while the payload, starting to lose speed, continues to gain altitude. If the missile is sufficiently long-range, its payload rises above the Earth's atmosphere during cruise phase, where it jettisons its aerodynamic protective shroud and arcs under the influence of gravity. The payload may be a single cone-shaped warhead or a flat "bus" with several warheads attached to it like upside-down ice-cream cones arranged circularly on a plate.

Individual warheads are not propelled downward toward their targets on the ground, but follow ballistic paths determined by gravity and **aerodynamics**, gaining speed as they lose altitude. Modern reentry vehicles usually feature small external fins or other steering devices that enable them to control their course, within limits, as they fall through the atmosphere; though such maneuverable reentry vehicles (MARVs) are not, strictly speaking, ballistic objects, missiles delivering them are still termed "ballistic" missiles for convenience. Maneuverability increases **accuracy**; a modern MARV delivered by ICBM or SLBM can land within a few hundred feet of its target after a journey of thousands of miles. Warheads may explode in the air high above their targets, on the surface, or under the surface after striking into the ground.

Boosters

The booster rockets of early ballistic missiles were powered by liquid fuels. A liquid-fuel rocket carries fuel (hydrazine, liquid **hydrogen**, or other) and liquid **oxygen** in tanks. Pressurized streams of fuel and oxygen are mixed and ignited at the top of a bell-shaped chamber: hot, expanding gases rush out of the open end of the bell, imparting **momentum** to the rocket in the opposite direction. Liquid fuels are unwieldy, as they must be maintained at low temperatures and may leak fuel or oxygen from tanks, pipes, valves, or pumps. Early U.S. ICBMs such as the *Atlas* and *Titan I* required several hours of above-ground preparation, including fueling, before they could be launched.

Since the late 1950s, ballistic-missile design has concentrated on solid-fuel boosters, which require less maintenance and launch preparation time and are more reliable because they contain fewer moving parts. Solid-fuel rockets contain long, hollow-core casts of a fuel mixture that, once ignited, burn from the inside out in an orderly way, forcing gases out the rear of the rocket. Starting in the early 1960s, liquid-fuel ballistic missiles were gradually phased out of the U.S. and Russian arsenals in favor of solid-fuel missiles. The first U.S. solid-fuel ICBM was the *Minuteman I* missile (so-called be-

cause of its near-instant response time), which was deployed to underground silos in the Midwest starting in 1962. Today, the ballistic-missile fleet of the United States consists almost entirely of solid-fuel rocket boosters. The *Minuteman III*, for example, like the *Minuteman I and II* it replaces, has a three-stage solid-fuel booster and a range of over 7000 miles (11,265 km). (*Stages* are independent rockets that are stacked to form a single, combined rocket. The stages are burned from the bottom up; each is dropped as it is used up, and the stage above it is ignited. The advantage of staging is that the booster lightens more rapidly as it gains speed and altitude. There are single-stage, two-stage, and three-stage ballistic missiles; the greater the number of stages, the longer the range of the missile.)

Payloads, warheads, and MIRV

As mentioned above, the payload of a ballistic missile may be either a single warhead or a bus bearing several warheads which can each be sent to a different target in the same general area (e.g., the eastern United States). Such a payload is termed a *multiple independently targetable reentry vehicle* (MIRV) system, and missiles bearing multiple independently targetable warheads are said to be MIRVed. The first MIRVed missiles were deployed the U.S. in 1970; only long-range ballistic missiles (ICBMs and SLBMs) are MIRVed. After a MIRV bus detaches from the burnt-out upper stage of its booster, it arcs through space in its cruise phase. It may possess a low-power propulsion system that enables it to impart slightly different velocities to each of its warheads, which it releases at different times. (Slight differences between individual warhead trajectories in space can translate to relatively large differences between trajectories later on, when the individual warheads are approaching their targets.) The U.S. *Minuteman III* ICBM is a modern MIRVed missile carrying up to three warheads; other MIRVed missiles, such as the MX, have been capable of carrying up to ten warheads.

Regional or approximate targeting for each MIRVed warhead is achieved by bus maneuvering and release timing during cruise phase. During descent phase, the warhead may steer itself to its precise target by means of **inertial guidance**, **radar**, or a combination of the two. Inertial guidance is based on the principle that every change in an object's **velocity** can be sensed by that object as an **acceleration**. By knowing its exact prelaunch location and state of **motion** (e.g., by consulting the **Global Positioning System**) and by precisely measuring all accelerations during and after launch, an inertial guidance system can calculate its location at all times without needing to make further observations of the outside world. Ballistic-missile payloads rely primarily on inertial guidance to strike their targets; MARVs may refine their final course by consulting the Global Positioning System (as is done, for example, by the Chinese CSS-6 SRBM) or by using radar to guide themselves during final approach (as was done, for example, by the Pershing II IRBM deployed the U.S. in **Europe** during the 1980s).

The nuclear warheads mounted on modern long-range ballistic missiles are usually thermonuclear warheads having yields in the range of several hundred kilotons to several megatons. (One kiloton equals the explosive power of one thousand tons of the chemical explosive TNT; one megaton is equivalent to a million tons of TNT.) Those nations that do not possess nuclear weapons mount conventional-explosive warheads on their ballistic missiles.

History

The world's first ballistic missile was the V-2, developed by Nazi Germany during World War II. The V-2, which was first test-launched on October 3, 1942, could deliver a 1,650-lb (750-kg) warhead to a target 225 miles away. Germany launched approximately 3,000 V-2s during the war, but with little military effect; the V-2, lacking the sophisticated guidance computers of later ballistic missiles, were inaccurate. Only 50% of V-2s aimed at a given point would, on average, land within 11 mi (17 km) of that point. The V-2 was therefore not aimed at military installations but, like its predecessor the V-1 (the first cruise missile, also developed by Nazi Germany), at the city of London. Some 518 V-2s struck London during the final years of World War II, killing over 20,000 people and making the V-2 the deadliest ballistic missile in history—so far. (The "V" in V-1 and V-2 stands for *Vergeltungswaffe*, German for "retaliation weapon," reflecting the fact that the V-2's primary purpose was not victory but vengeance.)

The United States and Soviet Union were far behind Germany in the design of large rockets during World War II, but both captured V-2 technicians and information at the end of the war and used them to accelerate their own missile programs. The U.S. began by experimenting with captured V-2s, and during the late 1940s built several new rockets of its own based on the V-2. During the 1950s both the Soviet Union and the United States turned their attention to the development of ballistic-missile boosters that could reach the other country's heartland from anywhere in the world. The Soviet Union flight-tested the world's first ICBM, the R-7, in August, 1957. Two months later the R-7 was soon used to launch the world's first artificial **satellite**, *Sputnik I*, and four years later launched the world's first orbital manned

space flight. The U.S. was not far behind, and by 1959 had deployed its own ICBMs, the liquid-fueled Atlas and Titan missiles. The Americans also used their ICBMs for early space-flight efforts; the first manned U.S. space flights (Mercury and Gemini programs) used the *Redstone, Atlas*, and *Titan II* missile boosters.

See also Nuclear weapons.

Resources:

Books

Cimbala, Stephen J. *Nuclear Strategy in the Twenty-First Century.* Westport, CT: Praeger, 2000.

Cochran, Thomas B., William M. Arkin, and Milton M. Hoenig. *Nuclear Weapons Databook: Vol. I, U.S. Nuclear Forces and Capabilities.* Cambridge, MA: Ballinger Publishing Company, 1984.

Other

"Ballistic Missile Threats." Centre for Defense and International Security Studies, Lancaster University, UK. Aug. 10, 2001 [cited March 3, 2003]. <http://www.cdiss.org/bm threat.htm>.

Daniel Smith. "A Brief History of 'Missiles' and Ballistic Missile Defense." Center for Defense Information. 2000 [cited March 3, 2003]. <http://www.cdi.org/hotspots/issuebrief/ch2/>.

Larry Gilman

Ballistics

Ballistics is the study of projectile **motion**. A projectile is an object that has been launched, shot, hurled, thrown, or by other means projected, and continues in motion due to its own inertia. The path of the projectile is determined by its initial **velocity** (direction and speed) and the forces of gravity and air resistance. For objects projected close to **Earth** and with negligible air resistance, the flight path is a **parabola**. When air resistance is significant, however, the shape and **rotation** of the object are important and determining the flight path is more complicated. Ballistics influences many fields of study ranging from analyzing a **curve** ball to developing missile guidance systems.

Free-falling bodies

In order to understand projectile motion it is first necessary to understand the motion of free-falling bodies—objects that are simply dropped from a certain height above Earth. For the simplest case, when air resistance is negligible and when objects are close to the earth's surface, Galileo Galilei (1564-1642) was able to show that two objects fall the same **distance** in the same amount of **time**, regardless of their weights. It is also true that the speed of a falling object will increase by equal increments in equal time periods. For example, a ball dropped from the top of a building will start from rest and increase to a speed of 32 ft (9.8 m) per second after one second, to a speed of 64 ft (19.5 m) per second after two seconds, to a speed of 96 ft (29.4 m) per second after three seconds, and so on. Thus, the *change* in speed for each one second time interval is always 32 ft per second. The change in speed per time interval is known as the **acceleration** and is constant. This acceleration is equal to 1 g, which stands for the acceleration due to the **force** of gravity. By comparison, a pilot in a supersonic jet pulling out of a nose dive may experience an acceleration as high as 9 g (of course, a jet is not in free fall but is being accelerated by its engines and gravity).

The acceleration of gravity, g, becomes smaller as the distance from Earth increases. However, for most earthbound applications, the value of g can be considered constant (it only changes by 0.5% due to a 10 mi [16 km] altitude change). Air resistance, on the other hand, can vary greatly depending on altitude, **wind**, and the properties and velocity of the projectile itself. It is well know that sky divers may change their altitude relative to other sky divers by simply changing the shape of their body. Also, it is obvious that a rock will fall more quickly than a feather. Therefore, when treating problems in ballistics, it is necessary to separate the effects due to gravity, which are fairly simple, and the effects due to air resistance, which are more complicated.

Projectile motion without air resistance

The motion of projectiles without air resistance, can be separated into two components. Motion in the vertical direction where the force of gravity is present, and horizontal motion where the force of gravity is **zero**. As Isaac Newton (1642-1727) proposed, an object in motion will remain in motion unless acted upon by an external force. Therefore, a projectile in motion will remain with the same horizontal velocity throughout its flight, since no force exists in the horizontal direction, but its velocity will change in the vertical direction due to the force of gravity. For example, a cannon ball is fired in the horizontal direction. The velocity of the cannon ball will remain constant, in the horizontal direction, but the ball will accelerate toward the Earth, in the vertical direction, with an acceleration of 1 g. The combination of these two effects produces a path which describes a parabola. Since the vertical motion is determined by the same acceleration that describes the motion of objects in free fall, a second cannon ball that is dropped at precisely the same instant as the first cannon ball is fired, will reach the ground at precisely the same instant. Therefore, the motion in the horizontal direction does not affect the motion in the vertical direction. This fact can be confirmed by knocking one coin off the edge of the desk with a good horizontal whack, while a second coin is simultaneously knocked off the desk with a gentle nudge. Both coins will reach the ground at the same time.

By increasing the amount of gun powder behind the cannon ball, one could increase the horizontal velocity of the cannon ball as it leaves the cannon and cause the cannon ball to land at a greater distance. If it were possible to increase the horizontal velocity to very high values, there would come a point at which the cannon ball would continue in its path without ever touching the ground, similar to an orbiting **satellite**. To attain this orbiting situation close to the earth's surface, the cannon ball would have to be fired with a speed of 17,700 MPH (28,500 km/h)! In most instances, projectiles, like cannon balls, are fired at some upward angle to the earth's surface. As before, the flight paths are described by parabolas. (The maximum range is achieved by aiming the projectile at a 45° angle above the horizontal.) Angles equally greater or less than 45° will produce flight paths with the same range (for example 30° and 60°).

Projectile motion with air resistance

If projectiles were only launched from the surface of the **moon** where there is no atmosphere, then the effects of gravity, as described in the previous section, would be sufficient to determine the flight path. On Earth, however, the atmosphere will influence the motion of projectiles. As opposed to the situation due to purely gravitational effects, projectile motion with air resistance will be dependent on the weight and shape of the object. As one would suspect, lighter objects are more strongly affected by air resistance. In many cases, air resistance will produce a drag force which is proportional to the velocity squared. The effects of increased air drag on an object such as a cannon ball will cause it to fall short of its normal range without air resistance. This effect may be significant. In World War I, it was realized that cannon balls would travel farther distances if aimed at higher elevations, due to the decreased air **density** and decreased drag.

More subtle effects of air resistance on projectile motion are related to the shape and rotation of the object. Clearly, the shape of an object can have an effect on its projectile motion, as anyone has experienced by wadding up a piece of **paper** before tossing it into the waste can. The rotation of an object is important also. For example, a good quarterback always puts a spin on a football when making a pass. By contrast, to produce an erratic flight, a knuckle ball pitcher in baseball puts little or no spin on

KEY TERMS

Acceleration of gravity—The vertical downward acceleration equal to 32 ft (9.8 m) per second per second experienced by objects in flight close to Earth.

Air resistance—The drag force on an object in flight due to the interaction with air molecules.

Free falling body—A falling object in one dimensional motion, influenced by gravity when air resistance is negligible.

Gyroscope—A device similar to a top, which maintains rotation about an axis while maintaining a constant orientation of that axis in space.

Inertia—The tendency of an object in motion to remain in motion in a constant direction and at a constant speed, and the tendency of an object at rest to remain at rest.

Projectile—An object that is projected close to Earth and whose flight path is determined by gravity, air resistance, and inertia.

the ball. The physical property that tends to keep spinning objects spinning is the conservation of angular **momentum**. Not only do spinning objects tend to keep spinning but the orientation of the spin axis tends to remain constant. This property is utilized in the design of rifle barrels that have **spiral** grooves to put a spin on the bullet. The spinning of the bullet around its long axis will keep the bullet from tumbling and will increase the **accuracy** of the rifle. This property is also utilized in designing guidance systems for missiles. These guidance systems consist of a small spinning device called a **gyroscope**, which keeps a constant axis orientation and thus helps to orient the missile. Small deviations of the missile with respect to the orientation of the gyroscope can be measured and corrections in the flight path can be made.

See also Conservation laws.

Resources

Books

Armenti, Angelo Jr. *The Physics of Sports.* New York: American Institute of Physics, 1992.

Hewitt, Paul. *Conceptual Physics.* Englewood Cliffs, NJ: Prentice Hall, 2001.

Munson, Bruce, et al. *Fundamentals of Mechanics.* 4th ed. New York: John Wiley and Sons, 2002.

Young, Hugh. *University Physics.* Reading, MA: Addison-Wesley, 1999.

Kurt Vandervoort

Balloon

A balloon is a nonsteerable **aircraft** consisting of a thin envelope inflated with any gas lighter than the surrounding air. The balloon rises from the ground similar to a gas bubble in a **glass** of soda. The physical principle underlying this ability to ascend is Archimedes' law, according to which any immersed body is pushed upward by a **force** equal to the weight of the displaced fluid. If this force is greater than the weight of the body itself, the body rises. The lighter the balloon is in comparison with air of the same **volume**, the more load (envelope, people, instruments) it can lift. The approximate lifting capacity of some lighter-than-air gases in a 1000 cu m balloon at 32°F (0°C) is shown below (in pounds):

Hydrogen	Helium	Methane	Neon	Nitrogen
1203	1115	576	393	42

For example, a balloon filled with **nitrogen** possesses only about 1/30 of the lifting capacity of the same balloon filled with **hydrogen**. As a matter of fact, only hydrogen, helium, and hot air are of practical importance. Hydrogen, the lightest existing gas, would be ideal for the balloon inflation if it had not one serious demerit: inflammability. Helium is 7% less efficient than hydrogen. It is absolutely safe in usage, however, but it is not easily available and its production is not cheap. Hot air is safe and easy to obtain, making it the most often used for common manned flights. But to get from hot air a lifting power equal to at least 40% of that of helium it would be necessary to **heat** it to about 570°F (299°C). The ascensional force of a hot-air balloon is difficult to control, since it is very unstable, sharply reacting to any variations of the inside air **temperature**. There is always an element of uncertainty in the balloon flight. Once airborne, it floats freely in air currents, leaving a man the possibility to regulate only the vertical **motion**.

For 123 years, since the first flight of a bag filled with smoke publicly launched by the Montgolfier brothers in Paris in 1783 till the first flight of the practical powered airplane of the Wright brothers in 1905, a balloon and its later modification, an **airship**, remained the only means of aerial navigation. This period was full of exciting achievements of courageous aeronauts. The crossing of the English Channel (1785, by Blanchard, France and Jeffries, USA), the parachute descent from the balloon (1797, by Garnerin, France), the crossing of the Irish Sea (1817, by Windham Sadler, England), and the long-distance flight from London to Nassau (1836, by Green, England) are a few of the milestones in the balloon's early history.

The suitability of balloons for making observations and for reaching inaccessible areas was soon generally recognized. The first air force in the world was created by France in 1794, and by the end of the nineteenth century balloon corps, whose main function was reconnaissance, were the common feature of European and American armies. With the introduction of a heavier-than-air craft military interest in balloons faded. However, the most challenging pioneering and scouting missions via balloons were yet to come.

Balloons and the exploration of the unknown

In 1863, the first remarkable high-altitude ascent was made by Glaisher and Coxwell in England. The purpose of this flight was purely a scientific one: to observe and record the properties of the upper atmosphere. The explorers rose to over 33,000 ft (10,000 m). The attempt almost cost them their lives, but fortunately they survived to describe the unique experience. This outstanding attempt was followed by many others, and high-altitude scientific ascents continued until the early 1960s. A specific layer of the atmosphere between 35,000 and 130,000 ft (11,000 and 40,000 m), which is called stratosphere, for some time became a new challenge to human spirit and **engineering** art. The mark of 72,395 ft (22,066 m), achieved by Stevens and Anderson in 1935, was a tremendous success for that time and was surpassed only twenty years later, when the United States resumed the manned stratosphere ballooning. The last in the series was the flight of Ross and Prather, who attained the altitude of 113,740 ft (34,467 m) in 1961. The technology developed to secure man's survival in extreme conditions became a germ of future **space** life-support systems.

The introduction of new lightweight and very strong plastic materials made it possible to build extremely big balloons able to take aloft huge payloads. Loaded with sophisticated instruments, such balloons began to carry out complex studies of the atmosphere, biomedical and geographical research, and astronomical observations.

Each day, thousands of balloons measure all possible characteristics of the atmosphere around the entire globe, contributing to the worldwide meteorological database. This information is needed for understanding the laws of air-mass movement and for accurate **weather forecasting**.

Balloon **astronomy** takes advantage of making observations in the clarity of the upper air, away from dust, **water** vapor, and smoke. Telescopes with a diameter of up to 3.3 ft (1 m) are placed on platforms, which are supported by mammoth balloons, as high as an eight-story building, at elevations of up to 66,000-120,000 ft (20,000-35,000 m).

The Russian mission to **Venus** in 1985 used two helium balloons to examine the motion of the Venusian at-

mosphere. For 46 hours, they floated above Venus with an attached package of scientific equipment that analyzed the environment and transmitted the information directly to **Earth**. For comparison, a landing module in the mission functioned for only 21 minutes.

The success of balloons on Venus may be possibly continued on **Mars**. To carry a multipurpose research probe above the Martian surface, American scientists suggested an original device consisting of a big hot-air balloon and a much smaller helium-filled balloon connected together. During the day, the air-balloon, heated by the **sun**, would drift in the Martian atmosphere with a payload of instruments. At night, the air-balloon would cool and descend to the ground, where it would stay, supported in the upright position by the smaller gas-balloon. Thus, the same probe would perform the on-ground experiments at night and the atmospheric experiments during the day, travelling from one location to another.

See also Aerodynamics; Buoyancy, principle of.

Resources

Books

The Cambridge Encyclopedia of Space. Cambridge University Press, 2002.
Curtis, A. R. *Space Almanac.* Arcsoft Publishers, 1990.
DeVorkin, D. H. *Race to the Stratosphere.* Springer-Verlag, 1989.
Jackson, D. D. *The Aeronauts.* Time-Life Books, Inc., 1980.

Elena V. Ryzhov

Balsa *see* **Silk cotton family (Bombacaceae)**
Bamboo *see* **Grasses**

Banana

Bananas, plantains, and their relatives are various **species** of plants in the family Musaceae. There are about 40 species in this family, divided among only two genera. The most diverse genus is *Musa*, containing 35 species of bananas and plantains, followed by *Ensete*, the Abyssinian bananas. The natural range of bananas and plantains is the tropics and subtropics of the Old World, but agricultural and horticultural species and varieties are now cultivated in suitable climates all over the world.

Biology of bananas

Plants in the banana family are superficially tree-like in appearance. However, they are actually tall, erect, perennial herbs, because after they **flower** and set fruit, they die

Young green bananas growing in clusters. *Photography by Nigel Cattlin. The National Audubon Society Collection/Photo Researchers, Inc. Reproduced by permission.*

back to the ground surface. Their perennating structure is a large, underground, branched **rhizome** or **corm**.

Bananas and their relatives have a pseudostem, so-called because it has the appearance of a **tree** trunk. However, the banana stem is actually herbaceous, and is comprised of the densely overlapping sheath and petiole bases of their spirally arranged leaves. The pseudostem contains no woody tissues, but its fibers are very strong and flexible, and can easily support the erect **plant**, which is among the tallest of any herbaceous plants.

Bananas can grow as tall as 19.7-23 ft (6-7 m), and typically have a crown of leaves at the top of their greenish stem. The leaves of bananas are large and simple, with a petiole, a stout mid-rib, and a long, expanded, roughly oval, **leaf** blade, which can reach several meters in length. The leaf blade has an entire (smooth) margin, although it often becomes frayed by the **wind**, and may develop lobe-like ingrowths along its edge.

The flowers of bananas are finger-shaped, with three petals and sepals, and are subtended by large, fleshy, bright reddish-colored scales, which fall off as the fruit matures. The flowers are imperfect (that is, unisexual), and the plants are monoecious, meaning **individual** plants contain both female and male flowers. The flow-

ers are arranged in a group, in an elongate structure known as a raceme, with male flowers occurring at the tip of the structure, and female flowers below. Only one inflorescence develops per plant. The flowering stalk develops from the underground rhizome or corm, and pushes up through the pseudostem of the plant, to emerge at the apex. The flowering stalk eventually curves downwards, under the weight of the developing **fruits**. The central axis of the raceme continues to elongate during development, so that older, riper fruits occur lower down, while flowers and younger fruit occur closer to the elongating tip. The same is true of the male flowers, with spent flowers occurring lower down, and pollen-producing ones at the tip of the inflorescence.

The flowers of bananas are strongly scented, and produce large quantities of **nectar**. These attract **birds** and **bats**, which feed on the nectar, and pollinate the flowers. The mature fruits are a type of multi-seeded berry, with a leathery outer coat known as an exocarp, and a fleshy, edible interior with numerous **seeds** embedded.

Bananas and people

Various species in the banana family are cultivated as agricultural **crops**, with a world production of about 66 million tons (60 million tonnes). The best-known species is the banana (*Musa paradisiaca*; sometimes known as *M. sapientum*). The cultivated banana is a sterile triploid, and does not produce viable seeds. This banana is believed to be derived from crosses of *Musa acuminata* and *M. balbisiana*, likely occurring in India or Southeast **Asia** at some prehistoric time. The banana is an ancient fruit, thought to have been cultivated for at least 3,000 years in southern Asia.

Bunches of banana fruits can be quite large, weighing as much as 110 lb (50 kg). Each bunch consists of clusters of fruits, known as hands; each hand contains 10-20 individual fruits, or fingers. After the fruits of a banana plant are harvested, the plant dies back, or is cut down, and a new stalk regenerates from the same, perennating rhizome or corm.

Bananas intended for export to temperate countries are generally harvested while the fruits are still green. As they ripen, the skin turns yellowish or reddish, depending on the variety. When dark blotches begin to appear on the skin, the fruits are especially tasty and ready for eating. Bananas are a highly nutritious food.

The cultivated banana occurs in hundreds of varieties, or cultivars, which vary greatly in the size, **color**, and **taste** of their fruits. The variety most familiar to people living in temperate regions has a rather large, long, yellow fruit. This variety is most commonly exported to temperate countries because it ripens slowly,

KEY TERMS

Berry—A soft, multi-seeded fruit, developed from a single, compound ovary.

Imperfect—In the botanical sense, this refers to flowers that are unisexual, containing either male or female reproductive parts, but not both.

Monoecious—A plant breeding system in which male and female reproductive structures are present on the same plant, although not necessarily in the same flowers.

Rhizome—This is a modified stem that grows horizontally in the soil and from which roots and upward-growing shoots develop at the stem nodes.

Triploid—An organism having three sets of chromosomes. In plants, triploids develop from crosses between a diploid parent, having two sets of chromosomes, and a tetraploid parent, with four sets.

and travels well without spoiling. However, this variety of banana has proven to be susceptible to a recently emerged, lethal fungal **disease**. The long, yellow banana will soon be largely replaced in the temperate marketplace by another variety, which has a smaller, reddish, apple-tasting fruit.

Most varieties of the cultivated banana occur in tropical countries, especially in southern Asia, where many may be grown in any particular locality. These varieties range greatly in their size, taste, and other characteristics, some being preferred for eating as a fresh fruit, and others for cooking by frying or baking. Plantains or platanos are a group of about 75 varieties of cultivated bananas that are only eaten after they are cooked or processed into chips or flour. Like bananas, plantains are a highly nutritious food.

Another important economic product is manila **hemp** or abaca, manufactured from the fibers of the large, sheathing leaf-stalks of the species *Musa textilis*, as well as from some other species of bananas and plantains. *Musa textilis* is native to the Philippines and the Moluccas of Southeast Asia, and most manila hemp comes from that general region, although it has also been introduced elsewhere, for example, to Central America. The fibers of manila hemp are tough, flexible, and elastic, and are mostly woven into rope. Because it is resistant to **salt water**, this cordage is especially useful on boats and ships, although its use has now been largely supplanted by synthetic materials, such as polypropylene. The fibers of manila hemp can also been woven into a cloth, used to make bags, hats, twine, and other goods.

Bananas are also sometimes cultivated as ornamental plants, in gardens and parks in warm climates, and in greenhouses in cooler climates. Some taxonomic treatments include the genus *Strelitzia* in the banana family. The best-known species in this group is the bird-of-paradise plant (*Strelitzia reginae*), a beautiful and well-known ornamental plant that is cultivated in tropical and sub-tropical climates, and in greenhouses in more temperate places.

Resources

Books

Brucher, H. *Useful Plants of Neotropical Origin and Their Wild Relatives.* New York: Springer-Verlag, 1989.

Hvass, E. *Plants That Serve and Feed Us.* New York: Hippocrene Books, 1975.

Judd, Walter S., Christopher Campbell, Elizabeth A. Kellogg, Michael J. Donoghue, and Peter Stevens. *Plant Systematics: A Phylogenetic Approach.* 2nd ed. with CD-ROM. Suderland, MD: Sinauer, 2002.

Klein, R. M. *The Green World. An Introduction to Plants and People.* New York: Harper and Row, 1987.

Bill Freedman

Bandicoots

Australian **wildlife** holds many surprises, but few as intriguing as the widely distributed bandicoots. These small, rabbit-sized **marsupials** have a thick set body, short limbs, a pointed muzzle, short neck and short hairy tail. Their teeth are similar to those of insect- and flesh-eating **mammals**, but their hind feet resemble those of kangaroos and possums. The hindfeet are not only considerably longer than the front pair, which gives most bandicoots a bounding gait, but the second and third toes of the hind foot are fused together, with only the top joints and claws being free. The forefeet, in contrast, have three prominent toes, with strong claws used for digging and searching for insect **prey**. The fused toes of the hind limbs not only provide a strong base for a hopping **animal**, but also make a highly effective comb for grooming dirt and **parasites** from the fur. One **species** now thought to be extinct, the pig-footed bandicoot (*Chaeropus ecaudatus*), had a slightly different morphology, with only two toes on their forefeet and one on the hindfoot—adaptations for running, as this was a species of the open plains.

The taxonomic status of bandicoots is still uncertain, although two families are now recognized—the Peramelidae, with some 13 species, and the Thylacomyidae, with just one species represented, the greater bilby

Stuffed specimen of a pig-footed bandicoot. *Photograph by Tom McHugh. The National Audubon Society Collection/Photo Researchers, Inc. Reproduced by permission.*

(*Macrotis lagotis*). All of these species are unique to the Australian region, specifically mainland **Australia**, Tasmania, New Guinea and several offshore islands. Within this range, however, bandicoots have adapted to a wide range of habitats, including arid and semi-arid regions, coastal and sub-coastal **habitat**, savannah and lowland and mid-montane **rainforest**. On New Guinea, some species have been recorded at an altitude of 5,000 ft (1,500 m). The animals themselves may also vary considerably in size and appearance. The smallest species *Microperoryctes murina* weighs less than 4 oz (100 g), while one of the largest *Peroryctes broadbenti* may weigh more than 11 lb (5 kg).

Bandicoots are terrestrial and nocturnal species, constructing shallow burrows and surface nests beneath vegetation. The greater bilby is the only species that constructs a large burrow system, which may extend 7 ft (2 m) underground. Bandicoots are normally solitary, with

males occupying a larger home range (4.2–12.8 acres, or 1.7-5.2 ha) than females (2.2–5.2 acres, or 0.9-2.1 ha). Despite their genteel appearance, males, in particular, can be very aggressive towards other males. Little is known of the social **behavior** of most species, but they are thought to defend a central part of their ranges, especially the area surrounding their nest, against other animals. Males do not cooperate with bringing up the litter.

In the wild, bandicoots feed mainly on **insects** and their larvae but they are opportunistic feeders and will also consume fruit, berries, **seeds**, and **fungi**. Prey is either dug out of the **soil** or gleaned from the surface; their long, pointed snout is probably an **adaptation** for poking into tiny crevices or foraging under **leaf** litter for insects. Bandicoots have a keen sense of **smell** which is probably the main way in which they locate food at night. Most species also have prominent ears which may also assist with locating moving prey. Some species, such as the greater bilby have long naked ears that probably help with thermoregulation, as this species is adapted to living in arid conditions.

As with all marsupials the young are born at a very early stage of development, usually after a gestation period of just 12 days—one of the shortest periods of any mammal. When they are born, young bandicoots measure less than an inch (about 1 cm) and weigh a fraction of an ounce (0.2 g). Following **birth** the infant, which has well developed forelimbs, crawls its way into the mother's pouch where it attaches to a nipple and where it will remain for much of its pouch life. The average litter size is four; litters of seven young have been recorded. As the young grow, the mother's pouch increases in size to accommodate her developing family. Juveniles remain in the pouch for about 50 days, after which the mother begins to wean them. By the time they are seven weeks old, they are covered with short hair and the eyes are open.

Where climate and food conditions are favorable, bandicoots may breed throughout the year—females of some species may even become pregnant before its current litter is fully weaned. Bandicoots therefore have a very specialized pattern of breeding behavior among the marsupials, with a pattern of producing many young in a short period of time and with little parental investment.

Despite these adaptations, many species are now threatened as a result of human activities. Bandicoots have proven to be highly vulnerable to habitat modification and predation from introduced predators. In Australia, large areas of former brush habitat have been converted into rough pasture for **sheep** and cattle grazing, whose close cropping feeding actions have had a considerable effect on the soil microhabitat. Overgrazing by rabbits has had a similar effect. Introduced predators, especially foxes, **cats** and dogs have also had a major impact on populations in some areas. In New Guinea, many species are trapped for their fur and meat, but current levels of exploitation are unlikely to pose a significant threat to most species.

Bar code

Almost everyone is familiar with the striped bars found on grocery and retail store items. These are bar codes, or more specifically, the Universal Product Code (UPC). UPC codes first appeared in stores in 1973 and have since revolutionized the sales industry.

The UPC code consists of ten pairs of thick and thin vertical bars that represent the manufacturer's identity,

The parts of the Universal Product Code (UPC). © *Kelly A. Quin. Reproduced by permission.*

product size and name. Price information, which is not part of the bar code, is determined by the store. Bar codes are read by hand-held wand readers or fixed scanners linked to point of sale (POS) terminals.

Bar codes are also used for non-retail purposes. One of the earliest uses for bar codes was as an identifier on railroad cars. Organizers of sporting events also take advantage of bar code technology. For example, as runners of the Boston Marathon complete the 26 mi (42 km) course, they turn over a bar code tag that allows race officials to quickly tabulate results.

From 1965 through 1982, the United States Post Office experimented with optical character recognition (OCR) and bar code technology to speed up mail delivery. The Post Office now utilizes another type of bar code called the POSTNET bar code. Consisting of full and half height bars representing the zip-code and delivery address, the bar code allows mail to be sorted automatically at speeds of up to 700 pieces per minute.

Barberry

Barberries are about 600 **species** of plants in the genus *Berberis*, family Berberidaceae, occurring throughout the Northern Hemisphere and **South America**. Most species of barberry are shrubs or small trees, and many of these have persistent, evergreen leaves. The flowers are small, arranged in clusters, and insect pollinated. The **fruits** of barberries are multiple-seeded berries.

Barberry hybrids are often cultivated as attractive, ornamental shrubs. Some of these commonly cultivated species include the Japanese barberry (*Berberis thunbergii*), common barberry (*B. vulgaris*), Oregon grape (*Mahonia aquifolium*), and heavenly bamboo (*Nandina domestica*).

The common barberry is a native of **Europe**, with attractive foliage and bright-red berries. The common barberry has been widely planted in **North America** as a garden shrub, and it has escaped to natural habitats, where it can maintain viable populations.

The presence of wild populations of common barberry is considered a significant agricultural problem, because this species is an alternate host in the life cycle of the **wheat** rust (*Puccinia graminis*). This fungus is a pathogen of wheat (*Triticum aestivum*), one of the most important food-producing plants in North America, and the world. The control of populations of common barberry is critical to control of the wheat rust.

The inner **bark** and roots of the common barberry are bright yellow, and were once used to make a natural dye. Practitioners of folk medicine thought that this yellow **color** indicated that common barberry could be useful in the treatment of **jaundice**. However, this barberry has not proven to be useful for this purpose.

There are also native species of barberry in North America, for example, the American barberry (*Berberis canadensis*) of the eastern United States. Other native species in the barberry family include various shrubs known as Oregon **grapes** (for example, *Mahonia repens*). Several species of spring-flowering, herbaceous perennials occur in the understory of hardwood **forests** in eastern North America, including blue cohosh (*Caulophyllum thalictroides*), mayapple (*Podophyllum peltatum*), and twin-leaf (*Jeffersonia diphylla*). All of these genera of herbaceous plants of eastern North America have counterparts in the same genus is eastern **Asia**, but not in western North America. This represents a biogeographically interesting, disjunct pattern of distribution.

See also Rusts and smuts.

Barbets

Barbets are about 76 **species** of medium-sized **birds**, divided among 13 genera. These comprise the family Capitonidae, in the order Piciformes, which also contains the **woodpeckers**, **toucans**, and their allies. Barbets are birds of tropical **forests**, occurring in Central and **South America**, **Africa**, and **Asia** as far south as Indonesia. However, none of the species of barbets occur in more than one **continent**. The usual **habitat** of barbets is tropical forests and savannas.

Barbets are thick-set birds, with species ranging in body length from 3.5-13 in (9-33 cm). Barbets have short, rounded wings, a short tail, a large head on a short neck, and a stout, sharp-tipped bill. These birds have a distinctive "barbet" of bristles around the base of their beak. Like other birds in the order Piciformes, the barbets have short but strong legs, and feet on which two toes point backwards and two forwards. This arrangement of the toes is adaptive to clinging to **bark** on the sides of trees, while also using the tail feathers as a prop for additional support.

Barbets are brightly colored birds, many of them beautifully so. The basal coloration of many species is bright green, often with white, red, yellow, or blue markings, especially around the head. South American barbets have a different plumage for each sex, but the African and Asian species are not usually dimorphic in this way.

Barbets mostly eat **fruits** of various sorts. Although they are not migratory, barbets will move about in search of their food, especially wild figs (*Ficus* spp.), which tend to fruit seasonally. The strong bill of barbets is used to crush the hard fruits that these birds commonly feed upon. Barbets also feed on **insects**, and the young birds are raised almost exclusively on that high-protein diet.

Barbets nest in cavities that they excavate themselves in heart-rotted stems of trees, in termite mounds, or in earthen banks. Incubation of the two to four eggs and raising of the young is shared by both sexes of the pair. The young are dependent on their parents for a relatively long time. In some cases, older youngsters will help their parents raise a subsequent brood of babies. Barbets are noisy birds, but they are not particularly musical.

The many-colored or red-crowned barbet (*Megalaima rafflesii*) occurs over much of Southeast Asia. This species has a body length of 10 in (25 cm), and is a particularly beautiful bird, with a bright-green back, wings, and belly, bright-red cap and throat, yellow on the forehead, blue on the cheeks, and a black line through the **eye**. Most other species of barbets are similarly attractive in the colors and patterns of their plumage.

Barbiturates

Barbiturates are in the group of medicines known as central **nervous system** (CNS) depressants. Also known as sedative-hypnotic drugs, barbiturates make people very relaxed, calm, and sleepy. These drugs are sometimes used to help patients relax before **surgery**. Some may also be used to control seizures (convulsions). Although barbiturates have been used to treat nervousness and **sleep** problems, they have generally been replaced by other medicines for these purposes.

These medicines may become habit-forming and should not be used to relieve everyday **anxiety** and tension or to treat sleeplessness over long periods.

Description

Barbiturates are available only with a physician's prescription and are sold in capsule, tablet, liquid, and injectable forms. Some commonly used barbiturates are phenobarbital (Barbita) and secobarbital (Seconal).

Recommended dosage

Recommended dosage depends on the type of barbiturate and other factors such as the patient's age and the condition for which the medicine is being taken. Check with the physician who prescribed the drug or the pharmacist who filled the prescription for the correct dosage.

Always take barbiturates exactly as directed. Never take larger or more frequent doses, and do not take the drug for longer than directed. If the medicine does not seem to be working, even after taking it for several weeks, do not increase the dosage. Instead, check with the physician who prescribed the medicine.

Do not stop taking this medicine suddenly without first checking with the physician who prescribed it. It may be necessary to taper down gradually to reduce the chance of withdrawal symptoms. If it is necessary to stop taking the drug, check with the physician for instructions on how to stop.

Precautions

See a physician regularly while taking barbiturates. The physician will check to make sure the medicine is working as it should and will note unwanted side effects.

Because barbiturates work on the central nervous system, they may add to the effects of **alcohol** and other drugs that slow the central nervous system, such as **antihistamines**, cold medicine, **allergy** medicine, sleep aids, medicine for seizures, **tranquilizers**, some **pain** relievers, and **muscle relaxants**. They may also add to the effects of anesthetics, including those used for dental procedures. The combined effects of barbiturates and alcohol or other CNS depressants (drugs that slow the central nervous system) can be very dangerous, leading to unconsciousness or even death. Anyone taking barbiturates should not drink alcohol and should check with his or her physician before taking any medicines classified as CNS depressants.

Taking an overdose of barbiturates or combining barbiturates with alcohol or other central nervous system depressants can cause unconsciousness and even death. Anyone who shows signs of an overdose or a reaction to combining barbiturates with alcohol or other drugs should get emergency medical help immediately. Signs include:

- Severe drowsiness
- Breathing problems
- Slurred **speech**
- Staggering
- Slow heartbeat
- Severe confusion
- Severe weakness.

Barbiturates may change the results of certain medical tests. Before having medical tests, anyone taking this medicine should alert the health care professional in charge.

People may feel drowsy, dizzy, lightheaded, or less alert when using these drugs. These effects may even occur the morning after taking a barbiturate at bedtime. Because of these possible effects, anyone who takes these drugs should not drive, use machines or do anything else that might be dangerous until they have found out how the drugs affect them.

Barbiturates may cause physical or mental dependence when taken over long periods. Anyone who shows these signs of dependence should check with his or her physician right away:

- The need to take larger and larger doses of the medicine to get the same effect
- A strong desire to keep taking the medicine
- Withdrawal symptoms, such as anxiety, nausea or vomiting, convulsions, trembling, or sleep problems, when the medicine is stopped.

Children may be especially sensitive to barbiturates. This may increase the chance of side effects such as unusual excitement.

Older people may also be more sensitive that others to the effects of this medicine. In older people, barbiturates may be more likely to cause confusion, **depression**, and unusual excitement. These effects are also more likely in people who are very ill.

Special conditions

People with certain medical conditions or who are taking certain other medicines can have problems if they take barbiturates. Before taking these drugs, be sure to let the physician know about any of these conditions.

Allergies

Anyone who has had unusual reactions to barbiturates in the past should let his or her physician know before taking the drugs again. The physician should also be told about any allergies to foods, dyes, preservatives, or other substances.

Pregnancy

Taking barbiturates during pregnancy increases the chance of **birth defects** and may cause other problems such as prolonged labor and withdrawal effects in the baby after **birth**. Pregnant women who must take barbiturates for serious or life-threatening conditions should thoroughly discuss with their physicians the benefits and risks of taking this medicine.

Breastfeeding

Barbiturates pass into breast milk and may cause problems such as drowsiness, breathing problems, or slow heartbeat in nursing babies whose mothers take the medicine. Women who are breastfeeding should check with their physicians before using barbiturates.

Other medical conditions

Before using barbiturates, people with any of these medical problems should make sure their physicians are aware of their conditions:

- Alcohol or drug abuse
- Depression
- Hyperactivity (in children)
- Pain
- Kidney **disease**
- Liver disease
- Diabetes
- Overactive thyroid
- Underactive adrenal gland
- Chronic lung diseases such as **asthma** or **emphysema**
- Severe **anemia**
- Porphyria

Use of certain medicines

Taking barbiturates with certain other drugs may affect the way the drugs work or may increase the chance of side effects.

Side effects

The most common side effects are dizziness, lightheadedness, drowsiness, and clumsiness or unsteadiness.

These problems usually go away as the body adjusts to the drug and do not require medical treatment unless they persist or interfere with normal activities.

More serious side effects are not common, but may occur. If any of the following side effects occur, check with the physician who prescribed the medicine immediately:

- Fever
- Muscle or joint pain
- Sore throat
- Chest pain or tightness in the chest
- Wheezing
- Skin problems, such as rash, hives, or red, thickened, or scaly skin
- Bleeding sores on the lips
- Sores or painful white spots in the mouth
- Swollen eyelids, face, or lips

In addition, check with a physician as soon as possible if confusion, depression, or unusual excitement occur after taking barbiturates.

Patients who take barbiturates for a long time or at high doses may notice side effects for some time after they stop taking the drug. These effects usually appear within 8-16 hours after the patient stops taking the medicine. Check with a physician if these or other troublesome symptoms occur after stopping treatment with barbiturates:

- Dizziness, lightheadedness or faintness
- Anxiety or restlessness
- Hallucinations
- Vision problems
- Nausea and vomiting
- Seizures (convulsions)
- Muscle twitches or trembling hands
- Weakness
- Sleep problems, nightmares, or increased dreaming

Other side effects may occur. Anyone who has unusual symptoms during or after treatment with barbiturates should get in touch with his or her physician.

Interactions

Birth control pills may not work properly when taken while barbiturates are being taken. To prevent pregnancy, use additional or additional methods of birth control while taking barbiturates.

Barbiturates may also interact with other medicines. When this happens, the effects of one or both of the drugs may change or the risk of side effects may be

KEY TERMS

Adrenal glands—Two glands located next to the kidneys. The adrenal glands produce the hormones epinephrine and norepinephrine and the corticosteroid (cortisone-like) hormones.

Anemia—A condition in which the level of hemoglobin falls below normal values due to a shortage of mature red blood cells. Common symptoms include pallor, fatigue, and shortness of breath.

Central nervous system—The brain and spinal cord components of the nervous system that control the activities of internal organs, movements, perceptions, thoughts, and emotions.

Hallucination—A sensory experience of something that does not exist outside the mind. A person can experience a hallucination in any of the five senses. Hallucinations usually result from drugs or mental disorders.

Hypnotic—A medicine that causes sleep.

Porphyria—A disorder in which porphyrins build up in the blood and urine.

Porphyrin—A type of pigment found in living things, such as chlorophyll which makes plants green and hemoglobin which makes blood red.

Sedative—Medicine that has a calming effect and may be used to treat nervousness or restlessness.

Seizure—A sudden attack, spasm, or convulsion.

Withdrawal symptoms—A group of physical or mental symptoms that may occur when a person suddenly stops using a drug to which he or she has become dependent.

greater. Anyone who takes barbiturates should let the physician know all other medicines he or she is taking. Among the drugs that may interact with barbiturates are:

• Other central nervous system (CNS) depressants such as medicine for allergies, colds, hay fever, and asthma; sedatives; tranquilizers; prescription pain medicine; muscle relaxants; medicine for seizures; sleep aids; barbiturates; and anesthetics.

• **Blood** thinners.

• Adrenocorticoids (cortisone-like medicines).

• Antiseizure medicines such as valproic acid (Depakote and Depakene), and carbamazepine (Tegretol).

The list above does not include every drug that may interact with barbiturates. Be sure to check with a physician or pharmacist before combining barbiturates with any other prescription or nonprescription (over-the-counter) medicine.

Resources

Books

Klaassen, Curtis D. *Casarett and Doull's Toxicology* 6th ed. Columbus: McGraw-Hill, Inc., 2001.

Miller, Norman S. "Sedative-Hypnotics: pharmacology and use." *Journal of Family Practice.* 29 (December 1989):665.

O'Neil, Maryadele J. *Merck Index: An Encyclopedia of Chemicals, Drugs, & Biologicals.* 13th ed. Whitehouse Station, NJ: Merck & Co., 2001.

Periodicals

Kiefer, D.M. "Chemistry Chronicles: Miracle Medicines." *Today's Chemist* 10, no. 6 (June 2001): 59-60.

Nancy Ross-Flanigan

Bariatrics

Recent reports estimate that the proportion of individuals who are overweight in the United States surpasses 50% of the adult population. Such a staggering statistic has associated with it profound ramifications. Excess weight is a major contributor to serious health conditions that affect millions of people and can result in early death. Aside from tangible diseases, **obesity** is the root of much psychological distress, which adds to the negative impact of being overweight. As the social pressure to be thin climbs, the number of obese individuals also continues to rise, creating a potentially devastating emotional and physical dilemma for many. Additionally, the overall cost of health care is increased by obesity in this country and in other post-industrialized nations. As health care resources become more limited, then, potentially preventable conditions such as obesity are being pushed into the spotlight of managed care concerns. As a result, many different facets of bariatric medicine have emerged that try to reduce obesity effectively and safely.

Bariatrics is the field of medicine that is concerned with the causes, prevention, and treatment of obesity. Just as cardiac surgeons treat **heart** conditions, and podiatrists manage diseases of the feet, bariatric physicians are concerned with weight loss. Bariatrics utilizes varied techniques to reduce body **fat**. Among these are surgical procedures like gastric (or stomach) bypass, cosmetic procedures like liposuction, or pharmaceutical therapies that use drugs to reduce weight. Bariatrics also includes special dieting techniques and **exercise** regimens, often tailored to meet **individual** needs. While even a small reduction in weight is known to have beneficial impacts on health, some bariatric methods are controversial because they have significant health risks of their own.

The problem of obesity

Obesity can be defined as a body weight that is in excess of 20-30% of one's ideal weight. In order to fit this definition, though, the excess weight must be from fatty **tissue**, or adipose tissue. Muscle **mass** is not counted in obesity weight measurements. Obesity is detrimental because it can cause or contribute to the development of many other very serious health problems. Obesity, and simply being overweight, has been linked to coronary artery **disease** and congestive heart failure, non-insulin dependent **diabetes mellitus** (also called adult-onset diabetes), gout and gall stones, some forms of **cancer**, and **arthritis**. Some experts believe that obesity has reached **epidemic** proportions in the U.S. The increase in obesity, and associated illness, might be the result of the stresses of modern lifestyles where time is limited, high **calorie** food is abundant, and physical activity is reduced.

Obesity is a persistent, continuing disease caused by many factors. However, a useful and precise definition of obesity has been elusive. This is generally due to the fact that many factors can influence the weight of a person. Diet, gender, exercise frequency, **genetics** factors, and environment all contribute to weight. Regardless of the lack of a distinct definition, obesity is charged with over 300,000 preventable deaths in the United States each year. Second only to the effects of cigarette smoking and other tobacco use, obesity costs are estimated to exceed $100 billion per year, and continues to increase. As a whole, the portion of American society considered to be obese rose from 25% to over 38% from 1991-2002. Therefore, the demand for bariatric services has also increased proportionally.

The tools of bariatric medicine

Just as there are many factors contributing to obesity, bariatrics utilizes many techniques to control weight and induce weight loss. Although bariatric medicine is concerned with all methods of weight loss including **nutrition** and exercise, the term is used almost synonymously with surgical procedures. Such techniques include cosmetic procedures such as liposuction and somewhat radical operations like gastric resections. In addition, drug therapy is often employed in the fight for weight loss. Pharmaceuticals are used alone or in combination to reduce appetite or inhibit the **metabolism** of ingested fats. The ease of taking pills to reduce body fat make pharmaceuticals attractive to many. However, side effects and the potential for abuse of medications can be significant.

Restriction operations

The surgeries most often used to cause weight loss are restriction operations. Restriction operations limit or restrict food intake by creating a small pouch at the top of the stomach where the food enters from the esophagus. The reduced outlet delays the passage into the stomach and results in a sensation of fullness after eating very little food. The effect is like creating a tiny stomach on top of the existing stomach. The patient, then, perceives only the filling of the tiny stomach. After an operation, the person usually can eat only one half to one whole cup of food without discomfort or nausea. Most people cannot eat normal quantities of food. As a result, patients lose weight.

Restriction operations include gastric banding and vertical banded gastroplasty. Both operations serve only to restrict food intake, maintaining normal digestive processes. In gastric banding, a band made of surgical material restricts food movement near the upper portion of the stomach leaving a small passage for food to enter the rest of the stomach. Vertical banded gastroplasty is more prevalent and uses surgical staples to limit access of food to the stomach. Restrictive operations induce weight loss in nearly all patients. Unfortunately, some patients regain the weight that was lost. Because some patients are unable to modify their eating habits, restriction operations have limited success. Only about 30% of people reach normal weights after vertical banded gastroplasty. However, nearly 80% achieve some degree of weight loss.

Bypass surgery

Gastric bypass operations combine the creation of small stomach pouches to restrict food intake and the construction of paths for food to bypass the small intestine to cause malabsorption, or a lack of adequate nutrient and **vitamin** absorption into the body. Some gastric bypass surgeries are extensive, involving major portions of the stomach. Because they completely bypass major regions of food absorption, gastric bypass surgeries are more effective in causing weight loss than restrictive operations. Bypass operations generally result in a loss of two-thirds of one's body weight within two years. However, the weight loss comes with severe drawbacks. Because essential vitamins and **minerals** are obtained from food in the stomach and small intestine, gastric bypass prevents their absorption. Therefore, major vitamin deficiencies can be induced. For instance, **anemia** may result from insufficient absorption of vitamin B12 and **iron**. Accordingly, patients are required to take nutritional supplements. Also, the more extensive the bypass operation, the greater is the risk for complications, sometimes requiring the life-long use of special foods and medications.

Liposuction

Lipectomy is the surgical removal of fatty tissue from the body. Liposuction, a specific kind of lipectomy, is the surgical removal of fat from beneath the surface of the skin using suction or **vacuum** techniques. Most often, it is used to reduce adipose tissue from limited areas of the body, and therefore is considered to be a cosmetic procedure, or one that primarily improves appearance. One method of liposuction in weight control is to inject large volumes of **salt water**, or saline, containing a local anesthetic into the patient. The **solution** also contains adrenaline. The anesthetic numbs the procedure area, and the adrenaline constricts **blood** vessels to minimize bleeding, bruising, and swelling during the procedure. Adipose tissue is then removed by suctioning. Generally, only people who are slightly overweight benefit from liposuction. Because liposuction is surgical, there are some risks involved even though the procedure is considered very safe. While uncommon occurrences, **infection**, excessive bleeding, and occasional nerve damage can occur.

Drug therapy in bariatrics

Most of the medications that are used in bariatric treatment are appetite suppressants. These pharmaceuticals promote weight loss by decreasing appetite or increasing a sensation of fullness. Often, these medications work by increasing the action of neurotransmitters, substances that nerve cells release to chemically communicate with one another. Two **brain** neurotransmitters that can affect appetite are serotonin and the catecholamines. Because of their potency and wide physiological effects, most appetite suppressants are approved for short-term use only. An exception is Sibutramine, an appetite suppressant which can be used to treat obese individuals for extended periods of time. In general, appetite suppressant medications used in bariatric treatment have only limited effectiveness. Average weight loss using such medications ranges from 5-22 lb (2.3-10 kg). Some obese individuals, however, may lose up to 10% of their body mass. Amphetamine drugs can result in greater weight reduction, but their use is highly regulated because of their potential for abuse and dependence. In some cases, bariatric physicians will utilize a combination of drugs to maximize weight loss. An example is the concurrent use of fenfluramine and phentermine, called Fen/Phen. Fen/Phen was implicated in a number of serious adverse reactions and is therefore no longer used. (In March 2003, the death of Baltimore Orioles pitcher, Steve Bechler, was linked to use of the diet drug, ephedra.)

A multiple approach to weight management

Often, bariatric treatment uses multiple approaches to manage weight loss. For instance, liposuction may ad-

KEY TERMS

Adipose—Fatty tissue.

Bariatrics—The field of medicine concerned with the causes, prevention, and treatment of obesity.

Liposuction—The surgical removal of adipose tissue using suction techniques.

Obesity—A metabolic condition of excess weight defined by some as a body weight over 30% of an ideal value.

Restrictive operations—Bariatric procedures designed to promote weight loss by restriction the volume of food that can physically pass through the stomach. Stomach stapling is an example of a restrictive operation.

dress cosmetic concerns and reduce some bulk weight, while diet and behavioral regimens are taught to the same individual to reduce caloric intake. Furthermore, drug therapy may be used to reduce the absorption of fats from the diet and psychotherapy to address emotional issues may also be administered at the same time. In this multiple approach, many techniques are used simultaneously to induce and maintain weight loss for individuals who find weight loss to be very difficult. Often perceived as the result of a lack of willpower, weakness, or a lifestyle choice of overeating, obesity is in reality a chronic condition that is the result of many interacting factors that include genetics, environment, and **behavior**. As weight problems persist and increase in the United States, the need for new and effective bariatric treatments will continue to rise.

Resources

Books

Ackerman, Norman B. *Fat No More: The Answer for the Dangerously Overweight*. Prometheus Books, 1999.

Terry Watkins

Barium

Although pure barium is rarely used outside the laboratory, barium's many compounds have a number of practical applications. Perhaps the most familiar is the barium enema. When doctors need to examine a patient's **digestive system**, a mixture containing *barium sulfate* is used to coat the inner lining of the intestines. Similarly, to enhance examination of the stomach and esophagus, the patient drinks a chalky **barium sulfate** liquid. When the patient is x rayed, the barium coating inside the digestive tract absorbs a large proportion of the **radiation**. This highlights the black-and-white contrast of the x-ray photograph, so that doctors can better diagnose digestive problems.

Barium sulfate ($BaSO_4$) is safe to use for this purpose because it doesn't dissolve in **water** or other body fluids. However, barium and all of its soluble compounds are poisonous. Because pure barium reacts immediately with **oxygen** and water vapor to produce barium oxide when exposed to air, barium does not naturally occur in an uncombined state. Barium compounds are found primarily in two mineral ores—*barite*, which contains the sulfate compound, and *witherite*, which contains barium carbonate.

Like other metals, barium (Ba) is a good conductor of **heat** and **electricity**. It is silvery white and relatively malleable. Chemically, it resembles **calcium** and strontium, which are fellow members of the alkaline-earth family of metals. The **metal** gets its name from the Greek word for "heavy," *barys*, which was first used to name the mineral barite, or heavy spar. Barium's **atomic number** is 56 and it has seven stable isotopes.

During the 1700s, chemists thought that barium oxide and **calcium oxide** were the same substance. In 1774, Carl Wilhelm Scheele showed that barium oxide is a distinct compound, pointing the way toward discovery of the element. In the early 1800s, after electric batteries had been invented, chemists began using electric currents to break compounds apart. Humphry Davy, who pioneered this technique of **electrolysis**, discovered barium in 1808. Davy produced the metal for the first time by passing an **electric current** through molten barium hydroxide. He also used electrolysis to isolate potassium, **sodium**, calcium, **magnesium**, and strontium.

Although pure barium metal can be used to remove undesirable gases from electronic **vacuum** tubes, barium's compounds are much more important to industry. Barium sulfate is a component of lithopone, a white pigment used in paints. Barium carbonate is used in the production of optical **glass**, **ceramics**, glazed pottery, and specialty glassware; it is also an ingredient in oil drilling "muds" or slurries that lubricate the drill bit. The bright yellow-green colors in fireworks and flares come from barium nitrate. Motor oil detergents, which keep engines clean, contain barium oxide and barium hydroxide.

Barium sulfate

Barium sulfate ($BaSO_4$) is a white or yellow powder or crystalline **salt** with no **taste** or odor. Its **density** is

4.24-4.5 and it melts at 2,876°F (1,580°C), decomposing above that **temperature**. The compound is insoluble in **water**, but dissolves in hot concentrated **sulfuric acid**.

Barium sulfate occurs in nature as the mineral barite, or baryte, which is mined in Canada and Mexico and, in the United States in Arkansas, Missouri, Georgia, and Nevada. It is also prepared synthetically either by treating a **solution** of a barium salt with **sodium** sulfate or as a by-product in the manufacture of **hydrogen peroxide**.

Barium sulfate is used in diagnostic **radiology** of the **digestive system**. A suspension of barium sulfate in water is administered either orally or via an enema, which coats the lining of the upper or lower digestive tract. Because barium is a heavy **metal**, the compound is opaque to **x rays**, and the shapes of the coated organs can be clearly seen. Although barium in solutions is highly toxic, the sulfate is so insoluble that the suspension is harmless.

Barium sulfate is also used as a filler to products such as rubber, linoleum, oil cloth, **plastics**, **paper**, and lithographic inks. The compound is also used in paints and pigments, especially in the manufacture of colored papers and wall papers. Recently, barium sulfate has become popular as a substitute for natural ivory.

Bark

Bark is a protective, outer **tissue** that occurs on older stems and roots of woody coniferous and **angiosperm** plants. Bark is generally considered to occur on the outside of the tissue known as **wood**, or the water-conducting xylem tissues of woody plants. The inner cells of bark, known as phloem, grow by the division of outer cells in a generative layer called the vascular cambium, located between the bark and wood (inner cells of this cambium produce xylem cells). The outer cells of bark, known as **cork**, grow through cellular division in the cork cambium, present outside of the phloem. The outer part of the bark is a layer of dead cells, which can be as thick as several inches or more, and serves to protect the internal living tissues from injury, **heat**, and desiccation.

The macroscopic structure of bark varies greatly among **species** of woody plants. For example, the bark of American beech (*Fagus grandifolia*) is distinctively grey and smooth. In contrast, many species have a deeply fissured, rough bark, as in the cases of sugar maple (*Acer saccharum*) and spruces (*Picea* spp.). The **color** and pattern of fissuring and scaling of bark can often be used to identify species of trees and shrubs.

Harvesting barley with a combine tractor. *Photograph by Holt Studios Limited Ltd. Photo Researchers, Inc. Reproduced by permission.*

Some types of bark have specific uses to humans. The young, brownish, inner bark of young shoots of the cinnamon **tree** (*Cinnamomum zeylanicum*), native to Sri Lanka but now widely cultivated in the humid tropics, is collected, dried, and used whole or powdered as an aromatic flavoring of drinks and stews. Some barks have medicinal properties, such as that of the cinchona tree (*Cinchona calisaya*), from which **quinine** has long been extracted and used to reduce the fevers associated with **malaria**. More recently, taxol, an anti-cancer chemical, has been identified in the bark of the Pacific **yew** (*Taxus brevifolia*). Taxol is now used to treat ovarian **cancer**, and is also effective against some other cancers.

The bark of some species of trees contains large concentrations of a group of organic chemicals known as tannins, which can be reacted with **animal** skins to create a tough, flexible, and very useful material known as leather. Tannins will also react with certain **metal** salts to form dark pigments, which are used in **printing** and dyeing. Major sources of tannins in **North America** are the barks of hemlock trees (especially *Tsuga canadensis* and *T. heterophylla*), and **oaks**, especially **chestnut** oak (*Quercus prinus*) and tanbark oak (*Lithocarpus densiflo-*

ra). Eurasian oaks and hemlocks are also used, as are several tropical species, such as red mangrove (e.g., *Rhizophora mangle*) and wattle (e.g., *Acacia decurrens*). The thick, outer bark of the cork oak (*Quercus suber*) of **Europe** is collected and used to manufacture bottle corks, flotation devices, insulation, and **composite materials** such as parquet flooring. The bark of some conifers is used as a mulch in landscaping, for example, that of Douglas fir (*Pseudotsuga menziesii*) and redwood (*Sequoia sempervirens*).

Bill Freedman

Barley

Barley is one of the world's major cultivated **crops**. It is a member of the grass family (Poaceae). In 1999, approximately 142 million acres (57.5 million ha) of barley were grown worldwide and the total production was 147.0 million tons of grain (133.6 million tonnes).

The most widely cultivated **species** is six-rowed barley (*Hordeum vulgare*; also known as *Hordeum sativum*), which has its **seeds** arranged in six vertical rows on the flowering axis. The natural range of the wild barley from which the cultivated species is derived is not known for certain. Most likely, it was in southwestern **Asia** or northeastern **Africa**, now Ethiopia. The modern six-rowed barley probably has a polyphyletic origin, meaning several species in the genus *Hordeum* contributed genes to its **evolution** and domestication. The primary progenitor, however, is thought to have been a wild barley known as *Hordeum spontaneum*, a grass with only two rows of seeds on its flowering axis. Other contributing barleys may include *Hordeum distichon* and *Hordeum deficiens*.

There is archaeological evidence suggesting barley was gathered as a wild food 17,000 years ago. Barley has been cultivated in the Middle East for more than 8,000 years, and was domesticated at least 2,000 years before that. Within the last 5,000 years, barley has been widely cultivated throughout Eurasia. Some of the major advantages of barley as a crop are that it can be cultivated in a relatively cool, dry climate, in infertile **soil**, and can mature in as few as 60-80 days after planting. In fact, six-rowed barley is routinely grown further north than any other grain crop derived from a grass, such as the family Poaceae. Most barley is sown in the springtime and used primarily as **livestock** fodder or in the production of malt for **brewing** beer. In the past, barley was primarily used to make cakes and bread, as a gruel, or as a thickening ingredient in soup. These are now minor uses of the world production of barley.

There are also numerous non-crop plants in the same genus as the economically important barley species. One of these is the foxtail barley, *Hordeum jubatum*, a weedy **plant** native to marine shores having very long, attractive awns attached to its seeds.

Bill Freedman

Barnacles

The rocky shores of most coastlines are liberally dotted with clusters of barnacles (phylum Arthropoda, class **Crustacea**). Few people take any notice of these animals, despite their common occurrence. Barnacles are exclusively marine animals: some 900 **species** have been identified worldwide. Many are tiny organisms measuring just a few centimeters in diameter, while others such as the South American *Balanus psittacus* may reach a height of 9 in (23 cm) and a diameter of 3 in (8 cm).

Some of the smallest barnacles are parasitic, burrowing into mollusc shells and corals. The majority, however, are free-living animals that occur in distinct parts of the shoreline: while most species live within the intertidal range, some are limited to the low tide mark, while others are adapted to living in the spray zone, which only the highest **tides** can reach. A few species are even adapted to living in deep **water**.

There are two main types of barnacles—acorn and goose. Acorn barnacles are generally recognized by their squat, limpet-like appearance and extremely tough outer covering made up of five calcareous plates, which surround and protect the soft body cavity. With muscular contractions these plates can be opened or closed, depending on the state of the tide: at full tide, the plates are pushed outwards to allow the barnacle to feed, but as the tide withdraws, the barnacle closes its shell once again, leaving just a tiny opening for **oxygen** to enter. Thus enclosed in their shells, barnacles can resist drying at low tide.

Goose, or stalked, barnacles differ in appearance by having a long stalk (peduncle), the base of which is attached to the substratum and the main part of the body (capitulum) poised at the other end. The latter is enclosed in a toughened carapace, similar to that of acorn barnacles, while the peduncle is muscular and capable of movement.

Rocks are not the only substrate that attract barnacles. Some species attach to intertidal **grasses**, while others fix onto the shells of **crabs** or other molluscs such as clams, where they may help camouflage the host **animal**, and some even become attached to active-swimming species such as marine **turtles** or even the fins or other body parts of whales. Floating timber and flotsam, marine buoys, piers and ship's keels are also convenient anchoring points for many barnacles.

Adult barnacles remain in the same position for their lifetime. Being literally stuck in one place might prove an obstacle to many species, but barnacles have overcome this problem by having a larval dispersal phase. Most barnacles are hermaphroditic—each **individual** having both male and female reproductive organs and, while self-fertilization may take place in some instances, the normal pattern is for cross **fertilization**. Barnacles have extremely long male reproductive organs, some of which measure more than 30 times the length of the animal's body. The advantage of this is that although barnacles usually live in crowded conditions, they may also be able to reach other, more distant animals and fertilize them. Using this means to reach other barnacles, sperm are deposited in neighboring animals and the eggs brooded for about four months in a special sac within the mantle cavity. When they hatch, the tiny larvae will be released to the **ocean** where they drift with the **currents**. As many as 13,000 larvae may be released by a single individual. These larvae feed and mature through a series of six stages, following which they are ready to settle—a critical time in the life of the barnacle. As the larvae settles, it attaches itself to some substrate by means of cement **glands** located in the base of the first antennae. It then undergoes a period of **metamorphosis** in which the existing larval carapace becomes covered with interlocking calcareous plates.

Locked in that position for the remainder of its life, the barnacle has evolved a simple, but effective means of feeding. When covered with water, the barnacle extends six pairs of curved, hairy legs (cirri) from the body cavity into the water column. Here they are able to trap tiny **plankton** and small crustaceans directly from the water. As food is captured, the cirri are withdrawn into the mouth where particles are cleaned off and the cirri unfolded once again to continue feeding. Rhythmic beating of the cirri also creates a gentle flow of water down towards the mouth, further enhancing the chances of obtaining additional food.

In spite of their small size barnacles are of considerable economic importance, particularly for shipping, as high densities of barnacles on a ship's keel can reduce its speed by as much as one-third. Similar agglomerations may form on the legs of oil rigs, piers and other semi-permanent features, causing considerable damage. A great deal of research and money has been invested in the design of anti-fouling paints, which would deter barnacles from settling in the first place. Many of these products, however, have had a negative effect on the marine environment, causing poisoning among some species.

See also Arthropods.

Barometer

A barometer is an instrument for measuring **atmospheric pressure**. Two kinds of barometers are in common use, a mercury barometer and an aneroid barometer. The first makes use of a long narrow **glass** tube filled with mercury supported in a container of mercury, and the second makes use of a diaphragm whose size changes as a result of air **pressure**.

Mercury barometers

The principle of the mercury barometer was discovered by the Italian physicist Evangelista Torricelli in about 1643. That principle can be illustrated in the following manner. A long glass tube is sealed at one end and then filled with liquid mercury **metal**. The filled tube is then inverted and its open end inserted into a bowl of mercury. When this happens, a small amount of mercury metal runs out of the tube, leaving a **vacuum** at the top of the tube.

Under normal circumstances, the column of mercury in the glass tube stands at a height of about 30 in (76 cm). The column is sustained because air pressure pushes down on the surface of the mercury in the bowl at the bottom of the barometer. At the same time, the vacuum at the top of the glass tube exerts essentially no pressure on the column of mercury. The height of the mercury column in the glass tube, then, reflects the total pressure exerted by the atmosphere at the moment of measurement.

In theory, a barometer could be made of any liquid whatsoever. Mercury is chosen, however, for a number of reasons. In the first place, it is so dense that the column sustained by air pressure is of practicable height. A similar barometer made of **water**, in comparison, would have to be more than 34 ft (100 m) high. Also, mercury has a low **vapor pressure** and does not, therefore, evaporate easily. In a water barometer, the situation would be very different. Water has a much greater vapor pressure, and one would have to take into consideration the pressure exerted by water vapor at the top of the barometer, a factor of almost no consequence with a mercury barometer.

Two important additions needed to increase the **accuracy** of a barometer are a vernier scale and a **thermometer**. The vernier allows one to make an even more accurate measurement than is possible by reading the scale itself. The thermometer is needed because the **density** of mercury and other materials used in the construction of a barometer change with **temperature**. Most barometers come equipped with thermometers attached to them, therefore, along with conversion charts that permit one to correct barometer readings for a range of actual temperatures.

Modifications to the mercury barometer

The barometer described above is adequate for making rough measurements of atmospheric pressure. When more accurate readings are needed, however, modifications in the basic design of the barometer must be made. The most important factor to be considered in making such modifications is changes that take place in the mercury reservoir at the bottom of the barometer as a result of changes in atmospheric pressure.

When the atmospheric pressure decreases, for example, air pressure is able to sustain a slightly smaller column of mercury, and some mercury flows out of the glass tube into the reservoir. One might hope to find the

An antique aneroid barometer. *Photograph by Diana Calder. The Stock Market. Reproduced by permission.*

new pressure by reading the new level of the mercury in the glass tube. However, the level of the mercury in the glass tube must be compared to the level of the mercury in the reservoir, and the latter has changed also as a result of a new atmospheric pressure.

This problem is dealt with in one of two ways. In one instrument, the English Kew barometer, no modification is made in the mercury reservoir itself. Instead, changes that take place in the mercury level in the reservoir as a result of changes in atmospheric pressures are compensated for by making small changes in the measuring scale mounted to the glass tube. As one moves upward along the scale, the graduations between markings become slightly smaller to correct for the changing level of the mercury in the reservoir.

A second type of barometer, the Fortin barometer, contains a flexible bag that holds an extra supply of mercury metal. The flow of mercury into and out of that bag and then out of and into the glass tube is controlled by an adjustable screw whose point is moved so as just to touch the surface of the mercury in the reservoir. As atmospheric pressure and mercury levels change, modifications of the adjustable screw keep the mercury level at a constant height.

Aneroid barometer

A major disadvantage of the mercury barometer is its bulkiness and fragility. The long glass tube can break easily, and mercury levels may be difficult to read under unsteady conditions, as on board a ship at sea. To resolve these difficulties, the French physicist Lucien Vidie invented the aneroid ("without liquid") barometer in 1843.

An aneroid barometer can be compared to a coffee can whose sides are flexible, like the bellows on an accordion. Attached to one end of the coffee can (aneroid barometer) is a pointer. As atmospheric pressure increases or decreases, the barometer contracts or expands. The movement of the barometer is reflected in the **motion** of the pointer, which rides up and down with changes in atmospheric pressure.

One way to observe the motion of the pointer is to attach it to the hand on a dial that moves around a circular scale, from low pressure to high pressure. The simple clock-like aneroid barometer hanging on the wall of many homes operates on this basis. Another way to observe the movement of the pointer is to have it rest on the side of a rotating cylinder wrapped with graph **paper**. As the cylinder rotates on its own axis, the pen makes a tracing on the paper that reflects increases and decreases in pressure. A recording barometer of this design is known as a barograph.

The altimeter

An important application of the aneroid barometer is the altimeter, an instrument used to measure one's **distance** above **sea level**. Atmospheric pressure is a function of altitude. The farther one is above sea level, the less the atmospheric pressure, and the closer one is to sea level, the greater the atmospheric pressure. A simple aneroid barometer can be used to confirm these differences. If the barometer is now mounted in an airplane, a **balloon**, or some other device that travels up and down in the atmosphere, one's height above the ground (or above sea level) can be found by noting changes in atmospheric pressure.

Resources

Books

Banfield, Edwin. *Barometers: Aneroid and Barographs*. Trowbridge, Wiltshire, England: Baros Books, 1985.

KEY TERMS

. . . .

Altimeter—An aneroid barometer used to measure altitude.

Barograph—An aneroid barometer modified to give a continuous reading of atmospheric pressures on graph paper.

English Kew barometer—A mercury barometer with a contracting measuring scale and a constant reservoir of mercury.

Fortin barometer—A mercury barometer with a fixed measuring scale and an adjustable reservoir of mercury.

Vapor pressure—The amount of pressure exerted by liquid molecules in the vapor state.

Vernier scale—A movable scale for measuring a fractional part of one division on a fixed scale.

Brombacher, W. G. *Mercury Barometers and Manometers.* Washington, DC: U.S. Department of Commerce, National Bureau of Standards, 1960.

The Illustrated Science and Invention Encyclopedia. Westport, CT: H. S. Stuttman, 1982, vol. 2: 238-40.

Middleton, W. E. Knowles. *The History of the Barometer.* Baltimore: Johns Hopkins Press, 1964.

Periodicals

Caristi, Anthony J. "Build a Portable Barometer." *Popular Electronics* (January 1994):31-6.

Walker, Jearl. "Making a Barometer that Works with Water in Place of Mercury," *Scientific American.* (April 1987):122-27.

David E. Newton

Barracuda

A barracuda is a long, cylindrical, silvery **fish**. It has two widely separated dorsal fins, in roughly the same location as the two fins on its belly, and a forked tail. The largest **species**, the great barracuda, seldom grows longer than 6.5 ft (2 m) and is an aggressive fearsome **predator** of other fish. All barracudas have an underhung jaw that houses long, incredibly sharp teeth; their teeth are conically shaped, are larger in the front, like fangs, and their horizontal mouths can open very wide. In general, the barracuda inhabits tropical and warmer temperate waters throughout the world, specifically in the Atlantic, Pacific, and Indian Oceans. Different species of barracuda

thrive in a variety of specific habitats but they are common over reefs and near continental shelves. Barracudas have been known to attack humans.

Barracudas are classified in the order Perciformes, an incredibly diverse group, containing 18 suborders and nearly 7,000 species of fish. Barracudas are the only fish in the suborder Sphyraenoidei and in the family Sphyraenidae. Within their family, there is one genus, *Sphyraena*, with 20 species.

Predatory behavior

Barracudas usually swim actively in clear **water** searching for schools of plankton-feeding fish. Their silver coloring and elongated bodies make them difficult for **prey** to detect, especially when viewing them head-on. Barracudas depend heavily on their sense of sight when they hunt, noticing everything that has an unusual **color**, reflection, or movement. Once a barracuda sights an intended victim, its long tail and matching anal and dorsal fins enable it to move with incredibly swift bursts of speed to catch its prey before it can escape. Barracudas generally assault schools of fish, rushing at them head first and snapping their strong jaws right and left.

When barracudas are mature, they usually swim alone, however, there are circumstances when they tend to school. Two such instances are while they are young and when they are spawning. Additionally, to feed more easily, barracudas sometimes swim in groups. In this case, they can herd schools of fish into densely populated areas or chase them into shallow water; when the barracudas accomplish this, they can eat practically all the fish they want at leisure.

The great barracuda (*Sphyraena barracuda*)

As its name implies, the great barracuda is most notable because of its size. Like all species of barracuda, this species has a long, silvery body and very sharp teeth. It generally appears silvery with green or gray on its back and black blotches on its belly. However, this fish can change color to match its background environment. While individuals in most species of barracuda rarely grow longer than 5.5 ft (1.7 m), some specimens of the great barracuda have been reported to reach 10 ft (3 m) long. (Usually, members of this species average about 3 ft [0.9 m] long.) An aggressive hunter, the great barracuda inhabits temperate and tropical waters all over the world, except in certain parts of the Pacific Ocean and Mediterranean Sea. Specifically, it is found in the west Atlantic from Brazil to New England and in the Gulf of Mexico. Also, it is one of three barracuda species found

A barracuda in its reef habitat. The creature can move with great speed to slash an intended victim to ribbons with its teeth. *Photograph by Gregory Ochocki/Eric Haucke. Photo Researchers, Inc. Reproduced by permission.*

in the Caribbean. The other two species are *Sphyraena guachancho* and *Sphyraena picudilla*. These two species, much smaller and more rare, often swim in the company of **jacks** and are commonly named sennets.

The Pacific barracuda

A great deal is known about the Pacific barracuda (*Sphyraena argentea*). It spends its winters off Mexico and joins a school in the spring to swim up the coast and spawn. Males are sexually mature when they are two or three years old; females mature one year later. The female Pacific barracuda lays her eggs at intervals, and the eggs float freely in the water. Measuring up to 4.9 ft (1.5 m) long, Pacific barracuda eat **sardines**. They are caught throughout the year off the Mexican coast and in the summertime off the California coast. Their meat is reportedly very good.

Human fear of barracudas

People who dive in tropical regions are often quite afraid of barracudas, fearing them even more than **sharks**. Often, divers in tropical waters report feeling as if there is "someone watching them" when they are submerged. With barracudas, this is probably the case. Barracudas are curious animals and commonly follow and watch divers, noticing any strange movements or colors. Unlike sharks, barracudas only attack their prey one time, usually with one massive bite.

Barracudas are not as dangerous as many people think. Unless they are provoked, they rarely attack. Barracuda attacks usually take place under certain circumstances, including: (1) when the water is very murky; (2) when a diver is carrying a shiny reflecting object, like jewelry; (3) when the barracuda is provoked; or (4) when a diver is carrying a wounded fish. Attacks can also be caused by excessive splashing or other irregular movements in the water, especially in murky conditions. Also, some people think that the likelihood of a barracuda attack depends on location. For instance, while the great barracuda seldom attacks humans in Hawaii, it is considered more dangerous in the West Indies.

Resources

Books

Grzimek, H. C. Bernard, ed. *Grzimek's Animal Life Encyclopedia.* New York: Van Nostrand Reinhold, 1993.

Hauser, Hillary. *Book of Fishes.* New York: Pisces Books, 1992.

The Illustrated Encyclopedia of Wildlife. London: Grey Castle Press, 1991.

MacMillan Illustrated Animal Encyclopedia. New York: MacMillan Publishing, 1992.

Moyle, Peter B., Joseph Cech. *Fishes: An Introduction to Ichthyology.* 4th ed. New York: Prentice Hall, 1999.

Nelson, Joseph S. *Fishes of the World.* 3rd ed. New York: John Wiley & Sons, 1994.

The New Larousse Encyclopedia of Animal Life. New York: Bonanza Books, 1987.

Whiteman, Kate. *World Encyclopedia of Fish & Shellfish.* New York: Lorenz Books, 2000.

Wilson, Josleen. *The National Audubon Society Collection Nature Series, North American Fish.* New York: Gramercy Books, 1991.

Kathryn Snavely

Barrier islands

A barrier **island** is a long, thin, sandy stretch of land oriented **parallel** to the mainland coast, which protects the coast from the full force of powerful **storm** waves. Between the barrier island and the mainland is a lagoon or bay. Barrier islands are dynamic systems that migrate under the influence of changing sea levels, storms, waves, **tides**, and longshore **currents**. Approximately 2,100 barrier islands are known to exist around the world. In the United States, barrier islands occur along gently sloping sandy coastlines such as those along the Gulf Coast and the Atlantic Coast as far north as Long Island, New York. They are, in contrast, absent along the generally steep and rocky Pacific Coast. Some of the better known barrier islands along the coast of the United States are Padre Island, Texas, the world's longest barrier island; Sanibel and Captiva Islands, Florida; Cape Hatteras, North Carolina; and Assateague Island, Maryland.

Residential and recreational development on barrier islands has become a contentious issue in recent years. Although the islands serve as buffers against the sea by

constantly shifting and changing their locations, the owners of homes, stores, and hotels on barrier islands often try to stabilize the shifting **sand** to protect their property. This is accomplished by beach hardening (the construction of engineered barriers) or **beach nourishment** (the continual replacement of sand that has been washed away during storms). In either case, the result is an interruption of the natural processes that formed the islands. Construction that prevents the naturally occurring **erosion** of sand from the seaward side of a barrier island, for example, may result in erosion problems when the landward side is consequently starved of sand. Beach nourishment projects typically require ever increasing amounts of sand to maintain a static beachfront, and are therefore economically viable for only short periods of time. Although coastal management activities have long been directed towards beach hardening and nourishment, current scientific thinking suggests that the islands are more appropriately viewed as geologically transient features rather than permanent shorelines suitable for development.

Barrier island origins

Currently existing barrier islands are geologically young features that formed during the Holocene epoch (approximately the last 10,000 years). During that time, the rapid rise in **sea level** associated with melting **glaciers** slowed significantly. Although the exact mechanisms of barrier island formation are not fully understood, the decreasing **rate** of sea level rise allowed the islands to form.

Several conditions must be met in order for barrier islands to form. First, there must be a source of sand to build the island. The sand may come from coastal deposits or offshore deposits called shoals. In either case, the sand originated from the **weathering** and erosion of rock and was transported to the coast by **rivers**. In the United States, much of the sand composing barrier islands along Florida and the East Coast came from the Appalachian Mountains. Second, the topography of the coastline must have a broad, gentle slope. This condition occurs from the coastal plains of the mainland to the edge of the **continental shelf** along the Atlantic and Gulf Coasts. Finally, the waves, tides and currents in the area of a future barrier island must be strong enough to move the sand. Waves must be the dominant of these three **water** movement mechanisms.

Several explanations for barrier island development have been proposed. According to one theory, coastal sand is transported shoreward as sea level rises and, once sea level stabilizes, waves and tides work the sand into a barrier island. Another possibility is that sand is transported from shoals. Some barrier islands may form when

A barrier island in the Cape Hatteras National Seashore Park, off the coast of North Carolina. *JLM Visuals. Reproduced by permission.*

low-lying areas of spits, extensions of beaches that protrude into a bay, are breached by the sea. Finally, barrier islands may form from sandy coastal ridges that become isolated and form islands as sea level rises.

Once formed, barrier islands do not persist as static landforms. They are dynamic features that are continuously reworked by **wind** and waves. Changes in sea level also affect barrier islands. Most scientists agree that sea level has been gradually rising over the last thousand years, and this rise may be accelerating today due to **global warming**. Rising sea levels cause existing islands to migrate shoreward, and barrier islands off the Carolina coast are thought to have migrated 40-50 mi (64-80 km) during the Holocene epoch.

Barrier island zonation

Individual barrier islands do not stand alone; instead, systems of islands develop along favorable coastlines. The formation of an island allows other landforms to develop, each characterized by its dominant sediment type and by the water that shapes them. For example, each barrier island has a shoreline that

faces the sea and receives the full force of waves, tides, and currents. This shoreline is often called the beach. The beach zone extends from slightly offshore to the high water line. Coarse sand and gravel are deposited along the beach, with finer sand and silts carried farther offshore.

Behind the beach are sand dunes. Wind and plants such as sea oats form dunes, but occasionally dunes are inundated by high water and may be reworked by storm surges and waves. On wide barrier islands, the landscape behind the foredunes gently rolls as dunes alternate with low-lying swales. If the dunes and swales are well developed, distinct parallel lines of **dune** ridges and swales can be seen from above. These differences in topography allow some **soil** to develop and **nutrients** to accumulate despite the porous sandy base. Consequently, many medium to large barrier islands are host to trees (which are often stunted), bushes, and herbaceous plants. Smaller or younger barrier islands may be little more than loose sand with few plants.

The back-barrier lies on the shoreward side of the island. Unlike the beach, this zone does not bear the full force of waves. Instead, the back-barrier region

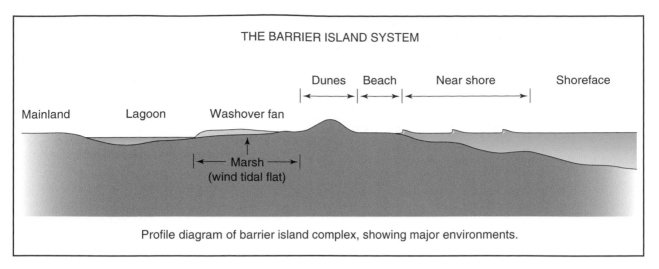

THE BARRIER ISLAND SYSTEM

Profile diagram of barrier island complex, showing major environments.

Diagram of the barrier island system. *Photo Researchers, Inc. Reproduced by permission.*

consists of a protected shoreline and lagoon, which is influenced by tides more than waves. Water may occasionally rush over the island during storms, carrying beach and dune sand and depositing it in the lagoon. This process, called rolling over, is vital to the existence of barrier islands and is the method by which a barrier island migrates landward. Sand washover fans in the lagoons are evidence of rolling over. **Salt** marsh, sea grass, and mudflat communities develop along the sheltered back-barrier. These communities teem with **plant** and **animal** life and their sediments are rich in organic **matter**.

Finally, barrier islands are associated with tidal inlets and tidal deltas. Tidal inlets allow water to move into and out of bays and lagoons with rising and falling tides. Tidal inlets also provide a path for high water during storms and hurricanes. As water moves through an inlet, sand is deposited at both ends of the inlet, forming tidal deltas. Longshore currents may also **deposit** sand at the **delta**. Eventually, the growing deltas close the inlet and a new inlet appears elsewhere on the island, usually at a low-lying spot. The size and shape of the inlet are determined by various factors, including the size of the associated lagoon and the tidal range (the vertical height between high and low tide for the area). A large tidal range promotes the formation of many inlets, thereby creating shorter and wider barrier islands referred to as drumsticks. In addition, the larger the lagoon and the greater the tidal range, the deeper and wider the inlet due to the large quantity of water moving from **ocean** to lagoon and back. Deep, wide inlets occur where the main source of **energy** shaping the coastal area is tides, or tides in conjunction with waves. In contrast, wave-dominated areas form long barrier islands with narrow bays and narrow, shallow inlets.

Can humans and barrier islands coexist?

Barrier islands bear the full force of coastal storms and hurricanes, buffering the mainland coast. This often occurs at the expense of the island. Although the processes creating and maintaining barrier islands have been occurring for thousands of years, they have only become of concern in the last few decades. Billions of dollars worth of real estate development on barrier islands is now threatened by migrating beaches as sand continues to be reworked and transported by natural forces. Cities such as Miami Beach and Atlantic City are on barrier islands. **Engineering** efforts to stop erosion through beach nourishment projects, seawalls, and other means are merely temporary fixes against the powerful forces of nature. Some engineered structures can actually accelerate the rate of erosion.

Laws in some coastal states prohibit building between the sea and the dunes closest to the sea. Some laws prohibit the rebuilding of structures lost or damaged due to storms and erosion. Preservation may be the best long-term solution to ensure the future of these islands, but for many people the desire for beach resorts is a more immediate concern.

See also Coast and beach; Dune.

Resources

Books

Bascom, W. *Waves and Beaches.* Garden City, New York: Anchor Press/Doubleday, 1980.

Carter, R. W. G. *Coastal Environments.* San Diego: Academic Press, 1991.

Davis, R. A. *Depositional Systems.* New Jersey: Prentice-Hall, 1983.

Kaufman, W., and O. Pilkey. *The Beaches Are Moving.* Garden City, New York: Anchor Press/ Doubleday, 1979.

Leeder, M. *Sedimentology and Sedimentary Basins.* London: Blackwell Science, 1999.

Periodicals

Bennett, D. "Paying for Sand." *Audubon* (September/October 1993): 132.

Hawes, E. "Castles in the Sand." *The New York Times Magazine* (July 1993): 24–32.

Overby, P. "Beachfront Bailout." *Common Cause* (Summer 1993): 12–17.

Stuller, J. "On the Beach." *Sea Frontiers* (December 1994): 28–34.

Tibbetts, J. "On Shifting Sands." *E: The Environmental Magazine* (July/August 1993): 19–21.

Other

Earth & Sky. "Transcript—Barrier Islands." July 18, 2002 [cited October 19, 2002]. <www.earthsky.com/2002/es 020718.html>.

List, J. "USGS Fact Sheet: Louisiana Barrier Islands." 1995 [cited October 19,2002]. <www.marine.usgs.gov/factsheets/Barrier/barrier.html>.

Pilkey, O. H. "Engineered Barrier Islands: Lifeless Piles of Sand." 2002 [cited October 19, 2002]. <www.gsa.confex.com/gsa/2002AM/finalprogram/abstract_40202.htm>.

Elaine L. Martin

Barrier reef *see* **Coral and coral reef**

Baryons *see* **Subatomic particles**

Base *see* **Acids and bases**

Base (numeration) *see* **Numeration systems**

Basin

Within the broad field of **geology**, the term basin can be used to represent a number of features. These include topographic or drainage basins, structural basins, and sedimentary basins. In some cases, a single basin can include aspects of more than one of these types of basins.

A topographic, or drainage, basin is a sloping or depressed area from which runoff collects and flows into a channel, stream, or **lake**. The flow of any river or stream is collected within that area of land comprising its particular drainage basin. These basins can vary greatly in size from millions of square miles to a few square feet. The Amazon River's drainage basin covers more than 2.7 million sq mi (3.22 million km). A drainage divide lies at the top of a topographically elevated ridge that separates each drainage basin from the neighboring basins. The divide encircles the entire basin. Closed drainage basins have no outlet. Any **water** entering one of these cannot exit by overland flow. Death Valley is a closed basin and the floor of the valley is covered with salts that have been precipitated from water that was later lost to **evaporation**.

An area in which the **rocks** dip or tilt toward the center of the structure is known as a structural basin. A structural basin is analogous to a set of progressively smaller bowls that fit neatly within each other. Each of the bowls represents a different layer of rock, all symmetrically arranged around a common center. The rocks that make up such a basin have been displaced downward at the center of the feature, leaving the flanking rocks with a relatively equal dip in all directions. Most structural basins are from a few miles to hundreds of miles in width. In some cases, the rocks that form the outer limbs of the structural basin might be removed by **erosion**, leaving the center of the basin intact and higher than the outside of the basin. Such a feature would be a topographic high but have the internal structure of a basin. The entire lower **peninsula** of Michigan is an example of this type of structural basin.

A sedimentary basin is a depressed area in the earth's crust in which sediments tend to accumulate over a long period of time. Sedimentary basins form when the crust of the **earth** subsides, or becomes depressed, as a result of changes within the earth resulting from tectonic events or other internal changes, or downward warping as sediments are loaded onto the crust. Sedimentary basins can be shaped like bowls or elongate troughs, or have irregular shapes. They can be tens to hundreds of miles in diameter and can contain layers of sediment as thick as 49,000 ft (15,000 m). Much of our interest in these basins results from the fact that some sedimentary basins contain economically significant accumulations of **petroleum**. There are approximately 600 sedimentary basins on Earth, of which about 150 contain petroleum.

Sedimentary basins often form beneath large deltas. Deltas form in areas where **rivers** drain continents and deposit sediment. The load of sediment can cause the crust to sag or warp, forming a basin. The Gulf of Mexico is an area of tremendous sediment deposition by the Mississippi River and also contains significant petroleum reserves.

See also Fold; Tectonics.

Bass

Bass is the common name for a number of popular **freshwater** and **saltwater fish**, which include the wide mouth bass, the striped bass, groupers, jewfish, and wreckfish, which are some of the finest sports and food fish in the world.

Fish known as bass actually belong to different families and are distributed worldwide in tropical and temperate waters.

The freshwater family Centrarchidae includes the black basses, crappies, and sunfish. All **species** have plump, oval bodies with rough scales with a comblike edge. There are two dorsal fins which are connected, the first fin with thick spines. Heavy spines are also found in the anal fin. Freshwater basses include the predatory largemouth bass, *Micropterus salmoides,* which is generally found in shallow waters with plenty of vegetation such as slow moving streams with muddy bottoms and lakes. The largemouth bass is widespread in the United States and southern Canada and is an important sport fish. The average size of largemouth bass varies; bass from southern regions are heavier than those from the north. In southern Georgia a specimen was caught by rod and reel weighing 22.25 lb (10 kg). The species spawns in late March in the south but in June in the north. Largemouth bass demonstrate characteristic parental care of the young. When females are about to lay their eggs males will prepare the nest. After a female has laid her eggs, the male will deposit sperm (milt) over them. He will guard the nest from potential enemies. In the course of their development the eggs are fanned to aerate them and to keep off mud and silt. After the eggs develop into fry the male guards the young for several weeks.

The largemouth bass, so called because of its large mouth compared to the other basses, is greenish with a lateral stripe on each side, the **color** depending on size, and **water temperature** and **chemistry**.

The smallmouth bass, *M. dolomieui,* is found in **North America** farther north than the largemouth bass. The smallmouth bass is found in cold, moving water, and not generally on muddy bottoms. It is a smaller species

than the largemouth bass, on average weighing under a pound. It is somewhat more aggressive and provides more sport for the fisherman.

The spotted bass, *M. punctulatus,* is found between Ohio and Florida and in parts of Kansas and Texas. The spotted bass may reach up to 4 lb (1.8 kg). This fish is found in cool, fast flowing waters like the smallmouth bass, but may also be found in deep water (100 ft, 33 m). Three other species resemble the spotted bass. These are the Guadalupe bass, *M. treculi,* reaching a weight of 1 lb (0.45 kg), the Suwanee bass, *M. notius,* of northern Florida weighing under a pound, and the redeye bass, *M. coosae,* of the Alabama River and nearby areas, weighing up to 2 lb (0.9 kg).

The saltwater basses are distinct from the freshwater basses, and most are included in the family Serranidae. Although designated as saltwater fishes, some of these bass may be found in fresh or **brackish** water. Some of these species are prized sport fishes especially for the hook and line enthusiast. Saltwater bass have the characteristic basslike or perchlike body shape with a thick spine on the first dorsal fin and a soft rayed second dorsal fin. Saltwater bass vary in size from massive forms of about 1,000 lb (454 kg) to tiny individuals under 1 in (2.5 cm) long, some of which are aquarium fish.

The belted sandfish, *Serranus subligarius,* matures when it reaches 2 in (5 cm) in length, and grows to a maximum of about 6 in (15 cm) long. The belted sandfish is found in the tropical Atlantic and Caribbean and may descend to deeper waters of 60 ft (20 m).

Sea basses in the genus *Morone* are occasionally found in fresh water. Some taxonomists group these fish with the wreckfish (*Polyprion americanus*) and the giant sea bass in the family Percichthyidae rather than the Serranidae. The wreckfish reaches lengths of over 6 ft (2 m), weighs up to 100 lb (45 kg), and descends to depths of 300 ft (1,000 m).

The striped bass, *Morone saxatilis,* also called rockfish, rock bass, or striper, inhabits the waters along the mid-Atlantic coasts from the St. Lawrence River to northern Florida. Some specimens of striped bass transferred to the Pacific Ocean off California in 1879 have established themselves well and are now abundant. The striped bass may reach 100 lb (45.4 kg), although most average approximately 10 lb (4.54 kg) and are highly prized for sport and food. Striped bass spawn in fresh waters in early summer; some have become landlocked in the Santee-Cooper Reservoir in South Carolina.

Groupers (genus *Epinephelus*) frequently gather on rocky shores. Groupers are popular sport fish, caught by

Two striped bass on the ocean floor. *Photograph by Andrew J. Martinez. The National Audubon Society Collection/Photo Researchers, Inc. Reproduced by permission.*

rod and reel as well as by hand line, and highly prized for their superb flavor. Examples are the red grouper, *E. mario,* the Nassau grouper, *E. striatus,* the red hind, *E. guttatus,* and the rock hind, *E. adscensionis.*

In **Australia** the Queensland grouper, *E. lanceolatus,* reaches half a ton in weight and has a huge appetite. There are reports that this giant fish stalks pearl and shell divers, and there are unconfirmed tales of divers being swallowed by groupers.

A close relative of the Queensland grouper is the giant jewfish (*Epinephelus itajara*), which frequents the Caribbean up to Florida and also the Pacific Ocean in the Gulf of California, Mexico, and Panama. Although the jewfish's weight averages about 20 lb (9 kg), some may weigh up to 100 lb (45.4 kg) or more. A rod and reel catch of a 680 lb (309 kg) specimen has been reported. To catch the larger fish, shark hooks and ropes may be used. The large fish are cut into steaks and fillets.

The taste of the smaller fish is preferable to that of the larger specimens, which are edible but have a strong flavor.

Several species of sea bass in the Serranidae are hermaphrodites, where both male and female organs are found in the same **individual**. Cross-fertilization is the rule since fish do not fertilize their own eggs. Examples of hermaphroditic species include *Paralabrax clathratus,* *P. hepatus,* and *Diplodus vulgaris* of the Mediterranean Sea. Most species of true **hermaphrodite** fish among the teleosts are in the Serranidae, the sea bass family.

Many species of grouper and other sea basses are fish that function as females at first, then transform into males and function as males for the rest of their lives. This explains why large specimens are all males. An example is the Atlantic or black sea bass (*Centropristis striatus*), which demonstrates this condition. It is found from Massachusetts to North Carolina with some individuals straying to Florida.

The black sea bass demonstrates a distinct form of hermaphroditism characteristic of hermaphroditic groupers, where the sexes are physically distinct. Functional males and females are easily distinguished; the female has a more pointed snout, and is a darker or duller blue color. In July, the post breeding period, some females may be brown or almost completely white. Males at spawning time in May and June may be bright blue, especially around the eyes, with an adipose (fatty) hump behind the head which is most prominent at spawning time. Early in life there is a sex change from female to male, the changeover reaching its summit at about five years of age. By the eighth year there are no females in the population, and the males continue to grow for several more years. As with the groupers this accounts for the fact that the large black sea bass are all males.

Resources

Books

Whiteman, Kate. *World Encyclopedia of Fish & Shellfish.* New York: Lorenz Books, 2000.

Periodicals

Cousteau, J. "The Act of Life." In *The Ocean World of Jacques Cousteau* Vol. 2. Canada: Prentice-Hall of Canada, 1975.
Hildebrand, S. F., and W. C. Schroeder. "Fishes of Chesapeake Bay." *Bulletin of the United States Bureau of Fisheries* Vol XLIII, Part I, 1927.

Nathan Lavenda

Basswood

Basswoods are about 30 **species** of trees in the genus *Tilia*, in the linden family Tiliaceae. In **North America**, these trees are generally known as basswoods in **forestry**, and as lindens in **horticulture**.

Basswoods have simple, long-petioled, coarsely toothed, broadly heart-shaped leaves, arranged alternately on their twigs. The flowers occur in clusters, and emerge from a specialized **leaf** known as a bract. The flowers produce relatively large amounts of **nectar**, and are insect pollinated. The ripe **fruits** are grey, hard, and nut-like, and each contains one or several **seeds**.

Mature basswood trees issue sprouts from their roots, which develop as shoots around the **tree**. In addition, after a mature basswood tree is harvested, numerous sprouts arise from the surviving roots and stump, and these can develop into a new tree.

Basswoods produce a relatively light, clear, strong, and durable **wood** that can be used as lumber to manufacture boxes and crates, furniture, picture frames, and

for carving. Basswood honey is another economic product, as is a herbal tea made from the dried flowers. North American species of basswood and the related European linden (*Tilia cordata*) are commonly planted as shade trees in residential areas of towns and cities.

Four species of basswood are native to North America. The most widely distributed species is the American basswood (*Tilia americana*), a tree of temperate, hardwood **forests** of the northeastern United States and southeastern Canada. This is a relatively tall-growing species, which can reach a height of up to 82 ft (25 m). The leaves of this species are largest in relatively shaded parts of the crown, where they can achieve a length of more than 7.8 in (20 cm) and a width of 3.9 in (10 cm).

The white basswood (*T. heterophylla*) occurs in the eastern United States, while the ranges of the Carolina basswood (*T. caroliniana*) and Florida basswood (*T. floridana*) are the southeastern coastal plain of the United States.

Bill Freedman

Bathysphere

Throughout human history, the sea has yielded an abundance of resources for man's existence and provided efficient routes for exploration and transportation. In return, it has exacted a toll in terms of human life and property. The fear and respect that it earned from those who ventured out upon its surface was itself a deterrent to **learning** more about its mysteries. The physical restrictions of penetrating the sea made sub-surface exploration nearly impossible.

The most immediate restriction was air supply. Only the most disciplined divers could stay under for more than a few minutes. As scientists began to recognize the sea as a realm to be explored, this was their first hurdle.

In 1716, the English astronomer, Edmund Halley, whose interests in the universe included the **earth**, invented a wooden diving bell which was open at the bottom. The significance of Halley's bell was the system developed with it to provide its occupants air. The trapped air in the bell was resupplied by sending down weighted barrels of fresh air. Divers could also venture from the bell with the aid of leather helmets and leather air hoses.

The next major restriction to be surmounted was the tremendous **water pressure** which increases with depth. While some sea creatures have adapted to this environment, an unprotected human can only dive to about 330 ft (100 m) even with an air supply. The diving suits, bells, and submarines invented in the eighteenth and nineteenth centuries were not constructed well enough to protect divers very far below the surface.

The collaboration in the late 1920s of two scientists, William Beebe, a naturalist from Columbia University in New York, and Otis Barton, an engineer at Harvard University in Boston, led to the invention of the bathysphere, the first ever deep sea exploration vessel.

The **sphere** was nearly 5 ft (1.5 m) in diameter, had **steel** walls and weighed 5,400 lb (2,451 kg). It had a circular manhole at the top and three windows made of thick fused quartz. Air was released from two tanks and chemicals were used to absorb moisture and **carbon dioxide**. It was equipped with a **telephone** and searchlight. The sphere was tethered to the surface vessel by a steel cable.

After two unmanned test dives, Beebe and Barton made the first manned descent on June 6, 1930, to a world-record depth of 800 ft (244 m). Continued dives led to a record of 3,028 ft (923 m) in the Atlantic Ocean off Bermuda.

The primary purpose of the bathysphere was to explore rather than to set records. The explorations led to discovery of deep sea **plant** and **animal species** and observation of already known species. It also gave scientists new knowledge of submarine topography, **geology**, and geomorphology.

After some improvements to his diving vessel, Barton set a final diving record of 4,501 ft (1,372 m) in the Pacific Ocean off of Southern California in 1949. But by this time efforts led by Swiss physicist Auguste Piccard were being made to develop the successor to the bathysphere, the self-propelled bathyscaphe.

The bathysphere was limited to vertical travel. It was constantly dependent on its umbilical connection to its mothership. Besides visual observation, other exploration activities, such as specimen collection, were difficult with the sphere. Increased water pressures and potential problems with the support line, including the shear weight of the steel cable, made the deeper dives ever more risky.

The development of the bathyscaphe in the early 1950s would allow scientists to overcome the third, and perhaps final, obstacle to deep sea exploration, the need for freedom and independence for meaningful research.

Bats

Bats are one of the most diverse and widely distributed groups of **mammals** on **Earth**, second only to ro-

dents in the number of **species**. More than 900 species of bats have been described. They occur in most terrestrial biomes, except for the high Arctic and all of **Antarctica**. Bats are the only truly flying mammals, and are distinct from the flying **lemurs** and flying **squirrels**, which actually glide. Bats make up the order Chiroptera, named from the Greek words *cheir* (hand) and *pteron* (wing); this is an appropriate name, since the wing is formed by modified bones of the hand.

The bat order Chiroptera is divided into two suborders. The Megachiroptera is composed of a single family that is restricted to the tropics, and includes the largest species of bats, such as fruit bats and flying foxes. The megachiropterans are characterized by large eyes, simple ears, and a dog-like face. The Microchiroptera are made up of 17 families, and feature small eyes, complex ears, and an ability to find their **prey** and navigate by **echolocation**. Certain differences between the two suborders in flight and sensory capabilities have led some biologists to propose that they evolved from separate ancestral lineages, and that the megachiropteran bats are more closely related to **primates**. This idea is highly controversial, however, and other morphological data and DNA evidence support the hypothesis that all of the bats evolved from a common ancestor. Most bat scientists believe that bats evolved from tree-dwelling, shrew-like ancestors that scampered along branches and fed on **insects**. The proto-bat likely had long fingers supporting webs of skin attached to the body, which it used to glide while in pursuit of its insect prey.

Basic body plan

The body of bats is well designed for flight. However, they achieve flight differently from **birds**. Like birds, the bones of bats are light-weight and delicate. However, bats have a short neck compared to birds, and they lack a deeply keeled sternum, or breastbone, where the flight muscles attach in birds. Instead, three shallow pairs of muscles on the breast power the downstroke of the wing during bat flight, while the upstroke is provided by three pairs of muscles on the back. Because they do not have a well-developed breastbone, bats have a flat profile through the chest, and so they can squeeze through small openings and roost in narrow crevices.

The wing structure of birds and bats is also different. The skin and feathers of the wings birds are supported mainly by their second and third "fingers." In comparison, the wings of bats are formed by thin, elastic skin extending from the sides of the body to the tips of all four elongated "finger" bones. Their much-reduced thumb remains free of the wing **membrane** and is used to manipulate food, and as a hook when the bat climbs and clings to surfaces or vegetation. The wing membranes are also supported by the hind legs, and in species with a tail, it is entirely or partially enclosed by wing membrane stretching between the hind legs. The hind legs of bats are unique among mammals in being rotated 180°, so that the knees point backward, allowing the leg to flex in a reverse fashion. This is believed to assist in steering during flight, and in taking off from the characteristic head-down roosting position of bats.

The Megachiroptera have a dog-like face. However, the face of the Microchiroptera is often striking and weird-looking, with fleshy embellishments that form complicated dimples, wrinkles, and horseshoe- or leaf-like structures. Some species have tubular nostrils. Biologists have suggested that these facial embellishments function in the projection of sounds produced for echolocation, like megaphones or acoustic lenses. While Megachiroptera generally have simple ears, there is huge variation in the size, shape, and elaboration of ears among microchiropteran bats. Depending on the species, their ears may feature special folds and ridges that are thought to play a role in sound **perception**. For instance, many of these bats possess a large tragus, a fleshy projection on the bottom front edge of the **ear** opening, and believed to aid an echolocating bat in determining the horizontal position of a target.

Like many birds, bats pass the food they have eaten fairly quickly through their digestive tract, so as to reduce the amount of time they must carry the extra weight of undigested food. Total output time is as little as 20 minutes in some smaller bat species, which is similar to birds of the same size.

Diet

The dietary diversity of bats is unmatched among living mammals. Most bats living in temperate areas, about 265 species, eat insects. Fruit bats, restricted to tropical areas, eat fruit and leaves, which they chew, swallowing the juice and spitting out the pulp. The long-tongued fruit bats (*Macroglossus* species) specialize in a diet of pollen and **nectar**, which they acquire using their elongated snout and unusually long tongue (up to one-third their body length). The fisherman bats (*Noctilio* species) of Central and **South America** catch small **fish**, while the frog-eating bat (*Trachops cirrhosus*) uses the calls of **frogs** to locate this prey, and can distinguish between the calls of poisonous and edible frogs. The large slit-faced bat (*Nycteris grandis*) of **Africa** eats small birds and even other bats, which are caught on the wing.

The infamous vampire bats (three genera in the subfamily Desmodontinae) dine on the **blood** of other mammals, such as domestic **livestock**, by making a shallow

cut with their incisors and lapping the blood that flows from the wound; their saliva contains an anticoagulant that keeps the blood from clotting. These bats are quite agile on the ground, typically landing beside their sleeping victim, and crawling gingerly onto them to feed. Vampire bats are dietary specialists, but most other species consume several food items, varying their consumption to get enough protein and other **nutrients**.

Sensory systems and echolocation

Contrary to popular myth, bats are not blind. In fact, the large eyes of many species suggest that they have well-developed **vision**. Like most mammals, they have keen senses of **taste** and **smell**, the latter being useful in locating food items, and in identifying roost sites and other bats, including family members. Bats also have excellent **hearing**. Many species use a wide range of vocalizations to communicate with one another. Some species hunt for food by listening to the sounds of their prey moving about.

The most remarkable sensory **adaptation** of bats is their capacity for echolocation. This sensory ability allows bats to maneuver in total darkness, using echoes of their ultrasonic calls to detect objects in their vicinity. Efforts to understand how bats can fly in complete darkness date back to the late eighteenth century, when the Italian scientist Lazarro Spallanzani (1729-1799) conducted experiments that included denying bats the use of their senses of smell, **touch**, and vision. He observed that bats lost their way only if their head was covered by a small sack, and concluded that bats must have a sixth sense, not shared with humans.

A Swiss scientist named Charles Jurine reported in 1794 that if a bat's ears are blocked, it cannot maneuver. Spallanzani heard this report, and taking the experiment a step further, showed that bats with brass tubes inserted into their ears can only navigate when the tubes are open. He then concluded that bats must somehow "see" with their ears. How this could occur was not explained until the 1930s, when the echolocation pioneer Donald Griffen (then an undergraduate at Harvard University) detected ultrasonic signals produced by bats in the lab, using a microphone capable of picking up sounds above 20 kHz-ultrasound. In a series of experiments, Griffen showed that bats used the echoes of their calls to locate obstacles, and he coined the term "echolocation" to describe this sensory ability.

It is now known that some other animals are also able to echolocate, including whales, dolphins, **shrews**, and some birds such as **cave** swiftlets. It is also known that some bats, including all but one of the flying foxes, are not able to echolocate. (The only megachiropteran genus that can echolocate is *Rosettus*, whose sounds are produced by tongue clicks.) Evidence indicates that bats can echolocate using reflected sound as effectively as we see with our eyes, using reflected **light**. Bats do this by sensing the time elapsed between the production of the sound (by means of their larynx) and the return of its echo, thus gauging the **distance** of objects near them. Of course, they do not perform these computations in a conscious way, any more than a person willfully determines the **frequency** of incoming light waves to perceive an object as blue or green. Rather, the bat **brain** carries out the necessary functions in a split second, providing them with a continuously updated "picture" of their surroundings.

What bats "see" in this way might best be imagined as something like what a human visitor might see in a darkened disco, where a strobe light is flickering to illuminate dancers and other objects in a pulse-like fashion; every time the strobe light flashes, the observer gets a brief update on the position of nearby things. A faster pulse **rate** in the strobe means that more information about these objects can be conveyed to the observer. The same is true, more or less, for bats, which vary their calling rate depending on what they are doing. The rate for a bat on a routine cruise is 5-10 calls per second; as it locates and closes in on a flying insect the call rate increases, and it finally accelerates briefly to more than 200 per second in a terminal "feeding buzz" that pinpoints the location of the food item. However, such a high rate of calling is only suitable for near targets; if an object is too far away, the outgoing signal gets mixed up with those returning from other distant objects. In addition, echolocation takes considerable **energy**; the intensity of the calls of some bats, if they were of a frequency audible to humans, would make them as loud as the beeping alarm of a home smoke detector. Expending the energy required for the feeding buzz does not pay off during an ordinary commute, and so the calling rate at this time is relatively low.

The brain of an echolocating bat carries out some astonishing perceptual feats, and solves problems similar to those facing engineers during the early development of **radar** technology. For example, there is the problem of signal attenuation: sounds progressively degrade as they get farther away from their source. Pulses must be loud enough to survive degradation, but a loud sound can overwhelm the sensitive receiver structures needed to pick up the return signal. Through their **evolution**, bats solved this problem in a similar way as the early radar engineers: they are equipped with a send/receive switch that momentarily disconnects the receiver function just as the loud outgoing pulse is produced; the receiver is then reconnected in time to receive the echo. The switching is accomplished by muscles that attach to the bones

of the inner ear; when the muscles contract, these bones do not transmit sound well. When the muscles relax, the ear returns to its normal sensitivity. Further signal attenuation happens within the brain itself, as neurons responsible for sound perception block the transmission of messages to higher regions of the brain at the moment that a bat vocalizes. In addition, special echo-detector cells in the brain respond more intensely to the *second* of two separate sound pulses, which is an excellent way of picking up the echo of a vocalization.

There is also the problem of sorting out echoes returning from near, medium, and distant objects; how can a bat tell the distance between itself and its next meal? They accomplish this by means of frequency modulation, so the sorting can be done by differences in pitch (frequency). If a bat utters a downward-sweeping whistle call, an echo returning from a more distant object will be older and thus higher in pitch compared with echoes from closer objects. The bat thus has a standard for comparison: lower pitch means close by, higher pitch means farther away. In addition, bats can measure the speed of a moving target by means of the **Doppler effect**; this is the phenomenon responsible for the change in pitch of an ambulance siren as it moves toward and then passes a stationary observer. To do this, the bat's brain compares the pitch of an echo with that of the original call; they can do this reliably even in the midst of hundreds of echolocating colleagues engaged in a midnight feeding frenzy. All of this is accomplished automatically and instantaneously, with no more conscious effort than a person might exert while watching images on **television**.

Of course, the echolocation system does not guarantee a perfect success rate while hunting. Some flying insects, **mice**, and other potential prey can detect the echolocation signals of bats and then take evasive action.

Bats also use their senses of sight and smell to find food. Their other senses are also important in recognizing other bats, including their offspring, and perhaps also in identifying roost sites.

Roosting

Most bats rest during the day and disperse around dusk to feed. Bats typically spend more than half of their lives in their roost environment, which may be in a cave, mine, crevice in **rocks**, cavity in a **tree**, in dense foliage (sometimes rolled up into a tent), or in a human structure. Many bats roost communally, often for brooding of young or for **hibernation**; such colonies range in size from a few individuals to several millions in a large cave. Females of some colonial bat species will share food and nursing of the young during the breeding season. For a hibernating bat in temperate areas, a communal living

Looking up at fruit bats in Tasavo National Park, Kenya. *Photograph by Don Mason. Stock Market. Reproduced by permission.*

arrangement offers a relatively stable microhabitat in which their body **temperature** may drop to within a few degrees of the ambient temperature, thus permitting the **conservation** of critical reserves of body **fat**.

Reproduction and social organization

Most bats have a breeding season, which is in the spring for species living in a temperate climate. Bats may have one to three litters in a season, depending on the species and on environmental conditions such as the availability of food and roost sites. Females generally have one offspring at a time; this is probably a result of the mother's need to fly to feed while pregnant. Female bats nurse their youngster until it has grown nearly to adult size; this is because a young bat cannot forage on its own until its wings have assumed adult dimensions.

Female bats use a variety of strategies to control the timing of pregnancy and the **birth** of young, so as to make delivery coincide with maximum food ability and other ecological factors. Females of some species have delayed **fertilization**, in which sperm are stored in the reproductive tract for several months after mating; in many such cases, mating occurs in the fall, but fertilization does not occur until the following spring. Other species exhibit delayed implantation, in which the egg is fertilized after mating, but remains free in the reproductive tract until external conditions become favorable for giving birth and caring for the offspring. In yet another strategy, fertilization and implantation both occur but development of the fetus is delayed until favorable conditions prevail. All of these adaptations result in the pup being born during a time of high local production of fruit or insects.

Bats exhibit every kind of breeding system that has been described for mammals. Some species are monogamous, with a particular male and female mating only with each other. Other species are polygynous, which means that one male may mate with several females. In these species, males may fight for control of preferred female roosting sites, or over aggregations of females, called harems. In hammer-headed fruit bats (*Hypsignathus monstrosus*) of Africa, the males assemble into leks, which are aggregations of displaying males at traditional sites that females visit for the purpose of selecting a mate. The males display by vigorous wing flapping, erecting patches of hair, and loud vocalizations. Still other bats have a promiscuous mating system, in which both males and females mate with more than one other **individual**.

Primary cells

Primary cells are designed to be discharged only once. This is despite the fact that all electrodes must theoretically participate in a reversible reaction when current is generated. The reason that the primary cell reaction is not reversible has to do with reactions that prevent or limit the efficiency of recharging. For example, a **magnesium** anode decomposes to produce magnesium ions and electrons. The magnesium ions react with **water** to produce magnesium hydroxide, which causes the cell to swell, and hydrogen gas. Any attempt to recharge the cell would only generate more hydrogen gas at the oxide surface, because the voltage required to generate hydrogen is less than that required to redeposit the magnesium.

Moderate energy primary cells

Zinc/manganese dioxide systems

The cell developed by Georges LeClanche in 1866 used inexpensive, readily available ingredients. It therefore quickly became a commercial success. The anode is a zinc **alloy** sheet or cup (the alloy contains small amounts of lead, cadmium, and mercury). The electrolyte is an aqueous solution of zinc chloride with solid ammonium chloride present. The cathode is manganese dioxide blended with either graphite or acetylene black to conduct electrons to the oxide. The system is relatively tolerant of many impurities. These cells are used in barricade flashers, flashlights, garage door openers, lanterns, pen lights, radios, small lighted toys and novelties, and in others.

The zinc chloride cell without ammonium chloride was patented in 1899, but the technology from commercially producing such cells did not prove practical until about 70 years later. Currently zinc chloride cells deliver more than seven times the energy **density** of the original LeClanche cell. This cell is used in same applications as the LeClanche cell.

Zinc/manganese dioxide alkaline cells

The zinc/manganese dioxide alkaline cell's anode consists of finely divided zinc. The cathode is a highly compacted mixture of very pure manganese dioxide and graphite. The cells operate with higher efficiency than the zinc chloride or LeClanche cells at temperatures below 32°F (0°C). Manganese/manganese dioxide cells have much higher energy densities than zinc chloride systems. Cylindrical batteries are used in radios, shavers, electronic flash, movie cameras, tape recorders, **television** sets, cassette players, clocks, and camera motor drives. Miniature batteries are used in calculators, toys, clocks, watches, and cameras.

Medium to high energy primary cells

Mercuric oxide/zinc cells

Mercuric oxide/zinc cells use alkaline electrolytes and are frequently used in small button cells. The cell has about five to eight times the energy density available in the LeClanche cell and four times that in an alkaline manganese dioxide/zinc cell. The cell provides a very reliable voltage, and is used as a standard reference cell. These cells are used for walkie-talkies, **hearing** aids, watches, calculators, microphones, and cameras.

Silver oxide/zinc cells

Silver oxide/zinc cells use cellophane separators to keep the silver from dissolving and the cells from self discharging. The system is very popular with makers of hearing aids and watches because the high conductivity of the silver cathode reaction product gives the cell a very constant voltage to the end of its life. These cells are also used for reference voltage sources, cameras, instruments, watches, and calculators.

Lithium (nonaqueous electrolyte) cells

Lithium/iron sulfide cells take advantage of the high electrochemical potential of **lithium** and low cost of iron sulfide. The high reactivity of lithium with water requires that the cells use a nonaqueous electrolyte from which water is removed to levels of 50 ppm.

Lithium/manganese dioxide cells are slowly increasing in commercial importance. The voltage provides a high energy density, and the materials are readily available and relatively inexpensive.

Lithium/copper monofluoride cells are used extensively in cameras and smaller devices. They provide high voltage, high power density, long shelf life, and good low **temperature** performance.

Lithium/thionyl chloride cells have very high energy densities and power densities. The cells also function better at lower temperatures than do other common cells.

Lithium/sulphur cells are used for cold **weather** use and in emergency power units.

Air-depolarized cells

Zinc/air cells are high energy can be obtained in a galvanic cell by using the **oxygen** of air as a "liquid" cathode material with an anode such as zinc. If the oxygen is reduced in the part of the cell designed for that purpose and prevented from reaching the anode, the cell can hold much more anode and electrolyte **volume**.

Aluminum/air cells have difficulty protecting the **aluminum** from the electrolyte during storage. Despite much research on this type of cell, aluminum/air cells are not in much current use.

Secondary cells

Secondary cells are designed so that the power withdrawn can be replaced by connecting the cell to an outside source of direct current power. The **chemical reactions** are reversed by suitably applying voltage and current in the direction opposite to the original discharge.

Moderate energy storage cells

Lead secondary cells

The lead/acid rechargeable battery system has been in use since the mid-1950s. It is the most widely used rechargeable portable power source. Reasons for the success of this system have included: great flexibility in delivery currents; good cycle life with high reliability over hundreds of cycles; low cost; relatively good shelf life; high cell voltages; ease of casting, **welding**, and recovery of lead.

The chief disadvantage of this battery is its high weight.

Nickel electrode cells with alkaline electrolytes

Nickel/cadmium cells provide portable rechargeable power sources for garden, household tools, and appliance use. The system carries exceptionally high currents at relatively constant voltage. The cells are, however, relatively expensive. These cells are used for portable **hand tools** and appliances, shavers, toothbrushes, photoflash equipment, tape recorders, radios, television sets, cassette players and recorders, calculators, personal pagers, and laptop computers.

Alkaline zinc/manganese dioxide cells

Alkaline zinc/manganese dioxide systems been developed and used as special batteries for television sets and certain portable tools or radios.

High energy storage batteries

Silver/zinc cells

Silver/zinc cells are expensive. They are chiefly used when high power density, good cycling efficiency, and low weight and volume are critical, and where poorer cycle life and cost can be tolerated. They are used in primarily four areas: under water, on the ground, in the atmosphere, and in **space**.

KEY TERMS

Anode—A positively charged electrode.

Battery—A battery is a container, or group of containers, holding electrodes and an electrolyte for producing electric current by chemical reaction and storing energy. The individual containers are called "cells". Batteries produce direct current (DC).

Cathode—A negatively charged electrode.

Direct current (DC)—Electrical current that always flows in the same direction.

Electrode—The conductor by which electricity enters or leaves a galvanic cell.

Electrolyte—The medium of ion transfer between anode and cathode within the cell. Usually liquid or paste that is either acidic or basic.

Galvanic cell—Combination of electrodes separated by electrolyte capable of producing electric energy by electrochemical action.

Primary cell—A galvanic cell designed to deliver its rated capacity once and then be discarded.

Secondary cell—A galvanic cell designed for reconstitution of power by accepting electrical power from an outside source.

Lithium secondary cells

Lithium secondary cells are attractive because of their high energy densities.

Sodium/sulfur systems

Sodium/sulfur systems are high-temperature batteries that operate well even at 177°F (80.6°C).

See also Cell, electrochemical; Electricity; Electrical conductivity; Electric conductor.

Resources

Books

Macaulay, David. *The New Way Things Work.* Boston: Houghton Mifflin Company, 1998.

Meyers, Robert A., *Encyclopedia of Physics Science and Technology.* New York, NY: Academic Press, Inc., 1992.

Bayberry *see* **Sweet gale family (Myricaceae)**

Beach *see* **Coast and beach**

ton (about 900 kg). Most male bears, called boars, are considerably larger than the females, or sows. This is especially important when boars compete for females. Sun bears and sloth bears (*Melursus ursinus*), however, take only one mate and so the sexes are nearly the same size.

Bears from cold regions do not truly hibernate. Instead, during the coldest part of the winter when food is not readily available, they enter a long period of lethargy. However, their body **temperature** and **heart rate** do not drop much, as would occur during true **hibernation**. During this time bears **sleep** a great deal and do not eat; instead, they live off **energy** stored in their body **fat**. The exception to this pattern is the polar bear, which continues to hunt for seals during winter.

Bears have an amazing ability to adjust physiologically to seasonal ecological changes. Non-tropical bears mate during the spring or summer, but the fertilized eggs float free in the uterus, rather than immediately implanting and developing. Later, in early winter, the sow finds or creates a den in which to sleep away the winter. At that time, the eggs implant and their gestation starts. The cubs are poorly developed when born, being blind, nearly hairless, almost helpless, and extremely small. For example, a female brown bear weighing 450 lb (205 kg) produces cubs weighing less than a pound (about 450 g). The cubs are born during the winter, and are so small that they are incapable of regulating their body temperature. The warm den in which the mother winters provides a snug place for them to nurse and grow until they can maintain their body temperature. The sow's milk, which is extremely rich in fat, sustains the rapid growth of the cubs during the winter denning. By the time the sow is ready to leave her den, the cubs have grown enough to follow her.

All bears are thickly furred, often with a coat of a single **color**. With the exception of the sun bear, all species have fur on the bottom of their feet, around the pads of the soles. This is especially important for the polar bear, which spends most of its life walking on snow and **ice**. The sun bear, on the other hand, has furless feet as an aid in climbing trees. The feet of bears are well-armed with heavy claws. Bears walk in plantigrade fashion, meaning they walk on the heel and sole of the foot. Most other **mammals** walk on their toes. Although they may look rather lumpish and clumsy, bears can run for short distances at speeds up to 40 MPH (64 km/h).

Bears have a reputation for having poor eyesight, but their sight is actually quite good. Their sense of **smell**, however, is extremely acute. Bears can identify the odor of animals that passed by as much as several days before. Unlike other carnivores, bears do not have their lips attached to their gums. This means that they can make facial expressions and use their lips to suck in food, such as **insects** or honey.

Grizzly and other brown bears

Brown or grizzly bears (*Ursus arctos*) live in forest, **tundra**, and grassland across the top of the Northern Hemisphere, including both **North America** and Eurasia. Some biologists separate them into three subspecies: the Eurasian brown bear (*U. a. arctos*) of much of temperate and subarctic Eurasia; the grizzly bear (*U. a. horribilis*) of Canada, Russia, and the United States; and the Kodiak bear (*U. a. middendorffi*) of Kodiak Island and two smaller islands in the Bering Sea off Alaska. These various brown bears may vary in color from white to cinnamon to black, but their fur is most commonly brown. Some brown bears have white tips on the end of the hairs, a coloring called grizzled. A grizzly bear can be distinguished from other bears in North America by the profile of its body. The outline of its head as seen from the side is concave, or scooped inward, and its shoulders are high due to a thick layer of muscle and fat, which makes the back appear to slope downward.

Grizzlies that live near **rivers** accessible to the **ocean** feed on migrating **salmon** as much as they can. Those bears that feed well can be huge in size. Even with their great size, they like to frolic in the **water** and are adept at catching **fish**, which they carry to land to strip the flesh from the bones. They may also hunt **rodents**, and will opportunistically predate large **prey** such as **moose**, **caribou**, and even black bears. When meat is not available, grizzlies feed primarily on roots, sedge leaves, and berries.

Grizzly bears mate in the late spring. In the autumn, the sow finds a den in a cave or hollow **tree**, and settles in for her winter lethargy. Two or three cubs are born, usually in February. The cubs stay close to their mother for at least two years, continuing to nurse during most of that time. The mother is exceedingly protective, and teaches the cubs to climb trees to escape danger. Most attacks on humans are made by sow grizzlies protecting their cubs. One of the major enemies from which she must defend her cubs is male grizzlies, which will kill and eat them.

After a young female leaves its mother, it may continue to share the same feeding range. A male, however, will go off on its own, traveling up to 100 mi (160 km) before finding a place to settle down. It may have difficulty finding a suitable **habitat** because older males will fight to keep new ones out of their territory.

The range of the grizzly bear originally extended from Alaska to Mexico, as far east as Hudson Bay, through the **prairie** region, and even extending into **desert** habitat. Grizzlies have now disappeared from most of the western United States, with only small numbers surviving in Wyoming, Montana, Idaho, and Washington. There are larger numbers in Alberta, British Columbia, Yukon, Alaska, and the Northwest Territories. The populations of grizzlies in the United States (outside of Alaska) have declined mostly because of excessive hunting and habitat loss. These are still important problems for the species.

Eurasian brown bears vary greatly in size. The few remaining in Spain rarely weigh more than 250 lb (114 kg). Those in Siberia may rival the huge Kodiak bear in size. Brown bears are still found throughout much of **Europe** and **Asia**, but in rapidly decreasing numbers because of excessive hunting and habitat loss. Those that survive live primarily in hardwood forest in mountainous regions. Fewer than 35,000 brown bears are thought to live in Eurasia.

Polar bear

Polar bears are huge, whitish or yellowish, marine bears that are well adapted to life in and around the icy Arctic Ocean. They are relatively slender bears, with a longer neck and head than brown bears. The individual hairs of their thick fur are transparent; it is reflected sunlight that makes their coat appear white. Oddly, their skin is black. Any sunlight that gets through its fur is absorbed by the black skin, helping to keep the animal warm. Polar bears swim by paddling their furry, slightly webbed front feet and steering with the back feet.

Polar bears hunt by wandering extensively over the sea ice, often moving from ice floe to floe, and sometimes swimming for hours in the cold water. They do this to find good places for hunting their favorite food, the ringed seal (*Phoca hispida*). Their claws are longer than those of other bears, and are used to grasp their seal prey as it rises out of the water at a breathing hole or along the edge of an ice floe. During the spring, polar bears break into the snowy dens where female seals have given **birth** to their young.

Polar bears may congregate in areas where ice floes move freely in the **wind**, because seals are more easily obtained in that habitat. Polar bears will tolerate each other if they find a stranded whale carcass, or if a walrus has been killed. During the summer, when the ice is gone from the mainland coast, polar bears may move onto the land and feed on berries, or they may fast. They may also be attracted to garbage dumps near towns, where there can be dangerous encounters with people.

Polar bears are usually solitary animals, coming together only to mate in late spring (March-June). One to three cubs are born in December or January in a snow den constructed by the mother. The cubs average about

A polar bear. *Photograph by G. Bell. Stock Market/Zefa Germany. Reproduced by permission.*

23 oz (650 g) at birth, but weigh 20 lb (9 kg) or more when they emerge from the den in April. This rapid weight gain is possible because the milk of the sow contains more than 30% fat. The cubs stay with their mother for at least two years, **learning** to hunt and defend themselves. At that time the mother will mate again. Young females become sexually mature at five years of age. Polar bears live to be 20 or more years old.

Five countries have populations of polar bears: the United States (Alaska), Canada, Denmark (in Greenland), Norway, and Russia. These nations are cooperating in the management of their polar bears, allowing only a tightly controlled hunt. The hunt is mostly carried out by aboriginal people, who eat the bear meat and sell the valuable hides.

American black bear

The American black bear (*Ursus americanus*) occurs largely in forested habitat, but also in grassy meadows and tundra. It ranges across most of North America, from Alaska to Newfoundland, and south into Mexico. Although its populations are depleted in many parts of its range, it is still a widespread species. The black bear is usually black-colored, but may also be brown, light tan, or even white. Different colors may occur within a litter of cubs.

Black bears have larger ears than other bears. Size varies among populations; the overall range of adult weight is about 125-600 lb (57-272 kg). Males typically weigh about one-third more than females. The weight of individuals varies considerably during the year, depending on the amount of nourishing food available. Because they are much smaller than grizzlies, black bears try to stay out of sight when territories of the two species overlap.

Each black bear has a territory where it forages for berries, nuts, honey, insect grubs, fish, rodents, and carrion. A male's territory typically overlaps those of several sows. Black bears mate during the summer, with each female being visited several times by nearby males. The fertilized egg does not implant until the autumn. However, if the female bear is poorly nourished, the fertilized eggs do not implant at all. One to four cubs are born in January, after a gestation of 8-10 weeks.

Other black-colored bears

The Asiatic black bear (*Selenarctos thibetanus*) is a black-colored animal with a white crescent on its chest; an alternative name is moon bear. The hair around its neck is considerably longer than elsewhere on the body. A male may weigh up to 350 lb (159 kg), while females usually weigh less than 200 lb (91 kg). This bear inhabits mountain **forests** from Afghanistan, across China to Japan, and south to Southeast Asia.

The Asiatic black bear is generally nocturnal, but will sometimes venture out in the daytime to feed on sun-warmed fruit. It is also an opportunistic **predator**, and can kill fairly large animals by breaking the neck. Northern populations of Asiatic black bears sleep away the winter, but in warmer parts of their range they remain active year-round. They mate in the autumn, and the cubs are born 3-4 months later. The cubs are weaned by four months of age, much earlier than in other species of bears.

The sun bear (*Helarctos malayanus*) is the smallest species of bear, with some adults weighing less than 100 lb (45 kg). It has black fur, a whitish snout, and often a whitish or orange, U-shaped mark on the chest. It has a much longer tongue than other bears, which is used to lap honey, **bees**, or **termites** from their nests. They also eat a wide variety of tropical **fruits**. The sun bear is a tropical species, found in **rainforest** from southern China to Borneo and Sumatra. Sun bears climb trees

well, and often build nests by breaking and bending branches together. Cubs are born at any time of the year, after a gestation of 14 weeks.

The sloth bear (*Melursus ursinus*) of India and other countries in South Asia has the longest hair in the bear family, although its belly is almost hairless. Otherwise, it looks much like the Asiatic black bear. However, it has several distinctive behavioral traits: it carries its young on its back, and the male remains with the female to help raise the cubs. The sloth bear has long, strong claws, which are used to break open termite nests during feeding.

The spectacled or Andean bear (*Tremarctos ornatus*) is the only species in South America. It occurs in the Andean region, even living above 14,000 ft (4,300 m). It has a whitish, eyeglasses-shaped pattern around its eyes and a band of white on its neck and chest. It climbs trees well and may sleep in them. Male spectacled bears may reach a head-body length of almost 6 ft (1.8 m) and a weight of 400 lb (182 kg). In profile, they have a shorter nose than other bears.

The spectacled bear is primarily nocturnal, sleeping during the day in an excavated cavity or under tree roots. At night, it often climbs high into trees to feed on fruit. If it finds a good supply and decides to stay a while, it will build a platform of branches as a nest. They are an **endangered species**, with fewer than 2,000 remaining in the wild. An international program has succeeded in breeding them in captivity.

Bears and humans

All bears except polar bears are regarded as therapeutic in traditional medicine in Asia. In particular, fluid from the gall bladder is thought to have many health-enhancing qualities. Many Asiatic black bears are kept in captivity, where they have tubes implanted into their gall bladder from which bile fluid is continuously withdrawn without killing the animal. It is also thought that an aphrodisiac, or love potion, can be made from the gall bladders of bears. The flesh of bear paws is regarded as a gourmet food in eastern Asia. Because of these uses, the Asiatic black bear has long been intensively hunted, and is now an endangered species. Many American black bears and Eurasian brown bears are also killed so that their gall bladder can be harvested and exported to eastern Asia.

Even the extremely rare spectacled bear of South America is hunted by indigenous people, who believe their fat is useful in the treatment of **arthritis**.

In addition to the problem of excessive hunting, all bears are being affected by habitat destruction, mostly to develop agricultural land. When humans move into natural habitat, bears are often the first animals to be eliminated. All species of bears have declining populations, and some are endangered. If their populations are not better conserved, it is possible that some species of bears will become extinct.

See also Pandas.

Resources

Books

Bailey, Jill. *Polar Bear Rescue.* Austin, TX: Raintree Steck-Vaughn Publishers, 1991.

Bauer, E.A., and P. Bauer. *Bears: Biology, Ecology, and Conservation.* Voyageur Press, 1997.

Bears. Zoobooks Series. San Diego, CA: Wildlife Education, 1982.

Brown, Gary. *The Great Bear Almanac.* New York: Lyons & Burford, 1993.

Bruemmer, Fred. *World of the Polar Bear.* Minocqua, WI: NorthWord Press, 1989.

Caras, Roger A. *North American Mammals: Fur-Bearing Animals of the United States and Canada.* New York: Meredith Press, 1967.

Domico, Terry. *Bears of the World.* New York: Facts on File, 1988.

Elman, Robert. *Bears.* Stamford, CT: Longmeadow Press, 1992.

Hunt, Joni P. *Bears.* San Luis Obispo, CA: Blake Publishing, 1993.

Nowak, Ronald M. *Walker's Mammals of the World.* 5th ed. Baltimore: Johns Hopkins University Press, 1991.

O'Toole, Christopher, and John Stidworthy. *Mammals: The Hunters.* New York: Facts on File, 1988.

Polar Bears. Zoobooks Series. San Diego, CA: Wildlife Education, 1991.

Stirling, Ian. *Bears: Majestic Creatures of the Wild.* Emmaus, PA: Rodale Press, 1993.

Van Wormer, Joe. *The World of the Black Bear.* Philadelphia: J. B. Lippincott Co., 1966.

Ward, P., and S. Kynaston. *Bears of the World.* Blandford Press, 1997.

Wison, Don E., and Sue Ruff, eds. *The Smithsonian Book of North American Mammals.* Washington, DC: Smithsonian Institution Press, 1999.

Jean F. Blashfield

Beavers

The true beavers are robust, aquatic herbivores in the family Castoridae, order Rodentia. Many taxonomists believe that two, closely related **species** of true beavers exist—the American beaver (*Castor canadensis*) and the Eurasian beaver (*C. fiber*). Other taxonomists, however, classify these as closely related variants of the same species, under the name *Castor fiber*.

A few other **rodents** are also called beavers, such as the mountain beaver (*Aplodontia rufa*) of western **North America**, and the swamp beaver or nutria (*Myocastor coypu*) of **South America**. However, these two species of rodents are not in the family Castoridae, and are not true beavers.

The true beavers are large animals, weighing as much as about 88 lb (40 kg), and they are the largest rodents to occur in Eurasia and North America. Only the **capybaras** of South America (family Hydrochoeridae), which can weigh as much as 110 lb (50 kg), are larger rodents. However, a now-extinct species of giant beaver in the genus *Castoroides*, which occurred in North America as recently as about 10,000 years ago at the end of the most recent **ice** age, is estimated to have weighed several hundred pounds. This enormous rodent was similar in size to a black bear (*Ursus americanus*).

One of the most distinctive features of beavers is their scaly, naked, paddle-like tail. The flattened beaver tail is used as a rudder while the **animal** swims, using its webbed hind feet to propel itself through the **water**. The unusual tail is also used as a support while the beaver is standing and as a brace while the animal is dragging logs to the water. If danger is perceived, the tail is energetically splashed onto the water surface to warn other beavers of the threat. However, contrary to what some people believe, the tail is not used as a trowel to daub mud onto the **dams** that beavers often build.

The American beaver

The American beaver (*Castor canadensis*) is widespread in North America, ranging from the limits of the boreal forest in the north, through almost all of the United States, except for the Florida **peninsula** and parts of the southwestern states. The American beaver has also been introduced beyond its natural range, for example, into some regions in **Europe**. As a result, some hybridization has occurred with the European beaver, suggesting a close evolutionary relationship between the two species.

The beaver is a large animal, with the biggest animals reaching a weight of about 88 lb (40 kg), but more typically being 33-77 lb (15-35 kg). The beaver has a robust body, a broad and blunt head, and a short neck and limbs. Beavers have very large, continuously growing incisor teeth and large cheek teeth used for chewing their food of **plant** materials. The incisor teeth meet outside of the closed lips of the mouth, enabling the animal to feed easily underwater. The nostrils and ears have skin flaps that serve as valves to keep water out when the animal is submerged. The forepaws have long fingers, useful for dextrous handling of branches and twigs while feeding and building lodges and dams. The hind feet have two serrated claws that are used in preening and oiling the fur, a task in which beavers are commonly engaged. The other three claws on the hind feet are blunt and flat. The pelage of this animal is thick and lustrous, with a dense, brown underfur and longer, coarser guard hairs.

Beavers are social animals, with the basic unit being the family, which forms a colony with a hierarchical structure among the individuals. The oldest female is the central **individual** in the group. She establishes the colony, and, if she is killed and no daughter exists to take over the matriarchal role, the site is abandoned. The average colony size is about six animals. All of the animals in the colony work cooperatively, especially in building and maintaining the group's dams and lodge.

Beavers are famous for their industriousness and **engineering** skills. If an open-water wetland such as a pond is not available locally, beavers will construct one by building a dam of logs, sticks, stones, and mud-plaster across a stream, causing the water to back up. Beavers maintain their dams assiduously, and they seem to be constantly working on improving these structures. This is necessary, of course, because the beaver pond provides essential local **habitat** for the species, yet it is in some respects artificial, having been created by the animals themselves. Of course, many other species of **wildlife** benefit greatly from the habitat-creating enterprise of beavers. The dams are generally constructed to create a pool that is 6.5-10 ft (2-3 m) deep in some places, so that in the wintertime unfrozen water will occur beneath the ice, which can be as much as 3 ft (1 m) thick. Some dams can be hundreds of feet long, and several feet high.

Beavers also build lodges of sticks and mud. Beaver lodges are commonly located in shallow, open water, and their tops project as high as several feet above the water surface. Lodges may also be located near the edge of the beaver pond, rather than in open water. The lodge has a hollow, gnawed-out core, in which the family lives. The roof of the lodge is relatively thin and porous, allowing fresh air to circulate. The interior of the lodge is reached by several underwater passageways. Beavers also burrow into mud banks, and animals living on large **rivers** or lakes will do this instead of building a lodge.

Beavers are crepuscular, meaning they are most active between sunset and sunrise. They have a slow lumbering gait on land, but are skilled swimmers. Beavers mostly eat the inner **bark** and cambium of trees and shrubs, as well as the buds, leaves, and flowers of woody plants. They also supplement this diet with aquatic plants and herbaceous terrestrial vegetation during the summer and autumn. Beavers sometimes fell quite large trees (up to 16 in [40 cm] in diameter) to get access to the relatively nutritious branches and twigs of the canopy. Trembling aspen (*Populus tremuloides*) is the food species of choice, with willows, birches, poplars, and alders also being favored.

Beavers fell trees by gnawing them at the base, often while standing erect, propped by their tail, and gripping the **tree** with their forepaws. The stumps of beaver-felled trees have a distinctive, conical top, with clear evidence of the large cuts made by the chisel-like incisors of the animal, which remove substantial chips of **wood**. The beavers do not gnaw right through the trunk—they leave a central core intact, and rely on rocking motions from a later **wind** to actually cause the final felling of the tree. Beavers seem unable to plan the direction of the eventual tree-fall, and they are sometimes killed when this actually happens. Beavers occasionally construct canals in wet terrain in order to make their logging areas easier to reach, and they often develop wide, well-trodden paths to facilitate the dragging of branches to their pond.

Beavers do not hibernate—they remain active in their lodges and beneath the ice of their pond. During winter, these animals mostly feed on underwater piles of branches and twigs accumulated for this purpose during the previous summer and autumn. However, beavers will sometimes emerge above the ice and snow to feed if they run short of their stored winter food.

Many animals **prey** on beavers. When predators are relatively abundant, beavers are wary and do not like to forage or fell trees very far from the safety of their pond.

Beavers and the fur trade

The pelts of American beavers are valuable in the fur trade and are largely used in making coats and hats. During the first several centuries of the European colonization of North America, beaver pelts were one of the most important natural resources to be exported from the northern regions of that **continent**. The most important markets were in Europe, where the pelts were used to make gentlemen's hats, also known as "beavers." In fact, most of the initial exploration and settlement of the interior of North America was undertaken by fur traders, and these intrepid men were most enthusiastically searching for beaver pelts. For many years in vast regions of North America, beaver pelts were the measure of wealth, and

An American beaver (*Castor canadensis.*) dragging a willow branch over his dam in Denali National Park, Alaska.
Photograph by Ron Sanford. Stock Market. Reproduced by permission.

were even a common unit of currency. In view of the great importance of the beaver in the early colonial history of Canada, this animal has become a national symbol of that country. However, beavers were similarly important in the northeastern and central United States.

The extraordinary overharvesting of beavers for their pelts caused great reductions in the abundance of these animals, and they were widely extirpated from much of their original range in North America. Moreover, the American beaver will not breed in captivity, so fur-farming is not possible. Fortunately, the implementation of **conservation** measures after about the 1940s has allowed a substantial rebound in the populations of American beavers. These animals are now re-occupying much of their former range, as long as the habitat has remained suitable for their purposes. Beavers are sometimes hunted for meat, although their use in this way is usually secondary to the taking of their pelts.

Beavers can be viewed as a nuisance, their constructions **flooding** roads, culverts, railroads, lawns, and agricultural land. Beavers also may cut down valuable ornamental trees in some places where they are living in proximity to humans. As a result, many states and

provinces will live-trap problem beavers for relocation to less built-up areas.

The Eurasian beaver

The natural range of the Eurasian beaver (*Castor fiber*) extends through most of Europe and through much of northwestern **Asia**. However, the modern range of the Eurasian beaver has become rather restricted and fragmented, and the species is now much less abundant than it used to be. The population declines are due to the conversion of the species' natural **wetlands** habitat into agricultural landscapes as well as overhunting. The European beaver has long been exploited as a natural resource, for both its thick, lustrous fur, and also a musky oil called castoreum, which is extracted from the anal or castor gland of this animal.

The mountain beaver

The mountain beaver (*Aplodontia rufa*, family Aplodontidae) occurs in the Cascade Mountains from east-central California north through Oregon and Washington to southwestern British Columbia. Other than being a large rodent, the mountain beaver is not particu-

KEY TERMS

Crepuscular—Refers to animals that are most active in the dim light of dawn and dusk, and sometimes at night as well.

Overharvesting—The unsustainable exploitation of a potentially renewable natural resource, such as hunted animals or trees. Overharvesting eventually leads to a collapse in the abundance of the resource.

larly closely related to the true beavers, which are in the family Castoridae. In fact, mountain beavers are the only species in their family, and they not closely related to any other rodents. Because of their ancient evolutionary history, mountain beavers are sometimes considered to be living fossils. In body form, the mountain beaver looks like a tailless **muskrat** or large vole, with small ears, short legs, and grizzled, brownish fur. Mountain beavers are terrestrial animals, digging long, complex burrows in moist, workable **soil** near streams, with numerous entrances and exits located in concealed places. These animals live in loose colonies, but they are not very social animals, preferring to avoid frequent, direct contact with each other. Mountain beavers eat a wide range of plant foods, including herbaceous plants and **fruits**, young twigs of woody plants, and **conifer** foliage and shoots in the wintertime. Mountain beavers store food for the winter in underground haystacks.

Resources

Books

Banfield, A.W.F. *The Mammals of Canada.* Toronto: University of Toronto Press, 1974.
Grzimek, B., ed. *Grzimek's Encyclopedia of Mammals.* London: McGraw-Hill, 1990.
Hall, E.R. *The Mammals of North America.* 2nd ed. New York: Wiley & Sons, 1981.
Nowak, R.M., ed. *Walker's Mammals of the World.* 5th ed. Baltimore: John Hopkins University Press, 1991.
Ryden, H. *The Beaver.* London: Lyons & Burford, 1992.
Wilson, D.E., and D. Reeder. *Mammal Species of the World.* 2nd ed. Washington, DC: Smithsonian Institution Press, 1993.

Bill Freedman

Bedrock

Bedrock is the solid rock that is exposed at the earth's surface, or buried beneath one or more layers of loose sediment. It is of igneous, sedimentary or metamorphic origin and forms the upper surface of the rocky foundation that composes the earth's crust.

Bedrock exposures

A surface exposure of bedrock is called an **outcrop**. Bedrock is only rarely exposed, or crops out, where sediment accumulates rapidly, for example, in the bottom of stream valleys and at the base of hills or **mountains**. Outcrops are common where **erosion** is rapid, for example, along the sides of steep stream channels and on steep hill or mountain slopes. Deserts and mountain tops above the treeline also host good bedrock exposures due to the scarcity of vegetation, and resulting rapid erosion. Man-made outcrops are common where roadways cut through mountains or hilltops, in quarries, and in mines.

Generally, the more rock resists erosion, the more likely it is to crop out. Granite and sandstone commonly form well-exposed outcrops. Natural exposures of shale and claystone, both soft, fine-grained **rocks**, are rare—especially in humid climates.

Bedrock features

In addition to the occasional mineral **crystal** or fossil that attracts rockhounds, all outcrops contain through-going fractures called joints. These form during the application of stresses to bedrock on a regional scale, for example during mountain-building. Even greater stresses may cause faulting movement of the rock on the sides of a fracture. An example is the large-scale bedrock movement that occurs along the San Andreas Fault in California. When stresses cause plastic rather than brittle deformation of bedrock, it folds rather than faulting.

Bedrock distribution

Bedrock is distributed in a fairly predictable pattern. Generally in the central area of a **continent** you will find very ancient (one billion years or more) mountain chains, consisting of igneous and **metamorphic rock**, eroded to an almost flat surface. This area, called a continental shield, typically contains the oldest continental bedrock. Shields have experienced multiple episodes of deformation so they are intensely folded and faulted. These ancient igneous and metamorphic rocks, called basement rocks, compose much of the continental crust. However, on the shield margins, thick sequences of relatively undeformed, sedimentary rocks cover the basement rocks. These deposits, called the continental platform, commonly exceed 1 mi (1.6 km) in thickness and 100 million years in age.

Together, the shield and platform make up the bedrock area known as the continental craton. The craton is considered more or less stable, that is, it is not currently experiencing significant deformation. On the margins of the craton, there may be areas of geologically-active bedrock, called orogens, from the Greek word for mountain. Orogens are relatively young mountain belts where **uplift**, folding, faulting, or volcanism are occuring. The bedrock here varies in age from lava flows that may be only days old to igneous, sedimentary, and metamorphic rock that are hundreds of millions of years old. All bedrock belongs to either the continental shield, platform, or the orogens.

Bee-eaters

Bee-eaters are 24 **species** of **birds** that make up the family Meropidae. Bee-eaters occur in open habitats and savannas of the south-temperate and tropical zones, ranging through **Africa**, southern **Europe**, southern **Asia**, Southeast Asia, and many Pacific Islands. Species that breed in temperate habitats migrate to the tropics for the winter.

Bee-eaters have large, pointed wings and a long tail, usually with the two central feathers quite extended. The bill is long, slender, down-curved, and pointed. Their feet and legs are rather small and weak and are only used for perching. Bee-eaters are brightly colored, most commonly with a basal hue of green, and have bold markings of yellow, blue, red, brown, black, and white. All species have a black stripe running through the **eye**, known as a "mask." Both sexes are similarly colored and patterned, as are the juvenile birds.

Bee-eaters tend to occur in groups, often perched in the open. They commonly feed by pursuing and catching **insects** in the air, a foraging strategy known as "hawking." True to their name, the principal food of most species of bee-eaters is **bees** and **wasps**. However, a wide diversity of flying insects is taken, depending on their local and seasonal availability. After a bee or wasp is captured in the bill, its abdomen is forcefully wiped against a branch, causing the venom to be discharged.

Bee-eaters nest in a burrow dug into an earthen bank or **sand** cliff. The tunnels are as long as several meters, and have a nesting chamber at the end, in which two to six eggs are laid. Nesting sites are generally colonial, with large numbers of pairs breeding in the same vicinity, commonly near **water**. Both sexes share in the incubation of the eggs and care of the young.

The European bee-eater (*Merops apiaster*) is a blue-bellied, cinnamon-backed, yellow-throated species of

A common green bee-eater with its catch. *Photograph by E. Hanumantha Rao. Photo Researchers, Inc. Reproduced by permission.*

southern Europe and western Asia, wintering in sub-Saharan Africa and India.

The blue-cheeked bee-eater (*Merops persicus*) is a widespread species, occurring in Africa and Madagascar through to western Asia. This species has a lime-green body, with a bluish breast, a yellow and chestnut-brown throat, and a black eye-line.

As its name implies, the rainbow-bird (*Merops ornatus*) of **Australia** is an especially lovely and multi-hued bee-eater. This species migrates north to New Guinea after its breeding season in temperate Australia.

Clay Harris

Beech family (Fagaceae)

The beech family is an important group of flowering plants that includes the beeches, **oaks**, and sweet chestnuts. Most members of the family are deciduous or evergreen trees or shrubs. The leaves are arranged alternately along branches, are leathery in texture, often strongly ribbed, and have margins that are entire, toothed, or deeply lobed. The flowers are unisexual. Male flowers are usually arranged in catkins (round heads in *Fagus*) whereas female flowers are in few-flowered clusters. The fruit is a one-seeded **nut** that is partially or completely covered by a cupule of scales or a spiny bur.

The family includes eight genera and about 1,000 **species**. The Fagaceae are widely distributed and most abundant in temperate and subtropical regions of the Northern Hemisphere, although there is one tropical and one south temperate (*Nothofagus*) genus. Beech (*Fagus*) and sweet **chestnut** are, or were, in the case of American

Sweet chestnuts. The thorny husks split to release the nuts.
*Photograph by Geoff Bryant. Photo Researchers, Inc.
Reproduced by permisison.*

A beech forest. *P. Berger/National Audubon Society/Photo Researchers, Inc. Reproduced by permission.*

chestnut (*Castanea dentata*), prominent components of mature deciduous and mixed **forests** of **North America** and Eurasia. Much of central **Europe** was covered by forests of beech and oak until cleared by people for agriculture. Similar forests in the Southern Hemisphere are dominated by southern beeches in the southern Andes, eastern **Australia**, and New Zealand. The American chestnut was formerly a dominant species of deciduous forest in eastern North America. Oaks (*Quercus*) are also important in rich temperate deciduous forests as well as on droughty soils such as the **sand** plains of the eastern North American seaboard, where oaks grow abundantly with **pines**. Evergreen oaks are especially important in the arid regions of the Gulf of Mexico, southern China, and southern Japan. Evergreen species of both oaks and southern beeches are prominent in the mixed mountain forests of Southeast **Asia**. Given their prominence as the producers of deciduous and mixed forests, the Fagaceae are extremely important ecologically. Furthermore, their nuts are important sources of food for a variety of **insects**, **birds**, and **mammals**.

The beech family is among the most valuable sources of hardwood timber in the world. There are over 300 species of oak worldwide and the characteristics of their **wood** varies. Nevertheless, most produce strongly grained, durable wood that polishes well. The white oaks of North America, in particular, produce fine wood that is commonly used for furniture, panelling, and flooring. The **bark** of oaks is often rich in tannin, which is used in tanning leather. Oak wood is used in the construction of barrels for the storage of whiskey, wine, and sherry. Oak is also valued as firewood and for making charcoal. Tropical members of the family, such as *Castanopsis* and

Lithocarpus (called chinkapin and tanoak respectively in North America), also produce high quality wood, but they have not as yet been heavily exploited.

The bark of the Mediterranean **cork** oak (*Quercus suber*) is the principle source of commercial cork. Cork bark is stripped in summer by making a circular cut at the base of the trunk and another just below the first branches. A lengthwise slit is then made between the two circular cuts and the bark is carefully removed with special hatchets so as not to damage the conducting **tissue** of the inner bark. The stripped bark is stacked and allowed to season for a few weeks, then boiled in tanks of **water**, followed by the removal of the rough outer bark. The cork is then dried and ready for use. The bark is stripped from trees once every 8-10 years. Aside from its role in stoppering bottles, cork is used for gaskets, floats, non-slip walkways, corkboard, and flooring. The main cork-producing areas are Portugal and southwest Spain.

Chestnuts are especially prized for their large, sweet, edible nuts. Although there are about 10 species of sweet chestnuts, the most widely grown is *Castanea sativa*, commonly called the Italian, Spanish, or European sweet chestnut, which is a native of southern Europe. The nuts can be roasted over an open fire, used whole or sliced in stews and stuffings, or pureed into the exquisite French dessert called *marron glacé*. The American chestnut was once an important timber and nut **tree**, and it is said that its nuts were sweeter and tastier than those of European species. Unfortunately, this once-dominant member of

KEY TERMS

. . . .

Catkin—An elongate, spikelike cluster of unisexual flowers, often drooping at maturity.

Evergreen—Plants whose leaves persist and function for two or more years.

the deciduous forest of eastern North America is now a rarity, struck down by chestnut blight.

Chestnut blight was introduced into eastern North America from abroad in about 1904. By 1940, the American chestnut had disappeared from most of its range, clinging in places as sprouts from the root collars of trees whose trunks had died, but although these sprouts may grow for 40 years, they do not produce fruit, and so the tree is condemned to die without leaving offspring. The **disease** is caused by the fungus *Endothia parasitica*, which produces spores that stick onto the beaks and feet of bark-feeding birds and on the bodies of bark **beetles**. The spores are carried by the birds and insects to healthy trees that become infected when spores are accidentally deposited into wounds caused by the feeding of the birds and beetles. The spores germinate and the growing fungal threads enter vital cells of the inner bark, killing them. There are no adequate control measures for chestnut blight.

Beeches (*Fagus* and *Nothofagus*) both produce valuable wood. European beeches have proved to be extremely useful in bentwood furniture, which is made by steaming laminated pieces of wood and then pressing them against forms until dry. Beech nuts were once commonly eaten by people, but now the nuts, along with acorns, are mostly fed to **pigs**, especially in Europe. In many parts of eastern North America, the American beech has been afflicted by a canker disease that disfigures the lovely, smooth, gray bark and obstructs the conducting tissue causing reduced rates of growth.

Resources

Books

Heywood, Vernon H. ed. *Flowering Plants of the World.* New York: Oxford University Press, 1993.

Mitchell, A. *The Guide to Trees of Canada and North America.* Surrey, U.K.: Dragon's World Ltd., 1987.

Les C. Cwynar

Bees

Bees belong to the insect order Hymenoptera, which includes **wasps** and **ants**. Its name is derived from Greek, meaning "winged membrane," and it is the third largest group of **insects** with more than a hundred thousand **species** in the order. Ants and bees play vital roles in agriculture, ants being useful in aerating **soil** and bees in pollinating plants. Wasps play an important part as predators to other insect **pests** and bees are the source of honey and wax, which have been highly valued by human beings since antiquity.

Hymenoptera are distinguished by having two pair of wings that are veined in cross angles creating a cell-like pattern. The rear wings are smaller than the front ones, and wing **color** ranges from brown with yellow markings to red, white, blue, or green marks. Male Hymenoptera have 13 segments in their antennae, while females have only 12. Most Hymenopterons have chewing mouthparts with a pair of mandibles, but bees have a long tongue (proboscis) to lap **nectar**. Bees have a complete, four-stage **metamorphosis** from egg, larva, pupa, to adult. Some species of bees, as well as ants and some wasps, form colonies under a *caste system*, while other species are solitary.

Bee families

The more than 20,000 species of bees are assigned to the superfamily Apoidea, which includes eight families. The diversity of bees includes the yellow-faced, plasterer, oxaeid, andrenid, sweat, melittid, leafcutting, mason, cuckoo, digger, carpenter, bumble, and honey bees. The latter two are the most common and both belong to the family Apidae. Bees are characterized by the vein pattern on their wings and by the size of their tongues. Some have a short tongue and others a long, slender one. Bees are able to chew as well as suck with their mouthparts.

Bees mainly eat nectar and pollen, which they also store in their hives or nests for their larvae to eat. A segment of the rear legs of bees is enlarged and somewhat flattened and serves as a carrying device for the pollen they collect. Male bees have seven segments in the abdominal region, while females have only six. Hairlike setae densely cover the bodies of bees. Plants that bees pollinate include most **fruits**, numerous **vegetables**, and field **crops** like **cotton**, tobacco, and clover. While the bees that are most beneficial for commercial production of honey are social bees, many families of bees are solitary in nature. Bees are also diurnal, that is, they are active in the daytime.

Solitary bees

Among the *solitary bees*, where each queen bee builds her own nest, there is sometimes evidence of a division of labor. Some of the bee families are more sociable than others and build their nests close to one another and may even share the same entrance to the nests. In such cases, a bee might stand guard at the entrance of the group of nests to protect them from predators. This is not the same social organization as the caste system established by true social bees, where there is only one queen bee laying eggs. Some species of solitary bees build nests, while some scavenge and use the nests of other bees or convenient crevices for laying their eggs. Nest building patterns among solitary bees vary from species to species.

Plasterer bees, members of the family Colletidae, get their name from a secretion they use to plaster the sides of their mud nests, which may be in the ground or in crevices of stones and bricks. Plasterer bees are black with light-colored body hairs. Yellow-faced bees, which belong to the same family as the plasterer bee, build nests in **plant** stalks and insect burrows. Yellow-faced bees feed their larvae on a mixture of pollen and nectar which is stored in their nests.

There are over 1,200 species of the Andrenid bee family found in **North America**. These yellow, white, or black bees make their nests underground in tunnels, which may include many branches and may house large groups of bees. Over 500 species of the "sweat bee" can be found in North America. Their sting is not painful, although sweat bees have a reputation for stinging persons who are sweating. They nest in clay and **sand** banks of streams. Some have metallic blue or green colorations, but they are mostly black or brown.

The leafcutting bee gets its name from its habit of cutting pieces of leaves to use as a nest. It places a ball of pollen on the cut **leaf** and then lays its eggs on top. It locates its nests in **wood**, under loose **bark**, or in the ground. It is closely related to the mason bee, a shiny, blue-green insect, that builds its nest under stones, where it builds clusters of small cells. Mason bees also like empty snail shells and the empty nests of other bees.

Digger, cuckoo, and carpenter bees belong to the same subfamily, Apidae, as the honey bee and bumble bee, but they are not social. The larger carpenter bees nest in open spaces in wood, while the smaller ones use the stems of bushes in which to build their nests. They are robust in build, as are the digger and cuckoo bees. The large ones look like bumble bees. Digger bees are much more hairy than other members of the bee family and they build their nests in burrows in the ground. Cuckoo bees look like small wasps and lay their eggs in the nests of other bees.

Social bees

Honey bees and bumble bees are two of the 500 species of bees which are *social*. Their colonies or hives range in size from several hundred to as many as 80,000 inhabitants. They are organized within a rigid caste system, where members of a caste carry out specific tasks. The social system consists of a queen bee, male drones, and worker bees. The queen bee is responsible for laying eggs, which the drones have fertilized, and the worker bees build the nest and care for the fertilized eggs and larvae.

The female worker bees differ in structure from the queen bee. They have pollen sacs in their rear legs, which the queen bee does not have. Workers also have wax **glands** and other differences in their head structure. Their life span, usually a season, is shorter than that of the queen bee, who lives on the average for several years. Worker bees are not capable of mating, but if a queen bee is not present in the hive, their ovaries do develop and they become capable of laying eggs that become drones. Drones are not constructed for collecting pollen. They are hatched from unfertilized eggs that the queen lays by withholding sperm. While the queen bee has a life span of several years, the drone dies when he has impregnated the queen.

Bumble bees are social bees and are characteristically black, yellow, and hairy. After mating in the fall, a queen will hibernate over the winter, while the workers and drones from the past season's colony die before winter. In the following spring, the queen begins a new nest, lays her eggs, and spends her time protecting them and sipping from a honey pot. She is often compared to a mother hen hatching her eggs. A favorite nesting place for bumble bees is an abandoned mouse nest.

When the larvae mature in about 10 days, they construct a cocoon for their pupal stage. After several weeks, female workers leave their cocoons to take up the work of building the nest. Males and potential queens are hatched later in the season. Bumble bees are important for the **pollination** of red clover, which is an important field crop in agriculture. Plants that bloom eight to nine weeks before clover are planted near clover fields to lure bumble bees to the area with a supply of food before the red clover comes into bloom. This ensures the bumble bees will be present when it is time to pollinate the crop.

The stingless bee flourishes in Central and **South America**. Before European honey bees were introduced in the Western Hemisphere, these bees supplied honey and wax to communities in these regions. The wax was also used as casting material for the molding of gold jewelry. There are several hundred species of stingless bees in tropical regions and their colonies range from several hundred to as many as 80,000 **individual** bees. They use a blend of wax, resin, and mud to build their nests, which may have walls as thick as eight inches. Eggs are laid in the nest with a store of food and the cells are sealed. New nests are created in preparation for the departure of a new, young queen from the old nest, and workers and males follow to join her.

Honey bees

The social structure of honey bees is the caste system of queen, drones, and workers. Unlike the stingless bee, the honey bee queen is the one to leave the old colony to form a new one. The move to a new nest begins with a swarming of bees and ends when a suitable place, such as the hollow in a **tree**, is found to establish a new colony. A young queen will take over the old colony.

Of interest to entomologists is the so-called "dance language" of honey bees. A worker can communicate the location of a food source, how far away it is, and the type of **flower** that will be found. Some of this information is transmitted by the scent the flower has left on the messenger's body, but there are other features to this communication. One is a **circle** dance that communicates sources of the nectar. The other dance involves wagging the tail and the abdominal region, which indicates the **distance**. The tail wagging is accompanied by wing vibrations produced at the same rate. The closer the source of the food, the more wags. Different species of honey bees follow different dance tempos.

Direction to the food source is shown by the angle of the bee to the **sun** when it is wagging its tail. Besides this "dance communication," bees seem to know when flowers have a supply of nectar available. This built-in biological clock is not as well understood as their "dance language." The person responsible for unraveling the dance language of honey bees was Karl von Frisch, who received a Nobel Prize in 1973 for this work.

Honey bees are susceptible to debilitation of their honey production by bee **mites**, a parasite that reduces their natural pollinating and honey-making activities. In the mid 1980s, 150 million honey bees had to be destroyed in several parts of the United States to eliminate the infestation of these mites. Other diseases that honey bees are susceptible to include foulbrood, which attacks larvae or pupae, **stress** diseases, such as sacbrood and nosema, which can shorten the lives of adult bees, and acarine **disease**, another mite disease. **Animal** predators that are dangerous to bees are **mice**, **birds**, **bears**, **squirrels**, **skunks**, **raccoons**, and **opossums**. The first line of defense of a bee is of course its sting.

Beekeeping

References to bees and honey can be found in early civilizations from the Sumerians, Babylonians, Egyptians, Hindus, Greeks, Romans, and Mayans in the warmer climates to Celts, Slavs, and Northern **Europe** in colder climates. Honey as a sweetener was valued even in areas where sugar was available. A number of these early civilizations held the bee and its honey in high regard, using the bee as a symbol for royalty and honey for anointing their kings and for embalming the dead. Besides using honey for a sweetener in food, it was also used medicinally during the Middle Ages and Renaissance in Europe, where beeswax was also used for candle making and for molds to cast statues.

There is evidence from an Egyptian tomb dating to 2400 B.C. that this culture had learned to raise bees in man-made hives and no longer had to rely on raiding beehives for their honey. In southern climates, a round type of beehive was constructed from hollow tubes made of mud or clay and baked in the sun. The bees then built their honeycombs in these early beehives. In the northern climates of Europe, a horizontal hive was developed that was made of wicker or straw. Other materials used were cane in China, **cork** in Spain, and hollow tree trunks in eastern Europe. The bees were smoked out from the hive in order to collect the honey, as they frequently are today, especially with bee colonies that are aggressive in nature.

A vertical beehive was invented by François Huber in 1792 that was made of wooden frames, hinged like a book, and with **glass** covering the end leaves so that the bees' activities could be observed. Many other similar hives were developed, but they all shared the problem of becoming gummed together by the beeswax that was produced along with the honey. In 1851, a Pennsylvania minister solved the problem by establishing the correct measurement for *bee space* needed around the frames and other movable parts of man-made hives. This measurement is one-quarter to three-eighths of an inch or six to ten millimeters.

A subsequent innovation in beehive construction was the introduction of a fabricated wax honeycomb foundation on which the bees could accelerate the production of honey, since they did not have to spend time building the honeycomb. Further improvement in honey production was made after the introduction of a mechanical honey extractor.

Beekeeping is carried out by large-scale commercial beekeepers and by thousands of hobby beekeepers. It is estimated that the annual worldwide production of honey exceeds a million metric tons. Besides marketing honey as a product, beekeepers serve agricultural businesses by supplying bees for pollinating at least 90 commercially valuable crops, such as fruits, nuts, and field crops like alfalfa. Beekeepers often migrate from northern locations in the summer to southern ones in the winter. The largest honey-producing beekeepers are found in California,

Florida, and Minnesota, but New York, Ohio, Michigan, and Illinois also have commercial beekeepers. While a hobbyist might have only a dozen or so hives to tend, a commercial beekeeper often has thousands of hives.

Killer bees

During the mid-1950s, a **hybrid** African honey bee was accidentally released in Brazil. This bee was more aggressive than the European honey bee and by the mid-1960s had gained the name of "killer bee." The African bee was introduced by Warwick Kerr in Brazil in an attempt to find a bee that was more suitable to the climate. This bee was found to be more productive than other bees and many beekeepers in South America use them for the production of honey. Because these bees are more aggressive, beekeepers must wear more protective clothing. By the late 1980s, the "killer bees" had migrated across the Rio Grande. While some entomologists fear that the killer bees will replace the European honey bee and upset honey production in the United States, others feel this will not happen.

Resources

Books

Arnett, Ross H. *American Insects.* New York: CRC Publishing, 2000.

Hubbell, Sue. *Broadsides from the Other Orders: A Book of Bugs.* New York: Random House, 1993.

Imes, Rick. *The Practical Entomologist.* New York: Simon & Schuster, 1992.

Morse, Roger. *The ABC and XYZ of Bee Culture.* Medina, Ohio: A.I. Root Co., 1990.

Style, Sue. *Honey: From Hive to Honeypot.* San Francisco: Chronicle Books, 1992.

Winston, Mark L. *Killer Bees: The Africanized Honey Bee in the Americas.* Cambridge, MA: Harvard University Press, 1992.

Vita Richman

Beet

Beet belongs to the genus *Beta* in the goosefoot family, Chenopodiaceae. There are several varieties of beet and all are used as food for either animals or humans. Most **species** of beet are biennial and are harvested after the first growing season when the roots are most nutritious.

The wild beet, *Beta maritima*, is thought to be the species from which cultivated beets (*Beta vulgaris*), originate. Wild beet is found on the Mediterranean and Atlantic European coasts. Although beets are native to these temperate areas, they are now cultivated in many parts of the world for food, fodder, and as a source of sugar.

The cultivated beet has several commonly used varieties. Probably the most recognizable agricultural beet (*Beta vulgaris escuelenta*) is the bulbous, reddish-purple root that shows up on the dinner table. Although these beets can be successfully stored during winter, most of the **crops** of red garden beets and table beets in the United States are canned, pickled or frozen. The red beetroot is often dried, made into a powder and used for food coloring and fabric dyes. Beet plants can have round globular, or long conical roots with several stems growing above ground. The leaves on the stems vary in size and from green to purple in **color**.

Another important variety of the cultivated beet is *B. vulgaris crassa*, the **sugar beet**. Sugar beets are quite large, with green leaves and white roots weighing about 2.2 lb (1 kg). The root is shredded, mixed into **water**, and heated. The impurities are removed and the remaining sugary liquid is concentrated and crystallized. The sugar beet has been cultivated for centuries, but its use as a primary source of sugar dates back to the beginning of the nineteenth century. France is a leading contributor to the world's sugar beet stores, which today maintains half of the world's sugar supply.

Although beets are usually grown for the root part of the **plant**, one type of common beet is grown for its greens. This beet is known as Swiss chard (variety *B. cicla*), or **spinach** beet, and has large stems, fleshy red and green curly leaves, and small, branched roots. The

Mangel-Wurzel beet (variety *B. macrorhiza*) has large roots and is grown for **livestock** feed. Most of the **nutrients** in beets are found in the tops, which are used as greens. Swiss chard, for example, is a good source of vitamins A, B$_1$, and B$_2$, as well as **calcium** and **iron**.

Beetles

Beetles make up the large, extremely diverse order Coleoptera of the class Insecta, and comprise the largest single group of animals on **Earth**. There are at least 250,000 **species** of beetles, compared to the 5,000 known species of **mammals**. The weevil family of beetles alone contains about 50,000 species, and is the largest family in the **animal** kingdom. Thus, the order Coleoptera, representing about 40% of the known insect species, contributes greatly to making the **insects** the largest class of the largest phylum—Arthropoda. **Arthropods** are thought to have first evolved as long as 500 million years ago in Precambrian times, while the most primitive insect fossils date to the **rocks** of the Middle Devonian period about 350 million years ago. Coleoptera are thought to have evolved in the early Permian period about 225-280 million years ago, and were common even before the age of **reptiles**.

Beetles are found in virtually all climates and latitudes throughout the world except at very high altitudes or in regions with extreme temperatures, e.g. the Antarctic. Most species of beetles occur in the tropics, but fewer individuals of a particular species are generally found in tropical regions rather than in temperate areas.

The success of the beetles is due to at least three important characteristics. First, they undergo complete **metamorphosis** (egg, larva, pupa, adult), with larval and adult stages usually living in different places and eating different food. This division greatly expands the number of ecological niches and food available to these insects. Second, the front pair of wings is modified into a hard cover (the elytra) that protects the soft body underneath. Third, most beetles have mouth parts capable of chewing a wide variety of solid foods. Some beetles, however, have mouth parts modified for sipping sap and **nectar**.

The front pair of wings, modified into horny covers (elytra), hide the rear pair of wings and abdomen, and their inner edges appose each other, creating a straight line down the back of the insect. The elytra form a rigid, closely interlocking sheath that covers the mesothorax and metathorax, and most of the abdomen. (The name Coleoptera is derived from the Greek word *koleos*, meaning sheath.) The perfect alignment of the edges of the elytra form the characteristic, straight line that seems to split the back of the beetle, and gives these insects their common name (beetle from the German word *bheid*, meaning to split).

Beetles are found on vegetation, under **bark**, stones, and other objects, as well as almost anywhere on or in the **soil**, rotting vegetation, dung, and carrion. They vary widely in size and appearance, and many have noteworthy **behavior**. Some beetles (e.g., Lampyridae) produce **light**, while others (Cerambycidae) can stridulate, that is, they can produce sound. Large beetles usually make a loud noise during flight, and some, such as the scarab beetles, have a bizarre physical form.

Varieties of beetles

The Coleoptera includes the largest and smallest insects in the world, ranging from the giant, 6.3-in (16-cm) Longhorn beetle (*Titanus giganteus*) of the Amazon region to the dot-sized, fringed ant beetle (*Nanosella fungi*) of **North America**, which reaches only 0.25 mm in length—smaller than a large protozoan.

Beetles are economically important in agriculture, either feeding directly on **crops** and trees, or preying on other species that harm **plant** crops. For example, the ground beetles (Carabidae) and the rove beetles (Staphylinidae) feed on caterpillars and other larvae as well as on many soft-bodied insects and insect eggs. Many of the adult and larval forms of the ladybugs or ladybird beetles (Coccinellidae) feed on plant-sucking insects (Homoptera) such as **aphids** and **scale insects**, while only a few of the Coccinellidae themselves (e.g., *Epilachna*) feed on plants.

Many other beetles, however, do feed on plants. Among the most important of these beetles are the **leaf** beetles (Chrysomelidae) and the **weevils** and their relatives (Curculionoidea). The larvae of leaf beetles feed on leaves, stems, or roots, while most adults chew on leaves; the larvae of weevils feed on almost every part of plants. For example, larvae and adult forms of bark beetles (Scolytidae) attack **tree tissue** beneath the bark.

The scarab beetles (Scarabaeidae) are important **pests** of crops, lawns, and pastures. One of these insects, the dung beetle, was an important religious symbol to ancient Egyptians, who considered its life cycle to be a reflection of the cyclical processes of nature, especially the "rebirth" of the **sun** each morning. Glazed steatite (soapstone) and other ancient Egyptian ceramic or stone representations of the beetle, called scarabs, were a symbol of the soul and used as talismans.

Many beetles act as scavengers, breaking down organic material such as **wood** and dead plant and animal **mat-**

ter. The larvae of some beetles, such as the wedge-shaped beetles, are parasitic on **wasps**, **bees**, and **cockroaches**. The European **elm** bark beetle *(Scolytus multistriatus)* transmits the fungus that causes Dutch elm **disease**.

The vast array of forms and colors of Coleoptera ranges from the black, furry Brazilian beetle, with creamy-white and orange spots, to the squat tortoise beetle, the long-snouted Peruvian beetle, the stag beetle with its two threatening "horns," and the whirligig beetle, often found gyrating rapidly on the surface of ponds.

Click beetles (family Elateridae) are named for the sharp noise they make. When turned onto its back, a click beetle will bend its head and the upper part of its body backward, then suddenly straighten. This movement produces a click and propels the beetle into the air. This maneuver is repeated until the beetle lands right side up.

Lightning bugs or fireflies (family Lampyridae) produce light; some species produce flashes, while others produce continuous **luminescence**. These insect light shows, common in spring and summer, are a mating ritual through which the opposite sexes find each other.

The classification of beetles established by R. A. Crowson in 1955 *(The Natural Classification of the Families of Coleoptera)*, divides the order into four suborders: Archostemata (rarely found beetles), Adephaga (the tiger beetles and various **water** beetles), Myxophaga (the minute bog beetles and skiff beetles), and Polyphaga (the majority of beetles, such as carrion beetles, scarab beetles, ladybugs, and long-horned beetles). The Polyphaga is the largest suborder, with 18 superfamilies. In all, there are about 135 known families of beetles, of which 120 are found in the Western Hemisphere.

Beetle anatomy and physiology

As insects, beetles share common traits with all other arthropods. The legs are jointed, and there is an external skeleton called the exoskeleton, an inert compound made mostly of a **carbohydrate** called chitin (polyacetylglucosamine). Those sections of the exoskeleton that do not need to be flexible to allow for movement are further strengthened by sclerotin, a hard, proteinaceous substance similar in composition to human fingernails. The exoskeleton serves in both protection and in muscle attachment. A superficial layer of wax secreted on the outside of the exoskeleton prevents water loss through **evaporation**.

Beetles share with all insects the body form that differentiates them from other arthropods. The body of insects is divided into three main sections: head, thorax and abdomen. In Coleoptera, however, two of the three segments of the thorax (mesothorax and metathorax) are attached to the abdomen, while the third one (prothorax) is isolated between the head and trunk and is covered by a dorsal plate called the pronotum. The insect thorax usually has three pairs of legs and two pairs of wings. This body section also contains the powerful muscles that operate both the wings and legs. The abdomen has nine or ten segments, some not externally visible, each bearing a pair of spiracles, or respiratory openings, which direct air through the exoskeleton into the body.

Beetles can fly from hostile environments, escape enemies, and seek mates over wide areas. The first pair of wings of beetles, which arise from the mesothorax, is modified as the elytra-forming the protective cover for the hind wings and abdomen. This is a particular advantage for these insects, because they spend so much time on the ground rummaging through decaying plant matter, wood, and soil. The hind wings are membranous and usually fold beneath the elytra—when not in use. When the beetle flies, the elytra are held open at an angle, providing additional stability and lift as the back wings beat.

Beetles have three pairs of legs that are usually well-developed, with a strong femur and tibia, and five or fewer tarsal (end) segments tipped with a paired claw. The front pair of legs arises from cavities under the pronotum, with a spiracle positioned just to the rear of the base of each of the front legs. The mesothorax bears the second pair of legs, while the third pair of legs arises from the metathorax.

The legs of beetles may be modified for running, swimming, jumping, digging, or clasping, depending on the species. For example, the hind legs of some species of water beetles are long, flattened, and covered with long, matted hairs that serve as paddles for swimming. The water strider has slender legs, which, together with a lightweight body covered with tiny hairs that buoy it up, permit it to skitter over the surface of the water.

The head bears a pair of compound eyes, a pair of antennae (usually with 11 segments), and the mouthparts. The eyes consist of many tiny individual units (facets), which together resemble a honeycomb. Under each facet is a group of six or seven retinal cells surrounding a rod-like light-receptive zone (rhabdom). Each of these tiny, individual "eyes" has its own nerve, which together with the nerves of the other eyes, form the optic nerve.

The beetle **eye**, like that of other insects, does not move, and its lenses cannot focus. Instead, each individual eye contributes a tiny bit of the image; these combine to form a crude mosaic of the scene rather than a clear, continuous picture. In addition, insects can't close their eyes, and can see well only to a **distance** of a few feet (about 90 cm). The whirligig beetle, which is found on the surface of bodies of water, has eyes divided into an upper part, with which the insect observes the surface environment, and a lower part, for underwater viewing.

The antennae (feelers) are sense organs that gather information about the **touch**, sound, **taste**, **smell**, **temperature** and **humidity** of the beetle's environment. The maxillae hold a pair of lobed sense organs, called palps, which may detect smells. The beetle's mouth is a simple hole that lacks jaws, but is surrounded by specialized structures for grasping and grinding. Behind the upper "lip," or labrum, a pair of jawlike appendages (called mandibles) serves as pincers. Behind the mandibles are a pair of bladelike appendages (called maxillae), followed by a second pair of maxillae that are fused in the midline to form the lower lip, or labium.

While most beetles have mouth parts designed for chewing solid food, many of the beetles of the superfamily Curculionoidea have a distinct snout that can bore into wood and suck sap. The snout has mouthparts at its end and is used for penetration and feeding, and for boring holes for egg-laying. These beetles are mostly plant feeders and are economically important pests of crops. For example, the 30,000 species of weevils in the family Curculionidae include many insect pests, such as the **cotton** boll weevil, the apple blossom weevil, and the **rice** weevil. The Curculionidae are also called true weevils, or snout weevils.

The chewed food is passed into the mouth (which secretes the digestive **enzyme** amylase), then into the muscular pharynx, and then to the esophagus. From there food enters the midgut, where digestive enzymes break it down further. Attached to the end of the midgut are the malpighian tubules, the insect's kidney-like organs of excretion that empty into the hindgut (located just past the midgut). The hindgut is followed by the rectum, which ends in the anus. Digested food enters the hemocoele, or body cavity, and is transported to the organs by means of the circulatory fluid, or hemolymph.

Beetles have an open **circulatory system**, that is, they lack an extensive system of **arteries** and **veins** and their hemolymph bathes their tissues directly. A tube-like "heart" in the abdomen pumps the hemolymph forward through a dorsal tube ("aorta") in the thorax to the head. Tiny pumps send the hemolymph to the wings, antennae and legs, after which the fluid flows back passively to the **heart** in the abdomen. The hemolymph transports **nutrients** throughout the body, and carries waste products from the organs to the malpighian tubules. Free cells called hemocytes travel in the hemolymph and serve to devour foreign **microorganisms**. Unlike the **blood** of the **vertebrates**, the hemolymph is not involved in **oxygen** transport; that function is performed by the spiracles.

Life cycle

The mouth parts, which allow beetles to utilize a wide variety of solid foods in their environment, and the elytra, which protect the hindwings, give beetles great survival advantages. Another factor that contributes to the enormous success of beetles is the fact that they undergo complete metamorphosis. Beetles pass through three distinct developmental stages—egg, larva (grub), and pupa—before becoming adults.

Beetles reproduce sexually, although a few species consist of females only and **parthenogenesis** sometimes occurs. The male reproductive **organ** is the aedeagus, a hard, tubelike structure that is inserted into the tip of the female's abdomen through the bursa copulatrix during mating. The female stores sperm in a saclike structure called the spermatheca until they are used to fertilize eggs.

The beetle larva hatches from an egg and feeds, growing until its burgeoning body splits the skin (cuticle). The larva crawls out of the old skin and forms a new one, a process called molting. This occurs several times, until the larva is mature.

Beetle larvae are always very different from adults in both form and habits. They usually have only chewing mouth parts even if as adults they develop siphoning or piercing mouth parts. Wings develop internally and are not evident until the pupal stage. Because larvae, pupas, and adults live in different places and eat different foods, they do not compete with each other.

The different larval forms of beetles reflect a wide variety of feeding habits and habitats. The predatory larvae of water beetles (dytiscids) and ground beetles (carabids) are slender or have gradually tapered bodies and long legs adapted to chasing **prey**, and large, slender mandibles for holding food.

The larva of the tiger beetle (Cicindelidae) lives in the ground, digging a burrow up to 2 ft (0.6 m) deep to avoid high temperatures in the subtropical and tropical environments. The head of the tiger beetle is large and is bent at right angles to the body. When the larva is poised vertically within the burrow waiting for passing prey, its lidlike head acts like a living plug flush with the surface. When a potential meal nears the burrow, the beetle springs out like a jack-in-the-box doing a partial back somersault to catch the prey in its jaws. Two barbed spines on the tiger beetle's back hook into the burrow wall and prevent a strongly struggling victim from pulling the beetle out of its burrow.

The eggs of the European stag beetle (Lucanidae) hatch in the decaying heartwood of old trees, slowly developing into plump larvae, which remain in the tree while they develop into a pupa. A month later, the adult form emerges from the pupa and searches along the forest floor for prey. The large branched jaws of the adult resemble the antlers of a stag. The larvae of ambrosia beetles feed on fungus gardens cultivated by adults in the sapwood of trees.

Following the larval stage, the beetle enters the pupal stage. The pupa develops beneath the skin of the final larval stage, then emerges when the skin splits. The pupa is a soft, pale image of the adult it is to become. The pre-adult appendages are curled or loosely attached to the body and the wings are in flat bags called wing pads. After the pupa sheds its thin skin, the adult emerges, the wings stretch out to full size, and the outer skeleton hardens. The beetle has undergone complete metamorphosis—from egg to larva, to pupa, to adult.

Defense

Beetles produce a variety of noxious chemicals to protect themselves against predators. For example, members of the genus *Meloe* release an oily substance from the joints of their legs that can raise blisters on human skin. In addition, members of the genus *Eleodes* emit an offensive black fluid when disturbed. However, the bombardier beetle displays one of the most dramatic repellent devices. This beetle shoots a boiling hot mixture of liquid and vapor from a "turret" at the rear of its abdomen. Able to fire repeatedly, the beetle has been observed to shoot 29 times in rapid succession within four minutes. The spray protects the beetle from **ants**, **frogs**, spiders, and praying mantids.

Parasitic beetles

Throughout the world, beetles have evolved parasitic relationships with a wide variety of animals, feeding on epidermal secretions and the hair of vertebrate hosts. The small beetle *Leptinus testaceus* of Britain is sometimes found living on the fur of **voles** and **mice**, and has also been found in bees' nests. The American species of this beetle, *L. americanus,* is also found on small **rodents**. Another North American member of this genus, *L. validus,* is an ectoparasite of the common beaver; and *L. aplodontiae* is an ectoparasite of the mountain beaver. Two South American species, *Uroxys gorgon* and *Trichillium brachyporum,* in the Scarabaeidae family live in the fur of the three-toed sloth, while the Australian genus *Macropocopris* lives in the fur of kangaroos.

Beetles and humans

The vast number and variety of beetles have inevitably had an important impact on the human populations that share environments with these insects. Beetles,

like some other insects, pose a threat to agriculture, feeding on crops and wood, both harvested and stored. For example, the dermestid beetles of the family Dermestidae are widely distributed and feed on cereal products, grains, stored food, rugs and carpets, upholstery, and fur coats. Although the adults of some species may be destructive, usually it is beetle larvae that do the most damage.

The grain weevil and the rice weevil are particularly destructive, having evolved a snout that can penetrate food plants and also bore holes to deposit eggs. The boll weevil *(Anthonomus grandis)* is a major cotton crop pest in North America. The boll weevil deposits up to 300 eggs at a time in cotton buds or fruit. The larvae live within the cotton boll, destroying the **seeds** and the surrounding fibers. The Colorado **potato** beetle *(Leptinotarsa decemlineata)* attacks the leaves of potato plants; it became a major pest in the United States during the late nineteenth century.

Beetles, like other insects, also eat a variety of plants that are not of agricultural value, but may be of aesthetic value to humans. For example, two forms of ladybird beetles, the Mexican bean beetle and the squash beetle, are voracious garden pests; and some blister beetles commonly parasitize eggs or larvae of bees.

Not all beetle activity is destructive, however. Beetles, along with other insects, also help to pollinate flowers, which then produce **fruits** and seeds. Beetles crawling over flowers brush up against pollen-bearing organs, and carry the pollen dust to another **flower** of the same species.

In addition, some beetles keep gardens from being overrun by plant pests. For example, most species of ladybird beetles feed as adults and larvae on aphids, scale insects, **mites**, and other insect pests. These highly predatory beetles are an important factor in keeping populations of plant-feeding pests such as leaf beetle larvae and other insects from reaching plague levels. And the larvae of wedge-shaped beetles are parasitic on cockroaches.

An Australian ladybird beetle, the vedalia beetle (*Rodolia cardinalis*), is used throughout the world to control crop pests, such as the coconut scale, **sugarcane** mealy bug, potato aphid, and fir aphid. In addition, certain pollen or sap beetles of the family Nitidulidae prey on the eggs, nymphs, and adult stage of a variety of whitefly and aphid species, among other insects.

Resources

Books

Johnson, Sylvia A. *Beetles*. Minneapolis: Lerner, 1982.
White, Richard E. *A Field Guide to Beetles*. Boston: Houghton Mifflin, 1983.

Christine Miner Minderovic
Marc Kusinitz

Begonia

Begonias (genus *Begonia*) are attractive perennial herbs with soft, succulent stems, and white, pink, red, orange, or yellow flowers. Begonias are members of the begonia family, Begoniaceae, order Violales, subclass Dilleniidae, class Magnoliopsida (dicotyledons), division Magnoliophyta (flowering plants). The begonia family consists of five genera and 920 true **species**, the majority of which belong to the genus *Begonia*. Begonia **taxonomy** can be ambiguous, mainly due to the enormous number of horticultural varieties and hybrids, which many gardeners treat as species. These horticultural varieties of Begonia number in the thousands.

Begonia flowers are either staminate (male) or pistillate (female), and occur on the same **plant**, the plants being monoecious. Wild type flowers have four or five sepals, no petals, numerous stamens in males and an inferior ovary with three fused carpels in females. The colorful begonia sepals resemble petals, and **plant breeding** has produced many showy **flower** varieties. The begonia's fruit is a dry, winged capsule that splits lengthwise to release the **seeds**. Most begonias sprout easily from seeds and can also be propagated from leaves and stems. Leaves are simple and have wavy or serrated margins. **Leaf** arrangement on the stem is alternate. Two fleshy stipules occur at the base of the leaf petiole.

Horticulturists classify begonias into three categories based on rootstock: tuberous, fibrous, and rhizomatous. Unlike the tuberous and fibrous rooted begonias, which are cultivated for their flowers, rhizomatous

Imperfect begonia flowers (*Begonia tuberosa*). © A. Gurmankin 1987/Phototake NYC. Reproduced with permission.

begonias are grown for their large, attractive foliage. The cultivated rex begonia, *Begonia x rex-cultorum*, is a rhizomatous begonia. This horticultural variety with beautiful foliage was developed in England from *Begonia rex* of India. Since most rhizomatous begonias originate from Brazil and Mexico, some people speculate that *Begonia rex* was also a cultivar. Rhizomatous begonias have striking foliage that takes many forms. Leaves can be hairy, fuzzy, or smooth, and are flecked with colorful patterns. Beefsteak begonia, *Begonia feastii*, is another example of a rhizomatous-rooted begonia.

The popular wax begonia, *Begonia semperflorens*, is a fibrous rooted begonia. Wax begonias are outdoor bedding plants that have smooth leaves and an abundance of flowers, hence the scientific name *semperflorens*, which means always flowering. Like the rex begonia, many colorful varieties of the wax begonia have been developed. The angel wing begonia, *Begonia coccinea*, with its thick, jointed stems, is another popular fibrous rooted begonia. Angel wing begonias have cane-like stems. Many cane begonias develop woody **tissue** in their stems.

Tuberous begonias such as *Begonia x tuberhybrida*, are best known for their showy flowers. They originate from South American begonias with large pink (*Begonia boliviensis* and *Begonia veitchii*) and yellow (*Begonia pearcei*) flowers. Cultivated tuberous begonias may resemble other popular flowers such as carnations and daffodils, and come in a range of sizes, including some varieties with large showy blossoms.

Begonias are indigenous to tropical and subtropical regions; no species is native to the United States. They occur primarily in Central and **South America**, **Asia**, and sub-Saharan **Africa**. The natural **habitat** of many begonias are moist, cool **forests** and tropical rainforests, but some begonias are adapted to dryer climates. Tuberous begonias are adapted to cool mountain habitats such

KEY TERMS

. . . .

Alternate—Leaves that occur one at a time on alternating sides of the stem.

Capsule—A dry, dehiscing fruit derived from two or more carpels.

Carpel—Female reproductive organ of flowers which is composed of the stigma, style, and ovary.

Hybrid—Offspring produced from the sexual union of two different species.

Inferior ovary—An ovary embedded within a flower, below the other flower parts.

Monoecious—Plants which have separate male and female flowers on the same plant.

Perennial—Plants which live for several years, often bearing fruits and flowers each year.

Rhizome—This is a modified stem that grows horizontally in the soil and from which roots and upward-growing shoots develop at the stem nodes.

Stipule—An appendage found at the base of a leaf where it joins a branch or stem.

---

as the Andes Mountains of Peru, where many horticultural varieties originate.

Although begonias are herbaceous perennials, they are susceptible to frost, and many varieties planted in the United States are treated as annuals. Begonias are easy to grow, both outdoors and in containers. They like bright **light**, not direct **sun**, a humid environment, and rich, aerated **soil**. Bright, indirect light is required to bring out the colorful patterns on rex begonia leaves. Begonias do best in mild temperatures (above 65°F [18°C] but can tolerate hot **weather** if they are kept in cool, shady places. Regular **fertilization** keeps plants lush and healthy.

Begonias are as easy to propagate as they are to grow. Plant seeds in rich, well-drained soil, such as African violet soil, and keep them protected. Many growers propagate plants from stem and leaf cuttings. Leaves are cut into wedges, each wedge with a central vein. The wedges are dusted with rooting hormone and planted in builders **sand**. The developing leaf wedges are given high **humidity**, bright indirect light, and occasional waterings. New shoot growth appears in two to three months. Cane and rhizomatous begonias may also be propagated from stem cuttings.

Begonias are susceptible to mealybugs and **aphids**, controllable with insecticidal soaps. Rex begonias may be infected with nematodes, soil-dwelling plant **parasites**

that are more difficult to treat. Many garden shops carry products to control nematodes. A home remedy for these **pests** is mothballs. Watering the plants with mothballs on the soil surface will help eliminate the nematodes.

Because of their success and popularity as ornamental bedding and container plants, begonias are economically important.

Resources

Books

The American Horticultural Society. *The American Horticultural Society Encyclopedia of Plants and Flowers.* New York: DK Publishing, 2002.

Heywood, Vernon H. ed. *Flowering Plants of the World.* New York: Oxford University Press, 1993.

Periodicals

Carlquist, S. "Wood Anatomy of Begoniaceae, with Comments on Raylessness, Paedomorphosis, Relationships, Vessel Diameter, and Ecology." *Bulletin of the Torrey Botanical Club* 1985.

Martin, T. "Rex Begonias." *Horticulture* (January 1988).

Neuman, L. "Top-notch Tuberous Begonias." *Horticulture* (July 1988):18-23.

Elaine L. Martin

Behavior

Behavior is the way that living things respond to their environment. A behavior consists of a response to a **stimulus** or factor in an individual's internal or external environment. Stimuli include chemicals, **heat**, **light**, **pressure**, and gravity. All living things exhibit behavior. When dust irritates our throats, for example, we respond with coughing behavior. Plants respond with growth behavior when light stimulates their leaves. Generally, behavior helps organisms survive. Behavior can be categorized as either innate or learned, but the distinction is frequently unclear. Learned behavior often has innate or inborn components. Behavior is considered innate when it is present and complete without the need for experience. Babies, even blind ones, at about four weeks of age smile spontaneously at a pleasing stimulus. Such innate behavior is stereotyped (always the same) and, as a result, quite predictable. Plants, protists, and animals that lack a well developed **nervous system** rely on innate behavior. Higher animals use both innate and learned behavior.

Behavior in plants

The innate behavior of plants depends mainly on growth in a given direction or movement due to changes

Among brown bears, the highest ranking animals are the large adult males. These bears are fighting to establish dominance. Overt fighting is usually brief, and serious wounds are not usually inflicted. *Photograph by Ron Sanford. Stock Market. Reproduced by permission.*

in **water** content. **Plant** behavior in which a plant **organ** grows toward or away from a stimulus is known as a tropism. A positive tropism is growth toward a stimulus, while a negative tropism is growth away from a stimulus. During positive **phototropism**, stems and leaves grow in the direction of a source of light. Roots exhibit positive gravitropism, growth toward gravity, while stems demonstrate negative gravitropism. Since roots grow toward water, they are said to behave with positive hydrotropism. **Touch** stimulates positive thigmotropism, such as vines growing on supporting surfaces.

Sometimes a part of a plant moves in a specific way regardless of the direction of the stimulus. These movements are temporary, reversible, and due to changes in the water pressure inside the plant organ. The leaves of peas and beans open in the morning and close up at night. In light, ion channels open within the pulvinus, a gland present at the **leaf** base. Ions and water enter the leaf, causing it to open. The reverse occurs in darkness. When pollinating **insects** contact cornflowers, the male stamens respond by shortening rapidly, thereby releasing pollen onto the exposed female style. In the Venus flytrap, a carnivorous plant, the touch of an insect on the leaf stimu-

lates the triggering hairs, causing the hinged lobes of the leaf to close quickly around the unsuspecting **prey**.

Animal behavior

The study of **animal** behavior is known as ethology. Ethologists investigate the mechanisms and **evolution** of behavior. Charles Darwin founded the scientific study of behavior, and showed by many examples that behavior, as well as morphology and **physiology**, is an **adaptation** to environmental demands, and can increase the chances of **species** survival.

Between 1930 and 1950, the Austrian naturalist Konrad Lorenz and the Dutch ethologist Niko Tinbergen found that certain animals show fixed-action patterns of behavior (FAPs), which are strong responses to specific stimuli. For example, male stickleback **fish** attack other breeding males that enter their territory. The defending male recognizes intruders by a red stripe on their underside. Tinbergen found that the male **sticklebacks** he was studying were so attuned to the red stripe that they would try to attack passing red British mail trucks visible through the **glass** of their tanks. Tinbergen termed the red stripe a behavioral releaser, a simple stimulus that brings about an FAP.

Once an FAP is initiated, it continues to completion even if circumstances change. If an egg rolls out of a goose's nest, the goose stretches her neck until the underside of her bill touches the egg. Then she rolls the egg back to the nest. If someone takes the egg away while she is reaching for it, the goose goes through the motions anyway without an egg. FAPs have innate components.

Complex programmed behavior involves several steps and is more complicated than FAP. When **birds** build nests and **beavers** build **dams** they are exhibiting complex programmed behavior.

Reflexes are also innate. A **reflex** is a simple, inborn, automatic response by a part of the body to a stimulus. At its simplest, a reflex involves receptor and sensory neurons and an effector organ, for example, when certain coelenterates withdraw their tentacles. More complex reflexes include processing interneurons between the sensory and motor neurons as well as specialized receptors. Complex reflexes occur when food in the mouth stimulates the salivary **glands** to produce saliva, or when a hand is pulled away rapidly from a hot object. Reflexes help animals respond quickly to a stimulus, thus protecting them from harm. Learned behavior results from experience, and enables animals to adjust to new situations. Unless an animal exhibits a behavior at **birth**, however, it is often difficult to determine if the behavior is learned or innate. For example, pecking, an innate behavior in chicks, gets more accurate as the chicks get older. The improvement in pecking aim does not occur because the chicks learn and correct their errors, but is due to a natural maturing of muscles and eyes. Scientific studies have shown that pecking is entirely innate.

The interaction of heredity and **learning** can be observed in a learning program known as **imprinting**, seen frequently in birds. Imprinting is the learning of a behavior at a critical period early in life that becomes permanent. Such behavior was studied in the 1930s by Lorenz. Newly hatched **geese** are able to walk at birth. They survive because they follow their parents. How do young geese recognize their parents from all the objects in the environment? Lorenz found that if he removed the parents from view the first day after hatching and he walked in front of the young geese, they would follow him. This tactic did not work if he waited until the third day after hatching. Lorenz concluded that during a critical period, the goslings follow their parents' movement and learn enough about their parents to recognize them. Since Lorenz found that young geese will follow any moving object, he determined that movement is their releaser for parental imprinting.

Habituation is a type of behavior in which an animal learns to ignore a stimulus that is repeated over and over. A snail will pull its head back into its shell when touched. When touched repeatedly with no subsequent harm, however, the withdrawal response ceases. Apparently, the snail's nervous system "learns" that the stimulus is not threatening and stops the reflex.

In classical **conditioning**, an animal's reflexes are trained to respond to a new stimulus. Ivan Pavlov, a Russian physiologist working in the early twentieth century, was the first to demonstrate this type of learning behavior. He placed powdered meat in a dog's mouth and observed that by reflex saliva flowed into the mouth. Then Pavlov rang a bell before he gave the dog its food. After doing this for a few times, the dog salivated merely at the sound of the bell. Many experiments of this type demonstrate that an innate behavior can be modified.

Further information about behavior modification came in the 1940s and 1950s with the work of B.F. Skinner, an American physiologist. He demonstrated operant conditioning, the training of certain behaviors by environmental rewards. This type of learning is also known as trial-and-error. During operant conditioning, a **random** behavior is rewarded and subsequently retained by an animal. If we want to train a dog to sit on command, all we have to do is wait until the dog sits. Then say "sit" and give the dog a biscuit. After a few times the dog will sit on command. Apparently, the reward reinforces the behavior and fosters its repetition.

Operant conditioning also occurs in nature. By watching their parents, young chimps learn to prepare a stick by stripping a twig and then using it to pick up **termites** from rotten logs. Their behavior is rewarded by the meal of termites, a preferred food. Operant conditioning lets animals add behaviors that are not inherited to their repertory.

Reasoning is a way to solve problems without trial-and-error. This is accomplished by thinking. Using reasoning or insight, we apply memories of past experiences to new situations to help find answers. **Memory** is the storing and retrieving of learned material. The two types of memory are short-term memory, the memory of recent events, and long-term memory, the memory of events that occurred in the past. When given a phone number, we quickly forget it. This is typical of short-term memory which is temporary. Long-term memory lasts longer, days or even a lifetime. Humans use reasoning more than other animals, but **primates** and others have been observed to solve problems by thought processes. Researchers recently discovered that black-capped chickadees develop new **brain** cells to improve their memory, which helps the birds locate buried **seeds** in winter.

In much of their behavior, animals interact with each other. In order to do this they communicate with each other, using their sense organs. Birds hear each other sing, a dog sees and hears the spit and hiss of a cornered cat, **ants** lay down scent signals (**pheromones**), to mark a trail that leads to food.

There are many kinds of interactive behavior. One of them is **courtship** behavior that usually takes place at the start of the mating season. During courtship, some animals leap and dance, others sing, still others ruffle their feathers or puff up pouches. The male peacock displays his glorious plumage to the female. Humpback whales advertise their presence under the sea by singing a song that can be heard hundreds of miles away. Courtship behavior enables an animal to find, identify, attract, and arouse a mate. During courtship, animals use rituals, a series of behaviors for communication that is performed the same way by all the males or females in a species. Territorial behavior is also interactive. Here, animals use signals such as pheromones and visual displays to claim and defend a territory.

Some animals live together in groups and display social behavior. The group helps protect individuals from predators, and allows cooperation and division of labor.

Insects, such as **bees**, ants, and termites live in complex groups in which some individuals find food, some defend the colony, and some tend to the offspring. A method of reducing fighting in a group is accomplished by a dominance hierarchy or ranking system. Chickens, for example, have a peck-order from the dominant to the most submissive. Each **individual** knows its place in the peck-order and does not challenge individuals of higher rank, thereby reducing the chances of fighting. Interactions among group members gets more complex with more intelligent species such as **apes**.

See also Brain; Geotropism; Nervous system; Territoriality.

Resources

Books

Attenborough, David. *The Trials of Life.* Boston: Little, Brown and Co., 1990.

Morris, Desmond. *Animal Watching.* New York: Crown Publishers, Inc, 1990.

Segal, Nancy L. *Entwined Lives: Twins and What They Tell Us About Human Behavior.* New York: Plume, 2000.

Periodicals

"Leaves with Clocks." *Discover* (September 1993).

Maestripieri, Dario. "Evolutionary Theory And Primate Behavior." *International Journal of Primatology* 23, no. 4 (2002): 703-705.

KEY TERMS

Classical conditioning—A procedure involving pairing a stimulus that naturally elicits a response with one that does not until the second stimulus elicits a response like the first.

Fixed-action pattern (FAP)—A strong response by animals to specific stimuli known as releasers.

Habituation—Behavior in which an animal learns to ignore an often repeated stimulus.

Operant conditioning—Trial and error learning in which a random behavior is rewarded and subsequently retained.

Tropism—Orientation of an organism in response to an external stimulus such as light, gravity, wind, or other stimuli, in which the stimulus determines the orientation of the movement.

Matsumoto, Oda. "Behavioral Seasonality In Mahale Chimpanzees." *Primates* 43, no. 2 (2002): 103-117.

Simons, Paul. "Touchy Flowers Work on Elastic." *New Scientist* (July 10, 1993).

Bernice Essenfeld

Belladonna *see* **Nightshade**

Bennettites

The bennettites are an extinct group of gymnosperms—seed-bearing plants whose **seeds** are exposed to the air, not enclosed in the ovary of a **flower**. Botanists hypothesize that bennettites are related to the **cycads**, an extant group of gymnosperms, and paleobotanists believe the bennettites originated from the **seed ferns** (Pteridospermales) about 220 million years ago during the Triassic period. Bennettites became extinct in the Upper Cretaceous period, about 100 million years ago.

Bennettites had palm-like leaves with stems that were thin and branched in some **species**, and stout and trunk-like in others. Most species had stems with a large central pith. The bennettites are also distinguished by certain microscopic features of their guard cells. Guard cells are specialized cells on the surface of a **leaf** that regulate opening of stomata (pores) in the leaf for photosynthetic gas exchange. The bennettites had guard cells with a large amount of cutin, a naturally occurring **plant** wax.

The best-known genus of the bennettites is *Cycadeoidea*. Knowledge about this genus has been gleaned mostly from fossils found in the Black Hills of South Dakota. Many of these fossils were collected and studied in the early 1900s by George R. Wieland. Wieland proposed that the strobili (reproductive structures) of *Cycadeoidea* functioned like flowers and thus that this genus was a close ancestor of the angiosperms, the flowering plants. More recent evaluation of these and other fossil strobili of the bennettites, however, indicates that they differed significantly from angiosperms. It has been shown instead that bennettite strobili were bisporangiate (containing male and female reproductive organs in the same structure) and that they probably relied on self-pollination to reproduce.

See also Paleobotany.

Bentgrass *see* **Grasses**

Benzene

Benzene is an aromatic organic compound with the **molecular formula** C_6H_6. Credit for its discovery and identification in 1825 is usually given to the English chemist and physicist Michael Faraday.

Benzene is a clear, colorless, highly flammable liquid with a pronounced characteristic odor. It has a freezing point of 41.9°F (5.5°C), a **boiling point** of 176.2°F (80.1°C), and a **density** of 0.8787 g/mL. It is only slightly soluble in **water** (0.18 g/100 mL at 77°F [25°C]), but is completely miscible with **alcohol**, **chloroform**, **ether**, **carbon** disulfide, **carbon tetrachloride**, and other organic solvents. Benzene is not to be confused with benzine, which is not a pure chemical compound but a mixture of **petroleum** hydrocarbons used as a solvent and a fuel.

Structure

The structure of the benzene **molecule** proved to be a challenge for chemists for more than 40 years after the compound's discovery by Faraday. Its formula suggests the existence of multiple double and/or triple carbon-carbon bonds, because there are too few **hydrogen atoms** for six single-bonded carbon atoms. However, benzene exhibits none of the chemical properties associated with such a structure, the property of addition, for example. That problem was largely solved in 1865 by the German chemist Friedrich August Kekulé. Kekulé's own story is that he fell asleep in front of his fireplace and dreamed of a snake with its tail in its mouth. He awoke to the realization that the benzene molecule might be a ring con-

sisting of six carbon atoms, with one hydrogen atom attached to each carbon atom. That general structure is still accepted today, although the concept of **resonance** has replaced that of simple single and double bonds between adjacent carbon atoms in the benzene ring.

Properties

The most common chemical property of benzene is that it undergoes substitution reactions. Substitution is a reaction in which an atom or group of atoms replaces a hydrogen atom in an organic molecule. The **halogens**, **nitric acid**, **sulfuric acid**, and alkyl halides all react with benzene to form substituted derivatives. Two, three, or more substitutions can occur on the same benzene molecule, although the ease and location on the benzene ring of these substitutions varies depending on the earlier substitutions.

Benzene derivatives

A number of the substituted benzene derivatives are well known and commercially important compounds. For example, the substitution of a single methyl, hydroxyl, or amino group in benzene results in the formation, respectively, of toluene ($C_6H_5CH_3$), phenol (C_6H_5OH), or aniline ($C_6H_5NH_2$). Probably the best known disubstituted products are the xylenes, $C_6H_4(CH_3)_2$. Three different xylene molecules are possible depending on whether the methyl groups are adjacent to each other on the benzene ring (ortho-xylene), separated by one carbon atom (meta-xylene), or opposite each other on the ring (para-xylene). The removal of one hydrogen atom from the benzene molecule results in a radical known as the **phenyl group**.

Benzene occurs so abundantly in and is obtained so easily from **coal** tar and petroleum that there is virtually no reason to make it synthetically. Although benzene had been recognized as a component of petroleum for many years, it was not produced commercially from that source until the beginning of World War II.

Uses

Benzene is used as a solvent in many commercial, industrial, and research operations. It has long been of interest as a fuel because of its high octane number. Some manufacturers, particularly in **Europe**, have used it as a gasoline additive to increase engine efficiency and to improve starting qualities.

By far the most important use of benzene, however, is in the production of other aromatic compounds. The word *aromatic* was originally applied to benzene because of its distinctive odor, but it later took on a broader meaning, referring to any compound whose molecular structure includes one or more benzene rings. The largest volume of compounds made from benzene goes toward the production of commercially valuable polymers, such as polystyrene, nylon, and synthetic rubber.

The benzene derivative produced in largest quantity is ethylbenzene ($C_6H_5C_2H_5$). Ethylbenzene is converted to styrene ($C_6H_5CH=CH_2$) which, in turn, is polymerized to form polystyrene. Nearly half of all benzene used in chemical synthesis is used for this process.

In another example, benzene is treated with propylene to form cumene ($C_6H_5CH[CH_3]_2$). The cumene thus formed is then oxidized to produce phenol. Phenol is the starting point for a large number of polymers known as phenolic **resins**.

Synthetic fibers are produced by yet a third kind of benzene substitution sequence. The addition of hydrogen to benzene converts it to cyclohexane (C_6H_{12}), which is then oxidized to adipic acid ($COOH[CH_2]_4$-COOH) The acid can then be treated with hexamethylene diamine to form nylon.

Health issues

The health risks associated with exposure to benzene have been known for many years. The compound has both chronic and acute effects whether ingested by mouth, taken in through the **respiratory system**, or absorbed through the skin. Acute effects resulting from inhalation include irritation of the mucous membranes, headache, instability, euphoria, convulsions, excitement or **depression**, and unconsciousness.

The ingestion of benzene has been associated with the development of **bronchitis** and **pneumonia**, while exposure through the skin can cause drying, blistering, and erythema (redness). Death can result from exposure to high concentrations of benzene. Chronic effects resulting from benzene exposure include reduced white and red **blood cell** counts, aplasia, and more rarely, **leukemia**.

See also Hydrocarbon.

Resources

Books

Browning, E. *Toxicity and Metabolism of Industrial Solvents.* New York: Elsevier, 1965, pp. 3-65.

Carey, Francis A. *Organic Chemistry.* New York: McGraw-Hill, 2002.

Graham, John D., Laura C. Green, and Marc J. Roberts. *In Search of Safety: Chemicals and Cancer Risk.* Cambridge, MA: Harvard University Press, 1988.

KEY TERMS

. . . .

Acute—A medical condition that arises over a relatively brief period of time, reaches some crisis, and then may be resolved.

Aromatic—In organic chemistry, a compound whose molecular structure includes some variation of the benzene ring.

Chronic—A disease or condition that devlops slowly and exists over a long period of time.

Hydrogenation—A chemical reaction in which hydrogen is added to a compound.

Polymer—A molecule that consists of a few small units repeated over and over again many times.

Substitution—In organic chemistry, a chemical reaction in which an atom or group of atoms substitutes for a hydrogen atom in a molecule.

Synthesis—A chemical process by which some new substance is produced by reacting other substances with each other.

Purcell, William P. "Benzene." *Kirk-Othmer Encyclopedia of Chemical Technology.* 4th ed. Suppl. New York: John Wiley & Sons, 1998.

Solomons, T. W. Graham. *Organic Chemistry.* 2nd edition. New York: John Wiley, 1980, Chapter 11.

David E. Newton

Benzoic acid

Benzoic acid is a derivative of **benzene** with the chemical formula C_6H_5COOH. It consists of a **carboxyl group** attached to a **phenyl group**, and is thus the simplest aromatic carboxylic acid. It is also known as carboxybenzene, benzene carboxylic acid, and phenylformic acid.

In its pure form, benzoic acid exists as white needles or scales with a strong characteristic odor. It melts at 252.3°F (122.4°C), although it may also sublime at temperatures around 212°F (100°C). It dissolves only sparingly in cold **water** [0.4 g/100 g at 77°F (25°C)], but more completely in hot water [6.8 g/100 g at 203°F (95°C)].

Benzoic acid occurs naturally in gum benzoin, also known as benzoin resin or Benjamin gum, a brown resin found in the benzoin **tree** of Southeast **Asia**. It is also found naturally in many kinds of berries, where its **concentration** may reach 0.05%.

One of the most common uses of benzoic acid is as a food preservative. Both the acid and its **sodium salt, sodium benzoate** (usually listed on labels as benzoate of soda), are used to preserve many different kinds of foods, including fruit juices, soft drinks, pickles, and salad dressings. In fact, it is the acid rather than the sodium salt that is toxic to **bacteria**. Thus, the two additives can be used only in acidic solutions, where the sodium salt is converted to the acid form.

The acid and the sodium salt are both considered to be safe for human consumption in limited amounts. In the United States, foods may contain a maximum of 0.1% benzoic acid or sodium benzoate, although the limit in other nations may be as high as 1.25% in some types of prepared foods. Benzoic acid is also used in the manufacture of artificial flavors and perfumes and for the flavoring of tobacco.

Berkelium *see* **Element, transuranium**

Bernoulli's principle

Bernoulli's principle states that flowing fluids like air and **water** press less than still fluids and that **pressure** decreases quadratically with speed; i.e., with speed squared.

History

One quarter of a millennium ago, Daniel Bernoulli pioneered the use of kinetic theory that molecules moved and bumped things. He also knew that flowing fluids pressed less, but he did not connect these ideas logically. In *Hydrodynamica*, Daniel's logic that flow reduced pressure was obscure, and his formula was awkward. Daniel's father Johann, amid controversy, improved his son's insight and presentation in *Hydraulica*. This research was centered in St. Petersburg where Leonhard Euler, a colleague of Daniel and a student of Johann, generalized a rate-of-change dependence of pressure and **density** on speed of flow. Bernoulli's principle for liquids was then formulated in modern form for the first time.

In this same group of scientists was d'Alembert, who found paradoxically that fluids stopped ahead of obstacles, so frictionless flow did not push.

Progress then seems to have halted for about a century and a half until Ludwig Prandtl or one of his students solved Euler's equation for smooth streams of air in order to have a mathematical model of flowing air for

designing wings. Here, speed lowers pressure more than it lowers density because expanding air cools, and the **ratio** of density times degrees-kelvin divided by pressure is constant for an ideal gas.

More turbulent flow, as in atmospheric winds, requires an alternative solution of Euler's equation because mixing keeps air-temperature fixed.

Applying Bernoulli's principle

Bernoulli's principle is regarded by many as a paradox because currents and winds upset things, but standing a stick in a stream of water helps to clarify the enigma. You will see calm, smooth, level water ahead of the stick and a cavity of reduced pressure behind it. Calm water pushes the stick, as lower pressure downstream fails to balance the upsetting **force**.

Bernoulli's principle never acts alone; it also comes with molecular entrainment. Molecules in the lower pressure of faster flow aspirate and whisk away molecules from the higher pressure of slower flow. Solid obstacles such as airfoils carry a very thin stagnant layer of air with them. A swift low-pressure airstream takes some molecules from this boundary layer and reduces molecular impacts on that surface of the wing across which the airstream moves faster.

For Bernoulli's principle to dominate a dynamic situation, **friction** must be less dominant. Elastic molecular impacts are frictionless—no heating. Molecules of dry air, even more than those of water, collide elastically; so Bernoulli's principle with its molecular-entrainment agent is dominate for windy air.

See also Aerodynamics.

Beryllium *see* **Alkaline earth metals**

Beta-blockers

Beta-blockers are medications used primarily for treating high **blood pressure**. The usefulness of these medications rests on their ability to block the effects of a **nervous system** transmitter chemical known as norepinephrine and the related "fight-or-flight" hormone epinephrine. Beta-blockers are also used to treat heart-related chest **pain** (*angina pectoris*, or simply angina), abnormalities of **heart** rhythm, and certain other conditions.

Adrenergic receptors

Like all nervous system transmitter chemicals and many **hormones**, norepinephrine and epinephrine exert

TABLE 1 FREQUENTLY USED BETA-BLOCKERS	
Generic Name	*Trade Names*
Pindolol	Visken
Propranolol	Inderal
Timolol	Blocarden
Enter Brain Poorly	
Atenolol	Tenormin
Nadolol	Corgard
Selective for Beta1 Receptors	
Metoprolol	Lopressor
Acebutolol	Sectral
Block Both Alpha and Beta Receptors	
Labetalol	Normodyne, Trandate

their effects by interacting with **proteins** on the target cell's outer surface. Scientists refer to the ones on which epinephrine and norepinephrine act as adrenergic receptors, and group them into two major classes. These classes are formally known as a- and b-adrenergic receptors. However, many medical articles use the short forms "alpha receptors" and "beta receptors," respectively.

The most fundamental distinction between alpha and beta receptors is their response or lack of response to specific synthetic chemicals. They also respond differently to their natural stimuli: alpha receptors are more responsive to norepinephrine than to epinephrine, while beta receptors respond equally to both.

Some **cell** types carry both alpha and beta receptors, while others carry only one, or neither. The two classes of receptors often have opposite effects. This allows the body to "fine-tune" its response by varying the relative amounts of circulating epinephrine and locally released norepinephrine in different tissues. In the **circulatory system**, however, both alpha and beta receptors raise blood pressure. Nevertheless, they do so in different ways: alpha receptors by constricting the blood vessels, beta receptors by increasing the **force** and **rate** of the heartbeat.

Mechanism of action

Beta-blockers are not general blood-pressure-lowering drugs, that is, they do not cause already normal blood pressure to go still lower. Nor do they usually affect the

heartbeat of a person at rest, although they do limit the ability of **exercise** or emotion to make the heart beat more quickly and strongly.

Indeed, exactly how beta-blockers combat elevated blood pressure remains unclear. One important aspect is their ability to relax small **arteries**, thus allowing blood to flow more easily and with less pressure behind it. No one knows how beta-blockers do this, however, it would not be expected from their known actions. Furthermore, since relaxation occurs only after several days of beta-blocker use, it is very likely to be an indirect effect.

Scientists also know that beta-blockers reduce the kidney's release of renin, an **enzyme** essential for production of the hormone angiotensin II. Since angiotensin II raises blood pressure in several ways, there can be little doubt that renin plays a significant role in regulating blood pressure. Unfortunately, researchers have found little relationship between blood pressure levels and the amount of renin circulating in the blood. This leaves them uncertain whether beta-blockers' ability to lower blood pressure is tied to their effect on renin release.

By contrast, reasons for beta-blockers' ability to relieve angina are obvious. This condition results from fatty deposits narrowing the arteries that carry blood to the heart muscle. As a result, the heart muscle does not get enough blood to meet its needs—especially when those needs increase because it is beating harder and faster than usual. Since beta-blockers limit the effects of exercise and emotion on the heartbeat, the gap between the amount of blood the heart receives and what it needs will be smaller. As a result, the patient will experience less pain.

Side effects

Different parts of the body contain different beta receptor subtypes, designated beta$_1$ and beta$_2$. Receptors in the circulatory system belong to the beta$_1$ subclass, while those on cells lining the small airways of the lung are of the beta$_2$ subclass. Beta$_2$ receptors help relax these small airways and therefore make breathing easier-indeed, patients with **asthma** and other obstructive lung diseases often inhale beta$_2$—stimulating medications to help them breath more easily. Thus, patients with such diseases should not take medications that block beta$_2$ receptors. Fortunately, several blood pressure medications that selectively block only beta$_1$ receptors are now available.

Beta-blockers are also probably not the best choice of treatment for people who have diabetes along with their high blood pressure or angina. In a hypoglycemic crisis (where blood sugar drops too low), the body pours out large amounts of epinephrine to stimulate release of stored sugar into the blood stream. This epinephrine also causes a rapid, pounding heartbeat that is often the diabetic's first indication something is wrong. Beta-blockers blunt both responses, leading to a crisis that is worse and longer-lasting than it would be otherwise.

These medications may likewise not be the best choice for people with poor circulation in their hands or feet, since beta-blockers sometimes make circulation in the extremities even worse.

About 10% of patients treated with beta-blockers may become dizzy or light-headed. More seriously, about 5% may become clinically depressed, with feelings of helplessness and hopelessness that sometimes lead to suicide. As might be expected, all such reactions are less common with beta-blockers that do not enter the **brain** readily.

Other moderately common side effects of beta-blockers include diarrhea, rash, slow heartbeat, and impotence or loss of sexual drive.

An additional concern with beta-blockers is their effect on blood **cholesterol**, they lower the amount of "good" (HDL) cholesterol, while increasing the amount of "bad" (LDL) cholesterol. They also raise the amounts of fatty materials known as **triglycerides** in the bloodstream; some scientists believe triglycerides may increase the risk of a heart attack to almost the same extent as cholesterol. Nevertheless, there is no concrete evidence that people treated with beta-blockers are more likely to have heart attacks than those treated with other blood-pressure medications.

KEY TERMS

Alpha receptors (BETA-adrenergic receptors)— Proteins on the surface of target cells through which epinephrine and norepinephrine exert their effects.

Angina pectoris (angina)—Chest pain that occurs when blood flow to the heart is reduced, causing a shortage of oxygen. The pain is marked by a suffocating feeling.

Beta receptors (BETA-adrenergic receptors)— Proteins on the surface of target cells through which epinephrine and norepinephrine exert their effects; beta receptors respond to the two substances to approximately the same extent.

Epinephrine (adrenaline)—The "flight-or-fight" hormone synthesized by the adrenal gland.

Norepinephrine (noradrenaline)—A substance that certain nerve cells release in order to produce their effects.

Recent studies indicate that beta-blockers are actually underused. While clinical evidence shows that beta-blockers are in fact very effective medications, other blood pressure drugs are prescribed more often. One reason they are under prescribed may be that physicians are wary of potential side effects when patients are given recommended doses. A new study, however, has provided preliminary evidence that beta-blocker medications can still provide positive clinical outcomes at lower doses, making the risk of side effects lower. Some beta-blocker medications are, however, commonly used. Inderal (Propranolol), Lopressor (Metoprolol), and Ternormin (Atenolol) are widely used beta-adrenergic blocking agents. In general, the generic names for this class of drugs end with the suffix -olol, as in Propranolol.

Summary

Beta-blockers are highly useful and relatively inexpensive medications. They remain among the most commonly used treatments for high blood pressure, although other types of medication have become more popular in recent years. Popularity of these newer medications rests almost entirely on their lower frequency of side effects: they have not been shown to treat the condition any more effectively. In fact, beta-blockers remain one of the two types of medication that have actually been shown to extend the life of people with high blood pressure.

See also Heart diseases; Hypertension.

Resources

Books

Edelson, Edward. *The ABCs of Prescription Drugs.* Garden City, NY: Doubleday & Co., 1987.

Hoffman, Brian B., and Robert J. Lefkowitz. *"Adrenergic Receptor Antagonists." Goodman and Gilman's The Pharmacological Basis of Therapeutics.* 8th ed. New York: McGraw-Hill, 1990.

Oppenheim, Mike. *100 Drugs that Work.* Los Angeles: Lowell House, 1994.

W. A. Thomasson

Beta particle *see* **Radioactive decay**

BHA *see* **Butylated hydroxyanisole**

BHT *see* **Butylated hydroxytoluene**

Big bang theory

The big bang theory is the conceptual and mathematical model that scientists use to describe the origin of the Universe. It states that the Universe began as a tiny, violent explosion about 15 billion years ago. That event produced all of the **matter** and **energy** in the universe, including its **hydrogen** and helium. Some of these light **atoms** were forged in the cores of stars, over billions of years, into atoms of the heavier elements that exist today, including the atoms of which we ourselves are made. One consequence of the big bang is that today the Universe, which is of finite size and contains a finite amount of matter, is expanding; in fact, the occurrence of the big bang was originally deduced from the fact of the Universe's expansion. In recent years astronomers have made many observations that verify predictions of the big bang theory.

Studying the Universe

Since ancient times, people have wondered about the origin of the Universe. Questions about how and when **Earth** and the heavens formed have been pondered by philosophers, theologians, and scientists. The modern, scientific study of the origin and structure of the Universe is known as the science of **cosmology**.

For many centuries cosmological thought was limited mostly to speculation. For example, it was not obvious to the ancients that the **Sun** and the stars are objects of the same sort. Today it is known that the Sun is a **star** much like the other stars, and is only brighter because it is closer. This simple fact took centuries to determine because it is difficult to determine the **distance** to most of the objects seen in the night sky. Early astronomers of the scientific era, although they knew that the stars are also suns, assumed that all stars have the same intrinsic brightness and thus, that only their distance from Earth determines their apparent brightness. This is now accepted as untrue—enormous variations in brightness among individual stars do exist. Examination of binary stars (paired stars that **orbit** each other) demonstrated these differences. When **binary star** systems in which the two stars did not have the same brightness were observed, it became clear that the amount of light received from any given star is dependent on more than just its distance. Until the elementary task of measuring the position of astronomical objects could be pursued systematically, larger questions about the structure and history of the Universe as a whole could not even begin to be answered.

Measurement techniques

All measurements of the stars must necessarily be made from the neighborhood of the Earth, since the distances involved are enormous. The nearest star other than the Sun is more than four light-years away, and most objects seen from earth, even with the naked **eye**, are much farther off. (A **light-year** is the distance that light travels in one year: 5.88 trillion miles [9.46 trillion kilometers], or about 60,000 times the distance from the Earth to the Sun.)

There are two fairly direct ways to determine the distance to the nearest stars. The first is to measure their *parallax* or apparent change in position during the year. As the Earth circles the Sun, stars are seen from a shifting vantage point. The furthest objects do not appear to move because the Earth's change in position is too small to affect our view, but the nearest stars seem to move back and forth slightly during the course of a year. **Parallax** can be seen by holding up a finger a few inches before your eyes and closing first one eye, then the other, thus repeatedly shifting your point of view by the distance between your eyes: the finger seems to jump back and forth dramatically, while objects across the room move much less. The shift in the finger's apparent position is its parallax. By measuring the parallax of a nearby star over a six-month period (during which the Earth moves from one side of its orbit to the other), and knowing the radius of the Earth's orbit, it is a matter of straightforward **trigonometry** to determine the distance to that star.

Another technique to determine stellar distances is to measure the proper **motion** of a star. This is the apparent motion of a star with respect to other stars caused by the star's actual motion through the sky. (All stars are moving, including the Sun.) Although the motion of distant stars is too small to detect, closer stars can be seen changing position with respect to more distant stars over the years.

Such techniques are only applicable to a few of the nearest stars, however, and disclose nothing about the large-scale structure of the Universe. More sophisticated methods had to be developed for this task, requiring different astronomical observations. One such method depends on the examination of a star's (or other celestial object's) **spectrum**, that is, the intensity of its **radiation** (including, but not limited to, its visible light) at various wavelengths. If the light from a star is divided into its component wavelengths using a **prism**, a continuous spread of wavelengths punctuated by a number of dark lines can be seen. (The visible part of our Sun's spectrum is the rainbow.) These absorption lines are caused by elements in the star's outer atmosphere that absorb light at specific wavelengths. Each dark line in a star's spectrum corresponds to a specific element; the absorption lines in a star's spectrum thus give a catalogue of the substances in its outer layers. Furthermore, these lines can reveal how fast the star is approaching the Earth or receding from it using the **Doppler effect**, a fundamental property of all traveling waves (including light waves). The absorption lines in the spectra of objects moving *away* from the Earth are shifted to longer wavelengths, while absorption lines in the spectra of objects moving *toward* the Earth are shifted to shorter wavelengths. A shift to longer wavelengths is called a **redshift** because red light appears near the long-wavelength end of the visible spectrum, while a shift to shorter wavelengths is called a blueshift. Measurement of the Doppler shifting of **spectral lines** has made it possible to **map** the large-scale structure of the cosmos, and it is this structure that the theory of the Big Bang and the theory of general relativity explain.

Historical background

In 1905, Danish astronomer Ejnar Hertzsprung (1873–1967) compared the width of various stars' absorption lines to the absolute luminosity, or brightness of the stars as determined from proper motion measurements. Hertzsprung found that that wider lines correspond to larger and brighter stars. This provided a way to determine the absolute brightness of a star from its spectrum. Knowing its absolute brightness, he could then determine its distance from Earth. This method applied to stars at any distance, as opposed to the parallax and absolute-motion methods, which applied only to stars quite near the Sun, but was limited in **accuracy**.

In 1908, U.S. astronomer Henrietta Swan Leavitt (1868–1921) discovered that Cepheid variables, a type of star in which the brightness changes in a regular manner, showing a well-defined relationship between period (time

required for brightness to wax and wane through a full cycle) and absolute luminosity. Brighter Cepheid **variable stars** have longer periods while dimmer ones have shorter periods. Leavitt calculated a simple relationship between brightness and period. This discovery had a profound effect on stellar distance measurements. Now, any time a Cepheid could be found—in, say, a distant galaxy—the distance to it could be determined accurately.

The spiral nebulae

In the early twentieth century, there was a debate among astronomers over the nature of the **spiral** nebulae, the diffuse spiral-shaped structures visible (through telescopes) in most parts of the sky. Some believed these were nearby objects that were part of our **Milky Way Galaxy**, while others thought that they were much further away, and in fact were "island universes," or separate galaxies. If the distances to these objects could be measured, then the debate could be settled, and important knowledge gained about the structure of the Universe.

In 1914, U.S. astronomer Vesto M. Slipher (1875–1969) presented figures for the velocities of 14 spiral nebulae, obtained by measuring the Doppler shifts of their spectral lines as described above. Slipher found that most of the nebulae were, surprisingly, moving away from earth. If the nebulae's motions were **random**, just as many of them would be expected to move toward earth as were moving away. Why should the sun just happen to be at the center of an organized pattern of motion? Another puzzling finding was the large velocities at which these objects were receding. The nearby Andromeda nebula, for example, was speeding toward earth at 180 miles per second (300 km per second). Many astronomers interpreted this to mean that the nebulae must be outside our galaxy.

In 1923, U.S. astronomer Edwin Hubble (1889–1953), using the 60-in and 100-in (152-cm and 254-cm) telescopes at Mount Wilson Observatory, succeeded in identifying Cepheid variables in the outer regions of two nebulae, M31 and M33. By measuring the periods of these Cepheids and using the formula developed by Leavitt several years earlier, he calculated that they are about 930,000 light years distant. From these distances and the observed sizes of the nebulae, their actual sizes could be calculated. These turned out to be similar to that of Earth's galaxy, strongly supporting the idea that the nebulae are galaxies in their own right.

In 1929, Hubble plotted data on a number of galaxies on a graph. He plotted the distance to the galaxy along the horizontal axis and the **velocity** of the galaxy's recession along the vertical axis. From this limited data, it was clear that a simple, linear relationship between the two quantities existed; on average, the velocity of a galaxy's recession was proportional to the distance to the galaxy. (The Andromeda galaxy is a member of our local galactic group and does not obey this general rule.) The constant of proportionality between distance and velocity, now called the *Hubble's constant* (H), was given by the slope of the line. From those data, Hubble's constant was estimated to be 310 miles per second per megaparsec (500 km/s/Mpc)—that is, a galaxy 1 megaparsec (one million parsecs, where one parsec = 3.26 light-years) away would be moving away from our galaxy at 310 mi/s (500 km/s), while a galaxy ten times further away would be moving ten times as fast. Modern values for H are much smaller than Hubble's estimate of 500 km/s/Mpc. The relationship between speed and distance governed by Hubble's constant is termed Hubble's law.

Implications of Hubble's law

Galaxies are moving away from the Earth in all directions because all galaxies are receding from each other, that is, every galaxy is getting further away from every other galaxy. There is a simple way to visualize this effect. Imagine a partially blown up **balloon**, on the surface of which a number of spots (representing galaxies) are drawn. As the balloon inflates, each spot gets farther away from its neighbors; furthermore, the distance from any one spot to any other on the same side of the balloon (as measured on the surface of the balloon) grows faster the farther away from each other the two spots become. This can be seen by imagining three spots in a row (A, B, and C), with the balloon's surface expanding evenly between them. The distance from A to B is growing at x inches per second, and so is the distance from B to C; the distance from A to C must therefore be growing at $2x$ in/s ($x + x = 2x$). Thus no matter where an observer is located on the balloon, they will observe that all the other spots on the balloon's surface obey Hubble's law: the farther away they are, the faster they recede. There is no preferred direction, no preferred position, and no "center" of the balloon's surface. The behavior of galaxies in our spatially three-dimensional universe is analogous to that of the spots on the two-dimensional surface of the balloon.

There is an important implication to this model. If all galaxies are moving away from each other with a velocity proportional to their separation, then at all earlier times the galaxies were closer together. If one goes back far enough, there must have been a time when all were at the same position—that is, there must have been a beginning to the Universe. In fact, if the expansion has been constant for all time, the **age of the Universe** is simply the inverse of Hubble's constant. The situation is not quite so simple, as scientists expect Hubble's constant to change over time (the gravitational attraction between the galaxies has a tendency to slow the **rate** of expansion, while recent observations show that the expansion of the Universe is actually accelerating under the influence of a still-mysterious force that acts opposite to gravity).

Hubble's original measurements gave an age of the Universe of about two billion years. This immediately caused dispute, since it was known from measurements of **radioactive decay** that the age of the **solar system** is more than twice this value. How could the solar system have been formed before the Universe itself? It is now known that Hubble's original measurements were in **error**. Current measurements put Hubble's constant in the range of 50–100, giving an age of 10–20 billion years.

Other developments

When Einstein developed his general theory of relativity he added a term, the cosmological constant, to his equations in order to permit a static universe which was neither expanding nor contracting. (He later came to regret this, calling it one of the worst mistakes he ever made; however, the recent discovery that the expansion of the Universe is actually accelerating has brought Einstein's cosmological constant back into favor). Despite his use of this extra term, however, Russian mathematician Alexander Friedmann (1888–1925) and Belgian astronomer Georges LeMaître (1894–1966) found solutions (in 1922 and 1927, respectively) to Einstein's equations that permitted an expanding universe. After Hubble's 1929 discovery there was a great deal of interest in these models, which could be used to explain the observations.

A big bang seemed an obvious implication of the new data, but steady-state models that avoided the embarrassment of a universe with a definite beginning had adherents for decades after Hubble's measurements were made. One of the more promising models, "constant creation," postulated that new hydrogen atoms form constantly and spontaneously throughout **space**, out of nothing, providing the material for new galaxies as the older galaxies move apart. On this theory, the Universe has always been expanding, had no beginning, has always looked as it does now, and will always look as it does now. This theory predicted that nearby galaxies will look similar to those far away, but it was found that distant galaxies are in fact different from nearby ones, which agrees with the big bang's claim that the Universe is not in a steady state. It was one of the originators of this **steady-state theory**, British astronomer Fred Hoyle (1915–2001), who coined the term "big bang," now used to describe the expanding-universe model based on Hubble's observations. Hoyle chose the expression to ridicule the theory, but the name stuck.

The evolution of the Universe

The current picture of the big bang can be described briefly as follows. Because current formulations of the laws of **physics** break down very close to the big bang itself, the account will start one second after the event occurs. At this time, the **temperature** was 10,000,000,000K. This was too hot for atoms to exist, so their elementary particle constituents (electrons, protons, and neutrons) existed separately, along with photons (particles of light), and various exotic particles. Over the next 100 seconds, the temperature dropped by a factor of 10, enough to allow the nuclei of light elements such as **deuterium** (an **isotope** of hydrogen) and helium to form. As further cooling took place, these nuclei combined with electrons to form atoms.

At this point it should be stressed that the expansion of the Universe means that space itself is expanding. This differs fundamentally from an ordinary explosion, in which matter expands into a surrounding **volume** of space. The expansion of space itself can be compared to the increase of the surface area of the inflating balloon mentioned earlier; as the balloon expands, its surface area grows, but not by expanding into any larger, surrounding surface, as a circular ripple expands across the surface of a pond. Similarly, our universe is not expanding into any larger, surrounding volume of space. The expansion of space has important cosmological implications. One is that as space expands, the average temperature of the Universe drops with it. This cooling has an important effect on the **cosmic background radiation**.

Early in the history of the Universe, when its **density** was extremely high, particles and radiation were in equilibrium, meaning that there was a very uniform temperature distribution. Such a distribution gives rise to radiation with a particular spectrum, a blackbody spectrum, which has a well-defined peak wavelength. Radiation of this type currently pervades all space in the form of microwave radiation, the afterglow of the big bang. Due to expansion of the Universe, the peak of this radiation's spectrum—its temperature—has by now been shifted to below 3K ($-454°F$ [$-270°C$]), or three degrees above **absolute zero**, despite its initial high temperature. The cosmic background radiation was first detected by U.S. astrophysicist Arno Penzias (1933–) and U.S. **radio** astronomer Robert Wilson (1936–) in 1965. Measurements from the COBE spacecraft have shown that the spectrum is a nearly perfect blackbody at 2.73K ($-454.5°F$ [$-270.27°C$]), as predicted by the big bang theory.

As described above, only the lightest elements were created in the big bang itself. As the Universe expanded, inhomogeneities eventually developed, and regions of more-dense and less-dense gas formed. Gravity eventu-

ally caused the high-density areas to coalesce into galaxies and eventually stars, which became luminous due to nuclear reactions in their cores. These reactions take hydrogen and helium and create some of the heavier elements. Once its light-element nuclear fuels are exhausted, a star may explode in a **supernova**, creating still heavier elements in the process. It is these heavier elements from which the solar system, the earth, and humans are made. Every atom in the human body (every element other than hydrogen) was created in the core of an exploding star billions of years ago.

What will be the ultimate fate of the Universe? Will it continue to expand forever, or will it eventually contract in a "big crunch?" To understand this question, the analogy of a projectile being launched from the surface of the earth can be used. If a projectile is launched with enough velocity, it will escape the Earth's gravity and travel on forever. If it is too slow, however, gravity will pull it back to the ground. This same effect is at work in the Universe today. If there is enough **mass** in the Universe, the force of gravity acting between all matter will eventually cause the expansion to slow, stop, and reverse, and the Universe will become smaller and smaller until it ultimately collapses.

This does not seem likely. Astronomers have made estimates of the mass in the Universe based on the luminous objects they see, and calculated a total mass much less than that required to "close" the Universe, that is, to keep it from expanding forever. From other measurements, they know that there is a large amount of unseen mass in the Universe as well, called **dark matter**. The amount of this dark matter is not known precisely, but greatly exceeds that of all the stars in the Universe. The nature of dark matter is being intensely researched and debated by astronomers.

In 1998, astronomers studying a certain group of supernovas discovered that the older objects were receding at a speed about the same as the younger objects. According to the theory of a closed universe, the expansion of the Universe should slow down as it ages, and older supernovas should be receding more rapidly than the younger supernovas. The fact that observations have shown the opposite has led many scientists to believe that the Universe is, in fact, open. Other theorists hold that the Universe is flat—that is, that it will neither collapse nor expand forever, but will maintain a gravitational balance between the two and remain in a coasting expansion. In the last few years, various observations have indicated—to astronomers' astonishment—that the expansion of the Universe is actually accelerating. If this is true, then the fate of the Universe will be to expand without end. Eventually, all sources of energy will exhaust themselves, and after many trillions of years even the protons and neutrons of which ordinary matter is con-

KEY TERMS

. . . .

Cepheid variable star—A class of young stars that cyclically brighten and dim. From the period of its brightness variation, the absolute brightness of a Cepheid variable can be determined. Cepheid variables in distant galaxies give a measure of the absolute distance to those galaxies.

Hubble constant—The constant of proportionality in Hubble's Law which relates the recessional velocity and distance of remote objects in the universe whose motion is determined by the general expansion of the universe.

Light-year—The distance that light travels in one year, equal to 5.87 million miles 9.46 million km.

Parsec—3.26 light-years.

Resources

Books

Hawking, Stephen W. *A Brief History of Time*. Toronto: Bantam Books, 1988.

Silk, Joseph. *A Short History of the Universe*. New York: Scientific American Library, 1994.

Weinberg, Steven. *The First Three Minutes: A Modern View of the Origin of the Universe*. New York: Basic Books, 1977.

Periodicals

Glanz, James, "Photo Gives Weight to Einstein's Thesis of Negative Gravity." *New York Times*. April 3, 2001.

Overbye, Dennis, "Radio Telescope Proves a Big Bang Prediction." *New York Times*. September 20, 2002.

Peebles, P. James, David N. Schramm, Edwin L. Turner, and Richard G. Kron. "The Evolution of the Universe." *Scientific American* (October 1994): 53–57.

Larry Gilman

structed will break down. If this vision is correct, the Universe will end up as a diffuse, eternally expanding gas of **subatomic particles** at a uniform temperature.

Future work

Although the big bang model has done a good job of explaining what is seen in the Universe, there are still many unanswered questions. There is still disagreement about the exact value of the Hubble constant by approximately a factor of two. The **Hubble Space Telescope** is making observations similar to those made by Edwin Hubble in order to try to measure this quantity more accurately. Preliminary results have been announced, but it will be some time before a value can be accurately determined. These measurements are very difficult to make, since they are at the limits of the telescope's ability to observe.

Another open question is how galaxies actually formed from what was very close to a uniform, homogeneous medium in the early universe. From the uniformity of the microwave background radiation, it is known that this uniformity was better than one part in a thousand. Just by looking at the sky, however, a great deal of structure in the Universe is seen up to very large-size scales of clusters of galaxies and beyond. There must have been some type of clumping which occurred to start the process (with gravity helping the process along), but what started it? Physicists are seeking the answer in **quantum mechanics**, the science of very small events. The answer resides there because at the moment of the big bang, the Universe was subatomic in size.

See also Blackbody radiation; Elements, formation of; Redshift.

Binary star

Binary stars, often called double stars, refer to pairs of stars sufficiently close to each other in **space** to be gravitationally bound together. Following the laws of gravitation, each of the components revolves around the common center of **mass** of the system. At least 50% of stars are found to exist as binary systems, according to conservative **statistics**. There seems to be no obvious preference for particular combinations of brightness, size, or mass differences and a wide range in periods of revolution from less than a day to thousands of years. Likewise, there is a large range in separations from those stars in contact to those separated by thousands of times the **Earth** to **Sun distance**. Historically, visual binaries, those that appear as double stars when seen through a **telescope**, were discovered to be gravitationally bound by William Herschel around 1800.

Techniques of observation

There are a number of telescopic techniques used to discover and study binary stars. No one telescopic method can be used because of the wide range in the separations exhibited in the systems. The desired information about the orbital **motion** and the physical quantities of the stars themselves must come from different ways of observing. Hence, there are descriptive classifications of binary stars as determined by the various modes of study discussed below.

Importance

The importance of binaries lies in a number of areas: 1) The analysis of a visual binary (where the two

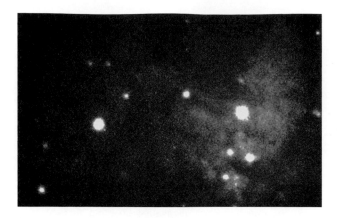

An x-ray image of the x-ray binary star system LMC X-1 in the Large Magellanic Cloud (LMC). LMC X-1 is seen as the two bright objects at center left and center right. The Tarantula Nebula is at the top of the cloud at left. The picture was taken by the ROSAT x-ray astronomy satellite. ROSAT discovered 15 new x-ray sources in the LMC, which is 163,000 light years from our galactic center and 33,000 light years in diameter. ROSAT was launched on June 1, 1990, and is a collaboration between Germany, UK, and USA. *© Max-Planck-Institut Fur Physik Und Astrophysik. Photo Researchers, Inc. Reproduced by permission.*

components can be seen visually through a telescope) leads to the only direct method for the evaluation of stellar mass, one of the most important parameters of the physical universe. In some cases the mass of the system is found, in other situations the mass **ratio** of the components, and in some the individual masses can be determined; 2) Stellar duplicity plays an important role in the study of the physical aspects of stars, such as relative diameters, surface brightness, the genesis and evolution of stars, and the study of stellar-mass loss.

Visual binaries

Visual binaries are those stellar systems which appear as two stars in the eyepiece of a telescope. They are traditionally observed by manually measuring, at the eyepiece of a telescope, the angular separation of the two components. The angle the fainter stars make with respect to the brighter component is also taken into account, and refers to the north direction on the **plane** of the sky. If the measurements of each component are made and photographically compared to background stars, the masses of each **star** can be found. Many observations are needed over a period of time commensurate to the period of revolution of the pair.

Study of orbital motion

The first goal generally is to determine the period of revolution, then when feasible, the geometric elements (relating the apparent **orbit** on the plane of the sky to that on the true plane of the orbit) and the dynamical orbital elements of the system as far as possible, which will lead to the physical characteristics of the stellar components. The simple laws which govern the dynamics of orbital motion of double stars stem from the three Kepler laws which were originally formulated by Johann Kepler to describe the motions in our **solar system**. They had far-reaching implications of gravitational forces explained later by Newton. The laws of Kepler, as used in binary star analysis, are: 1) The orbit described by the fainter component (often called B) around the brighter (A component) is an **ellipse** with the A component at one of the foci; 2) The component sweeps over equal areas, throughout its orbital path, in equal lengths of time; 3) The sum of the masses of the two components (in units of the solar mass) is equal to the scale or semi-major axis cubed (in units of the earth-sun distance) divided by the period squared (in unit of years). The mass of the binary system from Kepler's third law is the only direct way stellar mass can be determined.

Astrometric binaries

Astrometric binaries are double star systems visible on astrometric photographs as single stars. They have a telltale wavy motion across the sky indicating that the visible star is revolving around the center of mass of the visible star and its invisible companion, and thus its motion over an interval of time is analyzed for gravitational orbital motion. This process is a slightly modified form of the method for visual binaries. Generally, the companion star is either too faint to be seen or too close to the primary star to be resolved as two stars. The largest ground telescopes and also the **Hubble space telescope** are used to try and "see" the fainter component which might turn out to be a **brown dwarf** or a **planet**. The star Sirius is a fine example of a visual binary, discovered first as an astrometric binary in 1844 by the German astronomer F. Bessel.

Spectroscopic binaries

Spectroscopic binaries are pairs which are too close to each other as seen from the earth to be resolved into two stars. However, when the **light** from the star is analyzed with a spectrograph, which spreads the light into a continuous **spectrum** of colors with dark absorption lines superimposed, the **spectral lines** are alternately shortened or lengthened indicating Doppler motion, a to-and-fro motion as seen from Earth. This shift in the wavelengths results from the periodic motion, in the line of sight, of the visible star revolving around the center of mass of the system. When only the brighter component has sufficient light to show on the spectrogram the system

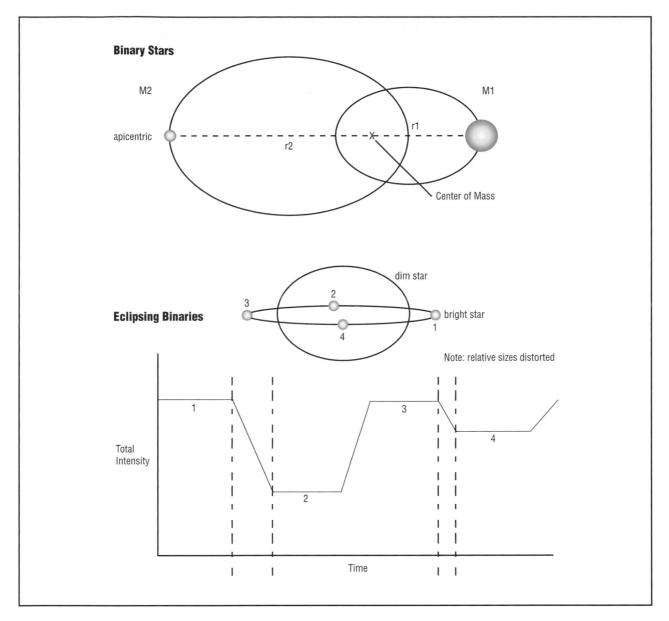

Binary Stars

M2

apicentric

r2

X

r1

M1

Center of Mass

Eclipsing Binaries

dim star

3

2

4

1 bright star

Note: relative sizes distorted

Total Intensity

1

2

3

4

Time

Binary star systems contain two stars held by mutual gravitational attraction. The stars orbit a common center of mass.
Illustration by K. Lee Lerner, with Argosy. The Gale Group.

is known as a single-lined spectroscopic binary. When the spectra of the fainter component is also recorded the name double-line spectroscopic binary is used.

Eclipsing binaries

Eclipsing binaries are those systems, seen as a single star, which show periodic changes in brightness. This occurs when one component **eclipses** the other during their orbital motions around the center of mass of the system. The plane of the orbital motion must necessarily be close to **perpendicular** to our line of sight; eclipses are further facilitated when their separation is small. By

analyzing the brightness with the passage of time, the resulting light **curve** can indicate some geometric and dynamical components of orbital motion. Astrophysical properties can be determined such as the relative brightness of the two components, their relative diameters, and some aspects of their atmospheres.

Mass exchange binaries

Mass exchange binaries are short period pairs whose components are virtually in contact with each other so that they interact with each other; their large gaseous atmospheres may touch forming a figure eight in three di-

KEY TERMS

Astrometric photographs—Photos taken with "telephoto type" telescopes yielding large scale portrayal suitable for accurate measurements of positions of stellar images.

Dark absorption lines—Part of the stellar spectrum coming from different atomic elements in the atmosphere of the star.

Doppler shift—The change in frequency or wavelength resulting from the relative motion of the source of radiation and the observer. A motion of approach between the two will result in a compression of the waves as they pass the observer and a rise in "pitch" in the frequency of the wave and a shortening of the relative wavelength called a "blue shift." A relative motion of recession leads to a lowering of the "pitch" and a shift to longer "redder" wavelengths.

Dynamical orbital elements—Used in equations to describe the true orbital path of a binary star component in the plane of the orbit.

Geometrical orbital elements—Used in equations to describe the orbital path of a binary star component as seen on the plane of the sky.

Mass—The quantity of matter in the star as exhibited by its gravitational pull on another object. Stellar mass is usually measured in units of the sun's mass.

mensions. Sometimes the atmospheres overlap to such an extent that they make an envelope around the entire system. Mass from one component flows into the other resulting in mass loss and exchange. This in turn affects a change in the period of revolution of the pair. Generally, the components are at different stages in their evolutionary track. X-ray binaries described below are also binaries that exchange mass. Much remains to be done to understand the details of the **physics** of the interaction.

X-ray binaries

X-ray binaries are discovered through space telescopes, which focus on very short-wave **energy radiation** sources. The International Explorer and the Einstein X-Ray Observatory and other satellites have been used. Some semi-detached pairs emit x–rays provided by mass transfer in a common atmospheric envelope. Close pairs with one component, a **neutron star** or a **black hole**, are likely indicated from enormous energy output in the

form of ultraviolet and x-rays which are generated around the massive star as gas from the companion, un-evolved star, is sucked toward the massive central degenerate component. This type of binary may have a period of revolution around two days or less.

See also Brown dwarf; Gravity and gravitation; X-ray astronomy.

Resources

Books

Couteau, Paul. *Observing Double Stars.* Cambridge, MA: The MIT Press, 1981.

Introduction to Astronomy and Astrophysics. 4th ed. New York: Harcourt Brace, 1997.

Zeilik, M., and J. Gaustad. *Astronomy, the Cosmic Perspective.* New York: John Wiley, 1990.

Periodicals

Degirmenci, L. "Formation, Structure And Evolution Of Stars." *Astronomy And Astrophysics* 363, no. 1 (2000): 244-252.

"Determination of the Ages of Close Binary Stars on the Main Sequence" *Astrophysics* 45, no. 3 (2002) 342-357.

Sincell, M. " Profile: Twin Stars Of Astrophysics Make Room For Two." *Science* 293, no. 5532 (2001): 1040-1041.

Other

McAlister, H.A., and Wm. I. Hartkopf, eds. *Complementary Approaches to Double and Multiple Star Research, IAV Colloquium 135.* ASP Conference Series, Vol. 32, 1992. Published by the Astronomical Society of the Pacific.

Sarah Lee Lippincott

Binary system *see* **Numeration systems**

Binocular

In 1823, a new optical instrument began to appear in French opera houses that allowed patrons in the distant (and less expensive) seats to view the opera as if they were in the front row. Called opera glasses, the device combined **telescope** lenses with stereoscopic prisms to provide a magnified, three-dimensional view. After many years (but relatively few modifications), opera glasses have evolved into the binocular.

In their simplest form, binoculars are a pair of small refracting telescope lenses, one for each **eye**. The **brain** assembles the two views, one from each **lens**, into a single picture. Because each eye sees its own view, the final image has depth; this is not so with conventional telescopes, which possess only one eyepiece and, therefore, a two-dimensional image.

While some simple binoculars can be found, most quality binoculars possess a more intricate design. In more complex binoculars, there is a system of prisms between the large front lens, called the objective lens, and the smaller eyepiece. These prisms serve two important functions. First, they bend the **light** so that the final image is both upright and nonreversed. In a common telescopic view, the image is both reversed and upside-down. Second, the bending produced by the prisms lengthens the overall light path, which allows for much greater magnification while staying within the binocular's short tube. Without prisms, average-powered binoculars would need to be more than one foot (0.305 m) in length.

In order to enhance the stereoscopic, or three-dimensional, effect, the binocular's two objective lenses are placed further apart than the viewer's eyes. When the two views are then assembled by the brain, a greater impression of depth and clarity results.

Many different factors influence the quality of a binocular. For the majority of users, the most important of these is magnifying power. Binocular magnification usually ranges from six to twenty times—that is, the object appears six to twenty times larger in the binoculars than it would with the unaided eye. Magnification is usually expressed as "X," whereby six times magnification would be written 6 X.

Another factor governing a binocular's quality is the size of the two object lenses, called the aperture. Larger aperture sizes are valued, because they collect a greater amount of light. This is crucial, because an image becomes fainter as the magnification increases. Thus, high magnifications are usually coupled with wide apertures. While object lenses of 30-80 mm (1-3 in) are common, apertures as large as 150 mm (6 in) have been designed; these, however, are used chiefly in military reconnaissance. Binocular makers generally express the quality of their instruments in terms of both magnification and aperture size. A common rating is 7×50, which describes a magnification of seven power and an aperture of 50 mm (2 in).

Professional users of binoculars, such as astronomers and military personnel, also consider the size of the light beam that exits the eyepiece, called the exit pupil. The closer this light beam is to the width of the viewer's own pupil, the more efficient the binocular. This is a tricky factor because the size of a human pupil varies: in bright light (such as daylight) the pupil is only 0.078 in (2 mm) across, while in dim light (such as moonlight) it opens to almost 0.273 in (7 mm). Thus, binoculars with a small exit pupil are best for daytime use, while those with a wider exit pupil are essential for nighttime observing.

Yet another factor affecting the quality of a binocular is the straightness of its beams. In a perfectly adjusted instrument, the two beams entering each eye will be **parallel**. If these beams are offset even slightly, a doubled image will be produced. Such a poorly adjusted view is uncomfortable and bad for the eyes. In order to fix the image, the binoculars should be collimated so that the beams are parallel.

Although they are inferior to telescopes in magnification, binoculars are often better devices for viewing the heavens. Because of their wider object lenses, binoculars can collect more light than telescopes; this makes objects such as distant stars or planetary satellites appear much brighter than they would in many telescopes. Even household binoculars are sufficient for viewing the **Moon** and the visible planets, and they are usually much less expensive than a telescope.

Binomial theorem

The binomial **theorem** provides a simple method for determining the coefficients of each term in the expansion of a binomial with the general equation $(A + B)^n$. Developed by Isaac Newton, this theorem has been used extensively in the areas of probability and **statistics**. The main argument in this theorem is the use of the combination formula to calculate the desired coefficients.

The question of expanding an equation with two unknown variables called a binomial was posed early in the history of **mathematics**. One solution, known as **Pascal's triangle**, was determined in China as early as the thirteenth century by the mathematician Yang Hui. His solution was independently discovered in **Europe** 300 years later by Blaise Pascal whose name has been permanently associated with it since. The binomial theorem, a simpler and more efficient solution to the problem, was first suggested by Isaac Newton. He developed the theorem as an undergraduate at Cambridge and first published it in a letter written for Gottfried Leibniz, a German mathematician.

Expanding an equation like $(A + B)^n$ just means multiplying it out. By using standard **algebra** the equation $(A + B)^2$ can be expanded into the form $A^2 + 2AB + B^2$. Similarly, $(A + B)^4$ can be written $A^4 + 4A^3B + 6A^2B^2 + 4AB^3 + B^4$. Notice that the terms for A and B follow the general pattern A^nB^0, $A^{n-1}B^1$, $A^{n-2}B^2$, $A^{n-3}B^3$,...,A^1B^{n-1}, A^0B^n. Also observe that as the value of n increases, the number of terms increases. This makes finding the coefficients for individual terms in an equation with a large n value tedious. For instance, it would be cumbersome to find the **coefficient** for the term A^4B^3 in the expansion of $(A + B)^7$ if we used this algebraic approach. The inconvenience of this method led to the development of other solutions for the problem of expanding a binomial.

One solution, known as Pascal's triangle, uses an array of numbers (shown below) to determine the coefficients of each term.

$(A+B)^0$ 1
$(A+B)^1$ 1   1
$(A+B)^2$ 1   2   1
$(A+B)^3$ 1   3   3   1
$(A+B)^4$ 1   4   6   4   1
$(A+B)^5$ 1   5   10   10   5   1

Pascal's Triangle

This triangle of numbers is created by following a simple rule of **addition**. Numbers in one row are equal to the sum of two numbers in the row directly above it. In the fifth row the second term, 4 is equal to the sum of the two numbers above it, namely 3 + 1. Each row represents the terms for the expansion of the binomial on the left. For example, the terms for $(A+B)^3$ are $A^3 + 3A^2B + 3AB^2 + B^3$. Obviously, the coefficient for the terms A^3 and B^3 is 1. Pascal's triangle works more efficiently than the algebraic approach, however, it also becomes tedious to create this triangle for binomials with a large n value.

The binomial theorem provides an easier and more efficient method for expanding binomials which have large n values. Using this theorem the coefficients for each term are found with the combination formula. The combination formula is

$$_nC_r = \frac{n!}{r!(n-r)!}$$

The notation n! is read "n factorial" and means multiplying n by every positive whole integer which is smaller than it. So, 4! would be equal to $4 \times 3 \times 2 \times 1 = 24$. Applying the combination formula to a binomial expansion $(A + B)^n$, n represents the power to which the formula is expanded, and r represents the power of B in each term. For example, for the term A^4B^3 in the expansion of $(A + B)^7$, n is equal to 7 and r is equal to 3. By substituting these values into the combination formula we get $7! / (3! \times 4!) = 35$, which is the coefficient for this term. The complete binomial theorem can be stated as the following:

$$(A + B)^n = \Sigma_nC_r A^{n-r} B^r$$

See also Factorial.

Resources

Books

Dunham, William. *Journey Through Genius.* New York: John Wiley & Sons, 1990.

Eves, Howard Whitley. *Foundations and Fundamental Concepts of Mathematics.* New York: Dover, 1997.
Larson, Ron. *Precalculus.* 5th ed. New York: Houghton Mifflin College, 2000.

Perry Romanowski

Bioaccumulation

Bioaccumulation is the gradual build up over **time** of a chemical in a living **organism**. This occurs either because the chemical is taken up faster than it can be used, or because the chemical cannot be broken down for use by the organism (that is, the chemical cannot be metabolized).

Bioaccumulation need not be a concern if the accumulated compound is not harmful. Compounds that are harmful to health, such as mercury, however, can accumulate in living tissues.

Chemical pollutants that are bioaccumulated come from many sources. **Pesticides** are an example of a contaminant that bioaccumulates in organisms. Rain can wash freshly sprayed pesticides into creeks, where they will eventually make their way to **rivers**, estuaries, and the **ocean**. Anther major source of toxic contaminants is the presence of compounds from industrial smokestacks and **automobile** emissions that return to the ground in rainfall. Deliberate discharge of compounds into **water** is another source of chemical pollutants.

Once a toxic pollutant is in the water or **soil**, it can easily enter the food chain. For example, in the water, pollutants adsorb or stick to small particles, including a tiny living organism called **phytoplankton**. Because there is so little pollutant stuck to each phytoplankton, the pollutant does not cause much damage at this level of the food web. However, a small **animal** such as a **zooplankton** might then consume the particle. One zooplankton that has eaten ten phytoplanktons would have ten times the pollutant level as the phytoplankton. As the zooplankton may be slow to metabolize or excrete the pollutant, the pollutant may build up or bioaccumulate within the organism. A small **fish** might then eat ten zooplankton. The fish would have 100 times the level of toxic pollutant as the phytoplankton. This multiplication would continue throughout the food web until high levels of contaminants have biomagnified in the top **predator**. While the amount of pollutant might have been small enough not to cause any damage in the lowest levels of the food web, the biomagnified amount might cause serious damage to organisms higher in the food web. This phenomenon is known as **biomagnification**.

Mercury **contamination** is a good example of the bioaccumulation process. Typically, mercury (or a chemical version called methylmercury) is taken up by **bacteria** and phytoplankton. Small fish eat the bacteria and phytoplankton and accumulate the mercury. The small fish are in turn eaten by larger fish, which can become food for humans and animals. The result can be the build up (biomagnification) of large concentrations of mercury in human and animal **tissue**.

One of the classic examples of bioaccumulation that resulted in biomagnification occurred with an insecticide called dichlorodiphenyltrichloroethane (DDT). DDT is an insecticide that was sprayed in the United States prior to 1972 to help control **mosquitoes** and other **insects**. Rain washed the DDT into creeks, where it eventually found its way into lakes and the ocean. The toxic pollutant bioaccumulated within each organism and then biomagnified through the food web to very high levels in predatory **birds** such as bald **eagles**, osprey, peregrine **falcons** and brown **pelicans** that ate the fish. Levels of DDT were high enough that the birds' eggshells became abnormally thin. As a result, the adult birds broke the shells of their unhatched offspring and the baby birds died. The population of these birds plummeted. DDT was finally banned in the United States in 1972, and since that time there have been dramatic increases in the populations of many predatory birds.

The bioaccumulation and biomagnification of toxic contaminants also can put human health at risk. When humans eat organisms that are relatively high in the food web, we can get high doses of some harmful chemicals. For example, marine fish such as **swordfish**, shark, and **tuna** often have bioaccumulated levels of mercury, and bluefish and striped **bass** sometimes have high concentrations of **polychlorinated biphenyls (PCBs)**. The federal government and some states have issued advisories against eating too much of certain types of fish because of bioaccumulated and biomagnified levels of toxic pollutants.

Advances are being made in efforts to lessen the bioaccumulation of toxic compounds. Legislation banning the disposal of certain compounds in water helps to reduce the level of toxic compounds in the environment that are capable of being accumulated in the food chain. As well, **microorganisms** are being genetically engineered so as to be capable of using a toxic material such as mercury as a food source. Such bacteria can directly remove the compound from the environment.

See also Bivalves; Ecosystem.

Resources

Books

Beek, B.O. *Bioaccumulation New Aspects and Developments.* New York: Springer Verlag, 1999.

Neff, J.M. *Bioaccumulation in Marine Organisms: Effect of Contaminants from Oil Well Produced Water.* Amsterdam: Elsevier Science Publishers, 2002.

Periodicals

Bae, W., R.K. Mehra, A. Mulchandani, et al., "Genetic Engineering of *Escherichia coli* for Enhanced Uptake and Bioaccumulation of Mercury." *Applied and Environmental Microbiology* 67 (November 2001): 5335–5338.

Brian Hoyle

Bioassay

A bioassay is the use of a living **organism** to test for the presence of a compound or to determine the amount of the compound that is present in a **sample**. The organism used is sensitive to the compound for which the test is conducted. Thus, the effect observed is typically the death or deteriorated health of the test organism. Depending on the test organism, **soil**, air, or liquid samples can be assayed.

The classic historical example of a bioassay was the use of canaries by miners in past centuries. Because canaries are more sensitive than humans to noxious gases like methane, they reacted quickly to even small amounts of the gas. This would give the miners time to escape.

Today's bioassays are more sophisticated than the canary. The ASTM (formerly known as the American Society for the Testing of Materials) has catalogued over 70 different bioassays. These are used to analyze soil, **freshwater**, and the sediment at the bottom of watercourses like streams and **rivers**, **saltwater**, and air.

Plants can be used as indicators of the presence of toxic compounds in the soil. In this bioassay, **seeds** or the mature **plant** is introduced into the soil of a site that is suspected of being contaminated. Failure of the seeds to germinate, or failure of the mature plant to thrive, can be evidence of **contamination**. If the assay is done in a controlled manner with the use of standards to provide reference points, then the geographical area of contamination can be determined.

Some **species** of plants can also be used to accomplish bioassays in the **water**. More commonly, however, the test organisms are single-celled organisms such as **algae**, water **fleas** (in particular a species called *Daphnia magna*, or **fish** (in particular the fathead minnow).

Bacteria can be used in bioassays. For example, the use of bacteria to detect and determine the amount of **antibiotics** or compounds that might be carcinogens in a sample has been practiced for decades. Another particularly useful bacterial bioassay involves the use of bacteria that have been designed to fluoresce (to emit **light**). If the bacteria are harmed by a toxic compound in the test sample, then they fail to fluoresce. The decrease in **fluorescence** of bacterial populations is measured in a device called a spectrophotometer. The degree of decrease can be compared to standards in order to determine the **concentration** of the toxic compound.

With the wide variety of bioassays available, the investigator must choose the assay with care. The test organism used needs to be appropriate for the task, and must provide a result that is readily apparent. Being able to determine the level of the toxic agent (in other words being able to quantify the agent) can be very useful.

Bioassays continue to be developed. For example, in the mid-1990s, a bioassay was introduced to detect the presence of some human **hormones** in the environment.

See also Ames test; Bioluminescence; Ecological monitoring; Poisons and toxins.

Resources

Books

Haynes, K., and J. Millar. *Methods in Chemical Ecology, Volume II: Bioassay Methods.* New York: Kluwer Academic Publishers, 1998.

Periodicals

Pauwels, A., et al. "Comparison of Chemical-Activated Luciferase Gene Expression Bioassay and Gas Chromatography for PCB Determination in Human Serum and Follicular Fluid." *Environmental Health Perspective* 108 (June 2000): 553–557.

Traunspurger, W., et al. "Ecotoxicological Assessment of Aquatic Sediments with *Caenorhabditis elegans* (nematoda)—A Method for Testing in Liquid Medium and Whole Sediment Samples." *Environmental Toxicology and Chemistry* 16 (1997): 245–250.

Brian Hoyle

Biochemical oxygen demand

Oxygen helps liberate biochemical **energy** from food by acting as the **electron** acceptor for the reaction that metabolizes **adenosine triphosphate**, ATP, one of the body's major chemical energy sources. Metabolic processes that require oxygen are called **aerobic**. Naturally occurring oxygen is in the form of molecular oxygen, O_2. Atmospheric oxygen is obtained by the body in the lungs in tiny air sacs called alveoli. Within the alveoli, red **blood** cells (rbc's) in narrow blood vessels absorb oxygen and carry it to cells throughout the body. **Respiration**, inhaling and exhaling of air, is subconsciously controlled by the **brain** in response to fluctuations in **carbon dioxide** levels. The average human body of 139 lb (63 kg) consumes 250 ml of O_2 each minute. The major single-organ oxygen consumers are the liver, brain, and **heart** (consuming 20.4%, 18.4%, and 11.6%, respectively), while the sum total of all the body's skeletal muscles consume about 20%. In addition, the kidneys use up about 7.2%, and the skin uses 4.8% The rest of the body consumes the remaining 17.6% of the oxygen. Oxygen use can also be measured per 100 gm of an **organ** to indicate concentrations of use; as such, heart usage is highest, followed by the kidneys, then the brain, and then the liver. During **exercise**, the biochemical oxygen demand increases for active tissues including the heart and skeletal muscles.

Oxygen is the **molecule** used by animals as a final electron acceptor for **metabolism**. Two electrons (one at a time) from metabolic products can chemically bind each oxygen molecule. While numerous molecules combine with oxygen in the human body, one of the major **chemical reactions** involving oxygen is the synthesis of the high-energy phosphate bonds in ATP. ATP is the cell's currency for generating muscle contractions and driving certain ions through membrane-bound ion channels. Oxygen facilitates aerobic ATP production in mitochondria of cells throughout the body. Aerobic production of 36 molecules of ATP from one glucose molecules occurs in the **citric acid** metabolic cycle. About 1 L of oxygen can release the chemical energy stored in 1 g of food.

Oxygen is carried through the body in a number of chemical forms including simple O, **water** (H_2O), **carbon** dioxide (CO_2), and oxyhemoglobin. Unbound oxygen radicals can be highly toxic to cells. Allowing

random oxidation reactions to occur throughout the **cell**, these radicals can be very destructive, and cellular defenses have evolved to combat them. In fact, the oxygen radical H_2O_2 is highly toxic to cells and can be used as a bactericidal agent. H_2O and CO_2 are end products for several aerobic reactions. And oxyhemoglobin is the oxygen shuttle complex that carries oxygen to needy cells. One rbc contains around 350 million hemoglobin molecules. Hence, one rbc can carry about 1.5 billion oxygen molecules.

Hemoglobin is a large globular protein made up of four polypeptide chains (two alpha and two beta hemoglobins, in adults) that each contain one heme complex. Heme complexes are sophisticated ring structures that contain a central ferrous **iron** atom. The iron **atoms** can each bind one O_2 molecule. Hence, one hemoglobin molecule can bind four O_2 molecules. This is conventionally represented as Hb_4O_8. The four Hb components can alter their orientation to favor uptake or release of the oxygen. When the Hb bonds are relaxed, they favor uptake, and when they are tense they favor release. Affinity is the chemical term used to indicate how eager multiple units are to interact with one another. The chemical affinity for the first oxygen to bind is lower than the affinity for the later oxygens to bind. In other words, once one oxygen has bound to the hemoglobin, the binding of the other three oxygen molecules is more favorable. In addition, the amount of oxygen bound or released depends on the **concentration** of oxygen in two locations (where it is coming from and where it is being absorbed).

Oxygen flow is greatly determined by local partial **pressure** gradient. Just like it is more difficult to push water up a waterfall, so it is difficult to absorb oxygen into an area that already has more oxygen than the place it is coming from. In both cases, a certain amount of pressure is causing something to flow one way. As rbcs travel through **arteries** and **veins** around the human body, they collect oxygen in the alveoli where the partial pressure of oxygen is higher than it is in the rbcs. Usual average **atmospheric pressure** is measured as 1 atmosphere (1 atm) or 14.7 pounds per square inch (14.7 psi). Since oxygen makes up about 21% of atmospheric air, the partial pressure of oxygen is 0.21 atm or 3.09 psi. Because venous oxygen partial pressure is less than 3.09 psi, oxygen is driven into the blood in the lungs. Although a small amount of O_2 gas dissolves into the **plasma** (the fluid surrounding the blood cells), most is bound by hemoglobin. The reverse process occurs as **capillaries** supply tissues with oxygen. The partial pressure of oxygen in the **tissue** is lower than in the blood, so oxygen flows into the tissue. CO_2 travels a reverse course where high tissue partial pressures push CO_2 out into the veins that carry it to the lungs for release into the atmosphere. The CO_2 partial

---

KEY TERMS

Hemoglobin—An iron-containing, protein complex carried in red blood cells that binds oxygen for transport to other areas of the body.

Hypoxia—State of deficient oxygen supply.

---

pressure of the atmosphere is significantly lower than that of body tissues. The relationship of gaseous absorption to atmospheric pressure makes it crucial for mountain climbers and scuba divers to calculate their expected partial pressure gaseous exposure before climbing or diving. Miscalculations could lead to death.

Body tissues vary in their oxygen dependency. Hypoxia is the condition of existing with a lowered oxygen supply. The brain and heart are the two most hypoxia sensitive organs. A severe drop in available oxygen can cause brain death in five minutes. Less severe hypoxia can lead to other mental problems such as dizziness, headache, disorientation, drowsiness, or impaired judgment. Although basic brain functions may recover fully from a short hypoxic period, higher neural functions can be severely impaired.

During rigorous exercise, oxygen demand may increase to up to 15 times the normal demand. As muscles deplete their oxygen supplies, the lowered muscular oxygen partial pressure steepens the pressure gradient, so that even more oxygen leaves the blood and enters the muscles. If aerobic metabolism is unable to supply enough ATP to muscle cells, then **anaerobic** metabolism can provide some ATP. However, anaerobic metabolism temporarily adds **lactic acid** to the muscle creating an oxygen debt whereby the muscle fatigues and requires a recovery period to get rid of the lactic acid. An initial oxygen debt is also always present for about the first 30 seconds of exercise until circulation can accelerate to provide additional oxygen.

Various life forms are classified on the basis of their tolerance or requirement of oxygen. Different types of **bacteria** are aerobic, facultatively aerobic, or anaerobic. Aerobes use oxygen to generate energy. Facultative aerobes can use oxygen but survive without it. Oxygen is highly toxic to anaerobes which die rapidly when exposed to it.

See also Cellular respiration; Circulatory system.

Resources

Books

Guyton & Hall. *Textbook of Medical Physiology.* 10th ed. New York: W. B. Saunders Company, 2000.

Rhoads R., and R. Pflanzer, eds. *Physiology*. 2nd ed. New York: Saunders College Publishing, 1992.

Louise Dickerson

Biochemistry

Biochemistry is the study of the molecular basis of life. The study of biochemistry includes the knowledge of the structure and function of molecules found in the biological world and an understanding of the precise biochemical pathways by which organic molecules are either put together or broken down.

Biochemistry seeks to describe the structure, organization, and functions of living **matter** in molecular terms. Essentially two factors have contributed to the excitement in the field today and have enhanced the impact of research and advances in biochemistry on other life sciences. Firstly, it is now generally accepted that the physical elements of living matter obey the same fundamental laws that govern all matter, both living and non-living. Therefore, the full potential of modern chemical and physical theory can be brought in to solve certain biological problems. Secondly, incredibly powerful new research techniques, notably those developing from the fields of **biophysics** and **molecular biology**, are permitting scientists to ask questions about the basic process of life.

Biochemistry draws on its major themes from many disciplines. For example from organic **chemistry**, which describes the properties of biomolecules; from biophysics, which applies the techniques of **physics** to study the structures of biomolecules; from medical research, which increasingly seeks to understand **disease** states in molecular terms and also from **nutrition**, microbiology, **physiology**, **cell biology,** and **genetics**.

Origins of and development of biochemistry

Biochemistry draws strength from all of these disciplines but is also a distinct discipline, with its own identity. It is distinctive in its emphasis on the structures and relations of biomolecules, particularly enzymes and biological catalysis; on the elucidation of metabolic pathways and their control; and on the principle that life processes can, at least on the physical level, be understood through the laws of chemistry. It has its origins as a distinct field of study in the early nineteenth century, with the pioneering work of Friedrich Wöhler. Prior to Wöhler's time it was believed that the substance of living matter was somehow quantitatively different from

that of nonliving matter and did not behave according to the known laws of physics and chemistry. In 1828 Wöhler showed that **urea**, a substance of biological origin excreted by humans and many animals as a product of **nitrogen** metabolism, could be synthesized in the laboratory from the inorganic compound ammonium cyanate. As Wöhler phrased it in a letter to a colleague, "I must tell you that I can prepare urea without requiring a kidney or an **animal**, either man or dog." This was a shocking statement at the time, for it breached the presumed barrier between the living and the nonliving. Later, in 1897, two German brothers, Eduard and Hans Buchner, found that extracts from broken and thoroughly dead cells from **yeast**, could nevertheless carry out the entire process of **fermentation** of sugar into **ethanol**. This discovery opened the door to analysis of biochemical reactions and processes in vitro (Latin "in glass"), meaning in the test tube rather than in vivo, in living matter. In succeeding decades many other metabolic reactions and reaction pathways were reproduced in vitro, allowing identification of reactants and products and of enzymes, or biological catalysts, that promoted each biochemical reaction.

Until 1926, the structures of enzymes (or "ferments") were thought to be far too complex to be described in chemical terms. But in 1926 J.B. Sumner showed that the protein urease, an **enzyme** from jack beans, could be crystallized like other organic compounds. Although **proteins** have large and complex structures, they are also organic compounds and their physical structures can be determined by chemical methods.

Modern biochemistry

Today, the study of biochemistry can be broadly divided into three principal areas: (1) the structural chemistry of the components of living matter and the relationships of biological function to chemical structure; (2) metabolism, the totality of **chemical reactions** that occur in living matter; and (3) the chemistry of processes and substances that store and transmit biological information. The third area is also the province of molecular genetics, a field that seeks to understand heredity and the expression of genetic information in molecular terms.

Biochemistry is having a profound influence in the field of medicine. The molecular mechanisms of many diseases, such as **sickle cell anemia** and numerous errors of **metabolism**, have been elucidated. Assays of enzyme activity are today indispensable in clinical **diagnosis**. To cite just one example, liver disease is now routinely diagnosed and monitered by measurements of **blood** levels of enzymes called transaminases and of a hemoglobin breakdown product called bilirubin. **Deoxyribonucleic** acid (DNA) probes are coming into play in diagnosis of **genetic disorders**, infectious diseases, and cancers. Genetically engineered strains of **bacteria** containing **recombinant DNA** are producing valuable proteins such as **insulin** and growth hormone. Furthermore, biochemistry is a basis for the rational design of new drugs. Also the rapid development of powerful biochemical concepts and techniques in recent years has enabled investigators to tackle some of the most challenging and fundamental problems in medicine and physiology. For example in **embryology**, the mechanisms by which the fertilized embryo gives rise to cells as different as muscle, the **brain** and liver are being intensively investigated. Also, in **anatomy**, the question of how cells find each other in order to form a complex **organ**, such as the liver or brain, are being tackled in biochemical terms. The impact of biochemistry is being felt in so many areas of human life, through this kind of research and the discoveries are fuelling the growth of the life sciences as a whole.

The biochemistry of digestion, for example, includes the study of the pathways involving changes in molecular structure, and all enzyme interactions that take place when large food molecules (proteins, lipids, or carbohydrates) are broken down into smaller molecules capable of uptake and use by the cells of a living body.

Fundamental advances in biochemistry have also allowed important advances in genetics and molecular biology. The discovery of the composition—and subsequently the double-helical structure—of DNA provided a critical and highly demonstrable link between biochemical form and function. The structure of the DNA **molecule** is remarkably well suited to its ability to make more copies of itself. The **hydrogen** bonds between the base pairs of the DNA **double helix** enable the two-stranded double helix molecule to unzip easily so that each piece can act as a pattern to build new DNA molecules.

Another area of interest to biochemists involves the flow of genetic information from DNA to RNA to protein is the same in all organisms. ATP (**adenosine triphosphate**), which is the universal currency of **energy** in biological systems, is generated in similar ways by all forms of life. Furthermore, biochemists were able to unravel some of the central metabolic pathways and energy-conversion mechanisms. The determination of the three-dimensional structure and the mechanism of action of many protein molecules were also significant achievements in the field of biochemistry.

Identifying and **sequencing** genes responsible for causing diseases, and genetic cloning technology has led to remarkable progress in understanding the relationships of genes and proteins. In addition, the molecular bases of several diseases, such as sickle cell **anemia**, and numerous inborn errors of metabolism are now known. Biochemical assays for enzyme activities have become indispensable in clinical diagnosis.

Indeed, the field of biochemistry, which investigates the relationship between molecular structure and function of living things at a molecular level, has been profoundly transformed by recombinant **DNA technology**. This has led to the integration of molecular genetics and protein biochemistry. The intricate interplay of the genetic make-up (genotype) and how the molecular structure influences function and the various physical traits (phenotype) is now being unraveled at the molecular level.

See also Amino acid; Hormones; Photosynthesis.

Resources

Books

Boyer, Rodney. *Concepts in Biochemistry.* Pacific Grove, CA: Brooks/Cole Publishing Company, 1999.

Jorde, L.B., J.C. Carey, M.J. Bamshad, and R.L. White. *Medical Genetics,* 2nd ed. St. Louis: Mosby-Year Book, Inc., 2000.

Lodish, H., et. al. *Molecular Cell Biology,* 4th ed. New York: W. H. Freeman & Co., 2000.

Nelson, David L., David L. Nelson, and Michael M. Cox. *Lehninger Principles of Biochemistry,* 3rd edition. Worth Publishing, 2000.

Judyth Sassoon

Biodegradable substances

The term *biodegradable* is used to describe materials that decompose through the actions of **bacteria**, **fungi**, and other living organisms. **Temperature** and sunlight may also play roles in the **decomposition** of biodegradable **plastics** and other substances. If such materials are not biodegradable, they remain in the environment for a long time, and, if these same substances are toxic, they may pollute the **soil** and **water**. Some nonbiodegradable pollutants may be capable of causing harm to organisms in the environment.

Common, everyday substances that are biodegradable include food refuse, **tree** leaves, and grass clippings. Many communities now encourage people to compost these materials and use them as **humus** (an organic-rich material in soil) for gardening. Because **plant** materials are biodegradable, **composting** is one way to reduce amounts of solid waste that towns and cities otherwise have to dispose in landfills.

In many cases, scientists can come up with biodegradable alternatives to nonbiodegradable products.

For example, when household detergents were developed and came into wide use, foam began to clog streams and **sewage treatment** plants. The foam was caused by the presence of a complex phosphate, **sodium** tripolyphosphate, an ingredient in the detergent that reacted with, and removed dirt from, the surfaces of clothes. These complex phosphates, collectively called surfactants for their actions on material surfaces, were not biodegradable, and appeared to be harming plants and **fish** in streams. Detergent manufacturers responded to the problem by replacing phosphates with enzymes like protease and amylase, which are biodegradable.

Nonbiodegradable plastics are a particular problem, because they take up so much room in landfills or require special handling at waste incinerators. Most plastics are petroleum-based, meaning they are made from oil and other **petroleum** products. Until recently, plastics have been nonbiodegradable. Today, however, various techniques for producing biodegradable plastics are being explored, developed, and marketed. In some cases, organic compounds like sugar, corn starch, silk, and bamboo are being incorporated into the plastic production process. This allows large pieces of plastic to break down into smaller units, but on a molecular level, many of these plastics remain nonbiodegradable. Other researchers have come up with non-petroleum based plastics, using bioengineered organisms, such as bacteria, to produce plastic. In some cases, enzymes produced by the same **organism** can be used to break down the biologically produced plastic. Currently, these plastics are expensive to produce, but as the technology becomes more readily available, they are likely to become much more common.

Governments and industries have taken various measures to replace nonbiodegradable materials with those that will degrade or decompose. For example, the plastic rings that bind six-packs of soda and beer are required by law to be biodegradable in Oregon and Alaska. Italy has banned all nonbiodegradable plastics. The packaging industry continues to experiment with biodegradable packaging for food and fast food. Several coalitions have been formed to address biodegradable products in the oil and plastics industries, and to evaluate the benefits of **recycling** stable but nonbiodegradable materials versus developing biodegradable substances that may be costly for both industry and the consumer. The Council for Solid Waste Solutions and the Council on Plastics and Packaging in the Environment are action groups led by industry. Environmental groups like Keep America Beautiful also advocate recycling out of concern that biodegradability tells consumers littering is acceptable, but really, toxic chemicals that may leach out of biodegradable substances can poison **groundwater**. Interestingly, grain growers and processors strongly favor biodegradable plastics be-

cause in some cases, corn starch is used to replace some of the plastic resin during manufacture.

Successful moves toward biodegradable substances have been made in some markets. Europeans have used degradable plastic shopping bags as mulch to cover new **crops** in the spring since 1975. Lawn bags that degrade would benefit the composting business because nonbiodegradable bags have to be removed before yard waste can be composted. In landfills, where bagged yard waste occupies approximately 20% of the space, decomposing waste and degradable bags produce methane gas that can be recovered and sold for power generation. Marine and coastal environments can benefit from the use of biodegradable plastics in the fishing and boating industries; public outrage over the killing of dolphins, game fish, whales, and sea **turtles** fuels interest in these industries. In fact, the public is ultimately the driving force behind the development of biodegradable substances because litter on beaches, roadsides, and parks is an eyesore with apparent potential to harm the environment.

See also Aerobic; Hazardous wastes; Landfill; Waste management.

Biodiversity

Biodiversity is the total richness of biological variation. The scope of biodiversity is usually considered to range from the genetic variation of **individual** organisms within and among populations of a **species**, to different species occurring together in ecological communities. Some definitions of biodiversity also include the spatial patterns and temporal dynamics of populations and communities on the landscape. The geographical scales at which biodiversity can be considered range from local to regional, state or provincial, national, continental, and ultimately to global.

Biodiversity at all scales is severely threatened by human activities; this is one of the most important aspects of the global environmental crisis. Humans have already caused permanent losses of biodiversity through the **extinction** of many species and extensive losses of distinctive, natural ecosystems. Ecologists predict that unless there are substantial changes in the ways that humans affect ecosystems, there will be much larger losses of biodiversity in the near future.

Species richness of the biosphere

About 1.7 million of Earth's species have been identified and designated with a scientific name. About 6% of the identified species live in boreal or polar latitudes,

59% in the temperate zones, and the remaining 35% in the tropics. However, the knowledge of Earth's species is highly incomplete, especially for tropical countries. According to some estimates, there could be as many as 30-50 million species on **Earth**, 90% of them occurring in tropical ecosystems. Tropical ecosystems are much richer in species than are those at higher latitudes.

Most of the described species on Earth are **invertebrates**, particularly **insects**, and most of the insects are **beetles** (order Coleoptera). A famous scientist, J.B.S. Haldane, was once asked by a theologian to briefly explain what his knowledge of **biology** told of God's purpose. Haldane reputedly said that God has "an inordinate fondness of beetles," reflecting the fact these insects are so much richer in species than any other group of creatures on Earth. Some biologists believe that beetles account for most of the undescribed tropical insects.

The suggestion of enormous numbers of undescribed insects in tropical **forests** initially emerged from the research of Terry Erwin. This entomologist performed experiments in which tropical-forest canopies in Amazonia were treated with an insecticide, and the subsequent "rain" of dead **arthropods** was collected using ground-level sampling devices. This innovative sampling procedure indicated that: (1) a large fraction of the insect species of tropical forests is unknown to science; (2) most insect species are confined to a single type of tropical forest, or even to a particular **tree** species, which may itself have only a local distribution; and (3) most species of tropical-forest insects have little ability to disperse very far. Erwin's studies of tropical **rainforest** found that beetles accounted for most of the insect species and that most of the beetles are narrowly **endemic**, that is, they have a local distribution and are found nowhere else. For example, the tree *Luehea seemanii* has more than 1,100 species of beetle in its canopy, of which 15% are specific to that **plant**. The emerging conclusion from this and other descriptive research is that there is an enormous abundance of undescribed species of insects and other invertebrates living in tropical forests.

Compared with invertebrates, the numbers of species (that is, the species richness) of other groups of tropical-forest organisms are better known. Although it is extremely difficult to do so, the numbers of species of vascular plants have been described for a few tropical forests. For example: a plot of only 0.0004 sq mi (0.1 ha) in a moist forest in Ecuador had 365 species of vascular plants; there were 98 species of large trees in 0.006 sq mi (1.5 ha) of forest in Sarawak, Malaysia; there were 90 tree species in 0.0032 sq mi (0.8 ha) of forest in Papua New Guinea; 742 woody species occurred in 0.012 sq mi (3 ha) of forest in Sarawak, with 50% of the species recorded as single individuals; and more than 300 species of woody plants were

discovered on a 2-sq mi (50-ha) area of forest in Panama. These tropical forests are much richer than temperate forests, stands of which typically support fewer than 12-15 species of trees. The Great Smokey **Mountains** of the eastern United States have some of the richest temperate forests in the world, and they typically contain 30-35 species, far fewer than tropical rainforests.

It is extraordinarily difficult to determine the numbers of **birds** in tropical forest because the dense foliage and darkness of the understory make it inconvenient to see small animals, even if they are brightly colored. As a result, few studies have been made of the birds of tropical rainforest. However, one study in Peru discovered 245 resident and 74 transient species in 0.4 sq mi (97 ha) of Amazonian forest. Another study of rainforest in French Guiana recorded 239 species of birds, and another found 151 species in a forest in Sumatra. In comparison, stands of temperate forest in **North America** typically support only 15-30 species of birds.

Almost no systematic surveys have been made of all of the species of tropical ecosystems. In one case, a 42 sq mi (108 sq km) reserve of dry forest in Costa Rica was estimated to support about 700 plant species, 400 vertebrate species, and 13,000 species of insects, including 3,140 species of **moths** and **butterflies**.

Why is biodiversity important?

Biodiversity is valuable for the following classes of reasons:

(1) Intrinsic Value. Biodiversity has its own intrinsic value, regardless of its worth in terms of human needs. Because of its intrinsic merit, there are ethical considerations to any degradation of biodiversity. For example, do humans have the "right" to diminish or exterminate elements of biodiversity, all of which are unique and irretrievable? Is the human existence itself diminished by losses of biodiversity? Ethical issues cannot be resolved through science, but enlightened people would mourn any loss of species, or of natural ecosystems.

(2) Utilitarian Value. Humans have an absolute requirement for the products of other species. Because of this need, wild and domesticated species and their communities are exploited in many ways to provide food, materials, **energy**, and other goods and services. This fact can be illustrated in many ways. In the United States, for example, about one-quarter of prescription drugs have active ingredients obtained from higher plants, and these uses contribute about $14 billion per year to the U.S. economy, and $40 billion per year worldwide. Potentially, harvests of biodiversity can be conducted in ways that foster their renewal. Unfortunately, potentially renewable biodiversity resources are often harvested too intensively

or inadequate attention is paid to regeneration, so the resource is degraded or becomes extinct.

(3) Provision of Ecological Services. Biodiversity provides many ecological services that are directly or indirectly important to human welfare. Examples of these services include biological productivity, nutrient cycling, cleaning **water** and air of pollutants, control of **erosion**, provision of atmospheric **oxygen**, removal of **carbon dioxide**, and other functions related to the integrity of ecosystems. According to the biologist Peter Raven: "Biodiversity keeps the **planet** habitable and ecosystems functional."

There are many cases of the discovery, through research on previously unexploited plants and animals, of bio-products useful to humans as food, medicine, or for other purposes. Consider the case of the rosy periwinkle (*Catharantus roseus*), a small plant native to the tropical **island** of Madagascar. During an extensive screening of wild plants for anti-cancer chemicals, an extract of rosy periwinkle was observed to inhibit the growth of cancerous cells. The active biochemicals are several alkaloids in foliage of the plant, which probably serve to deter herbivores. These natural substances are now used to prepare the drugs vincristine and vinblastine, which can be successfully used to treat childhood **leukemia** and a lymphatic **cancer** known as **Hodgkin's disease**. In this case, a species of wild plant known only to a few botanists has proven to be of great benefit to humans by treating previously incurable diseases, in the process sustaining a large pharmaceutical economy. There is a tremendous undiscovered wealth of other biological products useful to humans in unexplored biodiversity.

Biodiversity and extinction

Extinction refers to the loss of some species or other taxonomic unit (e.g., subspecies, genus, family, etc.; each is known as a taxon), occurring over all of its range on Earth. (Extirpation refers to a more local disappearance, with the taxon still surviving elsewhere.) The extinction of any species is an irrevocable loss of part of the biological richness of Earth, the only place in the universe known to support living creatures. Extinction can be a natural occurrence caused by an unpredictable catastrophe, chronic environmental stress, or ecological interactions such as **competition**, **disease**, or predation. However, there have been dramatic increases in extinction rates since humans have become Earth's dominant large **animal** and the cause of global environmental change.

Extinction has always occurred naturally. Almost all species that have ever lived on Earth have become extinct. Perhaps they could not cope with changes occurring in their environment, such as climate changes, the intensity of predation, or disease. Alternatively, many extinctions may have occurred simultaneously as a result of unpredictable catastrophes. From the geological record it is known that species, families, and even phyla have appeared and disappeared over **time**. For example, numerous phyla of invertebrates proliferated during an evolutionary radiation occurring at the beginning of the Cambrian era about 570 million years ago, but most of these animals are now extinct. The 15-20 extinct phyla from that period are known from the Burgess Shale of British Columbia, and they represent unique experiments in invertebrate form and function. Similarly, entire divisions of plants have appeared, radiated, and disappeared, such as the **seed ferns** Pteridospermales, the cycad-like Cycadeoidea, and woody plants known as Cordaites. Of the 12 orders within the class Reptilia, only three survive today: crocodilians, **turtles**, and snakes/lizards. Clearly, the fossil record displays a great deal of evidence of natural extinctions.

Overall, the geological record suggests that there have been long periods of time characterized by uniform rates of extinction, but punctuated by about nine catastrophic episodes of **mass extinction**. The most intense extinction event occurred at the end of the Permian period some 245 million years ago, when 54% of marine families, 84% of genera, and 96% of species are estimated to have become extinct.

Another famous, apparently synchronous, extinction of vertebrate animals occurred about 65 million years ago at the end of the Cretaceous period. The most renowned extinctions were of the last of the reptilian dinosaurs and pterosaurs, but many plants and invertebrates also became extinct at that time. In total, perhaps 76% of species and 47% of genera became extinct in the end-of-Cretaceous crisis. One hypothesis to explain the cause of this mass extinction involves a meteorite impacting the Earth, causing huge quantities of fine dust to be spewed into the atmosphere, and resulting in a climatic deterioration that most large animals could not tolerate. However, some scientists believe that the extinctions of the last dinosaurs were more gradual.

However, humans have been responsible for almost all of Earth's recent extinctions. These extinctions are occurring so quickly that they represent a modern mass extinction of similar intensity to those documented in the geological record. Examples of recent extinctions caused by humans include such well-known cases as the dodo, passenger pigeon, and great auk. Many other high-profile species have been taken to the brink of extinction, including the plains **bison**, whooping crane, ivory-billed woodpecker, right whale, and other marine **mammals**. These losses have been caused by insatiable overhunting and intense disturbance or conversion of natural habitats.

Beyond these well-known and tragic cases involving large animals, Earth's biodiversity is experiencing an even larger loss. This ruin is mostly being caused by extensive conversions of tropical ecosystems, particularly rainforest, into agricultural habitats that sustain few of the original species. As was described previously, tropical ecosystems sustain extremely large numbers of species, most of which have restricted distributions. The conversion of tropical forest into other habitats inevitably causes the loss of most of the locally endemic biota. This is a great tragedy, and the lost species will never occur again.

The most important human influences causing the extinction or endangerment of species are: (1) excessive exploitation, (2) effects of introduced predators, competitors, and diseases, and (3) **habitat** disturbance and conversion. These stressors can result in small and fragmented populations that experience the deleterious effects of inbreeding and population instability and then decline further, ultimately to extirpation or extinction.

The increased **rate** of extinction and endangerment of biodiversity during the past several centuries is best documented for **vertebrates** because, as noted previously, most invertebrate species, particularly insects, have not yet been described by scientists. During the last four centuries there have been more than 700 known extinctions globally, including about 100 species of mammals and 160 species of birds, all because of human influences.

A much larger number of species is facing imminent extinction; they are endangered. For example, more than 1,000 species of birds are considered to be threatened with extinction. Of this total, 46% live only on oceanic islands, a situation in which species are especially vulnerable to extinction caused by stresses associated with human activity. Birds of tropical forests account for 43% of the threatened bird species, wetland species for 21%, grassland and savanna species 19%, and other habitats 17%. Only 1.5% of the threatened species are North American, 4.2% are European or Russian, 33% Central and South American, 18% African, 30% Asian, and 14% from **Australia** and the Pacific.

Protection of endangered biodiversity

Biodiversity can be protected in ecological reserves. These are protected areas established for the **conservation** of natural values, usually the known habitat of **endangered species**, threatened ecosystems, or representative examples of widespread communities. In the early 1990s there were about 7,000 protected areas globally, with an area of 2.5 million sq mi (651 million ha.) Of this total, about 2,400 sites comprising 1.5 million sq mi (379 million ha) were fully protected and could be considered to be true ecological reserves.

Ideally, the design of a national system of ecological reserves would provide for the longer-term protection of all native species and their natural communities, including terrestrial, **freshwater**, and marine ecosystems. So far, however, no country has implemented a comprehensive system of ecological reserves to fully protect its natural biodiversity. Moreover, in many cases the existing reserves are relatively small and threatened by eenvironmental change and other stressors, such as illegal poaching of animals and plants and sometimes excessive tourism.

The World Conservation Union, World Resources Institute, and United Nations Environment Program are three important agencies whose mandates center on the conservation of the world's biodiversity. These agencies have developed the *Global Biodiversity Strategy,* an international program to help protect biodiversity. The broad objectives are to: (1) preserve biodiversity; (2) maintain Earth's ecological processes and life-support systems; and (3) ensure that biodiversity resources are used in a sustainable manner. As such, the *Global Biodiversity Strategy* is a mechanism by which countries and peoples can initiate meaningful actions to protect biodiversity for the benefit of present and future generations of people, and also for its intrinsic value. Because it only began in the late 1970s, it is too early to evaluate the success of this program. However, the existence of this comprehensive international effort is encouraging, as is the participation of most of Earth's countries, representing all stages of socioeconomic development.

Another important international effort is the *Convention on Biological Diversity*, negotiated under the auspices of the United Nations Environment Program and signed by many countries at a major conference at Rio De Janeiro in 1992. This international treaty requires signatory nations to take measures to systematically catalogue their indigenous biodiversity, and to take action to ensure that it is conserved.

Numerous other agencies are working to preserve biodiversity. In the United States, the World Wildlife Fund and Nature Conservancy are important organizations at the national level. There are numerous other national and local groups. The lead federal agency is the U.S. Fish and Wildlife Service, and all states have similar agencies.

Important progress is being made, and the progressive worldwide development of activities intended to identify, conserve, and preserve biodiversity will hopefully come to be regarded as an ecological "success story."

See also Biological community; Ecosystem.

Resources

Books

Becher, A. *Biodiversity: A Reference Handbook.* Abo-Clio Pub, 1998.

Wilson, E.O., ed. *BioDiversity*. Washington, DC: National Academy Press, 1988.

Periodicals

Caballero A., and M.A. Toro. "Interrelations Between Effective Population Size and Other Pedigree Tools for the Management of Conserved Populations." *Genet Res* 75(no. 3) (June 2000): 331–43.

Bill Freedman

Bioenergy

Bioenergy is **energy** derived using organic material, especially plant **matter**, as fuel. The material burned or processed to produce bioenergy (the "feedstock") is called **biomass**. Biomass has been an energy source for as long as humans have used **wood** fires to warm themselves and cook food. Wood is still the most commonly used biomass fuel. In some developing countries, crop and logging residues, and a dried mixture of straw and **animal** dung are also common biomass fuels.

Unlike most other sources of energy, biomass is a potentially renewable resource. By definition, bioenergy sources also include **coal**, **petroleum**, and **natural gas**, because these **fossil fuels** are derived from ancient **plant** biomass laid down in early geologic ages. However, these are non-renewable sources of energy and by convention are omitted when discussing bioenergy. In fact, the big advantage of bioenergy is that its use reduces the use of such non-renewable energy sources.

Primary ways of using bioenergy

There are three major ways in which the energy in plants is utilized: direct burning, conversion to gas, and conversion to **alcohol**.

Direct burning

Biomass materials can be burned directly as fuel, as when wood logs are put on a fire. A major environmental problem in many developing countries is that **forests** are disappearing as people use up nearby trees as fuel for cooking. Without a forest cover to hold **soil**, the land can erode and become unproductive, making it difficult to raise **crops** or regenerate another forest. **Deforestation** has contributed to severe environmental damage, including widespread famine, in large regions of **Africa** and elsewhere. The direct burning of biomass also releases many of the same pollutants into the air as does the burning of fossil fuels.

Less damaging than deforestation of natural forest is the use of biomass grown specifically for that purpose.

Dobson, A.P. *Conservation and Biodiversity*. W.H. Freeman and Co., 1998.

Freedman, B. *Environmental Ecology*. 2nd edition. San Diego: Academic Press, 1994.

Gaston, K.J., and J.I. Spicer. *Biodiversity: An Introduction*. Blackwell Science Inc., 1998.

Hamblin, W.K., and E.H. Christiansen. *Earth's Dynamic Systems*. 9th ed. Upper Saddle River: Prentice Hall, 2001.

Levin, Simon A., ed. *Encyclopedia of Biodiversity*. San Diego, CA: Academic Press, 2000.

Myers, Judith, and Dawn Bazely. *Ecology and Control of Introduced Plants*. Cambridge: Cambridge University Press, 2003.

Nebel, Bernard J., and Richard T. Wright. *Environmental Science: Toward a Sustainable Future*. 8th ed. Englewood Cliffs, NJ: Prentice Hall, 2002.

Schneiderman, Jill S. *The Earth Around Us: Maintaining a Livable Planet*. New York: W.H. Freeman & Co., 2000.

Wilson, E.O. *The Diversity of Life*. Cambridge: Harvard University Press, 1992.

As fossil-fuel resources become scarce, such bioenergy crops will undoubtedly be much more important sources of energy in the future. The organic material in municipal and industrial solid waste is also a biomass fuel, and can be combusted in "waste-to-energy" incinerators, which generate **electricity** for local use.

Conversion to gas

Biomass can also be converted into methane gas, also called biogas, which can be burned as a source of energy. When **bacteria** digest organic materials in the absence of **oxygen** (anaerobic digestion), the gas produced is about two-thirds methane, CH_4, which is the main component in natural gas. Methane is also produced in landfills as organic waste is digested anaerobically. The gas can be collected and piped out of landfills and used as a fuel. Biogas can also be obtained from the **anaerobic** digestion of sewage sludge.

In China, many farmers use small closed pits as anaerobic digesters. Agricultural waste and sewage is placed in the pit, and the biogas given off is used as a fuel for cooking. On a larger scale, some dairy farmers in the United States have begun to produce biogas from cow manure. The gas is used to run electrical generators. In addition, the **heat** given off by the generators can be fed back into the manure digesters to speed up the process, as well as into the barns for space heating.

Conversion to alcohol

About one-fourth of all energy use in the United States is for transportation. Biomass can be converted by microbial **fermentation** into liquid alcohol, which can be used to fuel vehicle engines. The fermentation is essentially the same natural process that has been used to make alcoholic beverages since civilization began: **yeast** feeds on sugars and starches in the plant biomass, producing ethyl alcohol, or **ethanol**, (C_2H_6O; also written as CH_3CH_2OH). A **ratio** of 1 part ethanol to 9 parts gasoline is used to produce a fuel known as gasohol, which can be burned in a standard **automobile** engine. Ethanol can also be used alone in engines that have been slightly modified. Brazil, which makes ethanol from **sugarcane**, has many cars that run on gasohol. Ethanol has a higher octane rating and produces less **carbon monoxide** than gasoline.

Most ethanol in the United States has been by a fermentation of corn and **sorghum** grain. This can be done in the U.S., where there is an excess of farmland, but in most countries the agricultural land is needed to grow food. Researchers are working on solving the problems of converting cornstalks, instead of grain, into ethanol. The industrial fermentation results in some emissions of pollutants to the atmosphere from the fossil fuel used in the **distillation** process. However, the burning of ethanol in vehicles can reduce the amount of **carbon** monoxide emitted.

The fermentation of organic material also gives off carbon monoxide and **hydrogen** gas, which can be heated in the presence of a catalyst to synthesize a poisonous but energy-efficient alcohol called *methanol*, or wood alcohol (CH_4O or CH_3OH). Methanol can also be substituted for gasoline in motor vehicles and other machinery. However, compared with the production of ethanol, that of methanol is not so efficient, and its wider use as a fuel is not yet commercially realistic. Also, emissions given off by burning methanol include toxic formaldehyde, although this could be controlled by **pollution control** equipment on vehicles.

Sources of biomass

There are four major sources of biomass: agricultural and **forestry** residues, municipal solid waste, industrial waste, and specifically grown bioenergy crops.

Agricultural and forest residues

In some respects, the conversion of crop-quality biomass to energy is a dubious strategy. It is less efficient to produce biofuel than food on Earth's limited arable land. However, many people in developing countries use crop residues and woody debris left after logging as a fuel source. Although they may seem a viable energy resource, crop and forest residues are often parts of plants containing high concentrations of **nutrients**. If the residues are harvested as biofuel, the nutrients are lost to the field or forest, and synthetic chemical **fertilizers** may have to be applied later. However, the burning of secondary wastes from processing, such as **rice** hulls, sawdust, or **paper** sludge, is an efficient use of material that might otherwise have been wasted.

Municipal solid waste

Municipal solid waste (MSW), much of which is organic, is being incinerated in some cities both as a means of keeping the waste out of landfills, and also as a way to acquire relatively inexpensive fuel to generate electrical power. On average, the burning of one ton of typical MSW produces as much heat as one barrel of oil. More than 70 waste-to-energy plants are now installed in the United States, and about that many others are being planned.

However, unexpected problems can arise in cities that do such a good job of **recycling** that not enough high-energy organic waste is available for the waste-to-energy incinerator. **Plastics**, for example, are a high-energy waste that can be efficiently recycled, while food waste can be composted. Another problem is that the

KEY TERMS

Biogas—Methane derived from the anaerobic digestion of biomass.

Biomass—Any biological material used to produce energy.

Cogeneration—The generation of electricity from the heat derived as a "waste" from an industrial process. The electricity generated is used to carry out the industrial processes.

Digester—A sealed, enclosed volume in which anaerobic digestion of biological material is carried out.

Emission—Any by-product of combustion, especially from industrial processes and vehicle engines.

Fermentation—A process by which complex organic molecules are enzymatically broken down to simpler molecules. For example, in alcohol fermentation glucose is reduced to ethyl alcohol and carbon dioxide.

Gasohol—A mixture of ethanol or methanol and gasoline, used to increase the octane rating of the gasoline while reducing emissions of carbon dioxide.

Methane—A biogas resulting from the anaerobic digestion of organic matter by bacteria; CH_4.

Methanol—Methyl alcohol, or CH_4O (also written CH_3OH).

Municipal solid waste—The entire waste of a municipality, its homes, and businesses, but not including those of industry or sewage; also called MSW.

Octane—A number indicating the ability of a petroleum compound to burn smoothly in an engine, rather than with power-losing explosions. The higher the octane number, the smoother the engine burns.

Organic—Refers to material made of the material of organisms. In pure chemistry, however, organic refers to compounds that include carbon, whether or not they are biological in origin.

content of MSW is usually not known. It often contains materials that after **incineration** can send toxic pollutants into the air, or that make the final residue toxic, so it must be treated as hazardous waste.

Industrial waste

Some industrial wastes are a good source of bioenergy, especially if its use involves a **cogeneration** facility. Cogeneration is the use of heat or material left over from manufacturing, often in combination with a conventional fuel such as oil, to produce electricity. The electricity can be used to run the factory, with any power left over put into the regional transmission grid. The U.S. pulp and paper industry satisfies about 8% of its energy needs by the cogeneration of wood waste. In South Florida's Palm Beach County, an extremely large cogeneration facility produces enough electricity to power 46,000 homes. It burns *bagasse*, a byproduct of milling sugar, plus wood waste obtained from building and demolition firms.

Bioenergy crops

Wood and other high-biomass crops can be grown specifically for use as bioenergy sources. The growing of annual crops such as corn may not be efficient enough in the long run for this purpose; too much labor and energy are required to grow this annual plant. However, perennial **grasses** such as switchgrass can be more efficiently cultivated for bioenergy. The grass biomass can be compressed into dense pellets, and efficiently transported and burned in this form. Woody crops, such as high-yield **hybrid** poplar trees, can be grown in plantations with the biomass harvested on a 3-to-10-year cycle, and then regenerated from stump sprouts. New trees might have to be replanted every 15-20 years.

The best biomass crops grow fast, use nutrients and **water** efficiently, grow densely, are hardy, are perennial, and regenerate easily after harvesting. Ideally, they would also have nitrogen-fixing capability, which would limit the need for fertilizer application. In the future, bioengineering may increase the nitrogen-fixing capability of certain plants and increase their **rate** of productivity. Bioengineering has already improved the bioenergy qualities of such trees as black cottonwood and hybrid poplar. Other trees with good potential as bioenergy sources include eucalyptus, sweetgum, and black locust.

Advantages and disadvantages of bioenergy

Although the burning or conversion of biomass does not fully relieve **pollution** of the atmosphere, it does have several major benefits. In many regions, biomass is more reliable than solar or **wind** energy. This is because the energy in plants is captured and stored, while in solar and wind energy this must be done by manufactured technology. Another advantage of bioenergy is that it can be produced using organic waste material that might otherwise be discarded; this saves the environmental and economic costs of their disposal. Used in mass quantities, bioenergy could boost the economy of any nation that must now import fossil fuels. If crops grown for their biomass increase the biomass of growing plants on the **planet**, this would reduce the amount of **carbon dioxide** in the atmosphere. Perhaps the most significant advantage of bioenergy is that it is a potentially renewable natural resource that would help supply energy needs indefinitely.

However, there are some disadvantages to using bioenergy. Biomass has a smaller energy content for its bulk than fossil fuels. Therefore the costs of labor, transportation, and storage are higher. Water and nutrients, which are in short supply in many areas, must be used to grow biomass crops.

Perhaps the major difficulty with bioenergy, however, is the same problem that has arisen with recycling. People will not demand bioenergy until there is a considerable cost saving in doing so, but there will not be much savings until there is a much larger demand for bioenergy, or the non-renewable sources become significantly more expensive.

See also Alternative energy sources; Hazardous wastes; Hydrocarbon; Landfill.

Resources

Books

Blashfield, Jean F., and Wallace B. Black. *Recycling.* Saving Planet Earth series. Chicago: Childrens Press, 1991.

Chartier, P. *Biomass for Energy and the Environment.* Pergamon Press, 1997.

Klass, D.L. *Biomass for Renewable Energy, Fuels, and Chemicals.* Academic Press, 1998.

Miller, Alan. *Growing Power: Bioenergy for Development and Industry.* Washington, DC: World Resources Institute, 1986.

Pack, Janet. *Fueling the Future.* Saving Planet Earth series. Chicago: Childrens Press, 1992.

Rickard, Graham. *Bioenergy.* Alternative Energy series. Milwaukee, WI: Gareth Stevens, 1991.

Jean F. Blashfield

Biofeedback

Biofeedback is a means by which a person can mentally influence a natural physiologic process that may or may not be consciously regulated under normal conditions. This could include lowering **blood pressure**, regulating the **heart rate**, or influencing the skin **temperature**.

Deliberate control of bodily functions is not a new accomplishment. Many historical accounts exist of Indian yogis who controlled their body temperature with such precision that they could make the palm of one hand warmer or cooler on one side than the other. Biofeedback also played a role in the performances of the famous escape artist Harry Houdini who consciously suppressed his gag **reflex** to suspend a key in his throat so as to regurgitate it later to unlock his elaborate bindings.

Development of modern biofeedback methods

As a form of therapy, biofeedback is a relatively recent development that has begun to gain acceptance among some members of the medical community. Research in the topic began in the 1960s, and by the end of the decade a number of research projects had demonstrated its effectiveness. While some early studies indicated that physiologic processes not usually under conscious control could not be influenced by biofeedback, subsequent research soon disproved this assumption.

During the 1970s biofeedback developed a devout following and as a result an entire industry was created for the manufacture and marketing of the instruments used to measure biofeedback alterations and alpha rhythms in the **brain**. Alpha rhythms are electrical waves formed by the brain at the rate of 8-13 cycles per second. Because they are associated with the state of meditation attained by practitioners of yoga or transcendental meditation, they were accepted as the optimal state of biofeedback. Instruments to detect and measure alpha rhythms soon became readily available.

Biofeedback training

Biofeedback training must begin with an auditory or visual signal to measure the activity of the **organ** being influenced and to indicate any changes that take place in it. Heart rate, for example, can be signaled by a beeping sound that occurs with each heartbeat. A subject can detect any increase or decrease in the number of heartbeats per minute by the increasing or decreasing rapidity of the beeping. A visual signal also could be devised—for example, spikes in a horizontal line that appear closer together or farther apart as the heartbeat changes. The ultimate objective is to develop a such a high level of consciousness that such changes can be determined without the signal device.

Biofeedback can be separated into three processes or steps. The first involves detecting the biological process being measured and amplifying it so as to be seen or heard. The second step is to convert the electrical signal into an easily understood form from which its alterations can be read. The third step is to make this signal available to the subject as soon as possible after the event being measured has occurred.

Uses of biofeedback

Clinical applications for biofeedback include the control of blood pressure for patients with **hypertension**, relief or control of migraine headaches, easing of muscle cramps, and relief of **insomnia**.

Biofeedback training begins with the basic control of heart rate or other readily accessible and controllable functions. The subject is provided an auditory or visual signal at first and is gradually weaned from it as he becomes more skillful at the practice. Once the basic skill has been learned he or she can then shift concentration to a specific problem. Ideally, patients will continue to practice biofeedback techniques and so increase their effectiveness over **time**.

Biofeedback has gained acceptance in the United States as its clinical use has increased. Most major cities have a biofeedback association, and practitioners can be certified by a national certification institute. Certification standards are rigorous to assure that the practitioner has a thorough understanding of **physiology** and **psychology** to better apply the methodology.

See also Alternative medicine.

Resources

Books

Burton Goldberg Group. *Alternative Medicine: The Definitive Guide.* Puyallup, WA: Future Medicine Publishing, 1993.

Periodicals

Fugh-Berman, A. "The Case for 'Natural' Medicine." *The Nation* 257 (September 1993):240-244.
Morrow, J., and R. Wolff, "Wired for Wonders." *Readers Digest* 140 (May 1992):105-108.

Larry Blaser

Biofilms

A biofilm is a population of **bacteria**, **algae**, **yeast**, or **fungi** that is growing attached to a surface. The surface can be living or nonliving. Examples of living surfaces where biofilms may grow include the teeth, gums, and the cells that line the intestinal and vaginal tracts. Examples of nonliving surfaces include

rocks in watercourses, and implanted medical devices such as **catheters**.

Rudimentary knowledge of the presence of biofilms has been known for centuries. For example, the bacterium *Acetobacter aceti* attached to **wood** chips has been used to manufacture vinegar since the nineteenth century. Despite this history, biofilms were viewed as more of a curiosity until the 1980s. Indeed, much of what is known about **microorganisms** and about specific areas such as bacterial antibiotic resistance has resulted from the use of bacteria growing as floating (planktonic) populations in liquid growth sources.

Beginning in the 1980s, evidence accumulated that has led to the recognition that the floating form of bacterial growth is artificial, and that the biofilm form of growth is the natural and preferred **mode** of growth for microbes. Now, it is accepted that virtually every surface that comes into contact with microorganisms is capable of sustaining biofilm formation.

Much of what is known about biofilms has come from the study of bacteria. Typically, the biofilm studied in the laboratory consists of one bacterial type. Observation of only one growing bacteria makes study of the formation and behavior of the biofilm easier to accomplish. In a natural setting, however, a biofilm is often comprised of a variety of bacteria. Dental plaque is a good example. Hundreds of **species** of bacteria can be present in the biofilm that forms on the surface of the teeth and gums.

The formation of a biofilm begins when floating bacteria encounter a surface. Attachment can occur nonspecifically or specifically. Specific attachment involves the recognition of a surface **molecule** by another molecule on the surface of the microorganism. Bacterial attachment can be aided by appendages such as **flagella**, cilia, or the holdfast of *Caulobacter crescentus*.

Attachment is followed by a more long-lasting association with the surface. For bacteria, this association involves structural and genetic changes. Genes are expressed following surface attachment. A particularly distinctive result of this preferential genetic activity is the production of a large amount of a sugary material known as the glycocalyx or the exopolysaccharide. The sugar layer buries the bacterial population, creating the biofilm.

As times passes, a biofilm can become thicker. An older, more mature, biofilm differs from a younger biofilm. Studies using instruments that can probe into a biofilm without physically disturbing its structure have demonstrated that the bacteria deeper within a biofilm stop producing the exopolysaccharide and slow their growth **rate** to become almost dormant. In contrast, the bacteria at the edge of the biofilm grow faster and produce large amounts of exopolysaccharide. These activities occur at the same time and indeed are coordinated. The bacteria can chemically communicate with one another. This phenomenon, which is called quorum sensing, allows a biofilm to grow and encourages bacteria to leave a biofilm and form new biofilms elsewhere.

Another difference in biofilms that develop over time concerns their three-dimensional structure. A young biofilm is fairly uniform in structure, with the bacteria arranged evenly throughout the biofilm. In contrast, a well-established biofilm consists of bacteria clustered together in microcolonies, with surrounding regions of exopolysaccharide and open channels of **water** that allow food to easily reach the bacteria and waste material to easily pass out of the biofilm.

Bacterial biofilms are important in the establishment and treatment of infections. Within the biofilm, bacteria are very resistant to chemicals like **antibiotics** that would otherwise kill the bacteria. Antibiotic resistant biofilms occur on inert surfaces such as artificial **heart** valves and urinary catheters, and on living surfaces, such as gallstones and in the lungs of those afflicted with **cystic fibrosis**. In cystic fibrosis, the biofilm formed by bacteria, mainly *Pseudomonas aeruginosa*, protects the bacteria from the host's **immune system**. The immune response may persist for years, which irritates and damages the lung **tissue**.

Resources

Books

Doyle, R.J. *Biofilms (Methods in Enzymology, Volume 310)*. New York: Academic Press, 1999.

Periodicals

Davies, D.G., M.R. Parek, J.P. Pearson, et al., "The Involvement of Cell-to-Cell Signals in the Development of a Bacterial Biofilm." *Science* (April 1998): 3486–3490.

Donlan, R.M., "Biofilms: Microbial Life on Surfaces." *Emerging Infectious Diseases* (September 2002): 881–890.

Murga, R., T.S. Forster, E. Brown, et al. "The Role of Biofilms in the Survival of *Legionella pneumophila* in a Model Water System." *Microbiology* (November 2001): 3121–3126.

Bioinformatics and computational biology

Bioinformatics, or computational **biology**, refers to the development of new database methods to store genomic information, computational software programs, and methods to extract, process, and evaluate this information, and the refinement of existing techniques to acquire the genomic data. Finding genes and determining their function, predicting the structure of **proteins** and RNA sequences from the available DNA sequence, and determining the evolutionary relationship of proteins and DNA sequences are also part of bioinformatics.

The **genome** sequences of some **bacteria**, **yeast**, a nematode, the fruitfly *Drosophila,* and several plants have been obtained during the past decade, with many more sequences nearing completion. Although work still continues in order to refine the data, the initial **sequencing** of the human genome was completed in 2000. In addition to this accumulation of nucleotide sequence data, elucidation of the three-dimensional structure of proteins coded for by the genes has been accelerating. The result is a vast ever-increasing amount of databases and genetic information The efficient and productive use of this information requires the specialized computational techniques and software. Bioinformatics has developed and grown from the need to extract and analyze the reams of information pertaining to genomic information like nucleotide sequences and protein structure.

Bioinformatics utilizes statistical analysis, stepwise computational analysis and database management tools in order to search databases of DNA or protein sequences to filter out background from useful data and enable comparison of data from diverse databases. This sort of analysis is on-going. The exploding number of databases, and the various experimental methods used to acquire the data, can make comparisons tedious to achieve. However, the benefits can be enormous. The immense size and network of biological databases provides a resource to answer biological questions about mapping, **gene** expression patterns, molecular modeling, molecular **evolution**, and to assist in the structural-based design of therapeutic drugs.

Obtaining information is a multi-step process. Databases are examined, or browsed, by posing complex computational questions. Researchers who have derived a DNA or protein sequence can submit the sequence to public repositories of such information to see if there is a match or similarity with their sequence. If so, further analysis may reveal a putative structure for the protein coded for by the sequence as well as a putative function for that protein. Four primary databases, those containing one type of information (only DNA sequence data or only protein sequence data), currently available for these purposes are the European Molecular Biology DNA Sequence Database (EMBL), GenBank, SwissProt and the Protein Identification Resource (PIR). Secondary databases contain information derived from other databases. Specialist databases, or knowledge databases, are collections of sequence information, expert commentary and reference literature. Finally, integrated databases are collections (amalgamations) of primary and secondary databases.

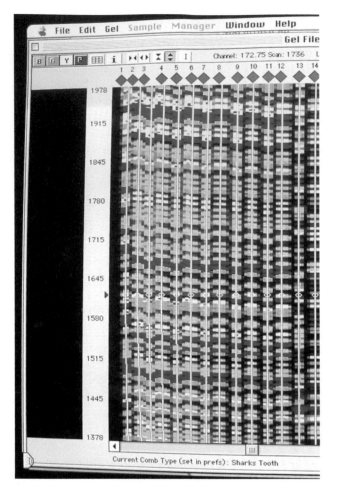

Computer monitor of automated DNA sequencer with gel image. *Photograph by T. Bannor. Custom Medical Stock Photo.*

The area of bioinformatics concerned with the derivation of protein sequences makes it conceivable to predict three-dimensional structures of the protein molecules, by use of computer graphics and by comparison with similar proteins, which have been obtained as a **crystal**. Knowledge of structure allows the site(s) critical for the function of the protein to be determined. Subsequently, drugs active against the site can be designed, or the protein can be utilized to enhance commercial production processes, such as in pharmaceutical bioinformatics.

Bioinformatics also encompasses the field of comparative genomics. This is the comparison of functionally equivalent genes across **species**. A yeast gene is likely to have the same function as a worm protein with the same **amino acid**. Alternately, genes having similar sequence may have divergent functions. Such similarities and differences will be revealed by the sequence information. Practically, such knowledge aids in the selection and design of genes to instill a specific function in a product to enhance its commercial appeal.

The most widely known example of a bioinformatics driven endeavor is the **Human Genome Project**. Work related to the Human Genome Project has allowed dramatic improvements in molecular biological techniques and improved computational tools for studying genomic function.

See also Chromosome mapping; Deoxyribonucleic acid (DNA); Genetic engineering; Genetic testing; Genome; Molecular biology; Proteomics; Ribonucleic acid (RNA).

Biological community

In **biology**, the term **species** refers to all organisms of the same kind that are potentially capable, under natural conditions, of breeding and producing fertile offspring. The members of a species living in a given area at the same time constitute a population. All the populations living and interacting within a particular geographic area make up a biological (or biotic) community. The living organisms in a community together with their nonliving or abiotic environment make up an **ecosystem**. In theory, an ecosystem (and the biological community that forms its living component) can be as small as a few mosquito larvae living in a rain puddle or as large as **prairie** stretching across thousands of kilometers.

A very large, general biotic community such as the boreal forest is called a **biome**.

It often is difficult, however, to define where one community or ecosystem stops and another starts. Organisms may spend part of their lives in one area and part in another. **Water**, **nutrients**, sediment, and other abiotic factors are carried from place to place by geologic forces and migrating organisms. While it might seem that a **lake** and the dry land surrounding it, for instance, are distinctly different in their environmental conditions and biological communities, there can be a great deal of exchange of materials and organisms from one to the other. **Insects** fall into the lake and are eaten by **fish**. **Amphibians** leave the lake to hunt on shore. **Soil** erodes from the land and fertilizes the water. Water evaporated from the lake surface falls back on the land as rain that nourishes **plant** life. Every biological community requires a more or less constant influx of **energy** to maintain living processes.

Several important ecological categories and processes characterize every biological community. Productivity describes the amount of **biomass** produced by green plants as they capture sunlight and create new organic compounds. A tropical **rainforest** or a Midwest corn field can have very high rates of productivity, while deserts and arctic **tundra** tend to be very unproductive. **Trophic levels** describe the methods used by members of the biological community in obtaining food. Primary producers are green plants that depend on **photosynthesis** for their nourishment. Primary consumers are the herbivores that eat plants. Secondary consumers are the carnivores who feed on herbivores. Top carnivores are large, fierce animals who occupy the highest level on the food chain or food web. Nobody eats the top carnivores except the scavengers (like **vultures** and hyenas) and decomposers (like **fungi** and **bacteria**) that consume dead organisms and recycle their bodies back into the abiotic component of the ecosystem. Because of the second law of **thermodynamics**, a majority of the energy in each trophic level is unavailable to organisms in the next higher level. This means that each successive trophic level generally has far fewer members than the **prey** on which they feed. While there might be thousands of primary producers in a particular community, there might be only a few top predators.

Abundance is an expression of the total number of organisms in a biological community, while diversity is a measure of the number of different species in that community. The arctic tundra of Alaska has vast **clouds** of insects, enormous flocks of migratory **birds**, and great herds of a few species of **mammals** during the brief summer growing season. Thus, it has high abundance but very little diversity. The tropical rainforest, on the other hand, might have several thousand different **tree** species and an even larger number of insect species in only a few hectares, but there may be only a few individuals representing each of those species in that area. Thus, the forest could have extremely high diversity but low abundance of any particular species. Complexity is a description of the variety of ecological processes or the number of ecological niches (ways of making a living) within a biological community. The tropical rainforest is likely to be highly complex, while the arctic tundra has relatively low complexity.

Biological communities generally undergo a series of developmental changes over time known as **succession**. The first species to colonize a newly exposed land surface, for example, are known as pioneers. Organisms such as **lichens**, **grasses**, and weedy flowering plants with a high tolerance for harsh conditions tend to fall in this category. Over time, the pioneers trap sediments, build soil, and retain moisture. They provide shelter and create conditions that allow other species like shrubs and small trees to take root and flourish. Larger plants accumulate soil faster than do pioneer species. They also provide shade, shelter, higher **humidity**, protection from **sun** and **wind**, and living **space** for organisms that could not survive on open ground. Eventually these successional processes result in a community very different from the one first established by the original pioneers,

most of whom are forced to move on to other newly disturbed land. It was once thought that every area would have a climax community such as an oak forest or prairie grassland determined by climate, topography, and mineral composition. Given enough time and freedom from disturbance, it was believed, every community would inevitably progress to its climax state. It is now recognized, however, that some ecosystems experience continuous disturbance. Certain biological assemblages such as **conifer forests** that we once thought were stable climax communities, we now recognize as chance associations in an ever changing mosaic of regularly disturbed and constantly changing landscapes.

Many biological communities are relatively stable over long periods of time and are able to withstand many kinds of disturbance and change. An oak forest, for example, tends to remain an oak forest because the species that make it up have self-perpetuating mechanisms. When a tree falls, others grow to replace it. The ability to repair damage and resist change is termed "resilience." For many years there has been an on-going debate between theoretical and field ecologists about whether complexity and diversity in a biological community increase resilience. Theoretical models suggest that a population of a few very hardy, weedy species, such as dandelions and box elder bugs, might be more resistant to change than a more highly specialized and more diverse community such as a tropical forest. Recent empirical evidence suggests that in at least some communities, such as prairies, higher diversity does impart greater resistance to change and a better ability to repair damage after stress or disturbance.

Biological rhythms

Biological rhythms are often referred to as biological clocks, since they operate on **time** schedules on a daily, monthly, seasonal, or annual basis. Some biological rhythms even occur on the basis of fractions of seconds. These internal clocks operate independent of the environment, but they are controlled by environmental conditions in changing situations. During times of change, such as seasonal decreases of **light**, or those caused by travel in an east-west or west-east direction, human beings are able to reset their biological clocks to become synchronized with the environment.

History

Daily rhythms in plants and animals have been noticed since early times. As early as the fourth century B.C., Alexander the Great's scribe Androsthenes noted that the leaves of certain trees opened during the day and closed at night. Two centuries before modern gardeners noticed their day lilies closed at night, the famous taxonomist Carolus Linnaeus discovered the petals of many **flower species** opened and closed at regular times. He even created a garden with flowers that opened at various times so that he could tell the time of day by looking in his garden.

Karl von Frisch observed that **bees** visited flowers only at specific times. He and Ingeborg Beling trained bees to visit a **nectar** feeding station between 4 and 6 P.M. The bees did not visit at other times, and they still visited even when the nectar was removed. When outside cues such as light were removed in laboratory trials, the bees still fed at prescribed times. Although von Frisch did not know it, the bees were operating on an internal clock. It wasn't until the 1950s that Gustav Kramer and Klaus Hoffmann proved the existence of a biological clock. With an ingenious apparatus, Kramer demonstrated that **starlings** used the **sun** as a compass to migrate even though the sun itself moves throughout the day. That is, the bird's internal clock reorients it in the direction of the moving sun. Hoffmann showed that the clock persisted in dim light and thus is endogenous to the **animal**. He showed that the animal's clock was synchronized to local time by the influence of the local environment.

Research in the area of biological clocks did not begin with humans. Early research was done on a variety of animals including **rats**, **hamsters**, sparrows, lizards, marine **snails**, and fruit **flies**. The choice of an animal depended on many considerations—size, **rate** of reproduction, expense to maintain, and availability—as well as behavioral issues, such as whether it entrained easily and what environmental factors seemed to affect its **behavior**.

In 1920, a landmark **paper** was written by W.W. Garner and H.A. Allard in which they showed that tobacco plants would flower only if exposed to a certain number of hours of light. The term "photoperiodism" was used to designate the response of organisms to relative length of day and night. The ability to sense day length is an important ability for plants so that they grow, reproduce, and develop during favorable times of the year. The changing times of dawn and dusk contain seasonal information as well as time of day information so that the organisms have, in effect, an internal clock and calendar. In plants, a photoperiodic clock not only controls flowering, but also induction and termination of dormancy in buds and bulbs, seed **germination**, and daily rhythms such as **leaf** movements, petal movements, and nectar secretion.

The study of biological rhythms is called chronobiology, a relatively new scientific specialty that began in the 1950s when researchers used new **heart** and lung monitors to answer some of the basic questions that

chronobiologists ask. Some of these questions deal with sleep/wake cycles, time of day **energy** and productivity levels, and mood changes. Other important areas of study in chronobiology deal with growth patterns, hormone secretions, and menstruation.

Types of internal clocks

Some biological rhythms occur more than once a day and are called ultradian rhythms. The release of **hormones** from the male pituitary gland of **mammals** occurs about every one to two hours during the day. **Sleep** cycles, that is, the cycle from drowsiness to REM (rapid **eye** movement, dream sleep) to dozing, then light and deep sleep, and finally slow-wave sleep, is a 90-minute sleep cycle that repeats itself during a night's sleep. Constant breathing and the beating of our hearts are also ultradian rhythms, but these activities are also affected by the daily sleep/wake cycle, which is a circadian rhythm. Heart rate and breathing both slow during sleep.

Circadian rhythms are those that occur once a day and relate to the sun. The one that dominates our activities is the sleep/wake cycle people experience every 24 hours. Body **temperature**, response to medications, **alcohol** blood level, alertness, and fatigue all have a daily up and down cycle. Circadian rhythms also control such daily activities as eating. Chronobiologists explain the synchronizing of circadian rhythms to sunlight. This external **factor** helps the body regulate its daily activities. When people travel in an east or west direction through many time zones, they may suffer from jet lag and will need several days to adjust their sleep/wake cycle to the daylight and darkness in their new environment.

The clock in humans is located in the suprachiasmatic nucleus (SCN), a distinct group of cells found within the hypothalamus. The SCN is only one part of the mechanism by which "time" is kept. There are light receptors found in the retina that have a pathway, called the retinohypothalamic tract, leading to the SCN. The pineal gland is a pea-like structure found behind the hypothalamus in humans. The pineal gland receives information indirectly from the SCN. It appears that the SCN takes the information on day length from the retina, interprets it, and passes it on to the pineal gland, which secretes the hormone melatonin in response to this message. Nighttime causes melatonin secretion to rise, while daylight inhibits it. Even when light cues are absent, melatonin is still released in a cyclic manner; yet if the SCN is destroyed, circadian rhythms disappear entirely.

Infradian cycles are monthly cycles, the most common of which is the monthly menstruation of females. Illness and death have been correlated to certain times within infradian cycles, with more deaths occurring in the second half of the **menstrual cycle**. Research with men has shown that weight fluctuations, hormone levels, growth of beards, body temperature, **pain** threshold, lung capacity, and physical strength all demonstrate monthly cyclical patterns. More men experience symptoms of prostate enlargement during the new **moon** period, while more deaths and accidents occur around the time of a full moon.

Lunar cycles are somewhat longer than circadian ones. They last about 24 hours and 50 minutes. Some marine **invertebrates**, such as the fiddler crab, synchronize daily activity to the **tides**, which are affected by the moon. The change in **water** levels during the tides influences the activity of many marine invertebrates (animals without a spinal column).

The longest cycle is of course the life cycle, which has the distinct stages of growth, maturation, decline, and death in all forms of life.

The circannual cycle is a yearly occurring one. Some people and animals periodically gain or lose weight at certain times of the year. Transplant patients respond better to certain drug treatments at certain times of the year than at others.

A disorder called SAD (seasonal affective disorder) afflicts many people during the winter months of the year when the days are shorter. The hormone called melatonin is secreted from the pineal gland, which is located just behind the hypothalamus in the **brain**. Melatonin affects our moods. Its greatest production is during the night and when there is more darkness during a 24 hour period, as it is in winter, there is a tendency for some people to become depressed. An effective treatment is to expose the person to intense bright light.

Adaptations to time

The basic time **adaptation** in biological rhythms is entrained, which means it is influenced by external cues. Other aspects of our biological rhythms are so influential that most of us adapt our activities to the time of day during which we function best. Some people are morning people, while others are night people. Morning people (sometimes called larks) wake up with lots of energy and perform their best work in the early hours of the day. Night people are sluggish in the morning and do their best work late in the day or in the evening. They are often called **owls**. Work environments in our society are not adjusted to these differences, though, and many people have to work at times when they are least likely to be productive.

An experiment that took place in New Mexico in 1989, when a woman spent 130 days in isolation in a **cave** that had no natural light, demonstrated how external cues affect our biological clocks. After six weeks, she was functioning within a 44 hour cycle of sleeping and wakefulness. Her **perception** of time was also compressed to a considerable degree. In other experiments, volunteers drifted into a 25-hour day, while others have experienced 50-hour days or irregular cyclical patterns.

Problems due to circadian desynchrony

Diminished performance is not the only problem caused by being out of phase with one's environment. Worker and public safety also can be compromised. Researchers have found that the neural processes controlling alertness and sleep produce an increased sleep tendency and diminished capacity to function during certain early morning hours (circa 2-7 A.M.) and, to a lesser degree, during a period in the mid-afternoon (circa 2-5 P.M.), whether or not one has slept. Several studies of single-vehicle car accidents that have been judged to be fatigue-related have shown two peak times for accidents—a major one between midnight and 7 A.M., and a secondary peak between 1 and 4 P.M.

Shift work is a necessity in an industrially developed country. Manufacturing, transportation, and health care rely on shift workers to maintain operations around the clock. A shift worker can be defined as someone who works evenings, nights, rotating shifts, or extended shifts. Unlike jet lag, which is usually a temporary disruption, shift work schedules and their disruptions can last for years. Humans are diurnal creatures (active during the day), and shift workers are often out of sync with environmental and social cues. Circadian adjustment to a new shift is gradual (taking a week or more), and changing to a new shift before adjustment is complete can cause perpetual desynchronization. Research shows that shift workers suffer from sleep disruption and fatigue, domestic disturbances, and health problems such as gastrointestinal disorders and increased risk of cardiovascular **disease**. Task performance is also poorer at night. In order to minimize circadian disruptions, many large companies employ consulting firms to advise them on how to manage or prevent problems caused by shift work. Other than adjusting schedules, implementing educational and support programs regarding biological rhythms, sleep, and family counseling may be a good way to improve the coping ability of shift workers.

Compromised performance among health care workers can jeopardize patients' lives. The medical profession is notorious for having its doctors-in-training work very long hours. In the case of a young woman who died eight hours after being admitted to the emergency room, the intern and resident had each been on duty for 18 hours before treating her. The grand jury ex-

KEY TERMS

. . . .

Chronobiology—The study of biological clocks.

Circadian rhythm—The rhythmical biological cycle of sleep and waking which, in humans, usually occurs every 24 hours.

Entrainment—Regulation of the biological cycle to the environment.

Free-running clock—Response to internal clocks without any influence from the external world.

Infradian cycle—The monthly biological cycle.

Lunar rhythm—The regulation of the biological cycle to the movement of the moon.

Synchrony—The adjustment of biological rhythms to the environment.

Ultradian rhythm—Biological cycles of less than a day.

amining the case determined that the number of hours the residents and interns were required to work contributed to her death. Some improvements may be on the way in the medical profession as well. The Accreditation Council for Graduate Medical Education (ACGME), which sets standards for medical residency training programs, has begun within the past few years to introduce new standards for residents' work hours.

Medical uses

The work of chronobiologists in the area of biological rhythms has been useful to medical science in helping them diagnose illness more accurately. Twenty-four hour monitoring of heart rate and blood **pressure** gives the treating physician a better picture of health problems. Newborn infants of families with histories of heart disease can be monitored and abnormalities can be seen. Early detection of breast **cancer** can be made through the recording of skin temperature fluctuations over breasts. Noncancerous temperatures of the skin of breasts has a greater fluctuation cycle than temperatures recorded of the skin of cancerous breasts.

Research on biological clocks is shedding new light on standard prescribing practices for medication. For example, cortisone injections are given in the morning for the treatment of adrenal gland malfunctions. Hormones in various **glands** throughout the human body are released when they are needed by major organs and the chemical changes that occur manage our biological rhythms. As new ways of monitoring them are discov-

ered, early warning signs of disease will become more apparent to diagnosing physicians. Evidence shows that certain medical illnesses whose symptoms show a circadian rhythm respond better when drugs are coordinated with that rhythm. Medications for **asthma**, **epilepsy**, cancer, cardiovascular disease, and allergies all have shown better results with minimum side effects when given at particular times.

Researchers are currently trying to re-educate the medical profession about the limitations of traditional prescribing practices and the greater potential benefits of administering medication at the most appropriate time. In the future, other research will no doubt provide even stronger arguments for coordinating medical tests and procedures more closely with temporal information and with influences of the biological clock that have not yet been discovered.

See also Depression.

Resources

Books

Campbell, Jeremy. *Winston Churchill's Afternoon Nap.* New York: Simon and Schuster, 1986.

Dotto, Lydia. *Losing Sleep.* New York: William Morrow, 1990.

Hughes, Martin. *Body Clock.* New York: Facts on File, 1989.

Orlock, Carol. *Inner Time.* New York: Birch Lane Press, 1993.

Perry, Susan L. *The Secrets Our Body Clocks Reveal.* New York: Rawson Associates, 1988.

Shafii, Mohammad and Sharon Lee Shafii. *Biological Rhythms, Mood Disorders, Light Therapy, and the Pineal Gland.* Washington, DC: American Psychiatric Press, 1990.

Periodicals

Hardin, P.E. "From Biological Clock to Biological Rhythms." *Genome Biology* 1 (2000): 1023.1-1023.5.

Vita Richman

Biological warfare

Biological warfare is the use of living organisms (e.g., **bacteria**, **virus**) or biochemical agents (e.g., chemical neurotoxins) as strategic military weapons to cause harm in humans, animals, or plants. In contrast to **bioterrorism**, biological warfare is considered the government-sanctioned use of biological weapons to attack a clearly defined military force or civilian population. These agents can be classified into six distinct groups; bacteria, viruses, rickettsiae, chlamydia, **fungi**, and biological toxins.

Bacteria (by itself or genetically engineered) can be used to produce **disease** that often can be treated with **antibiotics**. These are small organisms that can be easily grown on solid supports or liquid media. Viruses, however, require a living host in which to reproduce and are dependent on the cells they infect. Viruses sometimes can be destroyed using antiviral compounds, however, there is a limited supply of these compounds and only a few types available. Rickettsiae are organisms with similar characteristics common to bacteria, as they are susceptible to antibiotics, require **oxygen**, and have **cell** membranes. Rickettsiae are also similar to viruses in that they require a host to grow and reproduce. Chlamydia are **parasites** that cannot produce their own **energy** sources. Fungi can live without oxygen, are non-photosynthetic **species**, utilize decaying vegetable **matter** as their food source, and form spores. Toxins can be synthetic (chemical agents) or derived from poisonous substances produced by living plants, animals, or **microorganisms**. Toxins have an important advantage over **pathogens** as they are not alive and, therefore, they are more stable and easier to produce and distribute.

The degree at which these biological weapons become threatening involves several characteristics such as the level of infectiousness, toxicity, virulence, pathogenicity (ability to cause infectious disease), stability, and how easily they are transmitted from one species to the next.

Historical perspective of biological warfare

Biological weapons have been used as part of an arsenal during warfare for centuries. For example, ancient battle records suggest that diseased human remains and cattle that had died of microbial diseases were used to poison wells. In the spring of 1346, after a Mongol army attempted an overtake of the Crimean city of Caffa for three years without success, the Mongols gathered a number of their own people who had died of the plague, laid them on their catapults, and hurled them into the town. Eventually, the plague spread through Caffa causing residents to flee. It is quite possible that the successful conquest of the town by the Mongols may have resulted in the most damaging attack using biological weapons in world history. In the 1760s, a rebellion among Native Americans was countered when Lord Jeffrey Amherst, then commander-in-chief of British forces, suggested grinding pox scabs (pustule scabs) into blankets intended for distribution to members of an Ohio tribe. The disease broke out among the Native Americans, ending the rebellion.

During the twentieth century, a trend of designer biological weapons targeted for larger, more defined military objectives became more prevalent. Research projects aimed at developing anthrax-based biological weapon programs during World War II were initiated by Canada, the United States, and Britain. Britain produced anthrax-based weapons at the United Kingdom Chemi-

cal and Biological Defense Establishment at the Porton Down facility that were intended to be dropped on Germany to infect their food supply (reportedly, they were never used). Prisoners in Nazi Germany concentration camps were infected with pathogens, such as **hepatitis** A, *Plasmodium falciparum,* as well as other types of a bacteria for unclear research objectives. Countries such as Japan also conducted extensive biological weapons research during World War II while occupying part of China. Prisoners were infected with a variety of bacterial pathogens, including *Neisseria meningitis, Bacillus anthracis* (**anthrax**), and *Yersinia pestis.* It has been estimated that over 10,000 prisoners died as a result of either **infection** or execution following infection. In addition, biological agents contaminated the food and **water** supply. It is also thought that approximately 15 million potentially plague-infected **fleas** were released from Japanese aircrafts, affecting many Chinese cities with an estimated 10,000 illnesses and 1,700 deaths.

The development of airborne biological weapons make biological warfare particularly dangerous. For example, British open-air testing of anthrax weapons in 1941 on Gruinard Island in Scotland rendered the island uninhabitable for five decades. The United States Army conducted a study in 1951–52 called *Operation Sea Spray* to study how **wind** currents might effect the dispersion of bioagents. As part of the project design, balloons were filled with *Serratia marcescens* (then thought to be relatively harmless) and exploded over San Francisco. Reportedly, there was a corresponding significant increase in the number of **pneumonia** and urinary tract infections. In 1979, an accidental release of anthrax spores (approximately a gram for several minutes) at a bioweapons facility near the Russian city of Sverdlovsk, infected 77 and killed 66 people that were approximately 2.5 mi (4 km) downwind of the facility. **Sheep** and cattle up to 31 mi (50 km) downwind also became ill. From these events, anthrax is considered one of the most deadly biological weapons. A dose of 10,000 spores, or one millionth of a gram, is fatal within days after exposure in 90–100% of the population. According to the U.S. Congress Office of Technology Assessment, an **aircraft** that drops 220 lb (100 kg) of anthrax over a city under normal **weather** conditions would be lethal for approximately one to three million people. Skin contact with spores can also produce a less lethal, but dangerous cutaneous anthrax infection. Without antibiotic treatment, the mortality **rate** for cutaneous anthrax is 10–20% and with treatment, the mortality rate falls to less than 1%.

The first diplomatic effort to reduce biological warfare was initiated by the international community in 1925 and called the Geneva Protocol, a treaty prohibiting the development and use of any form of biological weapon in war.

During the 1950s and 1960s, the United States constructed research facilities to develop antisera, vaccines, and equipment for protection against a possible biological attack as well as the use of microorganisms as offensive weapons. Since then, other initiatives to ban the use of biological warfare and/or destroy the stockpiles of biological weapons have been attempted. Despite the international prohibitions, the existence of biological weapons remains an impending and growing threat. In 1972, 87 nations, including the United States, signed the Biological Weapons Convention Treaty, which banned the development, testing, and storage of such weapons. For a time, it appeared that the world agreed to ban biological weapons. By the 1980s, the political mood had changed. In 1982, President Ronald Reagan declared that world politics justified research on biological and chemical weapons and that the United States would return to a more ambitious program.

Over the last decade, an emphasis on the development of chemical toxins and the development of genetically engineered biological weapons emerged due to international political instability in several countries, along with the uprising of extremist groups and disestablishmentarianism in the Middle East. During the Persian Gulf War, U.S. troops were exposed to an uncertain source of biological weapons leading to what are called the Gulf War Illnesses. In Iraq, five hidden laboratories were discovered that were designed to refine and stockpile several biological weapons including anthrax, **botulism**, and gas **gangrene** bacteria. In 1997, Secretary of Defense William Cohen pinpointed Libya, Iraq, Syria, and Iran as countries that were launching aggressive biological weapon development programs. In 1998, several scientists that were part of the former Soviet Union's biological and chemical weapons programs were reported to have been recruited to develop biological warfare in Libya. Not long after the terrorist attacks on the U.S. World Trade Center and Pentagon in 2001, the postal service was used by terrorists (possibly unrelated to World Trade Center and Pentagon attacks) to deliver anthrax by mail to several locations within the U.S. On October 4, 2001, a 63-year-old man was reported to have contracted the first known intentionally inflicted case of inhalation anthrax in the U.S. By the time the outbreak was contained, the U.S. Centers for Disease Control reported 23 cases of inhaled or cutaneous anthrax (19 confirmed, 4 suspected), including five deaths.

Genetically engineered weapons and other biological weapons

Since about 1960, the development of **genetic engineering** has greatly expanded the possibilities of biological warfare. Genetic **engineering** is the process by which an organism's genetic properties are manipulated so that they acquire desired characteristics. In the case of biological warfare, the desired characteristics are harmful to mankind. Several issues arise when tampering with the genetic makeup of microorganisms such as the development of new diseases for which there is no treatment available for either opponent. Genetic engineering can also be applied for the purposes of making existing pathogens more pathogenic (disease-causing).

Other known biological agents that are considered biological weapons and might be used during biological warfare are botulism, **brucellosis**, Q fever, smallpox, saxitoxin, ebola hemorrhagic fever, tularemia, and staphylococcus enterotoxin.

Like most potential biological weapons, brucellosis occurs naturally among domestic and wild animals. It is spread by one of three varieties of the *Brucella* bacterium. It is seldom fatal, but causes a long-term, debilitating illness characterized by fever, loss of weight, general lassitude, and **depression**. Brucellosis is a biological weapon that does not kill people, but renders them so ill that they are unable to fight effectively. The disease can be treated with tetracycline.

Q fever is a biological agent that is not particularly harmful but is highly infectious. It is caused by a bacterial strain of the rickettsial **organism** *Coxiella burnetii* and produces headache, fever, chills, sweats, and a loss of appetite. Scientists believe that the condition is often mistaken for the flu. Like brucellosis, the value of Q fever is not in its toxicity, but in its ability to be easily spread. No more than a dozen microbes are needed to initiate an infection. It represents a particularly noxious biological weapon in its potential to be modified to cause disease or death.

From time to time, coastal waters in various parts of the world develop a reddish tinge because of a unicellular organisms called *dinoflagellates*. These so-called *red tides* are a warning that seafood from these waters are potentially unsafe. The dinoflagellates release a variety of toxins including saxitoxin. Some of these toxins quickly cause a type of paralysis that can cause death in humans in less than an hour. U.S. scientists prepared and tested shellfish toxins, including saxitoxin, for possible use as a biological weapon after World War II.

Many forms of *Staphylococcus* exist, some harmless and some quite dangerous. They are often responsible for specific forms of **food poisoning** and, in rare instances, exposure to these microorganisms can lead to **toxic shock syndrome** within hours. A person infected with *Staphylococcus* contaminated food experiences severe nausea, vomiting, diarrhea, and is essentially incapacitated for two or three days. *Staphylococcus* has a practical advantage in that it can be dried and stored for up to a year without losing its toxicity.

Tularemia is a plague-like disease caused by the bacterium *Francisella tularensis*. After an incubation period of about a week, a person infected with the bacterium begins to develop a fever accompanied by chills and headaches. If the bacterium is inhaled, symptoms also include chest **pain** and difficulty in breathing. The death rate is relatively low when exposure occurs through the skin, but much higher when inhaled. Delivery of a genetically modified form of the bacteria by aerosolization would cause a disease that is expected to have a case fatality rate which may be higher compared to the 5–10% seen when disease is acquired naturally.

See also Anthrax; Poisons and toxins.

Resources

Books

Frist, William. *When Every Moment Counts: What You Need to Know About Bioterrorism from the Senate's Only Doctor.* Lanham, MD: Rowman and Littlefield, 2002.

McCuen, Gary E. *Poison in the Wind: The Spread of Chemical and Biological Weapons.* Hudson: GEM Publications, Inc., 1992.

Tucker, Jonathan B. *Scourge: The Once and Future Threat of Smallpox.* New York: Atlantic Monthly Publishers, 2001.

Periodicals

Drotman, D. Peter, et al. "Bioterrorism-Related Anthrax." *Emerging Infectious Diseases* 8, no. 10 (October, 2002).

Inglesby, T.V., D.T. Dennis, D.A. Henderson, et al., for the Working Group on Civilian Biodefense. "Plague as a Biological Weapon: Medical and Public Health Management." *Journal of the American Medical Association* 283 (2000): 2281–90.

Other

United States Centers for Disease Control. "Bioterrorism and Public Health Preparedness." [cited October 20, 2002]. <http://www.cdc.gov/od/oc/media/presskit/bio.htm>.

Brian R. Cobb

Biology

Biology is the scientific study of all forms of life, including plants, animals, and **microorganisms**.

Among the numerous fields in biology are microbiology, the study of microscopic organisms like **bacteria**; **cytology**, the study of cells; **embryology**, the study of development; **genetics**, the study of heredity; **biochemistry**, the study of the chemical structures in living things; morphology, the study of the **anatomy** of plants and animals; **taxonomy**, the identification, naming, and classification of organisms; and **physiology**, the study of how organic systems function and respond to stimulation. Biology often interfaces with subjects like **psychology**. For example, **animal** behaviorists would need to understand the biological nature of the animal they are studying in order to evaluate the animal's **behavior**.

Important discoveries in biological science

The history of biology begins with the careful observation of the external aspects of organisms and continues with investigations into the functions and interrelationships of living things.

The ancient Greek philosopher Aristotle is credited with establishing the importance of observation and analysis as the basic approach for scientific investigation. By A.D. 200, studies in biology were centered in the Arab world. Most of the investigations during this period were made in medicine and agriculture. Arab scientists continued this activity throughout the Middle Ages.

When ancient Greek and Roman writings were revived in **Europe** during the Renaissance, scientific investigations began to accelerate. Leonardo da Vinci and Michelangelo, Italian Renaissance artists, produced detailed anatomical drawings of human beings. At the same time others were dissecting cadavers (dead bodies) and describing internal anatomy. By the seventeenth century, formal experimentation was introduced into the study of biology. William Harvey, an English physician, demonstrated the circulation of the **blood** and so initiated the biological discipline of physiology.

So much work was being done in biological science during this period that academies of science and scientific journals were formed, the first of which being the Academy of the Lynx in Rome in 1603. In Massachusetts, the Boston Philosophical Society was founded nearly a hundred years before the American Revolution. The first scientific journals were established in 1665 with the *Journal des Savants* (France) and in Great Britain with the *Philosophical Transactions of the Royal Society.*

KEY TERMS

Genetic engineering—The manipulation of genetic material to produce specific results in an organism.

Germ theory of disease—The theory that disease is caused by germs.

Metabolism—The chemical changes within cells that produce energy for vital organism activity and the assimilation of nutrients.

Microorganisms—Living units that cannot be seen without magnification under a microscope.

Molecular biology—The study of the cellular structure of living units.

Prokaryote—A cell that does not have a distinct nucleus, such as bacterium or alga.

Spontaneous generation—The theory that disease was caused spontaneously, not from germs.

The invention of the light **microscope** opened the way for biologists to investigate living organisms at the cellular level, and ultimately at the molecular level. The first drawings of magnified life were made by Francesco Stelluti, an Italian who published drawings of a honeybee at a 10-times magnification in 1625.

During the eighteenth century, Carolus Linnaeus proposed a system for naming and classifying plants and animals which is still used today. In his book, *Species plantarum*, which was written in 1753, Linnaeus described 6,000 plants, each one assigned a binomial name—genus and **species**. For example, the binomial name for the wolf is *Canis lupus*, and for humans, *Homo sapiens*. In the nineteenth century, many explorers contributed to biological science by collecting **plant** and animal specimens from around the world. In 1859, Charles Darwin published *On the Origin of Species,* in which he outlined the theory of **evolution** by means of natural **selection**. This was an important discovery; it disproved the idea that organisms generated spontaneously. Later, French chemist Louis Pasteur confirmed Darwin's findings by the discovery of certain bacteria caused diseases. Pasteur also developed the first vaccines. By the end of the nineteenth century the **germ theory** of **disease** was established by Robert Koch, and by the early twentieth century, chemotherapy was developed. The use of **antibiotics** began with penicillin in 1928 and steroids were discovered in 1935.

From the nineteenth century until the present, the amount of research and discovery in biology has been voluminous. Two fields of rapid growth in biological science today are **molecular biology** and **genetic engineering**.

See also Biodiversity; Biological community; Biological rhythms; Botany; Ecology; Ecosystem; Evolution, convergent; Evolution, divergent; Evolution, evidence of; Evolution, parallel; Evolutionary change, rate of; Evolutionary mechanisms.

Resources

Books

Byatt, Andrew, et al. *Blue Planet.* New York: D.K. Publishing, 2002.

Campbell, Neil A., and Jane B. Reese. *Biology.* 6th ed. San Francisco: Benjamin/Cummings, 2001.

Perlman, Dan L., and Edward O. Wilson *Conserving Earth's Biodiversity.* Washington, DC: Island Press, 2000.

Purves, William K. *Life: The Science of Biology.* 6th ed. New York: W.H. Freeman, 2001.

Starr, Cecie, and Ralph Taggart. *Biology: The Unity and Diversity of Life.* Pacific Grove, CA: Brooks/Cole Pub Co, 2000.

Wilson, Edward O. *The Diversity of Life.* New York: W.W. Norton, 1999.

Other

WGBH Educational Foundation and Clear Blue Sky Productions. "Evolution: A Journey into Where We're From and Where We're Going." 2001 [cited January 15, 2003]. <http://www.pbs.org/wgbh/ evolution/>

Vita Richman

Bioluminescence

Bioluminescence is the production of **light** by living organisms. Some single-celled organisms (**bacteria** and **protista**) as well as many multicellular animals and **fungi** demonstrate bioluminescence.

Bioluminescence in nature

Marine environments support a number of bioluminescent organisms including **species** of bacteria, dinoflagellates, **jellyfish**, coral, **shrimp**, and **fish**. On any given night one can see the luminescent sparkle produced by the single-celled dinoflagellates when **water** is disturbed by a ship's bow or a swimmer's motions. Many multicellular marine organisms have specialized light emitting organs that project light in a particular direction or convey a unique shape to the light. The **anglerfish** has a light-emitting **organ** that projects from its head, which serves as a bait to attract smaller **prey** fish. The light emitted from this organ in the anglerfish is actually produced by bacteria, living in a symbiotic relationship in which both the fish and bacteria profit from their shared existence.

Bioluminescent organisms in the terrestrial environment include species of fungi and **insects**. The most familiar of these is the firefly, which can often be seen glowing during the warm summer months. In some instances organisms use bioluminescence to communicate, such as in fireflies, which use light to attract members of the opposite sex. Certain reef fish use light produced from organs under their eyes to illuminate the interior of crevices and caves. This not only helps the fish to navigate, but also allows it to locate prey. Organisms that are unpalatable or dangerous, such as jellyfish, use bioluminescence as a signal to warn off attacks by predators. A newly-discovered deep sea **octopus** has bioluminescent organs in place of suckers.

Biochemical mechanism

Light is produced by most bioluminescent organisms when a chemical called luciferin reacts with **oxygen** to produce light and oxyluciferin. The reaction between luciferin and oxygen is catalyzed by the **enzyme** luciferase. Luciferases, like luciferins, usually have different chemical structures in different organisms.

In addition to luciferin, oxygen, and luciferase, other molecules (called cofactors) must be present for the bioluminescent reaction to proceed. Cofactors are molecules required by an enzyme (in this case luciferase) to perform its catalytic function. Common cofactors required for bioluminescent reactions are **calcium** and ATP, a **molecule** used to store and release **energy** that is found in all organisms.

The terms luciferin and luciferase were first introduced in 1885. The German scientist Emil Du Bois-Reymond obtained two different extracts from bioluminescent clams and **beetles**. When Du Bois mixed these extracts they produced light. He also found that if one of these extracts was first heated, no light would be produced upon mixing. Heating the other extract had no effect on the reaction, so Reymond concluded that there were at least two components to the reaction. Reymond hypothesized that the heat-resistant chemical undergoes a chemical change during the reaction, and called this compound luciferin. The **heat** sensitive chemical, Reymond concluded, was an enzyme which he called luciferase.

Bioluminescence as a research tool

The two basic components needed to produce a bioluminescent reaction, luciferin and luciferase, can be isolated from the organisms that produce them. When they are mixed in the presence of oxygen and the appropriate cofactors, these components will produce light with an intensity dependent on the quantity of luciferin and lu-

Fireflies have a bioluminescent organ in their abdomen that they use to attract mates. Enzymes within the organ react with oxygen to produce light. The insect controls the flashes by regulating the flow of oxygen. *Photograph by Roy Morsch. Stock Market. Reproduced by permission.*

ciferase added as well as the oxygen and cofactor concentrations. Luciferases isolated from fireflies and other beetles are commonly used in research.

Scientists have used isolated luciferin and luciferase to determine the concentrations of important biological molecules such as ATP and calcium. After adding a known amount of luciferin and luciferase to a **blood** or **tissue sample**, the cofactor concentrations may be determined from the intensity of the light emitted. Scientists have also found numerous other uses for the bioluminescent reaction such as using it to quantify specific molecules that do not directly participate in the bioluminescence reaction. To do this, scientists attach luciferase to antibodies—molecules produced by the **immune system** that bind to specific molecules called antigens. The antibody-luciferase complex is added to a sample where it binds to the molecule to be quantified. Following washing to remove unbound antibodies, the molecule of interest can be quantified indirectly by adding luciferin and measuring the light emitted. Methods used to quantify particular compounds in biological samples such as the ones described here are called assays.

Luciferase is often used as a "reporter gene" to study how individual genes are activated to produce protein or repressed to stop producing protein. Most genes are turned on and off by DNA located in front of the part of the **gene** that codes for protein. This region is called the gene promoter. A specific gene promoter can be attached to the DNA that codes for firefly luciferase and introduced into an **organism**. The activity of the gene promoter can then be studied by measuring the bioluminescence produced in the luciferase reaction. Thus, the luciferase gene can be used to "report" the activity of a promoter for another gene.

KEY TERMS

Antigen—A molecule, usually a protein, that the body identifies as foreign and toward which it directs an immune response.

Assay—Method used to quantify a biological compound.

ATP—Adenosine triphosphate; a high energy molecule that cells use to drive energy-requiring processes such as biosynthesis, transport, growth, and movement.

Cofactor—Molecule required by an enzyme to perform its catalytic function.

Dinoflagellate—Chloroplast containing protists that primarily inhabit marine environments.

DNA—Deoxyribonucleic acid; the genetic material in a cell.

Enzyme—Biological molecule, usually a protein,

which promotes a biochemical reaction but is not consumed by the reaction.

Extract—Solution from a biological material that contains an active compound.

Gene promoter—Regions of DNA that control gene activity.

Luciferase—An enzyme that catalyzes the reaction between oxygen and luciferin.

Luciferin—Complex carbon molecules that produce light when oxidized.

Oxidation—The process where a molecule loses one or more electrons.

Oxyluciferin—An oxidized luciferin molecule which is the product of a bioluminescent reaction.

Protista—Kingdom composed of single-celled organisms whose DNA is enclosed by a nucleus.

In recent studies, luciferase has been used to study viral and bacterial infections in living animals and to detect bacterial contaminants in food. The luciferase reaction also is used to determine DNA sequences, the order of the four types of molecules that comprise DNA and code for **proteins**.

Resources

Books

Herring, Peter, Anthony Campbell, Michael Whitfield, and Linda Maddock, eds. *Light and Life in the Sea*. Cambridge, England: Cambridge University Press, 1990.

Herring, Peter, ed. *Bioluminescence in Action*. New York: Academic Press, 1978.

Purves, William, Gordon Orians, and H. Heller. *Life: The Science of Biology*. 3rd ed. Sunderland, Massachusetts: Sinaur Associates, Inc., 1992.

Smith, D.C., and A.E. Douglas. *The Biology of Symbiosis*. Baltimore: Edward Arnold, 1987.

Steven MacKenzie

Biomagnification

Biomagnification (or **bioaccumulation**) refers to the ability of living organisms to accumulate certain chemicals to a **concentration** larger than that occurring in their inorganic, non-living environment, or in the case of animals, in the food that they eat. Of course, organ-

isms accumulate any chemical needed for their **nutrition**. In environmental science, however, the major focus of biomagnification is the accumulation of certain nonessential chemicals, especially certain **chlorinated hydrocarbons** that are persistent in the environment, insoluble in **water**, but highly soluble in fats. Because almost all fats within ecosystems occur in the living bodies of organisms, chlorinated hydrocarbons such as DDT and PCBs tend to selectively accumulate in organisms. This can lead to ecotoxicological problems, especially for top predators at the summit of ecological food webs.

Biomagnification and food-web accumulation

Organisms are exposed to a myriad of chemicals in their environment. Some of these chemicals occur in trace concentrations in the environment, and yet they may be selectively accumulated by organisms to much larger concentrations that can cause toxicity. This tendency is referred to as biomagnification, or bioaccumulation.

Some of the biomagnified chemicals are elements such as selenium, mercury, or nickel, or organic compounds of these such as methylmercury. Diverse others are in the class of chemicals known as chlorinated hydrocarbons (or organochlorines). These are extremely insoluble in water, but are freely soluble in organic solvents, including **animal** fats and **plant** oils (these are collectively known as lipids). Many of the chlorinated hydrocarbons are also very persistent in the environment, because they are not easily broken down to simpler chemicals through the **metabolism** of **microorganisms**, or by ultraviolet **radiation** or other inorganic processes. Common examples of bioaccumulating chlorinated hydrocarbons are the **insecticides** DDT and dieldrin, and a class of industrial chemicals known as PCBs.

Food-web accumulation is a special case of biomagnification, in which certain chemicals occur in their largest ecological concentration in predators at the top of the food web. An ecological food web is a complex of **species** that are linked through their trophic interactions, that is, their feeding relationships. In terms of **energy** flow, food webs are supported by inputs of solar energy, which is fixed by green plants through **photosynthesis**. Some of this fixed energy is used by the plants in their own **respiration**, and the rest, as plant **biomass**, is available to be passed along to animals, which are incapable of metabolizing any other type of energy. Within the food web, animals that eat plants are known as herbivores. These are eaten by first-level carnivores, which in turn may be eaten by higher-level carnivores. Top predators occur at the summit of the food web. In general, food webs have a pyramidal structure, with plant productivity being much greater than that of herbivores, and these being more productive than their predators. Top predators are usually quite uncommon. Within food webs, biomagnifying chemicals such as DDT, dieldrin, and PCBs have their largest concentrations, and cause the greatest damage, in top predators.

Biomagnification of some inorganic chemicals

All of the naturally occurring elements occur in the environment in at least trace concentrations. This ubiquitous **contamination** is always detectable, as long as the analytical **chemistry** methodology has detection limits that are small enough. About 25 of the elements are required by plants and/or animals, including the micronutrients **copper**, **iron**, molybdenum, zinc, and rarely, **aluminum**, nickel, and selenium. However, under certain ecological conditions these micronutrients can biomagnify to very large concentrations, and even cause toxicity to organisms.

One such case refers to serpentine **soil** and the vegetation that grows in it. Serpentine **minerals** contain relatively large concentrations of nickel, cobalt, chromium, and iron, and soils derived from this mineral can be toxic to plants. However, some plants grown on serpentine soils are physiologically tolerant of these metals, and can bioaccumulate them to very large concentrations. For example, the normal concentration of nickel in plants is about 1–5 ppm (parts per million, a concentration equiv-

alent to mg/kg). However, on sites with serpentine soils much larger concentrations of nickel occur in plant foliage and other tissues. Nickel concentrations as large as 16% occur in tissues of a plant in the mustard family, *Streptanthus polygaloides,* in California, and 11-25% nickel occurs in the blue-colored latex of *Sebertia acuminata* on the **island** of New Caledonia in the Pacific **Ocean**. It is common for plants growing on serpentine soils to have nickel concentrations of thousands of parts per million, which is usually considerably larger than the concentration in soil.

Another case of biomagnification occurs on some sites in semiarid regions in which the soil is contaminated by selenium, which may then be hyperaccumulated (i.e., extremely accumulated) by specialized species of plants. These plants are poisonous to grazing **livestock** and other large animals, causing a toxic reaction called "blind staggers." The most important selenium-accumulating plants in **North America** are milk vetches in the genus *Astragalus,* in the legume family. There are 500 species of *Astragalus* in North America, of which 25 are accumulators of selenium. The foliage of these plants can contain thousands of ppm of selenium, to a maximum of about 15,000 ppm, much larger than the concentration in soil. Sometimes, accumulator and non-accumulator *Astragalus* species grow together, as in the case of a place in Nebraska with 5 ppm selenium in soil, and 5,560 ppm in *Astragalus bisulcatus,* but only 25 ppm in *A. missouriensis.*

Mercury can also be biomagnified from trace concentrations in the environment. In this case, trace concentrations of mercury in water can result in large contaminations of **fish** and other predators. For example, fish species known to bioaccumulate mercury in offshore waters of North America include Atlantic **swordfish**, Pacific blue marlin, tunas, and halibut, among others. These fish can accumulate mercury from trace concentrations in seawater (less than 0.1 ppm, equivalent to 0.0001 ppb or 0.1 mg/L) to concentrations in flesh that commonly exceed 0.5 ppm fresh weight (f.w.), the maximum acceptable concentration in fish for human consumption. The contamination of oceanic fish by mercury is probably natural, and is not only a modern phenomenon. Studies have found no difference in mercury contaminations of modern **tuna** and museum specimens collected before 1909, or concentrations in feathers of pre-1930 and post-1980 seabirds collected from islands in the northeast Atlantic Ocean. In this phenomenon of mercury biomagnification, there is a tendency for larger, older fish to have relatively large concentrations. In a study of Atlantic swordfish, for example, the average mercury concentration of animals smaller than 51 lb (23 kg) was 0.55 ppm, compared with 0.86 ppm for those 51–99 lb (23–45 kg) in weight, and 1.1 ppm for those heavier than 99 lb (45 kg). Large concentrations of mercury also occur in fish-eating marine **mammals** and **birds** that are predators at or near the top of the marine food web.

Mercury biomagnification has also been observed in many **freshwater** ecosystems, even in some relatively remote lakes where this may be a natural occurrence. Mercury concentrations exceeding 0.5 ppm are often measured in fish from lakes and **rivers** in many parts of the United States and Canada. For example, about three-quarters of 1,500 lakes monitored in Ontario had at least some fish that exceeded 0.5 ppm f.w. in flesh. In one remote **lake** in northern Manitoba, the mercury concentration averaged 2 ppm f.w. in northern **pike**, and one **individual** had 5 ppm. As in the marine case, fresh-water fish that are top predators have the largest concentrations of mercury, and older and larger individuals are the most contaminated. Because of the common occurrence of large mercury concentrations in fish in certain regions, some governments have developed fish-consumption advisories and/or restrictions. In Sweden about 250 lakes have been "blacklisted" in terms of fish consumption, and 9,400 other lakes are candidates for that status. In Ontario about 1,200 lakes have restrictions on fish consumption. Some fish-eating **wildlife**, such as **loons** and **mink**, may also be affected by mercury in their food.

Biomagnificaiton of some chlorinated hydrocarbons

Chlorinated hydrocarbons such as the insecticides DDT, DDD, dieldrin, and methoxychlor, the dielectric fluids known as PCBs, and the chlorinated **dioxin**, TCDD, have a very sparse **solubility** in water. As a result, these chemicals cannot be "diluted" into this ubiquitous solvent, which is so abundant on the surface of **Earth** and in organisms. Therefore, even situations considered to be highly contaminated by chlorinated hydrocarbons have very small concentrations of these chemicals in water, typically less than 1 ppb, and in the case of the dioxin TCDD smaller than 1 ppt (i.e., 0.001 ppb). However, chlorinated hydrocarbons are highly soluble in lipids. Because most lipids within ecosystems occur in biological tissues, the chlorinated hydrocarbons have a strong affinity for living organisms, and they tend to biomagnify by many orders of magnitude from vanishingly small aqueous concentrations. Furthermore, because chlorinated hydrocarbons are persistent in the environment, they accumulate progressively as organisms grow older, and they accumulate into especially large concentrations in top predators, as described previously. In some cases, older individuals of top-predator animals such as raptorial birds and fish-eating marine mammals have been found to have thousands of ppm of DDT and PCBs in their fatty tissues. The toxicity caused by these animals accumulated exposures to DDT, PCBs, and other chlorinated hydrocarbons is a well-recognized environmental problem.

The biomagnification and food-web accumulation characteristics of DDT are especially well known. Typically, DDT has extremely small concentrations in air and water, and, to a lesser degree in soil. However, concentrations are much larger in organisms, especially in animals at or near the top of their food web, such as humans and predatory birds. The food-web biomagnification of DDT can be illustrated by the case of Lake Kariba, Zimbabwe. Although banned in most industrialized countries since the early 1970s, DDT is still used in many tropical countries for agriculture purposes and to control insect vectors of human diseases. The use of DDT in agriculture was banned in Zimbabwe in 1982, but DDT continues to be used to control **mosquitoes** and tsetse **flies**, **insects** that spread **malaria** and diseases of livestock. The concentration of DDT in the water of Lake Kariba was less than 0.002 ppb, but concentrations in sediment were 0.4 ppm (because sediment contains a relatively large concentration of organic **matter**, it contains much more DDT than the overlying water). Planktonic **algae** contained 2.5 ppm. A filter-feeding mussel had 10 ppm (values for animal tissues are for DDT in **fat**), while two species of plant-eating fish contained 2 ppm, and a bottom-feeding fish contained 6 ppm. A predatory fish and a fish-eating bird, the great cormorant, contained 5-10 ppm. The Nile crocodile is the top **predator** in Lake Kariba (other than humans), and it had 34 ppm. Therefore, the data for Lake Kariba illustrates a substantial biomagnification of DDT from water, and to a lesser degree from sediment, as well as a marked food-web accumulation from herbivores to top carnivores.

The widespread occurrence of food-web biomagnification of DDT and other chlorinated hydrocarbons caused chronic, ecotoxicological damage to birds and mammals of many species, even in habitats remote from sprayed sites. In some species, effects on predatory birds were severe enough to cause large declines in abundance beginning in the early 1950s, and resulting in local or regional losses of populations. Prominent examples of North American birds that suffered population decreases because of exposure to chlorinated hydrocarbons include bald eagle, golden eagle, **peregrine falcon**, osprey, brown pelican, and double-crested cormorant, among others. However, since the banning of the use of DDT in North America in the early 1970s, these birds have increased in abundance. In the case of the peregrine falcon, this increase was enhanced by a captive-breeding and release program over much of its former range in eastern North America.

See also Food chain/web.

KEY TERMS

. . . .

Biomagnification—Tendency of organisms to accumulate certain chemicals to a concentration larger than that occurring in their inorganic, non-living environment, such as soil or water, or in the case of animals, larger than in their food.

Ecotoxicology—The study of the effects of toxic chemicals on organisms and ecosystems. Ecotoxicology considers both direct effects of toxic substances and also the indirect effects caused, for example, by changes in habitat structure or the abundance of food.

Food-web accumulation—Tendency of certain chemicals to occur in their largest concentration in predators at the top of the ecological food web. As such, chemicals such as DDT, PCBs, and mercury in the aquatic environment have their largest concentrations in predators, in comparison with the non-living environment, or with plants and herbivores.

Hyperaccumulation—A syndrome in which a chemical is bioaccumulated to an extraordinary degree.

Resources

Books

Freedman, B. *Environmental Ecology.* 2nd ed. San Diego: Academic Press, 1994.

Moriarty, F. *Ecotoxicology,* 2nd ed. London: Academic Press, 1988.

Smith, R.P. *A Primer of Environmental Toxicology.* Philadelphia: Lea & Febiger, 1992.

Bill Freedman

Biomass

Biomass consists of living organisms, or parts of living organisms, as well as waste products and incompletely decomposed remains of living organisms. The term is quite encompassing and includes plants (referred to as phytomass), microbes, and **animal** material, or zoomass. Biomass density is a distinguishing feature of ecological systems and is usually presented as the amount of dry biomass per unit area. To insure a uniform basis for comparison, biomass samples are dried at 221°F (105°C) until they reach a constant weight.

In most settings, phytomass is by far the most important component. A square yard (0.84 m^2) of the planet's land area has, on average, about 18–22 lb (8.4–10 kg) of phytomass, although values may vary widely depending on the type of **biome**. Tropical rain **forests** contain four or five times the average while **desert** biomes may have a value near **zero**. The global average for non-plant biomass is approximately 1% of the total. Organic compounds typically constitute about 95% by weight of biomass, and inorganic compounds account for the remaining 5%. An exception occurs in **species** that incorporate large amounts of inorganic elements such as silicon or **calcium**, in which case the inorganic portion may be several times higher.

Photosynthesis is the principle agent for biomass production. **Light energy** is used by **chlorophyll** containing green plants to remove (or fix) **carbon dioxide** from the atmosphere and convert it to energy rich organic compounds or biomass. It has been estimated that on the face of the **earth** approximately 200 billion tons of **carbon** dioxide are converted to biomass each year. Carbohydrates are usually the primary constituent of biomass, and **cellulose** is the single most importan t component. Starches are also important and predominate in storage organs such as tubers and rhizomes. Sugars reach high levels in **fruits** and in plants such as **sugar-cane** and **sugar beet**. Lignin is a very significant non-carbohydrate constituent of woody **plant** biomass.

Biome

A biome is a major, geographically extensive **ecosystem**, structurally characterized by its dominant life forms. Terrestrial biomes are usually distinguished on the basis of the major components of their mature or climax vegetation, while aquatic biomes, especially marine ones, are often characterized by their dominant animals. Most of the oceans are considered part of a single biome, although areas with particularly unusual or unique physical characteristics or inhabitants may be considered as separate biomes.

Similar biomes can occur in widely divergent places as long as the environmental conditions are appropriate for their development. Some environmental conditions affecting the location of biomes include climate, latitude, topography, and fire. Often, different **species** having similar, convergent growth forms will dominate at different places within the same biome. For example, the boreal coniferous forest occurs in suitable environments of northern **North America** and Eurasia. In northeastern North America this biome is dominated by stands of black **spruce**, while in the northwest white spruce is dominant. Norway spruce is most important in this biome in northwestern **Europe**, while in parts of Siberia species of pine and larch are dominant. Because biomes are described according to the structural characteristics of their dominant organisms (in this example, coniferous trees growing under a particular climatic regime) all of these different forest types are considered convergent ecosystems within the same biome, the boreal coniferous forest.

Major biomes and their characteristics

Most biomes are delineated on the basis of their naturally occurring communities of plants, animals, and **microorganisms**. Exceptions are the so-called anthropogenic biomes, which are strongly influenced by humans and their activities, as in the case of cities and agroecosystems. However, it should be remembered that to some degree, all of Earth's biomes have been influenced by human activities. All organisms, for example, contain trace contaminations of certain organochlorine chemicals of human manufacture, such as DDT and PCBs.

Ecologists have used a number of systems to divide the **biosphere** into major biomes. A classification of global biomes, modified from one proposed by E.P. Odum, is described below:

Terrestrial biomes

Tundra

Tundra is a treeless biome occurring in areas with cold climates and a short growing season. Alpine tundra occurs at high altitudes on **mountains**, while arctic tundra occurs at high latitudes. Most tundras receive very small inputs of **water** as **precipitation**, but nevertheless their **soil** may be moist or wet because there is little **evaporation** in such cold climates, and deep drainage may be prevented by frozen soil. The coldest, most northern, high-arctic tundras are very unproductive and dominated by long-lived but short-statured plants, typically less than 1.97-3.94 in (5-10 cm) tall. Low-arctic tundras are dominated by shrubs as tall as 3.28 ft (1 m), while wet sites develop relatively productive meadows of sedge, cotton-grass, and grass. In North America, arctic tundras can support small densities of mammalian herbivores such as **caribou** and muskox (although during **migration** these animals can occur in locally large densities), and even smaller numbers of their predators, such as wolves.

Boreal coniferous forest

The boreal coniferous forest, or taiga, is an extensive northern biome occurring in moist climates with cold winters. The boreal forest is dominated by coniferous trees, especially species of fir, larch, pine, and spruce. Some broad-leaved, **angiosperm** trees are also important in the boreal forest, especially species of aspen, birch, poplar, and willow. Usually, particular stands of boreal forest are dominated by only one or several species of trees. Most regions of boreal forest are subject to periodic events of catastrophic disturbance, most commonly caused by **wildfire** and sometimes by **insects**, such as spruce budworm, that kill trees through intensive defoliation. Montane **forests**, also dominated by conifers and similar in structure to the boreal forest, can occur at sub-alpine altitudes on mountains in southerly latitudes.

Temperate deciduous forest

Forests dominated by species-rich mixtures of broad-leaved trees occur in relatively moist, temperate climates. Because these forests occur in places where the winters can be cold, the foliage of most species is seasonally deciduous, meaning that all leaves are shed each autumn and re-grown in the springtime. Common trees of this forest biome in North America are species of ash, **basswood**, birch, cherry, **chestnut**, dogwood, **elm**, hickory, **magnolia**, maple, oak, tulip-tree, and walnut, among others. These various **tree** species segregate into intergrading communities on the basis of site variations of soil moisture, fertility, and air **temperature**.

Temperate rainforest

Temperate rainforests develop under climatic regimes characterized by mild winters and an abundance of precipitation. Because these systems are too moist to support regular, catastrophic wildfires, they often develop into **old-growth forests**, dominated by coniferous trees of mixed age and species composition. **Individual** trees can be extremely large, and in extreme cases can be more than 1,000 years old. Common trees of this biome are species of Douglas fir, hemlock, cedar, redwood, spruce, and yellow cypress. In North America, temperate rainforests are best developed on the humid west coast.

Temperate grassland

These **grasslands** occur under temperate climatic regimes that are intermediate to those that support forest and **desert**. In the temperate zones, grasslands typically occur where rainfall is 9.9-24 in (25-60 cm) per year. Grasslands in North America are called **prairie** (they are often called steppe in Eurasia), and this biome occupies vast regions in the interior. The prairie is often divided into three types according to height of the dominant vegetation—tall-grass, mixed-grass, and short-grass. The once extensive tall-grass prairie is dominated by various

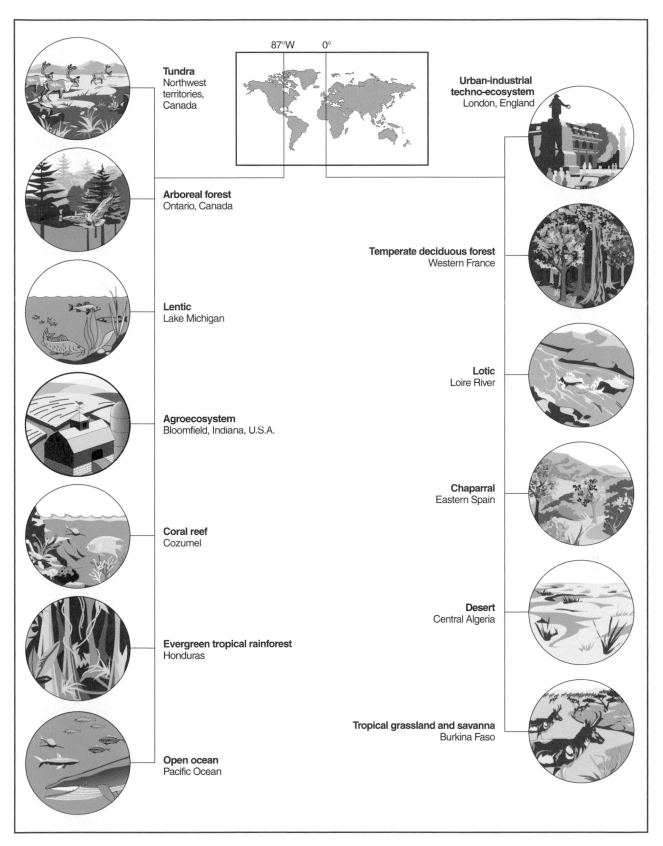

Tundra
Northwest territories, Canada

Arboreal forest
Ontario, Canada

Lentic
Lake Michigan

Agroecosystem
Bloomfield, Indiana, U.S.A.

Coral reef
Cozumel

Evergreen tropical rainforest
Honduras

Open ocean
Pacific Ocean

87°W   0°

Urban-industrial techno-ecosystem
London, England

Temperate deciduous forest
Western France

Lotic
Loire River

Chaparral
Eastern Spain

Desert
Central Algeria

Tropical grassland and savanna
Burkina Faso

Biomes along 87 degrees west longitude and along 0 degrees longitude. *Illustration by Hans & Cassidy. Courtesy of Gale Group.*

species of **grasses** and herbaceous broad leaved plants such as sunflowers and blazing stars, some as tall as 9.8-13.1 ft (3-4 m). Fire was an important natural factor that prevented much of the tall-grass prairie from developing into an open forest. The tall-grass prairie is now an endangered natural ecosystem, because it has been almost entirely converted to agriculture. The mixed-grass prairie occurs where rainfall is less, and it supports shorter species of grasses and other herbaceous plants. The short-grass prairie has even less precipitation, and is subject to unpredictable years of severe **drought**.

Tropical grassland and savanna

Tropical grasslands can occur in regions with as much as 47.2 in (120 cm) of rainfall per year, but under highly seasonal conditions with a pronounced dry season. Savannas are dominated by grasses and other herbaceous plants, but they also have scattered shrub and tree-sized woody plants, which form a very open canopy. Tropical grasslands and savannas can support a great seasonal abundance of large, migratory **mammals**, as well as substantial populations of resident animals. This is especially true of savannas in **Africa**, where this biome supports a very diverse assemblage of large mammals, including **gazelles** and other antelopes, rhinos, **elephant**, hippopotamus, and buffalo, and some of their predators, such as lion, cheetah, wild dog, and **hyena**.

Chaparral

Chaparral is a temperate biome that develops in environments with seasonally extreme moisture gradients, characterized by a so-called Mediterranean climate with winter rains and summer drought. The chaparral biome is typically composed of dwarf forest and shrubs, and interspersed herbaceous vegetation. Chaparral is highly prone to events of catastrophic wildfire. In North America, chaparral is best developed in parts of the southwest, especially coastal southern California.

Desert

Desert is a temperate or tropical biome, commonly occurring in the center of continents, and in the rain shadows of mountains. The distribution of this biome is determined by the availability of water, generally occurring where there is less than 9.9 in (25 cm) of precipitation per year. Not surprisingly, the productivity of desert ecosystems is strongly influenced by the availability of water. The driest deserts support almost no **plant** productivity, while less-dry situations may support communities of herbaceous, succulent, and annual plants, and somewhat moister places will allow a shrub-dominated ecosystem to develop.

Semi-evergreen tropical forest

This type of tropical forest develops when there is a seasonality of water availability due to the occurrence of pronounced wet and drier **seasons** during the year. Because of this seasonality, most of the trees and shrubs of this biome are seasonally deciduous, meaning that they shed their foliage in anticipation of the drier season. This biome supports a great richness of plant and **animal** species, though somewhat less than in tropical rainforests.

Evergreen tropical rainforest

This biome occurs under tropical climates with abundant precipitation and no seasonal drought. Because wildfire and other types of catastrophic disturbance are uncommon in this sort of climate regime, tropical rainforests usually develop into old-growth forests. As such, they contain a diverse size range of trees, a great richness of species of trees and other plants, as well as an extraordinary diversity of animals and microorganisms. Many ecologists consider this biome to represent the epitome of ecosystem development on land, because of the enormous variety of species that are supported under relatively benign climatic conditions in old-growth tropical rainforests.

Freshwater biomes

Lentic ecosystems, such as lakes and ponds, occur in basins containing standing water. Depending on the rates of input and output of water, the flushing time of these ecosystems can range from days, in the case of small pools, to centuries, in the case of the largest lakes. The biological character of lentic ecosystems is strongly influenced by water **chemistry**, especially nutrient **concentration** and water transparency. Waters with a large nutrient supply are highly productive, or eutrophic, while infertile waters are unproductive, or oligotrophic. Commonly, shallow water bodies are much more productive than deeper water bodies of the same surface area. However, plant productivity is also influenced by how far **light** can penetrate into the water column. This factor is restricted in waters with large concentrations of turbidity associated with silt, or with a brown **color** caused by dissolved organic **matter**. In such cases, primary productivity is smaller than might be expected on the basis of nutrient supply. The ecosystems of lentic waters are usually highly zonal in two dimensions. Horizontal zonation is associated with variations of water depth, usually related to slope and length of the shoreline. Vertical zonation of deeper waters is determined by light availability, water temperature, and nutrient and **oxygen** concentrations. There are also distinct, benthic ecosystems in the sediment of lentic ecosystems.

Lotic

The character of this running-water biome is determined by physical factors, especially the quantity, **velocity**, and seasonal variations of water flow. These hydrologic characteristics influence other important characteristics of lotic ecosystems. For example, the bottom tends to be muddy in places with calm water flows where silt is deposited, and rocky in more vigorous places where fine particles are selectively eroded from the bottom. Similarly, turbidity is great during high water flows, and this interferes with the penetration of light, restricting plant productivity. Although they sustain some primary productivity of aquatic plants, the common lotic ecosystems such as **rivers**, streams, and brooks are not usually self-supporting in terms of fixed **energy**. These ecosystems typically rely on inputs of organic matter from their surrounding, terrestrial **watershed**, or from upstream lakes to support much of their productivity of aquatic **invertebrates** and **fish**.

Wetlands

Freshwater wetlands (or mires) occur in shallow waters, usually having pronounced seasonal variations of water depth, sometimes including dry periods during which water does not occur at the surface. The four major wetland types are: marsh, swamp forest, bog, and fen. Marshes are the most productive wetlands, and are typically dominated by relatively tall, emergent species of angiosperm plants such as reed, cat-tail, and bulrush, and by floating-leaved plants such as water lily and lotus. Swamps are forested wetlands, seasonally or permanently flooded, and in North America, dominated by tree-sized plants such as bald cypress or silvermaple. Bogs are acidic, unproductive wetlands that develop in relatively cool but wet climates. Bogs depend on atmospheric inputs for their supply of **nutrients**, and are typically dominated by species of sphagnum **moss**. Fens also develop in cool and wet climates, but they have a better nutrient supply than bogs, and are consequently less acidic and more productive.

Marine biomes

Open ocean

The character of the open-water, or pelagic oceanic biome is determined by physical and chemical environmental factors, particularly waves, **tides**, **currents**, salinity, temperature, light intensity, and nutrient concentration. Primary productivity of this biome is small, and comparable to some of the least productive terrestrial biomes, such as deserts. Primary production in the open **ocean** is carried out by **phytoplankton** of diverse species, ranging in size from extremely small photosyn-

thetic **bacteria**, to larger but microscopic unicellular and colonial **algae**. The phytoplankton are grazed by small crustaceans known as **zooplankton**, and these are eaten in turn by small fish. At the top of the pelagic food web are very large predators such as bluefin **tuna**, **sharks**, **squid**, and whales. The deep benthic ecosystems of this biome are supported by a sparse rain of dead **biomass** from its surface waters. The benthic ecosystems are not well known, but they appear to be extremely stable, rich in species, and low in productivity.

Continental shelf waters

Near-shore waters of the oceans are relatively shallow, because they overlie continental shelves. Compared with the open ocean, waters over continental shelves are relatively warm, and are well supplied with nutrients. The nutrients originate with inputs from rivers, and from occasional movements of deeper, more fertile waters to the surface, stirred from the bottom by **turbulence** caused by storms. Mostly as a result of the nutrient inputs, the phytoplankton of these waters are relatively productive, and they support a larger biomass of animals than occurs in the open ocean. Some of the world's most important pelagic and benthic fisheries are supported by the **continental shelf** biome, for example, those in the North Sea and Barents Sea of western Europe, the Grand Banks and other shallow waters of northeastern North America, the Gulf of Mexico, and inshore waters of much of western North America.

Upwelling regions

In certain places or regions, oceanographic conditions favor upwellings to the surface of relatively deep, nutrient rich waters. Because of the enhanced nutrient supply, **upwelling** areas are relatively fertile, and they sustain this highly productive, open-ocean biome. Because of their large foundation of primary production, upwelling regions support sizable populations of animals, including large species of fish and shark, marine mammals, and seabirds. Some of Earth's most productive fisheries occur in upwelling areas, such as that off the west coast of Peru and elsewhere off **South America**, and large regions of the Antarctic Ocean.

Estuaries

Estuaries are a complex group of coastal ecosystems that are semi-enclosed, but open to the sea. Estuaries display characteristics of both marine and freshwater biomes, because they typically have substantial inflows of fresh water from the nearby land, along with large fluctuations of **salt** water resulting from tidal cycles. Examples of estuaries include coastal bays, sounds, river mouths, salt marshes, and tropical mangrove forests. Because their large water-borne inputs of terrestrial nutrients are partially retained by their semi-enclosed water circulation, estuaries are highly productive ecosystems. Estuaries provide important **habitat** for juvenile stages of many commercially important species of fish, shellfish, and crustaceans, and they are often characterized as "nursery" habitat for these species.

Seashores

The seashore biome is a complex of ecosystems occurring at the interface of terrestrial and oceanic biomes. The local character of the seashore biome is determined by environmental factors, such as the intensity of wave action, the frequency of major events of disturbance, and bottom type. In temperate waters, hard-rock and cobble bottoms often develop ecosystems dominated by large species of macroalgae, broadly known as seaweeds, or kelp. In some cases, so-called kelp "forests" can develop. These are highly productive ecosystems, which maintain large quantities of biomass, mostly of macroalgae. In situations characterized by softer bottoms of **sand** or mud, ecosystems are typically dominated by benthic invertebrates, especially **mollusks**, echinoderms, crustaceans, and marine worms.

Coral reefs

Coral reefs are a distinctive marine biome of tropical seas, occurring locally in shallow but relatively infertile areas close to land. The physical structure of coral reefs is provided by the **calcium carbonate** exoskeletons of dead coral polyps. This structure supports a species-rich population of living coral, crustose algae, invertebrates, and fish. This biome is dominated by corals, a diverse group of coelenterate animals, living in **symbiosis** with unicellular algae. Because this symbiosis is highly efficient in the acquisition and recycling of nutrients, coral reefs typically sustain a high productivity, even though they occur in nutrient-poor waters.

Human-dominated biomes

Urban-industrial techno-ecosystems

This anthropogenic biome consists of large metropolitan districts, dominated by humans, human dwellings, businesses, factories, and other types of infrastructure. This biome supports many species in addition to humans, but, with few exceptions, these are non-native plants and animals that have been introduced from other places, and that cannot live independently outside of this biome, unless they are returned to their native biome.

Rural techno-ecosystems

This is another anthropogenic biome, occurring outside of intensively built-up areas, and consisting of certain components of the extensive technological infrastructure

---

KEY TERMS

Anthropogenic—A situation that occurs because of, or is influenced by, the activities of humans.

Biome—A geographically extensive ecosystem, usually characterized by its dominant life forms.

Community—In ecology, a community is an assemblage of populations of different species that occur together in the same place and at the same time.

Convergence—An evolutionary pattern by which unrelated species that fill similar ecological niches tend to develop similar morphologies and behavior. Convergence occurs in response to similar selection pressures.

Monoculture—An ecosystem dominated by a single species.

Species richness—The number of species occurring in a community, a landscape, or some other defined area.

---

of civilization. This biome is comprised of transportation corridors (such as highways, railways, transmission corridors, and aqueducts), small towns, and industries involved in the extraction, processing, and manufacturing of products from natural resources. Typically, this biome supports mixtures of **introduced species** and those native species that are tolerant of the disturbances and other stresses associated with human activities.

Agroecosystems

This biome consists of ecosystems that are managed and harvested for human use. The components of this biome are uneven in their anthropogenic influence. The most intensively managed agroecosystems typically involve monocultures of non-native crop species of plants or animals in agriculture, aquaculture, or **forestry**, and are not favorable to native **wildlife**. The management objective is to cultivate the economically valuable species under conditions that ensure optimal growth. Less-intensively managed agroecosystems involve mixtures of species, or polycultures, and these may provide habitat for some native wildlife species.

See also Continental shelf; Desert; Ecosystem; Forests; Grasslands; Lake; Ocean; Savanna; Tundra; Wetlands.

Resources

Books

Barbour, M.G., and W.D. Billings. *North American Terrestrial Vegetation.* New York: Cambridge University Press, 1988.

Begon, M., J.L. Harper, and C.R. Townsend. *Ecology. Individuals, Populations and Communities.* 2nd ed. London: Blackwell Sci. Pub., 1990.

Odum, E.P. *Ecology and Our Endangered Life-Support Systems.* Sunderland, MA: Sinauer Assoc. Ltd., 1993.

Walter, H. *Vegetation of the Earth.* New York: Springer-Verlag, 1977.

Periodicals

Clark, B.C. "Planetary Interchange of Bioactive Material: Probability Factors and Implications." *Origins of Life and Evolution of the Biosphere* no. 31 (2001): 185-197.

Bill Freedman

Biophysics

Biophysics is the integration and application of the principles of **physics** to explain and explore the form and function of living things. The most familiar examples of the role of physics in **biology** are the use of lenses to correct visual defects and the use of **x rays** to reveal the structure of bones. Principles of physics have been used to explain some of the most basic processes in biology such as **osmosis**, **diffusion** of gases, and the function of the **lens** of the **eye** in focusing **light** on the retina.

The understanding that living organisms obey the laws of physics as non-living systems do has had profound effects on the study of biology. The discovery of the relationship between **electricity** and muscle contraction by Luigi Galvani, an eighteenth-century physician, initiated a field of research that had continued to give information about the nature of muscle contraction and nerve impulses. It has led to the development of such instruments and devises as the electrocardiograph, electroencephalograph, and cardiac **pacemaker**. Medical technology in particular has benefited from the association of physics and biology. Medical imaging with 3-D diagnostic techniques such as computer tomographic (CAT) scanning, magnetic **resonance** imaging and positron emission tomography have permitted researchers to look inside living things without disrupting life processes. Today, lasers and x rays are routinely used in medical treatments.

The use of non-invasive imaging traces it roots to advances in the understanding of the fundamentals and biophysical interactions of **electromagnetism** during the nineteenth century. By 1900, physicist Wilhelm Konrad Roentgen's (1845–1923) discovery of high **energy** electromagnetic **radiation** in the form of x rays found use in medical **diagnosis**. Developments in **radiology** progressed throughout the first half of the twentieth century, finding extensive use in the treatment of soldiers during World War II.

Although nuclear medicine—heavily based upon advances in biophysics—traces its clinical origins to the 1930s, the invention of the scintillation camera in the 1950s brought nuclear medical imaging to the forefront of diagnostics.

The use of a wide array of instruments and techniques futhered by discoveries in physics, especially **electronics**, has helped biology to change from a descriptive science to an analytical one. An example of this is one of the most important events of this century, deciphering the structure of the DNA **molecule** using x-ray **diffraction**, a technique which has also been used to determine the structure of hemoglobin, viruses, and a variety of other biological molecules and **microorganisms**. The ability to apply information discovered in physics to the study of living things led to the development and use of the **electron microscope** and **ultracentrifuge**, instruments that have revealed much information about **cell** structure and function. Other applications have been sensors for **heat** and **pressure** detection that give information about body functions under a variety of conditions which have been of great importance in the **space** program.

See also Computerized axial tomography; Laser; Microscopy; Nanotechnology.

Bioremediation

Bioremediation is a type of **biotechnology** in which living organisms or ecological processes are utilized to deal with some environmental problem. The most common use of bioremediation is to metabolically break down or otherwise remove toxic chemicals before or after they have been discharged into the environment. In such uses, bioremediation takes advantage of the fact that certain **microorganisms** can utilize toxic chemicals as metabolic substrates, in the process rendering them into simpler, less toxic compounds. Bioremediation is a relatively new and actively developing technology.

In general, bioremediation methodologies focus on: (1) enhancing the abundance of certain **species** or groups of microorganisms that can metabolize toxic chemicals (this is also known as *bioaugmentation*) and/or (2) optimizing environmental conditions for the actions of these organisms (also known as *biostimulation*). Bioaugmentation may involve the deliberate addition of strains or species of microorganisms that are specifically effective at treating particular toxic chemicals, but are not indigenous to or abundant in the treatment area. Biostimulation usually involves **fertilization**, aeration, or **irrigation** in order to decrease the importance of environmental factors in limiting the activity of microorganisms. Biostimulation focuses on rapidly increasing the abundance of naturally occurring, ubiquitous microorganisms capable of dealing with certain types of environmental problems.

Bioremediation of spilled hydrocarbons

Accidental spills of **petroleum** or other hydrocarbons on land and **water** are regrettable but frequent occurrences. Such spills can range in size from a few gallons that may be spilled during refueling to enormous spillages of millions of tons as occurred to both the sea and land during the Gulf War of 1991. Once spilled, petroleum and its various refined products can be persistent environmental contaminants. However, these organic chemicals can also be metabolized by certain microorganisms, whose processes transform the toxins into simpler compounds, ultimately to **carbon dioxide**, water, and other inorganic chemicals.

Numerous attempts have been made to increase the rates by which microorganisms break down spilled hydrocarbons. In some cases, specially prepared concentrates of **bacteria** that are highly efficient at metabolizing hydrocarbons have been "seeded" into spill areas in an attempt to increase the **rate** of degradation of the spill residues. Although this technique has sometimes been effective, it commonly is not. This is because the indigenous microbial communities of soils and aquatic sediments contain many species of bacteria and **fungi** that are capable of utilizing hydrocarbons as a metabolic substrate. After a spill, the occurrence of large concentrations of hydrocarbons in **soil** or sediment stimulates rapid growth of those microorganisms. Consequently, seeding of microorganisms that are metabolically specific to hydrocarbons does not always make much of a difference to the overall rate of degradation.

More important, however, is the fact that the environmental conditions under which spill residues occur are almost always highly sub-optimal for their degradation by microorganisms. Most commonly, the rate of microbial breakdown of spilled hydrocarbons is limited by the availability of **oxygen** or of certain **nutrients** such as nitrate and phosphate. Therefore, the microbial breakdown of spilled hydrocarbons on land can be greatly enhanced by occasionally tilling the soil to keep conditions aerated and by fertilizing with **nitrogen** and **phosphorus** while keeping conditions moist but not wet. Thus, bioremediation systems for dealing with soils contaminated by spilled gasoline or petroleum can be based on simple tillage and fertilization.

Similarly, petroleum refineries may utilize a bioremediation process called land farming, in which oily wastes are spread onto land, which is then tilled and fertilized until microbes reduce the residue concentrations to an acceptable level.

After some petroleum spills, more innovative approaches may prove to be useful. For example, it is difficult to fertilize aquatic habitats, because the nutrients simply wash away and are therefore not effective for very long. In the case of the *Exxon Valdez* spill in Alaska in 1989, research demonstrated that nutrients could be applied to soiled beaches as an oleophilic (that is, oil-seeking), nitrogen and phosphorus-containing fertilizer. Because of its oleophilic nature, the fertilizer adhered to the petroleum residues and was able to significantly enhance the rate of oil degradation by the naturally occurring community of microorganisms. This treatment was applied to about 73 mi (118 km) of oiled beach and proved to be successful in speeding up the process of degradation of the residues by increasing the rate of oxidation by about 50%. No attempts were made in this case to "seed" the microbial community with species specifically adapted to metabolizing hydrocarbons. It was believed that hydrocarbon-specific microbes were naturally present in the beach sediment and that their activity and that of species with broader substrate tolerances only had to be enhanced by making the ecological conditions more favorable, that is, by fertilizing.

Bioremediation of metal pollution

Metals are common pollutants of water and land because they are emitted by many industrial, agricultural, and domestic sources. In some situations, organisms or ecological processes can be successfully utilized to concentrate metals that are dispersed in the environment, especially in water. The metals can then be removed from the system by harvesting the organisms. For example, **metal** polluted waste waters can be treated by encouraging the vigorous growth of certain types of **algae**, fungi, or vascular plants, usually by fertilizing the water within some sort of constructed lagoon. This bioremediation system works because the growing plants and microorganisms absorb metals from the water (acting as so-called biosorbents), and thereby reduce their concentrations to a more tolerable range. The plants can then be harvested to remove the metals from the bioremediation system. In some cases, the **plant biomass** may even be processed to yield metal products of economic value.

Bioremediation of acidification

In some situations, artificial **wetlands** can be engineered to treat acidic waters associated with **coal mining** or other sources of acidity. Coal mining disturbs soil and fractures **rocks** exposing large quantities of pyritic **sulfur** to atmospheric oxygen. Under such conditions, certain species of bacteria oxidize the sulfide of the mineral pyrites to sulfate, generating large quantities of acidity in the process known as *acid mine drainage*. The resulting acidity is often treated by adding large quantities of acid-neutralizing chemicals such as lime or limestone. However, it has also been recently demonstrated that natural, acid-consuming, ecological processes operate in wetlands. These processes can be taken advantage of in constructed wetlands to decrease much of the initial acidity of acid mine drainages, and thereby reduce the costs of conventional treatments with acid-neutralizing chemicals. The microbial processes that consume acidity are various, but they include: (1) the chemical reduction of sulfate to sulfide at the oxygen-poor interface between the sediment and the water column and around plant roots, (2) the reduction of ferric **iron** to ferrous in the same anoxic microhabitats, as well as (3) the primary productivity of **phytoplankton**, which also consumes some acidity.

A less intensive type of bioremediation can be used to mitigate some of the deleterious ecological effects associated with the acidification of surface waters, such as lakes and ponds. In almost any fresh waters, fertilization with phosphate will greatly increase the primary productivity of algae and vascular plants. In acidic waters, this process can be taken advantage of to reduce the acidity somewhat, but the most important ecological benefit occurs through enhancement of the **habitat** of certain aquatic animals. **Ducks** and **muskrat**, for example, can breed very successfully in fertilized acidic lakes, because their habitat is improved through the vigorous growth of vegetation and of aquatic **insects** and crustaceans. However, the productive but still acidic habitat remains toxic to **fish**. In this case, manipulation of the **ecosystem** by fertilization mitigates some but not all of the negative effects of acidification.

Bioremediation of sewage

Sewage is a very complex mixture of wastes, usually dominated by fecal materials but also containing toxic chemicals that have been dumped into the disposal system by industries and home owners. Many advanced sewage-treatment technologies utilize microbial processes to both oxidize the organic **matter** associated with fecal wastes and to decrease the concentrations of soluble compounds or ions of metals, **pesticides**, and other toxic chemicals. The latter effect, decreasing the aqueous concentrations of toxic chemicals, is accomplished by a combination of chemical adsorption as well as microbial biodegradation of complex chemicals into their simpler, inorganic constituents. Microbial processes are relied

---

KEY TERMS

Bioaugmentation—Increasing the abundance of microorganisms that are specifically effective at bioremediation.

Biostimulation—Optimizing environmental conditions for the actions of microorganisms important in bioremediation, usually by fertilizing or aerating.

---

upon in many **sewage treatment** systems including activated sludges, aerated lagoons, **anaerobic** digestion, trickling filters, waste stabilization ponds, **composting**, and disposal on land.

Bioremediation of soils contaminated with toxins

Many of the chemicals commonly used in industry or agriculture, or produced as by-products of industrial processes are persistent poisons, such as DDT and other organochlorines that are dangerous to humans and **wildlife** and that do not readily breakdown under natural processes. In most cases, contaminated soils would be burned and then hauled away for storage as hazardous waste. However, new bioremediation techniques that use bacteria to break down the toxins are being tested around the world. These techniques involve many varieties of bacteria, working in a carefully orchestrated sequence. By controlling environmental conditions around the contaminated site—for example, by enclosing the site in a large tent-like construction and controlling **heat** and moisture inputs as well as oxygen levels—clean-up experts believe they can encourage the proper series of bacterial relationships required to break down the contaminates. The bacteria used in such projects are not genetically altered or specially bred strains. If successful, this approach to the bioremediation of contaminated sites will offer a cheaper, less environmentally damaging alternative to traditional clean-up technologies.

See also Hazardous wastes; Oil spills.

Resources

Books
Freedman, B. *Environmental Ecology.* 2nd ed. San Diego: Academic Press, 1994.

Periodicals
Frederick, R.J., and M. Egan. "Environmentally Compatible Applications of Biotechnology." *BioScience,* 44 (1994): 529-535.

Bill Freedman

Biosphere

The biosphere is the space on and near the earth's surface that contains and supports living organisms and ecosystems. It is typically subdivided into the **lithosphere**, atmosphere, and **hydrosphere**. The lithosphere is the earth's surrounding layer composed of solid **soil** and rock, the atmosphere is the surrounding gaseous envelope, and the hydrosphere refers to liquid environments such as lakes and oceans, occurring between the lithosphere and atmosphere. The biosphere's creation and continuous **evolution** result from physical, chemical, and biological processes. To study these processes a multi-disciplinary effort has been employed by scientists from such fields as **chemistry**, **biology**, **geology**, and **ecology**.

History

The Austrian geologist Eduard Suess (1831–1914) first used the term biosphere in 1875 to describe the space on **Earth** that contains life. The concept introduced by Suess had little impact on the scientific community until it was resurrected by the Russian scientist Vladimir Vernadsky (1863–1945) in 1926 in his book, *La biosphere*. In that work Vernadsky extensively developed the modern concepts that recognize the interplay between geology, chemistry, and biology in biospheric processes.

Requirements for life

For organisms to live, appropriate environmental conditions must exist in terms of **temperature**, moisture, **energy** supply, and nutrient availability.

Energy is needed to drive the functions that organisms perform, such as growth, movement, waste removal, and reproduction. Ultimately, this energy is supplied from a source outside the biosphere, in the form of visible **radiation** received from the **Sun**. This electromagnetic radiation is captured and stored by plants through the process of **photosynthesis**. Photosynthesis involves a light-induced, enzymatic reaction between **carbon dioxide** and **water**, which produces **oxygen** and glucose, an organic compound. The glucose is used, through an immense diversity of biochemical reactions, to manufacture the huge range of other organic compounds found in organisms. Potential energy is stored in the chemical bonds of organic molecules and can be released through the process of **respiration**; this involves enzymatic reactions between organic molecules and oxygen to form **carbon** dioxide, water, and energy. The growth of organisms is achieved by the accumulation of organic **matter**, also known as **biomass**. Plants and some **microorganisms** are the only organisms that can form organic molecules by photosynthesis. Heterotrophic organisms, including humans, ultimately rely on photosynthetic organisms to supply their energy needs.

The major elements that comprise the chemical building blocks of organisms are carbon, oxygen, **nitrogen**, **phosphorus**, **sulfur**, **calcium**, and **magnesium**. Organisms can only acquire these elements if they occur in chemical forms that can be assimilated from the environment; these are termed available **nutrients**. Nutrients contained in dead organisms and biological wastes are transformed by **decomposition** into compounds that organisms can reutilize. In addition, organisms can utilize some mineral sources of nutrients. All of the uptake, excretion, and transformation reactions are aspects of nutrient cycling.

The various chemical forms in which carbon occurs can be used to illustrate nutrient cycling. Carbon occurs as the gaseous **molecule** carbon dioxide, and in the immense diversity of organic compounds that make up living organisms and dead biomass. Gaseous carbon dioxide is transformed to solid organic compounds (simple sugars) by the process of photosynthesis, as mentioned previously. As organisms grow they deplete the atmosphere of carbon dioxide. If this were to continue without carbon dioxide being replenished at the same **rate** as the consumption, the atmosphere would eventually be depleted of this crucial nutrient. However, carbon dioxide is returned to the atmosphere at about the same rate that it is consumed, as organisms respire their organic molecules, and microorganisms decompose dead biomass, or when **wildfire** occurs.

Evolution of the biosphere

During the long history of life on Earth (about 3.8 billion years), organisms have drastically altered the chemical composition of the biosphere. At the same time, the biosphere's chemical composition has influenced which life forms could inhabit its environments. Rates of nutrient transformation have not always been in balance, resulting in changes in the chemical composition of the biosphere. For example, when life first evolved, the atmospheric **concentration** of carbon dioxide was much greater than today, and there was almost no free oxygen. After the evolution of photosynthesis there was a large decrease in atmospheric carbon dioxide and an increase in oxygen. Much of carbon once present in the atmosphere as carbon dioxide now occurs in fossil fuel deposits and limestone rock.

The increase in atmospheric oxygen concentration had an enormous influence on the evolution of life. It was not until oxygen reached similar concentrations to what occurs today (about 21%, by **volume**) that multicellular organisms were able to evolve. Such organisms require high oxygen concentrations to accommodate their high rate of respiration.

Current research

Most research investigating the biosphere is aimed at determining the effects that human activities are having on its environments and ecosystems. **Pollution**, fertilizer application, changes in **land use**, fuel consumption, and other human activities affect nutrient cycles and damage functional components of the biosphere, such as the **ozone** layer that protects organisms from intense exposure to solar ultraviolet radiation, and the **greenhouse effect** that moderates the surface temperature of the **planet**.

For example, fertilizer application increases the amounts of nitrogen, phosphorus, and other nutrients that organisms can use for growth. An excess nutrient availability can damage lakes through algal blooms and **fish** kills. Fuel consumption and land clearing increases the concentration of carbon dioxide in the atmosphere, and may cause **global warming** by intensifying the planet's greenhouse effect.

Recent interest in long-term, manned space operations has spawned research into the development of artificial biospheres. Extended missions in space require that nutrients are cycled in a volume no larger than a building. The Biosphere 2 project, which received a great deal of popular attention in the early 1990s, has provided insight into the difficulty of managing such small, artificial biospheres. Human civilization is also finding that it is difficult to sustainably manage the much larger biosphere of planet Earth.

See also Lithosphere.

Resources

Books

Allen, John. *Biosphere 2: The Human Experiment.* New York: Viking, 1991.

Bradbury, I. K. *The Biosphere.* London/New York: Bellhaven Press, 1991.

Hamblin, W.K., and E.H. Christiansen. *Earth's Dynamic Systems.* 9th ed. Upper Saddle River: Prentice Hall, 2001.

Levin, Simon A., ed. *Encyclopedia of Biodiversity.* San Diego, CA: Academic Press, 2000.

Odum, Eugene. *Ecology and Our Endangered Life-Support Systems.* 2nd ed. Sunderland, MA: Sinauer Associates, 1993.

Smil, V. *Cycles of Life: Civilization and the Biosphere.* W.H. Freeman and Co., 1997.

Tudge, Colin. *Global Ecology.* New York: Oxford University Press, 1991.

KEY TERMS

· ·

Decomposition—The breakdown of the complex molecules composing dead organisms into simple nutrients that can be reutilized by living organisms.

Energy—The ability to do work. Energy occurs in various forms. The most important ones in biospheric processes are solar (electromagnetic), kinetic, heat (or thermal), and chemical-bond energies.

Global warming—Atmospheric warming caused by an increase in the concentration of greenhouse gases, which absorb infrared energy emitted by Earth's surface, thereby slowing its rate of cooling. Carbon dioxide and water vapor are particularly important in this respect.

Nutrient cycle—The cycling of biologically important elements from one molecular form to another, and eventually back to the original form.

Nutrients—The molecules organisms obtain from their environment and are used for growth, energy, and other metabolic processes.

Photosynthesis—Enzymatic, sunlight-induced reaction between carbon dioxide and water, which produces oxygen and organic molecules. Plants, algae, and certain bacteria are photosynthetic organisms.

Respiration—Enzymatic chemical reactions between organic molecules and oxygen, which result in the production of carbon dioxide, water, and energy.

Periodicals

Clark, B. C. "Planetary Interchange of Bioactive Material: Probability Factors and Implications." *Origins of Life and Evolution of the Biosphere* no. 31 (2001): 185–197.

Huggett, R. J. "Ecosphere, Biosphere, Or Gaia? What To Call The Global Ecosystem." *Global Ecology And Biogeography* 8, no. 6 (1999): 425-432.

Salthe, S.N. "The Evolution of the Biosphere: Towards a New Mythology." *World Futures.* 30 (1990): 53-67.

Steven MacKenzie

Biosphere Project

The **Biosphere** 2 Project is an experiment in which scientists, engineers, and some intrepid "biospherians" (or dwellers) within the Biosphere have recreated several of the main terrains and habitats of our **planet** and attempted to co-habit with these environments to the environments' benefit. Many environmentalists see planet **Earth** as the first, original, and only known biosphere endangered by **pollution**, **acid rain**, **global warming**, the destruction of tropical rainforests, and a thousand other manmade ills. The Biosphere 2 Project is an experiment to prove that humans can live in the most precious terrains on Earth successfully and nondestructively.

The physical structure

Biosphere 2 is located in the Sonora Desert at the foot of the Santa Catalina Mountains not far from Tucson, Arizona. It is one of the most spectacular structures ever built. It is the world's largest greenhouse, made of tubular **steel** and **glass**, covering an area of three football fields (137,416 sq ft [12,766 sq m]), and rising above the **desert** floor to a height of 85 ft (25.9 m). Within the structure, there is a human **habitat** and a farm for the Biospherians to work to provide their own food. There are five other wild areas representing the **savanna** (extensive **grasslands**), a **rainforest**, a marsh, a desert, and the **ocean**. These five areas plus the Human Habitat and the farm are called "biomes," which might be interpreted to mean biological homes or dwellings for life forms.

Biosphere 2 is completely sealed so no moisture or air can flow in or out. Beyond the Biosphere, two large white domes also dominate the landscape and capture the imagination. These are balloon-like structures that operate like a pair of lungs for Biosphere 2 in maintaining air **pressure** inside the biosphere. Only sunlight and **electricity** are provided from outside.

The residents

Within the biosphere, four women and four men from three countries lived in the Human Habitat during a two year experiment. During this time, they ran the farm and grew their own food in the company of some **pigs**, **goats**, and many chickens. They shared the other biomes with over 3,800 **species** of animals and plants, which are native to those habitats. It was the responsibility of the humans, scientists, and environmentalists to make sure the model of the planet in miniature survived and thrived. In 1991, the doors to Biosphere 2 were sealed for the two-year-long initial program of survival and experimentation.

Scientific objectives

The idea behind Biosphere 2 was to establish a planet in miniature where the inhabitants not only survived but learned to live cooperatively and happily together. The resident scientists observed the interactions of plants and animals, their reactions to change, and their unique methods of living. In the real world, scientists still know little about many of these relationships and how Earth achieves a balance or regains balance after some disruption. The residents also had the assignment of experimenting with new methods of cleaning air and **water**. Lessons learned from Biosphere 2 may help engineers to design workable living environments and life-support systems for **space** stations and settlements on other planets. Other biospheres may also be adaptable to less hospitable parts of our own planet, they may be used to house **endangered species** or environments, they may provide recreational areas of vastly different terrains near cities and, perhaps most importantly, they may be used as living classrooms to educate future generations about preservation of the original biosphere, Earth.

Earlier Biosphere experiments

Scientists have struggled for generations to understand the complex interrelationships of life forms and the atmospheric and hydrologic cycles that provide life on Earth with its essentials. A Russian scientist, Dr. Vladimir Vernadsky (1863-1943), was the first to understand that the life systems of Earth are perfectly balanced and able to self-correct; Vernadsky is considered the "Father of Biospherics." Efforts to recreate these water, air, and life cycles began in 1968 when an American experimenter in Hawaii, Dr. Clair Folsome, accidentally discovered that microbes, seawater, and **algae** trapped and sealed in a glass bottle did not die but created their own tiny environment in which the materials recycled naturally. Dr. Folsome's sealed bottles survived for some years, and, even after he added **shrimp**, the shrimp were able to live although they did not reproduce.

In the former Soviet Union, scientists sealed humans into small buildings to study methods of creating life-support systems in space. These experiments were known as BIOS-3, but the dwellers only had to produce half their food, and waste was disposed outside the buildings. When the Biosphere 2 Project was conceived, its creators met with Dr. Folsome, the Soviet scientists who had worked on the BIOS-3 experiments, and experts at the National Atmospheric and Space Administration (NASA) whose spacecraft and lunar module designs had allowed Americans to reach the **moon**.

Designing Biosphere 2

The design of the Biosphere habitats was the work of an international team of hundreds of engineers, scientists, and specialists in agriculture and diverse life species. The site in the Arizona desert was chosen because of the relative ease, in the United States, of obtaining building permits and of importing the species that would reside in the Biosphere. The project was funded by Ed Bass, a billionaire from Texas, and initiated by John Allen, a leading ecologist. The name was chosen in deference to Earth; there was no Biosphere 1 Project because our home planet is the original biosphere. The five wild zones, the human "microcity," and the farm were selected to show diversity representing planet Earth but also to improve survival of as many species as possible. Roof heights for each zone had to be selected to allow air and rain to rise and fall as it does in nature, sunlight was essential to the survival of all zones, and the habitats had to be somewhat separated so desert life would not relocate to the rainforest.

To be perfectly isolated, Biosphere 2 needed a floor separating it from the supporting ground. After the groundbreaking in 1987, construction of this sealed floor began. To support the extensive glass, steel spaceframes consisting of 5-ft (1.5-m) lengths of strong but lightweight tubular steel were constructed. During building, the glass panes were lifted by cranes into place and sealed with liquid silicone to retain interior air and moisture and keep out the exterior. The surface of Biosphere 2 consists of 170,000 sq ft (15,794 sq m) of glass. Because of the purpose of the project and the huge expanse of glass, Biosphere 2 was nicknamed the "Glass Ark". The dome-like lungs were needed to equalize air pressure as the air inside Biosphere expands under the **heat** of the **sun**; the domes do not provide or replenish any of Biosphere's air, they only provide room for heated air to evacuate.

A smaller Biosphere module was made as a proving ground. The test module was only 1/400th of the size of Biosphere. The test module was sealed to make sure it was leakproof, then a living test was performed using plants. A plants-only habitat was installed, along with the correct moisture and air balance, and the module was sealed for a month. Opening the door and smelling the air immediately indicated whether or not the test had succeeded; if the plants had not adjusted to this environment, they would have produced gas that smelled. The test was a success with healthy, growing plants.

In 1988, the first of three humans moved into the test module for a three-day period. John Allen had his health monitored constantly as he lived exclusively on a small tropical garden and in the fully sealed environment. The second scientist occupied the module for a five-day-long test, and the third lived in the module for three weeks. These three early experimenters were all amazed at the cause-and-effect relationships they witnessed in the module, which were much more obvious than those in the larger world.

Mechanical devices were needed to regulate the **temperature** and to provide rain, but the moisture supply remained constant in Biosphere 2. Soil-bed reactors were constructed to use the properties of **soil** and its microbes to clean the air. An artificial mountain was built to provide streamflow by gravity to add moisture to the rainforest, flow past the savanna, enter the marsh, and empty in the ocean.

The **animal** residents of the biomes were chosen by the captain of each **biome**. No large predators were allowed, and many species were rejected because their needs were greater than the scale of Biosphere 2 could support. **Hummingbirds** were an easy choice because they are surprisingly hardy and are excellent pollinators. **Bees**, **bats**, **moths**, and **butterflies** were also brought in as pollinators. About 40 land dwelling species included **snakes**, **reptiles**, and **turtles**. The resident **mammals** were bats and bushbabies. Of Earth's thousands of species of **insects**, **ants**, **termites**, and **cockroaches** were chosen to share the habitat because they break down dead plants and animals into recyclable materials that benefit the Biosphere. Oysters and **crabs** were brought in to populate the replica ocean. Even a small coral reef was incorporated in the Biosphere 2 ocean. The plants were equally important and included medicinal plants, the agave, the jojoba, rubber trees, mosses, **ferns**, and trees that produce gums and soaps. Plants and animals that were imported into the United States were quarantined before being allowed in Biosphere 2.

The Human Habitat or microcity was also given special consideration to prevent the scientists from experiencing "cabin fever" or any feeling of being trapped. Windows look out over all other parts of the Biosphere, and each Biospherian had a private apartment with an upstairs bedroom and a downstairs sitting room. The bathrooms were equipped with showers and toilets; but shower time was limited to conserve water, and toilet **paper** was forbidden. Drinking water was collected from moisture produced by the plants in the Biosphere. The Habitat also includes laboratory space, a medical clinic, an **exercise** room, recreation facilities, and a communications center. Kitchen, cooking, and cleanup duties were all shared.

The farm for the humans' food consisted of eighteen separate garden plots designed to produce three **crops** per year. Beans, potatoes, and peanuts for protein were key elements of the farm. Oats, **barley**, and **rice** were grown; and the grains were used by the Biospherians to bake bread. **Fruits** included pineapple, guava, apples, bananas, **grapes**, strawberries, oranges, and papayas; these produced not only fruit but juice and jam. Sugar cane was grown as a sweetener. The Biospherians all received special training before they were finally selected for their assignments. Physical fitness courses, training in emergencies, skin diving classes (to care for the ocean), and special training in cooking were all provided.

KEY TERMS

Biome—A geographically extensive ecosystem, usually characterized by its dominant life forms.

Biosphere—Life forms that not only live together, but provide functions to maintain their environment.

Soil-bed reactor—A bed of soil that, when air is passed through it, works like a filter to clean the air.

Spaceframe—Steel tubes that form lightweight frames to shape and support walls, floors, and ceilings.

The pulse of Biosphere was monitored constantly by extensive instrumentation. The functions of the limited external systems were carefully controlled, and internal health including moisture, temperature, and other vitals were also monitored both by the Biospherians and scientists working beyond the containment. It was up to the Biospherians to adjust living conditions for all the Biosphere's residents, but external monitoring provided a final check on the safety of the Biosphere population.

Epilogue

The eight human residents of Biosphere 2 lived inside the containment from September 26, 1991, to September 26, 1993, the longest period on record that humans have lived in an "isolated confined environment." The Biospherians experienced many difficulties, including an unusually cloudy year in the Arizona desert that stunted food crops, proliferation of some ant species, and unusual **behavior** by bees fooled by the glass walls. Columbia University took over the operation of the facility in 1996, a visitors' center was opened later in 1996, and Biosphere 2 has been maintained for study but without human inhabitants. The future of Biosphere 2 remains uncertain, but environmentalists and scientists hope to restart the project and to continue answering the questions and testing the environments envisioned initially.

Resources

Books

A Living Laboratory for Earth & the Environment. New York: Columbia University Press, 2001.

Periodicals

Elliott, Caroline. "My Holiday in a Giant Greenhouse." *Focus,* (February 1998): 106-109.
Stover, Dawn. "Second Chance for Biosphere." *Popular Science* (April 1997):56-59.

Other

Alper, Joseph. "Biosphere II: Out of Oxygen." [cited March 2003]. <http://www.chemistry.org/portal/Chemistry?PID=acs display.html&DOC=vc2%5C2my%5Cmy2_biosphere.html>.
Biosphere II Center, Project Home [cited March 2003]. <http://www.bio2.edu/>.

Gillian S. Holmes

Biotechnology

Biotechnology is the use of any technique involving living organisms to manufacture or change products, to improve the desired characteristics of a **plant** or **animal**, or to alter **microorganisms** for a purpose.

Biotechnology has a long history. For example, **yeast** microorganisms were harnessed to prepare wine by Egyptians some 4,000 years before the **birth** of Christ. In 1865, Gregor Mendel presented his laws of heredity, which he deduced by the careful observation of the results of breeding different types of pea plants. Although he did not realize it at the time, Mendel was observing the results of the exchange and altered expression of genetic material.

The modern day conception of biotechnology, with the deliberate experimental manipulation of genetic material, had its roots in the mid years of the twentieth century. In 1940, **deoxyribonucleic acid (DNA)** was isolated by Oswald Avery. Thirteen years later, James Watson and Francis Crick described the **double helix** structure of DNA, a feat that earned them a Nobel Prize just a few years later. The modern age of biotechnology began in 1973, when Stanley Cohen and Herbert Boyer devised **recombinant DNA** technology; the deliberate introduction of DNA from one **species** into another. Their work made possible feats such as the production of human **insulin** by the bacterium *Escherichia coli*. This genetically engineered human insulin was, in fact, the first genetically engineered product approved for sale in the United States in 1982.

The latter decades of the twentieth century saw an explosion in the experimental and commercial use of biotechnology.

The basic concept of biotechnology involves recombination, or the process where genetic material from just about any living **organism** can be isolated, cut up into pieces using special enzymes, and the pieces encouraged to recombine. The recombination can be between genetic material from the same organism, or between genetic material from different organisms. Differences between

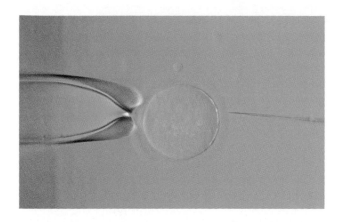

An animal cell being microinjected with foreign genetic material. *Photograph by M. Baret/RAPHU. National Audubon Society Collection/Photo Researchers, Inc. Reproduced by permission.*

the organization of the genetic material of organisms like **bacteria** and "higher" organisms such as humans, and the difference in how the genetic traits coded for by the material are expressed, has complicated the advances in biotechnology. But, increasingly, such species differences are being understood.

Applications of biotechnology are numerous. For example, foods are being genetically altered to engineer in more nutritional compounds. The nutraceutical industry is growing to become a potent economic force, generating billions of dollars in sales each year in the United States alone. Genetic manipulation can also help preserve foods longer, allowing a fresher product to reach the supermarket shelves.

An aspect of biotechnology that has garnered much attention since the 1990s is cloning. Until 1997, a fully developed organism could not be cloned. But, in early 1997, the first success at producing live animals by embryo cloning occurred in Edinburgh, Scotland. The procedure that produced Dolly the **sheep** was reported in the March 6, 1997 edition of *Nature*.

While embryo cloning is still a "hit or miss" procedure, the consensus among researchers involved in embryo cloning is that cloning animal embryos will be perfected. The resulting ease of genetic tailoring could produce higher yielding and disease-resistant **livestock**.

Cloning embryos is similar to what happens naturally when identical twins are created in the womb. All human embryos begin as a single **cell**. Normally, millions of rounds of division and the formation of cells that differ in structure and function from other cells gives rise to a human. With identical twins, as the cell divides it separates into two separate, **individual** cells. The two separate, individual cells then divide and differentiate in-

dependently. The result is two embryos that are identical in the composition of their genetic material.

In embryo cloning, a cell is mechanically encouraged to divide into two separate, individual cells. These grow and develop separately, creating identical twins.There is continuing debate around the moral and ethical limits on cloning human embryos. Currently, it is illegal to use federal research funds in the United States to clone human embryos.In November of 2001, the **human cloning** debate was raised from a theoretical discussion to a concrete discussion. Then, a company in suburban Boston announced that a human cell had been cloned to provide **stem cells** for research. While the experiment was carried on for only a few cell divisions, the technology required to develop a cloned human being may be almost in place.

The prospects offered by biotechnology have not been greeted with unanimous enthusiasm by everyone. Many scientists and laypersons assert that the hope of curing or avoiding **genetic disorders** through biotechnology is a positive advance. Some hold that the genetically derived nutritional enhancement of foods, such as the nutritional supplementation of **rice** grown in developing countries, is a worthy aim. Others oppose all forms of **genetic engineering**, or warn of the dangers of having such technology as the commercial property of a few large companies. There are also concerns about genetic privacy, the effects of transgenic organisms on other organisms and the environment, and animal rights.

As the technology available for genetic **engineering** continues to improve, debates over the use of these techniques in practical settings are almost certainly going to continue and escalate in the future.

See also DNA technology; Human Genome Project; Ribonucleic acid (RNA).

Resources

Books

Charles, D. *Lords of the Harvest: Biotech, Big Money, and the Future of Food.* Cambridge, MA: Perseus Books, 2001.

Wilmut, I., K. Campbell, and C. Tudge. *The Second Creation: Dolly and the Age of Biological Control.* New York: Farrar, Straus and Giroux, 2000.

Periodicals

Lerner, J., and R.P. Merges, 1998, "The Control of Technology Alliances: An Empirical Analysis of the Biotechnology Industry." *Journal of Industrial Economics* 66 (June 1998): 125–156.

Martin, G.B., S.H. Brommonschenkel, J. Chunwongse, et al., "Map-based Cloning of a Protein Kinase Gene Conferring Disease Resistance in Tomato." *Science* 262 (1993): 1432–1436.

Brian Hoyle

Bioterrorism

Bioterrorism is the use of a biological weapon against a civilian population. As with any form of terrorism, its purposes include the undermining of morale, creating chaos, or achieving political goals. Biological weapons use **microorganisms** and toxins to produce **disease** and death in humans, **livestock**, and **crops**.

Biological, chemical, and **nuclear weapons** can all be used to achieve similar destructive goals, but unlike chemical and nuclear technologies that are expensive to create, biological weapons are relatively inexpensive. They are easy to transport and resist detection by standard security systems. In general, chemical weapons act acutely, causing illness in minutes to hours at the scene of release. For example, the release of **Sarin gas** by the religious sect Aum Shinrikyo in the Tokyo subway in 1995 killed 12 and hospitalized 5,000 people. In contrast, the damage from biological weapons may not become evident until weeks after an attack. If the pathogenic (disease causing) agent is transmissible, a bioterrorist attack could eventually kill thousands over a much larger area than the initial area of attack.

Bioterrorism can also be enigmatic, destructive, and costly even when targeted at a relatively few number of individuals. Starting in September 2001, bioterrorist attacks with **anthrax** causing **bacteria** distributed through the mail, targeted only a few U.S. government leaders, media representatives, and seemingly **random** private citizens. As of May 2003, these attacks remain unsolved. Regardless, in addition to the tragic deaths of five people, the terrorist attacks cost the United States millions of dollars and caused widespread concern. These attacks also exemplified the fact that bioterrorism can also strike at the political and economic infrastructure of a targeted country.

Although the deliberate production and stockpiling of biological weapons is prohibited by the 1972 Biological Weapons Convention (BWC)—the United States stopping formal weapons programs in 1969—unintended byproducts or deliberate misuse of emerging technologies offer potential bioterrorists opportunities to prepare or refine biogenic weapons. **Genetic engineering** technologies can be used to produce a wide variety of bioweapons including organisms that produce toxins or that are more weaponizable because they are easier to aerosolize (suspend as droplets in the air). More conventional laboratory technologies can also produce organisms resistant to **antibiotics**, routine vaccines, and therapeutics. Both technologies can produce organisms that cannot be detected by antibody-based sensor systems.

Among the most serious of protential bioterrorist weapons are those that use **smallpox** (caused by the **Variola virus**), anthrax (caused by *Bacillus anthracis*), and plague (caused by *Yersinia pestis*). During naturally occurring epidemics throughout the ages, these organisms have killed significant portions of afflicted populations. With the advent of vaccines and antibiotics, few U.S. physicians now have the experience to readily recognize these diseases, any of which could cause catastrophic numbers of deaths.

Although the last case of smallpox was reported in Somalia in 1977, experts suspect that smallpox viruses may be in the biowarfare laboratories of many nations around the world. At present, only two facilities—one in the United States and one in Russia—are authorized to store the **virus**. As recently as 1992, United States intelligence agencies learned that Russia had the ability to launch missiles containing weapons-grade smallpox at major cities in the U.S. A number of terrorist organizations—including the radical Islamist Al Qaeda terrorist organization—actively seek the acquisition of state-sponsored research into weapons technology and **pathogens**.

There are many reasons behind the spread of biowarfare technology. Prominent among them are economic incentives; some governments may resort to selling bits of scientific information that can be pieced together by the buyer to create biological weapons. In addition, scientists in politically repressive or unstable countries may be forced to participate in research that eventually ends up in the hands of terrorists.

A biological weapon may ultimately prove more powerful than a conventional weapon because it's effects can be far-reaching and uncontrollable. In 1979, after an accident involving *B. anthracis* in the Soviet Union, doctors reported civilians dying of anthrax **pneumonia** (i.e., inhalation anthrax). Death from anthrax pneumonia is usually swift. The bacilli multiply rapidly and produce a toxin that causes breathing to stop. While antibiotics can combat this bacillus, supplies adequate to meet the treatment needs following an attack on a large urban population would need to be delivered and distributed within 24 to 48 hours of exposure. The National Pharmaceutical Stockpile Program (NPS) is designed to enable such a response to a bioterrorist attack.

Preparing a strategy to defend against these types of organisms, whether in a natural or genetically modified state, is difficult. Some of the strategies include the use of bacterial RNA based on structural templates to identify pathogens; increased abilities for rapid **genetic identification of microorganisms**; developing a database of virtual pathogenic molecules; and development of antibacterial molecules that attach to pathogens but do not harm humans or animals. Each of these is an attempt to increase—and make more flexible—identification capabilities.

Researchers are also working to counter potential attacks using several innovative technological strategies. For example, promising research with biorobots or microchip-mechanized **insects** with computerized artificial systems that mimic biological processes such as neural networks, and can test responses to substances of biological or chemical origin. These insects can, in a single operation, process DNA, screen **blood** samples, scan for disease genes, and monitor genetic **cell** activity. The **robotics** program of the Defense Advanced Research Project (DARPA) works to rapidly identify bio-responses to pathogens, and for designs to effectively and rapidly treat them.

Biosensor technology is the driving force in the development of biochips for detection of biological and chemical contaminants. **Bees**, **beetles**, and other insects outfitted with sensors are used to collect real-time information about the presence of toxins or similar threats. Using **fiber optics** or electrochemical devices, biosensors have detected microorganisms in chemicals and foods, and off the promise of rapid identification of biogenic agents following a bioterrorist attack. The early accurate identification of biogenic agents is critical to implementing effective response and treatment protocols.

To combat biological agents, bioindustries are developing a wide range of antibiotics and vaccines. In addition, advances in bioinformatics (i.e., the computerization of information acquired during, for example, genetic screening) also increases flexibility in the development of effective counters to biogenic weapons.

In addition to detecting and neutralizing attempts to weaponize biogenic agents (i.e., attempts to develop bombs or other instruments that could effectively disburse a bacterium or virus), the major problem in developing effective counter strategies to bioterrorist attacks

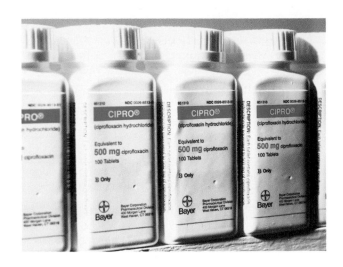

Cipro is used to treat anthrax and as of late 2002 was produced in limited supply. © FRI/Corbis Sygma. Reproduced by permission.

involves the breadth of organisms used in **biological warfare**. For example, researchers are analyzing many pathogens in an effort to identify common genetic and cellular components. One strategy is to look for common areas or vulnerabilities in specific sites of DNA, RNA, or **proteins**. Regardless whether the pathogens evolve naturally or are engineered, the identification of common traits will assist in developing counter measures (i.e., specific vaccines or antibiotics).

See also Contamination; Genetic identification of microorganisms.

Resources

Books

Drexler, Madeline. *Secret Agents: The Menace of Emerging Infections.* Joseph Henry Press, 2002.

Koehler, T.M. *Anthrax* New York: Springer-Verlag, 2002.

Tucker, Jonathan. *Toxic Terror: Assessing Terrorist Use of Chemical and Biological Weapons.* Cambridge, MA: MIT Press, 2002.

Periodicals

Glass, Thomas, and Monica Schoch-Spana. "Bioterrorism and the People: How to Vaccinate a City against Panic" *Clinical Infectious Diseases* (January 15, 2002): 34 (2).

Inglesby, Thomas V, et. Al. "Anthrax as a Biological Weapon, 2002-Updated Recommendations for Management " *Journal of the American Medical Association* (May 1,2002):

Organizations

United States Food and Drug Administration Center for Food Safety and Applied Nutrition. "Food Safety and Terrorism" [cited February 5, 2003]. <http://www.cfsan.fda.gov/~dms/fsterr.html>.

Other

Unites States Department of Homeland Security. "President Discusses Measures to Protect Homeland from Bioterrorism" [cited February, 5, 2003]. <http://www.whitehouse.gov/homeland/>.

Brian Hoyle

Birch family (Betulaceae)

The birch family is a group of flowering plants of **tree** or shrub form that includes the birches (*Betula*), alders (*Alnus*), hornbeams (*Carpinus*), and hazels (*Corylus*). Members of the birch family have simple and alternate leaves that bear appendages (stipules) where they join the branch. The leaves are also deciduous, generally thin and often doubly toothed along the margin. The flowers are densely borne on elongate, spike-like structures called catkins. Each catkin bears flowers of only one sex but male and female catkins occur on the same **plant**. Female catkins are stiffer and fewer-flowered than male catkins. The flowers lack petals or sepals, although some **species** have small scale-like appendages that represent reduced perianth parts. **Pollination** occurs in the spring by the **wind**. The fruit is a one-seeded **nut** or nutlet which is often winged and enclosed or surrounded at the base by leaf-like appendages called bracts.

The family includes six genera and about 170 species worldwide. The Betulaceae occur throughout the temperate and boreal regions of the Northern Hemisphere, where they are prominent members of forest communities. Some alders and dwarf birches extend into the southern Arctic where they are common shrubs. Members of the family occur only in a few regions of the Southern Hemisphere, notably in the tropical **mountains** of the northern Andes of Columbia and in Argentina.

Ecologically, this an important family. Many members of the Betulaceae play pivotal roles in early successional stages following disturbance. In **North America**, for example, the **paper** or white birch (*Betula papyrifera*) is a transcontinental species that rapidly recolonizes areas disturbed by fire or logging, thus stabilizing the **soil** and providing suitable conditions for recolonization by other species. Paper birch is able to invade disturbed areas by dispersal of its winged **seeds**. In addition, birches typically resprout from the base when the main trunk is killed. Birches grow fast and die young; paper birch, for example, reaches maturity in 60-75 years and seldom lives longer than 140 years.

Alders are also important early successional species. In northern regions, cold soils limit the activity of nitro-

A birch tree (*Betula pendula*). © *K. G. Vock/Okapia, National Audubon Society Collection/Photo Researchers, Inc. Reproduced with permission.*

gen-producing soil **bacteria**, thus, northern soils frequently lack adequate quantities of **nitrogen** for vigorous plant growth. Alders are among the few plants that form symbiotic relationships with nitrogen-fixing bacteria. In alders, filamentous bacteria known as actinomycetes form nodules on the roots. The nitrogen compounds produced by the actinomycetes are in a form directly usable by the host alder plant. In addition, some of the nitrogen compounds are leached from the nodules or released when nodule-bearing roots die. These compounds accumulate in the soil where they remain available for use by other plants of later successional stages that cannot produce their own nitrogenous compounds.

In the United States there are five genera and about 25 species of Betulaceae. Yellow birch (*Betula alleghaniensis*) and paper birch are common in the northern hardwood **forests** of northern New England. The American hornbeam, also called blue beech, is the only North American member of the genus *Carpinus* (*C. carolini-*

KEY TERMS

Perianth—A collective term for the calyx and corolla that generally form the most obvious, showy part of a flower

Succession—A process of ecological change, involving the progressive replacement of earlier communities with others over time, and generally beginning with the disturbance of a previous type of ecosystem.

ana). It grows in moist woods as a tall shrub or small tree reaching heights of 39 ft (10 m) and with a smooth, gray, ridged trunk that is often described as fluted. American hornbeam has a curious range that covers much of eastern North America, parts of Texas, and then skips to southern Mexico, Guatemala, and Honduras. Eastern hop hornbeam (*Ostrya virginiana*) has a similar distribution to American hornbeam. However, eastern hop hornbeam is taller, reaching heights of 65 ft (20 m), and has scaly **bark** that breaks off in short strips. The **wood** of eastern hop hornbeam is extremely dense, making it difficult to drive a nail into it, hence its other common name of ironwood. A variety of alder species occur as shrubs in moist ground throughout North America. In the Pacific Northwest, however, red alder (*A. rubra*) is an abundant, fast-growing tree of disturbed areas that attains heights of 131 ft (40 m). **Hazel** (*Corylus*) is a widespread shrub (3-10 ft [1-3m] tall) in eastern North America and parts of the Midwest, with the beaked hazel (*C. cornuta*) reaching Oregon.

The birch family is economically valuable. In North America, yellow and sweet birch are important sources of wood for cabinet-making, furniture of various kinds, floors, and doors. Paper birch is excellent as firewood and is used in the making of plywood and boxes. The bark of paper birch was once used by indigenous North Americans for making canoes. In northern **Europe** and especially in Russia, birch switches are traditionally used to beat one's skin during sauna baths. The sap of birches is sweet and can be collected and condensed into syrup. Hazels produce edible nuts that are sometimes called filberts or cobnuts. Store-bought hazelnuts generally come from cultivated European species, most commonly *Corylus avellana*, *C. colurna*, and *C. maxima*. Turkey is the largest producer of hazelnuts at 300,000 metric tons annually, followed by Spain and Italy. In the United States, hazelnuts are produced mostly in Oregon with annual harvests of about 10,000 metric tons.

Most species of alder are shrubs of no commercial value, but red alder of the Pacific Northwest and a number of European species reach tree-size and are valuable sources of wood. Alder wood produces a superior charcoal that imparts a delicious flavor to meat. Charcoal made from red alder is preferred for smoking **salmon** on the west coast of North America.

Resources

Books

Heywood, Vernon H., ed. *Flowering Plants of the World.* New York: Oxford University Press, 1993.

Raven, Peter, R.F. Evert, and Susan Eichhorn. *Biology of Plants.* 6th ed. New York: Worth Publishers Inc., 1998.

Les C. Cwynar

Birds

Birds are vertebrate animals in the class Aves. There are approximately 8,800 **species** of birds, divided among 28 living orders. Of these, slightly more than 900 species are found in **North America**. There has been considerable disagreement among ornithologists about the appropriate level for differentiating species, leading to multiple classification schemes. But however one distinguishes between species, each species belongs to a larger group, called a genus; each genus belongs to a family; and each family belongs to an order. One order, the Passeriformes or perching birds, accounts for more than one-half of the living species of birds.

Birds are believed to have evolved from saurischian dinosaurs, about 150 million years ago. The first truly bird-like **animal** was *Archaeopteryx lithographica*, which lived during the Jurassic period, about 130 million years ago. This 3 ft (1 m) long animal is considered to be an evolutionary link between the birds and the dinosaurs. *Archaeopteryx* had teeth and other dinosaurian characters, but it also had a feathered body and could fly.

Characterization

Birds can often be characterized by their habitats. Species are more likely to be found in a single **habitat** in the tropics, than in more temperate regions (with variable climates) where they may have to accommodate several environments. During times of **migration**, birds may show up in very non-traditional places. Major habitats that serve as homes to birds include polar regions; **tundra**; alpine regions; coniferous **forests**; deciduous forests; tropical rainforests; **grasslands**; deserts; **freshwater** lakes, ponds, and streams; shore and marshes; and seas.

Most species of birds will defend their territories, particularly in nesting season, against other birds that happen to stray into the area, and are perceived to be a threat to the nest. Territories generally fall into the following categories: mating, nesting, and feeding; mating and nesting; mating; narrowly restricted nesting; feeding; wintering; roosting; and group territory.

Some birds prefer to live apart from other birds, while others are more social. Certain mated **woodpeckers**, for example, seem to constantly resent each other, and go out of their way to avoid each other's presence. The Australian wood swallows, on the other hand, feed, bathe, roost, attack predators, breed, preen, perform aerial acrobatics, and feed each other in a communal fashion. Assembly into a flock (consisting of either a single species or several different species) offers the advantage of more eyes and ears for spotting predators and finding new food sources.

Most birds have distinctive calls. Males may sing to warn off rivals, or to attract a mate. Some species have separate calls for **courtship** and for communication. Bird calls serve several functions including: reproductive functions (to proclaim the sex of an **individual**, advertise for a mate, or establish territorial sovereignty); social functions (space out birds in a given environment, teach the young their species' song, or convey information about food and enemies); or individual functions (provide emotional release, identify individuals to each other, or simply to perfect a song).

All birds have feathers, and in this way are unlike all other types of animals. Most birds loose their old feathers each year, these being replaced by new ones (molting). Usually, molting occurs when the bird is neither nesting nor migrating. Some birds will undergo a complete molt in the summer or fall, and then a partial molt in the spring (replacing feathers of the body and head). The chief functions of feathers are to protect the body and to promote flight.

Because birds have four limbs, they are referred to as tetrapods. However, the fore limbs are highly modified into structures known as wings, which are primarily used for flying or gliding. The hind limbs are mostly used for walking or hopping, so these animals are bipedal when moving on the ground. Some species of birds are flightless, but most can fly or soar well.

The bones of all flying birds are modified for flight, and are relatively light with many hollow regions. The flying birds have a strongly keeled breastbone or sternum, to which the flight muscles are attached. Their pelvic bones are fused into a structure known as a synsacrum. Birds have a relatively long neck and their mandibles are modified into a keratinous beak. They do not have teeth. Birds have a four-chambered **heart** and a double circulation of the **blood**, with complete separation of oxygenated and de-oxygenated blood.

Approximately 90% of all birds are monogamous, which is to say that one male bonds with one female. Bonds may last for one nesting period (e.g., house **wrens**), an entire breeding season (most species), several **seasons** (e.g., American **robins**), or life (e.g., **geese**, **swans**, **hawks**, **albatrosses**, and **loons**). Current thinking is that most monogamous birds probably mate outside of the primary pair bond, but the pair contributes substantially only to the raising of young in their own nest. Most birds whose young are reared in a nest form monogamous pairs. But monogamous versus polygamous relationships are not rigidly fixed in some species. Swans, hawks, doves, **finches**, and **thrushes**, for example, normally show monogamous **behavior** but occasionally form polygamous bonds.

Situations in which one male mates with more than one female, while each female mates with only one male, are referred to as polygyny. Scientists have noted that when male birds hold territories that vary greatly in the quality of resources, females tend to choose males with high-quality territories as mates. Given the choice between a poor nesting territory and becoming the second mate of a male with a high quality territory, many females will opt for the latter course. In other cases, where **territoriality** does not appear to be a factor, an overabundance of resources may also lead to polygyny. In North America, red-winged **blackbirds** and yellow-headed blackbirds sometimes form polygynous relationships.

There are also rare mating systems in which one female mates with more than one male, while each male mates with only one female (known as polyandry). Two forms of polyandry are known. In one case, the female holds a large territory that includes the nesting territories of two or more males who care for the eggs and the young, such as the spotted **sandpipers** and red phalaropes of North America. In an even rarer type of polyandry, more than one male may mate with a single female with the resultant offspring reared collectively by the female and her mates. The North American acorn woodpecker occasionally exhibits this type of cooperative polyandry.

Courtship behavior may include dances, feeding rituals, flights, posturing, and cries. Frequently, courtship displays highlight some distinctive feature of the male bird's plumage. In addition to singing, male birds may vibrate their wings, fluff their body feathers, raise their bills, thrust their head forwards, or run by taking short steps. But other birds may advertise their skills rather than their coloring to attract a mate. In its courtship behavior, for example, the male tern finds it sufficient to display a freshly caught **fish** to a potential mate.

Female birds are fertilized internally. They are **oviparous**, laying relatively large, hard-shelled eggs, with a discrete yolk. The eggs are incubated by one or both parents. The young of almost all species of birds are cared for by their parents.

Birds tend to build nests to cradle their eggs, rather than as residences. (Compared with nesting behavior, the night-time roosting behavior of most birds is poorly understood.) The construction of nests can range from very basic to highly complex. Some birds will use a nest repeatedly, while others will build a new nest for each new brood. Although each type of bird typically has a preferred site for building its nest, some birds may be very accommodating in their choice of nesting sites. It is usually the female that builds the nest, though the males of some species may assist. The materials used in the building of a nest may be of animal, vegetable, or mineral origin, but **plant** materials are most commonly used.

The number of eggs laid by the female varies widely from species to species. This number may depend on the **temperature** of the environment, the time in the season, or the size of the food supply. For the most part, only birds incubate eggs by sitting on them and keeping them warm. During the brooding period, some birds may develop a brooding patch on their bellies where feathers and extra blood vessels develop to keep the eggs warm. The number of broods raised in a single year depends on the length of the available season, and the time it takes for one brood to reach independence. There is some evidence that birds living under crowded conditions are less likely to lay extra clutches in a given year.

Upon hatching, young birds may be active and wide-eyed (precocial), or blind and immobile (altricial). Some precocial birds are able to find their own food immediately after being hatched. Most altricial young remain in the nest for several weeks after they learn to leave it. Among precocial species, young birds may be completely independent of their parents, as in the megapodes; or they may follow their parents but find their own food (e.g., **ducks** and shorebirds). Others may follow their parents, who point out food to them, for example, **quail** and chickens. Still others, like **grebes** and **rails**, may follow their parents and be fed by them.

The young of semi-precocial species are active and wide-eyed at **birth**. They remain in the nest even though they are able to walk, so they can be fed by their parents (e.g., **gulls** and **terns**). Semi-altricial young are unable to leave the nest, and may be born with their eyes open (e.g., **herons** and hawks), or with their eyes closed (e.g., **owls**). The passerines are examples of altricial species.

Although few birds are completely vegetarian, most eat plant material. Some birds are seed-eaters, others feed on fruit and berries, and still others feed on buds and green shoots of plants. Waterfowl may eat various parts of aquatic plants and aquatic creatures such as marine worms. But, without **insects** and other **arthropods** such as spiders, most birds would starve. Some birds may employ a variety of foraging methods to find their food, while others may have a much narrower range of techniques. In addition to food, birds must have **water**, which they either drink or obtain from the foods they eat.

Birds require large and dependable sources of rich food to sustain their high metabolisms. Although some birds are able to sustain themselves by only moving a few miles from where they were hatched, most birds, at least in the temperate zones, move from place to place with the change in seasons. Geese and **cranes** appear to learn their migration routes by accompanying their elders, but other birds, including many songbirds, seem to reach their wintering grounds by pure **instinct**. Clues to migratory routes may be provided by the position of the **sun**, the positions of the stars, the **earth's magnetic field**, and the sound of the **ocean** crashing against the shore.

Man's fascination with birds dates back thousands of years. Old World **cave** drawings show ostriches, **auks**, **grouse**, passerines, snowy owls, swans, ducks, **eagles**, and others. Men were even painted with birds' heads. Ancient myths suggest that early man viewed birds not only as food sources, but as colorful mystical creatures capable of flight and disappearance, and the gift of song.

Although there is little doubt that most birds are beneficial, or at least not harmful, to the interests of man, they are frequently perceived as **pests**. While it is true, birds have wreaked economic havoc on some of man's agricultural activities, it should not be forgotten that birds also feed on insects, **rodents**, and other life forms that destroy **crops**.

Both intentionally and unintentionally, man has destroyed many bird species. Two hundred years ago, birds were considered such an inexhaustible resource that the wholesale slaughter of a species, leading to its **extinction**, hardly raised an eyebrow. But the greatest impact man has had on birds occurred through his expansion into their natural habitats with the construction of farms, cities, roads, and industrial buildings. A by-product of industrial development has been widespread environmental **pollution**, including **pesticides** and other noxious species. Intended to rid fields of such insects as the fire ant, pesticides have accumulated in many places traditionally frequented by birds, where they have been subsequently ingested by them. **Oil spills** have also taken their toll on bird populations. It is not surprising, therefore, that many species have disappeared as a result of man's encroachment on the natural environment. According to one estimate, 85 species of birds, representing 27 families, have become extinct since 1600.

Sibley, David Allen. *The Sibley Guide to Birds*. New York: Knopf, 2000.

KEY TERMS

Brood—To sit or hatch eggs.

Molt—To periodically shed all or part of a bird's feathers.

Monogamy—The practice of having only one mate.

Ornithology—The branch of zoology concerned with the study of birds.

Oviparous—Producing eggs that hatch outside of the body.

Passeriformes—An order of birds whose members have four toes, three directed forward and one backward, and all joining the foot at the same level.

Passerines—Members of the order Passeriformes, which includes perching birds and songbirds such as jays, blackbirds, warblers, and sparrows.

Polyandry—A mating pattern in which a female mates with more than one male in a single breeding season.

Polygyny—A mating pattern in which a male mates with more than one female in a single breeding season.

The best approach to limiting man's impact on birds is the simplest one: if man would limit his destruction of the natural environment, many of the habitats of threatened species could be preserved. Interventions that could protect bird populations include checks on unbridled commercial and residential development, and the dumping of environmental pollutants. But for man to adopt these measures would require a fundamental change in his attitude toward the environment, which is paradoxically only likely to occur through his gaining greater first-hand knowledge of nature in its unspoiled state.

See also Birds of prey; Chordates; Shore birds; Song birds.

Resources

Books

Ehrlich, Paul R., David S. Dobkin, and Darryl Wheye. *The Birder's Handbook*. New York: Simon & Schuster Inc., 1988.

Forshaw, Joseph. *Encyclopedia of Birds*. New York: Academic Press, 1998.

Mitchell, A. *The Guide to Trees of Canada and North America*. Surrey, U.K.: Dragon's World Ltd., 1987.

Peterson, Roger Tory. *North American Birds*. Houghton Mifflin Interactive (CD-ROM). Somerville, MA: Houghton Miflin, 1995.

Birds of paradise

The birds of paradise are some of the most fascinating **birds** in the world. This is due to the striking coloration of the males of most **species**, and the wide range of behaviors demonstrated in the group. Researchers of **animal** behavior are particularly interested in the elaborate mating displays performed by male birds of paradise.

Birds of paradise are members of the family Paradisaeidae, which probably evolved on the **island** of New Guinea. The family is comprised of 43 species, 38 of which are found mainly or entirely on New Guinea. Two species are found only in the Moluccan Islands to the west of New Guinea, and four others are found mainly or entirely in northeastern **Australia**. Included within the family are such birds as astrapias, manucodes, paradisaeas, parotias, riflebirds, and sicklebills.

Description

The birds of paradise have a crow-like body shape, with strong feet and bill. However, in some species this basic pattern has been modified substantially. For example, the sicklebills have evolved a long, curved beak used to probe for **insects** in thick **moss** and **tree bark**. In many species the plumage of the males is modified with fantastic plumes, streamers, and wiry head and tail extensions. Although the body of most of the birds of paradise is 10-17 in (25-45 cm) long, the head plumes may reach 16 in (40 cm) in length, and the tail feathers up to 27 in (70 cm) long.

The females of most species are colored drab buff to black, with patterning that helps them remain hidden in the forest canopy while sitting on a nest. (This type of coloration is called cryptic.) Nests of most species are cup-shaped, and are built in forks of trees using leaves, twigs, and other **plant** material. Females lay one or two eggs, which average 1.4 in (37 mm) long and 1 in (26 mm) wide. Incubation periods are 17-21 days, and young birds remain in the nest 17-30 days.

Habitat and diet

New Guinea is an extremely mountainous island. Its equatorial location results in a tropical climate near **sea level**, but cooler conditions higher in the **mountains**. In fact, the highest peaks have **glaciers** present. In addition, the prevailing oceanic winds carry moisture-laden air over the island, resulting in as much as 27 ft (8.5 m) of rain per year in some places. Sites on the lee side of mountains, however, may be quite dry. The great variations of climate in New Guinea result in numerous different habitats occurring. The various species of birds of paradise are rather specific to particular kinds of **habitat**. For example, the crested bird of paradise is only found in upper montane forest and subalpine shrubland, while the trumpet manucode is found only in lowland and lower mountain **forests**, and the blue bird of paradise prefers mid-montane forest.

In addition to inhabiting different ecological zones in New Guinea, the various birds of paradise use different food resources. The two basic kinds of foods eaten are **fruits** and insects. There are also two groups of fruits: simple fruits rich in **carbohydrate**, such as figs, and complex fruits with high levels of **fat** and protein, such as those of **mahogany** and **nutmeg**. Species of birds of paradise tend to eat mainly simple fruits (e.g., the trumpet manucode), mainly complex fruits (e.g., the raggiana bird of paradise), or complex fruits plus significant quantities of insects (e.g., the magnificent bird of paradise).

When animals eat tree fruits, they may also digest the **seeds**, or the seeds may pass through the **digestive system** intact. If seeds are not digested, a tree seedling may sprout from them, helping the forest regenerate. In most forest habitats worldwide, the main fruit-dispersing animals are **mammals**. In New Guinea, however, this role is largely played by birds of paradise, which eat fruits and distribute the seeds, helping to ensure the dispersal of important species of forest trees.

Mating behavior

Polygynous birds of paradise

As mentioned above, many species of birds of paradise are sexually dimorphic, meaning the males and females have different appearances. The males have elaborate plumage patterns, which are used in their mating displays. The females of these species are drab and cryptic. Sexually dimorphic species are usually also polygynous, meaning the males may mate with more than one female. To attract a female, a male may perform a mating dance on the ground while conspicuously displaying its bright plumage and calling loudly, or it may display while perched on a shrub, or while hanging upside down on a tree branch. Males may perform these displays alone, or in competitive groups in a place called a lek. The females watch the displays and choose which male to mate with. The female choice appears to be based on the vigor of the display of the male, and the condition and **color** of his feathers. By choosing a vigorous mate, the female presumably ensures that her offspring will also be relatively healthy. Therefore, the strongest, most brightly-feathered males have a better chance of being

An **Emperor of Germany's bird of paradise in Baiyer River Sanctuary, New Guinea.** *Photograph by Tom McHugh. National Audubon Society Collection/Photo Researchers, Inc. Reproduced by permission.*

chosen as a mate by females, while less attractive males may be passed over. The elaborate plumage of the males is thought to have evolved through this evolutionary process of sexual **selection** (i.e., females choosing mates on the basis of their desirable behavioral and anatomic traits, including color). After mating, the female returns to her nest and raises her offspring alone.

Researchers have noticed a relationship between the mating system and diet in the birds of paradise. Polygynous bird of paradise species that display in leks (such as the raggiana bird of paradise) tend also to eat mainly complex fruits. This is thought to be because females searching for these fruits fly long distances in the forest, and thus are likely to encounter groups of males displaying together. Polygynous species in which solitary males display (such as the magnificent bird of paradise) tend to eat insects plus complex fruits. To get insects, females need not fly long distances, so a male is more likely to be seen and chosen as a mate if he displays alone near the small home range of the female.

Interestingly, the polygynous birds of paradise also show sexual bimaturism. This means that males and females become sexually mature at different ages. Females of these species are thought to begin to breed when 2-3 years old, while males do not acquire mature plumage (and do not breed) until age 4-7 years. However, males of these species will grow adult plumage at a younger age when kept alone in captivity. This suggests that the delay in male maturation in the wild is due to hormonal suppression related to the presence of already-mature adult males.

Monogamous birds of paradise

Nine species of birds of paradise, including the manucodes, are sexually monomorphic. The males and females have coloring that is the same or nearly so (they tend to be brown or black), and both lack the elaborate plumage that characterizes most other birds of paradise. These species are monogamous, meaning the males and females mate with only one partner at a time, and in some species pair for life. As in most monogamous species, the males help the females raise the young.

These species typically feed mainly on simple fruits, such as figs. This type of fruit is relatively low in **nutrients**, compared to complex fruits and insects. Scientists think that the monogamous mating system may have developed in these species because two parents are necessary to provide enough **nutrition** to raise the young. Thus, in the birds of paradise, it appears that the diet of a species has influenced the **evolution** of its social system.

Habitat loss

Although much of New Guinea is still covered with **rainforest**, extensive areas are being logged or converted to agriculture. Moreover, because of population growth and economic development the habitat destruction by **deforestation** will increase in the future. Some species of birds of paradise are found in highly limited ranges, so deforestation of their local habitat could result in their **extinction**. Other species are found throughout New Guinea, but only within a particular altitudinal range. For example, the blue bird of paradise occurs only between 4,200 and 5,900 ft (1,300-1,800 m). This species is under **pressure** from habitat loss associated with human colonization at these altitudes.

In addition to habitat loss, many species are threatened by overhunting. After Europeans discovered the birds, the demand for their plumage to use as decoration increased, so that by 1900 the populations of many species were greatly reduced. At present, the importation of bird of paradise feathers into the United States and most of **Europe** is illegal. However, bird of paradise feathers and skins continue to be of great cultural importance for the indigenous highlanders of New Guinea, who use the feathers in headdresses and other decorations.

Six of 24 species of New Guinea birds thought to require urgent **conservation** action are birds of paradise. To conserve the birds of paradise of New Guinea, a network of large rainforest reserves must be designated. The network of protected areas should be designed to include large areas of the habitat of all New Guinea birds of paradise. The reserves would also provide habitat for many other rare species of New Guinea **wildlife**. If these reserves are to be successfully established, they must be managed in a manner that also provides sustainable livelihoods for the indigenous people of the area. One of the great challenges of the future will be to balance human development with environmental conservation.

KEY TERMS

Cryptic—Drab, usually brownish coloration that makes an organism difficult to see in its natural habitat and allows it to hide from predators.

Lek—A central area in which many males of a species perform mating displays simultaneously.

Montane—Habitat on relatively cool, moist mountain slopes below the tree line.

Sexual bimaturism—A condition in which the males and females of a species become sexually mature at different ages.

Sexual dimorphism—The occurrence of marked differences in coloration, size, or shape between males and females of the same species.

Subalpine—Habitat on high mountain slopes, above the treeline.

Resources

Books

Beehler, Bruce M. *A Naturalist in New Guinea.* Austin, TX: University of Texas Press, 1991.

Beehler, Bruce M., Thane K. Pratt, and Dale A. Zimmerman. *Birds of New Guinea.* Princeton, NJ: Princeton University Press, 1986.

Frith, C.B., and B.M. Beehler. *The Birds of Paradise: Paradisaeidae.* Oxford University Press, 1998.

Perrins, C.M., and A.L.A. Middleton, eds. *The Encyclopedia of Birds.* New York: Facts on File, 1985.

Raven, Peter, R. F. Evert, and Susan Eichhorn. *Biology of Plants.* 6th ed. New York: Worth Publishers Inc., 1998.

Periodicals

Beehler, Bruce M. "The Birds of Paradise." *Scientific American* 261 (December 1989): 116-123.

Amy Kenyon-Campbell

Birds of prey

Birds of **prey** are predators that catch and eat other animals. These birds are called **raptors** (from the Latin *rapere*, meaning to snatch), a reference to their specialized, powerful feet, which are used to seize their prey. Raptorial birds eat birds, small **mammals**, **reptiles**, **amphibians**, **fish**, and large **insects**.

Birds of prey are members of five avian families within the order Falconiformes. The Accipitridae includes the **hawks**, **eagles**, kites, harriers, and Eurasian **vultures**.

Some widespread North American **species** in this family are the bald eagle (*Haliaeetus leucocephalus*), the red-tailed hawk (*Buteo jamaicensis*), the sharp-shinned hawk (*Accipiter striatus*), and the northern harrier (*Circus cyaneus*). The Falconidae includes the **falcons** and caracaras, such as the **peregrine falcon** (*Falco peregrinus)* and the merlin (*F. columbarius*). The Pandionidae includes only one species, the osprey (*Pandion haliaetus*), which has an almost worldwide distribution. The Sagittariidae of **Africa** includes only the **secretary bird** (*Sagittarius serpentarius*). Finally, the Cathartidae (American vultures) live only in North, Central, and **South America** and include the turkey vulture (*Cathartes aura*), the black vulture (*Coragyps atratus*), and the endangered California condor (*Gymnogyps californianus*).

Owls are also birds of prey. Their order Strigiformes includes two families, the Tytonidae (barn owls), represented in **North America** by the barn owl (*Tyto alba*), and the Strigidae, including the eastern screech-owl (*Otus asio*), the great horned owl (*Bubo virginianus*), the snowy owl (*Nyctea scandiaca*), and the tiny elf owl (*Micrathene whitneyi*), which is only 5 in (13 cm) long.

As a group, birds of prey eat large numbers of small mammals that might otherwise be **pests** of agriculture or homes. Many people find raptors fascinating because of the grace, speed, ferocity, and power that most species display as predators. However, some people dislike birds of prey because certain species hunt game birds and songbirds, and in the past, raptors have been killed in large numbers. Large numbers of raptors have also been indirectly poisoned by the use of **insecticides**. For example, the widespread agricultural use of persistent, bioaccumulating, chlorinated-hydrocarbon insecticides such as DDT and dieldrin caused the collapse of populations of peregrine falcons, bald eagles, and other species of birds. The use of these **pesticides** is now restricted, and their effect on raptors is becoming less significant. The peregrine falcon, for example, has recovered somewhat in abundance and is no longer considered an **endangered species** in the United States. Nevertheless, populations of many birds of prey remain perilously small.

See also Condors.

Birth

Birth, or parturition, in **mammals** is the process in which a fully developed fetus is expelled from the mother's uterus by the force of strong, rhythmic muscle contractions. The birth of live offspring is a reproductive feature shared by mammals, some fishes, and selected invertebrates (such as scorpions), as well as some **reptiles** and **amphibians**. Animals who give birth to live offspring are called viviparous (meaning "live birth").

In contrast to viviparous animals, other animals give birth to eggs; these animals are called **oviparous** (meaning "egg birth"). Some oviparous **species**, such as **birds**, retain their eggs inside their bodies for long periods of **time**; in these animals, the eggs are laid at an advanced stage of development. Other animals, such as **frogs**, give birth to less developed eggs, which undergo development outside the mother's body.

Viviparous animals

In both viviparous animals and oviparous animals, **fertilization** of the mother's egg with the father's sperm takes place inside the mother's body. One of the advantages to giving birth to live young is that the mother protects the fetus inside her body as it develops. The developing fetus derives **nutrients** from the mother's body, and so is assured of receiving all the nourishment it needs to complete development.

The length of time between fertilization and birth in viviparous animals is called the gestation period. The length of the gestation period varies according to species. The gestation period of **mice** is 21 days, of rabbits is 30-36 days, and of dogs and **cats** is 60 days. The largest mammal, the baleen whale, has a gestation period of 12 months—only three months longer than the gestation period of humans. Elephants have one of the longest gestation periods of all animals, 22 months.

Some viviparous animals such as humans, **horses**, and cows, give birth to only one offspring at a time, although occasionally these animals produce twins or triplets. Other animals give birth to many offspring at a time. Usually, the multiple offspring in a litter are each derived from a separate egg, but the armadillo gives birth to four identical offspring that are derived from the same fertilized egg.

How does birth begin?

At the end of the gestation period, the mother's uterus begins to contract rhythmically, a process called labor. The initiation of labor leading up to birth is the result of a number of **hormones**, notably oxytocin.

Maternal progesterone

Shortly after fertilization the hormone progesterone increases and is maintained at high levels in the mother's bloodstream. The high levels of progesterone prevent the uterus from contracting. The progesterone prepares the

A Holstein calf being born. Unlike a newborn human child it will walk almost immediately. *Photograph by Cynthia Matthews. Stock Market. Reproduced by permission.*

lining of the uterus (the endonestrium) for its supporting role in nurturing the developing fetus, and helps form the placenta. Maternal progesterone levels begin to drop during the last weeks of gestation, while the levels of estrogen begin to rise. When progesterone levels drop to very low levels and estrogen levels are high, the uterus begins to contract.

Oxytocin

Oxytocin is a hormone released from the pituitary gland in the **brain**, which stimulates uterine contractions and also controls the production of milk in the mammary **glands** of the breast (a process called lactation). Synthetic oxytocin is sometimes given to women in labor to induce labor.

The mechanism that prompts the secretion of oxytocin from the pituitary during labor is thought to be initiated by the **pressure** of the fetus's head against the cervix, the opening of the uterus. As the fetus's head presses against the cervix, the uterus stretches, and relays a message along nerves to the pituitary, which responds by releasing oxytocin. The more the uterus stretches, the more oxytocin is released.

Fetal endocrine control

Fetal hormones are also thought to play a role in initiating labor. At the end of gestation, the fetal adrenal glands secrete steroid hormones called cortico steroids, which cause the hormone-like substances known as prostaglandins. Prostaglandins contribute to the contraction of the uterus during labor.

Birth in humans

Labor culminating in birth in humans begins with the rhythmic contractions of the uterus, which dilate the cervix. This causes the fetus to move down the birth canal and be expelled together with the placenta, which had supplied the developing fetus with nutrients from the mother. Ususally, the entire birth process takes about 16 hours, but it can range anywhere from less than one hour to 48 hours.

The first stage: dilation of the cervix

In order for the fetus to leave the uterus and to enter the birth canal, it must pass through the cervix, the opening of the uterus. The cervix is normally tightly closed, and is sealed with a plug of mucus during gestation to protect the fetus from invading **microorganisms**. During the first stages of labor, the contractions of the uterus dilate the cervix, which widens to about 4 in (10 cm), to accommodate the passage of the fetal head.

In the last weeks of pregnancy, before labor begins, the uterus undergoes irregular contractions, which serve to **exercise** the muscles of the uterus and may even dilate the cervix; it's not unusual for a woman to go into active labor with a cervix that is already dilated to 1 or 2 cm. During the last weeks of pregnancy, the cervix also thins out (or effaces), which makes dilation easier.

In preparation for birth, the fetus moves further down into the mother's pelvis. When labor begins, the fetus is usually positioned with its head engaged with the top of the cervix. This engagement is called "lightening" or "dropping." When labor begins, the contractions loosen the mucus plug in the cervix causing small **capillaries** in the cervix to break, and the mucus and **blood** are discharged from the vagina. This discharge is sometimes called "bloody show" and signals the onset of labor.

Another sign that may signal the beginning of labor is the rupturing of the amniotic sac. In the uterus the fetus is encased in a **membrane** (the amniotic sac) and literally floats in amniotic fluid. When uterine contractions begin, this sac ruptures and the amniotic fluid can leak from the uterus. Not all women experience an abrupt rupturing of the amniotic sac; in some, the amniotic fluid gradually leaks out as labor progresses. Once the amniotic sac has ruptured, or the amniotic fluid begins to leak, labor usually progresses more rapidly. During the first stage of labor, the cervix dilates about 0.5-0.6 in (1.2-1.5 cm) an hour. The uterine contractions are 5-30 minutes apart, and last for 15-40 seconds. The end of the first stage of labor is associated with the strongest uterine contractions. Contractions are two to five minutes apart, and last for 45-60 seconds. The cervix opens rapidly at this point. This period of labor, sometimes called transition, is usually the most difficult for the mother. The contractions are very strong and close together, and nausea and vomiting are common. After the cervix has dilated to its full width of 4 in (10 cm), the contractions slow down somewhat to about three to five minutes apart. The fetus is then ready to be born, and the second stage of labor begins.

The second stage: birth

During the second stage, lasting about one to two hours, the mother uses her abdominal muscles to push the fetus through and out of the birth canal.

The pushing is actually a **reflex** action, but if a woman can help the reflex by actively using her muscles,

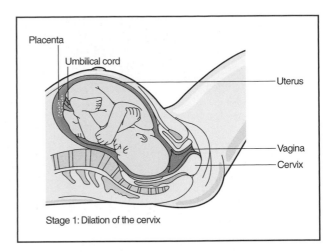

Stage 1: Dilation of the cervix. *Illustration by Hans & Cassidy. Courtesy of Gale Group.*

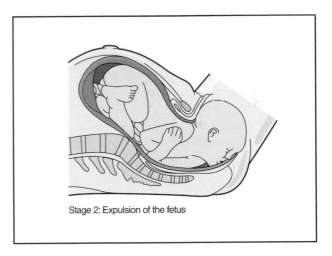

Stage 2: Expulsion of the fetus. *Illustration by Hans & Cassidy. Courtesy of Gale Group.*

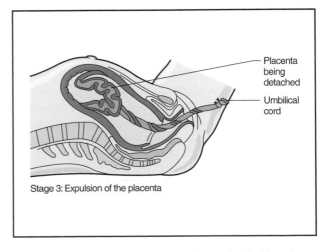

Stage 3: Expulsion of the placenta. *Illustration by Hans & Cassidy. Courtesy of Gale Group.*

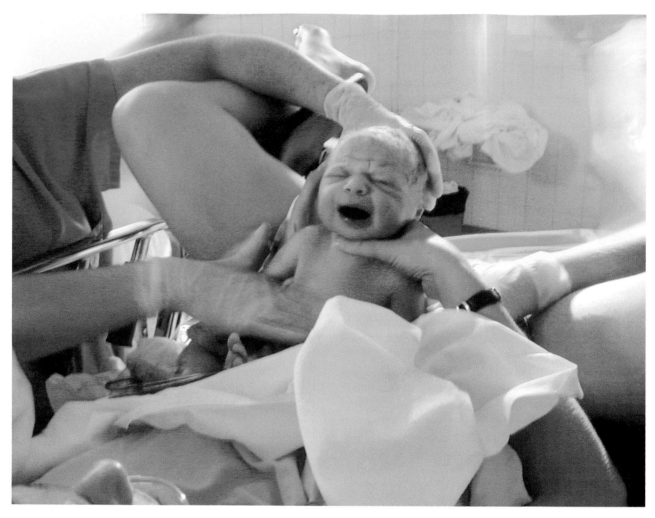

Natural childbirth. *© Jules Perrier/Corbis. Reproduced by permission.*

birth goes much faster. As the fetus moves down the birth canal to the vaginal opening, the head begins to appear. The appearance of the head at the opening of the vagina is called crowning. After the head is delivered, first one shoulder is delivered, then the other. The rest of the body follows.

After the baby is born, the umbilical cord that has attached the fetus to the placenta is clamped. The clamping cuts off the circulation of the cord, which eventually stops pulsing due to the interruption of its blood supply. The baby now must breathe air through its own lungs.

The third stage: delivery of the placenta

Before delivery, the placenta separates from the wall of the uterus. Since the placenta contains many blood vessels, its separation from the wall of the uterus causes bleeding. This bleeding, if not excessive, is normal. After the placenta separates from the uterine wall, it moves into the birth canal and is expelled from the vagina. The uterus continues to contract even after the placenta is delivered, and it is thought that these contractions serve to control bleeding.

History of childbirth

Until the twentieth century, childbirth was the province of women. A woman giving birth was attended by her female relatives and perhaps a woman in the community known for her midwifing skills. As the scientific revolution of the seventeenth century took place, concerned physicians noticed that childbirth was sometimes a dangerous, often fatal, process. Infections, injury to the baby and mother, and death occurred when unskilled midwives attempted to manage complications. Gradually, childbirth changed from an entirely female-centered activity to a medical process overseen by predominately male physicians. By the early twentieth century, childbirth moved from the home to the hospital. By the mid-twentieth century, childbirth had become a completely medical process, attended by physicians and managed by medical equipment and procedures, such as fetal monitors, **anesthesia**, and surgical interventions.

Later in the twentieth century, some women became dissatisfied with this medical approach to birth. Many felt that the medical establishment had taken control of a natural biological process. Women wanted more control over labor and birth and new ways of giving birth that sought to reduce or eliminate the medical interventions became popular. With the increasing concern about the effect of anesthesia on the fetus, many women refused artificial means of controlling **pain**, and instead relied on breathing and relaxation techniques. Fathers, once banished from labor and delivery rooms, were now welcomed as partners in the birth process.

Today, women have many options for labor and birth. Some women deliver in a hospital with doctors and nurses close by to supervise the birth process. Others choose a nurse-midwife, a person who has been trained to deliver babies but who is not a doctor. Still others choose home birth, attended either by a doctor or midwife, or sometimes both. Whatever option a woman chooses, it is important to get good medical care throughout pregnancy. Periodic prenatal checkups are one of the best ways to avoid birth complications.

Types of childbirth preparation

Many childbirth experts believe that the more a mother knows about the birth process, the less fear and apprehension she will feel giving birth. Many childbirth preparation methods prepare both mother and father for the birth experience and teach relaxation and breathing techniques. The Read method, for instance (named after its founder, British physician Grantley Dick-Read), is based on the notion that fear leads to pain. The Read method includes childbirth education, exercises to improve muscle tone, and relaxation techniques. The Lamaze method (named for Dr. Ferdinand Lamaze) takes a psychological approach to managing labor. The Lamaze method teaches women to relax and breathe in response to pain, the theory being that this substitution of favorable activity for negative sensations reduces pain. The Bradley method focuses on deep relaxation and slow, deep breathing, and ascribes an important role to the father.

Types of anesthesia

Two types of anesthesia are commonly used during labor and birth. In general anesthesia, the mother is given drugs that put her to **sleep**, but this type of anesthesia is rarely used today, since the drugs can depress the fetal **heart** beat. In regional anesthesia, drugs are inject-

---

KEY TERMS

Amniotic fluid—The fluid in which the fetus "floats" in the uterus.

Amniotic sac—The sac that contains the amniotic fluid.

Gestation—The period of carrying developing offspring in the uterus after conception; pregnancy.

Labor—The strong, rhythmic contractions of the uterus leading to birth.

Placenta—The organ that develops during gestation through which a fetus receives nourishment from the mother.

Prostaglandins—A substance released by uterine cells that cause contraction of the uterus.

Umbilical cord—The cord that attaches the fetus to the placenta.

---

ed to deaden sensation around the spinal nerves that carry sensations from the pelvic region. Controversy about whether these drugs affect the fetus is ongoing, although some kinds of regional anesthesia affect the fetus less than others.

See also Embryo and embryonic development; Reproductive system; Sexual reproduction; Viviparity.

Resources

Books

Bean, Constance A. *Methods of Childbirth.* 2nd ed. Garden City, NY: Doubleday, 1990.

Karmel, Marjorie. *Thank You, Dr. Lamaze.* New York: Harper and Row, 1993.

Knobil, Ernst, and Jimmy D. Neill, eds. *The Physiology of Reproduction.* 2nd ed. New York: Raven Press, 1994.

Korte, Diana. *The VBAC Companion: The Expectant Mother's Guide to Vaginal Birth After Cesarean.* Cambridge, MA: Harvard Common Press, 1999.

Mitford, Jessica. *The American Way of Birth.* New York: Dutton, 1992.

Moore, Michele, and Caroline De Costa. *Cesarean Section: Understanding and Celebrating Your Baby's Birth.* Baltimore, MD: Johns Hopkins Medical Press, 2003.

Periodicals

"Deciding to Be Born." *Discover* 13 (May 10, 1992).

Fischman, Joshua. "Putting a New Spin on the Birth of Human Birth." *Science* 264 (20 May 1994): 1082.

Ventura, S.J. "Births: Final Data for 1999." *Service Today* 49, no. (2001): 1-100.

Kathleen Scogna

Birth control *see* **Contraception**

Birth defects

Birth defects or **congenital** defects are present at birth. They result from heredity, environmental influences, or maternal illness. Such defects range from the very minor, such as a dark spot or birthmark that may appear anywhere on the body, to more serious conditions that may result in marked disfigurement, impaired functioning, or decreased lifespan.

A number of factors individually or in combination may cause birth defects. Heredity plays a major role in passing birth defects from one generation to the next. Inherited conditions are passed on when a baby receives a flawed **gene** from one or both parents. Conditions such as **sickle cell anemia**, **color blindness**, deafness, and extra digits on the hands or feet are hereditary. The condition may not appear in every generation, but the defective gene usually is passed on. A classification of structural defect can be as follows: Malformation (poor formation), deformation (due to fetal constraint that can result in damage (e.g., central **nervous system** damage or limb reduction) and disruption of previous normally formed structures (due to vascular damage, vascular exchange of necrotic debris).

Causes of defects

Low birth weight deriving from a fetal growth restriction (FGR) is the most common birth defect, with one in every 15 babies being born at less than their ideal weight. A baby whose weight lies in lowest 10% of the normal population is designated as having a FGR. At term of pregnancy, a baby who weighs 5 lb, 8 oz (2,500 g) at birth has a low birth weight. One who is born weighing 3 lb, 5 oz (1,500 g) has a very low birth weight. A low birth weight baby born after a normal gestation period is called a small-for-date or small-for-gestational-age baby.

Exposure of the mother to chemicals such as mercury or to **radiation** during the first three months of pregnancy may result in an abnormal alteration in the growth or development of the fetus. The mother's diet may also be a factor in her baby's birth defect. A balanced and healthy diet is essential to the proper formation of the fetus because the developing baby receives all of its **nutrition** from the mother.

Prenatal development of the fetus may also be affected by **disease** that the mother contracts, especially those that occur during the first trimester (three months)

of pregnancy. For example, if a pregnant woman catches rubella, the **virus** crosses the placenta and infects the fetus. In the fetus, the virus interferes with normal **metabolism** and **cell** movement and can cause blindness (from cataracts), deafness, **heart** malformations, and mental retardation. The risk of the fetal damage resulting from maternal rubella **infection** is greatest during the first month of pregnancy (50%) and declines with each succeeding month.

It is especially important that the mother not smoke, consume **alcohol,** or take drugs while she is pregnant. Drinking alcohol heavily can result in **fetal alcohol syndrome** (FAS), a condition that is physically apparent. FAS newborns have small eyes and a short, upturned nose that is broad across the bridge, making the eyes appear farther apart than normal. These babies also are underweight at birth and do not catch up as **time** passes. They often have some degree of mental retardation and may exhibit **behavior** problems. A mother who continues to take illicit drugs such as heroin, crack, or **cocaine** will have a baby who is already addicted. The **addiction** may not be fatal, but the newborn may undergo severe withdrawal, unless the addiction is revealed and carefully treated. Furthermore, some behavior problems/cognitive deficits are suspected to be associated with fetal drug exposure and addiction.

Some therapeutic drugs taken by pregnant women have also been shown to produce birth defects. The most notorious example is **thalidomide**, a mild sedative-hypnotic agent. During the 1950s women in more than 20 countries who had taken this drug gave birth to more than 7,000 severely deformed babies. The pattern of malformation seen in affected infants included phocomelia, polydactyly, syndactyly, facial capillar hemangiomas, **hydrocephalus**, renal anomalies, cadiovascular anomalies, **ear** and **eye** defects, and intestinal anomalies. The principal defect these children suffered is phocomelia, characterized by extremely short limbs often with no fingers or toes.

Physical birth defects

Clubfoot

Approximately one newborn out of every 735 has a form of clubfoot. In the most serious form, known as equinovarus, the foot is twisted inward and downward and the foot itself is cupped or flexed. If both feet are clubbed in this manner the toes point to each other rather than straight ahead. Often the heel cord or Achilles tendon is taut so that the foot cannot be straightened without **surgery**.

A milder and more common type of clubfoot is called calcaneal valgus, in which the foot is bent upward and outward in the same way that you would flex your foot at the ankle. Still other forms include the talipes cavus in which the instep is abnormally elevated; talipes valgus in which the heel is turned outward, and talipes varus in which the heel is turned inward.

The seriously deformed clubfoot requires surgery to realign the bones and ligaments. The milder forms often can be cured by fitting the baby with corrective shoes to gradually move the bones back into alignment.

Cleft lip and cleft palate

Approximately 7,000 newborns (one of every 930 births) are born with cleft lip and/or cleft palate each year in the United States. Cleft lip and palate describe a condition in which a split remains in the lip and roof of the mouth. Although cleft lip and palate are two distinct anomalies, they frequently occur together. Cleft lip with or without cleft palate occurs in 60-75% of the cases. twenty-five to forty percent are isolated cleft palate. During growth *in utero* (in the womb) the lip or palate, which develop from the edges toward the middle, fail to grow together. Such a failure is a consequence of the abnormal **migration** and proliferation of facial embryonic tissues called mesenchyme. The defect occurs most often among Asians and certain Native American groups, less frequently among whites, and least often among African Americans.

Approximately 25% of infants born with cleft palate have inherited the trait from one or both parents. The cause for the other 75% remains unknown, but may be a combination of heredity, poor nutrition, use of drugs, or a disease the mother contracted while pregnant. Maternal smoking represents the most controversial association. The cleft may involve only the upper lip, may extend into the palate, or may be located on the back of the palate.

Surgery is especially important to correct the defect in the palate. Feeding a baby with cleft palate is difficult because the food can pass through the palate into the nasal cavity and may be inhaled and cause choking. In the newborn, whose bones have not completely hardened, surgery is relatively simple. As the child ages, however, surgical correction is more difficult and the child will require **speech** therapy.

Spina bifida

Spina bifida or open spine occurs once in 2,000 births in the United States. It belongs to a group of defects known as neural tube defects that are the second most prevalent neonatal anomaly in the United States after cardiac malformations. It occurs when the edges of the spine that should grow around the spinal cord do not meet. An open area remains, which can mean that an

area of the spinal cord (or the entire spinal cord, in the most severe cases) are unprotected. The mildest form of spina bifida may be so slight that the defect does not have any effect on the child and is discovered by accident, usually when an x ray is taken for another reason. The term spinal bifida means the spine is cleft, having an opening or space, in two parts.

Spina bifida may present itself as a cyst, ranging in size from a walnut to a grapefruit, in which some parts of the meninges (layers of **connective tissue** covering the spinal cord), spinal cord, or both are contained. The lump can be removed surgically. In the most serious form, the lump or cyst has little skin or covering so spinal fluid may leak from it. Roots of the spinal nerves are contained within the cyst and the cyst may be covered with sores. Infection is a serious risk until surgery has been performed and the area has healed. Unfortunately, this condition may leave the child's legs partially or completely paralyzed and without feeling. Other associated problems may include control of the bowels and bladder.

Newborns with spina bifida often have an associated condition called hydrocephalus, which literally means **water** in the head. In this condition, cerebrospinal fluid collects in and around the **brain** and will not drain. Mental retardation can result if the fluid is not drained regularly. This can be accomplished by implanting a special tube (called a shunt) leading from the brain down into a vein in the child's neck or into the child's chest to allow the fluid to drain harmlessly. Hydrocephaly also can occur in infants who do not have spina bifida. The cause of spina bifida is not known, nor is any means of prevention. It can be diagnosed before birth by **amniocentesis** (by dosing the intra-amniotic alpha feto protein) or ultrasound. The risk of having a baby with spina bifida or other associated defects seems to be reduced if a woman takes at least 400 mg of folic acid just before and throughout pregnancy.

Heart defects

Congenital heart defects occur in one of every 115 births in the United States. The defect may be so mild that it is not detected for some years or it may be fatal. A baby with a heart defect may be born showing a bluish tinge around the lips and on the fingers. This condition, called cyanosis, is a signal that the body is not receiving enough **oxygen**. The blue **color** may disappear shortly after birth, indicating that all is normal, or it may persist, indicating that further testing is needed to determine the nature of the heart defect.

A normal heart has four chambers; two upper, called the atria (singular: atrium) and two lower called the ventricles. The right heart receives the **blood** that is returning from the body, and has been depleted of oxygen. This oxygen-poor blood arrives in the right atrium, where it is pumped into the right ventricle. The right ventricle sends oxygen-poor blood to the lungs, where it is exposed to and picks up plenty of oxygen again. This oxygen-rich blood enters the left atrium and is then pumped into the left ventricle. The left ventricle pumps oxygen-rich blood through the aorta to all the organs and tissues of the body.

During fetal development, blood circulation occurs differently, because the fetus' blood does not need to flow through its lungs. It receives its oxygen from the mother through the placenta via the umbilical cord. Since the atria communicate during fetal life, blood rich in oxygen coming from inferior vena cava crosses the foramen ovale and into the left atrium bypassing the lungs (eventually the foramen ovale is closed from the higher **pressure** generated at the left side after the lunds expand at birth). Another special shunt, the arterious duct connects the main pulmonary artery to the aorta. In such a way, the blood flow that does enter the right atrium enters the right ventricle, then the main pulmonary artery, then enters the ductus arteriosus which connects to the aorta. In this way the vast majority of blood flow bypasses the lungs during development of the fetus. Normally the shunts should close at birth. After birth, blood should begin to circulate through the lungs for the first time, because the newborn baby's lungs are now responsible for delivering oxygen to the blood. Sometimes, however, the shunt does not close properly, and blood is not appropriately circulated through the lungs. When this occurs, surgery is required to close the shunt and restore normal circulation.

If it is undetected at birth, a heart defect may impair the growth of a child. He will be unable to exert the **energy** that other children do at play because he cannot supply sufficient oxygen to his body. He may become breathless at small amounts of exertion and may squat frequently because it is easier to breathe in that position.

Some minor defects may disappear over time as the child grows. A small hole in the wall between the left and right sides of the heart, which causes symptoms by allowing the mixing of oxygen-poor and oxygen-rich blood, for example, may spontaneously close over time. A larger defect will require surgical patching.

Some newborns may have only one upper chamber or only a single lower chamber of the heart. The aorta, where it begins at the heart, may be narrowed (stenosed) and impair the flow of blood from the heart. Some of the heart valves may not function correctly and occasionally the vessels of the heart may be transposed so that the aorta leads from the right side of the heart, delivering oxygen-poor blood to the organs and tissues.

These are only a few of the heart anomalies that can be present in the newborn. The heart is a complicated **organ** and its formation can be influenced by hereditary factors as well as by alcohol consumption or smoking. Fortunately, most heart defects correct themselves over time or can be corrected with surgery.

Other physical deformities

Physical defects in newborns are common. They can affect any of the bones or muscles in the body and may or may not be correctable. Among the more common are the presence of extra fingers or toes (polydactyly), which presents no health threat and can be corrected surgically. Similarly, webbed fingers and toes, a genetic disorder, seen in approximately one of every 1,700 to 2,000 births, can be treated surgically to resemble a normal appendage.

A more serious, though relatively rare, condition is called achondroplasia; this term means without cartilage formation and refers to the supposed lack of cartilage growth plates near the ends of a child's bones. In fact, the plates are present, but grow poorly. Achondroplasia is a type of dwarfism. This genetic disorder of bone growth is seen in one in 20,000 births and is one of the oldest known birth defects. Ancient Egyptian art shows individuals with this condition.

The cause of achondroplasia is not known, nor is there a cure. The child who has this condition will be slow at walking and sitting because of his short arms and legs, and this may be interpreted as mental retardation. However, these individuals have normal intelligence.

Hereditary diseases and syndromes

In addition to physical deformities, certain diseases and syndromes also are passed to the infant through the parents' genes. Some of these conditions can be controlled or treated while others are untreatable and fatal.

Sickle cell anemia

Sickle cell **anemia** is an inherited disease of the blood cells that occurs in about one of every 400 African Americans. An **individual** can be a carrier of sickle cell anemia, in which case he or she has the gene but does not show any active signs of the disease. If two carriers become parents, however, some of their children may have sickle cell anemia.

The disease gets its name because certain red blood cells assume a sickle shape and lodge in small blood vessels. This altered shape is a function of the hemoglobin **molecule** present in red blood cells. Two forms of hemoglobin make up these cells: hemoglobin A (Hb A) and hemoglobin B (Hb B). In individuals with sickle cell anemia, Hb B is instead produced as Hb S, a form of hemoglobin with a rigid, sickle shape that deforms the red blood cell. When the cell becomes wedged in a small blood vessel it prevents the flow of blood through the vessel and can initiate what is called a sickle cell crisis. The lack of blood flow to the tissues being blocked causes **pain** and **inflammation** of the oxygen-deprived **tissue**.

Abnormal red blood cells are removed from the **circulatory system** by the spleen, but removal of large numbers of such cells can lead to anemia, a lack of adequate numbers red blood cells. Unfortunately, the breakdown of abnormal red blood cells can in itself cause a serious condition in which excess **iron**, scavenged from the hemoglobin molecule, is deposited in tissues such as the heart and liver. So, although replacement of the destroyed red blood cells could be achieved with blood transfusion, the replacement cells will only add to the iron content of blood. There is no cure for sickle cell anemia, though scientists are **learning** how to better control it to prevent sickling of the blood cells.

Tay-Sachs disease

Tay-Sachs disease affects Jews of eastern European origin, the Ashkenazi Jews, and is a condition that is fatal at an early age. A carrier of the disease will have a gene for Tay-Sachs disease and another gene that is normal. If two carriers have children, every pregnancy will have a 25% chance of producing a completely normal child; a 50% chance of producing a child who will carry the trait, but reveal no symptoms; and a 25% chance of producing a child who actually suffers from the disease.

The newborn Tay-Sachs child lacks a blood **enzyme** called hexosaminidase A, which breaks down certain fats in the brain and nerve cells. When first born, the baby appears totally normal. However, over a short period of time, the brain cells become clogged with fatty deposits, and the child begins to lose functioning. As the disease progresses, the child will no longer be able to smile, crawl, or turn over, and will ultimately become blind and unaware of his surroundings. Usually the child dies by the age of three or four years.

There is currently no cure for Tay-Sachs disease, although carriers can be detected by a simple blood test that measures the amount of hexosaminidase A. A carrier will have half the amount of the enzyme as a normal person, and two carriers can be counseled to explain the probability of producing an offspring with Tay-Sachs disease. Researchers are trying to find a way to provide sufficient levels of the missing enzyme in the newborn, or to find a suitable substitute that could be supplied as the child ages, much like **insulin** is used to treat diabetes. A more technologically advanced line of research is examining the possibility of transplanting a normal gene to replace the defective one in carriers.

KEY TERMS

Hemoglobin—The substance within red blood cells that gives blood its characteristic red color and carries oxygen to the cells of the body.

In utero—While in the uterus, prior to birth.

Placenta—The flat, plate-like organ of exchange between the blood of the mother and that of the embryo. It attaches to the wall of the uterus and provides nutrients and oxygen for the embryo and removes wastes from the embryo.

Down syndrome

One in every 800-1,000 babies is born with **Down syndrome**. Down **syndrome** babies may have eyes that slant upward, small ears that may turn over at the top, a small mouth and nose that also is flattened between the eyes (at the bridge). Mental retardation is present in varying degrees, but most Down's syndrome children have only mild to moderate retardation. Generally these children walk, talk, dress themselves, and are toilet trained later than children with normal intelligence.

Down syndrome results when either the egg or the sperm that fertilizes it has an extra **chromosome**. Normally a human has 23 pairs of chromosomes, for a total of 46. An extra chromosome, specifically an extra number 21 chromosome, present when the egg is fertilized, leads to a baby with Down syndrome. Of course, if either parent has Down syndrome, the probability of passing the condition on to the offspring is increased. Also, parents who have had one Down syndrome child and mothers older than 35 years of age are at increased risk of having a Down syndrome baby. There is no cure, though many of these children can go on to attend school and hold jobs as do unaffected individuals.

It should be apparent from this small **sample**, that some birth defects are hereditary, passed from parents to offspring; little can now be done to prevent or cure these conditions, but genetic therapy offers hope that this situation may change in the future. Other birth defects result from infections of the mother during pregnancy, or from maternal consumption of alcohol or drugs, use of tobacco, or exposure to radiation or chemicals during pregnancy. In some cases, these birth defects can be prevented through education or improved prenatal care.

See also Embryo and embryonic development; Genetics.

Resources

Books

Nussbaum, R.L., Roderick R. McInnes, Huntington F. Willard. *Genetics in Medicine*. Philadelphia: Saunders, 2001.

Rimoin, D.L. *Emery and Rimoin's Principles and Practice of Medical Genetics*. London; New York: Churchill Livingstone, 2002.

Sadler, T.W. and Jan Langman. *Langman's Medical Embryology,,* 8th ed. Lippincott Williams & Wilkins Publishers, 2000.

Larry Blaser

Bismuth *see* **Element, chemical**

Bison

The American bison (*Bison bison*) is a large, herbivorous land mammal native to the **grasslands** and open **forests** of **North America**. It is a member of the family Bovidae, which also includes cattle, **sheep**, and **goats**. When French explorers first saw these large, shaggy, cow-like animals, they called them *boeufs,* the French word for "cattle." This later became anglicized into the word "buffalo," a name still applied to the bison, despite the fact that there are other bovids in **Africa** and **Asia** more properly known as buffalo.

There are two subspecies of American bison, the plains bison (*B. b. bison*) and the wood bison (*B. b. athabascae*). The wood bison lives west and north of the plains bison, and is larger and darker in **color**. Because its **habitat** is open woodland and muskeg, it does not live in such huge herds as the plains bison once did. However, some taxonomists do not recognize wood bison and plains bison as separate subspecies.

Bison probably came to North America from Eurasia during the most recent ice age. They did this perhaps 25,000 years ago, by travelling along a land-bridge across the present-day Bering Strait. The land-bridge existed because **sea level** was lower then, owing to so much of Earth's **water** being on land in the form of glacial **ice**. The only Eurasian remnant is the European bison or wisent (*B. bonasus*), which now only survives in protected forest on the borderland of Poland and Belarus. Another population of wisents in the Caucasus Mountains became extinct in 1925. The wisent is somewhat smaller than the American bison and does not have as large a hump.

America's largest mammal

The male (or bull) bison can reach 6.5 ft (2 m) in height and measure up to 12 ft (3.7 m) long, and is twice as large as the female (or cow). Male bison commonly weigh 2,000 lb (900 kg), but animals twice as large have been reported. American bison have a huge hump across the shoulders that rises 1 ft (30 cm) or more above the top of their head. The dark brown hair growing on the hump is long and shaggy, and the thick body fur allows bison to tolerate air temperatures as low as -49°F (-45°C). In the spring, bison molt their winter coat and great quantities of their hair falls off. The back half of their body has short, lighter-colored hair. The head is covered with a helmet of thick black hair that terminates under the chin as a "goatee." Extremely rare white bison are revered by Native Americans of the Plains. When one was born in 1994, people traveled long distances to Wisconsin to see it. As the calf grew older, however, its fur turned darker.

Bison have large, curved, hollow horns growing sideways from their big, heavy head, which are never shed. The horns may reach a length of 24-26 in (60-65 cm), but are usually much shorter. The horns of females curve back toward the head more than that of males. Bison have no front teeth, and eat grass and other herbaceous plants by wrapping them around their tongue and pulling to break them off the rootstock. Bison are ruminant animals. As such, they have a four-chambered stomach and chew a cud.

Their short tail is used both as a flyswatter and as an indicator of excitement, for the normally drooping tail rises into the air when a bison is angry or otherwise excited. Bison are subject to attack by numerous biting **flies** and ticks, and they have a habit of rolling in dirt or mud to relieve their itchiness. Because of their hump they cannot roll over completely, so they rock and kick first on one side and then on the other. This maneuver is called wallowing, and it also helps get rid of molting hair.

Life in the herd

Estimates of the prehistoric numbers of bison on the American plains, before the arrival of Europeans, vary from 30-80 million. At the time, bison were a "pantry on the hoof" for the Plains tribes of Native Americans, who used the animals for food, clothing, fuel, and shelter, killing only what they needed to survive.

Bison herds move around constantly in search of food and water, and typically travel several miles a day. Unlike many migrating animals, bison do not follow set paths each year, and even minor stimuli can change their direction of travel. This might be a strange scent in the **wind**, an unfamiliar **animal** crossing their path, or a dried-up water hole. The only set pattern to their **migration** is that they meander north to find a good place to raise their young in the summer, and then south to ride out the winter. The herd follows experienced, lead animals. Bison in herds continually make noises to each other, using roars, grunts, sneezing, snorts, and bawling to communicate different meanings.

The least hint of danger can set a herd stampeding across the **prairie**. The stampede speed may reach 35 mph (56 kph), though they cannot keep that speed up for more than about half an hour. If running does not shake their pursuers, bison may turn abruptly and charge. The animal's skull in the forehead is double-thick, protecting the **brain** from damage during impact.

The continuing generations

In the early summer, bulls and cows gather for the rut. During the rut, bulls challenge each other for the right to mate with cows. Cows first mate when about two years old, but bulls do not do so until they are older and strong enough to challenge dominant males. A few loud roars and an enthusiastic demonstration of kicking and wallowing is usually enough to convince a lower-ranked bull to look elsewhere for a mate. Only rarely does an actual fight occur, when bulls lock horns and charge each other.

When a dominant bull selects a mate, he bonds with her by grazing side by side for some hours, away from the rest of the herd. After a brief nocturnal mating, the companionship may continue for a brief time, but then the male departs, looking for another female. The females usually mate only once.

The gestation period lasts nine-and-a-half months, and new calves are born in the spring. This begins about mid-April, though births may continue into the early autumn. Newborn calves are cinnamon-colored, weigh about 50 lb (23 kg), and can walk and nurse within 2-3 hours of **birth**. The calves start to eat grass within about 15 days, and at about two months of age their hump and horns begin to show and their coat darkens to the adult color. Bison typically live to an average age of about 20 years in the wild, but can reach 40 years in captivity.

The disappearing bison

The enormous bison herds of North America shrank rapidly in the face of relentless over-hunting during the westward migration of European settlement. The presence of bison conflicted with the aspirations of people looking for land to settle and farm, and with the aims of government, which wanted to subdue the native tribes of Plains Indians. As the railroads were built through the Central Plains, travelers were encouraged to shoot bison from the windows of the train for fun and excitement. The carcasses were often left to rot. More important, however, was the huge market hunt for the plains bison,

Bison (*Bison bison*) in Golden Gate Park, California.
Photograph by Robert J. Huffman. Field Mark Publications. Reproduced by permission.

with trainloads of butchered carcasses and hides being shipped to markets in American cities. By 1905, the excessive hunting resulted in only an endangered 500 bison left surviving on their range in the United States. A herd of endangered wood bison, discovered in northwestern Canada in 1957, is protected in Wood Buffalo National Park and nearby areas.

Although bison are still a threatened **species**, there has been some recovery and there are now substantial wild herds in some places, such as Yellowstone National Park. Some ranchers are raising semi-domestic animals and sell buffalo meat, which has less **cholesterol** than beef. It has even been suggested that herds of bison might once again be allowed to roam free on parts of the North American prairie, creating a tourist attraction to bring money to economically depressed areas. However, not much suitable habitat is left in a natural condition, and the enormous bison herds of the past will never again be seen.

KEY TERMS

Bovid—An animal of the Bovidae, or cow family, characterized by a grazing habit and having hollow horns.

Molt—To lose one type of hair in preparation for a new type to grow in. Bison molt twice a year as the seasons change.

Ruminant—A cud-chewing animal with a four-chambered stomach and even-toed hooves.

Rut—The period during which males challenge each other to acquire access to females.

Wallowing—Rolling and kicking in the dust to eliminate insect pests and scratch itchy skin.

Resources

Books

Berman, Ruth. *American Bison.* Minneapolis: Carolrhoda Books, 1992.

Caras, Roger A. *North American Mammals: Fur-Bearing Animals of the United States and Canada.* New York: Meredith Press, 1967.

Geist, V. *Buffalo Nation: History and Legend of the North American Bison.* Voyageur Press, 1998.

Green, Carl R., and William R. Sanford. *The Bison.* New York: Crestwood House, 1985.

MacDonald, David, and Sasha Norris, eds. *Encyclopedia of Mammals.* New York: Facts on File, 2001.

Stidworthy, John. *Mammals: The Large Plant-Eaters.* New York: Facts On File, 1988.

Time-Life Books, eds. *Lords of the Plains.* Alexandria, VA: Time-Life Books, 1993.

Jean F. Blashfield

Biting lice *see* **Lice**

Bitterns

Bitterns are about 12 **species** of wading **birds** in the subfamily Botaurinae of the family Ardeidae, which also includes **herons** and egrets. There are two genera: four species of the relatively large and stocky true bitterns (*Botaurus* spp.), and eight species of the much smaller and more slender, least bitterns (*Ixobrychus*).

Bitterns have brown-and-black, vertically streaked plumage, which renders them well camouflaged in their marshy or reed-fringed habitats. Male and female bitterns have identical plumage. When a potential **predator** is in the vicinity, bitterns will try to blend in with their surroundings by extending their neck and bill upright, compressing their brownish-streaked breast plumage, and facing the intruder. Bitterns may also sinuously wave their body to emulate the movements of the surrounding, wind-blown reeds and bulrushes.

Bitterns are unobtrusive animals, and many people are unaware of the presence of these animals, even in marshes where they are breeding. Bitterns fly with a slow wingbeat, low over the tops of the marsh vegetation, and then suddenly drop down out of sight to land.

Bitterns mostly eat **fish**, but they also take aquatic **invertebrates**, **snakes**, **frogs**, baby birds, and small **mammals** when these are available. Bitterns slowly and deliberately stalk their **prey**, and then capture their victim by quickly spearing it with a rapid thrust of the beak.

Bitterns usually nest on rough platforms that they construct out of sticks or marsh vegetation. The nest may be placed in a concealed place on the ground, or in a shrub or low **tree**. Bitterns lay three to six eggs, which are incubated by the female, who is also mostly or totally responsible for rearing the brood.

Species of bitterns

The American bittern (*Botaurus lentiginosus*) breeds in **freshwater** and **brackish** marshes over most of the temperate zone of **North America**, and as far south as central Mexico. Most North American populations migrate to the southern United States and Central America to spend their non-breeding season. American bitterns mostly eat fish, but they will also predate on other appropriately sized, aquatic prey. Male American bitterns have a distinctive song in the springtime, when they are establishing a breeding territory and attempting to attract a mate. This booming call can be heard over a **distance** of several miles, and more or less sounds like a pumping, "ong-ka-chonk." Some local names of the American bittern reflect its call: "thunder pump" and "stake driver." Other species of *Botaurus* occur in non-overlapping ranges on other continents. These species are all rather similar, and although considered to be taxonomically distinct, they form a closely-related, "super-species." The other species of bitterns include the Eurasian bittern (*Botaurus stellaris*) of **Europe** and **Asia**, the Australian bittern (*B. poiciloptilus*), and the South American bittern (*B. pinnatus*).

The least bittern (*Ixobrychus exilis*) is the smallest species of heron in North America. This secretive species breeds widely in freshwater and coastal marshes in the eastern United States and southeastern Canada. Least bitterns also breed in a disjunct, western range in California

An American bittern in Florida. *JLM Visuals. Reproduced by permission.*

and Oregon, and south to Brazil, Paraguay, and northern Argentina. Northern populations of the least bittern migrate to northern Mexico and Baja California for the winter. The least bittern mostly feeds on small fish, which it stalks patiently and then spears with its bill.

The various species of *Ixobrychus* also have non-overlapping ranges, and also form a closely-related "superspecies." The species include the little bittern (*I. minutus*) of Europe, Asia, **Africa**, and **Australia**, and the Chinese little bittern (*I. sinensis*) of eastern Asia.

Conservation of bitterns

Habitat losses associated with the drainage of **wetlands** for agricultural and residential developments are the most important threats to bitterns in North America and elsewhere. **Pollution** may also be significant in degrading habitat in some regions.

As a result of these and other stressors, the populations of both American bitterns and least bitterns are widely acknowledged as having declined substantially in North America. There is significant concern about the population status of both species in most parts of their ranges in the United States and Canada.

Like so many other species that require wetlands as habitat, the survival of bitterns can only be ensured by caring for the ecosystems of which they are an integral part. In this case, the key is the **conservation** and protection of wetlands.

Resources

Books

Bird Families of the World. Oxford: Oxford University Press, 1998.

Ehrlich, P. R., D. S. Dobkin, and D. Wheye. *Birds in Jeopardy.* Stanford, CA: Stanford University Press, 1992.

Forshaw, Joseph. *Encyclopedia of Birds.* New York: Academic Press, 1998.

Marquis, M. *Herons.* London: Colin Baxter, 1993.

Bill Freedman

Bivalves

Bivalve molluscks belong to the class Bivalvia (or Lamellibranchia) of the phylum Mollusca. Known by such common names as clams, mussels, cockles, oysters, and scallops, bivalves are among the most familiar aquatic **invertebrates**. They occur in large numbers in marine, estu-

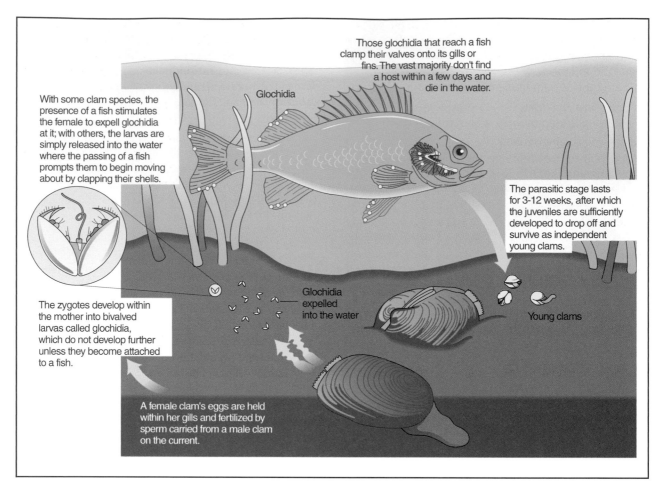

Those glochidia that reach a fish clamp their valves onto its gills or fins. The vast majority don't find a host within a few days and die in the water.

Glochidia

With some clam species, the presence of a fish stimulates the female to expel glochidia at it; with others, the larvas are simply released into the water where the passing of a fish prompts them to begin moving about by clapping their shells.

The parasitic stage lasts for 3-12 weeks, after which the juveniles are sufficiently developed to drop off and survive as independent young clams.

The zygotes develop within the mother into bivalved larvas called glochidia, which do not develop further unless they become attached to a fish.

Glochidia expelled into the water

Young clams

A female clam's eggs are held within her gills and fertilized by sperm carried from a male clam on the current.

The life cycle of a typical freshwater clam. For species that do not have the parasitic larval stage, the fertilized eggs develop into young clams within the gills of the mother. *Illustration by Hans & Cassidy. Courtesy of Gale Group.*

arine, and **freshwater** habitats all over the world. More than 30,000 living **species** of bivalves have been described. The main divisions of the Bivalvia are the Protobranchia (the primitive nutshells), the Filibranchia (the mussels, scallops, and oysters), and the Enamellibranchia (the cockles, clams, venus shells, razor shells, and shipworms).

The name bivalve refers to the limy shell, which consists of two pieces or valves, right and left, held closed by a pair of adductor muscles, and joined together by an elastic hinge ligament. The internal, compressed body is completely enclosed by the shell valves, except for a hatchet-shaped, muscular foot, which can be extended between the lower edges of the shell so that the mollusk can burrow in soft **sand** or mud. The two halves of the shell are secreted by the two lobes of the body wall (the mantle), and consist of layers of **calcium carbonate** crystals embedded in a protein matrix.

The innermost shell layer, which is often shiny and iridescent, is called mother of pearl. If a grain of sand or other hard foreign matter gets lodged between the mantle

and the shell, a layer of this pearly material is secreted around it, forming, in some species, a pearl. The space between the body wall and the mantle is known as the mantle cavity. This cavity contains a pair of large, perforated, plate-like gills that have a ciliated surface and function in both **respiration** and feeding. The posterior edges of the mantle lobes join to form two tubes, or siphons. The beating of the gill cilia causes **water** to be drawn into the mantle cavity through the lower incurrent siphon; after passing across the gills where **oxygen** is extracted, the water is expelled by the excurrent siphon. Bivalves lack a well-developed head, and so their sense organs (such as eyes) are located on the fringe of the mantle.

Bivalves are filter feeders, using their perforated gills as a sieve, which collects minute **algae** and other food particles suspended in the incoming respiratory water. These particles are trapped in strings of mucus secreted by the gills and conveyed to the mouth by cilia. Marine bivalves reproduce by releasing prodigious numbers of eggs and sperm into the water, where external **fertiliza-**

tion occurs. The fertilized eggs then float in the surface **plankton**. Within 48 hours after fertilization, the embryo develops into a minute, planktonic, trochophore larvae. This stage is followed by another larval form, the veliger, which settles to the seabed and transforms into an adult. In freshwater bivalves, the eggs are retained in the gill chambers of the female, where they undergo fertilization and develop into a peculiar larval form, the glochidium. Upon its release, the larva attaches to passing **fish**, and lives as an ectoparasite for several weeks before settling.

Mussels and oysters do not burrow, but remain permanently fixed to a hard substrate. Mussels are attached to **rocks** by clumps of byssus threads. In oysters, only the left valve is cemented to the rock. Scallops are able to swim by clapping their valves and ejecting water through an opening near the hinge area to produce a jet action. Scallops also have rows of eyes on the lower edges of the mantle. Other bivalves are able to bore into limestone, clay, or **wood**.

Bivalves range in size from the fingernail-sized "nut shells" of the Atlantic coast of **North America** to the giant clam of the Indo-Pacific, which measures up to 4.9 ft (1.5 m) in length and weighs more than 495 lb (225 kg).

Bivalves are of great economic importance as a food source, and as a source of valuable products such as pearls. Some bivalves cause important economic damage. Shipworms bore into and destroy the wooden hulls of ships and wharf pilings. The zebra mussel recently colonized inland waters of North America by hitch-hiking from Eurasia in ships' ballast water. This prolific species is causing extensive damage by clogging water pipes and displacing native species of bivalve molluscs.

BL Lacertae object

BL Lacertae objects, abbreviated BL Lac, are one subclass of **active galactic nuclei** (AGN), the extremely energetic nuclei of active galaxies. Roughly 40 BL Lac objects are known. Perhaps the most obvious property of BL Lac objects is that they look like stars. Astronomers originally thought the prototype, BL Lac, was a **star**. In fact, BL Lacertae is normally a variable star designation, two letters followed by a **constellation** name, because astronomers originally thought that BL Lac was a star whose brightness varies, a variable star. BL Lac objects do however have properties that are clearly not stellar.

Unlike most stars, BL Lac objects are very strong sources of **radio** and infrared **emission**. This emission, which is called synchrotron emission, arises from elec-

trons traveling near the speed of **light** in **spiral** paths in strong magnetic fields. Synchrotron emission generally is polarized, so BL Lac objects have polarized emission. When light or other electromagnetic **radiation** is polarized, the directions of the **oscillations** are the same. The amount of polarization and the brightness of BL Lac objects is highly variable. This variability is very rapid and erratic. They can change very significantly in times as short as 24 hours or less. The rapid variability tells us that the **energy** source is very small. Nothing can travel faster than the speed of light, including whatever signal or mechanism causes the BL Lac object to change its brightness. Therefore, if a BL Lac object changes its brightness significantly in a day its energy source must be less than one light day in radius.

The **spectrum** of a BL Lac object contains very few if any absorption or emission lines, which are caused by interstellar gas. Their essentially featureless spectra tell us that there is very little interstellar gas around BL Lac objects. There is evidence for a faint fuzziness in some pictures of BL Lac objects. This fuzziness is most likely the host **galaxy**, of which the BL Lac object is the active nucleus.

Currently, the most popular explanation of BL Lac objects is that they are the central very energetic nuclei of galaxies. The small central energy source in the nucleus is probably a supermassive **black hole**. However, astronomers are still very uncertain of their nature. We need to continue studying BL Lacertae objects to better understand them.

Black hole

A sufficiently intense gravitational field can prevent the escape not only of **matter**, but even of **light**. Such gravitational fields are produced by the bodies known as black holes.

The maximum intensity of a spherical object's gravitational field is a function both of the amount of matter it contains and of its **volume**. The more matter is contained in an object and the smaller its volume—in other words, the higher its density—the more intense the gravitational field at its surface will be. If the **earth** were compacted so that it had the same **mass**, but half its present radius, the **force** of gravity at its surface would be four times as great as it is now; if it were compacted further, a **density** would eventually be reached at which its constituent **subatomic particles** would be unable to support their own weight and would collapse to a state of (theoretically) infinite density, producing a black hole.

Black holes can (and some do) contain very large amounts of matter—millions or billions of times the mass of the Sun—but may be formed by even a small amount of matter sufficiently compressed.

The idea of black holes is not new; the French mathematician Pierre Simon Laplace (1749–1847) reasoned in 1795 that if the corpuscular theory of light proposed by English physicist Isaac Newton (1642–1727) were correct, there could exist massive objects from which light could not escape. The theory of general relativity, put forward by German physicist Albert Einstein (1679–1959) in 1915 and today basic to physicists' understanding of the Universe, also predicts the existence of black holes, though on from rather different reasoning. In recent decades, much observational evidence has been gathered to support the existence of black holes; there is no debate among astronomers today about whether black holes exist, only regarding their precise properties.

The event horizon

According to general relativity, the path taken by a beam of light is the shortest **distance** between two points; such a path is called a **geodesic**. Furthermore, gravity warps **space**, bending geodesics; the stronger a gravitational field is in a certain region, the more bent the geodesics are in that region. Within a certain radius of a black hole, all geodesics are so warped that a **photon** of light cannot escape to another part of the Universe; essentially, there are no straight lines connecting any point that is within a certain radius of a black hole (which, in theory, has no dimension) to any point that is farther away. The spherical surface defined by this radius is termed the **event horizon** of the black hole because events inside the event horizon can have no effect on events outside it. Whatever is inside the event horizon is sealed off forever from the space-time of outside observers.

The event horizon thus imposes a form of censorship on the makeup of a black hole; the only properties of a black hole that can be ascertained from the outside are its mass, net charge, and **rate** of spin. No internal time-dependent processes can be detected in the external environment, for that would involve sending signals from inside the black hole to the outside—which is impossible, for not even light can escape. This "censorship" is what is responsible for the fewness of a black hole's measurable properties: mass, spin, and charge.

Although there are complications in defining the "size" of a black hole, due to the fact that our everyday concept of "size" assumes Euclidean three-dimensional space and such space does not exist even approximately in the near vicinity of a black hole, one can uniquely specify a black hole's circumference and thus. its radius

as the circumference divided by 2PI. This value is known as the Schwarzschild radius (R_s) after German astronomer Karl Schwarzschild (1873–1916), who first defined it as $R_s = 2GM/c^2$ where G is the gravitational constant, M is the mass of the black hole, and c is the speed of light.

R_s, however, cannot be interpreted as the radius of a Euclidean sphere—that is, as the distance from a spherical surface (the event horizon) to its center (the black hole). As mentioned above, the **geometry** of space-time in the interior of the black hole is so warped that Euclidean notions of distance no longer apply. Nevertheless, R_s does provide a measure of the space around a black hole of mass M. R_s for an object having the mass of the **Sun** is about 3 km. Thus, in order to turn the Sun into a black hole, one would have to compress it from a **sphere** with a radius of 696,000 km to a sphere with a radius of just 3 km. Squeezing any mass into a volume dictated by its Schwarzschild radius presents a serious assembly problem; in fact, the only processes that might lead to the formation of a sizable black hole are the explosive death of a moderately massive **star** or the formation of a supermassive star by sheer accumulation. Physicists also speculate that extremely small black holes might be created by the collision of subatomic particles at high energies. In fact, they estimate that as many as 100 subatomic-size black holes may be produced in the atmosphere of the Earth every year by cosmic rays. The European Laboratory for Particle Physics (CERN, for Conseil Européen pour la Recherche Nucléaire) hopes to produce such microscopic black holes on demand in its new Large Hadron Collider, due to begin operation in 2006.

Very small black holes are predicted by theory to be short-lived, however, due to a quantum phenomenon termed "evaporation." Only large black holes are long-lived enough to have cosmic effects—to swallow millions of suns' worth of mass, to squeeze sufficient **energy** from the matter approaching their event horizons to outshine entire galaxies, to organize the orbits of billions of stars into well-defined galaxies, and so forth. Large black holes are thought to form primarily from exploding stars or by direct gravitational accumulation of large quantities of matter. A black hole may be produced by an exploding star (**nova**) as follows: An older star eventually exhausts the nuclear fuel that enables it to produce energy at its core, thus supporting its own weight (and shining steadily for many millions of years). It then begins a rapid collapse. The crushing **pressure** of the collapsing matter may be sufficient to form a black hole with the mass of several times that of the Sun. Such black holes would have Schwarzschild radii of several to a few tens of kilometers. Considering the amount of mass filling that space, such objects are truly tiny.

Detection of black holes

In their near vicinity, black holes produce bizarre effects; from a distance, however, they are well-behaved. If one were to replace the Sun with a black hole of the same mass as the Sun, there would be a region of space a few kilometers in size, located where the center of the Sun currently resides, in which space would be extremely warped. The gravitational field of this object, measured at the distance of the earth, would be exactly that of the present-day Sun. The earth and planets would continue in their orbits and the **solar system** would continue much as it does today—only in the dark.

Normally, an observer must get within a few Schwarzschild radii in order to feel the distinctive effects of the black hole. Indeed, one of the observational tests for the presence of a black hole in binary systems is to look for the characteristic radiations of matter being heated as it is squeezed during its final plunge toward the black hole's event horizon. Such matter will emit fluctuating **x rays** as a result of being squeezed. The rate of fluctuation is tied to the size of the emitting region. Astronomers find that in such systems the x rays come from a volume of space only a few kilometers in diameter. In several instances, analysis of the orbital motions in a **binary star** system with only one visible member (a conventionally shining star) indicates that the dark, unseen member of the binary system is much more massive than the Sun. A dark stellar component more massive than the Sun confined to a volume smaller than a few kilometers is a prime candidate for a black hole.

There is at least one other situation in which astronomers suspect the existence of a black hole. Because a black hole that is not actively swallowing large amounts of matter does not radiate significantly, we must detect it indirectly, through the effect of its gravitational field on neighboring objects. In the centers of many galaxies, the stars, gas, and dust of the **galaxy** are moving at very high speeds, suggesting they are orbiting some very massive, comparatively small object. If the object was a tightly packed collection of massive stars, it would shine so brightly as to dominate the light from the galactic center. The absence of light from the massive, central object suggests it is a black hole. Recent astronomical observations have confirmed many galaxies, including our own, have supermassive black holes at their centers. Scientists believe that all galaxies may be organized around such black holes; after the Big Bang, supermassive black holes may have formed first, then gathered the galaxies around them.

In one galaxy, the **Hubble Space Telescope** (HST) has photographed a spiraling disk of matter that appear be accreting onto a central massive dark object that is likely to be a black hole. Recently a large team of astronomers reported the results of a worldwide study involving the HST, the **International Ultraviolet Explorer satellite**, and many ground based telescopes. Instruments were able to detect light that was emitted by the accreting matter as it spiraled into the black hole that was subsequently absorbed and re-emitted by the orbiting **clouds** just a few light-days away from the central source. Mass estimates of the central source determined from the **motion** of these clouds suggests that the object has a mass of at least several million times the mass of the Sun. So much material contained in a volume of space no larger than a few light-days in diameter provides some of the clearest evidence yet for the existence of a black hole at the center of any galaxy.

In another study with the HST and a ground-based **telescope** in Hawaii, scientists were able to observe a black hole in a two-star system in the **constellation** Cygnus. This black hole is sucking material from its companion star in a swirling disk of material and hot gases, swallowing nearly 100 times as much energy as it radiates. The material being pulled in toward the black hole stores its energy as **heat** until the critical moment. The observations show gas at temperatures of over a million degrees falling toward the event horizon of the black hole.

Centerpiece of the galaxy

The concept of massive black holes at the centers of some galaxies is supported by theoretical investigations of the formation of very massive stars. Stars of more than about a hundred times (and less than about a million times) the mass of the Sun cannot form because they will explode from nuclear energy released during their contraction before the star can shrink enough for its self-gravity to hold it together. However, if a collapsing cloud of interstellar material contains more than about a million times the mass of the Sun, the collapse will occur so fast the nuclear processes initiated by the collapse will not be able to stop the collapse. The collapse will continue unrestrained until the object forms a black hole.

Such objects appear to be required to understand the observed behavior of the material in the center of some galaxies. Indeed, it is now certain that black holes reside at the centers of many normal galaxies, including our own **Milky Way**. Evidence comes from the motion of gas clouds near the galactic center and from the detection of x-rays bursts from the galactic center such as would typically be produced by a supermassive black hole swallowing matter.

Quantum physics and black holes

All that has been said so far involves black holes as described by the general theory of relativity. However,

KEY TERMS

Binary system—Any system of two stellar-like objects that orbit one another under the influence of their combined gravity.

Galaxy—A large collection of stars and clusters of stars, containing anywhere from a few million to a few trillion stars.

General relativity—A theory of gravity put forth by Albert Einstein in 1915 that basically describes gravity as a distortion of space-time by the presence of matter.

Interstellar material—Any material that resides between the stars. It makes up the material from which new stars form.

Perfect radiator—Also known as a black body (not to be confused with a black hole). Any object that absorbs all radiant energy that falls upon it and subsequently re-radiates that energy. The radiated energy can be characterized by a single dependant variable, the temperature.

Quantum gravity—A theory resulting from the application of quantum principles to interaction between two objects normally attributed to gravity.

Quantum mechanics—The theory that has been developed from Max Planck's quantum principle to describe the physics of the very small. The quantum principle basically states that energy only comes in certain indivisible amounts designated as quanta. Any physical interaction in which energy is exchanged can only exchange integral numbers of quanta.

in the realm of the very small, **quantum mechanics** has proved to be the proper theory to describe the physical world. To date, no one has successfully combined general relativity with quantum mechanics to produce a fully consistent theory of quantum gravity; however, in 1974, British physicist Stephen Hawking (1942–) suggested that quantum principles showed that a black hole should radiate energy like a perfect radiator having a **temperature** inversely proportional to its mass. This radiation—termed Hawking radiation—does not come about by the conventional departure of photons from the black hole's surface—which is impossible—but as a result of certain effects predicted by quantum **physics.** While the amount of **radiation** for any astrophysical black hole is very small (e.g., the radiation temperature for a black hole with the mass of the Sun would be 10^{-7}K), the suggestion that loss of energy from a black

hole was possible at all was revolutionary. It suggested a link between quantum theory and general relativity, and has spawned a host of new ideas expanding the relationship between the two theories. It is the ability of a black hole to lose mass via Hawking radiation (i.e., to evaporate) that prevents microscopic black holes, such as those that physicists hope to produce at CERN, from swallowing up the earth. These black holes evaporate faster than they can grow.

See also Relativity, general; Stellar evolution; Supernova.

George W. Collins, II

Resources

Books

Hawking, Stephen. W. *The Illustrated A Brief History of Time.* 2nd ed. New York: Bantam Books, 2001.

Periodicals

Cowan, John. "Supernova Birth for a Black Hole." *Nature.* (September 9, 1999): 124–125.

Glanz, James. "Evidence Points to Black Hole At Center of the Milky Way." *New York Times.* October 17, 2002.

Irion, Robert. "Galaxies, Black Holes Shared Their Youths." *Science.* (June 16, 2000): 1946–1947.

Johnson, George. "Physicists Strive to Build a Black Hole." *New York Times.* September 11, 2001.

Black smoker *see* **Hydrothermal vents**

Blackberry *see* **Rose family (Rosaceae)**

Blackbirds

The blackbird family (Icteridae) consists of 94 medium-sized **species** of **birds** that occur only in the Americas. Blackbirds are found in widespread habitats, ranging from **wetlands**, to prairies, to **forests**. The most common members of the family are various species of blackbirds, grackles, cowbirds, **orioles**, meadowlarks, bobolink, and others.

Biology of blackbirds

Blackbirds tend to have conical shaped, pointed beaks. Most species are sexually dimorphic, particularly the relatively northern, migratory species. Males often have brilliant hues in their plumage and are commonly iridescent, while female blackbirds are usually relatively drab and cryptically marked. Some male blackbirds incorporate splendid yellow, orange, and red colors in their plumage.

Some species of the blackbirds are accomplished vocalists, displaying a complex repertoire of loud and clear whistles and calls. Orioles are among the most musical of the blackbirds, producing rather pleasing, flute-like melodies.

Species of blackbirds

Blackbirds exploit a wide range of habitats. Most species occur in tropical forests of various sorts, but virtually all types of terrestrial and wetland habitats are utilized by some species. In **North America**, the northern oriole (*Icterus galbula*) is a species of open forests, where it builds its characteristic, pendulous nests, often in **elm** trees. The bobolink (*Dolichonyx oryzivorus*) and meadowlarks (*Sturnella* spp.) are more typical of open **grasslands** and prairies, while the red-winged blackbird (*Agelaius phoeniceus*) and yellow-headed blackbird (*Xanthocephalus xanthocephalus*) are typical of marshes and some other wet habitats.

The most widespread species is the red-winged blackbird, which ranges from the subarctic to Central America. This common and familiar bird breeds in tall marshes and other wet places. The male red-winged blackbird is colored as its name implies, with a jet-black body and richly red epaulets on the shoulders. Female red blackbirds have a streaky, brown plumage, and look much like large sparrows. After the breeding season, red-winged blackbirds aggregate into large flocks that forage widely for small grains, and can cause agricultural damage. Red-winged blackbirds generally spend the winter in these flocks, mostly in southern parts of their range, where they forage during the day and roost communally at night in woodlands and marshes.

The most northerly blackbird is the rusty blackbird (*Euphagus carolinus*), which in some places breeds as far as the limits of the mainland in the western Arctic. The most southerly species is the red-breasted blackbird (*Leistes militaris*) of the Falkland Islands.

Blackbirds that breed in the north are migratory. The longest migrations are undertaken by the bobolink. This species breeds in prairies and hayfields as far north as southern Canada, and winters in pampas and other grasslands as far south as Argentina. Tropical members of the blackbird family are not migratory, but they may undertake local, seasonal movements.

Cowbirds, such as the brown-headed cowbird (*Molothrus ater*) of North America, are species of open habitats. As their name implies, these birds often associate with grazing **livestock**, feeding on **insects** that these animals flush as they move through vegetation. Cowbirds have an unusual breeding strategy. Instead of building their own nest and raising their young, cowbirds are

A male brown-headed cowbird in Kirtland's warbler breeding territory near Mio, Michigan. Because the cowbird lays its eggs in the nests of other birds, where its aggressive hatchlings often out-compete their nest mates for food, the cowbird has been identified as a factor in the decline of the Kirtland's warbler, an endangered species known only to nest in north-central Michigan. *Photograph by Robert J. Huffman. Field Mark Publications. Reproduced by permission.*

nest **parasites**. They surreptitiously lay single eggs in the nests of other species, and sometimes remove eggs of the host bird. If the host birds do not recognize the cowbird egg as being alien, they will brood it and care for the hatchling until it fledges. Many of the approximately 200 species of birds known to be parasitized by the brown-headed cowbird are relatively small, such as **vireos**, **warblers**, and **thrushes**. The nestlings of these birds suffer as a result of the disproportionate demands placed by the voracious cowbird chick on its foster parents, and in many cases this causes the reproductive effort of the host birds to fail. The cowbird chick develops rapidly, and can fly in only nine or ten days after hatching.

Outside of the Americas, the blackbird (*Turdus merula*) of **Europe**, is actually a member of the thrush family, Turdidae.

Blackbirds and humans

Cowbirds are considered to be an important pest in those parts of North America to which the species has expanded its range as a result of the fragmentation of the initially forested landscape by humans. Birds in those regions tend not to be well adapted to cowbird parasitism, and the reproductive success of their populations can be markedly reduced by this relationship. In some cases, cowbirds are sought out and killed by **conservation** biologists, in order to reduce the negative impact of parasitism on rare and **endangered species** of birds, such as the Kirtland's warbler (*Dendroica kirtlandii*) of Michigan.

Some species of blackbirds are highly gregarious, especially during autumn and winter when they aggregate into flocks that can contain millions of birds. Winter flocks of the red-winged blackbird are sometimes regarded as an agricultural nuisance because of damages caused to fields of winter **wheat** and some other **crops**. Sometimes, pest-control actions are mounted against these flocks, and millions of these native birds may be killed when they are sprayed with chemicals at their communal roost sites.

Resources

Books

Brooke, M., and T. Birkhead. *The Cambridge Encyclopedia of Ornithology.* Cambridge: Cambridge University Press, 1991. U.K.

Forshaw, Joseph. *Encyclopedia of Birds.* New York: Academic Press, 1998.

Orians, G. H. *Blackbirds of the Americas.* Seattle: Washington University Press, 1986.

Bill Freedman

Blackbody radiation

The term blackbody radiation refers to electromagnetic **radiation** emitted by a completely opaque object. Such an object is referred to as a blackbody since it absorbs all of the radiation that falls on it and thus appears to be colorless, or black. According to Kirchoff's law, any object that qualifies as a blackbody must also be a perfect emitter of radiation.

In fact, no real object fits the definition of a blackbody since even the most opaque of materials reflects some small fraction of the **light** incident upon it. Soot, **carbon** black, platinum black, and carborundum are among the materials that come closest to a blackbody in the real world. the concept of an idealized blackbody is still of major importance in **physics**. It serves as a stan-

dard for **heat** and **temperature** measurements just as the wavelength of light emitted by krypton-86 **atoms** is a standard for measurements of length.

For research purposes, physicists replicate the principle of a blackbody with a device known as a cavity radiator. A cavity radiator is a hollow **sphere** with a small hole through which radiation can enter and leave. Radiation that enters the hole is reflected continuously within the sphere until it is completely absorbed (as would be the case with a blackbody). It follows, then, that any radiation emitted by the cavity radiator corresponds to the definition of blackbody radiation.

The study of blackbody radiation was of considerable interest to physicists in the late 1800s. Experiments showed that for any given temperature, the intensity (brightness) of blackbody radiation is a maximum for a relatively narrow range of wavelengths, dropping off sharply at shorter and longer wavelengths. A number of attempts were made to use classical electromagnetic theory to derive a mathematical formula that would describe the intensity/wavelength relationship, but all failed for one or another part of the **curve**. Finally, in 1900, the German physicist Max Planck solved the problem. By assuming that radiation travels not in continuous waves, but in discrete "packages" (called quanta), Planck was able to derive a formula for the blackbody radiation curve. That formula is:

$$I = (\hbar c^2/\lambda^5)[\exp{(\hbar c/\lambda kT - 1)}]^{-1}$$

where $\hbar$ is **Planck's constant**, k is Boltzmann's constant, λ is the wavelength of the radiation, and T is the absolute temperature.

Bleach

Bleaches are substances that whiten **textiles** and **paper** by chemical reaction. These reactions usually involve processes that degrade **color**. They may destroy or modify chemical bonds or groups that give fabrics their characteristic colors. This process degrades color bodies into smaller, more soluble units that are easily removed in laundering. Conventional bleaching agents, include two types: chlorine-based bleaches, such as **sodium hypochlorite**, and peroxygen bleaching agents such as **hydrogen peroxide** and **sodium** perborate.

Textile bleaching

The bleaching of textiles appears to have been known as early as 300 B.C. when soda ash was prepared from burned seaweed and used to clean cloth. Then the cloth was treated with soured milk to reduce its alkalinity. The bleaching process was completed when the cloth was exposed to the **Sun**. This type of sun bleaching typically took several weeks.

A Swedish chemist discovered **chlorine** gas in 1784 and succeeded in demonstrating its use for decolorizing vegetable dyes. Fifteen years later a patent was awarded for a bleaching powder formed by the absorption of chlorine gas into dry hydrate of lime. Following World War I the technology for shipping liquid chlorine was developed. This allowed for on-site production of sodium hypochlorite in textile mills and led to the development of other chlorine-based bleaches. In 1928, the first dry **calcium** hypochlorite bleach containing 70% available chlorine was produced in the United States. This material largely replaced bleaching powder in commercial bleaching.

Hydrogen peroxide was prepared as early as 1818 but did not find use in the bleaching of textiles until much later. By 1930, the prices of peroxides had dropped sufficiently to allow the use of hydrogen peroxide in the bleaching of **cotton**, wool, and silk. By 1940, 65% of all cotton bleaching was done with hydrogen peroxide.

Pulp bleaching

There are many parallels in the histories of pulp and textile bleaching because early paper was commonly made from rags. In the 1700s sunlight was used to bleach paper. After 1800, bleaching powder was used to whiten the rags used to make paper. In the early 1800s **wood** came into use as a source of paper and calcium hypochlorite was used as the bleaching agent.

After World War I chlorine bleaching came into use in paper production because compressed chlorine gas became available. By the 1950s, chlorine dioxide had become the principal pulp bleaching agent. More recently, peroxygens such as hydrogen peroxide have been used.

Household and commercial laundering

Before the twentieth century, home laundry bleaching in the United States was done by the same method used by the Romans and Gauls in ancient times: clothes were first laundered in a mildly alkaline bath then subjected to sunlight. In 1910, 20 sodium hypochlorite solutions were developed and distributed regionally in the United States. By the mid-1930s these solutions had become available nationwide. In the 1950s, dry sources of hypochlorite were introduced but these products had disappeared by the late 1960s because consumers preferred liquid hypochlorites.

In **Europe** sodium perborate was first used as a bleaching agent in the early 1900s. The perborate dis-

solves during bleaching to release hydrogen peroxide. Sodium perborate continues to be used in European laundering because their laundering temperatures tend to be higher than those used in America.

See also Sodium chloride.

Resources

Books

Kirk-Othmer Encyclopedia of Chemical Technology. 4th ed. Suppl. New York: John Wiley & Sons, 1998.

Blennies

Blennies are small, primarily tropical and subtropical marine **fish**. They are elongated and often eel-like in shape, with a dorsal fin running from the back of the head almost to the tail fin, and small abdominal fins; the pelvic fin is often completely absent. Many **species** also lack scales. The blenny's **anatomy** is well suited for hiding in cracks and crevices along shallow, rocky shorelines, their preferred **habitat**. Some species, however, dwell in deeper waters. Living and foraging close to the **ocean** floor, blennies are either carnivorous (meat eaters) or omnivorous (eating both meat and vegetation). A wide variety of body shapes, colors, patterns, and behaviors are displayed by the more than 732 species in the six families belonging to the suborder Blennioidei.

Many species of scaleless blennies (family Blenniidae) are distributed throughout the world. Scaleless blennies are often found in tide pools and complete their entire life cycle in one general location. These species are divided into two groups—the subfamily Blenniinae, which have immovable teeth firmly rooted in the jaw, and the subfamily Salariinae, which have moveable teeth rooted in the gums.

Scaled blennies belong to the very large family Clinidae, or clinids. They inhabit temperate oceans primarily south of the Equator. Dazzling and varied colors and markings differentiate the species. The largest clinid, one of the many pointy-headed blennies, is the 24-in (61 cm) kelpfish (*Heterostichus rostratus*), which inhabits the Pacific shoreline from British Columbia to Southern California, while the 8-in (20-cm), blunt-headed, hairy blenny (*Labrisomus nuchipinnis*) lives in the tropical waters off both Atlantic coastlines. The pike blenny (*Chaenopsis ocellata*) is a tube-dwelling species found in Florida. Male pike blennies jealously defend their territories from other intruding males by aggressively displaying a stiffly raised dorsal fin and a widely gaping mouth. Two males may literally face off, gaping mouths touching, until one snaps its mouth shut on the other. Some of the smallest blennies are also found in the Clinidae family—the female of the species *Tripterygion nanus* found in the Marshall Islands, is fully grown at less than 0.75 in (1.9 cm) in length.

The largest blennies, often reaching 9 ft (2.7 m) in length, are found among the nine species of wolf fish and wolf eels belonging to the family Anarhichadidae. These cold **water** fish are found in the northern hemisphere. They have prominent canine teeth in the front of their jaws and massive grinding teeth in the back of their mouths. Two species, the Atlantic wolf fish (*Anarhichas lupus*) and the spotted wolf fish (*A. minor*), are fished commercially along the European coasts.

Blindness and visual impairments

Blindness is usually considered as an inability to see or a complete loss of **vision**, although legally, a blind person may retain some vision. In contrast, visual impairment indicates a loss of vision such that there is an impact on daily living, which usually implies partial loss of vision.

There are many causes of visual impairment or blindness, and all parts of the **eye** (cornea, retina, **lens**, optic nerve) can be affected. The causes can be genetic (inherited eye diseases affecting both eyes), accidental (mechanical injury to the eyeball), **inflammation** of the eye tissues (uveitis), acute or extended exposure to harmful chemicals or **radiation** (acids, alkali, tobacco smoke, UV radiation), dietary imbalance (lack of **vitamin** A), medication (corticosteroids), systemic diseases (diabetes, renal failure), or simply an aging process.

The majority of visual impairments do not lead to blindness and are related to the refractive power of the lens and cornea. However, they are often troublesome and possibly restrictive in one's choice of job. A large number of people have problems focusing due to a variety of conditions. These can include near-sightedness (myopia), far-sightedness (hyperopia), astigmatism (inability to obtain a sharp focus), presbyopia (difficulty in

accommodation), animetropia (unequal vision in each eye), and finally, aniseikonia can develop as a result of **surgery**, resulting in images that are perceived by the eyes as different sizes and shapes.

Keratoconus, which arises from the thinning of the central stromal layer of the cornea possibly due to abnormalities in **collagen metabolism**, affects the cornea and usually causes some impairment of vision, but can be treated.

Cataracts are the leading cause of blindness in developing countries and result from increased opacity of the lens, which interferes with vision. In developed countries, cataracts are mainly age-related or arise as a diabetic complication. They can also result from an environmental trauma (toxic substance exposure, radiation, mechanical or electrical injury), and a small proportion of cataracts are **congenital**, resulting from the over-proliferation of lens epithelial cells. Most cataracts can be removed by surgery, although in rare cases, post-operative bacterial **infection** (endophthalmitis) develops, which can compromise newly restored vision.

The eye tissues are all interconnected and a problem with one can cause a problem with another. The best example is the vitreous, which in addition to the accumulation of **calcium** and **cholesterol** leading to decreased transparency and subsequent impairment of vision, can shrink, leading to vitreal or retinal detachment. If the macular region is affected, some loss of visual acuity can follow and in any case floaters or flushes' appear in the visual field.

Disorders and changes affecting the retina are the leading cause of blindness in developed countries. The abnormalities in the central retina can affect retinal pigment epithelium leading to blurry vision, or can affect the macular region (photoreceptors) leading to **color** misperception. **Color blindness** can also originate from the lack of one or more type of cones. Total color blindness (monochromatic vision) is very rare; most commonly various levels of single color deficits are found. The central vision can also be destroyed by hemorrhages of the neovascular vessels developing in the retina as a result of the aging process or diabetic retinopathy.

Irreversible loss of vision occurs due to optic nerve damage resulting from glaucoma. Glaucoma is caused by an increase in the intraocular **pressure** (IOP), which develops in the aqueous and is transmitted to the back of the eye, damaging the optic nerve and consequently causing severe reduction of visual field and loss of peripheral vision.

In the older population, complete or partial blindness is caused mainly by the aging process. Changes that lead to the destruction of vision are the non-enzymatic modifications in **proteins**, lipids, and DNA, which affect their structure, composition, and function. Glycosylation, carbamylation and deamination of the proteins, oxidation of proteins and lipids, UV induced damage to proteins and to DNA are the main culprits. An accumulation of these changes leads to decreased transparency of the lens (cataracts) and retinal degeneration (age-related macular degeneration, AMD) both resulting in blindness or severe visual impairment. Most of the age-related changes are non-reversible, with the exception of the cataracts that can be surgically treated.

Research into the causes of blindness, especially glaucoma and AMD, is being undertaken by many groups in order to develop preventative measures and new treatment methods.

See also Nerve impulses and conduction of impulses; Nervous system; Neuron; Neuroscience; Neurosurgery.

Agnieszka M. Lichanska

Resources

Books

Berman, E.R. *Biochemistry of the Eye. Perspectives in Vision Research.* New York: Plenum Press, 1991.

Guyton, Arthur C., and John E. Hall. *Textbook of Medical Physiology.* 10th ed. Philadelphia: W.B. Saunders Co., 2000.

Kandel, E.R., J.H. Schwartz, and T.M. Jessell, eds. *Principles of Neural Science.* 4th ed. New York: Elsevier, 2000.

Other

Karolinska Institutet. "Eye diseases" [cited January 28, 2002]. <http://www.mic.ki.se/Diseases/c11.html>.

Blindsnakes

These tiny, primitive burrowers live underground and forage for **ants**, **termites**, soft-bodied **insects**, and insect larvae. The eyes of most blindsnakes are degenerate; they are covered by scales and do not function. However, the eyes do have light-sensitive cells (rods), so these **snakes** may not be completely blind. The head is large and the mouth, like a shark's, is below and behind the snout, which is blunt or hooked. There is a tiny spine under the snake's stubby tail which anchors the **animal** while burrowing. The body is covered with smooth, shiny, tough scales, which even cover the belly, making it difficult for the snake to slither on solid surfaces. The body is cylindrical, either thin or thick. Blindsnakes are commonly black or brown, although some **species** completely lack any pigment.

The blindsnakes are classified in three families—the Typhlopidae (wormsnakes), the Leptotyphlopidae (threadsnakes), and the Anolmalepidae. The Typhlopidae and Leptotyphlopidae are so similar that most herpetolo-

gists include both in the Scolecophidia. Both families have a remnant pelvic girdle, one lung, and one oviduct. Threadsnakes have large teeth in the lower jaw and none in the upper jaw, while wormsnakes have teeth only on the upper jaw. The 15 species of threadlike Anomalepidae of tropical **South America** are also included in the Scolecophidia, but this family has teeth on both jaws and no vestigial pelvic girdle.

The 80 species of threadsnakes are found in tropical South America, **Africa**, and the southern United States, and range in length from 6 to 16 in (15 to 41 cm). Some species release foul-smelling excretions to ward off ravaging predators such as army ants. The Texas threadsnake incubates her eggs by muscular shivering to raise its body **temperature**, a strategy found in only one or two species of python.

The 200 species of wormsnakes occur in tropical and temperate South America, Africa, Madagascar, southern **Europe**, **Asia**, and **Australia**. Wormsnakes range in length from 4.5 in (11.5 cm) to about 3 ft (91.5 cm). The Brahminy wormsnake is parthenogenic, the only species of snake to reproduce without mating, and every specimen found so far is female. This tiny species lives among **plant** roots and is transported by unsuspecting humans carrying potted plants from place to place. A single Brahminy wormsnake can populate an entirely new region.

Blindsnakes are harmless to humans, except for a species from India that is reputed to crawl into the ears of people sleeping on the ground.

See also Reptiles.

Resources

Books

Bellairs, Angus. *The Life of Reptiles.* Vols. I and II. New York: Universe Books, 1970.

Cogger, Harold G., David Kirshner, and Richard Zweifel. *Encyclopedia of Reptiles and Amphibians.* 2nd ed. San Diego, CA: Academic Press, 1998.

Mattison, Christopher. *Snakes of the World.* New York/Oxford: Facts on File Publications, 1986.

Zug, George R., Laurie J. Vitt, and Janalee P. Caldwell. *Herpetology: An Introductory Biology of Amphibians and Reptiles.* 2nd ed. New York: Academic Press, 2001.

Marie L. Thompson

Blood

Blood is a liquid **connective tissue** that performs many functions in the body, including transport of **oxy-**

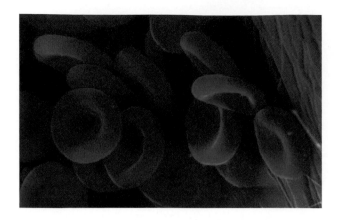

Red blood cells alongside the wall of the blood vessel.
Photograph by Dennis Kunkel. Phototake NYC. Reproduced by permission.

gen, **carbon dioxide**, **nutrients**, waste products, and **hormones**; clotting; and defense against **microorganisms**. Blood consists of formed elements, or blood cells suspended in **plasma**, a watery liquid that contains **proteins**, salts, and other substances. When a blood **sample** is placed in a test-tube and spun rapidly (a process called centrifugation), the heavier blood cells sink to the bottom of the test tube, while the straw-colored plasma floats on top.

Kinds of blood found in the animal kingdom

All **vertebrates** circulate blood within blood vessels. Because blood is enclosed within blood vessels, the circulatory systems of vertebrates are called closed circulatory systems. Some animals without vertebrae, called **invertebrates**, have circulatory systems that do not contain blood vessels. In these open circulatory systems, the fluid analogous to blood is called hemolymph (Greek, *hemo*, blood + *lympha*, **water**). Examples of animals that circulate hemolymph include **insects** and aquatic **arthropods** such as **lobsters** and crawfish. Like blood, hemolymph transports oxygen and **carbon** dioxide and has a limited clotting ability. Unlike blood, hemolymph is colorless. Other invertebrates have no true **circulatory system**. In these animals, it is not possible to distinguish blood or hemolymph from the watery fluid that bathes the tissues. This fluid contains a few defensive cells, proteins, and salts. However, oxygen and carbon dioxide are not transported in this fluid.

The composition of human blood

The human body contains about 4-6.3 qt (4-6 L) of blood. Men have more blood than women, due to the presence of higher levels of testosterone, a hormone that regulates sex characteristics and function and also stimulates blood formation. Plasma makes up 55% of the blood, while the blood cells constitute the other 45%.

Plasma

Plasma contains mostly water, which accounts for 91.5% of the plasma content. The water acts as a solvent for carrying other substances.

Proteins account for 7% of plasma. The most prevalent of these proteins in plasma is albumin, a protein also found in egg white. Albumin **concentration** is four times higher in the blood than in the interstitial fluid (the watery fluid that bathes tissues, but is located outside and between cells). This high concentration of albumin in plasma serves an important osmotic function. The higher concentration of protein in blood prevents water from moving from the blood into the interstitial fluid. Without this osmotic protection, water would move from the interstitial fluid into the blood, diluting the plasma and swelling the blood **volume**. A high blood volume could have disastrous consequences, because the circulatory system can only pump so much blood before it becomes overloaded.

Other proteins that are present in plasma are immunoglobins and fibrinogen. Immunoglobins, also called antibodies, are proteins that function in the immune response. Antibodies attach to invading **bacteria** and other microorganisms, marking them for destruction by other immune cells. Fibrinogen is a protein that functions in a complex series of reactions that leads to the formation of blood clots.

The other components of plasma are salts, nutrients, enzymes, hormones, and nitrogenous waste products. Together, these substances account for 1.5% of plasma. The salts present in plasma include **sodium**, potassium, **calcium**, **magnesium**, chloride, and bicarbonate. These salts function in many important body processes. For instance, calcium functions in muscle contraction; sodium, chloride, and potassium function in nerve impulse transmission in nerve cells; and bicarbonate regulates **pH**. These salts are also called electrolytes. An imbalance of electrolytes, which can be caused by dehydration, can be a serious medical condition. Many gastrointestinal illnesses, such as **cholera**, cause a loss of electrolytes through severe diarrhea. When electrolytes are lost, they must be replaced with intravenous solutions of water and salts or by having the patient drink solutions of salts and water.

The remaining substances present in plasma are elements that the plasma is transporting from one place to another. For instance, plasma contains nutrients that nourish tissues. The nutrients found in plasma include amino acids, the building blocks of proteins; glucose, or

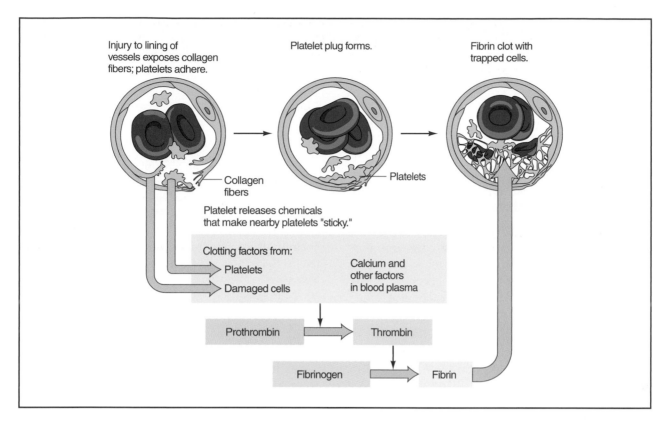

Injury to lining of vessels exposes collagen fibers; platelets adhere.

Platelet plug forms.

Fibrin clot with trapped cells.

Collagen fibers

Platelets

Platelet releases chemicals that make nearby platelets "sticky."

Clotting factors from:

Platelets

Damaged cells

Calcium and other factors in blood plasma

Prothrombin → Thrombin

Fibrinogen → Fibrin

Figure 1. The clotting process. *Illustration by Hans & Cassidy. Courtesy of Gale Group.*

sugars; and **fatty acids** and **glycerol**, the components of lipids (fats). In addition to nutrients, plasma also contains enzymes, or small proteins that function in **chemical reactions**, and hormones, which are transported from **glands** to body tissues. Waste products from the breakdown of proteins are also found in plasma. These waste products include creatinine, uric acid, and ammonium salts. Blood transports these waste products from the body tissues to the kidneys, where they are filtered from the blood and excreted in the urine.

Formed elements, or blood cells

Blood cells make up 45% of the total composition of blood. The various types of blood cells are erythrocytes, or red blood cells; leukocytes (also spelled leucocytes), or white blood cells; and platelets.

Red blood cells

The human body contains an estimated 25 trillion red blood cells; approximately 4.8-5.4 million are found in every microliter of blood. The structure of a red blood **cell** is eminently suited to its primary function, the transport of oxygen from the lungs to body tissues. Red blood cells are very small (about 6 nanometers wide), disk-shaped, and contain a small **depression** on either side. Their small size allows them to squeeze through the tiniest blood vessels, called **capillaries**. In addition, the small size of red blood cells allows a greater **diffusion** of oxygen across the blood cells' plasma membranes than if the cells were larger. Because blood contains so many of these small cells, the combined surface area of these many blood cells translates into an extremely large amount of surface area for the diffusion of oxygen. The disk shape and the depressions on either side also contribute to a greater surface area.

Red blood cells are unusual in that they do not contain nuclei or mitochondria, the cellular organelle in which **aerobic metabolism** (the breakdown of nutrients that requires oxygen) is carried out. Instead, red blood cells acquire **energy** through metabolic processes that do not require oxygen. The lack of nuclei and mitochondria therefore allow the red blood cell to function without depleting its cargo of oxygen, leaving more oxygen for the body tissues.

The **molecule** that binds oxygen in red blood cells is called hemoglobin. Hemoglobin is a large, globular protein consisting of four protein chains surrounding an **iron** core. Hemoglobin is densely packed inside the red blood cell; in fact, hemoglobin accounts for a third of the weight of the

KEY TERMS

ABO blood groups—Blood types established by the A and B antigens present on the plasma membrane of red blood cells; ABO blood groups include A, B, AB, and O.

Aerobic metabolism—Metabolic processes that require oxygen.

Agranular leukocyte—A white blood cell without granules in its cytoplasm; these white blood cells include the monocytes and lymphocytes.

Albumin—A protein found in plasma.

Antibody—A molecule created by the immune system in response to the presence of an antigen (a foreign substance or particle). It marks foreign microorganisms in the body for destruction by other immune cells.

Antigen—A molecule, usually a protein, that the body identifies as foreign and toward which it directs an immune response.

B lymphocyte—Immune system white blood cell that produces antibodies.

Basophil—A type of white blood cell; functions in the inflammatory response by releasing histamines and other chemicals that have specific effects on tissues.

Capillary—The smallest blood vessel; it connects artery to vein.

Centrifugation—A laboratory procedure in which a test tube of blood or other liquid is spun at a high speed.

Circulatory system—The body system that circulates blood pumped by the heart through the blood vessels to the body tissues.

Clotting factor—A set of substances released by platelets that function in the clotting mechanism.

Cytotoxic T lymphocyte—A type of white blood cell that attacks and kills cells infected by a foreign microorganism.

Electrolytes—The salts and other substances present in the plasma that function in crucial body processes.

Eosinophil—A type of white blood cell that counteracts the effects of histamine and other inflammatory chemicals; also phagocytizes bacteria tagged by antibodies.

Erythrocyte—A red blood cell.

Fibrin—A protein that functions in the clotting mechanism; forms mesh-like threads that trap red blood cells.

Fibrinogen—The inactive form of fibrin present in plasma; activated by clotting factors released by platelets.

Formed elements—The cells present in blood.

Granular leukocyte—A white blood cell that contains granules in its cytoplasm; includes basophils, eosinophils, and neutrophils.

Helper T lymphocyte—The "lynch pin" of specific immune responses; helper T cells bind to APCs (antigen-presenting cells), activating both the antibody and cell-mediated immune responses.

Hemoglobin—An iron-containing, protein com-

---

entire red blood cell. Each red blood cell contains about 250 molecules of hemoglobin. In the lungs, oxygen diffuses across the red blood cell **membrane** and binds to hemoglobin. As blood circulates to the tissues, oxygen diffuses out of the red blood cells and enters tissues. The waste product of aerobic metabolism, carbon dioxide, then diffuses across red blood cells and binds to hemoglobin. Once circulated back to the lungs, the red blood cells discharge their load of carbon dioxide, which is then breathed out of the lungs. However, only 7% of carbon dioxide generated from metabolism is transported back to the lungs for exhalation by red blood cells; the majority is transported in the form of bicarbonate, a component of plasma.

The complexity of blood is apparent. Still, researchers hope to create synthetic blood substitutes, which will ease the burden of dwindling donations to meet the demand for surgeries, transfusions, and emergency use. Currently under development is an artificial blood that uses *perfluorocarbons* to carry oxygen to tissues, replacing the function of hemoglobin. Perfluorocarbons are long, fatty **hydrocarbon** chains containing fluorine that have the ability to pick up oxygen in lungs, and release it into tissues. The artificial blood made with these molecules is a mixture of the perfluorocarbons with saline (physiological **salt** water) using surfactants, substances that allow the mixing of oil and water. The **solution** then can be administered to patients. Over **time**, as the artificial blood helps deliver oxygen to tissues, the perflourocarbon molecules are exhaled from the body. Strictly, this substance is not a whole blood substitute since it only has the ability to carry oxygen and cannot

plex carried in red blood cells that binds oxygen for transport to other areas of the body.

Hemolymph—The blood-like liquid present in the open circulatory systems of certain invertebrates.

Hemophilia—A genetic disorder in which one or more clotting factors are not released by the platelets; causes severe bleeding from even minor cuts and bruises.

Hemopoiesis—The process of red blood cell formation in the bone marrow.

Histamine—A chemical released by basophils during the inflammatory response; causes blood vessels to dilate.

Human leukocyte antigen (HLA)—A type of antigen present on white blood cells; divided into several distinct classes; each individual has one of these distinct classes present on their white blood cells.

Immunoglobulin—The protein molecule that serves as the primary building block of antibodies.

Inflammatory response—A type of non-specific immune response; involves the release of chemicals from basophils that increase blood circulation and white blood cell migration to the affected area.

Interstitial fluid—The fluid that bathes cells.

Leukocyte—A white blood cell.

Lymph node—A small structure located at several points in the body; consists of lymphatic tissue that filters blood and removes microorganisms.

Lymphocyte—A type of white blood cell; includes B and T lymphocytes.

Lymphoid stem cell—The cell from which B and T lymphocytes are derived.

Lysozyme—An enzyme released by neutrophils that kills cells.

Macrophage—A type of phagocytic cell derived from monocytes.

Monocyte—A type of white blood cell that phagocytizes foreign microorganisms.

Neutrophil—A type of white blood cell that phagocytizes foreign microorganisms; also releases lysozyme.

Phagocytize—To engulf and digest a cell.

Plasma—The straw-colored liquid portion of blood that contains water, proteins, salts, nutrients, hormones, and metabolic wastes.

Plasma cell—The cell derived from the B lymphocyte, which secretes antibodies.

Platelet—A piece of a cell that contains clotting factors.

Pluripotent stem cell—The type of stem cell from which red blood cells and more white blood cells are derived in the bone marrow.

Sickle cell anemia—A genetic disorder caused by a defect in one of hemoglobin's four protein chains; causes red blood cells to be sickle-shaped.

T cells—Immune-system white blood cells that enable antibody production, suppress antibody production, or kill other cells.

Thymus—The organ in which T cells undergo further development and maturation.

replace the other important functions of blood. However, it is valuable because it eliminates the risk of transmitting **disease** during transfusions as well as preventing accidental blood type mismatches.

Sickle cell anemia is an inherited disorder caused by a defect in one of hemoglobin's four protein chains. The sickle hemoglobin distorts the shape of the red blood cells and injures the red blood cell membrane. Water and potassium leak from the cells, causing the red blood cells to become "sickle-shaped." The cells also become inflexible and rigid. As a result of these changes, oxygen transport is severely interrupted and circulation of the blood through the blood vessels can become blocked. These irregular blood cells do not carry as much oxygen as their normally-shaped counterparts.

Sickle cell **anemia** is invariably fatal; most people with the disease die in early adulthood.

Red blood cells are formed in red bone marrow from precursor cells called pluripotent **stem cells**. The process of red blood cell formation is called hemopoiesis, or hematopoiesis. In adults, hemopoiesis takes place in the marrow of ribs, vertebrae, breast bone, and pelvis. On average, a red blood cell lives only 3-4 months. Constant wear and tear on the red blood cell membrane, caused by squeezing through tiny capillaries, contribute to the red blood cell's short life span. Worn out red blood cells are destroyed by phagocytic cells (cells that engulf and digest other cells) in the liver. Parts of red blood cells are recycled for use in other red blood cells, such as the iron component of hemoglobin.

An interesting aspect of red blood cells is that they carry certain proteins, called antigens, on their plasma membranes. These antigens are responsible for the various blood groups known as A, B, AB, and O. A person with A antigens is type A; a person with B antigens is type B; a person with both antigens is type AB; and a person with none of the antigens is type O. A individuals have antibodies to B antigens; B individuals have antibodies to A antigens; AB individuals do not have antibodies to the antigens, and O individuals have antibodies to both A and B antigens. These combinations are necessary to know for blood transfusions. For instance, if a type A **individual** donates blood to a type B individual, the A antibodies in the recipient's B blood will react with the A antigens of the donor's A blood. This reaction, called the agglutination reaction, causes the blood cells to clump together. Agglutination can be fatal. Until blood typing was worked out early in this century, many deaths from blood transfusions occurred due to incompatibility of antigens and antibodies.

White blood cells

White blood cells are less numerous than red blood cells in the human body; each microliter of blood contains 5,000-10,000 white blood cells. The number of white blood cells increases, however, when the body is fighting off **infection**. White blood cells, therefore, are maintained at a stable number until the **immune system** detects the presence of a foreign invader. When the immune system is activated, chemicals called lymphokines stimulate the production of more white blood cells.

White blood cells function in the body's defense against invasion and are key components of the immune system. They usually do not circulate in the blood vessels, and are instead found in the interstitial fluid and in lymph nodes. Lymph nodes are composed of lymphatic **tissue** and are located at strategic places in the body. Blood filters through the lymph nodes, and the white cells present in the nodes attack and destroy any foreign invaders.

The human body contains five types of white blood cells: monocytes, neutrophils, basophils, eosinophils, and lymphocytes. Each type of white blood cell plays a specific role in the body's immune defense system.

Under a **microscope**, three kinds of white blood cells appear to contain granules within their cytoplasm. These three types are the neutrophils, basophils, and eosinophils. Together, these three types of white blood cells are called the granular leukocytes. The granules are specific chemicals released by these white blood cells during the immune response. The other two types of white blood cells, the monocytes and lymphocytes, do not contain granules. These types are known as the agranular leukocytes.

Monocytes, which comprise 3-8% of the white blood cells, and neutrophils, which comprise 60-70% of white blood cells, are phagocytic cells. They ingest and digest cells, including foreign microorganisms such as bacteria. Monocytes differentiate into cells called macrophages. Macrophages can be fixed in one place, such as the **brain** and lymph nodes, or can "wander" to areas where they are needed, such as the site of an infection. Neutrophils have an additional defensive property: they release granules of lysozyme, an **enzyme** that destroys cells.

Basophils comprise 0.5-1% of the total composition of white blood cells and function in the body's inflammatory response. Allergies are caused by an inflammatory response to relatively harmless substances, such as pollen or dust, in sensitive individuals. When activated in the inflammatory response, basophils release various chemicals that cause the characteristic symptoms of allergies. Histamines, for instance, cause the runny nose and watery eyes associated with allergic reactions; heparin is an anticoagulant that slows blood clotting and encourages the flow of blood to the site of **inflammation**, inducing swelling.

Eosinophils, which comprise 2-4% of the total composition of white blood cells, are believed to counteract the effects of **histamine** and other inflammatory chemicals. They also phagocytize bacteria tagged by antibodies.

Lymphocytes, which comprise 20-25% of the total composition of white blood cells, are divided into two types: B lymphocytes and T lymphocytes. The names of these lymphocytes are derived from their origin. T lymphocytes are named for the thymus, an **organ** located in the upper chest region where these cells mature; and B lymphocytes are named for the bursa of Fabricus, an organ in **birds** where these cells were discovered. T lymphocytes play key roles in the immune response. One type of T lymphocyte, the helper T lymphocyte, activates the immune response when it encounters a macrophage that has ingested a foreign microorganism. Another kind of T lymphocyte, called a cytotoxic T lymphocyte, kills cells infected by foreign microorganisms. B lymphocytes, when activated by helper T lymphocytes, become plasma cells, which in turn secrete large amounts of antibodies.

All white blood cells arise in the red bone marrow. However, the cells destined to become lymphocytes are first differentiated into lymphoid stem cells in the red bone marrow; from the red bone marrow, these stem cells undergo further development and maturation in the spleen, tonsils, thymus, adenoids, and lymph nodes.

HIV, the **virus** that causes Acquired Immune Deficiency **syndrome** (**AIDS**), attacks and kills T lymphocytes. This disease cripples the immune system and leaves the body helpless to stave off infections. As AIDS progresses, the number of helper T lymphocytes drops from a normal 1,000 to **zero**.

Like red blood cells, the plasma membranes of white blood cells also contain antigens. These surface antigens are called the human leukocyte associated (HLA) antigens. Like the red blood cell types, these HLA antigens represent different white blood cell "groups." When a person receives an organ transplanted from a donor, the recipient and the donor must have the same HLA antigen group for the transplant to be successful. If the donor and recipient are two different HLA antigen groups, the recipient's body will "reject" the organ; in other words, the recipient's immune system will be activated by the foreign cells of the organ and initiate an immune response against the organ.

Platelets

Platelets are not cells; they are fragments of cells that function in blood clotting. Platelets number about 250,000-400,000 per liter of blood. Blood clotting is a complex process that involves a cascade of reactions that leads to the formation of a blood clot. Platelets contain chemicals called clotting factors. These clotting factors first combine with a protein called prothrombin. This reaction converts prothrombin to thrombin. Thrombin, in turn, converts fibrinogen (present in plasma) to fibrin. Fibrin is a thread-like protein that traps red blood cells as they leak out of a cut in the skin. As the clot hardens, it forms a seal over the cut. This process works for relatively small cuts in the skin. When a cut is large, or if an artery is severed, blood loss is so severe that the physical **pressure** of the blood leaving the body prevents clots from forming. In addition, in the inherited disorder called **hemophilia**, one or more clotting factors are lacking in the platelets. This disorder causes severe bleeding from even the most minor cuts and bruises.

Platelets have a short life span; they survive for only 5-9 days before being replaced. Platelets are produced in red bone marrow and are broken off from other red blood cells.

See also Anemia; Anticoagulants; Blood gas analysis; Blood supply; Heart; Hematology; Respiratory system.

Resources

Books

Agre, Peter C., and Jean-Pierre Cartron, eds. *Protein Blood Group Antigens of the Human Red Cell: Structure, Function, and Clinical Significance.* Baltimore: Johns Hopkins University Press, 1992.

Belcher, Anne E. *Blood Disorders.* St. Louis: Mosby Year-Book, 1993.

Kapff, Carola R. *Blood: Atlas and Sourcebook of Hematology.* 2nd edition. Boston: Little, Brown. 1991.

Long, Michael W., and Max S. Wicha, eds. *The Hematopoietic Microenvironment: The Functional and Structural Basis of Blood Cell Development.* Baltimore: Johns Hopkins University Press, 1993.

Periodicals

Roush, Wade. "An 'Off-Switch' for Red Blood Cells." *Science.* 268 (April 7, 1995): 27.

Ware, Anthony J., and Donald D. Heistad. "Platelet-Endothelium Interactions." *New England Journal of Medicine,,* 328 (March 4, 1993): 628.

Weller, Peter F. "The Immunobiology of Eosinophils." *New England Journal of Medicine* 324 (April 18, 1991): 110.

Kathleen Scogna

Blood gas analysis

Blood gas analysis is a means of determining the amount of **oxygen** or **carbon dioxide** being carried in the blood, and in some cases, of discovering the identity of a toxic gas, such as **carbon monoxide**, that may be present. Also, the determination can be made as to whether the blood is too acidic or too alkaline, which may help the physician in his **diagnosis**.

Among other functions, blood carries oxygen to the body's tissues and removes **carbon** dioxide. The blood laden with carbon dioxide passes through the right side of the **heart** into the lungs and exchanges the carbon dioxide for fresh oxygen. The oxygenated blood is then pumped by the left side of the heart out into the body to repeat the cycle.

The red blood cells, or erythrocytes, carry the blood gases. Hemoglobin, the substance that gives blood its red **color**, is the molecular substance in the erythrocyte that attaches to oxygen and exchanges it for carbon dioxide.

Carbon monoxide, a colorless, very toxic gas, can displace oxygen in the bloodstream. Hemoglobin has approximately 12 times the affinity for carbon monoxide as it does for oxygen, so it will pick up carbon monoxide if both gases are present. This means that the body does not get the oxygen it needs, and eventually death will occur.

Testing blood gases is a means to determine whether an acid-base (biochemical) disturbance is of respiratory or metabolic origin. Respiratory conditions such as **pneumonia**, **bronchitis**, **emphysema**, or severe **asthma** can cause the blood to become more acid. Respiratory conditions such as aspirin toxicity, strenuous **exercise**, fever, or overactive thyroid can cause the blood to be more alkaline. Kidney failure, burns, heart attack, or starvation are the metabolic reasons for the blood to become acid, and liver failure, vomiting, ulcer, or **cystic fibrosis** cause metabolic alkaline blood.

To determine blood gases, the blood specimen must be taken from an artery (usual blood specimens are taken from a vein) and the blood specimen is placed in **ice** to prevent any changes in blood gases, and rushed to the laboratory for analysis.

See also Bronchitis; Circulatory system.

Blood supply

Blood supply refers to the blood resources in blood banks and hospitals that are critical to the health care community. The blood supply consists of donated blood units (in pints) that are used to replace blood lost during **surgery** or from trauma.

Blood transfusions were attempted as early as 1667 when Jean-Baptiste Denis, a French physician, transfused 12 fl oz (355 ml) of lamb's blood into a 15-year-old male patient. While Denis's subject improved immediately, later attempts to transfuse blood met with mixed results. Prior to the end of the nineteenth century, some patients who received blood from another person improved while others died quickly. In 1900, Austrian physician Karl Landsteiner discovered the four types of human blood, A, B, O, and AB, and the rules that govern their compatibility. Type O can be given to any recipient and is called the universal donor. Type A can be given to type A and AB recipients, and type B to type B and AB recipients. Transfusing blood of a type not compatible with the patient's blood can be fatal.

Donating blood

Blood units are collected in the United States by the American Red Cross and by blood banks at local hospitals. Donations are always needed due to the constant demands of hospitals and their trauma units as well as to the brief shelf-life of stored blood. Some 12 million units are used every year in the U.S.

Blood is collected by simply inserting a needle into a vein and allowing the blood to flow into a plastic bag that has been specially treated to prevent the blood from clotting. The average adult human has approximately 6 qt (5.7 l) of blood in their body, so the loss of 1 pt (0.5 l) will have little effect on them. The liquid portion of the blood is quickly replaced from fluid the donor drinks afterwards and the blood cells are regenerated from the bone marrow. Healthy donors can make blood donations about every six to eight weeks without suffering any ill effects.

Once the blood has been collected in the bag, other small specimens are collected for testing. Blood is tested for **hepatitis** viruses, syphilis, **AIDS**, or other diseases and classified as to type. Technicians also run a series of antigen tests to provide information on the presence of any factor that may provoke a reaction in the recipient.

After the blood is tested and found satisfactory it is refrigerated for use. Blood in blood banks is distributed to medical facilities in the region of the blood bank. Patients undergoing surgery have blood tests to determine their blood type and antigen structure so that compatible blood can be selected to replace that which is lost during surgery. Patients who arrive in the emergency room following a trauma such as an **automobile** accident also are tested for possible blood replacement. Loss of between 1-2 qt (1-2 l) of blood will result in shock and require the immediate transfusion of blood. Loss of up to 3 qt (2.8 l), half the blood supply in the human body, can be fatal.

Blood components

Not all of the collected blood is used as whole-blood transfusions. Some of the supply is broken down into its components. Blood **plasma**, the liquid part of blood that remains once the blood has coagulated (clotted), can be dried into a powder and used as a replacement for blood **volume** lost from wounds. Blood plasma need not be refrigerated so it is useful in situations such as battlefields and areas that lack proper refrigeration. The plasma powder is reconstituted by mixing it with sterile **water**. It is then infused into the wounded or injured patient.

Thousands of units of blood can be combined and the clotting factor (Factor VIII) refined from the mixture. Factor VIII is used by hemophiliacs to provide the ingredient they lack to make their blood clot. Without the clotting factor a hemophiliac can bleed uncontrollably.

Blood cells also can be separated and used. Concentrated red blood cells can improve the blood count of an anemic patient. White blood cells and platelets also are isolated and transfused into needy patients.

AIDs and the blood supply

When Acquired Immune Deficiency (AIDS) entered the U.S. population in the late 1970s and early 1980s, much was unknown about the **disease** and no test existed to detect the **virus** in blood. While initially confined to populations demonstrating high risk **behavior**, such as homosexuals and intravenous drug users, AIDS began to infect hemophiliacs and surgical patients who did not fit into such high risk categories. Members of the medical community soon determined that these patients had contracted the disease from donated blood and that the process used to break blood into its components did not kill the AIDS virus. Because of its minute size the virus passed through the filters used in the extraction process.

In 1985 a test was developed to detect the HIV virus in blood, and immediately every unit of donated blood was tested and those tainted with the virus removed from the blood supply. Donors are now carefully screened to eliminate any who might be at high risk of contracting AIDS. To eliminate the possibility of contracting this or any other blood-borne disease, patients who are scheduled to undergo surgery are urged to donate blood ahead of time so that their own blood can be transfused into them if needed.

It is important to know that a blood donor cannot contract AIDS or any other disease by donating blood. The equipment used to collect donated blood is used only the once and then discarded.

Larry Blaser

Blotting analysis

Blotting analysis describes a series of techniques used to determine and describe protein and **nucleic acid** (e.g., DNA, RNA) sequences.

Blotting analysis allows scientists to transfer electrophoretically separated components from a gel to a solid support. This support may then be used for probing with reagents specific for particular sequences of amino acids or nucleotides. In this way, the size and/or quantity of the **proteins** or nucleic acids under study can be gauged.

Blotting analysis was originally developed in 1975 by British molecular biologist E. M. Southern while he was on a leave of absence from his Edinburgh lab to do research in Zurich. This process has since been referred to as Southern blotting. Southern's method was designed to transfer fragments of **deoxyribonucleic acid (DNA)** from the agarose gel in which they had been separated onto **cellulose** nitrate filters. The DNA fragments are deposited onto a filter laid over the gel as a result of **capillary action**, which is established and maintained by the flow of **buffer** from underneath the gel to a stack of dry **paper** towels placed on the filter.

In addition to the capillary transfer method described above, additional methods have been developed for the transfer of DNA. Electrophoretic transfer can be performed by mounting the gel and **membrane** between porous pads aligned between **parallel** electrodes in a tank containing buffer of high ionic strength. The **electric current** drives the transfer. However, resulting high temperatures require that the tank be cooled. This

method is most often used with gels made from polyacrylamide, not agarose, since polyacrylamide has a higher melting **temperature**. A third method involves the use of a **vacuum**. The gel is placed in contact with the membrane on a porous screen above a vacuum chamber, and a buffer elutes the DNA from the gel onto the membrane.

After the DNA is deposited onto the solid support, the filter is usually dried at a high temperature in order to bind the nucleic acid strongly to the membrane. Alternatively, the DNA can be covalently attached to the filter by cross linking with low doses of ultraviolet **radiation**. The process of attaching the nucleic acids allows the filter to be sequentially hybridized to several different probes with little loss in sensitivity.

Originally, the most common solid supports used were nitrocellulose filters. However, the filters become brittle when dried, and as a result, they can not be used for more than one or two cycles of hybridization. Currently, charged nylon membranes are a popular choice, as the DNA binds irreversibly and the membrane is more durable to withstand multiple hybridizations.

Once the DNA is bound to the membrane, the process of hybridization may begin. Typically, the membrane is first incubated with a buffer designed to limit the amount of non-specific binding of the chosen probe to parts of the membrane not covered by DNA. Next, a radiolabeled probe is added. This probe is complementary to sequences of the DNA of interest and therefore binds those sequences with a given affinity. The sort of probe used can vary greatly, from purified **ribonucleic acid (RNA)** to cloned cDNAs (DNA copies of RNA molecules) to short synthetic oligonucleotides. Unbound probe is then washed away from the membrane. Autoradiography of the labeled blot results in visualization of bands that correspond to the probe sequence. The size of the bands can be determined by their placement along the length of the gel in comparison with markers of known size. Additionally, the strength of the band seen can also be used to quantify the amount of DNA present.

Southern blotting has many uses in the field of **molecular biology**. Genomic Southern blots can provide a physical **map** of restriction sites within a **gene** in a **chromosome** through the analysis of fragments produced after digestion of genomic DNA with one or more restriction enzymes. Other uses include detecting major gene rearrangements and deletions involved in **disease**, identifying structurally related genes within the same **species** or homologous genes in other species, and more recently, screening genomes of various mutagenized lines in order to identify a mutant in a particular gene under investigation.

After the development of Southern blotting, similar procedures were created to analyze RNA as well as proteins. Named northern and western blotting, respectively, in reference to their progenitor, these techniques have many similarities to Southern blotting but require several modifications.

Northern blotting refers to the transfer of RNA to a solid support from a denaturing agarose gel. The membranes and transfer apparatus used are similar to those used in Southern blotting. The RNA run in the gel may be total RNA isolated from particular samples. Alternatively, RNA containing poly(A) tails, a characteristic of messenger RNA (mRNA), which is involved in the transfer of genetic information from DNA to protein, may be purified from the total RNA **sample** in order to analyze specifically mRNA molecules. As with Southern blotting, radiolabeled RNA, DNA, or oligonucleotides are used as probes, and the blots are hybridized and autoradiographed in a similar manner. More recently, additional methods of detection that do not require radioactivity are commonly used for both northern and Southern blotting. Northerns can be used to determine the size or amount of specific mRNAs, to examine in what tissues or under what conditions certain genes are expressed (i.e., are copied in the form of mRNA), and to study conditions that alter the level of particular mRNAs.

The analysis of proteins is accomplished by western blotting. Generally, proteins are put in **solution** (solubilized) with detergents and reducing agents, separated in polyacrylamide gels, and transferred to a nitrocellulose filter or polyvinylidene fluoride membrane, where they become covalently bound to the support. The crucial difference between westerns and Southerns is the probe. For western blots, specific unlabeled antibodies, which typically must be produced for each individual protein and which react specifically with their target, are used to recognize the protein of interest from within the background of other cellular proteins. Antibodies, unlike their nucleic acid probe counterparts, do not bind with predictable rates or specificities, and therefore require increased levels of optimization in order to extract the most useful information. The bound antibody is detected on the blot by one of several secondary reagents that is either radiolabeled or coupled to an **enzyme**. Activity of the enzyme in the presence of its substrate allows the detection of antibodies bound to the protein of interest. Westerns can be used to search for the presence of certain proteins in specific samples, tissues, or treatments. They may also be used for quantitative analyses or to determine the apparent **molecular weight** of a protein.

See also Antibody and antigen; Chromosome mapping; DNA synthesis; DNA technology; Electrophoresis.

Resources

Books

Griffiths, A., et al. *Introduction to Genetic Analysis*. 7th ed. New York: W.H. Freeman and Co., 2000.

Jorde, L.B., J. C. Carey, M. J. Bamshad, and R. L. White. *Medical Genetics*. 2nd ed. Mosby-Year Book, Inc., 2000.

Klug, W., and M. Cummings. *Concepts of Genetics*. 6th ed. Upper Saddle River: Prentice Hall, 2000.

Watson, J.D., et al. *Molecular Biology of the Gene*. 4th ed. Menlo Park, CA: The Benjamin/Cummings Publishing Company, Inc., 1987.

Watson, J.D., et al. *Recombinant DNA*. 2nd ed. New York: Scientific American Books, 1992.

Nicole D. LeBrasseur

Blue-green algae *see* **Algae**

Blue revolution (aquaculture)

The term "blue revolution" refers to the remarkable emergence of aquaculture as an important and highly productive agricultural activity. Aquaculture refers to all forms of active culturing of aquatic animals and plants, occurring in marine, **brackish**, or fresh waters.

Aquaculture has long been practiced in China and other places in eastern **Asia**, where **freshwater** fish have been grown as food in managed ponds for thousands of years. In recent decades, however, the practice of aquaculture has spread around the world. Many **species** of freshwater and marine organisms are being cultivated as highly productive and nutritious **crops** for consumption by humans. The tremendous growth of aquaculture has been stimulated by knowledge that there are intrinsic limitations to the productivity of the wild, unmanaged aquatic ecosystems that humans have traditionally exploited as sources of **fish**, aquatic **invertebrates**, and seaweeds. Moreover, in a depressingly large number of cases, the usable productivity of natural aquatic ecosystems has been overexploited or otherwise degraded by humans, and the harvested yields have declined substantially.

In many cases, however, the productivity of valuable aquatic species can be greatly increased under managed conditions, and also by genetic **selection** for varieties having desirable traits, such as higher productivity. The principal goal of aquaculture science is to develop systems by which aquatic organisms can be grown and harvested at high but sustainable rates, while not causing unacceptable environmental damage.

Aquaculture production

By far, the greatest aquacultural activity occurs in Asia, which realized 16 million tons of production in 1995, compared with 1.5 million tons in **Europe**, 0.46 million tons in **North America**, and 0.60 million tons in the rest of the world. Of the global production of 18.6 million tons, about 60% was freshwater fish, 9.5% was anadromous and marine fish, and 31% was **mollusks** and crustaceans.

In comparison, the global landings of marine fish in 1989 were about 84 million tons, while 13 million tons were produced by the world's freshwater fisheries. Clearly, aquaculture is an extremely large and rapidly growing enterprise.

Fish farming

Fish farming is a relatively intensive enterprise. It commonly involves the management of all stages in the life cycle of the cultivated fish, from the production of eggs and larvae, through to growth and eventual harvest of high-quality, market-sized fish. In this sense, fish farming is different from fish ranching, which is a less-intensive enterprise that usually involves the confinement and feeding of captured wild fish in order to increase their market value. This is done, for example, with bluefin **tuna** (*Thunnus thynnus*) in some places.

Fish have many potential benefits as cultivated species. Because they are cold-blooded (or *poikilothermic*), fish divert little **energy** to maintaining their body **temperature**, and can therefore convert a relatively large proportion of their food into their growing **biomass**. In other words, populations of fish can be very productive, especially under conditions where the animals are well-fed and the rates of mortality from **disease** and predation

Fish farming. *JLM Visuals. Reproduced by permission.*

are kept small. Moreover, fish are a tasty and highly nutritious food for humans. Consequently, the economic value of fish is great, as are the potential profits gained from cultivating them in large quantities.

Various species of fish are grown in aquaculture, using a variety of cultivation systems. The systems most commonly involve confinement in artificial ponds, or in cages set into larger bodies of **water**, including the **ocean**. The fish are fed with a nutritious diet, sometimes to excess so that their growth **rate** is maximized. When the fish are economically mature, they are carefully harvested and processed so that the highest-value economic products can be delivered to consumers.

The oldest fish-farming systems were developed in eastern Asia, and involved several species of freshwater fish. The first writings about methods of fish culture are dated from about 2,500 years ago and were written by Fan Lei, a wealthy Chinese fish farmer. The first species to be grown in aquaculture was probably the common **carp** (*Cyprinus carpio*), a species native to China but

known in **Australia** and New Zealand for excellent eating quality and sweet **taste**. Similarly popular for eating is the long-snouted boarfish (*Pentaceropsis recurvirostris*) from the waters off southern Australia.

Boas

Boas are a group of nonvenomous, constricting **snakes** (family Boidae), most of which are found in tropical America and in Madagascar. Boas bear live young, and in this way they differ from the Old World **pythons**, which lay eggs. Boas are of ancient derivation, retaining some of the features of their lizard-like ancestors, such as paired lungs (modern snakes have only one), tiny remnants of hind limbs (often called spurs), and a characteristic bone in their lower jaw, the coronoid, which is not found in advanced snakes. Boas have no poison fangs, and they kill their **prey** by squeezing, though their prey is often bitten first.

Boas, pythons, and wood snakes are classified in the family Boidae, whose members have small nostrils and small eyes with elliptical, vertical pupils. Boas are among the most ancient groups of living snakes, having been present in the Cretaceous period, 200 million years ago, when dinosaurs still stalked the **earth**. Although boas made up a major part of the snake **fauna** in the past, today only about 40 **species** of boas are known. Two quite distinct subfamilies are recognized, the true boas, which are mainly tropical **rainforest** animals from **South America** and Madagascar, and the sandboas from the deserts and other arid regions of the northern hemisphere. The reticulated python (*Reticulatus*) of **Asia** grows up to 33 ft (11 m) and is one the world's largest snakes.

True boas

All primitive snakes (other than the burrowing **blindsnakes**) have eyes with vertically elliptic pupils (cat-eyes) that can open widely in the dark. Boas have a stout body, short tail, and a green, brown, or yellow body with either blotches or **diamond** patterns. They tend to be most active at night, whether they inhabit deserts or rainforests. Beyond this general similarity, however, the South American boas are highly varied in their habits as well as in their overall appearance.

The boa constrictor, (*Constrictor constrictor*) and the West Indian boas (*Epicrates*) are primarily ground-dwellers, although they may also climb trees, but they show no specializations for a particular life-style. As in other boas, the young feed on small animals such as lizards, whereas the large adults tend to feed on larger **mammals** and **birds**.

The West Indian boas are found on many Caribbean islands. In general, each **island** has a single, unique species. The exception is the island of Hispaniola (Haiti and the Dominican Republic) which has three species; *Epicrates fordi* and *E. gracilis*, each only about 3 ft (1 m) long, and *E. striatus*, a much larger snake that reaches a length of 8 ft (2.5 m) or more. The latter species is also found on a number of islands in the Bahamas. On several Caribbean islands boas gather at **cave** entrances at night, snatching **bats** out of the air as they exit or enter the cave.

Probably the best known representative of the true boas is the boa constrictor of tropical America, from Mexico to Argentina. Although often depicted as a giant, man-eating snake in lurid stories or movies, this boa seldom grows to a length of more than 10 ft (3 m); the record is 16 ft (5 m). Boa constrictors are gentle and easy to care for, and have become one of the favorite pet snakes in recent years.

The green anaconda (*Eunectes murinus*) of South America may be the largest snake in the world. It has fairly reliably been reported to grow to 35 ft (11 m) long, although 18–20 ft (5.5–6 m) is the maximum seen in recent times. These dark green snakes with black marking are river-dwellers, and are highly adapted for an aquatic existence. The anaconda's eyes and nostrils are positioned on top of its head to allow it to see and breathe while the rest of its body is completely submerged. Anacondas lie submerged in the **water** at night waiting for **peccaries** (**pigs**) to come down to drink. Besides feeding on mammals, anacondas are also known to eat birds, crocodilians (caimans), and **turtles**. Like other boas, anacondas give **birth** to 40 or more live young.

The **tree** boas (*Corallus* and relatives) are highly modified for a life in the trees. Tree boas have slender bodies and their prehensile tails make them excellent climbers. They also have large eyes for nocturnal foraging, and long teeth for catching sleeping birds and other tree-dwellers, such as lizards, **rodents**, and **opossums**. Tree boas, like all boas, have heat-sensitive grooves between the labial scales under their nostrils that locate warm-blooded prey, even in total darkness. The emerald tree boa (*Corallus caninus*) and the green tree boa (*Boa canina*) are especially well-adapted for an arboreal life, with green, white-marked coloration making them almost invisible in the trees.

The West Indian boas (*Epicrates*) are common on those islands lacking the snake-eating Indian mongoose, which was introduced on several islands in the 1800s to control the **rats** in the **sugarcane** fields. However, boas are almost extinct on most islands where the mongoose

An anaconda in Venezuela, swimming. *Photograph by Stephen Green-Armytage. Stock Market. Reproduced by permission.*

occurs. As is the case with so many attempts to introduce exotic animals for a particular purpose, the mongoose did not do the job that was expected. Instead of eating rats, the mongoose preferred chickens. and also ate ground-nesting birds and terrestrial lizards and snakes, some of which have been driven to **extinction**.

The distribution of the true boas is rather odd. Although most of them inhabit tropical America, there is one group of three species, the Pacific boas (*Candoia*), that are found on the other side of the Pacific Ocean, in the Fiji islands and on other islands north and east of New Guinea. Although they appear to be closely related to the American species, these Pacific boas live in an area more than 4,000 mi (6,440 km) away. It is not easy to explain this disjunct distribution, but a parallel case is found in the Fiji **iguanas** (*Brachylophus*), whose closest relatives are also in tropical America.

Sandboas (family Erycidae)

Most sandboas are relatively small snakes, less than 3 ft (1 m) in length, currently found in southern Asia and northern **Africa**, and in western **North America**. In most ways the sandboas are very much like the true boas of South America, but because most of them live in relatively treeless areas, they are more adapted to burrowing in the **sand** than to climbing. Like the South American boas, they feed on small animals, such as lizards and rodents, which they kill by constriction. Because of their subterranean habits, however, the sandboas tends to have small, compact heads that can be pushed through the **soil**, and short, stubby tails that can act as "pushers." Their tail vertebrae are specialized and can be recognized in the fossil members of this family. Such sandboa fossils are known from many localities in western **Europe** and eastern North America that are very distant from the areas sandboas currently inhabit. The two American sandboas are the rubber boa of the dry pine **forests** of the western states of Washington and Oregon, and the rosy boa of the southwestern **desert** regions. The latter is a handsome snake that is a favorite of pet owners.

As in the true boas, there is a strange situation in sandboa distribution and relationship. On the island of Madagascar, off the southeast coast of Africa, there are two boas that resemble those of South America. One of these is specialized as a tree boa, the other is so similar in appearance to the South American boa constrictor that some experts have placed it in the same genus. Recent research, however, suggests that despite their superficial similarities the Madagascan boas are related to the sandboas, and probably represent an ancient division of that group.

Resources

Books

Cogger, Harold G., David Kirshner, and Richard Zweifel. *Encyclopedia of Reptiles and Amphibians.* 2nd ed. San Diego, CA: Academic Press, 1998.

Ross, R.A., and G. Marzec. *The Reproductive Husbandry of Pythons and Boas.* Stanford, CA: Institute for Herpetological Research, 1990.

Tolson, P.J., and R.W. Henderson. *The Natural History of West Indian Boas.* Excelsior, MN: R & A Publ., 1994.

Herndon G. Dowling

Bobcat *see* **Cats**

Bohr model

The Bohr model of atomic structure was developed by Danish physicist and Nobel laureate Niels Bohr (1885–1962). Published in 1913, Bohr's model improved the classical **atomic models** of physicists J. J. Thomson and Ernest Rutherford by incorporating quantum theory. While working on his doctoral dissertation at Copenhagen University, Bohr studied physicist Max Planck's quantum theory of **radiation**. After graduation, Bohr worked in England with Thomson and subsequently with Rutherford. During this time Bohr developed his model of atomic structure.

Before Bohr, the classical model of the atom was similar to the Copernican model of the **solar system** where, just as planets **orbit** the **Sun**, electrically **negative** electrons moved in orbits about a relatively massive, positively charged nucleus. The classical model of the atom allowed electrons to orbit at any **distance** from the nucleus. This predicted that when, for example, a **hydrogen** atom was heated, it should produce a continuous **spectrum** of colors as it cooled because its **electron**, moved away from the nucleus by the **heat energy**, would gradually give up that energy as it spiraled back

closer to the nucleus. Spectroscopic experiments, however, showed that hydrogen **atoms** produced only certain colors when heated. In addition, physicist James Clark Maxwell's influential studies on electromagnetic radiation (**light**) predicted that an electron orbiting around the nucleus according to Newton's laws would continuously lose energy and eventually fall into the nucleus. To account for the observed properties of hydrogen, Bohr proposed that electrons existed only in certain orbits and that, instead of traveling between orbits, electrons made instantaneous quantum leaps or jumps between allowed orbits.

In the Bohr model, the most stable, lowest energy level is found in the innermost orbit. This first orbital forms a shell around the nucleus and is assigned a principal **quantum number** (n) of n=1. Additional orbital shells are assigned values n=2, n=3, n=4, etc. The orbital shells are not spaced at equal distances from the nucleus, and the radius of each shell increases rapidly as the square of n. Increasing numbers of electrons can fit into these orbital shells according to the formula $2n^2$. The first shell can hold up to two electrons, the second shell (n=2) up to eight electrons, and the third shell (n=3) up to 18 electrons. Subshells or suborbitals (designated *s,p,d,* and *f*) with differing shapes and orientations allow each element a unique electron configuration.

As electrons move farther away from the nucleus, they gain potential energy and become less stable. Atoms with electrons in their lowest energy orbits are in a "ground" state, and those with electrons jumped to higher energy orbits are in an "excited" state. Atoms may acquire energy that excites electrons by **random thermal collisions, collisions with subatomic particles**, or by absorbing a **photon**. Of all the photons (quantum packets of light energy) that an atom can absorb, only those having an energy equal to the energy difference between allowed electron orbits will be absorbed. Atoms give up excess internal energy by giving off photons as electrons return to lower energy (inner) orbits.

The electron quantum leaps between orbits proposed by the Bohr model accounted for Plank's observations that atoms emit or absorb electromagnetic radiation only in certain units called quanta. Bohr's model also explained many important properties of the **photoelectric effect** described by Albert Einstein.

According to the Bohr model, when an electron is excited by energy it jumps from its ground state to an excited state (i.e., a higher energy orbital). The excited atom can then emit energy only in certain (quantized) amounts as its electrons jump back to lower energy or-

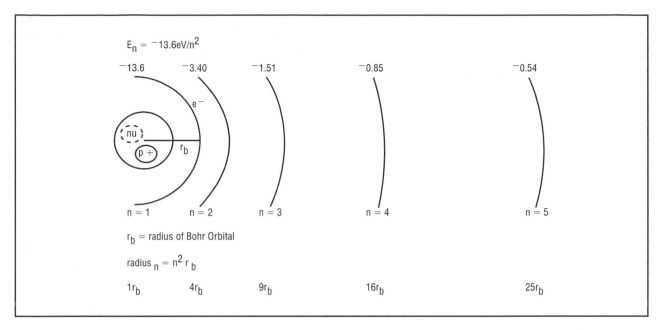

$$E_n = {}^-13.6eV/n^2$$

${}^-13.6$ ${}^-3.40$ ${}^-1.51$ ${}^-0.85$ ${}^-0.54$

e⁻

nu

p + r_b

n = 1 n = 2 n = 3 n = 4 n = 5

r_b = radius of Bohr Orbital

radius $_n$ = $n^2 r_b$

$1r_b$ $4r_b$ $9r_b$ $16r_b$ $25r_b$

The Bohr model accounted for quantum phenomena by encountered in chemistry and physics by restricting a hydrogen electron's transitions to instantaneous changes between allowed orbits (energy levels). *Illustration by Argosy. The Gale Group.*

bits located closer to the nucleus. This excess energy is emitted in quanta of electromagnetic radiation (photons of light) that have exactly same energy as the difference in energy between the orbits jumped by the electron. For hydrogen, when an electron returns to the second orbital (n=2) it emits a photon with energy that corresponds to a particular **color** or spectral line found in the Balmer series of lines located in the visible portion of the electromagnetic (light) spectrum. The particular color in the series depends on the higher orbital from which the electron jumped. When the electron returns all the way to the innermost orbital (n=1), the photon emitted has more energy and forms a line in the Lyman series found in the higher energy, ultraviolet portion of the spectrum. When the electron returns to the third quantum shell (n=3), it retains more energy and, therefore, the photon emitted is correspondingly lower in energy and forms a line in the Paschen series found in the lower energy, infrared portion of the spectrum.

Because electrons are moving charged particles, they also generate a magnetic field. Just as an ampere is a unit of **electric current**, a magneton is a unit of magnetic **dipole** moment. The orbital magnetic moment for hydrogen atom is called the Bohr magneton.

Bohr's work earned a Nobel Prize in 1922. Subsequently, more mathematically complex models based on the work of French physicist Louis Victor de Broglie (1892–1987) and Austrian physicist Erwin Schrödinger (1887–1961) that depicted the particle and wave nature of electrons proved more useful to describe atoms with

more than one electron. The **standard model** incorporating quark particles further refines the Bohr model. Regardless, Bohr's model remains fundamental to the study of **chemistry**, especially the **valence** shell concept used to predict an element's reactive properties.

The Bohr model remains a landmark in scientific thought that poses profound questions for scientists and philosophers. The concept that electrons make quantum leaps from one orbit to another, as opposed to simply moving between orbits, seems counter-intuitive, that is, outside the human experience with nature. Bohr said, "Anyone who is not shocked by quantum theory has not understood it." Like much of quantum theory, the proofs of how nature works at the atomic level are mathematical.

See also Atomic number; Atomic spectroscopy; Atomic theory; Atomic weight; Electromagnetic field; Electromagnetic induction; Electromagnetic spectrum; Quantum mechanics.

Resources

Books

Bohr, Niels. *The Unity of Knowledge.* New York: Doubleday & Co., 1955.

Feynman, Richard P. *QED: The Strange Theory of Light and Matter.* New Jersey: Princeton University Press, 1985.

Feynman, Richard P. *The Character of Physical Law.* MIT Press, 1965.

Griffiths, Robert B. *Consistent Quantum Theory.* Cambridge, MA: Harvard University Press, 2002.

Omnes, Roland. *Understanding Quantum Mechanics*. Princeton, NJ: Princeton University Press, 1999.

Pasachoff, Naomi. *Niels Bohr: Physicist and Humanitarian*. Enslow Publishers, 2003.

Silverman, Mark. *Probing the Atom*. Princeton, NJ: Princeton University Press, 2000.

Other

Kansas State University. "Visual Quantum Mechanics." (February 5, 2003).<http:// phys.educ.ksu.edu/>.

K. Lee Lerner

Boiling *see* **States of matter**

Boiling point

The boiling point of a liquid substance is the **temperature** at which the **vapor pressure** of the liquid equals the external **pressure** on the liquid. Vapor bubbles form in the liquid, rise to the surface and burst, causing the liquid to boil.

At room temperature, in a closed system, there is an equilibrium between the liquid and its vapor phase. For example, if a glass of **water** is left open, the water will eventually evaporate, although it may take a few days. On the other hand, if the glass is kept covered, when the cover is removed, there is a large amount of water on the bottom of the cover. The molecules of the liquid in the glass move into the vapor phase, but when they encounter the cold surface of the cover, they condense back into liquid form and fall back into the liquid. When the **rate** of movement of molecules into the gas phase from the liquid equals the rate of movement of molecules into the liquid phase from the gas phase, the two phases are in equilibrium.

When the temperature is raised, more molecules have enough **energy** to enter the gas phase. If the glass were left uncovered, the water would disappear more quickly. If the glass was kept covered, a new equilibrium would be established more quickly. As the temperature increases, more molecules enter the gas phase and the vapor pressure of the liquid increases.

At some point in time, as the temperature increases, the vapor pressure of the liquid will increase to the point where it equals the external or **atmospheric pressure**. At this point, bubbles begin to form and rise in the liquid and it is said to be boiling. Before this temperature, bubbles cannot form because the external pressure is greater than the pressure in the bubble and it collapses. The temperature at which the external pressure and the

vapor pressure are equal is called the normal boiling point. For water, this temperature is 212°F (100°C). In general, a liquid with high vapor pressures and a low normal boiling point is said to be volatile. Such liquids usually have strong smells. Liquids with low vapor pressures and high normal boiling points are non-volatile and have little or no odor.

If the external pressure is less than one atmosphere, liquids will boil at lower temperatures than their normal boiling points. At high elevations, the atmospheric pressure is much lower than one atmosphere. At the top of Mount Everest, where the atmospheric pressure is about 5 psi (260 mm Hg), the boiling point of water is only 160°F (71°C). At such high elevations, it is often necessary to follow special instructions for cooking and baking, as the water temperature is not high enough to cook food. Conversely, if the atmospheric pressure is greater than one atmosphere, liquids will boil at higher temperatures than their normal boiling points. We can use this to our advantage. In a pressure cooker, we increase the pressure so that it is greater than one atmosphere. As a result, water boils at a higher temperature and food cooks faster. We can also raise the boiling point of a liquid by adding a non-volatile solute to it. For example, adding **salt** to water raises its boiling point.

Rashmi Venkateswaran

Bond energy

Bond **energy** is the strength of a **chemical bond** between **atoms**, expressed as the amount of energy required to break it apart. It is as if the bonded atoms were glued together: the stronger the glue is, the more energy would be needed to break them apart. A higher bond energy, therefore, means a stronger bond.

Bond energies are usually expressed in kilojoules per **mole** (kJ/mol): the number of kilojoules of energy that it would take to break apart exactly one mole of those bonds is 6.02×10^{23}. There are several kinds of "glues," or attractions, by which atoms and molecules can stick together. Table 1 shows the approximate ranges of their strengths, from the strongest to the weakest.

Notice that ionic bonds are stronger than covalent bonds. Among covalent bonds, triple bonds are stronger than double bonds and double are stronger than single bonds. **Hydrogen** bonds are weaker than all, but they play a big role in determining the properties of important compounds such as **proteins** and **water**.

TABLE 1. TYPICAL BOND ENERGIES

Type of bond or attraction	Range of bond energies, kJ/mol
Ionic bonds	700-4000
Covalent triple bonds	800-1000
Covalent double bonds	500-700
Covalent single bonds	200-500
Dipole attractions between molecules	40-400
Hydrogen bonds	10-40

TABLE 2. AVERAGE BOND ENERGIES OF COMMON BONDS

Bond	Bond energy, kJ/mol
C–C	347
C=C	615
C≡C	812
C–O	360
C=O	728
F–F	158
Cl–Cl	244
C–H	414
H–H	436
H–O	464
O=O	498

Bonds of the same type can vary quite a bit in their strengths. The bond energies of several specific bonds are shown in this table.

Bond energies between certain pairs of atoms vary somewhat, depending on the particular **molecule** they are part of, because adjacent atoms can affect their bonding slightly. The values in Table II are average bond energies for the listed bonds.

See also Dipole.

Robert L. Wolke

Bony fish

Bony **fish** (Osteichthyes) are distinguished from other fish **species** that have a cartilaginous skeleton (Chondrichthyes—sharks, **rays** and **chimaeras**, for example) by the presence of true bone—a mixture of **calcium** phosphates and carbonates—in their skeletons. Other differences between the two groups are modifications in the structure and arrangement of the scales and fins and the presence of more specialized teeth in bony fish. When feeding, bony fish display a far wider range of adaptations than cartilaginous species: the former may be either carnivorous (like most cartilaginous species), plant-eating, or both. Combined, these features have helped them to exploit a much wider range of feeding and living habitats.

Fish breathe and feed in order to obtain sufficient **energy** to meet their daily physical requirements. Much of this energy is required for swimming, which can range from simple short distances to much greater seasonal migrations. However, considerable amounts of energy are also required for finding food and mates, as well as for avoiding predators. To conserve energy, bony fish have evolved a special swim bladder which is a gas-filled chamber that provides buoyancy and helps keeps them weightless in the **water** column. Through this system they may remain at the same level of water for several hours without expending too much valuable energy.

With few exceptions, all bony fish require a constant source of **oxygen** for **respiration**. Dissolved oxygen is freely available in **salt** and **freshwater** and to extract this, fish pass the water through specialised gill chambers that are richly supplied with **blood** vessels. As the water passes over the highly convoluted surface of the gills, the oxygen passes across the thin membranes and enters the red hemoglobin cells of the blood stream.

The classification of bony fish is complex and outside the scope of this present account. Basically there are two main groups of bony fish: the Crossopterygii and Actinopterygii. The latter contains some additional small groups of distinct fish such as the Polypteridae (bichirs), Acipenseridae (**sturgeons**), Polyodontidae (paddle fish) and Lepisosteidae (gar or pikes). It also includes the teleost fish, which are by far the most numerous of all fishes, with more than 20,000 species identified from a wide range of habitats—aquatic, marine, and terrestrial. Although these species vary considerably in size, appearance and structure, they are all much lighter than primitive species, largely through the loss of heavy body armor and thickened scales. The smallest known teleost fish is the Philippine goby (*Pandaka pygmaea*) which reaches a length of just 0.5 in (12 mm); the largest is the arapamia or pirarucu (*Arapamia gigas*) of the Amazonian waterways, which has been known to measure 16 ft (5 m).

In addition to the wide range of modern bony fish, a few primitive representatives still survive and, for taxonomic purposes, have been grouped together in the Crossopterygii. One of these is the **coelacanth** (*Latimera chalumnae*), a large blue-gray fish that may reach a length of 6.5 ft (2 m). Known only from the deep **ocean** trenches off the tiny Comoros Archipelago around northwest Madagascar, this sluggish species lives in complete darkness and preys on other fish. Many of its features are similar to fossil species, the most notable of which are its arrangement of fins. The rays of the second dorsal fin, the anal fin, and the paired fins rest on a muscular, scale-covered lobe, while the powerful tail fin is symmetrical in appearance. Almost nothing is known about the **ecology** of this species. A recent discovery showed that coelacanths actually give **birth** to live young.

Other unusual species of this same group are the **lungfish** of **Africa**, **Australia,** and **South America**, the only living representatives of a widespread group of fish that lived on **Earth** some 350 million years ago. In some of these, the dorsal, anal, and caudal fins are fused and modified to form a continuous median fin. Some species come to the surface to gulp in air, while others have developed a gill system that enables them to breathe when submerged. During periods of dry **weather**, the African and South American species dig deeply into the mud at the base of lakes and swamps and envelop themselves in a thick coat of mucus. As this dries out it provides a protective covering for the fish, enabling it to survive periods of **drought** in a state of aestivation. When the rains return, the mucus coating is softened and the fish reemerges.

Features such as these enabled some species to withstand periods of adverse weather or avoid excessive predation through the development of toughened skins that were often reinforced with bulky scales. As the world's climate changed and new species continued to evolve, most of these primitive features were lost and an explosion of new life forms spread throughout the seas and freshwater ecosystems. Witness to this is the present staggering diversity and numbers of bony fish that occur in almost every aquatic **habitat** on Earth, from the warm tropical waters to the frozen seas of the poles and from the margins of the **tides** to the deepest oceanic trenches.

See also Cartilaginous fish; Fish.

Boobies and gannets

Boobies and gannets are nine **species** of marine **birds** that make up the family Sulidae, in the order Pelecaniformes, which also includes the **pelicans**, **cormorants**, anhingas, **tropic birds**, and **frigate birds**.

Boobies and gannets have a narrow, cigar-shaped body, a longish, pointed tail, and long, narrow wings. Their feet are fully webbed, and are used in swimming. The beak is strong, pointed, has a serrated edge for gripping slippery **prey**, and is brightly colored in some species. Unlike some of the other groups in the order Pelecaniformes, boobies and gannets have fully waterproof plumage.

Gannets and boobies feed on **fish**, which they find by flying over the surface of the **ocean** at an altitude of up to 98 ft (30 m). These birds then catch their prey by spectacular, head-long, angled-winged plunges into the surface of the sea, seizing their quarry in their beak, and swallowing it underwater. During the breeding season, gannets and boobies are found in near-coastal waters. In their non-breeding season, however, these birds may occur far out to sea. Almost all species of gannets and boobies are colonial nesters.

Gannets (*Morus* spp.) are birds of temperate and subarctic oceans, and they breed in colonies on rocky cliffs and ledges. Both birds of a mated pair incubate their single egg, which they cover with their webbed feet before snuggling down to brood. After the chick develops its flight feathers and is ready to fledge, it is abandoned by its parents. It soon leaps into the sea from its cliff-top nest and begins to fish for itself.

The six species of boobies (*Sula* spp.) are all tropical and subtropical birds. Boobies breed in nests built on near-shore shrubs, or on coastal cliffs.

Species of gannets

The northern gannet (*Morus bassana*) breeds in north-temperate and subarctic waters on both sides of the Atlantic Ocean. In **North America**, the largest colonies of these birds occur at Cape Saint Mary's on Newfoundland and on Bonaventure Island in the Gulf of the Saint Lawrence River. There are another four smaller colonies of northern gannets in the western Atlantic Ocean and another 28 in the eastern Atlantic.

Adult northern gannets have a white body, with black wing-tips. The head is a bright lemon-yellow. During the first year after **birth** gannets are a dark-brown **color**, while older sub-adults have a dirty-white plumage and lack the yellow head of the sexually mature adults. The tail of gannets is pointed, as is the profile of their head, giving the bird a double-ended shape in flight.

The populations of northern gannets in some of their breeding colonies can be quite large. These birds are aggressively territorial, and their nests are therefore spaced at about twice the **distance** that a sharp beak can be thrust towards a neighbor. Other displays involve birds engaging in ritualized posturings to impress their neighbors, or to infatuate a potential mate.

In healthy colonies, all of the suitable **space** may be covered with nests. At Cape Saint Mary's, population growth in recent decades has resulted in all of the prime nesting **habitat** on cliffs to be fully utilized. This has forced many birds to nest on adjacent coastal meadows, an accessible habitat in which they are vulnerable to land-borne predators.

After their breeding season, northern gannets occur widely in waters of the continental shelves. During the winter, gannets range as far south as the northern Gulf of Mexico and the southeastern United States.

Other species of gannets include the Cape gannet (*Morus capensis*) of South **Africa**, and the Australian gannet (*M. serrator*) of **Australia**. These species are rather similar to the northern gannet, and some taxonomists consider all of these taxa to be subspecies of *Morus bassana*.

Species of boobies

Three species of boobies are relatively widespread in tropical waters. The brown booby (*Sula leucogaster*)

Blue-footed boobies in the Galápagos Islands. *Photograph by Anthony Wolff. PhototakeNYC. Reproduced by permission.*

is the most common species, breeding in all of the tropical oceans. This species has dark-brown upper parts and breast, and a white belly. Male birds have a dark-blue face, while that of females is yellow. Immature birds are more uniformly brown.

The blue-faced or masked booby (*Sula dactylatra*) is the largest species. This species breeds in the tropics of the Pacific and Indian Oceans. The blue-faced booby has a mostly white body, but the flight feathers are all black, resulting in a black stripe running the length of the back of the wings. The "mask" is an area of black feathers just behind the beak.

The red-footed booby (*Sula sula*) is a species of the tropical Pacific, Indian, and Atlantic Oceans. This species is named after its bright-red feet and legs. The red-footed booby nests in shrubs and trees.

Other, less widespread species are the Peruvian booby (*Sula variegata*) of offshore islands and coastal headlands of Peru, the blue-footed booby (*S. nebouxii*) of western Mexican and Central and South American waters, and Abbott's booby (*S. abbotti*), which only occurs in the vicinities of Assumption and Christmas Islands in the Indian Ocean. Peruvian boobies can nest in huge colonies, which can contain as many as one million pairs of birds.

Boobies do not breed in North America, but several species are regular visitors to coastal waters during their non-breeding season. The blue-faced booby occurs most frequently in the vicinity of the Dry Tortugas off extreme southern Florida, and also in the Caribbean Sea, the Gulf of Mexico, and Baha, California, as well as farther south of all of those places. The brown booby and blue-footed booby are also occasional visitors to the extreme southern United States.

Boobies, gannets, and people

Guano is a commercially important product obtained by digging the surface of the huge colonies of Peruvian boobies and other seabirds off northern **South America**, and at colonies of Cape Gannets off South Africa. Guano is a natural, phosphorus-rich compound derived from the excrement of seabirds, and is used as a fertilizer.

For many years, gannets, and to a much lesser degree boobies, were considered to be serious competitors with humans for commercially important marine fish. For this reason, gannets were often killed, and only a few decades ago their numbers were perilously small. This sort of indiscriminate killing is not much of a problem anymore, except in a few remote places.

In some regions, gannets and boobies may be killed for their meat and feathers, and where they are accessible, their eggs may be collected for eating.

Boobies and gannets are also vulnerable to collapses in the populations of the fish that they feed upon. For example, Peruvian boobies and other seabirds have suffered precipitous population declines when their most important prey of anchovies collapsed as a result of oceanographic changes associated with El Niño. El Niño is a warm-water phenomenon that impedes nutrient cycling, greatly reducing the productivity of **phytoplankton**, and ultimately, causing a collapse of fish stocks.

Since the beginning of the 1990s, there has also been a collapse of many fish stocks in coastal waters off eastern Canada. The reasons for this ecological change are not known for certain, but the leading hypotheses include the effects of overfishing and climate change. The collapse of the fisheries of the northwest Atlantic has led to severe economic hardship for many people who are dependent on that natural resource for their livelihood. However, there have also been severe effects on northern gannets and other seabirds, which depend on those fish stocks as a source of food, particularly when they are raising their young. Consequently, these birds have experienced unsuccessful reproduction, and this may pose a threat to the longer-term health of their populations in that region.

Resources

Books

Ehrlich, P.R., D.S. Dobkin, and D. Wheye. *Birds in Jeopardy.* Stanford, CA: Stanford University Press, 1992.

Forshaw, Joseph. *Encyclopedia of Birds.* New York: Academic Press, 1998.

Harrison, P. *Seabirds: An Identification Guide.* U.K.: Croom Helm, Beckenham, 1983.

Nelson, B. *The Sulidae: Gannets to Boobies.* London: Harrell Bokks, 1978.

Bill Freedman

Boolean algebra

Boolean **algebra** is often referred to as the algebra of logic, because the English mathematician George Boole, who is largely responsible for its beginnings, was the first to apply algebraic techniques to logical methodology. Boole showed that logical propositions and their connectives could be expressed in the language of **set theory**. Thus, Boolean algebra is also the algebra of sets. Algebra, in general, is the language of **mathematics**, together with the rules for manipulating that language. Beginning with the members of a specific set (called the universal set), together with one or more binary operations defined on that set, procedures are derived for manipulating the members of the set using the defined operations, and combinations of those operations. Both the language and the rules of manipulation vary, depending on the properties of elements in the universal set. For instance, the algebra of **real numbers** differs from the algebra of **complex numbers**, because real numbers and complex numbers are defined differently, leading to differing definitions for the binary operations of **addition** and **multiplication**, and resulting in different rules for manipulating the two types of numbers. Boolean algebra consists of the rules for manipulating the subsets of any universal set, independent of the particular properties associated with individual members of that set. It depends, instead, on the properties of sets. The universal set may be any set, including the set of real numbers or the set of complex numbers, because the elements of interest, in Boolean algebra, are not the individual members of the universal set, but all possible subsets of the universal set.

Properties of sets

A set is a collection of objects, called members or elements. The members of a set can be physical objects, such as people, stars, or red roses, or they can be abstract objects, such as ideas, numbers, or even other sets. A set is referred to as the universal set (usually called I) if it contains all the elements under consideration. A set, S, not equal to I, is called a proper subset of I, if every element of S is contained in I. This is written and read "S is contained in I." (see Figure 1)

If S equals I, then S is called an improper subset of I, that is, I is an improper subset of itself (note that two sets are equal if and only if they both contain exactly the same elements). The special symbol is given to the set with no elements, called the empty set or null set. The null set is a subset of every set.

When dealing with sets there are three important operations. Two of these operations are binary (that is, they involve combining sets two at a time), and the third involves only one set at a time. The two binary operations

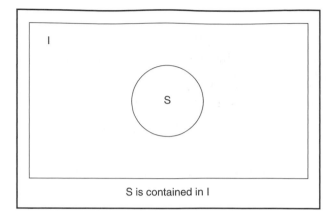

S is contained in I

Figure 1. *Illustration by Hans & Cassidy. Courtesy of Gale Group.*

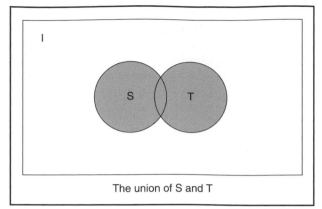

The union of S and T

Figure 2. *Illustration by Hans & Cassidy. Courtesy of Gale Group.*

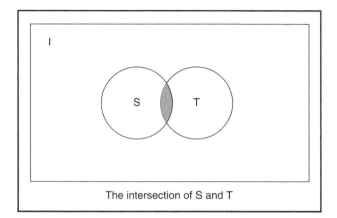

The intersection of S and T

Figure 3. *Illustration by Hans & Cassidy. Courtesy of Gale Group.*

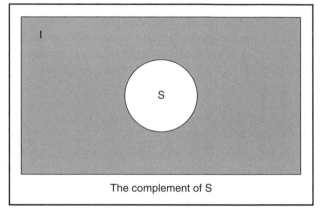

The complement of S

Figure 4. *Illustration by Hans & Cassidy. Courtesy of Gale Group.*

are union and intersection. The third operation is complementation. The union of two sets S and T is the collection of those members that belong to either S or T or both. (see Figure 2)

The intersection of the sets S and T is the collection of those members that belong to both S and T. (see Figure 3)

The complement of a subset, S, is that part of I not contained in S, and is written S'. (see Figure 4)

Properties of Boolean algebra

The properties of Boolean algebra can be summarized in four basic rules.

(1) Both binary operations have the property of commutativity, that is, order doesn't matter.

$$S \cap T = T \cap S, \text{ and } S \cup T = T \cup S.$$

(2) Each binary operation has an **identity element** associated with it. The universal set is the identity element for the operation of intersection, and the null set is the identity element for the operation of union.

$$S \cap I = S, \text{ and } S \cup \emptyset = S.$$

(3) each operation is distributive over the other.

$$S \cup (T \cap V) = (S \cup T) \cap (S \cup V), \text{ and}$$
$$S \cap (T \cup V) = (S \cap T) \cup (S \cap V).$$

This differs from the algebra of real numbers, for which multiplication is distributive over addition, $a(b+c) = ab + ac$, but addition is not distributive over multiplication, $a+(bc)$ not equal $(a+b)(a+c)$.

(4) each element has associated with it a second element, such that the union or intersection of the two results in the identity element of the other operation.

$$A \cup A' = I, \text{ and } A \cap A' = \emptyset.$$

This also differs from the algebra of real numbers. Each real number has two others associated with it, such that its sum with one of them is the identity element for addition, and its product with the other is the identity element for multiplication. That is, $a + (-a) = 0$, and $a(1/a) = 1$.

KEY TERMS

. . . .

Binary operation—A binary operation is a method of combining the elements of a set, two at a time, in such a way that their combination is also a member of the set.

Complement—The complement of a set, S, written S', is the set containing those members of the universal set that are not contained in S.

Element—Any member of a set. An object in a set.

Intersection—The intersection of two sets is itself a set comprised of all the elements common to both sets.

Set—A set is a collection of things called members or elements of the set. In mathematics, the members of a set will often be numbers.

Set theory—Set theory is the study of the properties of sets and subsets, especially those properties that are independent of the particular elements in a set.

Subset—A set, S, is called a subset of another set, I, if every member of S is contained in I.

Union—The union of two sets is the set that contains all the elements found in one or the other of the two sets.

Universal set—The universal set is the set containing all the elements being considered.

Applications

The usefulness of Boolean algebra comes from the fact that its rules can be shown to apply to logical statements. A logical statement, or proposition, can either be true or false, just as an equation with real numbers can be true or false depending on the value of the **variable**. In Boolean algebra, however, variables do not represent the values that make a statement true, instead they represent the truth or falsity of the statement. That is, a Boolean variable can only have one of two values. In the context of **symbolic logic** these values are true and false. Boolean algebra is also extremely useful in the field of electrical **engineering**. In particular, by taking the variables to represent values of on and off (or 0 and 1), Boolean algebra is used to design and analyze digital switching circuitry, such as that found in personal computers, pocket calculators, cd players, cellular telephones, and a host of other electronic products.

See also Computer, digital.

Resources

Books

Christian, Robert R. *Introduction to Logic and Sets.* Waltham, MA: Blaisdell Publishing Co., 1965.

Garfunkel, Soloman A., ed. *For All Practical Purposes: Introduction to Contemporary Mathematics.* New York: W. H. Freeman, 1988.

Hoernes, Gerhard E., and Melvin F. Heilweil. *Boolean Algebra and Logic Design.* New York: McGraw Hill, 1964.

Ryan, Ray, and Lisa A. Doyle. *Basic Digital Electronics, 2nd ed.* Blue Ridge Summit, PA: Tab Books, 1990.

J.R. Maddocks

Boric acid

Boric acid, also known as boracic acid and arthoboric acid, is a very weak acid with the formula H_3BO_3, often used as a mild antiseptic. Chemically, it acts as a tribasic acid—an acid that can dissociate successively to produce three **hydrogen** ions in **solution**. However, because it dissociates to such a small extent, it is a very weak acid that is actually used in **water** solution as an **eye** wash. Pure boric acid is a colorless, odorless, white powder or transparent crystals that melt at about 340°F (171°C). Boric acid loses water as it is heated, changing first into metaboric acid (HBO_2) and then into pyroboric acid ($H_2B_4O_7$). The three acids can be thought of as hydrates of boric oxide (B_2O_3). Orthoboric acid is fairly soluble in water (especially hot water), **alcohol**, and glycerine.

Boric acid has a wide variety of industrial applications. It is used in the manufacture of heat-resistant borosilicate **glass** and other **ceramics**, such as crockery, porcelains, enamels, and artificial gemstones. It also used in waterproofing **wood** and fireproofing **textiles**. It also finds application as an insecticide for **cockroaches** and black carpet **beetles** and as an **fungicide** on citrus **fruits**.

Its use in the last of these applications is carefully monitored, however, because of the compound's toxicity. When swallowed, boric acid can cause nausea, vomiting, diarrhea, and other intestinal problems. In large doses, it can cause **coma** and death. The toxic level of boric acid in infants can be less than 0.2 oz (5 g) and in adults, from 0.2 oz (5 g) to 0.7 oz (20 g).

Boring machine *see* **Machine tools**

Boron *see* **Element, chemical**

Bosons *see* **Subatomic particles**

Botany

Botany is the study of plants. It is one of the major fields of **biology**, together with zoology (the study of animals) and microbiology (the study of **bacteria** and viruses). Specializations within the field of botany include the study of mosses, **algae**, **lichens**, **ferns**, and **fungi**. Other specialties in botany include **plant** physiology, the study of the vital processes of plants, such as **photosynthesis**, **respiration**, and plant **nutrition**. Biochemists study the effects of **soil**, **temperature**, and **light** on plants, while plant morphologists study of the **evolution** and development of leaves, roots, and stems with a focus on the tissues at the tips of stems where the cells have the ability to divide.

Plant **pathology** studies the causes and control of **plant diseases**. Pathologists may work with a specific group of plants, such as forest trees, vegetable **crops**, grain, or ornamental plants, and they may concentrate on the interactions between host plants and **pathogens**, the carriers of **disease**. Economic botanists study the economic impact of plants as they relate to human needs for food, clothing, and shelter, while plant geneticists investigate the structure and behavior of genes in plants and plant heredity in order to develop crops that are resistant to diseases and **pests**. **Paleobotany** deals with the biology and evolution of plants by studying the fossil record in order to reconstruct the 600 million year history of plant life on this **planet**.

The relationship between plants and animals is one of interdependence. Without the evolution of plants, animals would not have been able to subsist. Animals, in turn, contribute to plant distribution, plant **pollination**, and every other aspect of plant growth and development. It is through this interdependence that plants continue to adapt and change. Human intervention in the cultivation of plants has contributed equally to plant development. Today, the study of botany is only one aspect of **ecology**, the study of the environment. Plant ecologists are concerned with the effects of the environment on plants.

History of botany

Aristotle and Theophrastus, living in ancient Greece about the fourth century B.C., were both involved in identifying plants and describing them. Theophrastus is called the "father of botany," because of his two surviving works on plant studies. While Aristotle also wrote about plants, he received more recognition for his studies of animals.

The early study of plants was not limited to Western cultures. The Chinese developed the study of botany along lines similar to the ancient Greeks at about the same time. In A.D. 60, another Greek, Dioscorides, wrote *De Materia Medica*, a work that described a thousand medicines, 60% of which came from plants. It remained the guidebook on medicines in the Western world for 1,500 years until the compound **microscope** was invented in the late sixteenth century, opening the way to the careful study of plant **anatomy**.

During the seventeenth century progress was made in experimenting with plants. Johannes van Helmont measured the uptake of **water** in a **tree** during the 1640s, and in 1727 Stephen Hales, an Englishman who is credited with establishing plant **physiology** as a science, published his experiments dealing with the nutrition and respiration of plants in a work entitled *Vegetable Staticks*. He developed techniques to measure area, **volume**, **mass**, **pressure**, gravity, and temperature in plants. In the latter part of the eighteenth century, Joseph Priestley laid the foundation for the chemical analysis of plant **metabolism**.

During the nineteenth century advances were made in the study of plant diseases because of the **potato** blight that killed potato crops in Ireland in the 1840s, an event that led to a mass migration of Irish to America. The study of plant diseases developed rapidly after this event. When the work in **genetics** by Gregor Mendel, an Austrian monk, was applied after 1900 to **plant breeding**, the development of modern plant genetics began. During the early part of the nineteenth century, progress in the study of plant fossils was made, and ecology began to develop as a science in the late nineteenth and early twentieth centuries.

Technology has helped specialists in botany to see and understand the three-dimensional nature of cells, and **genetic engineering** of plants has improved agricultural output. The study of plants continues as botanists try to both understand the structure, behavior, and cellular activities of plants in order to develop better crops, find new medicines, and explore ways of maintaining an ecological balance on **Earth** to continue to sustain both plant and **animal** life.

See also Taxonomy.

Resources

Books

Campbell, N., J. Reece, and L. Mitchell. *Biology*. 5th ed. Menlo Park: Benjamin Cummings, Inc. 2000.

Evans, Howard Ensign. *Pioneer Naturalists*. New York: Holt, 1993.

Heiser, Charles B. *Of Plants and People*. Norman: University of Oklahoma Press, 1985.

Morton, A.G. *History of Botanical Science*. London: Academic Press, 1981.

Roth, Charles E. *The Plant Observer's Guidebook*. Englewood Cliffs, NJ: Prentice-Hall, 1984.

Vita Richman

Botulism

Botulism is an extremely serious **disease** caused by the bacterium *Clostridium botulinum*. *C. botulinum* release one of the most potent toxins known—one gram of botulinum toxin theoretically can kill one million people. The toxin is swift-acting. It kills by binding to nerve cells, thereby causing paralysis of the muscles used in breathing.

First coined in the 1870s, the term botulism comes from the Latin word for sausage, *botulus*, since botulism used to be associated with eating sausage. Although botulism is still commonly associated with food **contamination** in the United States, it is more likely to occur from eating plants, not meat.

The canning connection

Plant foods associated with botulism are canned **vegetables**. In a typical scenario, vegetables contaminated with *C. botulinum* from the **soil** are not washed adequately and subjected to temperatures inadequate for killing the **bacteria**. As the vegetable sits on the shelf, botulinum toxin is released into the can. Because the toxin is odorless and colorless, the unsuspecting person eats the contaminated vegetable. Vegetables having neutral **pH** are most likely to harbor botulinum bacteria. Examples are beans, peas, and corn. Canned vegetables with low pH, such as canned tomatoes, are resistant to the growth of *C. botulinum* because of the acidic environment.

Fortunately, this scenario is rare as modern commercial canning techniques have virtually eliminated the risk of botulism. However, many home canners do not know the proper prevention techniques. About 10 outbreaks of botulism still occur each year in the United States. Most of these outbreaks are traced to **food poisoning**. Less frequently, botulism in humans stems from wound infections, or even more rarely a gastrointestinal **infection** of newborns. In animals, botulism can be traced to the eating of contaminated **animal** carcasses, or hay or grass that has been contaminated by a dead, toxic animal. Animals that characteristically feed on dead animals, such as **vultures**, are apparently resistant to the effects of botulinum toxin.

Clostridium botulinum

Clostridium botulinum has been classified into eight different strains. Each strain releases the deadly toxin but in slightly different forms. Humans are susceptible to four of these eight toxins; the other four are deadly in cattle, **sheep**, and **horses**. *C. botulinum* is a strict anaerobe, meaning it survives only in conditions completely lacking **oxygen**. In fact, the presence of oxygen kills *C. botulinum*. However, the bacteria can survive for long periods of time by producing endospores. Endospores are small, protective capsules that surround the bacteria. They can withstand incredible extremes of **temperature**. The *C. botulinum* endospore can survive several hours at 212°F (100°C, the **boiling point** of **water**) and 10 minutes at 248°F (120°C). *C. botulinum* endospores also can survive at -374°F (-226°C). The endospores can even resist **radiation**. The botulism endospore is one of the most hardy organisms on **Earth**.

The botulinum toxin is in fact a group of seven closely related poisons produced by the bacteria. The fatal toxin the bacteria produces is a neurotoxin—it binds to nerve cells. Once bound, the toxin prevents the release of a **neurotransmitter** called **acetylcholine** from the nerve cells. Since nerve cells use the release of acetylcholine to transmit nerve impulses, preventing the release of acetylcholine stops the transmission of nerve impulses. Muscle paralysis eventually results.

The toxin targets nerve cells of the peripheral **nervous system** which govern the muscles associated with breathing. The muscles that control the tongue, pharynx, and ribs succumb swiftly to the toxin, becoming paralyzed within hours of ingestion of contaminated food. If rapid **diagnosis** is not made and treatment with an antitoxin (a substance that blocks the binding of the toxin to nerve cells) is not started, death can result quickly.

Symptoms

Symptoms can occur anywhere from 12 to 36 hours to eight days after eating toxin-contaminated food. The early symptoms of botulism are mild. Dizziness, fatigue, and weakness are common complaints. None of the early symptoms indicate the seriousness of the disease. Later, neurological symptoms develop, including difficulty in speaking and swallowing, and double **vision**. Fever rarely is present. Abdominal distension can further complicate the diagnosis leading to the incorrect conclusion of appendicitis. As more toxin binds to more nerve cells, paralysis sets in. Weakness of the muscle groups in the neck, extremities, and respiratory muscles is followed by complete paralysis.

Treatment

Because botulism is caused by a toxin and not the effects of the bacteria, **antibiotics** are not usually prescribed. The treatment for botulism is the antitoxin that neutralizes the toxin in the body. The antitoxin must be given early in the course of the disease to be effective. The antitoxin is useless if given too late, for too many nerve cells already are affected by the toxin. A botulism patient also must receive intensive, supportive care, such as a **respirator**, to assist breathing. Sometimes **dialysis**, a process that artificially cleanses the **blood**, is used to assist the kidneys in removing the toxin from the bloodstream.

Prevention

The risk of botulism has been virtually eliminated from the commercial canning industry, which uses sterilization techniques to kill the *C. botulinum* spores. For the home canner it is essential to follow recommended guidelines to prevent the growth of *C. botulinum*. These guidelines are:

(1) All non-acidic foods, such as green beans and corn, must be canned using a pressure cooker. Acidic foods, such as tomatoes and citrus **fruits**, contain natural acids that kill the botulism bacteria.

(2) To can non-acidic foods, cook them at 10 pounds of pressure at a temperature of 248°F (120°C) for 80 minutes.

Obviously, canning non-acidic foods requires special equipment. If you are not sure about the origin or safety of any home-canned food do not eat it. Dispose of the can safely and be sure to wash hands and touched surfaces thoroughly with **bleach** or **ammonia**.

The toxin can also be inactivated if the canned vegetable is cooked at 176°F (80°C) for five minutes or boiled for one minute before eating.

It is also wise to be wary of uncooked **fish** and meats. Sushi, a popular Japanese dish of raw fish, and venison have been known to spread botulism. All meats should be cooked thoroughly to kill the botulism endospores.

Because of its potent effects upon the human nervous system, botulinum toxin has been investigated for use in the treatment of various neurological diseases. The purified toxin used as a drug, called Botox, is used to treat conditions in which the nervous system cannot adequately control muscles, resulting in debilitating spasms. By its nature, the toxin induces muscular paralysis and is therefore useful in alleviating spasms in disorders such as cerebral palsy, spasmotic dysphonia (vocal spasms), facial spasms, and strabismis (squinting spasms of the eyelids). More controversially, Botox is also used by plastic surgeons in the temporary elimination of facial wrinkles. Administered as an injection, the toxin creates a temporary loss of muscle tone in areas of wrinkling. The result is wrinkle removal. Having reportedly few risks and complications, cosmetic Botox treatment is

---

KEY TERMS

Antitoxin—A antidote to a toxin that neutralizes its poisonous effects.

Endospore—A small, protective capsule surrounding a bacterium.

Toxin—A poisonous substance.

---

temporary. Injections initially last only three to four months, eventually requiring one to two injections per year for sustained effects on wrinkles. Botox is most used to eliminate the vertical wrinkles of the forehead in between the eyebrows (called glabellar frownlines) and so-called "crow's feet," wrinkles at the corners of eyes. The value of the cosmetic use of such a potent biological toxin should be weighed carefully against any potentially harmful effects of treatment.

See also Poisons and toxins.

Resources

Books

Francis, Frederick. *Wiley Encyclopedia of Food Science and Technology.* New York: Wiley, 1999.

Houschild, Andreas H.W., and Karen L. Dodds, eds. *Clostridium botulinum: Ecology and Control in Foods.* New York: M. Dekker, 1993.

Lance, Simpson L., ed. *Botulinum Neurotoxin and Tetanus Toxin.* San Diego: Academic Press, 1989.

Periodicals

Binder, W.J. "Botulinum Toxin Type A (Botox) For Treatment Of Migraine Headache." *Otolaryngology And Head And Neck Surgery* 123, no. 6 (2000): 669–676.

Jankovic, Joseph, and Mitchell F. Brin. "The Therapeutic Uses of Botulinum Toxin." *New England Journal of Medicine* 324 (April 25, 1991): 1186.

Morse, Dale L., et. al. "Garlic-in-Oil Associated Botulism: Episode Leads to Product Modification." *American Journal of Public Health* 80 (November 1990): 1372.

Nebel, Diane. "Case Study: Botulism in Home-Canned Food." *Journal of Environmental Health* 54 (July-August 1991): 9.

"Preventing Food Poisoning." *Professional Nurse* 18, no. 4 (2002): 185-186.

Kathleen Scogna

Bowen's reaction series

As hot **magma** cools, it undergoes specific reactions. Bowen's reaction series describes the **tempera-**

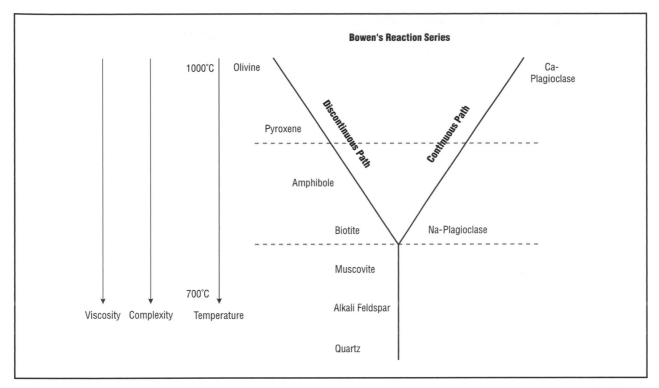

Bowen's Reaction Series

Bowen's reaction series depicts mineral formation in a cooling magma. The discontinuous side depicts mineral formation at decreasing temperatures. The continuous side depicts a solid solution series. As the magma cools there is a trend towards molecular complexity. As the temperature cools, viscosity increases. *Illustration by Argosy. The Gale Group.*

ture dependent formation of **minerals** as magma cools. **Rocks** formed from magma are **igneous rocks**, and minerals crystallize as magma cools. The temperature of the magma and the **rate** of cooling determine which minerals are stable (i.e., which minerals can form) and the size of the mineral crystals formed (i.e., texture). The slower a magma cools, the larger crystals can grow.

Named after geologist Norman L. Bowen (1887–1956), Bowen's reaction series allows geologists to predict chemical composition and texture based upon the temperature of a cooling magma.

Bowen's reaction series is usually diagramed as a "Y" with horizontal lines drawn across the "Y." The first horizontal line—usually placed just above the top of the "Y"—represents a temperature of 3272°F (1800°C). The next horizontal line, represents a temperature of 2012°F (1100°C) and is located one-third of the way between the top of the "Y" and the point where the two arms join the base. A third line representing a temperature of 1652°F (900°C) is located two-thirds of the way from the top of the "Y" to juncture of the upper arms. A fourth horizontal line—representing a temperature of 1112°F (600°C) intersects the triple point junction where the upper arms of the "Y" meet the base portion.

The horizontal temperature lines divide the "Y" into four compositional sections. Mineral formation is not possible above 3272°F (1800°C). Between 2012°F (1100°C) and 3272°F (1800°C), rocks are ultramafic in composition. Between 1652°F (900°C) and 2012°F (1100°C), rocks are mafic in composition. Between 1112°F (600°C) and 1652°F (900°C), rocks are intermediate in composition. Below 1112°F (600°C), felsic rocks form.

The upper arms of the "Y" represent two different formation pathways. By convention, the left upper arm represents the discontinuous arm or pathway. The upper right arm represents the continuous arm or continuous path of formation. The discontinuous arm represents mineral formations rich in **iron** and **magnesium**. The first mineral to form is olivine—it is the only mineral stable at or just below 3272°F (1800°C). As the temperature decreases, pyroxene becomes stable. The general chemical compositional formula—used throughout this article and not to be confused with a balanced molecular or empirical chemical formula—at the highest temperatures includes iron, magnesium, silicon and **oxygen** (FeMgSiO, but no quartz). At approximately 2012°F (1100°C), **calcium** containing minerals (CaFeMgSiO) become stable. As the temperature lowers to 1652°F (900°C), amphibole (CaFeMgSiOOH) forms. As the magmas cools to 1112°F (600°C), biotite (KFeMgSiOOH) formation is stable.

The continuous arm of Bowen's reaction series represents the formation of feldspar (plagioclase) in a continuous and gradual series that starts with calcium rich feldspar (Ca-feldspar, CaAlSiO) and continues with a gradual increase in the formation of **sodium** containing feldspar (Ca-Na-feldspar, CaNaAlSiO) until an equilibrium is established at approximately 1652°F (900°C). As the magmas cool and the calcium ions are depleted, the feldspar formation becomes predominantly sodium feldspar (Na-feldspar, NaAlSiO). At 1112°F (600°C), the feldspar formation is nearly 100% sodium feldspar (Na-feldspar, NaAlSiO).

At or just below 1112°F (600°C), the upper arms of the "Y" join the base. At this point in the magma cooling, K-feldspar or orthoclase (KAlSiO) forms and as the temperature begins to cool further, muscovite (KAlSiOOH) becomes stable. Just above the base of the "Y," the temperature is just above the point where the magma completely solidifies. At these coolest depicted temperatures (just above 392° F [200°C]), quartz (SiO) forms.

The time that the magma is allowed to cool will then determine whether the rock will be pegmatite (produced by extremely slow cooling producing very large crystals), phaneritic (produced by slow cooling that produces visible crystals), aphanitic (intermediate cooling times that produce microscopic crystals), or glassy in texture (a product of rapid cooling without **crystal** formation). When magmas experience differential cooling conditions, they produce porphyritic rock, a mixture of crystal sizes and exhibit either a phaneritic or aphanitic groundmass.

Although the above temperature and percentage composition data are approximate, simplified (e.g., the formation of hornblende has been omitted), and idealized, Bowen's reaction series allows the prediction of mineral content in rock and—by examination of rock—allows the reverse determination of the conditions under which the magma cooled and igneous rock formed.

See also Chemical bond; Earth's interior; Igneous rocks; Lithosphere; Mineralogy; Volcano.

Resources

Books

Hamblin, W.K., and E.H. Christiansen. *Earth's Dynamic Systems*. 9th ed. Upper Saddle River: Prentice Hall, 2001.

Hancock, P.L., and B.J. Skinner, eds. *The Oxford Companion to the Earth*. New York: Oxford University Press, 2000.

Klein, C. *The Manual of Mineral Science*. 22nd ed. New York: John Wiley & Sons, Inc., 2002.

Press, F., and R. Siever. *Understanding Earth*. 3rd ed. New York: W.H Freeman and Company, 2001.

Tarbuck, Edward. D., Frederick K. Lutgens, and Tasa Dennis. *Earth: An Introduction to Physical Geology*. 7th ed. Upper Saddle River, NJ: Prentice Hall, 2002.

Periodicals

Hellfrich, George, and Bernard Wood. "The Earth's Mantle." *Nature*. (August 2, 2001): 501–507.

Other

James Madison University, Department of Geology and Environmental Science. "Bowen's Reaction Series and The Igneous Rock Forming Minerals." August 17, 2000 [cited January 22, 2003]. <http://csmres.jmu.edu/geollab/Fichter/RockMin/RockMin.html>.

K. Lee Lerner

Bowerbirds

The 18 **species** of bowerbirds are unique in that the males build and decorate a bower, a structure of sticks or grass on the ground, for the purpose of attracting and courting females. Members of the bowerbird family (Ptilonorhynchidae) are found in **Australia** and New Guinea, and are related to **lyrebirds** and **birds of paradise**. Most bowerbirds are about the size of a blue jay or grackle, and as a group they show a wide variety of plumage characteristics, vocal **behavior**, and bower-building styles. The bowers of some species are quite large (3.3-6.6 ft/1-2 m) in length, and are decorated with a variety of objects, making them some of the most remarkable examples of **animal** architecture.

Naturalists have been fascinated by bowerbirds for decades. Early observers believed that the bower was a nest; however, in 1865, the ornithologist John Gould suggested that the bower was used for sexual display and mating. Not all bowerbirds build a bower; some, such as the toothbilled bowerbird of eastern Australia, simply clear a display court on the ground, and decorate it with leaves.

Bowers occur in several forms, each built by species that appear to share closer genealogical relationships with each other than with those species that build a different type of bower. Avenue bowers (constructed by the satin bowerbird) have two vertical walls running **parallel** to each other, with one end opening onto a display area where most of the decorations are arranged. Maypole bowers consist of sticks woven around a central pole, formed by a sapling or fern, surrounded by a circular, raised court. Two species, the striped gardener bowerbird and Vogelkop's bowerbird, build massive hutlike structures (up to 6.6 ft/2 m across) around the central maypole, opening onto a cleared exhibition area. The golden bowerbird places sticks against adjacent saplings which are joined by a crossbranch; this he uses as a display perch. A variety of objects are used to decorate the bowers of different species of bowerbird, including **fruits**,

A female satin bowerbird (*Ptilonorhynchus violaceus*) visiting a bower. © *Tom McHugh, National Audubon Society Collection/Photo Researchers, Inc. Reproduced with permission.*

flowers, feathers, **moss**, snail shells, colored stones and **bark**; the recent presence of humans has added coins, bottle tops, pieces of **glass**, teaspoons, nails, and screws to the bowers.

Female bowerbirds visit the bowers of numerous males in the process of selecting a mate. When a female arrives, she will take up a position within the bower, while the owner launches into a stereotyped display that is unique for each species. The male emits varied chirps, whistles, buzzes, and mechanical sounds, while performing a series of dance-like movements that have been described as "rooster walks" and "penguin walks." Males of some species pick up and hold decorations in their beak during their display, bobbing their heads or tossing the items with considerable vigor. One especially vigorous species, the spotted bowerbird, actually rushes at the visitor watching from within the bower, crashing bodily into the wall that separates the two **birds**. When the visiting female is ready to mate, she crouches low in the bower, lifting her tail, and the male approaches from the rear of the bower to mount her and copulate for just one or two seconds.

It is clear that a bower can be extremely valuable to its owner, for bowerbirds spend hours constructing, deco-

rating, and maintaining their bowers, and no male has ever been reported to successfully court a female without one. However, success is not guaranteed. Males in some species are highly competitive, and are observed to steal decorations from the bowers of other males, and to destroy rival bowers. Many bowerbird males simply fail to mate with even one visiting female during a breeding season, even though the males have been tending bowers.

Researchers have sought to understand why some individuals enjoy high mating success, while most others do not, and the functional role of the bower in female mate choice. Elaborate display traits, such as bowers, suggest the influence of sexual selection—the process of evolutionary change due to **competition** between members of one sex (usually males), and selective mate choice by the other sex (usually females). In species where males give little or no parental care to their offspring (as is the case for most bowerbirds), researchers have attempted to address the question: What is the female choosing when she selects a mate? Bowerbird observers have attempted to ascertain whether certain features of bowers or of male display are reliable predictors of a male's success in obtaining mates.

Gerald Borgia and his collaborators have used an ingenious method to investigate this question in the field. A video camera, positioned a short **distance** in front of the bower, was outfitted with a motion-sensitive infrared detector, which turned on the camera whenever there was movement at the bower. Such continuous monitoring compiled a comprehensive record of activity at the bower, and allowed Borgia to measure and compare the **courtship** of many individuals. The results showed that female bowerbirds differentiated among males, at least in part on the basis of the quality of the bowers and the display. In satin bowerbirds, for instance, the number of decorations, especially snail shells and blue feathers, as well as the degree of bower **symmetry** and the density of the sticks used to construct it, were excellent predictors of male mating success. The more decorations, and the more symmetrical and densely constructed the bower, the more matings were achieved by the bower owner. The importance of decorations was underscored when the researchers experimentally removed decorations from some bowers. The owners of the manipulated bowers had far less success in attracting females. Clearly, female choice could exert a potent effect on the **evolution** of male display in bowerbirds, helping shape the elaborate courtship structures and behaviors observed in these animals.

Resources

Periodicals

Borgia, Gerald. "Sexual Selection in Bowerbirds." *Scientific American* 254 (1986): 92-101.

Bowfin

The bowfin is a **bony fish** (*Amia calva*, family Amiidae) found in eastern **North America**. It is a relic species—the sole living representative of the order Amiiformes, which first appeared in the Triassic period more than 200 million years ago.

Members of this family were common in **Europe** and **Asia**, as well as North America, during the Cretaceous and the early part of the Cenozoic. Fossil **species** of the genus *Amia* occurred in Europe as recently as the Miocene (about 20 million years ago) but this single species in North America is the only one of its group still living.

The bowfin is a relatively common **fish** in swamps and slow-moving streams in the southeastern United States, and is often caught by fishermen. Often called a grennel or mudfish, it is very bony and seldom used as food if there is an alternative. In appearance it is a rather stout cylindrical fish, with a distinctive bony head and a wide mouth with many sharp-pointed teeth. It differs from most **freshwater** fishes in having a very long dorsal fin and widely separated pectoral and pelvic fins.

The bowfin preys on other fishes, and because its swim bladder is connected with its esophagus, it can gulp air in situations when the **oxygen** in the **water** is depleted. Thus, it can survive in warm isolated pools during a **drought** and feed on the other fishes that cannot withstand these conditions. The male bowfin builds a circular nest in shallow water and guards the eggs until they hatch. He continues to guard the young for some time. They can be seen following him in shallow water like so many baby chicks following their mother hen.

Boxfish

Boxfish, also called trunkfish or cowfish, are a small group of shallow-water, marine **fish** in the family Ostraciontidae (order Tetraodontiformes). The family includes the genera *Lactoria, Ostracion,* and *Tetrosomus* and is closely related to the poisonous **puffer fish** of the family Tetraodontidae. To avoid confusion with these poisonous relatives, some people avoid eating boxfish despite their being good for food. Boxfish are generally oval in shape when seen from the side; while being viewed from one end, different **species** of boxfish resemble triangles, squares, or pentagons.

Boxfish take their shared common name from the hard shell or carapace, composed of strongly joined plates corresponding to the scales of other fish, which surrounds their bodies. Only the eyes, the low-set mouth, the fins, and the tail are not covered by this rigid shell. Boxfish reach lengths of up to 2 ft (61 cm). They inhabit warm waters from the Pacific Ocean to the Caribbean. Boxfish usually prefer shallow waters, and are often found around coral reefs. In some areas the boxfish are dried and used as decorations.

Boxfish possess an intriguing method of defense against predators, such as **sharks**. Like other species of fish, when disturbed boxfish can secrete molecules known as icthyocrinotoxins or ichthyotoxins—literally, fish poisons—from their skin. The toxins of two Pacific boxfish species, *Ostracion immaculatus* from Japan and *O. lentiginosus* from Hawaii, have been characterized by biochemists. These toxins turn out to be esters of choline chloride. Choline is a **vitamin** in the B complex that makes up part of many of the **fatty acids** found in the membranes of human cells.

Similar to the ichthyotoxic compounds of fish from other families, boxfish toxins are surfactants. In general they act as biological detergents, promoting the dissolu-

A male Pacific boxfish (*Ostracion meleagris*) near Cocos island, Costa Rica. *© Fred McConnaughey, National Audubon Society Collection/Photo Researchers, Inc. Reproduced with permission.*

tion of fatty acids in **water**. This occurs because surfactants, like all detergents, include parts that can interact with water and parts that can interact with fats. Thus, fats and water can mix, something that normally does not occur. By this mechanism, these ichthyotoxins cause hemolysis in the laboratory by dissolving **cell** membranes. When these membranes dissolve, cells break down and die. This irritates and deters predators, saving the boxfish from being eaten.

Boyle's law *see* **Gases, properties of**

Brachiopods

Brachiopods, or lampshells, are a phylum of small marine animals with a two-valved shell that, at first glance, resemble bivalved **mollusks** such as clams. The resemblance, however, is quite superficial. The orientation of the shells of brachiopods is very different from that of bivalved mollusks, and brachiopods have two additional structures virtually unique to them, the lophophore (a cili-

ated feeding apparatus) and pedicle (a muscular stalk). The valves of the brachiopod shell are symmetrical—being dorsal and ventral and covering the upper and lower parts of the body, with the hinge at the posterior end. In contrast, shells of a clam are on the right and left sides of the body, with the hinge dorsal.

The lophophore, which occupies much of the space in the anterior portion of the shell, resembles a circle of small tentacles surrounding the mouth. Each tentacle bears large numbers of tiny cilia that, when beating, create a **water** current that draws in water and suspended food particles down toward the mouth. Once trapped in the lophophore, tiny **plankton** are passed along special grooves that lead to the mouth, and from there, to a stomach and intestine. The water current also maintains a steady supply of **oxygen** to the **animal**. The lophophore is a complex feeding apparatus found in only a few other groups of marine and **freshwater** animals, chiefly the Bryozoa and Phoronida. Collectively, these three groups of marine **invertebrates** are sometimes referred to as lophophorates.

Brachiopods can be divided into two major groups, articulate and inarticulate, based on their use of the pedi-

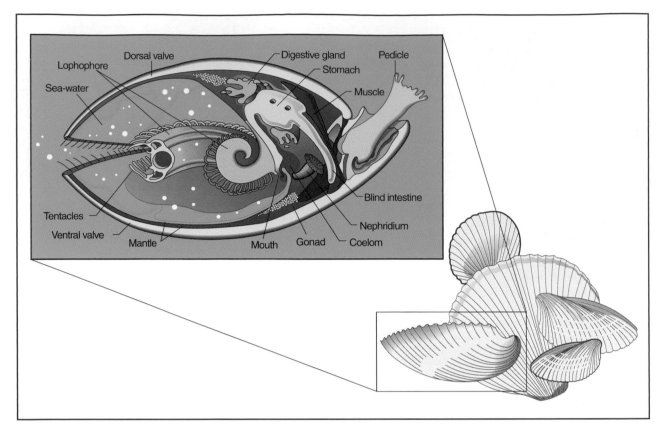

The anatomy of an articulate brachiopod. *Illustration by Hans & Cassidy. Courtesy of Gale Group.*

cle. Articulate brachiopods are fixed directly to a hard substrate by the pedicle, a short piece of **connective tissue** at the posterior end of the shell. The brachiopod has a very limited range of **motion** and remains, for the most part, sessile. The inarticulate brachiopods are not fixed to one location. Instead, they use their specialized muscular pedicles to burrow through **sand** and other soft sediments. At the distal end of the pedicle a sticky substance is secreted that forms a sand anchor, enabling them to withdraw deeper into the sediment by contracting the muscular pedicle when threatened. The pedicle ranges from about 0.1 in (2.5 mm) in some **species** to more than 7.9 in (20 cm) in others. Some bivalved mollusks, such as oysters, also attach themselves to the substratum, but they lack a pedicle.

Brachiopods first appeared about 500 million years ago during the Paleozoic era, as shown by their common occurrence as fossils in many parts of the world. This accounts for their great interest to geologists. Over 30,000 species are believed to have evolved over the years. Today, roughly 300 living species are know to exist. Most lampshells are found in deepwater, so their shells are not commonly found on the beach. In the Pacific, *Lingula unguis* is commonly found living in vertical bur-

rows in sand and mud. In Maine, *Terebratulina septentrionalis* is sometimes exposed to the air at low tide. Five species of brachiopods are readily collected from **rocks** off the shores of the South Island of New Zealand.

Most lampshells are dioecious, producing either male or female gametes. In the majority of species, when the gonads have ripened, the gametes are released into the coelom and pass through the nephridia to the sea. **Fertilization** takes place outside the body. Some species, however, brood their young by retaining their eggs and awaiting the arrival of male gametes with the incoming flow of water. The fertilized eggs develop into free-swimming larvae that are capable of feeding. Further development of the larvae depends on the species: in most articulate brachiopods, larvae undergo a transformation of the body shape and structure before settling, while the larvae of inarticulate brachiopods already resembles the final adult stage apart from its diminutive size. As the shell develops in the latter species, the larvae are encouraged to settle on the seabed.

Since brachiopods are fixed in position after the larval phase, there is little observable **behavior** beyond the opening and closing of valves. The valves may remain closed for periods of 5-22 hours, during which there is no feeding and little or no oxygen uptake. Lampshells obtain oxygen from the water that passes over the lophophore, which functions as a respiratory surface although brachiopods tolerate lack of oxygen (anoxia) well. Oxygen consumption rates of brachiopods are generally lower than those of **bivalves** of similar size.

The food of brachiopods is mainly algal cells of the **phytoplankton**, which are strained out of water **currents** passing over the lophophore in a process called filter-feeding. Clearance rates of several species are in the same range or a little lower than rates shown by bivalve mollusks. Research on the **physiology** of brachiopods is heavily slanted toward comparisons with mussels and clams because of the obvious similarity in protective shell and **mode** of feeding. The fossil record indicates the rise of one group followed by the decline of the other, as if they might have been competitors. Unlike mollusks, brachiopods are not significantly preyed upon or harvested. Some observations suggest that **fish** find lampshells distasteful.

Resources

Books

Pearse, V., et al. *Living Invertebrates.* Palo Alto, CA: Blackwell, 1987.

Prothero, Donald R. *Bringing Fossils To Life: An Introduction To Paleobiology.* Columbus: McGraw-Hill Science/Engineering/Math, 1997.

Periodicals

Hammen, C. S. "Brachiopod Metabolism and Enzymes." *American Zoologist* 17 (1977): 141-147.

Morris, S. C. "The Fossil Record and the Early Evolution of the Metazoa." *Nature* 361 (1993): 219-225.

Rhodes, M. C., and R. J. Thompson. "Comparative Physiology of Suspension-Feeding in Living Brachiopods and Bivalves: Evolutionary Implications. *Paleobiology* 19 (1993): 322-334.

Wilson, E. O. 1994. "Biodiversity: Challenge, Science, Opportunity." *American Zoologist* 34 (1994): 5-11.

C. S. Hammen

Brackish

Brackish refers to **water** with a salinity intermediate to that of fresh water and sea water (the latter has a **salt** concentration of about 3.5%, or 35 parts per thousand). Brackish waters originate by the mixing of sea water and **freshwater**, and are most common near the coasts of the oceans.

Brackish waters can occur as enclosed systems such as lakes and ponds that receive occasional inputs of oceanic water during severe storms. Brackish waters also occur as coastal estuaries or salt marshes that are more frequently flooded with saline water as a result of tidal cycles. Sometimes, brackish waters can occur far inland, for example, in parts of the prairies of **North America** where saline ponds and **wetlands** have variable salt concentrations depending on the diluting effects of recent rains or snowmelt.

The salt **concentration** of water is highly influential on the transport of ions across cellular membranes, the availability of **nutrients** in **soil**, and for other reasons. Most **species** can tolerate either **saltywater** or freshwater, but not both. However, organisms that live in brackish habitats must be tolerant of a wide range of salt concentrations (such species are known as *euryhaline*). For example, the small **fish** known as **killifish** (*Fundulus* spp.) are common residents of brackish coastal habitats known as estuaries, where within any day the salt concentration in tidal pools and creeks can vary from that typical of freshwater to that of the open **ocean**. Other fish such as **salmon** (e.g., *Salmo* spp.) and eels (*Anguilla* spp.) move to or from marine waters during their spawning migrations, in the process moving from environments characterized by the salt concentration of full seawater, through brackish, to freshwater. Animals that live in or move through estuaries must be tolerant of the physiological stresses associated with such large and rapid changes in salinity, as must the plants of those habitats, such as the aquatic eelgrass (*Zostera* spp.) and the cord grass of salt marshes (*Spartina* spp.).

The environmental conditions of brackish waters are highly stressful for organisms that cannot tolerate such wide swings of salinity. However, for those relatively few species that are tolerant of such difficult environmental conditions, the environmental conditions of brackish habitats represent a relatively uncompetitive, ecological opportunity to be exploited as a livelihood.

See also Saltwater.

Brain

The brain is a mass of nerve **tissue** located in an animal's head that controls the body's functions. In simple animals, the brain functions like a switchboard picking up signals from sense organs and passing information to muscles. The brain is also responsible for a variety of involuntary **behavior**, including keeping the **heart** beating, and maintaining **blood** pressure and **temperature**. In more advanced forms, particularly **vertebrates**, a more analytical brain coordinates complex behaviors. In higher vetebrates, the brain coordinates thinking, **memory**, **learning**, and emotions. The brain is part of an animal's central **nervous system**, which receives and transmits impulses. It works with the peripheral nervous system, which carries impulses to and from the brain and spinal cord via nerves running throughout the body.

Invertebrate brain

Nematodes (**roundworms**) have a simple brain and nervous system consisting of approximately 300 nerve cells, or neurons. Sensory neurons located in the head end of the **animal** detect stimuli from the environment and pass messages to the brain. The brain then sends out impulses through a ventral nerve cord to muscles which respond to the **stimulus**. The way that the interneurons of the brain process the data determines the response.

The earthworm and other annelids, as well as **insects** and other **arthropods**, have more complex nervous systems. In these animals, there are paired ventral nerve cords that run from head to tail on the animal's underside. **Cell** bodies of neurons in the cords form pairs of ganglia in each body segment. Four of the most anterior ganglia fuse in the head to form a brain. As a result, the brain ganglia are larger than the segmental ganglia, and also contain a larger proportion of sensory motor neurons. The brain ganglia have some dominance over the segment ganglia. The ventral nerve cords, brain, and segmental ganglia comprise the central nervous system. Neuronal fibers in the cords, bundled into nerves that carry communications between ganglia, make up the peripheral nervous system.

The earthworm's brain consists of paired ganglia in the head end. An impulse, such as **touch**, **light**, or moisture, is detected by receptor cells in the skin. A pair of nerves in each of the earthworm's segments carries the signal to the brain and smaller ganglia in each segment, where the signals are analyzed. The central nervous system then transmits impulses on nerves that coordinate muscle action, causing the earthworm to move.

In insects, specialized sense organs detect information from the environment and transmit it to the central nervous system. Such sense organs include simple and compound eyes, sound receptors on the thorax or in the legs, and **taste** receptors. The brain of an insect consists of a ganglion in the head. Some of the segmental ganglia

are fused, allowing better communication between the segments. The information that insects use for behaviors such as walking, flying, mating, and stinging is stored in the segmental ganglia. In experiments in which heads are cut off of **cockroaches** and **flies**, these insects continue to learn.

Vertebrate brain

The central nervous system of vertebrates consists of a single spinal cord, which runs in a dorsal position along the back, and a highly developed brain. The brain is the dominant structure of the nervous system. It is the master controller of all body functions, and the analyzer and interpreter of complex information and behavior patterns. We can think of the brain as a powerful neural computer. The peripheral nervous system, composed of nerves that run to all parts of the body, transmits information to and from the central nervous system.

The vertebrate brain is divided into three main divisions: the forebrain, the midbrain, and the hindbrain. The hindbrain connects the brain to the spinal cord, and a portion of it, called the medulla oblongata, controls important body functions such as the breathing **rate** and the heart rate. Also in the hindbrain, the cerebellum controls balance. The forebrain consists of the cerebrum, thalamus, and hypothalamus. The forebrain controls, among other things, the sense of **smell** in vertebrates.

During the first few weeks of development, the brain of a vertebrate looks like a series of bulges in the neural tube. It is hard to see a difference when we examine the early embryonic brains of **fish**, **amphibians**, **reptiles**, **birds**, and **mammals**. As the brain develops, the bulges enlarge, and each type of vertebrate acquires its own specific adult brain that helps it survive in its environment. In the forebrain of fish, the olfactory (smell) sense is well developed, whereas the cerebrum serves merely as a relay station for impulses. In mammals, on the other hand, the olfactory division is included in the limbic system, which also controls emotions, and the cerebrum is highly developed, operating as a complex processing center for information. Optic lobes are well-developed in the midbrain of nonmammalian vertebrates, whereas in mammals the **vision** centers are mainly in the forebrain. In addition, a bird's cerebellum is large compared to the rest of its brain, since it controls coordination and balance in flying.

Human brain

The living human brain is a soft, shiny, grayish white, mushroom-shaped structure. Encased within the skull, it is a 3 lb (1.4 kg) mass of nerve tissue that keeps us alive and

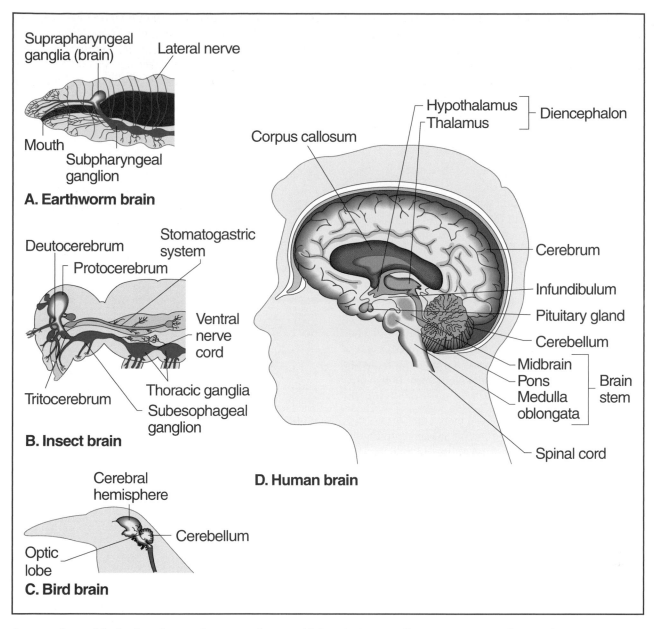

A. Earthworm brain

- Suprapharyngeal ganglia (brain)
- Lateral nerve
- Mouth
- Subpharyngeal ganglion

B. Insect brain

- Deutocerebrum
- Stomatogastric system
- Protocerebrum
- Ventral nerve cord
- Tritocerebrum
- Thoracic ganglia
- Subesophageal ganglion

C. Bird brain

- Cerebral hemisphere
- Cerebellum
- Optic lobe

D. Human brain

- Corpus callosum
- Hypothalamus
- Thalamus
- Diencephalon
- Cerebrum
- Infundibulum
- Pituitary gland
- Cerebellum
- Midbrain
- Pons
- Medulla oblongata
- Brain stem
- Spinal cord

A comparison of the brains of an earthworm, an insect, a bird, and a human. *Illustration by Hans & Cassidy. Courtesy of Gale Group.*

functioning. On average, the brain weighs 13.7 oz (390 g) at **birth**, and by age 15 grows to approximately 46 oz (1,315 g). The human brain is composed of up to one trillion nerve cells—100 billion of them are neurons, and the remainder are supporting (glial) cells. Neurons receive, process, and transmit impulses, while glial cells (neuroglia) protect, support, and assist neurons. The brain is protected by the skull and by three membranes called the meninges—the outermost the dura mater, the middle the arachnoid, and the innermost the pia mater. Also protecting the brain is cerebrospinal fluid, a liquid that circulates between the arachnoid and pia mater in the subarachnoid space. Many bright red **arteries** and bluish **veins** on the surface of the brain penetrate inward. Glucose, **oxygen**, and certain ions pass easily from the blood into the brain, whereas other substances, such as **antibiotics**, do not. The capillary walls are believed to create a blood-brain barrier that protects the brain from a number of biochemicals circulating in the blood.

The parts of the brain can be studied in terms of structure and function. Four principal sections of the human brain are the brain stem (the hindbrain and midbrain), the diencephalon, the cerebrum, and the cerebellum.

The brain stem

The brain stem is the stalk of the brain, and is continuous with the spinal cord. It consists of the medulla oblongata, pons, and midbrain. A part of the brain stem, the medulla oblongata is a continuation of the spinal cord. All the messages that are transmitted between the brain and spinal cord pass through the medulla via fibers in the white matter. The fibers on the right side of the medulla cross to the left and those on the left cross to the right. The result is that each side of the brain controls the opposite side of the body. There are three vital centers in the medulla which control the heartbeat, the rate of breathing, and the diameter of the blood vessels. Centers that help coordinate swallowing, vomiting, hiccoughing, coughing, and sneezing are also located in the medulla. The reticular formation occurs partially in the medulla and in other parts of the central nervous system. The reticular formation operates in maintaining our conscious state. The pons (meaning bridge) conducts messages between the spinal cord and the rest of the brain, and between the different parts of the brain. The midbrain conveys impulses from the cerebral cortex to the pons and spinal cord. It also contains visual and audio **reflex** centers involving the movement of eyeballs and head.

Twelve pairs of cranial nerves originate in the underside of the brain, mostly from the brain stem. They leave the skull through openings and extend as peripheral nerves to their destinations. Cranial nerves include the olfactory nerve that brings messages about smell from the nose and the optic nerve that conducts visual information from the eyes.

The diencephalon

The diencephalon lies above the brain stem, and embodies the thalamus and hypothalamus. In the diencephalon, the thalamus is an important relay station for sensory information for the cerebral cortex from other parts of the brain. The thalamus also interprets sensations of **pain**, **pressure**, temperature, and touch, and is concerned with some of our emotions and memory. It receives information from the environment in the form of sound, smell, and taste. The hypothalamus performs numerous important functions. These include the control of the autonomic nervous system (a branch of the nervous system involved with control of a number of body functions, such as heartbeat rate and digestion). The hypothalamus helps regulate the **endocrine system** and controls normal body temperature. It tells us when we are hungry, full, and thirsty. It helps regulate **sleep** and wakefulness, and is involved when we feel angry and aggressive.

The cerebrum

The cerebrum, constituting about 87.5% of the brain weight, spreads over the diencephalon. The cerebral cortex is the outer layer of the brain and is composed of gray matter made up of nerve cell bodies. It is about 0.08 in (2 mm) thick and its surface area is about 5 sq ft (1.5 sq m)—around half the size of an office desk. White matter, composed of nerve fibers covered with myelin sheaths, lies beneath the gray matter. With the rapid growth of the brain during embryonic development, the gray matter grows faster than the white matter and folds on itself. The folds are called convolutions or gyri, and the grooves between them are known as sulci. A deep longitudinal fissure separates the cerebrum into a left and right hemisphere. Each cerebral hemisphere is divided into frontal, temporal, parietal, and occipital lobes. The corpus callosum, a large bundle of fibers, connects the two cerebral hemispheres. The thalamus and subcortical nuclei, or basal ganglia, are areas of gray matter that exist below the white matter.

Sensory areas of the cerebrum interpret sensory impulses. Spoken and written language are transmitted to a part of the cerebrum called Wernicke's area where meaning is extracted, and sent to Broca's area, one of the motor areas of the cerebrum. Motor areas of the cerebrum control muscle movements. Within Broca's area, thoughts are translated into **speech**, and muscles are coordinated for speaking. Impulses from other motor areas direct our hand muscles when we write, and our **eye** muscles when we scan the page for information.

Association areas of the cerebrum are concerned with emotions and intellectual processes, by connecting sensory and motor functions. In our association areas, innumerable impulses are processed that result in memory, emotions, judgment, personality, and intelligence.

Certain structures in the cerebrum and diencephalon make up the limbic system. These regions function in memory and emotions, and are associated with pain and pleasure.

By studying patients whose corpus callosum were destroyed, scientists realized that differences existed between the left and right sides of the cerebral cortex. The left side of the brain functions mainly in speech, logic, writing, and **arithmetic**. The right side of the brain, on the other hand, is more concerned with imagination, art, symbols, and spatial relations.

The cerebellum

The cerebellum is located below the cerebrum and behind the brain stem, and is shaped like a butterfly. The "wings" are the cerebellar hemispheres, and each consists of lobes that have distinct grooves or fissures. The cerebellum controls the movements of our **muscular system** needed for balance, posture, and maintaining posture.

Studying the brain

At the end of the nineteenth century, Santiago Ramon y Cajal, a Spanish scientist, studied neurons using stain developed by Camillo Golgi. Cajal realized that the brain was made up of individual units and not a continuous net as was believed at the time. His studies uncovered a large variety of neurons that differed in size and shape. He explained that neurons received signals on dendrites and transmitted impulses on axons. Since his work, researchers have learned that neurons carry information in the form of brief electrical impulses called action potentials that result when positively charged **sodium** ions travel across the axon **membrane** from the fluid outside to the cytoplasm inside. When a nerve impulse reaches the end of an axon, neurotransmitters are released at junctions called synapses. The neurotransmitters are chemicals that bind to receptors on the receiving neurons, triggering the continuation of the impulse. Fifty different neurotransmitters have been discovered since the first one was identified in 1920. By studying the chemical effects of neurotransmitters in the brain, scientists have made advances in finding medicines for the treatment of mental disorders, and determining the actions of drugs on the brain.

Researchers today are able to trace various molecules that are transported along axons during action potentials. Microelectrodes are used to detect the currents that cross synapses. Using this information, wiring diagrams are created that model the patterns of information flow within the brain.

Considerable knowledge about the human brain has been obtained during brain **surgery** by stimulating specific areas with a mild **electric current**, and from the observation of patients with brain damage. In the 1920s, a Canadian neurosurgeon named Wilder Penfield electrically stimulated different parts of the brains of some of his patients. He found this caused them to remember specific events from the past. For example, one patient heard someone from the past singing a particular song. From this and other studies, scientists realized that specific functions are localized in specific parts of the brain. Recently, scientists observed the behavior of a woman whose amygdala (an almond-shaped group of cells in the cerebrum) was destroyed. The amygdala plays a role in emotions and social relationships. The researcher realized that without an amygdala, the patient could not read facial expressions. As a result, she couldn't judge the intentions of others, and often made poor social decisions.

Until recently, scientists believed that brain cells do not regenerate, thereby making brain injuries and brain diseases untreatable. Researchers are now trying to help such patients with **neuron** transplants, introducing nerve tissue into the brain. They are also studying substances, such as **nerve growth factor** (NGF), that someday may be used to help regrow nerve tissue.

Since the 1950s, scientists have begun to understand the process of sleep. They find that sleep occurs in different stages. One stage is called rapid eye movement (REM) sleep. We dream during REM sleep, a period when there is a lot of brain activity and eye movements and the body is inactive. The pons, an area of the brainstem, sets off REM sleep and dreaming. During REM sleep, the brain emits characteristic brain waves. Non-REM sleep usually comes first, takes up about 75% of our sleep, and is much quieter. Its stages get deeper and deeper. Non-REM sleep also has its own particular brain waves. The two types of sleep alternate during the night. Scientists are beginning to understand the factors that control sleep and wakefulness. These include a biological clock, a group of about 10,000 neurons in the hypothalamus that trigger off waking up; **homeostasis**, the body's tendency to maintain equilibrium in physiological systems; and changes in the level of norepinephrine and serotonin, neurotransmitters in the brain.

Where and how does memory occur? This is another question that has puzzled scientists for decades. Recent information suggests that memory is not stored in a single brain center, but instead is part of numerous processing systems in the cerebral cortex. Scientists believe that memory involves chemical and structural changes in neurons, as well as changes in the strength of synapses.

Technology provides useful tools for researching the brain and helping patients with brain disorders. An **electroencephalogram (EEG)** is a record of brain waves, electrical activity generated in the brain. An EEG is obtained by positioning electrodes on the head and amplifying the waves with an electroencephalograph, and is valuable in diagnosing brain diseases such as **epilepsy** and tumors.

Scientists use three different techniques that involve scans to study and understand the brain and diagnose disorders:

(1) **Magnetic resonance imaging (MRI)** depends on the use of a magnetic field to display the living brain at various depths as if in slices. Not blocked by bone, MRI allows the viewer to zoom in on any region and obtain reliable pictures of brain tissue.

(2) **Positron emission tomography (PET)** results in **color** images of the brain displayed on the screen of a monitor. During this test, a technician injects a small

KEY TERMS

Broca's area—Area in the cerebrum that organizes thought and coordinates muscles for speech.

Ganglion—A cluster of nerve cell bodies, usually found outside of the central nervous system.

Glial cells—Nerve cells (other than neurons) located in the brain that protect, support, and assist neurons.

REM sleep—Rapid eye movement sleep that is characterized by dreaming, active brain activity, and numerous eye movements.

Wernicke's area—The portion of the left side of the brain that stores and retrieves words and word patterns.

amount of a substance, such as glucose, that is marked with a radioactive tag. The marked substance shows where glucose is consumed in the brain. PET is used to study the **chemistry** and activity of the normal brain and to diagnose abnormalities such as tumors.

(3) Magnetoencephalography (MEG) measures the electromagnetic fields created between neurons as electrochemical information is passed along. When under the machine, if the subject is told, "wiggle your toes," the readout is an instant picture of the brain at work. Concentric colored rings appear on the computer screen that pinpoint the brain signals even before the toes are actually wiggled.

Using an MRI along with MEG, physicians and scientists can look into the brain without using surgery. They hope to use these techniques for the early **diagnosis** of disorders such as Alzheimer and **Parkinson disease**. They foresee that these techniques could help paralysis victims move by supplying information on how to stimulate their muscles, or indicating the signals needed to control an artificial limb. Furthermore, by understanding what areas of the brain are active while performing particular tasks, and by mapping information regarding increased or decreased **metabolism** in particular regions of the brain, a variety of questions regarding various **disease** states, as well as a variety of questions regarding the phenomenal potential of the human brain, may be answered

Researchers are studying any number of issues regarding brain functioning. Fascinating research revealing the presence of tiny magnetic bits within the brain has suggested that humans (like some insects and birds) have the potential to navigate via interactions with the **earth's**

magnetic field. Technological research into computers and artifical intelligence are furthered by a good understanding of the amazing, computer-like intelligence of the human brain; conversely, studies of the human brain have been furthered by efforts to create artifical intelligence. Research into the human brain is actually thought to be in its infancy; myriad topics of investigation must be explored in order to grasp the complexities and intricacies of the human brain.

See also Electrocardiogram (ECG); Neurotransmitter; Nuclear medicine.

Resources

Books

Jackson, Carolyn, ed. *How Things Work: The Brain*. Alexandria, VA: Time-Life Books, 1990.

Rumbaugh, Duane M., and D.A. Shaw. *Intelligence of Apes and Other Rational Beings (Current Perspectives in Psychology)*. New Haven, CT: Yale University Press, 2003.

Periodicals

Carey, Joseph, ed. *Brain Facts* Washington, DC: Society for Neuroscience, 1993.

Friedrich, M.J. "A Bit of Culture for Children: Probiotics May Improve Health and Fight Disease." *Journal of the American Medical Association* no. 284 (September 2000): 1365-1366.

Golden, Frederic. "Mental Illness: Probing the Chemistry of the Brain." *Time* 157 (January 2001).

"Mind and Brain." *Scientific American* (September, 1992).

Other

The Nature of the Nerve Impulse. Films for the Humanities and Sciences, 1994-95. Videocassette.

Bernice Essenfeld

Brewing

Brewing is a multi-stage process during which the brewer encourages a grain such as **barley** to germinate briefly, steeps the grain in **water** to release its sugars, and adds **yeast** to the mixture, which ferments the sugar, turning it into **alcohol** and **carbon dioxide**. Around the world, people have brewed grains and starchy **vegetables** for thousands of years from ingredients as varied as **rice**, corn, cassava, pumpkins, **sorghum**, and millet, and brewed beverages are staples in the diets of many cultures.

History

Archaeologists have turned up evidence that the Sumerian people in the Middle East were brewing barley as long as 8,000 years ago. These early brewers may have discovered the fundamental processes of brewing as they observed—or tasted—what happened when they left fruit juices or cereal extracts exposed to the wild yeasts that naturally float in the air. Native Americans made a beer from corn, which they softened by chewing and mixing into a pulpy mass with saliva. They then set this masticated corn out in a vessel to ferment, and enjoyed the resulting drink.

Throughout **Europe**, breweries sprang up where there was good water for brewing. During the Middle Ages, monasteries became the centers for brewing, and the monks originated brewing techniques and created many of the beers still popular today. Early European settlers in **North America** brought with them from Europe a **taste** for beer, but followed the Native Americans' example and initially made beer from corn and pumpkins, which they flavored with such local additives as the tops of **spruce** trees.

With the spread of the British empire and rise of industrialization over the past 150 years, traditional European beer began to move from its local markets to towns, pubs, and countries far away from its place of origin. This emigration was aided with the introduction of bottled beer in 1875 by the Joseph Schlitz Brewing Company in Milwaukee, Wisconsin, and the later advent of canned beer in the 1930s.

Brewing process

The brewer's first step in making a beer is to wet the kernels of grain, promoting their **germination**, or sprouting. During this process, the germ or embryo of the grain breaks down slightly, releasing the enzymes that turn starches, complex sugars, into simple sugars. The brewer allows the grain to germinate for a short period and dries it quickly in a kiln. Dried, malted grain is very durable and easily stored.

The brewer next mixes the dried malt with water to create a porridgy substance called mash. During this phase, the mash is brought to a **temperature** of around 150°F (66°C) and is kept there a number of hours. The starchy components of the grain break down during mashing, and are converted into simple sugars, which yeast will be able to digest during the **fermentation** phase.

The brewer now adds more water to the mash and rinses out a watery sugar **solution** called wort, which is heated to a boil. At this point, the brewer adds hops, the **flower** of the *Humulus lupulus* vine, which give beer its characteristic bitter flavor and aroma.

When the wort has cooled to the temperature most friendly to yeast (50–60°F [10–16°C]), these organisms are added and go to work consuming the sugars in the wort, growing to be as much as five to ten times their

A brewmeister and fellow worker inspect the current brew of a local beer in a brewery in the Dominican Republic. *Photograph by John Olson. Stock Market. Reproduced by permission.*

original weight. They fuel their growth with the sugar and in the process, change it from **carbohydrate** to **ethanol**, a form of alcohol, and **carbon** dioxide, the gas that is responsible for beer's foam and bubbles.

Types of beer

Most familiar beers are barley-based beverages. In Europe and the United States, beer is usually one of two types, ale or lager. Ale is a traditional British beer that, until the latter part of the nineteenth century, was not flavored with hops. Ale is made with a variety of yeast that rises to the top of the fermentation tank, and that produces a higher alcohol content than lagers. Ales range from pale ales such as Kolsch (from a district around Cologne, Germany) which is strongly hopped and not sweet, to strong, dark, ales such as porter and stout. Lager (which means cellar in German) originated in the Bavarian region of Germany. Lager is made with bottom-fermenting yeast, and is lower in alcohol content than ales. Lagers include: pilsner, a pale beer with a distinctive hop flavor; bock, a strong dark beer; weiss, a pale beer made with **wheat** that is usually served with slices of lemon or flavored with fruit juices.

Future developments

Current research in brewing technology is focusing on events at the cellular level. For example, brewery researchers are working on methods that will rapidly and accurately detect the presence and identity of unwanted yeasts or other **microorganisms** that find their way into the beer during brewing, and that change the flavor of the beer. (Potentially problematic organisms include *Lactobacillus, Pediococcus,* and *Obesumbacterium.*) Some of the methods already available or soon to be

KEY TERMS

Ale—A top-fermented beer that, until the latter part of the nineteenth century, was not flavored with hops; traditionally British ale has a higher alcohol content than lagers.

Fermentation—The process during which yeast consumes the sugars in the wort and releases alcohol and carbon dioxide as byproducts.

Hops—Comes from the cone of the flower of the vine *Humulus lupulus,* which is cultivated throughout temperate regions for use in brewing. The addition of hops gives beer its characteristic bitter flavor and aroma.

Lager—Lager (which means cellar in German) is a traditional Bavarian beer made with bottom-fermenting yeast. These beers are lighter in color and lower in alcohol content than ales. Over the past few decades, lager surpassed ale in popularity.

Malt—Grain that has germinated, or sprouted, for a short period and is then dried. During germination, the enzymes in the germ of the grain are released, which will make the sugars in the grain available to the yeast.

Wort—The sugar water solution made when malted barley is steeped in water and its complex sugars break down into simple sugars.

Yeast—A microorganism of the fungus family that promotes alcoholic fermentation, and is also used as a leavening in baking.

available include those using DNA probes, and protein and **chromosome** fingerprinting. Some of these techniques are already used in tests designed for at-home pregnancy testing, drug screening, **AIDS** testing, and tests for the presence of environmental contaminants such as **pesticides**.

Breweries are also using biotechnological techniques and **genetic engineering** to select and propagate choice characteristics in barley, rice, corn, yeast, and hops. Some of the characteristics that brewery researchers are working toward include **disease** resistance in the **crops**, and in yeast, the ability to resist **contamination** and to ferment carbohydrates that yeasts have been incapable of fermenting. Micropropagation is a new and experimental technique that involves making any number of genetically identical copies of a **plant**, by removing minute quantities of its growth points and placing them in a medium that encourages the rapid growth of shoots. This process is then repeated again and

again. **Mutation** breeding is a process for creating mutations, some of which may be desirable, by exposing the plant to **x rays** or chemical mutagens; and transformation technologies, which allow the breeder to make a change in a single **gene** by adding or deleting it.

See also Fermentation; Yeast.

Resources

Books

Briggs D.E. and J.S. Hough. *Malting and Brewing Science.* Volumes 1 & 2. London: Chapman and Hall, 1981.
Gourish and Wilson. *The British Brewing Industry, 1830-1980.* Cambridge: Cambridge University Press, 1994.
Hough, J.S. *Biotechnology of Brewing.* Cambridge: Cambridge University Press, 1985.

Other

Gump, Barry, ed. *Beer and Wine Production: Analyses, Characterization, and Technological Advances.* American Chemical Society Symposium Series 536, American Chemical Society, Washington D.C., 1993.

Beth Hanson

Brick

Bricks are one of the oldest types of building blocks. They are an ideal building material because they are relatively cheap to make, very durable, and require little maintenance. Bricks are usually made of kiln-baked mixtures of clay. In ancient times, bricks were made of mud and dried in the **sun**; modern bricks are made from **concrete**, **sand** and lime, and **glass**. The physical and chemical characteristics of the raw materials used to make bricks, along with the **temperature** at which they are baked, determine the **color** and hardness of the finished product. Bricks are made in standard sizes, are usually twice as long as they are wide and, since most bricklaying is done manually, are made small enough to fit in the hand. Bricklayers use a trowel to cover each brick with mortar—a mixture of cement, sand, and **water**. The mortar hardens when dry and keeps the bricks in place. Bricks are arranged in various patterns, called bonds, for strength.

History

Archaeologists have found bricks in the Middle East dating 10,000 years ago. Scientists suggest that these bricks were made from mud left after the **rivers** in that area flooded. The bricks were molded by hand and left in the sun to dry. Structures were built by layering the bricks using mud and tar as mortar. The ancient city of

Ur (modern Iraq) was built with mud bricks around 4,000 B.C. The Bible (Exodus 1:14; 5:4-19) provides the earliest written documentation of brick production—the Israelites made bricks for their Egyptian rulers. These bricks were made of clay dug from the **earth**, mixed with straw, and baked in crude ovens or burned in a fire. Many ancient structures made of bricks, such as the Great Wall of China and remnants of Roman buildings, are still standing today. The Romans further developed kiln-baked bricks and spread the art of brickmaking throughout **Europe**.

The oldest type of brick in the Western Hemisphere is the adobe brick. Adobe bricks are made from adobe **soil**, comprised of clay, quartz, and other **minerals**, and baked in the sun. Adobe soil can be found in dry regions throughout the world, but most notably in Central America, Mexico, and the southwestern United States. The Pyramid of the Sun was built of adobe bricks by the Aztecs in the fifteenth century and is still standing. In **North America**, bricks were used as early as the seventeenth century. Bricks were used extensively for building new factories and homes during the **Industrial Revolution**. Until the nineteenth century, raw materials for bricks were mined and mixed, and bricks were formed, by manual labor. The first brickmaking machines were steam powered, and the bricks were fired with **wood** or **coal** as fuel. Modern brickmaking equipment is powered by gas and **electricity**. Some manufacturers still produce bricks by hand, but the majority are machine made.

Brick manufacturing

The manufacture of bricks entails several steps and starts with obtaining the raw materials. Clays are mined from open pits or underground mines. Storage areas are located at the **mining** site so that portions from various "digs" can be blended. The clay mixture goes through a process called primary crushing, where the clay is put through giant rollers that break the clay into small chunks. This mixture is transported to the manufacturing site, where the clay mixture is pulverized and screened to remove impurities. Further blending of materials may take place at this time.

There are three methods of forming bricks. The most common is the stiff-mud process where the clay blend is put into a machine called a pug mill that mixes the clay with water (12-15% by weight), kneads the mixture, removes trapped air, and transfers the mixture to an auger machine. The auger forces or extrudes the wet clay through a die that forms a continuous rectangle-shaped column. The column is cut with **steel** wires into desired lengths. The newly formed bricks are place on drying racks for a few days and then fired in a kiln. The soft-

mud process is used when the mined clay is naturally too wet (20-30% by weight) to undergo the stiff-mud process. The clay is mixed, extruded, and placed in lubricated molds. Each mold makes six to eight bricks. The drying process takes more time than with stiff mud, but the firing procedure is the same. The third method is the dry-press process, which is most commonly used when making refractory bricks. The clay has minimal water content (up to 10% by weight) and is exposed to high **pressure** (in a hydraulic or mechanical press) while in the molds. The bricks are dried and fired. While still damp and moldable, textures, designs, or functional grooves can be pressed into the brick. Special glazes can be applied for decorative and for functional purposes.

Firing or burning the bricks takes two to five days. The most common type of kiln used to fire bricks is the tunnel kiln, where the bricks, stacked on cars, move slowly though a long chamber or tunnel. Many changes in the physical properties occur during the firing process. During firing, any residual water evaporates, some minerals melt, blend, and fuse, and organic **matter** oxidizes. The hardness of the brick increases and the color develops. The whole process of making bricks takes 10-12 days.

With handmade bricks, the clay is kneaded and put into molds. Excess clay is skimmed off the top of the mold, and the brick is then dumped out, dried, and fired. Handmade bricks are usually more expensive than machine-made bricks. They are often used in special projects, such as historical restoration.

Types of brick

Some bricks are made for specific purposes and are made of certain raw materials, formed in a particular shape, or with added special textures or glazes. Common brick is the everyday building brick. They are not made of special materials, and do not have special marks, color, or texture. Common brick is typically red and sometimes used as a "backup" brick, depending on the quality. Face brick is often applied on top of common backup brick. Face brick can be obtained in a variety of colors, has a uniform surface appearance and color, is more durable, and is graded according to its ability to withstand freezing temperatures and moisture. Refractory bricks are made from fireclays—clays with a high alumina or silica content or nonclay minerals such as bauxite, zircon, silicon carbide, or dolomite. Fireclays are **heat** resistant and are used in various types of furnaces, kilns, and fireplaces. **Calcium** silicate bricks are often made in areas where clay is not readily available. Glazed bricks are made primarily for walls in buildings such as dairies, hospitals, and laboratories, where easy cleaning is necessary.

See also Stone and masonry.

A bowstring arch bridge in Arizona. The deck is supported from the arch by hangers. The load is transmitted to the abutments by the outward thrust of the arch. *JLM Visuals. Reproduced by permission.*

of the bridge, and the inner piers that anchor the fixed end of the cantilever.

The two longest cantilever bridges in the world are the Forth Railway Bridge in Scotland, completed in 1890, and the Quebec Bridge in Canada, built in 1917. The former is 1,700 ft (520 m) in length and the latter, 1,800 ft (550 m) long.

Trusses

The strength of a cantilever bridge can be increased by the use of trusses. A truss is structure that consists of a number of triangles joined to each other. The triangle is an important component of many kinds of structures because it is the only geometric figure that can not be pulled or pushed out of shape without actually changing the length of one of its sides. By combining a number of triangles into a single unit, the unit is given a great deal of strength.

The cantilever beam, end beams, and joining beams in a cantilever bridge are often strengthened by adding trusses to them. The trusses act somewhat like an extra panel of **iron** or steel, adding strength to the bridge with relatively little additional weight. The open structure of a truss also allows the wind to blow through them, preventing additional stress on the bridge from this factor.

Trusses are used not only in cantilever bridges, but in all other kinds of bridges also. In fact, you have probably noticed the complex pattern of intersecting triangles on bridges over which you have passed. These truss patterns are one of the most efficient ways of adding strength to any type of bridge an engineer designs.

Arch bridges

As its name suggests, the main supporting structure in an arch bridge is one or more curved elements.

The dead and live forces that act on the arch bridge are transmitted along the curved line of the arch into abutments at either end. These abutments are sunk deep into the **earth**, into bedrock if at all possible. They are, therefore, essentially immovable and able to withstand very large forces exerted on the bridge itself. This structure is so stable that piers are generally unnecessary in an arch bridge.

The deck of an arch bridge can be placed anywhere with relationship to the arch: on top of it, beneath it, or somewhere within the arch. The deck is attached to the arch by vertical posts (ribs and columns) if the deck is above the arch, by ropes or cables (suspendors) if the deck is below the arch, and by some combination of the two if the deck is somewhere within the arch.

Most arch bridges today are made either of steel or of reinforced concrete. The longest existing steel arch bridge is the New River Gorge Bridge in Fayetteville, West Virginia, built in 1977. It is 1,700 ft (518 m) long. The longest reinforced concrete bridge is the Jesse H. Jones Memorial Bridge at the Houston Ship Channel, Texas, with a length of 1,500 ft (455 m).

Suspension bridges

The longest bridges in the world are all suspension bridges. Some examples are the Humber Bridge in Hull, England, with a length of 4,626 ft (1,410 m), the Verrazano-Narrows Bridge in lower New York Bay (4,620 ft [1,298 m]); the Golden Gate Bridge over the entrance to San Francisco Bay (4,200 ft [1,280 m]); and the Mackinac Straits Bridge connecting the Upper and Lower Peninsulas of Michigan (3,800 ft [1,158 m]).

In a suspension bridge, the dead and live loads are carried by thick wire cables that run across the top of at least two towers and are anchored to the shorelines within heavy abutments. In some cases, the bridge deck is supported directly by suspendors from the cables, while in other cases, the suspendors are attached to a truss, on top of which the deck is laid. In either case, the dead and light load of the bridge are transmitted to the cables which, in turn, exert stress on the abutments. That stress is counteracted by attaching the abutments to bedrock.

The towers in a suspension bridge typically rest on massive foundations sunk deep into the river bed or sea bed beneath the bridge itself. The wire cables that carry the weight of the bridge and its traffic are made of parallel strands of steel wire woven together to make a single cable. Such cables typically range in diameter from about 15 in (38 cm) to as much as 36 in (91 cm). Smaller cables can be ordered from a factory, while thicker cables may have to be assembled on the construction site itself.

An interesting hybrid of the cantilever and suspension bridge is known as the cable-stayed bridge. The 1,200 ft (366 m) Sunshine Skyway across the entrance to Tampa Bay in Florida is one of the most beautiful examples of the cable-stayed bridge. In a cable-stayed bridge, the deck is cantilevered outward in both directions from a central tower. The deck is then attached to the tower by a series of cables, similar to those in a suspension bridge. Often, a cable-stayed bridge will make use of two towers. In that case, the cantilevered sections extending towards each other in the middle of the bridge can be joined together, producing an unusually long central span. The advantage of the cable-stayed bridge is that support for dead and live loads come from three distinct places: the towers, the cables, and the abutments to which the bridge is attached at each end.

Pontoon bridges

Pontoon bridges are bridges that float on water. They find use primarily in two situations. First, they find application during wars when engineers need to construct a simple bridge quickly and easily. In such instances, they can be assembled from inflatable rubber or plastic and put into place in a matter of hours. Second, they can be used in rivers and lakes where the river bottom makes it very difficult or impossible to install piers. Lake Washington, in the state if Washington, for example, once had three floating bridges. All were made of large hollow concrete blocks tied to each other.

Movable bridges

Traditionally, three kinds of movable bridges have been constructed. In one, the swing bridge, the deck is rotated around a central span, a large, heavy pier sunk into the river bottom. The swing bridge has one serious disadvantage. The central pier, on which the bridge rotates, is usually located in the deepest part of the waterway. Ships with significant drafts may, therefore, have difficulty passing through such bridges. The swing bridge also has one important advantage. Since it never moves upward in a vertical direction, it will not interfere with air traffic that might be present in the area.

In the second type of movable bridge, the bascule bridge, the deck is raised, either at one end or at two ends. The bascule bridge acts, therefore, something like a cantilever in which the free end is raised to permit passage of seagoing vessels.

In the third type of movable bridge, the vertical-lift bridge, the whole central portion of the bridge is raised straight up by means of steel ropes. One disadvantage of the vertical-lift bridge, of course, is that it can not open

---

KEY TERMS

Abutment—Heavy supporting structures usually attached to bedrock and supporting bridge piers.

Cable-stayed bridge—A type of bridge that is a mix of cantilever and suspension bridge, in which the deck is supported both by one or more central towers and cables suspended from the tower(s).

Dead load—The force exerted by a bridge as a result of its own weight.

Dynamic load—The force exerted on a bridge as a result of unusual environmental factors, such as earthquakes or strong gusts of wind.

Live load—The force exerted on a bridge as a result of the traffic moving across the bridge.

Piers—Vertical columns, usually made of reinforced concrete or some other strong material, on which bridges rest.

Suspendors—Ropes or steel wires from which the deck of a bridge is suspended.

Truss—A very light, yet extremely strong structural form consisting of triangular elements, usually made of iron, steel, or wood.

---

entirely above the waterway, but can only be raised to a given maximum height.

Resources

Books

Brash, Sarah, Matthew Cope, Charles Foran, Dónal Kevin Gordon, and Peter Pocock. *How Things Work: Structures*. Alexandria, VA: Time-Life Books, 1991.

"Bridge," *McGraw-Hill Encyclopedia of Science & Technology*, 7th ed. New York: McGraw-Hill Book Company, 1992, vol. 2.

Corbett, Scott. *Bridges*. New York: Four Winds Press, 1978.

DeLony, Eric. *Landmark American Bridges*. Boston: Little, Brown and Company, 1993.

MacGregor, Anne, and Scott MacGregor. *Bridges: A Project Book*. New York: Lothrop, Lee & Shephard Books, 1980.

Trefil, James. *Encyclopedia of Science and Technology*. The Reference Works, Inc., 2001.

David E. Newton

Bristletails

Bristletails are about 300-400 **species** of small, elongate, terrestrial **insects** in the order Thysanura. Bristletails have an ancient evolutionary lineage, and they are believed to be relatively primitive, that is, similar in form and function to the most early evolved insects.

Bristletails have a simple **metamorphosis**, with three life-history stages: egg, nymph, and adult. Both the nymphal and adult stages are wingless, and they are rather similar in their physical appearance, the main difference being in size and sexual maturity. The adults are large and mature.

Bristletails are easily distinguished by the three thread-like appendages that emerge from the end of their abdomen, by their long, backward-pointing antennae, and by their body covering of glistening scales. Bristletails have chewing mouth parts, and they feed on a wide range of types of soft, usually decaying organic **matter**. Bristletails hide during the day, and if they are disturbed they run quickly away to seek a new hiding place. Otherwise, bristletails are active at night. Bristletails are unusual in that they continue to grow and molt even after they have become sexually mature adults. Almost all other insects stop growing after they become breeding adults.

The most familiar bristletails are those in the family Lepismatidae, including the light-colored, common silverfish (*Lepisma saccharina*) and the darker-colored firebrat (*Thermobia domestica*), which often occurs in warm, moist places, for example, near hearths, stoves, and furnaces. Both of these species can be common in moist places in buildings, where they feed on a wide variety of starchy foods, and can sometimes cause significant damages to books, wallpaper, fabrics, and stored foods.

The jumping bristletails (family Machilidae) are widespread in **North America**, living under organic debris in various types of natural, terrestrial habitats. The primitive bristletails (Lepidotrichidae) and nicoletiid bristletails (Nicoletiidae) are all rather rare in North America.

Brittle star

Brittle stars are starfishlike echinoderms (phylum Echinodermata) in the family Ophiodermatidae, whose star-shaped bodies are radially symmetrical and are supported by a hard endoskeleton made of **calcium** salts. Brittle stars are closely related to basket stars, and more distantly related to **starfish**, **sand dollars**, and **sea urchins**. Brittle stars are named for the ease by which their arms fall off when touched; these animals, known collectively as ophioroids, are also called serpent stars (*ophis* means snake in Greek) because their long arms resemble serpents. The brittle stars comprise the largest

number of **species** (about 1,000) of echinoderms and are found on the seabed in all of the world's oceans. Brittle stars also inhabit the dark high-pressure environments on the floor of the abyssal zone, the deepest part of the **ocean** where few other living things can survive. Some species of brittle star can swim, but most species simply crawl along the ocean floor. Brittle stars in shallow seas tend to avoid **light** and prefer to hide in dark crevices, becoming more active at night, or they inhabit the ocean depths where it is always dark. Brittle stars usually have five long, thin, jointed arms covered with spines; sometimes they have six or seven arms. Brittle stars can be as large as 2 ft (0.6 m) in diameter or as small as a few millimeters. Ophioroids are usually a drab green, grey, or brown, but some have variegated **color** patterns. Some species of brittle stars glow in the dark-a bright green **luminescence** appears if they are disturbed. Others can change their color.

The arms of brittle stars are attached to a central disklike body that houses on the underside the mouth and jaws, stomach, and saclike body cavities called bursae, which are peculiar to ophioroids. The mouth is surrounded by five moveable jaw segments, making the mouth opening look like a star. Some brittle stars are carnivorous, while others feed on small particles of **plankton**. The food enters the mouth and goes directly into the stomach, and there is no intestine and no anus; absorption and excretion are carried out instead by the five pairs of bursae located at the base of each of the arms. **Water** circulates in the bursae and there is an exchange of respiratory gases and excretion of wastes. The coelomic walls of the bursae contain the gonads that discharge sex cells in the water for **fertilization**. The larva, which develops from a fertilized egg, is called an ophiopluteus and is free swimming in the plankton until it transforms into the juvenile stage when it settles on the bottom of the ocean. Brittle stars are either male or female, although a few individuals are hermaphrodites. Brittle stars spawn at the end of summer, and most species release their eggs into the plankton and invest no parental care thereafter. They can also reproduce asexually; if an arm breaks off and it still has a small piece of the central disk attached, the arm can regenerate into a whole new brittle star.

Brittle stars are among the most active of the echinoderms and, unlike starfish and sea urchins, can move easily and quickly. The arms of a brittle star reach out in pairs to pull the **animal** along. Each arm is supported by a central internal skeletal support (ossicle). Like starfish, brittle stars have tube feet, but those of brittle stars lack suckers. The tube feet help more with feeding than with locomotion. Like other echinoderms, brittle stars can regenerate lost parts. If an arm is broken off, a new one grows in its place within months. If an arm is injured, the star can cast off the injured arm (autotomy) and then eventually grow a new one.

Bromeliad family (Bromeliaceae)

The pineapple family (Bromeliaceae) consists of about 1,500 **species** of flowering plants. Most species are medium-sized herbs with tightly packed, thick, stiff, spi-ralling leaves that usually have spiny margins. Some species are semi-woody, and a few rare ones, such as *Puya raimondii,* are trees that can reach 33 ft (10 m) in height. Most species of bromeliads are epiphytes in rainforests, while others are terrestrial, inhabiting dry habitats in mountainous and coastal regions. Bromeliads are native to tropical and subtropical regions of North and **South America**. The only exception may be *Pitcairnia feliciana* of coastal west **Africa**; this species appears to represent a remarkable example of long-distance, natural dispersal, although it may have been introduced into Africa by hu-mans. The range of the bromeliads extends from Tierra del Fuego in southernmost South America, northward into Mexico, then through the southeastern United States to their northernmost limit in southern Maryland.

The most commercially valuable member of this family is the common pineapple, *Ananus comosus.* A na-tive of South America, the pineapple has been widely planted as a cash crop throughout tropical and subtropical regions, especially in the Hawaiian Islands, the Philip-pines, South Africa, and southern **Asia**. The pineapple is a biennial **plant**, that is, it usually lives for only two years. In the first year this species produces a dense growth of sharp-pointed, overlapping leaves. During the second year a short stalk with many flowers is produced. Each **flower** gives rise to **fruits** that are technically berries, but then the stalk bearing the fruit begins to swell. This results in the development of a thick, sweet, fleshy mass, within are embedded the fruits. This whole thing is the "pineapple," which is therefore an aggrega-tion of fruits within an accessory structure (the thickened stem). When you buy a pineapple at the market you no-tice at the top of the "fruit" a tuft of prickly, reduced leaves or bracts, and a cleanly cut-off base. This may lead you to believe that the pineapple was cut off from the plant at ground level, but this was not the case. Pineap-ples grow at the end of shoots, two to four per plant, so they are cut off from an underlying, large-leafed shoot.

Bromeliads are xerophytes and possess many of the usual, water-conserving adaptations of such plants: a thick epidermis covered with wax, water-storage cells

Single swamp cypress covered with spanish moss.
Photograph by Chris Sharp. Photo Researchers, Inc. Reproduced by permission.

that cause the leaves to appear succulent (that is thick and fleshy), and sheathing **leaf** bases. One of the notable characteristics of bromeliads is the distinctive, water-ab-sorbing scales on their leaves and stems. These thin scales occur in grooves in the epidermis, and they resem-ble opened umbrellas. Their thinness and large surface area make the scales ideal for rapidly absorbing **water**.

Equally important in terms of water economy are the shape and arrangement of the leaves of bromeliads. In many species the leaves are wide and deeply U-shaped where they join the stem, forming a series of vessel-like compartments. When it rains, water flows down the leaves and pools in the compartments, where it can be ab-sorbed by the umbrella scales. Especially remarkable are the "tank plants," such as *Nidularium* and *Billbergia.* In these species the stem is greatly reduced and the densely packed leaves have broad, overlapping bases, resulting in a pitcher or vase-like center-the tank. Rainwater fills the tank, where some of the moisture is absorbed by the um-

A bromeliad growing on the bole of a cypress tree in Everglades National Park, Florida. *Photograph by E.R. Degginger. Photo Researchers, Inc. Reproduced by permission.*

brella scales. Because the tank is shaded by the dense crown of leaves around it, the water does not evaporate quickly and can persist, enabling the plant to survive periods of **drought**. Interestingly, some species of **mosquitoes** breed nowhere else but in these tanks.

Another important **adaptation** of some bromeliads to drought-prone environments is seen in their stomata. Stomata allow gaseous **carbon dioxide**, a necessary ingredient for **photosynthesis**, into the leaf. In most plants the stomata are open during daylight hours, because that is when **light** is available to drive photosynthesis. When open during the day, however, stomata also lose water, which is a disadvantage for plants growing in environments where water is scarce. Many plants, including the bromeliads pineapple and Spanish **moss**, have evolved stomata that are closed during the **heat** of the day but open at night when temperatures are cooler. These plants trap the **carbon** dioxide that enters through their stomata

and store the gas until daylight, when light **energy** is available for further processing through photosynthesis.

Why are some plants of wet tropical and subtropical regions, such as bromeliads, specially adapted to dry conditions? The key is that most bromeliads are epiphytic, occurring high in the crowns of forest trees. When it is not raining and the **sun** is out, these plants can dry out rapidly because their roots are not in **soil**. In addition most bromeliads that are terrestrial generally occur in rocky habitats, where rainwater rapidly percolates through the soil, leaving it dry.

Spanish moss (*Tillandsia usneoides*) is not a true moss but rather is one of the interesting bromeliads known as "air plants." Most of these species grow as epiphytes high in the crowns of trees, where of course there is no soil. In order to obtain the **nutrients** necessary for growth, air-plants absorb some directly from the atmosphere, such as gases of **sulfur** and **nitrogen**. Other nutrients must be absorbed from rainwater or atmospheric dusts. Air-plants are commonly available in flower shops. These attractive plants need to be misted occasionally with a sprayer, because their foliar scales must be moistened to absorb the nutrients. Because they derive most of their nutrients directly or indirectly from the atmosphere, these bromeliads are called air plants. Spanish moss commonly grows in the southeastern United States, where it often forms large and beautiful drapes in **tree** canopies. The roots of Spanish moss serve to hold the plant to the tree on which it is growing, and do not function in obtaining water and nutrients.

Bromeliads have become increasingly popular as indoor plants. Air plants are appealing novelty items, but many of the other, leafier bromeliads are also now prized. These have attractive, usually bright-red flowers, designed to attract **birds** as pollinators, and often surrounded by brightly colored modified leaves. Furthermore, as xerophytic plants, bromeliads can withstand the benign neglect of forgetful watering.

Resources

Books

Dahlgren, R.T., H.T. Clifford, and P.F. Yeo. *The Families of the Monocotyledons: Structure, Evolution, and Taxonomy.* Berlin: Springer-Verlag, 1985.

Les C. Cwynar

Bromine *see* **Halogens**

Bronchitis

Bronchitis is the **inflammation** of the bronchi and is a commonly seen winter condition. The bronchi (the air passages leading into the lungs) are formed by the division of the trachea (the main windpipe leading from the larynx [Adam's apple] down through the neck into the chest). The trachea branches left and right into the bronchi which branch to supply lung lobe with the means for air to pass in and out.

Like the trachea, the bronchi are formed of cartilage rings overlain with muscle. One layer of muscle runs lengthwise along the tube; the other layer is circular. These muscles regulate the diameter of the air passages.

A common cold or extended exposure to cold temperatures or **air pollution** over **time** may lead to bronchitis (the suffix "-itis" means inflammation). A cough and sore throat are the primary symptoms but difficulty in breathing and development of a fever are also charac-

teristics. This sudden and short-lived bronchitis is called acute bronchitis. This is easily cured with aspirin in more serious cases an antibiotic. Acute bronchitis usually causes no long-term problems.

Chronic bronchitis is a more serious condition. Chronic means it is a condition that persists over a long period of time. Chronic bronchitis is a **disease** of cigarette smokers and is often accompanied by **emphysema**. Emphysema is a condition in which lung **tissue** is destroyed and the capacity to breathe is seriously impaired.

The form of bronchitis associated with emphysema is called chronic obstructive pulmonary disease (COPD) and is difficult to cure. It is often difficult to tell whether the respiratory problems associated with COPD are the result of emphysema or chronic bronchitis. Some physicians consider the two conditions to be synonymous: if one has chronic bronchitis one also has emphysema.

See also Respiratory diseases; Respiratory system.

Brown dwarf

A brown dwarf is a pseudostar; a body of gas not massive enough for the gravitational **pressure** in its core to ignite the hydrogen-fusion reaction that powers true stars. The name "brown dwarf" is a play on the name of the smallest class of true stars, "red dwarf," but while red dwarfs are actually red, brown dwarfs are not brown, but purple or magenta. Objects ranging in **mass** between 13 and 75 times the mass of Jupiter—between 1.2% and 7% the mass of the Sun—are generally considered brown dwarfs. Clear rules for distinguishing large planets from brown dwarfs, however, are lacking. Some astronomers consider objects down to seven or eight **Jupiter** masses to be brown dwarfs, while others reserve this term for objects heavy enough to initiate **deuterium** fusion in their cores, that is, objects of 13 Jupiter masses or more. (Deuterium is a relatively uncommon form of **hydrogen** that has both a **neutron** and a **proton** in its nucleus; deuterium fusion is a minor reaction in true stars and persists for only a few million years even in brown dwarfs.) In 2001, an international committee declared that objects heavier than 13 Jupiter masses should be labeled brown dwarfs regardless of whether they **orbit** true stars, while objects below this threshold should be labeled as planets if they are orbiting true stars and as sub-brown dwarfs if they are not.

Until recently, astronomers could only theorize that brown dwarfs were common in the Universe. They observed that the less massive stars are far more common than the more massive stars, a trend that would suggest that brown dwarfs should be still more numerous. Brown

la cause **infection** by actually entering host cells. As the bacteria cross the host cell **membrane**, they are engulfed by host cell vacuoles called phagosomes. The presence of *Brucella* within host cell phagosomes initiates a characteristic immune response, in which infected cells begin to stick together and form aggregations called granulomas.

Brucella species

Three **species** of *Brucella* cause brucellosis in humans: *Brucella melitensis*, which infects **goats**; *B. abortis*, which infects cattle and, if the animal is pregnant, causes the spontaneous abortion of the fetus; and *B. suis*, which infects **pigs**. In animals, brucellosis is a self-limiting disease, and usually no treatment is necessary for the resolution of the disease. However, for a period of time from a few days to several weeks, infected animals may continue to excrete brucella into their urine and milk. Under warm, moist conditions, the bacteria may survive for months in **soil**, milk, and even seawater.

Because the bacteria are so hardy, humans may become infected with *Brucella* by direct contact with the bacteria. Handling or cleaning up after infected animals may put a person in contact with the bacteria. *Brucella* are extremely efficient in crossing the human skin barrier through cuts or breaks in the skin.

Symptoms and treatment of brucellosis

The incubation period of *Brucella*—the time from exposure to the bacteria to the start of symptoms—is typically about three weeks. The primary complaints are weakness and fatigue. An infected person may also experience muscle aches, fever, and chills.

The course of the disease reflects the location of the *Brucella* bacteria within the human host. Soon after the *Brucella* are introduced into the bloodstream, the bacteria seek out the nearest lymph nodes and invade the lymph node cells. From the initial lymph node, the *Brucella* spread out to other **organ** targets, including the spleen, bone marrow, and liver. Inside these organs, the infected cells form granulomas.

Diagnosing brucellosis involves culturing the **blood**, liver, or bone marrow for *Brucella* organisms. A positive culture alone does not signify brucellosis, since persons who have been treated for the disease may continue to harbor *Brucella* bacteria for several months. Confirmation of brucellosis, therefore, includes a culture positive for *Brucella* bacteria as well as evidence of the characteristic symptoms and a history of possible contact with infected milk or other animal products.

In humans, brucellosis caused by *B. abortus* is a mild disease that resolves itself without treatment. Bru-

KEY TERMS

Granuloma—An immune response in which cells infected with bacteria clump together. Granuloma formation is typical of tuberculosis and brucellosis.

Lymph nodes—Small, bean-shaped structures located along the lymphatic vessels of the body. Lymph nodes function in the immune response to protect the body against foreign cells.

Pasteurization—The process in which milk or either dairy products are heated to a high temperature in order to kill disease-causing bacteria.

cellosis caused by *B. melitensis* and *B. suis*, however, is chronic and severe. Brucellosis is treated with administration of an antibiotic that penetrates host cells to destroy the invasive bacteria.

Prevention

Since the invention of an animal **vaccine** for brucellosis in the 1970s, the disease has become somewhat rare in the United States. Yet the vaccine cannot prevent all incidence of brucellosis. In 1989, the Centers for Disease Control reported only 95 total cases in the United States. Most of these were reported in persons who worked in the meat processing industry. Brucellosis remains a risk for those who work in close contact with animals, including veterinarians, farmers, and dairy workers.

Brucellosis also remains a risk when animal products from foreign countries are imported into the United States. Outbreaks of brucellosis have been linked to unpasteurized feta and goat cheeses from the Mediterranean region and **Europe**. In the 1960s, brucellosis was linked to bongo drums imported from **Africa**: drums made with infected animal skins can harbor *Brucella* bacteria, which can be transmitted to humans through cuts and scrapes in the human skin surface.

In the United States, preventive measures include a rigorous vaccination program that involves all animals in the meat processing industry. On an individual level, people can avoid the disease by not eating animal products imported from other countries. If this is not possible or desirable, make sure that imported cheeses have been made with pasteurized milk. If the package does not indicate pasteurization, do not eat the cheese.

Resources

Books

Prescott, L., J. Harley, and D. Klein. *Microbiology.* 5th ed. New York: McGraw-Hill, 2002.

Thimm, Bernhard M. *Brucellosos: Distribution in Man, Domestic, and Wild Animals.* Berlin: Springer-Verlag, 1982.

Periodicals

Kiel, Frank W., and M. Yousouf Khan. "Brucellosis in Saudi Arabia." *Social Science and Medicine* 29 (1989): 999-1001.

Olle-Goig, Jaime E., and Jaume Canela-Soler. "An Outbreak of *Brucella melitensis.* Infection by Airborne Transmission Among Laboratory Workers." *American Journal of Public Health* 77 (March 8, 1987): 335-38.

Wright, Paul. "Brucellosis." *American Family Physician* 35 (May 1987): 155-59.

Kathleen Scogna

Bryophyte

Bryophytes include the mosses, liverworts, and hornworts. Bryophytes are the simplest of plants (excluding the **algae**, which are not considered plants by most botanists). Bryophytes are small, seldom exceeding 6-8 in (15-20 cm) in height, and usually much smaller. They are attached to the substrate (ground, rock, or **bark**) by *rhizoids*, which are one or a few-celled, root-like threads that serve only for anchoring and are not capable of absorbing **water** and **nutrients** from the substrate. Brypohytes lack *vascular tissue* (the specialized cells grouped together to pipe water and nutrients to various parts of the body), or in the rare cases when this **tissue** is present, it is not well differentiated. The leaves of bryophytes are technically not true leaves, because in most **species** they lack vascular tissue. However, they are functionally equivalent to leaves, containing chlorophylls a and b for **photosynthesis**. Leaves are usually one-cell thick, except for the midrib, which may be up to 15 cells thick. Bryophytes satisfy their nutritional requirements by absorbing **minerals** from dust, rainfall, and water running over their surface.

The life cycle of bryophytes is characterized by an alternation of generations, one of which is a multicellular, diploid **individual** called a *sporophyte*, having two of each type of **chromosome** per **cell**. This stage alternates with multicellular, haploid individual called the *gametophyte*, with only one of each type of chromosome per cell, as is also the case with **animal** sperm. Bryophytes are unique among plants in that the dominant, conspicuous generation is the haploid gametophyte. In all other plants, the dominant stage is the diploid sporophyte.

Most reproduction of bryophytes is asexual, occurring by fragmentation of body parts, and by the production of specialized vegetative units called *gemmae*. Gemmae may be produced as microscopic plates (in the genus *Tetraphis*), as bulbils in the axils of leaves (in *Pohlia*), or as microscopic filaments (in *Ulota*). When **sexual reproduction** occurs, it always involves a flagellated sperm (produced in a specialized **organ** called an *antheridium*) that must swim through water to reach an egg located in a specialized, flask-shaped organ (the *archegonium*). The antheridia and archegonia are surrounded by a layer of sterile cells, which protects the sex organs from mechanical damage and desiccation.

The union of the sperm and egg results in a diploid zygote, i.e., a new sporophyte. This is nourished by the gametophyte and grows on it in a parasitic fashion, although the sporophytes of some bryophytes photosynthesize and make some contribution to their own growth. Initially, as the young sporophyte grows, the archegonium also enlarges. However, it ultimately fails to keep pace with the growth of the sporophyte and becomes detached from its base, forming a cap-like structure called a *calyptra*.

Classification, characteristics, and habitats of bryophytes

The classification of bryophytes has been controversial among botanists. Traditionally, the division Bryophyta has included the true mosses, liverworts, and hornworts. However, some scientists consider each of these groups sufficiently distinct to deserve their own division: Bryophyta for the mosses, Hepatophyta for the liverworts, and Anthoceratophyta for the hornworts. The latter view is followed here, although the bryophyte is used as a collective term for all of these.

About 15,000 species of bryophytes have been described. They are distributed throughout the world, and are especially abundant in arctic and boreal regions, where they often dominate the ground vegetation. Bryophytes also occur in humid tropical regions where they commonly grow on other plants, especially in higher-elevation **forests**. Bryophytes are considered the **amphibians** of the **plant** world, because they require abundant moisture to grow. This requirement for water results from a number of their characteristic features. Their stems and leaves are thin, and either lack a cuticle (that is, a waxy surface layer) or have a very thin one, making them prone to drying out. Because bryophytes lack roots and a vascular system, they cannot obtain water from the **soil** and transport it to above-ground tissues; for this same reason, bryophytes are necessarily small. In addition, their sperm require free water in order to swim from their parent plant to the egg on another plant.

Hepatophyta (division liverworts)

Hepatophyta means "liver plant" and refers to the body of some common species of liverworts, whose lob-

KEY TERMS

Antheridium—A sperm-producing organ consisting of sperm-producing tissue, surrounded by a sterile layer of cells.

Archegonium—An egg-producing organ, often flask-shaped, with an outer layer of sterile cells.

Calyptra—An enlarged and modified archegonium that forms a cap around the capsule of the developing sporophyte.

Gametophyte—Individual plant containing only one set of chromosomes per cell that produces gametes i.e. reproductive cells that must fuse with other reproductive cells to produce a new individual.

Sporophyte—The diploid, spore-producing generation in a plant's life cycle.

Zygote—The cell resulting from the fusion of male sperm and the female egg. Normally the zygote has double the chromosome number of either gamete, and gives rise to a new embryo.

spores, mosses play a vital role in being among the first colonizers of disturbed sites. They stabilize the soil surface, thereby reducing **erosion**, while at the same time reducing the **evaporation** of water, making more available for succeeding plants. Mosses are not an important source of food for vertebrate herbivores. Peat mosses are the dominant plants of extensive northern wetland areas, and are largely responsible for the development of bogs.

Most species of mosses are not of any direct economic importance, and none are a food source for humans. Peat mosses are economically the most important mosses. Peat mosses are an important source of fuel in some countries. Peat is abundant in northern regions and represents a vast reservoir of potential **energy**. In northern **Europe**, peat has historically been dried, and in some cases compressed into briquettes for use in fireplaces and stoves. In Ireland, peat is still extensively used for cooking. One great advantage of peat as a fuel is that it burns very cleanly. About 95% of peat harvested in Ireland is burned to generate **electricity**. Peat is also highly valued as a conditioner of inorganic soils. Because it absorbs large amounts of water readily, peat improves the water-holding capacity of soil. Peat mosses are characteristically acidic which prevents the growth of most **bacteria**. They have therefore been used by indigenous peoples for diapers, and during the World Wars, when bandages were in short supply, peat mosses were a commonly used antiseptic dressing for wounds.

In recent years, mosses have become important in monitoring the health of ecosystems, especially in **relation** to atmospheric **contamination**. Because bryophytes lack roots, many of their nutritional requirements are met by nutrients deposited from the atmosphere. Thus, they are sensitive indicators of atmospheric pollutants. Changes in the distributions of mosses (and **lichens**) are therefore an early-warning signal of serious effects of atmospheric **pollution**.

See also Liverwort; Symbiosis; Wetlands.

Resources

Books

Longton, R.E. *The Biology of Polar Bryophytes and Lichens.* U.K.: Cambridge University Press, 1988.

Richardson, D.H.S. *The Biology of Mosses.* New York: John Wiley and Sons, Inc., 1981.

Schofield, W.B. *Introduction to Bryology.* New York: Macmillan Publishing Co., 1985.

Les C. Cwynar

Bubble chambers *see* **Particle detectors**

Bubonic plague

Bubonic plague is a contagious, deadly **disease** caused by the bacterium *Yersinia pestis.* Sometimes referred to simply as "the plague," this disease has played a major role in world history. Because plague is highly contagious, it is easily transmitted from one person to another. Worldwide epidemics of the disease (called pandemics) have decimated populations since A.D. 542, when the first evidence of plague was recorded. Plague is rarely seen in the United States today, but is still **epidemic** in Southeast **Asia**.

Transmission of bubonic plague bacteria

Yersinia pestis is an intracellular parasite. In contrast to other kinds of **bacteria**, it enters cells. The bacterium that causes **tuberculosis** is also an intracellular parasite, as is the bacterium that causes chlamydia, a sexually transmitted disease.

Humans are not the "first choice" of host for *Yersinia pestis.* The *Yersinia pestis* bacterium infects the bloodstream of **rats** and other wild **rodents** such as **squirrels** and prairie dogs. Humans become infected only through the bite of a flea that has ingested **blood** from an infected rodent. Another route of transmission is through person-to-person contact. If a person's lungs are infected with the bacteria, the disease can be transmitted easily to another person through a cough or a sneeze. This form of transmission is extremely quick: cases have been recorded of persons dying from the disease within 24 hours of exposure to an infected person.

Symptoms of bubonic plague

In humans, plague can take two forms. One form, called the bubonic form, usually results from a flea bite and is characterized by a sore called a bubo. The bubo is actually the infected lymph node that drains the area through which the bacteria was introduced by the infected flea. The lymph node enlarges and turns black. Other symptoms of this form of plague include fever and congestion of the blood vessels of the **eye**. As the disease progresses, the bacteria spread to other parts of the body, resulting in septicemia, or widespread **infection**. The fatality **rate** of the bubonic plague is 15%.

In another form of plague, called the pneumonic form, the bacteria infect the lungs. This form of plague can follow the bubonic form, as the bacteria spreads to the lungs. Or, a person may simply contract the pneumonic form only, and show no evidence of a bubo. The pneumonic form is highly contagious and especially virulent: the average length of time from the first appearance to symptoms to death is less than two days.

For both types of plague, **antibiotics** can cure the disease. A **vaccine** is also available to protect those who are at risk of contracting plague. People who work with *Yersinia pestis* in laboratories and in environments where wild rodents are infected with the bacteria usually are vaccinated against plague. United States soldiers who fought in the Vietnam war in the 1960s and 1970s were vaccinated against plague. However, the vaccine only protects against the bubonic form, not the pneumonic form. People who are exposed to the pneumonic disease should take antibiotics as a precautionary measure.

Plague pandemics

Plague has played a major role in world history. Some evidence exists that a plague pandemic took place about 2,000 years ago, but the first recorded pandemic of plague occurred in A.D. 542 in Egypt and Ethiopia. This pandemic killed 100 million people.

The next great plague pandemic occurred in the fourteenth century in **Europe**, Central Asia, the North East, India, and China. In this pandemic, trading ships from China carried infected rats to Europe. About 25 million people in Europe alone died from plague; some experts estimate that this number constitutes a third of the European population. Because so many people died,

The Plague of 1665. *Mary Evans Picture Library/Photo Researchers, Inc. Reproduced by permission.*

the plague had a major impact on the economy and political structure of Europe. The scarcity of workers led to a scarcity of food; workers, previously given little compensation for their labors, began to demand higher wages. Some historians feel that the unrest of workers and the middle class in Europe that culminated in the beheading of King Charles I in England in the seventeenth century and the beheading of Louis XVI in France in the eighteenth century had its roots in the economic aftermath of the fourteenth plague pandemic.

The third plague pandemic began in Burma in 1894; from there, the plague spread to China and through Hong Kong to **North America**. One hundred million people in India died from plague over a period of 20 years. During this pandemic, the United States saw its first case of plague in 1900 in San Francisco. In 1907, 167 cases of plague in San Francisco were recorded. As a result of the pandemic, rats and other wild rodents in the areas around San Francisco became reservoirs of *Yersinia pestis*. Today, isolated cases of plague are still found in Kansas, Oklahoma, and Texas. The majority of cases worldwide (90%) occur in Southeast Asia: Burma, South Vietnam, Nepal, and Indonesia. Brazil also has a high number of plague cases. A recent outbreak in Surat, India, in 1994, killed 56 people and caused widespread panic.

Prevention

Plague pandemics can be prevented by the disinfection of ships, **aircraft**, and persons who are known to have the plague. The classic route of transmission that leads to pandemics is the transportation of infected rodents aboard transcontinental vehicles. Since many countries have instituted rigorous disinfection practices for ships and planes, plague cases have dropped dramatically.

If a person is diagnosed with plague, most countries, including the United States, require that the governmental health agency be notified. The person is usually kept under strict quarantine until the disease is brought under control with antibiotics.

Another way to prevent plague is to control rodent and flea populations in cities. **Fleas** are easier to control than rodents, since most homes can be easily decontaminated. Many cities, especially in the United States, have instituted rodent-control programs aimed at decreasing the numbers of rodents that roam the streets. Since rodents also carry **rabies** and other deadly diseases, controlling their numbers makes sense for a variety of reasons.

Resources

Books

Cantor, Norman F. *In the Wake of the Plague: The Black Death and the World It Made.* New York: Perennial, 2002.

Nelson, K.E., C.M. Williams, and N.M.H. Graham. *Infectious Disease Epidemiology: Theory and Practice.* Gaithersburg: Aspen Publishers, 2001.

Periodicals

Epstein, Richard. "A Persistent Pestilence." *Geographical Magazine* 63 (April 1994): 18.

Jayaramen, K.S. "Indian Plague Poses Enigma to Researchers." *Nature* (October 13, 1994): 547.

Mee, Charles L. "How a Mysterious Disease Laid Low Europe's Masses." *Smithsonian* (February 1990): 66.

Richardson, Sarah. "The Return of the Plague." *Discover* 16 (January 1995): 69.

Kathleen Scogna

Buckeye *see* **Horse chestnut**

Buckminsterfullerene

As recently as 1984, **carbon** was thought to exist in only two solid forms. There was graphite, in which the carbon **atoms** arranged themselves as layered sheets of hexagonally bonded atoms, and there was **diamond**, in which the carbon atoms formed octahedral structures in which each carbon atom had four nearest neighbors.

Then, in 1985, chemists R. E. Smalley, R. F. Curl, J. R. Heath, and S. O'Brien at Rice University, and H. W. Kroto of the University of Sussex in England observed that a hollow truncated icosahedron, similar in shape to a soccer ball, and consisting of 60 carbon atoms, tends to form spontaneously when carbon vapor condenses. In 1990, physicists D. R. Huffman and L. Lamb of the University of Arizona, working with W. Kratschmer and K. Fostiropoulos of the Max Planck Institute in Germany, discovered a way to make bulk quantities of this C_{60} **molecule**, which investigations using high resolution **electron** microscopes have shown to have sizes of about one billionth of a meter.

As C_{60} has the same structure as the **geodesic dome** developed by American engineer and philosopher R. Buckminster Fuller, these molecules were christened buckminsterfullerenes by the group at Rice University. The Swiss mathematician Leonhard Euler had proved that a **geodesic** structure must contain 12 pentagons to close into a spheroid, although the number of hexagons may vary. Later research by Smalley and his colleagues showed that there should exist an entire family of these geodesic-dome-shaped carbon clusters. Thus, C_{60} has 20 hexagons; whereas its rugby-ball shaped cousin C_{70} has 25. Research has since shown that **laser** vaporization of graphite produces clusters of carbon atoms whose sizes range from two to thousands of atoms. These molecules are now known as fullerenes. All the even numbered species between C_3 and C_{600} are hollow fullerenes, but below C_{32}, the fullerene cage is too brittle to remain stable. Helical microtubules of graphitic carbon have also been found.

Although many examples are known of five-membered carbon rings attached to six-membered rings in stable organic compounds (for example, the nucleic acids adenine and guanine), only a few occur whose two five-membered rings share an edge. The smallest fullerene in which the pentagons need not share an edge is C_{60}; the next is C_{70}. C_{72} and all larger fullerenes adopt structures in which the five-membered carbon rings are well separated, but the pentagons in these larger fullerenes occupy strained positions. This makes the carbon atoms at such sites particularly vulnerable to chemical attack. Thus, it turns out that the truncated icosahedral structure of C_{60} distributes the strain of closure equally, producing a molecule of great strength and stability. This molecule will, however, react with certain free radicals.

When compressed to 70% of its initial **volume**, the buckminsterfullerene is expected to become harder than diamond. After the **pressure** has been released, the molecule would be expected to take up its original volume. Experiments in which these molecules were thrown against **steel** surfaces at about 17,000 MPH (25,744 km/h) showed them to just bounce back.

Fullerenes are, in fact, the only pure, finite form of carbon. Diamond and graphite both form infinite networks of carbon atoms. Under normal circumstances, when a diamond is cut, the surfaces are instantly covered with **hydrogen**, which tie up the unattached surface bonds. The same is true of graphite. Because of their **symmetry**, fullerenes need no other atoms to satisfy their surface chemical bonding requirements.

The buckminsterfullerene seems to have an incredible range of electrical properties. It is currently thought that it may alternately exist in insulating, conducting, semiconducting, or superconducting forms.

Fullerenes with **metal** atoms trapped inside the carbon cage have also been studied. These are referred to as endohedral metallofullerenes. Reports of **uranium**, lanthanum, and **yttrium** metallofullerenes have appeared in the literature. It has been exceptionally difficult to isolate pure samples of these shrink-wrapped metal atoms, however.

Production of fullerenes

For reasons that are not yet fully understood, C_{60} seems to be the inevitable result of condensing carbon slowly at high temperatures. At high temperatures when carbon is vaporized, most of the atoms initially coalesce into clusters of 2-15 atoms. Small clusters from chains, but clusters containing at least 10 atoms commonly form monocyclic rings. Although these rings are favored at low temperatures, at very high temperatures they break open to form linear chains of up to 25 carbon atoms. These carbon chains may then link together at high temperatures to form graphite sheets, which somehow manage to form the geodesic fullerenes. One theory has it that the carbon sheets, when heated sufficiently, close in on themselves to form fullerenes.

Kratschmer and his coworkers in Germany managed to prepare the first concentrated **solution** of fullerenes in 1990 by mixing a few drops of **benzene** with specially prepared carbon soot.

Scientists later demonstrated that fullerenes can be conveniently generated by setting up an **electric arc** between two graphite electrodes. In their method, the tips of the electrodes are screwed toward each other as fast as the graphite is evaporated to maintain a constant gap. The process has been found to work best in a helium at-

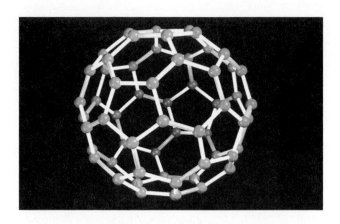

A supercomputer simulation of the atomic structure of a molecule of buckminsterfullerene. Carbon molecules appear as small spheres; double bonds between them are darker than single bonds. *Photograph by J. Bernholc et al, North Carolina State University. Photo Researchers, Inc. Reproduced by permission.*

mosphere in which other gases such as hydrogen and **water** vapor have been eliminated.

Fullerenes have been reported to occur naturally in certain coals, as well as in structures produced by lightening known as fulgurites, and in the soot of many flames.

Uses

Fullerenes have so far failed to realize their commercial potential. This is partly for reasons of cost and partly because it has proven difficult to isolate large quantities of sought-after types. At the beginning of 1994, fullerenes were actively being studied for the following applications: optical devices, hardening agents for carbides, chemical sensors, gas separation devices, thermal insulation, diamonds, batteries, catalysts, hydrogen storage media, polymers and **polymer** additives, and medical applications.

It has been predicted the first large-scale applications for fullerenes will not be found until manufacturing cost are close to those of **aluminum** (a few dollars per pound).

See also Platonic solids; Polyhedron.

Resources

Books

Baggott, J. *Perfect Symmetry.* New York: Oxford University Press, 1994.

Fuller, Buckminster. *Ideas and Integrities.* Toronto: Collier Books, 1963.

Porter, Roy, and Marilyn Ogilvie, eds. *The Biographical Dictionary of Scientists.* Vol. 2, Oxford: Oxford University Press, 2000.

KEY TERMS

Electric arc—A discharge of electricity through a gas.

Geodesic dome—A dome constructed of many light, straight structural elements in tension, arranged in a framework of triangles to reduce stress and weight.

Icosahedron—A 20–sided polyhedron.

Organic—Containing carbon atoms, when used in the conventional chemical sense. Originally, the term was used to describe materials of living origin.

Polymer—A substance, usually organic, composed of very large molecular chains that consist of recurring structural units.

Stewart, Ian, and Martin Golubitsky. *Fearful Symmetry: Is God a Geometer?* Oxford: Blackwell, 1992.

Periodicals

Curl, Robert F., and Richard E. Smalley. "Fullerenes." *Scientific American* (October 1991).

Randall Frost

Buckthorn

Buckthorns are various **species** of shrubs and small trees in the family Rhamnaceae, a mostly tropical and subtropical family of about 600 species. Most of the buckthorns are in the genus *Rhamnus*.

Buckthorns have a few economic applications, although none of these are very important. A dye known as sap green is made from the **fruits** of the European buckthorn (*Rhamnus cathartica*). Another pigment known as Chinese green is made from the **bark** of several Chinese species of buckthorn (*Rhamnus globosus* and *R. utilis*). A laxative and tonic known as cascara is made from the bark of the western buckthorn (*R. purshiana*) of the United States, and from a European species, the alder buckthorn (*R. frangula*). The **wood** of some buckthorns may also be carved, for example, as pipe stems.

There are a number of native species of buckthorns in **North America**. The Carolina buckthorn (*Rhamnus caroliniana*) occurs widely in the eastern United States. Cascara buckthorn (*R. purshiana*) occurs relatively broadly in forested areas of the west coast. Western

species of more restricted distributions in the southwestern United States include birchleaf buckthorn (*R. betulaefolia*) and California buckthorn (*R. californica*).

The European buckthorn (*Rhamnus cathartica*) and glossy buckthorn (*R. frangula*) are European species that were introduced to North America through **horticulture**, and have now become invasive weeds in some areas. These shrubs can form dense stands that exclude native species of shrubs and other plants, and that represent severely degraded ecosystems. The European buckthorn is also an alternate host of oat rust, a fungus that causes an economically important **disease** of oats (*Avena sativa*).

Bill Freedman

Buckwheat

Buckwheat, *Fagopyrum esculentum*, is not really a **wheat** at all—it belongs to the family Polygonaceae, and hence is a dicotyledonous **plant**, not a monocotyledonous **species**. However, the starchy **seeds** of buckwheat are utilized in much the same way as the cereal grains of cultivated **grasses**, such as wheat (*Triticum aestivum*).

The seeds of buckwheat can be used directly as poultry or **animal** feed. Processed, the seeds can be cooked as porridge for humans, or they can be milled to yield a nutritious flour that can be made into a variety of foods, such as pancakes and biscuits. Technically, the seeds of buckwheat are achenes (simple, dry, one-celled, one-seeded **fruits**), as they are surrounded by dry, brown fruit coats, and are slightly winged.

Fagopyrum esculentum was probably derived from the wild species *F. cymosum*, a perennial species with rhizomes (underground storage organs) that occurs naturally in China and northern India. Buckwheat has been cultivated in China for about 1,500 years, and was introduced to **Europe** (via Germany) in the fifteenth century, and arrived in England about A.D. 1600. From Europe it was taken to the American colonies and to **Africa**. The production of buckwheat has been declining in countries where it has been popular in the recent past, such as the former Soviet Union, France and the United States, but against this trend, production has increased in Canada since the 1960s.

Cultivated buckwheat is an annual plant that grows well in poor soils, reaching a height of about 24 in (60 cm). Another attractive feature to farmers is the excellent resistance of buckwheat to many insect **pests** and diseases. Possession of such resistance is fortunate, since

Buckwheat (*Fagopyrum esculentum*) flowering. *© John Kaprielian, National Audubon Society Collection/Photo Researchers, Inc. Reproduced with permission.*

breeding for improvements by conventional methods has proved to be difficult.

See also Crops.

Buds and budding

Bud is a term used to refer to three different types of undeveloped forms described in this article.

Plant buds

Plant buds, such as the buds of flowers, trees, and scrubs, are small, rounded, incompletely developed, dormant parts of a plant consisting of cells capable of rapid **cell division** when conditions are right for growth. They first appear in the spring when sap starts to flow, causing the buds to swell, which makes them more noticeable. These buds are first formed by the plant in late summer and early fall but remain small over the winter. When spring arrives, the structures for new shoot and floral growth are already formed and are tightly packaged and ready for quick growth when days lengthen and temperatures rise. The delicate, immature structures of these plant buds are covered with tough protective scales formed from modified leaves that enable the tender structures to get through winter in a dormant, resting state.

Plant buds can be classified in two ways, either according to their location on the plant, or according to the type of tiny immature structures that are contained within the bud. Buds on the tip of stems are called terminal buds, those at the sides of stems are called lateral buds, while those formed in the **angle** the **leaf** makes with the plant stem are known as axillary buds. When buds are classified according to their internal structures, those that contain only the beginning of a **flower** are called flower buds. Those that contain only immature leaves are called leaf buds, while those buds containing both flowers and leaves in the earliest stages of development are termed mixed buds. Flower buds on herbaceous plants and on woody plants are made up of undeveloped and tightly packed groups of cells that are the precursors of the various floral parts—petals, stamens, and pistils—with a whorl of sepals or outer leaf bracts covering and protecting the inner parts of the flower bud. Some small buds can produce a surprising amount of growth when attached to, and growing out from, a piece of the parent plant that is put into suitable **soil**. People familiar with growing potatoes by planting small sections of **potato** that contain "eyes" (sunken buds) know that the sprouts that grow from these buds will develop into whole new potato plants. Because of this remarkable ability for vegetative reproduction, these fertile pieces of potato are referred to as seed potatoes. The little sunken bud (the potato "eye") draws its nourishment to sprout from the stored starchy food contained in the piece of parent potato **tuber**.

The buds that woody plants such as shrubs, woody vines, and trees produce contain miniature shoots with a short stem and small, undeveloped, tightly packed leaves or immature floral structures covered with tough protective, overlapping scales. A terminal bud on a woody twig overwinters and grows out during the next spring and summer into a whole new shoot that extends the length of the twig and may also produce flowers. Apple and cherry blossoms are well known examples of this type of bud growth. Growth from a lateral bud will produce either a branch, or just a leaf on the side of the twig depending upon the nature of the precursor cells that were packaged in the bud during the previous fall.

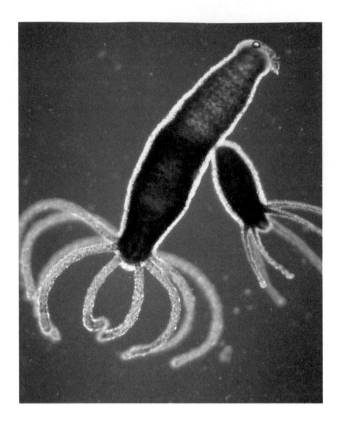

A hydra, budding. © Biophoto Associates/Science Source/Photo Researchers, Inc.

Animal buds

Buds and budding are also found in the **asexual reproduction** (involving only one parent) of some animals, such as the **freshwater** hydra and **species** of marine colonial **jellyfish**, where a single parent gives rise to one or more new individuals. When a single **hydra** reaches maturity and is well fed, outpocketings of the animal's body wall begin to form a rounded growth projecting from the tube-like section or stalk of the adult's body. This growth, called a bud, develops in time into a miniature hydra whose body layers and inner body cavity, the digestive cavity, are continuous with that of the parent **individual**. Food captured and gathered in by the adult parent also supports the growth of the bud. Early in this budding process tiny tentacles appear on the free end of the hydra bud. It is not unusual to find two or more buds on an adult hydra in different stages of growth and development. An adult hydra may have its body and tentacles fully extended while its bud may have its whole form contracted into a rounded mass. Conversely the bud may be stretched out, while the adult is contracted. The bud will, however, sometimes contract soon after the adult contracts as the nerve net in the mesoglea (middle jelly layer) of the two individuals is continuous. When a newly budded hydra offspring is fully formed and suffi-

ciently developed to take up an independent existence, the base of the new hydra seals off and thus allows the new individual to break off from the parent hydra.

Buds and budding also refers to the extensions of microscopic **yeast** cells and some types of bacterial cells produced during asexual reproduction, forming the beginning of daughter cells. The **taste** buds of the mammalian tongue, are so called because of their small size and bud-like shapes, but bear no relationship to the buds of plants and animals discussed above.

Resources

Books

Campbell, N., J. Reece, and L. Mitchell. *Biology.* 5th ed. Menlo Park: Benjamin Cummings, Inc. 2000.

Raven, Peter, R.F. Evert, and Susan Eichhorn. *Biology of Plants.* 6th ed. New York: Worth Publishers Inc., 1998.

Julia M. Van Denack

Buffer

In **chemistry**, a buffer is a system, usually an aqueous (**water**) **solution**, that resists having its **pH** changed when an acid or a base is added to it.

Normally, the addition of acid to a solution will lower its pH and the addition of a base will raise its pH. If the solution is a buffer, however, its pH will be changed to a much lesser extent than would be expected from the amounts of acid or base that are added. So-called "buffered aspirin" is not really a buffer, because it does not resist **acids and bases**. It is simply aspirin combined with a basic compound, such as **magnesium** carbonate or **aluminum hydroxide**, which neutralizes some stomach acid.

Almost all **chemical reactions** that take place in aqueous solution—meaning almost all chemical reactions—are sensitive to the concentrations of **hydrogen** ions and hydroxide ions, that is, to the pH of the solution. This is because hydrogen and hydroxide ions are the ions of water itself. In particular, many biochemical processes essential to life are quite sensitive to the acidities of various body fluids. A variety of natural buffer systems keep the body's pH values within the limits that are necessary for health. For example, a system of several buffers holds the pH of human **blood** between 7.33 and 7.43 in a healthy person. A blood pH below 7.0 or above 7.8 can be fatal.

How buffers work

There are two common kinds of buffer solutions: solutions that contain a weak acid plus one of its salts (e.g., **acetic acid** plus **sodium** acetate) and solutions that contain a weak base plus one of its salts (e.g., **ammonia** plus ammonium chloride). Their workings can be understood in terms of LeChâtelier's principle.

Weak acid buffers

When a weak acid is dissolved in water, only a few of its molecules dissociate to form only a few hydrogen ions; the rest of the acid molecules remain as undissociated, neutral molecules that do not affect the pH. For example, whenever acetic acid is added to water, the following three species will be in the solution:

$$HC_2H_3O_2 \longleftrightarrow H^+ + C_2H_3O_2^-$$

many	a few	a few
acetic	hydrogen	acetate
acid	ions	ions
molecules		

To make a buffer solution out of this system, we can add many more acetate ions to the solution in the form of sodium acetate, which is a strong **electrolyte** and dissociates completely. Ignoring the sodium ions that come along with the sodium acetate because they do not affect the acidity at all, we then have:

Buffer solution

$$HC_2H_3O_2 \longleftrightarrow H^+ + C_2H_3O_2^-$$

many	a few	many
acetic	hydrogen	acetate
acid	ions	ions
molecules		

This solution will resist having its hydrogen ion **concentration** changed. To see how that works, first consider what would happen if we were to add some acid-some extra hydrogen ions-to this solution. According to LeChâtelier's principle, the equilibrium will be shifted to the left. That is, the added hydrogen ions will react with some of the acetate ions to form more acetic acid molecules. The result is that almost all of the added

hydrogen ions are used up to form "harmless" neutral molecules; they therefore are not available to increase the acidity of the solution. The solution has resisted having its pH lowered more than a little bit.

What if we were to add some base-hydroxide ions to the buffer solution? Hydroxide ions to react with hydrogen ions, because the resulting **molecule**, H_2O, is so stable.

$$OH^- + H^+ \rightarrow H_2O$$

| hydroxide | hydrogen | water |
| ion | ion | |

Therefore, the added hydroxide ions will quickly remove hydrogen ions from the buffer solution, which according to LeChâtelier's principle will then shift its equilibrium to the right, making more acetate ions out of acetic acid molecules. Thus, the added hydroxide ions will have been used up, and only "harmless" acetate ions will have been formed. (Acetate ions are slightly basic, however, so the pH of the buffer solution does increase slightly.)

Weak base buffers

When ammonia gas is added to water, it forms a solution of a weak base whose equilibrium can be represented as follows:

$$NH_3 + H_2O \longleftrightarrow NH_4^+ + OH^-$$

many	water	a few	a few
ammonia		ammonium	hydroxide
molecules		ions	ions

To make a buffer solution out of this system, we can add many more ammonium ions to the solution in the form of ammonium chloride, which is a strong electrolyte and dissociates completely. Ignoring the chloride ions that come along with the ammonium chloride because they do not affect the acidity at all, we will then have:

Buffer solution

$$NH_3 + H_2O \longleftrightarrow NH_4^+ + OH^-$$

many	water	many	a few
ammonia		ammonium	hydroxide
molecules		ions	ions

If we add hydrogen ions to this buffer solution, they will be neutralized by reacting with the hydroxide ions to form water. According to LeChâtelier's principle, this removal of hydroxide ions will shift the equilibrium to the right, producing more ammonium ions, which does not affect the pH (but ammonium ions are slightly acidic). If we add hydroxide ions to the ammonia buffer solution, they will shift the equilibrium to the left, which uses up the added hydroxide ions and forms more whole NH_3 molecules, which does not affect the pH.

Some important buffers

Common weak acid buffer systems are based upon carbonic acid (H_2CO_3), **citric acid** ($H_3C_6H_5O$), and **phosphoric acid** (H_3PO_4). Common weak base buffer systems are based upon amines (organic bases) or amino acids, which can act as both acids and bases.

A buffer solution based upon a dibasic or tribasic acid (an acid that can produce two or three hydrogen ions per molecule) may be made from two different ions of the same acid, rather than from the acid itself and one **salt** ion. For example, the phosphoric acid ions $H_2PO_4^-$ and HPO_4^{2-} can form what is known as the phosphate buffer system, which is one of the buffers that control the pH of human blood. In this system the $H_2PO_4^-$ ion plays the role of the weak acid and the HPO_4^{2-} ion plays the role of its salt. The relevant equilibrium is

$$H_2PO_4^- \longleftrightarrow H^+ + HPO_4^{2-}$$

dihydrogen   hydrogen   monohydrogen
phosphate   ion   phosphate
ion

The main buffer that is involved in controlling blood pH, however, is the carbonate system, which is based on the following equilibrium:

$$H_2CO_3 \longleftrightarrow H^+ + HCO_3^-$$

carbonic   hydrogen   bicarbonate
acid   ion   ion

The carbonic acid in the blood comes from dissolved **carbon dioxide**:

$$CO_2 + H_2O \longleftrightarrow H_2CO_3$$

carbon   water   carbonic acid
dioxide

Our breathing (**oxygen** in, **carbon** dioxide out) controls the amount of carbon dioxide that is available to dissolve in the bloodstream. Therefore, our lungs also play an important part in controlling our blood's pH through the carbonate buffer system.

Resources

Books

Brown, William H., and Elizabeth Rogers. *General, Organic and Biochemistry*. Boston: Willard Grant, 1980.

Creedy, John. *A Laboratory Manual for Schools and Colleges*. London: Heinemann, 1977.

Lide, D.R., ed. *CRC Handbook of Chemistry and Physics*. Boca Raton: CRC Press, 2001.

Ouellette, Robert. *Chemistry: An Introduction to General, Organic, and Biological Chemistry*. Prentice Hall, 1994.

Robert L. Wolke

Building design/architecture

Architects design buildings, but architecture is more than just building design and more than just art on a massive scale. Architecture is about **light** and **space**. It is about stimulating emotions in the people who see and inhabit the structure. Architecture creates an environment, whether it is the uplifted spirituality of the Chartres Cathedral, the drama and anticipation of the Schauspielhaus auditorium, or tranquil serenity of Fallingwater. It is experiential. When you approach or enter a building you move through the space, the scale changes, the proportions shift around you. This is what makes architecture glorious; the way it manipulates light.

Structure is fundamental to architecture. For a design to move from the mind of the architect to reality, there must be a method of building it. The development of structural forms has thus been a driving force in architecture, every bit as much as changes in social order and historical events. Through history there have been relatively quiet periods of stylistic development, interspersed with almost muscular leaps of structural innovation such as the Roman period, the Gothic period or the **Industrial Revolution**.

Prehistory

The architecture of prehistory is largely one of tombs and temples, though mudbrick Neolithic settlements have been discovered. Houses in these settlements were one story, rectangular structures with a hole in the roof that served as both chimney and entry. The doorway, with its horizontal lintel and vertical posts, was developed somewhat later and constitutes the first significant leap forward in architecture, making all future styles possible. The architectural term for post-and-lintel construction is trabeation. Trabeated passages, some involving enormous stones, were built into huge mounds of earth to form burial tombs called barrows.

The thought of the ancient Egyptians immediately brings the image of the pyramids to mind, but architecturally speaking the advances they made with materials and design were far more important. The Egyptians built structures as we know them, with walls, trabeated doorways and small window openings. They eventually developed the freestanding column, which allowed them to build enormous halls with trabeated roofs, structures that essentially consisted of **parallel** rows of post-and-lintel constructions.

The emphasis of ancient Egyptian architecture was on mass rather than space, typified by the Hypostyle hall where some of the columns are 11 or 12 ft (3.5-4 m) in diameter. The structural strength of the lintel in a trabeated roof limits the expanse of open space that can be spanned by this method. When the lintel is too long, the load carried by the stone is greater than its strength, and it fails. Thus, the open space in Egyptian halls was very limited, though the ceilings soared to heights of almost 70 ft (21 m) and the actual halls were hundreds of feet wide.

The Egyptians were the first structural designers with an identifiable visual style. Their temples and tombs had such a strong identity and coherence that architects still echo their designs in modern structures. They drew some of their inspiration from nature, carving columns to look like **palms**, or plants crowned with papyrus or lotus blossoms. Structures were also designed to elicit emotions in the viewers, such as the temples interiors that progressed from the bright, relatively open spaces allowed to the public, to the dim, confined spaces of the inner sanctum, accessible only to the priests and rulers. Both the use of nature in architectural decoration and the use of design to control emotion were themes that would be repeated over and over in coming centuries.

The architecture of the Near-Eastern civilizations that coexisted with the Egyptians evolved distinct identities. The Sumerians, for example, had little access to stone. Their available building material was mudbrick, a structurally weak material that could not produce the lintels required for trabeated roofing. To solve the problem of roofing, the Sumerians are believed to have developed the curved arch and tunnel vault to enclose narrow interior spaces. The ancient Persians had access to a variety of building materials, and they were influenced by the architecture of the Egyptians, the Greeks, and the other civilizations that inhabited their enormous empire. With such rich source material, they were able to develop a unique, fanciful architectural style, and structurally refine the approach of the ancient Egyptians. In the royal audience hall in Persopolis, the ancient capital, the pillars were half the diameter of the Egyptian columns, with a significantly wider spacing than the Egyptian version. The ceiling was still trabeated, but the effect is of a much lighter, much more open space.

Classical architecture

The Greeks

Ancient Greek architecture was a miracle of style, balance, and harmony, with a powerful simplicity whose influence persists in architecture to this day. It is distinctive and immediately recognizable, a graceful massing that creates a sense of dignity, wisdom, and timelessness. The Greeks believed in human intellect and rational thought, in community, and the achievements of the living, rather than the cult of death featured in earlier civilizations. Accordingly, their architectural focus was on public buildings: temples, theaters, civic structures. They were the pioneers of city planning.

Greek society was built upon democracy, stressing the involvement of the individual in government and culture. This was reflected in their architecture. Structures were rarely aligned with one another but were instead set at angles to enhance the individuality of the structures and draw the viewer into participating in the process. The viewer's viewpoint was not strictly regimented, as we will see in later eras, but instead allowed to form naturally.

The elements of classical Greek structures are simple and few, yet carefully organized to create an overall effect. The primary structural form was trabeation—stone columns supporting lintels. There were three styles, or orders, of Greek architecture: the Doric, the Ionic, distinct styles that originated in different parts of the country, and the Corinthian, which is essentially a spin-off of the Ionic order. The structural elements, proportions, and composition of a building were defined by its order. Doric structures, in particular, followed a rigidly prescribed design, evolving toward the ideal Doric form expressed in the Parthenon. **Symmetry** of proportion runs through the structures, with set ratios of building length to width, and column height, diameter, and spacing. The Ionic style was less rigidly defined than the Doric, though proportion was still important, and the Corinthian order differed from the Ionic only in column design. In fact, the names of the orders are most often associated with the capitals or decorative tops of the columns used in the buildings. Doric columns are simple angled tops, Ionic columns are delicate scrolls, and Corinthian columns are crowned with a delicate, leafy capital.

The Romans

Roman philosophy and engineering fundamentally changed the way people thought about architecture. The Romans took previously invented structural elements, such as the arch and vault, to the limits of their potential. The Greeks, for example, dabbled with **concrete** and in later periods occasionally used arches, but only as freestanding decorative elements. In the hands of the Romans, these same two concepts were used to build the Colosseum.

The development of the rounded arch was critical to Roman architecture. In our earlier discussion of trabeation, we pointed out the lintel as the weak element of the structure. The tensile strength of the lintel limits the size of the opening and the load that can be carried, because the load on the lintel is only supported at the ends. In the non-supported portion of the lintel, the lines of force are aimed through the lintel to the ground. In an arch, on the other hand, the lines of force from the load are directed through the **curve** of the arch to the supporting piers, and from there to the ground. The supporting pier strength is limited by the compressive strength of the stone, which is far greater than its tensile strength. Moreover, the form of the arch and arch elements is such that the structure actually becomes stronger with a uniform load placed on top of it.

Rounded arches can be placed in a row to make a barrel vault, which means that a significant amount of space can be roofed over. The Romans developed a number of variations on the barrel vault, changing the way that space could be enclosed. The disadvantage of the barrel vault is that it requires continuous support along the sides, limiting the number of window openings permitted in the sidewalls. One answer to this was the groin vault, a structure consisting of two intersecting barrel vaults. Whereas the weight and force of the barrel vault is carried all along the line of the supporting wall, the groin vault concentrates weight and force in the corners of the structure, permitting a more open space and windows. The hemispherical dome, which can be thought of as an arch in three dimensions, was another variation on roofing. This form appears time and time again in Roman architecture, most notably in the Pantheon.

Another important Roman development was that of improved concrete. For several centuries, a concrete made of lime, **sand**, and **water** had been used sporadically by various builders. The Romans added a volcanic ash called pozzolana to their concrete, obtaining a stronger mortar that had the added advantage of setting up in water, allowing underwater construction. The Romans mixed this concrete with gravel or chips of stone and molded it into blocks or even arches and vaults, simplifying construction methods.

The Romans considered architecture differently than the Greeks. The Greeks emphasized structure and form, the Romans emphasized space. Whereas Greek architecture was exterior, focused on the outside, on creating an experience for the viewer, Roman architecture was interior, based on the idea of creating an environment for the inhabitants. The Romans were more pragmatic than spiritual. Rather than focusing on temples, they built sumptuous bath houses, theaters, and other public spaces. Great administrators, they created the basilica to house government offices. The interiors of their buildings focused on space, on creating spaces to serve man, spaces that were emotionally and functionally pleasing. The Romans were also sophisticated urban planners, designing freeform civic centers, or forums, and rigidly styled *castrum*, a combination military outpost/colonial settlement in newly conquered territory.

Byzantine

After the fall of the western Roman Empire, the creative focus of architecture moved to Constantinople, located at the site of the Hellenic city of Byzantium. Most of the surviving structures from this period are churches. Rather than the long, axial floor plan characteristic of Roman barrel vaulted structures, Byzantine designs tended toward a centralized, vertically focused structure crowned with a dome that flooded the interior with light. Light, form, and structure in these churches were orchestrated to express the spiritual ideas of the new Christian religion, to exalt the worshippers.

Byzantine churches achieved the floating effect of the central dome by the use of pendentives and pendentive domes. A hemispherical dome requires a circular base to support it fully, like that on the Pantheon. Interior spaces, however, tend to be **square**. Ancient architects wishing to top a non-circular space by a spherical dome added diagonal elements across the corners, called squinches, which helped support the structure. The Romans developed a more sophisticated method of support called the pendentive. A pendentive is a spherical, inverted triangle that rises from its point and curves downward. It is essentially a section of a domed surface and as such can be constructed to fully support a hemispherical dome. More important than the full support, though, is the nature of that support. The structural support of a dome on a cylinder or on squinches is visually dense, giving a sense of massing and of enclosure. A dome on pendentives appears to rise from only four points, giving the impression of floating overhead, adding to the other-worldly effect of the church interior.

The spiritual effect of the floating domes was enhanced by the use of light and ornamentation. Hagia Sophia, the most spectacular of the Byzantine churches, is flooded with light from a multitude of wall openings. A row of windows around the base of the dome adds to the impression of a magically hovering surface, as do the additional half domes and colonnades in the lower levels. To look up from the floor of the building is to look into a gleaming, billowing surface that appears to follow no known rules of structure, generating an exaltation of religion in its intensity.

Medieval architecture—Romanesque and Gothic

In the medieval period, European architects were profoundly influenced by the Roman structures dotting the countryside. A style known as Romanesque emerged, featuring the Roman hallmarks of rounded arches and barrel vaults. Because barrel vaults must be supported continuously along the sides, these structures had few windows. The focus was on an almost claustrophobic massing: thick walls, small windows, heavy, ponderous piers.

The lines of force in an arch are designed to run down into the supporting column or piers. When the load is too extreme, as in the case of the large vaults of the Romanesque cathedrals, the lines of force are shifted laterally, with the result that the base of the arch tends to push outward from its supporting column. To prevent this, the Romanesque designers added buttresses, massive piers of masonry built against the walls at critical points to resist the lateral stress. Structurally it was effective, but it only added to the oppressive feel of the buildings.

Romanesque was primarily an adaptation of existing ideas, but the Gothic period was one of profound innovation. Elements were developed that allowed interior space to be approached in a way it had not been before. In contrast to the heavy inertness of the Romanesque structures, the Gothic cathedrals were open and buoyant, with a dynamic use of space. If Byzantine churches like Hagia Sophia gave an impression of the otherworldly, to the medieval peasant unused to any but the smallest interior spaces, the soaring lines of the Gothic cathedrals with their brilliantly colored walls of **glass** and traceries of stone must have felt like heaven.

The rounded arch is a powerful structural element but it has its limitations. Force applied to the rounded arch is carried along the full **arc** and driven into the supports of the arch. When the applied load is too high, however, the lines of force move outside of the structure of the arch; in such a case, the arch fails. The arch developed by the Gothic cathedral builders was pointed, rather than round. Its lines described a catenary arc rather than a hemisphere, keeping the lines of force within the structure of the arch so that the load applied to the arch was deflected straight down through the arch supports to the ground. The Gothic arch is more stable and can be thinner than the same size rounded arch. This allowed the cathedral designers to build larger, lighter looking structures.

A second important development of the Gothic period was the rib vault. Romanesque naves were roofed with barrel vaults, essentially a line of rounded arches. Load applied to the vault was carried all along the wall holding up the arch. This limited the number of openings possible in the wall, leading to the claustrophobically small and infrequent windows of the Romanesque churches. Groin vaults made from the intersection of two barrel vaults permitted a somewhat wider open space, but again the load on the vault was carried by the arches. Groin vaults carried the load in the corners, allowing more openings in the walls, but they were difficult to build and could only span a square area. The rib vault represented a new concept in structure.

The rib vault consists of six arches: the arches on each side and a pair of transverse arches. The roofing of the vault is just a thin, relatively lightweight layer of stone webbed over this supporting structure. Load is minimized and carried at the corners of the vault. Construction is dramatically simplified. Design of the rib vault is facilitated by the pointed arch, which allows the vault to be any variety of **rectangle**, as opposed to the square vault dictated by use of rounded arches. The development of the rib vault allowed the cathedral architects to roof over enormous spaces. Because the vaults carried the load to the corners of the vault, the need for massive load-bearing walls was gone. The Gothic builders were able to open enormous holes in the walls without compromising the structure, and the cathedrals became traceries of stone filled with stained glass.

The third major development of Gothic architecture was the flying buttress. Romanesque churches used massive piers to support their sidewalls, the inertial bulk resisting the side forces created by the load of the ceiling. The flying buttress is a half arch that connects to a massive pier. It allowed the Gothic architects to apply resistive force to the ceiling loads at the point needed, rather than simply building huge piers and hoping they would not fail. The flying buttresses allowed the cathedrals to soar to heights well over 100 ft (30 m). Moreover, they lightened the feel of the cathedral exteriors, making them appear like lacework, fragile and airy.

The Renaissance and the Baroque

Renaissance architecture was a response to the ornamentation of the Gothic period and a homage to the new values of symmetry, balance, logic, and order. Man was once again exerting control over his environment, and the structures of this time reflected the dedication to mathematical forms and the ordering of space. The innovations of Renaissance architecture are less structural than paradigmatical. Leon Battista Alberti (1404-1472) codified what was known of architecture, emphasizing the theoretical aspects, turning it from a trade into a profession. More importantly, he developed perspective drawing techniques that allowed architects to draw accurate architectural renderings, techniques that are still used today.

Like the Renaissance, the Baroque period in architecture was marked by design rather than structural innovation. In response to the bareness of Renaissance architecture, Baroque buildings were lavishly decorated. Rather than the approach of form following function, designers of this period approached architecture as theater, emphasizing effects, interpretations, and movement. The Spanish Steps in Rome, for instance, were designed to mimic the movements of a popular dance of the time, with sidewalls that moved in, moved out, met, separated, and met again. The design elements became more emotional and dynamic, and the focus was on **energy** versus balance. At the same time, underlying the ornateness of the Baroque was the logic of Renaissance design.

One notable aspect of Baroque architecture was its use of controlled viewing to emphasize a building. The approach to St. Peter's in Rome is a classic example. It was carefully orchestrated, beginning some blocks away with the Ponte S. Angelo, progressing through a tangle of narrow streets that heighten anticipation by providing only glimpses of St. Peter's. Finally, the pilgrim emerged into an enormous oval piazza that permitted a clear view of St. Peter's for the first time, framed by the curved colonnade. The sequence was designed to enhance the religious experience of St. Peter's by intensifying the emotions—creating a sense of anticipation and delayed gratification—felt during the approach.

The Industrial Revolution—new materials

After the Baroque faded slowly away, eighteenth-century architecture consisted primarily of revivals of previous periods. This time was to be the calm before the storm, for the approaching Industrial Revolution was to change everything about the world as it was then, including architecture. Previously, building materials had been restricted to a few manmade materials along with those available in nature: timber, stone, timber, lime mortar, and concrete. Metals were not available in sufficient quantity or consistent quality to be used as anything more than ornamentation. Structure was limited by the capabilities of natural materials. The Industrial Revolution changed this situation dramatically.

In 1800, the worldwide tonnage of **iron** produced was 825,000 tons. By 1900, with the Industrial Revolution in full swing, worldwide production stood at 40 million tons, almost 50 times as much. Iron was available in three forms. The least processed form, cast iron, was brittle due to a high percentage of impurities. It still displayed impressive compressive strength, however. Wrought iron was a more refined form of iron, malleable, though with low tensile strength. **Steel** was the strongest, most versatile form of iron. Through a conversion process, all of the impurities were burned out of the iron **ore**, then precise amounts of **carbon** were added for hardness. Steel had tensile and compressive strength greater than any material previously available, and its capabilities would revolutionize architecture.

This change did not happen over night. Prior to the introduction of bulk iron, architecture relied on compressive strength to hold buildings up. Even great structures like the Chartres Cathedral or the Parthenon were essentially orderly piles of stone. Architects were accustomed to thinking of certain ways of creating structure, and though they glimpsed some of the possibilities of the new materials, the first applications were made using the old ideas.

The explosion in the development of iron and steel structures was driven initially by the advance of the railroads. **Bridges** were required to span gorges and **rivers**. In 1779, the first iron bridge was built across the Severn River in Coalbrookdale, England. It was not an iron bridge as we might conceive of it today, but rather a traditional arch made of iron instead of stone. The compressive strength of limestone is 20 tons per square foot. The compressive strength of cast iron is 10 tons per square inch, 72 times as high, permitting significantly larger spans. Later, the truss, long used in timber roofs, became the primary element of bridge building. A triangle is the strongest structural element known, and applied force only makes it more stable. When a diagonal is added to a square, the form can be viewed as two triangles sharing a side, the fundamental element of a truss. Trusses were used to build bridges of unprecedented strength throughout the nineteenth century, including **cantilever** bridges consisting of truss complexes balanced on supporting piers. A third, more attractive type of steel bridge was the suspension bridge, in which the roadway is hung from steel cables strung from supporting towers in giant catenary arcs.

As with bridges, some of the first structural advances using steel were prompted by the railroads. Trains required bridges and rails to get them where they were going, but once there, they required a depot and storage sheds. These sheds had to be of an unprecedented scale, large enough to enclose several tracks and high enough to allow smoke and fumes to dissipate. Trusses spanned the open area of the tracks, creating a steel skeleton hung with steel-framed glass panes. The structures were extraordinarily light and open. Some of the sheds were huge, such as St. Pancras Station, London, England. To the people of the nineteenth century these sheds were breathtaking, the largest contiguous enclosed space the world had ever seen.

At this point the capabilities of iron and steel had been proven and it was natural to extend the idea to another utilitarian application—factories. The first iron frame factory was built in 1796-97 in Shrewsbury, England, followed rapidly by a seven story **cotton** mill with cast iron columns and ceiling beams. Wrought iron beams were developed in 1850, a significant advance over brittle cast iron versions.

The new materials were not just used as skeletal elements. In the 1850s, 1860s, and 1870s, cast iron was used as a facade treatment, especially in the Soho district of New York City. Buildings such as the Milan Galleria, an indoor shopping area, and the Bibliotheque Nationale in Paris used iron as an internal structural and decorative element. In 1851, the Crystal Palace was built for the London Exposition, truly the Chartres Cathedral of its time. In 1889, Gustav Eiffel built the Eiffel Tower for the Exposition Universelle in Paris, initially the target of harsh criticism and now the symbol of Paris.

The Industrial Revolution provided more than just ferrous building materials. A stronger, more durable and fire resistant type of cement called Portland Cement was developed in 1824. The new material was still limited by low tensile strength, however, and could not be used in many structural applications. By a stroke of good fortune, the **thermal expansion** properties of the new cement were almost identical to those of iron and steel. In a creative leap, nineteenth century builders came up with the idea of reinforced concrete. Though expensive, iron and steel had high tensile strength and could be easily formed into long, thin bars. Enclosed in cheap, easily formed concrete, the bars were protected from fire and **weather**. The result was a strong, economical, easily produced structural member that could take almost any form imaginable, including columns, beams, arches, vaults, and decorative elements. It is still one of the most common building materials used today.

The modern era

In the mid-nineteenth century, Viollet-le-Duc published a series of tracts on architecture. He decried the fact that with the notable exception of engineers like Eiffel, new structures were being built with old methods. In the face of the new material, architects had either substituted the new material for the old (the Coalbrookdale Bridge), or adapted old methods to the new material (the truss). What architecture needed, according to Viollet, was to uncover new methods that tapped the potential of the new materials.

Some of the most significant advances in architecture at this point were made by a group of architects collectively known as the Chicago School, developers of the modern skyscraper. The skyscraper was developed in response to the rising price of city land. To maximize use of ground space, it was necessary to construct buildings 16 or more stories tall. Initial attempts included iron frame construction with heavy masonry walls requiring massive, space consuming piers. This was an extension of the traditional methods in which the exterior walls of a building added structural support, unnecessary in the face of iron/steel framing. The approach was merely an adaptation, at a time when a completely new approach was needed.

The solution, developed in Chicago, was to separate the load-bearing frame of the building from a non-structural facade. The facade became a curtain wall that was supported by the steel frame story by story. Because the facade material on a given story supported only itself, it could be very light and thin. The further evolution of this approach by architects of the Chicago School led to the modern skyscraper, with its fireproof steel frame, curtain wall facade, and internal **wind** bracing.

Much of the architecture of the twentieth century has been a process of refinement, of **learning** to work with the multitude of new materials and techniques presented by the Industrial Revolution. The various movements, such as art nouveau, art deco, modernism, and postmodernism have all been about design and ornamentation rather than major structural innovation. Certainly architects during this period have explored the possibilities of the new materials and construction techniques. Most recently, however, architecture has been working through a revival period in which old styles are being reinterpreted, similar to the eighteenth century. Perhaps, like the eighteenth century, we are poised at the start of a new period of innovation.

KEY TERMS

Barrel vault—A vault made of a series of rounded arches.

Buttress—A strong stone pier built against a wall to give additional strength. (Flying buttresses extend quite far out from the wall.) A buttress must be bonded to the wall.

Curtain wall—A non-loadbearing facade that is supported by the steel frame of the building.

Flying buttress—A buttress incorporating arches that applies force at the arch/column interface.

Groin vault—A vault made of the intersection of two barrel vaults.

Nave—The long central portion of a church or cathedral.

Pendentive—An inverted concave masonry triangle used to support a hemispherical dome.

Piazza—Italian for plaza or square.

Pier—A vertical support element, usually extremely massive.

Post and lintel—A structural form consisting of a horizontal element (lintel) resting on two support posts. Also called trabeation.

Squinch—Diagonal elements crossing the corners of a rectangular room to supply additional support for a dome.

Trabeation—A structural form consisting of a horizontal element (lintel) resting on two support posts. Also called post and lintel construction.

Truss—A very light, yet extremely strong structural form consisting of triangular elements, usually made of iron, steel, or wood.

Vault—A roof made using various types of arches.

This entry is only a brief discussion of the role of technical advances in the history of architecture. It is by no means a thorough discussion of architecture itself. Technology permits architecture, but architecture is not about technology. Architecture can perhaps best be expressed in the words of Le Corbusier, one of the most influential architects of the twentieth century: "You employ stone, **wood**, concrete, and with these materials you build houses and palaces. This is construction. Ingenuity is at work. But suddenly you touch my **heart**, you do me good, I am happy and I say 'This is beautiful.' That is Architecture. Art enters in."

See also Brick; Cranes; Stone and masonry.

Resources

Books

Allen, Edward. *The Architect's Studio Companion.* 3rd ed. New York: John Wiley & Sons, 2001.

Allen, Edward. *Fundamentals of Building Construction.* 3rd ed. New York: John Wiley & Sons, 1998.

Ching, Francis D. *Architecture: Space, Form, and Order.* New York: Van Nostrand Reinhold, 1979.

McCoy, Esther. *Case Study Houses, 1945-1962.* Santa Monica, CA: Hennessey & Dyalls Inc., 1977.

Trachtenberg, Marvin, and Isabelle Hyman, *Architecture from Prehistory to Postmodern.* Englewood Cliffs, NJ: Prentice Hall Inc., 1986.

Kristin Lewotsky
Stephen K. Lewotsky

Bulbuls

Bulbuls are about 120 **species** of medium-sized, perching **birds**, distributed among 15 genera, and making up the family Pyncnontidae. The most diverse genus is *Pycnonotus*, with about 50 species. Bulbuls are mostly tropical and subtropical birds, occurring in **Africa**, **Asia**, and Southeast Asia. Some relatively northern species are migratory, but most species of bulbuls are local birds.

Bulbuls have rather short, rounded wings, a long tail, small, relatively delicate legs and feet, a small, slender bill, and prominent bristles about the base of the top mandible (these are known as rictal bristles). The body size of bulbuls ranges from 6-11 in (15-28 cm).

The coloration of bulbul species is commonly black or grey, often with reddish markings, and sometimes with a distinctive crest on the top of the head. Male and female birds look very similar, as do juvenile birds, although their coloration is more subdued than that of adults.

Bulbuls build a cup-shaped nest in a bush or **tree**, and lay two to four eggs. Both parents share in the incubation and care of the young.

Species of bulbuls occur in diverse tropical habitats, but not in deserts. They may occur in dense vegetation in tropical **forests** or in more open habitats, such as gardens in towns or even city parks. Some species of bulbuls are accomplished singers, and they are among the more pleasing avian vocalists in tropical towns and parks where **habitat** is available for these birds.

Most species of bulbuls eat small **fruits**, but they may also feed on **insects**, particularly when they are raising babies, which require high-protein foods. Both parents feed and care for the young, which typically fledge about two weeks after hatching. During the non-breeding season, bulbuls often occur in mixed-species flocks with other bulbuls, and sometimes with birds of other families.

The black bulbul (*Hypsipetes madagascariensis*) is a wide-ranging species, occurring in young forests and other disturbed habitats from Madagascar to Southeast Asia. This species has an all-black plumage, but red legs, feet, bill, and eyes.

The red-whiskered bulbul (*Pycnonotus jocosus*) is a common and familiar species of open habitats from India, through south China, to mainland Southeast Asia. This distinctively head-crested species has also been introduced to **Australia**, Mauritius, Fiji, and southern Florida.

The yellow-crowned bulbul (*Pycnonotus zeylanicus*) of Malaya, Borneo, Sumatra, and Java is an especially accomplished singer, and is sometimes kept in that region as a caged songbird.

Bulimia *see* **Eating disorders**

Bunsen burner

Named after the German chemist Robert Wilhelm Bunsen, who contributed to its development, the Bunsen burner was already known to Michael Faraday, who may have created the first design. The idea behind the Bunsen burner is to reduce the considerable loss in **heat energy** typical in ordinary gas burners. This reduction of energy waste is accomplished by using a mixture of gas and air, the optimal proportion being three volumes of air to one of gas, instead of pure gas. As a result, **combustion** is intensified, producing a nonluminous but remarkably hot flame.

The Bunsen burner consists essentially of a long **metal** tube set on a flat base. Gas enters the burner through a hole in the bottom of the tube. Some burners have a gas adjustment screw that allows one to control the amount of gas entering the tube. With burners lacking a gas adjustment screw, gas flow can be controlled only at the supply valve. A second opening at the bottom of the metal tube allows air to enter and mix with the gas. The air inlet may be the bottom opening of the tube itself, or it may be a pair of holes cut into the tube near the base. The amount of air entering the tube in the

former design was controlled by a flat piece of metal that can be slid across the hole to allow more or less air to enter. Some burners have threaded bases that allow the air supply to be controlled by turning the tube. In the second design described above, air supply is controlled by a collar that covers the hole in the tube. The collar can be rotated to allow more or less air to enter the tube.

The gas-air mixture is ignited at the top of the barrel. The flame produced at this point commonly consists of two cones. The outer cone is blue, while the inner remains quite pale, almost invisible. The hottest part of the burner flame is at the tip of the inner cone, where a rich supply of air ensures the nearly total combustion of the gas. The **temperature** at this point may be in excess of 3,272°F (1,800°C) in an inexpensive laboratory burner.

Beyond the laboratory, the the principle of Bunsen combustion is widely used in industry, in gas furnaces, and in everyday life, as exemplified by the kitchen gas range.

Buntings *see* **Sparrows and buntings**

Buoyancy, principle of

The principle of buoyancy is called Archimedes' Principle, since it was discovered by this Greek mathematician in the third century B.C. The principle states that the buoyant **force** acting on an object placed in a fluid is equal to the weight of the fluid displaced by the object. An object completely immersed in a fluid (liquid or gas) displaces a **volume** of fluid exactly equal to the volume of the object. The weight of that volume of displaced fluid is the buoyant force acting on the object.

Fluids such as **water** or air exert **pressure** in all directions and the amount of pressure depends on the depth of the fluid. The pressure on the bottom of an object immersed in a fluid will be greater than the pressure on the top of the object. The imbalance of pressure acting on the object creates an upward force called the buoyant force. If the buoyant force is greater than the weight of the object, the object will float. If the buoyant force is less than the weight of the object, the object will sink in the fluid.

The **density** of a fluid is its weight per unit of volume. Liquids and gases exhibit widely different densities. The buoyant force, or the weight of the volume of displaced fluid, will depend on the density of the fluid as well as the displaced volume. Fresh water has a density of

The molecular structure of water begins to expand once it cools beyond 39.4°F (4°C) and continues to expand until it becomes ice. For this reason, ice is less dense than water, floats on the surface, and retards further cooling of deeper water, which accounts for the survival of freshwater plant and animal life through the winter. *Illustration by Hans & Cassidy. Courtesy of Gale Group.*

62.4 lb per cubic foot (pcf), **saltwater** density is, on average, 64 pcf. Air at **sea level** has a density of 0.08 pcf and at 10,000 ft (3,050 m), 0.06 pcf. Saltwater is denser than **freshwater** because of its salt content, and, as a result, a swimmer is more buoyant in the **ocean** than in a freshwater **lake**. The density of saltwater depends on its salinity and varies around the world. The molecular structure of water expands when it freezes, therefore, **ice** is less dense than liquid water. As a result, ice cubes float and lakes freeze from the top down rather than the bottom up.

Boats, bladders, and blimps

Steel has a density of 487 pcf, about eight times that of water. Steel boats float, however, because they are hollow and shaped to displace a volume of water that weighs more than the boat's weight. Ships are often rated by their displacement. Displacement is measured in units called tons, which are the weight of the water displaced by the ship. As a ship is loaded with cargo, it settles deeper into the water. This displaces an additional volume of water and produces the greater buoyant force required to support the added load. Plimsoll marks are painted onto the hull of cargo ships to indicate the depth to which the ship could be loaded. The different marks refer to fresh and saltwater and to the various **seasons** where **temperature** also effect water density.

Fish can alter their buoyancy by changing the volume of their internal swim bladder. Scuba divers can inflate their external buoyancy compensator vest to change its volume. Both of these changes alter the amount of displaced water and, thus, the buoyant force acting on the body. With this control, divers and fish can ascend or descend at will as they observe each other at play.

The principle of buoyancy applies to all fluids, including gases. A blimp is filled with very light helium gas with a density of 0.01 pcf. As a result, the weight of the blimp is less than the weight of the air that it displaces and the blimp will float in air. By dropping ballast or venting helium, the blimp can control its buoyancy and, thus, its altitude. A hot air **balloon** gets it buoyancy because hot air is less dense than cold air. The density of air at 200°F (93°C) at sea level is 0.06 pcf and serves the same function as the light helium gas in the blimp. Although highly flammable, **hydrogen** gas is much less dense than helium and was used for lift in dirigibles up until 1937, when the German **airship** *Hindenberg* burned and crashed.

Richard A. Jeryan

Buret

A buret (also spelled burette) is a long **glass** tube open at both ends, that is used to measure out precise volumes of liquids or gases. Most burets are about 0.04 in (1 mm) in diameter and 30 in (75 cm) long. The bottom of a buret is tapered so that its diameter is only about 0.1 mm in diameter. Burets are most commonly designed to hold volumes of 1 ml or less.

Fluid is dispensed form a buret through a glass stopcock at the lower end of the glass tube. The stopcock consists of an inner piece of ground glass that fits tightly into the glass tube and that can be rotated in a tightly fitting casing. The stopcock allows a fluid to be released in very small, precise amounts. Commercially available burets can usually be read with an **accuracy** of ±0.01 ml.

Probably the most familiar application of a buret is in the process known as titration. In this process, accurately measured amounts of two solutions are allowed to react with each other in order to determine the **concentration** of one. Burets have far more uses, however. A single buret can be used, for example, to release a known **volume** of a **solution** of known concentration in order to determine the **mass** of an unknown solid. Oxidation-reduction reactions can also be studied quantitatively using burets.

Gas-dispensing burets consist of arrangements in which some gas is contained by and forced out of a graduated cylindrical tube by means of some liquid, such as mercury. In their appearance, gas burets look something like an upside-down version of their liquid counterparts.

Burets became necessary in chemical research only with the development of relatively precise analytical techniques in the eighteenth century. Credit for their invention is usually given to the French chemist Joseph Louis Gay-Lussac, who first developed them for the purpose of assaying silver.

Burn

A burn is damage to the skin. Depending on the type of and severity of the burn, skin may be only superficially damaged, or damage may extend deep within the layers of the skin.

Burns can be caused by extreme **heat**, extreme cold, chemicals, **electricity**, or radiant **energy** (i.e., ultraviolet rays from the **sun** or an artificial source, and **x rays**).

The degree of damage of a burn can be classified in two ways. The first, and more traditional way, uses the terms first degree, second degree, and third-degree burns. The second means of classification refers to partial thickness and full thickness burns.

A first-degree burn is one that affects only the uppermost layer of the skin, which is called the epidermis. This type of burn is the most common. Touching a hot stove element or scalding of skin by steam are examples of first-degree burns. Despite the **pain** associated with these burns, the first-degree burn is the least damaging. Within a few days healing is complete. A first degree burn is also a partial burn.

A second-degree burn extends through the epidermis to the underlying layer of skin, which is called the dermis. Redness and blistering of the skin are characteristics of a second-degree burn. Healing takes longer than with a first-degree burn and some scarring of the healed area might result. A second degree burn is also a partial burn.

A third-degree burn is the most serious type of burn. Here, skin damage extends all the way through the epidermal and dermal layers. These are full thickness burns. The patient may not feel as much pain with these burns, as the nerve endings in the burned area have been destroyed. Skin grafts are usually necessary to repair the damage of third-degree burns, and scarring is routine.

Chemical burns differ from radiant burns in that the skin has no protective mechanism to prevent them. With **radiation**, the melanin cells spread melanin in the skin to block ultraviolet **light** from penetrating. With chemical burns no such protective measure exists.

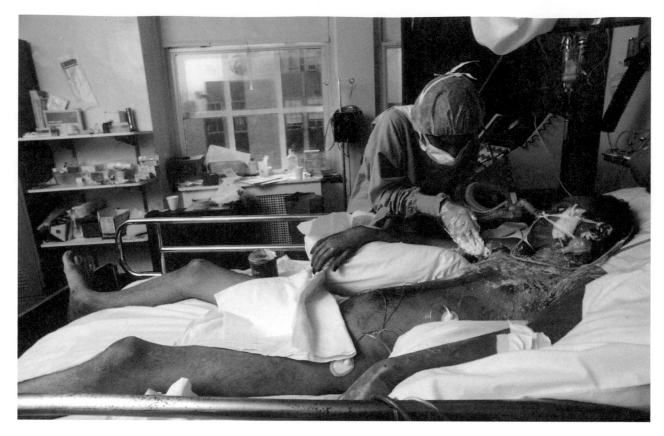

A patient is treated for burn wounds. *Photograph by Ken Sherman. Phototake.*

Chemical burns can occur with the application of acids, strong alkali (such as lye), or other agents. Some 25,000 industrial chemicals (of about 300,000 in use) can produce chemical burns, either internal or external.

Eyes are also vulnerable to chemical burns. Ideally the worker who is using dangerous materials wears goggles and other protective gear, but the home craftsman may not. Alkalis burn into the eyes rapidly and deeply. Acids burn rapidly, but usually are neutralized by the tears before they burn deeply. Initially, chemical burns may appear to be mild, but during the following day or so the injured **tissue** may slough off and the extent of the injury will be revealed.

Chemical burns may occur from unexpected sources. Dry cement, for example, because of its lime content, is capable of causing burns if one's skin is exposed to it for hours. Gasoline can penetrate skin and cause a burn after several hours of exposure as well. Never use fuel to clean the hands or use it in any other way that would result in long-term exposure.

Surprisingly, air bags in automobiles have burned some accident victims. The bags inflate explosively upon impact to cushion the car's occupants. However, the gas that inflates the bag is hot when it is released from the cylinder. Several burns from contact with inflating air bags have been reported.

Household chemicals can be as dangerous as industrial ones. Drain openers, for example, are based on lye with other additives and can be responsible for serious burns. Lawn fertilizer should never be handled with bare hands, and any that gets on the skin should be rinsed off immediately. Chemicals present a ready source for both internal and external burns. It is important that these materials be stored out of reach of young people and that they be used with great care. Protective clothing, gloves, and goggles should be worn whenever working with such chemicals. Spills and splashes should be cleaned up immediately and any chemical that contacts the skin should be rinsed off quickly. Excess chemicals or empty containers should be disposed of with care and in accordance with existing regulations. Empty containers should not be saved for reuse.

See also Physiology.

Resources

Books

Munster, A. M., and S. Burns. *A Family Guide to Medical and Emotional Recovery.* Baltimore: Johns Hopkins University Press, 1993.

KEY TERMS

Dermis—The internal layer of skin lying below the epidermis. It contains the sweat and oil glands, hair follicles, and provides replacement cells for those that are shed from the outer layer.

Epidermis—The outer layer of the skin consisting of dead cells. It is the primary protective barrier against sunlight, chemicals, and other possible harmful agents. The epidermal cells are constantly being shed and replenished.

Graft—The attachment of skin to an injured area. The new skin, natural or artificial, will prevent the loss of fluids and provide the means for a new, intact layer of skin to form.

Hydration—Restoring fluids to the body. Deep and extensive burns allow the escape of fluids needed for proper body functions.

KEY TERMS

Conversion—In the ecological context, this usually refers to a managed change of a natural ecosystem to one dominated by a human purpose, such as agriculture or an urbanized land-use. Losses of habitat associated with these sorts of conversion are among the most important causes of extinction and endangerment.

Dimorphic—This refers to the occurrence of two different shapes or color forms within the species, usually occurring as sexual dimorphism between the males and females.

Extirpated—The condition in which a species is eliminated from a specific geographic area of its habitat.

Over-hunting—Harvesting of wild animals at a rate that exceeds their capacity for regeneration, causing their population to collapse.

Wardrope, J., and J. A. Edhouse. *The Management of Wounds and Burns (Oxford Handbooks in Emergency Medicine)*. Oxford: Oxford University Press, 1999.

Wolf, S. E., and D. N. Herdon. *Burn Care*. Georgetown, TX: Landes Bioscience, 1999.

Brian Hoyle

Bushbabies *see* **Lorises**

Bustards

Bustards are 22 **species** of tall **birds** that make up the family Otidae. Bustards occur in relatively open habitats in **Africa**, central and southern **Europe** and **Asia**, Southeast Asia, and **Australia**. Most species, however, are African.

Bustards are large birds, with species ranging in body length from 14.5-52 in (37-132 cm), and in weight from 1-48 lbs (0.6-22 kg). Bustards have a stocky body, a long neck, and stout legs and feet, with three toes pointing forward, and no hind toe. The wings are broad, and the tail is short. The bill is stout, flattened, and blunt.

Bustards are colored in various subdued hues and patterns of brown, buff, gray, black, and white. Bustards are sexually dimorphic, with males being larger and more brightly colored than females.

Some species of bustards occur during the non-breeding season in flocks of various size. Bustards walk while feeding, and although they can fly, they tend to run to escape from predators. Bustards have keen **vision**, and are wary and difficult to approach closely on foot. Bustards are omnivores, eating a wide range of **plant** and **animal** foods. Bustards predate on large **insects** such as **grasshoppers** and **beetles**, as well as on small **reptiles** and nestling birds.

Bustards nest on the ground, and lay one to five eggs. The female incubates the eggs and cares for the young birds, which are precocious and can leave their nest soon after hatching.

Species of bustards

The great bustard (*Otis tarda*) occurs in scattered populations in Eurasia. Its present distribution is greatly reduced compared with several centuries ago because of overhunting and conversions of its natural **habitat** to agriculture. However, this species is still abundant in some places where its seasonal flocks can contain as many as 500 birds. The male great bustard has a spectacular **courtship** display in which internal air sacs are expanded to greatly puff out the chest while white plumes are erected on the wings, tail, breast, and head. This strutting display is generally performed on slightly raised ground, in front of a hopefully appreciative audience of as many as six female birds.

The little bustard (*Otis tetrax*) is another Eurasian species with a similarly wide distribution as the great bustard. This and the preceding species undertake seasonal

A Kori bustard (*Ardeotis kori*) in courtship display. *Photograph by Nigel J. Dennis. Photo Researchers, Inc. Reproduced by permission.*

migrations, flying south from the northern parts of their range. The Houbara or MacQueen's bustard (*Chlamydotis undulata*) occurs from the Canary Islands off western Africa, through North Africa, as far as southwestern Asia.

The Australian bustard (*Ardeotis australis*) is the only species to occur on that **continent**. This species utilizes rather dense, shrubby habitat, in contrast to the open spaces preferred by other species of bustards.

The smallest bustards are the lesser florican (*Sypheotides indica*) and the Bengal florican (*Houbaropsis bengalensis*) of India. The world's heaviest flying bird is the great bustard, which can achieve a weight of 48 lb (22 kg). Other large species include those in the genus *Ardeotis*, such as the Kori bustard (*A. kori*) of Southern Africa.

Bustards and humans

Bustards are large, palatable birds, and they are hunted for sport or as food in most parts of their range.

Some species of bustards have become endangered through the combined effects of overhunting and conversions of their habitat to agricultural land uses. Other species, while not endangered, have had their breeding ranges significantly reduced. The great bustard, for example, was extirpated in England in 1832.

Resources

Books

Bird Families of the World. Oxford: Oxford University Press, 1998.

Brooke, M., and T. Birkhead. *The Cambridge Encyclopedia of Ornithology.* Cambridge, UK: Cambridge University Press, 1991.

Johnsgard, P.A. *Bustards, Hemipodes, and Sandgrouse.* Oxford: Oxford University Press, 1991.

Bill Freedman

Butane *see* **Hydrocarbon**

Buttercup

Buttercups and crowfoots are about 275 **species** of plants in the genus *Ranunculus*, family Ranunculaceae. Buttercups mostly occur in cool and temperate regions of both hemispheres of the world, including **mountains** in tropical latitudes.

Buttercups are annual or perennial, and they are herbaceous plants, dying back to the ground surface before the winter. The leaves of terrestrial species are simple or compound. However, the underwater leaves of aquatic buttercups can be very finely divided. Some of the aquatic buttercups have dimorphic foliage, with delicately divided leaves in the **water**, and distinctly broader leaves in the atmosphere.

The flowers of buttercups have numerous stamens and pistils, arranged in a **spiral** fashion on a central axis. The flowers of most species of buttercups are radially symmetric and showy, owing to their large, yellow petals. However, some species have red or white petals. The petals secrete **nectar**, important in attracting the **insects** that are the pollinators of most buttercups. There are usually five sepals, but these generally fall off the **flower** relatively soon. The **fruits** are loose heads of one-seeded fruits called achenes.

Many species of buttercups are native to **North America**. The wood buttercup (*Ranunculus abortivus*) is a widespread species of rich, temperate **forests**. The yellow water-crowfoot (*R. gmelini*) is a widespread species of **freshwater** marshes and shores, while the seashore-buttercup (*R. cymbalaria*) occurs in **salt** marshes and estuaries. Many native species of buttercups occur in alpine and arctic tundras, for example, the Lapland buttercup (*R. lapponicus*) and snow buttercup (*R. nivalis*).

Several species of Eurasian buttercups have been introduced to North America where they have become widespread weeds of lawns, fields, and other disturbed places. Some of the more familiar **introduced species** are the tall or meadow buttercup (*R. acris*), the creeping buttercup (*R. repens*), and the corn crowfoot or hungerweed (*R. arvensis*).

A few species of buttercups are used in **horticulture**. The most commonly used species for this purpose is the garden buttercup (*R. asiaticus*), available in varieties with white, red, or yellow-colored flowers. Aquatic buttercups, such as the water crowfoot (*R. aquatilis*), are sometimes cultivated in garden pools. Various alpine species of buttercups can be planted in rock gardens.

Meadow buttercups (*Ranunculus acris*). *© John Buitenkant 1993, National Audubon Society Collection/Photo Researchers, Inc. Reproduced with permission.*

Butterflies

Butterflies are **insects** in the order Lepidoptera, which also includes the **moths**. Butterflies at rest fold their wings vertically over their head, whereas moths hold their wings horizontally. Most butterflies are active during daylight, while moths are mostly nocturnal. Butterflies undergo complete **metamorphosis**, that is, their egg hatches to a larva (or caterpillar), which pupates in a chrysalis, from which emerges the adult butterfly (or imago).

Evolution

Butterflies probably first evolved about 150 million years ago, appearing at about the same time as the flowering (or **angiosperm**) plants. Of the 220,000 **species** of Lepidoptera, about 45,000 species are butterflies, which probably evolved from moths. Butterflies are found throughout the world, except in **Antarctica**, and are especially numerous in the tropics. They fall into eight families: Papilionidae (swallowtail butterflies), Pieridae (whites), Danaidae (**milkweeds**), Satyridae (browns), Morphidae (morphos), Nymphalidae (nymphalids), Lycaenidae (blues), and Hesperidae (skippers).

Development and life cycle

The adult life of butterflies is only one week for some species, but up to 10 months for others. Adult butterflies spend much of their time courting, mating, and (for females) laying eggs. The females lay eggs on the specific food **plant** upon which the caterpillar must feed. The eggs are typically laid on leaves, flowers, or stems, but sometimes on **tree bark** or even stones, leaving the newly hatched caterpillars to find their necessary food plant. A few species of butterfly lay their eggs while in flight, the eggs attaching to food plants as they fall.

The egg

By the time most female butterflies emerge from the chrysalis, their eggs are fully mature, permitting immediate **fertilization**. Minute pores in the egg allow entry of the male sperm for fertilization. Eggs may be laid singly, in rows, in clusters, or in rings around a plant stem. The eggs are highly vulnerable to predators and **parasites**. Females lay up to 600 eggs at a time. A sticky substance produced during egg laying glues the eggs to the host plant. Most eggs hatch within a few days, although those of some species remain dormant over winter and hatch in the spring.

The caterpillar

The larval stage of butterflies, known as a caterpillar, emerges from the egg fully formed. Caterpillars see by means of groups of tiny eyes on each side of their head. The body of many species is protected by bristly hairs. Caterpillars have as many as eight pairs of appendages. Three pairs are true legs, and become the legs of the adult butterfly. These are located on the thorax behind the head. Another four pairs of appendages occur on the abdomen, and are known as false, or prolegs. There is also a single pair of claspers at the tip of the abdomen, which grip tightly to the food plant. Prolegs and claspers also carry tiny hooks, which catch a silk thread spun by the larva as it moves about its food plant and help keep it from falling. The silk thread is produced as a thick liquid, excreted through **glands** near the mouth, which is then spun into a thread by a spinneret located behind the jaws.

Caterpillars spend most of their time eating. Their large jaws (the mandibles) move sideways while shred-ding their plant diet. Caterpillars munch their way through life: first eating their egg shell, from which important **nutrients** are retrieved, and then continuously consuming its food plants. A very small number of butterfly species have caterpillars that are predators that feed on other insects. Caterpillars typically eat more than twice their own weight each day, pausing only to shed their old skin, in order that they may grow larger. Skin shedding (molting, or ecdysis) occurs at least four times before the caterpillar is fully grown. Growth **rate** depends on **temperature**; it is faster in warm weather, and slower when cool. Caterpillars seek a protected place in which to pupate, forming a protective shell and becoming dormant. Some caterpillars travel rather long distances (330 ft or 100 m) when seeking a sheltered place to pupate.

The chrysalis

The protective case surrounding the pupating caterpillar can take many shapes. It is usually brownish green in **color**, and may be speckled to aid in camouflage. With few exceptions, butterfly larvae pupate above ground, usually attached to a **leaf** or stem by a silken thread. Inside the chrysalis the pupated caterpillar gradually transforms into a butterfly. The process takes two weeks for some species, and as long as several years for others. When fully developed, the butterfly inside the chrysalis swallows air, inflating its body and splitting the pupal skin. An adult butterfly struggles out shortly after dawn, its moist wings hanging limply from its thorax. Crawling to a place where it can hang by its legs, the newly emerged butterfly pumps up its wings by swallowing air, increasing its internal **pressure** and forcing **blood** through tiny **veins** in the wings. Within several hours the expanded wings dry and harden. At about the same time the butterfly excretes waste products accumulated during pupation. The butterfly is now free to begin the adult part of its life cycle.

The adult (or imago)

Butterflies, like all insects, have an external skeleton (or exoskeleton) to which muscles are attached. The exoskeleton provides the butterfly's body with support and reduces **water** loss through **evaporation**. The **respiratory system** does not have a pumping mechanism. The sides of the thorax and abdomen have tiny pores (or spiracles) through which air enters and leaves the body via tubes (tracheae). The insect's blood (hemolymph) is pumped as it passes through a long, thin **heart**, and bathes the organs inside the body cavity.

Two large compound eyes, made up of hundreds of tiny units (ommatidia), cover much of the butterfly's head and allow a wide field of **vision**, including partially backwards. However, the eyes cannot distinguish much detail or determine **distance**, although they can readily identify color and movement, both of which are vital for survival. Colors aid in the identification of flowers, larval food plants, and the opposite sex of the species, while detecting movement may save butterflies from attacks by their predators.

The long, coiled proboscis (tongue) of the butterfly is projected into the center of flowers while searching for **nectar** (a liquid, sugar-rich food). Above the eyes and on either side of the head are two antennae covered with microscopic sense organs. The antennae are often incorrectly called "feelers," but "smellers" would be more accurate because it is through these organs that the butterfly sniffs out its favorite foods and potential mate. The antennae can detect **pheromones**, which are specific chemical signals released by the opposite sex that are detectable over a great distance.

The thorax of a butterfly is divided into three segments, each having one pair of legs. Each leg ends in a claw, enabling the butterfly to hold on while feeding and egg laying. Sensory receptors on the leg just above the claw detect the chemical makeup of appropriate food plants.

The wings are a spectacular part of the butterfly, and are extremely large compared to its body. The wings attach at the two rear segments of the thorax. Tough veins provide a framework which supports the wing **membrane**, which is covered by millions of microscopic scales arranged in rows like shingles on a roof. It is these scales (about 99,000 per sq in [15,000 per sq cm]) which form the color and pattern of the wing. Brown, orange, and black tones are due to the presence of chemical pigments. However, the brilliant, iridescent blues and greens are parts of the **light spectrum** separated by tiny facets on the surface of the scales, creating colors in the fashion of a **prism**. The wing colors provide camouflage against predators and aid in identifying mates of the correct species. Scales also cover the segmented abdomen, in which the digestive and reproductive organs are located.

Reproduction

Reproduction in butterflies begins with **courtship**, during which the male vigorously flaps its wings, releasing a dust of microscopic scales carrying pheromones above the female's antennae. These male pheromones act as a sexual stimulant to the female. Some males release additional pheromones from "hair pencils" under the abdomen. Female butterflies that are ready to mate dispense with courtship. Some species, however, perform complicated courtship maneuvers, probably to find a mate strong enough to endure the rigorous rituals, there-

A monarch butterfly (*Danaus plexippus*) in Livonia, Michigan. The larvae of this butterfly feed on the leaves of the milkweed, ingesting substances that make them toxic to birds and other predators. *Photograph by Robert J. Huffman. Field Mark Publications. Reproduced by permission.*

by increasing the chance of producing healthy offspring. Males usually must wait one or two days after emerging from the chrysalis before they can mate, but then they may mate many times. Females can mate immediately after emerging, some species mating several times. However, it is the last male to mate that fertilizes the eggs. Females of some species mate once only.

Migration

Most butterfly species live and die within a narrow geographic range, because their brief life span allows little time for wandering. However, the monarch butterfly migrates thousands of miles from southern Canada and northern parts of the United States to Mexico and southern coastal California. The round trip takes two or three generations to complete, because of the short **individual** life span. Each butterfly must find its own way, having no living guide, yet the species follows the same broad migratory route year after year, century after century. Monarchs and some other butterflies migrate in massive numbers, forming **clouds** of color sometimes a mile wide and a mile long; millions perish during the journey.

Vulnerability and defense

The survival of butterflies is influenced by many factors, and their populations may increase or decrease quite rapidly, and from year to year. Brilliantly colored butterflies are often toxic when eaten, so predators learn to leave them alone. Occasionally, a non-poisonous species evolves that mimics the appearance of a poisonous species, thus being less susceptible to predators. Dark colors, displayed by butterflies in cooler climates (such as alpine or arctic **tundra**), readily absorb sunlight so the cold-blooded insect can warm up more quickly.

Some butterflies are well camouflaged and almost undetectable among the flowers of their particular food plant. A variety of spots resembling eyes on the outer edges of the wings of certain species may startle or distract predators, drawing the point of attack of a swooping bird away from the butterfly's soft body. Some butterflies do a 180-degree turn before landing, so the eye-spots on the wings are positioned where a **predator** may think the head should be. Butterflies surviving bird attacks can often fly well even with a piece of wing missing.

The underpart of the wing is usually duller than the surface, so that when the butterfly assumes its resting position with wings folded above its head, it is cryptically colored, blending into the background on which it lands. Some butterflies use the club on the end of the antennae to knock small predators off their body.

The hairy spines on many caterpillars deter predators, and their colors often blend in with the leaves of their food plant. Some pupae develop hornlike appendages on the head and rear of the chrysalis, which appear to point menacingly at predators. In other cases, a chrysalis hanging from a twig may look like a dead leaf.

Conservation

Habitat loss to agriculture, **deforestation**, urbanization, draining of **wetlands**, and other changes in land-use is the foremost threat to butterfly populations. Although **pollution**, **pesticides**, and specimen collection pose serious threats to some species, none of these is as damaging as habitat loss. The short life span of butterflies usually makes it impossible for displaced populations to find another appropriate habitat. Although no species of butterfly is known to have been made extinct through human actions, some subspecies have been rendered extinct, and

---

KEY TERMS

Chrysalis—A soft casing, shell, or cocoon protecting the dormant pupa of insects during metamorphosis.

Metamorphosis—A complete change of form, structure, or function in the process of development, shown by insects.

Pheromones—Chemical substances, secreted by most animals, that stimulate a response from others of that species (e.g., sex hormones for mating) or other species (e.g., a predator sensing chemicals produced by its prey's fear).

Pupa—An insect in the nonfeeding stage during which the larva develops into the adult.

---

some rare species are endangered. Protection of habitat is the most effective way to prevent major reductions in populations and endangerment of butterflies, and of other wild animals and plants.

Resources

Books

Eid, A., and M. Viard. *Butterflies and Moths of the World.* Book Sales Pubs., 1997.

Feltwell, John. *The Natural History of Butterflies.* New York: Facts on File, 1986.

Pollard, Ernest. *Monitoring Butterflies for Ecology and Conservation.* London: Chapman & Hall, 1993.

Sbordoni, V., and S. Forestiero. *Butterflies of the World.* Firefly Books, 1998.

Scott, James A. *The Butterflies of North America.* Stanford: Stanford University Press, 2001.

Smart, Paul. *The Illustrated Encyclopedia of the Butterfly World.* New York: Random House, 1996.

Periodicals

Barbour, Spider. "Overnight Sensation." *Natural History* (May 1989): 24-28.

Boppre, Michael. "Sex, Drugs, and Butterflies." *Natural History* (January 1994): 28-33.

Marie L. Thompson

Butterfly fish

Butterfly **fish** (family Chaetodontidae) are some of the most colorful and varied fish of the oceans, the majority of which live on or close to coral reefs. Most **species** measure from 5-9.5 in (13-24 cm) in length and have deep, flattened bodies that are frequently adorned by extended fins. In some species these may form a large arc over the body. In addition to refinements in the body shape, the colors and patterns of most butterfly fish are quite enthralling. The head and body is usually a dark background **color** which is broken up by a series of stripes and other patterns. Coloration varies considerably but often includes patches of yellow, orange, blue, and white. A "false eye" is commonly seen on some butterfly fish; this is usually located towards the back of the fish or even on a fin, the objective being to distract a striking **predator** from the butterfly fish's own head.

Like the parrotfish and wrasses, butterfly fish swim by synchronous rowing strokes of the pectoral fins, while the tail fin is used as a rudder for direction and balance. They are capable of very rapid movement and rely largely on their agility to avoid capture from other larger species. Butterfly fish are strictly diurnal, exploiting the diversity of feeding and living spaces in and around coral reefs and atolls. As dusk approaches, however, most begin to seek out a safe hiding place where they will rest for the night.

Butterfly fish are specialist feeders: many species have the mouth placed at the end of a short, tubular snout that facilitates the animals poking into tiny crevices in coralline reefs and extracting **prey** from seemingly inaccessible places. Their diet may either be restricted to living polyps plucked from just a few coral species, while other members of the family spare the polyps themselves, preferring to graze on **algae** growing on the corals. Some even actively pursue small shrimps and **copepods** that lurk within the many crevices of the reef face.

The social **behavior** of butterfly fish varies according to the particular species. Many species are solitary, but some do form stable monogamous relationships with a member of the opposite sex. In the latter case, the two fish commonly patrol and defend a particular patch of coral against other members of the same species. A series of threat and aggressive gestures including color changes have evolved in these species which help prevent aggressive encounters from developing. Spawning frequently occurs at dusk—a strategy that may have evolved to increase the survival rates of young butterfly fish, as many of the tiny **plankton** feeders are less active in the evening.

Butyl group

Butyl group consists of the group of **atoms** C_4H_9. It is derived by removing one **hydrogen** atom from either

---

KEY TERMS

Aldehydes—A class of organic chemical compounds that contain a -CHO group.

Alkyl group—A paraffinic hydrocarbon group that may be obtained from an alkane by removing a hydrogen atom from the latter.

Antioxidant—Any substance that prevents oxidation from occurring.

Copolymer—A compound with high molecular weight formed in the reaction between two different raw materials.

Esters—A family of organic compounds formed in the reaction between an alcohol and an organic acid.

Food additive—A substance added to prepared foods to keep them from spoiling, add flavor or odor, increase their nutritional value, or make some other commercially desirable change in the food.

Herbicide—A chemical that kills entire plants, often selectively.

Inhibitor—Any substance that prevents some form of chemical reaction from taking place.

Isomers—Two molecules in which the number of atoms and the types of atoms are identical, but their arrangement in space is different, resulting in different chemical and physical properties.

Ketones—A family of organic compounds characterized by the presence of the C=O group appearing anywhere except at the end of the molecule.

Polymerization—A chemical reaction in which small molecules react with each other over and over many times, forming very large product molecules.

Silicone—A large group of organic compounds whose molecules consist of organic groups attached to silicon atoms.

---

of the isomers of butane, C_4H_{10}, and exists in four isomeric forms. Two isomers of butane exist, n-butane, also called 1–butane ($CH_3CH_2CH_2CH_3$), and iso-butane, also called 2-methylpropane ($CH_3CH(CH_3)CH_3$).

Removing a hydrogen atom from the first **isomer** can result in the formation of two different butyl groups, one designated as n-butyl, and the other as sec-butyl (secondary butyl). The structure of the n-butyl group is $CH_3CH_2CH_2CH_2^-$; the structure of the sec-butyl group is $CH_3CH_2CH(CH_3)^-$. The difference is that the n-butyl group

TABLE 1. BUTYL COMPOUNDS AND USES	
Compound	*Use*
n–butyl acetate	manufacture of photographic film, safety glass, artificial leather, perfumes, and flavoring agents
N–butyl pthalate	insect repellant
n–butyltrichlorosilane	production of silicones
4–tert–butyl catechol	used to slow down or stop polymerization reactions in the production of certain plastics and synthetic rubbers
butylate (bis [2–methyl–propyl] carbamothoic acid S–ethyl ester) herbicide t–butyl acetate	gasoline additive
4–tert–butylphenyl salicylate	light absorber in plastic food wrappings
n–butylamine	raw material in the synthesis of many products, including dyes, pharmaceuticals, insecticides, and rubber chemicals
sec–butylamine	prevents the growth of fungi
n–butyl citrate	anti–foaming agent; plasticizer; manufacture of inks and polishes
1–butyl–3–metanilyurea	treatment of hypoglycemia (an abnormally low blood sugar level)
butylmethoxydibenzoyl–methane	screens out ultraviolet light
t–butyl nitrate	jet propellant
n–butyl strearate	used to soften cosmetics, plastics, textiles, and other types of polymers

bonds to other atoms via an end **carbon** atom, while the sec-butyl group bonds via an "inner" carbon atom.

Removal of a hydrogen atom from the iso-butane isomer can result in the formation of two different butyl groups. One of these groups is commonly known as tert-butyl or 1, 1-dimethylethyl [$(CH_3)_3C^-$]. The tert- prefix stands for tertiary, indicating that the open bond is from a carbon atom that is attached to three other carbon atoms. The other butyl group that can be formed from iso-butane is the isobutyl or 2-methylpropyl group [$(CH_3)_2CHCH_2^-$].

Butyl compounds

A large number of compounds containing the butyl group exist. Many of them are relatively simple and can be represented by the formula C_4H_9X, where X stands for a halogen, a hydroxyl group, an amine, or some other group. The following sections review some of the most important of these compounds.

Butyl alcohols

Four butyl alcohols exist, each formed by the addition of a hydroxyl group (OH) to one of the four butyl isomers discussed above. Their names and structures are as follows: n-butyl **alcohol** (or 1-butanol) $CH_3CH_2CH_2CH_2OH$; iso-butyl alcohol (or 2-methyl-1-propanol) $(CH_3)_2CH\ CH_2OH$; sec-butyl alcohol (or 2-butanol) $CH_3CHOHCH_2\ CH_3$; tert-butyl alcohol (or 2-methyl-2-propanol) $(CH_3)_3\ COH$.

The boiling points of the butyl alcohols decrease regularly in moving down the above list, from 244°F (118°C) for n-butyl alcohol to 226°F (108°C) for iso-butyl alcohol to 212°F (100°C) for sec-butyl alcohol to 180°F (82°C) for tert-butyl alcohol. A similar pattern exists for **solubility** of the alcohols, increasing from 8 g per 100 g of **water** for n-butyl alcohol to 10 g per 100 g of water and 12.5 g per 100 g of water for the next two forms to complete **miscibility** for tert-butyl alcohol.

The four butyl alcohols undergo very different reactions in many instances. As an example, n-butyl and iso-butyl alcohol can be oxidized rather easily to yield **aldehydes**. Oxidation of sec-butyl alcohol, however, results in the formation of a ketone. Oxidation of tert-butyl alcohol occurs only under the most extreme conditions, resulting in the complete oxidation of the compound to **carbon dioxide** and water.

Of the four butyl alcohols, n-butyl alcohol is in the greatest demand commercially. It is used as a solvent for fats, waxes, gums, shellac, varnish, and other materials in many industrial processes. It is also used as the starting **point** in the preparation of other butyl compounds. All of the butyl alcohols are of some interest in the synthetic flavoring industry since they can react with organic acids to make pleasant smelling esters. For example, n-butyl butanoate has the odor of pineapple; 2-methylpropyl propanoate smells like rum; 2-methylpropyl methanoate, like apple; n-butyl methanoate, like **banana**; and n-butyl ehtanoate, like strawberry.

The accompanying table summarizes some butyl compounds and their most important uses.

See also Isomer; Oxidation-reduction reaction.

Resources

Books

Budavari, Susan, ed. *The Merck Index.* 11th ed. Rahway, NJ: Merck and Company, 1989. pp. 236-242.

Carey, Francis A. *Organic Chemistry.* New York: McGraw-Hill, 2002.

Hawley, Gessner G., ed. *The Condensed Chemical Dictionary.* 9th edition. New York: Van Nostrand Reinhold, 1977, pp. 133-142.

Loudon, G. Mark. *Organic Chemistry.* Oxford: Oxford University Press, 2002.

Butylated hydroxyanisole

Butylated hydroxyanisole is a food additive much more widely known by its abbreviation, BHA. BHA is an aromatic organic compound with the chemical names of 2- and 3-tert-butyl-4-methoxyphenol. It can exist in either of the two isomeric forms or as a mixture of the two isomers. In its pure form, BHA is a waxy white or pale yellow solid with a melting point of 118.4–131°F (48–55°C) and a **boiling point** of 507.2–518°F (264–270°C). It is normally insoluble in **water**, but can be treated in order to make it so.

The chemical property of BHA that is of greatest commercial interest is its tendency to reduce the **rate** at which other substances undergo oxidation. It has long been used as a preservative in foods containing fats, which turn rancid by oxidation. First used as an antioxidant in 1947, it is now added to a wide variety of foods, including beverages, ice cream, candy, baked goods, instant mashed potatoes, edible fats and oils, breakfast cereals, dry **yeast**, and sausages. The compound is some-

times used in conjunction with a related antioxidant, **butylated hydroxytoluene** (BHT).

Some studies have found that BHA can produce allergic reactions and, in larger doses, affect liver and kidney functions. The Select Committee on GRAS (Generally Regarded as Safe) Substances of the U.S. Food and Drug Administration (FDA) reported in 1980 that no evidence exists to indicate that BHA is a health hazard. However, it recommended caution in its use and suggested additional studies on possible risks to human health. Currently FDA regulations limit the **concentration** of BHA in commercial foods to 0.02% in products containing fats and oils and to somewhat higher concentrations in other food products. In spite of current FDA regulations, some **nutrition** experts have recommended that BHA be banned from use in foods on the grounds that safer antioxidant alternatives are available.

Butylated hydroxytoluene

Butylated hydroxytoluene (BHT) is a derivative of cresol, an aromatic organic compound in which two additional **hydrogen atoms** in the **benzene** ring are replaced by tertiary butyl groups. Its technical name is 2,6-di-tert-butyl-p-cresol. In its pure form BHT is a white crystalline solid with a melting point of 158°F (70°C) and a **boiling point** of 509°F (265°C). It is normally insoluble in **water**, but for commercial applications, it can be converted to a soluble form.

BHT was first used as an antioxidant food additive in 1954. An antioxidant is a substance that prevents the oxidation of materials with which it occurs. BHT, therefore, prevents the spoilage of food to which it is added.

BHT has grown to be very popular among food processors and is now used in a great range of products that include breakfast cereals, chewing gum, dried **potato** flakes, enriched **rice**, potato chips, candy, sausages, freeze-dried meats, and other foods containing fats and oils. BHT is sometimes used in conjunction with a related compound, **butylated hydroxyanisole** (BHA) as a food additive.

Some evidence exists that BHT may be harmful to human health. Studies suggest that the compound may damage the liver and kidneys. However, the U.S. Food and Drug Administration has deemed that BHT is safe enough when used in limited concentrations. It currently permits its use in concentrations of about 0.01% to 0.02% in most foods. As an **emulsion** stabilizer in shortening, it may be used in a somewhat higher **concentration**, 200 parts per million. Some authorities suggest that

BHT poses too large a health risk and that it should be banned in foods. That policy has been adopted in some other nations, such as England and **Australia**, where its use is permitted as a food additive only in special cases.

BHT does have other commercial uses, as in **animal** feeds and in the manufacture of synthetic rubber and **plastics**, where it also acts as an antioxidant.

See also Food preservation.

Buzzards

The true buzzards are diurnal **birds of prey** in the genus *Buteo*, sub-family Buteonidae, family Accipitridae. In **North America**, buzzards are also commonly known as **hawks**, although other genera in the family Accipitridae are also given this common name, for example, the *Accipiter* hawks. There are 25 **species** of buzzards.

Buzzards are in the order Falconiformes, which also includes other types of hawks, **eagles**, osprey, **falcons**, and **vultures**. All of these **birds** have strong, grasping (or raptorial) talons, a hooked beak, extremely good **vision**, and a fierce demeanor. However, buzzards can be distinguished by their relatively large size, wide, rounded tail, broad wings, and their soaring flight. The usual **color** of the feathers is a barred pattern of browns and black, with some buff or red. The sexes are colored similarly, but females are substantially larger than males.

Buzzards occur on all of the continents, except for **Antarctica**. However, most species of buzzards occur in the Americas. Buzzards are most common seen in relatively open habitats, such as prairies, savannas, and forest edges.

Buzzards are mostly predators of small **mammals**, rabbits and hares, and to a lesser degree, **snakes**, lizards, birds, and larger **insects**, such as **grasshoppers**. Buzzards commonly soar in huge circles at great heights, looking for **prey** in the open, using their extremely acute vision. If prey is seen, an attempt may be made to catch it by undertaking a steep dive, known as a stoop. Some species also hunt regularly from perches in trees or on posts. The prey is generally killed by the powerful, sharp-clawed, grasping talons of these birds.

Migrating buzzards also soar during their long-distance movements, utilizing the lift obtained from high thermals during sunny days to achieve a relatively effortless flight. Some species migrate in large groups, and occasionally thousands of individuals can be seen at one time. These birds seem to fill the sky as they soar to great heights on one thermal and then glide slowly to pick up the next thermal along their path of travel, using the presence of other birds to identify the otherwise invisible **habitat** of rising, warm air.

Buzzards defend a territory during their breeding season. The territory is proclaimed by aerial displays and by loud, harsh screams. Buzzards nest in trees. Often the nest was built by another species, such as crows or ravens, and is then appropriated by the buzzard. However, buzzards will also build their own nests. The same pair may continue to utilize a nest for several **seasons**.

Species of buzzards

The largest, most widespread and familiar species of buzzard in North America is the red-tailed hawk (*Buteo jamaicensis*). This species breeds in almost all regions below the arctic **tundra**, and as far south as Panama and the West Indies. The red-tailed hawk nests in trees at or near the edge of woodlands, but feeds in open country. The plumage of the red-tailed hawk is quite variable, but adults have a reddish top of their tail. Northern populations migrate to the south in winter, although they will stay quite far north if an abundance of their prey of small mammals is available.

The red-shouldered hawk (*B. lineatus*) is a common species of the eastern United States and southeastern Canada, with a separate, disjunct population in coastal California and Oregon. This species commonly hunts from perches. Northern populations winter in the southeastern states.

The broad-winged hawk (*B. platypterus*) is a relatively small and common woodland species of the eastern United States and southeastern Canada. This species usually hunts for small mammals, **reptiles**, and insects from a perch in a **tree**. During the autumn the broad-winged hawk migrates in spectacular flocks, which occur as large groups riding thermals in a southerly direction. This species winters from southern Mexico to northern **South America**.

The rough-legged hawk (*B. lagopus*) breeds in the northern tundra of Canada and Alaska and winters in open habitats of the United States. This species also breeds throughout the tundra of northern Eurasia, from Scandinavia to eastern Siberia. The rough-legged hawk commonly hunts while hovering in the air.

The ferruginous hawk (*B. regalis*) is a buzzard of prairies and other open habitats of western North America, wintering in the southwestern States and Mexico. Swainson's hawk (*B. swainsoni*) is another western species of open habitats, breeding from central Alaska to northern Mexico. This species migrates in flocks, and winters in Argentina.

A common buzzard (*Buteo buteo*) perched in a spruce.
Photograph by H. Reinhard/Okapia. National Audubon Society Collection/Photo Researchers, Inc. Reproduced by permission.

Other species of buzzards in North America are relatively uncommon and localized in their distributions. These include the Harlan's hawk (*B. harlani*), Harris' hawk (*B. unicinctus*), and the zone-tailed hawk (*B. albonotatus*).

The common buzzard (*Buteo buteo*) breeds widely in **Europe** and northern **Asia**. Northern populations of this species are migratory, but southern populations are sedentary, as long as there is sufficient prey of small mammals available to support their needs. The long-legged buzzard (*B. rufinus*) has a more southern Eurasian distribution.

Buzzards and humans

Some people consider all hawks to be **pests**, believing that they eat game birds such as **grouse** and

ducks, or that they kill **song birds**. For these reasons, buzzards and other hawks have been killed in large numbers in some regions. Fortunately, however, this is rarely the case today, and few people now seek to kill these predators.

To some degree, buzzards have also been detrimentally affected by the toxic effects of insecticide use in agriculture and **forestry**. However, these birds have been somewhat less damaged by **pesticides** than some other types of **raptors**, such as falcons and eagles.

In fact, because they eat large numbers of small mammals, which can cause serious agricultural damages, buzzards provide a useful service to humans.

Because of buzzards' large size and fierce demeanor, many bird-watchers avidly seek out quality sightings of individuals, which can represent a highlight of a day's field expedition.

Resources

Books

Clark, W.S. and B.K. Wheeler. *A Field Guide to the Hawks of North America.* Boston: Houghton-Mifflin, 1987.

Forshaw, Joseph. *Encyclopedia of Birds.* New York: Academic Press, 1998.

Freedman, B. *Environmental Ecology.* 2nd ed. San Diego: Academic Press, 1995.

Johnsgard, P. A. *Hawks, Eagles, and Falcons of North America. Biology and Natural History.* Washington, DC: Smithsonian Press, 1990.

Scholz, F. *Birds of Prey.* Harrisburg, PA: Stackpole Books, 1993.

Bill Freedman

Cabbage *see* **Mustard family (Brassicaceae)**

Cactus

The cactus family or Cactaceae is made up of about 2,000 **species** of perennial plants with succulent stems, most of which are well-armed with sharp spines. The natural distribution of most cacti is American, ranging from southern British Columbia and southern Ontario in Canada, through much of the United States, to the tip of southern **South America**. One genus, *Rhipsalis*, occurs in **Africa**, Madagascar, and India, and is probably native there. Cacti usually inhabit deserts and other dry, open places. The major use of cacti by humans is as attractive, ornamental plants in gardens, or as indoor house plants. A few species produce edible **fruits**, and one yields peyote, a hallucinogenic drug.

Biology of cacti

Cacti are perennial plants. Their stems are fleshy or succulent, and are cylindrical or flattened in shape. The stems are green-colored, and are photosynthetic, usually performing this function instead of leaves, which are greatly reduced in abundance or even absent in most mature cacti. Most species of cactus are well-protected by sharp bristles and spines, which serve to deter most herbivores.

The stems of cactus plants have numerous cushion- or pit-like structures known as areoles on their surface, from which usually emerge clusters of spines. In terms of developmental **biology**, areoles are usually interpreted as being incompletely developed, axillary stem branches. The spines are actually modified leaves. The areoles may also be protected by hook-like barbs known as glochidia. The roots of cacti are shallow and may be widely spread in the **soil**.

The flowers of cacti are usually perfect (bisexual), containing both male reproductive organs (stamens) and female parts (a pistil). The flowers occur singly, rather than in groups, although many discrete flowers may be present on a cactus at the same time. The flowers of most species of cacti are large and showy, and they can be colored white, red, pink, orange, or yellow, but not blue. The sepals of the calyx are petal-like in shape and **color**, and they combine with the numerous petals to form an attractive, often richly scented, nectar-producing **flower**, designed to lure such pollinators as hawk-moths, **bees**, **bats**, and **birds**, especially **hummingbirds** and small doves. The fruit is a many-seeded berry.

Cacti are xerophytic plants, meaning they are physiologically and morphologically adapted to coping with the extreme **water** deficiencies of dry habitats, such as deserts. The xerophytic adaptations of cacti include: (1) their succulent, water-retaining stems, (2) a thick, waxy cuticle and few or no leaves to greatly reduce the losses of water through **transpiration**, (3) stems that are photosynthetic, so leaves are not required to execute this function, (4) stems that are cylindrical or spherical in shape, which reduces the surface to **volume** ratio, and helps to preserve moisture, (5) tolerance of high **tissue** temperatures, (6) protection of the **biomass** and moisture reserves from herbivores by an armament of stout spines, (7) a physiological tolerance of long periods of **drought**, and (8) a periodic pattern of growth, productivity, and flowering, which takes advantage of the availability of moisture during the brief, rainy season, while the **plant** remains dormant at drier times of the year.

Cacti have a so-called crassulacean-acid **metabolism**, in which atmospheric **carbon dioxide** is only taken up during the night, when the stomates are open. The **carbon** dioxide is fixed into four-carbon, organic acids, and can later be released within the plant, to be fixed into sugars by **photosynthesis** when the **sun** is shining during the daylight hours. Because this system allows stomates to be kept tightly closed during the day, crassulacean-acid metabolism is an efficient way of conserving water in dry environments.

A prickly pear cactus in Big Bend National Park, Texas. The prickly pear is the only widespread eastern cactus in the United States. It can be found as far north as southern Ontario. *Photograph by Robert J. Huffman. Field Mark Publications. Reproduced by permission.*

Some plant species of dry habitats that are not related to cacti are nevertheless remarkably similar in appearance (at least, apart from their flowers and fruits, which are always distinctive among plant families). This is the result of convergent evolution, the similar evolutionary development of unrelated species or families that are subjected to comparable types of environmental selective pressures. Some species of spurges (family *Euphorbiaceae*) that grow in dry habitats are commonly thought by non-botanists to be cacti, even though they are quite unrelated.

Species of cacti in North America

Species of cacti are prominent in many arid and semiarid habitats in the Americas. Cacti provide important elements of the **habitat** for many species of animals, especially larger species such as saguaro and candelabra cacti.

One of the most familiar groups of cacti are the prickly-pears, beaver-tails, or chollas (*Opuntia* spp.), of which there are about 300 species. These species have flattened, succulent, segmented stems (sometimes known as stem-joints), and are usually well-armed with spines of various sizes. *Opuntia lindheimeri* is a red- or yellow-flowered species that grows in Louisiana, Texas, and northeastern Mexico. This plant can reach a height of almost 13 ft (4 m) and can sometimes form dense thickets. *Opuntia macrorhiza* is a yellow-flowered species that grows in dry prairies from Kansas and Missouri to Texas. *Opuntia imbricata* has cylindrical instead of flattened stems, grows as tall as 6.6 ft (2 m), has red- or purple-colored flowers, and is commonly known as the **tree** or candelabra cactus. *Opuntia compressa* or the beaver-tail is a low-growing, yellow-flowered, eastern species that ranges from Massachusetts to Georgia. *Opuntia fulgida* or cholla occurs in the Sonora and other deserts of the southwestern United States and Mexico.

The pin-cushion cacti (*Mammillaria* spp.) are about 300 species of relatively small cacti that have spherical stems, with numerous, small, spiny, nipple-like protuberances on their surface. *Mammillaria microcarpa* and *M. thornberi* are species native to the southwestern states and Mexico.

The hedge or candelabra cacti (*Cereus* spp.) are made up of about 40 species. The barbed-wire cactus (*Cereus pentagonus*) is an arching, sometimes climbing species that grows in southern Florida, while the organ-pipe cactus (*C. thurberi*) is an erect, multi-stemmed species of deserts of Arizona and Mexico, which can achieve a height greater than 39 ft (12 m). The **desert** night-blooming cereus (*C. greggii*) occurs in deserts of the southwestern United States and Mexico. The odorous, nectar-rich, white flowers of this species open synchronously on only a few nights each year, and are pollinated by bats and hawk **moths**.

The saguaro, giant, or tall cactus (*Carnegiea giganteus*, sometimes known as *Cereus giganteus*) is a spectacular, multi-columnar species that dominates the landscape of deserts of Arizona and down into Mexico. This candelabra-like species can grow as tall as 49 ft (15 m) and has showy flowers that are pollinated by bats, birds, moths, and bees. The saguaro is an important component of the habitat of many species of animals. The gila woodpecker (*Centurus uropygalis*) and gilded flicker (*Colaptes chrysoides*) excavate nesting cavities in the saguaro cactus, and when these are abandoned they may be used secondarily by elf **owls** (*Micrathene whitneyi*) and other species of birds. The cactus wren (*Campylorhynchus brunneicapillus*) is another prominent species in saguaro-dominated deserts. In addition, many species of animals feed on the **nectar** of the saguaro, and on the bright-red, juicy pulp of its ripened fruits.

The barrel cacti (*Echinocactus* spp.) are seven species with stout, rotund, barrel-like stems. The barrel cactus (*Echinocactus polycephalus*) is a relatively large species of the southwestern states and Mexico, while the horse crippler (*E. texensis*) and star cactus (*E. asterias*) are smaller species. The hedgehog cacti (*Echinocereus* spp.) are 70 species with relatively small, densely aggregated, spiny stems. The red-flowered hedgehog cactus (*E. triglochidiatus*) occurs widely in arid habitats of the southwestern United States and Mexico. The organ-pipe cacti (*Lemaireocereus* spp.) are 25 species of tall, multi-stemmed, columnar cacti, including the candebobe (*L. weberi*) of Mexico. The barrel cacti (*Ferocactus* spp.) are 35 species of stout, short-columnar species, including *F. acanthodes*, *F. wislizenii*, and *F. covillei* of the southwestern states and Mexico.

Economic importance of cacti

Many species of cacti are highly prized by horticulturalists as botanical oddities and ornamental plants. These may be cultivated for their beautiful flowers, the aesthetics of their stems and spines, or merely because the plants have a strange-looking appearance. In addition, many people like to grow cacti because they are relatively easy to maintain—it does not matter much if you forget to water your cacti for a few days, or even a few weeks or more. In fact, over-watering is usually the greatest risk to most cacti that are kept as house plants, because too much moisture will pre-dispose these drought-adapted plants to developing fungal and bacterial diseases, such as soft-rot.

Virtually any of the native species of cacti of **North America** may be used in **horticulture**, as are many of the species of Central and South America. The genera *Mammillaria* and *Opuntia* are most commonly grown, but virtually any species may be found in cultivation around or in homes and greenhouses. One of the most common and familiar species is the Christmas cactus (*Zygocactus elegans*), a flat-stemmed, red-, pink-, or white-flowered species that is grown as a garden and house plant. This species blooms during the winter, and florists often induce this plant to bloom around Christmas-time, when it is commonly sold as a living ornament to brighten homes during that festive season. The candelabra cactus (*Cereus peruvianus*) is a tree-sized species native to South America that is commonly cultivated outdoors in hot climates, or in greenhouses in colder climates.

Many species of cacti can be rather easily transplanted from natural habitats into the vicinities of homes and businesses, where they may be used as central com-

---

KEY TERMS

. . . .

Berry—A soft, multi-seeded fruit, developed from a single, compound ovary.

Cuticle—A waxy, superficial layer that covers the foliage of vascular plants, and the stems of cacti.

Monoecious—This is a plant breeding system in which male and female reproductive structures are present on the same plant, and in the case of cacti, in the same flowers.

Perfect—In the botanical sense, this refers to flowers that are bisexual, containing both male and female reproductive parts.

Stomate—These are microscopic pores in the leaf or stem cuticle, bordered by guard cells which control opening or closing of the pore.

Succulent—Having thick, fleshy leaves or stems that conserve moisture.

Xerophyte—A plant adapted to dry or drought prone habitats.

---

ponents of low-maintenance gardens in places where rainfall is sparse, and the development of grassy lawns would require an excessive use of scarce and expensive water. Wild cacti are also collected to grow in or around the home, and to develop private collections of these interesting plants.

Unfortunately, most species of cacti re-colonize disturbed sites very slowly and infrequently. Extensive losses of cactus habitat to industrial and residential developments, coupled with excessive collections of wild plants, have resulted in the populations of some species of cacti becoming endangered. In some areas, populations of wild cacti must be guarded against illegal, often nocturnal collecting of valuable plants for horticultural purposes. Unfortunately, it is difficult to protect many endangered cacti from poaching. This is because of the extensive areas that must be patrolled, in the face of multi-million-dollar profits that can potentially be made in the illicit cactus trade. Some species of cactus are now critically endangered in the wild because of excessive, illegal collecting, and this represents an important ecological problem in many areas.

The most commonly edible cactus fruit is that of *Opuntia* species, especially *O. ficus-indica*. The fruits of prickly-pears, sometimes known as apples or tunas, can be eaten directly or used to make a jelly. Prickly-pear fruits are considered to be a delicacy around Christmas time in some regions.

Peyote or mescal buttons (*Lophophora williamsii*) is a cactus containing several alkaloids in its tissues that are used as a hallucinogen and folk medicine. Peyote is important in the culture of some tribes of native Amerindians in the southwestern United States and Mexico, especially in the vicinity of the Rio Grande River. These aboriginal peoples use peyote to induce religious experiences and revelations. Peyote is also commonly used as a recreational drug by many people, and by several religious cults.

Some species of spiny cacti, such as *Opuntia*s, are used as living fences, for example, to keep **livestock** out of gardens. The long, sharp spines of other cacti were used as needles in some of the earliest types of phonographs. The "wood" of the saguaro cactus has long been used by Amerindian peoples, and is still utilized to make crafts and novelty furniture.

A few species of cacti have become **pests**, or weeds, when they escaped from cultivation in places where they were not native, and were not controlled by diseases or herbivores. The best known example is that of a prickly pear cactus (*Opuntia* spp.) that was imported to **Australia** from North America for use as an ornamental plant and living fence, but became invasive and a serious weed of rangelands. This pest has now been almost completely controlled through the introduction of one of its natural herbivores, the moth *Cactoblastis cactorum*, whose larvae feed on the cactus.

See also Hallucinogens; Spurge family.

Resources

Books

Benson, L. *The Cacti of the United States and Canada.* Stanford, CA: Stanford University Press, 1982.

Judd, Walter S., Christopher Campbell, Elizabeth A. Kellogg, Michael J. Donoghue, and Peter Stevens. *Plant Systematics: A Phylogenetic Approach.* 2nd ed. with CD-ROM. Suderland, MD: Sinauer, 2002.

Klein, R.M. *The Green World. An Introduction to Plants and People.* New York: Harper and Row, 1987.

Bill Freedman

CAD/CAM/CIM

CAD/CAM is an acronym for computer-aided design and computer-aided manufacturing. The use of

A CAD system used for Boeing airplanes. © *Ed Kashi/Phototake NYC. Reproduced by permission.*

computers in design and manufacturing applications makes it possible to remove much of the tedium and manual labor involved. For example, the many design specifications, blueprints, material lists, and other documents needed to build complex machines can require thousands of highly technical and accurate drawings and charts. If the engineers decide structural components need to be changed, all of these plans and drawings must be changed. Prior to CAD/CAM, human designers and draftspersons had to change them manually, a time consuming and error-prone process. When a CAD system is used, the computer can automatically evaluate and change all corresponding documents instantly. In addition, by using interactive graphics workstations, designers, engineers, and architects can create models or drawings, increase or decrease sizes, rotate or change them at will, and see results instantly on screen.

CAD is particularly valuable in space programs, where many unknown design variables are involved. Previously, engineers depended upon trial-and-error testing and modification, a time consuming and possibly life-threatening process. However, when aided by computer simulation and testing, a great deal of time, money, and possibly lives can be saved. Besides its use in the military, CAD is also used in civil aeronautics, automotive, and data processing industries.

CAM, commonly utilized in conjunction with CAD, uses computers to communicate instructions to automated machinery. CAM techniques are especially suited for manufacturing plants, where tasks are repetitive, tedious, or dangerous for human workers.

Computer integrated manufacturing (CIM), a term popularized by Joseph Harrington in 1975, is also known as autofacturing. CIM is a programmable manufacturing method designed to link CAD, CAM, industrial **robotics**, and machine manufacturing using unattended processing workstations. CIM offers uninterrupted operation from raw materials to finished product, with the added benefits of quality assurance and automated assembly.

CAE (computer aided **engineering**), which appeared in the late 1970s, combines software, hardware, graphics, automated analysis, simulated operation, and physical testing to improve **accuracy**, effectiveness, and productivity.

Caddisflies

North America's streams, **rivers**, and lakes are home to more than 1,200 different **species** of caddisflies, which are aquatic **insects** in the order *Trichoptera*. Adaptations to different **water** conditions and food types allow this group of insects to populate a variety of habitats in America's waters.

Caddisflies are best known and most easily identified in their larval stages. Most caddisfly larvae either spin shelters of silk or build tubular cases. The type of shelter can be used to assign caddisflies to their different families. Some species make shelters from the hollow stems of **grasses**. Others inhabit shelters constructed from rock fragments, pieces of **bark**, or other available materials. Some species of caddisfly carry their shelters with them as they graze on the **algae** on **rocks**; others remain anchored to a rock. Caddisflies that spin silk shelters also spin nets that filter out food particles from the flowing water.

Immature caddisflies are aquatic and must obtain **oxygen** from the water. Mobile caddisfly larvae move water through their gills. Sedentary caddisfly larvae make undulating movements to move water across their gills. The larval cases of sedentary caddisflies restrict or direct flow in some essential way, for if the cases are removed, the larvae usually die.

Like many other insects, caddisflies undergo complete **metamorphosis**, from egg to larva to pupa to adult. The aquatic larvae eventually spin a cocoon or pupal case and become dormant. At the end of the pupation period, adult caddisflies break free of their cocoons and swim to the surface. There, the new adults dry their wings and begin their short adult lives as active, sexually mature air-breathing insects.

Most adult caddisflies live less than a month. During that time, they are inactive during the day, and active at night. Adult caddisflies feed on **plant nectar**, or other plant liquids. After the females have mated they lay their eggs. Where they lay their eggs depends on their species: One species of caddisfly remains underwater for more than 15 minutes as she lays her eggs. Another species deposits her eggs on plants above water, while another species lays her eggs on the water surface—the mass of eggs then absorbs water, sinks, and adheres to an underwater rock or other surface.

As an order, caddisflies are associated with a variety of aquatic habitats: rushing mountain streams, ephemeral spring seeps, slow moving rivers and tranquil lakes. A single **habitat** such as a stream, can support several different species of caddisflies as part of a complex aquatic food web. The larvae of a species grazes on algae on rocks, another feeds on leaves and other plant parts that fall into the water, shredding the material into fine particles. Another species filters food from the fast moving rapids, while another species catches of food items that flow by in slower-moving waters.

Caddisflies specialize in how they acquire food rather than in the type of food they ingest. These special-

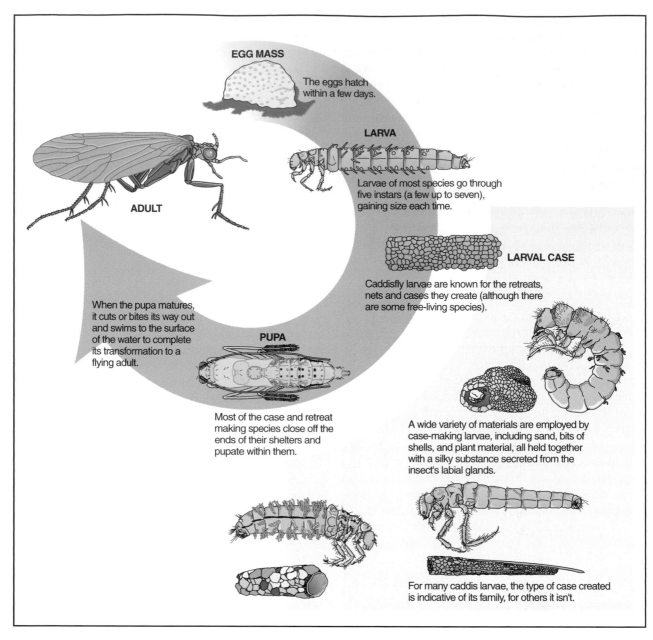

EGG MASS

The eggs hatch within a few days.

LARVA

Larvae of most species go through five instars (a few up to seven), gaining size each time.

LARVAL CASE

Caddisfly larvae are known for the retreats, nets and cases they create (although there are some free-living species).

ADULT

When the pupa matures, it cuts or bites its way out and swims to the surface of the water to complete its transformation to a flying adult.

PUPA

Most of the case and retreat making species close off the ends of their shelters and pupate within them.

A wide variety of materials are employed by case-making larvae, including sand, bits of shells, and plant material, all held together with a silky substance secreted from the insect's labial glands.

For many caddis larvae, the type of case created is indicative of its family, for others it isn't.

The life cycle of a caddisfly. *Illustration by Hans & Cassidy. Courtesy of Gale Group.*

izations make caddisflies one of the most varied and abundant species of aquatic insects in **North America**.

Cadmium *see* **Element, chemical**

Caecilians

Caecilians are long, worm-like legless **amphibians** in the order Gymnophiona (sometimes known as Apoda, meaning without legs). There are 163 **species** of caecilians, in 35 genera. Little is known about these animals, and few species have common names. Most of the caecilians are tropical or sub-tropical, and occur in Central and **South America**, **Africa**, and south and southeast **Asia**.

Caecilians grow up to 5 ft (1.5 m) in length in the case of *Caecilia thompsoni* of Colombia. Caecilians are virtually all body, with almost no tail. Caecilians do not even have rudimentary leg or girdle bones, and have probably been a distinct amphibian lineage for a very long time. However, caecilians have almost no fossil record, so little is known about their evolutionary history.

Caecilians are generally a uniformly or mottled gray in **color**, and somewhat lighter beneath. They have a small mouth, nostrils, and small eyes incapable of movement, covered by skin, and are probably only able to sense changes in **light** intensity. Caecilians have numerous ring-like, segmental grooves along their body, which enhance their superficial resemblance to earthworms.

Most caecilians live secretively in burrows made in moist **soil** or in forest litter, often near streams and **wetlands**. Some species occur in aquatic habitats, where they also burrow in soft substrates. Caecilians feed on **invertebrates**, some species specializing in earthworms or **termites**. The skull of caecilians is heavily boned, and the skin adheres to the skull—both of these are adaptations to the chisel-like burrowing methods of these animals.

Caecilians have internal **fertilization**, a relatively uncommon trait among amphibians. About one-half of caecilian species lay eggs that are guarded by the female, which coils around them until hatching occurs. The other species of caecilians are viviparous, meaning the eggs are retained within the reproductive tract of the female, where they develop and hatch into miniature adults. The larvae feed on their egg yolk, on a rich maternal secretion known as "uterine milk," and by scraping nutritious material from the lining of the reproductive tract of their mother. After a relatively long gestation period of 9-11 months, the baby caecilians emerge as fully metamorphosed but miniature replicas of the adults.

Caffeine

Caffeine is an **alkaloid** found in coffee, tea, chocolate, and other natural foods. It is also a component of cola soft drinks. Caffeine has been a part of the human diet for many centuries and is one of the most widely used central **nervous system** stimulants worldwide. In recent years, research has raised questions about possible deleterious health effects of caffeine, but no definitive conclusions have been reached about the harmfulness of moderate amounts.

Chemistry of caffeine

Caffeine's chemical name is 3,7-dihydro-1,3,7-trimethyl-1H-purine-2,6-dione. It is also known as theine, methyl theobromine, and 1,3,7-trimethylxanthine. Its **molecular formula** is $C_8H_{10}N_4O_2 \cdot H_2O$, and it consists of bicyclic molecules derived from the purine ring system.

In its pure form, caffeine is a fleecy white solid or long silky crystals. It is odorless, but has a distinctive bit-ter **taste**. When heated, caffeine loses **water** at 176°F (80°C), sublimes at 352.4°F (178°C), and/or melts at 458.2°F (236.8°C). It is only slightly soluble in water and **alcohol**, but dissolves readily in **chloroform**. Water solutions of caffeine are essentially neutral (**pH** = 6.9).

Caffeine is a member of the alkaloid family, a group of compounds obtained from plants whose molecules consist of nitrogen-containing rings. In general, alkaloids tend to have identifiable physiological effects on the human body, although these effects vary greatly from compound to compound.

History

The pleasures of coffee, tea, and chocolate drinking have been known to humans for centuries, but the isolation of caffeine from these beverages was accomplished only in the early 1800s. During the 1820s, researchers identified the active agents in tea and chocolate and gave them a variety of names such as guaranin. In 1840, T. Martins and D. Berthemot independently showed that these compounds are all identical with caffeine. Caffeine itself was originally called cofeine or caffein and only in the late 1820s was given the name by which we know it today.

Much of the work leading to the full characterization of caffeine's molecular structure was completed by the German chemist Emil Fischer (1852-1919). Fischer first synthesized the compound from raw materials in 1895, and two years later derived its precise structural formula.

Sources

Of all the commercial sources of caffeine, guarana paste has the highest **concentration** of the pure compound, about 4%. Guarana paste is made from the seed of the Paullinia **tree**, found primarily in Brazil. More common sources of caffeine contain lower concentrations of the compound: 1.1-2.2% in coffee beans; 3.5% in tea leaves; and 1.5% in **kola** nuts. Other less common sources of caffeine include maté leaves, obtained from the Ilex **plant** (less than 0.7% caffeine), and yoco **bark**, obtained from the Paullinia yoco tree (2.7% caffeine).

Because of the way in which these foods are prepared, the above data do not give an accurate picture of the amount of caffeine that people consume. The average cup coffee, for example, contains approximately 100-150 mg of caffeine; the average cup of tea, 50 mg of caffeine; and the average cup of cocoa, about 5 mg of caffeine. Cola drinks tend to contain 35-55 mg of caffeine, and the average chocolate bar contains about 20 mg of the compound. In 1997, more products were introduced that contained caffeine including bottled water and chewing gum.

Pharmacological effects

The most important physiological effect of caffeine is that it stimulates nerve cells, particularly those in the **brain**. It appears that caffeine molecules bind to **neurotransmitter** receptor sites in nerve cells, causing the continual stimulation of those cells. This property explains the most common clinical symptoms of caffeine ingestion: wakefulness, excitability, increased mental awareness, and restlessness.

Caffeine affects nerve **tissue** in the brain much more quickly than it does nerve tissue anywhere else in the body. As a result, it will bring about muscular changes such as convulsions only with very high doses of the drug—10 g or more, the equivalent of drinking 70-100 cups of coffee in a short time. Death from caffeine overdose is, therefore, extremely unlikely.

Teratogenic and mutagenic effects

Concerns about the possible health effects of consuming caffeine have been expressed for well over a hundred years. Recent concern about its physiological effects tend to focus on mutagenic and teratogenic effects. Mutagenic effects are those that change the reproductive genes, producing mutations in subsequent generations. Scientific reports have also appeared connecting the consumption of large doses of caffeine with particular types of **cancer**. However, the significance of such findings to the average coffee or tea drinker is unclear.

The situation with teratogenic effects—those that affect the fetus while it is still in the womb—is somewhat clearer. Caffeine passes readily across the placental lining, exposing the fetus to concentrations of the stimulant that are comparable to those in the mother's **blood**. Since the developing nervous system of the fetus is more likely to be

affected by the drug than is the mother's, a reduction in caffeine intake is often recommended for pregnant women.

Resources

Books

Dews, Peter B., ed. *Caffeine: Perspectives from Research.* Berlin: Springer-Verlag, 1984.

Gilbert, Richard J. *Caffeine: The Most Popular Stimulant.* New York: Chelsea House, 1986.

Selinger, Ben. *Chemistry in the Marketplace.* 4th ed. Sydney: Harcourt Brace Jovanovich, 1989.

David E. Newton

Caimans *see* **Crocodiles**

Caisson

A caisson is a hollow structure made of **concrete**, **steel**, or other materials that can be sunk into the earth. It used as the substructure for a bridge, a building, or other large structures. Caissons come in many sizes and shapes depending on their future use. The one shared feature is that their bottom edges are sharp so they easily can be sunk into the ground. These sharp edges are known as the cutting edges of the caisson.

General principle

The purpose of using a caisson in construction is to provide a temporary structure from which earth, **water**, and other materials can be removed and into which concrete or some other fill material can be placed. For example in the construction of a bridge it may be necessary to burrow into the **soil** at the bottom of a river until **bedrock** is reached. One way of doing this is to sink a caisson filled with compressed air into the river until it reaches the river bottom. Workers then can go into the caisson and dig soil out of the river bed until they come to bedrock. As they remove soil it can be transported upward out through the caisson. During this process the caisson continues to sink more deeply into the river bed until it reaches bedrock. At that point concrete may be poured into the caisson to form the lowest section of the new bridgepier.

Caissons may consist of a single unit looking like a tin can with both ends cut out. Or they may be subdivided into a number of compartments similar to a honeycomb. One factor in determining the shape of the caisson is the area it must cover. The larger the size of the caisson the more necessary it may be to subdivide it into smaller compartments.

Types of caissons

All caissons feature the shape of a tube, often with a cylindrical contour but it may also be rectangular, elliptical, or some other form. Some caissons are open at both ends, some are open only at the top, and some are open only at the bottom. It depends on the way each type of caisson is to be used.

A caisson open at both ends might be used to lay down a pier for a new skyscraper. The caisson would be driven into the ground to a certain depth and the earthy material inside the caisson would be scooped out. Depending on the depth of the pier required one long open cylindrical caisson could be used or a sequence of shorter caissons could be laid down one on top of the other. When the caisson(s) have been inserted to the desired depth and all the soil within them removed they might be filled with concrete. The decision as to whether to remove the caissons themselves before adding concrete would depend on the surrounding soil's nature. If the soils were too unstable to hold their shape the caisson would be left in place. With stable soils the caisson could be removed.

A caisson closed at the bottom and open at the top is a floating caisson. This type of caisson often is used in the construction of bridgepiers. The caisson is constructed on land of concrete, steel, **wood**, or some other material and floated to its intended position in a river, **lake**, or other body of water. The caisson then is filled with gravel, concrete, or some other material and allowed to sink to the river bed. The filled caisson then becomes the lowest portion of the new bridgepier. A floating caisson can be used only if engineers can be assured that the soil beneath and around the filled caisson will not wash away.

One interesting application of the floating caisson is in the reclamation of land from the North Sea around the Netherlands. In the first stage of this process a series of floating caissons are moved into the **ocean** where they are arranged to form a new dike system. Ocean water trapped within the line of caissons is pumped out to form new farmland.

A caisson closed at the top and open at the bottom is a pneumatic caisson. This type of caisson generally is used in underwater construction projects. It can be used only if air is pumped in to produce a **pressure** greater than water pressure outside. Workers entering a pneumatic caisson must first pass through an intermediate chamber that allows their bodies to adjust from normal **atmospheric pressure** to the higher pressure within the caisson or vice versa. Pneumatic caissons can not be used at a depth of more than 120 ft (36.6 m). Beyond that point the air pressure needed inside the caisson to keep out water is too great for the human body to withstand.

See also Bridges.

Resources

Books

How Things Work: Structures. Alexandria: Time-Life Books, 1991.

Trefil, James. *Encyclopedia of Science and Technology.* The Reference Works, Inc., 2001.

David E. Newton

Calcium

Calcium is a chemical element, a member of the alkaline-earth metals group, represented by the atomic symbol Ca and the **atomic number** 20. It has an **atomic weight** of 40.08. In its pure form, calcium is a silvery-white **metal**, although it is never found in this free state naturally. It is, however, one of the most abundant substances on **Earth**, comprising approximately 3.64% of the earth's crust.

Pure calcium metal has a melting point of 1547.6°F (842°C) and boils at 2703°F (1484°C). It consists of six stable isotopes with **mass** numbers between 40 and 48. By far the most abundant, however, is ^{40}Ca, which constitutes 96.941% of all the calcium **atoms** that are found in nature. Calcium is used in the production of many products including **glass**, batteries, and **steel**. It also combines readily with many other elements, and these compounds are used as well for a variety of purposes.

Calcium (from the Latin *calx*, meaning lime) was not known as an element until the early 1800s. During this time, chemists who were trying to prove the existence of unknown metals in natural compounds, began using the newly discovered phenomenon of **electricity** to break them apart. The English chemist Humphry Davy

(1778-1829), who was a pioneer in the field of electro-chemistry, first isolated elemental calcium in 1808 by electrolyzing a mixture of lime and mercuric oxide. Today, calcium metal is obtained by electrolyzing molten calcium chloride ($CaCl_2$) or by reducing **calcium oxide** with **aluminum** metal.

Calcium is the fifth most abundant element (after **oxygen**, silicon, aluminum and **iron**), in the earth's crust, making up 3.63% of the crust by weight. It occurs in the form of **minerals** such as limestone (**calcium carbonate**, $CaCO_3$), gypsum (**calcium sulfate**, $CaSO_4 \cdot 2H_2O$), and fluorite (calcium fluoride, CaF_2).

In living things, calcium is a component of leaves, bones, teeth, shells, and coral. Calcium plays a crucial role in good health, although its biological significance came to be understood only during the late nineteenth century. It is the most abundant metallic element in the human body, comprising about 1.4% of body weight. This makes it even more prevalent even than iron. Ninety-nine **percent** of the body's calcium is stored in the skeleton and teeth. Bones are 70% calcium by weight, which gives them their strength and rigidity. The remaining 1% circulates in the bloodstream, where, as American biochemist Elmer McCollum proved in the early 1900s, it is essential for muscle contractions. Calcium helps regulate contractions of the most important muscle in the body—the **heart**. This was discovered in 1882, when British physician Sydney Ringer (1835-1910) showed that a heart would continue to beat in a **solution** of **salt**, calcium, and other chemicals.

Among its many other functions, calcium plays a role in the transmission of nerve impulses and aids **blood** clotting. Too little calcium in the diet can cause **osteoporosis**, a progressive weakening of the bones. Rickets can occur if there is insufficient **vitamin** D to aid calcium **metabolism**. Natural food sources of calcium include milk and dairy products, leafy green **vegetables**, and canned **sardines**. Calcium supplements are often recommended to prevent these diseases in older people, mostly women.

Calcium is a very active metal and is never found uncombined in nature. It tarnishes quickly when exposed to air and burns with a bright yellowish red flame, forming mostly calcium nitride (Ca_3N_2). It reacts directly with **water** to form calcium hydroxide [$Ca(OH)_2$] and **hydrogen** gas. Because of its strong reducing power it is used to produce other metals such as thorium and **uranium** by reducing their compounds, and to purify various alloys by removing oxides and sulfides. Calcium forms useful alloys with aluminum, **copper**, and **lead**.

Many calcium compounds have important uses. Calcium oxide or lime is widely used to make cement (lime+clay), mortar (cement+sand+water) and **concrete** (cement+sand+gravel+water). It is also used in the manufacture of glass. When water is added to calcium carbide (CaC_2) the highly flammable gas acetylene (C_2H_2) is produced; it is used in lamps and **welding** torches, and as a starting material in the synthesis of many organic compounds. Calcium chloride ($CaCl_2$) is used as a drying agent, because it is a *deliquescent* solid: it can absorb so much water from the air that it turns into a liquid. It is also used as a more effective and less corrosive substitute for common salt (NaCl) for melting **ice** on roads in the winter. Calcium hypochlorite [$Ca(OCl)_2$] is used as a **bleach**. Calcium phosphate [$Ca_3(PO_4)_2$] and calcium cyanamide [$Ca(CN)_2$] are used in the production of **fertilizers**. Other calcium compounds include the minerals fluorspar, phosphorite, gypsum, and apatite. Calcium acetate is used in the production of **plastics**, and calcium hypochlorite is a bleaching agent and disinfectant.

See also Alkaline earth metals; Alloy; Calcium propionate; Element, chemical; Hard water.

Calcium carbonate

Calcium carbonate, $CaCO_3$, is one of the most common compounds on **Earth**, making up about 7% of Earth's crust. It occurs in a wide variety of mineral forms, including limestone, marble, travertine, and chalk. Calcium carbonate also occurs combined with **magnesium** as the mineral dolomite, $CaMg(CO_3)_2$. **Stalactites and stalagmites** in caves are made of calcium carbonate. A variety of **animal** products are also made primarily of calcium carbonate, notably coral, sea shells, egg shells, and pearls.

Calcium carbonate has two major crystalline forms-two different geometric arrangements of the calcium ions and carbonate ions that make up the compound. These two forms are called aragonite and calcite. All calcium carbonate **minerals** are conglomerations of various-sized crystals of these two forms, packed together in different ways and containing various impurities. The large, transparent crystals known as Iceland spar, however, are pure calcite.

In its pure form, calcium carbonate is a white powder with a specific gravity of 2.71 in the calcite form or 2.93 in the aragonite form. When heated, it decomposes into **calcium oxide** (CaO) and **carbon dioxide** gas (CO_2). It also reacts vigorously with acids to release a froth of **carbon** dioxide bubbles. It is said that Cleopatra, to show her extravagance, dissolved pearls in vinegar (**acetic acid**).

Every year in the United States alone, tens of millions of tons of limestone are dug, cut, or blasted out of huge deposits in Indiana and elsewhere. It is used mostly for buildings and highways and in the manufacture of **steel**, where it is used to remove silica (silicon dioxide) and other impurities in the **iron** ore; the calcium carbonate decomposes to calcium oxide in the **heat** of the furnace, and the calcium oxide reacts with the silica to form calcium silicates (slag), which float on the molten iron and can be skimmed off.

Deposits of calcium carbonate can be formed in the oceans when calcium ions dissolved from other minerals react with dissolved carbon dioxide (carbonic acid, H_2CO_3). The resulting calcium carbonate is quite insoluble in **water** and sinks to the bottom.

However, most of the calcium carbonate deposits that we find today were formed by sea creatures millions of years ago when oceans covered much of what is now land. From the calcium ions and carbon dioxide in the oceans, they manufactured shells and skeletons of calcium carbonate, just as clams, oysters, and corals still do today. When these animals die, their shells settle on the sea floor where, long after the seas have gone, we now find them compressed into thick deposits of limestone. The White Cliffs of Dover in England are chalk, a soft, white porous form of limestone made from the shells of microscopic sea creatures called Foraminifera that lived about 136 million years ago. Blackboard "chalk" isn't made of chalk; it is mostly gypsum, $CaSO_4$.

In pearls—which **mollusks** make out of their shell-building material when they are irritated by a foreign body in their flesh—and in sea shells, the individual $CaCO_3$ crystals are invisibly small, even under a **microscope**. But they are laid down in such a perfect order that the result is smooth, hard, shiny, and sometimes even iridescent, as in the rainbow colors of abalone shells. In many cases, the mollusk makes its shell by laying down alternating layers: calcite, aragonite, calcite, aragonite, and so on. This gives the shell great strength, as in a sheet of plywood where the grain of the alternating **wood** layers runs in crossed directions.

Calcium oxide

Calcium oxide (CaO), more commonly known as lime or quick lime, has been studied by scholars as far back as the pre-Christian era. In his book *Historia Naturalis,* for example, Pliny the Elder discussed the preparation, properties, and uses of lime. Probably the first scientific paper on the substance was Dr. Joseph Black's "Experiments Upon Magnesia, Alba, Quick-lime, and Some Other Alkaline Substances," written in 1755.

Lime does not occur naturally since it reacts so readily with **water** (to form hydrated lime) and **carbon dioxide** (to form limestone). It is produced in very large quantities synthetically, however, by the heating of limestone. For many years, calcium oxide has ranked among the top ten chemicals in the United States in terms of production. Other common names by which the compound is known include burnt lime, unslaked lime, fluxing lime, and calx.

In its pure form, calcium oxide occurs as white crystals, white or gray lumps, or a white granular powder. It has a very high melting point of 4,662°F (2,572°C) and a **boiling point** of 5,162°F (2,850°C). It dissolves in and reacts with water to form calcium hydroxide and is soluble in acids and some organic solvents.

Like other calcium compounds, calcium oxide is used for many construction purposes, as in the manufacture of bricks, mortar, plaster, and stucco. Its high melting point makes it attractive as a refractory material, as in the lining of furnaces. The compound is also used in the manufacture of various types of **glass**. Common soda-lime glass, for example, contains about 12% calcium oxide, while high-melting aluminosilicate glass contains about 20% calcium oxide. One of the new forms of glass used to coat surgical implants contains an even higher **ratio** of calcium oxide, about 24% of the compound.

Among the many other applications of calcium oxide are its uses in the production of pulp and paper, in the removal of hair from **animal** hides, in clarifying cane and **beet** sugar, in poultry feeds, and as a drilling fluid.

Calcium propionate

Calcium propionate is an organic **salt** formed by the reaction of calcium hydroxide with propionic acid (also known as propanoic acid). Its chemical formula is $Ca(OOCCH_2CH_3)_2$. The compound occurs in either crystalline or powder form. It is soluble in **water** and only very slightly soluble in **alcohol**.

Calcium propionate is used as a food preservative in breads and other baked goods because of its ability to inhibit the growth of molds and other **microorganisms**. It is not toxic to these organisms, but does prevent them from reproducing and posing a health risk to humans.

Propionic acid occurs naturally in some foods and acts as a preservative in those foods. Some types of cheese, for example, contain as much as 1% natural propionic acid.

Studies indicate that calcium propionate is one of the safest food additives used by the food industry. **Rats** fed a diet containing nearly 4% calcium propionate for a year showed no ill effects. As a result, the U.S. Food and Drug Administration has placed no limitations on its use in foods. In addition to baked goods, it is commonly used as a preservative in chocolate products, processed cheeses, and fruit preserves. The tobacco industry has also used calcium propionate as a preservative in some of its products.

Beyond its role as a food additive, calcium propionate finds some application in the manufacture of butyl rubber. Adding it to the raw product makes it easier to process the rubber and protects the rubber from scorching during manufacture.

Calcium sulfate

Calcium sulfate ($CaSO_4$) occurs in nature in both the anhydrous and the hydrated form. The former is found primarily as the mineral known as anhydrite, while the latter is probably best known as alabaster or gypsum. Calcium sulfate also occurs in forms known as selenite, terra alba, satinite, satin spar, and light spar.

Gypsum was well known to ancient cultures. Theophrastus of Eresus (about 300 B.C.), for example, described its natural occurrence, its properties, and its uses. The Persian pharmacist Abu Mansur Muwaffaq is believed to have first described one of the major products of gypsum, plaster of Paris, around 975 A.D.

Calcium sulfate occurs as a white odorless powder or as crystals that may be tinged with **color** by impurities. It has a melting point of 2,642°F (1,450°C) and is only slightly soluble in **water**. When heated, the hydrated forms of calcium sulfate lose 1.5 molecules of water and form the hemihydrate, $CaSO_4 \cdot 1/2H_2O$, commonly known as plaster of Paris. When added to water, plaster of Paris forms a hard mass used in making plaster casts, quick-setting cements, molds, wall plasters and wall board, and inexpensive art objects. Neither the anhydrous nor the hydrated calcium sulfate will react with water as does the hemihydrate.

Among the many other uses of calcium sulfate are as a pigment in white paints, as a **soil** conditioner, in Portland cement, as a sizer, filler, and coating agent in papers, in the manufacture of **sulfuric acid** and **sulfur**, in the **metallurgy** of zinc ores, and as a drying agent in many laboratory and commercial processes.

See also Calcium.

Calculator

The calculator is a computing machine. Its purpose is to do **mathematics**; basic calculators do the basic mathematical functions (**addition**, **subtraction**, **division**, and **multiplication**) while the more advanced ones, which are relatively new in the history of computing machines, do advanced calculations such as solving **polynomials**. The odometer, or mileage counter, in your car is a counting machine as is the calculator in your backpack and the computer on your desk. They may have different ability levels, but they all tally numbers.

The first calculators

Perhaps the earliest calculating machine was the Babylonian rod numerals. Not only was it a notational device, but administrators carried rods of bamboo, ivory, or **iron** in bags to help with their calculations. Rod numerals used nine digits (Figure 1).

The next invention in counting machines was the **abacus**. A counting board dating from approximately 500 B.C. now resides in the National Museum in Athens. It is not an abacus, but the precursor of the abacus. Oriental cultures have documents discussing the abacus (the Chinese call it a *suan phan* and the Japanese the *soroban*) as early as the 1500s A.D.; however, they were using the abacus at least a thousand years earlier. While experts disagree on the origin of the name (from either the Semitic *abq*, or dust, or the Greek *abax*, or **sand** tray), they do agree the word is based upon the idea of a sand tray which was used for counting.

A simple abacus has rows (or wires) and each row has ten beads. Each row represents a unit ten greater than the previous. Thus, the first stands for units of one; the second, units of ten; the third, units of hundred; and so on. The appropriate number of beads are moved from left to right on the wire representing the unit. When all ten beads on a row have been moved, they are returned to their original place and one bead on the next row is moved. The soroban divides the wires into two unequal parts. The beads along the lower, or larger, part represent units, tens, hundreds, and so on. The bead at the top represent five, fifty, five hundred, and so on. These beads are stored away from the central divider, and as needed are moved toward it.

Finger reckoning must not be ignored as a basic calculator. Nicolaus Rhabda of Smyrna and the Venerable Bede (both of the eighth century A.D.) wrote, in detail, how this system using both hands could represent numbers up to one million. The numbers from 1 to 99 are created by the left and the numbers from 100 to 9,900 by the right hand. By bending the fingers at their various joints and using the index finger and thumb to represent the multiples of ten, combinations of numbers can be represented. Similar systems were devised much earlier than the eighth century, probably by merchants and traders who could not speak each other's languages but needed a system to communicate: their fingers. Multiplication using the fingers came much later; well into the fifteenth century such complex issues, as multiplication, were left to university students, who were forced to learn a different finger reckoning system to accommodate multiplication.

Early calculators

Schickard, a German professor and Protestant minister, seems to have been the first to create an adding machine in the 1620s. It performed addition, subtraction, and carrying through the use of **gears** and preset multiplication tables. The machine only computed numbers up to six digits. Once the operator surpassed this limit, he was required to put a brass ring around his finger to remind him how many carries he had done. Schickard made sure to include a bell, so that the user would not forget to add his rings. His drawings of the machine were lost until the mid-1960s when a scrap of **paper** was located inside a friend's book which had a drawing of his machine. With this drawing and Schickard's letters, a reconstruction was made in 1971 to honor the adding machine's 350th anniversary.

Finished in 1642, Blaise Pascal's calculating machine also automatically carried tens and was limited to six digit numbers. The digits from 0 to 9 were represented on dials; when a one was added to a nine the gear turned to show a **zero** and the next gear, representing the next higher tens unit, automatically turned. Over 50 of these machines were made. A few remain in existence.

Gottfried Wilhelm Leibniz, a German, created a machine in 1671 which did addition and multiplication. For over 200 years this machine was lost; then it was discovered by some workmen in the attic of one of the Göttingen University's buildings.

Charles Xavier Thomas de Colmar, in 1820, devised a machine which added subtraction and division to a Leibniz type calculator. It was the first mass produced calculator and became a common sight in business offices.

Difference engine

The great English mathematician Charles Babbage tried to build a machine which would calculate mathematical tables to 26 significant figures; he called it the Difference Engine. However, in the early 1820s his plans were stalled when the British government pulled its funding. His second attempt, in the 1830s, failed because some of the tools he needed had yet to be invented. Despite this, Babbage did complete part of this second machine, called the Analytical Engine, in the early 1840s. It is considered to be the first modern calculating machine. The difference between the Difference and Analytical Engines is that the first only performed a certain number of functions, which were built into the machine, while the second could be programmed to solve almost any algebraic equation.

By 1840 the first difference engine was finally built by the Swedish father and son team of George and Edvard Scheutz. They based their machine on Babbage's 1834 publication about his experiments. The Scheutz's three machines produced the first automatically created calculation tables. In one 80 hour experiment, the Scheutz's machine produced the **logarithms** of 1 to 10,000; this included time to reset the machine for the 20 polynomials needed to do the calculations.

Patents

The first patent for a calculating machine was granted to the American Frank Stephen Baldwin in 1875. Baldwin's machine did all four basic mathematical functions and did not need to be reset after each computation. The second patent was given in 1878 to Willgodt Theophile Odhner from Sweden for a machine of similar design to Baldwin's. The modern electronic calculators are based on Baldwin's design.

In 1910, Babbage's son, Henry P. Babbage, built the first hand held **printing** calculator based on his father's Analytical Engine. With this machine, he was able to calculate and then print multiples of π to 29 decimal places.

In 1936, a German student, Konrad Zuse, built the first automatic calculating machine. Without any knowledge of previous calculating machines, Zuse built the Z1 in his parents' living room. He theorized that the machine had to be able to do the mathematical fundamentals. To do this, he turned to binary mathematics, something no other scientist or mathematician had contemplated. (The mathematics we use in every day life, decimal, is based on 10 digits. Binary mathematics uses two digits: 0 and 1.) By using binary mathematics, the calculating machine became a series of switches rather than gears, because a switch has two options: on (closed or 1) or off (open or 0). He then connected these switches into logic gates, a combination of which can be selected to do addition or subtraction. Zuse's third model (finished in 1941) was not only programmable, but hand held; it added, subtracted, multiplied, divided, discovered square roots, and converted from decimal to binary and back again. However, to

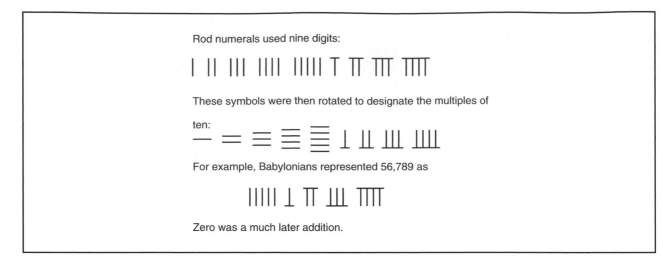

Rod numerals used nine digits:

These symbols were then rotated to designate the multiples of

ten:

For example, Babylonians represented 56,789 as

Zero was a much later addition.

Figure 1. *Illustration by Hans & Cassidy. Courtesy of Gale Group.*

add it took a third of a second, and to subtract an additional three to five seconds. By changing his relays into **vacuum** tubes he believed he could speed his machine up 1,000 times, but he could not get the funding from the Third Reich government to rebuild his machine.

Electronic predecessor to computer

In 1938, the International Business Machine Corporation (IBM) team of George Stibitz and S. B. Williams began building the Complex Number Calculator. It could add, subtract, multiply, and divide **complex numbers**. They completed the project in 1940 and until 1949 these calculators were used by Bell Laboratories. It was the first machine to use remote stations (terminals, or input units, not next to the computer) and to allow more than one terminal to be used. The operator typed the request onto a teletype machine and the response was sent back to that teletype. The relays inside the machine were basic **telephone** relays.

The IBM Automatic Sequence Controlled Calculator (ASCC) was based upon Babbage's ideas. This machine, completed in 1944, was built for the United States Navy by Harvard University and IBM. It weighed about 5 tons (10,000 kg) and was 51 ft (15.5 m) long and 8 ft (2.4 m) high. A second, more useful version, was completed in 1948.

The Electronic Numerical Integrator and Computer (ENIAC), also based on Babbage's concepts, was completed in 1945 at the University of Pennsylvania for the United States Army. It weighted over 30 tons (27,240 kg) and filled a 30 by 55 ft (9 by 17 m) room. In one second it could do 5,000 additions, 357 multiplications, or 38 divisions. However, to reprogram ENIAC meant rewiring it, which caused up to two days in delays.

The first electronic calculator was suggested by the Hungarian turned American John von Neumann. Von Neumann introduced the idea of a stored memory for a computer which allowed the program and the data to be inputted to the machine. This resolved problems with rewiring computers and permitted the computer to move directly from one calculation to another. This type of machine architecture is called "von Neumann." The first completed computer to use von Neumann architecture was the English Electronic Delay Storage Automatic Calculator (EDSAC) finished in 1949. An added feature of the EDSAC was that it could be programmed in a type of shorthand, which it then converted into binary code, rather than its precursors which demanded the programmer actually write the program in binary code, a laborious process.

Inside calculators

The early counting machines, items like the car's odometer, work with a set of gears and wheel. A certain number of wheels are divided in 10 equal parts on each of which one of the 10 digits appears. (Windows are placed on top of these wheels so that only one digit appears at a time.) These wheels are then attached to gears which as they turn rotate the wheel so that the digit being displayed changes. When the right most wheel changes from 9 to 0, a mechanism is set in **motion** which turns the wheel to the left one unit, so that the digit it displays changes. This is the carry sequence. When this second wheel changes from 9 to 0, it too has a mechanism to carry to the next left wheel.

Electronic calculators have the four major units van Neumann created: input, processing, memory, and output. The input unit accepts the numbers keyed in, or sent through the reader in the case of the punch card, by the operator. The processing unit performs the calculations.

KEY TERMS

. . . .

Binary—The base 2 system of counting using two digits: 0 and 1. Each unit is a2 to the n+1 power. For example, the first unit is 2^0 or 1; the second 2^1 or 2; the third 2^2 or 4; the forth 2^3 or 8; and so forth. Thus 221 in binary is 11011101 or 1 hundreds twenty-eight (2^7), 1 sixty-four (2^6), no thirty-twos (2^5), 1 sixteen (2^4), 1 eight (2^3), 1 four (2^2), no twos (2^1), and 1 one (2^0). This unit of counting was invented by Gottfried Leibniz in the 1600s.

Decimal—The base 10 system of counting using ten digits: 0, 1, 2, 3, 4, 5, 6, 7, 8, 9. Each unit is 10 to the n+1 power. For example, the first unit is 10^0 or 1; the second 10^1 or 10; the third 10^2 or 100; and so forth. Thus, 221 is 2 hundreds (10^2), 2 tens (10^1), and 1 one (10^0).

Punch cards—Made of a heavy cardboard, these rectangular cards have holes punched in them. Each hole is placed in a designated area which the computer then translates into a binary code. A series of such punched cards contain a sequence of events, or a program. The first punch cards were invented by Jacquard, a weaver, who wanted to automate the creation of patterns in his fabric. Thus, his loom was the first machine to use these cards.

When the processing unit encounters a complex calculation, it uses the memory unit to store intermediary results or to locate **arithmetic** instructions. At the completion of the calculation, the final answer is sent to the output unit which informs the operator of the result; this may be through a display, paper, or a combination. Since the calculator thinks in binary, the output unit must convert the result into decimal units.

Modern advances

At the end of 1947, the **transistor** was invented, eventually making the **vacuum tube** obsolete. This tiny creation, composed of semiconductors, were much faster and less **energy** consumptive than the tubes. Problems arose with the connections between the components with size, speed, and reliability as more complex machines needed more complex circuitry which in turn required more components soldered to more boards (the actual board to which the pieces were attached). The next breakthrough came with the invention of the **integrated circuit** (IC) in 1959 by Texas Instruments (TI) and Fairchild (a semiconductor manufacturing company). The integrated-circuit is akin to a solid mass of transistors, resistors, and

capacitors. Again, the speed of computation increased (since the resistance in the circuit was reduced) and the energy required by the machine was decreased. Finally, a computer could fit on a spaceship (they were part of the Apollo computer) or missile. In 1959, an IC cost over $1,000 but by 1965 they were under $10.

Ted Hoff, an electrical engineer for Intel conceived of a radical new concept-the microprocessor. This incorporated the circuitry of the integrated circuit and the programs used in a computer onto a single chip, or piece of silicon. This model of this microprocessor was finished in 1970. The idea of a disposable piece of a calculator was revolutionary. Its compactness and speed changed the face of the computing industry.

The creation of ICs allowed calculators to become much faster and smaller. By the 1960s, they were hand held and affordable. By the late 1980s, calculators were found on watches. However, once again engineers are being blocked by the size factor. The limits are now the size of the chip which in turn limits the speed and programmability of the calculator, or computer.

See also Computer, analog; Computer, digital.

Resources

Books

Macaulay, David. *The New Way Things Work.* Boston: Houghton Mifflin Company, 1998.
Palfreman, Jon, and Doron Swade. *The Dream Machine: Exploring the Computer Age.* London: BBC Books, 1991.
Williams, Michael R. *A History of Computing Technology.* Englewood Cliffs, New Jersey: Prentice-Hall, 1985.

Mara W. Cohen

Calculus

Calculus is the branch of **mathematics** that deals with rates of change and **motion**. It grew out of a desire to understand various physical phenomena, such as the orbits of planets, and the effects of gravity. The immediate success of calculus in formulating physical laws and predicting their consequences led to development of a new **division** in mathematics called analysis, of which calculus remains a large part. Today, calculus is the essential language of science and **engineering**, providing the means by which physical laws are expressed in mathematical terms. As a scientific tool it is invaluable in the further analysis of physical laws, in predicting the behavior of electrical and mechanical systems governed by those laws, and in discovering new laws.

Calculus divides naturally into two parts, differential calculus and **integral** calculus. Differential calculus is concerned with finding the instantaneous **rate** at which one quantity changes with respect to another, called the **derivative** of the first quantity with respect to the second. For example, determining the speed of a falling body at a particular instant of time, say that of a skydiver or bungi jumper, is equivalent to calculating the instantaneous rate of change in his or her position with respect to time. In general, evaluating the derivative of a **function**, f(x), involves finding another function, f'(x), such that f'(x) is equal to the slope of the tangent to the graph of f(x) at each x. This is accomplished, for each 2, by determining the slope of an approximating line segment in the **limit** that its length approaches **zero**.

Integral calculus deals with the inverse of the derivative, namely, finding a function when its rate of change is known. For example, if a skydiver's **velocity** is a known function of time, then we may ask what is his or her position at any given time after jumping. Finding the original function, given its derivative, is called integration, and the function is called the indefinite integral. Evaluating the indefinite integral of any function between specific limits to definition of the definite integral, which is equal to the area under the graph of the function between the specified limits. The latter is developed as a natural consequence of approximating an area by summing the areas of a number of inscribed rectangles. The **approximation** becomes exact in the limit that the number of rectangles approaches **infinity**. Thus, both differential and integral calculus are based on the theory of limits.

The usefulness of calculus is indicated by its widespread application. For example, it is used in the design of navigation systems, particle **accelerators**, and synchrotron **light** sources. It is used to predict rocket trajectories, and the orbits of communications satellites. Calculus is the mathematical tool used to test theories about the origins of the universe, the development of tornadoes and hurricanes, and **salt** fingering in the oceans. It has even found extensive application in business, where it is used, among other things, to optimize production.

History

Calculus was invented, more or less simultaneously, by Isaac Newton and Gottfried Leibniz. Some of the essential ingredients, however, had their beginnings in ancient Greece. In the fourth century B.C., Eudoxus invented the so called method of exhaustion, in order to furnish proofs of certain geometric theorems without having to resort to arguments involving the infinite. Approximately a century later, Archimedes used the same method to find a formula for the area of a **circle**. Archimedes' method con-

sisted of inscribing a polygon with n sides inside a circle, and circumscribing a similar polygon, again with n sides, outside the circle. Then, allowing n, in other words the number of sides, to get very large, he was able to show that the area of the circle was always greater than the area of the inscribed polygon and less than the area of the circumscribed polygon. As n grew very large, the areas of the two **polygons** tended to become equal, thus leading him to the area of a circle. The method of Archimedes persisted from the third century B.C. until the beginning of the seventeenth century A.D.when the work of Johannes Kepler, a German Astronomer, led to the discovery of general principles for the calculation of areas and volumes. Kepler's contribution was the notion of infinitesimals. He envisioned the inscribed polygons of Archimedes as being a collection of infinitely many, vanishingly small triangles. Thus, the area of a circle could be calculated by summing the areas of these triangles. While, Eudoxus and Archimedes had worked hard to avoid the infinite, Kepler embraced it. Simultaneous work on infinite **sequences** and sums of infinite sequences led the French mathematician Fermat, and others, to discover general methods for evaluating areas and volumesas the sums of infinite sequences rather than the sums of areas of common geometric figures. Finally, around the middle of the seventeenth century, Isaac Newton, in attempting to develop a universal theory of gravitation, discovered the derivative, a general method for determining the instantaneous rate of change of a function, based on the notion of infinitesimals. Though he did not explicitly define the integral at the time, Newton did recognize the need to solve differential equations. As a result, he invented methods of evaluating indefinite integrals very soon after introducing the derivative. It was Leibniz, however, whose work postdated that of Newton by some 10 years, who recognized and formulated the definite integral as an infinite sum of "lines," that is, as an area calculated by summing an infinite number of infinitely narrow rectangles.

Differential calculus

Differential calculus involves the analysis of functions, specifically, determining their instantaneous rates of change. An important feature of any function is its rate of change. Geometrically, rate of change is associated with the graph of a function. The rate of change of a straight line (the simplest kind of real valued function) is the slope of the line. The slope is defined as the **ratio** of the vertical change, or "rise," to the horizontal change, or "run," that occurs between any two points on the line. Because the slope is the same between any two points, the rate of change of such a function is said to be constant. In general, however, any function whose graph is not a straight line has a varying rate of change. The rate of change in the vicinity of a particular **point** on the graph of **curve** can be approximated by drawing a straight line through two points in the neighborhood of that point, and determining the slope of the line. Suppose we are interested in the rate of change at the point (x,f(x)). First, choose a second nearby point, say (x+h, f(x+h)). Then the slope of the line segment connecting these two points is [f(x+h) - f(x)] 4 [(x+h) - x]. The shorter the approximating line segment becomes, the more accurate the approximation of the rate of change at the point (x, f(x)) becomes. In the limit that h approaches zero the slope of the approximating line segment becomes exactly the rate of change of the function at the point (x,f(x)). Thus, the instantaneous rate of change of a function, called the derivative of the function, is defined by:

$$f'(x) \quad = \quad \frac{df(x)}{dx} \quad = \quad \lim_{h \to 0} \frac{f(x + h) - f(x)}{h}$$

where the notation df(x) is intended to indicate that the derivative is the ratio of an infinitesimal change in f(x) (the rise) to the corresponding infinitesimal change in x (the run). The derivative of a function is itself a function, and so may also have a derivative. Often times the derivative of the derivative is an important quantity. Called the second derivative of the original function f, it is denoted by f''(x).

An important application of differential calculus involves using information about the first and second derivatives, and the appropriate geometric interpretations, to graph functions. For example, the first derivative of a function is its rate of change. The value of the first derivative at a given point is equal to the slope of the tangent to the graph of the function at that point. When the derivative is positive, the function is said to be increasing (the value of the function increases with increasing x). When the derivative is **negative**, the function is said to be decreasing (the value of the function decreases with increasing x). When the value of the derivative is zero at a point, the tangent is horizontal, and the function changes from increasing to decreasing, or vice versa, depending on the sign of the second derivative. The second derivative is the rate of change of the rate of change, and thus contains information about the curvature of the function. When the second derivative is positive, the function is concave upward (as though it would hold **water**). When the second derivative is negative, the function is concave downward. With this knowledge, and a few points in the function, a reasonable graph can be drawn without having to plot hundreds of points.

Other applications of differential calculus include the solution of rate problems and optimization problems. In general, a rate is the ratio of change in one quantity to

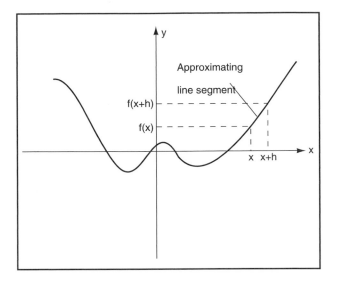

Figure 1. The rate of change in the vicinity of a particular point on the graph of a curve can be approximated by drawing a straight line through two points in the neighborhood of that point, and determining the slope of the line. *Illustration by Hans & Cassidy. Courtesy of Gale Group.*

the simultaneous change in a second quantity. Thus, the derivative, being an instantaneous rate, is applicable to any problem in which the rate of change of one quantity with respect to another is of interest. Innumerable applications in engineering and science affect our daily lives. For instance, the instantaneous velocity of an orbiting communications **satellite** is calculated from knowledge of its position as a function of time. The **acceleration** of a falling body is calculated from knowledge of its velocity as a function of time, which in turn is calculated from knowledge of its position as a function of time. The **force** required to deliver **natural gas** through a pipeline, over large distances, is calculated using the derived of the gas **pressure** with respect to **distance**.

Optimization problems are problems that require knowledge of maximum or minimum values of functional relationships. For example, it can be shown that a **sphere** has the least surface area for a given **volume** of any geometric solid. Thus, the optimum shape for a raindrop is spherical because this shape contains the most water, but has the least amount of surface area, hence the least surface **energy** (a measure of the work required to form the drop).

Integral calculus

Integral calculus is the study of integration and methods for evaluating integrals. Integrals come in two kinds, definite and indefinite. The definite integral of a function, interpreted geometrically, corresponds to the area under the curve of the function between any two

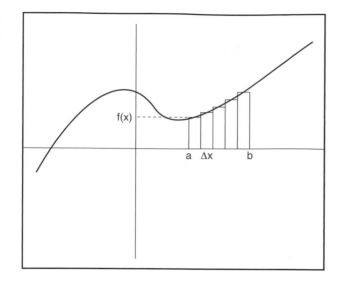

Figure 2. Consider approximating the area under the graph of a function f(x) by drawing a series of rectangles, and summing their areas to arrive at the total area. *Illustration by Hans & Cassidy. Courtesy of Gale Group.*

limits. Thus, it has a definite value depending on the limits chosen, The indefinite integral of a function is the inverse of the derivative of that function. That is, integrating (finding the integral) undoes differentiating (finding the derivative f'(x)). Integrating the derivative of a function returns the original function.

Indefinite Integral

The indefinite integral is the inverse of the derivative, that is, the integral of the derivative of a function is the original function. From the definition of derivative, $f'(x) = df(x)/dx$, we find $f'(x)dx = df(x)$. To obtain the original function from this second equation, we "integrate" both sides, and write, $\int f'(x)dx = \int df(x) = f(x) + C$. The integral sign, $\int$, is intended to symbolize the summing, integrating, or putting together of the infinitesimal pieces df(x) to obtain the original function f(x). The constant, C, arises because functions that differ only by a constant are "parallel" to one another, and so have the same derivative (or slope) at each value of x. Defined in this manner, integrating amounts to guessing original functions based on prior knowledge of their derivatives. For example, if $f(x) = x^2$ then $f'(x) = 2x$. Thus, if asked to integrate the function g(x) = 2x, it is apparent that $\int 2xdx = x^2 + C$. In order to determine the value of C it is necessary to have an additional piece of information. Such information is referred to as an initial condition or a boundary condition, and is sufficient to determine which of the **parallel** curves is the desired one.

The primary application of indefinite integrals is in the solution of differential equations. A differential equa-

tion is any equation that contains at least one derived. The equation $f'(x) = ax^2+bx+c$ is an example of a differential equation. Many natural relationships are described by differential equations. For instance, **heat** conduction is related to the derivative of the **temperature** with respect to distance; the velocity of a fluid flowing through a pipe is related to the derivative of the pressure with respect to length of pipe; and the force on any massive body is related to the derivative of its **momentum** with respect to time.

Definite Integral

The definite integral corresponds to the area under the graph of a function, above the x-axis and between two vertical lines called the limits of integration. Consider approximating the area under the graph of a function f(x) by drawing a series of rectangles, and summing their areas to arrive at the total area, A(x) (see Figure 2). The height of each **rectangle** is the value of the function at x, namely f(x). The width of each rectangle is $\Delta x = (b-a)/n$, where n is the number of rectangles chosen. If we wish to know the area between x=a and x=b, then the area is given by the sum

$$A(x) = \sum_{k=1}^{n} f(x_k)\,\Delta x$$

(the Greek letter Σ (sigma) is used to indicate that the n products $f(x_k)\,\Delta x$ corresponding to the n rectangles are to be summed). In the limit that n approaches infinity, Δx approaches 0, and the sum is exactly equal to the are. Since Δx approaches 0, it represents an infinitesimal change in the **variable** x, so the same notation used in defining the derivative is used to replace Δx with d. The product $f(x)\,\Delta x$ becomes f(x)dx, and corresponds to an infinitesimal area, dA(x). The total area, then, is the sum of an infinite number of an infinite number of infinitesimal areas. Thus, the area A(x) between a and b is equal to the integral of f(x)dx, written,

$$A(x) = \int_a^b dA(x) = \int_a^b f(x)dx$$

The limits included above and below the integral sign indicate that the indefinite integral of f(x)d is to be evaluated at a and b and the values subtracted, that is,

$$\int_a^b f(x)dx = A(b) - A(a)$$

This is interpreted as the area under the curve to the left of b minus the area under the curve to the left of a. Comparing this with the form of the indefinite integral we see that a function f(x) is the derivative of its "Area func-

KEY TERMS

Derivative—The derivative of a function is the instantaneous rate of change of the function's dependent variable with respect to its independent variable, interpreted geometrically as the slope of the tangent to the graph of the function.

Function—A set of ordered pairs for which each of the first and second elements are related, no two first elements being equal.

Integral—The integral is the inverse of the derivative, interpreted geometrically as the area under the graph of the function.

Limit—A limit is a boundary which the value of a variable may come infinitely close to, but never reach.

tion," the constant C being evaluated by use of boundary conditions, namely the values a and b. There are many applications of definite integrals, among the most common are the determination of areas and volumes of revolution.

Resources

Books

Abbot, P., and M.E. Wardle. *Teach Yourself Calculus.* Lincolnwood, IL: NTC Publishing, 1992.

Larson, Ron. *Calculus With Analytic Geometry.* Boston: Houghton Mifflin College, 2002.

Silverman, Richard A. *Essential Calculus With Applications.* New York: Dover, 1989.

Swokowski, Earl W. *Pre Calculus, Functions, and Graphs.* 6th. ed. Boston, MA: PWS-KENT Publishing Co., 1990.

Weisstein, Eric W. *The CRC Concise Encyclopedia of Mathematics.* New York: CRC Press, 1998.

Periodicals

Moore, A.W. "A Brief History of Infinity." *Scientific American* 272 (1995): 112-116.

J. R. Maddocks

Calendars

There are three units of **time** which have a direct basis in **astronomy**: the day, which is the period of time it takes for the **Earth** to make one **rotation** around its axis; the month, which is the period of time it takes for the **Moon** to revolve around the earth; and the year, which is the period of time it takes for the earth to make one revolution around the **Sun**.

The week has an indirect basis in astronomy—the seven days of the week probably were named for the seven objects which the ancients saw moving on the zodiac, which were the Sun, Moon, and the five planets that can be seen with the naked **eye**.

A calendar is a system for measuring long units of time, usually in terms of days, weeks, months, and years. The year is the most important time unit in most calendars, since the cycle of **seasons**, which are associated with change of climate in the earth's temperate and frigid zones, repeat in a yearly cycle with the change in the Sun's apparent position on the ecliptic as Earth revolves around the Sun.

Types of calendars

There are three main types of calendars. One type of calendar is the lunar calendar, which is based on the month (Moon). A lunar calendar year is 12 synodic months long, where a synodic month is the time interval in which the phases of the Moon repeat (from one full moon to the next), and averages 29.53 days. Thus, a lunar calendar year averages 354.37 days long. Because Earth takes slightly longer than 365 days to revolve completely around the Sun, a lunar calendar soon gets out of phase with the seasons. Thus, most lunar calendars have died out over the centuries. The main exception is the Muslim calendar, which is used in Islamic countries, most of which are in or near Earth's torrid zone, where seasonal variation of climate is slight or non-existent, and the climate is usually consistently hot.

The second type of calendar is the luni-solar calendar, in which most years are 12 synodic months long, but a thirteenth month is inserted every few years to keep the calendar in phase with the seasons. There are two important surviving luni-solar calendars: the Hebrew (Jewish) calendar, which is used by the Jewish religion, and the Chinese calendar, which is used extensively in eastern **Asia**.

The third type of calendar is the solar calendar, which is based on the length of the year. Our present calendar is of this type; however, it evolved from the ancient Roman calendar, which passed through the stage of being a luni-solar calendar.

In the first centuries after Rome was founded (753 B.C.), the Roman calendar consisted of ten synodic months; the year began near the start of spring with March and ended with December (the tenth month). The remaining 70 winter days were not counted in the calendar. Some centuries later two more months, January, named for Janus, the two-faced Roman god of gates and doorways, and February, named for the Roman festival of purification, were added between December and

March. An occasional thirteenth month was later inserted into the calendar; at this stage the Roman calendar was luni-solar calendar. It was quite complicated and somewhat inaccurate even by the year 45 B.C.

That year, Julius Caesar (100-44 B.C.) commissioned the Greek astronomer Sosigenes (c. 50 B.C.) from Alexandria to plan a sweeping reform of the Roman Calendar. The calendar Sosigenes devised and Caesar installed for the Roman Empire had the following main features:

The months January, March, May, July, August, October, and December each have 31 days. The months April, June, September, and November each have 30 days. February has 28 days in ordinary years, which have 365 days.

Every fourth year is a Leap year with 366 days. The 366th day appears in the calendar as February 29th.

The calendar year begins on January 1 instead of March 1. January 1 is set by the time of year when the Sun seems to set about half an hour later than its earliest setting seen in Rome, which occurs in early December.

This calendar was named the Julian calendar for Julius Caesar. He also had the month Quintilis (the fifth month) renamed July for himself. Augustus Caesar (63 B.C.-A.D 14.) clarified the Julian calendar rule for leap year by decreeing that only years evenly divisible by four would be leap years. He also renamed the month Sextilis (the sixth month) August for himself.

The average length of the Julian calendar year over a century or more is 365.25 days. This time interval is between the lengths of two important astronomical years. The shorter one is the tropical year, or the year of the seasons. The tropical year is defined as the time interval between successive crossings of the Vernal **Equinox** by the Sun (which marks the beginning of spring in the earth's northern hemisphere) and averages 365.2422 days long. The sidereal year, which is defined as the time interval needed for Earth to make a complete 360° orbital revolution around the Sun, is slightly longer, being 365.25636 days long. The small difference between the lengths of the sidereal and tropical years arises because the **earth's rotation** axis is not fixed in **space** but describes a cone around the line passing through the earth's center that is **perpendicular** to the earth's **orbit plane** (the ecliptic). The rotation axis describes a complete cone in 25,800 years.

This phenomenon is called precession, and it causes the equinoxes (the intersections of the celestial equator and ecliptic) to shift westward on the ecliptic by 50.″2 (O°0139) each year and also the celestial poles to describe small circles around the ecliptic poles. Because

the Sun appears to move eastward on the ecliptic at an average **rate** of 0°.9856/day, the Sun moves only 359°.9861 eastward along the ecliptic in an average tropical year, whereas it moves 360° eastward in a sidereal year, making the tropical year about 20 minutes shorter than the sidereal year. Precession is caused by stronger gravitational pulls of the Sun and Moon on the closer parts of the earth's equatorial bulge than on its more distant parts. This effect tries to turn Earth's rotation axis towards the line perpendicular to the ecliptic, but because the Earth rotates, Earth precesses like a rapidly spinning top, producing the effects described above.

An astronomer wants to make the average length of the calendar year equal to the length of the tropical year in order to keep the calendar in phase with the seasons. Sosigenes knew that **precession of the equinoxes** existed; it had been discovered by his predecessor Hipparchus (c. 166-125 B.C.). From his observations and records of earlier observations, Sosigenes allowed for precession of the equinoxes by making the average length of the Julian calendar year slightly shorter (0.00636 day, or about nine minutes) than the length of the sidereal year. But he did not know the physical cause of precession (a gravitational tidal effect), so he could not calculate what the annual rate of the precession of the equinoxes should be. The crude astronomical observations existing at that time may have led Sosigenes to believe that the rate of precession of the equinoxes was about half its true value, and therefore, that the 365.25 day average length of the Julian calendar year was an adequate match to the length of the tropical year. Unfortunately, this is not true for a calendar intended for use over time intervals of many centuries.

The development of our present (Gregorian) calendar

The Sun appeared to reach the vernal equinox about March 25 in the years immediately after the Roman Empire adopted the Julian calendar. It continued to be the official Roman calendar for the rest of the empire's existence. The Roman Catholic Church adopted the Julian calendar as its official calendar at the Council of Nicaea in A.D.325, soon after the conversion of the emperor Constantine I, who then made Christianity the Roman Empire's official religion. By that time, the Sun was reaching the Vernal equinox about March 21; the fact that the tropical year is 0.0078 day shorter than the average length of the Julian calendar year had accumulated a difference of three to four days from the time when the Julian calendar was first adopted. The Council of Nicaea also renumbered the calendar years; the numbers of the Roman years were replaced by a new numbering system in an effort to have in accord with Christian tradition and beliefs Christ's **birth** occur in the year A.D.1. (*Anno Domini*). This effort was somewhat unsuccessful; the best historical evidence indicates that Christ probably was born sometime between 7 B.C. (before Christ) and 4 B.C. Another feature of this modified Julian calendar is that it has no year **zero**; the 1 B.C. is followed by the year A.D.1.

This Julian calendar remained the official calendar of the Roman Catholic Church for the next 1,250 years. By the year 1575, the Sun was reaching the Vernal Equinox about March 11. This caused concern among both church and secular officials because, if this trend continued, by the year 11,690, Christmas would have become an early spring holiday instead of an early winter one, and would be occurring near Easter.

This prompted Pope Gregory XIII to commission the astronomer Clavius to reform the calendar. Clavius studied the problem, then he made several recommendations. The rate of the precession of the equinoxes was known much more precisely in the time of Clavius than it had been in the time of Sosigenes. The calendar which resulted from the study by Clavius is known as the Gregorian calendar; it was adopted in 1583 in predominantly Roman Catholic countries. It distinguished between century years, that is, years such as 1600, 1700, 1800, 1900, 2000, etc., and all other years, which are non-century years. The Gregorian calendar has the following main features:

All non-century years evenly divisible by four, such as 1988, 1992, and 1996 are leap years with February 29th as the 366th day. All other non-century years are ordinary years with 365 days.

Only century years evenly divisible by 400 are leap years; all other century years are ordinary years. Thus, 1600 and 2000 were leap years with 366 days, while 1700, 1800, and 1900 had only 365 days.

The Gregorian calendar was reset so that the Sun reaches the Vernal Equinox about March 21. To accomplish this, ten days were dropped from the Julian calendar; in the year 1582 in the Gregorian calendar, October 4 was followed by October 15.

The Gregorian calendar is the official calendar of the modern world. From the rules for the Gregorian calendar shown above, one finds that, in any 400-year interval, there are 97 leap years and 303 ordinary years, and the average length of the Gregorian calendar year is 365.2425 days. This is only 0.0003 day longer than the tropical year. This will lead to a discrepancy of a day in about the year 5000. Therefore, the Sun has usually reached the Vernal Equinox and northern hemisphere Spring has begun about March 21 according to the Gregorian calendar.

The Gregorian calendar was not immediately adopted beyond the Catholic countries. For example, the British Empire (including the American colonies) did not adopt the Gregorian calendar until 1752, when 11 days had to be dropped from the Julian calendar, and the conversion to the Gregorian calendar did not occur in Russia until 1917, when 13 days had to be dropped.

One feature of the Gregorian calendar is that February is the shortest month (with 28 or 29 days), while the summer months July and August have 31 days each. This disparity becomes understandable when one learns that Earth's orbit is slightly elliptical with eccentricity 0.0167, and the earth is closest to the Sun (at perihelion) in early January, while it is most distant from the Sun (at aphelion) in early July. It follows from Kepler's Second Law that Earth, moving fastest in its orbit at perihelion and slowest at aphelion, causes the Sun to seem to move fastest on the ecliptic in January and slowest in July. The fact that the Gregorian calendar months January, February, and March have 89 or 90 days, while July, August, and September have 92 days makes some allowance for this.

Possible future calendar reform and additions

Although the Gregorian calendar partially allows for the eccentricity of the earth's orbit and for the dates of perihelion and aphelion, the shortness of February introduces slight inconveniences into daily life. An example is that a person usually pays the same rent for the 28 days of February as is paid for the 31 days of March. Also, the same date falls on different days of the week in different years. These and other examples have led to several suggestions for calendar reform.

Perhaps the best suggestion for a new calendar is the World Calendar, recommended by the Association for World Calendar Reform. This calendar is divided into four equal quarters that are 91 days (13 weeks) long. Each quarter begins on a Sunday on January 1, April 1, July 1, and October 1. These four months are each 31 days long; the remaining eight months all have 30 days. The last day of the year, a World Holiday (W-Day), comes after Saturday December 30 and before January 1 (Sunday) of the next year; it is the 365th day of ordinary years and the 366th day of leap years. The extra day in leap years appears as a second World Holiday (Leap year or L-Day) between Saturday June 30 and Sunday July 1. The Gregorian calendar rules for ordinary, leap, century, and non-century years would remain unchanged for the foreseeable future.

The most recent, and much-discussed, calendrical confusion concerned the so-called *Y2K*, the rollover of the calendar from 1999 to 2000. Debates ensued as to whether the new millennium would begin on January 1,

KEY TERMS

Eccentricity—Measurement of the earth's deviation from a circular orbit around the Sun, based on its actual elliptical orbit.

Ecliptic—The plane of Earth's orbit about the Sun as projected on the sky. The Sun always appears to lie directly on the ecliptic; the Moon and planets lie near it but not necessarily on it, as their orbital planes are all oriented slightly differently from Earth's.

Equinox—The days of the year when the Sun appears to lie directly on the celestial equator, meaning it appears to rise due east and set due west. This happens twice per year, on or about March 21 (the spring or vernal equinox) and September 22 (the fall equinox), and on these dates the day and night are each 12 hours long. The word equinox is derived from Latin words meaning "equal" and "night."

Precession—The wobbling motion of Earth's rotational axis, much like a spinning top wobbles about its axis of rotation.

Sidereal year—The time interval needed for the earth to make a complete 360° orbital revolution around the Sun, 365.25636 days.

Synodic month—The time interval in which the phases of the Moon repeat (from one Full Moon to the next), and averages 29.53 days.

Tropical year—The time interval between successive crossings of the Vernal Equinox by the Sun, 365.2422 days.

Zodiac—The zone 9° on each side of the ecliptic where a geocentric observer always finds the Sun, Moon, and all the planets except Pluto.

endar has been suggested, but much more must be done before an official Martian calendar is adopted.

The Julian day calendar

This calendar is extensively used in astronomy, **oceanography**, and other sciences. It must not be confused with the Julian civil calendar.

This calendar was devised in 1582 by Josephus Justus Scaliger; the Julian date for a given calendar date is the number of days that have elapsed for that date since noon (by Universal Time [U.T.]) on January 1, 4713 B.C. It is based on a time interval 7,980 years long, which Scaliger called the Julian period. For example, noon (12:00 U.T.) on January 1, 1996 is Julian Day J.D. 2,450,084.0 = 1.5 January 1996 U.T.

Resources

Books

Branley, Franklyn M., Mark R. Chartrand III, and Helmut K. Wimmer. *Astronomy.* New York, Thomas Y. Crowell Co., 1975, pp. 407–415.

Oriti, Ronald A., and William B., Starbird. *Introduction to Astronomy.* Encino, CA: Glencoe Press, 1977, pp. 45-51.

Frederick R. West

Calibration

Calibration is the process of checking the performance of a measuring instrument or device against some commonly accepted standard. A watch, for example, has to be calibrated so that it keeps correct **time**, agreeing with the international standard. The dials on a **radio** must also be calibrated so that the correct **frequency** or station is actually being received. Calibration provides consistency in a variety of applications. Because rulers and calipers are calibrated, for instance, a 0.39 in (10 mm) nut made by one factory will fit a 0.39 in (10 mm) screw machined by another halfway around the world. Without calibration, such standardization and interchangeability would not be possible.

Laboratories exist that provide official calibration of various instruments. In addition to clocks, such instruments and tools as electrical meters, **laser** beam power analyzers, **torque** wrenches, thermometers, and surveyors' theodolites all need calibration to an accepted standard to be useful. Calibration is performed by comparing the results of the instrument or device being tested (the value you actually get) to the accepted standard (the

2000, or January 1, 2001. Technically, the first day of the new millennium is January 1, 2001, because there was no year zero. The first year of the first millennium A.D. began at the start of the year 1, so the first year of the next two millennia must begin at the start of the years 1001 and 2001. However, a reasonable case can be made that the change of digits to the even year 2000 makes it more significant in human reckoning than the minor change from 2000 to 2001.

A future **Mars** calendar for the human colonization of Mars in future centuries poses interesting problems. There are about 668.6 sols (mean Martian solar days, which average 24 hours 39 minutes 35.2 seconds of mean solar time long) in a Martian sidereal year. At least one Martian cal-

value you should get), and adjusting the instrument/device being tested until the two agree.

Frequency of calibration varies according to the device being calibrated and the applications. A clock or a common ruler for home use, for example, will only be calibrated at the time it is manufactured. A torque wrench on a NASA project, on the other hand, may require calibration every year. Some sophisticated electronic instruments for such projects may require calibration every few months.

In the United States, calibration of instruments or devices for high precision applications is generally traceable to standards established by the National Institute of Standards and Technology (NIST). In other words, if a laboratory is calibrating a meter stick for **distance** measurement, they need to prove that the standard they are measuring against has also been calibrated against the NIST definition of the meter. NIST keeps standard definitions of **mass**, length, **temperature**, etc. Historically, standards have been based on a "magic measure" such as the platinum-iridium bar initially used as the standard for the meter. The trend now is away from physical expressions of standards and toward standards based on some physical constant. One of the official definitions of the meter, for example, is based on the wavelength of **light** emitted by **calcium atoms** under certain specific conditions. Such a definition can be recreated at need and does not depend on the physical existence of a slab of **metal**. International standards have been agreed upon, simplifying international trade and science.

Traceability to NIST is not in and of itself a guarantee of **accuracy**, however. All measurements have some uncertainty associated with them. Common sense should thus be used when measurement accuracy requirements for an instrument or device are formulated. Accuracy is limited by calibration, and even calibration has its limits.

See also Caliper.

Californium *see* **Element, transuranium**

Caliper

A caliper is an instrument used for measuring linear dimensions that are not easily measured by devices such as meter sticks or rulers. Two examples of such measurements include the outer dimensions of a pipe or the internal diameter of a **glass** tube.

Although many kinds of calipers exist, they are all designed on a common principle: two legs are hinged at

one end to allow movement of the free ends of the legs both towards and away from each other. A caliper looks like a common pair of tweezers. The **distance** between the free ends of the two legs is the linear dimension measured by the caliper.

The crescent-shaped legs of an outside caliper are curved inward toward each other. When the ends of the caliper are placed on the outside of some object, the distance between the ends of the legs can be read on a scale (usually incised on the pivot end of the caliper), giving the outside diameter of the object. The legs of an inside caliper, on the other hand, are curved away from each other, like an hourglass. To find an interior diameter, the caliper is placed inside an object and opened. Again, the distance between the ends of the caliper can be read on a scale.

The micrometer is one of the most common devices using the caliper principle. The micrometer consists of a **metal** handle around which a movable cylinder called the thimble is attached. As the thimble is rotated, a spindle connected to the handle moves toward or away from a fixed anvil. The dimensions of an object can be measured by placing the object between the anvil and the spindle and slowly rotating the thimble. When the object is in firm contact with the anvil on one side and the spindle on the other, its linear dimensions can be read on the scale located on the micrometer handle.

The micrometer caliper can provide precise measurements relatively easily. A turn of the thimble advances the spindle only a small distance. A single **rotation** of the thimble in most micrometers advances the spindle a distance of 0.025 in (0.064 cm). The thimble itself is divided into 25 segments, which enables the micrometer to measure distances as small as 0.001 in (0.0025 cm). The addition of a vernier scale to the micrometer can further improve the precision of a measurement by a factor of 10.

Calomel *see* **Mercurous chloride**

Calorie

A calorie is the amount of **energy** required to raise the **temperature** of 1g of pure **water** by 34°F (1°C) under standard conditions. These conditions include an **atmospheric pressure** of one atmosphere, and a temperature change from 60° to 62°F (15.5 to 16.5°C).

The calorie is also sometimes designated as a gram-calorie or small calorie (abbreviated: cal), to distinguish it from the calorie of dieticians (abbreviated: Cal), also known as a large calorie, or kilocalorie (kcal), which is equal to 1,000 (small) calories.

One calorie is equivalent to 3.968 British thermal units (btu), a non-metric measure of energy content. A calorie is also equivalent to 4.187 joules (also known as an International Table calorie), which is now the unit of energy that is most commonly used in science.

Scientists are often interested in the energy contents of organic materials. These data are usually obtained by completely oxidizing (burning) a known quantity of a substance by igniting it in an oxygen-rich atmosphere inside of a device known as a bomb-calorimeter. The quantity of energy released is determined by measuring the increase in temperature of a known quantity of water contained within the bomb.

Dieticians are interested in the calorie contents of foods of various sorts. The potential energy of food is utilized metabolically by animals to drive their physiological processes, and to achieve growth and reproduction. Foods vary tremendously in their energy contents, so careful planning of food intake requires an understanding of the balance of the **nutrients**, such as vitamins and amino acids.

On average, pure carbohydrates have a calorific content of about 4,600 cal/g (or 4.6 Cal/g), while **proteins** contain about 4,800 cal/g, and fats or lipids about 6,000-9,000 cal/g. Because fats are so energy-dense, they are commonly used by organisms as a compact material in which to store potential energy for future use. Of course, some of us store more of this potential energy of **fat** than others.

Engineers are often concerned with the energy contents of **petroleum**, **coal**, and **natural gas**, and of distillates or synthetic materials refined from any of these **fossil fuels**. Knowledge of the amounts of energy that are liberated through the complete oxidation of these materials is important in the design of engines, fossil-fueled generating stations, and other machines that we use to achieve mechanical work. In order to maximize the amount of useful work that is achieved per unit of fuel consumed, that is, the energy-conversion efficiency, engineers are constantly re-designing machines of these sorts, and tuning their operating parameters, such as fuel-oxygen ratios.

Ecologists are also interested in the energy contents of organic materials, and how these change over **time**. Although ecologists commonly measure **biomass** and productivity in terms of weight, these are often converted into energy units, in order to account for the greatly varying calorific contents of different sorts of biomass, as was described above for carbohydrates, proteins, and fats. In parallel with the interests of engineers, ecologists

are concerned with the efficiency of ecosystems in converting solar energy into **plant** productivity, as well as the transfers of the energy of plants to herbivores and carnivores. These efficiencies are best determined through knowledge of the amounts and transfers of energy, as expressed in calorific units.

Calorimeter *see* **Calorimetry**

Calorimetry

Calorimetry is the measurement of the amount of **heat** gained or lost during some particular physical or chemical change. Heats of fusion or vaporization, heats of **solution**, and heats of reaction are examples of the kinds of determination that can be made in calorimetry. The term itself derives the Latin word for heat, caloric, as is the name of the instrument used to make these determinations, the calorimeter.

History

Little productive work on the measurement of heat changes was accomplished prior to the mid-nineteenth century for two reasons. First, the exact nature of heat itself was not well understood. Until the work of the Scottish chemist Joseph Black in the late eighteenth century, the distinction between **temperature** and heat was not at all clear. It then took until about 1845 before the nature of heat as a form of **energy** and not of **matter** was made clear in the experiments of James Joule and others.

Secondly, given such uncertainties, it is hardly surprising that appropriate equipment for the measurement of heat changes was not available until after the 1850s. Lavoisier and Laplace had made use of a primitive **ice** calorimeter to measure the heats of formations of compounds in 1780, but their work was largely ignored by their colleagues in **chemistry**.

In fact, credit for the development of modern techniques of calorimetry should probably be given to the French chemist Pierre Eugène Berthelot (1827-1907). In the 1860s, Berthelot became interested in the problems of heat measurement. He constructed what was probably the first modern calorimeter and invented the terms **endothermic** and exothermic to describe reactions in which heat is taken up or given off, respectively.

The calorimeter

In essence, a calorimeter is any device in which the temperature before and after some kind of change can be accurately measured. Probably the simplest of such devices is the coffee cup calorimeter so-called because it is made of a styrofoam cup such as the ones in which coffee is commonly served. A styrofoam cup is used because styrofoam is a relatively good insulating material. Heat given off within it as a result of some physical or chemical change will not be lost to the surrounding environment. To use the coffee cup calorimeter, one simply carries out the reaction to be studied inside the coffee cup, measures the temperature changes that take place, and then calculates the amount of heat lost or gained during the change.

The type of calorimeter more commonly used for precise work is called the bomb calorimeter. A bomb calorimeter designed to measure heat of **combustion**, as an example, consists of a strong-walled **metal** container set inside another container filled with water. The inner container is fitted with an opening through which **oxygen** can be introduced and with electrical leads to which a source of **electricity** can be connected.

The object to be studied is then placed in a combustion crucible within the bomb and ignited. The reaction occurs so quickly within the reaction chamber that it is similar to the explosion of a bomb. Hence the instrument's name. Surrounding the bomb in this arrangement is a jacket filled with (usually) water. Heat given off or absorbed within the bomb heats up the **water** in the jacket, a change that can readily be measured with a **thermometer** inserted into the water.

Many variations in the basic design described here are possible. For example, the use of liquids other than water in the insulating jacket can permit the study of heat changes at higher temperatures than the **boiling point** of water 212°F (100°C). Aneroid (without liquid) calorimeters are also used for special purposes, such as the measurement of heat changes over very large temperature ranges. Such calorimeters use metals with a high **coefficient** of thermal conductivity, like **copper**, to measure the gain or loss of heat in some type of change.

Calorimetry theory

Suppose that a cube of sugar is burned completely within the bomb of a calorimeter. How can an experimenter determine the heat released in that reaction?

To answer that question the assumption is made that all of the heat produced in the reaction is used to raise the temperature of the water in the surrounding jacket and the metal walls of the bomb itself. The heat absorbed by each is equal to its **mass** multiplied by its specific heat multiplied by the temperature change (DT). Using a word equation to express this fact: heat released in reac-

tion = (mass of water × specific heat of water × DT) + (mass of bomb × specific heat of bomb × DT). The last part of this equation, (mass of bomb × specific heat of bomb × DT), is the same for any given calorimeter. Once measured, it is known as a constant value and, therefore, is given the name of calorimeter constant.

Resources

Books

Asimov, Isaac. *Asimov's Biographical Encyclopedia of Science & Technology.* 2nd revised edition. Garden City, NY: Doubleday & Company, Inc., 1982, pp. 443-444.

Masterson, William L., Emil J. Slowinski, and Conrad L. Stanitski. *Chemical Principles.* Philadelphia: Saunders, 1983.

David E. Newton

Camels

Camels and their relatives, the llamas, are long-legged, hoofed **mammals** in the family Camelidae in order Artiodactyla, whose members have an even number of toes. All camels have a cleft in their upper lip, and all have the ability to withstand great **heat** and great cold.

Camels evolved in **North America** and spread into **South America**, **Asia**, and **Africa**. Camels in Asia and Africa today have been domesticated for up to 3,500 years. The two smaller South American **species** of camel have been bred to develop two purely domestic species, the llama and the alpaca.

Like cattle, camels are cud-chewing animals, or ruminants. However, unlike other ruminants, which have four chambers to their stomachs, camels have only three. Both male and female camels have the same number of

---

KEY TERMS

Heat—The transfer of thermal energy that occurs between two objects when they are at different temperatures.

Insulator—An object or material that does not conduct heat or electricity well.

Specific heat—The amount of heat needed to increase the temperature of a mass of material by one degree.

Temperature—A measure of the average kinetic energy of all the elementary particles in a sample of matter.

A dromedary camel. © *Leonard Lee Rue, III/The National Audubon Society Collection/Photo Researchers, Inc. Reproduced by permission.*

teeth for feeding, but the front incisor teeth of the males are large and sharp, making them useful for fighting. Camels have oval shaped red **blood** cells, whereas all other mammals have round red blood cells.

Both New World and Old World camels communicate in a variety of ways, including whistling, humming, and spitting. In zoos, camels have been known to spit the contents of their first stomach at annoying visitors.

Old world camels

The Bactrian, or two-humped, camel (*Camelus ferus*) is the largest species, native to the rocky deserts in Asia. These wild camels were the ancestors of the domestic Bactrian camel, *C. bactrianus.* These animals are named for the Baktria region of ancient Persia (now Iran), and can withstand severe cold as well as extreme heat (up to 122°F [50°C]. Bactrian camels have a thick and shaggy coat, with very long hair growing downward from their necks. Domesticated bactrians have longer hair than the wild species.

The Bactrian camel stands about 6.5 ft (2 m) high at the shoulder and weighs up to 1,500 lb (680 kg) and can run up to 40 mph (65 kph). Bactrian camels can carry loads of up to 1,000 lb (454 kg), about twice as much as a dromedary can carry. Although Bactrian camels breed well in zoos, they are almost extinct in the wild, with probably only a few hundred left in the Gobi Desert of Asia.

The most common camel is the one-humped Arabian, or dromedary camel, which is known today only as a domesticated species, *C. dromedarius.* Although the name "dromedary" has now been given to all one-humped camels, it originated with a special breed developed for great speed in racing. Racing camels can also run over great distances—covering more than 100 mi (160 km) in a day.

Dromedary camels are taller but lighter than Bactrian camels, reaching 7 ft (2.1 m) at the shoulder, and an average weight of about 1,200 lb (550 kg). The animals' hair can vary in **color** from dark brown to white, though most are the tan color referred to as "camel's hair." The Arabian camel has long been extinct in the wild, though feral populations (domestic animals living in the wild) occur in various parts of the world, including central **Australia** where the herds number up to 50,000.

Although one-humped and two-humped camels are given separate species names, they can interbreed fairly easily and are probably varieties of a single species. The product of interbreeding, called a tulu, usually has two humps. This is not surprising in view of the fact that the one-humped camel actually has another hump that lies unnoticeably in the shoulder region.

Both species are used primarily as pack animals in **desert** countries, where they travel at a leisurely pace of about 25 mi (40 km) a day, carrying both goods and people, mounted in saddles that fit over the single hump or between the two humps. A camel's hump contains 80 lb (36 kg) of **fat**, not **water**, which provides the **animal** with **energy** when no food is available. The hump shrinks and becomes flabby as the fat supply is used up, but it firms up again when the animal eats plants and drinks water. A camel can go for a week and even travel 100 mi (160 km) or more in a desert summer without drinking. Camels can withstand a great deal of dehydration and can lose more than 40% of their body weight without harm. In winter, camels can go for many weeks without drinking. When dehydrated camels do reach water, they make up for any previous lack by drinking as much as possible very quickly. They have been known to drink as much as 30-40 gal (114-150 l) in a single session to rehydrate.

Other adaptations of camels for dealing with desert conditions include a reduced number of sweat **glands** which only function in extreme heat or exertion. At heat stresses that would cause most animals to sweat to cool their bodies, and thus use up water, the camel's body **temperature** can temporarily rise several degrees, a strategy known as heat storage. The thick coat on the back of the camel prevents heat from the **sun** from being absorbed. The hoofed feet have broad, thick pads that provide a solid base on shifting sands. The thick, bushy eyebrows and double rows of eyelashes keep **sand** out of their eyes. Any sand that does enter the **eye** is dislodged by a transparent third eyelid that slides across the eye. Hair inside a camel's ears prevents sand from easily blowing in the **ear** canal. In addition, camels can voluntarily squeeze shut both their slit-like nostrils and their mouth to prevent sand from entering. The camel's mouth has a thick, leathery lining that prevents the thorny desert plants from damaging the mouth. The round, leathery

A domesticated llama in Peru. *JLM Visuals. Reproduced by permission.*

kneepads of camels protect their knees when they kneel on the hot sand or on hard rocky ground.

Camels are central to the survival and culture of the nomadic peoples of the old world deserts. Camel hair, shed in large clumps, is woven into clothing and tents, while camel milk and meat provide nourishment, particularly on special occasions. Nomadic people often let their camels loose in the desert for several months at a time, which includes the mating season. Camels have a gestation period of about 14 months, after which the mother camel gives **birth** to a single 80 lb (36 kg) offspring, every other year. The long-legged calves, become independent at about four years, and domesticated camels can live up to about 50 years.

New world camels

New World camels are native to the Andes **Mountains** on the western side of South America. The wild New World camels are the vicuña (*Vicugna vicugna*) and the guanaco *(Lama guanicoe)* while the llama and al-paca are domestic animals.

Wild South American camels live primarily at high altitudes in both open **grasslands** and **forests**. Their family groups may include a male and half a dozen or so females, each with a single young. Young males are chased from the group when they are between a year and a year and a half old, then they join a bachelor herd until they can later form their own family groups. Young females join a new group and mate, producing young after a gestation period of about 11 to 12 months.

Vicuñas are extremely rare in the wild, and are small animals, often standing no more than 3 ft (90 cm) at the shoulder and weighing no more than about 110 lb (50 kg). Vicuñas, are alone among the camels to have bottom incisor teeth that keep growing and enamel only on the outer surface. This characteristic has led taxonomists to assign them to a separate genus. Vicuñas are fast runners that can readily cover the dry, open grasslands where they live at altitudes between 11,500 and 18,700 ft (3,500-5,700 m). Male vicuñas defend both a grazing territory and a sleeping territory.

Vicuña fur can be woven into one of the softest **textiles** known and was for many centuries worn only by the Inca kings. After the Incan empire fell, Vicuñas were no longer protected, and were hunted for their meat and skins until they were close to **extinction**, with fewer than 10,000 animals left in 1967. Vicuñas are now protected in several Andean national parks, and their numbers are climbing once again.

The South American larger guanaco lives primarily in dry open country, from the coastal plains to the high mountains. Guanaco hair is cinnamon colored on their backs and white on their under parts. Unlike the smaller vicuñas, they have dark faces. Guanacos stand about 6 ft (less than 2 m) tall, and are the tallest South American camels, but are very light compared to camels, weighing only about 250 lb (113 kg). Most guanacos now live in Patagonia, a temperate large grassland in southern Argentina and Chile, and are found from **sea level** to about 14,000 ft (4,200 m). Male guanacos will mark their territories with piles of dung. Female guanacos give birth every two years with their newborn being called a chulengo, which can run within minutes of being born. Young guanacos often make a playful prance in which they lift all four feet off the ground at once, and guanacos also like playing in the running water of streams.

Starting about 4,000 years ago, natives of the Andes Mountains bred the guanaco to develop two other domesticated camels, the sure-footed llama *(Lama glama)*, bred for its strength, endurance, and its ability to carry

great loads over steep mountains, and the long-haired alpaca *(Lama pacos)*.

Llamas stand only about 4 ft (1.2 m) at the shoulder. Usually only male llamas are used for the long pack trains, while the females are kept for breeding. An average male llama weighs about 200 lb (90 kg) but can carry a load weighing two-thirds that amount on its back, for about 15-20 mi (24-32 km) a day across mountain terrain. Llamas were used by the Incas to transport silver from their mountain mines.

The alpaca, the other domesticated breed, has long hair valued for warm blankets and clothing because it is soft, lightweight, and waterproof. Some breeds of alpaca have hair that almost reaches the ground before it is sheared. Llama hair is not used for weaving because it is too coarse. Llamas and alpacas are often crossed to get an animal that produces hair that is both sturdier and softer than either of the parents' hair.

Resources

Books

Arnold, Caroline. *Camel.* New York: Morrow Junior books, 1992.
Camels. Zoobooks series. San Diego, CA: Wildlife Education, 1984.
Green, Carl R., and William R. Sanford. *The Camel.* Wildlife Habits & Habitats series. New York: Crestwood House, 1988.
LaBonte, Gail. *The Llama.* Remarkable Animals series. Minneapolis: Dillon Press, 1989.
Lavine, Sigmund A. *Wonders of Camels.* New York: Dodd, Mead & Company, 1979.
Perry, Roger. *Wonders of Llamas.* New York: Dodd, Mead & Co., 1977.
Stidworthy, John. *Mammals: The Large Plant-Eaters.* Encyclopedia of the Animal World. New York: Facts on File, 1988.

Jean F. Blashfield

Canal

A canal is a man-made waterway or channel that is built for navigation, **irrigation**, drainage, or **water** supply. When the word is used today however, it is usually in the context of transport or navigation by boats. Canal transport should not be confused with navigating on a river, because a canal is entirely artificial (although canals are in many cases connected with a natural body of water).

There are two major types of transport canals. One is an inland waterway, or water route, that either follows the lay of the land or has locks. The other is a canal that is built to shorten a sea route. Examples of the latter are the Suez Canal, which connects the Mediterranean and the Red Sea, and the Panama Canal, which shortens the voyage from **Europe** to America's west coast by 3,000 m (4,800 km). From the earliest times, canals were built because they were the simplest and cheapest way of moving heavy goods.

Canals can be among the most complicated **engineering** projects of their times. A canal across the Isthmus of Panama, for example, was first envisioned by Spanish explorers in 1513. The French began construction of a canal across the **isthmus** in 1880, but abandoned work in 1889. Excavation of the canal was resumed by the United States in 1904 and the Panama Canal finally opened in 1914. As many as 40,000 workers were employed at various times during the construction of the Panama Canal, which required the construction of towns to house workers, railroads to move supplies and enormous amounts of excavated **soil** and rock, and the invention of new kinds of construction equipment. Perhaps the most prominent engineering accomplishment during construction of the Panama Canal was the Culebra Cut, a 8.75-mi (14-km) excavation across the continental divide that was plagued by landslides. The total amount of soil and rock excavated along the Culebra Cut was estimated to have been about 100,000,000 cubic yards (76,500,000 cubic meters), or the equivalent of more than 8 million large dump truck loads. Monitoring of landslides along the Culebra Cut persists to this day.

Many long-abandoned inland canals in the United States, British Isles, and Europe are currently the focus of restoration projects. This work, often undertaken by volunteer groups, is being done to repair the effects of decades of neglect, educate the public about the historical significance of canals, and provide new recreational opportunities.

History

The earliest canals were built by Middle Eastern civilizations primarily to provide water for drinking and for irrigating **crops**. In 510 B.C. Darius I, King of Persia, ordered the building of a canal that linked the Nile River to the Red Sea; this canal was a forerunner of the modern Suez Canal. The Chinese were perhaps the greatest canal builders of the ancient world, having linked their major **rivers** with a series of canals dating back to the third century B.C. Their most impressive project was the famous Grand Canal, the first section of which opened around A.D. 610. With a total length of 1,114 ft (1,795 km), it is the longest canal in the world. Canals were employed by the highly practical Romans, but were neglected for the centuries after the fall of the Roman Empire. The commercial expansion of the twelfth century spurred the re-

Canal in Delft, the Netherlands. The first network of canals was built in the thirteenth and fourteenth centuries in the Netherlands. *Photograph by Porterfield/Chickering. The National Audubon Society Collection/Photo Researchers, Inc. Reproduced by permission.*

vival of canals, and it is estimated that as much as 85% of the transport in medieval Europe was by canal.

Many early canals were called contour canals because they followed the lay of the land and simply went around anything in their way. Major changes in ground and water levels have always presented canal builders with their greatest engineering problem; at first, boats were simply towed or dragged over slipways to the next level. The invention of the modern **lock** in China solved this problem at once, causing the full development of canals. The modern two-gate lock evolved from the slow and unsafe Chinese "flash lock" that had only one gate; the "flash lock" eventually made its way to Renaissance Europe, where it was modified with a second gate. The development of the lock heralded a period of extensive canal construction across Europe, and it is not surprising that each nation responded in its own particular way. The

Naviglio Grande Canale in Italy (1179–1209) and the Stecknitz Canal in Germany (1391–1398) are two that made important contributions to waterway technology. Also in China, a 684 mi (1,100 km) branch of the Grand Canal was finished in 1293. Over the next few centuries, France built the pioneering Briare Canal (completed in 1642) and the famous Canal du Midi (1681), which joined the Mediterranean and the Atlantic and would serve the world as an example of complex civil engineering at its best. This remarkable French canal stimulated the era of British canal construction that began with the completion of the Bridgewater Canal (1761). In Germany, the Friedrich Wilhelm Summit Canal was completed in 1669, and in other nations, extensive waterway systems were developed. With industrial production steadily growing in the nineteenth century, transport by canal became essential to the movement of raw materials and goods throughout Europe.

The United States has a shorter tradition of canal building. Its first major canal was the Erie Canal, constructed during the beginning of the nineteenth century. Completed in 1824, the 364-mi (586 km) canal provided a water route that brought grain from the Great Lakes region to New York and the markets of the East. With the construction of railroads in the 1830s, the U.S. quickly abandoned its canals in the belief that rail would be the best method for every transport task. The Europeans did not react the same way to railroads, maintaining their canal systems as a complementary system not in competition with railroads. Today, inland waterways or canals play a major transportation role in the United States and the rest of the world, for it has been realized that canals are perfectly suited for carrying low-value and high-bulk cargoes over long distances.

Sea canals, the great canals that shorten sea routes, are glamorous and highly visible engineering achievements. Three well-known examples are the Kiel Canal connecting the North and Baltic Seas (1895), the Suez Canal linking the Mediterranean and Red Seas (1869), and the Panama Canal between the Atlantic and Pacific Oceans (1914). Sea-to-sea ship canals face the problem of obsolescence: some new ships are too large for old canals. Both the Kiel and the Suez canals have been enlarged, but the Panama canal is not large enough to accommodate the world's largest ships.

Construction and operation

Engineers who are designing a canal must take several things into consideration. They must formulate the dimensions of the canal to accommodate the numbers and sizes of ships that are predicted to use the waterway. Natural obstacles in the path of the canal, such as rock

KEY TERMS

Contour canal—Usually early canals that followed the meandering natural contours of the earth.

Flash lock—A simple wooden gate that was placed across a moving body of water to hold it back until it had become deep; the sudden withdrawal of the gate would cause a "flash" of water that would carry a boat downstream and over the shallows below.

Inland waterway—An artificial waterway or channel that is cut through land to carry water and is used for transportation.

Lock—A compartment in a canal separated from the main stream by watertight gates at each end; as water fills or drains it, boats are raised or lowered from one water level to another.

Slipway—An inclined path or road leading into a body of water over which, in ancient times, boats were dragged or rolled from one body of water to another.

formations, must also be modified, removed, or avoided. There must be adequate vertical clearance above the canal and the clearance afforded by pre-existing **bridges** must be able to accommodate the vessels that will use the canal. Finally, engineers must decide the scale and location of associated structures, such as bridges, tunnels, and locks.

The paths of most canals are affected by variations in the levels of terrain. Engineers compensate for these variations with either locks or inclined planes. A lock is a segment of the waterway that is closed off by gates at either end. When a vessel enters the lock, the front gate is already closed. The back gate is then closed behind the vessel, and the water level within the lock is raised or lowered to the level of the water on the outside of the front gate. Valves on the gates control the level of the water. Whereas locks are the most common means of compensating for elevation changes, the procedure is slow and uses a large amount of water. Inclined planes can be used to elevate and lower smaller vessels; they use no water and often allow more rapid passage of vessels. When vessels reach certain stations along the waterway, they are pulled out of the water and moved on trucks up or down the **plane**.

Canal operators must monitor the canal's supply of water. If the natural supply of water at the upper end of the canal is deficient, it must be supplemented by water pumped into the reservoirs. If nature supplies the reservoir with excess amounts of water, some of the water must be diverted from the canal. Otherwise, excess water may strengthen the current and disrupt canal operations.

Although canals are among the oldest civil works, they play a major role in commerce because they are by far the least expensive form of inland transportation yet devised.

Resources

Books

Hadfield, Charles. *World Canals: Inland Navigation Past and Present.* New York: Facts On File, 1986.

Payne, P. S. Robert. *The Canal Builders: The Story of Canal Engineers Through the Ages.* New York: Macmillan Co., 1959.

Spangenburg, Ray, and Diane K. Moser. *The Story of America's Canals.* New York: Facts On File, 1992.

Other

Panama Canal Authority. *The Panama Canal.* (October 17, 2002). <www.pancanal.com/eng/index.html.>

Leonard C. Bruno

Cancel

Cancel refers to an operation used in term **mathematics** to remove terms from an expression leaving it in a simpler form. For example, in the fraction 6/8, the **factor** 2 can be removed from both the numerator and the denominator leaving the irreducible fraction 3/4. In this instance the 2 is said to be canceled out of the expression. Canceling is particularly useful for solving algebraic equations. The solution to the equation x - 7 = 4 is obtained by adding 7 to each side of the equation resulting in x = 11. When we add 7 to the left side of the equation, we cancel the -7 and put the equation in a simpler form. Typically, canceling is performed by using inverse operations. These are operations such as **multiplication** and **division** or **addition** and **subtraction** which "undo" one another.

Resources

Books

Bittinger, Marvin L., and Davic Ellenbogen. *Intermediate Algebra: Concepts and Applications.* 6th ed. Reading, MA: Addison-Wesley Publishing, 2001.

Blitzer, Robert. *Algebra and Trigonometry.* 2nd ed. Englewood Cliffs, NJ: Prentice Hall, 2003.

Cancer

Cancer is not just one **disease**, but a large group of almost 100 diseases. Its two main characteristics are un-

controlled growth of the cells in the human body and the ability of these cells to migrate from the original site and spread to distant sites. If the spread is not controlled, cancer can result in death.

One out of every four deaths in the United States is from cancer. It is second only to **heart** disease as a cause of death in the states. About 1.2 million Americans are expected to be diagnosed with cancer in 1998, of which, more than 500,000 are expected to die.

Cancer can attack anyone. Since the occurrence of cancer increases as individuals age, most of the cases are seen in adults, middle-aged or older. The most common cancers are skin cancer, lung cancer, colon cancer, breast cancer (in women), and prostate cancer (in men). In addition, cancer of the kidneys, ovaries, uterus, pancreas, bladder, rectum, and **blood** and lymph node cancer (**leukemia** and lymphomas) are also included among the 12 major cancers that affect most Americans.

The history of cancer as a known disease

The term cancer derives from the observation by Hippocrates in 400 B.C. that the **veins** radiating from a breast cancer resembled the legs of a crab, hence *karkinoma* in Greek and *cancer* in Latin. Cancer is not a single disease, but is many different diseases that all share common biological and pathological characteristics. In most western societies, cancer is a leading cause of death. The disease may develop in any body **tissue** or **organ** and over one hundred different types of cancer can occur in adults. Cancer also occurs in children and may even be present at **birth**.

The first clues to the cause of cancer came over two hundred years ago from an observation by Percivall Pott, a London doctor, who in 1775 found a high incidence of scrotal cancer in men who had worked as chimney sweeps. Later, **radiation** was found to cause skin cancer and tragically Marie Curie (1867–1934) the discoverer of **x rays**, died of a cancer caused by prolonged exposure to radiation. During the second half of the twentieth century, epidemiologists (those who study disease in populations) linked exposure to certain environmental toxins and particular types of cancer. Most notably, cigarette smoking and lung cancer, sunlight and skin cancer, and certain industrial chemicals to the cause of bladder and liver cancer. Finally several viruses were also been implicated in causing cancer, such as the **hepatitis** B **virus** and cancer of the liver, the Epstein Barr virus and lymphoma, and the human papilloma virus and cancer of the cervix. These important observations all suggested that specific external environmental agents could cause specific cancers.

How then could a diverse range of external agents such as chemicals, radiation and viruses, all lead to the development of cancer? The answer to this question has come over the last 25 years from two different lines of investigation; studies on cancer causing viruses and research into the **genetics** of some rare cancers in children.

In 1910, Frances Peyton Rous (1879–1970) isolated a virus from a cancer in chickens (a sarcoma) that caused new sarcomas to develop when infected into healthy chickens. Rous's work languished for over 50 years until he was awarded a Nobel Prize in 1966. By this time, methods for the study of viruses and cancer had improved considerably and many new **animal** derived viruses were found to cause cancer in a range of **species**. These viruses could also induce cancer-like changes when introduced into normal cells grown in the laboratory. A genetic study of these cancer causing viruses identified a small number of genes termed viral oncogenes (v-oncogenes) which, when introduced into cells, could transform the normal cells into malignant cells.

The presence of viral oncogenes led to the search for endogenous cellular oncogenes, which might cause cancer. In a crucial experiment in the late 1970s, DNA from mouse cells which had been transformed by a chemical **carcinogen**, was transfected into normal mouse cells. The normal mouse cells became malignant suggesting that a **gene** within the cancer (a proto-oncogene) had been mutated by exposure to the chemical and was able to induce cancer. Surprisingly, when these endogenous cellular oncogenes were eventually isolated they were found to be homologous to virally derived oncogenes.

In the early 1970s the American pediatrician and scientist, Alfred Knudson at the Fox Chase Cancer Center studied retinoblastoma, a rare childhood **eye** cancer that is sometimes inherited but is most often sporadic. He observed that children who had inherited retinoblastoma often had the cancer at birth, and were at high risk of developing multiple cancers in both eyes. Children with later onset retinoblastoma usually had no family history and developed isolated tumors. Knudson reasoned that children with inherited retinoblastoma had a germline **mutation** in one allele of a recessive cancer gene. The germline mutation was the first of two hits in knocking out a recessive cancer gene. This is known as Knudson's two hit hypothesis. Later genetic studies found the first hit in children with inherited retinoblastoma to be a partial deletion of the long arm of **chromosome** 13 causing loss of the **tumor** suppressor gene, RB1.

These two directions of study independently identified two different classes of cancer gene, the oncogene and tumor suppressor gene, that when mutated in a given **cell** can set in train the sequence of events leading to the development of a cancer.

The genetics of cancer

Cancer, by definition, is a disease of the genes. Cancer is also our most common genetic disease, but only rarely is it inherited. A gene is a small part of DNA, which is the master **molecule** of the cell. Genes make "proteins," which are the ultimate workhorses of the cells. It is these **proteins** that allow our bodies to carry out all the many processes that permit us to breathe, think, move, etc.

Throughout people's lives, the cells in their bodies are growing, dividing, and replacing themselves. Many genes produce proteins that are involved in controlling the processes of cell growth and division. An alteration (mutation) to the DNA molecule can disrupt the genes and produce faulty proteins. This causes the cell to become abnormal and lose its restraints on growth. The abnormal cell begins to divide uncontrollably and eventually forms a new growth known as a "tumor" or neoplasm (medical term for cancer meaning "new growth").

In a healthy individual, the **immune system** can recognize the neoplastic cells and destroy them before they get a chance to divide. However, some mutant cells may escape immune detection and survive to become tumors or cancers.

Tumors are of two types, benign or malignant. A benign tumor is slow growing, does not spread or invade surrounding tissue, and once it is removed, it doesn't usually recur. A malignant tumor, on the other hand, invades surrounding tissue and spreads to other parts of the body. The hallmark of a malignant cancer is the uncontrolled clonal proliferation and spread of abnormal cancer cells. If the cancer cells have spread to the surrounding tissues, then, even after the malignant tumor is removed, it generally recurs.

A majority of cancers are caused by changes in the cell's DNA because of damage due to the environment. Environmental factors that are responsible for causing the initial mutation in the DNA are called carcinogens, and there are many types.

There are some cancers that have a genetic basis. In other words, an individual could inherit faulty DNA from his parents, which could predispose him to getting cancer. While there is scientific evidence that both factors (environmental and genetic) play a role, less than 10% of all cancers are purely hereditary. Cancers that are known to have a hereditary link are breast cancer, colon cancer, ovarian cancer, and uterine cancer. Besides genes, certain physiological traits could be inherited and could contribute to cancers. For example, inheriting fair skin makes a person more likely to develop skin cancer, but only if they also have prolonged exposure to intensive sunlight.

Most cancers are sporadic and arise in a particular tissue such as the colon, breast, lung, or skin when normal cells acquire mutations in one or more oncogenes or tumor suppressor genes. The acquisition of multiple new genetic changes is what sets the cancer cell apart from the normal cells in its surrounding tissues.

The cancer cell develops when a normal cell in an organ or tissue acquires the capacity to divide in an uncontrolled fashion. Over time the developing cancer cell starts to multiply in a clonal fashion, begins to appear different (anaplastic or undifferentiated), and progressively acquires other characteristics, such as the capacity metastasise while losing cell-to-cell adhesion. The continued acquisition of new biologic characteristics is the key to many aggressive cancers evading the host defenses, and to the resisting some treatments such chemotherapy and radiotherapy.

It is important to appreciate that oncogenes and tumor suppressor genes are in fact normal cellular genes with vital functions within normal cells. It is only when they are mutated in some way that these genes become cancer causing.

The Ha-ras gene is a good example of an oncogene. Located on chromosome 11 at the normal cellular Ha-ras gene is one of a family of ras genes and encodes a small protein that is involved in intracellular signaling. Mutations in the ras oncogenes disrupt processing of cell signals and contribute to cell transformation. Mutations in ras oncogenes are found in approximately 10% of cancers especially cancer of the colon and lung.

The most important tumor suppressor gene is the p53 gene. This gene which is known as the guardian of the **genome** encodes for a protein with multiple intracellular functions related to the detection of DNA damage. When DNA is damaged by exposure to a **mutagen** such as UV irradiation the p53 gene is expressed. The p53 protein causes the cell to stop dividing so DNA mismatch repair genes can repair the DNA. If the DNA is successfully repaired, the cell resumes normal cell functions and the p53 gene is down regulated. However, if the DNA damage is beyond repair the p53 protein switches on a process called apoptosis (programmed **cell death**) leading to the death of the cell. For example, sunburn to the skin causes UV induced DNA damage, which often cannot be repaired. Expression of the p53 gene induces apoptosis the skin cells die and peel off.

Mutations in the p53 gene occur in approximately 50% of all cancers—particularly cancer of the breast, colon, lung, and **brain**. The mutant p53 protein is unable to stop uncontrolled **cell division** or switch on apoptosis, and can no longer protect the cell from acquiring additional mutation in other genes. The result is an unstable

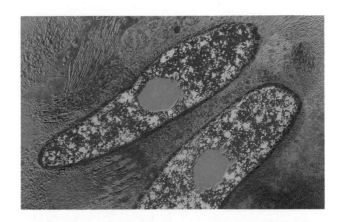

A transmission electron microscopy (TEM) of two spindle cell nuclei from a human sarcoma. Sarcomas are cancers of the connective tissue (bone, nerves, smooth muscle). *Photograph by Dr. Brian Eyden. National Audubon Society Collection/Photo Researchers, Inc. Reproduced by permission.*

cell genome liable to further progressive DNA damage. The inherited cancer condition, Li-Fraumeni **syndrome**, is an autosomal dominant disorder caused by inherited mutations in the p53 gene. Individuals affected with Li-Fraumeni syndrome may develop breast cancer, brain tumors, leukemia, prostate cancer and various sarcomas at a young age.

Mismatch repair genes are another class of cancer gene contributing to instability of the cancer cell genome. Damaged DNA is repaired by an active DNA mismatch repair mechanism that identifies damaged DNA, then cuts out and repairs the the damaged DNA bases. Mutations in these repair genes are common in cancer cancers of the colon.

Oncogenes, tumor suppressor genes and other cancer causing genes can become mutated in any number of different ways. Most oncogenes become activated by specific mutations within their DNA sequence that causes the gene protein to function abnormally. Some oncogenes such MYCN are activated by DNA amplification. Oncogene amplification occurs commonly in neuroblastoma an aggressive cancer in children. These tumors can acquire hundreds of copies of this gene by DNA amplification making the cancer very resistant to treatment. Another means of oncogene activation is by its translocation from one chromosome to another. In the Burkitt lymphoma the c-myc oncogene is translocated from chromosome 8 to chromosome 14 where it becomes activated by an immunoglobulin gene. Only one allele of an oncogenes need to be activated for it to participate in cell transformation.

Tumor suppressor genes on the other hand are recessive and normally act to suppress cell replication. Cell transformation occurs when both gene **alleles** are inactivated (knocked out). Most commonly, inactivation of one gene allele occurs by a chromosome deletion. The second event may be an inactivating **gene mutation**, a second deletion or methylation of the genes promoter.

Regardless of the actual mutations involved a crucial concept in the development of most cancer is that more than one gene is usually involved in the process. Indeed in the development of cancer of the colon at least six or more separate oncogenes and tumor suppressor genes are involved in a progressive multi-step process to transform a normal colon cell into an aggressive, self replicating and invading cancer.

More recently, the application of gene expression arrays (microarrays) to the study of cancer has found that in addition to multiple gene mutations, the expression of many hundreds of non-mutant genes is affected in the process of cell transformation.

Microarray analysis of cancers of the breast and soft tissues has also identified distinctive patterns of gene expression which can be used to aid **diagnosis** and predict the clinical behavior of individual tumors.

This type of genetic analysis will also aid the development of new cancer therapies directed specifically at the **molecular biology** of the cancer.

Types of cancers

There are several different types of cancers:

- Carcinomas are cancers that arise in the epithelium (the layers of cells covering the body's surface and lining the internal organs and various **glands**). Ninety **percent** of human cancers fall into this category. Carcinomas can be subdivided into two subtypes: adenocarcinomas and squamous cell carcinomas. Adenocarcinomas are cancers that develop in an organ or a gland, while squamous cell carcinomas refer to cancers that originate in the skin.

- Melanomas also originate in the skin, usually in the pigment cells (melanocytes).

- Sarcomas are cancers of the supporting tissues of the body, such as bone, muscle and blood vessels.

- Cancers of the blood and lymph glands are called leukemia and lymphomas respectively.

- Gliomas are cancers of the nerve tissue.

Causes and symptoms

The major risk factors for cancer are: tobacco, **alcohol**, diet, sexual and reproductive behavior, infectious agents, family history, occupation, environment and **pollution**.

According to the estimates of the American Cancer Society (ACS), approximately 40% of the cancer deaths in 1998 were due to tobacco and excessive alcohol use. An additional one-third of the deaths were related to diet and **nutrition**. Many of the one million skin cancers that are expected to be diagnosed each year are due to overexposure to ultraviolet **light** from the sun's rays.

Tobacco

Eighty to ninety percent of the lung cancer cases occur in smokers. Smoking has also been shown to be a contributory factor in cancers of upper respiratory tract, esophagus, larynx, bladder, pancreas, and probably liver, stomach, and kidney as well. Recently, scientists have also shown that second-hand smoke (or passive smoking) can increase one's risk of developing cancer.

Alcohol

Excessive consumption of alcohol is a risk factor in certain cancers, such as liver cancer. Alcohol, in combination with tobacco, significantly increases the chances that an individual will develop mouth, pharynx, larynx, and esophageal cancers.

Diet

Thirty five percent of all cancers are due to dietary causes. Excessive intake of **fat** leading to **obesity** has been associated with cancers of the breast, colon, rectum, pancreas, prostate, gall bladder, ovaries, and uterus.

Sexual and reproductive behavior

The human papilloma virus, which is sexually transmitted has been shown to cause cancer of the cervix. Having too many sex partners and becoming sexually active early has been shown to increase one's chances of contracting this disease. In addition, it has also been shown that women who do not have children or have children late in life, have an increased risk for both ovarian and breast cancer.

Infectious agents

In the later decades of the twentieth century, scientists have obtained evidence to show that approximately 15% of the world's cancer deaths can be traced to viruses, **bacteria**, or **parasites**. The most common cancer-causing **pathogens** and the cancers associated with them are shown in table form.

Family history

Certain cancers like breast, colon, ovarian and uterine cancer, recur generation after generation in some families. A few cancers, such as the eye cancer "retinoblastoma," a type of colon cancer, and a type of breast cancer known as "early-onset breast cancer," have been shown to be linked to certain genes that can be tracked within a family. It is therefore possible that inheriting particular genes makes a person susceptible to certain cancers.

Occupational hazards

There is evidence to prove that certain occupational hazards account for 4% of all cancer deaths. For example, **asbestos** workers have an increased incidence of lung cancer. Similarly, a higher likelihood of getting bladder cancer is associated with dye, rubber and gas workers; skin and lung cancer with smelters, gold miners, and arsenic workers; leukemia with glue and varnish workers; liver cancer with PVC manufacturers; and lung, bone, and bone marrow cancer with radiologists and **uranium** miners.

Environment

Radiation is believed to cause 1-2% of all cancer deaths. Ultra-violet radiation from the **sun** accounts for a majority of melanoma deaths. Other sources of radiation are x rays, **radon** gas, and **ionizing radiation** from nuclear material.

Pollution

Several studies have shown that there is a well-established link between asbestos and cancer. **Chlorination** of **water** may account for a small rise in cancer risk. However, the main danger from pollution occurs when dangerous chemicals from the industries escape into the surrounding environment. It has been estimated that 1% of cancer deaths are due to air, land and **water pollution**.

Cancer is a progressive disease, and goes through several stages. Each stage may produce a number of symptoms. Some symptoms are produced early and may occur due to a tumor that is growing within an organ or a gland. As the tumor grows, it may press on the nearby nerves, organs and blood vessels. This causes **pain** and some pressure which may be the earliest warning signs of cancer.

Despite the fact that there are several hundred different types of cancers, producing very different symptoms, the ACS has established the following seven symptoms as possible warning signals of cancer:

• changes in the size, **color**, or shape of a wart or a mole
• a sore that does not heal
• persistent cough, hoarseness, or sore throat
• a lump or thickening in the breast or elsewhere
• unusual bleeding or discharge
• chronic indigestion or difficulty in swallowing
• any change in bowel or bladder habits.

Many other diseases, besides cancer, could produce the same symptoms. However, it is important to have these symptoms checked, as soon as possible, especially if they linger. The earlier a cancer is diagnosed and treated, the better the chance of it being cured. Many cancers such as breast cancer may not have any early symptoms. Therefore, it is important to undergo routine screening tests such as breast self-exams and mammograms.

Diagnosis

Diagnosis begins with a thorough physical examination and a complete medical history. The doctor will observe, feel, and palpate (apply pressure by **touch**) different parts of the body in order to identify any variations from the normal size, feel and texture of the organ or tissue.

As part of the physical exam, the doctor will inspect the oral cavity or the mouth. By focusing a light into the mouth, he will look for abnormalities in color, moisture, surface texture, or presence of any thickening or sore in the lips, tongue, gums, the hard palate on the roof of the mouth, and the throat. To detect thyroid cancer, the doctor will observe the front of the neck for swelling. He may gently manipulate the neck and palpate the front and side surfaces of the thyroid gland (located at the base of the neck) to detect any nodules or tenderness. As part of the physical examination, the doctor will also palpate the lymph nodes in the neck, under the arms and in the groin. Many illnesses and cancers cause a swelling of the lymph nodes.

The doctor may conduct a thorough examination of the skin to look for sores that have been present for more than three weeks and that bleed, ooze, or crust; irritated patches that may itch or hurt, and any change in the size of a wart or a mole.

Examination of the female pelvis is used to detect cancers of the ovaries, uterus, cervix, and vagina. In the visual examination, the doctor looks for abnormal discharges or the presence of sores. Then, using gloved hands the physician palpates the internal pelvic organs such as the uterus and ovaries to detect any abnormal masses. Breast examination includes visual observation where the doctor looks for any discharge, unevenness, discoloration, or scaling. The doctor palpates both breasts to feel for masses or lumps.

For males, inspection of the rectum and the prostate is also included in the physical examination. The doctor inserts a gloved finger into the rectum and rotates it slowly to feel for any growths, tumors, or other abnormalities. The doctor also conducts an examination of the testis, where the doctor observes the genital area and looks for swelling or other abnormalities. The testicles are palpated to identify any lumps, thickening or differences in the size, weight, and firmness.

If the doctor detects an abnormality on physical examination, or the patient has some symptom that could be indicative of cancer, the doctor may order diagnostic tests.

Laboratory studies of sputum (sputum **cytology**), blood, urine, and stool can detect abnormalities that may indicate cancer. Sputum cytology is a test where the phlegm that is coughed up from the lungs is microscopically examined. It is often used to detect lung cancer. A blood test for cancer is easy to perform, usually inexpensive and risk-free. The blood **sample** is obtained by a lab technician or a doctor, by inserting a needle into a vein and is relatively painless. Blood tests can be either specific or non-specific. Often times, in certain cancers, the cancer cells release particular proteins (called tumor markers) and blood tests can be used to detect the presence of these tumor markers. However, with a few exceptions, tumor markers are not used for routine screening of cancers, because several non-cancerous conditions also produce positive results. Blood tests are generally more useful in monitoring the effectiveness of the treatment, or in following the course of the disease and detecting recurrent disease.

Imaging tests such as computed tomography scans (CT scans), **magnetic resonance imaging (MRI)**, ultrasound, and fiberoptic scope examinations help the doctors determine the location of the tumor even if it is deep within the body. Conventional x rays are often used for initial evaluation, because they are relatively cheap, painless and easily accessible. In order to increase the information obtained from a conventional x ray, air or a dye (such as **barium** or iodine) may be used as a contrast medium to outline or highlight parts of the body.

The most definitive diagnostic test is the biopsy, wherein a piece of tissue is surgically removed for **microscope** examination. Besides, confirming a cancer, the biopsy also provides information about the type of cancer, the stage it has reached, the aggressiveness of the cancer and the extent of its spread. Since a biopsy provides the most accurate analysis, it is considered the gold standard of diagnostic tests.

Screening examinations, conducted regularly by healthcare professionals can result in the detection of cancers of the breast, colon, rectum, cervix, prostate, testis, tongue, mouth, and skin at early stages, when treatment is more likely to be successful. Some of the routine screening tests recommended by the ACS are sigmoidoscopy (for colorectal cancer), mammography (for breast cancer), pap smear (for cervical cancer), and the PSA test (for prostate cancer). Self-examinations for cancers of the breast, testes, mouth, and skin can also help in detecting the tumors before the symptoms become serious.

A recent revolution in molecular **biology** and cancer genetics has contributed a great deal to the development of several tests designed to assess one's risk of getting cancers. These new techniques include **genetic testing**, where molecular probes are used to identify mutations in certain genes that have been linked to particular cancers.

Treatment

The aim of cancer treatment is to remove all or as much of the tumor as possible and to prevent the recurrence or spread of the primary tumor. While devising a treatment plan for cancer, the likelihood of curing the cancer has to be weighed against the side effects of the treatment. If the cancer is very aggressive and a cure is not possible, then the treatment should be aimed at relieving the symptoms and controlling the cancer for as long as possible.

Cancer treatment can take many different forms, and it is always tailored to the individual patient. The decision on which type of treatment is the most appropriate depends on the type and location of cancer, the extent to which it has already spread, the patient's age, sex, general health status and personal treatment preferences. The major types of treatment are: **surgery**, radiation, chemotherapy, immunotherapy, hormone therapy, and bone-marrow transplantation.

Surgery

Surgery is the removal of a visible tumor and is the most frequently used cancer treatment. It is most effective when a cancer is small and confined to one area of the body.

Surgery can be used for many purposes.

• Treatment: Treatment of cancer by surgery involves removal of the tumor to cure the disease. This is typically done when the cancer is localized to a discrete area. Along with the cancer, some part of the normal surrounding tissue is also removed to ensure that no cancer cells remain in the area. Since cancer usually spreads via the **lymphatic system**, adjoining lymph nodes may be examined and sometimes they are removed as well.

• Preventive surgery: Preventive or prophylactic surgery involves removal of an abnormal looking area that is likely to become malignant over time. For example, about 40% of the people with a colon disease known as ulcerative colitis, ultimately die of colon cancer. Rather than live with the fear of developing colon cancer, these people may choose to have their colons removed and reduce the risk significantly.

- Diagnostic purposes: The most definitive tool for diagnosing cancer is a biopsy. Sometimes a biopsy can be performed by inserting a needle through the skin. However, at other times, the only way to obtain some tissue sample for biopsy, is by performing a surgical operation.

- Cytoreductive surgery: is a procedure where the doctor removes as much of the cancer as possible, and then treats the remaining with radiation therapy or chemotherapy or both.

- Palliative surgery: is aimed at curing the symptoms, not the cancer. Usually, in such cases, the tumor is so large or has spread so much, that removing the entire tumor is not an option. For example, a tumor in the abdomen may be so large, that it may press on and block a portion of the intestine, interfering with digestion and causing pain and vomiting. "Debulking surgery" may remove a part of the blockage and relieve the symptoms. In tumors that are dependent on **hormones**, removal of the organs that secrete the hormones is an option. For example, in prostate cancer, the release of testosterone by the testicles stimulates the growth of cancerous cells. Hence, a man may undergo an "orchiectomy" (removal of testicles) to slow the progress of the disease. Similarly, in a type of aggressive breast cancer, removal of the ovaries (oophorectomy) will stop the synthesis of hormones from the ovaries and slow the progression of the cancer.

Radiation

Radiation kills cells. Radiation is used alone in cases where a tumor is unsuitable for surgery. More often, it is used in conjunction with surgery and chemotherapy. Radiation can be either external or internal. In the external form, the radiation is aimed at the tumor from outside the body. In internal radiation (also known as brachytherapy), a radioactive substance, in the form of pellets or liquid is placed at the cancerous site by means of a pill, injection, or insertion in a sealed container.

Chemotherapy

Chemotherapy is the use of drugs to more specifically kill cancer cells. It destroys the hard-to-detect cancer cells that have spread and are circulating in the body. Chemotherapeutic drugs can be taken either orally (by mouth) or intravenously, and may be given alone or in conjunction with surgery, radiation, or both.

When chemotherapy is used before surgery or radiation, it is known as primary chemotherapy or "neoadjuvant chemotherapy." An advantage of neoadjuvant chemotherapy is that since the cancer cells have not been exposed to anti-cancer drugs, they are especially vulnerable. It can therefore be used effectively to reduce the size of the tumor for surgery or target it for radiation.

However, the toxic effects of neoadjuvant chemotherapy are severe. In addition, it may make the body less tolerant to the side effects of other treatments that follow such as radiation therapy. The more common use of chemotherapy is adjuvant therapy, which is given to enhance the effectiveness of other treatments For example, after surgery, adjuvant chemotherapy is given to destroy any cancerous cells that still remain in the body.

Immunotherapy

Immunotherapy uses the body's own immune system to destroy cancer cells. This form of treatment is being intensively studied in clinical trials and is not yet widely available to most cancer patients. The various immunological agents being tested include substances produced by the body (such as the **interferons**, interleukins, and growth factors), monoclonal antibodies, and vaccines. Unlike traditional vaccines, cancer vaccines do not prevent cancer. Instead, they are designed to treat people who already have the disease. Cancer vaccines work by boosting the body's immune system and training the immune cells to specifically destroy cancer cells.

Hormone therapy

Hormone therapy is standard treatment for some types of cancers that are hormone-dependent and grow faster in the presence of particular hormones. These include cancer of the prostate, breast, and uterus. Hormone therapy involves blocking the production or action of these hormones. As a result the growth of the tumor slows down and survival may be extended for several months or years.

Bone marrow transplantation

The bone marrow is the tissue within the bone cavities that contains blood-forming cells. Healthy bone marrow tissue constantly replenishes the blood supply and is essential to life. Sometimes, the amount of drugs or radiation needed to destroy cancer cells also destroys bone marrow. Replacing the bone marrow with healthy cells counteracts this adverse effect. A bone marrow transplant is the removal of marrow from one person and the transplant of the blood-forming cells either to the same person or to some one else. Bone-marrow transplantation, while not a therapy in itself, is often used to "rescue" a patient, by allowing those with cancer to undergo very aggressive therapy.

Many different specialists generally work together as a team to treat cancer patients. An oncologist is a physician who specializes in cancer care. The oncologist provides chemotherapy, hormone therapy, and any other non-surgical treatment that does not involve radiation.

The oncologist often serves as the primary physician and co-ordinates the patient's treatment plan.

The radiation oncologist specializes in using radiation to treat cancer, while the surgical oncologist performs the operations needed to diagnose or treat cancer. Gynecologist-oncologists and pediatric-oncologists, as their titles suggest, are physicians involved with treating women's and children's cancers respectively. Many other specialists may also be involved in the care of a cancer patient. For example, radiologists specialize in the use of x rays, ultrasounds, computed tomography scans (CT scans), MRI imaging, and other techniques that are used to diagnose cancer. Hematologists specialize in disorders of the blood and are consulted in case of blood cancers and bone marrow cancers. The samples that are removed for biopsy are sent to a laboratory, where a pathologist examines them to determine the type of cancer and extent of the disease. Only some of the specialists who are involved with cancer care have been mentioned above. There are many other specialties, and virtually any type of medical or surgical specialist may become involved with care of the cancer patient should it become necessary.

Alternative treatment

There are a multitude of alternative treatments available to help the person with cancer. They can be used in conjunction with, or separate from, surgery, chemotherapy, and radiation therapy. Alternative treatment of cancer is a complicated arena and a trained health practitioner should be consulted.

The effectiveness of complementary therapies such as **acupuncture** in alleviating cancer pain has not been clinically proven. Bodywork therapies such as massage and reflexology ease muscle tension and may alleviate the side effects such as nausea and vomiting. Homeopathy and herbal remedies used in Chinese traditional **herbal medicine** have also been reported to alleviate some of the side effects of radiation and chemotherapy and are being recommended by many doctors.

Certain foods including many **vegetables**, **fruits**, and grains are believed to offer protection against various cancers. However, isolation of the individual constituent of vegetables and fruits that are anti-cancer agents has proven difficult. In laboratory studies, vitamins such as A, C and E, as well as compounds such as isothiocyanates and dithiolthiones found in broccoli, cauliflower, and cabbage, and beta-carotene found in carrots have been shown to protect against cancer. Studies have shown that eating a diet rich in fiber as found in fruits and vegetables reduces the risk of colon cancer. **Exercise** and a low fat diet help control weight and reduce the risk of endometrial, breast, and colon cancer.

Certain drugs, which are currently being used for treatment, could also be suitable for prevention. For example, the drug tamoxifen (Nolvadex), that has been very effective against breast cancer, is currently being tested by the National Cancer Institute, for its ability to prevent cancer. Similarly, retinoids derived from **vitamin** A are being tested for their ability to slow the progression or prevent head and neck cancers. Certain studies have suggested that cancer incidence is lower in areas where **soil** and foods are rich in the mineral selenium. More trials are needed to explain these intriguing connections.

Prognosis

"Life-time risk" is the term that cancer researchers use to refer to the probability that an individual, over the course of a lifetime will develop cancer or die from it. In the United States, men have a one in two lifetime risk of developing cancer, and for women the risk is one in three. Overall, African-Americans are more likely to develop cancer than whites. African-Americans are also 30% more likely to die of cancer than whites.

Most cancers are curable if detected and treated at their early stages. A cancer patient's prognosis is affected by many factors, particularly the type of cancer the patient has, the stage of the cancer, the extent to which it has metastasized and the aggressiveness of the cancer. In addition, the patient's age, general health status, and the effectiveness of the treatment being pursued are also important factors.

To help predict the future course and outcome of the disease and the likelihood of recovery from the disease, doctors often use **statistics**. The five-year survival rates are the most common measures used. The number refers to the proportion of people with cancer who are expected to be alive, five years after initial diagnosis, compared with a similar population that is free of cancer. It is important to note that while statistics can give some information about the average survival experience of cancer patients in a given population, it cannot be used to indicate individual prognosis, because no two patients are exactly alike.

Prevention

According to nutritionists and epidemiologists from leading universities in the United States, a person can reduce the chances of getting cancer by following some simple guidelines:

• Eating plenty of vegetables and fruits
• Exercising regularly
• Avoiding excessive weight gain
• Avoiding tobacco (including second hand smoke)
• Avoiding excessive amounts of alcohol

KEY TERMS

Benign—A growth that does not spread to other parts of the body. Recovery is favorable with treatment.

Biopsy—The surgical removal of a small part of a tumor. The excised tissue is studied under the microscope to determine whether it is benign or malignant.

Bone marrow—A spongy tissue located in the hollow centers of certain bones, such as the skull and hip bones. Bone marrow is the site of blood cell generation.

Carcinogen—Any substance capable of causing cancer by mutating the cell's DNA.

Chemotherapy—Use of powerful drugs to kill cancer cells in the human body.

Epithelium—The layer of cells that covers external and internal surfaces of the body. The many types of epithelium range from flat cells to long cells to cubed cells.

Hormone therapy—Treatment of cancer by inhibiting the production of hormones, such as testosterone and estrogen.

Immunotherapy—Treatment of cancer by stimulating the body s immune defense system.

Malignant—A general term for cells that can dislodge from the original tumor, invade and destroy other tissues and organs.

Metastasis—The spread of cancer from one part of the body to another.

Radiation therapy—Treatment using high-energy radiation from x-ray machines, cobalt, radium, or other sources.

Sore—An open wound or a bruise or lesion on the skin.

Tumor—An uncontrolled growth of tissue, either benign (noncancerous) or malignant (cancerous).

X ray—Electromagnetic radiation of very short wavelength, and very high energy.

• Avoiding the midday sun (between 11 a.m. and 3 p.m.) when the suns rays are the strongest
• Avoiding risky sexual practices
• Avoiding known carcinogens in the environment or work place.

See also Gene therapy; Immunology; Nuclear medicine; Radioisotopes in medicine; Stem cells.

Resources

Books
Haskell, Charles M. *Cancer Treatment.* 5th. ed. Philadelphia: W.B. Saunders, 2001.
Rosenbaum, Ernst H. MD, et al. *Everyone's Guide to Cancer Therapy.* 4th. ed. Andrews McMeel Publishing, 2002.
Steingraber, Sandra. *Living Downstream: A Scientist's Personal Investigation of Cancer and the Environment.* Vintage Books, 1998.

Periodicals
Brookes, Anthony, "Rethinking Genetic Strategies to Study Complex Diseases," *Trends in Molecular Medicine* (November 2001): 512–6.

Other
National Institutes of Health. "National Cancer Institute." (February 5, 2003).<http://www.nci.nih.gov/>.

Lata Cherath
Micheal Sullivan

Canines

Canines are **species** in the **carnivore** family, Canidae, including the wolves, coyote, foxes, dingo, jackals, and several species of wild dog. The family also includes the domestic dog, which is believed to have descended from the wolf. The Canidae includes 10-14 genera with 30-35 species, depending on the taxonomic treatment.

Canines originated in **North America** during the Eocene era (38-54 million years ago), from there they spread throughout the world. The social **behavior** of canines varies from solitary habits to highly organized, cooperative packs. Canines range in size from the fennec fox, about 16.5 in (41 cm) long including the tail and weighing about 1 lb (0.5 kg), to the gray or timber wolf, which is more than 6 ft (2 m) in length and weighs up to 175 lb (87.5 kg).

Canid skulls have a long muzzle, well-developed jaws, and a dental formula of 42 teeth. Most canines species live in packs, which offer several benefits including group defense of territory, communal care of the young, and the ability to catch large **prey** species.

Wolves

Wolves are found in North America, **Europe**, and **Asia**. The gray wolf (*Canis lupus*) is the largest member of

A blonde dingo. *Photograph by Jim Steinberg. The National Audubon Society Collection/Photo Researchers, Inc. Reproduced by permission.*

the dog family, and is a widely distributed species. It lives in a variety of habitats, including forest, **prairie**, **mountains**, **tundra**, and **desert**. The red wolf (*Canis rufus*) is found only in southeastern Texas and southern Louisiana. The red wolf is smaller than the gray wolf, and it may be a **hybrid** between the gray wolf and coyote (*Canis latrans*).

The gray wolf lives in packs and is a territorial species. Territories are scent marked, and range from 50 to 5,000 sq mi (128-12,800 sq km) in area. Pack size is usually about eight members, consisting of a mature male and female, their offspring, and close relatives. A system of dominance hierarchy is established within the pack. The dominant male leader of the pack is called the alpha male, and the dominant female is the alpha female. Hierarchy is acknowledged among pack members through submissive facial expressions and body postures.

Only the dominant male and female breed. Gestation is about two months and the average litter size is four to seven pups, which are born blind. The young are weaned within five weeks and reach physical maturity within the year, but do not become sexually mature until the end of their second year. The non-breeding members of a pack will help to protect and feed the young. The prey species of the wolf include **deer**, **moose**, elk, **caribou**, and beaver.

Besides scent marking, wolves communicate by howling. It is believed that howling lets dispersed pack members know each other's position, and warns other packs off the territory. During the spring and summer, the wolf pack has a stationary phase and remains within its territory. It is during this period that the pups are raised. During the nomadic phase in autumn and winter, wolf packs travel widely, often following the **migration** of prey species.

Foxes

There are 21 species of foxes in four genera. Foxes range in size from the 3 lb (15 kg) fennec fox (*Fennecus*

The endangered red wolf (*Canis rufus*). *Photograph by Tim Davis. The National Audubon Society Collection/Photo Researchers, Inc. Reproduced by permission.*

cerda) to the 20 lb (10 kg) red fox (*Vulpes vulpes*). The gray fox (*Urocyon cinereargenteus*) and arctic fox (*Alopex lagopus*) are highly valued for their pelts. **Color** phases of the arctic fox include the silver fox and blue fox. Species found in the United States are the kit fox (*Vulpes macrotis*) and the swift fox (*V. velox*), which live on the western plains. Other species of Central or **South America** include the crab-eating fox (*Dusicyon azarae*), the **sand** fox (*Vulpes ruppelli*), and the Corsac fox (*Vulpes corsac*).

Foxes have a pointed muzzle, large ears, a slender skull, and a long bushy tail. Foxes are territorial and scent-mark their territories. They use stealth and dash-and-grab hunting techniques to catch their prey. Foxes are generally solitary hunters and most species feed on rabbits, **rodents**, and **birds**, as well as **beetles**, **grasshoppers**, and earthworms. Foxes mate in winter, having a litter of one to six pups after a gestation period of 50-60 days. Besides scent marking, foxes proclaim their territory by vocalizations such as yapping, howling, barking, whimpering, and screaming.

Foxes are heavily hunted for their pelts. They also may be killed to prevent the spread of the viral **disease**

rabies. Some efforts at oral vaccination for rabies have been successful in Switzerland and Canada.

Coyotes, jackals, the dingo, and species of wild dog comprise the rest of the canine family. The distribution of coyotes is from Alaska to Central America. Coyote populations have flourished as wolves have been eliminated. Coyotes have interbred with wolves and with domestic dogs. Coyotes prey on small animals, but will also feed on carrion, **insects**, and fruit. Coyotes reach maturity within a year and produce a litter of about six pups. While the basic social unit of coyotes is the breeding pair, some coyotes form packs similar to wolves and scent-mark territory. In the United States, coyotes have been responsible for considerable losses of **sheep**.

There are four species of jackals, which replace wolves and coyotes in warmer parts of the world. Jackals are found throughout **Africa**, southeastern Europe, and southern Asia as far east as Burma. The four species are the golden jackal (*Canis aureus*), the simien jackal (*C. simensis*), the black-backed jackal (*C. mesomelas*), and the side-striped jackal (*C. adustus*). The golden jackal prefers arid **grasslands**, and is the most widely distributed of the four species. The silver-backed jackal prefers brushy woodlands, the simien jackal the high mountains of Ethiopia, and the side-striped jackal moist woodland. Jackals have a varied diet of fruit, **reptiles**, birds, and small **mammals**.

Jackals are unusually stable in their breeding relationships, forming long-lasting partnerships. They also engage in cooperative hunting. Jackals are territorial and engage in scent marking, usually as a male and female pair that tends to remain monogamous. Jackals communicate by howling, barking, and yelping. In Ethiopia, the simien jackal is an **endangered species** because it has been overhunted for its fur.

Other wild canines include the Indian wild dog (*Cuon alpinus*) of southeast Asia and China, the maned wolf (*Hrysocyon subatis*) of Central Asia, the bush dog (*Speothes venaticus*) of Central America and northern South America, the dingo (*Canis dingo*) of **Australia**, and the raccoon dog (*Nyetereutes procyonoides*) of eastern Asia. In Africa, the cape hunting dog (*Lycaon pictus*) hunts in packs which can overpower large mammal species.

The domestic dog

Kennel societies in the United States recognize 130 breeds of domestic dog, while those in Britain recognize 170 breeds, and the Federation Cynologique Internationale (representing 65 countries) recognizes 335 breeds. The size range of domestic dogs is from about 4 lb (2 kg) to 200 lb (100 kg). Some breeds, such as the dachshund, have short legs, while others, such as the greyhound, have long legs.

For dog owners, these animals can serve a number of different purposes. Pet dogs provide companionship and protection, while working dogs may herd sheep or cattle, or work as sled dogs. Police use dogs to sniff out illegal drugs and to help apprehend criminals. Dogs are also used for hunting and for racing. Guide dogs help blind people find their way around.

Female dogs may reproduce at an age of 7-18 months. Gestation lasts about two months, and the size of a litter is from three to six puppies. Born unable to see, like other canines, domestic dog puppies develop all their senses by 21 days. Around the age of two months, puppies are less dependent on their mother and begin to relate more to other dogs and people. Typical vocalizations of domestic dogs include barking and yelping.

About 10,000 years ago human civilization changed from a gatherer-and-hunter society to a farming culture, and the domestication of the dog began. However, dogs probably also associated with humans before this time. It is believed that all breeds of domestic dogs, whether small or large, long-haired or short-haired, are descended from a wolf-like **animal**. Breeds of domestic dogs have

---

KEY TERMS

Alpha male or female—The dominant male or female in a pack of wolves.

Dominance hierarchy—Rank-ordering among animals, with dominant and submissive ranks.

Opportunistic predator—An animal that eats what is available, either killing its own prey, stealing food from other predators, or eating plant material, such as berries.

Stationary or nomadic phase—Seasonal periods in which animals may remain within a specific area, usually during the breeding season, as compared to nomadic periods when the group moves extensively to follow prey.

---

been produced through selective breeding. One distinguishing feature between domestic dogs and wolves is the orbital angle of the skull. Dogs have a larger angle, which is measured from lines at the top of the skull and at the side of the skull at the **eye** socket.

The dog, sometimes known as "man's best friend," is treasured by humans. Dog stories abound in children's literature, from Lassie to Rin Tin Tin, and many politicians, including Presidents Franklin D. Roosevelt and Richard Nixon, used dogs to enhance their personal image.

Excessive hunting for their fur or as **pests**, often coupled with the destruction of **habitat**, have endangered some species of canines. Their reputation as predators has added to efforts to eradicate them from areas where **livestock** is raised or where they live close to human settlements. Some of the rare wild dogs, for instance the jackal of Ethiopia, are very few in number. The maned wolf (*Chrysocyon jubatus*) of Argentina and Brazil has a population of only 1,000-2,000. A successful effort has been made to reintroduce the gray wolf into Yellowstone National Park in the United States, but this is opposed by many local ranchers.

Resources

Books

Carey, Alan. *Twilight Hunters: Wolves, Coyotes, and Foxes.* Flagstaff, AZ: Northland Press, 1987.

Olsen, Stanley John. *Origins of the Domestic Dog: The Fossil Record.* Tucson, AZ: University of Arizona Press, 1985.

Sheldon, Jennifer W. *Wild Dogs: The Natural History of the Non-Domestic Canidae.* San Diego: Academic Press, 1992.

Wolves. San Francisco: Sierra Club Books, 1990.

Vita Richman

Cantaloupe *see* **Gourd family (Cucurbitaceae)**

Cantilever

A cantilever, also called a fixed end beam, is a beam supported only at one end. The beam cannot rotate in any direction; thus it creates a solid support. The cantilever is considered the third of the three great structural methods, the other two being post-and-beam construction and arch construction. The cantilever thrusts down which is different from the thrust of an arch which is outward against its supports.

Cantilevers did not become popular in architecture until the invention of **steel** and its widespread adoption in construction, because the combined strength of steel and cement is needed to create an effective cantilever system. A building using cantilevers has an internal skeleton, from which the walls hang very much like curtains. Unlike more traditional building methods where the walls are used as support for the ceiling and walls, here they are dividers of **space**. This allows the interior of the building to be designed for purpose and creative architecture, rather than on where columns and other structural supports must be. The most famous architect to use the cantilever system was Frank Lloyd Wright. He first used it in the 1906 construction of the Robie House in Chicago. With the use of steel and **concrete**, Wright was able to extend the roof 20 ft (6 m) beyond its support. With the cantilever and Wright's belief in the use of the nature in which the building resided, an entire new school of architecture was created called the **Prairie** School.

Before cantilevers were used in buildings, they were used to create **bridges**. The first cantilever bridge was built in the late 1800s by Heinrich Gerber in Germany. He based his ideas on ancient Chinese bridges which, much earlier, used the concept of the cantilever. By using the cantilever, bridges would no longer need supports in their middles and, thus, could span deep ravines or **rivers**. In addition, bridges could be built across extremely wide bodies of **water**, or valleys, because fewer supports are needed, and the supports which are used can be further apart. Thus, the incorporation of steel, cement, and cantilevers changed the world of architecture and civil **engineering**.

Capacitance

Capacitance is an electrical effect that opposes change in voltage between conducting surfaces separated by an insulator. Capacitance stores electrical **energy** when electrons are attracted to nearby but separate surfaces. The voltage across an unchanging capacitance value will stay constant unless the quantity of charge stored is changed.

The Farad, the unit of capacitance

The unit of capacitance is the Farad, in honor of Michael Faraday's work with electrostatics. When a 1-Farad capacitance store 1 **Coulomb** the result will be 1 volt. The Coulomb is the basic unit of electrical charge, equal to 6.2422×10^{18} charges the size carried by an **electron** or by a **proton**.

An electrical component that introduces capacitance is called a **capacitor**. Practical capacitors may have as small a value as a few trillionths of a Farad or as large as several Farads.

Energy storage in capacitors

Work is performed to accumulate charge in a capacitor. Each additional electron stored must overcome the repelling **force** caused by the charge previously stored. Energy storage increases as the square of the voltage across a capacitor. This often considerable energy can be used later.

Capacitors used as energy reservoirs can deliver powerful pulses of energy. A capacitor can discharge quickly then slowly recharge until the next power demand. The power source needs only to be large enough to supply the average energy. Inexpensive audio amplifiers often use large capacitors to provide high power peaks required by occasional loud sounds. Quiet intervals allow the capacitor to recharge before the next power burst.

Capacitance and alternating current

A capacitor effectively conducts alternating current even though electrons do not cross from one plate to other plate. Alternating current that appears to pass through a capacitor is actually, the charge and discharge current resulting from the constantly-changing voltage across the capacitor.

An uncharged capacitor always appears as a short circuit because its voltage must equal **zero** when its stored charge is zero. A capacitor carrying an alternating current continually charges and discharges, spending much of the time in a near-zero charge state. The resulting low voltage across its terminals means that it is often less significant in limiting circuit than other components in the circuit.

A capacitor's opposition to alternating current is called reactance. Higher capacitance introduces less reactance and higher frequencies result in lower reactance.

Capacitance and direct current

In a direct-current circuit a series capacitor will permit only a single pulse of charging current when the circuit voltage is changed. The charging current in quickly falls to almost zero as a capacitor charges from a constant-voltage source. Capacitors are sometimes used in circuits to oppose direct current. They may block direct current while simultaneously passing a superimposed alternating currents. A blocking capacitor is commonly used to separate alternating and direct current components.

Dielectrics

Dielectrics are the insulating materials used between the conducting plates of capacitors. Dielectrics increase capacitance or provide better insulation between the plates. Dielectrics materials exhibit very little ability to conduct **electric charge**. Mylar, **paper**, mica, and **ceramics** are commonly-used dielectrics. When extremely-high capacitance is required, a thin film of **aluminum** oxide on etched aluminum plates is used as a dielectric.

Dielectrics have a property called polarizability. A dielectric placed within an electric field appears to have electric charge on its surfaces even though the insulator remains electrically neutral. Each of the dielectric's molecules is stretched when the electric field causes its **negative** charges to be pulled toward the positive-charged capacitor plate and the molecule's positive charges are pulled toward the negative plate. This polarization strain causes each dielectric **molecule** to act as a voltage source. These voltages add in series aiding as do the voltage from several cells making up the **battery** in a flashlight. A phantom charge appears on each surface of the dielectric canceling much of the electric field produced by the real charges. The greater the polarization developed by a dielectric the larger the quantity of real charge the capacitor must store to develop a given voltage. The capacitance appears to increase as a result of dielectric polarization.

The capacitance multiplier for any dielectric is called its dielectric constant. The dielectric constant of a perfect **vacuum** is defined as exactly 1. Common dielectrics have dielectrics constants in the range of 2-4. Using a higher quality dielectric increases the capacitance by a factor equal to the dielectric constant.

Dielectric strength

Dielectric strength is the measure of a dielectric's ability to resist electric stress without losing its insulating capabilities. A high dielectric constant does not always correspond to high dielectric strength. Distilled **water** has a fairly high dielectric constant but it has poor

---

KEY TERMS

Alternating current—Electric current that flows first in one direction, then in the other; abbreviated AC.

Direct current (DC)—Electrical current that always flows in the same direction.

Electric field—The concept used to describe how one electric charge exerts force on another, distant electric charge.

Electron—A negatively charged particle, ordinarily occurring as part of an atom. The atom's electrons form a sort of cloud about the nucleus.

Farad—The unit of capacitance, equal to 1 Volt per Coulomb.

Neutral—No net charge, when positive and negative charges cancel.

Open circuit—A physical break in a circuit path that stops the current.

Polarizability—Possible asymmetrical charge distribution in a molecule.

Power supply—A source of electrical energy used to supply a circuit.

Proton—The positively-charged particle in atoms.

Short circuit—Unwanted bypass of the expected current path in a circuit.

Voltage—Ratio of electrical potential energy to the quantity of charge.

---

dielectric strength. Water, therefore, is not a useful dielectric for capacitors because it breaks down too easily. Some ceramics have dielectric constants as high as 10,000. These materials would be extremely valuable if they had better dielectric strength.

Working voltage

If the voltage across a capacitor is increased until charges jump from one plate to the other, the capacitor will probably fail, either momentarily or permanently. Capacitors are rated to specify the maximum continuous voltage that can be applied across the dielectric before the capacitor will fail.

Capacitors as a cause of electronics equipment failures

Failed capacitors are a common cause of electronic-equipment breakdowns. When a capacitor's dielectric is

---

destroyed the resulting short circuit may cause other components to fail. Capacitors also develop open circuits, causing the loss of the capacitance.

Electrolytic capacitors are generally less reliable than other types, a tradeoff made to secure very-high capacitance in a small package. They tend to fail if stored without a voltage across their terminals. The electrolytic paste may dry in time, causing a loss of capacitance. Experienced electronic technicians consider electrolytic capacitor failures as a likely cause of an equipment fault that is not otherwise immediately obvious.

The significance of capacitance

Capacitance, inductance, and resistance are the passive electrical properties affecting electrical circuits. Understanding capacitance is an essential part of the study of **electricity** and **electronics**.

Resources

Books

Asimov, Isaac. *Understanding Physics: Light, Magnetism, and Electricity.* Vol. II. Signet Books, The New American Library.

Bord, Donald J., and Vern J. Ostdiek. *Inquiry Into Physics.* 3rd ed. West Publishing Company, 1995.

Sear, Zemansky, and Young. *College Physics.* 6th ed. Addison-Wesley Publishing Company, 1985.

Donald Beaty

Capacitor

A capacitor stores electrical **energy**. It is charged by hooking into an electrical circuit. When the capacitor is fully charged a switch is opened and the electrical energy is stored until it is needed. When the energy is needed, the switch is closed and a burst of electrical energy is released.

A capacitor consists of two electrical conductors that are not in contact. The conductors are usually separated by a layer of insulating material, dielectric. The dielectric is not essential but it keeps the conductors from touching. When the capacitor is hooked into an **electric circuit** with a current, one conductor becomes positively charged and the other **negative**. The conductors are not in contact, so the current cannot flow across the capacitor. The capacitor is now charged up and the switch can be opened. The capacitor is storing electrical energy. When the energy is needed the capacitor is connected to the circuit needing the energy. The current flows rapidly in the opposite direction, discharging the capacitor in a burst of electrical energy.

Capacitors take many shapes, but the simplest is a **parallel** plate capacitor. It consists of two flat conductors placed parallel to each other. Larger plates can store more charge and hence more energy. Putting the plates close together also allows the capacitor to store more energy. The **capacitance** of a capacitor is the charge on the conductor divided by the voltage and is used to measure the ability of a capacitor to store energy. The capacitance of a parallel plate capacitor is proportional to the area of the plates divided by the **distance** between them. This number must then be multiplied by a constant which is a property of the dielectric between the plates. The dielectric has the effect of increasing the capacitance.

Capacitors come in a wide range of sizes. Banks of large capacitors can store and rapidly release large bursts of electrical energy. Among other uses, engineers can use such devices to test a circuit's performance when struck by a bolt of **lightning**. On an intermediate scale, a camera flash works by storing energy in a capacitor and then releasing it to cause a quick bright flash of **light**. Electronic circuits use large numbers of small capacitors. For example, a RAM (Random Access Memory) chip uses hundreds of thousands of very small capacitors coupled with switching transistors in a computer memory. Computer information is stored in a binary code of ones and zeros. A charged capacitor is a one, and an uncharged is a **zero**. These are just a few example of the many uses of capacitors.

Capillaries

Capillaries are microscopic **blood** vessels that connect small **arteries** (arterioles) and small **veins** (venules). Within the tissues, arterioles terminate into a network of microscopic capillaries. Substances move in and out of the capillary walls as the blood exchanges materials with the cells. Before leaving the tissues, capillaries unite into venules, which merge to form larger and larger veins that eventually return blood to the **heart**. Of all the blood vessels, only capillaries have walls thin enough to allow the exchange of materials between cells and the blood. Their extensive branching provides a sufficient surface area to pick up and deliver substances to all cells in the body.

Despite the fact that there are approximately 40 billion capillaries in the body, they hold only 5% of total blood **volume**. There are two reasons for this. First, the size of the capillaries is only 5–10 nm in diameter. Second, at any give time only a fraction (25%) of capillaries are fully filled with blood, especially in tissues at rest, as blood flow in microvessels is dependent on the metabolic activity of the **tissue** and is regulated at the sites of their origin by sphincter muscles.

Capillaries are essential for the delivery of **oxygen** to the tissues and the exchange of **nutrients** between blood and interstitial fluid surrounding the cells. This function is well supported by the **anatomy** of the vessels. The thin walls of the capillaries are composed of a single layer of endothelial cells. As a result, gasses such as oxygen and **carbon dioxide** can diffuse through their walls, as can **lipid** soluble substances. In contrast, an exchange of lipid-insoluble substances occurs via transcytosis, which involves formation of pinocytotic vesicles at one side of the endothelial **cell**, their transport across the cells, and release of contents from the other side of the cell.

Capillaries also play an important role in regulating the relative volume of the blood and interstitial fluid by allowing a bulk flow through their walls. This exchange of **water** and solutes occurs in response to the **pressure** gradient across the capillary wall.

Based on the structure of their endothelial cells, there are three types of capillaries. Continuous capillaries are a tube developed by the endothelial cells with no intercellular or intracellular gaps (**brain**, retina), or small intercellular gaps. In contrast, fenestrated endothelium has pores of 70–100 nm in size which allow some substances to pass through. Finally, there are discontinuous capillaries (or sinusoids) that are the largest capillaries and have little or no basal **membrane**, and large intercellular pores and fenestrations.

Capillaries form functional units known as capillary beds and these are not uniformly distributed among the different tissues. Sites of high metabolic activity (such as the liver and kidneys) contain numerous capillaries, while sites with little metabolic activity (such as the **lens** of the **eye**) are capillary-free.

See also Circulatory system.

Capillary action

Capillary action is the tendency of a liquid to rise in narrow tubes or to be drawn into small openings such as those between grains of a rock. Capillary action, also known as capillarity, is a result of the intermolecular attraction within the liquid and solid materials. A familiar example of capillary action is the tendency of a dry **paper** towel to absorb a liquid by drawing it into the narrow openings between the fibers.

The mutual attractive **force** that exists between like molecules of a particular liquid is called cohesion. This force is responsible for holding a raindrop together as a single unit. Cohesion produces the phenomenon known as **surface tension**, which may allow objects that are more dense than the liquid to be supported on the surface of the liquid without sinking. When an attractive force exists between two unlike materials, such as a liquid and a solid container, the attractive force is known as adhesion. Adhesion is the force that causes **water** to stick to the inside of a **glass**. If the adhesive force between the liquid and solid is greater than the cohesive force within the liquid, the liquid is said to wet the surface and the surface of the liquid near the edge of the container will **curve** upward. In cases where the cohesive force is greater than adhesion, the liquid is said to be nonwetting and the liquid surface will curve downward near the edge of the container.

The combination of the adhesive forces and the surface tension that arises from cohesion produces the characteristic upward curve in a wetting fluid. Capillarity is the result of cohesion of water molecules and adhesion of those molecules to the solid material forming the void. As the edges of the container are brought closer together, such as in a very narrow tube, the interaction of these phenomena causes the liquid to be drawn upward in the tube. The more narrow the tube, the greater the rise of the liquid. Greater surface tension and increased **ratio** of adhesion to cohesion also result in greater rise. However, increased **density** of the liquid will cause it to rise to a lesser **degree**.

The force with which water is held by capillary action varies with the quantity of water being held. Water entering a natural void, such as a pore within the **soil**, forms a film on the surface of the material surrounding the pore. The adhesion of the water molecules nearest the solid material is greatest. As water is added to the pore, the thickness of the film increases, the capillary force is reduced in magnitude, and water molecules on the outer portion of the film may begin to flow under the influence of gravity. As more water enters the pore the capillary force is reduced to **zero** when the pore is saturated. The movement of **groundwater** through the soil zone is controlled, in part, by capillary action. The transport of fluids within plants is also an example of capillary action. As the **plant** releases water from its leaves, water is drawn upward from the roots to replace it.

Caprimulgids

The frogmouths, oilbird, potoos, owlet frogmouths, and nightjars are five unusual families of **birds** that make up the order Caprimulgiformes, and are collectively referred to as caprimulgids.

Caprimulgids have a large head, with a short but wide beak that can open with an enormous gape, fringed by long, stiff bristles. This apparatus is used by caprimulgids to catch their food of **insects** in flight.

Caprimulgids have long, pointed wings, and short, weak legs and feet. Most of these birds are crepuscular, meaning they are active in the dim **light** of dusk. Some **species** are nocturnal, or active during the night. Caprimulgids have soft feathers and a subdued coloration, consisting of streaky patterns of brown, grey, and black. Caprimulgids are well camouflaged when they are at rest, and can be very difficult to detect when roosting or sitting on a nest.

Caprimulgids may nest on the ground, in a **tree** cavity, or in caves. They lay one to five eggs. The chicks are downy and helpless at first, and are fed and brooded by both parents.

The oilbird

The oilbird (*Steatornis caripensis*) of Trinidad and northern **South America** is the only species in the family Steatornithidae. This bird forages widely for its major food of oily palm nuts, and it roosts and nests in caves. The oilbird navigates inside of its pitch-black caves using **echolocation**, similar to **bats**. It rears two to four young, which are extremely fat, and at one stage of development are about 50% larger than their parents.

In the past, large numbers of fat, baby oilbirds were collected and boiled down (that is, rendered) as a source of oil for illumination and cooking. Dead young oilbirds were sometimes even impaled on a stick and used as a long-burning torch. Excessive exploitation soon threatened the oilbird, and it is now a protected species over most of its range. However, the forest **habitat** of oilbirds is not well protected, and **deforestation** represents an important threat to the species over much of its range.

Frogmouths

Frogmouths are 12 species occurring in lowland and secondary tropical **forests**, collectively making up the family Podargidae. Frogmouths occur from India, through Indochina, Southeast **Asia**, the Philippines, **Australia**, and many nearby Pacific Islands.

Frogmouths are rather large birds, with a body length of up to 20 in (50 cm). They have short, rounded wings and a long, pointed tail, and are relatively weak fliers. Their bill is very wide, flattened, and heavy. Unlike most caprimulgids, frogmouths do not feed aerially.

Rather, these nocturnal birds pounce on their **prey** of **invertebrates**, small **mammals**, birds, and other small animals on the ground and in tree branches. The cup-shaped or platform nest is usually built in a forked branch of a tree or on a horizontal branch, and depending on the species, contains one to four eggs.

The tawny frogmouth (*Podargus strigoides*) occurs in Australia and Tasmania. The Papuan frogmouth (*P. papuensis*) breeds in New Guinea. The large frogmouth (*Batrachostomus auritus*) occurs in lowland forests of Indochina, Sumatra, and Borneo.

Potoos

Potoos, or tree-nighthawks, are five species of birds that comprise the family Nyctibiidae. Potoos occur in open forests from southern Mexico and the West Indies to northern Argentina and Paraguay.

Potoos have long, pointed wings and a long tail. These birds have weak legs and feet, but long claws, and they perch in an upright, almost-invisible stance on tree limbs. Potoos are solitary birds, feeding nocturnally on insects in flycatcher-fashion, by making short sallies from a prominent perch. Potoos lay a single egg on a cup-like cavity atop a broken stub of a dead branch.

The common potoo (*Nyctibius griseus*) is a widespread species, occurring from southern Mexico to northern Argentina. The great potoo (*N. grandis*) occurs widely in forests of Central and northern South America.

Owlet frogmouths

Owlet frogmouths (or owlet nightjars) are eight species that make up the family Aegothelidae. These birds are Australasian, occurring in Australia, New Guinea, New Caledonia, and nearby islands. Their typical breeding habitat is open forests and brushlands.

Owlet frogmouths have long, pointed wings and a long, pointed tail. They are solitary, nocturnal animals that feed on insects in the air and on the ground. Owlet frogmouths lay their clutch of three to five eggs in tree cavities.

The owlet nightjar (*Aegotheles cristatus*) occurs in savannas and open woodlands in Tasmania, Australia, and New Guinea. The grey or mountain owlet frogmouth (*Aegotheles albertisi*) is widespread in mountain forests of New Guinea.

Goatsuckers and nighthawks

The **goatsuckers**, nightjars, and nighthawks are 70 species that make up the family Caprimulgidae. Most species in this family occur in **Africa** and Asia, but eight species breed in **North America**. These birds have extremely long, pointed wings, and are excellent fliers that feed aerially on flying insects.

The whip-poor-will (*Caprimulgus vociferous*) is a familiar species to forests in the eastern United States and southeastern Canada. Chuck-will's-widow (*C. carolinensis*) breeds in pine forests of the southeastern United States. The common poor-will (*Phalaenoptilus nuttallii*) occurs in the western United States. The common nighthawk (*Chordeiles minor*) is a familiar species over most of the United States and southern Canada, and sometimes nests on flat, graveled roofs in cities. The lesser nighthawk (*C. acutipennis*) is a smaller species of the southwestern United States. Like most species of caprimulgids, the North American species are declining because of habitat loss, and perhaps because of the effects of exposure to **pesticides**.

Resources

Books

Bird Families of the World. Oxford: Oxford University Press, 1998.

Brooke, M., and T. Birkhead. *The Cambridge Encyclopedia of Ornithology.* Cambridge, UK: Cambridge University Press, 1991.

Bill Freedman

Captive breeding and reintroduction

In 1973, the Endangered Species Act was passed in the United States to protect **species** that are rapidly declining due to human influences. Captive breeding and release is one of the tools available to halt or reverse the decline of some species in the wild. Such programs may be carried out by zoos, aquaria, botanical gardens, or **conservation** organizations. In some cases the efforts have met with a substantial degree of success, while in others the results have been limited. It takes a relatively deep understanding of the **biology** and **ecology** of an endangered species to implement a successful program of captive breeding and release.

Captive breeding

The primary goal of captive breeding, also known as ex situ conservation, is to develop a self-sustaining or increasing population of an **endangered species** in captivity, without the need to capture additional individuals

from the wild. Any surplus captive-bred individuals are available to support a program of release into the wild.

Another goal of captive breeding programs is to maintain an appropriate level of genetic diversity, which can allow the population be adaptable to conditions in the environment after release. Genetic diversity refers to the numerous **alleles** of genes in a population. (An allele is one of several forms of a **gene**, the latter being the unit inherited by offspring from their parents.) If all captive-bred individuals are offspring of the same parents, then the population is likely to have low genetic diversity because of the effects of inbreeding (or breeding between closely related individuals). This can also lead to a phenomenon known as inbreeding depression, a detrimental effect on offspring that can result from mating between close relatives. Inbreeding depression is due to an accumulation of deleterious recessive alleles, which can become expressed in a high frequency in inbred populations. Inbreeding depression can be manifested as lowered fecundity, smaller numbers of offspring produced, and decreased survival after **birth**.

If a highly inbred population was reintroduced into the wild, its chances of survival and reproduction are likely to be relatively low. In essence, genetic diversity helps to ensure that a released population will be able to survive and grow, despite natural **selection** against some of its individuals.

The size of released population is another important issue. A small population has a greater probability of **extinction** because of the potentially devastating effects of deaths caused by unpredictable environmental events or flaws in the reintroduction process. In addition, small populations may exhibit a phenomenon known as genetic drift, caused by the disappearance of certain alleles and fixation in the population of others. Genetic drift occurs readily in small populations, and results in a loss of genetic diversity.

It is also important that the alleles of the founder individuals, that is, the animals brought from the wild into the breeding program, are maintained so that the natural, "wild" alleles are not lost during years of captive breeding. Since the ultimate aim is to re-introduce animals back into a native **habitat**, maintenance of the original genetic diversity is crucial to the eventual survival of those individuals in the wild. Also, captive breeding over several generations may select for characteristics such as docility, which are not advantageous in the wild.

Many research programs are addressing these problems of captive breeding. At the Minnesota Zoo, for example, a program known as the International Species Inventory is keeping track of the pedigree of individual animals in zoos around the world. This information is used to help prevent mating among closely related individuals, and to thereby maintain the genetic diversity of captive populations.

Various methods are available for increasing the numbers of offspring that can be bred from a limited number of parents. One such method is artificial insemination, in which sperm is transferred to females by artificial means. This allows animals from different zoos to be mated without actually moving them from pace to place. Another enhancement technique involves the removal of eggs from nests of bird species that will subsequently lay replacement eggs. This allows more eggs to be produced by a female than would occur under natural conditions. Reproduction can also be enhanced by foster parenting the young of an endangered species by "parents" of a closely related one, thereby ensuring the rearing of young in a relatively natural, non-human environment. This method has been used to rear endangered whooping **cranes**, by foster-rearing captive-incubated nestlings by sandhill cranes.

Preparing for successful release

Captive breeding programs must address the issue of adequately preparing animals behaviorally for life in a wild environment. This is an especially formidable task with animals that have a complex social system, and whose behaviors for mating, communication, foraging, **predator** avoidance, offspring rearing, and **migration** are learned by observation of the parents or other experienced individuals. A captive environment does not adequately simulate natural conditions or ensure that exposure to appropriate **learning** opportunities occurs. To circumvent this important problem, training programs have been developed to teach survival skills to captive-bred animals before they are introduced to the wild. For example, red wolves have been taught to hunt and kill living **prey**, and golden lion tamarins to find and manipulate the kinds of fruit they depend on in the wild.

Another extremely important learned **behavior** is the fear of potential predators, including humans. Captive-reared individuals may be taught this essential behavior using realistic dummies in situations that frighten the animals, so they learn to associate fear with the model. **Imprinting** on humans is another potential problem, involving impressionable young animals learning to think that they are the same as humans, while not recognizing other individuals as their own species. Imprinting on people can be avoided by using a puppet of an adult of the proper species to "interact" with the young, including during feeding. For example, **peregrine falcon** chicks born in captivity are fed by people wearing puppets of adult **falcons** on their arms, while blocking the rest of their body from view with a partition. This prevents the falcon chicks from seeing the human caregiver, and helps it to imprint on an appropriate subject.

Perhaps the most difficult problem involves teaching captive-bred animals about the social hierarchy and other behavioral intricacies of their species. The most practical approach to this problem has been to keep wild-caught individuals together with captive-reared ones for some time, and to then release them together. This method has been somewhat successful in the reintroduction of the golden lion tamarin to tropical forest in Brazil.

Reintroduction

If a successful reintroduction of an endangered species is to occur, the factors causing its decline must be understood and managed. The most common cause of endangerment is habitat destruction or degradation. Obviously, it is crucial that the habitat of endangered species is conserved before captive-bred individuals are released into the wild. This is not necessarily an easily attained goal, because the causes of habitat destruction usually involve complex social, cultural, and economic factors. Controversy has, for example, accompanied reintroduction of the critically endangered California condor to the wild. The condor is a large scavenging bird that requires an extremely large range to survive, exceeding millions of acres per bird. Initially, the U.S. Fish and Wildlife Service failed to conserve enough habitat to support the highly endangered condor, resulting in controversy over the ultimate goal and likely success of the captive-breeding program. In 1986, however, an extremely large tract of suitable land was purchased for use as the base of the reintroduction of captive-reared **birds**, which has since begun.

After release, captive-reared animals must be monitored to determine whether they have been able to survive the stresses of living in a wild habitat. To ease the transition from captivity to the wild, the release may be somewhat gradual. For example, a "soft release" may involve the provision of food at the release point until animals learn to forage on their own. Moreover, if environmental conditions become particularly stressful, such as a **drought** making **water** and food scarce, it may be necessary to intervene temporarily until conditions improve. Monitoring the released population is necessary for the assessment of survival and the causes of mortality, so that future releases can attempt to avoid such pitfalls.

Although releases of captive-bred animals has received most of the public attention, there have also been attempts to reintroduce endangered plants to the wild. Many of the same issues are involved, but plants also present unique problems due to their lack of mobility and specific microhabitat requirements for establishment and growth. For example, the immediate environment in the **soil** surrounding a seed must have appropriate conditions of **light**, water, nutrient availability, and **temperature**, and must be free of seed predators and fungal **disease** spores. Moreover, the microhabitat requirements for **germination** often involve a specific disturbance regime, such as fire or canopy gaps created by tree-falls. Consequently, even in native habitats, only a very small percentage of **seeds** produced by a given **plant** can germinate and establish. In a successful reintroduction program, the habitat should be managed to allow these periodic disturbances to occur. Higher success rates in germination can be achieved in a greenhouse, after which the seedlings can be transplanted into the wild. This does not, however, dismiss the necessity of managing the land for future reproduction and survival of the plant in the wild; otherwise the reintroduction effort could fail.

A study was undertaken to evaluate 79 different reintroductions of birds and **mammals** in the United States It was found that certain reintroduction conditions had a higher probability of success than others. The highest probability of failure occurred when the species was a large **carnivore** requiring an extensive range, when the animals were released into marginal habitat, and when the released individuals were reared in captivity instead of being wild-caught and released within their lifetime. Any of these circumstances require particularly close attention if the reintroduction attempt is to be successful.

Programs of captive breeding and release can be extremely expensive, and their success may be limited because of difficulties in biology, ecology, and in addressing the ultimate cause of the species decline (such as habitat loss or excessive hunting). Moreover, reintroduction efforts should always be accompanied by a program of public education. The informed public has an influence on political decisions to attempt to reverse human-induced losses of biological diversity, and to avoid such ecological damage by preventing habitat loss, overhunting, and other destructive actions.

See also Condors.

Resources

Books

Primack, R.B. *Essentials of Conservation Biology.* Sunderland, MA: Sinauer, 1993.

Spellerberg, I.F., and S.R. Hardes. *Biological Conservation.* Cambridge: Cambridge University Press, 1992.

Periodicals

Frazer, N. B. "Sea Turtle Conservation and Halfway Technology." *Conservation Biology* 6 (1992): 179-184.

Griffith, B., et al. "Translocation as a Species Conservation Tool: Status and Strategy." *Science* 245 (1989): 477-480.

Kleiman, D.G. "Reintroduction of Captive Mammals for Conservation." *Bioscience.* 39 (1989): 152-161.

Puja Batra

Capuchins

Capuchins are **New World monkeys** characterized by a cap or crown patch of hair that resembles a hood, called a capuche, worn by Franciscan monks. Capuchins belong to the family Cebidae, which includes 31 **species** in 11 genera. The Cebidae is subdivided into seven subfamilies which include night **monkeys**, squirrel monkeys, titis, sakis, howlers, **spider monkeys**, and the capuchins.

Monkeys in the family Cebidae are thin animals with long legs and a prehensile tail, which is muscular and can be used to help the **animal** in climbing and swinging through trees. Most of these New World monkeys, including capuchins, are active during the day and **sleep** at night. Capuchins are medium-sized animals with a body and legs that are evenly proportioned, and have fingers and toes with nails. The nostrils of New World monkeys are round and set far apart while those of the Old World monkeys are set close together.

Physical characteristics

There are four species of capuchin monkeys found in **South America**. The brown capuchin (*Cebus apella*) lives in tropical and subtropical **forests** from Venezuela to Brazil. Capuchins are not found in the Andes Mountains along the western part of the **continent**. The brown capuchin has tufts

A white-fronted capuchin (*Cebus albifrons*). © *Renee Lynn, National Audubon Society Collection/Photo Researchers, Inc. Reproduced with permission.*

of hair on the forehead and a dark cap which extends downward on his forehead into a triangle. Other distinctive markings of the brown capuchin are black sideburns, and a coarse coat that is usually paler on the abdomen, with black limbs. The hair on the face is sparser than the rest of the fur, and the facial skin is pale. The average weight is 6 lb (3 kg) for females and 8 lb (4 kg) for males.

The white-faced capuchin (*C.capucinus*) is found in Central America from the southern region of Mexico, south into Colombia. White-faced capuchins live in dry or wet forests, and in mangroves. The **color** of their fur is pale cream to white on their bellies and the upper parts of their arms and legs, with black fur on their backs and lower limbs. They have white fur on their faces and a black cap. Many older white-faced capuchins have a ruff (fringe) of hair on their foreheads and crowns. The average weight for males is 7 lb (3.5 kg) and 5 lb (2.5 kg) for females.

Weeper capuchins (*C. nigrivittatus*) are found north of the Amazon and north and east of the Rio Negro in Brazil, the Guianas, and central Venezuela. Females weigh less than 5 lb (2.5 kg) and males weigh around 6 lb (3 kg). Their colorings are like the white-faced capuchins but there is less contrast between the dark and light colors.

They have a narrow crown patch that comes to a marked point on their foreheads. They also live in dry and wet forests and mangroves as do the white-faced capuchin.

The white-fronted capuchin (*C. albifrons*) is found in the moist forests of Venezuela, Brazil, Bolivia, Ecuador, Colombia, and on the **island** of Trinidad. This species is slightly smaller than other capuchin monkeys. The colors are similar to weeper and white-faced capuchins, with a pale and broad cap that covers most of the tops of their heads.

Capuchins are capable of running on two legs, as well as on all fours. They are very nimble and acrobatic in the treetops. The use of the tail as a fifth limb in capuchins is rather restricted. They do not spend all their time in trees, however, since they also find food on the forest floor. Fruit comprises 80% of capuchins' diet, the rest consisting mainly of leaves and **insects**. In laboratory studies capuchins will use tools to help them get food.

Social behavior

The social groups of capuchin monkeys vary in size from small groups with three members, to groups of 30 or more. There are usually more females in the group than males, and half of the members of these social groups are infants and adolescents. While there is a dominant male and female in each group, there is little evidence of any other hierarchy within the group, except that dominant males exhibit different degrees of tolerance among the various members of the group. This is particularly evident when the group is foraging for food.

The dominant male does not mingle much with other members of the group, but does play a role in defending the group from intruders. The dominant female establishes a special relationship with the dominant male and tries to keep others away from him.

Capuchins are polygamous, and it is the females who do the courting. Their methods of luring males include raising their eyebrows, gesturing, and making sounds. If a male is interested, he will mimic her gestures and sounds, follow her, and mate. Females give **birth** to one infant at a time, about every two years. Gestation is about five months, and infants are completely dependent on their mothers during the first three weeks of life.

Pastime activities among capuchins differ by age and gender. During the first few months, sisters especially take an interest in an infant sibling. After the third month of birth, the infant will also seek out the company of younger members of the group. A main social activity of male capuchins includes fighting games, while females spend a good deal of time sitting close together and in mutual grooming, particularly those parts of their bodies which are hard to reach or which they cannot see. Relationships among capuchins extend not only to siblings and their mothers, but to other relatives within the group as well.

Resources

Books

Loy, James, and Calvin B. Peters. *Understanding Behavior: What Primate Studies Tell Us about Human Behavior.* New York: Oxford University Press, 1991.

Mason, William A., and Sally P. Mendoza. *Primate Social Conflict.* Albany: State University of New York Press, 1993.

Vita Richman

Capybaras

Capybaras, also known as carpinchos or **water** hogs, are large South American **rodents** in the family Hydrochaeridae. *Hydrochaeris hydrochaeris* is the larger of the two **species** of capybaras, and is the world's largest rodent. It can reach a body weight of 110 lb (50 kg), a body length of 4.5 ft (1.3 m), and a height of 1.5 ft (50 cm). *Hydrochaeris isthmius* is about half this size. *H. hydrochaeris* has a wide distribution in **South America**, while *H. isthmius* has a relatively restricted distribution in Panama, western Colombia, and northwestern Venezuela. Some scientists, however, regard *H. hydrochaeris* and *H. isthmius* as a single species.

Capybaras have a large head with a blunt snout and small ears. The body is stout, robust, and almost tailless. The feet of capybaras are partially webbed and have four digits on the forefeet and three on the hind, and all have strong claws. The fur of capybaras is long, coarse, and rather sparse, so that naked skin can be seen. The body **color** is generally brownish or grayish. Capybaras have a strong physical resemblance to another group of much smaller South American rodents, the closely related guinea **pigs** (family Caviidae).

Capybaras are semi-aquatic animals, occurring in a wide range of terrestrial habitats in the vicinity of **freshwater**, including the forested edges of streams, **rivers**, ponds, lakes, swamps, and marshes. Capybaras can run easily on land, and when disturbed near water they generally swim and dive to escape. These animals can swim with only their eyes and nostrils exposed to the atmosphere, or they can swim completely submerged.

Capybaras are herbivores, eating a wide range of aquatic, near-shore, and riparian plants. Capybaras sometimes feed with cattle and other domestic herbivores, and they are known to raid gardens for **vegetables**, **fruits**, and grains. Capybaras feed most actively during the moderate temperatures of dawn and dusk, spending the **heat** of the day in a cool, underground excavation. However, in places where they are frequently disturbed or hunted by people, capybaras generally develop a nocturnal habit.

Capybaras are peaceful, social animals, living in extended family groups containing as many as several tens of animals. They give **birth** once a year to two to eight

A South American capybara (*Hydrochaeris hydrochaeris*). Photograph by Robert J. Huffman. Field Mark Publications. Reproduced by permission.

offspring. Capybaras are known to live as long as 10 years in the wild.

Capybaras are hunted by various species of natural predators, including jaguar and large caimans. They are also hunted by humans, because they are often regarded as agricultural **pests**. The meat of capybaras is sometimes consumed, although it is not regarded as one of the higher-quality game species. The capybara is still widespread and abundant over much of its natural range.

Bill Freedman

Caraway *see* **Carrot family (Apiaceae)**

Carbohydrate

Carbohydrates are naturally occurring compounds composed of **carbon**, **hydrogen**, and **oxygen**. The carbohydrate group includes sugars, starches, **cellulose**, and a number of other chemically related substances. For the most part, these carbohydrates are produced by green plants through the process known as **photosynthesis**. Countless varieties of plants use this process to synthesize a simple sugar (glucose, mostly) from the **light** energy absorbed by the **chlorophyll** in their leaves, **water** from the **soil**, and **carbon dioxide** from the air. Typically, plants use some of this simple sugar to form the more **complex** carbohydrate cellulose (which makes up the plant's supporting framework) and some to provide energy for its own metabolic needs; the rest is stored away for later use in the form of **seeds**, roots, or **fruits**.

Interestingly, the digestive and metabolic processes in animals and humans work almost in reverse fashion.

When a fruit is eaten, for instance, the complex carbohydrates are broken down in the digestive tract to simpler glucose units. The glucose is then used primarily to produce **energy** in a process which involves oxidation and the excretion of carbon dioxide and water as waste products. In the mid-1800s, German chemist Justus von Liebig was one of the first to recognize that the body derived energy from the oxidation of foods recently eaten, and also declared that it was carbohydrates and fats that served to fuel the oxidation-not carbon and hydrogen as Antoine-Laurent Lavoisier had thought.

Carbohydrates are usually divided into three main categories. The first category, the monosaccharides, are simple sugars that consist of a single carbohydrate unit that cannot be broken down into any simpler substances. The three most common sugars in this group are glucose (or dextrose), the most frequently seen sugar in fruits and **vegetables** (and, in digestion, the form of carbohydrate to which all others are eventually converted); fructose, associated with glucose in honey and in many fruits and vegetables; and galactose, derived from the more complex milk sugar, lactose. Each of these simple but nutritionally important sugars is a hexose, which means it contains six carbon **atoms**, 12 hydrogen atoms, and six oxygen atoms. All three require virtually no digestion but are readily absorbed into the bloodstream from the intestine.

Slightly more complex sugars are the disaccharides which contain two hexose units. The three most nutritionally important of these are sucrose (ordinary table sugar), maltose (derived from starch), and lactose, which is formed in the mammary **glands** and is the only sugar not found in plants. In the digestive tract, specific enzymes split all of these sugars into the more easily absorbed monosaccharides. If needed for future energy use, glucose units are typically squeezed together into larger, more slowly absorbed units and stored as polysaccharides, whose molecules often contain a hundred times the number of glucose units as do the simple sugars. These highly complex carbohydrates include dextrin, starch, cellulose, and glycogen. More efficient and more stable than the simple sugars, they are much easier to store. On the other hand, most of them need to be broken down by the digestive tract's enzymes before they can be absorbed. Some of them—cellulose, for instance—are almost impossible for humans to digest, but this indigestibility is useful since the colon needs a certain amount of bulk, or roughage, to perform at its best.

Glycogen is the form in which most of the body's excess glucose is stored. Both the liver and muscle are able to store glycogen, with muscle glycogen used primarily to fuel muscle contractions and liver glycogen used (when necessary) to replenish the bloodstream's dwindling supply of glucose.

Glycogen was named by French physiologist Claude Bernard, who in 1856 discovered a starchlike substance in the liver of **mammals**. This substance, he later showed, was not only built out of glucose taken from the **blood**, but could be broken down again into sugar whenever it was needed. In 1891, German physiologist Karl von Voit demonstrated that mammals could make glycogen even when fed sugars more complex than glucose. In 1919, Otto Meyerhof was able to show that glycogen is converted into **lactic acid** in working muscles. It was not until the 1930s, however, that the complicated process by which glycogen, stored in the liver and muscle, is broken down in the body and resynthesized was discovered by Czech-American biochemists Carl Cori and Gerty Cori. Building on their work, Fritz Lipmann was able a few years later to further clarify the way carbohydrates can be converted into the forms of chemical energy most usable by the body.

The chemical structure of the various sugars was worked out in great detail by German biochemist Emil Fischer, who began his Nobel Prize-winning work in 1884. Fischer not only was able to synthesize glucose and 30 other sugars, he also showed that the shape of their molecules was even more important than their chemical composition.

Carbon

Carbon is the non-metallic chemical element of **atomic number** 6 in Group 14 of the **periodic table**, symbol C, **atomic weight** 12.01, specific gravity as graphite 2.25, as diamond 3.51. Its stable isotopes are ^{12}C (98.90%) and ^{13}C (1.10%). The weight of the ^{12}C atom is the international standard on which atomic weights are based. It is defined as weighing exactly 12.00000 atomic **mass** units.

Carbon has been known since prehistoric times. It gets its name from *carbo*, the Latin word for charcoal, which is almost pure carbon. In various forms, carbon is found not only on **Earth**, but in the atmospheres of other planets, in the **Sun** and stars, in **comets,** and in some meteorites.

On Earth, carbon can be considered to be the most important of all the chemical elements, because it is the essential element in practically all of the chemical compounds in living things. Carbon compounds are what make the processes of life work. Beyond Earth, carbon-atom nuclei are an essential part of the **nuclear fusion** reactions that produce the **energy** of the Sun and of many other stars. Without carbon, the Sun would be cold and dark.

How carbon is found

In the form of chemical compounds, carbon is distributed throughout the world as **carbon dioxide** gas, CO_2, in the atmosphere and dissolved in all the **rivers**, lakes and oceans. In the form of carbonates, mostly **calcium carbonate** ($CaCO_3$), it occurs as huge rocky masses of limestone, marble, and chalk. In the form of hydrocarbons, it occurs as great deposits of **natural gas, petroleum,** and **coal**. Coal is important not only as a fuel, but because it is the source of the carbon that we dissolve in molten **iron** to make **steel**.

All plants and animals on Earth contain a substantial proportion of carbon. After **hydrogen** and **oxygen**, carbon is the most abundant element in the human body, making up 10.7% of all the body's **atoms**.

Carbon is found as the free (uncombined) element in three different allotropic forms-different geometrical arrangements of the atoms in the solid. The two crystalline forms (forms containing very definite atomic arrangements) are graphite and diamond. Graphite is one of the softest known materials, while diamond is one of the hardest.

There is also a shapeless, or amorphous, form of carbon in which the atoms have no particular geometric arrangement. Carbon black, a form of amorphous carbon obtained from smoky flames, is used to make rubber tires and inks black. Charcoal—wood or other **plant** material that has been heated in the absence of enough air to actually burn—is mostly amorphous carbon, but it retains some of the microscopic structure of the plant cells in the **wood** from which is was made. Activated charcoal is charcoal that has been steam-purified of all the gummy wood-decomposition products, leaving porous grains of pure carbon that have an enormous microscopic surface area. It is estimated that one cubic inch of activated charcoal contains 200,000 sq ft (18,580 m^2) of microscopic surface. This huge surface has a stickiness, called adsorption, for molecules of gases and solids; activated charcoal is therefore used to remove impurities from **water** and air, such as in home water purifiers and in gas masks.

Graphite

Graphite is a soft, shiny, dark gray or black, greasy-feeling mineral that is found in large masses throughout the world, including the United States, Brazil, England, western **Europe**, Siberia, and Sri Lanka. It is a good conductor of **electricity** and resists temperatures up to about 6,332°F (3,500°C), which makes it useful as brushes (conductors that slide along rotating parts) in electric motors and generators, and as electrodes in high-temperature electrolysiscells. Because of its slipperiness, it is used as a lubricant. For example, powdered graphite is used to lubricate locks, where oil might "gum up the works." The "lead" in pencils is actually a mixture of graphite, clay, and wax. It is called "lead" because the metallic element **lead** (Pb) leaves gray marks on **paper** and was used for writing in ancient times. When graphite-based pencils came into use, they were called "lead pencils."

The reason for graphite's slipperiness is its unusual crystalline structure. It consists of a stack of one-atom-thick sheets of carbon atoms, bonded tightly together into a hexagonal pattern in each sheet, but with only very weak attractions—much weaker than actual chemical bonds—holding the sheets together.

The sheets of carbon atoms can therefore slide easily over one another; graphite is slippery in the same way as layers of wet leaves on a sidewalk.

Diamond

Diamond, the other crystalline form of pure carbon, is the world's hardest natural material, and is used in industry as an abrasive and in drill tips for drilling through rock in oil fields and human teeth in dentists' offices. On a hardness scale of one to ten, which mineralogists refer to as the Moh scale of hardness, diamond is awarded a perfect ten. But that's not why diamonds are so expensive. They are the most expensive of all gems, and are kept that way by supply and demand. The supply is largely controlled by the De Beers Consolidated Mines, Inc. in South **Africa**, where most of the world's diamonds are mined, and the demand is kept high by the importance that is widely attributed to diamonds.

A diamond can be considered to be a single huge **molecule** consisting of nothing but carbon atoms that are strongly bonded to each other by covalent bonds, just as in other molecules. A one-carat diamond "molecule" contains 10^{22} carbon atoms.

The beauty of gem-quality diamonds—industrial diamonds are small, dark, and cloudy—comes from their **crystal** clarity, their high refractivity (ability to bend **light** rays), and their high dispersion—their ability to spread light of different colors apart, which makes the diamond's rainbow "fire." Skillful chipping of the gems into facets (flat faces) at carefully calculated angles makes the most of their sparkle. Even though diamonds are hard, meaning that they cannot be scratched by other materials, they are brittle—they can be cracked.

The chemistry of carbon

Carbon is unique among the elements because its atoms can form an endless variety of molecules with an endless variety of sizes, shapes, and chemical properties.

No other element can do that to anywhere near the degree that carbon can. In the **evolution** of life on Earth, Nature has always been able to "find" just the right carbon compound out of the millions available, to serve just about any required function in the complicated **chemistry** of living things.

Carbon-containing compounds are called organic compounds, and the study of their properties and reactions is called organic chemistry. The name organic was originally given to those substances that are found in living organisms-plants and animals. As we now know, almost all of the chemical substances in living things are carbon compounds (water and **minerals** are the obvious exceptions), and the name organic was eventually applied to the chemistry of all carbon compounds, regardless of where they come from.

Until the early nineteenth century, it was believed that organic substances contained a supernatural life force that made them special, and that they were not susceptible to chemical experimentation. But in 1828, a German chemist named Friedrich Wöhler (1800-1882) apparently broke down the mysterious barrier between living and non-living things. By simply heating a non-organic, non-living chemical called ammonium cyanate (NH_4OCN), he converted it into a chemical called **urea** (H_2N-CO-NH_2), which was known to be a waste product in the urine of **mammals** and was therefore an "organic" substance. As Wöhler put it, he was amazed to be able to create an organic substance "without benefit of a kidney, a bladder or a dog." What had happened in Wöhler's experiment was that the eight atoms in the ammonium cyanate molecule—two **nitrogen** atoms, four hydrogen atoms, one oxygen atom, and one carbon atom—simply rearranged themselves into a molecule having a different **geometry**. In chemical language, the two molecules are isomers of one another.

After Wöhler, chemists boldly synthesized (made artificially) many of the chemical compounds that formerly had been observed only in living things. Today, biochemistry—living chemistry—is one of the most active and productive fields of scientific research. It has taught us more about the processes of life than could ever have been imagined.

Why carbon is special

There are now more than ten million organic compounds known by chemists. Many more undoubtedly exist in nature, and organic chemists are continually creating (synthesizing) new ones. Carbon is the only element that can form so many different compounds because each carbon atom can form four chemical bonds to other atoms, and because the carbon atom is just the right, small size to fit in comfortably as parts of very large molecules.

Having the atomic number 6, every carbon atom has a total of six electrons. Two are in a completed inner **orbit**, while the other four are **valence** electrons—outer electrons that are available for forming bonds with other atoms.

The carbon atom's four valence electrons can be shared by other atoms that have electrons to share, thus forming covalent (shared-electron) bonds. They can even be shared by other carbon atoms, which in turn can share electrons with other carbon atoms and so on, forming long strings of carbon atoms, bonded to each other like links in a chain. Silicon (Si), another element in group 14 of the periodic table, also has four valence electrons and can make large molecules called silicones, but its atoms are too large to fit together into as great a variety of molecules as carbon atoms can.

Carbon's ability to form long carbon-to-carbon chains is the first of five reasons that there can be so many different carbon compounds; a molecule that differs by even one atom is, of course, a molecule of a different compound. The second reason for carbon's astounding compound-forming ability is that carbon atoms can bind to each other not only in straight chains, but in complex branchings, like the branches of a **tree**. They can even join "head-to-tail" to make rings of carbon atoms. There is practically no limit to the number or complexity of the branches or the number of rings that can be attached to them, and hence no limit to the number of different molecules that can be formed.

The third reason is that carbon atoms can share not only a single **electron** with another atom to form a single bond, but it can also share two or three electrons, forming a double or triple bond. This makes for a huge number of possible bond combinations at different places, making a huge number of different possible molecules. And a molecule that differs by even one atom or one bond position is a molecule of a different compound.

The fourth reason is that the same collection of atoms and bonds, but in a different geometrical arrangement within the molecule, makes a molecule with a different shape and hence different properties. These different molecules are called isomers.

The fifth reason is that all of the electrons that are not being used to bond carbon atoms together into chains and rings can be used to form bonds with atoms of several other elements. The most common other element is hydrogen, which makes the family of compounds known as hydrocarbons. But nitrogen, oxygen, **phosphorus**, **sulfur**, **halogens**, and several other kinds of atoms can also be attached as part of an organic molecule. There is

a huge number of ways in which they can be attached to the carbon-atom branches, and each variation makes a molecule of a different compound. It's just as if moving a Christmas tree ornament from one branch to another created a completely different tree.

Classes of carbon compounds

It is obviously impossible to summarize the properties of carbon's millions of compounds in one place. Introductory textbooks of organic chemistry generally run well over a thousand pages, and **biochemistry** textbooks run thousands more. But organic compounds can be classified into families that have similar properties, because they have certain groupings of atoms in common.

See also Alcohol; Aldehydes; Alkaloid; Amides; Amino acid; Barbiturates; Chemical bond; Carbohydrate; Carbon cycle; Carboxylic acids; Ester; Ether; Fat; Fatty acids; Glycol; Halide, organic; Hydrocarbon; Isomer; Lipid; Polymer; Proteins.

Resources

Books

Johnson, A. William. *Invitation to Organic Chemistry.* Boston: Jones & Bartlett, 1999.

Loudon, G. Marc. *Organic Chemistry.* Menlo Park, CA: Benjamin/Cummings,1988.

Parker, Sybil P., ed. *McGraw-Hill Encyclopedia of Chemistry.* 2nd ed. New York: McGraw-Hill, 1999.

Sherwood, Martin, and Christine Sutton, eds. *The Physical World.* New York: Oxford University Press, 1991.

Robert L. Wolke

Carbon cycle

The carbon cycle describes the movement of **carbon** in the atmosphere, where it is in the gaseous form **carbon dioxide**, through organisms, and then back into the atmosphere and the oceans. Carbon is a central element of the huge diversity of organic chemicals found in living things, such as the many kinds of carbohydrates, **proteins**, and fats. **Energy** is contained in the chemical bonds that hold the **atoms** of carbon and other elements together in these organic compounds. Organisms use chemical energy from organic compounds to carry out all the processes necessary to life.

How carbon is released into the atmosphere

Carbon is released into the atmosphere through three major processes: cellular respiration, the burning of **fossil**

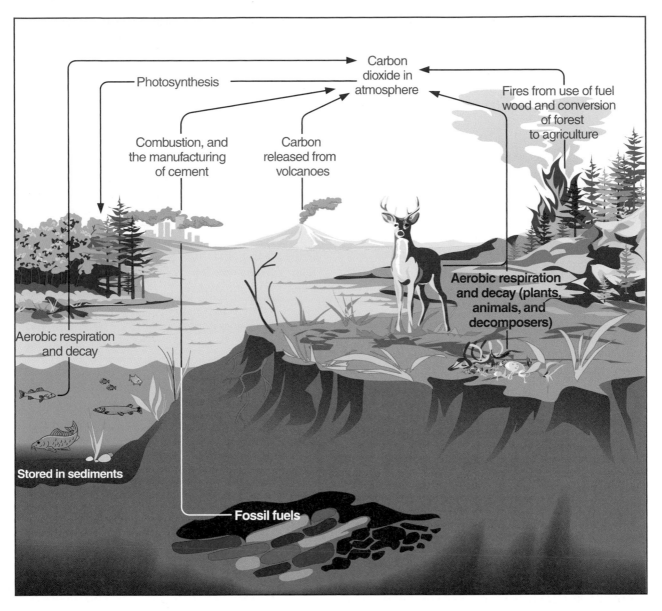

The carbon cycle. *Illustration by Hans & Cassidy. Courtesy of Gale Group.*

fuels, and volcanic eruptions. In each of these processes, carbon is returned to the atmosphere or to the **ocean**.

Cellular respiration

Plants convert the carbon in atmospheric carbon dioxide into carbon-containing organic compounds, such as sugars, fats, and proteins. Plants take in carbon dioxide through microscopic openings in their leaves, called stomata. They combine atmospheric carbon with **water** and manufacture organic compounds, using energy trapped from sunlight in a process called **photosynthesis**. The by-product of photosynthesis is **oxygen**, which plants release into the atmosphere through the stomata.

Animals that eat plants, or that eat other animals, incorporate the carbon in the sugars, fats, and proteins derived from the ingested **biomass** into their bodies. Inside their cells, energy is extracted from the food in a process called cellular respiration. Cellular **respiration** requires oxygen (which is the by-product of photosynthesis) and it produces carbon dioxide, which is used in photosynthesis. In this way, photosynthesis and cellular respiration are linked in the carbon cycle.

Photosynthesis requires atmospheric carbon, while cellular respiration returns carbon to the atmosphere, and vice versa for oxygen. The global rates of photosynthesis and cellular respiration influence the amount of carbon dioxide in the atmosphere. In the summer, the high **rate** of photosynthesis uses up much of the carbon dioxide in the atmosphere, and the amount of atmospheric carbon dioxide decreases. In the winter, when the rate of photosynthesis is low, the amount of atmospheric carbon dioxide increases.

Another way that cellular respiration releases carbon into the atmosphere is through the actions of decomposers. Decomposers, such as **bacteria** and **fungi**, derive their **nutrients** by feeding on the remains of plants and animals. The bacteria and fungi use cellular respiration to extract the energy contained in the chemical bonds of the decomposing organic **matter**, and so release carbon dioxide into the atmosphere.

In some ecosystems, such as tropical rainforests, **decomposition** is accomplished quickly, and carbon dioxide is returned to the atmosphere at a relatively fast rate. In other ecosystems, such as northern **forests** and **tundra**, decomposition proceeds more slowly. In some places, such as bogs and the deep ocean, the organic matter of plants and animals may accumulate in deep sediments, where decomposers cannot function well because of the lack of oxygen. Slowly, over millions of years, the carbon-rich materials are converted into carbon-rich fossil fuels, such as **petroleum**, **natural gas**, and **coal**. Also in marine environments, carbon-containing matter (such as **calcium carbonate**) is incorporated into the shells and other hard parts of aquatic organisms. When these organisms die, the carbon-rich hard parts sink to the ocean bed. There they become buried in sediment, and eventually densify into **rocks** such as limestone and dolomite.

The burning of fossil fuels

When fossil fuels are burned, their organic carbon is released into the atmosphere. During the past 100 years, fossil fuel consumption has increased dramatically, and this has led to a huge amount of carbon dioxide being released into the atmosphere. In addition, the widespread clearing of forests is resulting in the **emission** of huge quantities of carbon dioxide into the atmosphere (forests store large amounts of organic carbon in their biomass, most of which is emitted through decomposition and fire when **deforestation** occurs). The increasing concentrations of atmospheric carbon dioxide are a cause for concern, since they may be responsible for **global warming** and associated climatic and ecological disruptions. During the middle of the nineteenth century, the atmospheric **concentration** of carbon dioxide was about 270 ppm, but in 2000 it had increased to 365 ppm.

Volcanic eruption

Another way that carbon is released into the atmosphere is through volcanic eruptions. When a **volcano** erupts, it sends huge amounts of ash and soot high into the atmosphere. Some of this ash and soot is derived from ancient carbon-rich sediments, and the cloud of debris that results from a volcanic eruption returns large amounts of carbon to the atmosphere.

The carbon cycle in land and sea

The cycling of carbon takes place in oceans and other aquatic ecosystems as well as in terrestrial environments. The world's oceans contain about 50% more carbon than does the atmosphere. The oceans are able to absorb some of the carbon dioxide currently being released by the burning of fossil fuels, thus offsetting global warming. However, the oceans cannot do this as quickly as the carbon dioxide is being released to the atmosphere, and this time lag is resulting in increasing concentrations in the atmosphere.

In aquatic environments, carbon cycling is more complex because carbon interacts with water. When carbon dioxide is released by cellular respiration, it combines with water to form carbonic acid and bicarbonate. Therefore, in aquatic environments most inorganic carbon is in the form of bicarbonate rather than carbon dioxide. Carbon dioxide from the atmosphere readily diffuses into water, and it is quickly converted to bicarbonate.

Importance of the carbon cycle

The carbon cycle is important in ecosystems because it moves carbon, a life-sustaining element, from the atmosphere and oceans into organisms and back again to the atmosphere and oceans. If the balance between these latter two reservoirs is upset, serious consequences, such as global warming and climate disruption, may result. Scientists are currently looking into ways in which humans can use other, non-carbon containing fuels for energy. **Nuclear power**, solar power, **wind** power, and water power are a few **alternative energy sources** that are being investigated.

See also Greenhouse effect.

Resources

Books

Dunnette, David A., and Robert J. O'Brien, eds. *The Science of Global Change: The Impact of Human Activities on the Environment.* Washington, DC: American Chemical Society, 1992.

Hamblin, W.K., and E.H. Christiansen. *Earth's Dynamic Systems.* 9th ed. Upper Saddle River: Prentice Hall, 2001.

Levi, Barbara Gross, David Hafemeister, and Richard Scribner. *Global Warming: Physics and Facts.* New York: American Institute of Physics, 1992.

Matthews, John A., E. M. Bridges, and Christopher J. Caseldine. *The Encyclopaedic Dictionary of Environmental Change.* New York: Edward Arnold, 2001.

Tolbert, N. E., and Jack Preiss, eds. *Regulation of Atmospheric Carbon by Photosynthetic Carbon Metabolism.* New York: Oxford University Press, 1994.

Periodicals

Hileman, Bette. "New CO$_2$ Model Shows Whole Earth 'Breathing'" *Chemical and Engineering News* 72 (January 10, 1994): 6.

Vitousek, Peter M. "Beyond Global Warming: Ecology and Global Change." *Ecology* 75 (October 1994): 1861.

Volk, Tyler. "The Soil's Breath." *Natural History* 103 (November 1994): 48.

Zimmer, Carl. "The War Between Plants and Animals." *Discover* 14 (July 1993): 16.

Kathleen Scogna

Carbon dioxide

Carbon dioxide was the first gas to be distinguished from ordinary air, perhaps because it is so intimately connected with the cycles of **plant** and **animal** life. When we breathe air or when we burn **wood** and other fuels, carbon dioxide is released; when plants store **energy** in the form of food, they use up carbon dioxide. Early scientists were able to observe the effects of carbon dioxide long before they knew exactly what it was.

Around 1630, Flemish scientist Jan van Helmont discovered that certain vapors differed from air, which was then thought to be a single substance or element. Van Helmont coined the term gas to describe these vapors and collected the gas given off by burning wood, calling it gas sylvestre. Today we know this gas to be carbon dioxide, and van Helmont is credited with its discovery. He also recognized that carbon dioxide was produced by the **fermentation** of wine and from other natural processes. Before long, other scientists began to notice similarities between the processes of breathing (**respiration**) and burning (**combustion**), both of which use up and give off carbon dioxide. For example, a candle flame will eventually be extinguished when enclosed in a jar with a limited supply of air, as will the life of a bird or small animal.

Then in 1756, Joseph Black proved that carbon dioxide, which he called fixed air, is present in the atmosphere and that it combines with other chemicals to form new compounds. Black also identified carbon dioxide in exhaled breath, determined that the gas is heavier than air, and characterized its chemical behavior as that of a weak acid. The pioneering work of van Helmont and Black soon led to the discovery of other gases by Henry Cavendish, Antoine-Laurent Lavoisier, Carl Wilhelm Scheele, and other chemists. As a result, scientists began to realize that gases must be weighed and accounted for in the analysis of chemical compounds, just like solids and liquids.

The first practical use for carbon dioxide was invented by Joseph Priestley, an English chemist, in the mid 1700s. Priestley had duplicated Black's experiments using a gas produced by fermenting grain and showed that it had the same properties as Black's fixed air, or carbon dioxide. When he dissolved the gas in **water**, he found that it created a refreshing drink with a slightly tart flavor. This was the first artificially carbonated water, known as soda water or seltzer. Carbon dioxide is still used today to make colas and other soft drinks. In addition to supplying bubbles and zest, the gas acts as a preservative.

The early study of carbon dioxide also gave rise to the expression to be a guinea pig, meaning to subject oneself to an experiment. In 1783, French physicist Pierre Laplace used a guinea pig to demonstrate quantitatively that **oxygen** from the air is used to burn carbon stored in the body and produce carbon dioxide in exhaled breath. Around the same time, chemists began drawing the connection between carbon dioxide and plant life. Like animals, plants breathe, using up oxygen and releasing carbon dioxide. But plants also have the unique ability to store energy in the form of carbohydrates, our primary source of food. This energy-storing process, called **photosynthesis**, is essentially the reverse of respiration. It uses up carbon dioxide and releases oxygen in a complex series of reactions that also require sunlight and **chlorophyll** (the green substance that gives plants their **color**). In the 1770s, Dutch physiologist Jan Ingen Housz established the principles of photosynthesis, which helped explain the age-old superstition that plants purify air during the day and poison it at night.

Since these early discoveries, chemists have learned much more about carbon dioxide. English chemist John Dalton guessed in 1803 that the **molecule** contains one carbon atom and two oxygen **atoms** (CO$_2$); this was later proved to be true. The decay of all organic materials produces carbon dioxide very slowly, and the earth's atmosphere contains a small amount of the gas (about 0.033%). Spectroscopic analysis has shown that in our **solar system**, the planets of **Venus** and **Mars** have atmospheres very rich in carbon dioxide. The gas also exists in **ocean** water, where it plays a vital role in marine plant photosynthesis.

In modern life, carbon dioxide has many practical applications. For example, fire extinguishers use CO$_2$ to control electrical and oil fires, which cannot be put out with water. Because carbon dioxide is heavier than air, it spreads into a blanket and smothers the flames. Carbon dioxide is also a very effective refrigerant. In its solid form, known as dry **ice**, it is used to chill perishable food during transport. Many industrial processes are also cooled by carbon dioxide, which allows faster production rates. For these commercial purposes, carbon dioxide can be obtained from either **natural gas** wells, fermentation of organic material, or combustion of **fossil fuels**.

Recently, carbon dioxide has received negative attention as a **greenhouse effect** gas. When it accumulates in the upper atmosphere, it traps the earth's **heat**, which could eventually cause **global warming**. Since the beginning of the **Industrial Revolution** in the mid-1800s, factories and power plants have significantly increased the amount of carbon dioxide in the atmosphere by burning **coal** and other fossil fuels. This effect was first predicted by Svante August Arrhenius, a Swedish physicist, in the 1880s. Then in 1938, British physicist G. S. Callendar suggested that higher CO_2 levels had caused the warmer temperatures observed in America and **Europe** since Arrhenius's day. Modern scientists have confirmed these views and identified other causes of increasing carbon dioxide levels, such as the clearing of the world's **forests**. Because trees extract CO_2 from the air, their depletion has contributed to upsetting the delicate balance of gases in the atmosphere.

In very rare circumstances, carbon dioxide can endanger life. In 1986, a huge cloud of the gas exploded from Lake Nyos, a volcanic **lake** in northwestern Cameroon, and quickly suffocated more than 1,700 people and 8,000 animals. Scientists have attempted to control this phenomenon by slowly pumping the gas up from the bottom of the lake.

See also Air pollution; Carbon cycle; Planetary atmospheres.

Carbon monoxide

Carbon monoxide is a compound of carbon and **oxygen** with the chemical formula CO. It is a colorless, odorless, tasteless, toxic gas. It has a **density** of 1.250 g/L at 32°F (0°C) and 760 mm Hg **pressure**. **Carbon dioxide** can be converted into a liquid at its **boiling point** of -312.7°F (-191.5°C) and then to a solid at its freezing point of -337°F (-205°C).

History

The discovery of carbon monoxide is often credited to the work of the English chemist and theologian Joseph Priestley. In the period between 1772 and 1799, Priestley gradually recognized the nature of this compound and showed how it was different from carbon dioxide, with which it often appeared. None the less carbon monoxide had been well known and extensively studied in the centuries prior to Priestley's work. As early as the late 1200s, the Spanish alchemist Arnold of Villanova described a poisonous gas produced by the incomplete **combustion** of **wood** that was almost certainly carbon monoxide.

In the five centuries between the work of Arnold and that of Priestley, carbon monoxide was studied and described by a number of prominent alchemists and chemists. Many made special mention of the toxicity of the gas. Johann (or Jan) Baptista van Helmont in 1644 wrote that he nearly died from inhaling *gas carbonum*, apparently a mixture of carbon monoxide and carbon dioxide.

An important milestone in the history of carbon monoxide came in 1877 when the French physicist Louis Paul Cailletet found a method for liquefying the gas. Two decades later, a particularly interesting group of compounds made from carbon monoxide, the carbonyls, were discovered by the French chemist Paul Sabatier.

Sources

Carbon monoxide is the twelfth most abundant gas in the atmosphere. It makes up about $1.2 \times 10^{-5}\%$ of a **sample** of dry air in the lower atmosphere. The major natural source of carbon monoxide is the combustion of wood, **coal**, and other naturally occurring substances on the earth's surface. Huge quantities of carbon monoxide are produced, for example, during a forest fire or a volcanic eruption. The amount of carbon monoxide produced in such reactions depends on the availability of oxygen and the combustion **temperature**. High levels of oxygen and high temperatures tend to produce complete oxidation of carbon, with carbon dioxide as the final product. Lower levels of oxygen and lower temperatures result in the formation of higher percentages of carbon monoxide in the combustion mixture.

Commercial methods for producing carbon monoxide often depend on the direct oxidation of carbon under controlled conditions. For example, producer gas is made by blowing air across very hot coke (nearly pure carbon). The final product consists of three gases, carbon monoxide, carbon dioxide, and **nitrogen** in the **ratio** of 6 to 1 to 18. **Water** gas is made by a similar process, by passing steam over hot coke. The products in this case are **hydrogen** (50%), carbon monoxide (40%), carbon dioxide (5%) and other gases (5%). Other methods of preparation are also available. One of the most commonly used involves the partial oxidation of hydrocarbons obtained from **natural gas**.

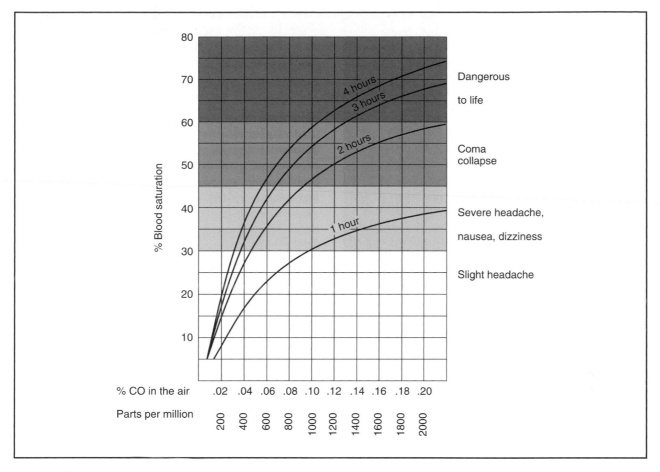

Figure 1. Effects of carbon monoxide on humans. *Illustration by Hans & Cassidy. Courtesy of Gale Group.*

Physiological effects

The toxic character of carbon monoxide has been well known for many centuries. At low concentrations, carbon monoxide may cause nausea, vomiting, restlessness, and euphoria. As exposure increases, a person may lose consciousness and go into convulsions. Death is a common final result. The U.S. Occupational Safety and Health Administration has established a limit of 35 ppm (parts per million) of carbon monoxide in workplaces where a person may be continually exposed to the gas.

The earliest explanation for the toxic effects of carbon monoxide was offered by the French physiologist Claude Bernard in the late 1850s. Bernard pointed out that carbon monoxide has a strong tendency to replace oxygen in the **respiratory system**. Someone exposed to high concentrations of carbon monoxide may actually begin to suffocate as his or her body is deprived of oxygen.

Today we have a fairly sophisticated understanding of the mechanism by which carbon monoxide poisoning occurs. Normally, oxygen is transported from the lungs to cells in red **blood** cells. This process occurs when oxygen **atoms** bond to an **iron** atom at the center of a complex protein **molecule** known as oxyhemoglobin. Oxyhemoglobin is a fairly unstable molecule that decomposes in the intercellular spaces to release free oxygen and hemoglobin. The oxygen is then available to carry out metabolic reactions in cells, reactions from which the body obtains **energy**.

If carbon monoxide is present in the lungs, this sequence is disrupted. Carbon monoxide bonds with iron in hemoglobin to form carbonmonoxyhemoglobin, a complex somewhat similar to oxyhemoglobin. Carbonmonoxyhemoglobin is, however, a more stable compound than is oxyhemoglobin. When it reaches cells, it has much less tendency to break down, but continues to circulate in the bloodstream in its bound form. As a result, cells are unable to obtain the oxygen they need for **metabolism** and energy production dramatically decreases. The clinical symptoms of carbon monoxide poisoning described above are manifestations of these changes.

Carbon monoxide poisoning—at least at moderate levels—is common in everyday life. Poorly vented charcoal fires, improperly installed gas appliances, and the exhaust from internal combustion vehicles are among the most common sources of the gas. In fact, levels of carbon monoxide in the air can become dangerously high in busy urban areas where automotive transportation is extensive. Cigarette smokers may also be exposed to dangerous levels of the gas. Studies have shown that the one to two pack-a-day smoker may have up to 7% of the hemoglobin in her or his body tied up in the form of carbonmonoxyhemoglobin.

Uses

Carbon monoxide is a very important industrial compound. In the form of producer gas or water gas, it is widely used as a fuel in industrial operations. The gas is also an effective reducing agent. For example, when carbon monoxide is passed over hot iron oxides, the oxides are reduced to metallic iron, while the carbon monoxide is oxidized to carbon dioxide.

In another application a mixture of metallic ores is heated to 122–176°F (50–80°C) in the presence of producer gas. All oxides except those of nickel are reduced to their metallic state. This process, known as the Mond process, is a way of separating nickel from other metals with which it commonly occurs.

Yet another use of the gas is in the Fischer-Tropsch process for the manufacture of hydrocarbons and their oxygen derivatives from a combination of hydrogen and carbon monoxide. Carbon monoxide also reacts with certain metals, especially iron, cobalt, and nickel, to form compounds known as carbonyls. Some of the carbonyls have unusual physical and chemical properties that make them useful in industry. The highly toxic nickel tetracarbonyl, for example, is used to produce very pure nickel coatings and powders.

See also Metallurgy.

David E. Newton

Resources

Books

Boikess, Robert S., and Edward Edelson. *Chemical Principles.* 2nd edition. New York: Harper & Row Publishers, 1981, pp. 672 - 673.

Brown, Theodore L., and H. Eugene LeMay Jr. *Chemistry: The Central Science.* 3rd edition. Englewood Cliffs, NJ: Prentice-Hall, 1985, pp. 390-392, 66 -669.

Budavari, Susan, ed. *The Merck Index.* 11th edition. Rahway, NJ: Merck and Company, 1989, pp. 1821.

Greenwood, N. N., and A. Earnshaw. *Chemistry of the Elements.* Oxford: Butterworth-Heinneman Press, 1997.

Lide, D.R., ed. *CRC Handbook of Chemistry and Physics.* Boca Raton: CRC Press, 2001.

Matthews, John A., E.M. Bridges, and Christopher J. Caseldine *The Encyclopaedic Dictionary of Environmental Change.* New York: Edward Arnold, 2001.

Partington, J.R. *A Short History of Chemistry.* 3rd edition. London: Macmillan & Company, 1957, pp. 49, 116, 142, 151.

Other

The United Nations. "The Conference and Kyoto Protocol." (March 2003). <http://unfccc.int/resource/convkp.html>.

Carbon tetrachloride

Carbon tetrachloride is an organic chemical that is commonly used as a solvent. It is also called tetra chloromethane and is composed of molecules that have one carbon atom and four **chlorine atoms** bonded together in the shape of a **tetrahedron**. It is made by combining elemental chlorine with simple carbon compounds like methane or carbon disulfide. It is a liquid at room **temperature**, with a freezing point of -9.4°F (-23°C) and a **boiling point** of 170.6°F (77°C). Carbon tetrachloride dissolves other organic materials such as oils, fats, and grease very well. This property makes carbon tetrachloride very useful for cleaning manufactured parts. Carbon tetrachloride was once used heavily in the dry-cleaning industry. Use in that industry has declined because it is toxic when inhaled or absorbed through the skin, and it is no longer used in products for the home. Since carbon tetrachloride is a good solvent, it is used to dissolve things like oils, fragrances, and colors from flowers and **seeds**. Carbon tetrachloride is not flammable, so it can be used in fire extinguishers or as an additive to make other chemicals nonflammable. It is also very useful as a raw material in synthesizing larger, more complicated organic compounds. Because of the health hazards of long-term exposure to carbon tetrachloride, it should only be used where there is adequate ventilation present.

Carbonyl group

A carbonyl group is a group of **atoms** that consists of a **carbon** atom covalently attached to an **oxygen** atom by a double bond: $C = O$. The carbon atom, to satisfy its **valence** of 4, must also be attached by covalent bonds to two other atoms. The simplest type of **molecule** that contains a carbonyl group is a ketone. Other types of molecules that contain carbonyl groups are **aldehydes**, acids, esters, and **amides**.

Ketones

A ketone is a compound whose molecules contain a carbonyl group and have two other groups attached to the carbon atom of the carbonyl group. There are many molecules that belong to this classification, but the simplest one is **acetone**. A condensed structural formula for acetone looks like this.

$$\overset{\displaystyle O}{\overset{\displaystyle \|}{CH_3 - C - CH_3}}$$

In this formula, the $C=O$ represents the carbonyl group, and the two CH_3 groups satisfy the carbon atom's valence of 4. In other molecules that contain the carbonyl group, the $C=O$ is still present, but the two CH_3 groups are traded for other atoms or groups of atoms.

Sometimes we need to talk about the entire class of possible ketone molecules, and then we use a structural formula that looks like this.

$$\overset{\displaystyle O}{\overset{\displaystyle \|}{R - C - R'}}$$

In this picture, R and R' can stand for any hydrocarbon-containing group and as CH_3-, C_2H_5-, etc.

Properties of the carbonyl group

The carbonyl group is somewhat polar. That means that one end of it (the carbon atom) has a slight positive **electric charge**, and one end of it (the oxygen atom) has

a slight **negative** charge. This makes the entire molecular a polar molecule.

The polar nature of the carbonyl part of the molecule affects the physical properties of the entire molecule. For instance, small ketone molecules, with fewer than six carbon atoms in all, are soluble in **water**, a very polar solvent. At the same time, small ketone molecules are themselves often good solvents for other compounds with polar groups. This is in contrast to small **hydrocarbon** molecules with no carbonyl group—they are insoluble in water, and they won't dissolve other polar molecules.

A carbonyl group in a molecule is often the most chemically reactive portion. When a molecule containing a carbonyl group undergoes a chemical reaction, it is often this polarity that controls which reaction will take place. Usually a chemical reaction in a molecule containing only a carbonyl group and hydrocarbon groups will take place at the carbonyl group.

Other molecules with carbonyl groups

In many molecules that contain a carbonyl group, the other two groups of atoms are not hydrocarbon groups. Molecules like this are so different chemically that they belong to entirely different classifications. There are four major classes of molecules like this. Again, R stands for any hydrocarbon group. There is more information about these kinds of carbonyl-containing molecules in their entries in this encyclopedia.

$$\begin{array}{c} O \\ \| \\ R - C - H \end{array}$$
aldehyde

$$\begin{array}{c} O \\ \| \\ R - C - OH \end{array}$$
acid

$$\begin{array}{c} O \\ \| \\ R - C - O - R' \end{array}$$
ester

$$\begin{array}{c} O \\ \| \\ R - C - NH_2 \end{array}$$
amide

See also Acids and bases; Amides; Ester.

Resources

Books

Mark, Herman F. *From Small Organic Chemicals to Large: a Century of Progress.* Washington DC: American Chemical Society, 1993.

Mauskopf, Seymour H. *Chemical Sciences in the Modern World.* Pennsylvania: University of Pennsylvania Press, 1993.

G. Lynn Carlson

Carboxyl group

A carboxyl group, also called a carboxy group, is a characteristic group of **atoms** found in organic molecules. Organic compounds that contain carboxyl groups are called **carboxylic acids**.

The carboxyl group occurs on the end or side of a **molecule**. The group consists of a **carbon** atom that forms two chemical bonds to one **oxygen** atom and one **chemical bond** to a second oxygen atom. This second oxygen is also bonded to a **hydrogen** atom. The arrangement is written -COOH or -C(O)OH (which emphasizes the different chemical bonding between the carbon atom and one of the oxygen atoms). The name "carboxyl" is actually a combination of the words "carbonyl" and" hydroxyl," because the carboxyl group itself can be considered as a combination of carbonyl (CO) and hydroxyl (OH) groups.

Carboxylic acids

Carboxylic acids are chemical compounds that contain a **carboxyl group**, which is -COOH. The carboxyl group is attached to another **hydrogen** atom or to one end of a larger **molecule**. Examples include formic acid, which is produced by some **ants** and causes their bites to sting. (In fact, the scientific name for ants, *Formica,* is what gives formic acid its name.) Another example is **acetic acid**, which is found in vinegar. Many carboxylic acids dissolve in **water**. Solutions of many carboxylic acids have a sour **taste** to them, a characteristic of many

acids. Carboxylic acids also react with alkalis, or bases. Generally, however, carboxylic acids are not as chemically active as the non-organic mineral acids such as hydrochloric acid or **sulfuric acid**.

Biological importance

Carboxylic acids are very important biologically. The drug aspirin is a carboxylic acid, and some people are sensitive to its acidity. The non-aspirin **pain** reliever ibuprofen is also a carboxylic acid. Carboxylic acids that have very long chains of **carbon** atoms attached to them are called **fatty acids**. As their name suggests, they are important in the formation of **fat** in the body. Many carboxylic acids are present in the foods and drinks we ingest, like malic acid (found in apples), **tartaric acid** (grape juice), **oxalic acid** (**spinach** and some parts of the **rhubarb plant**), and **lactic acid** (sour milk). Two other simple carboxylic acids are propionic acid and butyric acid. Propionic acid is partly responsible for the flavor and odor of Swiss cheese. Butyric acid is responsible not only for the **smell** of rancid butter, but also contributes to the odor of sweat. Lactic acid is generated in muscles of the body as the individual cells metabolize sugar and do **work**. A buildup of lactic acid, caused by overexertion, is responsible for the fatigue one feels in the muscles by such short-term use. When one rests, the lactic acid is gradually converted to water and **carbon dioxide**, and the feeling of fatigue passes. A form of **vitamin** C is called ascorbic acid and is a carboxylic acid.

A special form of carboxylic acids are the amino acids, which are carboxylic acids that also have a nitrogen-containing group called an amine group in the molecule. Aminoacids are very important because combinations of amino acids make up the **proteins**. Proteins are one of the three major components of the diet, the other two being fats and carbohydrates. Much of the human body, like skin, hair, and muscle, is composed of protein.

Industrial importance

Carboxylic acids are also very important industrially. Perhaps one of the most important industrial applications of compounds with carboxyl groups is the use of fatty acids (which are carboxyl groups attached to long carbon chains) in making soaps, detergents, and shampoos. In some such compounds, the hydrogen atom in the carboxyl group is replaced with some **metal cation**. The modified carboxyl group is soluble in water, while the long chain of carbons remains soluble in fats, oils, and greases. This double **solubility** allows water to wash out the fat- and oil-based dirt. Many shampoos are based on lauric, palmitic, and stearic acids, which have long chains of 12, 16, and 18 carbon atoms, respectively. To

KEY TERMS

Amino acid—An organic compound whose molecules contain both an amino group (-NH$_2$) and a carboxyl group (-COOH). One of the building blocks of a protein.

Carboxyl group—The —COOH group of atoms, whose presence defines a carboxylic acid.

Ester—A derivative of a carboxylic acid, where an organic group has been substituted for the hydrogen atom in the acid group. Esters contribute to tastes and smells.

Fatty acid—A carboxylic acid which is attached to a chain of at least 8 carbon atoms. Fatty acids are important components in fats, and are used to make soaps.

Lactic acid—A carboxylic acid formed during the metabolism of sugar in muscle cells. A buildup of lactic acid leads to a feeling of fatigue.

Mineral acid—An acid that is not organic. Examples include hydrochloric acid and sulfuric acid.

Saponification—A chemical reaction involving the breakdown of triglycerides to component fatty acids, and the conversion of these acids to soap.

make other cleansing agents, three molecules of fatty acid are combined with one molecule of a compound called glycerin in a reaction called saponification. This reaction also makes a **soap** molecule which has one end soluble in water and the other soluble in fat or grease or oil. Various fatty acids are used to make soaps and detergents that have different applications in society. Carboxylic acids are also important in the manufacture of greases, crayons, and **plastics**.

Compounds with carboxyl groups are relatively easily converted to compounds called esters, which have the hydrogen atom of the carboxyl group replaced with a group containing carbon and hydrogen **atoms**. Such esters are considered derivatives of carboxylic acids. Esters are important because many of them have characteristic tastes and odors. For example, methyl butyrate, a derivative of butyric acid, smells like apples. Benzyl acetate, from acetic acid, has a jasmine odor. Carboxylic acids are thus used commercially as raw materials for the production of synthetic odors and flavors. Other esters, derived from carboxylic acids, have different uses. For example, the **ester** ethyl acetate is a very good solvent and is a major component in nail polish remover.

See also Acetylsalicylic acid; Acids and bases.

Resources

Books

Loudon, G. Mark. *Organic Chemistry.* Oxford: Oxford University Press, 2002.
Snyder, Carl H. *The Extraordinary Chemistry of Ordinary Things.* New York: John Wiley & Sons, 1992.

Periodicals

Murray, Frank. "Hydroxycitric Acid." *Better Nutrition for Better Living* 56 (1994): 34–39.

David W. Ball

Carcinogen

A carcinogen is a substance that causes a normal **cell** to change into a cancerous cell. The word "carcinogen" is derived from Greek and means in English, cancer-causing. Carcinogens fall into two broad categories, naturally occurring substances that are found in food or **soil**, or artificial substances created by chemists for various industrial purposes. Although the way carcinogens cause **cancer** is still not completely understood, cancer researchers believe that humans and other animals must be exposed to a carcinogen for a certain period of time and at a high enough **concentration** for cancer to occur.

Cancer is a **group** of **diseases** in which cells grow abnormally. Like all living things, normal cells grow, reproduce, and die. These processes are controlled by chemicals and reactions within the cell, which are in turn controlled by the cell's genetic material within its nucleus. In a cancerous cell, the genetic material is altered, and the genes which encode and direct the **chemical reactions** within the cell are mutated, or changed. Cancerous cells grow uncontrollably, forming a large mass of cells called a **tumor**, which invades tissues and kills non-cancerous cells. Sometimes cancerous cells "break off" from a tumor and enter the bloodstream, traveling to other parts of the body and infecting other organs and tissues in a process known as metastasis. In this way, a cancer can spread from an isolated tumor to the entire body.

Several agents, such as viruses, medication, and synthetic carcinogens can cause mutations within a cell's genetic material. Some kinds of cancers are caused by viruses. For example, a special kind of **virus** called a **retrovirus** causes a rare form of **leukemia** (cancer of the white **blood** cells). **Radiation** from naturally-occurring radioactive substances (such as **uranium**) can disrupt a cell's genetic material and bring about cancer. Synthetic carcinogens are found in processed foods and industrial chemicals.

Carcinogens and cancer

For a carcinogen to cause cancer a person must be exposed for a certain length of time to the carcinogen and at a high enough concentration. Repeated exposure to a carcinogen over an extended period (such as 20 years) increases the likelihood that a normal cell's genetic material will mutate and initiate cancer. **Cigarette smoke** contains potent carcinogens; it can take many years and many repeated exposure to the carcinogens in smoke for smoking to cause cancer. Smoking-related lung cancers typically develop between 10 and 20 years of continuous smoking. In Hiroshima and Nagasaki in Japan, where atomic bombs were dropped in 1945, the leukemia rates in the surviving population increased dramatically some five years after the bombs were detonated.

Some carcinogens are more powerful, cancer-causing agents, than others. Powerful carcinogens are called tumor promoters, and can cause genetic mutations directly within cells. Less powerful carcinogens are called tumor initiators and can cause latent changes in the cell's genetic material. These changes are not enough to actually cause cancer, but sensitize the **tissue** for later exposures to tumor promoters. If a tumor initiator has already wrought some damage to the cell's genetic material, the likelihood that a tumor promotor will cause the cell to become cancerous is increased.

Cancerous tumors develop over many stages, and it is rare that exposure to a carcinogen is the sole cause of most cancers. Exposure to a carcinogen must usually be combined with other environmental factors for a cancer to develop. Some environmental risk factors are difficult to identify, while others such as prolonged, heavy cigarette smoking have been easy to identify. Heredity, such as whether close relatives have developed cancer, is another risk factor that is not as easily characterized.

Some carcinogens such as cigarette smoke can be avoided. Other factors, such as a diet, can be modified. Additional risk factors such as gender, immune status, metabolic **rate**, levels of certain enzymes, and age can neither be avoided nor modified.

Carcinogens used in industry

The idea that chemicals could cause cancer was first promoted in 1775 by Percivall Pott, a London physician. Dr. Pott noted that young chimney sweeps had a high incidence of scrotal cancer. Because most sweeps began their careers very early in life and seldom washed or changed clothes, the sweeps were exposed to soot re-

peatedly and for long periods of time, leading to scrotal cancer in young adulthood. Not until 150 years later were the actual carcinogenic substances in soot identified, but Dr. Pott made his case for more humane treatment and better working conditions for the chimney sweeps by noting the connections between cancer and this profession.

With the advent of industrial development in the nineteenth century, other connections between certain cancers and chemicals were noticed. Shale oil and **coal** and tar workers had a high incidence of skin cancer. Dyestuff workers developed bladder cancer. And a chemical called vinyl chloride, used in the manufacture of leather goods, caused a rare liver tumor.

In response to growing concern about cancer in industrial workers, the International Agency for Research on Cancer and the National **Toxicology** Program formulated a system to classify chemicals according to their cancer-causing risk. A chemical could be classified either as a probable carcinogen, or as a non-carcinogen.

A problem with this and other classification systems is that much of the research is based on experiments on animals. If cancer can be induced in experimental animals with high levels of a chemical, it is sometimes assumed that humans could also be at risk even with lower exposure levels over a long period of time. The science of risk assessment investigates the possibility of developing cancer from very low levels of exposure to chemicals that cause cancer at very high levels.

Dioxin is a case in point. In the early 1980s, dioxin was sprayed on the roads of Times Beach, Missouri, to seal the pavement. Dioxin had been classified by the Environmental Protection Agency (EPA) as a probable carcinogen. When the townspeople discovered that dioxin had been sprayed on their roads, the town was abandoned and lawsuits against the road contractor were initiated which were challenged by the defendants.

Carcinogens in food

Some foods contain naturally-occurring carcinogens. Safrole, found in sassafras root; estragole, found in the **herb** tarragon; allyl isothiocynate, found in mustard seed; and **benzene**, found in eggs, **fruits**, **vegetables**, cooked meats, and **fish** are all carcinogens. However, these substances must be consumed in large amounts, over a long period to initiate cancer.

Processed foods such as bacon, sausages, and canned meats contain the preservative nitrite. Frying the cured bacon can convert some non-carcinogenic substances in the nitrites into potent carcinogens. Browning meats such as hamburger can also cause carcinogenic chemicals to be produced. However, in both cases, the amount of carcinogen is extremely small. Interestingly, microwave cooking does not release the carcinogens in beef.

Foods that involve **fermentation** in their production, such as beer, wine, bread, and yogurt, all contain mildly carcinogenic substances. Again, these foods must be eaten in large amounts over decades to possibly cause the genetic mutations related to cancer.

Other carcinogens

Radioactive substances found in **rocks** and soil are also considered potentially carcinogenic. While most people do not come in contact with radioactive chemicals on a day-to-day basis, these substances can emit radioactive particles that can be dangerous. **Radon**, a radioactive substance emitted by uranium, can seep from rocks into buildings. In some areas of the country, radon can be emitted in relatively large amounts into buildings which should be tested for radon. If the levels are found to be high, changes should be made to the building's ventilation system to reduce the amount of radon in the building.

In 1993, the EPA designated second-hand smoke from cigarettes to be a known human carcinogen. It is estimated that 2,000 lung cancer deaths a year are caused by second-hand smoke, which has led to the designation of many public areas as smoke-free zones.

Avoiding carcinogens

It is recommended that people eat a varied diet high in fresh fruits and vegetable and avoid excess consumption of foods high in nitrites. While it is not possible to completely eliminate one's exposure to carcinogens, it is possible to avoid the concomitant risk factors that may lead to cancer. Avoiding smoking, eating a varied, balanced diet that includes fiber, and limiting **alcohol** consumption are all associated with a lowered cancer risk.

See also Mutagen; Mutation.

Resources

Periodicals

Ashby, John. "Change the Rules for Food Additives." *Nature* 368 (April 1994).

Begley, Susan. "Don't Drink the Dioxin." *Newsweek* 124 (September 1994): 57.

Boyle, Peter. "The Hazards of Passive-and Active-Smoking." *New England Journal of Medicine* 328 (June 1993): 1708.

Moolenaar, Robert J. "Overhauling Carcinogen Classification." *Issues in Science and Technology* 8 (Summer 1992): 70.

Nesnow, Stephen. "Breakthroughs in Cancer Risk Assessment." *EPA Journal* 19 (Jan/Mar 1993): 27.

"Second-hand Smoke Designated as a Known Human Carcinogen." *EPA Journal* 19 (2): 5. April/June 1993.

KEY TERMS
. . . .

Cancer—A disease in which cells grow abnormally.

Carcinogen—Any substance capable of causing cancer by mutating the cell's DNA.

Mutation—A change in the genetic material of a cell.

Risk assessment—The study of the risk of exposure to certain levels of an agent that may lead to the development of a disease, such as cancer.

Risk factor—Any habit, condition, or external force that renders an individual more susceptible to disease. Cigarette smoking, for example, is a significant risk factor for lung cancer and heart disease.

Yuspa, S. H. "Overview of Carcinogenesis: Past, Present and Future." *Carcinogenesis* 21 (2000): 341–344.

Kathleen Scogna

Cardiac cycle

The coordinated and rhythmic series of muscular contractions associated with the **heart** comprise the cardiac cycle.

In humans, the cardiac cycle can be subdivided into two major phases, the systolic phase and the diastolic phase. Systole occurs when the ventricles of the heart contract. Accordingly, systole results in the highest pressures within the systemic and pulmonary circulatory systems. Diastole is the period between ventricular contractions when the right and left ventricles relax and fill.

The cardiac cycle cannot be described as a linear series of events associated with the flow of **blood** through the four chambers. One can not accurately describe the cardiac cycle by simply tracing the path of blood from the right atrium, into the right ventricle, into the pulmonary circulation, the venous pulmonary return to the left atrium, and finally the ejection into the aorta and systemic circulation by the contraction of the left ventricle. In reality, the cardiac cycle is a coordinated series of events that take place simultaneously on both the right pulmonary circuit and left systemic circuit of the heart.

The cardiac cycle begins with a period of rapid ventricular filling. The right atrium fills with deoxygenated blood from the superior vena cava, the inferior vena cava, and the coronary venous return (e.g., the coronary sinus and smaller coronary **veins**). At the same time, the pulmonary veins return oxygenated blood from the lungs to the left atrium. During the early diastolic phase of the cardiac cycle, both ventricles relax and fill from their respective atrial sources. The atrio-ventricular valves (the tricuspid valve is located between the right atrium and right ventricle; the mitral valve is between the left atrium and left ventricle) open and allow blood to flow from the atria into the ventricles.

The flow of blood through the atrio-ventricular valves is unidirectional and as **volume** related **pressure** increases within the ventricles, the atrioventricular valves close to prevent backflow from the ventricles into the atria.

At the onset of the systolic phase, specialized cardiac muscle fibers within the sino-atrial node (S-A node) contract and send an electrical signal propagated throughout the heart. In a sweeping fashion, the right atrium contracts and forces the final volume of blood into the right ventricle. The left atrium contracts and contributes the final 20% of volume to the left ventricle.

The S-A node signal is delayed by the atrioventricular node to allow the full contraction of the atria that allows the ventricles to reach their maximum volume. A sweeping right to left wave of ventricular contraction then pumps blood into the pulmonary and systemic circulatory systems. The semilunar valves that separate the right ventricle from the pulmonary artery and the left ventricle from the aorta open shortly after the ventricles begin to contract. The opening of the semilunar valves ends a brief period of isometric (constant volume) ventricular contraction and initiates a period of rapid ventricular ejection.

As muscle fibers contract, they lose their ability to contract forcefully (i.e., the greatest **force** of muscular contraction in the ventricle occurs earlier in the contraction phase and decreases as contraction proceeds). When ventricular pressures fall below their respective attached arterial pressures, the semilunar pulmonary and aortic valves close. At the end of systole, the semilunar valves shut to prevent the backflow of blood into the ventricles.

After emptying, both ventricles collapse to undergo a period of repolarization and refilling. The receptivity of the ventricles to filling corresponding lowers atrial pressures and allows them to fill from their respective venous sources. At the outset, ventricular pressures remain greater than atrial pressures and the atrioventricular valves remain closed. Because the volume of blood in the ventricle is once again static—closed off by both the atrio-ventricular and semilunar valves—this period is described as isometric (same volume) relaxation.

The cardiac cycle is complete with the onset of another period of rapid ventricular filling that takes place when atrial pressures exceed ventricular pressures and the atrio-ventricular valves open to allow rapid filling.

It is the opening and snapping shut of the atrio-ventricular and semilunar pulmonary and aortic valves that creates the familiar pattern of sound associated with the cardiac cycle. Because the right to left contractions of the atria and ventricles are sweeping, the opening and closings of the right side and left side valves are separated by a short interval. The first heart sound results from the closure of the atrio-ventricular valves. The second heart sound results from the closure of the semilunar valves.

The electrical events associated with cardiac cycle are measured with the electrocardiogram (EKG).

See also Action potential; Circulatory system; Heart diseases; Heart, embryonic development and changes at birth; Heart, rhythm control and impulse conduction; Nerve impulses and conduction of impulses; Nervous system; Neuron.

Resources

Books

Gilbert, Scott F. *Developmental Biology.* 6th ed. Sunderland, MA: Sinauer Associates, Inc., 2000.

Guyton, Arthur C., and John E. Hall. *Textbook of Medical Physiology.* 10th ed. Philadelphia: W.B. Saunders Co., 2000.

Thibodeau, Gary A., and Kevin T. Patton. *Anatomy & Physiology.* 5th ed. St. Louis: Mosby, 2002.

Other

Klabunde, R. E. "Cardiac Cycle." Cardiovascular Physiology Concepts. January 17, 2003 [cited January 22, 2003] <http://www.cvphysiology. com/Heart%20Disease/HD002.htm>.

Brenda Wilmoth Lerner

Cardinal number

A measure of the number of elements in a group or a set. For example, the number of books on a shelf can be described by a single cardinal number. Similarly, the set can be assigned the cardinal number 3 because it has only three elements. Since cardinal numbers count the number of elements in a set, they are always positive whole **integers**. If the elements from two sets have a one-to-one relationship, namely each element can be paired together such that no elements are left over, then they can be represented by the same cardinal number.

Some sets have an infinite number of elements. However, not all infinite sets have a one-to-one relationship. Consider the following sets:

Set X Set Y Although both of these sets have an infinite number of elements, they can not be represented by the same cardinal number because set X contains all the elements of set Y, but it also contains additional elements. To solve this problem a nineteenth century mathematician named George Cantor (1845-1918) created a new numbering system to deal with infinite sets. He called these new numbers transfinite cardinal numbers and used the symbol $\aleph_0$ (aleph null) to represent the smallest one. He also developed an **arithmetic** system for manipulating these numbers.

Cardinals and grosbeaks

The cardinals and grosbeaks belong to the subfamilies Cardinalinae, of the finch family (Fringillidae), which is the largest of all North American bird families. (Some researchers include the cardinals and grosbeaks with the Emberizidae, the buntings and **tanagers**).

Cardinals and grosbeaks are New World **birds**, ranging from central Argentina as far north as central Canada. They live primarily in temperate zone woodlands, and have adapted to life around humans, whose help (in the form of birdseed) has helped cardinals extend their range north to Canada.

The name "grosbeak" is descriptive: these birds have thick, sturdy beaks, which help them crack open **seeds**. Their diet also includes blackberries, strawberries, **insects**, spiders, **bees**, corn, **snails**, **slugs**, and earthworms. Cardinals have been seen drinking maple sap from holes left by sapsuckers. The pine grosbeak has special throat pouches in which it transports food.

The males of cardinals and grosbeaks are brightly colored; females are duller. Pine grosbeaks (*Pinicola enucleator*) and evening grosbeaks (*Hesperiphona vespertina*) remain in their northern or high-mountain habitats year-round, but have been known to migrate out of these areas if food is in short supply. Three **species** of North American grosbeak—the rose-breasted grosbeak (*Pheuctinus ludovicanus*), black-headed, and blue grosbeak (*Guiraca caerulea*)—prefer southern areas, and the rose-breasted will migrate as far south as Venezuela and Peru come winter.

In some species, including the cardinal (*Cardinais cardinalis*), the female builds the nest and incubates the eggs without help from the male. In others, males and females share in these efforts. Between 2-5 eggs are laid. Young cardinals fledge quickly, leaving the nest when 10 or 11 days old. This rapid development allows the cardinal to raise multiple clutches in a season, up to as many

as four; the male cares for the hatchlings while the female incubates the next clutch.

Male cardinals and grosbeaks are renowned singers. The cardinal has at least 28 different songs. Male rose-breasted grosbeaks will compete for a female by hovering over her and singing a long, liquid, robin-like song; the winner of that **courtship** will sing while he is helping incubate the eggs.

Caribou

The caribou or reindeer (*Rangifer tarandus*) is a northern **species** of **deer** occurring in the boreal and arctic regions of **North America** and Eurasia. At one time, caribou and reindeer were considered to be separate species, but these animals are fully interfertile and are now considered to be the same species. In North America they are called caribou, whereas in Eurasia they are known as reindeer. However, there are well-differentiated, geographically distinct populations of these animals, which are designated as subspecies. Northern caribou are relatively small, while southern caribou are larger, with the biggest males (bucks) weighing up to 660 lb (300 kg).

Caribou are even-toed, hoofed **mammals** in the order Artiodactyla and suborder Ruminanta. They are in the family Cervidae, along with other species of deer. Like other deer and cattle, caribou have a four-chambered stomach capable of digesting the tough, fibrous **plant** and lichen materials containing **cellulose** that comprise most of their diet. Caribou ruminate, which means they re-chew forage that has previously fermented in the fore-pouches of their stomach.

Like other deer (family Cervidae), caribou have deciduous antlers, which are long, branching, bony outgrowths of the frontal bones of the skull. During their growth, antlers are covered with a heavily vascularized **tissue** called velvet, which eventually dries and is peeled or rubbed off, leaving the bare bone exposed. Unlike other species of deer, both sexes of caribou can develop antlers. However, the antlers of mature male animals are much larger and more elaborate than those of females. The males use their antlers for jousting during the rutting season, when they attempt to assemble a harem of does. The antlers of adult bucks grow most rapidly from May to July, and are at their largest size in August. By October the antlers are hard and velvet free, and are used in ritualized combat with other males, although this can escalate into real fighting. Soon after the rut, the joint between the antler and the skull weakens, and the antler is shed, usually by early December. Antler growth in female caribou

A reindeer bull in Finnish Lapland. *B&C Alexander/Photo Researchers, Inc. Reproduced by permission.*

starts later (June to September), and shedding is delayed until April or May when the calves are born.

The newborn calves of caribou are very precocious, and are able to stand within only one-half hour of **birth**. After a few days, calves are capable of running several kilometers an hour, and can keep up with the moving herd. This rapid development is, of course, an **adaptation** to reducing the predation **rate** of young calves, the stage with the highest risk of mortality.

In North America, the most northerly subspecies is the Peary caribou (*Rangifer tarandus pearyi*). This relatively small, whitish subspecies is a resident of the high-arctic islands of northern Canada. Peary caribou gain weight during the warmest two to three months of the year, foraging on **grasses** and forbs (broadleaf herbs) in relatively productive wet meadows and dwarf-shrub **tundra**. During most of the rest of the year, however, these animals must survive on the much sparser and less nutritious vegetation of upland, relatively snow-free ridges.

Farther to the south are woodland, or barren-ground caribou (these are mostly *R. t. caribou* in the east, and *R. t. groenlandicus* in the west). These relatively large, brown-colored caribou tend to undertake long-distance, seasonal migrations, which can exceed 600 mi (1,000

km) in their circuitous passage. The caribou often swim across large **rivers** during those journeys. During the growing season, these caribou move to open habitats such as tundra and muskeg (the latter is mostly peaty **wetlands** with tussock meadows and low woody vegetation), where they calve and feed on the lush growth of grasses, **sedges**, forbs, and young twigs of shrubs. At the end of the growing season, the barren-ground caribou migrate back to the boreal forest, where they feed largely on arboreal and ground **lichens** during the winter. More southerly caribou living in mountainous terrain undertake vertical migrations, to alpine tundra and meadows during the summer, and montane forest in winter.

In general, the winter diet of caribou consists of foods that are not very nutritious, such as lichens, twigs, and dried grasses and forbs. Caribou tend to slowly lose weight on this poor quality diet, during a time when there are great **energy** demands for thermoregulation in cold temperatures and windy conditions. During summer, a much wider range of more nutritious foods is available, and caribou put weight on at that time. Summer foods include grasses, sedges, forbs, new twigs and foliage of shrubs, **mushrooms**, and berries. Caribou will also opportunistically eat **lemmings** and **birds** eggs.

Caribou are rather social animals, tending to occur in groups of various size. These assemblies are loosely segregated by sex and age-class, and their size can vary seasonally. The density of animals in the groups also varies, being more compact when caribou are harassed by predators such as wolves or humans, or sometimes if the animals are being severely bothered by biting **flies**. During the autumn migrations and the rutting period, woodland caribou may occur in enormous herds of tens of thousands of animals, which disperse into much smaller herds at other times.

Caribou have an excellent sense of **smell**, but do less well visually. Sometimes, these animals can be closely approached from downwind. Caribou can be quite curious, and humans can occasionally approach these animals while walking directly towards them and holding their arms straight up, roughly simulating the silhouette of an oncoming caribou. When frightened, caribou usually run a short distance, **circle** around until they catch a confirming scent of the intruder, and then move to a safer distance.

Wolves are the most important natural predators of caribou, but grizzly bear, **wolverine**, and lynx also kill some animals. Caribou were also a staple food for aboriginal humans in North America, and they continue to be an important game species throughout their range. In some areas caribou have been overharvested, and they have been widely extirpated from most of the southern parts of their original North American range, for example in Maine, the

Maritime Provinces, parts of the southern boreal forest, and parts of the Rocky Mountains. In one unusual case, about 10,000 caribou drowned in northern Quebec when a hydroelectric authority released huge quantities of **water** into a river that the animals had to traverse.

Reindeer (*R. t. tarandus*) have long been domesticated by northern peoples of Eurasia, such as the Lapps of northern Scandinavia. Reindeer have also been introduced to the western Arctic of North America, and to subarctic South Georgia Island in the Southern Hemisphere, in attempts to develop commercial enterprises. Domestic reindeer are husbanded for their meat, hide, and milk. In recent years, a large export market has developed in China and elsewhere in eastern **Asia** for reindeer or caribou horn in velvet, which is made into a powder used in traditional medicine. Domestic reindeer have also been used to pull small sleighs and wagons.

Resources

Books

Banfield, A.W.F. *The Mammals of Canada.* Toronto: University of Toronto Press, 1974.

Grzimek, B., ed. *Grzimek's Encyclopedia of Mammals.* London: McGraw Hill, 1990.

Wilson, D.E., and D. Reeder, comp. *Mammal Species of the World.* 2nd ed. Washington, DC: Smithsonian Institution Press, 1993.

Bill Freedman

Carnivore

In the literal sense, a carnivore is any flesh-eating **organism**. However, in the ecological usage of the word, carnivores kill animals before eating them (that is, they are predators), as opposed to feeding on animals that are already dead (the latter are called scavengers or detritivores).

Trophic **ecology** deals with the feeding and nutritional relationships within ecosystems, and this field has developed some specialized terminology. Carnivores, for example, are heterotrophs, which means that they must ingest other organisms to obtain **energy** and **nutrition**. (In contrast, autotrophs such as green plants can fix their own energy and synthesize biochemicals utilizing diffuse sources such as sunlight and simple inorganic molecules.) Animals that feed on plants are herbivores (or primary consumers), while animals that eat herbivores are known as primary carnivores (or secondary consumers), and carnivores that feed upon other

A cougar (*Felis concolor*). *JLM Visuals. Reproduced by permission.*

carnivores are tertiary consumers. It is rare for an **ecosystem** to sustain carnivores of an order higher than tertiary. This is due to the pyramid-shaped structure of productivity in ecological food webs, which itself is caused by thermodynamic inefficiencies of **energy transfer** between levels. Therefore, the productivity of green plants is always much larger than that of herbivores, while carnivores sustain even less productivity. As a result of their trophic structure, ecosystems cannot sustain predators that feed upon, for example, lions, wolves, or killer whales.

Another consequence of the pyramidal structure of ecological webs is the tendency of top carnivores to bioconcentrate especially large residues of fat-soluble, persistent chemicals such as the **chlorinated hydrocarbons**, DDT, PCBs, and dioxins. This happens because organisms in successive levels of the trophic web absorb most of the chlorinated hydrocarbons that they ingest, storing these chemicals in fatty tissues. Consequently, top carnivores further concentrate the pre-concentrated residues of organisms lower in the ecological web. Therefore, the largest residues of these chemicals occur in peregrine **falcons**, polar **bears**, and **seals**, and these top predators have a disproportionate risk of being poisoned.

Almost all carnivores are animals. However, a few carnivores are specialized **species** of plants that trap, kill, and digest small animals, and then absorb some of their **nutrients**. Examples of these so-called **carnivorous plants** include Venus flytrap, sundews, and pitcher plants.

See also Food chain/web; Herbivore; Heterotroph; Omnivore; Predator; Scavenger; Trophic levels.

Carnivorous plants

Carnivorous plants are botanical oddities that supplement their requirement for **nutrients** by trapping, killing, and digesting small animals, mostly **insects**. Carnivorous plants are photosynthetic, and are therefore fundamentally autotrophic. Still, their feeding relationship with animals represents a reversal of the normal trophic connections between autotrophs and consumers.

Carnivorous plants have long been fascinating to humans. They have the subject of some captivating tales of science fiction, involving fantastic trees that consume large, unwary creatures in tropical **forests**. Tales have even been told about ritual sacrifices of humans to these awesome carnivores, presumably to appease evil, botanical spirits. Fortunately, fact involves much smaller predators than those of science fiction. Still, the few **species** of carnivorous plants that really exist are very curious variants on the usual form and function of plants. Scaled up, these carnivores would indeed be formidable predators.

All species of carnivorous plants are small, herbaceous plants, generally growing in nutrient poor habitats, such as acidic bogs and oligotrophic lakes. The usual **prey** of these green predators is not unwary **deer**, cattle, or humans, but insects and other small **invertebrates**, although a few of the larger species are capable of capturing tadpoles and small **fish**.

Ecology of carnivorous plants

Carnivorous plants are mostly herbaceous perennials with poorly developed root systems, and often propagate by vegetative means, such as stolons and rhizomes. Carnivorous plants are typically intolerant of competition, occurring in open, wet habitats subject to full sunlight. Carnivorous plants are often tolerant of a limited amount of disturbance, and in fact may benefit from a low intensity of trampling, which prepares a substrate suitable for the **germination** of their **seeds** and the establishment of new individuals. Some species are also tolerant of light fires, which also favor their reproduction.

Most carnivorous plants grow in acidic bogs, unproductive lakes, or sandy soils. These are all habitats that are poor in the nutrients that plants require for growth, particularly inorganic **nitrogen**, **phosphorus**, and **calcium**. The nutrients obtained through carnivory are important to these plants. In the absence of **animal** foods these plants grow less well, and they **flower** sparsely or not at all.

The types of traps

Contrary to some portrayals in science fiction, the flowers of carnivorous plants are not the organs that en-

A pitcher plant in Isle Royale National Park, Michigan.
Photograph by Robert J. Huffman. Field Mark Publications. Reproduced by permission.

snare their prey. Rather, in all cases the deadly traps are modified leaves and stems. There are three basic types of trapping organs: active, adhesive, and passive.

Active traps of carnivorous plants attract their mostly arthropod prey using various machinations, including **color**, scent, and **nectar**. Once a victim is suitably within, the trap rapidly closes, preventing the escape of the prey. The active trap of the Venus fly-trap (*Dionaea muscipula*) is modeled on a basic clamshell design. This species utilizes a fast-acting response to a mechanical **stimulus** caused when an insect triggers sensitive hairs in the trap, causing its clam-shell leaves to close. The fringing outer projectiles of the leaves rapidly enclose to form a barrier that prevents the trapped arthropod from escaping. At the same time, mechanical stimuli from the struggling victim trigger the synthesis and excretion of digestive enzymes onto the inner surface of the trap, which facilitate digestion of the prey.

Another design of active trap is based on a small, hollow chamber with a trap door. This design is utilized by the bladderworts (various species of *Utricularia*), small aquatic plants that form little bladders with diameters of several millimeters, that trap tiny aquatic invertebrates behind a rapidly closing trap door. The door of the bladderwort trap initially swings quickly into the bladder, triggered to respond in this way by **motion** sensed by fine, fringing bristles. The inward motion of the door develops a suction that can sweep invertebrates into the trap, where they are trapped by the re-closing door, and are digested for the nutrients they contain.

Adhesive, semi-active traps primarily rely on sticky, surface exudates to ensnare their prey. Once a victim is firmly entangled, the **leaf** slowly enfolds to seal the fate of the unlucky arthropod, and to facilitate the process of digestion. This manner of trap is typified by the most species-rich of the carnivorous plants, the genus of plants known as sundews (*Drosera* spp.). These plants develop relatively wide, modified leaves, that are densely covered with stalked **glands** that resemble tentacles several millimeters long. Each tentacle is tipped with a droplet of sticky mucilage. Unwary **arthropods**, lured by scent, color, and nectar, are caught by this gluey material and are then firmly entangled during their struggles. The leaf then slowly, almost imperceptibly, enfolds the prey, which is then digested by proteolytic enzymes secreted by special glands on the leaf surface.

Passive traps lie in deadly wait for their small victims, which are attracted by enticing scents, colors, and nectar. However, these seeming treats are located at the end of a fatal, usually one-way passage, from which the prey cannot easily exit. The passage terminates in a pit filled with **water** and digestive enzymes, where the victim drowns, or is attacked by predacious insects that live symbiotically with the carnivorous **plant**.

The ingenious design of the pitcher plant (*Sarracenia purpurea*) is a revealing example of passive traps. The pitcher plant has foliage modified into upright vessels, as much as 4-6 in (10-15 cm) tall. When mature, these are reddish-green in color, with ultraviolet nectar guides pointing into their interior, which also emits alluring scents. The fringing lip and upper part of the inside of the pitcher are rich in insect attracting nectaries (organs that secrete nectar), and are covered with stiff, downward pointing bristles. These bristles can be easily traversed by an insect walking into the trap, but they passively resist movement upwards and out of the trap. Beneath the zone of bristles is a very waxy, slippery zone, the surface of which is almost impossible for even the tiny feet of insects to grasp, so they fall to the bottom of the trap. There the victim encounters a pool of collected rainwater, replete with digestive enzymes and the float-

A Venus fly trap with its red jaws open waiting for prey. © Peter J. Aitken/The National Audubon Society Collection/Photo Researchers, Inc. Reproduced by permission.

ing corpses of drowned insects, in various stages of decay and digestion. The newest victim struggles for a while, then drowns, and is digested.

Interestingly, a few species of insects are capable of living happily in the water-filled vessels of the pitcher plant and related species, such as the cobra plant (*Darlingtonia californica*). These insects are resistant to the digestive enzymes of the carnivorous plants, and they utilize the pitchers as a micro-aquatic **habitat**. Some species of midges and **flies** that live in pitcher plants actually attack recently trapped insects, killing and feeding on them. Eventually, the carnivorous plant benefits from nutrients excreted by the symbiotic insects. These pitchers also support a rich microbial community, which are useful in the decay of trapped arthropods, helping to make nutrients available for uptake by the carnivorous plant.

Conservation and protection of carnivorous plants

Most species of carnivorous plants are rare, and many are endangered. The principle threats to these species are habitat destruction caused by the drainage

A slender-leaved sundew (*Drosera linearis*) in Bruce National Park, Ontario. *Photograph by Robert J. Huffman. Field Mark Publications. Reproduced by permission.*

and infilling of **wetlands** and bogs to develop housing, and ecological conversions associated with agriculture and **forestry**. The **mining** of bog peat for horticultural materials or as a source of **energy** is another threat to some species of carnivorous plants. In addition, some species of carnivorous plants are actively collected in the wild to supply the horticultural trade, and this can seriously threaten the populations of those species.

Venus flytrap is a famous North American example of a carnivorous plant that is endangered in the wild. The natural distribution of this species is restricted to a small area of the coastal plain of North and South Carolina, fringing inland as far as 124 mi (200 km) along about 186 mi (300 km) of the coast, on either side of Cape Fear. However, the Venus flytrap only occurs today in a few small, scattered remnants of its natural habitat, associated with open spots in acidic bogs and pine savannas. To some degree this species has been endangered in the wild by excessive collecting in the past, but the modern threat is mostly associated with habitat losses to urbanization, agriculture, and forestry.

Fortunately, the Venus flytrap and many other species of carnivorous plants are fairly easy to propagate by vegetative means, usually by sowing leaf fragments onto moist sphagnum peat. For these species, there is no need to collect plants from the wild to supply the economic demands of **horticulture**.

However, some other species of carnivorous plants cannot be easily propagated in greenhouses, and the demand for these species by aficionados of these charismatic carnivores must be satisfied by collecting wild plants. In some cases, these demands are resulting in unsustainable harvests that are endangering wild populations, for example, of some of the species of the tropical Eurasian pitcher plant, *Nepenthes*.

However, even species that can be propagated in greenhouses may be collected from the wild for sale to horticulturalists, because quick and easy profits can be made in this way. So, if you decide to try to grow carnivorous plants as unusual pets, ensure that you are obtaining stock that was cultivated in a greenhouse, and not collected from the wild.

Resources

Books

Juniper, B.E., R.J. Robins, and D.M. Joel. *The Carnivorous Plants*. San Diego: Academic Press, 1989.

Lecoufle, M. *Carnivorous Plants: Care and Cultivation*. Blandford, U.K.: Sterling Publishing Co., 1991.

Schwartz, R. *Carnivorous Plants*. New York: Avon Books, 1975.

Bill Freedman

Carp

Carp are **fish** species in the minnow family (Cyprinidae), one of the major groups of **freshwater** fish. The most familiar **species** are the common carp (*Cyprinus carpio*) and the closely related goldfish (*Carassius aureus*). The minnow family is characterized by having no teeth in the jaws, although well-developed teeth occur on the pharyngeal bones (located behind the gill chamber) and are used to grind food against a hard, rough pad in the roof of the pharynx.

The body of the common carp is covered with large scales. A single dorsal fin is present, with 17-22 branched rays and a strong, toothed spine in front. The carp has four barbels, which are fleshy outgrowths of the mouth that play a sensory role in the locating of food. There are two long barbels at the corners of the mouth and shorter ones on the upper lip. Carp mostly eat vegetable **matter**, but also feed on worms, crustaceans, aquatic **insects**, and smaller fish. Their food is mostly obtained by probing in the bottom mud of their aquatic **habitat**. The common carp is colored dull green-brown on the flanks, darkening on the back, and the underside may be golden-yellow. The fins are gray-green or dusky brown, with a reddish tinge.

The common carp is a hardy fish that can live and breed under difficult conditions. Its preferred habitat is lowland lakes and slow-flowing **rivers** with abundant vegetation for food and shelter. During periods of exceptionally cold weather, carp move into deep **water** and enter a resting phase in which their **metabolism** is greatly slowed. Breeding takes place during the spring, generally from May to June, when water **temperature** reaches about 68°F (20°C). Spawning takes place in shallow water, and the eggs are laid directly onto plants. When the eggs hatch the tiny fry remain in the shallows for several weeks, concealed amongst the vegetation. The growth **rate** of carp varies considerably according to local conditions. They can attain a large size: individuals weighing more than 66 lb (30 kg) have been recorded. In general, however, mature carp are about 20-25 in (50-60 cm) in length and weigh from 4.4-10 lb (2-4.5 kg). In their natural environment, carp are thought to live for as many as 15 years; in captivity, however, far greater ages have been recorded, with some being credited with a life span of more than 200 years.

Of the fish species that have been reared successfully in captivity from egg to maturity, the carp has probably been one of the most successful on a commercial basis. A number of domesticated varieties of common carp occur, including one that is scaleless. There is a long history of carp aquaculture in the Far East, and these fish are also grown in parts of **Europe**. Carp have also been released to freshwater habitats in **North America**, **Australia**, and New Zealand, both as a sport fish and as a commercial venture. Unfortunately, the release of carp into foreign aquatic ecosystems often causes intense ecological damage. This results from the physical disturbance caused by carp as they feed and spawn, as well as the intense **competition** and predation they can exert on native species of fish.

See also Minnows; Suckers.

Carpal tunnel syndrome

Carpal tunnel **syndrome** results from compression and irritation of the median nerve where it passes through the wrist. In the end, the median nerve is responsible for both sensation and movement. When the median nerve is compressed, an individual's hand will feel as if it has "gone to sleep." The individual will experience numbness, tingling, and a prickly pin like sensation over the

A carp (*Cyprinus carpio*). © *Y. Lanceau Jacana, National Audubon Society Collection/Photo Researchers, Inc. Reproduced with permission.*

palm surface of the hand, and the individual may begin to experience muscle weakness, making it difficult to open jars and hold objects with the affected hand. Eventually, the muscles of the hand served by the median nerve may begin to atrophy, or grow noticeably smaller.

Compression of the median nerve in the wrist can occur during a number of different conditions, particularly conditions that lead to changes in fluid accumulation throughout the body. Because the area of the wrist through which the median nerve passes is very narrow, any swelling in the area will lead to pressure on the median nerve, which will interfere with the nerve's ability to function normally. Pregnancy, **obesity**, **arthritis**, certain thyroid conditions, diabetes, and certain pituitary abnormalities all predispose to carpal tunnel syndrome. Furthermore, overuse syndrome, in which an individual's job requires repeated strong wrist motions (in particular, motions which bend the wrist inward toward the forearm) can also predispose to carpal tunnel syndrome.

Research conducted by the American Academy of Orthopaedic Surgeons has found that advanced carpal tunnel syndrome can be prevented in many cases. They concluded that by doing an uncomplicated set of wrist exercises consistently before work, during breaks, and after work, pressure on the median nerves that leads to carpal tunnel syndrome can be avoided. The exercises are simple, involving mild flexion and extension of the wrists. By stretching the associated tendons, trauma from repetitive exertion is made less likely by significantly lowering pressure within carpal tunnels. People most likely to benefit from such **exercise** are those who use computers and other electronic keyboard devices daily. Women are known to experience carpal tunnel syndrome more frequently than do men.

Carpal tunnel syndrome is initially treated by splinting, which prevents the wrist from flexing inward into the position that exacerbates median nerve compression. When carpal tunnel syndrome is more advanced, injection of steroids into the wrist to decrease **inflammation** may be necessary. The most severe cases of carpal tunnel syndrome may require **surgery** to decrease the compression of the median nerve and restore its normal function.

An often underestimated disorder, carpal tunnel syndrome affects significant numbers of workers. In some years, according to federal labor **statistics**, carpal tunnel syndrome exceeded lower back **pain** in its contribution to the duration of work absences. One estimate reports that as many as 5-10 workers per 10,000 will miss work for some length of time each year due to work-related carpal tunnel syndrome. Additionally, the affliction is not limited to those whose jobs involve long hours of typing. International epidemiological data indicate that the highest rates of the disorder also include occupations such as meat-packers, **automobile** and other assembly workers, and poultry processors. Also from these studies, strong evidence is presented which positively correlates carpal tunnel syndrome with multiple risk factors, rather than a single factor alone. It is believed that the risk of developing carpal tunnel syndrome is far greater when continual repetition of action is combined with increased **force** of the action, wrist vibration, and overall poor posture.

Carrier (genetics)

In **genetics**, the term carrier describes an **organism** that carries two different forms (**alleles**) of a recessive **gene** (alleles of a gene linked to a recessive trait) and is thus heterozygous for that the recessive gene. Although carriers may act to convey and maintain recessive genes within a population by passing them on to offspring, the carriers themselves are not affected by the recessive trait associated with the recessive gene.

Although a carrier's **genome** contains a particular mutant allele, another gene (e.g., a dominant gene), or series of genetic mechanisms act to prevent the observable expression of that mutant allele (phenotypic expression). If, for example, at the genetic level an organism had a genotype (T, t), with the capital letter "T" designating a completely dominant allele and the lowercase letter "t" representing the recessive allele, that organism would express the observed trait associated with "T" and be a carrier for the recessive gene designated by "t." In contrast, the human **blood** type AB presents an example of allele codominance because the allele IA and IA allele are both expressed and contribute to the phenotype (blood group AB).

Because heterozygous organisms carry contain different forms (alleles) of a particular gene, diploid carriers produce sex cells (gametes) by the process of **cell meiosis**. Accordingly, heterozygous organisms produce gametes that contain different copies of the genes for which they are heterozygous. With regard to a (T, t) genotype, such an diploid organism would produce equal numbers of gametes that carried a single "T" allele or a single "t" allele.

At the observable level, an individual may, for example, act to convey the sickle cell gene but remain unaffected by sickle cell **disease** that strikes those who are homozygous for the sickle cell gene (i.e., carry two copies of the recessive sickle cell allele).

Under some conditions, a carrier may actually be more fit for a particular environment. Carriers who benefit from this heterozygote superiority or advantage are able to pass on and maintain a particular recessive allele within a population. In the case of sickle cell, the heterzygote carrier has a greater resistance to some forms of **malaria**. Accordingly, in malaria stricken areas, carriers of sickle cell disease avoid (in greater numbers) the selective disadvantages of malaria.

Studies of patients of Ashkenazi Jewish heritage (Jews of Eastern European descent), indicate that as many as one in seven individuals acts as a carrier of at least one of several different genetic diseases. Although some of these diseases are potentially fatal, the carriers of these diseases remain observably healthy individuals and show no signs of being affected with the disease related to the particular gene they carry.

Geneticists and physicians have developed a number of screening tests (carrier screening) to identify individuals who may be carriers for a particular gene.

See also Chromosomal abnormalities; Chromosome mapping; Chromosome; Gene mutation; Germ cells and the germ cell line; Pedigree analysis; Sexually transmitted diseases.

Carrot family (Apiaceae)

The carrot family (Apiaceae, or Umbelliferae) is a diverse group of about 3,000 **species** of plants, occurring in all parts of the world.

Most Umbellifers are herbaceous, perennial plants, often with aromatic foliage. Some species have poisonous foliage or roots. The leaves are typically alternately arranged on the stem, and in many species they are compound and divided into lobes. The flowers are small and

Cow parsnip growing in a field. *Mary M. Thacker/Photo Researchers, Inc. Reproduced by permission.*

contain both female (pistillate) and male (staminate) organs. The individual flowers are aggregated into characteristic, flat-topped inflorescences (groups) called umbels, from which one of the scientific names of the family (Umbelliferae) is derived. The **fruits** are dry, two-seeded structures called schizocarps, which split at maturity into two one-seeded, vertically ribbed sub-fruits, known as mericarps.

Edible species in the carrot family

Various species of the Apiaceae are grown as food or as flavorings. The best known of the food **crops** is the carrot (*Daucus carota*), a biennial **plant** native to temperate Eurasia. The cultivated carrot develops a large, roughly conical, orange-yellow tap root, which is harvested at the end of one growing season, just before the ground freezes. The **color** of carrot roots is due to the pigment carotene, a metabolic precursor for the synthesis of **vitamin** A. Carotene is sometimes extracted from carrots and used to color other foods, such as cheddar cheese, and sometimes butter and margarine. Carrots can be eaten raw, cooked as a vegetable, or added to stews and soups.

The parsnip (*Pastinaca sativa*) is another biennial species in which the whitish tap root is harvested and eaten, usually as a cooked vegetable. Both the carrot and parsnip are ancient cultivated plants, being widely used as a food and medicine by the early Greeks and Romans.

The celery (*Apium graveolens*) is native to moist habitats in temperate regions of Eurasia. Wild celery plants are tough, distasteful, and even poisonous, but domesticated varieties are harvested for their crisp, edible petioles (stalks). The most commonly grown variety of celery has been bred to have long, crunchy, juicy petioles. Until this variety was developed, the flavor of cultivated celery was commonly improved by a technique known as blanching, in which the growing plant is partially covered by mulch, **paper**, or boards to reduce the amount of **chlorophyll** that it develops. This practice is still used to grow "celery hearts." The celeriac or celery-root is a variety in which the upper part of the root and the lower part of the stem, a **tissue** known as a hypocotyl, are swollen, and can be harvested and used in soups or cooked as a nutritious vegetable. Celery **seeds** are sometimes used as a savory garnish for cooked foods, and to manufacture celery **salt**.

Parsley (*Petroselinum crispum*) is one of the most common of the garden herbs, and is often used as a savory, edible foliage, rich in vitamin C and **iron**. Parsley is

Fennel (*Foeniculum vulgare*). © *Hans Reinhard/Okapia 1990, National Audubon Society Collection/Photo Researchers, Inc. Reproduced with permission.*

most commonly used as a pleasing, but not-to-be-eaten visual garnish for well-presented, epicurean foods. This dark-green plant can be a pleasant food in itself and is used to flavor tabouleh, a North African dish made with bulgar **wheat**, tomatoes, and lots of chopped parsley.

The foliage of dill (*Anethum graveolens*) is commonly used to flavor pickled cucumbers, gherkins, and tomatoes, and sometimes as a steamed garnish for **fish** or chicken. Chervil (*Anthriscus cerefolium*) leaves are also used as a garnish and in salads.

Other species in the Apiaceae are cultivated largely for their tasty, aromatic seeds. The economically most important of these savory seeds are those of caraway (*Carum carvi*), which are widely used to flavor bread and cheese. An aromatic oil extracted from caraway seeds is used in the preparation of medicine and perfume, and to flavor the liquors kummel and aqua-vitae. The anise or aniseed (*Pimpinella anisum*) is one of the oldest of the edible, aromatic seeds. An oil extracted from the seeds of anise is used to flavor candies, cough medicines, and a liquor known as anisette. The seeds of fennel (*Foeniculum vulgare*) also contain anise oil, and are used in the preparation of medicines and liquorice, while the foliage is sometimes used as a garnish. The seeds of angelica (*Angelica archangelica*) are also a source of an aromatic oil, used to flavor vermouth and other liquors. Coriander (*Coriander sativum*) seeds yield another aromatic oil, used to flavor candy, medicine, and liquors. The seeds of cumin (*Cuminum cyminum*) are used to flavor breads, cheese, candy, soup, and pickles.

Ornamental species

A few species in the carrot family are grown as ornamentals, usually as foliage plants, rather than for their flowers. A variegated variety of goutweed (*Aegopodium podagraria*) is often cultivated for this reason, as are some larger species, such as angelica (*Angelica sylvestris*).

Wild species occurring in North America

A number of species of wildflowers in the carrot family occur naturally in **North America**, or have been introduced from elsewhere and have spread to natural habitats.

Some of the more familiar and widespread native species of Apiaceae in North America include black snake-root (*Sanicula marylandica*), sweet-cicely (*Osmorrhiza claytoni, O. divaricata*), Scotch or sea lovage (*Ligusticum scothicum*), golden Alexanders (*Zizia aurea*), marsh-pennywort (*Hydrocotyle umbellata*), and **water** hemlock (*Sium suave*).

Some wild species in the Apiaceae are deadly poisonous. The poison hemlock (*Conium maculatum*) is a native of Eurasia, but has spread in North America as an introduced weed. Poison hemlock may be the most poisonous of the temperate plants, and it can be a deadly forage for cattle. The famous Greek philosopher, Socrates, is thought to have been executed by being condemned in the courts to drink a fatal infusion prepared from the poison hemlock. Native species are similarly poisonous, for example, the water hemlock or cowbane (*Cicuta maculata*), and the bulb-bearing water hemlock (*Cicuta bulbifera*). Most cases of poisoning by these plants involve cattle or people eating the roots or the seeds, which, while apparently tasty, are deadly toxic.

Some other wild species, while not deadly, can cause a severe dermatitis in exposed people. These include wild parsnip (*Pastinaca sativa*) and cow-parsnip (*Heracleum lanatum*).

The wild carrot, also known as Queen Anne's-lace, or bird's-nest plant (*Daucus carota*) is a common, **introduced species** in North America. This is a wild variety of the cultivated carrot, but it has small, fibrous tap roots, and is not edible. The Queen Anne's-lace probably escaped into wild habitats in North America from cultivation. However, other Eurasian species in the Apiaceae appear to have been introduced through the dumping of

KEY TERMS

Biennial—A plant that requires at least two growing seasons to complete its life cycle.

Inflorescence—A grouping or arrangement of florets or flowers into a composite structure.

Umbel—An arrangement of flowers, whereby each flower stalk arises from the same level of the stem, as in onions.

Weed—Any plant that is growing abundantly in a place where humans do not want it to be.

ships' ballast. This happened when ships sailing from **Europe** to America carried incomplete loads of cargo, so they had to take on **soil** to serve as a stabilizing ballast at sea. The soil ballast was usually dumped at an American port, serving as a means of entry for many species of Eurasian weeds, which had viable seeds in the material. Species of Apiaceae that are believed to have spread to North America in this way include the knotted hedgeparsley (*Torilis nodosa*), Venus-comb or shepherd's-needle (*Scandix pecten-veneris*), and cow-parsnip (*Heracleum sphondylium*).

See also Herb.

Resources

Books

Conger, R.H.M., and G. D. Hill. *Agricultural Plants*. 2nd ed. Cambridge: Cambridge University Press, 1991.

Hartmann, H.T., et al. *Plant Science. Growth, Development, and Utilization of Cultivated Plants*. Englewood Cliffs, NJ: Prentice-Hall, 1988.

Hvass, E. *Plants That Feed and Serve Us*. New York: Hippocrene Books, 1975.

Judd, Walter S., Christopher Campbell, Elizabeth A. Kellogg, Michael J. Donoghue, and Peter Stevens. *Plant Systematics: A Phylogenetic Approach*. 2nd ed. with CD-ROM. Suderland, MD: Sinauer, 2002.

Klein, R. M. *The Green World. An Introduction to Plants and People*. New York: Harper & Row, 1987.

Bill Freedman

Carrying capacity

Carrying capacity refers to the maximum abundance of a **species** that can be sustained within a given area of **habitat**. When an ideal population is at equilibrium with the carrying capacity of its environment, the **birth** and death rates are equal, and size of the population does not change. Populations larger than the carrying capacity are not sustainable, and will degrade their habitat. In nature, however, neither carrying capacity or populations are ideal—both vary over **time** for reasons that may be complex, and in ways that may be difficult to predict. Nevertheless, the notion of carrying capacity is very useful because it highlights the ecological fact that, for all species, there are environmental limitations to the sizes of populations that can be sustained.

Carrying capacity is never static. It varies over time in response to gradual environmental changes, perhaps associated with climatic change or the successional development of ecosystems. More rapid changes in carrying capacity may be caused by disturbances of the habitat occurring because of a fire or windstorm, or because of a human influence such as timber harvesting, **pollution**, or the introduction of a non-native competitor, **predator**, or **disease**. Carrying capacity can also be damaged by overpopulation, which leads to excessive exploitation of resources and a degradation of the habitat's ability to support the species. Of course, birth and death rates of a species must respond to changes in carrying capacity along with changes in other factors, such as the intensities of disease or predation.

Carrying capacity for humans

Humans, like all organisms, can only sustain themselves and their populations by having access to the products and services of their environment, including those of other species and ecosystems. However, humans are clever at developing and using technologics; as a result they have an unparalleled ability to manipulate the carrying capacity of the environment in support of their own activities. When prehistoric humans first discovered that crude tools and weapons allowed greater effectiveness in gathering wild foods and hunting animals, they effectively increased the carrying capacity of the environment for their species. The subsequent development and improvement of agricultural systems has had a similar effect, as have discoveries in medicine and industrial technology.

Clearly, the cultural **evolution** of human socio-technological systems has allowed enormous increases to be achieved in carrying capacity for our species. This increased effectiveness of environmental exploitation has allowed a tremendous multiplying of the human population to occur. In prehistoric times (that is, more than 10,000 years ago) all humans were engaged in a primitive hunting and gathering lifestyle, and their global population probably amounted to several million individuals. In the year 2000, because humans have been so adept at increasing the carrying capacity of their environment, more than six billion individuals were sustained, and the global population is still increasing.

Humans have also increased the carrying capacity of the environment for a few other species, including those with which we live in a mutually beneficial **symbiosis**. Those companion species include more than about 20 billion domestic animals such as cows, **horses**, **pigs**, **sheep**, **goats**, dogs, **cats**, and chickens, as well as certain plants such as **wheat**, **rice**, **barley**, maize, tomato, and cabbage. Clearly, humans and their selected companions have benefited greatly through active management of Earth's carrying capacity.

However, an enormously greater number of Earth's species have not fared as well, having been displaced or made extinct as a consequence of ecological changes associated with the use and management of the environment by humans, especially through loss of their habitat and over harvesting. In general, any increase in the carrying capacity of the environment for one species will negatively affect other species.

In addition, there are increasingly powerful indications that the intensity of environmental exploitation required to sustain the large populations of humans and our symbionts is causing important degradations of carrying capacity. Symptoms of this environmental deterioration include the **extinction** crisis, decreased **soil** fertility, **desertification**, **deforestation**, fishery declines, pollution, and increased **competition** among nations for scarce resources. Many reputable scientists believe that the sustainable limits of Earth's carrying capacity for the human enterprise may already have been exceeded. This is a worrisome circumstance, especially because it is predicted that there will be additional large increases in the global population of humans. The degradation of Earth's carrying capacity for humans is associated with two integrated factors: (1) overpopulation and (2) the intensity of resource use and pollution. In recent decades human populations have been growing most quickly in poorer countries, but the most intense lifestyles occur in the richest countries.

If it is true that the human enterprise has exceeded Earth's carrying capacity for our species, then compensatory adjustments will either have to be made by the human economy, or they will occur naturally. Those managed or catastrophic changes will involve a combination of decreased per-capita use of environmental resources, decreased birth rates, and possibly, increased death rates.

See also Sustainable development.

Resources

Books

Begon, M., J.L. Harper, and C.R. Townsend. *Ecology: Individuals, Populations and Communities.* 2nd ed. London: Blackwell Sci. Pub., 1990.

Freedman, B. *Environmental Ecology.* 2nd ed. San Diego: Academic Press, 1995.

Ricklefs, R. E. *Ecology.* New York: W.H. Freeman and Co., 1990.

Bill Freedman

Cartesian coordinate plane

The Cartesian coordinate system is named after René Descartes (1596–1650), the noted French mathematician and philosopher, who was among the first to describe its properties. However, historical evidence shows that Pierre de Fermat (1601-1665), also a French mathematician and scholar, did more to develop the Cartesian system than did Descartes.

To best understand the nature of the Cartesian **plane**, it is desirable to start with the number line. Begin with line L and let L stand for a number axis (see Figure 1). On L choose a **point**, O, and let this point designate the zeropoint or origin. Let the **distance** to the right of O be considered as positive; to the left as **negative**. Now choose another point, A, to the right of O on L. Let this point correspond to the number 1. We can use this distance between O and A to serve as a unit with which we can locate B, C, D,... to correspond to the +1, +2, +3, +4,... Now we repeat this process to the left of O on L and call the points Q, R, S, T,... which can correspond to the numbers -1, -2, -3, -4,... Thus the points A, B, C, D,..., Q, R, S, T,... correspond to the set of the **integers** (see Figure 1). If we further subdivide the segment OA into d equal parts, the 1/d represents the length of each part. Also, if c is a positive integer, then c/d represents the length of c of these parts. In this way we can locate points to correspond to rational numbers between 0 and 1.

By constructing rectangles with their bases on the number line we are able to find points that correspond to some irrational numbers. For example, in Figure 1, **rectangle** OCPZ has a base of 3 and a segment of 2. Using the Theory of Pythagorus we know that the segment OP has a length equal to $\sqrt{13}$. Similarly, the length of segment OW is $\sqrt{10}$. The **real numbers** have the following property: to every real number there corresponds one and only one point on the number axis; and conversely, to every point on the number axis there corresponds one and only one real number.

What happens when two number lines, one horizontal and the other vertical, are introduced into the plane? In the rectangular Cartesian plane, the position of a point is determined with reference to two **perpendicular** line called coordinate axes. The intersection of

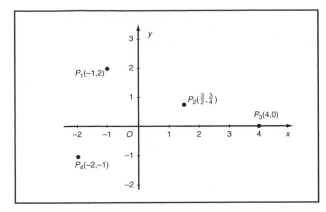

Figure 1. *Illustration by Hans & Cassidy. Courtesy of Gale Group.*

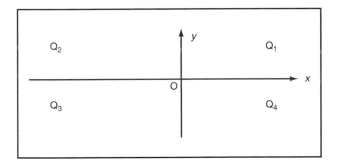

Figure 2. Cartesian Plane *Illustration by Hans & Cassidy. Courtesy of Gale Group.*

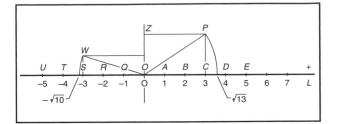

Figure 3. A plot of the points P $_1$(-1, 2); P $_2$(3/2,3/4); P $_3$(4,0); P $_4$(-2, -1). *Illustration by Hans & Cassidy. Courtesy of Gale Group.*

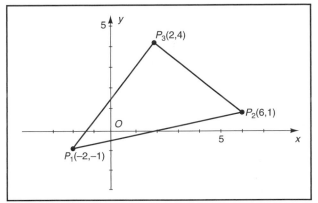

Figure 4. A graph of the triangle whose vertices are the points P $_1$(-2,-1), P $_2$(6,1), and P $_3$(2,4). *Illustration by Hans & Cassidy. Courtesy of Gale Group.*

these axes is called the origin, and the four sections into which the axis divide a given plane are called quadrants. The vertical axis is real numbers usually referred to as the y axis or functions axis; the horizontal axis is usually known as the x axis or axis of the independent **variable**. The direction to the right of the y axis along the x axis is taken as positive; to the left is taken as negative. The direction above the x axis along the y axis is taken as positive; below as negative. Ordinarily, the unit of measure along the coordinate axes is the same for both axis, but sometimes it is convenient to use different measures for each axis.

The symbol P_1 (x_1, y_1) is used to denote the fixed point P_1. Here x_1 represents the x coordinate (abscissa) and is the perpendicular distance from the y axis to P_1; y_1 represents the y coordinate (ordinate) and is the perpendicular distance from the axis to P_1. In the symbol P_1 (x_1,y_1), x_1 and y_1 are real numbers. No other kind of numbers would have meaning here. Thus, we observe that by means of a rectangular coordinate system we can show the correspondence between pairs of real numbers and points in a plane. For each pair of real numbers (x, y) there corresponds one and only one point (P), and conversely, to each point (P) there corresponds one and only pair of real numbers (x, y). We say there exists a

one-to-one **correspondence** between the points in a plane and the pair of all real numbers.

The introduction of a rectangular coordinate system had many uses, chief of which was the concept of a graph. By the graph of an equation in two variables, say x and y, we mean the collection of all points whose coordinates satisfy the given equation. By the graph of the **function** f(x) we mean the graph of the equation y= f(x). To plot the graph of an equation we substitute admissible values for one variable and solve for the corresponding values of the other variable. Each such pair of values represents a point which we locate in the coordinate system. When we have located a sufficient number of such points, we join them with a smooth **curve**.

In general, to draw the graph of an equation we do not depend merely upon the plotted points we have at our disposal. An inspection of the equation itself yields certain properties which are useful in sketching the curve like **symmetry**, asymptotes, and intercepts.

Cartilage *see* **Connective tissue**

Cartilaginous fish

Cartilaginous **fish** such as **sharks**, **skates**, and **rays** are **vertebrates** whose internal skeleton is made entirely of cartilage and contains no ossified bone. Cartilaginous fish are also known as Chondrichthyes and have one or two dorsal fins, a caudal fin, an anal fin, and ventral fins which are supported by girdles of the internal skeleton.

Placoid scales, or dermal teeth, are characteristic of the skin of both sharks and rays. The touch of shark's skin is similar to the feel of sandpaper and was used as such for many years. The tiny teeth that protrude from the skin vary in each **species** of shark. The tooth tip is dentine with an overlay of dental enamel, while the lower part of the tooth is made of bone, which anchors the tooth to the skin.

The skin of rays is naked in places, that is, without dermal teeth, but on the back of or upper tail surface, the dermal teeth have developed large, strong spines.

The jaw teeth of both sharks and rays are in fact modified dermal teeth, which are lost when they become worn, and are replaced by rows of new teeth from the space behind. In some species of sharks, the jaw looks like an assembly line, with new teeth filling spaces immediately.

Both sharks and rays breathe through gills and have an opening called a spiracle on both sides of the head behind the **eye**. The spiracle enables the rays, which often bury in the **sand**, and seabed-resting sharks to take in **water**, pump it through the gill chamber, and release it through the gill slits without taking in large amounts of mud and sand. These fish usually take respiratory water in through the mouth, extract the **oxygen** from the water in the gills, and pass it out through the gill slits.

Cartilaginous fish are divided into two subclasses on the basis of gill slits and other characteristics. The first is the Elasmobranchs, which have at least five gill slits and gills on each side, one spiracle behind each eye, dermal teeth on the upper body surface, a tooth jaw, and an upper jaw not firmly attached to the skull. Sharks (Selachii), rays, and skates (Rajiformes) belong to this group. The chimeras (Holocephali) have one gill opening on each side, tooth plates, and a skull with a firmly attached upper jaw.

Cartilaginous fish do not have swim bladders, so a swimming **motion** must be maintained continuously, even when sleeping, or they will sink to the bottom. The caudal fin of the shark provides the propellant **force** in swimming, the dorsal fin provides balance, and the pectoral fins are used for upward force and depth rudders.

The flattened body and the rear spine of the rays makes their swimming motion unique and completely different than that of sharks. The large flattened body of the rays has become fused with the pectoral fins, which produces vertical waves from front to rear, similar to that of a bird in flight.

The chimeras utilize their pectoral fins when swimming, beating these fins simultaneously for propulsion, or alternately, to change direction. This method is highly effective for this group of cartilaginous fish, but is seen most often in **bony fish**.

The pectoral fins in the male cartilaginous fish are also used for mating. The rear part of the pectoral fin is modified as a copulatory **organ**. All cartilaginous fish have internal **fertilization**. Some species are **oviparous**, or egg-layers, and some are **ovoviviparous**, hatching the eggs within the female and giving **birth** to live young. Still others may be viviparous, with the young developing in utero, similar to **mammals**, with the yolk sac developing into a yolk placenta providing **nutrients** to the embryo.

Only true rays, species of sharks which live near the sea bed, and the chimeras lay eggs. The eggs are often encased in a leathery shell with twisted tendrils which anchor the egg case to **rocks** or weeds. These leathery shells are known as the "mermaid's purse."

Cartilaginous fish are predatory, meaning that they feed on other animals, from zoo **plankton** to shellfish to whales. Cartilaginous fish themselves are sought after by humans as a food source. Shark meat, once marketed under the pseudonyms of "flake" and "steakfish" is now popular worldwide. Shark fins have long been popular in the Orient. Rays are considered delicacies in Great Britain and France, and thornback rays and flapper skates are often sold as sea trout.

F. C. Nicholson

Cartography

Cartography is the creation, production, and study of maps. Cartographers are often geographers who specialize in the combination of art, science, and technology to make and study maps. Some cartographers teach map-making skills and techniques, some design and produce maps, and some are curators of **map** libraries. All cartographers, however, focus on maps as the object of their study or livelihood. In other cases, biologists, economists, geologists, hydrologists, planners, and others can engage in cartography to summarize or analyze spatial data. Geologists, for example, produce highly specialized geologic maps to show the three-dimensional arrangement of rock types in an area.

A major change in cartography during the past decade has been the growing use of geographic information system (**GIS**) software to produce, store, and use maps. GIS software can be used to create custom maps that cover an area or portray features of specific interest to a user. For example, a map showing vegetation types can be placed over a shaded relief map of the earth's surface to illustrate the relationship between **biology** and topography. GIS software can also be linked to computer simulations of processes such as **flooding** or **earthquake** damage to help communities develop emergency response plans. Digital maps can also be widely distributed using internet map servers that allow users to interactively explore a large map by scrolling and zooming.

What is a map?

A map is a two-dimensional representation of the spatial distribution of phenomena or objects. For example, a map may show the location of cities, mountain ranges, **rivers**, or types of rock in a given region. Most maps are flat, making their production, storage, and handling relatively easy. Maps present their information to the viewer at a reduced scale. They are smaller than the area they represent and use mathematical relationships to maintain proportionally accurate geographic relationships between points. Maps portray information by using symbols that are identified in a legend.

The history of cartography

References to surveying and mapping are found in ancient Egyptian and Mesopotamian writings. The oldest known map is of an area in northern Mesopotamia. The baked clay tablet, found near Nuzi, Iraq, dates from approximately 3800 B.C. Fragments of clay maps nearly 4,000 years old have been found in other parts of Mesopotamia, some showing city plans and others showing parcels of land. Over 3,000 years ago, the ancient Egyptians were surveying the lands in the Nile Valley. They drew detailed maps on papyrus to use for taxation purposes.

Chinese cartographers produced maps as early as 227 B.C. Following the invention of **paper** about A.D. 100, cartography flourished throughout the Chinese empire. Chinese cartography continued to have its own distinctive style until the 1500s, when it began to be influenced by European cartography.

Although Chinese cartography followed certain standards, it was not based on the same scientific principles as European cartography. The Greeks developed many of the basic principles of modern cartography, including **latitude and longitude**, and map projections.

The maps of Ptolemy, a Greek astronomer and mathematician who lived in the first century A.D., are considered the high point of Greek cartography. Although his maps appear crude by current standards, they are amazingly accurate given the extent of geographic knowledge at the time.

Cartography came to a near-halt in **Europe** during the medieval period, when maps were little more than imaginative illustrations for theological texts. In Muslem countries, however, the science of cartography continued to grow, and various techniques were refined or improved by Arabic cartographers. Their knowledge and skills were introduced into Europe during the Renaissance.

The eras of exploration that followed the Renaissance supplied cartographers with a wealth of new information, which allowed them to produce maps and navigation charts of ever-increasing accuracy and detail. Europeans became fascinated with the idea of mapping the world. The French initiated the first national topographic survey during the 1700s, and soon other European countries followed suit. Today, most countries have an official organization devoted to cartographic research and production.

Types of maps

There are many different types of maps. So many, in fact, that it can be difficult to classify them into groups. A common classification system divides maps into two categories: general and thematic. General maps are maps that show spatial relationships between a variety of geographic features and phenomena, emphasizing their location relative to each other. Thematic maps illustrate the spatial variation of a single phenomenon or variable, or the spatial relationship between two particular phenomena or variables, emphasizing the pattern of the distribution.

Maps can be either general or thematic, depending on the intent of the cartographer. For example, a cartographer may produce a vegetation map showing the distribution of various **plant** communities. If the cartographer shows the location of various plant communities in relation to a number of other geographic features, the map is properly considered a general map. The map is more likely to be considered thematic if the cartographer uses it to focus on something about the relationship of the plant communities to each other, or to another particular phenomenon or feature, such as the differences in plant communities associated with changes in elevation or changes in **soil** type.

Some examples of general maps include topographic maps, planimetric maps, and charts. Topographic maps depict the form of the earth's surface, most commonly expressed as elevation above **sea level**, and are general maps if they also include features such as cities, rivers, and roads. Bathymetric maps depict underwater topography. Planimetric maps show features such as cities and roads without depicting elevations. Charts are used by the navigators of **aircraft** and seagoing vessels to plot positions and courses. World maps on a small or medium scale showing physical and cultural features, such as those in atlases, are also considered general maps.

Although the subject matter of thematic maps is nearly infinite, cartographers use common techniques involving points, lines, and areal units to illustrate the structure of spatial distribution. Isarithmic maps use lines to connect points of equal value; these lines are called isopleths, or isolines. Isopleths are used to show how certain quantities change with location and those used for a particular purpose may have a particular name. For example, isotherms connect points of equal **temperature**, **isobars** connect points of equal air **pressure**, and isohyets connect points of equal **precipitation**. Isopleths indicating differences in elevation are called topographic contour lines, and a topographic map that does not depict general features such as cities and roads would be a thematic map.

A topographic map is a good example of how isopleths are used to present information. Topographic maps use isopleths called elevation contour lines to indicate the topographic relief. Each contour line connects points of the same elevation, and the difference in elevation between each contour line is known as the contour **interval**. A contour interval of 20 ft (6.1 m) means that there is a 20 ft (6.1 m) difference in elevation between the points connected by one contour line and the points connected by an adjacent contour line. Closely spaced contour lines represent steep slopes and widely spaced contour lines represent gentle slopes. Closed contour lines, for example in the general shape of a **circle** or **ellipse**, represent hills.

Chloropleth maps are another type of thematic map. They use areas of graduated gray tones or colors to show spatial variations in the magnitude of a phenomenon.

Geographic illustrations

There are many portrayals of geographic relationships that do not qualify as maps as previously defined. Throughout human history, people have been illustrating geographic relationships between various elements of the physical and cultural environment. These geographic illustrations and representations are often beautiful, and can illustrate the world view of the culture that produced them. Some are extremely accurate in their representation of geographic relationships. Most geographic illustrations, however, are not considered true maps by modern cartographers because they do not use a scale based on **distance**. The development of the tools and techniques for accurately measuring distance requires a particular technical and scientific world view not shared by all cultures.

Many geographic illustrations or representations do not have a scale. Those that do, usually have a scale based on traveling times. Traveling times for the same distance can vary depending on the nature of the terrain, **weather** conditions, or other variables. For example, a 4-mi (6.4-km) journey across rugged **mountains** in a snow **storm** and a 12-mi (19.3-km) journey across a relatively smooth plain on a pleasant spring day may both take eight hours. A geographic illustration using a time-based scale would show two equal intervals; a distance-based scale would show the 12-mi (19.3-km) journey as three times longer than the rugged 4-mi (6.4-km) trek. Clearly, for a nomadic or migratory society, a geographic representation with a scale based on traveling times would be extremely useful, whereas one with a scale based on a distance would be of little or no use.

Map making

Cartographers traditionally obtained their information from navigators and surveyors. Explorations that expanded the geographical awareness of a map-making culture also resulted in increasingly sophisticated and accurate maps. Today, cartographers incorporate information from aerial photographs and **satellite** images in the maps they create.

Modern cartographers face three major design challenges when creating a map. First, they must figure out how to represent three-dimensional objects in two dimensions. Second, cartographers must represent geographic relationships at a reduced size while maintaining their proportional relationships. Third, they must select which pieces of information will be included in the map and develop a system of generalization that will make the information presented by the map useful and accessible to its readers. This includes the development of symbols that will effectively convey the subject of the map.

Showing three-dimensional relationships in two dimensions

When creating a flat map of a portion of the earth's surface, cartographers first locate their specific area of interest using latitude and longitude. They then use map projection techniques to represent the three-dimensional characteristics of that area in two dimensions. Finally, a grid, called a rectangular coordinate system, may be superimposed on the map, making it easier to use.

Latitude and longitude

Distance and direction are used to describe the location of an object in **space**. In conversation, terms like right and left, up and down, or here and there are used to indicate direction and distance. These terms are useful only if you know the location of the speaker; in other words, they are relative. Cartographers, however, need objective terms for describing location, because maps are intended for use by many individuals in many different situations. The system of latitude and longitude, a geographical coordinate system developed by the Greeks, is used by cartographers for describing location.

Earth is an oblate, rotating around an axis tilted approximately 23.5 degrees. The two points where the axis intersects Earth's surface are called the poles. The equator is an imaginary circle drawn around the center of Earth, equidistant from both poles. A **plane** that sliced through the earth at the equator would intersect the axis of Earth at a right **angle**. Lines drawn around the earth to the north and south of the equator and at right angles to Earth's axis are called parallels. Any point on Earth's surface is located on a **parallel**.

An **arc** is established when an angle is drawn from the equator to the axis and then north or south to a parallel. Latitude is the measurement of this arc in degrees. There are 90 degrees from the equator to each pole, and sixty minutes in each degree. Latitude is used to determine distance and direction north and south of the equator.

Meridians are lines running from the north pole to the south pole, dividing Earth's surface into sections, like those of an orange. Meridians intersect parallels at right angles, creating a grid. Just as the equator acts as the line from which to measure north or south, a particular meridian, called the prime meridian, acts as the line from which to measure east or west. There is no meridian that has a natural basis for being considered the prime meridian. The prime meridian is established by international agreement and currently runs through the Royal Observatory in Greenwich, England. Longitude is the measurement in degrees of the arc created by an angle drawn from the prime meridian Earth's axis and then east or west to a meridian. There are 180 degrees west of the prime meridian and 180 degrees east of it. The international dateline lies approximately where the 180th meridian passes through the Pacific Ocean.

Using the geographical coordinate system of latitude and longitude, any point on the earth's surface can be located with precision. For example, Buenos Aires, the capital of Argentina, is located 34 degrees 35 minutes south of the equator and 58 degrees 22 minutes west of the prime meridian. Anchorage, the capital Alaska, is located 61 degrees 10 minutes north of the equator and 149 degrees 45 minutes west of the prime meridian.

Map projections

After locating their area of interest using latitude and longitude, cartographers must determine how best to represent that particular portion of the Earth's surface in two dimensions. They must do this in such a way that minimal amounts of distortion affect the geographic information the map is designed to convey.

In order to understand the difficulty of such a task, imagine an orange with lines similar to parallels and meridians inked onto its surface. Now imagine removing the peel from the orange in one piece. If the peel of an orange is laid out flat on a tabletop, the peel will crack and break in various places. The cracks and breaks will distort the original shape of the orange, and the inked lines will no longer bear the same spatial relationship to each other as they did when the peel was on the orange. If the peel is arranged so that there are no cracks, breaks, or distortions in the relationships between the lines on its surface, the peel will assume the shape of a hollow **sphere**. There are only two choices: a spherical, distortion-free arrangement or a flat, distorted arrangement.

Cartographers have developed map projections to transform geographic information from a spherical surface onto a planar surface. A map projection is a method for representing a curved surface, such as the surface of Earth, on a flat surface, such as a piece of paper, so that each point on the curved surface corresponds to only one point on the flat surface.

There are many types of map projections. Some of them are based on **geometry**; others are based on mathematical formulas. None of them, however, can accurately represent all aspects of the earth's surface. Inevitably there will be some distortion in shape, distance, direction or area. Each type of map projection is intended to reduce the distortion of a particular spatial element. Some projections reduce directional distortion, while others try to present shapes or areas in as distortion-free a manner as possible. The cartographer must decide which of the many projections available will provide the most distortion-free presentation of the information to be mapped.

Rectangular coordinates

Although the geographical coordinate system is useful for large areas, it can be awkward to use for small areas. Many city maps use rectangular coordinate systems. After the map is complete, a grid is superimposed over it. The horizontal lines of the grid are assigned one set of numbers or letters, the vertical lines are assigned another set of numbers or letters. An index of place names is generated, which lists the horizontal and vertical coordinates for each place shown on the map. Used in conjunction with an index of place names, the rectangular coordinate system makes it simple for map readers to locate particular places. The Universal Transverse Mercator (UTM) system divides most of the earth's surface into zones, each of which has its own rectangular grid system. Most **global positioning system** (GPS) receivers can be set to show locations in UTM coordinates.

Reducing size while maintaining accurate proportions

Maps present geographical information at a reduced scale. In order for the information to be useful to the map user, the relative proportions of geographic features and spatial relationships must be kept as accurate as possible. Cartographers use various types of scales to keep those features and relationships in the correct proportion.

Scale is the mathematical relationship between a distance between two points on the map and the distance between two corresponding points on the ground. The relationship is expressed as a **ratio**, the first number being the distance between two points on the map and the second number being the actual distance represented. The number indicating map distance is always one. Thus, a map with a scale of 1:125,000 tells the map user that every unit of distance on the map equals 125,000 of the same units of distance on the ground. The units of distance used are not important as long as they are the same on both sides of the ratio. One centimeter on the map would equal 125,000 cm on the ground, 1 ft on the map would equal 125,000 ft on the ground, and 1 m on the map would equal 125,000 m on the ground.

Maps showing a large area are called small-scale maps. This is because the ratio between map distance and actual distance is a small number. The number is small because one distance unit on the map represents a large number of distance units on the ground. For example, a map showing **North America** at a scale of 1:40,000,000 would use one unit of map distance to depict 40,000,000 units of actual distance. One centimeter on these maps equals 40 km of actual distance. Such a map fits on a piece of paper only 9 in wide and 8.25 in high (23 cm by 21 cm).

Maps showing a small area are called large-scale maps. The ratio between map distance and actual distance is a large number. It is large because each unit of distance on the map represents a relatively small number of distance units on the ground. City maps are good example of large-scale maps. A city map of Portland, Oregon with a scale of 1:38,000 fits on a piece of paper 41.75 in by 35.5 in (106.5 cm by 90 cm). One centimeter

on this map equals 0.38 km of actual distance and one inch equals six tenths of a mile.

Every properly prepared map has a statement of its scale. This statement can take many forms, and many maps express scale in more than one way. The scale may be indicated by a ratio, such as 1:100,000 or 1/100,000 (the latter is less common). This ratio is called the representative fraction. Representative fractions are not particularly easy to use in everyday situations, so cartographers have developed other ways to communicate the scale of a map to its users.

Sometimes cartographers use a graphic scale, also called a bar scale. A graphic scale is a line or bar subdivided to show how many actual miles fit into a particular measurement on the map. In most parts of the world the graphic scale shows how many actual miles or kilometers are represented by a particular number of inches or centimeters on the map.

Two other means for expressing scale are the area scale and the verbal statement. Area scales are used for maps based on equal-area projections, that is, maps that present all areas shown in the same proportion to one another as they occur on Earth's surface. These scales tell the reader that one unit of area on the map represents a certain area on the ground. The scale can be written $1:250,000^2$, although 1:250,000 is more common. The latter expression assumes the reader is aware that the number represents a ratio of **square** units. A verbal statement of scale uses words, rather than numbers or graphic symbols. "One inch equals one mile" is a verbal statement of scale equivalent to the representative fraction 1:63,360 (there are 63,360 inches in one mile).

Presenting geographic information effectively

No single map can accurately show every feature on Earth's surface. There is simply too much spatial information at any particular point on Earth's surface for all of the information to be presented in a comprehensible, usable format. In addition, the process of reduction has certain visual effects on geographic features and spatial relationships. Because every feature is reduced by the ratio of the reduction, the distance between features is reduced, crowding them closer together and lessening the clarity of the image. The width and length of individual features are also reduced.

When designing a map, cartographers strive for clarity and effective communication. They use the technique of selection to determine which pieces of information to include and what kinds of symbols will most effectively portray that information.

A wide array of geographical information is available to mapmakers. When preparing a map, cartographers must choose only those pieces of information that are pertinent to the purpose of the map and then display those pieces of information in a way that effectively communicates their significance. Only information deemed significant or useful is selected for inclusion in the map.

Once cartographers have selected the information that will be portrayed on the map, the information must be displayed in an effective manner. Cartographers deal with this problem by applying the techniques of cartographic generalization. Both map geometry and map content are generalized.

Geometric generalization techniques change the placement and appearance of various map features in order to make the map easier to interpret and more pleasing to the **eye**. For example, not every twist and turn of a 15 mi stretch of river can be accurately portrayed at a 1:500,000 scale, where 1 in equals 7.89 mi. The path of the river is simplified, reducing excessive detail and angularity. A railroad running 50 ft from the river would appear to run in the riverbed when shown at a 1:500,000 scale. Using cartographic generalization, the cartographer displaces the railroad, showing it next to the river, avoiding graphic interference and increasing the readability of the map. The numerous right-angle bends in a highway following rural property boundaries along the river would be smoothed by the cartographer, reducing their angularity and thereby making the line of the highway easier for the eye to follow.

Linear and areal features can be generalized using the techniques of simplification, displacement, smoothing, and enhancement. Additionally, the techniques of dissolution, segmentation, and aggregation are applied to areal features. Point features are generalized by displacement, graphic association, and abbreviation.

Map content is generalized using the technique of classification, in which similar features to be grouped together are represented by a single symbol. Campgrounds, for example, are often represented by a tent-shaped symbol, even when the facilities can accommodate trailers or large recreational vehicles. Categorization is another form of classification. Many maps, for example, use one point symbol for population centers of 1,000–10,000. Another point symbol for population centers of more than 10,000 but less than 100,000, and a third point symbol for population centers of more than 100,000 but less than 500,000. Cartographers must carefully consider the implications of such classification schemes. The system described above implies that towns of 1,000 and towns of 9,000 have more in common than towns of 9,500 and towns of 10,500.

Cartographic production

For many centuries maps were produced entirely by hand. They were drawn or painted on paper, hide, parchment, clay tablets, and slabs of **wood**, among other things. Each map was an original work; the content may have been copied, but each map was executed by hand.

Once **printing** techniques were developed, many reproductions could be made from one original map. Chinese printmakers were producing maps on handmade paper using wood block printing techniques over 1,800 years ago. The Europeans developed the printing press and movable type in the 1400s, and maps became more common and more accessible. The paper they were printed on was still handmade, however, and any colored areas on the map had to be painted by hand.

The introduction of the lithographic printing method in the late 1800s allowed multi-colored maps to be produced by machine. Various photographic techniques were integrated into the printing process during the last 200 years, increasing the variety of scales at which maps were produced. Despite these production advances, each original map was still drawn by cartographers, using technical pens, various lettering devices, straight edges and razor knives, the traditional tools of the trade.

During the last two decades, however, the cartographer has acquired another production tool, the computer. Advanced computer-assisted design programs allow cartographers to set aside their technical pens and their straight edges. They use the computer to conjure and produce map images, but computer programs cannot replace cartographers. The various techniques for cartographic expression involve a sense of craft and artistry that has not yet been duplicated by electronic means.

See also Archeological mapping; Cartesian coordinate plane; Celestial coordinates; Earth science; Geographic and magnetic poles; Geologic map; Global Positioning System; Isobars; Latitude and longitude; Surveying instruments.

Resources

Books

Burrough, P. A., and R. A. McDonnell. *Principles of Geographical Information Systems.* Oxford, UK: Oxford University Press, 1998.

Hall, S. *Mapping the Next Millenium.* New York: Random House, 1992.

Harley, J. B., and D. Woodward, eds. *History of Cartography.* 6 vols. Chicago: University of Chicago Press, 1987.

Lobeck, A. K. *Things Maps Don't Tell Us: An Adventure into Map Interpretation.* Chicago: University of Chicago Press, 1984.

Monmonier, M. *How to Lie with Maps.* Chicago: University of Chicago Press, 1991.

Robinson, A., R. Sale, J. Morrison, and P. C. Muehrcke. *Elements of Cartography.* New York: John Wiley, 1994.

Other

U.C. Berkeley Library. *Maps and Cartography.* August 21, 2002 [cited January 3, 2003]. <www.lib.berkeley.edu/EART/MapCollections.html>.

Campbell, T. *Map History/History of Cartography.* January 1, 2003 [cited January 3, 2003]. <www.ihrinfo.ac.uk/maps/>.

Karen Lewotsky

Cashew family (Anacardiaceae)

The cashew family (Anacardiaceae) is a group of about 600 **species** of plants, most of which are tropical in distribution, although some occur in the temperate zone.

Maturing pistachio (*Pistacia vera*) nuts on a tree in California. © *Holt Studios International, National Audubon Society Collection/Photo Researchers, Inc. Reproduced with permission.*

Almost all members of the cashew family are trees or shrubs, though some are vines. Many species have foliage, **fruits,** or **bark** on the stems and roots that contain acrid, an often milky resin, and saps that are irritating or poisonous if touched or eaten. The leaves are typically compound, with at least three if not more leaflets per **leaf**. The flowers are small, five-parted, insect pollinated, and arranged in compact inflorescences. The fruits are either a one-seeded drupe or a many-seeded berry, and are generally eaten and dispersed by **birds** or small **mammals**.

The fruits of some species in the cashew family are an important source of food for people, while other species are used in **horticulture**. Many species are considered to be important weeds because they are poisonous, often causing a severe dermatitis (rash) in exposed people.

Edible species of the cashew family

Various nuts and other fruits are obtained from species in the cashew family.

The cashew (*Anacardium occidentale*) is the source of kidney-shaped cashew nuts. The cashew was originally from northeastern **South America**, but it is now planted widely throughout the humid tropics. The seedcoat of the fruit contains a toxic oil, and the raw cashew **nut** is also poisonous if eaten by people. However, the toxic chemical can be neutralized by roasting, and this richly delicious nut can be eaten safely.

The pistachio or green almond (*Pistacia vera*) is native to Syria, but is now widely cultivated in the Mediterranean region, the southern United States, and elsewhere. These fruits are prepared for eating by roasting and are usually salted by a brief soaking in a brine **solution**. The natural **color** of the pistachio's shell is white, but they are sometimes dyed red to make them more attractive to consumers.

The mango (*Mangifera indica*) is an evergreen tropical **tree** that grows up to 98 ft (30 m) and is native to southern **Asia**. Its fruit is known as a mango, possessing a yellow-red skin with a large, flat seed surrounded by a tasty, juicy pulp, which can be yellow, red, or orange in color. However, there are numerous cultivated varieties of mangos varying greatly in the size, shape, and color of their fruits. The flavor of the mango is an exotic blend of sweet and acidic tartness, with an aromatic undercurrent. Mangos are an ancient, cultivated fruit, having been grown in tropical Asia for as many as 6,000 years, and achieving sacred status in some Indian cultures. Most mangos are eaten as a fresh fruit, but this food is also used to prepare sauces, jams, and chutney.

Other, less-well known tropical fruits in the cashew family include the ogplum, Jamaica plum, Otaheite apple (obtained from species of *Spondias*), and the kaffir plum (*Harpephyllum caffrum*) of southern **Africa**.

Other useful species

The lacquer tree (*Rhus verniciflua*) occurs in China and Japan, where the viscous, milky sap of this **plant** has long been collected and applied as a natural varnish to fine **wood** carvings and furniture. The sap turns dark after oxidation in the atmosphere, providing an attractive, glossy coating to oriental lacquerware. A lacquer finish is resistant to **heat**, moisture, acid, alkali, and **alcohol**, and is therefore an excellent protection for fine works of art. Lacquering is an old art form, although it reached its greatest expression in China during the Ming Dynasty of 1368-1644 and in Japan during the seventeenth century. The finest pieces of lacquerware received as many as hundreds of individual coatings, applied over a period of several years. Other minor sources of lacquer are the Burmese lacquer tree (*Melanorrhoea usitata*) and an Indonesian sumac (*Rhus succedanea*).

The leaves of some species in the cashew family are dried and processed as a source of tannins, chemicals that are useful for preparing leather. The Sicilian sumac (*Rhus coriaria*) of southern Italy is especially useful, as its leaves can have a tannin **concentration** of 20-35%. This species is actually cultivated as a source of tannins, and it produces a superior, soft leather with a pale color, considered especially useful for fine gloves and bookcovers. The red quebracho (*Schinopsis lorentzii*) of South American temperate **forests** is another important source of tannins, which are obtained from the wood of this tree. The dried leaves of native sumacs (*Rhus* spp.) of **North America** have also been used as a minor source of tannins.

Poison ivy growing on a tree trunk. © *Carolina Biological Supply/Phototake NYC. Reproduced with permission.*

Chios mastic is a type of resin derived from the dried sap of *Pistacia lentiscus* of the Mediterranean region, while Bombay mastic is obtained from *P. cabulica* of southern Asia. These materials are used to manufacture a clear, high-grade varnish, which is sometimes used to coat metallic art and pictures, and in **lithography**.

The terebinth tree (*Pistacia terebinthus*) was the original source of artists' turpentine, but this solvent is now more commonly distilled from other types of trees. Another minor product is a yellow dye obtained from the twigs of the tropical South American tree, *Cotinus cuggygroa*.

Ornamental species

Various species of sumac are grown as ornamentals. The staghorn sumac (*Rhus typhina*) is cultivated for its attractive, purple-red foliage in the autumn and the interesting, reddish, horn-shaped fruiting inflorescences of female plants. This species is dioecious, meaning individual plants only bear female flowers (pistillate), or male flowers (staminate). The fragrant sumac (*R. aromatica*) is also commonly grown in horticulture.

The South American pepper-tree (*Schinus molle*) is also grown as an ornamental shrub. So are the smoke trees, *Cotinus obovatus* of North America, and the introduced *C. coggygria*, with their diffuse and fuzzy, smoke-like inflorescences, and attractive, purplish foliage in autumn.

Wild species occurring in North America

Various species in the cashew family are native to North America. One of the more familiar groups includes species of vines and shrubs in the genus *Toxicodendron*, many of which contain a toxic oil that causes a contact dermatitis in people exposed to crushed foliage, stems, or roots. It appears that some people develop an increased sensitivity to this toxic oil with increased exposure. Many people appear to not have been initially affected by contact with poison ivy and its relatives, but subsequent exposures then elicited sensitive responses. In contrast, others appear to progressively obtain an immunity to the toxic oil of these plants. Especially severe poisoning can be caused if smoke from the burning of *Toxicodendron* **biomass** is inadvertently inhaled—human deaths have been caused by this type of exposure resulting from severe blistering of the pharynx and lungs. The most widespread species is known as poison ivy (*T. radicans*, sometimes known as *Rhus radicans*), a plant with distinctive, shiny, compound leaves with three leaflets, and shiny, white berries. Poison ivy can grow as a perennial ground cover or as a vine that grows up trees. Other toxic species include poison oak (*T. toxicodendron*) and poison or swamp sumac (*T. vernix*), both of which are shrubs. The Florida poison tree or poisonwood (*Metopium toxiferum*) grows in southern Florida.

Various species of sumac (*Rhus* spp.) occur as shrubs in North America. One of the more familiar species is the staghorn sumac (*Rhus typhina*), the fruits of which are sometimes collected and used to prepare a lemonade-like drink. Other widespread species are the shining or mountain sumac (*Rhus copallina*), smooth or scarlet sumac (*R. glabra*), fragrant sumac (*R. aromatica*), and ill-scented sumac or skunkbush (*R. trilobata*).

The wild smoke-tree (*Cotinus obovatus*) occurs in the southeastern United States and is sometimes grown as an ornamental shrub.

See also Poisons and toxins.

Resources

Books

Conger, R.H.M., and G.D. Hill. *Agricultural Plants.* 2nd ed. Cambridge: Cambridge University Press, 1991.

Hartmann, H.T., et al. *Plant Science. Growth, Development, and Utilization of Cultivated Plants.* Englewood Cliffs, NJ: Prentice-Hall, 1988.

Judd, Walter S., Christopher Campbell, Elizabeth A. Kellogg, Michael J. Donoghue, and Peter Stevens. *Plant Systematics: A Phylogenetic Approach.* 2nd ed. with CD-ROM. Suderland, MD: Sinauer, 2002.

Kostermans, A.G.H., and J.M. Bompard. *The Mangoes. Their Botany, Nomenclature, Horticulture, and Utilization.* London: Academic Press, 1993.

Bill Freedman

Cassava *see* **Spurge family**

Cassini spacecraft

In the fall of 1997, the Cassini spacecraft began a seven-year, 2.175 billion mi (3.5 billion km) journey to **Saturn**. The 22.3 ft (6.8 m) robotic spacecraft is still functioning perfectly. When it arrives on July 1, 2004, it will spend four years probing the Saturnian system. Cassini, the first spacecraft to visit Saturn since *Voyager 2* swung past it in 1980 and the first spacecraft ever to take up **orbit** around Saturn, will observe Saturn's atmosphere, magnetic field, rings, and moons. Cassini will also release a small probe that will descend by parachute through the atmosphere of Saturn's **moon** Titan and impact on its surface.

Cassini bears a number of specialized cameras and other instruments. The **Magnetosphere** Imaging Instrument (MIMI), for example, designed by the Applied **Physics** Laboratory at Johns Hopkins University, will allow the first-ever imaging of a planet's magnetic field. MIMI will obtain images of the **plasma** and **radiation** surrounding Saturn and enveloping its moons and will observe the glow of Titan's exosphere (highest layer of atmosphere) caused by bombardment of high-speed protons trapped in Saturn's magnetic field.

Saturn's sixth moon, Titan—the second-largest moon in the **solar system** (larger than the **planet** Mercury) and one of only two moons in the solar system to have an atmosphere (the other is Neptune's moon Triton)—is of particular interest. Titan's atmosphere is so smoggy that observations of its surface have been limited, although images acquired by the Keck Observatory in Hawaii in 2001 at infrared wavelengths show that Titan has a blotchy surface. The dark parts, scientists theorize, may be pitch-black oceans of liquid hydrocarbons that are fed by methane rain falling on Titan's surface. Titan's surface **chemistry** may resemble that of a frigid, primordial **Earth**; if so, an understanding of Titan's atmosphere may help to understand the evolutionary **origin of life** on our planet.

A detailed study of Saturn's rings is another of the Cassini spacecraft's goals. Researchers hope that long-term, close-up observations of the planet's rings will help resolve whether they are material left over from Saturn's original formation or remnants of one or more moons shattered by comet or meteor strikes. This data should also help prove or disprove theories about the origin and evolution of the dust and gas from which the planets first formed. Researchers timed Cassini's arrival at Saturn so that the rings will be illuminated by sunlight. The tilt of the ring **plane** and resulting illumination **angle** will offer Cassini's instruments an excellent view of the ring disk.

The Cassini-Huygens mission, the result of an international collaboration between the U.S. National Aeronautics and Space Administration (NASA), the European Space Agency, the Italian Space Agency, and several other European academic and industrial partners, was named in honor of the seventeenth-century, French-Italian astronomer Jean Dominique Cassini. Cassini discovered the prominent gap in Saturn's main rings as well as the icy moons Iapetus, Rhea, Dione, and Tethys.

Cassini's launch was accompanied by political controversy and protest, as the probe carries 72 lb (34 kg) of plutonium with which to generate electric power during its seven years in **space**. All probes to the outer solar system have been powered by similar systems, as the dimness of the Sun's **light** at such distances makes the use of solar power difficult. Cassini caused unique worry because unlike earlier plutonium-carrying spacecraft, it was designed to follow a complex, looping path from the earth to **Venus**, the **Sun**, and past Earth again on its way to Saturn. There was concern that if the Cassini struck Earth by accident (as the Mars Climate Orbiter spacecraft struck Mars in 1999), its plutonium load might be vaporized in the atmosphere and constitute a global health hazard. (Plutonium has a **half-life** of 24,000 years and is highly carcinogenic; 72 lb [34kg], if divided into small particles and inhaled, would be sufficient to cause **cancer** in billions of people.) NASA disputed the protestors' claims. Whatever the merits of this controversy, Cassini did not crash into Earth during flyby and is now safely on its way to Saturn.

Cassini acquired valuable science data during its flyby of **Jupiter** on December 30, 2000, taking many photographs, measuring magnetic fields in collaboration with the Galileo spacecraft orbiting Jupiter, and taking advantage of Jupiter's gravity to boost itself toward Saturn at increased speed.

After the primary mission to study Saturn and its moons is finished in 2008, Cassini may fly closer to Saturn and pass inside the G ring while evading the regions known to contain a high **density** of potentially damaging ring particles. The spacecraft may also be sent into orbit around Titan to make a closer study.

Resources

Other

Cowing, Keith. *SpaceRef.com.* "Keck Observatory Provides Clearest Peek Yet of Titan's Surface." <http://www.space ref.com/news/viewnews.html?id=65>.

Jet Propulsion Laboratory, California Institute of Technology. *Cassini-Huygens Mission to Saturn and Titan.* <http://saturn.jpl.nasa.gov/index.cfm>.

Zarcelle, John. *CNN Interactive.* "Much Ado About Cassini's Plutonium." <http://www.cnn.com/TECH/9710/10/cassini.advancer/>.

Larry Gilman

Cassowaries *see* **Flightless birds**

Castor bean *see* **Spurge family**

Catabolism

Catabolism is the breakdown of large molecules into small molecules. Its opposite process is **anabolism**, the combination of small molecules into large molecules. These two cellular **chemical reactions** are together called **metabolism**. Cells use anabolic reactions to synthesize enzymes, **hormones**, sugars, and other molecules needed to sustain themselves, grow, and reproduce.

Energy released from organic **nutrients** during catabolism is stored within the **molecule adenosine triphosphate** (ATP), in the form of the high-energy chemical bonds between the second and third molecules of phosphate. The **cell** uses ATP for synthesizing cell components from simple precursors, for the mechanical **work** of contraction and **motion**, and for transport of substances across its **membrane**. ATP's energy is released when this bond is broken, turning ATP into **adenosine diphosphate** (ADP).

The cell uses the energy derived from catabolism to fuel anabolic reactions that synthesize cell components.

Although anabolism and catabolism occur simultaneously in the cell, their rates are controlled independently of each other. Cells separate these pathways because catabolism is a so-called "downhill" process during which energy is released, while anabolism is an energetically "uphill" process which requires the input of energy.

The different pathways also permit the cell to control the anabolic and catabolic pathways of specific molecules independently of each other. Moreover, some opposing anabolic and catabolic pathways occur in different parts of the same cell. For example, in the liver, the **fatty acids** are broken down to acetyl CoA inside mitochondria, while fatty acids are synthesized from acetyl CoA in the cytoplasm of the cell.

Both catabolism and anabolism share an important common sequence of reactions known collectively as the **citric acid** cycle, or **Krebs cycle**, which is part of a larger series of enzymatic reactions known as oxidative phosphorylation. Here, glucose is broken down to release energy, which is stored in the form of ATP (catabolism), while other molecules produced by the Krebs cycle are used as precursor molecules for anabolic reactions that build **proteins**, fats, and carbohydrates (anabolism).

Cells regulate the **rate** of catabolic pathways by means of allosteric enzymes, whose activity increases or decreases in response to the presence or absence of the end product of the series of reactions. For example, during the Krebs cycle, the activity of the **enzyme** citrate synthase is slowed by the buildup of succinyl CoA, a product formed later in the cycle.

Resources

Books

Alberts, Bruce, et al. *Molecular Biology of The Cell.* 2nd ed. New York: Garland Publishing, 1989.

Parker, Sybil, ed. *McGraw-Hill Encyclopedia of Chemistry.* 2nd ed. New York: McGraw Hill, 1999.

Catalyst and catalysis

Humans used the process known as catalysis long before, they understood what took place in that process. For example, soap-making, the **fermentation** of wine to vinegar, and the leavening of bread are all processes that involve catalysis. Ordinary people were using these procedures in their everyday lives without knowing that catalysis was involved.

The term catalysis was proposed in 1835 by the Swedish chemist Jons Berzelius. The term comes from the Greek words for down, *kata,* and loosen, *lyein.* Berzelius explained that he meant by the term catalysis the property of exerting on other bodies an action which is very different from chemical affinity. By means of this action, they produce **decomposition** in bodies, and form new compounds into the composition of which they do not enter.

One of the examples of catalysis familiar to Berzelius was the conversion of starch to sugar in the presence of strong acids. In 1812, the Russian chemist Gottlieb Sigismund Constantin Kirchhof had studied this reaction. He found that when a **water** suspension of starch is boiled, no change occurs in the starch.

However, when a few drops of concentrated **sulfuric acid** are added to the same suspension before boiling, the starch breaks down into a simple sugar called glucose. The acid can be recovered unchanged from the reaction. Kirchhof concluded that it had played a helping role in the breakdown of the starch, without itself having undergone any change.

Berzelius had been able to draw on many other examples of catalysis. For example, both Humphry Davy and Johann Dîbereiner had studied the effect of platinum **metal** on certain organic reactions. They had concluded that the metal increased the **rate** at which these reactions occurred without undergoing any change itself. Davy's most famous protégé, Michael Faraday, also demonstrated the ability of platinum to bring about the recombination of **hydrogen** and **oxygen** that had been obtained by the **electrolysis** of water.

When Berzelius first defined catalysis, he had in mind some kind of power or **force** by which an agent (the catalyst) acted on a reaction. He imagined, for example, that platinum might exert an electrical force on gases with which it came into contact in order to bring about a change in reaction rates.

This kind of explanation works well for heterogeneous catalysis, in which the catalyst and the reaction are in different phases. In a platinum catalyzed reaction, for example the platinum is in a solid state and the reaction in a gaseous or liquid state.

Homogeneous catalysis, in which catalyst and reaction are in the same state, requires a different explanation. How does sulfuric acid in the liquid state, for example, bring about the conversion of a starch suspension with which it is intermixed?

The solution to this problem came in the early 1850s. Alexander William Williamson was carrying out research on the preparation of ethers from **alcohol**. Chemists knew that concentrated sulfuric acid was an effective catalyst for this reaction, but they did not know why. Williamson was able to demonstrate that the catalyst does break down in the first stage of this reaction, but is regenerated in its original form at the conclusion of the reaction.

The role of catalysts in living systems was first recognized in 1833. Anselme Payen and Jean François Persoz isolated a material from malt that accelerated the conversion of starch to sugar. Payen called the substance diastase. A half century later, the German physiologist, Willy Kahne, suggested the name **enzyme** for catalysts that occur in living systems.

Toward the end of the nineteenth century, catalysts rapidly became important in a variety of industrial applications. The synthesis of indigo became a commercial possibility in 1897 when mercury was accidentally found to catalyze the reaction by which indigo was produced. Catalysis also made possible the commercial production of **ammonia** from its elements (the Haber-Bosch process), of **nitric acid** from ammonia (the Ostwald process) and of sulfuric acid from **sulfur** oxides (the contact process).

Catastrophism

Catastrophism is the doctrine that Earth's history has been dominated by cataclysmic events rather than gradual processes acting over long periods of time. For example, a catastrophist might conclude that the Rocky **Mountains** were created in a single rapid event such as a great **earthquake** rather than by imperceptibly slow **up-lift** and **erosion**.

Catastrophism developed in the seventeenth and eighteenth centuries. A prominent British theologian, Bishop James Ussher (1581–1656) added together the ages of people in the Bible and calculated that **Earth** must have been created in 4004 B.C. His calculation implied that all of the features of Earth's surface must be less than 6,000 years old and were therefore, formed as the result of violent upheavals or catastrophes. Current research, in contrast, suggests that the earth is about 4.5 billion years old. Baron Georges Cuvier (1769–1832), a French anatomist, tried to reconcile the fossil record with Biblical history. Cuvier stated that different groups of fossil organisms were created and then became extinct as the result of geologic catastrophes, the last of which was the great flood described in the Bible. Each catastrophe, according to Cuvier, killed the fossilized organisms and deposited the sediment that solidified into the rock surrounding the fossils.

A new concept, **uniformitarianism**, grew from the work of the Scottish geologist James Hutton (1726–1797) and eventually replaced catastrophism. Uniformitarianism is the doctrine that geologic processes operate at the same rates and with the same intensity now as they did in the past. Hutton suggested that Earth had a very long history that could be understood in terms of currently observable processes such as the **weathering** of **rocks** and erosion of sediment. Sandstone, for example, was formed by the same kinds of physical processes that form modern sandy beaches or deserts. Therefore, catastrophic events were not needed to explain Earth's history. Geologists often summarize the idea of uniformitarianism with the phrase, "The present is the key to the past."

The concept of catastrophism has been revived with the discovery of large meteorite impact structures and evidence of mass extinctions in the fossil record. The most notable of these events was the asteroid impact marking the boundary between the Cretaceous and Tertiary Periods about 65 million years ago, which coincides with the **extinction** of the dinosaurs. The resulting Chicxulub impact structure, located in the Gulf of Mexico near the Yucatan **Peninsula**, is approximately 120 mi (180 mi) wide and 1 mi (1.6 km) deep. It is thought that the crater was formed by the catastrophic impact of an asteroid 4–9 mi (6–15 km) in diameter. The impact produced a thin layer of clay that contains elements rare on Earth but abundant in meteorites and **minerals** that can only be formed under very high **pressure**. Soot within the clay also suggests that the impact triggered extensive wildfires, which may have acted with sulfate minerals pulverized in the impact, to slow **photosynthesis** and cause global cooling to occur. **Clouds** of dust may have darkened the atmosphere for weeks or months. Other large impact structures on Earth include a 100 mi (120 km) wide structure in Acraman, **Australia**, a 120 mi (200 km) wide structure near Sudbury, Ontario, Canada, and an 50 mi (85 km) wide structure in Chesapeake Bay, Virginia and Maryland. Meteor Crater, Arizona, is an example of a relatively small impact structure.

The recognition that Earth's history has been punctuated by rare but catastrophic events such as asteroid impacts has led most geologists to abandon strict uniformitarianism in favor of a doctrine known as actualism. Actualism states that the laws of nature do not change with time and much of Earth's history can be explained in terms of currently observable processes. It acknowledges, however, that rates of geologic change are not constant over long periods of time and there have been some catastrophic geologic events that are far beyond the range of human experience.

Catfish

Catfish include some 2,500 **species** of mostly **fresh-water** fish characterized by two to four pairs of whiskers or barbels around their mouth. Many species have spines on the dorsal fins and near the gills. In some species these spines may contain poison.

Catfish belong to the **bony fish** order Siluriformes, and are mainly freshwater forms with representatives throughout the world. Most species of catfish lack scales, although some species are covered with heavy plates of tough, armored skin. Catfish tend to be hardy and a few species can survive for some time out of **water** as long as their skin is kept moist by an external layer of mucus.

A brown bullhead catfish. *JLM Visuals. Reproduced by permission.*

A few species of catfish live in the oceans, such as the sea catfishes of the family Aridae which are found in tropical and subtropical seas, and in temperate waters during the summer. Catfish vary in size from the pygmy corydoras (*Corydoras hastatus*), about 0.8 in (2 cm) long, to the giant catfish (*Pangasianodon giga*) of southeastern **Asia**, which can exceed 7 ft (2.1 m) and weigh 250 lb (113 kg). This group also includes the glass catfish (*Physailla pellucida*), a popular aquarium fish.

North American freshwater catfish are found from Canada to Guatemala. They are often caught by rod and reel, have considerable commercial importance, and are also raised on **fish** farms. Catfish live in murky lakes and ponds, feeding on the bottom on both live and dead material. Catfish spawn around May and June. The parents prepare a nest in the mud or **sand**. After the eggs hatch the parents guard the nest and protect the young until they have developed enough to become independent.

The most abundant North American catfish are the bullheads, including: the black bullhead (*Ictalurus melas*), the brown bullhead (*I. nebulosus*), and the yellow bullhead (*I. natali*).

Bullheads are plentiful in streams and ponds of **North America**, from the Atlantic to the Pacific Oceans.

Bullheads are thought to have spread naturally from western North America to the east. Adhesive bullhead eggs may have stuck to the legs and feet of aquatic migratory **birds**, and later washed off when the birds traveled to another pond, thus establishing new populations of these fish.

Rivalling the bullheads in commercial importance in some areas are the channel catfish (*I. punctatus*), the blue catfish (*I. furcatus*), the white catfish (*I. catus*), and the flathead catfish (*Pylodictus olivaris*). These species may reach 150 lb (68 kg), although they are usually smaller. Catfish farming is an increasing enterprise in the southern United States, with more than 100 million lb (45.4 million kg) of fish being produced annually.

The diet of the ictalurids is varied since they eat almost anything, dead or alive. People fishing for catfish often use a use "stink bait." Such bait can lure catfishes over a wide expanse. Catfish can detect the bait using the extensive sensory surface of their body and their long barbels.

Also included in the North American catfish family are the madtoms in the genus *Noturus*. These are small fish under 5 in (13 cm) in length. Madtoms have **glands** associated with spines which can inflict extremely painful stab wounds.

The Eurasian catfish family Siluridae includes the wels (*Siluris glanis*), which grows over 12 ft (3.7 m) long and weighs hundreds of pounds. At the other extreme in this family is the glass catfish (*Kryptopterus bicirrhus*) of southeastern Asia, which is only 4 in (10 cm) long. The skin and muscles of the glass catfish are transparent enough to display its viscera.

The catfish family Clariidae includes labyrinthic fishes which have evolved a special air-breathing apparatus, found anterior to the gills and equipped with numerous **blood** vessels. These catfish can stay out of water for an extended period of time as long as their skin is kept moist with mucus. Air-breathing catfish can live in stagnant, low-oxygen water that would be lethal to other species of fish.

The walking catfish (*Clarius batrachus*) of southeast Asia "walks" on dry land by performing snakelike movements, using its pectoral fins as props. In times of severe **drought** these catfish try to move overland to ponds containing water, or they may dig into the bottom of a pool and wait there for the return of the rains.

The talking catfish (*Acanthodoras spinosissimus*, family Doradidae) makes a croaking sound, especially when captured. These sounds result from air forced in and out of the swim bladder due to changes in **pressure** when the pectoral fins flap.

In **Africa**, electric catfish (*Malapterurus electicus*, family Malapteruridae) range in size from 8 in (20 cm) to 4 ft (1.2 m) and reach a weight of 50 lb (23 kg). These fish can produce a 100-volt shock followed by lesser shocks, which can stun large fish. In addition to predation and defense, the electrical impulses are used to navigate in turbid water. The electric organs are found along the body and tail, and are derived from glandular cells in the epidermis, rather than from the muscles as occurs in other species of electric fish.

Nathan Lavenda

Catheters

Catheters are long, flexible tubes that are inserted into the body for various purposes, either to remove an unwanted substance or to instill nourishment or medication.

A relatively large catheter can be passed through the nose, down the throat and into the stomach to remove the contents of the stomach; for example, if someone has consumed a poisonous substance, a catheter can be used to remove a small **sample** of the stomach contents for laboratory testing. A catheter may also be used to pass liquid nourishment into the digestive tract, as in the case of someone unable to swallow for some reason. This catheter is called a nasogastric tube.

A smaller tube can be passed through the urethra into the bladder to empty its urine. Oftentimes after **surgery** or trauma an individual is unable to void and must have the bladder emptied. Sometimes these urinary catheters must be left in place; a small **balloon** near the end of the catheter is inflated to hold the catheter in the bladder.

Very long catheters are often passed through an incision in the thigh into an artery and into the **heart**; a doctor can then inject contrast agents into the patient to outline the coronary **arteries**. The physician can watch as the agent, which is visible on **x rays**, is injected and courses through the heart's arterial system. In this way a doctor can see a blockage and take measures to bypass it or remove it. These catheters have now been fitted with devices to open clogged arteries. Small balloons mash obstructions out of the way, and **laser** tips or whirring blades cut stubborn blockages from the arterial passage. Since there are no **pain** receptors inside **blood** vessels, passing the cardiac catheter is done under local anesthetic with the patient fully awake.

Still other catheters can be inserted through the trachea into the lungs to remove fluid or mucus. Some two-channeled catheters are used to induce a chemical into an **organ** and remove the organ's contents at the same time. Others can be used for wound drainage or for measuring blood **pressure** in any of the heart's four chambers.

See also Surgery.

Cathode

The cathode is one of the two electrodes that are present in any system in which **electricity** is entering and leaving a region; the other electrode is called the **anode**. The **electric current** enters through one of the electrodes and leaves through the other.

Two general kinds of systems employ electrodes: vacuum tubes (also called gas discharge tubes) and electrochemical cells.

In a **vacuum tube**, the cathode is the **negative** electrode-the electrode that carries a negative potential with respect to the other one. The cathode is often heated to drive out electrons, which then fly through the **vacuum** toward the positive electrode, the anode. These streams of electrons are referred to as cathode rays. Cathode ray tubes are vacuum tubes that are widely used as oscilloscopes, **television** tubes, and computer monitors.

Electrochemical cells are of two types: voltaic cells (also called galvanic cells) and electrolytic cells. In a galvanic cell, such as an **automobile battery**, an electric current is produced by a chemical **oxidation-reduction reaction**. In an electrolytic cell, such as a cell designed for the **electrolysis** of **water**, the chemical oxidation-reduction reaction is produced by an externally-supplied electric current. In either case, the cathode is defined as the electrode at which the chemical reduction process is taking place in the cell—that is, the electrode at which electrons are being taken up by **atoms**, molecules, or ions. The anode, on the other hand, is the electrode at which the oxidation process is taking place—that is, the electrode at which electrons are being given off by atoms, molecules, or ions.

See also Cathode ray tube.

Cathode ray tube

A **cathode** ray tube is a device that uses a beam of electrons in order to produce an image on a screen. Cathode ray tubes are also known commonly as CRTs. Cathode ray tubes are widely used in a number of electrical devices, such as computer screens, **television** sets, **radar** screens, and oscilloscopes used for scientific and medical purposes. A cathode ray tube consists of five major parts: an envelope or container, an **electron** gun, a focusing system, a deflection system, and a display screen.

Envelope or container

Most people have seen a cathode ray tube or pictures of one. The "picture tube" in a television set is perhaps the most familiar form of a cathode ray tube. The outer shell that gives a picture tube its characteristic shape is called the envelope of a cathode ray tube. The envelope is most commonly made of **glass**, although tubes of **metal** and ceramic can also be used for special purposes. The glass cathode ray tube consists of a cylindrical portion that holds the electron gun and the focusing and deflection systems. At the end of the cylindrical portion farthest from the electron gun, the tube widens out to form a conical shape. At the flat wide end of the cone is the display screen.

Air is pumped out of the cathode ray tube to produce a **vacuum** with a **pressure** in the range of 10^{-2} to 10^{6} pascal, the exact value depending on the use to which the tube will be put. A vacuum is necessary to pre-

vent electrons produced in the CRT from colliding with **atoms** and molecules within the tube.

Electron gun

An electron gun consists of three major parts. The first is the cathode, a piece of metal which, when heated, gives off electrons. One of the most common cathodes in use is made of cesium metal, a member of the alkali family that loses electrons very easily. When a cesium cathode is heated to a **temperature** of about 1,750°F (954°C), it begins to release a stream of electrons. These electrons are then accelerated by an **anode** (a positively charged electrode) placed a short **distance** away from the cathode. As the electrons are accelerated, they pass through a small hole in the anode into the center of the cathode ray tube.

The intensity of the electron beam entering the anode is controlled by the grid. The grid may consist of a cylindrical piece of metal to which a variable electrical charge can be applied. The amount of charge placed on the control grid determines the intensity of the electron beam that passes through it.

Focusing and deflection systems

Under normal circumstances, an electron beam produced by the electron gun described above would have a tendency to spread out to form a cone-shaped beam. However, the beam that strikes the display screen must be pencil-thin and clearly defined. In order to form the electron beam into the correct shape, an electrical or magnetic **lens**, similar to an optical lens, can be created adjacent to the accelerating electrode. The lens consists of some combination of electrical or magnetic fields that shapes the flow of electrons that pass through it, just as a glass lens shapes the **light** rays that pass through it.

The electron beam in a cathode ray tube also has to be moved about so that it can strike any part of the display screen. In general, two kinds of systems are available for controlling the path of the electron beam, an electrostatic system and a magnetic system. In the first case, negatively charged electrons are deflected by similar or opposite electrical charges and in the second case, they are deflected by magnetic fields.

In either case, two deflection systems are needed, one to move the electron beam in a horizontal direction, and the other to move it in a vertical direction. In a standard television tube, the electron beam completely scans the display screen about 25 times every second.

Display screen

The actual conversion of electrical to light **energy** takes place on the display screen when electrons strike a

Cathode rays *see* **Subatomic particles**

material known as a phosphor. A phosphor is a chemical that glows when exposed to electrical energy. A commonly used phosphor is the compound zinc sulfide. When pure zinc sulfide is struck by an electron beam, it gives off a greenish glow. The exact **color** given off by a phosphor also depends on the presence of small amounts of impurities. For example, zinc sulfide with silver metal as an impurity gives off a bluish glow and with **copper** metal as an impurity, a greenish glow.

The selection of phosphors to be used in a cathode ray tube is very important. Many different phosphors are known, and each has special characteristics. For example, the phosphor known as **yttrium** oxide gives off a red glow when struck by electrons, and yttrium silicate gives off a purplish blue glow.

The **rate** at which a phosphor responds to an electron beam is also of importance. In a color television set, for example, the glow produced by a phosphor has to last long enough, but not too long. Remember that the screen is being scanned 25 times every second. If the phosphor continues to glow too long, color will remain from the first scan when the second scan has begun, and the overall picture will become blurred. On the other hand, if the color from the first scan fades out before the second scan has begun, there will be a blank moment on the screen, and the picture will appear to flicker.

Cathode ray tubes differ in their details of construction depending on the use to which they will be put. In an **oscilloscope**, for example, the electron beam has to be able to move about on the screen very quickly and with high precision, although it needs to display only one color. Factors such as size and durability are also more important in an oscilloscope than they might be in a home television set.

In a commercial television set, on the other hand, color is obviously an important factor. In such a set, a combination of three electron guns is needed, one for each of the primary colors used in making the color picture.

Resources

Books

Keller, Peter A. *The Cathode-Ray Tube: Technology, History, and Applications.* Palisades Press, 1992.

Parr, Geoffrey, and O.H. Davie. *The Cathode-Ray Tube and Its Applications.* London: Chapman and Hall, 1959.

David E. Newton

Cation

A cation is any atom or group of **atoms** that has a net positive charge. While **matter** is electrically neutral overall, ionic compounds are matter that is composed of positively-charged and negatively-charged particles called ions. An ion is any atom or group of atoms with an overall electrical charge. According to the laws of **physics**, opposite charges attract, so the oppositely-charged ions attract each other to form compounds that are, overall, electrically neutral. In such compounds, the number of positive charges on the cations is equal to the number of **negative** charges. Species that have an overall negative charge are called anions. Compounds that are composed of cations and anions are called ionic compounds. Examples include table **salt (sodium chloride)** and potash (potassium carbonate).

Cations are formed when an atom or group of atom loses one or more electrons. The resulting species has more protons than electrons, so it has an overall positive charge. (Each **proton** has a +1 charge, and each **electron** has a -1 charge. In normal atoms, the number of protons equals the number of electrons, so a normal atom has an overall charge of **zero**. We say it is electrically neutral.) On the other hand, anions are formed when an atom or group of atoms accepts one or more electrons, so it has more negative charges than positive charges. Most metallic elements react chemically to form cations, losing electrons. Most nonmetallic elements react chemically to gain electrons, thereby forming anions.

See also Anion.

Cation-ratio dating *see* **Dating techniques**

Cats

Cats are **mammals** in the family Felidae of the order Carnivora, which includes all of the carnivores. The highly predatory instincts of **species** in the cat family are easily seen in domestic cats, for even well-fed individuals will aggressively hunt small mammals and **birds**.

The cat family includes both large species (jaguar, leopard, lion, and tiger) and small ones (bobcat, lynx, ocelot, and serval). Small species of cats purr but do not roar, whereas big cats roar but do not purr. The reason for this is that the tongue muscles of large cats are attached to a pliable cartilage at the base of the tongue, which allows roaring, while those of small cats are attached to the hyoid bone, which only allows purring.

Most cats have 30 teeth, including large canine and carnassal teeth, but few in their cheek. This arrangement is suited to crushing bones and tearing, cutting, and gripping the flesh of their **prey**. Their jaws are mostly adapted to vertical movement, and the chewing action is aided by sharp, backward projections on the tongue (known as papillae), which help to grip and manipulate food.

Members of the cat family occur naturally in all parts of the world, except **Antarctica**, **Australia**, and New Zealand (although domestic cats have been introduced and are now wild in the latter two places). There are 36 species of cats in four genera. The genus *Panthera* includes the jaguar, leopard, lion, and tiger. The cheetah is the sole member of the genus *Acinonyx*, while the clouded leopard is the only species of *Neofelis*. The puma, lynx, and smaller cats, including the domestic cat, are placed in the genus *Felis*.

Species of big cats

There are eight species of so-called "big cats," which includes the lion (*Panthera leo*), tiger (*P. tigris*), leopard (*P. pardus*), cheetah (*Acinonyx jubatus*), jaguar (*P. onca*), snow leopard (*Uncia uncia*), clouded leopard (*Neofelis nebulosa*), and cougar (*Puma concolor*).

The lion

Lions were once distributed over much of southern **Europe**, **Asia**, and **Africa**. Today, lions are found only in sub-Saharan Africa and in the Gir Forest, a **wildlife** sanctuary in India. Lions prefer open grassland and **savanna** to forest, and are also found in the Kalahari Desert. Adult male lions weigh from 300–500 lb (135–225 kg), while females weigh about 300 lb (135 kg). Lions are a light tawny **color** with black markings on the abdomen, legs, ears, and mane. Lions live up to 15 years, reaching sexual maturity in their third year. Male lions have been observed to kill cubs that they have not fathered.

Lions are the most social of the cats. They live in family groups called a pride, which consists of 4-12 related adult females, their young, and 1-6 adult males. The size of the pride usually reflects the amount of available food: where prey is abundant, lion prides tend to be larger, making them better able to protect their kills from hyenas and other scavengers. Most lion kills are made by the females, while the males defend the pride's territory, which may range from 8 sq mi (20 sq km) to more than 150 sq mi (385 sq km).

The tiger

The tiger is the largest member of the cat family, with males weighing from 400 to 600 lb (180-275 kg) and females 300-350 lb (135-160 kg). Tigers range from a pale yellow to a reddish orange background color (depending on **habitat**), overlain by vertical stripes. Tigers live in habitats with a dense cover of vegetation, commonly forest and swamps (or forested **wetlands**) on the Indian subcontinent, southern China, Southeast Asia, and Indonesia. A century ago, tigers commonly inhabited areas as far north as southern Siberia, all of India and Southeast Asia, and regions along the eastern part of China. Today, however, their range is much reduced and fragmented and all eight subspecies of tigers are endangered.

Tigers live a solitary life and systematically protect their territory by marking its boundary with urine, feces, glandular secretions, and scrape marks on trees. Tigers are solitary nocturnal hunters, approaching their prey stealthily in a semi-crouching position. When close enough, the tiger makes a sudden rush for the prey, attacking from the side or the rear. The tiger keeps its hind feet on the ground while using its front paws and jaws to seize its prey by the shoulder or neck. The tiger applies a throat bite that suffocates its victim, which is then carried into cover and consumed.

The leopard

Male leopards weigh about 200 lb (90 kg), with females weighing about half that amount. Leopards are found in sub-Saharan Africa, India, and Southeast Asia. There are also small populations in Arabia and North Africa. Leopards have a distinctive coloring of black spots over a pale brown coat. Their habitat includes tropical **rainforest**, dry savanna, and cooler mountainous areas.

Leopards feed on a variety of small and medium-sized prey, usually hunting at night by ambush. Leopards use trees as resting places and frequently drag their catch up there for feeding. The number of leopards is declining worldwide due to hunting and habitat destruction resulting from human population pressure.

The cheetah

Cheetahs can reach a speed of up to 70 MPH (112 km/h), and are the fastest animals on land. Cheetahs resemble leopards in having a black spotted pattern over a tawny coat, but are distinguished by a long, lithe body, large black "tear" stripes under their eyes, and a relatively small head. Cheetahs are the only members of the cat family that do not have retractable claws. Cheetahs are solitary hunters, feeding mostly on **gazelles** and impala. They hunt mainly in the morning and early afternoon, when other big cats are usually sleeping, thereby enabling them to share hunting areas with other large carnivores. Cheetahs are found in North and East Africa, in eastern parts of southern Africa, and in certain areas of the Middle East and South Asia. There is a considerable trade in cheetah skins, and hunting of these animals, together with the loss of habitat, threatens their survival in the wild.

Other big cats

Among the other large cats are the jaguar, snow leopard, and clouded leopard. These cats inhabit forested wilderness, and all are solitary, nocturnal predators. Jaguars are found in Central and **South America**, while the clouded leopard occurs in Southeast Asia, and the snow leopard at higher elevations in the Himalayas of Central Asia. The clouded leopard and the snow leopard have a rigid hyoid bone in their throat which prevents them from roaring. The black panther is a melanistic (or black) form of the jaguar; its spots are barely discernable within its dark coat. The cougar, also known as the puma or mountain lion, is about the size of a leopard and ranges from western Canada to Argentina. The cougar is found in **mountains**, plains, deserts, and **forests** and preys on **deer** and medium-sized mammals.

The smaller wild cats

Smaller wild cats are native to most areas of the world, except Australasia and Antarctica. Smaller cats are characterized by an inability to roar, retractable claws, and a hairless strip along the front of their nose. Examples of small wild cats are the Eurasian wild cat (*Felis sylvestris*), the African wild cat (*F. lybica*), the **sand** cat (*F. margarita*) of the Sahara, the African tiger cat (*Profelis aurata*) of tropical forests, the golden cat (*P. temminckii*), and Pallas' cat (*Otocolobus manul*) of central Asia. Medium-sized cats include the African serval (*Leptailurus serval*) and the caracal or desert lynx (*Caracal caracal*). Medium-sized cats of the New World include the ocelot (*Leopardus pardalis*) of South and Central America and the jaguarundi (*Herpailurus jaguarundi*).

The bobcat or wildcat (*Lynx rufus*) of **North America** is colored to blend into the rocky, densely vegetated background of its habitat. Bobcats rely more on **hearing** and **smell** than on sight to catch their prey. The lynx (*Lynx lynx*) lives in cold climates and has long legs and big feet to make trekking through deep snow easier. The Canada lynx (*L. canadensis*) has long hair and does not have a spotted coat.

The other 26 species of smaller cats live mainly in forest and feed on small prey, such as **squirrels** and other **rodents**, hares, small deer, birds, **snakes**, lizards, **fish**, and **insects**. Most species have a spotted or striped coat and usually a rounded head. Small wild cats are ei-

Two cheetahs in Kenya. *Photograph by Lew Eatherton. The National Audubon Society Collection/Photo Researchers, Inc. Reproduced by permission.*

ther solitary in habit or form small groups, depending on the abundance of the food supply. Most species are hunted for their spotted or striped skin and some are in danger of becoming extinct.

Senses

Cats have excellent **binocular vision**, which allows them to judge **distance** well. Cats can see well at night, using dim light to distinguish objects and prey. Their eyes have a special reflective layer behind the retina that aids in low-light **perception**. This tapetal layer makes their eyes appear to glow in the dark when a light is shone on them.

The senses of smell and **taste** in cats are closely connected, as they are in all mammals. Distinctive to cats is an avoidance of foods that taste sweet. Their taste buds are located along the front and side edges of their tongue. Their vomeronasal **organ**, also known as Jacobson's organ, is a saclike structure located in the roof of the mouth, that is believed to be involved in sensing chemical cues associated with sexual activity. When a male cat smells a female's urine, which contains **hormones** indicating sexual receptiveness, he may wrinkle his nose and

A leopard. *Photograph by Tom Brakerfield. Stock Market. Reproduced by permission.*

curl back his upper lip in a gesture known as flehmening. He will also raise his head and bare his teeth.

Cats have the ability to hear high-frequency sounds that humans are unable to perceive. This ability is particularly helpful when cats are stalking such prey as **mice**, since the cats can detect the high-frequency sounds emitted by these rodents. The external ears of cats are flexible and can turn as much as 180 degrees to locate sounds more precisely.

A cat's whiskers have a sensory function, helping to avoid objects in its path in the dimmest light. If a cat passes an object that touches its whiskers, it will blink, thus protecting the eyes from possible injury. Besides the long cheek whiskers, cats have thicker whiskers above their eyes. Cats use their nose to determine the **temperature**, as well as the smell, of food. The hairless pads of their feet are an important source of tactile information gained from investigating objects with their paws.

Behavior

In the wild, most forest-living members of the cat family tend to be solitary hunters. Some species live in pairs, while others, such as lions, live in family groups.

Cats engage in daily grooming which not only keeps their fur in good condition, but also helps regulate their body temperature (fur licking helps cool the cat through **evaporation**), and spreads oils to keep their coat waterproof.

Cats need a great deal of **sleep**, which is consistent with the large amounts of **energy** they expend when hunting. They typically sleep intermittently almost two-thirds of the day. Because of a slight fall in their body temperature when they sleep, cats often look for warm, sunny places for dozing.

Most cats are excellent climbers, great jumpers, and have remarkable balance. Except for the cheetah, cats have retractable claws that are curved, sharp, and sheathed. The claws are particularly useful when climbing trees. The bones of their feet (like those of dogs) are arranged in a digitigrade posture, meaning that only their toes make contact with the ground, which increases their speed of running. Cats have the a remarkable ability to right themselves when falling, when first the head, then the rest of the body turns toward the ground so that the cat lands on its feet.

Cats follow a well-defined hunting sequence that begins with the sighting or smelling of prey. The hunting skills that cats display are in some aspects instinctual and in others learned. Cats begin **learning** how to hunt through the play they engage in when young. Mother cats teach hunting skills to their young, first by bringing back dead prey, later by bringing back immobilized prey, allowing the young to kill it themselves. Still later, the mother cat will take the young on stalking and killing missions. Cats that do not have the opportunity to learn to hunt from their mother do not become good hunters.

Cats are territorial, marking their territory by spraying its boundary with urine. Cats also scratch and rub against fixed objects to mark their territory. Within the territorial boundary of a male cat, there may be several female territories. During mating, the male will seek out or be lured to nearby females that are in heat (estrous). Females may vocalize loudly when they are ready to mate, thus attracting males. Frequent scenting and rubbing against trees by the female cat also help the male know she is ready to mate. Frequent sexual intercourse during estrous is also important to ensure successful ovulation.

The gestation (pregnancy) period in cats depends upon their body size. Gestation ranges from slightly less than 60 days for smaller species to about 115 days for large cats, such as lions. The number in the litter varies from one to seven. The body size is not a consistent factor determining litter size; it may have more to do with the availability of food and the survival **rate** in the **ecosystem** the cat inhabits. With the exception of lions, the care and training of the cubs or kittens are left to the mother. Nursing continues until the young are weaned and learn to eat meat.

Evolution and history

The emergence of modern cats began about 25 million years ago during the Miocene epoch. Much more recently, the saber-toothed tiger (*Smilodon fatalis*) lived in Europe, Asia, Africa, and North America, and had extremely long, upper canine teeth for stabbing its prey. The remains of saber-toothed tigers have been found to be as recent as 13,000 years old.

Cats were probably first domesticated in ancient Egypt about 5,000 years ago. The Egyptians used cats to protect grain supplies from rodents. They also worshiped cats, and mummified large numbers of them and placed them in tombs so they could continue to serve their owners in their afterlife. Since that time, the domestic cat has been carried by people throughout most of the world. The breeding of cats into numerous specific pedigrees was particularly active beginning with the middle of the nineteenth century.

Domestic cats

The breeding of the domestic cat (*Felis cattus*) involves basic principles of heredity, with consideration of dominant and recessive traits. It was in England that this breeding first became serious enough that so-called "purebred" cats were displayed at shows and a system of authenticating the genetic lineage was begun by issuing pedigree certificates. Special associations were established to regulate the validation of cat pedigrees and to sponsor the shows.

Cat breeds can be categorized as either long-haired or short-haired. Within each group, the breeds are distinguished by their color and pattern, head and **ear** shape and size, body shape, hair color and length, **eye** color and shape, and special markings like stripes and color variations on the feet, tail, face, and neck.

More than a hundred different breeds of domestic cats are recognized, subdivided into five broad groups. One group includes Persian longhairs, another the rest of the long-haired cats, a third the British short-haired cats, a fourth the American short-haired cats, and a fifth the Oriental short-haired cats.

The Persian cat, highly prized among cat fanciers, has a rounded body, face, eyes, and head, with a short nose and legs. Its fur is long and woolly, and its tail is fluffy and bushy. Persian cats vary from black to white, cream, blue, red, blue-cream, cameo, tortoiseshell, smoke, silver, tabby, calico, pewter, chocolate, and lilac. Other popular long-haired cats include the Balinese, the ragdoll, the Turkish angora, and the Maine coon cat. Among the short-haired cats, the Manx, British shorthair, American shorthair, Abyssinian, Burmese, and Siamese are popular. One breed is hairless: the sphynx, bred from a mutant kitten in 1966, does not even have whiskers.

The domestic cat is rivaled only by the domestic dog as a household pet, and in recent years has outnumbered the dog in urban areas. Cats are more self-sufficient than dogs in that they self-groom, need little if any training to use a litter box, and don't have to be walked. Cats are generally quiet and aloof, but will display affection to their owners. They have a reputation of being fussy eaters, but will usually adapt quickly to a particular kind of food.

See also Marsupial cats.

Resources

Books

Alderton, David. *Wildcats of the World.* New York: Facts on File, 1993.

Bailey, Theodore N. *The African Leopard: Ecology and Behavior.* New York: Columbia University Press, 1993.

Bailey, Theodore. *Wild Cats of the World.* New York: Sterling Publications, 1998.

Loxton, Howard. *The Noble Cat: Aristocrat of the Animal World.* London: Merehurst, 1990.

Savage, R. J. G., and M. R. Long. *Mammal Evolution—An Illustrated Guide.* New York: Facts on File, 1986.

Taylor, David. *The Ultimate Cat Book.* New York: Simon and Schuster, 1989.

Turner, A. *The Big Cats and Their Fossil Relatives: An Illustrated Guide to Their Evolution and Natural History.* Columbia University Press, 1997.

Turner, Dennis C., and P. P. G. Bateson. *The Domestic Cat: the Biology of Its Behavior.* New York: Cambridge University Press, 1988.

Vita Richman

Cattails

Cattails or reedmaces are about 10 **species** of monocotyledonous plants in the genus *Typha*, comprising the family Typhaceae. Cattails are tall, herbaceous, aquatic plants, growing from stout rhizomes located in shallow sediments of **wetlands**. The leaves of cattails are long and strap-like, sheathing at the base of the **plant**, while the spike-like inflorescence is borne by a long cylindrical shoot. The typical **habitat** of cattails is productive marshes, the edges of shallow, fertile lakes, and ditches. Cattails occur in temperate and tropical regions but not in **Australia** or **South America**.

Cattail inflorescences are dense aggregations of numerous small separate female and male flowers, the latter occurring segregated at the top of the club-like flowering structure. **Pollination** is anemophilous, meaning

Great meadows cattails. *Photograph by Farrell Grehan. Photo Researchers, Inc. Reproduced by permission.*

the pollen is shed copiously to the **wind** which transports it to the stigmatic surfaces of female flowers. The small, mature **fruits** of cattails are shed late in the growing season or during the ensuing autumn or winter. The fruits have a white filamentous pappus that makes them aerodynamically buoyant so they can be easily dispersed by the wind.

Two familiar species of cattail in **North America** are the broad-leaved cattail (*Typha latifolia*) and the narrow-leaved cattail (*T. angustifolia*). Both occur commonly in a wide range of fertile, **freshwater** wetlands.

Cattail leaves are sometimes used for weaving mats, baskets, chair bottoms, and floor mats. Cattail pollen can be collected and used to make or include in protein-rich pancakes and breads. The rhizomes of cattails are sometimes collected and used as a starchy food. Sometimes the dried plants are collected, dried, and used for winter bouquets and other natural decorations.

The major importance of cattails, however, is ecological. These plants are a significant component of the vegetation of many productive marshes. As such, cattails help to provide **critical habitat** for many species of aquatic wild life such as waterfowl, waders, other species of **birds**, and **mammals** such as muskrats.

Because of their great productivity, cattails can take up large quantities of **nutrients** from **water** and sediment. This makes these plants rather efficient and useful at cleansing nutrients from both natural and waste waters. In this way cattails can help to alleviate **eutrophication**, an environmental problem associated with large rates of aquatic productivity caused by nutrient loading.

Cattle family (Bovidae)

The cattle family, Bovidae, is a widespread group of **mammals** which also includes the **goats**, **sheep**, **gazelles**, antelopes, and goat-antelopes. Of the 107 **species** currently recognized within this family, just 12 are wild cattle. Even the large muskox (*Ovibos moschatus*), which looks quite cow-like, is more closely related to the goats than to cattle. Cattle are generally characterized by their large size and the single pair of non-branching horns growing from their forehead. The horns, which are not shed, are largely hollow and differ considerably among species: some of the largest are those of the wild **yak** (*Bos mutus*) and African buffalo (*Syncerus caffer*), while the smallest are those of the anoas or dwarf water-buffalo (*Bubalus* species). Both males and females develop horns as they mature, but those of the male are usually much longer. Wild cattle vary considerably in appearance—from the brown-black colors of the anoas to the banteng (*Bos javanicus*), in which the females are reddish brown and the males shiny black. Both sexes are adorned with white stockings, a white rump patch, a white patch over the eyes, and a white band around the muzzle. The antithesis to these decorative cattle are wild yak, whose unkempt, shaggy appearance and unpredictable temperament are an example of how one species has evolved to withstand extreme conditions which few other animals could exploit.

The precise origins of wild cattle are still unclear, but it is thought that they arose from species resembling the small four-horned antelope (*Tetracerus quadricornis*) and larger nilgai (*Boselaphus tragocamelus*), which today survive only in India. The ancestors of the modern-day cow (*Bos taurus*) have been traced back to a species known as the auroch (*Bos primigenius*), an extinct wild ox of **Europe** and **Asia** that reached more than 7 ft (2m) at the shoulder. These were largely forest-dwelling cattle, feeding in open glades and around the fringes of woodlands. Adult males, or bulls, had long curving horns, a black coat with a white stripe down the middle of the back, and a patch of short tufts of white hair between the horns. Females, or cows, were smaller and a reddish brown **color**. The last known auroch died

in Poland in 1627. Domestic cows, of course, are extremely abundant in captivity, with a world-wide population exceeding a billion animals.

Wild cattle have evolved to survive in a wide range of habitats, from Arctic to tropical conditions. Wild yak are an example of a species that has adapted to living in a harsh climate, with their thick shaggy outer hair and densely matted undercoat providing insulation against the extreme cold in its native **habitat** in the high, **wind** and snow-swept plateaus of the Himalayas. Species that live in the tropics have other physiological problems to overcome, such as avoiding becoming too hot. These species are typically forest-dwelling, and have short hair which does not retain body **heat** very effectively. Some species, such as the kouprey (*Bos sauveli*), have extended dewlaps, large fat-filled folds of loose skin that hang below the neck and serve as heat-radiating surfaces. Other species, such as water buffaloes, cope with the heat and ever-present **flies** by immersing themselves in **water**. By frequent wallowing, buffalo cover themselves with a layer of mud that also helps protect them from the piercing bites of **insects**. Most species feed at dawn and dusk in order to avoid the midday heat. Some of the shyer species, such as gaur, anoa, and tamaraw (*Bubalus mindorensis*) may feed at nighttime in order to avoid detection from predators, including humans.

Over the centuries, several species of wild cattle have been domesticated, including the auroch, yak, gaur, banteng, and water buffalo. In some cultures, cattle are their owner's most important possessions. In **Africa**, cattle are an important symbol of wealth and are used for trade purposes, as well as for their milk and **blood**, both of which feature in the diet of certain tribes. Their meat, of course, is also of importance, as are their valuable hides, but cattle are only butchered on special occasions. Even their dung is of considerable importance as it is dried and stored as a source of cooking fuel in many parts of the **continent** where **tree** cover is sparse. Domesticated yak are equally important for the people of mountainous Nepal and Tibet in regions where roads are few and the climate extreme. Working at altitudes of up to 20,000 ft (6000 m), yak are capable of hauling heavy loads and surviving on a low-quality diet. In addition, they provide people with an essential supply of milk, meat, wool, and hides.

Cattle are grazing animals that feed mainly on **grasses**, herbs, and tree and shrub leaves. They feed by twisting grasses and stems around their tongue and cutting the vegetation off with the lower incisors, which protrude slightly forward. The jaw is designed to allow a circular grinding **motion**, which allows the food to be thoroughly crushed and masticated between the animal's large teeth. Their **plant** food, however, is largely com-

A Longhorn bull, heifer, and calf. *JLM Visuals. Reproduced by permission.*

posed of **cellulose**, a tough **carbohydrate** that few animals are able to digest. Cows, however, have evolved a means of overcoming this problem and are able to benefit from the relatively abundant supplies of available plant **biomass**. Cows belong to a group of animals known as ruminants, which have a specialized system of digestion that enables them to break down the cellulose fibers and extract the **energy** from the plants they eat. One of the most significant features of this system is a specialized stomach which, in ruminants, consists of four.distinct chambers: the rumen, reticulum, omasum and abomasum. Another is the presence of large numbers of specialized **bacteria** that live within a sort of liquid broth in the stomach, and are essential in digesting cellulose. Without these symbiotic bacteria, no amount of chewing would render the plant material in a state from which **nutrients** can be absorbed.

Cattle are prodigious eaters; at any one feeding, large quantities of grasses are consumed and pass directly into the large rumen. Here they are moistened and mixed with the bacteria which begin to attack the tough plant fibers. From the rumen, partly digested materials pass on to the reticulum, where a similar process continues. Most wild cattle have preferred feeding and resting grounds and, when not feeding or moving, they withdraw to protective cover to digest their food. At this time the **animal** will regurgitate this partly digested food once again into the mouth, where it is chewed a second time, mixed with salivary enzymes, and then swallowed again in a process commonly known as "chewing the cud." When it is swallowed again, the food passes through the upper stomach to the omasum and abomasum, where normal digestive enzymes take over the process of breaking down the plant materials and freeing the nutrients, which can then be absorbed and used by the cow. In total, food takes from 70 to 100 hours to pass through the **digestive system** of a cow—one of the slowest passage rates in the animal kingdom. Although this passage is slow, it is extremely effective. A thoroughly masticated and digested food base allows a large fraction of the **proteins** and other nutrients to be absorbed.

Wild cattle play an important ecological role through their grazing habits, in particular by keeping **grasslands** open from invading shrubs and coarse grasses, and creating a habitat favorable for other grazing animals, such as **deer** and antelope. Their dung, which is widely scattered across the grasslands or throughout the forest, serves as an important fertilizer and is broken down by a wide range of **beetles**, **fungi**, and bacteria that release essential nutrients and **minerals** that promote further plant growth.

Cattle are naturally social animals and form small herds, the composition of which varies according to the species. Some species, like the anoa, may be solitary or travel in groups of just two to four animals. At the other extreme, herds of thousands of **bison** once ranged across the vast fertile prairies of **North America**, each massive herd separated into sub-units of several hundred animals.

The African buffalo displays a relatively advanced system of social **behavior**. The entire herd not only feeds and moves around as a colossal single unit, but individual animals will also gather around an injured or sick animal if it is threatened by predators. This is a non-territorial species that forms herds of 50-2,000 animals, with an average of about 350 animals. Within the herd, a distinct social hierarchy is evident: dominant males are most successful in breeding with females. There is constant rivalry between these dominant bulls and transient males who do not belong to the local herd. Subordinate juvenile males are allowed to remain with the herd until they reach sexual maturity, at which time they are driven out and may remain solitary or form small bachelor groups with other males. Juvenile females remain with the herd and maintain a strong bond with their mother.

Breeding is a frantic time for all species of wild cattle, with adult males vying with other socially dominant males for the right to mate with receptive females. Dominant males must also keep a vigilant **eye** on sexually mature transient males, who may attempt to steal their cows and form their own herd. Adult bulls challenge others of similar status with loud roars and mock charges. On most occasions these displays of strength are sufficient to deter challenging males, but sometimes the challenging male refuses to back down and the situation escalates to fierce clashes in which one bull pitches his strength and agility against the power of another. The bull's horns are one focus of attention in such battles, as males interlock their horns and try to wrestle their opponent into a vulnerable position where they may strike other parts of the body with their pointed horns. The winner of such conflicts is almost certainly guaranteed the rights of dominant male within the herd, until he, in turn, is deposed by another challenging male.

The gestation period of wild cattle varies considerably, from about 270 days to over 400 days. As the calving time approaches, most cows leave the herd, returning to join it again 7-10 days later with their offspring. Most cows give **birth** to a single calf, which remains close to its mother until weaned. Following that, calves usually remain with the herd for a further three years until they reach sexual maturity. The fate of young cattle thereafter depends on the social system of the particular species, as described above.

Wild cattle have few natural predators, at least at the adult stage. Calves, however, are susceptible to predation from lions, tigers, leopards, and wild dogs. Predation is thought to have been one of the main reasons why wild cattle developed a herding life style, as the presence of large numbers of heavily armored animals is often enough to deter a **predator** from attacking. When feeding as a group, the animals are slightly spread apart and it is advantageous to have many eyes on the lookout - each animal takes its turn to scan the surrounding vegetation between feeding bouts. Through the centuries, however, wild cattle have suffered considerably at the hands of humans, as they have been hunted for their meat, hides, and as sport. Widespread herds of auroch were decimated by the sixteenth century; the European bison (*Bison bonasus*) suffered a similar fate during the nineteenth century; while the once vast herds of American bison suffered heavily at the hands of European settlers in their quest to open up the American West.

The European bison is an example where a species became extirpated in the wild, but then recovered somewhat through the breeding and release of captive animals. A sedentary, woodland-dwelling species, the European bison was reduced to a few scattered populations by the beginning of the twentieth century. Many of these were destroyed during World War I, and the last remaining wild herds died out in Lithuania and the Caucasus by the middle 1920s. Concentrated efforts by a few zoos led to the re-introduction of a small herd to natural habitat in the Bialowieza Forest in Poland. At first, these animals were retained in semi-captive conditions, but they were later released to form free-ranging herds. Although this species has been saved from **extinction**, present-day herds are closely inbred— they are all descended from just 17 animals. Additional **conservation** programs are underway to try and maximize the genetic exchange between breeding herds.

One of the least-known cattle species is the tamaraw, a small species that reaches just 3 ft (1 m) at the shoulder. This species inhabits forested parts of the **island** of Mindoro in the Philippines. It is thought to be nocturnal and quite aggressive, but almost nothing is known about its **ecology**. Overhunting, as well as loss of habitat and human encroachment on this species' habitat, has resulted in a serious population decline from an estimated 10,000 animals in the early 1900s, to fewer than 400 animals today. The majority of these free-living animals are confined to Mount Ilgo National Park. In view of the extent of **deforestation** on this island, the species is classified as endangered by conservation organizations. Several con-

A water buffalo during a monsoon in Nepal. *© George Turner, National Audubon Society Collection/Photo Researchers, Inc. Reproduced with permission.*

servation initiatives have been attempted in the past, but these have met with little success so far.

Like the tamaraw, the kouprey is another highly **endangered species**. Once wide-ranging throughout Indochina, it is now thought to be extinct in countries such as Thailand. The last remaining herds may survive in the forested border countries of Laos, Cambodia and Vietnam. A large species reaching a height of 6 ft (1.9 m) at the shoulder and measuring from 7-8 ft (2.1-2.3 m)in body length, males of this species are much larger than females. Young calves and females are generally a gray color (the species is locally known as the "gray ox"), with the undersides a lighter hue and the neck, chest and forelegs slightly darker. Mature males are a rich dark brown color, with white or gray coloring on the lower legs. The species may be recognized by its dewlap, which in some older males may even reach and drag along the ground. Apart from color differences, bulls are easily distinguished from cows by their horns: the latter generally have horns that **spiral** upwards, while those of a bull are more widely spaced and often frayed at the ends. The reason why these split at the ends is unknown, but some authorities believe kouprey use their horns for digging in the **earth**, perhaps in search of mineral salts.

Kouprey are animals of gently rolling hills in deciduous and semi-evergreen tropical **forests** that offer a wide range of open feeding and resting sites. They may, however, move to higher ground during the wettest periods of the year. Little is known about the ecology of this species. They are known to form small herds of cows and offspring, with separate bachelor herds of young males. Animals appear to be most active in the morning and again in late afternoon, and frequently travel at night. Mixed herds of kouprey, banteng, and feral water buffaloes have been reported. In 1949, there were thought to be around 1,000 kouprey surviving. This number is thought to have declined to 500 in 1951, and just 100 in 1969. No one is quite sure how many kouprey survive in the wild today, as this species' habitat is at the center of almost constant human warfare. Authorities fear for its survival in view of the heavy hunting pressure in the region. Major conservation efforts have been designed to undertake captive breeding programs in safe parts of Kampuchea and Laos if sufficient animals can be obtained from the wild.

Other species, which are perhaps equally threatened, are the wild anoas, of which two species have been recorded: the mountain (*Bubalus quartesi*) and lowland

(*B. depressicornis*) anoa. Both species are only found in Sulawesi, Indonesia. In appearance, anoas resemble dwarf water buffaloes, measuring just 2-3 ft (0.7-1 m) at the shoulder and 5-6 ft (1.6-1.7 m) in body length, and weighing about 300-600 lb (150-300 kg). Adults are usually a dull brown color, but the shading pattern may vary, with some animals having lighter undersides. Calves are covered with woolly yellow-brown hair, but this is lost as they mature. Little is known about the ecology of these secretive cattle as few observations have been made in the wild. It is known that they are widely hunted for their meat, but another serious threat is continuing loss of forest habitat, which affects both species.

Wild cattle are of enormous importance for our present civilization, just as they have been in our past. They not only play a vital role in the local ecology of their diverse environments, but also represent an important genetic reservoir which is important for breeding purposes. Hybrids of cattle-yak origin are already of great importance in many parts of Nepal and China. Elsewhere in Asia, farmers in Laos and Kampuchea used to drive their cows into the forest in the hope that they would breed with wild kouprey bulls, as they found that such offspring were stronger than if the cows were mated with domestic cattle. All wild cattle therefore, have considerable potential in the breeding arena, since these species are often far better adapted to local climatic and forage conditions, as well as being stronger and more resistant to diseases, many of which are debilitating for domestic breeds.

If we are to save these remaining species in the wild, however, there is not much time to lose. Most species are now severely threatened by hunting as well as habitat loss. As the habitat range of these species continues to shrink in the face of human encroachment, all wild populations of cattle risk becoming isolated and susceptible to **disease**, as well as inbreeding, as there will no longer be a possibility of genetic exchange between isolated populations. Future conservation efforts must continue to focus on preserving natural feeding and breeding ranges, as well as essential **migration** corridors of these species, in order to ensure the continued viability of the wild herds of cattle.

See also Livestock.

Cauterization

Cauterization is the application of **heat**, mechanically or chemically, to prevent or stop bleeding. It is widely used in **surgery** to hold bleeding to a minimum and speed the surgical process.

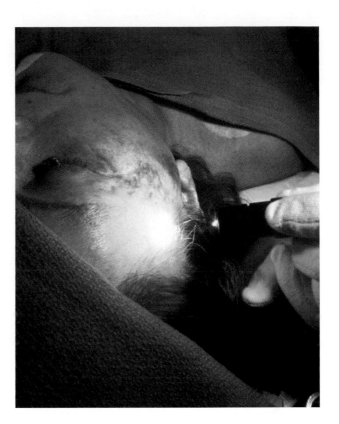

Birthmark removal by cauterization with an argon laser.
Photograph by Alexander Tsiaras. National Audubon Society Collection/Photo Researchers, Inc. Reproduced by permission.

History

Searing areas of bleeding with a hot instrument, a hot **iron** or other metallic object, was practiced for many years for the treatment of wounded soldiers. Even thousands of years ago, all wounded were treated by pouring boiling oil into the wound to arrest bleeding. Of course, in this case the cure was nearly as harmful as the original injury. Many of the wounded, already in shock from their trauma, were plunged into deeper shock and death by the oil.

As surgery progressed and **anesthesia** was introduced to quiet the patient and prevent his feeling **pain**, more care and more time could be devoted to preventing bleeding. In making an incision the surgeon would cut across small **blood** vessels such as **capillaries** and arterioles that would begin to ooze blood. The surgeon then had to locate each point of bleeding and apply a clamp to stop it, and then go back and tie a suture around each bleeder, a long and exacting process.

The electric cautery, a form of scalpel, then was invented and introduced into the surgical suite. Using this instrument the surgeon could make his incision and the cautery seared and sealed off all sites of bleeding except the largest ones. This considerably reduced the time the surgeon spent in stanching the flow of blood into the surgical field. It was also a benefit to the patient who spent less time under the anesthetic and reduced the amount of blood loss.

Currently the ubiquitous **laser** has been introduced as a scalpel. The powerful beam cuts into **tissue** and opens the incision while at the same time heat-searing the small blood vessels.

Other applications

Chemical cauterization also is used in limited circumstances. For example, one means of stopping a nosebleed that has defied all other means of cure is to use an applicator with silver nitrate on one end. The silver nitrate is applied directly to the bleeding area and cauterizes it.

See also Laser surgery.

Cave

A cave is a naturally occurring hollow area inside the earth. Most caves are formed by some type of erosional process. The most notable exception is hollow lava tubes such as those in Hawaii. The formation of caves depends upon geologic, topographic, and hydrologic factors. These factors determine where and how caves develop, as well as their structure and shape. The study of caves is called speleology. Some caves may be small hillside openings, while others consist of large chambers and interconnecting tunnels and mazes. Openings to the surface may be large gaping holes or small crevices.

Caves have provided shelter to prehistoric, ancient, and primitive contemporary people such as the Tasadays of the Philippine Islands. Human remains, artifacts, sculptures, and drawings found in caves have aided archeologists to learn about early humans. Caves are sites of many important archeological discoveries such as The Dead Sea Scrolls. Many religious traditions have regarded caves as sacred and have used them to perform rituals, ceremonies, and sacrifices. Some ancient traditions felt that caves led to the underworld. Caves have fascinated poets, artists, philosophers, and musicians.

Cave types

Caves hosted in **rocks** other than limestone are usually formed by **water** erosion. For example, **rivers** running through canyons with steep walls erode the rock at points where the current is strong. Such caves usually have large openings and are not too deep. Caves of this type can be found in the southwestern United States and were at one time inhabited by prehistoric American Indians known as Cliff Dwellers. Sea caves are formed by waves continually crashing against cliffs or steep walls. Often these caves can only be entered at low tide. **Ice** caves are also formed in **glaciers** and **icebergs** by meltwater that drains down crevices in the ice.

Lava caves, which are often several miles long, form when the exterior of a lava flow hardens and cools to form a roof, but lava below the surface flows out, leaving a hollow tube. **Wind** or aeolian caves usually form in sandstone cliffs as wind-blown **sand** abrades the cliff face. They are found in **desert** areas, and occur in a bottle neck shape with the entrance much smaller than the chamber. Talus caves are formed by boulders that have piled up on mountain slopes. The most common, largest, and spectacular caves are solution caves.

Solution caves

Solution caves form by chemical **weathering** of the surrounding **bedrock** as **groundwater** moves along fractures in the rock. These caves produce a particular type of terrain called karst. Karst terrain primarily forms in bedrock of **calcium carbonate**, or limestone, but can develop in any soluble **sedimentary rock** such as dolomite, rock gypsum, or rock **salt**. The host rock extends from near the earth's surface to below the water table. Several distinctive karst features make this terrain easy to identify. The most common are **sinkholes**, circular depressions where the underlying rock has been dissolved away. Disappearing streams and natural bridges are also common clues. Entrances to solution caves are not always obvious, and their discovery is sometimes quite by accident.

Formation of karst involves the chemical interaction of air, **soil**, water, and rock. As water flows over and drains into the earth's surface, it mixes with **carbon dioxide** from the air and soil to form carbonic acid (H_2CO_3). The groundwater becomes acidic and dissolves the **calcium** carbonate in the bedrock, and seeps or percolates through naturally occurring fractures in the rock. With continual water drainage, the fractures become established passageways. The passageways eventually enlarge and often connect, creating an underground drainage system. Over thousands, perhaps millions of years, these passages evolve into the caves we see today.

During heavy rain or **flooding** in a well-established karst terrain, very little water flows over the surface in stream channels. Most water drains into the ground through enlarged fractures and sinkholes. This underground drainage system sometimes carries large amounts of water, sand, and mud through the passageways and further erodes the bedrock. Sometimes ceilings fall and passageways collapse, creating new spaces and drainage routes.

Not all solution caves form due to dissolution by carbonic acid. Some caves form in areas where **hydrogen** sulfide gas is released from the earth's crust or from decaying organic material. **Sulfuric acid** forms when the hydrogen sulfide comes in contact with water. It chemically weathers the limestone, similar to **acid rain**.

Cave environment and formations

The deep cave environment is completely dark, has a stable atmosphere, and the **temperature** is rather constant, varying only a few degrees throughout the year. The **humidity** in limestone caves is usually near 100%. Many caves contain unique life forms, underground streams and lakes, and have unusual mineral formations called speleothems.

When groundwater seeps through the bedrock and reaches a chamber or tunnel, it meets a different atmosphere. Whatever mineral is in solution reacts with the surrounding atmosphere, precipitates out, and is deposited in the form of a **crystal** on the cave ceiling or walls. Calcite, and to a lesser degree, aragonite, are the most common **minerals** of speleothems. The amount of mineral that precipitates out depends upon how much gas was dissolved in the water. For example, water that must pass through a thick layer of soil becomes more saturated with **carbon** dioxide than water that passes through a thin layer. This charges the water with more carbonic acid and causes it to dissolve more limestone from the bedrock. Later, it will form a thicker mineral **deposit** in the cave interior as a result.

Water that makes its way to a cave ceiling hangs as a drop. When the drop of water gives off carbon dioxide and reaches chemical equilibrium with the cave atmosphere, calcite starts to precipitate out. Calcite deposited on the walls or floors in layers is called flowstone.

Sometimes water runs down the slope of a wall, and as the calcite is deposited, a low ridge is formed. Subsequent drops of water follow the ridge, adding more calcite. Constant buildup of calcite in this fashion results in the formation of a large sheet-like formation, called a curtain, hanging from the ceiling. Curtain formations often have waves and folds in them and have streaks of various shades of off-white and browns. The streakiness results from variations in the mineral and **iron** content of the precipitating solution.

Often, a hanging drop falls directly to the ground. Some calcite is deposited on the ceiling before the drop falls. When the drop falls, another takes its place. As

Stalactites and stalagmites in the Hams Caves, Spain.
Photograph by Pedro Coll. Stock Market. Reproduced by permission.

with a curtain formation, subsequent drops will follow a raised surface and a buildup of calcite in the form of a hanging drop develops. This process results in icicle-shaped speleothems called stalactites. The water that falls to the floor builds up in the same fashion, resembling an upside down icicle called a stalagmite.

Of course, there are variations in the shape of speleothems depending on how much water drips from the ceiling, the temperature of the cave interior, rates and directions of air flow in the cave, and how much dissolved limestone the water contains. Speleothems occur as tiered formations, cylinders, cones, some join together, and occasionally **stalactites and stalagmites** meet and form a tower. Sometimes, when a stalactite is forming, the calcite is initially deposited in a round ring. As calcite builds up on the rim and water drips through the center, a hollow tube called a straw develops. Straws are often transparent or opaque and their diameter may be only that of a drop of water.

Stalactites and stalagmites occur in most solution caves and usually, wherever a stalactite forms, there is also a stalagmite. In caves where there is a great deal of seepage, water may drip continuously. Speleothems formed under a steady drip of water are typically smooth. Those formed in caves where the water supply is seasonal may reveal growth rings similar to those of a **tree** trunk. Stalactites and stalagmites grow by only a fraction of an inch or centimeter in a year, and since some are many yards or meters long, one can appreciate the time it takes for these speleothems to develop.

The most bizarre of speleothems are called helictites. Helictites are hollow, cylindrical formations that grow and twist in a number of directions and are not simply oriented according to the gravitational pull of a water drop. Other influences such as, crystal growth patterns and air currents influence the direction in which these speleothems grow. Helictites grow out from the side of other speleothems and rarely grow larger than 4 in (8.5cm) in length.

Speleothems called anthodites are usually made of aragonite. Calcite and aragonite are both forms of calcium carbonate, but crystallize differently. Anthodites grow as radiating, delicate, needle-like crystals. Pools of seepage water that drain leave behind round formations called cave popcorn. Cave pearls are formed in seepage pools by grains of sand encrusted with calcite; flowing water moves the grains about and they gather concentric layers of calcite.

Cave life

There are three main groups of animals that inhabit caves. These animals are classified by their degree of dependence on specific cave conditions such as amount of **light**, temperature, atmospheric conditions, and water. Animals that commonly use caves but depend on the outside world for survival, are called trogloxenes. The best known trogloxenes are **bats**.

Other examples include **birds**, **bears**, and **crickets**. Troglophiles are **species** that live their entire life cycle within a cave, generally near the entrance, but are also found living outside caves. **Cockroaches**, **beetles**, and **millipedes** are some examples of troglophiles. Certain **fungi** and **algae** are also classified as troglophiles. The third classification are troglobites. Troglobites are permanent cave dwellers found deep within the cave system in total darkness, and consequently lack **color**. These species are either white, transparent, or slightly pinkish. Troglobites have no need for eyes. Accordingly, groups of organisms that once could see, have evolved into eyeless creatures, although some species have retained **eye** sockets. They rely on their sense of **touch** to get around.

Some examples of troglobites are **fish**, **shrimp**, **crayfish**, **salamanders**, worms, **snails**, **insects**, **bacteria**, fungi, and algae. Each cave has a self-contained **ecosystem**, and it is thought that some have not changed for millions of years. As new caves are discovered, speleobiologists regularly find new species of animals.

Resources

Books

Hamblin, W.K., and E.H. Christiansen. *Earth's Dynamic Systems*. 9th ed. Upper Saddle River: Prentice Hall, 2001.

Jacobson, Don, and Lee Stral. *Caves and Caving*. Harbor House, 1987.

Mohr, C.E., and T. Poulson. *Life of the Cave*. New York: McGraw-Hill, 1966.

Christine Miner Minderovic

Cave fish

Many **species** of **fish** have evolved to living under strange conditions, but few are more intriguing than those that have adapted to living in complete darkness. Some of these fish have developed a tendency to live at great depths in the **ocean** where no **light** penetrates, while others have found refuge in equally dim locations such as caves, wells, and subterranean streams. This specialization to living in complete darkness has arisen many times during the course of **evolution**, and the physical and behavioral attributes these species have developed are not confined to a single taxonomic group of fish. Some 32 species of fish have been observed to exhibit this cave-dwelling **behavior**. Most of these are small species, measuring some 3 in (7 cm) in length; the Kentucky blind fish (*Amblyopsis spelaea*), which lives in limestone caves, is one of the largest known cave-dwelling species, with adults reaching a length of almost 8 in (20 cm).

A few of these species have functional eyes but, for the vast majority, **vision** is of little use. Some of them are completely blind, with the vestiges of eyes still visible beneath a thin layer of skin. In many species, functional eyes are present in the young fry, but these either disappear or are covered up by a layer of skin as they grow and mature. Many cave fish have little or no skin pigment as these species have no need for body coloring, a feature widely used for communication purposes among other fish. Some cave fish do appear to be a pinkish **color**, but this is the result of the animal's **blood** vessels showing through the pale skin, rather than any pigmentation.

Apart from the specialized adaptations these fish have developed, what is interesting to scientists is the fact that so many species of different groups have evolved independently to living in perpetual darkness. In **Africa**, a group of Cyprinid fish (related to the **carp**) has adapted to living in underground streams and wells, while several species of **catfish** in Africa, the United States, and parts of **South America** have developed a similar habit. In Central America, the brotulid fish (Family Brotulidae) of Mexico and Cuba are typically cave-dwellers, lacking functional eyes and body pigment. Similar features have been detected among the Amblyopsidae, a group of five species of small **freshwater** fish that only occur in the United States. Some of these species actually live in slow-moving streams and swamps in the open air in southern Atlantic coastal plain, yet they are all blind and live amongst the rubble and vegetation in deeper **water**, well out of direct light.

To compensate for their lack of vision, all of these species have developed other means of locating food and finding a mate, two of the essential features for survival. Many species have developed sensory organs capable of detecting **prey** either through chemical means or **touch**. Most are thought to be sensitive to vibrations. Some of these species retain their body scales but, in others, these have been lost during the course of evolution, enabling them to develop additional sensory organs on the skin. In addition to this range of specialized adaptations, one feature that all of these fish have developed to a higher level than other ocean or freshwater species, is an improved lateral line system—a series of grooves or canals that run along each side of the body and extend over the head to the eyes, snout, and jaws. These lines are well equipped with sensory organs known as neuromasts, each of which consists of a group of sensory cells with fine hair-like projections that extend beyond the body wall. They function not only in orientation and balance, but also in helping the fish locate potential food sources.

While many of these adaptations assist with the detection and gathering of food, it is also likely that some nourishment is obtained on a chance basis. Very little is known about the feeding behavior of cave fish, but they are known to eat a wide range of **insects**, small crustaceans, smaller fish, and probably some detritus. Some species are even thought to feed on fecal droppings from overhead bat roosts.

The reasons why species should evolve to living in total darkness are not immediately obvious, and there is probably no single explanation for this phenomenon. Cave environments are known to provide a relatively stable **habitat** in terms of **temperature** fluctuations, but the species living in caves and wells are totally reliant on food being brought to them by underground streams. As such, they are highly vulnerable to external factors as subterranean aquifers are becoming increasingly tapped

for **irrigation** purposes, and many sites may be at risk from drying out either temporarily or permanently. These species are also at risk from water-borne pollutants in agricultural runoff and other waste products entering their underground watercourses. For these reasons, many species of cave fish are endangered. On the other hand, these fish experience a reduced level of feeding **competition** as so few species have succeeded in colonizing these habitats. They are also relatively safe from predators, which is an obvious advantage provided they can obtain enough food in their immediate habitat.

Scientists now believe that the colonization of caves and wells by these fish was a deliberate, rather than accidental, move. Many species are known to live in and around cave openings and pools near underground springs. It is therefore possible that these dim habitats were gradually explored by a few different species that may have already favored living in poorly lit conditions, such as beneath **rocks**, or in the murky depths of swamps and lakes. Through a gradual progression and corresponding behavioral and anatomical changes, these fish could have eventually moved further into the **cave** system, exploiting the untapped food resources of these habitats.

Cavies *see* **Guinea pigs and cavies**

Celestial coordinates

Celestial coordinates locate objects on the sky, which is considered to be an infinitely large (celestial) **sphere**. The four conventional celestial coordinate systems are defined.

Horizon coordinates

These refer to an observer on the Earth's surface at the center of an imagined celestial sphere.

Altitude h, an object's **arc** distance along a vertical **circle** from the horizon, positive for objects above the horizon [between it and the zenith $(h=+90°)$] and **negative** for objects below it [between it and the nadir $(h=+90°)$].

Azimuth A, the arc distance from the north point of the horizon $(h=0°, A=0°)$ eastward along it to where it meets the object's vertical circle, going from 0° to 360° from the north point to full circle back to the North Point. Azimuth is undefined at Earth's north and south poles, where the north and south celestial poles (NCP and SCP) coincide with the zenith and nadir, and the celestial meridian become undefined.

KEY TERMS

Celestial equator—The projection into space of the earth's equator.

Celestial merdian—The circle passing through the zenith, zadir, north celestial pole, and south celestial pole.

Ecliptic—The intersection of the earth's orbital plane with the celestial sphere.

Galactic equator—The circle through the middle of our Milky Way galaxy's disk which passes through the direction from the solar system to the center of the Milky Way.

Horizon—The circle on the celestial sphere 90° from the zenith and nadir.

Hour circles—Half circles from the north celestial pole to the south celestial pole.

Nadir—The point on the celestial sphere directly below (by downward extension of the plumb line through the earth) the observer.

North point of the horizon—The intersection of

the horizon and celestial meridian closer to the north celestial pole (NCP).

North (south) celestial poles (NCP, SCP)—The intersection(s) of the earth's rotation axis extended beyond the north (south) geographic poles, respectively, with the celestial sphere.

Secondaries to the ecliptic—Half circles from the north ecliptic pole to the south ecliptic pole.

Secondaries to the galactic equator—Half circles from the north galactic pole to the south galactic pole.

Vernal equinox—The intersection of the celestial equator and ecliptic which the Sun appears to reach on or about March 21.

Vertical circles—Half circles from the zenith to the nadir.

Zenith—The point on the celestial sphere directly above (by upward extension of the local direction of gravity (plumb line) the observer.

Equatorial coordinates

These are based on the **earth's rotation**, which produces an apparent westward **rotation** of the celestial sphere around the NCP and SCP.

Right ascension *a*

Measured eastward along the celestial equator from the vernal **equinox** to where an object's hour circle meets the celestial equator, usually measured full circle in **time** units from 0^h to 24^h.

Declination *d*

An object's arc distance along its hour circle from the celestial equator, positive north of the equator, negative south of it. $d=+90°$ for the NCP, and $d=+90°$ for the SCP.

Hour angle

An object's hour angle t is the arc distance westward along the celestial equator from its intersection with the celestial meridian above the horizon to where the celestial equator meets the object's hour circle; its value increases with time from 0^h to 24^h.

Right ascension and declination on the celestial sphere are analogous to geographic longitude and latitude, respectively, on **Earth**, but their measurement differs somewhat.

Ecliptic coordinate

The ecliptic is the basic circle and the vernal equinox is the **zero** point for these coordinates. The north (NEP) and south (SEP) ecliptic poles are $23^0.5$ from the NCP and SCP, respectively, and are everywhere 90° from the ecliptic.

Celestial longitude l

The arc distance eastward along the ecliptic from the vernal equinox to where an object's secondary to the ecliptic meets the ecliptic; it is expressed in arc units from 0° to 360°.

Celestial latitude *b*

An object's arc distance from the ecliptic along its secondary to the ecliptic; it is positive north of the ecliptic, negative south of it. $b=+90°$ at the NEP and $b=90°$ at the SEP. These coordinates may be either geocentric (Earth-centered) or heliocentric (Sun-centered).

Galactic coordinates 1, the galactic equator is the basic circle and the direction from the **solar system** to the center of our **Milky Way galaxy** is the zero point for them. The north (NGP) and south (SGP) galactic poles are 90° from the galactic equator and are 62.6° from the NCP and SCP, respectively.

Galactic longitude

The arc distance eastward along the galactic equator from the direction to the center of our Milky Way galaxy to where an object's secondary to the galactic equator crosses the galactic equator; it increases from 0° to 360°.

Galactic latitude b, the arc distance of an object from the galactic equator along its secondary to the galactic equator. Values of b vary from +90° at the NGP to -90° at the SGP. Galactic coordinates of the direction to the center of the Milky Way galaxy are 1=0°, b=0°. Galactic coordinates are usually heliocentric but can also be centered at the center of our Milky Way galaxy.

Resources

Books

Berg, Rebecca M., and Laurence W. Frederick. *Descriptive Astronomy.* New York: Van Nostrand, 1978.

Bernhard, Hubert J., Dorothy A. Bennett, and Hugh S. Rice. *New Handbook of the Heavens.* New York: McGraw-Hill, 1948.

Motz, Lloyd, and Anneta Duveen. *Essential of Astronomy.* Belmont, CA: Wadsworth, 1966.

Celestial mechanics

Modern celestial mechanics began with Isaac Newton's generalization of **Kepler's laws** published in his *Principia* in 1687. Newton used his three **laws of motion** and his law of universal gravitation to do this. The three generalized Kepler's law are:

1) The orbits of two bodies around their center of **mass** (barycenter) are **conic sections** (ellipses, circles, parabolas, or hyperbolas) with the center of mass at a focus of each conic sections; 2) The line joining the center of the two bodies sweeps out equal areas in their orbits in equal **time** intervals. Newton showed that this is a consequence of conservation of angular **momentum** of an isolated two-body system unperturbed by other forces (Newton's third law of motion); 3) From his law of universal gravitation, which states that Bodies 1 and 2 of masses M_1 and M_2 whose centers are separated by a **distance** r experience equal and opposite attractive gravitational forces F_g of magnitudes

$$F_g = \frac{G m_1 m_2}{r^2} \quad \text{(Eq. 1)}$$

where G is the Newtonian gravitations factor, and from his second law of motion, Newton derived the following general form of Kepler's third law for these bodies moving around the center of mass along elliptical or circular orbits:

$$P^2 = \left[\frac{4\pi^2}{G(m_1 + m_2)} \right] a^3 \quad \text{(Eq. 2)}$$

where P is the sidereal period of revolution of the bodies around the center of mass, π is the **ratio** of the circumference of a **circle** to its diameter, X, m_1 and m_2 are the same as in Equation 1 and a is the semi-major axis of the *relative* orbit of the center of the less massive Body 2 around the center of the more massive Body 1.

These three generalized Kepler's Law form the basis of the two-body problem of celestial mechanics. **Astrometry** is the branch of celestial mechanics which is concerned with making precise measurements of the positions of celestial bodies, then calculating precise orbits for them based on the observations. In theory, only three observations are needed to define the **orbit** of one celestial body relative to a second one. Actually, many observations are needed to obtain an accurate orbit.

However, for the most precise orbits and predictions, the vast majority of systems investigated are not strictly two-body systems but consist of many bodies (the **solar system**, planetary **satellite** systems, multiple **star** systems, star clusters, and galaxies).

Planetary perturbations

To a first **approximation**, the solar system consists of the **Sun** and eight major planets, a system much more complicated than a two-body problem. However, use of Equation 2 with reasonable values for the **astronomical unit** (a convenient unit of length for the solar system) and for G showed that the Sun is far more massive than even the most massive **planet** Jupiter (whose mass is 0.000955 the Sun's mass). This showed that the gravitational forces of the planets on each other are much weaker than the gravitational forces between the Sun and each of the planets, which enabled astronomers to consider the gravitational interactions of the planets as producing small changes with time perturbations) in the elliptical orbit of each planet around the center of mass of the solar system (which is always in or near the Sun). If the Sun and a planet (say the **Earth** or **Jupiter**) were alone in empty **space**, we would have an ideal two-body problem and we would expect the two-body problem as defined by the generalized Kepler's law to exactly describe their orbits around the systems center of mass. Then the seven orbital elements (of which a and y are two) of a planet's orbit should remain constant forever.

However, the gravitational forces of the other planets on a planet cause its orbit to change slightly over time; these changes can be accurately allowed for over limited time intervals by calculating the perturbations of its orbital elements over time that are caused by the gravitational forces of the other planets.

Historically, perturbation theory has been more useful than merely providing accurate predictions of future planetary positions. Only six major planets were known when Newton published his *Principia*. William Herschel (1738-1822) fortuitously discovered **Uranus**, the seventh major planet from the Sun, in March 1781. The initial orbital elements calculated for Uranus did not accurately allow prediction of its future position even after inclusion of the perturbations caused by the six other major planets. Before 1821, Uranus was consistently observed to be ahead of its predicted position in its orbit; afterwards, it lagged behind its predicted positions.

John Couch Adams (1819-1892) in England and Urbain Leverier (1811-1877) in France, hypothesized that Uranus had passed an undiscovered massive planet further than it was from the Sun in the year 1821. They both made detailed calculations to locate the position of the undiscovered planet perturbing the **motion** of Uranus. Johann Galle (1812-1910) in Berlin, Germany used Leverier's calculations to discover the unknown planet in September 1846, which was then named **Neptune**.

Further unexplained perturbations of the orbits of Uranus and Neptune led Percival Lowell (1855-1916) and several other astronomers to use them to calculate predicted positions for another undiscovered (trans-Neptunian) planet beyond Neptune's orbit. Lowell searched for the trans-Neptunian planets he predicted from 1906 until his death in November 1916 without finding it. The search for a trans-Neptunian planet was resumed in 1929 at Lowell Observatory, where Clyde Tombaugh (1905–1997) who discovered **Pluto** in February 1930.

Lowell had predicted that a planet more massive than Earth produced the unexplained perturbations. During the years following Pluto's discovery, however, detailed studies of its perturbations of the orbits of Uranus and Neptune showed that Pluto is considerably less massive than Earth. The discovery of Pluto's satellite, Charon, in 1978 allowed the determination of the total mass of Pluto and Charon from Equation 2 which is about 0.00237 Earth's mass (about 0.2 the mass of Earth's **moon**). There are two consequences of this discovery; Tombaugh's discovery of Pluto may have been fortuitous, and one may make the case that Pluto is not a major planet.

The discrepancy in mass between the masses predicted by Lowell and others for the trans-Neptunian planet and the mass of the Pluto-Charon double planet has led to a renewed search for one or more additional trans-Neptunian planet(s) that still continues. The opinion also exists that the unexplained perturbations of the orbits of Uranus and Neptune are caused by systematic errors in some early measurements of their positions and that no trans-Neptunian planets with masses on the order of Earth's mass exist.

Resonance phenomena

Ceres, the first asteroid or **minor planets**, was discovered to orbit the Sun between the orbits of liars and Jupiter in 1801. Thousands of other asteroids have been discovered in that part of interplanetary space, which is now called the Main asteroid Belt.

Daniel Kirkwood (1815-1895) noticed in 1866 that the periods of revolution of the asteroids around the Sun did not form a continuous distribution over the Main Asteroid Belt but showed gaps (now known as Kirkwood's gaps) at periods corresponding to 1/2, 1/3, and 2/5 Jupiter's period of revolution (11.86 sidereal years). This phenomenon can be explained by the fact that, if an asteroid is in one of Kirkwood's gaps, then every second, third, or fifth revolution around the Sun, it will experience a perturbation by Jupiter of the same direction and magnitude; over the course of millions of years, these perturbations move asteroids out of the Kirkwood's gaps. This is a **resonance** effect of planetary perturbations, and it is only one of several resonance phenomena found in the solar system.

Ratios between the periods of revolution of several planets around the Sun are another resonance phenomenon that is poorly understood. The periods of revolution of **Venus**, Earth, and **Mars** around the Sun are nearly in the ratio is 5:8:15. The periods of revolution of Jupiter and **Saturn** are nearly in a 2:5 ratio, and for Uranus, Neptune, and Pluto they are nearly in a 1:2:3 ratio. The 2:3 ratio between the periods of revolution of Neptune and Pluto makes Pluto's orbit more stable. Due to the ellipticity of its orbit, near perihelion (the point on its orbit closest to the Sun) Pluto comes closer to the Sun than Neptune. Pluto last reached perihelion in September 1989; it has been closer to the Sun than Neptune since 1979 and will continue to be closer until 1998, when it will resume its usual place as the Sun's most distant known planet. However, Neptune will be the Sun's most distant known planet from 1995 to 1998! Recent calculations showed that, because of the 2:3 ratio of the orbital periods, the orientation of Pluto's orbit, and of the positions of Neptune and Pluto in their orbits, Neptune and Pluto have never been closer than 2,500,000,000 km in the last 10,000,000 years. Without the 2:3 ratio of their orbital periods, Pluto probably would have had a close encounter with Neptune which could have ejected Pluto and Charon into separate orbits around the Sun that are drastically different from the systems present orbit.

Jupiter's inner three Gallean satellite, Io, Europa, and Ganymede, have orbital periods of revolution around Jupiter that are nearly in the ratio 1:2:4. Five of Saturn's closest satellites, Pandora, Mimas, Enceladus, Tethys, and Dione, have orbital periods of revolution around Saturn that are nearly in the ratio 4:6:9:12:18. Resonance effects produced by some of these satellites, especially the 1.2 resonance with Mimas' period of revolution around Saturn, seem to have produced the Cassini Division between Saturn's A and B rings, which is analogous to the 1:2 Kirkwood's gap in the Main Asteroid Belt. A 3:4 orbital period resonance seems to exist between Saturn's largest satellite Titan and its next satellite out Hyperion.

Other resonances between the satellites and ring systems of Jupiter, Uranus, and Neptune are not clear because these ring systems are far less developed than that of Saturn.

Tidal effects

A tidal effect is produced when the gravitational pull of one body on a second one is appreciably greater on the nearer part of the second one than on its center, and in turn, the first body's pull on the second one's center is greater than its pull on the second one's most distant part. Unlike the gravitational **force** F_g, which varies as the inverse square of the distance r between the centers of the two bodies ($1/r_2$); see Equation 1, the tidal effect varies as the inverse cube ($1/r^3$) of the distance between their centers. Both the Moon and the Sun raise **tides** in Earth's oceans, atmosphere, and solid body. The lag of the tides raised in the oceans behind the Moon's crossings of the celestial meridian causes a gravitational interaction between Earth and Moon which slows the **earth's rotation** and moves the Moon's orbit further from Earth.

Tides raised in the Moon's solid body by Earth have slowed its **rotation** until it has become tidally locked to Earth (the Moon keeps the same hemisphere turned towards Earth, and its periods of rotation and revolution around Earth are the same, 27.32 mean solar days). Eventually Earth's rotation will be slowed to where Earth will be tidally locked to the Moon, and the durations of the sidereal day and sidereal month will both equal about 47 present mean solar days.

Tidal evolution has forced most planetary satellites to become tidally locked to their planets. This includes all of Jupiter's Galilean satellites and its four small satellites closer to Jupiter than Io, probably most of Saturn's satellites out to Iasetus (Titan, Saturn's largest satellite, is probably tidally locked to Saturn), the satellites of Uranus and Neptune, and the Pluto-Charon double planet. Tidal action in Io's interior produced by Jupiter (and to a lesser degree by its next satellite out Europa) powers volcanism on Io, making it the most volcanically active body in the solar systems. Tidal effects also may have powered volcanic activity on Europa and Ganymede, Saturn's satellite Enceladus, and Uranus satellites Miranda and Ariel; all of them show some evidence of resurfacing. Some of Earth's internal **heat** may have been produced by the Moon's tidal action.

The Sun's tidal action on Mercury at perihelion has tidally locked Mercury's rotation to its angular **velocity** near perihelion, which is 1.5 times Mercury's average orbital angular velocity; Mercury's rotation period is 58.6 days, 2/3 of its 87.9 day period of revolution around the Sun.

When two bodies are very close together, tidal forces tending to disrupt a body can equal or exceed the attractive gravitational forces holding it together. If the tidal stresses exceed the yield limits of the body's material, the body will gradually disintegrate into many smaller bodies. The mathematician E. Roche (1820-1885) studied the limiting separation of two bodies where the tidal and gravitational forces are equal; it usually between 2-3 times the radius of the more massive body and depends on the relative densities of the bodies and their state of motion. If two bodies approach closer than this Roche limit, one (usually the smaller, less massive body) or both bodies will begin to disintegrate. The rings of some of the Jovian planets may have formed from the tidal disintegration of one or more of their close satellites. Theory predicts that after Earth's and Moon's rotations become tidally locked (see above), the Sun's tides raised on Earth will cause the Moon to approach Earth. If this effect lasts long enough, the Moon may get closer to Earth than its Roche limit, be disintegrated by Earth's tidal forces, and form a ring of small bodies which orbit Earth.

Tidal effects act on close double stars, distorting their shapes, changing their orbits, and sometimes tidally locking their rotations. In some cases, tidal effects cause streams of gas to flow in a double star system and can transfer **matter** from one star to the other or allow it to escape into interstellar space. Tidal effects even seem to act between galaxies, with one **galaxy** distorting the form of its neighbor.

Precession

Rapidly rotating planets and satellites have appreciable equatorial bulges as a consequence of Newton's First Law of Motion. If the rotation axis of such a body is not **perpendicular** to its orbit, other bodies in the system will exert stronger gravitational attractions on the near part of the bulge than its far part. The effect of this difference is to tend to turn the body's rotation axis perpendicular to the **plane** of its orbit. Because the body is rotating rapidly, however, this does not happen, and, like the rotation axis of a spinning top, the body's rotation axis describes a cone in space whose axis is the perpendicular to the body's orbit (in a two-body system). This phenomenon is called precession, and it is important for Earth, Mars, and the Jovian planets. For Earth, precession causes its celestial poles to describe small circles of 23.°5 **arc** radii around its ecliptic poles and the equinoxes to move westward on the ecliptic. They require 25,800 years to make one 360 circuit around the ecliptic poles and the ecliptic. For liars, the estimated period of precession is about 175,000 years.

Non-gravitational effects

Twentieth century **physics** have found that photons of **light** possess momentum which, when they are absorbed or reflected by material bodies, transfers momentum to the bodies, producing a light **pressure** effect. The interaction of **photon** velocity of light with the orbital velocities of bodies orbiting the Sun produces a retarding effect on their orbits known as the Poynting-Robertson Effect. These effects are insignificant for large solar system bodies, but are important for bodies smaller than 0.394 in (1 cm) in diameter. The Poynting-Robertson Effect causes such small interplanetary particles to **spiral** inwards towards the Sun and to eventually be vaporized by heating from its **radiation**. Much smaller (micron-sized) particles will be pushed out away from the Sun by light pressure which, along with electromagnetic forces, are the dominant mechanisms for the formation of comet tails.

The three-body problem

No closed general solution has been found for the problem of systems of three or more bodies whose motions are controlled by their mutual gravitational attractions in a form analogous to the generalized Kepler's Laws for the two-body problem.

However, in 1772 Joseph Lagrange (1736-1813) found a special stable solution known as the Restricted Three-Body Problem. If the second body in the three-body system has a mass M_2 less than $0.04M_1$ where M_1 is the mass of the most massive Body 1, then there are five stability points in the orbital plane of Bodies 1 and

2. Three of these points, L_1, L_2, and L_3 lie on the line joining Bodies 1 and 2. The stability of particles placed at these points is minimal; slight perturbations will cause them to move away from these points indefinitely. The points L_4 and L_5, respectively 60° ahead of and 60° behind Body 2 in its orbit around the system's center of mass, are more stable; particles placed there will, if slightly perturbed, go into orbits around these points.

Lagrange's solution became relevant to the solar system in 1906 when Max Wolf (1863–1952) discovered the asteroid Achilles in Jupiter's orbit but about 60° ahead of it (near the L_4 point of the solution). Several hundred such asteroids are now known; they are called the Trojan asteroids, since they are named for heroes of the Trojan War. Following the three-body problem, the Sun is Body 1, Jupiter is Body 2, and the asteroids Achilles, Agamemnon, Ajax, Diomedes, Odysseus, and other asteroids named after Greek heroes cluster around the L_4 point of Jupiter's orbit, forming the "Greek camp." The asteroids Anchises, Patroclus, Priam, Aneas, that are named for Trojan heroes cluster around the L_5 point (60° behind Jupiter in its orbit), forming the "Trojan camp." The L_4 and L_5 points of the orbits of Earth, Mars, and Saturn around the Sun and of the Moon's orbits around Earth have been searched for the presence of small bodies ranging in size from asteroids to interplanetary dust without confirmed success. In Saturn's satellite system, with Saturn as Body 1 and its satellite Dione as Body 2, Saturn's small satellite Helene orbits Saturn in Dione's orbit near the L_4 point; with Saturn's satellite Tethys as Body 2, Saturn's satellite Telesto orbits in Tethys' orbit close to the leading L_4 point and Calypso orbits close to the-following L_5 point.

The n-body problem

For systems of n gravitationally interacting bodies where n = 3 to thousands, that is, multiple stars and star clusters where the member stars are of comparable mass, the Virial **Theorem**, by working with a systems gravitational potential **energy** and the kinetic energies of the member stars, can give some insight into the system's stability and evolution. However, the theorem gives mainly information of a statistical nature about the system; it cannot define the space trajectory of a specific star in the system over an extended time interval, and therefore, it cannot predict close encounters of it with other stars nor whether or not this specific star will remain part of the system or will be ejected from it.

Recent developments

In the last 30 years high performance computers have been used to study the n-body problem (n = 3 to n = 10 or more) by stepwise integration of the orbits of the gravitationally interacting bodies. Earlier computers were incapable of performing such calculations over sufficiently long time intervals. The study of the stability of Pluto's orbits over the last 10,000,000 years mentioned above was made for n = 5 (the Sun and the Jovian planets) perturbing Pluto's orbit. Some other studies have treated the solar system as a n = 9 system (the Sun and the eight major planets) over time intervals of several million years.

However, the finite increments of space and time used in stepwise integrations introduce small uncertainties in the predicted positions of solar system objects; these uncertainties increase as the time interval covered by the calculations increases. This has led to the application to celestial mechanics of a new concept in science, **chaos**, which started to develop in the 1970s. Chaos studies indicate that, due to increasing inaccuracy of prediction from integration calculations and also due to incompleteness of the mathematical models integrated, meaningful predictions about the state or position of a system cannot be made beyond some finite time. One result is that Pluto's orbit is chaotic over times of about 800 million years, so that its orbit and position in the early solar system or billions of years from now cannot be specified. Also the rotation of Saturn's satellite Hyperion appears to be chaotic. Chaos is now being applied to studies of the stability of the solar system, a problem which celestial mechanics has considered for centuries without finding a definite answer.

Chaos has also been able to show how certain orbits of main belt asteroids can, over billions of years, evolve into orbits which cross the orbits of Mars and Earth, producing near-Earth asteroids (NEA), of which about 100 are now known. Computer predictions of NEA orbits are now being made to identify NEA which may collide with Earth in the future; such collisions would threaten the very existence of our civilization. The prediction of such Earth-impacting asteroids may allow them to be dejected past Earth or to be destroyed; the space technology to do this may be available soon.

High performance computers and the concept of chaos are now also being used to study the satellite systems of the Jovian planets. They have also been used to study the orbits of stars in multiple star systems and the trajectories of stars in star clusters and galaxies.

The search for planets around other stars is also a recent development. It uses the theory of the two-body problem, starting from earlier work on astrometric double stars. These are stars whose proper motions on the sky are not straight lines as are the case for single stars, but are wavelike curves with periods of some years. This indicates that they are actually double stars with the visible star moving around the system's center of mass (which has straight-line proper motion) with an unseen companion. The stellar companions of Airius A (Gliese 244A), Procyon A (Gliese 280A), Ross 614 A (Gliese 234A), and Mu Cassiopeiae (Gliese 53A) were first detected as astrometric double stars before being observed optically. Small departures of the proper motions of stars from straight lines have been used since 1940 to predict the presence of companions of substellar mass (less than 0.07 solar mass) around nearby stars.

Action of a star around a double star system's center of mass produces periodic variations of the Doppler shift of the star's **spectral lines** as the star first approaches Earth, then recedes from it as seen from the system's center of mass. Since 1980, very precise spectroscopic observations have allowed searches for companions of substellar mass of visible stars to be made at several observatories.

These methods have allowed several dozen companions of substellar mass (so-called "brown dwarfs" and bodies of Jovian planet mass) to be suspected near stars other than the Sun. Unfortunately, as of late 1994 none of the suspected bodies of planetary mass associated with other stars has been confirmed by consistent observations at two or more observatories. Surprisingly, the two or three most reliably established planets have been detected orbiting a **pulsar**, which is a **neutron star**, a star that has used up its nuclear energy sources and has almost completed its evolution. The planets have been detected by apparent periodic variations in the period of the **radio** pulses from this neutronstar pulsar PSR 1257 + 12, and moreover, they seem to have masses on the order of Earth's mass or less. The search for planets orbiting normal stars continues; this is closely associated with the Search for Extra-Terrestrial Intelligence (the **SETI** Project).

Since 1957, the Space Age has accelerated the development of the branch of celestial mechanics called astrodynamics, which is becoming increasingly important. In addition to the traditional gravitational interactions between celestial bodies, astrodynamics must also consider (rocket) propulsion effects that are necessary for inserting artificial satellites and other spacecraft into their necessary orbits and trajectories. Aerodynamic effects must sometimes be considered for planets and satellites with appreciable atmospheres (Venus, Earth, Mars, the Jovian planets, Io, Titan, Neptune's satellite Britons and Pluto). Trajectory building is a new part of astrodynamics; it consists of combining different conic section orbits and propulsion segments along with planet and planetary satellite flybys to increase spacecraft payload on missions requiring very large propellant expenditures. The spacecraft *Voyagers 1* and 2, *Magellan*, and *Galileo* have all used trajectory building, and future spacecraft such as the Cassini/Huygens mission to Saturn and Titan plan to use it to reach their destinations. Minor perturbations due to light pressure, the Poynting-Robertson Effect, and electromagnetic effects sometimes must also be considered. The solar sail is now being studied in spacecraft design as a way of using the light pressure from sunlight on solar sails to maneuver spacecraft and propel them through interplanetary space. Finally, the development of astrodynamics has increased the importance of hyperbolic orbits, since so far all flybys of planets and planetary satellites by spacecraft have occurred along hyperbolic orbits. The spacecraft Pioneers 10 and 11 Voyagers 1 and 2 are leaving the solar system along hyperbolic orbits with respect to the Sun that will take them into interstellar trajectories around the center of our **Milky Way** galaxy. Their hyperbolic orbits are being checked by intermittent radio signals from their transmitters as they leave the solar system for perturbations that could be produced by the gravitational attractions of undiscovered trans-Neptunian planets.

See also Brown dwarf; Celestial coordinates; Gravity and gravitation; Precession of the equinoxes; Solar system; Stellar evolution.

Resources

Books

Glelek, Jame. *Chaos: Making a New Science.* New York: Viking Penguin, Inc. 1988.

Motz, Lloyd, and Anneta Duveen. *Essentials of Astronomy.* Belmont, CA: Wadsworth, 1966.

Periodicals

"Pulsar's Planets Confirmed." *Sky and Telescope.*" 87 (1994).

Celestial sphere: The apparent motions of the Sun, Moon, planets, and stars

The celestial **sphere** is an imaginary projection of the **Sun**, **Moon**, planets, stars, and all astronomical bodies upon an imaginary sphere surrounding **Earth**. The celestial sphere is a useful mapping and tracking remnant of the **geocentric theory** of the ancient Greek astronomers.

Although originally developed as part of the ancient Greek concept of an Earth-centered universe (i.e., a geocentric model of the Universe), the hypothetical celestial sphere provides an important tool to astronomers for fixing the location and plotting movements of celestial objects. The celestial sphere describes an extension of the lines of **latitude and longitude**, and the plotting of all visible celestial objects on a hypothetical sphere surrounding the earth.

The ancient Greek astronomers actually envisioned concentric crystalline spheres, centered around Earth, upon which the Sun, Moon, planets, and stars moved. Although heliocentric (Sun-centered) models of the universe were also proposed by the Greeks, they were disregarded as "counter-intuitive" to the apparent motions of celestial bodies across the sky.

Early in the sixteenth century, Polish astronomer Nicolaus Copernicus (1473–1543) reasserted the **heliocentric theory** abandoned by the Ancient Greeks. Although sparking a revolution in **astronomy**, Copernicus' system was deeply flawed by the fact the Sun is certainly not the center of the Universe, and Copernicus insisted that planetary orbits were circular. Even so, the heliocentric model developed by Copernicus fit the observed data better than the ancient Greek concept. For example, the periodic "backward" **motion (retrograde motion)** in the sky of the planets **Mars**, **Jupiter**, and **Saturn** and the lack of such motion for Mercury and **Venus** was more readily explained by the fact that the former planets' orbits were outside of Earth's. Thus, the Earth "overtook" them as it circled the Sun. Planetary positions could also be predicted much more accurately using the Copernican model.

Danish astronomer Tycho Brahe's (1546–1601) precise observations of movements across the "celestial sphere" allowed German astronomer and mathematician Johannes Kepler (1571–1630) to formulate his laws of planetary motion that correctly described the elliptical orbits of the planets.

The modern celestial sphere is an extension of the latitude and longitude coordinate system used to fix terrestrial location. The concepts of latitude and longitude create a grid system for the unique expression of any location on Earth's surface. Latitudes—also known as parallels—mark and measure **distance** north or south from the equator. Earth's equator is designated 0° latitude. The north and south geographic poles respectively measure 90° north (N) and 90° south (S) from the equator. The **angle** of latitude is determined as the angle between a transverse **plane** cutting through Earth's equator and the right angle (90°) of the polar axis. Longitudes—also known as meridians—are great circles that run north and south, and converge at the north and south geographic poles.

On the celestial sphere, projections of lines of latitude and longitude are transformed into declination and right ascension. A direct extension of Earth's equator at 0° latitude is the celestial equator at 0° declination. Instead of longitude, right ascension is measured in hours.

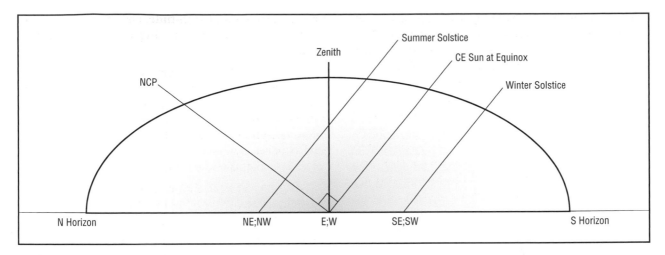

The celestial sphere, is a projected sphere surrounding Earth. The angle of the north celestial pole (NCP) with the horizon varies with latitude. The observer's zenith is directly overhead. *Illustration by K. Lee Lerner with Argosy. The Gale Group.*

Corresponding to Earth's **rotation**, right ascension is measured from **zero** hours to 24 hours around the celestial sphere. Accordingly, one hour represents 15 angular degrees of travel around the 360° celestial sphere.

Declination is further divided arcminutes and arcseconds. In 1° of declination, there are 60 arcminutes (60') and in one arcminute there are 60 arcseconds (60"). Right ascension hours are further subdivided into minutes and seconds of **time**.

On Earth's surface, the designation of 0° longitude is arbitrary, international convention, long held since the days of British sea superiority, establishes the 0° line of longitude—also known as the Prime Meridian—as the great **circle** that passes through the Royal National Observatory in Greenwich, England (United Kingdom). On the celestial sphere, zero hrs (0 h) right ascension is also arbitrarily defined by international convention as the line of right ascension where the ecliptic—the apparent movement of the Sun across the celestial sphere established by the plane of the earth's **orbit** around the Sun—intersects the celestial equator at the vernal **equinox**.

For any latitude on Earth's surface, the extended declination line crosses the observer's zenith. The zenith is the highest point on the celestial sphere directly above the observer. By international agreement and customary usage, declinations north of the celestial equator are designated as positive declinations (+) and declinations south of the celestial equator are designated as negative declinations (−) south.

Just as every point on Earth can be expressed with a unique set of latitude and longitude coordinates every object on the celestial sphere can be specified by declination and right ascension coordinates.

The polar axis is an imaginary line that extends through the north and south geographic poles. The earth rotates on its axis as it revolves around the Sun. Earth's axis is tilted approximately 23.5 degrees to the plane of the ecliptic (the plane of planetary orbits about the Sun or the apparent path of the Sun across the imaginary celestial sphere). The tilt of the polar axis is principally responsible for variations in solar illumination that result in the cyclic progressions of the **seasons**. The polar axis also establishes the principal axis about which the celestial sphere rotates. The projection of Earth's geographic poles upon the celestial sphere creates a north celestial pole and a south celestial pole. In the Northern Hemisphere, the **star** Polaris is currently within approximately one degree (1°) of the north celestial pole and thus, from the Northern Hemisphere, all stars and other celestial objects appear to rotate about Polaris and, depending on the latitude of observation, stars located near Polaris (circumpolar stars) may never "set."

For any observer, the angle between the north celestial pole and the terrestrial horizon equals and varies directly with latitude north of the equator. For example, at 30° N latitude an observer views Polaris at +30° declination, at the terrestrial North Pole (90° N), Polaris would be directly overhead (at the zenith) at +90° declination.

The celestial meridian is an imaginary **arc** from the north point on the terrestrial horizon through the north celestial pole and zenith that terminates on the south point of the terrestrial horizon.

Regardless of location on Earth, an observer's celestial equator passes through the east and west points of the terrestrial horizon. In the Northern Hemisphere, the celestial equator is displaced southward from the zenith (the point directly over the observer's head) by the number of degrees equal to the observer's latitude.

Rotation about the polar axis results in a diurnal cycle of night and day, and causes the apparent motion of the Sun across the imaginary celestial sphere. The earth rotates about the polar axis at approximately 15 angular degrees per hour and makes a complete rotation in 23.9 hours. This corresponds to the apparent rotation of the celestial sphere. Because the earth rotates eastward (from west to east), objects on the celestial sphere usually move along paths from east to west (i.e., the Sun "rises" in the east and "sets" in the west). One complete rotation of the celestial sphere comprises a diurnal cycle.

As the earth rotates on its polar axis, it makes a slightly elliptical orbital revolution about the Sun in 365.26 days. Earth's revolution about the Sun also corresponds to the cyclic and seasonal changes of observable stars and constellations on the celestial sphere. Although stars grouped in traditional constellations have no proximate spatial relationship to one another (i.e., they may be billions of **light** years apart) that do have an apparent relationship as a two-dimensional pattern of stars on the celestial sphere. Accordingly, in the modern sense, constellations establish regional location of stars on the celestial sphere.

A tropical year (i.e., a year of cyclic seasonal change), equals approximately 365.24 mean solar days. During this time, the Sun appears to travel completely around the celestial sphere on the ecliptic and return to the vernal equinox. In contrast, one orbital revolution of Earth about the Sun returns the Sun to the same backdrop of stars—and is measured as a sidereal year. On the celestial sphere, a sidereal day is defined as the time it takes for the vernal equinox—starting from an observer's celestial median—to rotate around with the celestial sphere and recross that same celestial median. The sidereal day is due to Earth's rotational period. Because of precession, a sidereal year is approximately 20 minutes and 24 seconds longer than a tropical year. Although the sidereal year more accurately measures the time it takes Earth to completely orbit the Sun, the use of the sidereal year would eventually cause large errors in **calendars** with regard to seasonal changes. For this reason the tropical year is the basis for modern Western calendar systems.

Seasons are tied to the apparent movements of the Sun and stars across the celestial sphere. In the Northern Hemisphere, summer begins at the summer **solstice** (approximately June 21) when the Sun is reaches its apparent maximum declination. Winter begins at the winter solstice (approximately December 21) when the Sun's highest point during the day is its minimum maximum daily declination. The changes result from a changing orientation of Earth's polar axis to the Sun that result in a change in the Sun's apparent declination. The vernal and autumnal equinox are denoted as the points where the celestial equator intersects the ecliptic.

The location of sunrise on the eastern horizon, and sunset on the western horizon also varies between a northern most maximum at the summer solstice to a southernmost maximum at the winter solstice. Only at the vernal and autumnal equinox does the Sun rise at a point due east or set at a point due west on the terrestrial horizon.

During the year, the moon and planets appear to move in a restricted region of the celestial sphere termed the zodiac. The zodiac is a region extending outward approximately 8° from each side of the ecliptic (the apparent path of the Sun on the celestial sphere). The modern celestial sphere is divided into twelve traditional zodiacal **constellation** patterns (corresponding to the pseudo-scientific astrological zodiacal signs) through which the Sun appears to travel by successive eastwards displacements throughout the year.

During revolution about the Sun, the earth's polar axis exhibits parallelism to Polaris (also known as the North Star). Although observing parallelism, the orientation of Earth's polar axis exhibits precession—a circular wobbling exhibited by gyroscopes—that results in a 28,000-year-long precessional cycle. Currently, Earth's polar axis points roughly in the direction of Polaris (the North Star). As a result of precession, over the next 11,00 years, Earth's axis will precess or wobble so that it assumes an orientation toward the star Vega.

Precession causes an objects **celestial coordinates** to change. As a result, celestial coordinates are usually accompanied by a date for which the coordinates are valid.

Corresponding to **Earth's rotation**, the celestial sphere rotates through 1° in about four minutes. Because of this, sunrise, sunset, moonrise, and moonset, all take approximately two minutes because both the Sun and Moon have the same apparent size on the celestial sphere (about 0.5°). The Sun is, of course, much larger, but the Moon is much closer. If measured at the same time of day, the Sun appears to be displaced eastward on the star field of the celestial sphere by approximately 1° per day. Because of this apparent displacement, the stars appear to "rise" approximately four minutes earlier each evening and set four minutes later each morning. Alternatively, the Sun appears to "rise" four minutes earlier each day and "set" four minutes earlier each day. A change of approximately four minutes a day corresponds to a 24-hour cycle of "rising" and "setting" times that comprise an annual cycle.

In contrast, if measured at the same time each day, the Moon appears to be displaced approximately 13°

eastward on the celestial sphere per day and therefore "rises" and "sets" almost one hour earlier each day.

Because the earth is revolving about the Sun, the displacement of the earth along it's orbital path causes the time it takes to complete a cycle of lunar phases—a synodic month—and return the Sun, Earth, and Moon to the same starting alignment is slightly longer than the sidereal month. The synodic month is approximately 29.5 days.

Earth rotates about its axis at approximately 15 angular degrees per hour. Rotation dictates the length of the diurnal cycle (i.e., the day/night cycle), creates "time zones" with differing local noons. Local noon occurs when the Sun is at the highest point during its daily skyward arch from east to west (i.e., when the Sun is at its zenith on the celestial meridian). With regard to the solar meridian, the Sun's location (and reference to local noon) is described in terms of being ante meridian (am)—east of the celestial meridian—or post meridian (pm) located west of the celestial meridian.

See also Astrolabe; Celestial mechanics; Cosmology; Geographic and magnetic poles; Precession of the equinoxes; Solar illumination: Seasonal and diurnal patterns; Zodiacal light.

Resources

Books

Hamblin, W. K., and E.H. Christiansen. *Earth's Dynamic Systems.* 9th ed. Upper Saddle River: Prentice Hall, 2001.
Hancock, P.L. and B.J., Skinner eds. *The Oxford Companion to the Earth.* New York: Oxford University Press, 2000.
Press, F., and R. Siever. *Understanding Earth.* 3rd ed. New York: W.H Freeman and Company, 2001.

K. Lee Lerner

Cell

The cell is the smallest living component of organisms and is the basic unit of life. In multicellular living things, a collection of cells that work together to perform similar functions is called a **tissue**; various tissues that perform coordinated functions form organs; and organs that work together to perform general processes form body systems. The human **digestive system**, for example, is composed of various organs including the stomach, pancreas, and the intestines. The tissue that lines the intestine is called epithelial tissue. Epithelial tissue, in turn, is composed of special cells called epithelial cells. In the small intestine, these epithelial cells are specialized for their absorptive function: each epithelial cell is covered with thousands of small projections called microvilli. The numerous microvilli greatly increase the surface area of the small intestine through which **nutrients** can be absorbed into the bloodstream.

Types of cells

Multicellular organisms contain a vast array of highly specialized cells. Plants contain root cells, **leaf** cells, and **stem cells**. Humans have skin cells, nerve cells, and sex cells. Each kind of cell is structured to perform a highly specialized function. Often, examining a cell's structure reveals much about its function in the **organism**. For instance, as we have already seen, epithelial cells in the small intestine are specialized for absorption due to the numerous microvilli that crowd their surfaces. Nerve cells, or neurons, are another kind of specialized cell whose form reflects function. Nerve cells consist of a cell body and long processes, called axons, that conduct nerve impulses. Dendrites are shorter processes that receive nerve impulses.

Sensory cells—the cells that detect sensory information from the outside environment and transmit this information to the brain—often have unusual shapes and structures that contribute to their function. The rod cells in the retina of the **eye**, for instance, look like no other cell in the human body. Shaped like a rod, these cells have a light-sensitive region that contains numerous membranous disks. Within each disk is embedded a special light-sensitive pigment that captures **light**. When the pigment receives light from the outside environment, nerve cells in the eye are triggered to send a nerve impulse in the **brain**. In this way, humans are able to detect light.

Cells, however, can also exist as single-celled organisms. The organisms called protists, for instance, are single-celled organisms. Examples of protists include the microscopic organism called *Paramecium* and the single-celled alga called *Chlamydomonas.*

Prokaryotes and eukaryotes

Two types of cells are recognized in living things. Prokaryotes (literally, "before the nucleus") are cells that have no distinct nucleus. Most prokaryotic organisms are single-celled, such as **bacteria** and **algae**. Eukaryotic (literally, "true nucleus") organisms, on the other hand, have a distinct nucleus and a highly organized internal structure. Distinct organelles, the small structures that each perform a specific set of functions, are present within eukaryotes. These organelles are bound by membranes. Prokaryotes, in addition to their lack of a nucleus, also lack these membrane-bound organelles.

Cell size and numbers

It is estimated that an adult human body contains about 60 trillion cells. Most of these cells, with some exceptions, are so small that a **microscope** is necessary to see them. The small size of cells fulfills a distinct purpose in the functioning of the body. If cells were larger, many of the processes that cells perform could not occur efficiently. To visualize this concept, think about the intestinal epithelial cells discussed earlier. What if the intestinal epithelium were composed of one, large cell instead of thousands of small cells? A large cell has a large **volume**, or contents. The surface area, or **membrane**, of this large cell is the site through which nutrients enter the small intestine for delivery to the bloodstream. Because the volume of this large cell is so large, the surface area, by comparison, is relatively small. Large cells, therefore, have a small surface area to volume **ratio**. Only so many nutrients can pass through the limited membrane area of this large cell. With a small surface area to volume ratio, the amount of substances passing into and out of the cell is severely restricted.

However, if the intestinal epithelium is divided into thousands of smaller cells, the volume stays the same, but the surface area—the number of cell membranes—greatly increases. Many more nutrients can pass through the intestinal epithelium cells. Small cells, therefore, have a large surface area to volume ratio. The large surface area to volume ratio of small cells makes the transport of substances into and out of cells extremely efficient.

Another reason for the small size of cells is that control of cellular processes is easier in a small cell than in a large cell. Cells are dynamic, living things. Cells transport substances from one place to another, reproduce themselves, and produce various enzymes and chemicals for export to the extracellular environment. All of these activities are accomplished under the direction of the nucleus, the control center of the cell. If the nucleus had to control a large cell, then this direction might break down. Substances transported from one place to another would have to traverse great distances to reach their destinations; reproduction of a large cell would be an extremely complicated endeavor; and products for export would not be as efficiently produced. Smaller cells, because of their more manageable size, are much more efficiently controlled than larger cells.

The structure and function of cells

The basic structure of all cells, whether **prokaryote** and eukaryote, is the same. All cells have a **plasma** membrane through which substances pass into and out of the cell. With the exception of a few minor differences, plasma membranes are the same in prokaryotes and eukaryotes. The interior of both kinds of cells is called the cytoplasm. Within the cytoplasm of eukaryotes are embedded the cellular organelles; the cytoplasm of prokaryotes contains no organelles. Finally, both types of cells contain small structures called **ribosomes** that function in protein synthesis. Composed of two protein subunits, ribosomes are not bounded by membranes; therefore, they are not considered organelles. In eukaryotes, ribosomes are either bound to an organelle, the endoplasmic reticulum, or exist as "free" ribosomes in the cytoplasm. Prokaryotes contain only free ribosomes.

The structure of prokaryotes

An example of a typical prokaryote is the bacterial cell. Bacterial cells can be shaped like rods, spheres, or corkscrews. All prokaryotes are bounded by a plasma membrane. Overlying this plasma membrane is a cell wall, and in some bacteria, a capsule consisting of a jelly-like material overlies the cell wall. Many bacteria that cause illness in animals have capsules. The capsule provides an extra layer of protection for the bacteria, and often pathogenic bacteria with capsules cause much more severe **disease** than those without capsules.

Within the cytoplasm of prokaryotes is a nucleoid, a region where the genetic material (DNA) resides. This nucleoid is not a true nucleus because it is not bounded by a membrane. Also within the cytoplasm are numerous ribosomes. These ribosomes are not attached to any structure and are thus called "free" ribosomes.

Attached to the cell wall of some bacteria are **flagella**, whip-like structures that provide for movement. Some bacteria also have pili, which are short, finger-like projections that assist the bacteria in attaching to tissues. Bacteria cannot cause disease if they cannot attach to tissues. Bacteria that cause **pneumonia**, for instance, attach to the tissues of the lung. Bacterial pili greatly facilitates this attachment to tissues, and thus, like capsules, bacteria with pili are often more virulent than those without.

The structure of eukaryotes

The organelles found in eukaryotes include the membrane system consisting of plasma membrane, endoplasmic reticulum, Golgi body, and vesicles; the nucleus; cytoskeleton; and mitochondria. In addition, **plant** cells have special organelles not found in animals cells. These organelles are the chloroplasts, cell wall, and vacuoles.

The membrane system

The membrane system of a cell performs many important functions. This system controls the entrance and exit of substances into and out of the cell, and also provides for the manufacture and packaging of substances

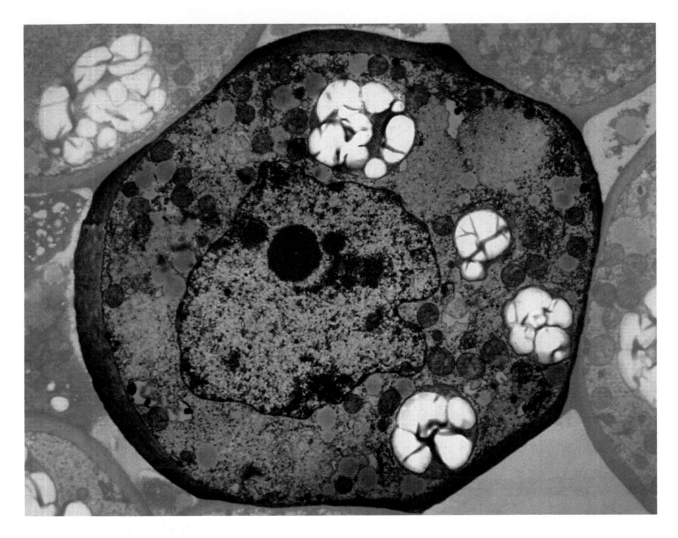

Figure 1. A plant cell. *Photograph by Dr. Dennis Kunkel. Phototake. Reproduced by permission.*

within the cell. The membrane system of the cell consists of the plasma membrane, which encloses the cell contents; the endoplasmic reticulum, which manufactures lipids and **proteins**; the Golgi body, which packages substances manufactured within the cell; and various vesicles, which perform different functions.

The plasma membrane

The plasma membrane of the cell is often described as "selectively permeable;" that is, the plasma membrane is designed so that only certain substances are allowed to traverse its borders. The plasma membrane is composed of two layers of molecules called phospholipids. Each phospholipid **molecule** consists of a phosphate "head" and two fatty acid chains that dangle from the head.

The orientation of these two sections of the phospholipid molecule is crucial to the function of the plasma membrane. The phosphate region is hydrophilic (literal-ly, "water-loving") and attracts **water**. The fatty acid region is hydrophobic (literally, "water-hating") and repels water. In the phospholipid bilayer of the plasma membrane, the phospholipid layers are arranged so that the two phosphate hydrophilic regions face outward, towards the watery extracellular environment, and inward, towards the cellular cytoplasm, which also contains water. The two hydrophobic fatty acid portions of the chains face each other, forming a water-tight shield. The plasma membrane, then, is both water-proof and water-attracting. It functions both as a boundary between the cell's contents and the external cellular environment, yet also allows the transport of water-containing and other substances across its boundaries.

Embedded within the plasma membranes of eukaryotes are various proteins. These proteins serve several distinct functions in the cell. Some proteins are pumps or channels for the import and export of substances. Other proteins, called antigens, serve as identification markers for the cell. Still other proteins help the cell form attachments with other cells. Because these membrane proteins often protrude out of the cell membrane into the extracellular environment, they too have hydrophobic and hydrophilic regions. Portions of the proteins that are embedded within the plasma membrane are hydrophobic, and portions of the proteins that extend outward into the extracellular environment are hydrophilic.

Scientists studying plasma membranes use the term "fluid-mosaic model" to describe the structure of plasma membranes. The "mosaic" portion of the model describes the way proteins are embedded within the plasma membrane. The "fluid" part of the model explains the fluid nature of plasma membranes. Rather than being fixed in one place within the plasma membrane, experiments have shown that the phospholipids exhibit some movement within the plasma membranes, sometimes moving laterally, sometimes (although rarely), flip-flopping from one phospholipid layer to another. The membrane proteins also move within the plasma membrane, albeit more slowly than the phospholipids.

Endoplasmic reticulum

The endoplasmic reticulum (meaning "within the cytoplasm" and "net") consists of flattened sheets, sacs, and tubes of membrane that cover the entire expanse of a eukaryotic cell's cytoplasm. This internal system of membrane is continuous with the double membrane that surrounds the cell's nucleus. Therefore, the encoded instructions that the nucleus sends out for the synthesis of proteins flow directly into the endoplasmic reticulum. Within the cell, the endoplasmic reticulum synthesizes lipids and proteins. The proteins that the endoplasmic reticulum synthesizes, such as enzymes, are exported from the cell to perform various functions in the body. Proteins that are made in the cell for use by the cell-for instance, as channels in the plasma membrane-are made by the free ribosomes that dot the cytoplasm.

Two types of endoplasmic reticulum are found in the eukaryotic cell. Rough endoplasmic reticulum is studded with ribosomes on its outer face. These ribosomes are the sites of protein synthesis. Once a protein is synthesized on a ribosome, it is enclosed within a vesicle, a small, membrane-bound "bubble." The vesicle travels to another organelle, the Golgi body. Within the Golgi body, the proteins within the vesicle are further modified before they are exported from the cell. Cells that specialize in protein secretion contain large amounts of rough endoplasmic reticulum. For instance, cells of the pancreas that produce the protein **insulin**, have abundant rough endoplasmic reticulum. Plasma cells, white **blood** cells that secrete immune proteins called antibodies, are so crowded with

rough endoplasmic reticulum it is difficult to distinguish other organelles within the cytoplasm.

The other type of endoplasmic reticulum is smooth endoplasmic reticulum. Smooth endoplasmic reticulum does not have ribosomes and is the site of **lipid metabolism**. Here, macromolecules containing lipids are broken down into their constituent parts. In addition, smooth endoplasmic reticulum functions in the synthesis of lipid-containing macromolecules. Smooth endoplasmic reticulum is not as common in cells as rough endoplasmic reticulum. Large amounts of smooth endoplasmic reticulum are found in cells that specialize in lipid metabolism. For instance, liver cells remove **alcohol** and drugs from the bloodstream. Liver cells have an impressive network of smooth endoplasmic reticulum. Similarly, cells of the ovaries and testes, which produce the lipid-containing **hormones** estrogen and testosterone, contain large amounts of smooth endoplasmic reticulum.

The Golgi body

Named for its discoverer, nineteenth century Italian scientist Camillo Golgi, the Golgi body is one of the most unusually shaped organelles. Looking somewhat like a stack of pancakes, the Golgi body consists of stacked, membrane-bounded, flattened sacs. Surrounding the Golgi body are numerous, small, membrane-bounded vesicles. The Golgi body and its vesicles function in the sorting, modifying, and packaging of macromolecules that are secreted by the cell or used within the cell for various functions.

The Golgi body can be compared to the shipping and receiving department of a large company. Each Golgi body within a cell has a *cis* face, which is analogous to the receiving division of the department. Here, the Golgi body receives macromolecules synthesized in the endoplasmic reticulum encased within vesicles. The *trans* face of the Golgi body is analogous to the shipping division of the department, and is the site from which modified and packaged macromolecules are transported to their destinations.

Within the Golgi body, various chemical groups are added to the macromolecules so ensure that they reach their proper destination. In this way, the Golgi body attaches an "address" to each macromolecule it receives. For example, cells called goblet cells in the lining of the intestine secrete mucous. The protein component of mucous, called mucin, is modified in the Golgi body by the addition of **carbohydrate** groups. From the Golgi body, the modified mucin is packaged within a vesicle. The vesicle containing its mucous cargo fuses with the plasma membrane of the goblet cell, and is released into the extracellular environment.

Vesicles

Vesicles are small, membrane-bounded spheres that contain various macromolecules. Some vesicles, as we have seen, are used to transport macromolecules from the endoplasmic reticulum to the Golgi body, and from the Golgi body to various destinations. Special kinds of vesicles perform other functions as well. Lysosomes are vesicles that contain enzymes involved in cellular digestion. Some protists, for instance, engulf other cells for food. In a process called phagocytosis, the protist surrounds a food particle and engulfs it within a vesicle. This food containing vesicle is transported within the protist's cytoplasm until it is contiguous with a lysosome. The food vesicle and lysosome merge, and the enzymes within the lysosome are released into the food vesicle. The enzymes break the food down into smaller parts for use by the protist.

Lysosomes, however, are found in all kinds of cells. In all cells, lysosomes digest old, worn-out organelles. They also play a role in the self-destruction of old cells. Although scientists do not understand the trigger mechanism of this self-destruction, cells that are not functioning properly due to old-age apparently self-digest by means of lysosomes. **Cell death** is also a component of normal **developmental processes**. For instance, a human fetus has web-like hands and feet. As development progresses, the cells that compose these webs slowly self-destruct, freeing the fingers.

Peroxisomes, as their name implies, contain **hydrogen peroxide**. Peroxisomes function in the oxidation of many materials, including fats. In oxidation, **oxygen** is added to a molecule. When oxygen is added to fats, **hydrogen** peroxide is formed. As anyone who has treated a cut with hydrogen peroxide knows, this substance is lethal to cells. Therefore, the oxidation of fats takes place within the membranes of peroxisomes so that the harmful chemical does not leak out into the cell's cytoplasm.

The nucleus

The nucleus is the control center of the cell. Under a microscope, the nucleus looks like a dark blob, with a darker region, called the nucleolus, centered within it. The nucleolus is the site where the subunits of ribosomes are manufactured. Surrounding the nucleus is a double membrane called the nuclear envelope. The nuclear envelope is studded all over with tiny openings called nuclear pores.

The nucleus directs all cellular activities by controlling the synthesis of proteins. The nucleus contains encoded instructions for the synthesis of proteins in a helical molecule called **deoxyribonucleic acid (DNA)**. The cell's DNA is packaged within the nucleus in a structural form called **chromatin**. Chromatin consists of DNA wound tightly around spherical proteins called histones. When the cell prepares to divide, the DNA unwinds from the histones and assumes the shape of chromosomes, the X-shaped structures visible within the nucleus prior to **cell division**. Chromatin packaging of DNA allows all of the cell's DNA to fit into the combined **space** of the nucleus. If DNA was not packaged into chromatin, it would spill out over a space about 100 times as large as the cell itself.

The first step in protein synthesis begins in the nucleus. Within the nucleus, DNA is translated into a molecule called messenger ribonucleic acid (mRNA). mRNA then leaves the nucleus through the nuclear pores. Once in the cytoplasm, mRNA attaches to ribosomes (either bound to endoplasmic reticulum or free in the cytoplasm) and initiates protein synthesis. Proteins made for export from the cell function as enzymes that participate in all the body's **chemical reactions**. Because enzymes are essential for all the body's chemical processes-from cellular respiration to digestion-direction of the synthesis of these enzymes in essence controls all the activities of the body. Therefore, the nucleus, which contains the instructions for the synthesis of these proteins, directs all cellular activities and thus all body processes.

The cytoskeleton

The cytoskeleton is the "skeletal" framework of the cell. Instead of bone, however, the cell's skeleton consists of three kinds of protein filaments that form networks. These networks give the cell shape and provide for cellular movement. The three types of cytoskeletal fibers are microtubules, actin filaments, and intermediate filaments.

Microtubules are 25 nanometers in diameter and consist of protein subunits called tubulin. Each microtubule is composed of eleven pairs of these tubulin subunits arranged in a ring. In **animal** cells, microtubules arise from a region of the cell called the microtubule organizing center (MTOC) located near the nucleus. From this center, microtubules fan out across the cell, forming a network of "tracks" over which various organelles move within the cell. Microtubules also form small, paired structures called centrioles within animal cells. These structures are not considered organelles because they are not bounded by membranes. Scientists once thought that centrioles formed the microtubules that pull the cell apart during cell division; now it is known that each centriole with the pair move apart during cell division and indicate the plan along which the cell divides.

KEY TERMS

Actin filament—A type of cytoskeletal filament that has contractile properties.

Amyloplast—A plant cell plastid that stores starch.

Centriole—Paired structures consisting of microtubules; in animal cells, directs the plane of cell division.

Chloroplast—Green organelle in higher plants and algae in which photosynthesis occurs.

Chromoplast—A plant cell plastid that contains yellow and orange pigments.

Cilia—Short projections consisting of microtubules that cover the surface of some cells and provide for movement.

Cisface—The side (or "face") of the Golgi body that receives vesicles containing macromolecules.

Crista—pl., cristae, the folds of the inner membrane of a mitochondrion.

Cytoplasm—All the protoplasm in a living cell that is located outside of the nucleus, as distinguished from *nucleoplasm,* which is the protoplasm in the nucleus.

Cytoskeleton—A network of assorted protein filaments attached to the cell membrane and to various organelles that makes up the framework for cell shape and movement.

Deoxyribonucleic acid (DNA)—The genetic material of a cell that contains encoded instructions for the synthesis of proteins

Endoplasmic reticulum—The network of membranes that extends throughout the cell; involved in protein synthesis and lipid metabolism.

Endosymbiotic theory—A theory that proposes that mitochondria, chloroplasts, and other eukaryotic organelles originally arose within cells by symbiosis between a single-celled prokaryote and another prokaryote.

Eukaryotic cell—A cell whose genetic material is carried on chromosomes inside a nucleus encased in a membrane. Eukaryotic cells also have organelles that perform specific metabolic tasks and are supported by a cytoskeleton which runs through the cytoplasm, giving the cell form and shape.

Flagellum—Thread-like appendage of certain cells, such as sperm cells, which controls their locomotion.

Fluid-mosaic model—The model that describes the nature of the plasma membrane; the "mosaic" portion describes the proteins embedded within the plasma membrane, and the "fluid" portion describes the fluidity of the plasma membrane.

Golgi body—The organelle that manufactures, sorts, and transports macromolecules within a cell.

Granum—Sacs within a chloroplast that contain photosynthetic enzymes.

Hydrophilic—"Water-loving;" describes the phosphate portion of a phospholipid.

Hydrophobic—"Water-hating;" describes the fatty acid portion of a phospholipid.

Intermediate filament—A type of cytoskeletal filament that anchors organelles.

Lysosome—A vesicle that contains digestive enzymes.

Some eukaryotic cells move about by means of microtubules attached to the exterior of the plasma membrane. These microtubules are called flagella and cilia. Flagella and cilia both have the same structure: a ring of nine tubulin triplets arranged around two tubulin subunits. The difference between flagella and cilia lies in their movement and numbers. Flagella are attached to the cell by a "crank"-like apparatus that allows the flagella to rotate. Usually, a flagellated cell has only one or two flagella. Cilia, on the other hand, are not attached with a "crank," and beat back and forth to provide movement. Ciliated cells usually have hundreds of these projections that cover their surfaces. For example, the protist *Paramecium* moves by means of a single flagellum, while the protist *Didinium* is covered with numerous

cilia. Ciliated cells also perform important functions in the human body. The airways of humans and other animals are lined with ciliated cells that sweep debris and bacteria upwards, out of the lungs and into the throat. There, the debris is either coughed from the throat or swallowed into the digestive tract, where digestive enzymes destroy harmful bacteria.

Actin filaments are 8 nanometers in diameter and consist of two strands of the protein actin that are wound around each other. Actin filaments are especially prominent in muscle cells, where they provide for the contraction of muscle tissue.

Intermediate filaments are 10 nanometers in diameter and are composed of fibrous proteins. Because of

Matrix—The inner space of a mitochondrion formed by cristae.

Microtubule—A type of cytoskeletal filament; the component of centrioles, flagella, and cilia.

Mitochondrion—The power-house of the cell; contains the enzymes necessary for the oxidation of food into energy.

Nuclear envelope—The double membrane that surrounds the nucleus.

Nuclear pore—Tiny openings that stud the nuclear envelope.

Nucleoid—The region in a prokaryote where the cell's DNA is located.

Nucleolus—The darker region within the nucleolus where ribosomal subunits are manufactured.

Nucleus—The control center of a cell; contains the DNA.

Organelle—A membrane-bounded cellular "organ" that performs a specific set of functions within a eukaryotic cell.

Peroxisome—A vesicle that oxidizes fats and other substances and stores hydrogen peroxide.

Phospholipid—A molecule consisting of a phosphate head and two fatty acid chains that dangle from the head; the component of the plasma membrane.

Phospholipid bilayer—The double layer of phospholipids that compose the plasma membrane.

Photosynthesis—In plants, the process in which carbon dioxide and water are converted to sugars.

Pili—Short projections that assist bacteria in attaching to tissues.

Plasma membrane—The membrane of a cell.

Plastid—A vesicle-like organelle found in plant cells.

Prokaryote—A cell without a true nucleus.

Protist—A single-celled eukaryotic organism.

Ribonucleic acid—RNA; the molecule translated from DNA in the nucleus that directs protein synthesis in the cytoplasm; it is also the genetic material of many viruses.

Ribosome—A protein composed of two subunits that functions in protein synthesis.

Stroma—The material that bathes the interior of chloroplasts in plant cells.

Surface area to volume ratio—The relationship between the surface area provided by the plasma membrane to the volume of the contents of a cell.

Thylakoid—A membranous structure that bisects the interior of a chloroplast.

Transface—The side (or "face") of a Golgi body that releases macromolecule-filled vesicles for transport.

Tubulin—A protein that comprises microtubules.

Vacuole—Membrane-enclosed structure within cells which store pigments, water, nutrients, and wastes.

Vesicle—A membrane-bound sphere that contains a variety of substances in cells.

their relative strength, they function mainly to anchor organelles in place within the cytoplasm.

Mitochondria

The mitochondria are the power plants of cells. Each sausage-shaped mitochondrion is covered by an outer membrane; the inner membrane of a mitochondrion is folded into compartments called cristae (meaning "box"). The matrix, or inner space created by the cristae, contains the enzymes necessary for the many chemical reactions that eventually transform food molecules into **energy**.

Cells contain hundreds to thousands of mitochondria. An interesting aspect of mitochondria is that they contain their own DNA sequences, although not in the profusion that the nucleus contains. The presence of this separate DNA, along with the resemblance of mitochondria to single-celled prokaryotes, has led to a theory of eukaryotic **evolution** called the endosymbiotic theory. This theory postulates that mitochondria were once separate prokaryotes that became engulfed within other prokaryotes. Instead of being digested, the mitochondrial prokaryotes remained within the engulfing cell and performed its energy-releasing functions. Over millions of years, this symbiotic relationship fostered the evolution of the eukaryotic cell.

Plant organelles

Plant cells have several organelles not found in animal cells. These are plastids, vacuoles, and a cell wall.

Plastids

Plastids are vesicle-type organelles that perform a variety of functions in plants. Amylopasts store starch, and chromoplasts store pigment molecules that give some plants their vibrant orange and yellow colors.

Chloroplasts are plastids that carry out **photosynthesis**, a process in which water and **carbon dioxide** are transformed into sugars. The interior of chloroplasts contains an elaborate membrane system. Thylakoids bisect the chloroplasts, and attached to these platforms are stacks of membranous sacs called grana. Each granum contains the enzymes necessary for photosynthesis. The membrane system within the chloroplasts is bathed in a fluid called stroma, which also contains enzymes.

Like mitochondria, chloroplasts resemble some ancient single-celled prokaryotes and also contain their own DNA sequences. Their origin within eukaryotes is thought to have arisen from the endosymbiotic relationship between a photosynthetic single-celled prokaryote that was engulfed and remained within another prokaryotic cell.

Vacuoles

Plant vacuoles are large vesicles bound by a single membrane. In many plant cells, they occupy about 90% of the cellular space. They perform a variety of functions in the cell, including storage of organic compounds, waste products, pigments, and poisonous compounds, as well as digestive functions.

Cell wall

All plant cells have a cell wall that overlies the plasma membrane. The cell wall of plants consists of a tough carbohydrate substance called **cellulose** laid down in a matrix or network of other carbohydrates. The cell wall provides an additional layer of protection between the contents of the cell and the outside environment. The crunchiness of an apple, for instance, is attributed to the presence of these cell walls.

See also Cellular respiration; Chloroplast; Chromosome; Enzyme; Eukaryotae; Flagella; Gene; Meiosis; Mitosis; Neuron; Nucleus, cellular; Organ; Ribonucleic acid (RNA); Tissue.

Resources

Books

Barritt, Greg J. *Communication within Animal Cells.* Oxford: Oxford University Press, 1992.

Bittar, F. Edward, ed. *Chemistry of the Living Cell.* Greenwich, CT: JAI Press, 1992.

Bray, Dennis. *Cell Movements.* New York: Garland Press, 1992.

Carroll, Mark. *Organelles.* New York: Guilford Press, 1989.

The Cell Surface. Plainview, NY: Cold Spring Harbor Laboratory Press, 1992.

Periodicals

Maddox, John. "Why Microtubules Grow and Shrink." *Nature* 362 (March 18, 1993): 201.

Pante, Nelly, and Ueli Aebi. "The Nuclear Pore Complex." *The Journal of Cell Biology* 122 (September 1993): 5-6.

Scott, J. D. and T. Pawson. "Cell Communication: The Inside Story." *Scientific American* 282 (June 2000): 54-61.

Shay, Jerry W., and Woodring E. Wright. "Hayflick, His Limit, and Cellular Aging." *Nature Reviews/ Molecular Cell Biology* (October 1, 2000): 72-76.

Kathleen Scogna

Cell death

Like all living things, the various types of cells in plants, animals, and the many different **cell** types in humans must eventually die. Cell death occurs in one of two ways. Cells can be killed by the effects of physical, biological, or chemical injury. Additionally, cells are induced to kill themselves. Cell suicide is also referred to as apoptosis (from the Greek words *apo,* meaning from, and *ptosis,* meaning to fall or to drop).

Cell death is important in **disease** and the aging process. Cellular suicide is also necessary in the fetal development of some organs and tissues.

Cell death that results from injury can be caused by mechanical damage such as tearing, or can be due to physical stresses such as **heat**. A third-degree sunburn, for example, results in the death of many skin cells. Exposure to toxic chemicals such as acids, corrosive bases, metabolic poisons, and other chemicals is also lethal to many types of cells. Excessive drinking of **alcohol (ethanol)** causes death of liver cells in humans.

Substances that dehydrate cells can also cause cell death. If the environment outside of a cell contains more **salt** than the interior of the cell, **water** flows out of the cell in an attempt to dilute the outside environment. The loss of water can disrupt the functioning of the cell to the point of death. This is called plasmolysis. Conversely, if the interior of a cell is saltier than the exterior environment, water flows into the cell. The cell can swell and burst. This phenomenon is called plasmoptisis.

Some diseases and infections cause chemical cell death. For example, **infection** of the upper respiratory cells with viruses that causes the common cold kills cells during the viral life cycle.

Causes of chemical or mechanical cell death are varied. Some agents act on the **membrane** that surrounds cells. The membrane can be dissolved or damaged. Other agents disrupt enzymes that the cell requires to sustain life. Still other agents can disrupt the genetic material inside the cell.

The process of programmed cell death, apoptosis, or suicide, is a necessary part of the functioning of an **individual** cell and, in multi-celled organisms such as humans, of the whole **organism**. For example, reabsorption of a tadpole's tail during the change from tadpole to frog involves apoptosis. Sloughing of uterine cells in women at the start of menstruation is due to apoptosis of the cells lining the uterine wall. Additionally, apoptosis of extraneous cells during development of a human fetus produces the distinct fingers and toes.

Apoptosis is also important as a means of dealing with threats to an organism. For example, the human **immune system** contains cells that can stimulate apoptosis of other cells that have been infected with a **virus**. Similarly, cells with damaged genetic material undergo cell death. Thus, apoptosis helps the entire organism function efficiently by eliminating cells that threaten the whole organism.

Programmed cell death occurs either by the withdrawal of a chemical signal that is required to continue living, or by exposure to a chemical signal that begins the death process. Once stimulated to die, apoptotic cells shrink, develop irregular cell surfaces, and show disintegration of genetic material within their nuclei. Eventually, these cells break into small membrane wrapped fragments that are engulfed by nearby cells. The apoptosis process is complex, and involves interactions between numbers of different biochemical compounds. This helps ensure that apoptosis does not initiate by accident, and that the process is limited only to specifically targeted cells.

Molecular biologists Sydney Brenner, Robert Horvitz, and John Sulston were awarded the 2002 Nobel Prize in Physiology or Medicine for their pioneering studies on the genetic regulation of programmed cell death. Their studies, which were carried out in the 1980s using a nematode worm as the model system, has since been shown to have relevance to the process of cell death in humans.

See also Deoxyribonucleic acid (DNA); Eukaryotae.

Resources

Periodicals

Wang, J.Y.J. "DNA Damage and Apopotosis." *Cell Death and Differentiation* 8 (November 2001): 1047–1049.

Zhivotovsky, B. "From the Nematode and Mammals Back to the Pine Tree: On the Diversity and Evolution of Programmed Cell Death." *Cell Death and Differentiation* 9 (September 2002): 867–70.

Cell division

Cell division is the process where a single living cell splits to become two or more distinct new cells. All cells divide at some point in their lives. Cell division occurs in single-celled organisms like **bacteria**, in which it is the major form of reproduction (binary fission), or in multicellular organisms like plants, animals, and **fungi**. Many cells continually divide, such as the cells that line the human digestive tract or the cells that make up human skin. Other cells divide only once.

There are two major ways in which biologists categorize cell division. The first, *mitosis*, is simple cell division that creates two daughter cells that are genetically identical to the original parent cell. The process varies slightly between prokaryotic and eukaryotic organisms. In eularyotes, **mitosis** begins with replication of the **deoxyribonucleic acid (DNA)** within the cell to form two copies of each **chromosome**. Once two copies are present, the cell splits to become two new cells by cytokinesis, or formation of a fissure. Mitosis occurs in most cells and is the major form of cell division.

The second process, called *meiosis* is the production of daughter cells having half the amount of genetic material as the original parent cell. Such daughter cells are said to be haploid. **Meiosis** occurs in human sperm and egg production in which four haploid sex cells are produced from a single parent precursor cell. In both mitosis and meiosis of nucleated cells, shuffling of chromosomes creates genetic variation in the new daughter cells. These very important shuffling processes are known as independent assortment and **random** segregation of chromosomes.

Cell division is stimulated by certain kinds of chemical compounds. Molecules called cytokines are secreted by some cells to stimulate others to begin cell division. Also, contact with adjacent cells can control cell division. The phenomenon of contact inhibition is a process where the physical contact between neighboring cells prevents cell division from occurring. When contact is interrupted, however, cell division is stimulated to close the gap between cells. Cell division is a major mechanism by which organisms grow, tissues and organs maintain themselves, and wound healing occurs. **Cancer** is potentially a deadly form of uncontrolled cell division.

Eukaryotic cell division

Although prokaryotes (i.e., non-nucleated unicellular organisms) divide through binary fission, eukaryotes undergo a more complex process of cell division because DNA is packed in several chromosomes located inside a

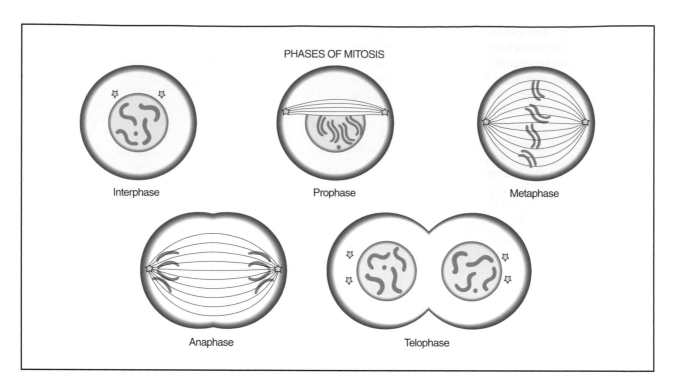

PHASES OF MITOSIS

Interphase

Prophase

Metaphase

Anaphase

Telophase

Cell dividing by mitosis. *Illustration by Hans & Cassidy. Courtesy of Gale Group.*

cell nucleus. In eukaryotes, cell division may take two different paths, in accordance with the cell type involved. Mitosis is a cellular division resulting in two identical nuclei is performed by somatic cells. The process of meiosis results in four nuclei, each containing half of the original number of chromosomes. Sex cells or gametes (ovum and spermatozoids) divide by meiosis. Both prokaryotes and eukaryotes undergo a final process, known as cytoplasmatic division, which divides the parental cell in new daughter cells.

The series of stages that a cell undergoes while progressing to division is known as cell cycle. Cells undergoing division are also termed competent cells. When a cell is not progressing to mitosis, it remains in phase G_0 ("G" zero). Therefore, the cell cycle is divided into two major phases: interphase and mitosis. Interphase includes the phases (or stages) G_1, S, and G_2 whereas mitosis is subdivided into prophase, metaphase, anaphase and telophase.

The cell cycle starts in G_1, with the active synthesis of RNA and **proteins**, which are necessary for young cells to grow and mature. The time G_1 lasts, varies greatly among eukaryotic cells of different **species** and from one **tissue** to another in the same **organism**. Tissues that require fast cellular renovation, such as mucosa and endometrial epithelia, have shorter G_1 periods than those tissues that do not require frequent renovation or repair, such as muscles or connective tissues.

The cell cycle is highly regulated by several enzymes, proteins, and cytokines in each of its phases, in order to ensure that the resulting daughter cells receive the appropriate amount of genetic information originally present in the parental cell. In the case of somatic cells, each of the two daughter cells must contain an exact copy of the original **genome** present in the parental cell. Cell cycle controls also regulate when and to what extent the cells of a given tissue must proliferate, in order to avoid abnormal cell proliferation that could lead to **dysplasia** or **tumor** development. Therefore, when one or more of such controls are lost or inhibited, abnormal overgrowth will occur and may lead to impairment of function and **disease**.

Cells are mainly induced into proliferation by growth factors or **hormones** that occupy specific receptors on the surface of the cell **membrane**, being also known as extra-cellular ligands. Examples of growth factors are as such: epidermal growth factor (EGF), fibroblastic growth factor (FGF), platelet-derived growth factor (PDGF), insulin-like growth factor (IGF), or by hormones. PDGF and FGF act by regulating the phase G_2 of the cell cycle and during mitosis. After mitosis, they act again stimulating the daughter cells to grow, thus leading them from G_0 to G_1. Therefore, FGF and PDGF are also termed competence factors, whereas EGF and IGF are termed progression factors, because they keep the process of cellular progression to mitosis going on.

Growth factors are also classified (along with other molecules that promote the cell cycle) as pro-mitotic signals. Hormones are also pro-mitotic signals. For example, thyrotrophic hormone, one of the hormones produced by the pituitary gland, induces the proliferation of thyroid gland's cells. Another pituitary hormone, known as growth hormone or somatotrophic hormone (STH), is responsible for body growth during childhood and early adolescence, inducing the lengthening of the long bones and protein synthesis. Estrogens are hormones that do not occupy a membrane receptor, but instead, penetrate the cell and the nucleus, binding directly to specific sites in the DNA, thus inducing the cell cycle.

Anti-mitotic signals may have several different origins, such as cell-to-cell adhesion, factors of adhesion to the extra-cellular matrix, or soluble factor such as TGF beta (tumor growth factor beta), which inhibits abnormal cell proliferation, proteins p53, p16, p21, APC, pRb, etc. These molecules are the products of a class of genes called tumor suppressor genes. Oncogenes, until recently also known as proto-oncogenes, synthesize proteins that enhance the stimuli started by growth factors, amplifying the mitotic signal to the nucleus, and/or promoting the accomplishment of a necessary step of the cell cycle. When each phase of the cell cycle is completed, the proteins involved in that phase are degraded, so that once the next phase starts, the cell is unable to go back to the previous one. Next to the end of phase G_1, the cycle is paused by tumor suppressor **gene** products, to allow verification and repair of DNA damage. When DNA damage is not repairable, these genes stimulate other intracellular pathways that induce the cell into suicide or apoptosis (also known as programmed **cell death**). To the end of phase G_2, before the transition to mitosis, the cycle is paused again for a new verification and "decision:" either mitosis or apoptosis.

Along each pro-mitotic and anti-mitotic intra-cellular signaling pathway, as well as along the apoptotic pathways, several gene products (proteins and enzymes) are involved in an orderly sequence of activation and inactivation, forming complex webs of signal transmission and signal amplification to the nucleus. The general goal of such cascades of signals is to achieve the orderly progression of each phase of the cell cycle.

Interphase is a phase of cell growth and metabolic activity, without cell nuclear division, comprised of several stages or phases. During Gap 1 or G_1 the cell resumes protein and RNA synthesis, which was interrupted during mitosis, thus allowing the growth and maturation of young cells to accomplish their physiologic function. Immediately following is a variable length pause for DNA checking and repair before cell cycle transition to phase S during which there is synthesis or semi-con-

servative replication or synthesis of DNA. During Gap 2 or G_2, there is increased RNA and protein synthesis, followed by a second pause for proofreading and eventual repairs in the newly synthesized DNA sequences before transition to Mitosis.

At the start of mitosis the chromosomes are already duplicated, with the sister-chromatids (identical chromosomes) clearly visible under a light **microscope**. Mitosis is subdivided into prophase, metaphase, anaphase and telophase.

During prophase there is a high condensation of chromatids, with the beginning of nucleolus disorganization and nuclear membrane disintegration, followed by the start of centrioles' migration to opposite cell poles. During metaphase the chromosomes organize at the equator of a spindle apparatus (microtubules), forming a structure termed metaphase plate. The sister-chromatids are separated and joined to different centromeres, while the microtubules forming the spindle are attached to a region of the centromere termed kinetochore. During anaphase there are spindles, running from each opposite kinetochore, that pull each set of chromosomes to their respective cell poles, thus ensuring that in the following phase each new cell will ultimately receive an equal division of chromosomes. During telophase, kinetochores and spindles disintegrate, the reorganization of nucleus begins, **chromatin** becomes less condensed, and the nucleus membrane start forming again around each set of chromosomes. The cytoskeleton is reorganized and the somatic cell has now doubled its **volume** and presents two organized nucleus.

Cytokinesis usually begins during telophase, and is the process of cytoplasmatic division. This process of division varies among species but in somatic cells, it occurs through the equal division of the cytoplasmatic content, with the **plasma** membrane forming inwardly a deep cleft that ultimately divides the parental cell in two new daughter cells.

Terry Watkins

Cell, electrochemical

Electrochemical cells are devices based on the principle that when a chemical **oxidation-reduction reaction** takes place, electrons are being transferred from one chemical species to another. In one type of electrochemical cell called a voltaic or galvanic cell, these electrons are deliberately taken outside the cell and made to flow through an **electric circuit** to operate some kind of electrical device. A flashlight **battery** is an example of a voltaic electrochemical cell.

In the other type of electrochemical cell, called an electrolytic cell, the reverse process is taking place: electrons in the form of an **electric current** are deliberately being pumped through the chemicals in the cell in order to **force** an oxidation-reduction reaction to take place. An example of an electrolytic cell is the setup that is used to decompose **water** into **hydrogen** and **oxygen** by **electrolysis**.

Thus, a voltaic cell produces **electricity** from a chemical reaction, while an electrolytic cell produces a chemical reaction from electricity. Voltaic and electrolytic cells are considered separately below, following a general discussion of the relationship between **chemistry** and electricity.

Chemistry and electricity

In order to understand the intimate relationship between **chemical reactions** and electricity, we can consider a very simple oxidation-reduction reaction: the spontaneous reaction between a **sodium** atom and a **chlorine** atom to form **sodium chloride**:

$$
\begin{array}{ccccc}
Na & + & Cl & \rightarrow & Na^+Cl^- & + & energy \\
\text{sodium} & & \text{chlorine} & & \text{sodium} & & \text{chloride} \\
\text{atom} & & \text{atom} & & \text{ion} & & \text{ion}
\end{array}
$$

What happens in this reaction is that an **electron** is passed from the sodium atom to the chlorine atom, leaving the sodium atom positively charged and the chlorine atom negatively charged. (Under normal conditions, the chlorine **atoms** are paired up into diatomic chlorine molecules, Cl_2; but that does not change the present argument.) When a large number of sodium atoms and chlorine atoms are mixed together and react, a large number of electrons move from sodium atoms to chlorine atoms. These moving electrons constitute a flow of electricity. The "push" or *potential* for this electron flow comes from the sodium atoms' eagerness to get rid of electrons and the chlorine atoms' relative eagerness to grab them.

Voltaic cells

The practical problem when large numbers of sodium and chlorine atoms react is that the electrons are flowing in every direction—wherever a sodium atom can find a chlorine atom. We therefore cannot harness the electron flow to do useful electrical **work**. In order to use the electricity to **light** up a bulb, for example, we must make the electrons flow in a single direction through a wire; then we can put a bulb in their path and they will have to push through the filament to get from the sodium atoms to the chlorine atoms, lighting the filament up in the process. In

other words, we must separate the sodium atoms from the chlorine atoms, so that they can only transfer their electrons on our terms: through the wire that we provide. Such an arrangement constitutes a voltaic or galvanic cell. It has the effect of converting chemical potential energy—a chemical push—into electrical potential energy—an electrical push: in other words, a voltage.

The sodium-plus-chlorine reaction is difficult to use in practice, because chlorine is a gas and sodium is a highly reactive **metal** that is nasty to handle. But many other chemical reactions can be used to make voltaic cells for generating electricity. All that is needed is a reaction between a substance (atoms, molecules, or ions) that wants to give up electrons and a substance that wants to grab onto electrons: in other words, an oxidation-reduction reaction. Then it is just a matter of arranging the substances so that the passing of electrons from one to the other must take place through an external wire. Strictly speaking, the resulting devices are voltaic cells, but people generally call them batteries.

As an illustration of how a voltaic cell works, we can choose the metallic elements silver (Ag) and **copper** (Cu) with their respective ions in **solution**, Ag^+ and Cu^{++}. Because copper atoms are more eager to give up electrons than silver atoms are, the copper atoms will tend to force the Ag^+ ions to take them. Or to say it the other way, Ag^+ ions are more eager to grab electrons than Cu^{++} ions are, so they will take them away from copper atoms to become neutral silver atoms. Thus, the spontaneous reaction that will take place when all four species are mixed together is

$$
\begin{array}{ccccc}
Cu & + & 2Ag^+ & \rightarrow & Cu^{++} & + & 2Ag \\
\text{copper} & & \text{silver} & & \text{copper} & & \text{silver} \\
\text{metal} & & \text{ions} & & \text{ion} & & \text{metal}
\end{array}
$$

This equation says that a piece of copper metal dipped into a solution containing silver ions will dissolve and become copper ions, while at the same time silver ions "plate out" as metallic silver. (This is not how silver plating is done, however, because the silver comes out as a rough and non-adhering coating on the copper. The silver plating of dinnerware and jewelry is done in an electrolytic cell.) To make a useful voltaic cell out of the copper-silver system, we must put the Cu and Cu^{++} in one container, the Ag and Ag^+ in a separate container, and then connect them with a wire. Bars of copper and silver metal should be dipped into solutions of copper nitrate, $Cu(NO_3)_2$, and silver nitrate, $AgNO_3$, respectively. A *salt bridge* should be added between the two containers. It is a tube filled with an electrolyte—a solution of an ionic **salt** such as **potassium nitrate** KNO_3, which allows ions to flow through it. Without the salt bridge, electrons would tend to build up in the silver container and the reaction would stop because the **negative** charge has no place to go. The salt bridge allows the negative charge, this time in the form of NO_3^- ions, to complete the circuit by crossing the bridge from the silver container back into the copper container. Now the circuit is complete and the reaction can proceed, producing a steady flow of electrons through the wire and keeping the bulb lit until something runs out—either the copper bar is all dissolved or the silver ions are all depleted: our "battery" is dead.

In principle, a voltaic cell can be made from the four constituents of any oxidation-reduction reaction: any two pairs of oxidizable and reducible atoms, ions, or molecules. For example, any two elements and their respective ions can be made into a voltaic cell. Examples: Ag/Ag^+ with Cu/Cu^{++} (as above), or Cu/Cu^{++} with Zn/Zn^{++} (zinc), or H_2/H^+ (hydrogen) with Fe/Fe^{+++} (**iron**), or Ni/Ni^{++} (nickel) with Cd/Cd^{++} (cadmium). The last cell is the basis for the rechargeable nickel-cadmium (nicad) batteries that are used to power many electrical devices from razors to computers. When voltaic cells are used for portable purposes, they are "dry cells:" instead of a liquid solution, they contain a non-spillable paste. The **lead** storage battery in automobiles, however, does contain a liquid: a **sulfuric acid** solution.

Electrolytic cells

There are many chemical reactions that, unlike the sodium-chlorine and copper-silver reactions above, simply will not occur spontaneously. One example is the breakup of water into hydrogen and oxygen:

$$2H_2O \ + \ energy \ \rightarrow \ 2H_2O \ + \ O_2$$
water   hydrogen   oxygen
                               gas   gas

This will not happen all by itself (that is, without the added **energy**) because water is an extremely stable compound. We can force this reaction to go, however, by pumping energy into the water in the form of an electric current. When we do this—passing an electric current through a chemical system in order to drive chemical reactions happen—creates an electrolytic cell.

Electrolytic cells are used for a variety of purposes other than the electrolysis of water. They are used for obtaining metals such as sodium, **magnesium**, and **aluminum** from their compounds; for refining copper; for producing important industrial chemicals such as **sodium hydroxide**, chlorine, and hydrogen, and for electroplating metals such as silver, gold, nickel, and chromium onto jewelry, tableware, and industrial machine parts.

Sea also Bond, chemical.

Resources

Books

Chang, Raymond. *Chemistry.* New York: McGraw-Hill, 1991.
Oxtoby, David W., et al. *The Principles of Modern Chemistry.* 5th ed. Pacific Grove, CA: Brooks/Cole, 2002.
Umland, Jean B. *General Chemistry.* St. Paul: West, 1993.

Robert L. Wolke

Cell membrane transport

The **cell** is bound by an outer **membrane** that, in accord with the fluid mosaic model, is comprised of a phospholipid **lipid** bilayer with proteins—molecules that also act as receptor sites—interspersed within the phospholipid bilayer. Varieties of channels exist within the membrane. There are a number of internal cellular membranes that partially partition the intercellular matrix, and that ultimately become continuous with the nuclear membrane.

There are three principal mechanisms of outer cellular membrane transport (i.e., means by which molecules can pass through the boundary cellular membrane). The transport mechanisms are passive, or gradient **diffusion**, facilitated diffusion, and active transport.

Diffusion is a process in which the **random** motions of molecules or other particles result in a net movement from a region of high **concentration** to a region of lower concentration. A familiar example of diffusion is the dissemination of floral perfumes from a bouquet to all parts of the motionless air of a room. The **rate** of flow of the diffusing substance is proportional to the concentration gradient for a given direction of diffusion. Thus, if the concentration of the diffusing substance is very high at the source, and is diffusing in a direction where little or none is found, the diffusion rate will be maximized. Several substances may diffuse more or less independently and simultaneously within a space or **volume** of liquid. Because lightweight molecules have higher average speeds than heavy molecules at the same **temperature**, they also tend to diffuse more rapidly. Molecules of the same weight move more rapidly at higher temperatures, increasing the rate of diffusion as the temperature rises.

Driven by concentration gradients, diffusion in the cell usually takes place through channels or pores lined by **proteins**. Size and electrical charge may inhibit or prohibit the passage of certain molecules or electrolytes (e.g., **sodium**, potassium, etc.).

Osmosis describes diffusion of **water** across cell membranes. Although water is a polar **molecule** (i.e., has overall partially positive and **negative** charges separated by its molecular structure), transmembrane proteins form hydrophilic (water loving) channels to through which water molecules may move.

Facilitated diffusion is the diffusion of a substance not moving against a concentration gradient (i.e., from a region of low concentration to high concentration) but which require the assistance of other molecules. These are not considered to be energetic reactions (i.e., **energy** in the form of use of **adenosine triphosphate** molecules (ATP) is not required. The facilitation or assistance—usually in physically turning or orienting a molecule so that it may more easily pass through a membrane—may be by other molecules undergoing their own random **motion**.

Transmembrane proteins establish pores through which ions and some small hydrophilic molecules are able to pass by diffusion. The channels open and close according to the physiological needs and state of the cell. Because they open and close transmembrane proteins are termed "gated" proteins. Control of the opening and closing mechanism may be via mechanical, electrical, or other types of membrane changes that may occur as various molecules bind to cell receptor sites.

Active transport is movement of molecules across a cell membrane or membrane of a cell organelle, from a region of low concentration to a region of high concentration. Since these molecules are being moved against a concentration gradient, cellular energy is required for active transport. Active transport allows a cell to maintain conditions different from the surrounding environment.

There are two main types of active transport; movement directly across the cell membrane with assistance from transport proteins, and endocytosis, the engulfing of materials into a cell using the processes of pinocytosis, phagocytosis, or receptor-mediated endocytosis.

Transport proteins found within the phospholipid bilayer of the cell membrane can move substances directly across the cell membrane, molecule by molecule. The sodium-potassium pump, which is found in many cells and helps nerve cells to pass their signals in the form of electrical impulses, is a well-studied example of active transport using transport proteins. The transport proteins that are an essential part of the sodium-potassium pump maintain a higher concentration of potassium ions inside the cells compared to outside, and a higher concentration of sodium ions outside of cells compared to inside. In order to carry the ions across the cell membrane and against the concentration gradient, the transport proteins have very specific shapes that only fit or bond well with sodium and potassium ions. Because the transport of these ions is against the concentration gradient, it requires a significant amount of energy.

Endocytosis is an infolding and then pinching in of the cell membrane so that materials are engulfed into a vacuole or vesicle within the cell. Pinocytosis is the process in which cells engulf liquids. The liquids may or may not contain dissolved materials. Phagocytosis is the process in which the materials that are taken into the cell are solid particles. With receptor-mediated endocytosis the substances that are to be transported into the cell first bind to specific sites or receptor proteins on the outside of the cell. The substances can then be engulfed into the cell. As the materials are being carried into the cell, the cell membrane pinches in forming a vacuole or other vesicle. The materials can then be used inside the cell. Because all types of endocytosis use energy, they are considered active transport.

See also Cell division; Cell staining; Cellular respiration; Nerve impulses and conduction of impulses.

Resources

Books

Cooper, Geoffrey M. *The Cell—A Molecular Approach.* 2nd ed. Sunderland, MA: Sinauer Associates, Inc., 2000.

Lodish, H., et. al. *Molecular Cell Biology.* 4th ed. New York: W. H. Freeman & Co., 2000.

Periodicals

Karow, Julia. "Cell Membrane Exclusivity." *Scientific American* 23 (October 2000).

Nicolson, Garth L., and S.J. Singer. "The Fluid Mosaic Model of the Structure of Cell Membranes." *Science* 18 (February 1972): 720– 730.

Other

Kimball, John W. "Transport Across Cell Membranes." *Kimball's Biology Pages.* August 17, 2002 (cited January 2003.) <http://users.rcn. com/jkimball.ma.ultranet/Biology Pages/D/Diffusion.html>.

K. Lee Lerner

Cell staining

Medical science depends on the staining of cells in tissues to make accurate diagnoses of a wide range of diseases from **cholera** to **sexually transmitted diseases**, to parasitic diseases and skin infections. Staining techniques performed routinely in microbiological laboratories include gram's stain, acid-fast stains, acridine orange, calcofluor white, toluidine blue, methylene blue, silver stains, and fluorescent stains. Stains are classified broadly as basic, acidic, or neutral stains. The chemical nature of the cells under examination determines which stain is selected for use.

Cell staining is important in the **diagnosis** of **microorganisms** because **bacteria** can be identified by the **color** differentiation of stains (dyes). Microscopic examination of stained cell samples allows examination of the size, shape, and arrangement of organelles, as well as external appendages such as the whip-like **flagella**, which are the cell's organs of **motion**. When sample cells are stained to show their chemical composition it is called differential staining.

Histochemistry is the specialty that studies the staining properties of cells. Histochemistry is used in other specialties such as histology (the study of tissues), **biochemistry** (the study of the chemical makeup of cells), **cytology** (the study of cells), and microbiology (the study of organisms that are too small to be seen without a **microscope**).

In 1880, Hans Christian Gram of Denmark noted the differences in the way bacteria react to stains. Those bacteria that retained a deep purple stain, even after they were washed, were termed "stain positive." Those that lost the stain and responded again to another stain, were termed "stain negative." Today, bacteria are classified as "gram-positive" or "gram-negative" to distinguish the two major groups of bacteria. This staining test highlights differences in the structure of the cell wall of the two types of bacteria.

Penicillin G is used to treat gram-positive infections, but it is ineffective against gram-negative bacteria. Other **antibiotics** are only effective against gram-negative bacteria. Chloromycetin, which was discovered in 1947, was the first antibiotic to be effective against both gram-positive and gram-negative bacteria.

Staining techniques

Bacteria are nearly colorless, so their features are difficult to distinguish when they are suspended in a fluid and viewed directly under a microscope. Stains are salts that color particular ions in the bacterial cell, and make more visible distinctions under the microscope. The chemical composition of the cell determines which stain is absorbed. Acidic parts of a cell absorb stains that are positively charged; alkaline parts of a cell combine with stains that are acidic or negatively charged.

Before tissues are stained, a thin layer of cells that have been sliced from the specimen (a smear) is prepared by fixing. Fixing a specimen that has been placed on a slide is done by either allowing it to dry at room **temperature**, or by passing the specimen quickly over a flame. Next the specimen is stained: either a simple stain, a differential stain, a negative or indirect stain, a stain for reserve materials, or for microbial structures is used. Most staining dyes are prepared from **coal** tar and those used in microbiology come from aniline, an oily liquid.

In simple (or direct) staining only one dye is used, which is washed away after 30–60 seconds, before drying and examination. Gentian violet, crystal violet, safranin, methylene blue, basic fuchsin, and others are the dyes used in this method. In differential staining, the gram stain and the acid-fast stain are used to distinguish different microorganisms.

There are four steps involved in the gram stain method, which is considered the most valuable cell-staining technique used in bacteriological cell analysis. In the first step, the specimen is stained with crystal violet or gentian, and one minute later, the second step is taken which involves washing the dye off and **flooding** the **solution** with iodine. The third step involves washing the iodine off 60 seconds after it is applied and then washing the slide with an ethyl **alcohol** solution of 95% or a 50:50 mixture of **acetone** and ethyl alcohol 15–30 seconds after this. The fourth and final step is to stain the slide for 30 seconds with a red or brown dye. The critical action in this process is the washing away of the stain, called decolorization stain, (sometimes called the Ziehl-

Neelsen technique) is particularly useful in identifying the **organism** that causes **tuberculosis**. When these microorganisms are stained with a red dye (carbol fuchsin), the color remains even though the slide is washed with a strong solution of acid alcohol. Most organisms, other than the ones responding to acid-fast staining, would decolorize from this wash. Methylene blue is then used to differentiate any other organisms present in the smear.

Negative (or indirect) techniques stain the background of cell smears, rather than the organisms directly. In this technique, a drop of the stain is placed on a slide and organisms are added to the stain. After the specimen is smeared over the slide, it is allowed to air dry and is then examined under the microscope. Negative or indirect staining procedures are useful when examining the size and shape of microorganisms.

Staining for reserve materials in cells isolates specific structures in the cells of microorganisms (such as granules or other reserve substances in bacteria that cause diseases such as **diphtheria**). In staining of microbial structures, the flagella, nuclear material of the cell, the cell wall, or capsule is stained for viewing under the microscope. These procedures use two or more stains.

Standardization of tests

Cell staining is one of a number of laboratory tests that are performed to aid in the analysis and diagnosis of **disease**. The work in these laboratories is performed for physicians as well as for government agencies involved in **water** purification and **sewage treatment**, and for industries such as the food industry involved in the manufacture of goods that need to adhere to strict health standards.

Standardization of these tests have been widely adopted throughout the microbiological laboratory community. The National Committee for Clinical Laboratory Standards (NCCLS), located in Villanova, Pennsylvania, continuously publishes standards for these laboratory tests. Among the factors that have been standardized in laboratory testing are temperature, **pH** (acidity or alkalinity), growth medium, antibiotics, quality control, and other factors.

Resources

Books

Hunt, Tim. *The Cell Cycle: An Introduction.* New York: Oxford University Press, 1993.

Keynes, Milton. *Handling Laboratory Microorganisms.* Philadelphia: Open University Press, 1991.

Koneman, Elmer W. *Color Atlas and Textbook of Diagnostic Microbiology.* 4th ed. Philadelphia: J. B. Lippincott, 1992.

KEY TERMS

Acidic stains—Stains that adhere to microorganisms having a high lipid (fatty) content.

Decolorization—Washing away of the staining medium.

Differential staining—Staining technique that uses more than one stain to differentiate the structure of the microorganism.

Fixing—Preparing a cell specimen on a slide for examination under a microscope.

Gram-negative—Those cells that lose the color of the stain after they are washed with a 95% alcohol solution during the staining process.

Gram-positive—Those cells that retain the color of the stain after they are washed with a 95% alcohol solution during the staining process.

Postgate, John R. *The Outer Reaches of Life.* Cambridge, England: Cambridge University Press, 1994.

Prescott, L., J. Harley, and D. Klein. *Microbiology.* 5th ed. New York: McGraw-Hill, 2002.

Vita Richman

Cellular respiration

Cellular respiration in the presence of **oxygen** (**aerobic respiration**) is the process by which energy-rich organic substrates are broken down into **carbon dioxide** and **water**, with the release of a considerable amount of **energy** in the form of **adenosine triphosphate** (ATP). **Anaerobic** respiration breaks down glucose in the absence of oxygen, and produces pyruvate, which is then reduced to lactate or to **ethanol** and CO_2. Anaerobic respiration releases only a small amount of energy (in the form of ATP) from the glucose **molecule**.

Respiration occurs in three stages. The first stage is **glycolysis**, which is a series of enzyme-controlled reactions that degrades glucose (a 6-carbon molecule) to pyruvate (a 3-carbon molecule) which is further oxidized to acetylcoenzyme A (acetyl CoA). Amino acids and **fatty acids** may also be oxidized to acetyl CoA as well as glucose.

In the second stage, acetyl CoA enters the **citric acid** (Krebs) cycle, where it is degraded to yield energy-rich **hydrogen** atoms which reduce the oxidized form of the coenzyme nicotinamide adenine dinucleotide (NAD^+) to NADH, and reduce the coenzyme flavin adenine dinucleotide (FAD) to $FADH_2$. (Reduction is the addition of electrons to a molecule, or the gain of hydrogen atoms, while oxidation is the loss of electrons or the addition of oxygen to a molecule.) Also in the second stage of cellular respiration, the **carbon** atoms of the intermediate metabolic products in the **Krebs cycle** are converted to carbon dioxide.

The third stage of cellular respiration occurs when the energy-rich hydrogen **atoms** are separated into protons [H^+] and energy-rich electrons in the **electron** transport chain. At the beginning of the electron transport chain, the energy-rich hydrogen on NADH is removed from NADH, producing the oxidized coenzyme, NAD^+ and a **proton** (H+) and two electrons (e-). The electrons are transferred along a chain of more than 15 different electron carrier molecules (known as the electron transport chain). These **proteins** are grouped into three large respiratory **enzyme** complexes, each of which contains proteins that span the mitochondrial **membrane**, securing the complexes into the inner membrane. Furthermore, each complex in the chain has a greater affinity for electrons than the complex before it. This increasing affinity drives the electrons down the chain until they are transferred all the way to the end where they meet the oxygen molecule, which has the greatest affinity of all for the electrons. The oxygen thus becomes reduced to H_2O in the presence of hydrogen ions (protons), which were originally obtained from nutrient molecules through the process of oxidation.

During electron transport, much of the energy represented by the electrons is conserved during a process called oxidative phosphorylation. This process uses the energy of the electrons to phosphorylate (add a phosphate group) **adenosine diphosphate** (ADP), to form the energy-rich molecule ATP.

Oxidative phosphorylation is driven by the energy released by the electrons as they pass from the hydrogens of the coenzymes down the respiratory chain in the inner membrane of the mitochondrion. This energy is used to pump protons (H^+) across the inner membrane from the matrix to the intermediate **space**. This sets up a **concentration** gradient along which substances flow from high to low concentration, while a simultaneous current of OH^- flows across the membrane in the opposite direction. The simultaneous opposite flow of positive and **negative** ions across the mitochondrial membrane sets up an electrochemical proton gradient. The flow of protons down this gradient drives a membrane-bound enzyme, ATP synthetase, which catalyzes the phosphorylation of ADP to ATP.

This highly efficient, energy conserving series of reactions would not be possible in eukaryotic cells without the organelles called mitochondria. Mitochondria are the "powerhouses" of the eukaryotic cells, and are bounded by two membranes, which create two separate compartments: an internal space and a narrow intermembrane space. The enzymes of the matrix include those that catalyze the conversion of pyruvate and fatty acids to acetyl CoA, as well as the enzymes of the Krebs cycle. The enzymes of the respiratory chain are embedded in the inner mitochondrial membrane, which is the site of oxidative phosphorylation and the production of ATP.

In the absence of mitochondria, **animal** cells would be limited to glycolysis for their energy needs, which releases only a small fraction of the energy potentially available from the glucose.

The reactions of glycolysis require the input of two ATP molecules and produce four ATP molecules for a net gain of only two molecules per molecule of glucose. These ATP molecules are formed when phosphate groups are removed from phosphorylated intermediate products of glycolysis and transferred to ADP, a process called substrate level phosphorylation (synthesis of ATP by direct transfer of a high-energy phosphate group from a molecule in a metabolic pathway to ADP).

In contrast, mitochondria supplied with oxygen produce about 36 molecules of ATP for each molecule of glucose oxidized. Procaryotic cells, such a **bacteria**, lack mitochondria as well as nuclear membranes. Fatty acids and amino acids when transported into the mitochondria are degraded into the two-carbon acetyl group on acetyl CoA, which then enters the Krebs cycle. In animals, the body stores fattyacids in the form of fats, and glucose in the form of glycogen in order to ensure a steady supply of these **nutrients** for respiration.

While the Krebs cycle is an integral part of aerobic **metabolism**, the production of NADH and $FADH_2$ is not dependent on oxygen. Rather, oxygen is used at the end of the electron transport chain to combine with electrons removed from NADH and $FADH_2$ and with hydrogen ions in the cytosol to produce water.

Although the production of water is necessary to keep the process of electron transport chain in **motion**, the energy used to make ATP is derived from a different process called chemiosmosis.

Chemiosmosis is a mechanism that uses the proton gradient across the membrane to generate ATP and is initiated by the activity of the electron transport chain. Chemiosmosis represents a link between the chemical and osmotic processes in the mitochondrion that occur during respiration.

The electrons that are transported down the respiratory chain on the mitochondrion's inner membrane release energy that is used to pump protons (H^+) across the inner membrane from the mitochondrial matrix into the intermembrane space. The resulting gradient of protons across the mitochondrial inner membrane creates a backflow of protons back across the membrane. This flow of electrons across the membrane, like a waterfall used to power an electric **turbine**, drives a membrane-bound enzyme, ATP synthetase. This enzyme catalyzes the phosphorylation of ADP to ATP, which completes the part of cellular respiration called oxidative phosphorylation. The protons, in turn, neutralize the negative charges created by the addition of electrons to oxygen molecules, with the resultant production of water.

Cellular respiration produces three molecules of ATP per pair of electrons in NADH, while the pair of electrons in $FADH_2$ generate two molecules of ATP. This means that 12 molecules of ATP are formed for each acetyl CoA molecule that enters the Krebs cycle; and since two acetyl CoA molecules are formed from each molecule of glucose, a total of 24 molecules of ATP are produced from each molecule of this sugar. When added to the energy conserved from the reactions occurring before acetyl CoA is formed, the complete oxidation of a glucose molecule gives a net yield of about 36 ATP molecules. When fats are burned, instead of glucose, the total yield from one molecule of palmitate, a 16-carbon fatty, is 129 ATP.

See also Catabolism; Respiration.

Resources

Books

Alberts, Bruce, et al. *Molecular Biology of The Cell.* 2nd ed. New York: Garland Publishing, 1989.
Lehninger, Albert L. *Principles of Biochemistry.* New York: Worth Publishers, 1982.

Marc Kusinitz

Cellular telephone

Cellular **telephone** technology is also called cellular **radio**. The cellular radio network became fully operational in **North America** in 1978. This technology relies on the distribution of what are called cell sites over a wide geographical area. Each cell site consists of a radio transceiver and a controller that sends and receives signals from the mobile phones in the area to a telephone switch. The signals can be beamed to a central **point** called the mobile telecommunications switching office. This office places calls from land based telephones to mobile telephones, and allows mobile phones to operate across the globe as the phone signal is relayed from one switching office to another.

Mobile telephone service was severely limited in availability until 1984. Until that year, only urban areas had mobile service, primarily for city services. Each city had a single **antenna** to transmit signals to and from the antennae of car phones. The Federal Communications Commission (FCC) assigned only 12–4 frequencies to an urban area. As a result, only one or two-dozen car phone calls could take place in the entire city at one time. The system was frustrating. Users had to wait up to 30 minutes to get a dial tone, and potential mobile phone customers were put on five to 10-year waiting lists.

Cellular phone technology changed all this. In a cellular system, each metropolitan area is divided into broadcasting zones or "cells." Each 6–10-mi^2 (15.5–25.9-km^2) cell has its own broadcast antenna or tower. As a car phone moves through the city, a computer automatically passes its **frequency** from one cell to the next. A single frequency can be used for multiple, nonadjacent cells; and, as the number of users increases, cells can be subdivided into any number of smaller cells, so the cellular system is capable of far greater usage than the old mobile service.

Rudimentary cellular technology was known as early as 1947. By the 1960s and 1970s, mobile phone service was overcrowded, and the need for a more efficient system led to a re-examination and refinement of cellular technology. Bell Laboratories, a research division within the American Telephone and Telegraph Company (AT&T), took the lead in this development. A prototype network had been developed by 1971. In 1978 the first experimental cellular service, which was called the Advanced Mobile Phone Service (AMPS), was operational in the Chicago, Illinois, area. AT&T was declared a monopoly and was reorganized in 1978. This opened the cellular phone market to competitors. Seven new regional phone companies began to pursue the cellular phone market.

In 1981, the FCC issued cellular phone regulations. In October 1983, Ameritech Mobile Communications (a subsidiary resulting from the Bell breakup) introduced the first American commercial cellular system in Chicago. Cellular service was also available by then in a number of other countries. The FCC also allowed one non-Bell service in each metropolitan area. For example, Cellular One began transmitting in Washington, D.C., in December 1983.

As the number of cellular phone systems and subscribers increased, the costs for equipment and service decreased. As of 2002, pocket-size personal telephones based on cellular technology are available, as are machines that combine a cellular phone, facsimile machine, voice and e-mail systems, answering machine, and pager.

The cellular technology that relied on ground-based antennae is giving way to **satellite** technology. Communication satellites are being used increasingly by the cellular services to provide uniform service as the telephone and its user travel through a number of cells. A new form of satellite technology called Global Mobile Personal Communications by Satellite (GMPCS) allows telecommunication virtually anywhere in the world.

Earlier communication satellites traveled high above **Earth** and orbited at the same speed as Earth rotates. The satellites stayed in one position relative to the ground. This geosynchronous **orbit** caused delays and loss of quality in signal transmission. GMPCS satellites have much lower orbits and can be used in clusters called constellations to transfer signals rapidly and with greater clarity. Developing countries and remote areas have access to cellular service, and, during natural disasters, emergency relief can be mobilized and coordinated when land-line telephones have been disabled or the region is remote.

The telephone industry is making advances toward a single telephone number per person. This is called a Personal Communications Network or PCN. The PCN can be used for both land-line networks (telephone service by cable that has traditionally been provided to homes and businesses) and cellular systems. Basically, this service is a form of call-forwarding in which the signal is transferred through the land-line network until a transceiver detects that the cell phone is within range. The call is then sent as a cellular signal.

Further development is expected to lead to digital transmissions that convert conversations into computer code that can be transferred by advanced cellular technology. Also microcellular technology that uses smaller and more closely spaced transceivers instead of cellular towers or by satellite is on the horizon. Microcells will relieve the pressure to provide enough access telephone numbers, and the FCC is working on modifying communications regulations to suit the new technologies and to free frequencies with less demand for cellular use. Radio frequencies may be converted to phone service, and dedicated land-lines for telephone service may be outmoded by delivering phone signals via cable **television**.

Telecommunications experts predict that all telephones will be wireless by 2010.

The ease of use and life saving potential of cellular telephones has not come without negative aspects. Foremost is the use of cellular telephones in motor vehicles.

KEY TERMS

Cellular signal—An analog or digital telephone signal that is transmitted on a specific frequency among areas or cells from cellular towers or by satellite.

Land-line network—A communications network that uses underground or overhead cables to carry signals.

Microcellular technology—Method of transmitting cellular telephone signals among smaller areas or microcells by transceivers or satellites.

Personal Communications Network—Also called a Personal Communications Service (PCS), a technology that uses one telephone access number assigned per person to transmit both land-line and cellular telephone calls.

The use of a phone can divert the driver's attention from the road. The increase in motor vehicle accidents and in injuries and death has been attributed to the use of cellular telephones. Increasingly, legislation is requiring the use of "hands off" cellular technology in motor vehicles, where the phone is positioned somewhere on the dashboard and a speaker is activated to carry on the conversation.

See also Fiber optics; Synthesizer, voice.

Resources

Books

Dodd, A.Z. *The Essential Guide to Telecommunications.* 3rd ed. Englewood Cliffs, NJ: Prentice-Hall, 2001.

Laino, J. *The Telephony Book-Understanding Systems and Services.* Gilroy, CA: CMP Books, 1999.

Noll, A.M. *Introduction to Telephones and Telephone Systems* Norwood. MA: Artech House, 1999.

Brian Hoyle

Cellulose

Cellulose is a substance found in the **cell** walls of plants. Although cellulose is not a component of the human body, it is nevertheless the most abundant organic macromolecule on **Earth**. The chemical structure of cellulose resembles that of starch, but unlike starch, cellulose is extremely rigid (Figure 1). This rigidity imparts great strength to the **plant** body and protection to the interiors of plant cells.

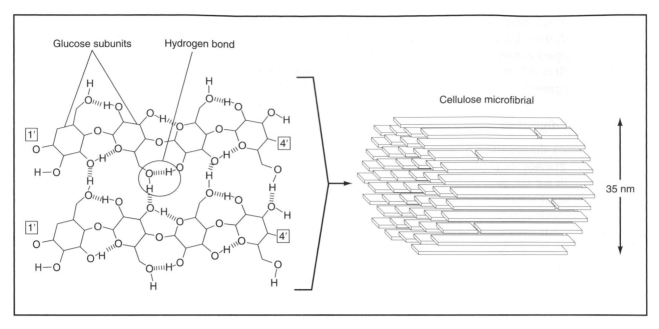

Glucose subunits   Hydrogen bond

Cellulose microfibrial

35 nm

Figure 1. The structure of cellulose. *Illustration by Hans & Cassidy. Courtesy of Gale Group.*

Structure of cellulose

Like starch, cellulose is composed of a long chain of at least 500 glucose molecules. Cellulose is thus a polysaccharide (Latin for "many sugars"). Several of these polysaccharide chains are arranged in **parallel** arrays to form cellulose microfibrils. The individual polysaccharide chains are bound together in the microfibrils by **hydrogen** bonds. The microfibrils, in turn, are bundled together to form macrofibrils (Figure 1).

The microfibrils of cellulose are extremely tough and inflexible due to the presence of hydrogen bonds. In fact, when describing the structure of cellulose microfibrils, chemists call their arrangement "crystalline," meaning that the microfibrils have crystal-like properties. Although starch has the same basic structure as cellulose—it is also a polysaccharide—the glucose subunits are bonded in such a way that allows the starch **molecule** to twist. In other words, the starch molecule is flexible, while the cellulose molecule is rigid.

How cellulose is arranged in plant cell walls

Like human bone, plant cell walls are composed of fibrils laid down in a matrix, or "background" material. In a cell wall, the fibrils are cellulose microfibrils, and the matrix is composed of other polysaccharides and **proteins**. One of these matrix polysaccharides in cell walls is pectin, the substance that, when heated, forms a gel. Pectin is the substance that cooks use to make jellies and jams.

The arrangement of cellulose microfibrils within the polysaccharide and protein matrix imparts great strength to plant cell walls. The cell walls of plants perform several functions, each related to the rigidity of the cell wall. The cell wall protects the interior of the plant cell, but also allows the circulation of fluids within and around the cell wall. The cell wall also binds the plant cell to its neighbors. This binding creates the tough, rigid skeleton of the plant body. Cell walls are the reason why plants are erect and rigid. Some plants have a secondary cell wall laid over the primary cell wall. The secondary cell wall is composed of yet another polysaccharide called lignin. Lignin is found in trees. The presence of both primary and secondary cell walls makes the **tree** even more rigid, penetrable only with sharp axes.

Unlike the other components of the cell wall, which are synthesized in the plant's Golgi body (an organelle that manufactures, sorts, and transports different macromolecules within the cell), cellulose is synthesized on the surface of the plant cell. Embedded within the plant's **plasma** membrane is an **enzyme**, called cellulose synthetase, which synthesizes cellulose. As cellulose is synthesized, it spontaneously forms microfibrils that are deposited on the cell's surface. Because the cellulose synthetase enzyme is located in the plasma **membrane**, the new cellulose microfibrils are deposited under older cellulose microfibrils. Thus, the oldest cellulose microfibrils are outermost on the cell wall, while the newer microfibrils are innermost on the cell wall.

As the plant cell grows, it must expand to accommodate the growing cell **volume**. However, because cellulose is so rigid, it cannot stretch or flex to allow this growth. Instead, the microfibrils of cellulose slide past each other or separate from adjacent microfibrils. In this way, the cellwall is able to expand when the cell volume enlarges during growth.

Cellulose digestion

Humans lack the enzyme necessary to digest cellulose. Hay and **grasses** are particularly abundant in cellulose, and both are indigestible by humans (although humans can digest starch). Animals such as **termites** and herbivores such as cows, **koalas**, and **horses** all digest cellulose, but even these animals do not themselves have an enzyme that digests this material. Instead, these animals harbor microbes that can digest cellulose.

The termite, for instance, contains protists (single-celled organisms) called mastigophorans in their guts that carry out cellulose digestion. The **species** of mastigophorans that performs this service for termites is called *Trichonympha*, which, interestingly, can cause a serious parasitic **infection** in humans.

Animals such as cows have **anaerobic bacteria** in their digestive tracts which digest cellulose. Cows are ruminants, or animals that chew their cud. Ruminants have several stomachs that break down plant materials with the help of enzymes and bacteria. The partially digested material is then regurgitated into the mouth, which is then chewed to break the material down even further. The bacterial digestion of cellulose by bacteria in the stomachs of ruminants is anaerobic, meaning that the process does not use **oxygen**. One of the by-products of anaerobic **metabolism** is methane, a notoriously foul-smelling gas. Ruminants give off large amounts of methane daily. In fact, many environmentalists are concerned about the production of methane by cows, because methane may contribute to the destruction of **ozone** in Earth's stratosphere.

Although cellulose is indigestible by humans, it does form a part of the human diet in the form of plant foods. Small amounts of cellulose found in **vegetables** and **fruits** pass through the human **digestive system** intact. Cellulose is part of the material called "fiber" that dieticians and nutritionists have identified as useful in moving food through the digestive tract quickly and efficiently. Diets high in fiber are thought to lower the risk of colon **cancer** because fiber reduces the time that waste products stay in contact with the walls of the colon (the terminal part of the digestive tract).

See also Rumination.

KEY TERMS

Anaerobic—Describes biological processes that take place in the absence of oxygen.

Cell wall—The tough, outer covering of plant cells composed of cellulose microfibrils held together in a matrix.

Cellulose synthetase—The enzyme embedded in the plasma membrane that synthesizes cellulose.

Colon—The terminal portion of the human digestive tract.

Golgi body—The organelle that manufactures, sorts, and transports macromolecules within a cell.

Lignin—A polysaccharide that forms the secondary cell wall in some plants.

Matrix—The material, composed of polysaccharides and protein, in which microfibrils of cellulose are embedded in plant cell walls.

Methane—A gas produced during the anaerobic digestion of cellulose by bacteria in certain animals.

Microfibril—Small fibrils of cellulose; consists of parallel arrays of cellulose chains.

Polysaccharide—A molecule composed of many glucose subunits arranged in a chain.

Ruminant—A cud-chewing animal with a four-chambered stomach and even-toed hooves.

Resources

Books

Brett, C.T. *Physiology and Biochemistry of Plant Cell Walls.* London: Unwin Hyman, 1990.

Van Soest, Peter J. *Nutritional Ecology of the Ruminant.* 2nd ed. Ithaca: Comstock Press, 1994.

Periodicals

Benedict, C. R., et al. "Crystalline Cellulose and Cotton Fiber Strength." *Crop Science* 24 (January-February 1994): 147.

Dunkle, Richard L. "Food Science Research: An Investment in Health." *Agricultural Research* 41 (December 1993): 2.

Dwyer, Johanna. "Dietary Fiber and Colorectal Cancer Risk." *Nutrition Reviews* 51 (May 1993): 147.

Kleiner, Susan M. "Fiber Facts: How to Fight Disease with a High-fiber Diet." *The Physician and Sportsmedicine* 18 (October 1990): 19.

Slavin, Joanne L. "Dietary Fiber: Mechanisms or Magic on Disease Prevention?" *Nutrition Today* 25 (December 1990): 6.

Young, Stephen. "How Plants Fight Back." *New Scientist* 130 (June 1, 1991): 41.

Kathleen Scogna

Celsius *see* **Temperature**

Centipedes

Centipedes (phylum Arthropoda, class Chilopoda) occur throughout the world in both temperate and tropical regions where they live in **soil** and **humus** and beneath fallen logs, **bark**, and stones. Because they lack a hard outer skeleton, centipedes are confined to moist environments in order to maintain **water** balance. Many **species** are therefore active only at night, remaining sheltered during the day. Most centipedes are active on the surface, but some of the more slender species are capable of burrowing in loose soils.

Four main orders of centipedes have been recognized with some 3,000 species described so far. Among these, there is considerable variation in size, **color**, and **behavior**. One of the largest species that has been recorded is *Scolopendra gigantea* from Latin America, which reaches a length of 10 in (26 cm). Most tropical species are distinguished by their bright colors (red, yellow, green, blue, or various combinations of these), while temperate-dwelling centipedes tend to be a reddish brown color. Many of these bold colors have evolved to deter potential predators. Such vivid yet simple colors advertise one of the following: that the animals can sting, inflict a painful or poisonous bite, produce a foul **taste** if eaten, or may cause an irritation to the skin. In the case of centipedes, all of these hold true: an inquisitive **animal** may receive a small injection of poison from special claws on the head, a painful pinch from the last pair of legs, or may be covered in foul acids produced from a series of **glands** along the body.

All centipedes are instantly recognizable by their segmented body, each segment of which bear a single pair of legs. The number of legs varies considerably according to species—from 15 to as many as 170 pairs. The legs, however, are not always of similar length: in some Scutigeromorpha species the posterior legs may be twice as long as those nearer the head. With so many legs, people have often wondered how centipedes manage to coordinate their movements, especially when running. But centipedes are well adapted for walking and running, as rhythmic waves of leg movements alternate on either side of the body. Thus at any one time, the feet on one side of the body may be clustered together in movement, while those on the opposite side are spread apart to provide balance. Some burrowing species, such as those of the Geophilomorpha, have a different form of locomotion, with each foot being able to move independently of the others. These centipedes usually have quite short feet that are used more as anchors in the soil rather than digging tools. The main digging force in these species is provided by the strong muscular body trunk, which pushes the body through the soil, much in the same manner as an earthworm.

The head betrays the highly predatory nature of these animals: extended antennae constantly move to detect potential **prey** which, once detected, is seized by the front pair of legs and firmly held by other, smaller pairs of claws. The front legs are not only sharply pointed but are also modified as poison claws and can deliver a lethal injection of paralyzing fluid produced from special glands. The sense of **vision** is limited in most species—probably to the level of being able to differentiate between **light** and dark. Many species, however, lack eyes, especially the burrowing and cave-dwelling centipedes. Prey consists of small **arthropods** as well as earthworms, **snails**, and nematodes. Some of the larger tropical species have been known to eat **frogs** and small **snakes**.

Male and female centipedes are quite similar on the outside and the sexes are difficult to tell apart. Tropical species may breed throughout the year but temperate-dwelling centipedes breed in the spring and summer months, becoming less active during the cold winter period. Most species have a simple **courtship** routine, after which the pair may mate. Reproduction takes place outside of the body, with the male constructing a shallow web of silk-like strands on which he deposits a single package known as a spermatophore, which contains his sperm cells. The female then moves over the web and collects the spermatophore which is transferred to the ovary, where **fertilization** occurs. After carrying the eggs for some time the female may deposit them one by one in a protected place in the ground, for example, under a stone or in a rotten log. These eggs are covered with a glutinous secretion which helps them to adhere to soil particles or other substances. Not all species lay their eggs in such a scattered fashion however: some females create a simple nest in an enlarged cavity in a fallen log, or similar suitable chamber, where she remains to guard her eggs and even the larvae once they have hatched. The young later disperse and grow through a series of molt stages to reach adult size and sexual maturity.

Centrifuge

A centrifuge is a device for separating two or more substances from each other by using centrifugal **force**. Centrifugal force is the tendency of an object traveling around a central **point** to continue in a linear **motion** and fly away from that central point.

Centrifugation can be used to separate substances from each other becausematerials with different masses experience different centrifugal forces when traveling at the same velocityand at the same **distance** from the common center. For example, if two balls of different **mass** are attached to strings and swung around a common point at the same **velocity**, the ball with the greater mass will experience a greater centrifugal force. If the two strings are cut simultaneously, the heavier ball will tend to fly farther from the common center than will the lighter ball.

Centrifuges can be considered devices for increasing the effects of the earth's gravitational pull. For example, if a spoonful of clay is mixed vigorously with a cup of **water** and then allowed to sit for a period of time, the clay will eventually settle out because it experiences a greater gravitational pull than does the water. If the same clay-water mixture is centrifuged, however, the separation will take place much more quickly.

Types of centrifuges

Centrifuges can be sub-divided into two major categories, stationary devices and rotating devices. Both types of centrifuge work on a common principle, however. A collection of particles of different mass is set into motion around a common center. The faster these particles move, the greater will be the difference with which they tend to escape from their common center, and the more easily they will be separated from each other.

In a stationary centrifuge, a fluid (a gas or liquid) consisting of two or more components is sprayed into a cylindrical or conical chamber at a high **rate** of speed. As the fluid travels around the inside of the chamber, it separates into its components, the heavier substance(s) traveling to the outside of the container, and the lighter substance(s) remaining closer to the center of the cylinder.

One application of the stationary centrifuge is in the separation of the isotopes of **uranium** isotope from each other. Naturally occurring uranium consists of a mixture of uranium-235, which will undergo fission, and uranium-238, which will not. A **sample** of uranium is first converted into the gaseous compound uranium hexafluoride and then injected into a stationary centrifuge. As the rapidly moving stream of uranium hexafluoride travels around inside the centrifuge, it begins to separate into two parts. The heavier uranium uranium-235-hexafluoride concentrates along the outer wall of the centrifuge, while the lighter uranium-238-hexafluoride is left toward the center of the stream. The heavier **isotope** can then be drawn out of the centrifuge, leaving behind a sample of uranium hexafluoride slightly richer in the desired uranium-235 isotope. This sample can then be re-centrifuged and made still richer in the lighter isotope.

A centrifuge. *Photograph by Charles D. Winters. National Audubon Society Collection/Photo Researchers, Inc. Reproduced by permission.*

Rotating centrifuges

Another type of centrifuge is one in which the fluid to be separated is introduced into a container, and the container is then set into rapid rotational motion. Most beginning **chemistry** students are familiar with this instrument. It is commonly used as a substitute for **filtration** in the separation of a solid precipitate from the liquid in which it is suspended.

In this kind of machine, hollow tubes about 2 in (5 cm) in length are attached to arms radiating from the center of the machine. When the machine is turned on, the arms are spun around the center at a speed of about 30,000 revolutions per minute. The gravitational force experienced by materials inside the tubes-about 25,000 times that of gravity-causes the separation of materials much more efficiently than would a conventional filtration system.

Laboratory centrifuges have become invaluable tools in many kinds of scientific research. For example, today a widely used method of studying cells is to break apart a **tissue** sample and then centrifuge the resulting fluid. In this way, the discrete components of the **cell** can be separated and identified.

Applications of the rotating centrifuge

The basic centrifuge design described above can be adapted for use in many different settings. Industrial centrifuges, for example, tend to be quite large, ranging in size from 4 in to 4 ft (10 cm to 1.2 m) in diameter, with rotational velocities from 1,000 to 15,000 revolutions per minute. They can be designed so as to remove separated portions continuously, all at once after the machine has been stopped, or intermittently.

Large-scale centrifugation has found a great variety of commercial and industrial uses. For example, the separation of cream from milk has been accomplished by this process for well over a hundred years. Today, centrifuges are used to remove water from oil and from jet fuel and in the removal of solid materials from waste water during the process of water purification.

A centrifuge for use with very small particles of similar weight—the ultracentrifuge—was first developed by the Swedish chemist Theodor Svedberg in about 1923. In the **ultracentrifuge**, containers no more than about 0.2 in (0.6 cm) in diameter are set into **rotation** at speeds of about 230,000 revolutions per minute. In this device, colloidal particles, not much larger than the size of molecules, can be separated from each other.

Centrifuge studies in the space sciences

Centrifuge studies have been very important in the development of manned **space** flight programs. Human volunteers are placed into very large centrifuges and then spun at high velocities. Inside the centrifuge, humans feel high gravitational velocities that correspond to high gravitational forces ("g forces") that occur during the launch of space vehicles. Such experiments help space scientists understand the limits of **acceleration** that humans can endure in such situations.

See also Gravity and gravitation.

Resources

Books

"Centrifugation." *McGraw-Hill Encyclopedia of Science & Technology.* 6th edition. New York: McGraw-Hill Book Company, 1987, volume 3, pp. 392—398.

Dufour, John W., and W. Ed Nelson. *Centrifugal Pump Sourcebook.* New York: McGraw-Hill, 1992.

Lobanoff, Val S., and Robert R. Ross. *Centrifugal Pumps.* 2nd edition. Houston: Gulf Publications, 1992.

KEY TERMS

. . . .

Centrifugal force—The tendency of an object traveling in a circle around a central point to escape from the center in a straight line.

Gravitation—The pull of the earth's mass on an object.

Revolutions per minute—The number of times per minute an object travels around some central point.

Rotation—The spinning of an object on its axis.

Trefil, James. *Encyclopedia of Science and Technology.* The Reference Works, Inc., 2001.

David E. Newton

Century plant *see* **Amaryllis family (Amaryllidaceae)**

Ceramics

Ceramic materials are usually understood to be compounds of metallic and nonmetallic elements, though some are actually ionic salts, and others are insulators. These materials can be very complicated, as are for example clays, spinels, and common window **glass**. Many ceramic compounds have very high melting points.

Ceramics have a wide range of applications. They have been used as refractories, **abrasives**, ferroelectrics, piezoelectric transducers, magnets, building materials, and surface finishes.

Unlike metals, there are really no **heat** treatments that can be used to modify the properties of ceramics, but their properties can be altered by changes in chemical composition. By carefully considering the choice of chemical composition, purity, particle size and uniformity and arrangement, and packing of **atoms**, high quality ceramics can be synthesized in a wide variety.

Traditional ceramics

Ceramics have been used by man since antiquity. The earliest ceramic articles were made from naturally occurring materials such as clay **minerals**. It was discovered in prehistoric times that clay materials become malleable when **water** is added to them, and that a molded object can then be dried in the **sun** and hardened in a high **temperature** fire. The word ceramic comes from

TABLE 1. TRADITIONAL CERAMICS	
Category	*Examples*
whitewares	dishes, plumbing materials, enamels, tiles
heavy clay products	brick, pottery, materials for the treatment and transport of sewage, water purification components
refractories	brick, cements, crucibles, molds
construction	brick, plaster, blocks, concrete, tile, glass, fiberglass

the Greek word for burnt material, *keramos*. Many of the same raw materials that were used by the ancients are still used today in the production of traditional ceramics. Traditional ceramic applications include whitewares, heavy clay products, refractories, construction materials, abrasive products, and glass.

Clay minerals are hydrated compounds of **aluminum** oxide and silica. These materials have layered structures. Examples include kaolinite, halloysite, pyrophillite, and montmorillonite. They are all formed by the **weathering** of **igneous rocks** under the influence of water, dissolved CO_2, and organic acids. The largest sources of these clays were formed when feldspar was eroded from granite and deposited in **lake** beds, where it became altered to a clay.

Silica is a major ingredient in glass, glazes, enamels, refractories, abrasives, and whiteware. Its major sources include quartz, which is made up primarily of **sand**, sandstone, and quartzite.

Feldspar is also used in the manufacture of glass, pottery, enamel, and other ceramic products. Other naturally occurring minerals used directly in ceramic production include talc, **asbestos**, wollastonite, and sillimanite.

Hydraulic cement

Hydraulic cements set by interaction with water. Portland cement, the most common hydraulic cement, is primarily a water-free **calcium** silicate. It is slightly soluble in water and sets by a combination of **solution** precipitation and chemical reaction with water to form a hydrated composition. The **ratio** of water to cement in the initial mix greatly influences the strength of the final **concrete**: the lower the water-to-cement ratio, the higher the strength.

Glass

Glass is a ceramic material consisting of uniformly dispersed mixtures of silica, soda ash, and lime, that is often combined with metallic oxides of calcium, **lead**,

lithium, cerium, etc. Glass is distinguished from solid ceramics by its lack of crystallinity. It is in fact a supercooled liquid. The atoms in glass remain disordered in the solid state, much as they are in the liquid state. Glasses are thus rigid structures whose atomic arrangements and properties depend on both composition and thermal history.

Group 16 elements (**oxygen**, sulphur, selenium, tellurium) are especially good candidates for glass formation. Oxygen is able to form stable bonds with silicon, boron, phosphorous, and arsenic and thereby form stable structures having oxygen atoms at the corners and one of the other atoms at the center. Pure oxide glasses are very stable because each oxygen atom is linked by **electron** bonds to two other atoms.

Various two-phase structures may exist as glasses. One such structure is a mixture of glass and crystal. These materials are converted into strong and durable ceramics that are part glass and part crystal by prolonged heat treatment.

Modern ceramics

In the twentieth century, scientists and engineers have acquired a much better understanding of ceramics and their properties. They have succeeded in producing ceramics with tailor-made properties. Modern ceramics include oxide ceramics, magnetic ceramics, ferroelectric ceramics, nuclear fuels, nitrides, carbides, and borides.

Aluminum oxide

Aluminum oxide (Al_2O_3) occurs naturally in the mineral corundum, which in gem-quality form is known as the precious stones ruby and sapphire. Ruby and sapphire are known for their chemical inertness and hardness. Al_2O_3 is produced in large quantities from the mineral bauxite. In the Bayer process, bauxite (primarily **aluminum hydroxide** mixed with **iron** hydroxide and other impurities) is selectively leached with caustic soda. Purified aluminum hydroxide is formed as a precipitate. This material is converted to aluminum oxide powder,

TABLE 2. MODERN CERAMICS	
Category	*Examples*
electronics	heating elements, dielectric materials, substrates, semiconductors, insulators, transducers, lasers, hermetic seals, igniters
aerospace and automotive	turbine components, heat exchangers, emission control
medical	prosthetics, controls
high-temperature structural	kiln furniture, braze fixtures, advanced refractories
nuclear	fuels, controls

which is used in the manufacture of aluminum-oxide-based ceramics. Aluminum oxide powder is used in the manufacture of porcelain, alumina laboratory ware, crucibles and **metal** casting molds, high temperature cements, wear-resistant parts (sleeves, tiles, seals), sandblast nozzles, etc.

Magnesium oxide

Magnesium oxide (MgO) occurs naturally in the mineral periclase, but not in sufficient quantity to meet commercial demand. Most MgO powder is produced from $MgCO_3$ or from seawater. MgO is extracted from sea water as a hydroxide, then converted to the oxide. MgO powder finds extensive use in high temperature electrical insulation and in refractory **brick**.

Silicon carbide

Silicon carbide (SiC) has been found to occur naturally only as small green hexagonal plates in metallic iron. The same form of silicon carbide has been manufactured synthetically, however. In this process, SiO_2 sand is mixed with coke in a large elongated mound in which large **carbon** electrodes have been placed at either end. As **electric current** is passed between the electrodes, the coke is heated to about 3,992°F (2,200°C). The coke reacts with the SiO_2 to produce SiC plus CO gas. Heating continues until the reaction has completed in the mound. After cooling, the mound is broken up, and the green hexagonal SIC crystals, which are low in impurities and suitable for electronic applications, are removed. The lower purity material is used for abrasives. The outer layer of the mound is reused in the next batch. SiC can be formed from almost any source of silicon and carbon. It has been produced in the laboratory from silicon metal powder and sugar, and from **rice** hulls. SiC is used for high-temperature kiln furniture, **electrical resistance** heating elements, grinding wheels and abrasives, wear-resistance applications, and incinerator linings.

Silicon nitride

Silicon nitride does not occur naturally. Most of the powder commercially available has been produced by reacting silicon metal powder with **nitrogen** at temperatures between 2,282°F (1,250°C) and 2,552°F (1,400°C). The powder that is removed from the furnace is not ready to use. It is loosely bonded and must be crushed and sized. The resulting powder contains impurities of Fe, Ca, and Al. Higher purity silicon nitride powder has been produced by reducing SiO_2 with carbon in a nitrogen environment, and by reaction of $SiCl_4$ with **ammonia**; these reactions produce a very fine powder. High purity silicon nitride powder has also been made by **laser** reactions in which a mixture of silane (SiH_4) and ammonia is exposed to laser **light** from a CO_2 laser. This produces spherical particles of silicon nitride of very fine size.

Processing

The raw materials for ceramics are chosen on the basis of desired purity, particle size distribution, reactivity, and form. Purity influences such high temperature properties as strength, stress rupture life, and oxidation resistance. Impurities may severely influence electrical, magnetic, and optical properties. Particle size distribution affects strength.

Binders may be added to the ceramic powder to add strength prior to densification. Lubricants reduce particle-tool **friction** during compaction. Other agents are added to promote flowability during shaping.

Forming processes

The ceramic powder along with suitable additives are placed in a die, to which **pressure** may be applied for compaction. Uniaxial pressing is often used for small shapes such as ceramics for electrical devices. Hydrostatic pressing (equivalent pressing from all sides) is often used for large objects.

Alternatively, the ceramic powder may be cast. Although molten ceramics may be cast into cooled metal plates and quenched to produce materials made up of very fine crystals with high material toughness, casting of ceramics is usually done at room temperature. The ceramic particles are first suspended in a liquid and then cast into a porous mold that removes the liquid, leaving a particulater compact in the mold.

Yet another method of shaping a ceramic involves plastic forming. In this process a mixture of ceramic powder and additives is deformed under pressure. In the case of pure oxides, carbides, and nitrides, an organic material is added in place of or in addition to water to make the ceramic mixture plastic. While forming the ceramic object, heat and pressure are usually applied simultaneously.

Sintering

Densification of the particulate ceramic compact is referred to as sintering. Sintering is essentially the removal of pores between particles, combined with particulate growth and strong bonding between adjacent particles. In order for sintering to occur, the particles must be able to flow, and there must be a source of **energy** to activate and sustain this material transport. Sintering can take place in the vapor, liquid, or solid phase, or in a reactive liquid.

Machining

The sintered material must frequently be machined to allow it to meet dimensional tolerances, to give it an improved surface finish, or to remove surface flaws. Machining must be done carefully to avoid brittle fracture. The machining tool must have a higher hardness than the ceramic. The ceramic material can be processed by mechanical, thermal, or chemical action.

Design considerations

When evaluating the suitability of a ceramic material for a particular application, it is first necessary to understand the requirements of the application. These requirements might typically be defined by the load that the material will experience, the stress distribution in the materi-

al, interface, frictional requirements, the chemical environment and range of temperatures that the material will experience, and restriction on the final cost of the materials. Usually, one or two material properties will dictate the choice of a material for a particular application.

Historically most ceramic designs have been developed by empirical, or trial-and-error investigation. Only since the advent of the digital computer has it been possible to predict the properties of a particular ceramic material prior to actually producing it.

Resources

Books

Richerson, David W. *Modern Ceramic Engineering.* New York, NY: Marcel Dekker, Inc., 1982.

Randall Frost

KEY TERMS

Ceramic—A hard, brittle substance produced by strongly heating a nonmetallic mineral or clay.

Glass—A ceramic material consisting of a uniformly dispersed mixture of silica, soda ash, and lime; and often combined with metallic oxides.

Refractory—Any substance with a very high melting point that is able to withstand very high temperatures.

Sintering—The bonding of adjacent surfaces of particles in a mass of metal powders by heating.

Cerenkov effect

The Cerenkov effect is the emission of **light** from a transparent substance like **water** or **glass** when a charged particle, such as an **electron**, travels through the material with a speed faster than the speed of light in that material.

The Cerenkov effect was discovered by Russian experimentalist P. A. Cerenkov in 1934 and explained by Russian theorists I. Y. Tamm and I. M. Frank. All three scientists received the Nobel prize in **physics** in 1958.

The electric field of a fast-moving charged particle shifts the electrons of the **atoms** of a nonconducting material as the particle passes through. When the particle travels at speeds faster than the speed of light in the material, the atoms respond by emitting light in a cone at an angle determined by the index of refraction of the material. The process can be compared to that of a shock wave of sound generated when an airplane exceeds the speed of sound in air.

The index of refraction is computed by dividing the speed of light in a **vacuum** (3×10^8 m/sec or 186,000 mi/sec) by the speed of light in the medium through which the particle passes. The speed of light in lucite or heavy **lead** glass, for example, is 2×10^8 m/sec; therefore the index of refraction for those substances is 1.5.

Cerenkov detectors use the properties of Cerenkov **radiation** in high **energy** physics and **cosmic ray** physics experiments. Since the radiation is only emitted when the **velocity** of the particle is above a predetermined speed, a "threshold" value for the particle velocity can be set on the detector to discriminate against slow particles. For a particle velocity above the threshold value, an angular measurement of the Cerenkov light relative to the particle direction determines the velocity of the particle.

James O'Connell

Cerium *see* **Lanthanides**
Cesium *see* **Alkali metals**

Cetaceans

Human contact with cetaceans—whales, dolphins, and porpoises—has a rich history, beginning with some of our earliest civilizations. Although ancient people believed cetaceans were **fish**, they are actually aquatic **mammals**, which means they bear live young, produce milk to feed their offspring, and have hair (albeit just a few sensory hairs). The Greek philosopher Aristotle (384-322 B.C.) was the first to record this fact; in his *Historia Animalium*, Aristotle noted that whales and dolphins breathe air through a blowhole, and therefore have lungs; and that instead of laying eggs like fishes, they deliver their offspring fully developed.

Modern biologists believe that life first appeared in the sea; from these marine beginnings, land-dwelling organisms such as mammals gradually evolved. Cetaceans have returned to the marine environment after an ancestral period on land. As evidence of their terrestrial pedigree, consider that a whale fetus possesses four limb buds, a pelvis, tail, and forelimbs with five fingers like any land mammal. Adult whales and dolphins have the streamlined, fish-like appearance befitting their watery existence, but they have maintained and modified key terrestrial features (e.g., a much-reduced pelvic girdle in the tail, and forelimbs now used as flippers for swimming). A blowhole atop the head (one in dolphins, two in whales) replaces the nostrils, and thus the passageways for food and air are completely separate, as opposed to the usual terrestrial condition, in which food and air partly share a common tube. Other anatomical changes in cetaceans include a reduced neck, sensory modifications, and the addition of a thick layer of blubber to insulate against the cold of the **ocean** depths and to provide extra **energy** stores.

The order Cetacea is divided into three suborders. The Archaeoceti are a group of extinct cetaceans with elongated bodies, and are known only from fossils that are still being discovered and described. The living cetaceans are the Mysticeti or baleen whales, ten **species** restricted to the ocean, and the Odontoceti or toothed whales, whose many species (including dolphins and porpoises) are found in diverse habitats ranging from deep oceans to **freshwater** rivers great distances from the sea. Cetologists (scientists who study cetaceans) still disagree about how many toothed whales may be distinguished, but at least 68 species are recognized.

Mysticeti: Baleen whales

Baleen whales include the great whales or rorquals, a word that comes from the Norse for "grooved whale," owing to the conspicuous grooves or pleats on the throat and belly of these huge animals. There are seven rorquals, including the blue whale (*Balaenoptera musculus*), the largest creature that has ever lived. The largest blue whale ever recorded, according to the *Guiness Book of World Records*, was a female caught in 1926 off the Shetland Islands, measuring 109 ft (33.3 m) long. The blue whale belongs to the family Balaenopteridae, a group that migrates from summer feeding grounds in cold polar waters to breed in warmer waters in the fall and winter. Additional species are the fin whale (*B. physalus*), sei whale (*B. borealis*), Bryde's whale (*B. edeni*), minke whale (*B. acutorostrata*), humpback whale (*Megaptera novaeangliae*), and gray whale (*Eschrictius robustus*).

Rounding out the baleen whales are the Balaenidae, three genera that include the right whale (*Balaena glacialis*), the bowhead whale (*B. mysticetus*), and their elusive cousin, the pygmy right whale (*Caperea marginata*). Many cetologists recognize three subspecific forms of the right whale, and some argue that the southernmost of these deserves recognition as a separate species, the southern right whale (*B. australis*).

Although baleen whales are very large, they subsist on some of the smallest creatures: tiny oceanic **plankton**, or krill. Mysticetes are filter feeders, straining the **water** to collect their microscopic food, swallowing them in vast numbers; one mouthful of water may net its owner tens or hundreds of thousands of these tiny **prey**, which are trapped on the baleen as the seawater rushes back out. The baleen are rows of flexible, horny plates suspended from the upper jaw, each one fringed with a mat of hairlike projections to create an effective filtering device. Baleen was used for many years in ladies' corsets and other fashions, and thus was given the misleading name "whalebone." It is not bone at all, but rather the tough, flexible protein called keratin. As a group, baleen whales are larger and slower moving than toothed whales, perhaps in part because they do not need to pursue their prey.

Odontoceti: Toothed whales

In contrast, the faster-moving, smaller-bodied toothed whales—including dolphins and porpoises—pursue **squid**, fishes of many sizes, and in the case of killer whales (*Orcinus orca*), sea **birds** and mammals, including other cetaceans. Many toothed whales travel in groups of from five to many dozens of animals, whose purpose seems to be in part to hunt cooperatively. Killer whales have been observed to gang up on and kill larger whales such as gray whales. They also collectively hunt **seals** resting on **ice** floes; once a seal is spotted, the whales will dive together, causing a great wave that upsets the ice, dumping the unfortunate seal into their midst. (Interestingly, there is no record of a killer whale having killed a human.) Cooperative **behavior** has also been suggested for several dolphin species; bottlenose dolphins (*Tursiops truncatus*) have been observed to circle a school of fish, causing them to group more tightly together, and then take turns lunging through the school, grabbing mouthfuls of fish as they pass. Perhaps in these species, individuals hunting together can catch more food than each would hunting alone. This is probably not true of the plankton-eating baleen whales.

There are many solitary species of odontocetes, however, each with innovative "solo" feeding strategies. Odontocete teeth are generally conical in shape, except in porpoises; these animals have spade-shaped teeth, and they lack the dolphin's protruding rostrum, or beak. However, the bouto (*Inia geoffrensis*) of the Amazon and Orinoco rivers in **South America**, has rear teeth that resemble molars, used for crushing the armored **catfish** which is among its favorite foods. The susu river dolphins (*Platanista minor* and *P. gangetica*) of the Indian subcontinent, have long, pincer-like jaws for grabbing prey out in front of them. The largest odontocete, the sperm whale (*Physeter catodon*), is also the deepest diver. These animals have been observed diving to 4,000 ft (1,220 m), and have been captured with squid inhabiting depths of 10,000 ft (3,050 m) within their stomachs; however, their average dives are probably around 1,100 ft (335 m).

Anatomy and physiology

The sperm whale's deep dives raise interesting questions about cetacean **anatomy** and **physiology**. Even the shallower dives performed regularly by many cetaceans would jeopardize the health of a human diver. One important issue for a diving **animal** is keeping warm in the cold depths. All cetaceans have a thick layer of blubber insulating them from the frigid water; in addition, a diving cetacean is aided by the increasing **pressure** of the greater depths, which reduces **blood** circulation automatically. Blood flow to peripheral body areas is further reduced by proximity to the cold water, keeping the warm blood circulating to the internal organs. It is still a mystery that the cetacean **brain** can function normally at great depths; the heartbeat drops during a long dive, but somehow the brain maintains its normal **temperature**.

Another important issue to a diving animal is **oxygen** deprivation. Many rorquals routinely stay under for 30 minutes before surfacing for a breath. Longer dives are always followed by a certain amount of panting at the surface, so the whale can restore its depleted blood oxygen levels. This is not the whole story, of course, since no land animal could hope to match this feat of breath-holding. Cetacean muscle **tissue** contains much greater amounts of myoglobin, the oxygen-binding protein found in the muscles of all mammals. This means that ounce for ounce, cetacean muscle is capable of storing more oxygen where it is needed most, even when new oxygen is not being provided via the lungs. In addition, cetaceans apparently have a high tolerance for the waste products (**lactic acid** and **carbon dioxide**) that accumulate in working muscle in the absence of sufficient oxygen.

Of special interest to people who dive for recreation is the dangerous phenomenon known as the bends. Human divers take to the depths with a tank of compressed air, whose pressure equals or exceeds that of the surrounding water. Otherwise, our relatively feeble chests would collapse under the pressure of the surrounding water. The bends occurs when **nitrogen** gas present in the compressed air dissolves into our blood and tissues, forming bubbles when the pressure is reduced too rapidly upon ascending. How do cetaceans avoid this deadly condition? Upon diving, their remarkably flexible chests and small, elastic lungs collapse; the tiny pouches that absorb oxygen within the lungs (the alveoli) are forced shut and gas exchange ceases. This means that little or no nitrogen is transferred into the bloodstream, and the dangerous bubbling-up of dissolved gas does not occur upon resurfacing.

In a swimming cetacean, the tail is the main source of forward propulsion, pushing the animal forward by an up-and-out movement against the water (rather than side-to-side, like a fish). The tail is a flexible extension of the last vertebra, which supports the muscular flukes. Rolling and changing position in the water is accomplished by the flippers; the flippers and dorsal fin (which is not present in all species) act together as stabilizers. Cetacean bones are heavier than water, but the body floats easily owing to buoyant blubber, oil in the bones, and air in the lungs. Some great whale species have been observed to **sleep** for hours, usually at night, their heads passively rising clear of the water to expose the blowhole. Both eyes are closed, and the body floats motionless with tail and flippers hanging limply, while the whale breathes once or twice per minute with a brief, snorting exhale. Amazingly, the smaller dolphins, who have more to fear from **sharks** and other predators, appear to sleep with only half of their brain at a time; one **eye** remains open, enabling the animal to rouse itself should danger arise.

Sensory perception

Cetologists have been able to infer a good deal about the sensory powers of whales and dolphins. Their **vision** is good, but is limited to about 45 ft (13.7 m) or so, even in the clearest water; the depths of the ocean are quite dark, and vision is of no use. Their sense of **hearing** is much more important, in part because water is such an excellent conductor of sound. In addition, many cetaceans navigate and find food using **echolocation**, or sonar.

An echolocating animal perceives objects in its path by listening for the reflected echoes of pulsed sounds that it produces. In keeping with this practice, the cetacean hearing range is much greater than ours; some species can hear sounds up to 180,000 Hz. The sound is produced in a complex chamber in the airway atop the head, and is conducted out through the melon, a waxy, lens-shaped structure in the forehead. The melon func-

tions to focus the beam of sound the way a magnifying **glass** focuses a beam of **light**. We know that around a dozen species of toothed whales and dolphins use sound to find food items; for instance, blindfolded bottlenose dolphins can find fish swimming in their tanks. Baleen whales lack the sophisticated structures for true echolocation, but may use echoes from lower-frequency sounds for a more rudimentary echonavigation.

Scientists once believed that cetaceans had no sense of **taste** or **smell**. More recently, bottlenose dolphins have been shown to distinguish the four basic taste stimuli (sour, sweet, salty, and bitter). In addition, beluga whales (*Delphinapteras leucas*) have been observed to show alarm when swimming through areas where other belugas have been killed and quantities of blood are present in the water. Several cetacean species are sensitive to substances found in mammalian urine and feces, which could provide information on the identity or status of other individuals. Most modern cetologists agree that taste and smell are important to many cetaceans.

The cetacean sense of **touch** is very keen, as becomes obvious to anyone who watches two familiar animals interacting: whales and dolphins large and small, rub up against each other, stroking and petting one another with flukes or flippers. Such touching clearly feels good to the recipient, and captive dolphins have been trained to do various tricks using touch alone as the positive reinforcement.

Scientists have suggested the existence of a cetacean magnetic sense; this would help to explain their remarkable navigational powers during long migrations in the otherwise featureless marine environment. In support of this possibility, the mineral magnetite has been found in the brains of some species (including common dolphins, Dall's porpoise, humpback whales and beaked whales). A magnetic sense could help explain the bizarre phenomenon of live stranding; although rare, stranding is generally fatal to the whale, which is ultimately crushed by its own weight out of the supporting water. A magnetic sense might not be fool-proof, and could be upset by various disturbances from on shore, leading the animals to beach themselves. Others have proposed that mass strandings of cetaceans are the result of the intense social bonds that form among members of some species. Perhaps the urge to avoid the dangers of separation is stronger than that to avoid the fatal risk of stranding along with a sick or injured comrade seeking shallow water.

Social behavior

Social behavior in cetaceans runs the gamut from species that are largely solitary to those that are highly social. Baleen whales are rarely seen in groups of more

Two Atlantic bottlenose dolphins. *JLM Visuals. Reproduced by permission.*

than two or three; however, gray whales and other rorquals may form transient groups from 5-50 animals during **migration**. Among toothed whales, river dolphins and narwhals (*Monodon menoceros*) are examples of species that appear to have mainly solitary habits.

The most highly social species are found among the toothed whales. In bottlenose dolphins and killer whales, for instance, individuals typically have social bonds with many others, which may last for life. In these species, females form the core of the society: long-term bonds between females and their adult daughters are important in many aspects of life, including foraging, fending off predators and aggressive dolphins, and delivering and raising the young. For example, sperm whale mothers often leave their infants in the company of a "babysitter," while they make the deep dives for squid where a baby could not follow. In bottlenose dolphins, social alliances between adult males are a prominent feature of society. Trios of adult

male bottlenose dolphins perform together in synchronous aggressive displays, herding a sexually receptive female, or attempting to dominate rival males from similar alliances.

Courtship and mating remains largely undescribed for many species, owing to the difficulty of observing events under water; systematic study of most species is simply lacking. The mating systems of many that have been studied are classified as promiscuous, meaning that individuals select a new mating partner each year, and both males and females may mate more than once in a given season.

Some cetacean species—for example, narwhals, killer whales, and sperm whales—exhibit rather pronounced sexual dimorphism in size, meaning that males are noticeably larger than females. Such a size difference is believed to indicate a relatively high degree of **competition** among males for a chance to mate. These species are assumed to be polygynous, meaning that a few males mate with most of the available females, excluding the rest of the males for mating opportunities. Male-male fights in the presence of a receptive female may become fierce, and adult males often bear the physical scars of such contests. Adult male narwhals have a long, **spiral** tusk made of ivory growing out from their head, which they use in jousting and sparring with their competitors.

Pregnancy and birth

Most baleen whales first mate when they are 4-10 years of age. Many toothed whales take longer to mature, and in sexually dimorphic species, males take longer still. Sperm whales require 7-12 years, killer whales 8-10, and false killer whales (*Pseudorca crassidens*) need up to 14 years to mature.

Gestation in mysticetes lasts 10-13 months on average. Many odontocetes have similar gestation times but some are longer: pilot, sperm, and killer whale pregnancies last up to 16 months. The **birth** of a baby has rarely been witnessed by humans—typically, a formerly pregnant female simply appears one day with her new infant at her side. On rare occasions, however, a human observer has been lucky enough to see a birth. One laboring gray whale spent the last 10 minutes of her labor hanging vertical in the sea, head down, with her flukes held 6 ft (1.8 m) out of the water. As she lowered her flukes to a horizontal position, the calf's snout was seen to be emerging from her belly. The mother shifted to a belly-up position, just at the surface, as the calf continued emerging; then the mother submerged, and the calf popped up to the surface, separate from its mother. Bottlenose dolphins are thought to usually deliver their calves tail-first, but head-first deliveries have been observed in captivity. In a captive situation, the dolphin mother typically uses her tail and rostrum (beak) to guide the baby to the surface for its first breath of air.

In all species, the baby soon takes up the "infant" position below and to her side, near her mammary slits. The nipple of a nursing mother protrudes from this slit on her belly, and the milk is ejected into the baby's mouth by her mammary muscles with no effort on the part of the calf. Nursing bouts are relatively brief, since the calf must surface to breathe. Even so, the baby gains weight steadily; blue whale infants gain 200 lb (90.8 kg) per day! Blue and gray whales nurse for a period of about seven months; bottlenose dolphins nurse for three to four years or more. The record must be for pilot and sperm whales, who are occasionally observed to be nursing at 10-15 years of age. Of course, these youngsters have been eating other foods as well.

Intelligence and communication

Lengthy juvenile periods are typical for animals of greater intelligence. In Greek myths and other ancient sources, whales and dolphins have been accorded the attributes of higher intelligence, congeniality, and kindness to humans. Observations of wild and captive dolphins supporting a dead baby dolphin at the surface for hours and days—probably an instinctive act by a naturally protective mother—are no doubt the source of long-standing anecdotes about dolphins saving injured human divers from drowning.

It is generally agreed that cetacean intelligence surpasses other non-human mammals, such as dogs, seals, and even many **primates**. Curiosity, affection, jealousy, self-control, sympathy, spite, and trick-playing are all common observations by human handlers of captive cetaceans. Their brains feature a sizeable and deeply convoluted cortex, suggesting considerable higher **learning** ability; dolphins in particular are known for their powers of innovation. An especially large supralimbic area of the brain explains their excellent powers of **memory** and social intelligence.

Greater communication skills often accompany greater powers of intelligence. An underwater listener in the vicinity of a group of cetaceans may be surprised by the great variety of sounds they produce, from the repetitive tonal pulses of fin whales, to the moans and knocks of gray whales, the whoops, purrs, and groans of bowhead whales, and the elaborate, eerie songs of humpbacks. Cetaceans do not have vocal cords; odontocete vocalizations are produced in a group of air sacs in the region below the blowhole atop the head, and are projected out through the melon. Mysticete sounds seem to come from an area off the lower side of the larynx, but the exact origin is unclear. Many of the social odontocetes emit whistles, and intense whistling seems to accompany excitement surrounding feeding, sex, and the joy of riding a wave before a speeding boat. Individuals may recognize one other from their unique whistles and other sounds.

Male humpback whales, who sing primarily during the breeding season, are the greatest balladeers of all cetaceans. In addition to their melodic, haunting songs, humpback whales produce a harsh gurgling noise, something like the sound a drowning person might make; their vocalizations, which may go on for hours, were audible through the wooden hulls of old whaling vessels. Superstitious sailors, hearing these voices out of the deep below their ship, believed they were the ghosts of drowned comrades, coming back to haunt their old vessel.

Commercial whaling and other threats

Cetaceans have been harvested on an individual basis by native peoples, including Inuit, in many parts of the world since before recorded time. Commercial whaling was underway in earnest by the twelfth century, when the Basque people of the French and Spanish coasts harvested whales harpooned from small boats, called shallops, in struggles lasting many hours. Such struggles were worth the risks, because a single whale yields enormous amounts of valuable commodities. The victorious whalers returned with huge quantities of meat and blubber, which was rendered down into valuable oil for fuel and lubricants. By the eighteenth century, commercial whaling was a burgeoning industry, notably for baleen, to be used in ladies' corsets.

Modern whaling methods are viewed by many outside the industry as grossly inhumane. Whales today are killed by a harpoon whose head has four claws and one or more grenades attached; when a whale is within range, the harpoon is fired into its body, where the head explodes, lacerating muscle and organs. The whale dives to escape, but is hauled to the surface with ropes and shot again. Death may not be swift, taking 15 minutes or more.

Commercial whaling inflicted a devastating blow on the world's baleen whale populations until 1986; most populations have yet to recover, and many populations and some species of whales are endangered. Public outcry resulted in reduced catch quotas during the 1970s, and finally in 1982 the International Whaling Commission set a moratorium on commercial whaling (to which all nations complied by 1989). Since then, however, Iceland, Norway, and Japan have demanded resumption of whaling; when the IWC refused, Norway announced plans to resume anyway, and Iceland left the IWC. Commercial whaling continues, although at a much reduced scale compared to former times.

Further dangers to cetacean populations include marine **pollution** and loss of food resources due to human activity. Drift nets hanging invisible in the water cause the death of cetaceans along with seals, seabirds, and fish; nets that get torn free during storms may drift at sea for many years. Thus, although the use of drift nets was halted by 1992, their effects persist. Meanwhile, purse-seine fishing methods for **tuna** have killed an estimated seven million dolphins since 1959.

Susan Andrew

CFCs *see* **Chlorofluorocarbons (CFCs)**

Chachalacas

Chachalacas, curassows, and guans are 42 **species** of **birds** that make up the family Cracidae. These birds are in the order Galliformes, which also includes the **grouse**, **pheasants**, **quail**, **guinea fowl**, and turkey. Curassows, chachalacas, and guans (or cracids) are believed to represent a relatively ancient and primitive lineage within this order. Fossil members of this family are known from deposits in **Europe**, but modern birds only occur in the Americas, ranging from the lower Rio Grande Valley of south Texas to Paraguay and northern Argentina.

The cracids are relatively large birds, ranging in body weight from about 1 lb (0.5 kg) for chachalacas to as much as 10.5 lb (4.8 kg) in the great curassow (*Crax rubra*). These birds have large, bare legs and large feet, well adapted for running and scratching for food in the forest floor. Cracids have a long tail, and a fowl-like body and head. The coloration of the plumage of cracids is generally a relatively plain brown or black, with little patterning. However, many species have colorful wattles and bare skin about the face, likely important in species recognition and **courtship**.

Most species of cracids occur in dense **forests** and thickets, although some species occur in more open forests. The curassows and guans mostly feed on the ground on **fruits**, **seeds**, and other **plant** materials, as well as **insects**, but guans mostly feed in the canopy. When disturbed, the ground-feeding birds typically fly up into the forest canopy. Cracids are not migratory, spending the entire year in a local environment.

Cracids build a simple nest of sticks, located on a **tree** branch. The clutch is two to five eggs, which are incubated by the female. The young are able to walk and run soon after **birth**, and are tended only by the female.

The chachalaca (*Ortalis vetula*) is the only species in **North America**, breeding in woodlands and thickets in extreme southeastern Texas and eastern Mexico. Like many other birds, this species is named after the sound that it makes.

All of the 10 species of chachalacas are in the genus *Ortalis*. The 20 species of guans occur in the genera *Pipile*, *Aburria*, *Chamaepetes*, and *Oreophasis*. The 13 species of curassows are in the genera *Pauxi*, *Mitu*, *Crax*, and *Nothocrax*.

Bill Freedman

Chameleons

Chameleons are small, strange-looking lizards in the family Chamaeleonidae. There are 86 **species** of chameleons, in four genera. The majority of species of

A chameleon catching an insect with its tongue. *JLM Visuals. Reproduced by permission.*

chameleon are found in the tropics of **Africa** and Madagascar, but some species live in southern Spain, Crete, India, and Sri Lanka. Most species of chameleons spend their lives in trees and shrubs, but some occur in herbaceous, grassy vegetation, and a few can be found on the ground.

Biology of chameleons

Most chameleons are green, yellow, or brown colored. However, these animals are famous for their ability to rapidly change the **color** and pattern of their skin pigmentation among these colors, and almost black or white shades can be achieved. This is done by varying the amount of pigment displayed by specialized cells in the skin, known as chromatophores. This visual **behavior** is primarily performed in response to changes in **temperature**, sunlight, and mood, especially when a chameleon is interacting socially with other chameleons. Chameleons may change the color of their skin to blend in better with their surroundings, as a type of opportunistic camouflaging. Although it is often believed that camouflage is the most common reason for the color changes of chameleons, the primary reasons are actually related to the mood or motivation of the **animal**.

Almost all chameleons are arboreal animals, moving slowly and deliberately in their **habitat** of trees and shrubs. The often imperceptible movements of these animals, coupled with their usual green or mottled color, makes chameleons difficult to detect among the foliage of their habitat. Chameleons have feet that grip twigs and branches well, with their toes fused in groups of two or three (a zygodactylous arrangement) that oppose each other to confer a strong grip.

The prehensile tail of most chameleons can be wrapped around twigs and other structures to give the animal stability while it is resting or moving around. In some respects, the tail serves as a fifth "leg" for these animals, because it is so useful for locomotion and securing their grip. The laterally compressed body of chameleons also appears to be an **adaptation** to a life in the trees, by conferring advantages related to the distribution of body weight, and perhaps in camouflage.

Chameleons have very unusual eyes, which extend rather far from the sides of the head within turret- or cone-like, fused eyelids. The eyes of chameleons can move and focus independently of each other. However, chameleons also have excellent **binocular vision**, which is necessary for sensing the **distance** of **prey** from the animal, so that it can be accurately snared by the long, unfurled tongue.

The tongue of chameleons is very important for the method of feeding. The tongue of chameleons is very long, and it can be rapidly extended by inflating it with **blood**. The tongue is accurately and quickly extruded from the mouth to catch prey up to a body length away from the animal. The sticky, club-like tip of the tongue snares the arthropod prey securely. Chameleons commonly feed on **insects** of all sorts, as well as on spiders, and scorpions.

Like other lizards, chameleons also use their tongue as a sense **organ**. The tongue is especially useful as a chemosensory organ which detects chemical signals from the air, ground, or food. The chemicals are transported to a sensory organ on the roof of the mouth, which analyzes the chemical signature.

Male chameleons aggressively defend a breeding territory, interacting with other males through drawn-out, ritualized displays. Some species have horn- and crest-like projections from their forehead which are used by the males as visual displays during their territorial disputes, and for jousting, during which the animals push at each other with their horns, until one combatant loses its grip on the branch, and falls to the ground. Most species of chameleons lay eggs, but a few are **ovoviviparous**, meaning the eggs are retained within the body of the female, where they hatch, so that the young are born as miniature replicas of the adults.

Some populations of chameleons consist entirely or mostly of female animals. This is an unusual trait in animal populations, and it may be indicative of **parthenogenesis** in these chameleon species, that is, the production of fertile eggs by females that have not mated with a male animal.

Species of chameleons

There are two genera of chameleons: *Chamaeleo* with 70 species, and *Brookesia* with 16 species. Species *Chamaeleo* occur in Africa, Madagascar, southern **Europe**, and southern and southeast **Asia**. Species of *Brookesia* occur only in East and West Africa and on the **island** of Madagascar.

The European chameleon (*Chamaeleo chameleon*) is represented by a number of subspecies in a few places in southern Europe, and much more widely in northern Africa, southern Arabia, and India. The African chameleon (*C. africanus*) occurs from West Africa through Somalia and Ethiopia. The common chameleon (*C. dilepis*) occurs throughout subsaharan Africa.

Some Madagascan and African chameleons have long projections on their snout which are used by male animals during their territorial jousts. Examples of these unusual,

horned chameleons are Fischer's chameleon (*C. fischeri*) and the mountain chameleon (*C. montium*). The most spectacular of the horned species is Owen's chameleon (*C. oweni*), which has three long, *Triceratops*-like horns. This species is only found in the lowlands of Cameroon.

Unlike *Chamaeleo*, species of *Brookesia* chameleons do not have a prehensile tail, and they do not undergo marked color changes. Examples of these stump-tailed chameleons are *Brookesia superciliaris*, *B. stumpfi*, and *B. tuberculata*, all found on Madagascar and several neighboring islands. *Brookesia spectrum* and *B. platyceps* occurs throughout subsaharan Africa.

Chameleons and people

In some regions local people have developed a fear of these unusual, bizarre-looking lizards, believing them to be poisonous or deadly in some other way. In other places, chameleons are believed to have medicinal value, and are sold dried for use in folk medicine.

Chameleons are striking and interesting animals, and they are sometimes kept as pets. Some chameleons are also sold internationally in the pet trade. However, chameleons are rather finicky creatures, and they usually do not survive very long in captivity. Not many people have the zoological skills needed to successfully keep chameleons alive.

As with so many other types of **wildlife**, the greatest threat to populations of chameleons is through the loss of their natural habitat. Most chameleon species do not adapt well to habitats that are intensively managed by humans. As a result, the populations of chameleons generally decline markedly when their natural habitats are converted to agricultural or residential land uses.

See also Anoles.

Resources

Books

Grzimek, H.C. Bernard, Dr., ed. *Grzimek's Animal Life Encyclopedia.* New York: Van Nostrand Reinhold Company, 1993.

Halliday, T.R., and K. Adler. *The Encyclopedia of Reptiles and Amphibians*. New York: Facts on File, 1986.

Bill Freedman

Chamomile *see* **Composite family**

Chaos

Chaos theory is used to model the overall behavior of complex systems. Despite its name, chaos theory is used to identify order in complex and otherwise seemingly unpredictable systems.

Chaos theory is used to understand explosions, complex **chemical reactions** (e.g., the Belousov-Zhabotinsky oscillating reaction that yields a red **solution** that turns blue at varying intervals of time), and many biological and biochemical systems. Chaos theory is now an important tool in the study of population trends and in helping to model the spread of **disease**. Epidemiologists use chaos theory to help predict the spread of epidemics.

Deterministic dynamical systems are those systems that are predictable based on accurate knowledge of the conditions of the system at any given time. When systems are, however, sensitive to their initial conditions they eventually become unpredictable. In particular, chaos theory deals with complex nonlinear dynamic (i.e., nonconstant, nonperiodic, etc.) systems. Nonlinear systems are those described by mathematical recursion and higher algorithms. Deterministic chaos, is mainly devoted to the study of systems the behavior of which can, in principle, be calculated exactly from equations of **motion**.

Chaos theory is the study of non-linear dynamic systems, that is, systems of activities (**weather, turbulence** in fluids, the stock market) that cannot be visualized in a graph with a straight line. Although dictionaries usually define "chaos" as "complete confusion," scientists who study chaos have discovered deep patterns that predict global stability in dynamic systems in spite of local instabilities.

Revising the Newtonian world view

Isaac Newton and the physicists of the eighteenth and nineteenth centuries who built upon his work showed that many natural phenomena could be accounted for in equations that would predict outcomes. If enough was known about the initial states of a dynamic system, then, all things being equal, the behavior of the system could be predicted with great **accuracy** for later periods, because small changes in initial states would result in small changes later on. For Newtonians, if a natural phenomenon seemed complex and chaotic, then it simply meant that scientists had to work harder to discover all the variables and the interconnected relationships involved in the physical behavior. Once these variables and their relationships were discovered, then the behavior of complex systems could be predicted.

But certain kinds of naturally-occurring behaviors resisted the explanations of Newtonian science. The weather is the most famous of these natural occurrences, but there are many others. The **orbit** of the **moon** around **Earth** is somewhat irregular, as is the orbit of the **planet Pluto** around the **sun**. Human heartbeats commonly exhibit minor irregularities, and the 24-hour human cycle of waking and sleeping is also irregular.

In 1961, Edward N. Lorenz discovered that one of the crucial assumptions of Newtonian science is unfounded. Small changes in initial states of some systems do not result in small changes later on. The contrary is sometimes true: small initial changes can result in large, completely **random** changes later. Lorenz's discovery is called the butterfly effect: a butterfly beating its wings in China creates small turbulences that eventually affect the weather in New York.

Lorenz, of MIT, made crucial discoveries in his research on the weather in the early 1960s. Lorenz had written a computer program to model the development of weather systems. He hoped to isolate variables that would allow him to forecast the weather. One day he introduced an extremely small change into the initial conditions of his weather prediction program: he changed one variable by one one-thousandth of a point. He found that his prediction program began to vary wildly in later stages for each tiny change in the initial state. This was the **birth** of the butterfly effect. Lorenz proved mathematically that long-term weather predictions based upon conditions at any one time would be impossible.

Mitchell Feigenbaum was one of several people who discovered order in chaos. He showed mathematically that many dynamic systems progress from order to chaos in a graduated series of steps known as scaling. In 1975 Feigenbaum discovered regularity even in orderly behavior so complex that it appeared to human senses as confused or chaotic. An example of this progression from order to chaos occurs if you drop pebbles in a calm pool of **water**. The first pebble that you drop makes a clear pattern of concentric circles. So do the second and third pebbles. But if the pool is bounded, then the waves bouncing back from the edge start overlapping and inter-

A head-on collision between two dipolar vortices entering a stratified fluid environment from the right and left sides of the picture. The original vortices have exchanged some of their substance to form two new mixed dipoles which are moving at roughly right angles to the original direction of travel (toward the top and bottom of the photo). Dipolar vortices are relevant to turbulence in large-scale geophysical systems like Earth's atmosphere or oceans. Turbulence within a fluid is an example of a chaotic system. *Photograph by G. Van Heijst and J. Flor. Photo Researchers, Inc. Reproduced by permission.*

fering with the waves created by the new pebbles that you drop in. Soon the clear concentric rings of waves created by dropping the first pebbles are replaced by a confusion of overlapping waves.

Feigenbaum and others located the order in chaos: apparently chaotic activities occur around some point, called an attractor because the activities seem attracted to it. Figure 1 illustrates an attractor operating in three-dimensional **space**. Even though none of the curving lines exactly fall one upon the other, each roughly circular set of curves to the left and right of the vertical line seems attracted to an orbit around the center of the set of circles. None of the curved lines in Figure 1 are perfectly regular, but there is a clear, visual structure to their disorder, which illustrates the structure of a simple chaotic system.

James Yorke applied the term "chaos" to non-linear dynamic systems in the early 1970s. But before Yorke gave non-linear dynamical systems their famous name, other scientists had been describing the phenomena now associated with chaos.

Current research

Chaos theory has a variety of applications. One of the most important of these is the stock market. Some researchers believe that they have found non-linear patterns in stock indexes, unemployment patterns, industrial production, and the price changes in Treasury bills. These researchers believe that they can reduce to six or seven the number of variables that determine some stock market trends. However, the researchers concede that if there are non-linear patterns in these financial areas, then anyone acting on those patterns to profit will change the market and introduce new variables which will make the market unpredictable.

Population **biology** illustrates the deep structure that underlies the apparent confusion in the surface behavior of chaotic systems. Some **animal** populations exhibit a boom-and-bust pattern in their numbers over a period of years. In some years there is rapid growth in a population of animals, followed by a bust created when the population consumes all of its food supply and most

KEY TERMS

Antibody—A molecule created by the immune system in response to the presence of an antigen (a foreign substance or particle). It marks foreign microorganisms in the body for destruction by other immune cells.

Boom-and-bust cycle—A recurring period of sharply rising activity (usually economic prosperity) which abruptly falls off.

Circadian rhythm—The rhythmical biological cycle of sleep and waking which, in humans, usually occurs every 24 hours.

Dynamics—The motion and equilibrium of systems which are influenced by forces, usually from the outside.

Electroencephalogram (EEG)—An electronic medical instrument used to measure brain activity in the form of waves printed on a sheet of paper.

Newtonian world view—The belief that actions in the physical world can be predicted (within a reasonable margin of error) according to physical laws, which only need to be discovered, combined appropriately, and applied accurately to determine what the future motions of objects will be.

Nonlinear—Something that cannot be represented by a straight line: jagged, erratic.

Population biology—The branch of biology that analyses the causes and (if necessary) solutions to fluctuations in biological populations.

Quantum—The amount of radiant energy in the different orbits of an electron around the nucleus of an atom.

Scaling—A regular series or progression of sizes, degrees, or steps.

members die from starvation. Soon the few remaining animals have an abundance of food because they have no **competition**. Since the food resources are so abundant, the few animals multiply rapidly, and some years later, the booming population turns bust again as the food supplies are exhausted from overfeeding. This pattern, however, can only be seen if many data have been gathered over many years. Yet this boom-and-bust pattern has been seen elsewhere, including disease epidemics. Large numbers of people may come down with measles, but in falling ill, they develop antibodies that protect them from future outbreaks. Thus, after years of rising cases of measles, the cases will suddenly decline sharply because so many people are naturally protected by their antibodies. After a period of reduced cases of measles, the outbreaks will rise again and the cycle will start over, unless a program of inoculation is begun.

Chaos theory can also be applied to human **biological rhythms**. The human body is governed by the rhythmical movements of many dynamical systems: the beating **heart**, the regular cycle of inhaling and exhaling air that makes up breathing, the circadian rhythm of waking and sleeping, the saccadic (jumping) movements of the **eye** that allow us to focus and process images in the visual field, the regularities and irregularities in the **brain** waves of mentally healthy and mentally impaired people as represented on electroencephalograms. None of these dynamic systems is perfect all the time, and when a period of chaotic behavior occurs, it is not necessarily bad. Healthy hearts often exhibit brief chaotic fluctuations, and sick hearts can have regular rhythms. Applying chaos theory to these human dynamic systems provides information about how to reduce **sleep disorders**, heart disease, and mental disease.

Chaos may depend on initial conditions and attractors

It is now understood that chaotic behavior may be characterized by sensitive dependence on initial conditions and attractors (including, but not limited to strange attractors). A particular attractor represents the behavior of the system at any given time. The actual state of any system (i.e., measured characteristics) depends upon earlier conditions. If initial conditions are changed even to a small degree the actual results for the original and altered systems become different (sometimes drastically different) over time even though the plot of the attractor for both the original and changed systems remains the same. In other words, although both systems yield different values as measured at any given time the plots of their respective attractors (i.e., the overall behavior of the system) look the same.

See also Mathematics; Quantum mechanics; Physics.

Resources

Books

Gleick, James. *Chaos: Making a New Science.* New York: Viking, 1987.

Prigogine, Ilya. *The End of Certainty: Time, Chaos, and the New Laws of Nature* New York: Free Press, 1998.

Other

Trump, Matthew A. The University of Texas. "What is Chaos?" <http://order.ph.utexas.edu/chaos/index.html> (February 5, 2003).

Patrick Moore